Materials Science for Dentistry

Related titles

Bone Response to Dental Implant Materials
(ISBN 978-0-08-100287-2)

Biomaterials for Oral and Dental Tissue Engineering
(ISBN 978-0-08-100961-1)

Biocompatibility of Dental Biomaterials
(ISBN 978-0-08-100884-3)

Woodhead Publishing Series in Biomaterials

Materials Science for Dentistry

B W Darvell *DSc CChem CSci FRSC FIM FSS FADM*
Honorary Professor, University of Birmingham, UK

10th edition

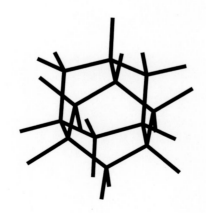

WOODHEAD
PUBLISHING
An imprint of Elsevier

Woodhead Publishing is an imprint of Elsevier
The Officers' Mess Business Centre, Royston Road, Duxford, CB22 4QH, United Kingdom
50 Hampshire Street, 5th Floor, Cambridge, MA 02139, United States
The Boulevard, Langford Lane, Kidlington, OX5 1GB, United Kingdom

Notices

Knowledge and best practice in this field are constantly changing. As new research and experience broaden our understanding, changes in research methods, professional practices, or medical treatment may become necessary.

Practitioners and researchers must always rely on their own experience and knowledge in evaluating and using any information, methods, compounds, or experiments described herein. In using such information or methods they should be mindful of their own safety and the safety of others, including parties for whom they have a professional responsibility.

To the fullest extent of the law, neither the Publisher nor the authors, contributors, or editors, assume any liability for any injury and/or damage to persons or property as a matter of products liability, negligence or otherwise, or from any use or operation of any methods, products, instructions, or ideas contained in the material herein.

Library of Congress Cataloging-in-Publication Data
A catalog record for this book is available from the Library of Congress

British Library Cataloguing-in-Publication Data
A catalogue record for this book is available from the British Library

ISBN: 978-0-08-101035-8 (print)
ISBN: 978-0-08-101032-7 (online)

For information on all Woodhead Publishing publications visit
our website at https://www.elsevier.com/books-and-journals

Working together
to grow libraries in
developing countries

www.elsevier.com • www.bookaid.org

Publisher: Matthew Deans
Acquisition Editor: Laura Overend
Editorial Project Manager: Natasha Welford
Production Project Manager: Joy Christel Neumarin Honest Thangiah
Cover Designer: Victoria Pearson

Typeset by B W Darvell and SPi Global, India

Contents

List of Tables

List of Key Data Figures

Preface

ἐπεὶ γὰρ τοῦ εἰδέναι χάριν ἡ πραγματεία,
εἰδέναι δὲ οὐ πρότερον οἰόμεθα ἕκαστον πρὶν ἂν λάβωμεν τὸ διὰ τί περι ἕκαστον

"Knowledge is the object of our inquiry,
and men do not think they know a thing till they have grasped the 'why' of it."
Aristotle (c. 350 BCE) *Physics*: Book II Chapter 3

Felix qui potuit rerum cognoscere causas

"Happy is he who gets to know the reasons for things."
Virgil (70-19 BCE) *Georgics ii 490*

As for previous editions, many changes have been made, arising from the notes and feedback gathered in various ways, to correct, revise, augment, extend and clarify wherever possible, as well as arising from my continued study. In adding some 150 pages, involving over 2500 changes or additions, including a new chapter on some further aspects of chemistry, and the beginnings of one on equipment, major new sections have again been included in a number of places, especially on wetting in practice, buckling, fluid mechanics and composite beams, all supported by 370 or so new references and a similar number of new figures. These continuing extensions are still a measure of the explanatory power of materials science in clinical dentistry, for ordinary processes and events, not in high theory or advanced materials, but in cumulative tiny insights into the way things work that tie ideas and phenomena together. However, much of the new material is in areas not covered elsewhere in accessible form and it is hoped that this will prove helpful. Indeed, the diversity of the subject continues to surprise. If I am able to convey any sense of the excitement of discovery, and contribute to another's understanding – and so advance the subject (and thereby patient care) – through this book, I will be happy indeed.

It remains my I hope that I have managed to maintain the right level and tone throughout, allowing an undergraduate access to every issue, whether or not it could be covered in a normal dental course. The availability of the explanations as a resource for use as required is the goal.

As ever, I would be grateful for a note of any errors, infelicities or problems as well as suggestions for new topics or subject matter – the field is by no means exhausted and much remains to be done. In addition, I invite questions and discussion through the online group at http://groups.yahoo.com/group/DentMatSci/. This is a live project that depends on feedback for its effectiveness as a teaching tool, both from the point of view of the teacher and the student. My thanks, again, to those who have helped.

Some 19 years on, I wonder whether those detractors (Preface to the 5th Edition) could sustain their argument? It is my view that the importance of the subject has never been higher, despite curriculum time being very low, and falling.

B.W.D. November 2017

Preface to the 5th Edition

Much of the success of clinical dentistry depends ultimately on the performance of the materials used. The correct selection and manipulation of those materials will itself rely on an understanding of their structures and properties, but the many types and formulations precludes simple rote learning of data as a means of supporting dentistry. This text provides the basis for study and the development of an understanding which will be of value indefinitely, irrespective of the changes in products that will inevitably occur with time. Even so, this is not a text which can or should stand alone: the subject is closely woven into the whole fabric of dentistry, and must be seen in that light if a full appreciation of its impact is to be gained. It is a key component of effective, practical, clinical dentistry.

This book arises from the undergraduate course in Dental Materials Science delivered at The University of Hong Kong and which has been developed over some 18 years. It has been written to complement existing textbooks dealing with dental materials. It thus avoids manipulation and processing instructions, whether for clinic or laboratory; tabulations of comparative data such as for composition and properties; long lists of product variants and alternatives; and discussion of treatment options and indications. Study of these other textbooks will reveal that the emphasis on manipulation necessarily entails much repetition and overlap of basic ideas, and that the expansion of the scope of the subject in recent years has been dealt with by the study of what may be called 'comparative composition', illustrated by tables of values. The essential unity of the subject is thereby obscured. It is the aim of this text to demonstrate that unity, as well as the general applicability of the fundamental principles by which all materials (not just dental) may be understood. This is done by chapters dealing with specific aspects of structure, behaviour or chemistry. Indeed, some topics are dealt with here for the first time, although they are inarguably within the ambit of the title. Thorough knowledge of a certain amount of basic chemistry and physics is assumed, as is the context of usage, general compositional details and handling or processing instructions for all types of material.

The subject is presented in a largely conceptual fashion. The mathematics present is used either by way of illustration of an approach or unavoidable statement of physical law. It is by being a conceptual approach that the wide range of relevant topics can be treated at the level which answers the question 'why?' in such a way as to maintain flexibility and permit the ideas to be carried forward to new developments for a broad and sound understanding of the field. More immediately, this approach permits the analysis of current problems and the handling of, and constraints on, current materials. What may superficially appear as dogma based on tradition can then be seen as material-dependent, but above all as rational and meaningful. Thus dental materials science can be seen to be in the service of dentistry.

This edition has undergone further thorough revision, with expansion of the discussion of many topics to explore the ramifications of ideas, as well as to improve the clarity of explanations. To this end many figures have been redrawn and many new ones introduced. As before, several suggestions for modifications have been incorporated. Again, some new auxiliary topics have been introduced. All of this remains with emphasis on, and always in the sense of, explaining actual dentistry.

In a sense, this is a work in progress as new topics, previously overlooked, will be developed for inclusion to suit new developments and to fill lacunae. However, and more importantly, development of the presentation of ideas continues in an effort to ease the comprehension and assimilation of ideas. I would be grateful therefore for feedback if passages are found wanting or errors of any kind are detected, even infelicities and gaps. I intend to produce revised editions as necessary.

I am indebted to many people for support and encouragement in producing this, including (in no particular order) Dr. Hugo Ladizesky, Dr. David Watts, Prof. George Nancollas, Prof. Ray Smallman, Dr. Vitus Leung, Prof. Peter Brockhurst and a good number of others. I must also thank Regina Chan for secretarial services, in particular for retyping the equations when the wordprocessor was changed, and my wife, Vivienne, for proofreading this and typing early editions. Nevertheless, any errors remain mine. I also owe a mention to all those who actively disparaged the project over many years claiming that it is irrelevant to dentistry, and to those innumerable publishers who turned it down on the grounds of lack of sales potential or because they "already have one" on their list. It was more of a spur than they would think.

B.W.D. Oct. 1998

Introduction

It is inevitable in preparing a book such as this that, because of the unusually wide scope of the subject and the interrelationships of the several aspects of any one material (and similar aspects in other classes of material), no one chapter order or sequence of ideas can be entirely logical or without anticipation of themes to be developed subsequently. This should not be taken as an indication of an unwieldy complexity and consequent difficulty of comprehension. Rather, it is a clear indication of the pervasiveness, the universality, of these themes. Time and again, reference will be made to an idea under different headings simply because that idea is a general principle. It will become apparent that there are relatively few major concepts embodied in this text, but that their applications are pretty much limitless.

The corollary to this is of course that it is quite impractical, if not actually impossible, to study each individual material in the context of its application and to analyse separately the contributions and interactions of each of its components. The welter of data would be nothing but indigestible. No student should be obliged to memorize data for its own sake. One implication of this is that pertinent comments are not written out in full at every opportunity. This is not to say that these are not relevant or important: fillers are fillers, with the same general effects, wherever and whenever they appear; surface energy considerations are relevant in almost every chapter, not just chapter 10, as they are in numerous other areas of dentistry which receive no mention here.

The student should aim to identify in any situation the relevant concepts, evaluate whether they are important in the context, and decide what their consequences might be. This procedure is one of analysis and synthesis, and can be recognized as a general approach to any problem. This book provides the wherewithal to do just that in respect of dental materials: a set of fundamental principles which may be viewed as a tool kit for dissecting a system, determining the role of each part, measuring its contribution and its interactions, and then understanding the behaviour of that system – as a whole. This is drawing a very careful distinction between just knowing and understanding[1] – compare Aristotle's remark, quoted in the Preface. Therein lies the key to this subject, as with many others: if its parts are treated independently, without reference to each other or the context, no real progress can be made; viewed in the round, the systems approach, the underlying patterns become much clearer. Basic concepts establish a framework, detail is built on to that.

This procedure, of building a mental model of a system, must be done gradually. Many concepts will be new, and the language used, the terminology, closely defined and specific. Learning the language will give a new view of the world, an alternative (and more explanatory) view of matter.

There are distinct parallels in other areas. It would be odd now, if once having grasped the concept of an atom or a cell, and thus their fundamental role in understanding the whole of chemistry and the organization of life respectively, that a new problem would require a reappraisal of that initially hard-won understanding. These ideas are second-nature, and are used as basic components of more complicated schemes. So it is with the concepts of materials science: rather than 'just' chemistry or physics, rather than 'just' observing behaviour, they are concerned with a composite of all of these, the intersection of all three major fields of Fig. 1.1. Dental materials science adds a further sphere, intersecting the first three, that of the clinical context, although this is more by way of changing the demands put on materials in service than on the underlying science.

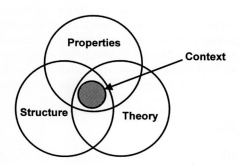

Fig. 1.1 Materials Science requires input from all aspects of a system.

An alternative approach is to consider the organizational hierarchy of matter in general (ignoring the sub-atomic!) (Fig. 1.2). The technology of dentistry deals with creating systems, primarily foreign objects in a biological environment, intended to be functional in some designed sense. Understanding of the system depends on that of the objects, which in turn depends on comprehending the constituent materials,[2] themselves derived from their component substances. This dependency provides one motivation for the organization of this book, and therefore in principle the manner of one's study of dental materials science within dentistry.

It is the existence of the clinical dental context – the motivation for the subject itself – that leads to the increasing use of the term 'biomaterials'. There is frequently confusion over how this term should be interpreted. The original use of the word, and the most intuitive, was as a contraction of 'biological materials', and referring to the materials of living organisms: skin, bone, shell, hair, cartilage and so on.[3] This usage is still to be preferred. Increasingly, however, foreign materials have been introduced into the body, primarily as prostheses (such as in hip replacements), but also surgical and other devices of various kinds (*e.g.* sutures, stents, pacemakers, boneplates and screws) where there is intimate contact with the internal environment, and these have, by association, been labelled biomaterials.[4] However, in dentistry, where contact may range from brief (impressions) through superficial semi-permanent (artificial dentures) to the intimate and invasive (transmucosal implants) *via* enamel-only restorations and pulp-capping, for some time the tendency has been more and more to apply the same term.[5] Where this is meant to leave investments and model materials, for instance, is unclear. Nevertheless, it is the application of the material which controls the properties of interest, whatever the context, whatever the label – the word is not the thing, the map is not the territory.[6]

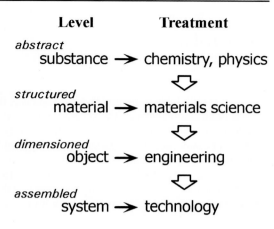

Fig. 1.2 The organizational hierarchy of matter. Understanding at one level depends on the level below in organizational complexity. Topics in the dental context form a subset at each level. Dentists are engineers using technologies.

With but a little thought the properties of a material which would be satisfactory to do a specific job can be identified. It is quite another matter to say what the ideal properties are, and one which reveals a major thread running through all aspects of real materials: compromise. The sad fact is that ideality simultaneously in several respects can never be approached because of the trade-off between one factor and another. The desired goals are on the two faces of the same coin; there is always a price to pay, and that price can be determined, in context, by reference to the ideas of this book.

There is yet another aspect of the subject which will be understood with advantage: "dental materials" includes all oral tissues. The tooth is part of the system which must be studied, as are the gingivae and other structures. They are certainly the contexts of many materials, but the interactions must be sought. Such an appreciation will be implicit in what follows.

If there is a message in this for the undergraduate it is probably that materials science is not so much an array of facts to be learned for their own sake, but more an attitude of mind. Necessarily the vocabulary and syntax of its language will need to be learned. But, as in any language, when these are mastered the number of sentences that can be constructed is unlimited. Some confidence can be taken from the thought that the ideas in this book are universal and enduring, unlike the commercial products themselves – or even the material itself – and the treatments. They are as valid now as they will be on the day you retire from practice.

References

[1]   Lipton P. Mathematical understanding. Chap. 5 in Meaning in Mathematics ed. Polkinghorne J. Oxford UP, Oxford, 2011. pp. 49 - 54.

[2]   Rand, A. Introduction to Objectivist Epistemology. 2nd ed, Binswanger H & Peikoff L (eds). Meridian, New York, 1990, pp. 16, 277.

[3]   Vincent, JFV. Structural Biomaterials. Macmillan, London, 1982.

[4]   see, for example, Cooke FW & Johnson JK (eds). Trans. Eleventh Int. Biomaterials Symposium, Clemson, South Carolina, Vol. III. Soc. for Biomaterials, Texas, 1979.

[5]   Burdairon, G. Abrégé de Biomateriaux Dentaires. Masson, Paris, 1981.

[6]   Korzybski A. Science and Sanity. p. 58 [5th ed. 1973; CD-ROM edition 1996] http://esgs.free.fr/uk/art/sands.htm

How To Use This Book

This book functions in several ways. Firstly, it is a reflection of the huge scope of materials in dentistry today, and thus the challenge to the dentist of tomorrow, no less than to those already in practice. There can be no reduction in this challenge as time goes on, rather will it increase, even if some materials fade away. In that light, this book is a toolkit of ideas that can be applied both in principle and in fact to any conceivable product in the future. This may seem like a grandiose claim, but if the identification of the fundamentals has been successful it must be true. That is the nature of materials science. Specific new polymers or composite combinations will almost certainly appear that have not been anticipated, but this should not matter. This is not meant to be an encyclopaedic treatise, but a foundation text that nevertheless serves that long-term purpose. It should permit answers to many questions not specifically addressed here to be worked out. Indeed, the focus on principles is for that very reason. It should be clear that whilst much work in a dental practice may be routine, it is the ability to recognize and deal with the non-routine events or conditions that is the mark of the professional. One must be prepared to think, but first observe.

Secondly, it provides material for an undergraduate-level course in the subject (although now the text has grown well beyond the needs of or timetable space for such a course). The level has, however, been chosen appropriately, and has been shown to be well within such a student's ability, and the 'reading age' set accordingly. That is not to say that an intellectual effort is not required – some topics are indeed challenging, but every effort has been made to explain rather than just declare: declaration is not teaching. Broadly speaking, though, more advanced ideas appear towards the ends of chapters and in the later chapters. Even so, everywhere the target in mind in writing has been the explanation in chemical and physical terms of the choices of material and procedure that are in use in everyday, clinical dentistry in a manner that empowers the student (and thus the practising dentist) to be in charge and control of what is done, rather than blindly following instructions with no capacity to recognize what is happening and intervene, adjust, or make independent, rational, scientific decisions, or indeed explain and justify treatment. This will be required more and more in respect of health insurance claims. In addition, in an increasingly litigious and regulated world, where accountability is paramount, a professional is no longer protected by a qualification and afforded unquestioning trust. As a professional, one may be called upon to defend a decision where something has gone wrong. It would seem essential that an appropriate knowledge and demonstrable understanding of the materials be part of the duty of care.[1] I have argued elsewhere that the ultimate competency in dentistry is this:

> To be able to defend competently in a court of law the selection, manner of use, and all aspects of the handling of products, instruments and devices used in the course of treatment and related processes as they affect patients, ancillary staff and others.

Thirdly, this is a source book, a reference for pursuing, as a first step, the ideas and equations needed for laboratory work, whether as part of undergraduate curriculum practical classes or further study, whether for a Master's degree or doctorate, or just research in general. At the undergraduate level, and in the vast majority of cases, it is not necessary to learn the equations or their derivation deliberately and explicitly. The practising dentist does not need them. It is therefore not appropriate that they be examinable as such (just as detailed compositions and property values are pointless). Nevertheless, it is important that the essence of these equations is understood: what variables are involved (*e.g.* stress, time), and in what sense (such as proportionally, inversely). The simpler equations, for example for Young's modulus, viscosity or surface tension, will in any case be absorbed almost automatically. The algebraic manipulations have deliberately been kept simple, but even if one's mathematical abilities are not up to reading and following these step by step, the text is intended to provide a parallel description of the process. Likewise, many diagrams are included to illustrate behaviour, structure and so on. These, too, should not be memorized. Rather, the nature of the trend or pattern of behaviour in broad terms should be identified. If the controlling factors are understood, the goal will have been achieved.

Fourthly, it is intended to demonstrate that instructions for use or clinical procedures are (or, at the very least, should be) justifiable, founded in reason, and logically traceable. Nothing should be taken on faith – including what is in this book![2] Be prepared to challenge dogmatic teaching or unsubstantiated claims. There must exist an explanation for every decision, every observation, every effect. Even so, in some cases, our understanding is incomplete (abrasion springs to mind), but the principle still stands. Therefore, identifying the path of the explanation, the chain of reasoning, should give confidence that dentistry has emerged from the mire of mediaeval magic, from the mystical arts of mere technical craftsmen, from the not-to-be questioned received wisdom of dogmatic practitioners, and from the era of the data handbook that lists simply what is rather than why. In that sense, nothing in this book is here just because it is interesting: it all has a purpose, a place in that chain intended to illuminate one idea or another. Thus, commonplace examples – or other non-dental topics – may be mentioned in order to underline the reality and broad applicability of the ideas, and to provide an image that may be better grasped by being familiar. Because, in places, the path is long, the student may not be able to see the ultimate goal or the part being played by a given idea. A little patience is required for things to fall into place – not everything can be said at once. Then again, the experience, background and knowledge of students do vary. It has been necessary in places to overlap what may have been taught prior to entry into university in order better to ensure that the path is complete for everyone (little has been assumed in this regard). In that case, the material will serve for revision. There are also many paths, which touch and cross each other in an elaborate network, so that cross-references are included liberally to assist in recognizing when other ideas are relevant, and as a guide to revision, if required.

This book does not deal with the operative techniques, instructions for use, laboratory procedures, or other matters of a direct treatment kind. These issues are dealt with in many other texts, and cross-reference should be made to those for the clinical background, handling instructions and so on. In addition, no attempt is made to deal with product comparisons or selection for purpose, but instead offers the means to support such judgements. It is emphasized that dental materials science is a foundation on which clinical dentistry totally and inescapably relies, whether or not it is recognized or acknowledged, and the implications and value of its contribution can only be found in the proper context. Study of dental materials science in isolation is not a meaningful proposition because clinical decisions should be in terms of and based on what it teaches. This may be understood better by reference to Fig. 1.3. The factors include not just the properties of the material but also the manner and conditions of use. These are seen to be acting on or through some kind of process, which then has an identifiable outcome. But the crucial point is then how this affects dentistry: what is the clinical implication of that outcome? How is treatment constrained? What limitations to performance in actual service might there be? Again, the ultimate purpose is patient well-being, and the present aim is to enable, facilitate and enhance just that.

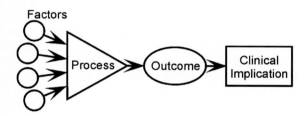

Fig. 1.3 Materials Science requires input from all aspects of a system, but end-point comprehension is the goal.

As indicated above, this is not an encyclopaedia. There is no attempt to cover all the many varieties of monomer, adhesives, and filler, for example. Such catalogues are available elsewhere (*e.g.* [3]) There are no tables of products, their recipes, and their property values – ephemera are pointless. Examples are used to illustrate principles, with such alternatives as are necessary to make a specific point, but always in a manner that may be applied elsewhere, including for products and systems yet to be invented. Even so, a deliberate attempt has been made to extend coverage into many areas not commonly dealt with (if ever), despite their evident importance. This is both a spur to thinking, and a resource for the curious, but all have emerged from questions put to me by students[1] or through observation of what dentistry uses or does (sometimes unwittingly). In that sense, the 'contained' undergraduate course needs to be created selectively by the teacher, omitting those many topics of lesser immediate need, according to the (ever diminishing) curriculum time available, yet allowing deeper discussion when the need arises.

So, to study. The terminology is of evident importance: knowing what the words mean and the context of their use allows comprehension of what is said, and the means of communication for reports and examinations (see the companion "Glossary"[4]). Much of this can be achieved by simply reading the appropriate portion of text once through, as for example in preparation for a lecture. At this stage one should not worry whether the chain of reasoning has been understood or a mechanism is clear – it is unimportant. What is more important is establishing a framework, a skeleton to be fleshed out in due course. If a lecture then follows, it will be far easier to understand. The relevant text can then be read through again, but now more carefully, this time identifying passages which cause some difficulty, but not dwelling unduly on them. There are many approaches to understanding concepts, and it may be that after reading other material, whether in other chapters of this book or in other dental materials texts, the ideas will be clearer. Still, if not, do not waste time struggling – ask for

---
[1] *sensu lato*: undergraduates, post-graduates, teachers, and others trying to understand.

help. In any case, if there is any doubt about the relevance or comprehension of a cross-reference, follow it up until a secure foundation is found. Then retrace your steps. Reinforcement will give you confidence. It is worth bearing in mind that total comprehension is not required to make coherent sense of system – missing pieces can be filled in later if there is a sufficient framework, indeed, it becomes easier as you proceed, just like a jigsaw puzzle (Fig. 1.4).

Try also to develop mental models, images in your mind of what is going on. Do not let the ideas remain just words on the page. Much of what is of interest here is dynamic: active processes of flow, diffusion, reaction and deformation. Ultimately, what is being described is observable behaviour, the macroscopic dependent on the microscopic. Verbal memory is not comprehension; it is not a viable route to understanding, as indicated in the Introduction.

Fig. 1.4 The full picture can be built slowly – it does not have to be complete for the essence to be grasped.

It will also be of value in consolidating the ideas, and in integrating them with clinical teaching, to begin to think about what is being done in the laboratory and on the clinic in terms of this subject. Try to relate your observations to the theory, and to answer clinical questions in these terms. This will give practice in using the tool kit, practice in thinking, and confidence in the meaning and utility of the subject.

Overall, a systematic, steady, integrated approach to study will yield a comprehension of the basis of much of modern dentistry in a manner that will become a second-nature, background, almost invisible part of your dentistry. It will be an investment for a long career, no matter what changes occur in the products and techniques.

References

[1]   Leung VWH & Darvell BW. The liability of dentists in the provision of dental materials. Hong Kong Law J 31 : 389 - 398, 2002.

[2]   'Nullius in verba'. http://royalsociety.org/page.asp?id=6186

[3]   [anon.] [Institut der Deutschen Zahnärzte] Das Dental Vademecum 2007/2008. Deutscher Zahnärzte Verlag, Köln. 9th ed. 2007.

[4]   https://www.academia.edu/1120368/Glossary_of_Terms_for_Dental_Materials_Science_12th_ed

Conventions

To facilitate the linking of ideas, copious cross-references are included in the text. Within a Chapter, the necessary section is indicated by the section mark thus: §4; if the section is in a different Chapter, the Chapter number is prefixed: 12§5. Figures, tables and equations are referred to the section always by the decimal form: equation 5.2 means equation 2 of §5, Fig. 3.12 means Fig. 12 in section 3, whether the reference is in the same section or not, but within the same Chapter. A figure, table or equation in a different Chapter is indicated by a prefixed Chapter number and section mark, thus: equation 1§2.3, Fig. 4§5.6. Ordinarily, it should not be necessary to cross-refer to these other locations to follow the sense of the text. The cross-references are to simplify the process, avoiding the index, when an idea is to be checked back. In many cases the cross-references indicate parallels in other systems that underline the general applicability of ideas.

References to selected primary and secondary sources in the literature are indicated by a superscript bracketed number, *e.g.*[1]. These references are listed at the end of each Chapter. Footnotes are indicated by a plain superscript number, *e.g.*[2]

SI units are used throughout except for a number of specialized uses where conversion of practice has yet to occur. These non-SI units are retained only for comparability with our sources.

A Note on References

In assembling a text of this sort, especially in view of the strongly multidisciplinary nature of the subject, a great many sources and influences will have been involved and their contribution to the narrative both diverse and intertwined, sometimes unconsciously. I acknowledge my indebtedness to all that have taught me, one way or another. Accordingly, many of the references given are of a general nature that may be relevant in a number of places in a chapter; I have given that reference on first use. In addition, many sources might now seem rather old. This only emphasises that the fundamental principles on which dental materials science is founded are universal and enduring, and well-established. On the other hand, no attempt is made either to be comprehensive or to track the latest information on all topics; this cannot be a key to the whole literature, which would be an impossible goal. Instead, in selected instances, some general reading or authority is cited to support the story being developed and to acknowledge sources. (I trust that there are no glaring oversights, but I apologize in advance for any that may be present. I should be obliged for advice of such omissions for correction in the next edition.) Even so, for material that is less commonly addressed, there are more given. For more advanced study, one's reading is researched in the usual way: cited instances of the references given may be a good start. However, I hope that I may be able to guide that study from firm foundations.

Greek Alphabet

Greek letters are used as symbols quite freely in many materials science contexts, and no less here. This list is given to facilitate reading. Upper case letters identical to the Roman are not used. (` indicates stress)

A	α	alpha	[`al-fuh]	N	ν	nu	[new]
B	β	beta	[`bee-tuh]	Ξ	ξ	xi	[zy]
Γ	γ	gamma	[`gam-uh]	O	o	omicron	[`o-mick-ron]
Δ	δ	delta	[`dell-tuh]	Π	π	pi	[pie]
E	ε, ε	epsilon	[`ep-sill-on]	P	ρ	rho	[row]
Z	ζ	zeta	[`zee-tuh]	Σ	σ	sigma	[`sigg-muh]
H	η	eta	[`ee-tuh]	T	τ	tau	[tore]
Θ	θ	theta	[`thee-tuh]	Y	υ	upsilon	[`up-sill-on]
I	ι	iota	[aye-`owe-tuh]	Φ	φ	phi	[fy]
K	κ	kappa	[`cap-uh]	X	χ	chi	[ky]
Λ	λ	lambda	[`lam-duh]	Ψ	ψ	psi	[sigh]
M	μ	mu	[mew]	Ω	ω	omega	[`ohm-egg-uh]

Chapter 1   Mechanical Testing

*One of the central requirements of any product in service is that the mechanical properties are suitable to the task. We may view mechanical testing as an attempt to understand the **response** of a material (deformation, failure) to a **challenge** experienced in service (loading). A range of relevant mechanical properties are described, introducing the terminology and interrelationships, as well as some of the tests themselves that are used to measure those properties in the laboratory. Much advertising reports the comparative merits of products in terms of these test data, and unless the tests are properly comprehended – and their limitations recognized – sensible buying and application decisions cannot be made in the clinic or dental laboratory.*

The challenges experienced by materials in dentistry extend beyond masticatory forces acting on restorations and prostheses. There are the deliberately applied forces such as in the deflection of a clasp as it moves over the tooth before seating, the seating of a crown, in rubber dam clamps and matrix bands, and in orthodontic appliances. In the laboratory, casting requires the hot metal to be forced into the mould, which must then be removed by mechanical means. In either context, shaping and finishing involves the application of forces through various tools, including the preparation of teeth and models.

*The responses of the materials of greatest concern include the **deformation**, both reversibly and irreversibly, and the outright **failure** in service or in preparation. Thus, reversible or elastic deformation is important in controlling the shape and continued functioning of a device whilst under load. Equally, whatever deformation occurs here should only be temporary: permanent deformation would ruin dimensionally accurate work. Undesired permanent deformation is a type of failure, but cracking and collapse is plainly not intended in many cases. Yet cutting, shaping and finishing, the debonding of orthodontic brackets and other procedures involve intentional breakage. These too must be understood to be controlled.*

*In normal service, except for dropping or traumatic events, loads are not instantaneous, one-off events – they have a duration, and are usually repetitive. This means that the **time scales** of loading and of the response mechanism must be considered, as well as the pattern of loading in the sense of **fatigue**.*

*The fact that a wide range of types of property are of interest to dentistry means that product selection must be based on a consideration of all of them. The problem is that not all can be optimal in the intended application, and some may be undesirable. The essence of this is that **compromise** is always involved, trading a bad point for a good, or putting up with a less than perfect behaviour in one respect to avoid a disaster in another or to ensure better performance in yet another. This theme recurs in all dental materials contexts.*

Materials Science for Dentistry
https://doi.org/10.1016/B978-0-08-101035-8.50001-8

The purpose of mechanical testing in the context of dental materials, as with all materials in any context, is to observe the properties of the materials themselves in an attempt to understand and predict service behaviour and performance. This information is necessary to help to identify suitable materials, compositions and designs.[1] It is also the most direct way in which the success or failure of improvements in composition, fabrication techniques or finishing procedure may be evaluated. The alternative would be to go directly to clinical trials which, apart from being very expensive, time-consuming, and demanding of large numbers of patients (of uncertain return rates for monitoring) in order that statistically-useful data be obtained, would provide the ethical problems of using people in tests of materials and devices which could possibly be to their detriment. Laboratory screening tests used at the stages of development and quality control are thus cheaper, easier, faster and (usually) without ethical problems. The results of such tests are generally the only information on which to base decisions. The comparison of products and procedures, to aid in the choices to be made at the chairside, are also informed by such data in publications in the scientific literature. It is therefore a prerequisite to understanding mechanical properties in general, and the basis of recommendations for clinical products and procedures in particular (making rational choices of materials and techniques), that the tests which are employed to study them, as well as their interpretation and implications, be understood thoroughly.

§1. Initial Ideas

We are all too well aware that things break (Fig. 1.1). Such breakages can be costly, even dangerous to patient or operator, and certainly an inconvenience, at the very least. Evidently, in such examples the forces applied were in excess of the objects' capacity to carry them. What needs to be understood, therefore, is what controls such behaviour: what is meant by strength, what determines it, how we can avoid exceeding it during use. Can we judge what is normal usage, what is abuse? Such damage does not arise spontaneously, but it depends on the forces acting during use. That is, the magnitudes, locations and directions of the loads applied, whether these were as intended by design, or inappropriate by accident or ignorance.

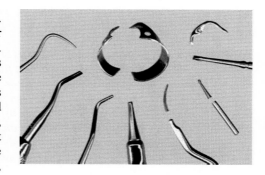

Fig. 1.1 Things break: probes, chisels, scalers, burs, clasps, rubber dam clamps.

There are several types of loading that an object or body may experience in practice: for example, tension, compression and shear (Fig. 1.2). We can therefore envisage that there will similarly be a number of ways or modes of testing which might be used in an attempt to understand the response of bodies to such loads in service. But even if such loading is externally realistic, and thus said to be modelling service conditions, it will be found that the internal conditions may be considerably

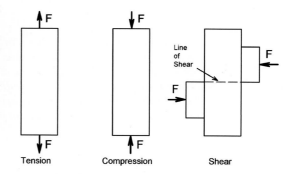

Fig. 1.2 Some of the principal kinds of applied load as might be used in laboratory tests. The applied forces are marked *F*.

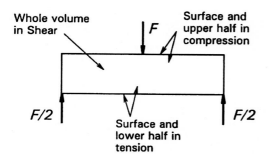

Fig. 1.3 The kinds of load acting in a body in a three-point bend test. The bending resulting from the application of the load makes the upper surface concave and the lower surface convex.

more complex. All three kinds of loading are typically present in any test, and they vary in intensity from place to place within the test piece. The interpretation of the results of such a test must therefore be done with care.

In a three-point bend test, for example (Fig. 1.3), we can identify a region where the principal result of the applied load is compression (extending from about the middle of the thickness of the beam to the concave surface), and a region in tension (similarly extending to the convex surface), but the entire region between the supports is in shear as well. This bend test thus gives us information about the performance of the material under those particular conditions of loading, but it may not say very much about any other circumstances. The deformation behaviour of the test piece certainly reflects contributions from all three aspects of the loading, although failure – meaning plastic deformation or crack initiation leading to collapse – may be attributable to one mode only. In the three-point bend beam case it is likely to be due to the tension on the lower, convex surface.

At the outset, then, we find that a major factor in testing materials is to match the conditions under which they will be put into service. Here, we are concerned about the mode of loading. This of course depends on an analysis being done of the service conditions themselves in order to determine what will occur and what will be relevant. It is this kind of thinking that underlies the selection of materials for particular purposes.

●1.1   Need to define properties

Now while we are perfectly capable (at least in principle) of determining the load that would cause a certain amount of deformation or even the collapse or failure in any sense of almost any given object, this in general is not a useful approach. There are an infinite number of possible shapes and sizes of object that might be put to some practical use, to say nothing of the variety of materials from which they could be made. The tabulation of such results would be impossibly cumbersome for even a small proportion of cases. Accordingly, a primary goal of materials science is to understand material behaviours in an abstract sense, that is, independent of shape and size. Thus, the underlying **postulate of consistency**[1] is that

a given material under chosen conditions will always behave in the same way, if subject to the same challenges

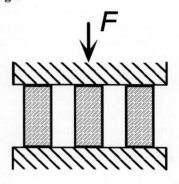

no matter what shape or size the object in which it is found. Notice that this is treating a material as having extent, as being a continuous but generalized 'body'. This is to distinguish **material properties** from the even more abstract sense in which the chemical and physical properties of substances are understood – the behaviour of matter itself – without any sense that we need be handling objects.

Fig. 1.4   A bearing capacity test on three identical objects simultaneously.

Consider an object that has an increasing compression load applied to it (as in Fig. 1.1, centre) until it collapses. We might term the ability of that object to carry a load without collapse the **bearing capacity**. Intuitively, this is what we need to know about any object that is meant to be load-bearing in any sense if we are to avoid overloading it. In other words, to determine if it is fit for duty under the conditions to which it will be exposed. Then, apply the same manner of loading to two, then three identical objects simultaneously (Fig. 1.3). Our elementary expectation is that the bearing capacity of the assemblage of objects will increase in a strictly linear fashion (Fig. 1.4). We naturally assume that each object carries its share of the load. We would therefore deduce that it is strictly unnecessary to test more than one such object at a time – there is no more *information* to be had in the multiple object test. Now, a little thought might suggest that perhaps in some sense it was the cross-sectional area of the set of objects being tested that was the underlying controlling variable. We should, for example, expect the same results whether we tested cylinders or rectangular objects – just so long as the cross-sections were the same.

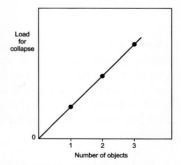

Fig. 1.5   The results of bearing capacity tests are expected to be strictly linear in the number of objects tested simultaneously.

Accordingly, our first efforts are directed to defining the **mechanical properties** of materials in just this size- and shape-free, abstract sense. We may therefore be in a better position to tabulate data in reference collections, having made the problem tractable. The intention is then that using such data we may calculate the expected behaviour of any

---

[1]   This, indeed, is a version of the underlying fundamental belief system on which the whole of science and 'scientific method' is founded.

arbitrary but real object, taking into account its shape and size, in order to determine its suitability or otherwise in a given application or circumstances. At the risk of being overly simplistic, **materials science** is about understanding such abstract properties as they affect the behaviour of objects, **engineering** is the design of real objects using that information. We shall return to consider the effects of shape in Chapter 23, indicating briefly how this is done.

Before proceeding, a distinction can be made with value. A mechanical property is limited to expressing the response of a material to externally applied forces in a scale-independent (that is, **intensive**) manner. In contrast, a **physical property** of a substance, compound or material, is an intensive character dependent on the spatial disposition of its matter (*e.g.* density), its response to a change in energy (*e.g.* thermal expansion coefficient), its effect on radiation (*e.g.* refractive index), or its effect on or response to fields (*e.g.* dielectric constant). This separation permits a clearer sense of the interactions occurring when we use or test materials.

§2. The Equations of Deformation

Perhaps the first inquiry to be made about the response of bodies to applied loads is the resulting deformation, the change in shape. In a great many aspects of dentistry it is the resistance to deformation – or the lack of it – that controls the suitability of a material for the application. Thus, the success of a partial or a full denture depends in part on its rigidity, but orthodontic devices must have readily-flexible components to do their intended job.

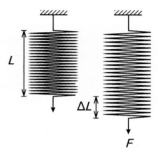

If a deformation is small enough then **Hooke's Law**[2] applies. This relates a deformation, for example the change in length of an object, L, to the force applied, F, through a **spring constant**, k (Fig. 2.1):

$$F = k \cdot \Delta L \qquad (2.1)$$

This force, for the moment, will be taken as acting in one axis only, *i.e.* the force is **uniaxial**. The spring constant here is relevant only to that particular test piece, and to no other, because it 'hides' the information about the material, length and cross-sectional area, even the shape, of the particular specimen.

Fig. 2.1 Hooke's Law: the extension under load of a spiral spring. Note the conventional representation of a fixed anchorage at the top.

●2.1 Stress and strain

To make things simple to start with, we first restrict the discussion to an object that is of uniform cross-section, that is, every section through the object perpendicular to the axis in which the load is applied is constant in both shape and area. We then imagine that the applied force is distributed uniformly over the entire cross-section at the end of the piece. We may now *define* **stress**, σ, as being force per unit area, A:

$$\sigma = F/A \qquad (2.2)$$

However, if the material from which the body is made is **homogeneous**, that is, uniform in composition and structure throughout, we have a reasonable expectation that the response of every part to the forces acting will be similar. Thus, the stress acting on each layer of the piece is expected to be identical, transmitted from layer to layer unchanged and uniform (Fig. 2.2). This may be called the **principle of uniformity**, and is a version of the postulate of consistency.

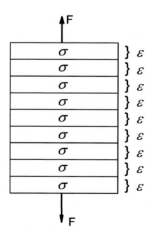

To define the response of a unit portion of the body to the applied stress, we define **strain**, ε, as the change in length per unit original length, L_o:

$$\varepsilon = \Delta L/L_o \qquad (2.3)$$

Fig. 2.2 The principle of uniformity – for all layers of a body.

---

[2]   First enunciated in 1678 by Robert Hooke (1635 - 1703).

This is on the basis of the principle of uniformity again: each layer is expected to respond to the applied stress in identical fashion. Similarly, this principle may be applied to the body divided into separate columns (parallel to the load axis). Not only do we anticipate that every unit area column within the body will behave identically, we have no reason to expect that any portion of such a column will behave any differently from any other portion (Fig. 2.3). To see this, consider their behaviour if the columns were not joined to each other: there can be no change. This is because the situation is then as depicted by Fig. 1.3.

The classification of the loading on a body includes the number of axes in which the forces are effectively acting. Along one axis only is **uniaxial**, along two mutually perpendicular axes is **biaxial**, while along all three Cartesian axes is called **triaxial**. The loads do not have to be equal, or even in the same sense. Thus, in the biaxial case, we may have tension in one axis and compression applied in the other. Similarly for triaxial loading we may have any combination. The special case here of all three stresses being equal is called **hydrostatic** loading.

These two views (Figs 2.2, 2.3) can be combined to show that all regions of a body subject to uniform loading, must behave in identical fashion (Fig. 2.4). Therefore, identical stresses and identical strains occur at every point.

It can now be seen that in defining stress and strain both the applied load and the resulting deformation have been scaled by the dimensions of the object, to make its actual shape and size irrelevant. Hooke's equation can therefore be reduced to a form which is, in principle, applicable to a test piece of any size and shape (but considering only regions of constant cross-sectional area and shape along the load axis). This version of Hooke's Law says that the stress (substituted for force) is proportional to the strain (instead of change of length overall), with E as the new constant of proportionality:

$$\sigma = E.\varepsilon \qquad (2.4\ a)$$

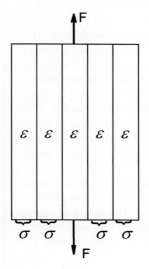

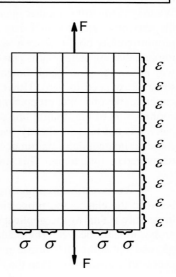

Fig. 2.3 The principle of uniformity – for all columns of a body.

Fig. 2.4 The principle of uniformity – for all infinitesimal regions.

Instead, therefore, of having to consider the behaviour of all kinds of shapes and sizes of bodies, we are now able to consider the behaviour of the material from which they are made, independently of size and shape. But, of course, from such data we can now *calculate* the response, if we so choose, for any other size and shape, not just that which was tested to get the data in the first place.

Care has to be taken to ensure that these defined terms are not confused as they are in ordinary, non-technical usage. The language of materials science, as in science in general, depends on the agreed definitions of such terms for effective communication.[3] The units of stress are N/m^2, or Pa, whereas strain is a dimensionless quantity, being the ratio of two lengths (in this example; there are also areal, volumetric and shear strains which could just as easily be considered in precisely the same manner).

The graphical expression of Hooke's Law is that the plot of stress against strain is a straight line (Fig. 2.5), the slope of which is the constant of proportionality between the two variables. The experimentally determined value of the slope of this line, E, is known as **Young's modulus** or the **modulus of elasticity**, which from equation 2.4 a is therefore defined by the ratio of stress to strain:

$$E = \sigma/\varepsilon \qquad (2.4\ b)$$

---

[3] There are, unfortunately, many instances in dentistry where terms are used loosely or incorrectly, without regard to their meaning outside the field. Sloppy terminology is unhelpful. Great care is suggested in reading some of the dental literature to ensure that usage is proper, dimensions match, and so on.

This quantity is therefore a measure of **stiffness** or resistance to deformation of the material.

Sometimes it is more convenient to think in terms of the reciprocal property, the **flexibility** of the material (sometimes called "springiness"), in other words the deformation obtained for unit applied stress. The term **compliance**, symbol J, is commonly used for this:

$$J = 1/E = \varepsilon/\sigma \qquad (2.4\ c)$$

If the stress were to be increased still further, we might be able to see the plot deviate appreciably from a straight line. We may then define another quantity from a stress-strain plot: the maximum stress which may be applied and still have the proportionality hold is known as the **proportional limit**. It is implied that the behaviour is **elastic** of course, that is, the piece will return to its original dimensions (zero strain) when the stress is reduced to zero. However, elastic deformation may continue beyond the proportional limit, it is simply not a continuation of the straight line plot. But, at some point, if the piece has not broken already, we will reach a stress at which some permanent deformation occurs, deformation which is not recovered on unloading. This stress is called the **elastic limit**.

We have now to emphasise the very important distinction between the elastic and proportional limits. The proportional limit only refers to the applicability of Hooke's Law: *strict* proportionality between stress and strain. If it is identifiable at all, and for some stiff brittle materials it might not be, it is necessarily always less than the elastic limit.

This essential *condition* – zero strain at zero stress after the temporary application of a stress – is a sufficient definition of **elasticity**. However, it says nothing about the kind of *deformation behaviour* shown by the piece under stress. There is no fundamental obligation for strict proportionality between stress and strain in any system. In fact, in some classes of material (**elastomers**, for example, Chap. 7) even a working approach to proportionality cannot be observed (see also §13). Even so, many other materials, such as metals and ceramics, show a substantially linear region in a plot of stress against strain and therefore may be considered to approach ideal Hookean behaviour sufficiently well. But some metals, for example, show a more obvious deviation from ideality: a region of proportional deformation is followed by a small, **non-Hookean** but still *elastic* deformation. We must therefore state that although under many circumstances the elastic limit is indistinguishable from the proportional limit, the elastic limit is always the higher of the two if there is a difference. Equally, the proportional limit may be zero, *i.e.* non-existent (which means not detectable), for some classes of material. The important point is the distinction between the definitions.

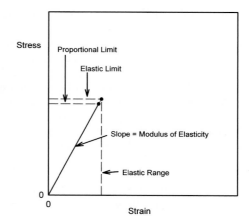

Fig. 2.5 A stress-strain diagram.

Modulus: In physics and mechanics, this term refers to a constant indicating the relation between the amount of a physical effect and that of the force producing it. In this kind of sense it was originally applied by T. Young (1807) to the quantity by means of which the amount of longitudinal extension or contraction of a bar of a given material, and the amount of the tension or pressure causing it, may be stated in terms of each other, *i.e.* the modulus of elasticity. On its own the word modulus does not convey any information, it must be qualified to indicate what it is a modulus *of*.

Proportional Limit: While it is very commonly mentioned in accounts of mechanical behaviour, as here, in practical terms it means very little. Essentially, its existence as a concept is merely a recognition and reminder that ideality is not to be found: reality is messier and less convenient. All it can indicate is when the value of E – the slope of the line – becomes *noticeably* different. It accounts for the comment for the applicability of Hooke's Law: "if the deformation is small enough". But, obviously it depends on the sensitivity of the detection: how far can the deviation be for it to count? Determination of a value is therefore not a precise matter, and there is no underlying physical event or condition change to mark it. Thus, it is not even a material property in any fundamental sense. This will become clearer in 10§3.6.

In the field of orthodontics it is common to characterize wires in terms of the strain ε_{max} observed at the elastic limit, σ_e, sometimes confusingly called the **maximum flexibility** or **springback**, and explicitly defined by:

$$\varepsilon_{max} = \sigma_e/E \qquad (2.4d)$$

A clearer term for this is **elastic range**. Thus, the strain at the elastic limit indicates how much deformation may be tolerated before the test piece will not return to zero strain when released. This needs to be known to avoid overloading a spring, for example. This quantity may also be found to be called the "range" (without further qualification). It can be seen that this assumes that the proportional limit and the elastic limit are the same, although for many metals the difference is not very great. These kinds of usage re-emphasize the need for very great care in reading and interpreting the dental literature, where technical definitions are often weak and terminology confused. This is not to say that the ideas themselves are not useful in certain contexts, but that the terms and their exact meanings may not be immediately obvious.

●**2.2   Poisson Ratio**
Experimentally it is commonly found that if a test piece is loaded in tension or compression, then, in directions perpendicular to the load axis, corresponding **lateral strains** will appear: a contraction if the load is tensile, an expansion if it is compressive (Fig. 2.6). This is known as **Poisson strain**. The behaviour under tension and compression is quite symmetrical, always so long as the deformations are small. The co-variation of lateral and axial dimensions is expressed through the **Poisson ratio**, ν. This material constant, also known as a modulus, is the ratio of the lateral strain to the axial strain, but given a negative sign because the strains themselves have opposite signs:

$$\nu = -\varepsilon_y/\varepsilon_x \qquad (2.5)$$

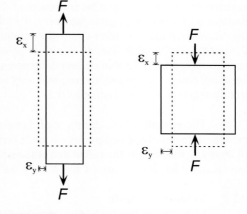

Fig. 2.6 Poisson strain is perpendicular to the load axis and (in most ordinary materials) of opposite sign to that of the axial strain.

This is adopting the convention that the *x*-direction is the load axis, and the *y*-direction is perpendicular to that. Values of ν for typical metals and ceramics lie in the range 0.2 ~ 0.4. On grounds of uniformity, as above, we expect the response will be similar in all directions perpendicular to the load axis, and in particular that:

$$\varepsilon_y = \varepsilon_z \qquad (2.6)$$

where the *z*-direction is, as usual, mutually perpendicular to the other two.

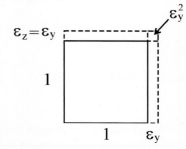

The definition of stress (equation 2.2) does not actually specify *when* the area is to be measured. We might assume, for example, that it is the original value, A, just prior to starting the test, that we should take. Yet it is plain from the mere existence of a non-zero Poisson Ratio that under stress the cross-sectional area of the test piece will change during the test, even if it remains within the (apparently) linear elastic range. We may explore the consequence of this changing area. The area A′ after deformation is given by:

Fig. 2.7 Poisson deformation of the cross-section of an axially-loaded (*i.e.*, in the x-axis) specimen under tension. The squared strain term can be ignored if the strain is small.

$$\begin{aligned}
A' &= A + \Delta A \\
&= A(1 + \varepsilon_y)(1 + \varepsilon_z) \\
&= A(1 + \varepsilon_y)^2 \\
&= A(1 + 2\varepsilon_y + \varepsilon_y^2)
\end{aligned} \qquad (2.7)$$

where ΔA is the change in area. However, $\varepsilon_y = -\nu.\varepsilon_x$ (from equation 2.5) and ε_y^2 may be ignored as it is very small (Fig. 2.7), so that we have:

$$A' = A(1 - 2\nu\varepsilon_x) \qquad (2.8)$$

as a good approximation. Therefore, at any point in the test, the actual stress, σ_x^*, is to be calculated from the cross-sectional area A' at that moment:

$$\sigma_x^* \ = \ \frac{F}{A'} \ = \ \frac{F}{A(1-2\nu\varepsilon_x)} \ = \ \frac{\sigma_x}{(1-2\nu\varepsilon_x)} \tag{2.9}$$

where σ_x is the stress calculated for the original cross-sectional area. Hence, the **true modulus of elasticity**, E_o, is given by:

$$E_o \ = \ \frac{\sigma_x}{\varepsilon_x} \cdot \frac{1}{(1-2\nu\varepsilon_x)} \tag{2.10}$$

It can be seen that E_o is very slightly larger than Young's Modulus (E) since this latter is determined experimentally only using a measurement of the original area. The Poisson Ratio is typically about 0.3 for a metal and so is not a negligible parameter if a full description of a material's behaviour is sought. Equation 2.10 also shows that the true stress-strain plot cannot in fact be exactly a straight line because of the extra term in the denominator. Nevertheless, it would in general be somewhat impractical to measure the lateral strain of specimens routinely (elaborate and delicate equipment is necessary), and it is usually sufficiently accurate – certainly easier – to employ the original area and just calculate Young's Modulus, using what is known as the **nominal stress**, *i.e.* from equation 2.2.

The remark "if the deformation is small enough" made for Fig. 2.1 needs a little amplification. In essence, it means that calculations are made on the assumption that the geometry remains unchanged in respect of anything that affects the calculated outcome. Generally, this may be effective as a working approximation, but clearly it has its limitations and these must be recognized – and checked – in all relevant contexts.

●2.3   Volumetric strain

From a knowledge of the magnitudes of the axial and lateral strains we can also calculate a volume strain, ε_v, that is, the relative change in the volume of the test piece. Thus, taking the new volume V' to be the original volume V plus the change in volume, ΔV, its value can be calculated from the usual expression for the volume of a cuboid in terms of the lengths of its sides. Thus,

$$V' \ = \ V \ + \ \Delta V \ = \ V(1 \ + \ \varepsilon_v) \tag{2.11}$$

but

$$1 \ + \ \varepsilon_v \ = \ (1 \ + \ \varepsilon_x)(1 \ + \ \varepsilon_y)(1 \ + \ \varepsilon_z) \tag{2.12}$$

so that from equation 2.6:

$$V'/V \ = \ (1 \ + \ \varepsilon_x)(1 \ + \ \varepsilon_y)^2$$

$$= \ 1 \ + \ \varepsilon_x \ + \ 2\varepsilon_y \ + \ \varepsilon_y^2 \ + \ 2\varepsilon_x\varepsilon_y \ + \ \varepsilon_y^2\varepsilon_x \tag{2.13}$$

Multiplied out, this has given a lengthy cubic expression, but because the strains themselves are typically very small in materials such as metals and ceramics, the quadratic and cubic terms can be considered negligible (see Fig. 2.7); that is, only the first three terms on the right need be considered. The error is only about the order of the square of the lateral strain. Subtracting the value one from each side we then have:

$$\varepsilon_v \ = \ \varepsilon_x \ + \ 2\varepsilon_y$$

$$= \ \varepsilon_x(1 \ + \ 2\varepsilon_y/\varepsilon_x)$$

$$= \ \varepsilon_x(1 \ - \ 2\nu) \tag{2.14}$$

We thus obtain an expression for the **volume strain** in terms of the axial strain and Poisson Ratio. What this means is that for values of $\nu < 0.5$, which is normally the case, there will be a change in volume of the piece when it is under load.

●2.4   True strain

Even the definition of ε_x itself needs refinement for it to be completely accurate, the point being that each increment of strain should be calculated in terms of the immediately prior value of the length of the specimen. In the limit, this requires the use of some calculus. So, considering the *increment of strain* at any moment, we can write

$$d\varepsilon \;=\; \frac{dL}{L} \tag{2.15}$$

which is the limiting version of equation 2.3. When the specimen has been deformed elastically to the new length L, the true total strain ε^* is obtained by integration:

$$\varepsilon^* = \int_{L_0}^{L} \frac{dL}{L} = \ln\!\left(\frac{L}{L_0}\right) \tag{2.16}$$

then, since $L = L_0 + \Delta L$,

$$\varepsilon^* \;=\; \ln(1 \,+\, \varepsilon) \tag{2.17}$$

This means that the true strain is slightly smaller than the **nominal strain** for tension, and slightly more negative for compression, as a few trials with a calculator will easily show. If the deformation is small enough it can be seen that $\varepsilon^* \approx \varepsilon$. In other words the approximation of equation 2.3 may be entirely adequate. However, to indicate that approximations are involved, σ_x and ε_x are sometimes referred to as **engineering stress** and **engineering strain**, to distinguish them from the true values. This also indicates that it is a matter of simple practicality in taking that approach. Most graphs of "stress" *vs.* "strain" are therefore in terms of these **nominal stress** and **nominal strain** values, unless they are explicitly labelled otherwise. We are prepared to compromise with the slightly less accurate values because of the expense and difficulty of obtaining the true values. However, one must not lose sight of the existence of Poisson strain.

•2.5   Shear

After the simple uniaxial tests discussed above, *i.e.* tension and compression, the next most important mode of testing is in **shear** (Fig. 2.8) where the layers of atoms or molecules of the material are envisaged as sliding over one another. The related mode of testing in **torsion** (Fig. 2.9) may be viewed as a particular case of shear. Shear is a common type of stress. For example, it is an aspect of the loading of beams in a three-point bend (Fig. 1.2). It is also relevant to the loading of interfaces such as between a bonded orthodontic bracket and a tooth, where the force is applied in the plane of the layer of cement. Endodontic files are ordinarily loaded in torsion in use, even though there is usually bending as well. There are many other examples.

In studying shear we are interested in the relative displacement of one layer sliding with respect to the next (Fig. 2.10). So, in a way analogous to that for direct uniaxial loading, we measure the length (L_0) over which the load is acting and the amount of displacement, Δs, at a given load, measured in the direction of load application. A version of Hooke's Law applies here also, where the shear force is related to the displacement by a shear spring constant, k_s:

$$F_s \;=\; k_s \cdot \Delta s \tag{2.18}$$

We can therefore define **shear stress**, τ, as the force per unit original cross-sectional area in the direction of shear:

$$\tau \;=\; F_s / A \tag{2.19}$$

and the **shear strain**, γ, as the displacement per unit original length over which the shear stress is applied,

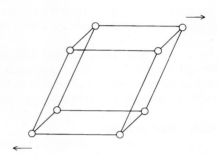

Fig. 2.8 The distortion resulting from the action of shear on a rectangular framework.

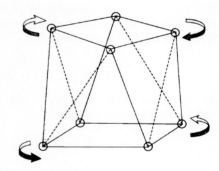

Fig. 2.9 The distortion resulting from the action of torsion on the same rectangular framework.

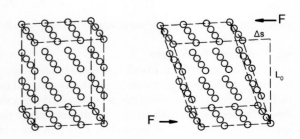

Fig. 2.10 Shear of materials involves the displacement of layers of atoms past each other.

that is, the depth, the distance between the top and bottom layers:

$$\gamma = \Delta s / L_o \tag{2.20}$$

Both of these are entirely analogous to the definitions for uniaxial loading. In particular, it can be seen that the shear stress on each layer is the same, and the relative displacement of each layer with respect to its neighbours must also be the same.

The **shear modulus of elasticity**, G, (sometimes called the **modulus of rigidity**) is then defined simply by:

$$G = \tau / \gamma \tag{2.21}$$

The reciprocal of this is known as the **shear compliance**, which is analogous to 'flexibility' (equation 2.4 c). It can also be shown that this shear modulus is related to Young's Modulus through the Poisson Ratio:

$$G = \frac{E}{2(1 + v)} \tag{2.22}$$

It is therefore not necessary to measure both moduli experimentally. To understand shear deformation we can use Young's modulus. Alternatively, it may be difficult to measure Young's modulus, yet the shear modulus might be easy to determine. However, either way it requires a knowledge of the Poisson ratio.

Two other points emerge from this. Firstly, the shear stress is the same on every layer, all the way through the specimen. The force acting over any layer is transmitted undiminished to the layer below. Secondly, also on the principle of uniformity (§2.1), the relative displacement of one layer with respect to the next must be the same for all layers in a homogeneous specimen.

●2.6   Bulk modulus

Using similar reasoning, the behaviour of materials under **hydrostatic loading** can be described with the **bulk modulus**, K [4] (again, it is understood that this is a type of elasticity). Under hydrostatic loading – in which the pressure (*i.e.* stress) on the sample is uniform in all directions – there is no change of shape, only of volume (Fig. 2.11). Again, the bulk modulus is related to the more easily measurable Young's Modulus:

$$K = \frac{\sigma}{(\Delta V / V)} = \frac{E}{3(1 - 2v)} \tag{2.23}$$

The reciprocal of this quantity is also known as the **compressibility**, κ, of a material:

$$\kappa = 1/K = \frac{(\Delta V / V)}{\sigma} \tag{2.24}$$

This is the bulk compliance.

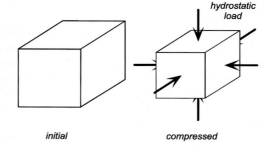

Fig. 2.11 Hydrostatic loading results in a change of volume but not of shape for an isotropic material.

If we combine equations 2.22 and 2.23 we get the following important relationship:

$$\frac{1}{E} = \frac{1}{3G} + \frac{1}{9K} \tag{2.25}$$

which illustrates the interdependence of the three moduli of elasticity.

●2.7   Effect of structure

We have so far assumed that the materials being tested are uniform, homogeneous and **isotropic**. While this may be true enough for an **amorphous** material (such as glass, which shows no long-range order at the molecular level), many materials are crystalline (thus showing long-range order) or heterogeneous. The latter class of materials, known generally as **composite** materials, are dealt with in later chapters as they have their own special characteristics. A crystalline structure necessarily has directionality, there can be no such thing as spherical symmetry in this case and the properties of the material must be **anisotropic**. That is to say, the

---
[4] The symbol B is also commonly used for bulk modulus.

mechanical properties vary according to the direction chosen for the load axis since the nature (*i.e.* the stiffness and strength) of the atomic or ionic bonds themselves vary with direction. Hydrostatic loading of single crystals will often involve some change of shape, since the three mutually perpendicular strain directions are unlikely to be equivalent if they are chemically or otherwise distinguishable. In fact, some 21 separate independent elastic moduli can be defined to express all of the possible directional variability in properties (for triclinic single crystals – see Table 11§3.1). However, other crystal types have more symmetry and so in practice there are rather fewer independent moduli, that is, the minimum number of moduli which are not algebraically derived from each other – for example, a cubic crystal has just three.

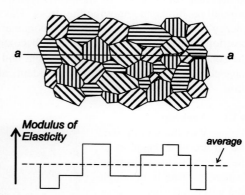

Even so, in dentistry we are very rarely, if ever, concerned with the properties of isolated single crystals. The materials we shall deal with are all **polycrystalline** (Fig. 2.12), such as metals, with usually random orientation of the individual crystals, or **amorphous**, such as glasses, when the local (at the atomic scale) anisotropies cancel out so that the overall effect is of an isotropic material. Because of these effects we can reduce the minimum number of moduli necessary to describe fully the behaviour of the material to just two: Young's Modulus and the Poisson Ratio, because the others can be calculated from these two values (see Table 2.1). (However, it can also be argued that the two fundamental independent moduli are G and B.) However, it must be emphasized that the validity of these properties in describing behaviour depends on the total deformations being small. Extremes lead to departures which require more complicated treatment.

Fig. 2.12 In a polycrystalline material local anisotropies in modulus of elasticity, i.e. at the level of the grain, and therefore both stresses and strains are averaged out, such as is shown for the section *a-a*.

Table 2.1 *Force and deformation moduli. The two most important are marked* *.

Modulus	Symbol	Units	relevant loading	associated with changes in
Elasticity	E_o	Pa	axial	shape + volume
Young's*	E	Pa	axial	shape + volume
Shear	G	Pa	lateral	shape
Bulk	K	Pa	hydrostatic	volume
Poisson*	ν	-	axial	shape + volume

§3. Plastic Deformation

So far we have dealt with the behaviour of materials for small deformations. This was to ensure that the geometry of the system was essentially unchanged, but also to keep the stress lower than the elastic limit. By definition, if a return to zero strain on removing the load is not obtained, there will have been **permanent deformation**, and thus the elastic limit will have been passed.

We may consider as a general example a specimen being tested in tension (Fig. 3.1), as this mode is somewhat easier to understand than compression (a tensile specimen will be assumed unless otherwise specified). The essential features of the test are that the specimen is gripped to apply the load, F, while a representative portion known as the **gauge length**, L (that is, the length of a representative portion before any load is applied), is monitored with some instrument (typically a **strain gauge extensometer**) to determine the change in length, ΔL for the load that is then

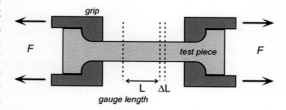

Fig. 3.1 A specimen undergoing a tensile test.

applied. From the measurements of the initial cross-section dimensions and gauge length, the stress and strain can be calculated. The behaviour of a such a specimen can now be examined in more detail.

Thus, after stressing past the elastic limit, the material may enter its **plastic** range (Fig. 3.2). The deformation has gone so far that some atoms or molecules cannot return to their original positions on removing the stress. Having gone past an energy maximum, they continue spontaneously into new positions, local energy minima, and so become stable. This is the source of the permanent deformation, and **yield** is said to have occurred. The **yield point** is thus identical with the elastic limit.

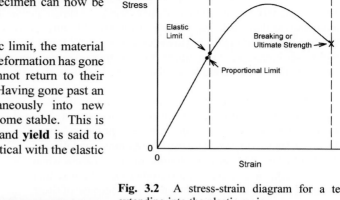

Fig. 3.2 A stress-strain diagram for a test extending into the plastic region.

●3.1   Necking

Past the yield point, the strain, or elongation of the specimen, may continue to increase steadily, although this will not normally occur uniformly over the entire length of the piece being tested but instead will be restricted to a short region called a **neck** (Fig. 3.3). No specimen could be entirely uniform and the unavoidable presence of microscopic defects of

Fig. 3.3 A neck in a tensile test specimen. Notice the rough surface over the neck (see 11§5.1).

one kind or another will mean that the yield point stress will be exceeded, and plastic deformation initiated, at one point only. The resulting plastic deformation takes the form of elongation in the axial direction. However, because the coherence of the material will tend to maintain the volume of the piece nearly constant, rather than let it increase indefinitely, this must result in a narrowing of the test piece in that region. This then is the neck. The cross-sectional area here is necessarily less than elsewhere in the piece, so that if a constant load were being applied, the stress at that location actually increases as necking proceeds (Fig. 3.4), and it is this local maximum stress that causes the ultimate failure of the test piece.

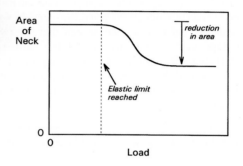

Fig. 3.4 The effect of necking on the cross-sectional area and true stress in a tensile specimen.

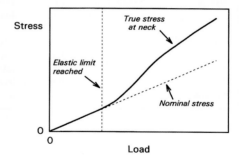

The experiment to achieve the above effect would be one in which the static load on the end of a wire is increased slowly and steadily, such as by adding lead shot into a bucket (Fig. 3.5). This is known as a **load-controlled** test, for the obvious reason: the resulting strain is the **dependent variable**. For any load giving a stress in the wire less than the elastic limit any extension is completely reversible but, most importantly, stable: nothing is expected to happen no matter for how long the stress is applied (but see Creep, §11). However, if the stress exceeds the elastic limit, even slightly, the corresponding reduction in cross-section due to necking results in a locally higher stress still (Fig. 3.4), causing immediate further deformation and, very rapidly, failure.

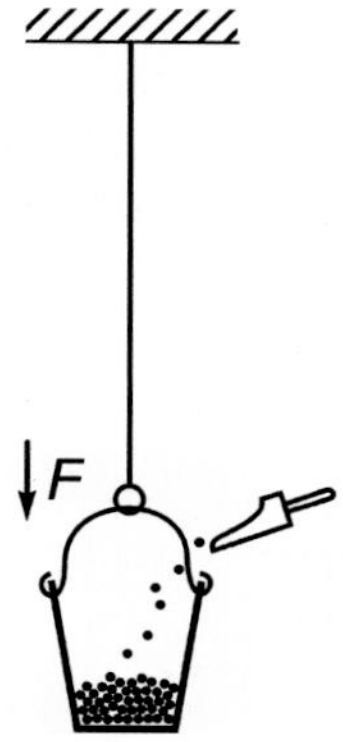

Fig. 3.5   The essential features of a load-control strength test.

On the other hand, the most common testing machines measure the load resulting from a particular applied specimen extension; this is therefore called a **strain-controlled** test and the load is now the dependent variable. This is done by a movable **crosshead** (Fig. 3.6), to which the test specimen is attached at one end, on a rigid frame to which the other end of the specimen is fixed. If desired, the crosshead can be stopped at any point in the test, even if plastic flow has occurred within the specimen, and the system will then be stable. Further motion of the crosshead then results in more deformation, and so on. It is during this stage of plastic deformation that the recorded load may actually level off or even fall (Fig. 3.2). This apparently strange behaviour needs to be explained.

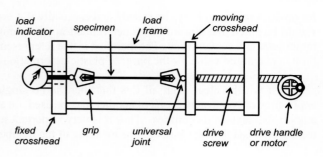

Fig. 3.6 The essential features of a machine for performing a strain-controlled tensile strength test.

•3.2 Types of strength

The **true stress**, that is having corrected for the cross-sectional area of the specimen at the narrowest point, is nevertheless always increasing (Figs 3.4, 3.7). The stress calculated on the basis of the original cross-section and the load at the moment when the test piece fails is called the **breaking strength**. This is quite distinct from the **tensile strength** or **ultimate**[5] **tensile strength** (**UTS**) which is calculated from the *maximum recorded* load and the original area (so-called 'compressive strength' is also calculated for maximum load and original cross-section; §6.3). The point here is that while the material is being deformed plastically the load may reduce, but without altering the total strain. In essence, the sample fails when the maximum strain that it can tolerate has been exceeded, which may be at an *apparent* stress lower than the recorded peak.

True breaking strength must be calculated from the actual cross-sectional area at the exact point of rupture (which is usually the cross-section minimum) and at the exact moment of rupture, and it is therefore very difficult to obtain. Instead, it is conventional (and much simpler) to record the apparent, "**nominal**" or **engineering** breaking strength, based on the original cross-sectional area and peak load. This may seem like cheating, but it is in fact a realistic measure of the true situation in service: if one asks what load will cause the object to fail, the answer corresponds to the peak observed during a test because we normally do not care about behaviour once past that point – the damage has been done.

The yield point sounds in principle as though it is a straightforward definition, but in practice it may be difficult if not impossible to identify on the stress-strain curve: the point of departure from elasticity cannot be seen at the time of the test even in a material that is Hookean up to the elastic limit in the same way that identifying a proportional limit is hard. Indeed, if a proportional limit exists, the problem is greater because deviation from linearity has already occurred. Permanent deformation can only be detected after unloading. The solution to all of this ambiguity is a compromise. A line is constructed parallel to the Hookean elastic region but offset to intersect the strain axis at some suitably large value; 0.2% strain is commonly used (Fig. 3.8). The (nominal) stress corresponding to where this line, when extended, intersects the stress-strain curve is then called the **proof stress**. The plastic **offset** used to identify this version of strength should,

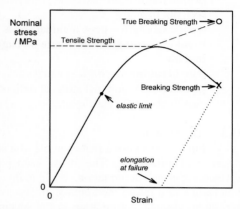

Fig. 3.7 The difference between the apparent and the true stress-strain curves. The solid line represents the stress calculated from the original cross-sectional area, even though the actual area is changing. The broken line represents the true stress, calculated for the actual cross-section at any moment.

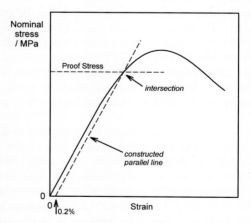

Fig. 3.8 The construction used to determine a proof stress. The offset used here is 0.2%, but other values, such as 0.1% or 0.5% may be encountered.

---

[5] Obviously, there is nothing 'ultimate' about such a value except it is a numerical artefact. It is an object behaviour, not a property.

for obvious reasons, always be stated,[6] and especially if it is called **yield strength**, as is sometimes the case. It should be noted that a proof stress determined in this manner is only an approximation, but one of practical importance and convenience. The intention is that it represents an upper bound for the value of the actual elastic limit, a maximum value for the onset of plastic deformation.

It will be clear from all this that there is no single definition of 'strength' and, depending on the application or convenience, one or another may be used. In addition, terminology and usage may vary between authors, contexts and countries. Thus it is very important to ensure that the definition or interpretation being used is checked to make sure that the meaning of the numbers is understood.

We can summarize the relevant terms, as defined here, as follows:

- **elastic limit** = **yield point** – stress beyond which plastic deformation occurs
- **proportional limit** – stress beyond which Hooke's Law can be seen not to apply
- **tensile strength** – peak nominal stress
- **breaking strength** = **ultimate strength** – nominal stress at rupture
- **true breaking strength** – actual stress at location of rupture
- **proof stress** – nominal stress at defined offset, intended to be clearly above the elastic limit.

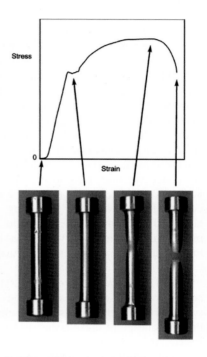

●3.3   An example

Figure 3.9 shows a real example of a stress-strain curve for a specimen of mild steel. The initial steep portion is very straight, representing elastic deformation. Yield in this case is very sharp, and even results in a fall in the nominal stress. Notice that at this point there is no obvious deformation in the specimen, just a loss of the originally shiny surface. There follows a long plastic region in which a pronounced neck develops, after which the nominal stress falls rapidly to the point of rupture.

It can be seen that the neck has caused a substantial change in length, *i.e.* the strain to failure is very large. Accordingly, the common measure of the ductility of metals in such tests is reported as the **elongation at failure** (Fig. 3.7). This is usually expressed as the **percentage elongation**, $\Delta L/L \times 100$, for the gauge length L (Fig. 3.1). However, it should be noted that this is not a material property in the strict sense as it depends on the gauge length (which is entirely arbitrary), while the necking is a localized event that has no 'knowledge' of the length of the specimen or the extensometer in use. Standardized procedures are used, and this requires complete reporting of conditions in order to be understood exactly.

Fig. 3.9 A tensile test result for mild steel, showing the deformation at various stages in the process of necking and failure.

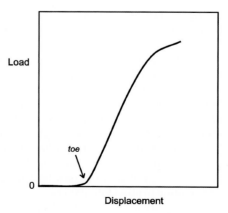

Another feature is of interest: the very beginning of the curve. Notice that the straight line elastic portion does not extrapolate through the origin. This is due to the imperfections in the testing machine and specimen grips and has nothing to do with the behaviour of the material itself. What may be termed the toe of the curve (Fig. 3.10) arises from the specimen bedding down into the grips, any slack in the joints or drive mechanism, or indeed in the load measuring system. Care must be taken to avoid mistaking the test system's imperfections for material behaviour in any experimental method.

Fig. 3.10 The effect on a load-displacement curve from a real tensile test in which there are imperfections in the load system.

---

[6] It is sometimes referred to as the **offset strength**, with the value of the offset stated, *e.g.* 0.2% offset strength.

§4. Work of Deformation

The stress-strain diagram also contains other information. It is easy to show that the area beneath the stress-strain curve is a measure of the work done in deforming the sample. Noting that the mechanical definition of work, W (J), is force times distance acted over, ΔL:

$$W = F \cdot \Delta L \qquad (4.1)$$

(Fig. 4.1, left). This is the area under the curve of the plot of force against distance. For a linearly varying force (Fig. 4.1, right), the corresponding area is that of the triangle beneath the line, that is

$$W = \tfrac{1}{2}F_{max} \cdot \Delta L_{max} \qquad (4.2)$$

This type of calculation also applies in the proportional elastic region of a stress-strain curve: the average force is one half of the maximum force applied. If this value of the work done is now scaled appropriately, by dividing the force by the cross-sectional area (*i.e.* stress) and the displacement by the original length (*i.e.* strain), effectively we have divided the total work done by the (original) volume of the specimen, V = A.L. This gives the work done per unit volume, U (J/m^3), at any point in the loading of the specimen, assuming linearity, *i.e.*:

$$U = \frac{F \cdot \Delta L}{2A \cdot L} = \frac{\sigma \cdot \varepsilon}{2} = \frac{\sigma^2}{2E} \qquad (4.3)$$

This quantity, U, is known as the **strain energy density**, and represents the *recoverable* stored work in the system.

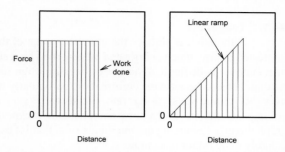

Fig. 4.1 Calculation of work done: (left) constant force, (right) linearly varying force. Distance means the displacement, ΔL, of the point of application of the load, F.

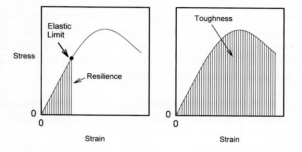

Fig. 4.2 The recoverable, elastic work of deformation called resilience (left) and the total work to failure called toughness (right). It is assumed here that the elastic limit is identical to the proportional limit.

It follows that the area beneath the curve up to the elastic limit is a measure of the maximum elastic energy which can be stored by, and which is therefore recoverable from, unit volume of the test piece: this is called the **modulus of resilience** or, simply, the **resilience** (Fig. 4.2). It is thus a material property rather than referring to a particular object. Similarly, the area under the curve up to the breaking point is a measure of the total energy which has been put into the specimen per unit volume up to the moment before rupture; this energy is known as the **toughness** (sometimes called the **modulus of toughness**), although this cannot be calculated easily. The value is easily expressed, however, as the integral of the stress-strain curve (which is effectively a normalized force × distance integration), and this can be obtained through suitable software or electronic hardware. Notice that the difference between the toughness and the resilience is the **plastic work** to failure, the *irrecoverable* part of the total work done (again, per unit volume).

The **modulus of resilience**, R (units of stress, Pa), is easily calculated as the area of a triangle, if the appropriate units are used, but *only* if the material is ideally Hookean:

$$R = \frac{\sigma_p \varepsilon_p}{2} = \frac{\sigma_p^2}{2E} = \frac{\varepsilon_p^2 E}{2} \qquad (4.4)$$

where the subscript p refers to the values of stress and strain at the proportional limit. Thus, this is true exactly only if the proportional limit and the elastic limit coincide. If this is not so then due allowance must be made for the non-linearity of the deformation. Since R represents the work done in deforming unit volume of the test material to the elastic limit, multiplying the modulus of resilience by the volume of the sample gives the total work done, again assuming uniformity of deformation. However, this roundabout route is not essential. It is easier just to measure the area under the force-deflection curve, whatever its shape.

For comparison, we can note that the area under the dotted line in Fig. 3.7 represents the elastic energy stored in the specimen and released when it breaks.

●4.1 Toughness

Although it is perhaps the more important of the two energy measures, because very frequently one is interested in how much energy can be absorbed by a structure before it fails, the calculation of the **toughness** of a material is difficult. This arises because often the shape of the stress-strain curve to failure is not very regular, but also because the deformation, particularly the plastic deformation, will not be uniformly distributed over the volume stressed, but rather tends to be very localized, *i.e.* in a neck. This makes stress, strain and volume calculations much more complicated. It can, of course, be estimated by measuring the area under the graph directly (assuming engineering stress, that is) based on the original cross-sectional area, but clearly this is likely to be rather inaccurate.

It will be seen that the above definition of toughness includes the resilience. In other words, when the specimen fails, the energy that went into plastic deformation is lost, but the elastic energy must now be accounted for. Of course there will be a recoil, just as a spring or elastic band will recoil on snapping. Often, the amount of energy stored in this way in strong materials can be considerable, and since this potential energy will be converted to kinetic energy, the velocity of the recoil may be a hazard to the operator of the testing machine. In fact, since the testing machine cannot be made perfectly rigid (*i.e.* there is no such thing as an infinite elastic modulus), much energy will also be stored in the frame of the testing machine and the specimen grips. Furthermore, if the test is of a material in compression, brittle failure – rapid cracking – may result in fragments of the test piece being ejected at very high velocity. Eye-protection is thus essential when performing such tests, and great care exercised at all times.

There are other definitions of toughness. The one just given is essentially a bulk property of an object, despite the scaling by volume, since the portion actually deforming is not controlled or measurable. It thus does not really meet our requirements for a specimen-free measure: it is not a material property (unlike resilience; *cf.* percentage elongation, §3.3). A second definition refers to the energy required to propagate a crack, and thus is measured in terms of the area of the crack that is formed. In principle this is far more reasonable, even if there is surrounding plastic deformation, since the crack area is measurable. We shall return to this idea below (§7) and several times later on. (The more commonly used but difficult calculation in engineering contexts is called **fracture toughness**; see 29§5.2.)

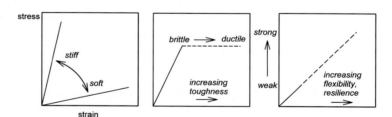

Fig. 4.3 Variation in the shape of stress-strain curves conveys much information about material behaviour.

●4.2 Stress-strain curves

The appearance or shape of the stress-strain curve can tell us much about a material (Fig. 4.3). Thus we may characterize its mechanical properties by taking note of
- **strength** – *e.g.* stress at failure (weak *vs.* strong)
- **stiffness** – slope of elastic portion (soft *vs.* stiff) or, conversely, the
- **maximum flexibility** – elastic range, maximum elastic strain (flexible *vs.* stiff or rigid)
- **ductility** – length of the plastic range (ductile *vs.* brittle), or its converse ...
- **brittleness** – the absence of plastic deformation (brittle *vs.* plastic)
- **resilience** – area beneath elastic portion
- **toughness** – total area beneath the curve (tough *vs.* brittle)

The difference between brittle and ductile behaviour may be seen clearly in Fig. 4.4.

Fig. 4.4 Examples of brittle failure (bottom) and ductile failure (top) in steel specimens.

It will be noticed that some terms do not have clear opposites, and this is in part due to the use of ordinary terms for technical concepts. Thus care is necessary in the use of these ideas to ensure that the correct meaning is conveyed. The relevant definition should be given clearly for any given usage.

§5. Modelling

It is frequently easier to discuss the behaviour of materials, when this is complicated, in terms of simpler 'ideal' components of behaviour. We may therefore construct convenient **mechanical analogues** for the basic types of idealized behaviour of materials as shown in their stress-strain curves. By a 'mechanical analogue' is meant a model which can mimic some aspect of the behaviour of a material; it is no more than a model, and should not be taken to imply anything about the structure of the material itself or the internal physical process producing the real behaviour.

There are two distinct kinds of model. A notional model may be built of conventional parts or **elements**, whose individual behaviour is well understood, and used to illustrate the kinds of process that may be considered to be occurring; this is what is represented by the mechanical analogue. Physical modelling, on the other hand, that is building a replica of the structure that will go into service (often on a smaller, sometimes on a larger scale), is a quite different proposition. We shall consider each in turn.

A further word of caution is appropriate before continuing. A model abstracts key features (or attempts to), often in an idealized or fitted fashion, for the purposes of study and an aid to thinking. They cannot be, and should not be taken as, an absolute representation of the real world. Thus, they cannot be pushed beyond their intended, usually illustrative, purposes to determine the behaviour of a system beyond the limits of the model: extrapolation is not possible. Bear in mind that "Essentially, all models are wrong, but some are useful." [2]

●5.1   Notional modelling
A simple spring, anchored at one end, may be taken to represent a perfectly elastic but brittle material. The corresponding stress-strain plot is a single straight line, as would be expected from Hooke's Law. The representation is meant to be that of a simple coiled spring, loaded in tension, reflecting perhaps the earliest experiments, but in fact it can stand for any loading system whatsoever – it is simply a conventional representation. The slope of the graph is, of course, the elastic modulus (Fig. 5.1a). The idea embodied here is that a spring returns to its unloaded state when the load is released.

A material which is perfectly rigid (E infinite) but with perfectly plastic behaviour at a definite yield point (σ) may be modelled by a **static friction** element (see §5.5), represented by a block resting on a horizontal surface (Fig. 5.1b) (the possibility that the **dynamic friction** may be different is simply ignored in this model). This is to be interpreted as showing continuous permanent deformation above the yield point, but no deformation, elastic or plastic, below that stress. The image used is that a block will move across a surface

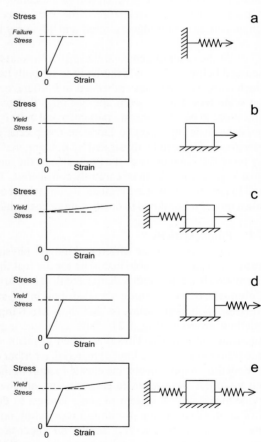

Fig. 5.1 The modelling of idealized stress-strain curves by mechanical analogues.
a: perfectly elastic, brittle
b: rigid, perfectly plastic
c: rigid, linear strain hardening
d: elastic perfectly plastic
e: elastic, linear strain hardening

when pulled, but it will not spontaneously return towards its starting position when the pulling stops.

The phenomenon of **strain hardening** (11§6.4), which is a very common behaviour of metals and polymers, is attributable to the atomic or molecular rearrangements which occur during plastic deformation. The rearrangements increase the resistance to further deformation. Such behaviour may be simulated by a static friction element which is restrained by being anchored through a spring (Fig. 5.1c). The spring cannot be stretched until the friction element block has started to move, but then the resistance to further displacement increases steadily as the spring is extended.

By other combinations of these basic elements such behaviour as elastic, perfectly plastic (Fig. 5.1d) and elastic, linear strain hardening (Fig. 5.1e) may be modelled. Indeed, more complex models are possible, and they can sometimes serve useful purposes of illustration, but a little caution is required in that real materials can be far from ideal in the sense of these models or, conversely and more to the point, that such models are far from ideal in representing many real materials. Non-linear behaviour, in particular, is very hard to represent like this.

●5.2   Physical modelling
It is a basic principle that material properties such as strength and deformation be expressed in terms that are independent of test specimen dimensions, thus permitting data to be tabulated in a convenient and universal form. This was the basis of the development of the ideas of stress and strain themselves (§2). Accordingly, it might be expected that any size of test piece would be satisfactory to determine a material's properties, whether the object to be built was a watch part or a locomotive. To a large extent, it is necessary to rely on this idea as being a satisfactory approximation to the truth, especially when the difficulty (and cost) of testing to destruction such objects as road bridges, aircraft and 40-storey buildings is contemplated.

At the other extreme similar difficulties arise. The problem is sometimes only that of the size effects discussed below (§7), but if the body to be built has a **composite structure** (6§1.13), for example concrete (which consists of pebbles embedded in Portland cement), it clearly would be meaningless to test a specimen on a scale less than (or even approaching) that of the pebbles. Thus, in dentistry, where the devices and restorations to be made are on a particular and quite small scale, from materials that are almost without exception composite, it is important to fabricate test pieces on a similar scale if service behaviour is to be properly understood. This can be illustrated by noticing that the width of a clasp on a cobalt-chromium partial denture may be of the same order as the grain size of the metal, and the idea of a polycrystalline test piece mentioned above (Fig. 2.12) needs to be carefully considered. Thus, it is possible to have specimens either too large or too small to give results that are realistic in the context of the service conditions. Even so, it is frequently useful to **scale** a model up or down to create a more easily handled model system.

●5.3   Photoelasticity
There is a rather different kind of physical modelling which substitutes readily available materials for the real thing in order to mimic deformation behaviour rather than strength. Thus, in using the phenomenon of **photoelasticity** to study the distribution of stresses in structures, transparent polymeric resins (which show **birefringence**) are used as the modelling materials (Fig. 5.2). This technique is particularly useful for complicated structures. However, when in practice the real structure is made from more than one kind of material, a further dimension is added to the modelling requirements: the elastic moduli of the modelling materials must be in the same *ratios* to each other as are the moduli of the actual service materials. A system of some interest in dentistry is that of the distribution of stress due to biting forces acting on an artificial denture. Here we have, say, acrylic acting on mucosa (which is far from homogeneous), itself overlying cortical bone with its core of cancellous bone. Not only must the dimensions of the denture be reproduced (or scaled), but also the relative thicknesses of the mucosa and bone cortex, as well as the elastic moduli of all four bodies. Even considering the simpler system of an amalgam restoration in a tooth presents a significant problem to the modeller if the results are to be meaningful.

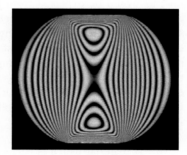

Fig. 5.2 Fringes in a photoelastic model indicate contours of equal shear stress, counting from zero at the left and right edges. This is a circular disc (of a polyurethane rubber) being compressed across a diameter between flat platens (*cf.* Fig. 6.3).

●5.4   Finite element analysis

The difficulties of physical modelling may be overcome (to a large extent) by calculating the stresses and strains in loaded structures. Essentially, this is based on the structure being divided up into a large number of discrete segments, each of which is effectively represented as spring and friction elements as in Fig. 5.1. Each is connected to its neighbours in a large network (Fig. 5.3). The advantages of this approach are that complicated shapes may be handled, and an indefinite number of elastic moduli may be represented. In addition, arbitrary non-linear behaviour may included by assigning appropriate equations to describe the behaviour of elements, while symmetry can simplify the model enormously by needing to deal only with one slice, as in Fig. 5.3. The disadvantages include the difficulties of deciding the interconnections and the resolution of the network (and thus computer time and power) that are necessary for an adequately precise and reliable answer, as well as knowing the values of the various properties sufficiently accurately. This approach can also be applied to heat and fluid flow problems in complicated systems, which are in fact somewhat easier to handle. These techniques have been applied to various dental problems, such as the deformations of teeth and bones as well as metal casting and cooling.

Fig. 5.3 Finite element analysis is based on the behaviour of a mesh, each segment of which is assigned a set of mechanical properties. This example represents a simplified radial section of a tooth and crown.

●5.5   Friction

For two bodies in contact, but stationary, there is a resistance to the sliding of one on the other (measuring in the plane of the common tangent interface). Thus, a body with a weight W (the 'normal' or perpendicular force) resting on a horizontal flat surface requires a force F to get it to move (Fig. 5.4 a). At the point of movement, the reaction force opposing F is the friction. This resistance may be due to the mutual interference or interlocking of roughness, the formation of chemical bonds, welding or other form of interaction, although its exact nature is not yet fully understood. Despite this uncertainty, there are two general '**laws of friction**' describing the relationship between the two forces:

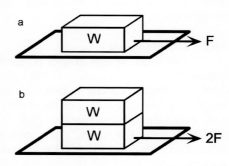

Fig. 5.4   Friction is the force opposing the sliding of the blocks.

1.   The force F is independent of (*i.e.* unaffected by) the area of contact – referring to the nominal or macroscopic, geometrical area, not the sum of the microscopic areas of actual physical contact. Thus for a block like a house brick, where all three edges are different in length and thus the faces are of different areas, it does not matter whether it is stood on end, on its side, or even balanced on an edge: F has the same value. We might say that spreading the load has no effect on F.

2.   F ∝ W. In other words, stacking blocks one on the other increases F by the same amount for each block added (Fig. 5.4 b). This proportionality is used to define the (dimensionless) **coefficient of friction** relevant to the interaction of the two materials:

$$\mu = \frac{F_\|}{W} \qquad\qquad (5.1)$$

where we introduce the distinction that $F_\|$ means the component of force acting in (*i.e.* parallel to) the direction of motion. For most systems, the value of μ lies in the range $0.1 \sim 1$ and only exceptionally will smaller values be encountered. As described, μ represents the coefficient of **static** friction, but friction still applies in a system in which the bodies are already in relative motion, and a **coefficient of dynamic friction** may be defined in similar terms.

Static friction represents the shear stress required for the collapse of a system otherwise in equilibrium and so provides a convenient conceptual model for the collapse occurring at a yield point (§5.1). On the other hand, dynamic friction represents work done in obtaining the relative movement of two bodies – force times distance. This work is ordinarily mostly dissipated at the interface as heat, hence frictional heating, but equally the dynamic system may be used as a model to illustrate the work of plastic deformation.

Friction is both good and bad. Without it we would not be able hold things in our hands, retain burs in chucks, or even stay in our chairs. However, in dentistry, friction and the consequent heating is the cause of many disadvantageous effects, whether the wear of moving parts or thermal changes in tissues and materials. Some of these issues are dealt with later, but for now we can simply note that the work done Q is force × distance acted over:

$$Q = F\| \times d \tag{5.2}$$

ignoring accelerations (including changes of gravitational height). This work must appear as heat at the interface. Thus, it is clear, substituting for $F\|$ from equation 5.1:

$$Q = \mu W \times d \tag{5.3}$$

that reducing μ reduces the work done, the heat generated, and the temperature rise caused, which is one reason why proper lubrication of moving parts is so important.

§6. Strength Testing

A variety of tests are available for gathering data on material behaviour. The advantages and limitations of each must be understood if the results obtained are to be interpretable.

•6.1 Direct tension

Direct tensile tests may be performed on wires, or on specially-shaped specimens. These latter may be either a dumb-bell shape of circular section, which is often particularly suited to testing cast alloys, or the similar 'dog-bone' shape with a rectangular cross-section (Fig. 6.1) which can be machined from flat sheet. The end regions are made substantially larger in cross-section to withstand the forces of gripping and to avoid failure occurring within the grips. The transition from the ends to the central portion is made gradual, with well-rounded corners that avoid stress-concentration effects that would otherwise cause failure there. Failure outside of the uniform central region of the specimen would provide no information except that the specimen or the gripping method was faulty. Gripping wires thus presents several problems which require careful technique to avoid.

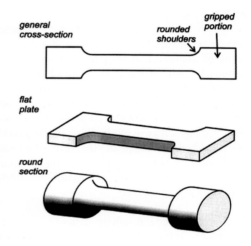

Fig. 5.5 Round and flat section test pieces for direct tensile tests (see also Fig. 4.4).

The calculation of the (nominal) tensile strength is straightforward for such a test, the ratio of the force acting at the moment of failure to the (nominal) cross-sectional area over which it is acting:

$$\sigma_t = F/A \tag{6.1}$$

•6.2 Indirect tension

Certain groups of materials are not amenable to being tested with direct-tension specimens such as these because of their extreme brittleness (when slight malalignment of specimen or machine would cause premature failure), difficulty of fabrication in these shapes, or merely their cost if large quantities are required. In an attempt to avoid these problems, the so-called **diametral compression** or **indirect tensile** test has been used. In this, a small cylinder of the material, which may be only a few millimetres across, is subjected to a compressive load across a diameter (Fig. 6.2).

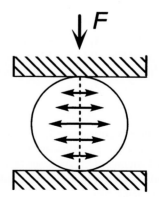

Fig. 6.1 How diametral compression is envisaged ideally: tension acting smoothly over the entire diameter, peak at the centre. This never happens.

A simple theoretical analysis[3] has been used to support the idea that this causes a failure of the piece in tension as if such a stress were applied across that diametral plane. This result can be qualitatively understood by noticing that for a direct axial load there is a complementary deformation defined by the Poisson ratio. Thus, there will be an expansion across the diametral compression specimen. This expansion is supposed to be entirely equivalent to that which would be produced by a tensile stress applied in the same direction. However, the true state of the piece under stress is rather more complicated. The initial line of contact top and bottom, being only a line and therefore of zero area, is subjected to a finite load, and so the stress there is infinite. This is expected to give immediate failure. Premature failure of very brittle materials, such as dental porcelain, occurs because of this.

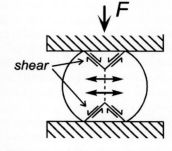

Fig. 6.2 Crushing at the points of contact in a diametral compression test redistributes the stress and causes failure by initiating cracking in shear. Note the two wedges, top and bottom. Cracking starts at their tips.

With less brittle specimens, such as dental silver amalgam, cements, or polymer-based materials, a flat area develops as the material deforms plastically (Fig. 6.3) and this leads to a serious deviation from the assumed pure tension across the diameter. A pair of **shear prisms** is developed, and these may crudely be envisaged as wedges driven into the sample to generate the tensile forces which operate across the remaining area of the diametral plane. However, it is at the apex of these regions that the shear stress is a maximum and thus becomes dominant. The location of these shear stress peaks can be seen in Fig. 5.2; they lie at the centres of the two loops.

It now appears that failure is actually initiated at these points, and in shear. It is therefore not true to say that this test measures tensile strength, it seems rather to be a measure of shear strength instead (even though it is not yet known how to calculate this).[4] While this makes interpretation difficult if not simply impossible, thus confounding one of the basic principles of testing, it is probably true enough that *comparative* measures of strength in the sense of load **bearing capacity** (§1.1) in this mode of loading are thereby obtained, although these can only be really useful within a single type of material such that the flattening along the contact lines is unchanging (which is, however, frankly impossible). The use of the indirect tensile test therefore in examining the effect of changing mixing conditions for a restorative material, for example, might just be acceptable, but the comparison of, say, glass ionomer and filled resin products is more dubious. Even so, the use of any soft material between the test piece and the platen changes the stress state even more and so is disallowed.

The equation used for "indirect tensile strength" is:

$$\sigma_{it} = \frac{2F}{\pi DL} \qquad (6.2)$$

where L is the length, and D the diameter of the specimen, although it is emphasized that this is not the stress causing failure. Broadly, there is no point to using this test.

●6.3 Compression

A similar analysis can be applied to the case of a cylindrical test piece loaded in compression across the end faces (Fig. 6.4). This is the commonly-used, so-called **compressive strength** test. It is not fundamentally different from the "indirect tensile strength" test. This time the flat contact areas between the specimen and the machine's anvils or platens are already present. The shear cones generated at either end may be envisaged as resulting in tensile forces acting outward from the central fibre of the sample. Again, the strength is ordinarily given by the load at failure divided by the cross-sectional area:

$$\sigma_c = F/A \qquad (6.3)$$

The problem is that no specimen can actually collapse directly under compression, unless of course it is a foam-like or open network structure. To understand this, consider hydrostatic compression (Fig. 2.11): clearly, no failure of the kind envisaged can occur. Essentially, there is no such thing as

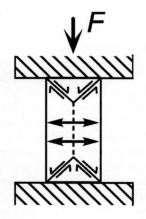

Fig. 6.3 The stresses and causes of failure in a cylindrical specimen loaded axially are no different from those in the diametral case except that the pattern is radially symmetrical.

compressive strength as a material property (the old term "crushing strength" would be preferable). Thus, in the axial loading case, cracks must form that permit the radial movement of fragments if the volume of the piece is to be preserved (even allowing for elasticity and Poisson strain). Thus, applying the same very simple analysis that was used to justify the indirect tensile strength test above, a cylinder in axial compression would also expected to fail in tension. Unfortunately, this too is wrong, and it is probable that failure occurs in shear in this case as well, making the two kinds of test essentially indistinguishable. This (not surprisingly) accounts for the usually very high correlation observed between the results of the two tests. More study of the meaning of such tests is very clearly urgently required. Meanwhile, published strength data derived from these tests must be used with caution except, as mentioned above, for comparison purposes within a type of material. In the sense of **bearing capacity**, 'compressive strength' has an interpretable meaning for this mode of loading in actual service, assuming that such an arrangement is ever used (which is probably not the case in dentistry). Even so, care is required: for a given cross-section the failure load depends on specimen height, which is therefore standardized for laboratory tests (again, this shows that it is not a material property). As above, any padding only complicates the system and its interpretation.

●6.4   Flexure

A rectangular bar-shaped or beam specimen may be tested as shown in Fig. 1.2. This is known as a **three-point bend** test (Fig. 6.5, top), and the corresponding **flexural strength** is given by

$$\sigma_{b3} = \frac{3FL}{2bh^2} \tag{6.4}$$

where L is the length of the specimen between the supports, h is its height (*i.e.* in the direction of the load axis) and b is its breadth (*i.e.* perpendicular to the page in Fig. 6.5). The maximum tensile stress is generated just opposite the load point. A similar **four-point bend** test may be conducted with the upper load divided equally at two points, each distance c from the supports (Fig. 6.5, bottom):

$$\sigma_{b4} = \frac{3Fc}{bh^2} \tag{6.5}$$

which has the advantage of testing the central portion in pure bending (23§2.5).[7] Both tests cause failure in tension, but in the four-point case a larger volume is involved. However, a difficulty with this general type of test is sensitivity to the surface conditions on the convex, tension face. That is, the presence of surface flaws such as porosity or scratches controls the observed strength. The larger effective test volume of the second example also means that a larger sample of flaws lies in the critical region. This brings us to the Griffith criterion.

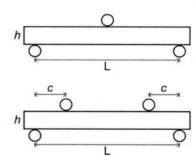

Fig. 6.4 The arrangement for flexural testing. The bar test pieces rest on rollers, with the load applied similarly through one on the centre line in three-point bend (top), or two equally spaced from the ends in four-point bend (bottom).

§7. Griffith Criterion

It has been established by many tests that the actual tensile strengths of all materials fall far short of the theoretical values, that is, taking into account electrostatic forces, molecular bond strengths, and so on. The factor is often about 1000 times between theory and practice. The reason is postulated to be structural imperfections. It is a matter of common experience that chipped or cracked cups are much easier to break, that scratches are used to facilitate the cutting to size of glass tubes and window panes, and that even a carrot is easier to snap if first a thumbnail has been used to make a notch; these are macroscopic demonstrations of the same effect.

It is an experimental as well as a theoretical result that for a crack in a brittle material, subjected to an increasing axial stress (Fig. 7.1), at a particular **critical stress** the crack will grow spontaneously and lead to

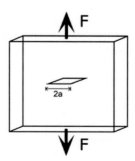

Fig. 7.1 A crack in a test piece in tension. Stress concentration at the crack tips is the cause of crack growth.

---

[7] The derivation of these equations is dealt with in 29§3.1.

the failure of the test piece, whatever its width.[5] The larger the crack, the smaller the stress that will cause it to grow spontaneously. These ideas are embodied in the **Griffith criterion**, which is expressed in the following equation:

$$\sigma_{crit} \cdot \sqrt{a} = \sqrt{2E\gamma/\pi} \qquad (7.1)$$

This relates the **critical stress** (σ_{crit}) for a given (flaw) crack size (a), the Young Modulus (E) and the **surface energy** (γ) of that material.[6][7]8 This latter factor is involved because in the propagation of a crack two new surfaces are being formed, which require work of formation to be expended (*cf.* 10§1). The amount of work is simply the product of the (specific) surface energy, in J/m^2, and the total new surface area created:

$$W = 2A\gamma \qquad (7.2)$$

assuming that A is the exact area of the cross-section used for calculating the stress, regardless of its irregularity or roughness, that is, there is no toughening due to any other energy-consuming process. The sudden, catastrophic growth of the crack can be understood in very simple terms: the amount of energy stored elastically in the specimen just exceeds that required for the creation of the (minimal) new surfaces.

It is emphasized that the Griffith equation as such, as an exact criterion for collapse, is applicable only to brittle materials. If the material has any appreciable ductility at the temperature of the test, there will be plastic deformation at the crack tip, and this process will therefore consume energy in excess of that required to create new surface at the fracture. In addition, the external surface area of the specimen is increased when there is plastic deformation (Figs. 3.3, 3.9, 4.4; 3§5.5) – in a complicated fashion, and the work which must be done to create this will be correspondingly hard to calculate. Of course, and structural aspect of the material that causes the crack to deviate and so increase its area also increases the work done. Thus, the surface energy term in equation 7.1 is inadequate on its own when any of these processes occur. Any process, in other words, that dissipates the elastically-stored energy tends to prevent catastrophic failure. This is a major reason for requiring that materials should be tough and not brittle if they are not to be unduly sensitive to defects of manufacture or preparation, or to accidental damage in service. This is also the basis of deliberately designing materials that encourage crack blunting in order to prevent brittle failure. To re-emphasize the point, notice the underlying meaning of the surface energy term: it is the work required to break the bonds between atoms and molecules that hold the material together. If there is any other means of consuming energy (*i.e.*, of doing work), the critical stress will differ from that indicated by equation 7.1. Nevertheless, even in plastic materials the presence and size of flaws is the controlling factor in determining failure. That is to say, the concept of a **critical flaw** causing collapse at some stress is of general applicability provided that the rate of release of stored energy is greater than the rate of consumption by all active processes, it is just that now we cannot calculate its size so easily, if at all. In other words, the **Griffith criterion** is a precise statement only for brittle materials that fail promptly on crack initiation in a strain-controlled (§3.1) context.

Because the right hand side of equation 7.1 is a constant for the material, it shows that the stress for crack propagation is inversely proportional to the square root of the (flaw) crack length (Fig. 7.2). Strength is therefore greatly sensitive to flaw size. The observed strength of glass corresponds to a "crack" size of about 1 μm. This is of a size that ought to be detectable, except that the width of the defect is presumed to be of the order of just a few atomic diameters at most, say 5 nm, but perhaps only one; it would be quite invisible except with an electron microscope, although not necessarily capable of being imaged even then since we are effectively talking about defects in the bonding structure. This then makes them 'virtual' or 'imaginary' cracks, but the structural defect is real.

It is this kind of thinking that leads to the idea of the limiting, theoretical tensile strength of a perfect single crystal of any material:

$$\sigma_{th} = \sqrt{E\gamma/a_0} \qquad (7.3)$$

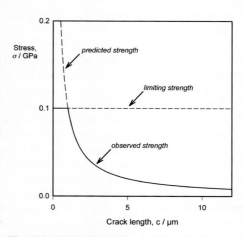

Fig. 7.2 Critical (experimental) crack size and critical stress (*i.e.* at failure) for a glass.

---

8   Griffith's original paper had errors; Sneddon corrects and develops the ideas. The core principle of eq. 7.1 is sound.

where a_0 is the interatomic spacing – a dimension which is clearly a lower limit (a_0 takes the place of the crack length a in equation 7.1; the factor $2/\pi$ there arises from geometrical considerations). Now it can be seen that any structural imperfection on a scale greater than a_0 inevitably reduces the strength of the material. Perfection is a seriously difficult proposition (perhaps even quite impossible), even in single crystals. How much more problematic, therefore, are the ordinary materials of commerce and those prepared or treated in the laboratory and clinic can perhaps now be appreciated.

Equations 7.1 and 7.3 are not intuitive. Let us examine a physical approach to the same issue by considering a tensile test on a cubic element as in Fig. 7.3. From equation 4.3 we can write down the specific elastic work done on a body, so that the total stored energy now is:

$$W = U.x^3 = \frac{\sigma^2}{2E}.x^3 \qquad (7.4)$$

(assuming it is Hookean). The minimum energy required to fracture the piece, if brittle, is given by equation 7.2. Hence, we can write that if the condition

$$\frac{\sigma^2}{2E}.x^3 \geq 2A\gamma = 2x^2\gamma \qquad (7.5)$$

is satisfied, failure must occur, but only if crack growth can be initiated. That is, if there is more stored energy than is required to 'pay' for the new surface that will be created, then the failure is unavoidable if a means for using that energy is available: if a failure path exists, it will be taken. Therefore, we have

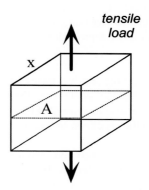

Fig. 7.3 A body, of dimension x, cross-sectional area A, under tensile load.

$$\sigma^2.x^3 \geq 4Ex^2\gamma$$

$$i.e. \quad \sigma^2.x \geq 4E\gamma$$

$$or \quad \sigma.\sqrt{x} \geq \sqrt{4E\gamma} \qquad (7.6)$$

Again, this differs in a detail from equation 7.1, and although not a rigorous treatment, it illustrates how a condition for brittle failure may be recognized: that there is sufficient energy ($U \times V$) stored in the system to allow the fracture surface to form if that crack can be propagated. The converse is perhaps clearer: unless there is enough energy to form the surface, it cannot be formed.

What the Griffith criterion for failure is about is the condition required for a pre-existing crack to grow spontaneously. Evidently, more energy can be stored in a test system (specimen and test machine) than is required to pay for the crack if that fundamental condition is not met. That is why many brittle specimens under test shatter into many pieces – often with high velocity: the excess elastic energy appears as extra surfaces formed and kinetic energy (and sometimes quite loud noises). Safety glasses are recommended when performing such tests.

Surface energy, upon which the critical crack size is dependent, is itself dependent on a number of things, not just of what the material is made. For example, with glass, atmospheric moisture and even oxygen are adsorbed onto the surface, which is enough to reduce the surface energy. But, more critically, the surface may be etched in the process, producing minute surface imperfections. Laboratory glassblowers take advantage of these effects when they wet the scratch prior to breaking the glass: the surface energy of the glass under the film of water or saliva (whose proteins will be adsorbed) is lower than when in air and the initiation of the crack is then very easy, requiring little force. In addition, the crack tends to follow the scratch because propagation through the scratch is easier than in the intact material at the side of it. Raising the temperature has a similar effect, generally lowering surface energy. Thus, with dental materials it is very important to test under conditions closely representative of actual service conditions, including such things as temperature, humidity, and the pH of, and chemical species present in the saliva or other fluids to which those materials may be exposed, as well as roughness and internal structure. Mechanical tests which fail to take these into account may not yield results interpretable in the service context. Even so, there may still be value in tests of a comparative nature when actual service conditions are very difficult to duplicate.

●7.1   Size effect

Just as in a chain the weakest link determines the test strength, the largest defect is the limiting factor to the observed strength of a test piece. In addition, there is a larger probability of finding a large defect in a large test piece simply because there is more room for it – it becomes *possible* (flaws larger than the specimen cannot occur). However, if the flaws in a specimen are considered to be obtained randomly from some parent distribution of flaws, so many per unit volume, then there is a greater chance of a flaw of a given size somewhere in the piece because that flaw sample is numerically larger.

Thus, it is a common observation that large test pieces tend to fail at lower (nominal) stresses (not loads) than their smaller counterparts. This is known as the **size effect**. For example, glass fibres show steadily *increasing* measured strength as their diameter is reduced, simply because of the impossibility of a large flaw being confined in a narrow fibre. Again, simulation of actual service conditions by selecting an appropriate sample size is necessary. The actual *number* of cracks within a sample is in fact of little concern: one critical crack in an appropriate place is all that is necessary for failure at a given load.

The implications of this idea can be explored for the system depicted in Fig. 1.3. It should now be obvious that the actual failure load for the three objects individually, were they to be tested separately, are likely to be different: there will be a distribution of failure loads for specimens of a given kind. Hence, the one specimen with the lowest individual failure load is expected to collapse first in the simultaneous test, effectively lowering the strength of the combined system to that of the weakest component. Now, in Fig. 2.4 there are the equivalent of many small specimens, the weakest one of which controls the outcome. Again, the larger the specimen, the lower on average the failure stress will be because the chances of low-strength elements increase.

The glass fibre model also illustrates another important point. Suppose that a long glass fibre is tested in tension until it fails and the load is noted. If either of the two pieces is then itself tested, the failure load must be higher than the first value. The load for the failure of both of those two pieces must again be higher, and so on (Fig. 7.4). The first test failed at the weakest point in the entire length of the fibre – it could not do otherwise. This corresponds to the flaw whose critical stress was lowest. The failure in the second test must then occur at the flaw in that piece which then had the lowest critical stress – but this must be higher than the first value otherwise it would have failed in the first test. This 'weakest link in the chain' argument shows that there is a distribution of flaw sizes, and thus of critical stresses. The strength observed for random (previously untested) short sections of fibre will then show a range of failure stresses. No single value can be obtained, therefore, for the 'strength' of any material, in any form of test. Repeated tests are essential, and 'strength' expressed in some appropriate statistical fashion. Most commonly this is done by reporting the mean, but the standard deviation for the sample data provides a measure of the spread or dispersion of the values to be

Fig. 7.4 The weakest link of a chain fails first – all others must then fail at successively higher loads.

It must be noted that the assumption in the "weakest link" argument is that each 'link' is independent. That is, the failure at one point has no effect on the properties anywhere of the rest of the material. In particular, we have to assume a completely brittle material such that the regions away from the point of failure have not suffered any change, *i.e.* that the yield point has not been exceeded, and therefore that no permanent deformations have occurred at flaw crack tips, whether growth or blunting.

expected in practice and should also be reported. This kind of behaviour must be taken into account when designing an object, such as an amalgam restoration or a partial denture, with due allowance – a safety margin – for the risk of a low-strength critical flaw in an area or volume under stress in service. Typically, the margin allowed may be some 25 ~ 50% of the nominal 'strength' of the material, taking into account the expected service stresses. Without that allowance the risk of failure is too high.

There is a further consideration that must be taken into account when testing non-homogeneous materials: the scale of structure. Thus, a specimen must be representative and allow for averaging such that all cross-sections are similar (*cf.* Fig. 2.12). Thus, in a specimen of concrete of a size suitable for laboratory testing a single pebble could represent an appreciable fraction of the width of the piece, yet be 'small' on the scale of a beam or pillar in a building. The properties of the pebble (and its interface) would have a disproportionate effect on the test result.

●7.2   Compressive *vs.* tensile strength

Brittle materials are well known to be much stronger in compression than in tension. This is because under a compressive load a transverse crack will tend to close up and so could not propagate. But defects such as cracks are likely to be randomly orientated with respect to the load axis, and because the stresses can be resolved into compressive and tensile components, regardless of the nominal mode of loading, and the presence of both stresses at a point is equivalent to a shear stress, there will always be some cracks in favourable orientations to propagate. In fact, the limit is set by the shear mode of failure (*cf.* §6), and the corresponding theoretical critical shear stress is about 8 times larger than the critical tensile stress, *i.e.* the material would appear about 8 times stronger in compression. This can be only an approximate guide because in practice other factors may play a part, depending on the material, but it appears to be a generally valid approximation.

Perfectly brittle materials are nevertheless very rare (if not actually impossible – see 29§2), the great majority of materials showing some appreciable plastic deformation at the crack tip even if generally appearing to be quite brittle. Of course, that plastic deformation itself requires work, and if the attempt were made to calculate surface energy from the total work of fracture, serious errors would be made. Put another way, the creation of new surface under most conditions and for most materials requires rather more than the theoretical quantity, and this work increases rapidly as the material becomes more obviously plastic or ductile.

The discussion above has been in terms of cracks. This is perhaps an unnecessarily restrictive term: *any* flaw or defect in the structure can act as a crack in the Griffith sense. Thus porosity, an unbonded interface between a filler particle and a resin matrix, a weak grain boundary in metals, or even a weak inclusion in a structure, and in the limit just one missing bond, can each act as a flaw in this sense. It does not have to be a real crack as it might ordinarily be understood, *i.e.* an open crevice resulting from the fracture of a body.

In conclusion, and taking into account remarks made in §6 about the uncertainty of mode of failure, we can see that strength tests in general of more or less brittle materials may be described as determining the nominal stress leading to a **critical flaw-dependent collapse**. What we mean, therefore, by the strength of a material must be carefully considered: it is not a substance strength (this would have no meaning, Introduction Fig. 1.2), but it refers to the effect of substance and structure in combination: the term 'material' must subsume all such aspects.

§8. Hardness

While strength itself is of great importance in determining a material's service performance, bulk properties are only part of the story. The surface behaviour of a restoration, for example, is relevant to considerations such as its ability to be polished, its retention of that smooth surface, *i.e.* withstand scratching, as well as to withstand the stress due to an opposing cusp – that is, avoid fracture or deformation under a concentrated load, and these aspects are not easily modelled by a simple axial loading test.

To obtain data about **hardness**, as it is called, a number of techniques are available (the word 'surface' as a qualifier for this is redundant as only the surface is accessible – there is no other kind of hardness). **Indentation hardness** is determined by applying an indenter of specified geometry to the surface under a predetermined load and, from a measurement of the width of the indentation (d) or its depth (t), its area is calculated (Fig. 8.1). From this is then calculated the stress which was present. Such a **hardness**, expressed in stress units, is effectively a measure of strength. It corresponds to the stress that the material could just support at equilibrium without further deformation, that is to say, a kind of yield point. Simply put, the smaller the indentation, the greater the hardness.[8] Indeed, there is a general association between hardness (H) values and the yield stress (Y) in tension of a material of the form

$$H = cY \qquad (8.1)$$

where c typically lies between 1 and 3, depending on whether the material has a structure that can collapse (porous, network, cellular) or be deformed at close to constant volume ($c = 3$). This

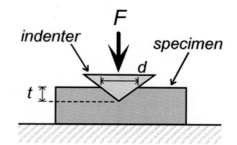

Fig. 8.1   Generalized principle of an indentation hardness test. The size of the indentation, *d*, or the depth, *t*, is determined for a given load, *F*.

then allows an estimate of the tensile yield strength of a material when a tensile specimen cannot readily be made. Obviously, the material under the indenter has to go somewhere when it is displaced. Commonly, in solid materials it appears all around and alongside the indentation as a small bulge as the displacement occurs sideways and upwards. In porous or cellular materials, collapse of the porosity or cells may accommodate some or all of the deformation. For example, gypsum products (Chap. 2) and casting investments (Chap. 17) are intrinsically porous, but built from very brittle ceramic materials. The fracture of crystals may be brittle, but the collapse or **densification** of the material under the indenter only appears to be plastic; it is therefore described as **quasi-plastic**. Some materials may undergo internal structural rearrangement to accommodate some deformation at the molecular level (such as polymers), or even a change of crystal structure to a denser form, but these latter are rare.

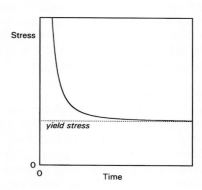

We may understand the nature of these tests by considering what happens as the indenter is lowered slowly into the specimen. Initially, there will be point contact. This corresponds to an infinite stress and the substrate deforms plastically. There is now a finite area of contact, but the stress is above the yield point; further deformation is unavoidable as long as this condition holds. The indenter thus sinks steadily into the specimen. Only when the area of contact has increased to the point where the actual stress is just identical to the yield stress will the movement stop (Fig. 8.2). Thus, we can see that this is a kind of strength test, but in a sense the inverse of the usual kind. In the place of steadily increasing the load or strain of a specimen of predetermined cross-sectional area until yield occurs, in a hardness test the *area* over which a fixed load is applied is allowed to vary until equilibrium is achieved.

Fig. 8.2 Form of the variation in stress beneath a hardness indenter as it sinks into the test substrate – flow takes time.

It is, of course, a necessary condition for such a test that the indenter is rather harder than the test piece, otherwise it would deform and invalidate the test, or even be destroyed. The pattern of stresses under such an indenter is very complicated, and very much dependent on its shape, but the key feature is that plastic deformation is plainly required to obtain a permanent deformation that can be measured. However, because of the complexity of the stress pattern, no hardness value[9] can be directly related to strength measured in any other way. Naturally enough, there will be some correlation, but not necessarily linear, and certainly not a scale factor that can be applied, even between the different hardness tests (see §8.6). What this means is that the type of test used for any mechanical property measurement must reflect as closely as possible the relevant service conditions if the data are to be interpretable.

An important practical point emerges from Fig. 8.2 and the accompanying discussion: it takes time for the equilibrium condition to be achieved. This is because it depends on the flow of the substrate under the applied stress, and flow is time-dependent (see Chap. 4). Hence, all such hardness indentation tests require that the load be applied for a specified duration or **dwell time**, typically 5 - 30 s, depending on the type of material.

A second point is that since there is necessarily sliding between the material and the indenter, of whatever its design, the friction at that interface is also part of the system, higher values limiting penetration more. Accordingly, measured hardness may vary according to the material of the indenter, but more importantly will be affected by its roughness as well as any contaminants on the specimen surface or remaining on the indenter from prior tests.

●8.1 Brinell

A method formerly in common use for metals uses the so-called **Brinell ball**. This requires, as its name implies, a spherical indenter. There are limitations to the Brinell method, however, because the relative geometry of the indentation varies with the load (Fig. 8.3): that is the steepness of the edge of the indentation increases with depth of penetration, and this in turn means that hardness values calculated for different loads or different diameter balls cannot be directly comparable: the apparent hardness increases with the relative depth of the indentation. Indeed, it is obvious that there is an upper limit to the size of indentation that is possible with any given ball, and an indentation diameter (d) greater than 60% of the ball diameter (D) is considered as unacceptable for a valid test result. Equally, if the indentation is too shallow, it is too difficult to measure and the result too imprecise; the diameter d is required to be greater than 25% of D. The **Brinell Hardness (H$_B$)** is

---

[9] Measurements of hardness used commonly to be called hardness *numbers*. Strictly speaking, these are not pure numbers, as would be implied by such a name, because they have the units of stress.

given by the equation:

$$H_B = \frac{2F}{\pi D[D - \sqrt{(D^2 - d^2)}]}$$ (8.2)

The denominator is the area of the spherical surface of the indentation obtained under the load, F, in units of kgf (*i.e.* the weight of 1 kg mass under standard gravity, $g_0 = 9.80665 \text{ ms}^{-2}$). If the projected area of the indentation is used $(4\pi/d^2)$, then the result is called the **Meyer Hardness** (H_M).

●8.2   Vickers

The problem of the continuously varying geometry of the Brinell indentation is overcome by using one of the pyramidal indenters. The **Vickers indenter** (Fig. 8.4) is a square pyramid, and the geometry of the indentation (under load) remains identical for all loads and hardnesses. Again, the surface area of the indentation is calculated, from the length of the diagonals, d, (averaged, if necessary, from $\{d_1^2 + d_2^2\}/2$), to obtain the **Vickers Hardness** or H_V:

$$H_V = \frac{2F}{d^2} \cdot \sin\left(\frac{136°}{2}\right) = \frac{1.854F}{d^2}$$ (8.3)

Here the angle 136° arises from the angle between opposing faces of the pyramid. Since F is again in kgf, H_V is expressed in kgf/mm².

The same test geometry is available with some automatic test machines which sense indentation depth through a precise transducer. This allows direct plotting of load *vs.* depth, observation of creep under load, and the calculation of modulus of elasticity, for example. This may be referred to as **Universal Hardness** (H_U), but is still defined in the same way as Vickers hardness. It avoids the operational difficulties arising from elastic recoil and optical measurement of the indentation.

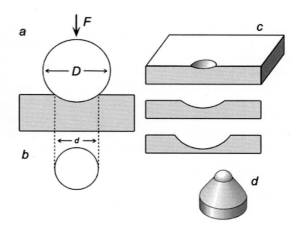

Fig. 8.3 The Brinell ball hardness test. Indentation shape varies with hardness. a: geometry of test, b: plan of indentation, c: indentation shape and depth of indentation, d: appearance of the indenter.

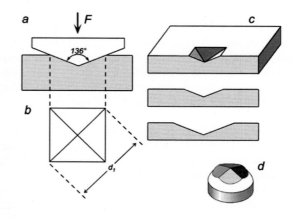

Fig. 8.4 The Vickers hardness test. For key, see Fig.8.3.

It is noteworthy that the area calculated for Vickers Hardness is that of the *surface* of the indentation itself as it is in the Brinell ball test (on which, in fact, its design was based). This approach is now recognized to be wrong in principle,[9] as for the calculation of the mean stress it should be the *projected* area, $d^2/2$, that is used. However, long usage means that it would now be difficult to rationalize this. Nevertheless, because the two areas are related only by a simple factor in the Vickers case (0.927 = 1.854/2; the actual stress is ~7 % greater), as a comparative measure it remains perfectly satisfactory. Clearly, Meyer Hardness is preferable to Brinell Hardness because of that, although the geometrical problems remain. (Indeed, this equation is exactly what is used for what is known as **Pfund Hardness**, a miniature version of the Brinell test.)

Neither the Brinell ball nor the Vickers indenter can be used effectively on materials with appreciably large flexibility, such as polymeric materials, because of the tendency of the test indentation to be obscured as the material springs back when the load is removed. Some substantial plastic deformation is obviously necessary for the indentation to persist. In the extreme case, materials such as rubbers have to be examined from different points of view. Problems also arise with these tests because the depth of penetration may cause extremely brittle materials to shatter under the test load if this is too large.

●8.3 Knoop

The **Knoop[10] pyramid** indenter[10] (Fig. 8.5) produces a very much longer and shallower indentation in relation to its width, as well as a more pronounced cutting action in the long diagonal direction. While the sides of the indentation may tend to spring back as a result of elasticity, the ends of the long diagonal usually remain quite visible, and it is this length which is used to calculate the area of the indentation, only this time it is the *projected* area which is calculated. The definition of the **Knoop Hardness (H$_K$)** is thus correct in principle, and still expressed in units of stress:

$$H_K = \frac{14.23F}{d^2} \qquad (8.4)$$

where d (mm) is the length of the long diagonal. H$_K$ is expressed in kgf/mm^2. The long diagonal is 30 times the depth, and 7 times the width of the indentation.

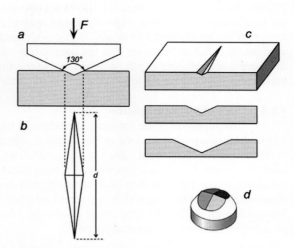

Fig. 8.5 The Knoop hardness test. For key, see Fig.8.3

●8.4 Units

All lengths and diameters are expressed in millimetres according to the definitions of these measures of hardness. Unfortunately, these definitions were made before the SI system was introduced so that all three have the load expressed in *kilograms-force*. Failure to convert from newtons (if the newton was the unit in which the force applied was measured) results in hardness values nearly ten times too high, so care must be taken. However, the interpretation of any of these hardness values depends on the application, because of the particular pattern of deformation each causes, but also upon the load that was used to perform the test. This latter problem arises from a size effect (§7.1) inasmuch as the indentation can range in scale from well above to well below that of any structure in the test piece, but also because the accompanying plastic deformation varies according to the size of the indentation, and therefore the properties of the deformed material, despite hardness being expressed in stress units. It is therefore usual to specify the test load when quoting a value, *e.g.* 235 H$_V$ 10 would mean a Vickers Hardness of 235 under a 10 kgf load. Clearly, the test methods will only rarely be a good approximation to service conditions, but they still provide valuable comparative data, rapid testing, and require little specimen preparation.

Despite the original definitions of these hardness measures, the tendency now is indeed to express Vickers and Knoop hardnesses in SI units, specifically in MPa or GPa. The equations are unchanged, but the input units are now newtons and metres. It is still correct to refer to hardness measures by these names as it is the geometry of loading that is important, but the two scale systems can be distinguished by the lack of a stated unit in the old style, and by the presence of a unit symbol in the new style. Thus, given that 1 kgf/mm^2 = 9.80665 MPa, a Vickers hardness stated as H$_V$ 235 ≅ 2305 MPa ≅ 2.305 GPa. Care is therefore required in reading any report to ensure that the values are understood correctly, in particular, equation 8.1 only works if the units are the same.

●8.5 Plastic zone

One point to notice in connection with all indentation hardness tests, and indeed scratches, whether deliberate in scratch tests or during a machining or abrading operation (Chap. 20), is that the plastically deformed zone extends outwards as much as twice its diagonal size, downwards from the indentation as much as 10 times its depth (Fig. 8.6). This has two main influences on these tests. Firstly, successive tests must be spaced out by more than 4 times their width to avoid testing already deformed and probably strain-hardened material. Secondly, the indentation must be far enough from an edge or other boundary that the deformed zone is fully contained.

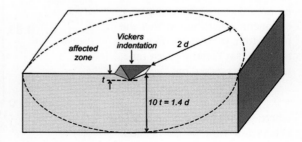

Fig. 8.6 Size of the affected zone around a Vickers indentation (the geometry gives *d/t* = 7.0006...).

---

[10] According to the family, this is pronounced with a voiced 'k', long 'oo': k'noop.

Thirdly, the test piece must be at least as thick as the indentation is wide if the supporting anvil is not to influence the result (there should be no sign of a bulge or other change on the reverse of the piece). This interference is known as the **anvil effect**. Similarly, if micro-indentations are made in individual grains in polished sections, care has to be taken to ensure that the indentation size is appropriately small in relation to the visible grain diameter – and possible depth, which is invisible. In transparent materials such as some polymers, viewing between crossed polarizers reveals the extent of the deformed zone very clearly.

●8.6   Lack of test equivalence

One important point: hardness values are not interchangeable or interconvertible even though there must be a high correlation between the values obtained with any pair of tests because the plastic deformation occurring is similar – but not identical. The varying geometries of these and other hardness tests means that a different mixture of a number of basic properties has been tested in each case – they are *different* mechanical tests. Their value lies in the fact that they are essentially non-destructive and provide cheap, rapid values which allow discrimination between materials and treatments in a way that correlates broadly with a range of meaningful engineering properties. Even so, there exist many "conversion" tables; these should be used with caution, and then only for the type of material specified.

§9. Dynamic Behaviour

Materials in service are not only subjected to static stresses, indeed it is rather unusual for any kind of structure to be given a single, unvarying load. This is particularly obvious in the mouth, where the forces of mastication can be large (the Inuit people, formerly with a diet that required a great deal of hard chewing, could generate up to 700 N between molars), and cyclic, as well as varying in direction. Dynamic conditions, that is, conditions where the magnitude or direction of a load can vary continuously (in any pattern and at any rate), automatically require dynamic tests for a full understanding of the properties of a material. Since it is obvious that in order to do a test of, say, compressive strength, the load must be changing during the course of the test, the question arises of the *rate* at which the loading is done for it to qualify as a static test. In fact, no such distinction can be made in any definite manner: what we must talk about properly is the **timescale** of the test, and then in relation to the processes going on inside the material, and especially those that are expected to be important in service. Typical considerations include atomic vibrations, heat diffusion, and atomic diffusion; there are others. If a test takes a shorter time to complete than one of these motions, that process (and all other slower processes) cannot possibly influence the outcome of the test. Conversely, for a test taking longer than the typical time of such a process, that process – and all faster processes – can be expected to dominate the observed behaviour. In no case is there a very precisely defined time (or, equivalently, a rate) associated with these processes: there is always a spread of values because such behaviour is based on random events. Such systems must be seen as essentially probabilistic: there is only a particular probability of something happening within a certain time, and near certainty if one waits long enough.

The statistical behaviour of such processes typically can be approximated by a negative exponential distribution: then the proportion, p, of atoms in a new position (say) after a given time, t, can be expressed by:

$$p = 1 - e^{-t/\theta} \qquad (9.1)$$

where θ is the characteristic **relaxation time** of the system (with the units of time). When the elapsed time is equal to the value of θ, then $t/\theta = 1$. This equation (9.1) is identical in form to the radioactive decay equation because that is a similar kind of

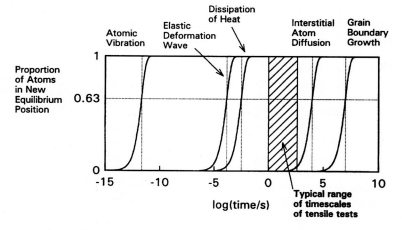

Fig. 9.1 The relaxation times for various processes cover a range of many orders of magnitude. (The value 0.63... corresponds to $1 - e^{-1}$, that is, after a time equal to θ has passed.) 10^5 s $\approx$ 1 day. NB: both diffusion and grain growth are temperature dependent; room temperature results are indicated here for 'typical' metals.

random process (11§2.6). An alternative view of such a process is that it indicates the 'typical' time required for the system to recover from a perturbation (that is, return to equilibrium). In mechanical testing the deformation resulting from an applied load would constitute such a perturbation. Thus, for example, on applying a load a wave of elastic deformation will travel through the specimen. This happens so fast (*i.e.* at the speed of sound c in the material) that our general perception is that elastic deformation in small specimens (length d) is instantaneous, *i.e.* d/c is very small. In many ordinary circumstances we may safely treat it as if it were, but care must always be taken to make sure that such hidden assumptions do not affect our interpretation of results. The value of p is therefore the proportion of completion of the process, and in that general sense can be applied to any time-dependent process.

The relative magnitudes of the timescales of some processes (particularly as applied to metals) are illustrated in Fig. 9.1. From this it is seen that typical 'static' tests, *i.e.* performed at ordinary rates and taking from about 1 to 100 s, would not be expected to be influenced by atomic diffusion.[11] It thus remains to enquire what might be expected at very high and very low test rates.

§10. Impact

Perhaps the most obvious dynamic stress that can be envisaged is that of **impact**. Rather difficult to define precisely, we have an intuitive idea of what constitutes an impact. There are several common kinds of example to consider. Thus, even if not intentional, there may be the impact of teeth on restorations, or merely accidentally dropping a denture on the floor. Random knocks on models occur in the dental laboratory, but a deliberate bang of the neck of a toothbrush on the edge of a sink is a quite common way used to clear water and debris from the bristles. We can therefore readily recognize the need for a material to be able to withstand some degree of impact. Even so, it is as well to appreciate that a considerable range of velocities falls within the scope of this kind of loading.

Two methods are commonly used in laboratory testing: one, the **Izod** test, supports the sample rigidly at one end, while the **Charpy** test supports both ends of the sample, but not rigidly (Fig. 10.1).[12] The specimen is then struck by a hammer in the form of a pendulum (Fig. 10.2). Before the test, the pendulum is held in an elevated position, and thus possesses potential energy by virtue of its position. When the pendulum is released, its free fall under the influence of the local gravity accelerates it to a maximum velocity at the lowest point of its travel. All the original

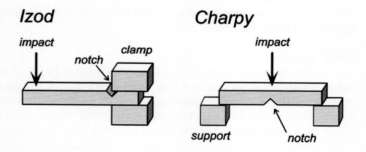

Fig. 10.1 The set up for the Izod (left) and the Charpy (right) impact tests. The notch is used when "notch sensitivity" is of interest.

potential energy has been converted to kinetic energy at that moment, and it is at that point that the specimen is positioned to be struck. If the specimen breaks, it absorbs some energy from the pendulum, which is therefore slowed down. The kinetic energy left, however, causes the pendulum to rise again on the far side to a height corresponding to the conversion of that remaining kinetic energy to potential energy once again. By determining how high the pendulum climbs after impact, the energy absorbed by the specimen may be calculated.

●10.1 Theory

The height of the pendulum in the start position (Fig. 10.2) above its lowest point (bottom dead centre, b.d.c., *i.e.* where it would hang naturally) is given by:

$$h = L(1 - \cos\alpha) \tag{10.1}$$

where L is its length to the **centre of mass**, and α is the angle of fall. Then, since work needs to be done to raise the pendulum from b.d.c. to the start position, against its weight due to gravity, gM (where M is its effective mass and g is the local acceleration due to gravity), the potential energy, E_p, of the pendulum at the start is given by

$$E_p = gMh = gML(1 - \cos\alpha) \tag{10.2}$$

All of this energy (ignoring friction) should be converted to kinetic energy at b.d.c, where contact occurs at the **centre of percussion**[13] (where impact produces no reaction force at the pivot)[11]:

$$E_k = \tfrac{1}{2}Mv_{bdc}^2 \qquad (10.3)$$

The pendulum velocity at that point, v_{bdc}, is given by:

$$v_{bdc} = \sqrt{2gL(1 - \cos\alpha)} \qquad (10.4)$$

Since the kinetic energy remaining after impact is converted back to potential energy, the energy absorbed, E_a, is simply calculated as

$$E_a = gML(\cos\beta - \cos\alpha) \qquad (10.5)$$

Ordinarily, of course, the test machine is calibrated in terms of energy absorbed, and no calculation of this kind is necessary. The equations are of interest in that they show how the impact conditions are affected by the machine that is used (and, indeed, affected by location in that a gravitational acceleration is involved, and this varies from place to place on the earth and with altitude) (see box).[14]

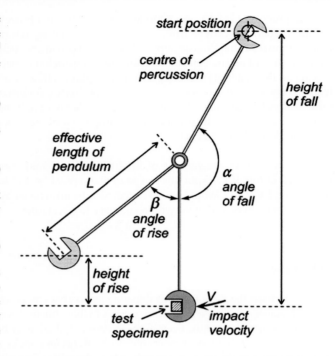

Fig. 10.2 Layout and terminology of the Charpy impact test machine. If hanging free, the pendulum's centre of percussion is just in contact with the specimen at the "bottom dead centre" of its travel.

Of course, no bearing is perfect, and unless the test was conducted in a vacuum air resistance is also a source of energy loss. Controlling for these factors is easy by using blank runs.

● **10.2 Notches**

In each type of test a standard shape of notch may be cut, if required, in the sample to assess the so-called **notch-sensitivity** of the material, a measure of the fracture resistance in the presence of a (macroscopic) flaw or, more to the point, the iystenfluence of complicated shapes on behaviour. We may cite the frenal notch in a denture base, or the sulcus carved at the junction of denture tooth and the base as examples of such features which are a part of the design and whose influence must be understood. In addition, there are many structures where changes of cross-section, holes and so on could have similarly important consequences. When a notch is used in the specimen in this way, the result of the test may be called the **notch toughness**. The importance of this lies also in relating the behaviour to the effects of the presence of flaws, whether porosity or cracks, in the bulk material.

● **10.3 Impact strength**

In impact tests it is not a failure stress that is measured but, rather, the energy absorbed in the course of the deformation and rupture of the specimen, and as such it is clearly related to the **modulus of toughness** (§4). This means that the common term **impact strength** is somewhat misleading as it is normally used. For brittle materials, the energy absorbed by the specimen from the hammer represents the work done in creating the two fracture surfaces and as such is dominated by the surface energy of the material. More ductile materials also have a plastic deformation component in the total. However, one or both pieces is accelerated as a result of the impact and this, of course, also absorbs some energy (both translational and rotational). The correction may be substantial for low impact strength (low fracture energy) materials. Unfortunately, the way that the pieces fly out is rather unpredictable, and the acceleration energy will vary a little from time to time and an average value should be used. It will also be influenced by the mass of the pieces (or, since standard sizes of specimen should be used, the density). Thus, the correction is expected to vary with gypsum products, for example, according to whether they are tested wet or dry, quite apart from the question of whether the true fracture energy changes. A control test may be performed by reassembling the pieces with a low-strength 'glue', such as petroleum jelly, then striking again.

---

[11] The design is such that the centre of mass lies just inside the centre of percussion, and on the same line to the pivot.

In a number of contexts, the acceleration due to gravity, g, appears in equations. In defined quantities, the standard exact value, $g_0 = 9.80665$ ms^{-2} is always used (which value was based, arbitrarily, on a kind of global sea-level average figure). However, if it is physically gravity that is involved, the local value must be used. The value for a non-rotating system of an object near a spherical body, radius R, mass M, is given by:

$$g = GM / (R + h)^2 \qquad (10.6)$$

where h is the height of the object above the surface, and G is the gravitational constant ($6.674\,28 \times 10^{-11}$ Nm^{2}kg^{-2}). However, the Earth is rotating, and the object in a given location is rotating with it. This generates the so-called centrifugal force (see 18§3.1), which is at a maximum at the equator, and zero at the (rotational) poles. On top of that, the rotation causes the Earth itself to bulge at the equator, which means that the effective value of R – the distance from the surface to the **centre of gravity** – also varies from pole to equator. The total effect is for a variation of about 0.5% (*i.e.* affecting the value in the 2nd decimal place) in an object's weight between the two surface locations at sea level. These effects are taken into account in standard gravity formulae such as

$$g = 9.780\,327\,(1 + 0.005\,302\,4 \sin^2 \varphi - 0.000\,005\,8 \sin^2 2\varphi) - 3.086 \times 10^{-6}\, h \text{ ms}^{-2} \qquad (10.7)$$

where φ is the latitude.

Altitude is therefore also very significant, affecting up to the 3rd decimal place, at about -0.003 ms^{-2}/1000 m, and clearly must be taken into account in even modestly high buildings. In addition, there are variations affecting up to the 4th decimal place due to variation in topography (mountains) and density (geology), so-called gravity anomalies. As might be imagined, the sun's and the moon's influences are also involved because they cause the tides. This, however, is at the level of a daily variation affecting up to the 6th decimal place.

It follows then that, for precise experimental work, where actual gravity is involved in generating a force, the location needs to be taken into account. Thus, for deadweight loads in force calibration (such as of load cells and electronic weighing machines), hardness and impact testing, mercury barometers (10§2) and the like, local gravity must be determined and used.

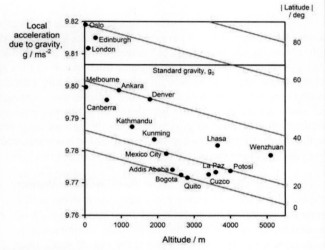

Fig. 10.3 Effect of altitude and latitude on the local value of the acceleration due to gravity.

Although a separate issue, the buoyancy of the air (affected by atmospheric pressure, temperature and humidity – altitude and weather!) must also be taken into account in any such 'weighing'. At sea level and ordinary conditions, this amounts to about 1.2 mg/mL. Clearly, local conditions (pressure, temperature, humidity) must be taken into account as well as the density of the material being weighed, including (and especially for) precise calibration masses.

It is necessary to note that conditions which affect surface energy, such as the temperature and the presence of fluid, must be controlled in such tests because of their influence on crack initiation and growth. Thus, denture base acrylic must be tested wet to mimic oral conditions, and ideally at 37 °C. Unfortunately, because of the complexity of the deformations usually occurring during a test, it is not possible to reduce impact strength data to the units of specific surface energy, *i.e.* J/m^2, as a scale-free material property, with any great reliability. There is a considerable variation in behaviour due to specimen shape and size (and, of course, impact velocity, according to the choice of testing machine, or, rather, the length of the pendulum). The generally desirable principle of making test results independent of cross-sectional area and so on (see §2) is not attainable in this case. What can be attempted though is the correlation of such test results under standardized conditions with the performance actually observed in service (if such observations are possible). Indeed, from what has gone before, this kind of remark applies with more or less emphasis to all kinds of test: the more complicated the service loading, the more difficult it becomes to interpret the data from laboratory tests. Given such difficulties, the tester must be prepared to accept that the results obtained sometimes may be only directly useful in a comparative rather than a predictive sense.

For convenience it is usual to employ rectangular bars as test specimens in impact testing. Obviously, the applications of materials in practical situations are unlikely to require regular rectangular bars, and the above remarks, as well as the expected influence of stress-concentrating features such as holes and changes in section, would need to be taken carefully into account in evaluating the results.

§11. Creep

At the other end of the time scale from impact we may consider the processes of **creep** and **flow**. The former term usually relates to the behaviour of solids, the latter to liquids, but as will be discussed later (Chap. 4) it may not be possible to draw very sharp lines between the two labels as it is sometimes difficult to decide on the nature of the material. For example, if glass rods or tubing are allowed to lean in storage, such as against a wall, very definite permanent curvature is soon evident. Indeed, flow has occurred, yet by 'normal' standards the material is a solid. If the glass is to stay straight, it must be stored horizontally, uniformly supported. For the moment the discussion will centre on creep in solids.

Reference to Fig. 9.1 shows that at long time scales atomic diffusion becomes important in controlling deformation under load. What is not shown there is that such deformation can, and does, occur at stresses well below those at which immediate fracture would be expected at 'normal' rates of loading (*i.e.* the 'strength' of the material under usual test conditions). Such materials thus appear to be plastic on long time scales at stresses which would otherwise be thought to be associated only with elastic behaviour. Dental silver amalgam is a good example of a material showing this kind of behaviour. Amalgam usually behaves in a brittle manner and does not show appreciable yield when tested on a timescale of a minute or so. However, under the stresses of mastication, which are usually well below the normal static test strength, creep is observed. This is thought to be the reason for the extrusion which is found to have occurred in old cervical restorations (14§6.5). As might be expected, creep strain rate depends on stress, but rarely linearly (Fig. 11.1).

Creep curves often show three roughly distinct stages. First, there is a very rapid permanent strain called **primary creep**. This is followed by a longer period of **secondary creep** in which the rate of strain accumulation is more or less steady. At some point, the rate of strain may start to increase again, and the test enters the **tertiary creep** stage.

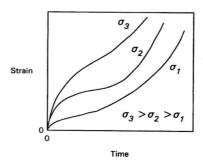

Temperature is also important because creep processes depend on diffusion, and thus require an **activation energy**. The conditions of any such tests must therefore be specified very carefully.

Fig. 11.1 The typical shapes of creep curves. Creep strain is measured after allowing for elastic strain.

§12. Fatigue

There is one other general kind of dynamic test, that of **fatigue**. Again, real examples of materials in service are not usually subjected to a single loading event. Stresses frequently vary cyclically, for example, as a result of chewing, and while the peak stresses (in tension, compression or shear) may be far below those at which a static test would give an immediate failure, failure may yet still occur eventually under a cyclic load. Fatigue testing in torsion and bending may also be used as appropriate to the service conditions.

The style of the variation of the load can be in any of several forms (Fig. 12.1). So, for a simple axial test, the peak stresses can be chosen independently to be zero, in tension, or in compression; similarly the mean stress can be chosen (depending on the first selection) to be zero, in tension, or in compression. In addition, the waveform for cycling can be one of several standard, convenient, forms (such as the sinusoidal function shown in Fig. 12.1) or a direct modelling of the load conditions shown by the real system. There are several examples of cyclic loading in dental applications such as denture bases and bridges as well as in teeth and supporting tissues. Notice that here there is an implicit sign convention in that tensile stress is treated as positive, compression stress as negative (*cf.* 23§1.1).

Fatigue test results are usually plotted as the peak stress *vs.* the number of cycles to failure, on a logarithmic scale (Fig. 12.2), often called the "S/N" plot. Notice that the number of cycles is frequently measured in millions. This means that such testing tends to be expensive and time consuming, and is normally only undertaken for critical applications. Most materials show a continuously declining stress for failure as the number of cycles increases, but steels in particular may show a stress below which no amount of load cycling produces a failure. In other words, the plot becomes horizontal at some point, which is then known as the **fatigue limit** or **endurance limit**. Notice that the one-cycle failure stress is just the 'static' strength.

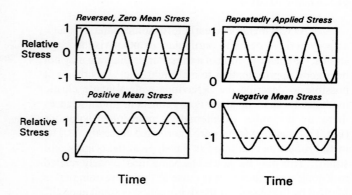

Fig. 12.1 Some examples of the loading conditions used in fatigue tests. A positive stress is tensile, negative is compressive.

Failure under fatigue conditions is due typically to crack propagation as a result of localized microscopic plastic deformation. This process has a very much longer timescale than a single static test, so that the failure occurs when the crack has grown to the critical size for the (peak) stress being applied (Griffith criterion, §7). Creep, on the other hand, which is a rather more macroscopic plastic deformation of materials, is explicable in terms of long time-scale diffusion processes.

As is indicated in Fig. 12.2, there may be two regimes of behaviour, these are called **low-cycle** and **high-cycle fatigue**. The principal distinction to be made is whether the stress exceeds the yield point of the material at any point in the cycle. If the yield point is exceeded, then plastic deformation of course occurs and the accumulation of damage is faster. The transition between the two regimes is typically at around 10^4 cycles.

One of the best examples of fatigue in a dental context is of the parts of the capsule-holder of an amalgam or cement oscillatory mixing machine. Commonly, the capsule is held in a spring-loaded fork (Fig. 15§1.3) whilst being shaken at ~70 Hz. The elements of this device are initially bent in order to exert a retaining force on the capsule, and so experience tension on the now more-convex surfaces, and then suffer further bending about this point when the machine is run. A positive mean stress (Fig. 12.1) is therefore relevant at such sites. Failures are known to occur in service, and a cover or protective enclosure therefore necessary to avoid injury. Likewise, endodontic machine files are rotated in curved root canals, and so suffer reversed stressing with a zero mean at every point on the curved portion. Failure here leads to the need to retrieve the tip fragment from the canal and is best avoided. Another context of relevance is that of chewing. If one envisages, say, 200 chewing cycles per meal, over 5 years for three meals per day, conservatively, a restoration may be subject to a million loading cycles. Clearly, materials and devices should be able to survive such exposure.

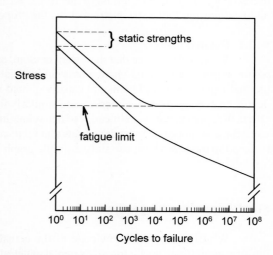

Fig. 12.2 Typical forms of fatigue test S/N plots. The upper plot shows a fatigue limit; the lower shows a transition between high- and low-cycle fatigue (below and above ~10^4 cycles, respectively).

§13. Elastic Moduli

As was mentioned earlier, for some materials (such as polymeric systems in particular) or otherwise at large strains, the elastic portion of a stress-strain diagram may not be a good approximation to a straight line. It is then not possible to describe the behaviour with a single elastic modulus because the slope is not constant. In these cases there are a number of alternative, convenient solutions (Fig. 13.1), as long as they are interpreted correctly.[15] The **initial modulus** is defined as the slope at the origin (zero strain). This may be viewed as being most closely related to Young's modulus as it is close to the zero strain that is implied by the phrase "if the deformation is small enough" (§2). This is a special case of the **tangent modulus**, E_t, which is the slope of the curve at a previously determined stress or strain of interest. This is obviously related to the differential form of the definition of strain (equation 2.15), and as such may be viewed as the "local" modulus, *i.e.* at a particular point in the elastic region of stress-strain curve:

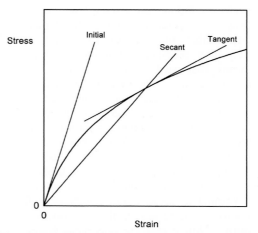

Fig. 13.1 Elastic modulus may be defined in a variety of ways for non-Hookean materials.

$$E_t = \frac{d\sigma}{d\varepsilon} \qquad (13.1)$$

It is this formulation that allows observation of whether the modulus of elasticity is increasing or decreasing during the deformation process.

The **secant** or **chord modulus**, on the other hand, corresponds to the slope of a straight line drawn between two chosen strain points (the first of which can be the origin) in order to give the **average modulus** over that interval. Which elastic modulus is selected depends on the application, but it must be remembered that these are *elastic* moduli, irrespective of the shape of the curve: the stored energy is fully recoverable, and a modulus of resilience is calculable, although this is less easily done than with more Hookean materials. When there is no linear portion to the stress-strain curve the proportional limit is zero.

●13.1 Related ideas

We can also notice that the secant or chord approach is taken for reporting the values of other properties where in practice the total change over an interval of the independent variable is determined. For example, thermal expansion coefficients are usually quoted with a temperature range such as "100 ~ 200 °C", *i.e.* per degree on average over that interval, and similarly for heat capacity. Note that the lower point need not be zero. Often, the approximation to linearity of behaviour implied by this is adequate, sometimes not at all. For detailed work the assumption may be dangerous and more complicated fitted equations (*e.g.* polynomials) may be used to describe the behaviour, certainly a simple graph is more informative.

§14. Comparative Properties

Whilst a personal knowledge of the actual values of mechanical properties is not of any particular intrinsic merit (there are far too many dental materials and far too many properties to make any deliberate attempt at memorizing rather silly, and many compilations exist in data books for easy reference), it is certainly of benefit to relate one class of materials to another. Figs 14.1-14.4 give an overview for metals, ceramics and polymers with respect to their elastic modulus, yield strength, crack toughness, and (for good measure) melting point, which has a great bearing on applications although not a mechanical property. There are two points in particular to be noted: (i) there is a rather definite upper limit to what can be achieved, determined primarily by the strength of interatomic forces; (ii) the relative ranks of the three classes of material change from plot to plot, supporting the idea that very often (indeed, nearly always) compromises have to be made in selecting a material to do a specified job.

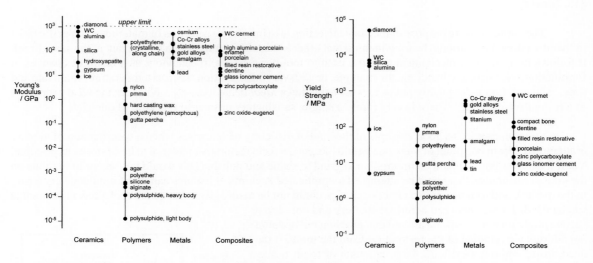

Fig. 14.1 Comparison of values of Young's modulus for some dental examples of the four major classes of materials. (WC: tungsten carbide; Co: cobalt; Cr: chromium)

Fig. 14.2 Comparison of values of the yield strength of some dental examples of the four major classes of materials.

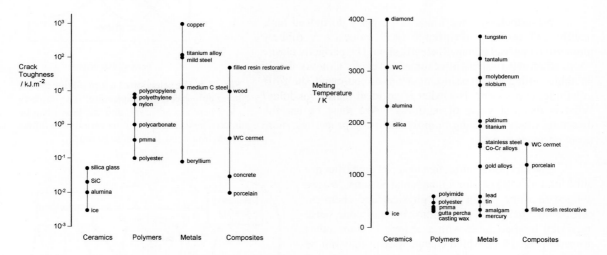

Fig. 14.3 Comparison of values of the crack toughness for some dental examples of the four major classes of materials.

Fig. 14.4 Comparison of values of the melting temperatures for some dental examples of the four major classes of materials.

Although much effort is made to obtain values for material mechanical properties as if they were characteristics of their substance and not their extent (despite the numerous difficulties in realizing this aim), in service it is necessarily the extent of the material, its thickness, length and width, which will control how it may be used. For example, knowing the load to which an object will be exposed will determine the minimum thickness, say, which will satisfactorily support it. If there are constraints on the thickness which may be used because of lack of space, then a minimum strength can be calculated in order to identify a suitable material from which to fabricate the object. This kind of reasoning is the essence of materials selection and object design: it must always be done in context.

§15. Ideality

The entire ultimate purpose of mechanical testing is to inform the selection of materials for specific tasks in specific (stress) conditions (bearing in mind that other kinds of factor may also be relevant at the same time). It follows from this that mechanical tests are used to monitor changes in behaviour as a result of changes in formulation, treatment and handling, for example, or with varying conditions such as temperature, strain rate and time. The question is frequently put as to what constitutes *ideal* behaviour, *e.g.*: What are the ideal properties of a restorative material? There is very rarely a single, satisfactory answer to this kind of question.

Strength provides a good example of the kind of difficulty which arises when a specification of ideality is attempted. From one point of view one would like, say, a resin restorative material to be as strong as possible. Yet if it is too strong it would be impossible to grind to shape and polish. This would put severe limitations on the kind of restoration that could be attempted (complicated anatomical features cannot be moulded very easily in the mouth), and routine adjustment of occlusion could not be made, given that a matter of a few micrometres can be the difference between being satisfactory and not. Even so, one might imagine working around that problem but it would not address the removal of the restoration from the tooth, if the need arose, without sacrificing large amounts of tooth tissue. This kind of irrevocability would not be acceptable. The key issue is that each property must be viewed in the context of all aspects of the material's use and handling, not merely from the narrow perspective of, say, just one aspect of final service. Selection of classes or types of material, even products, should be made in the light of this.

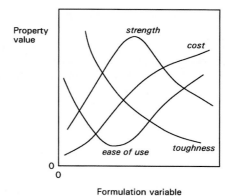

Nevertheless, some kinds of property do indeed have readily and usefully identifiable ideal values. A rubbery impression material (Chap. 7) clearly should be perfectly elastic – zero strain at zero stress after some load cycle. Colours should be stable, setting should be instantaneous on command, shelf-life indefinite, and so on. But what is an ideal elastic modulus? What is the best value of hardness? We must be careful, therefore, when discussing 'improvements' in properties; more is not necessarily better.

Fig. 15.1 The selection of a formulation almost invariably involves compromise. The property plots here are arbitrary, and the labels can be interchanged to explore how various conditions might affect the decision to be made.

The fact is, of course, that material properties are quite limited (see §14), for fundamental reasons. We are obliged to work within those physical and chemical constraints. In any case, an improvement in one sense may well be associated with a change in some other property (Fig. 15.1), which change may or may not be beneficial. The essential compromise in design and selection remains because an 'optimum' value cannot be attained in several variables simultaneously.

•15.1 Development

It may help to put biomedical materials design and selection in the context of the surrounding processes. First of all, a new material may arrive in the dental arena through one of three routes. Firstly, there is tradition or accident, when the origins of a use owe more to art and craft than science, although with a long enough history, much valuable information may have been accumulated. This is not to be underestimated, but on its own it is more or less stagnant, with no impetus for development. Secondly, borrowing, when a product is recognized to be valuable in a new application or field. Thirdly, deliberate design, from a knowledge of fundamental theory or observed behaviour, assembling

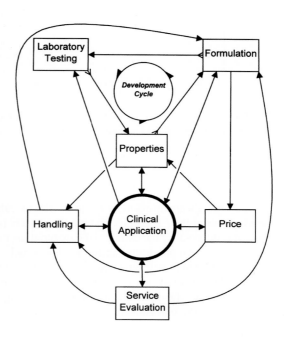

Fig. 15.2 The principal interrelationships in the development and evaluation of a product for clinical use.

components in new combinations or arrangements. In any case, the material now enters what may be termed the **development cycle** (Fig. 15.2), the process of laboratory testing, examination of properties and reconsideration of the formulation (this includes such things as composition, particle size, mixing ratio and manufacturing process details). It is the properties which feed directly into the determination of the clinical application, but such matters as the cost of the product or its use, and the handling characteristics interact with these other considerations. Ultimately, service evaluation – the only true test of the effectiveness of the material in the working context – will determine whether improvements need to be made in handling or some aspect of the formulation, when the development cycle is re-entered. There are other interactions, some of which are marked in Fig. 15.2, some of which may arise in special circumstances. The central issue is to suit the product to the purpose and to handle it so that it performs appropriately. This book is in large part about the interplay of factors relevant to making materials-related decisions.

References

[1]   Ashby MF & Jones DRH. Engineering Materials 1. Pergamon, Oxford, 1981.

[2]   Box GEP, *in*: Box GEP & Draper NR. Empirical Model-Building and Response Surfaces. Wiley, New York, 1987.

[3]   Hertz H. Gesammelte Werke, Vol. I, Ch. 15: 283 - 287, Leipzig, 1985.

[4]   Darvell BW. Uniaxial compression tests and the validity of indirect tensile strength. J Mater Sci 25: 757-780, 1990.

[5]   Rosenthal D & Asimov RM. Introduction to Properties of Materials. 2nd ed. Van Nostrand Reinhold, New York, 1971.

[6]   Griffith AA. The phenomena of rupture and flow in solids. Phil Trans Roy Soc 221A: 163 - 198, 1920.

[7]   Sneddon IN. The distribution of stress in the neighbourhood of a crack in an elastic solid. Proc Roy Soc London. Series A, Math Phys Sci 187 (1009): 229 - 260, 1946.

[8]   Tate DR. Hardness Characterization. *in:* Bever M (ed.) Encyclopedia of Materials Science and Engineering Vol 3: 2093 -2097, Pergamon, Oxford, 1986.

[9]   Tabor D. Chap. 7 *in* The Hardness of Metals. OUP, Oxford, 1951.

[10]   Knoop F, Peters CG & Emerson WB. A sensitive pyramidal-diamond tool for indentation measurements. J Res Nat Bur Stand 23: 39 - 61, 1939.

[11]   Guy AG. Introduction to Materials Science. McGraw-Hill, New York, 1972.

[12]   Hanks RW. Materials Engineering Science. An Introduction. Harcourt Brace & World, New York, 1970.

[13]   en.wikipedia.org/wiki/Center_of_percussion

[14]   earthobservatory.nasa.gov/Features/GRACE/page3.php

[15]   Wyatt OH & Dew-Hughes D. Metals, Ceramics and Polymers. An Introduction to the Structure and Properties of Engineering Materials. Cambridge UP, 1974.

Chapter 2   Gypsum Materials

Gypsum-based products play a major role in dental treatment procedures. To understand the behaviour and limitations of models, impression plaster and gypsum-bonded casting investments, the **setting** *reaction is of primary importance. However, much attention is focused on the dimensional accuracy of plaster objects since setting involves the dissolution of the powder and the growth of gypsum crystals. The* **fit** *of crowns, dentures, bridges and orthodontic appliances all depend on the fidelity of the model or mould. Accuracy is limited by the effects of crystal growth, and the means of control to limit it (in the case of models, say) or exaggerate it (in the case of investments). The relevant factors here are both chemical and physical.*

Plaster and so on are used as a mixture of the powder with water, and a key concern is the question of the **mixing volume**. *There is a compromise to be found in respect of strength and fluidity. These ideas are generally applicable to any type of product or system that is mixed from a liquid and a powder (whether by the dentist or in manufacture), they are subject to precisely the same constraints: for example, glass ionomer cement, amalgam, denture acrylic and resin restorative materials.*

Porosity *is also of great significance. It affects the accurate reproduction of surface detail, but mostly it needs to be controlled because of its strong effect in reducing the strength of the set material. Again, the principles involved here are also applicable to many other dental materials.*

Because of the bulk in which they are used, the **storage** *of dental plaster and stone may be under somewhat casual conditions. The deterioration which they may suffer may or may not be very noticeable, depending on how far the process has gone. Recognition of these effects and the means to avoid them is important in ensuring the expected outcome and avoiding waste of materials and effort.*

The rate of a setting reaction controls a material's handling and usage, expressed as **working and setting times**, *although neither is straightforward to explain or measure.*

The resistance of gypsum models to damage is important for the quality of work done on them. Various approaches have been used to combat a lack of hardness.

Many of the principles found to affect these materials also apply in several other dental systems.

Materials Science for Dentistry
https://doi.org/10.1016/B978-0-08-101035-8.50002-X

Gypsum products are probably the single most used class of material in the whole of dentistry, at least in terms of mass. This arises from the fact that models of the dentition provide the basis of the majority of restorative treatments: full and partial dentures, bridges, crowns, and inlays, as well as for orthodontic work and reconstructive surgery, both for study of the case and as a template on which to work for many laboratory procedures. Then there are impression materials and casting investments relying on the same chemistry. These materials provide important links in a chain of processes leading to the final treatment outcome. In view of this central position, gypsum products must be understood thoroughly if both their own use as well as the processes going on around them are to be effective. In fact, the design of some materials and procedures is based on the inadequacy of gypsum products in one respect or another. It is the low cost of the raw material that primarily contributes to its dominance, but this should not be considered as diminishing the value or importance of the resulting dental products.

§1. Basic Reactions

Gypsum is the common name of the mineral calcium sulphate dihydrate, which is mined in various places around the world and now is also being produced in a relatively pure form as a by-product of sulphur dioxide removal processes for power station exhaust gases. It has wide application, particularly in the building industry as plaster for finishing internal surfaces, but 'plaster casts' on broken limbs are possibly the most familiar use. **Plaster of Paris**, dental plasters and dental "artificial stones" are all manufactured by the dehydration of the dihydrate, using various processes, but they all have a common underlying chemistry.

•1.1 Dehydration

On heating, dehydration of the dihydrate occurs in two distinct stages. The fairly rapid loss of three quarters of the water of crystallization at about 130 °C gives the hemihydrate (Fig. 1.1). Further and now complete dehydration occurs at higher temperatures to give **anhydrite**. This has two crystal forms with differing reactivities with respect to water. The orthorhombic form, obtained by heating above 200 °C, is so slow in reacting with water that it is known as **dead-burnt** plaster: it does not form a usefully setting mixture with water. The crystal structures of the different forms are quite distinct from one other. The water molecules in the dihydrate are found in alternate double layers with the calcium ions and sulphate groups (Fig. 1.2),[1] and so its removal results in the destruction of that crystal. It may be deduced from the temperature required for dehydration that the process requires much energy, implying that at room temperature the dihydrate is a relatively stable form compared with the hemihydrate. It is this difference which provides the driving force for the setting reaction.

•1.2 Rehydration

Although all of the dehydration products will react with water to give the same reaction product (the dihydrate), the rates are considerably different. Dental

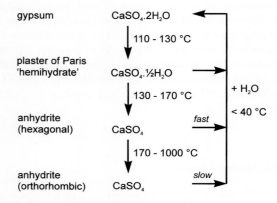

gypsum	$CaSO_4.2H_2O$
	↓ 110 - 130 °C
plaster of Paris 'hemihydrate'	$CaSO_4.\frac{1}{2}H_2O$
	↓ 130 - 170 °C
anhydrite (hexagonal)	$CaSO_4$ — fast
	↓ 170 - 1000 °C
anhydrite (orthorhombic)	$CaSO_4$ — slow

+ H_2O
< 40 °C

Fig. 1.1 Reactions of calcium sulphate and its hydrates. The orthorhombic form converts to the monoclinic at 1193 °C, which then melts at 1450 °C.

In the present context, the word **plaster** has two connotations in ordinary usage, firstly the hemihydrate powder, as in Plaster of Paris, and secondly the hardened mass of dihydrate resulting from the rehydration reaction of plaster powder. If there is any ambiguity, it is best to use phrases such as 'plaster powder' and 'set plaster' to make the distinction.

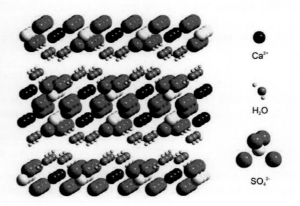

Ca²⁺

H_2O

SO_4^{2-}

Fig. 1.2 View of the layer structure of calcium sulphate dihydrate. Notice the clearly defined cleavage planes defined by the water molecules.

gypsum-based materials all rely on the fairly prompt reaction of the hemihydrate with water to produce the less-soluble dihydrate:

$$CaSO_4 \cdot \tfrac{1}{2}H_2O \ + \ 1\tfrac{1}{2}H_2O \qquad \Rightarrow \qquad CaSO_4 \cdot 2H_2O \qquad \text{-16.3 kJ/mol} \tag{1.1}$$

$$\begin{array}{lll} 8.8 \text{ g/L} & 2.1 \text{ g/L} & \text{(solubility @ 20 °C)} \\ 145.149 \qquad 1.5 \times 18.015 & 172.172 & \text{(molar mass)} \end{array}$$

The mechanism is straightforward: hemihydrate readily dissolves in water at low temperature, but the solution becomes supersaturated with respect to dihydrate well before it is saturated with respect to hemihydrate. As a result, dihydrate may commence to precipitate from the solution if it is appropriately nucleated, and it does so by the formation of needle-like crystals. This continuous precipitation then prevents the solution from becoming saturated with the hemihydrate, which therefore continues to dissolve. The process can continue until all hemihydrate is consumed, or other factors interfere, such as precipitated material completely occluding remaining hemihydrate, or insufficient water being present.

It can be deduced from this that there are a number of possibilities for intervening in the process and controlling the reaction, and so of modifying the setting rate of gypsum-based materials. These include control of the solubility of the reactant and product; the rate of solution of hemihydrate; the rate of nucleation of new product crystals; the crystal habit or morphology, and hence the rate of crystal growth; as well as modifying the amounts of reactants. To start, let us consider the unmodified reaction and its consequences. First of all, it is found that the amount of water required to mix a usably fluid **slurry** is rather greater than that calculated to be required for the complete stoichiometric reaction of the hemihydrate – this is shown by the fact that the set material is still wet. It is necessary to recognize that this apparent contradiction arises from two quite separate considerations.

§2. Mixing Liquid

Where a liquid is to be mixed with a powder or other granular material, the question arises of the amount of liquid needed for proper mixing. Although perhaps most obvious in the context of gypsum products, it is of direct relevance to many dental materials: acrylic, amalgam, cements, investments, filled resins, and so on. Exactly the same factors and compromises apply in all of these cases, and their handling should also be considered in the same way. The first issue is centred on the concept of **gauging liquid**.

●2.1   Volume fractions

For any particulate, non-fluid material ("powder") the amount that can occupy a given volume when it has been allowed to settle depends on the particle shapes and their relative size distributions. It does not depend on the overall or mean size of the powder particles. We are, of course, assuming that the container into which the powder is imagined to be put is very much larger than the particles, so that packing at the walls is not a consideration. Finer particles with the same shape and relative size distribution do not pack any better. Such packing is said to be **scale invariant**. This can be understood by considering a photograph of a cross-section of such a packed powder: the magnification does not affect the proportions of solid and pores.

The amount of powder in a given volume can be expressed by the **bulk density**:

$$\rho_{bulk} = \frac{\text{mass}}{\text{total volume}} \tag{2.1}$$

where the total volume is that defined by the **envelope** of the whole, the minimal surface that just encloses it. However, a more useful measure is the **volume fraction** of solid, to which we may give the symbol φ_s:

$$\varphi_s = \frac{\text{volume of solid}}{\text{volume of container}} \tag{2.2}$$

where the container is understood to be filled and the envelope for the powder then includes the boundary enclosing the free surface. Similarly, the volume fraction of pores in that bulk powder, φ_p, is given by:

$$\varphi_p = \frac{\text{volume of voids}}{\text{volume of container}} \tag{2.3}$$

This latter quantity properly must also include any porosity within the powder particles (as is the case with dental plaster, for example) as well as the interstices between particles. Hence, by definition,

$$\varphi_s + \varphi_p = 1 \tag{2.4}$$

●2.2   Gauging volume

In order that a fluid slurry be made from such a powder, a volume of liquid *at least* equivalent to φ_p must be added first. If this were not so, of course, the particles of solid would remain in contact with each other and thus inhibit flow markedly by frictional effects. Certainly, there would be that much porosity left in the final material if less liquid than the equivalent of φ_p were added; this porosity is, in general, undesirable because it reduces strength (§8). The volume fraction of pores in the bulk powder, φ_p, is commonly identified as the **gauging volume** of liquid for that powder. It is not necessarily (or even likely to be) the same as the volume of liquid required for the successful execution of the technique and due attention must be paid to the needs of the application. There are three principal reasons for this.

The gauging volume is *only* a measure of the **free space** in the packed powder, nothing else. Thus, and firstly, in defining gauging volume no account is, or should be, taken of the amount of liquid required for reaction: sand from the beach has an associated gauging volume, as does dental porcelain powder – and there are no setting reactions there. If it is a reacting system, we can determine only whether the gauging volume is greater or less than is required for complete reaction. Sometimes there will be excess liquid at the end of the reaction (as is usually the case with plaster and stone) and this remains detectable as liquid; sometimes there is none, when there is deliberately excess, unreacted, powder (as in amalgam and dental cements). Denture base acrylic requires at least sufficient monomer added to the powder to fill the spaces and avoid air porosity, but of course this monomer should all be converted to polymer ultimately.

Some care is required in considering gauging volume as it is not a fixed quantity: it depends on the state of consolidation or settling of the powder. That is, when a powder is poured into a container, the particles will settle in a random manner but limited by inter-particle friction. The rougher the particle, the worse the effect of this will be, leading to low density packing. Vibration, however, can overcome that friction and allow better settling, and greater density, with (of course) lower gauging volume. Depending on the nature of the powder, differences of 20 ~ 35% may be observed between the density in the freshly-poured state and the greatest attainable value (without crushing), the so-called **tap density** (that is, the density after tapping or vibrating the container to cause the particles to settle) (Fig. 2.1). This maximum density is, however, hard to approach without a lot of effort. There are implications for this variation of density in several areas (4§7.9, 15§4.1, 15§6.1, 25§1.2). The problem is one of reproducibility. In general, the as-poured condition is a better reference state. (This phenomenon also accounts for the appearance of many other powder or granular products sold by mass: their containers may not appear to be full in terms of volume because settling occurs in transit due to vibration.)

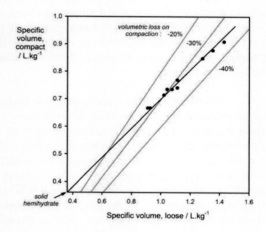

Fig. 2.1 Variation in volume of some commercial hemihydrate powders, showing the loss on attaining the fully-compact tap density.

●2.3   Dilatancy

Secondly, there is the phenomenon of **dilatancy**. When a powder is allowed to settle under gravity, each particle tends to fit as closely as possible into the available space, particularly if vibrated (this is commonly experienced with alginate impression material powders which are usually measured out by volume, 15§4). Most powders are irregular in shape, not spherical, so that if any particle were to be rotated the distance between its centre and that of an adjacent, contacting, particle must increase to allow that

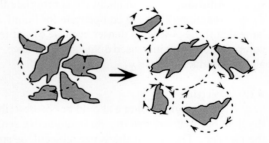

Fig. 2.2 The tumbling of particles during stirring requires sufficient space. If this is not present initially, the bulk must expand to permit it. Each particle has an associated "swept volume".

rotation to continue. Thus, when a powder is stirred, all particles will tend to rotate to some extent from their preferred (well-packed) positions (Fig. 2.2). If they move away from each other the bulk volume of the powder must increase, and therefore so must the gauging volume. This is dilatancy.

Beach sand shows the effect very well: if one walks across sand wet enough to have a smooth, glossy film of water at the surface, around one's foot when pressure is applied and while the sand is moving, a region appears to go dry or, at least, matte; it shows **loss of gloss**. This is because the locally-increased pore volume allows the surface water to drain away(or, rather, is pulled in) due to the small disturbance to each individual sand grain's position. So, to allow a slurry to be easily stirrable, somewhat more liquid must be added than the static gauging volume would suggest. This extra liquid is to allow for the dilatancy. Stirring with anything less would tend to introduce air bubbles into the mixture, again an unacceptable occurrence, at the very least because of the reduction in strength and stiffness that this causes.

Dilatant mixtures are easily recognized because they become harder to stir the faster one tries to stir them. It follows then that the liquid to be added to allow for the dilatancy of a given powder is a function of the stirring rate, at least up to the point where the powder particles are suspended in the liquid medium and capable of freely rotating without interfering with their neighbours. This limited amount of added liquid may be termed the **dilatancy allowance**. It may be viewed as the difference between the gauging volumes of the static, settled powder and that same powder when agitated – the *extra* gauging volume required by dilatancy.

●2.4   Minimum mixing volume

Beyond that point the fluidity of the mixture depends on the **dilution** of the slurry: the more liquid, the easier it will flow. It follows then that for a conveniently fluid mixture, for stirring, moulding and pouring, the actual liquid volume used must include an allowance, the **dilution volume**, along with those for gauging and dilatancy. However, the actual allowances made for dilution may vary somewhat with the application, even within one kind of product as, for example, when 'restoration' and 'luting' mixes of certain cements are defined with different powder-liquid ratios. We can therefore define the **minimum mixing volume** as the sum of the gauging volume and the dilatancy allowance, although recognizing that this might not (indeed, probably not) be usable in practice.

Where appreciable reaction occurs between the time of contact of liquid and powder in any such system (cements, amalgam and the like), the proportion of liquid actually present at the point of use will be different from that at the start, and other structures may be forming: the viscosity will rise and could make the material unworkable, damaging the structure that is forming by further manipulation. Thus, the liquid volume to be used for mixing must allow for that change. The actual **working mixing volume** therefore allows for this effect as well as the technique-dependent dilution volume, extra liquid is included to allow for loss by reaction – the **reaction allowance**. It is stressed that this is a purely physical concept because it relates to the physical (rheological) properties of the mixture in relation to its handling. The amount of liquid required for reaction, the chemical process, is an entirely separate consideration. The two should not be confused. However, it should be recognized that because of this allowance the initial viscosity, when freshly-mixed according to instructions, may appear to be too low: the temptation to thicken it up by adding more powder must be resisted.

Thus, the recommended working mixing volume can therefore be defined as involving allowances for
* **gauging**:   filling the spaces in the loosely settled powder
* **dilatancy**:   allowing for particle tumbling on stirring
* **dilution**:   to attain the appropriate fluidity for the application
* **reaction**:   to compensate for liquid volume fraction reduction as the material sets.

Note that if the powder is denser because of settling (§2.2), the decreased gauging volume will be completely compensated by an increased dilatancy allowance. That is, the minimum mixing volume is essentially fixed for a given product.

●2.5   Wetting

The gauging volume should not be described as the minimum liquid required to 'wet' the powder. In the limit one may say that the minimum in that sense is just a monolayer of the liquid over the entire surface of the powder. The size of the water molecule is such that about 0.5 mg or 30 μmol, just 1/400th of a drop, is required to cover 1 m², so that for a powder such as a dental stone, which may have a total surface area of greater than about 1 m²/g, only about 30 mg of water are required to 'wet' a 60 g portion of powder, substantially less than the 15 mL or so gauging volume. The fact is, this amount of water is probably present already, adsorbed from the atmosphere almost immediately upon exposure. Such a layer of adsorbed water is indeed likely to be present on almost every material's surface anyway.[2]

However, when mixing powder and liquid it is easy to trap air bubbles which contain dry powder, whether it is a gypsum product, alginate, investment, or cement, even though there may be well in excess of the gauging volume present. The mixing technique used, which always has some 'wiping' motion, is intended to break up and disperse such islands of dry powder and ensure that it all gets wetted. Certainly, with gypsum products and alginate, the method of adding powder to liquid (rather than the other way around) is intended to reduce the risk of that problem occurring and make correct mixing easier. (See also 15§6.)

One therefore has to distinguish carefully between the *sufficiency* of the liquid present to achieve wetting – a purely surface effect, and the *adequacy* of the mixing process, which refers to meeting the conditions for the entire bulk of material. We may therefore more usefully refer to the condition of the gauging requirement being met by saying that the bulk, undisturbed powder is **saturated**, that is, it cannot hold any more liquid in its porosity. A parallel common usage may be found when we speak of a sponge or a cloth being 'saturated' with water – it can hold no more. Sometimes the term **wetting out** is used in dentistry in the same sense of saturating a powder, but it is not as clear and distinct as a condition concept, although the process is clear enough.

§3. Dental Products

The amount of mixing water required for a given fluidity varies with the nature of the hemihydrate powder. The particles of **plaster** are very porous and, depending on the manufacturer's grinding process, often relatively large. In addition, being irregular the particles do not pack together very well, and being rough causes the friction between them to prevent settling. This material therefore requires a larger amount of gauging water than does dental stone, which is composed of regular, smooth and non-porous crystals that pack and settle well. These crystals also tend to be quite small. These variations are due to the different methods of commercial preparation.

●3.1   Manufacture

If dihydrate is heated in air at atmospheric pressure to above 130 °C or so, that is, **calcined** (Fig. 3.1), the crystal structure is rapidly disrupted as part of the water boils away, leaving a material with the composition of hemihydrate. This is **plaster**, and it has very poor crystallinity because of the lack of a medium through which the ions could diffuse to form good crystals. The calcined mass of plaster must be mechanically broken up and ground to a usable powder, but the particles remain irregular and porous. Commercially these days a 'kettle' batch process may be used, at 150 ~160 °C, which may give better control of the product.

Artificial stones, on the other hand, may be produced by **autoclaving** at high pressure, or by boiling dihydrate in a solution of calcium chloride (or other salt); both procedures raise the boiling point of the liquid present considerably. These conditions cause the decomposition of the dihydrate, because the temperature is high enough, but more importantly the presence of the liquid water permits the recrystallization of the hemihydrate. Such crystals are then well-formed, regular, and necessarily non-porous. Consequently, they pack rather better than plaster particles, hence the lower gauging water requirement.

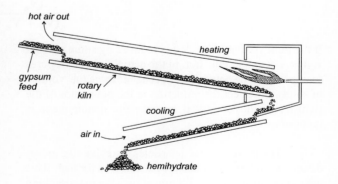

Fig. 3.1 Schematic of the type of rotary kiln formerly used for the production of calcium sulphate hemihydrate powder – plaster of Paris.

Note: the old but still commonly used labels α- and β- to identify the manufacturing process for hemihydrate have *no* crystal structure or compositional implications whatsoever in the case of dental gypsum products. To avoid confusion, therefore, they should not be used. These labels should be reserved for use only in other applications, principally for identifying different crystal structures in phase diagrams (see 12§4.1).

The distinction between the various grades of artificial stone depends essentially on the variation of the **crystal habit** of the hemihydrate powders, as controlled by the solution environment in which they are growing (*i.e.* the types and concentrations of other ions). The more **equiaxed** these crystals, the better they are able to pack in the bulk powder, the lower the mixing water requirement, the denser and harder the set material. The manufacturers' goals include finding the processing conditions that produce the highest packing density hemihydrate powder possible if a high strength, high hardness material is desired.

These particle properties are responsible in part for the choices made in the principal dental applications. Thus, **dental plaster** is used for the bases of models because it is cheap and easily shaped (*i.e.* weak), as well as because the setting expansion (§5) is not very important. The same kind of material is also used in **impression plaster** (see 27§2.4), but here primarily because it is weak enough not to run the risk of splinting the patient should there be an undercut and also to facilitate its removal from the model prepared in that impression (from artificial stone); however, anti-expansion solution is necessary here (§7.5). Should a plaster impression be broken, as is commonly the case, the fact that it is a rigid and brittle material means that it can be reassembled using sticky wax without appreciable loss of accuracy. Plaster powder is the reactive setting ingredient of **gypsum-bonded casting investments** (17§3.1), while powdered dihydrate – gypsum itself – is commonly used in **alginate impression materials** (7§9) as a source of cross-linking calcium ions because of its low but slow solubility (some alginate products may also use hemihydrate for this purpose). Autoclaved hemihydrate powders are used in **model stones**, as used for mouth models, and the harder **die stones**, which are used for the dies on which wax patterns for castings may be fabricated. The choice here is because their strength and hardness resist damage during handling, damage which would be detrimental to the accuracy and success of the final device. These different gypsum products may be variously coloured according to manufacturer and intended use, but this has no importance except for product identification and contrast with other materials.

●3.2   Minimum mixing volume (2)

As described in §3.1, the packing of the bulk powders is better when the particles are more regular, equiaxed and solid (*i.e.* non-porous). That is, the bulk density (§2.1) is higher. In other words, a given mass of artificial stone powder occupies a smaller volume overall than does the same mass of plaster powder (Fig. 3.2). Thus the gauging volume *per unit mass* of powder will automatically be much lower for the stone. Note that the volume *fraction* of pores does not change as rapidly as the bulk volume when the packing is better:

$$\varphi'_p = \frac{(V_p - x)}{(V_p + V_s - x)} \tag{3.1}$$

If the bulk volume ($V_p + V_s$) is reduced by some amount x, this is also the reduction in the actual volume of pores (compressing the powder cannot change the actual volume of solid). In effect the reduction in the numerator of the right-hand side is diluted because the denominator is reduced as well.

The more equiaxed particles also give rise to less dilatancy, so less water needs to be allowed for that. This also has an effect on the dilution allowance required for a sufficiently fluid to be easily workable. Thus the *working* liquid volumes can differ greatly. Typical values are around 60 mL water to 100 g plaster and 30 mL to 100 g for artificial stone. The 'improved' stones (which have even better crystal shapes) may require as little as 23 mL / 100 g.

Note that there would be no benefit whatsoever obtained from helping the powders pack better by vibration or compression (§2.2). Dilatancy ensures that the actual bulk volume when stirred depends on the particle shape, not on the original packing, even if proportionally the dilatancy allowance would seem to be greater. Any consolidation is undone on stirring so that the 'loose', as-poured, powder density is more relevant

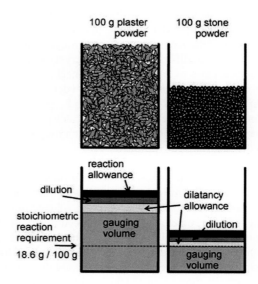

Fig. 3.2   The bulk volume of artificial stone powder is less than that of plaster because of the better packing. The mixing water requirements vary correspondingly but are unrelated to the reaction requirement. (Roughly to scale.)

(Fig. 2.1). Again, note the difference between minimum mixing volume and working mixing volume.

●3.3 Purity

As might be expected for a natural mined mineral, gypsum is not pure. Indeed, some commercially-used sources only amount to about 80% (by mass) gypsum, although some limited amounts are available at up to 95%.[3] The majority of the extraneous material is of three broad classes: silica and silicates, other sulphates (*e.g.* strontium), and carbonates (calcium, magnesium, magnesium+calcium – *i.e.* dolomite), although other ions must be present as well. While the silica would be essentially unchanged on calcining, some of the silicates may lose water. However, both would remain as unreactive 'filler' which would increase the setting expansion (§5, 17§3.2). Carbonates would remain as their decomposition temperatures are well above those for the manufacture of plaster (minimum 350 °C, for limestone), but clearly their presence must affect the pH of the slurry, to commonly around pH 8. Whether and how these substances get carried over in the manufacture of artificial stones is unknown (the solution-mediated process should in principle help), but it does seem likely that at least some will. It can also be noted that the loss of water in the formation of hemihydrate (*i.e.* ~16%) effectively increases the concentration of the contaminants in that powder. It is also likely that such compositional variation will affect the stability in storage of hemihydrate products (§10) by affecting water absorption.

> Gypsum is of a class of minerals called evaporites, that is, the result of the evaporation of ancient seas resulting in the fractional crystallization of the solutes therein. Such beds may be extensive and very thick, but contain many other substances.

For some applications, the purer grades are preferable (§13), but they still are unlikely to yield pure hemihydrate, even if gypsum from flue gas desulphurization or other commercial processes is used (in part at least because the limestone that is used will not be pure calcium carbonate). Chemical purification is unlikely to be cost-effective except for the most demanding applications. The point here is that the general assumption of purity in discussing gypsum product reactions and properties is an oversimplification at best. Variation in behaviour between products, and even over time within products (if sources vary), is to be expected. In some contexts this may not matter, in others it could be the source of complication, especially where precision is required. Batch checking is always worthwhile. This aspect will be put to one side in what follows, although clearly some cognizance at least ought to be taken in careful work. It follows that the assumption of purity in other dental systems is worth checking.

§4. Mixing Proportions

The hemihydrate mixtures used in dentistry are usually characterized by stating the **water : powder** ratio, expressed in units of mL/g.[1] This measure, however, whilst in general use for the practical handling of the material and entirely sensible in that context, is not very informative about the chemistry of the system, or very helpful in trying to understand mechanical properties. In particular, the scale is open-ended, there being no limit to the amount of water that can be taken. We shall therefore follow the lead given in §2, and convert to fractional forms. These have natural limits.

●4.1 Mass fractions

The water : powder ratio is numerically the same as the mass of water (W) added to unit mass of powder. Then, because the density of liquid water is very close to 1 g/mL at ordinary temperatures, the relative total mass of a mixture may be written as 1 + W, *i.e.* for each gram of powder there are W grams of water. The **mass fractions** of the two components are then simply given by the following:

$$A : \text{water} \quad \frac{W}{1 + W} \qquad B : \text{powder} \quad \frac{1}{1 + W} \qquad (4.1)$$

●4.2 Volume fractions

But, as remarked above (§2), volume fraction is a more informative measure. The definition of density is mass per unit volume, so the volume of a given mass of substance is obtained by dividing by its density. The individual volumes of the components are therefore $1/\rho_h$ and W/ρ_w for hemihydrate and water respectively. Expressions A and B can be converted to give the **volume fractions** of the components by first replacing each

---

[1] Fortunately, this is numerically the same as g/g for all practical purposes and hence may be treated as a dimensionless quantity.

term by the corresponding volume, then multiplying each term by ρ_h to avoid the presence of reciprocals, noting that $\rho_w = 1$:

$$C : \text{water} \quad \frac{\rho_h W}{1 + \rho_h W} \qquad\qquad D : \text{powder} \quad \frac{1}{1 + \rho_h W} \tag{4.2}$$

●4.3   Excess water

Knowing the stoichiometry of the reaction, the relative amount of water required for reaction is determined by the ratio of the **molar mass** of the extra water ($1{\cdot}5 \times 18{\cdot}02$) to the **molar mass** of the hemihydrate with which it will react, *i.e.* $27{\cdot}03/145{\cdot}15 = 0{\cdot}186$. By subtracting this quantity from the actual water added, the **excess water** is found from A and C respectively:

$$E : \text{by mass} \quad \frac{W - 0.186}{1 + W} \qquad\qquad F : \text{by volume} \quad \frac{\rho_h(W - 0.186)}{1 + \rho_h W} \tag{4.3}$$

It must be emphasized that this excess water is not the same as the gauging water, whose definition is fundamentally different (§2.2). The excess water is that remaining in the fully-set material (assuming no evaporation in the meantime). This is why fully-set gypsum products ordinarily are rather damp at first.

A negative value here means that there is insufficient water for full reaction of the hemihydrate (assuming that it is 100% pure to start with; see §3.3).

●4.4   Combined reactants

We may also enquire about the combined mass and volume fractions of the *reactants* only (by which is meant the amounts of material that will have been consumed when the setting reaction is complete), that is, as a fraction of the total amount of components in the mixture; these are given by:

$$G : \text{by mass} \quad \frac{1 + 0.186}{1 + W} \qquad\qquad H : \text{by volume} \quad \frac{1 + 0.186\rho_h}{1 + \rho_h W} \tag{4.4}$$

These expressions may be checked by noticing that the corresponding pairs sum to unity in each case, *i.e.* A + B, C + D, E + G, and F + H. Thus, *excess water* plus *total reactants* = 1, whether by mass or by volume. But now, if we compare the overall density of the reactants with the density of the reaction product, dihydrate, we identify an important characteristic of these materials, one with serious ramifications.

§5. Setting Dimensional Change

The mass proportions for the stoichiometric reaction of hemihydrate and water are 1 and 0.186 respectively (*i.e.* W = 0.186). Dividing these amounts by their densities ($\rho_h = 2.76$ Mg.m^{-3})[2] gives their individual volumes, so that the net reactant density, ρ_R, is given by:

$$\rho_R = \frac{1 + 0.186}{\rho_h^{-1} + 0.186} = 2.17 \tag{5.1}$$

The relative volume of the reaction product, V*, is defined by the ratio of the specific volumes (since the mass of material involved is unchanged):

$$V^* = V_d/V_R \tag{5.2}$$

where specific volume (*e.g.* mL/g) is defined by:

$$V = 1/\rho \tag{5.3}$$

V* is therefore in inverse proportion to the ratio of the two corresponding densities:

$$V^* = \rho_R/\rho_d = 0.935 \tag{5.4}$$

given that the density of dihydrate, ρ_d, is about 2.32 Mg.m^{-3}. There is therefore on reaction a volumetric

---

[2] Mg.m^{-3} = kg.dm^{-3} = g.cm^{-3} = mg.mm^{-3}

shrinkage, ΔV, of about 6.5 % ($\Delta V = 1 - V^*$); this corresponds to a linear shrinkage, ΔL, of about 2.2 % ($\Delta L = 1 - \sqrt[3]{V^*}$). Of course, any excess water that may be present undergoes no volume change. Hence, in the usual range of mixing proportions where there is water present in excess of the stoichiometric requirement of the reaction, the overall volumetric change would be somewhat less than this (it can, of course, be calculated in similar fashion if required).

●5.1 Crystal growth expansion

However, despite this expected shrinkage, in practice setting gypsum materials always show an overall expansion on setting, typically 0.2 ~ 0.6% (linear). The growing crystals, in radiating arrays called **spherulites** (Fig. 5.1), centred on some site of nucleation, must make contact with their neighbours and so exert a minute force on each other as they grow.

These contacts will not be geometrically perfect, so there will be many sites at which a new ion (or water molecule) could be fitted but some of these will not quite have the necessary space because of an adjacent surface (Fig. 5.2). However, there is a thermodynamic driving force for that next addition: the energy of the system is lower when it is in place than when it is not. This force then tends to squeeze it into place, that is it is now under a slight compression. It does not need to be much of a squeeze – thermal vibration might allow room from time to time. But that same thermal vibration also means that there will be a force acting to separate the two surfaces. As the two surfaces are forced apart so another site might become available for filling, and so on indefinitely, or at least until the force is insufficient against whatever external constraints may be operating to create any more space.

Summed over millions of such contact points (Fig. 5.3), the net effect is to generate an appreciable force, but it only becomes detectable when the reaction has proceeded sufficiently far for the growing crystals to contact their neighbours throughout the mass. This then is **crystal growth pressure**. We might predict that such forces would be generated in any system involving crystallization from a liquid.

If the setting mixture is floated on mercury, which provides no restraint to dimensional changes, an initial steady decline in length can be observed (Fig. 5.4) which is due to the reduction of reactant volume.[4] The next stage, a smaller increase in size, is due to crystal growth pressure. After this, a much slower decline is seen due to surface tension effects operating as the water is consumed and air is drawn into the mass. It is here that we will find another point of control for the system.

It should be borne in mind that whether used for models, dies, impressions, or as a component of an investment, dimensional considerations are paramount if the materials are to function satisfactorily in reproducing a structure. The expansions and contraction just described must be controlled if accuracy is to be achieved.

As long as the growing crystals of dihydrate are not

Fig. 5.1 Calcium sulphate dihydrate spherulite. Photograph courtesy of Barry Tse.

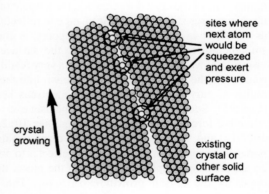

Fig. 5.2 Crystal growth pressure arises from the forces at contacts between crystals.

Fig. 5.3 Contacts between many crystals generates appreciable force overall.

in contact the mixture will remain fluid and flow under gravity, perhaps assisted by vibration, to conform to the container or mould (it is plainly the case that an accurate cast can only be made if this is true). The initial bulk shrinkage is therefore of no practical importance. When contact of crystal on crystal is made throughout the mass, as setting proceeds, such flow can no longer occur, and the shape and size of the cast is more or less determined. 'More or less', because if the mould material is of low modulus of elasticity (such as an alginate or other impression), it will not resist the expansion due to crystal growth pressure and the resulting cast will be oversize.

Even so, the net volume of the reactants and reaction product continues to decrease as the setting reaction proceeds: water is being consumed and occupies less and less space in the now continuous framework of the dihydrate crystals. The remaining water cannot be stretched to fill the same space but must instead be withdrawn from the free surface as it is consumed. This produces a quite sudden change from a wet and glossy surface to one which is matte and (relatively) dry-looking: this change is the **loss of gloss** (§2.3) which may be used as an indicator of 'initial' setting. Because at this point the growing crystals are thrusting against each other, causing an increase in the overall dimensions of the crystal framework, the amount of air drawn into the mass increases still further. The change in reflectivity of the surface, as the water is drawn down into the pores, is due to the originally smooth, mirror-like liquid surface – the surface tension and low viscosity of water means that this occurs very readily – becoming very rough. As the water surface is withdrawn into the pores, locally the radius of curvature of the resulting meniscus becomes very small (Fig. 5.5). Such small highly curved menisci exert large attractive forces on their surroundings (see 10§2, Fig. 10§2.12) and, again summed over the many millions of pores of the surface of the cast, the net effect is to provide an appreciable compressive stress on the piece, that is, opposing the crystal growth expansion. The extra pressure being exerted at crystal contact points means that crystal growth is now slightly more difficult, and the setting expansion is thus rather less than would otherwise be expected.

Notice that the point at which loss of gloss occurs cannot be associated in any fundamental way with the extent of conversion of hemihydrate to dihydrate because this depends on water : powder ratio (*i.e.* dilution), rate of nucleation and thus on mixing conditions (§6.1) and

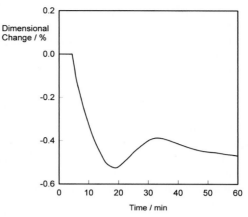

Fig. 5.4 Setting shrinkage of plaster floating on mercury.

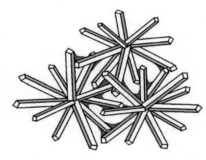

Fig. 5.5 Remnants of the mixing liquid are held by surface tension at crystal contacts and the small radii of curvature generate an appreciable force.

It is common to talk about the liquid phase in gypsum and similar systems, such as casting investments (Chap. 17), as just 'water'. It must be borne in mind that this is a convenience for brevity, because such liquid will always contain dissolved substances: saturated with calcium sulphate, here, obviously, but also impurities, additives and the residue of other reactions. If excess of a soluble solid is present, then the solution is necessarily saturated with respect to that substance (given enough time), otherwise all available such materials must dissolve completely. Evaporation of that solution must leave the solutes behind, and these must help cement the structure.

additives (§7), as well as crystal habit (additives again). It is a purely phenomenological issue, reflecting the point in the process when a continuous path of crystal contact has been achieved across the setting mass and sufficient expansion has occurred to disturb the flatness of the liquid surface noticeably.

It is worth noticing that one other factor may affect the timing of loss of gloss, although not under conditions normally relevant to dentistry. That is, if the slurry is sufficiently dilute (the water : powder ratio high enough), then some settling (4§9.3) of the hemihydrate may occur before reaction has proceeded very far. A layer of 'excess' water will therefore appear to form on the surface. Obviously, this would have to be consumed by reaction before loss of gloss could occur. Indeed, if the slurry were sufficiently dilute, it never would be consumed. But this reveals an experimental difficulty that might be encountered: once settling occurs in this way, what is being tested is the result of the setting of the settled material – with a more or less fixed water : powder ratio locally – and would not be representative of the overall, original mixing proportions. Such an effect would be exaggerated by vibration (shear thinning, 4§7.9), as is commonly used to ensure that a poured slurry properly fills an impression mould, that is, to displace trapped air and bubbles in slurry. In fact, even in normal use, this may well cause some densification of the solid component of the slurry at the bottom of the mould space (the cusps of the teeth?) and so lead to slightly different properties (including expansion) with depth.

●5.2   "Hygroscopic" expansion

However, the effect of the contraction force associated with loss of gloss may be overcome by providing extra water during setting. This may be done by immersing the setting mixture in water, or simply by adding a few drops to the surface, thus allowing water and not air to be drawn into the space created by the decrease in volume of the reactants. Since the liquid surface does not then enter the pores, the tension due to meniscus curvature is not generated, and the expansion due to crystal growth is thereby uninhibited. This may as much as double the expansion measured. This effect is ordinarily referred to in dentistry as **hygroscopic expansion**. It is purely physical in origin. Note that the proper interpretation of this effect is that the extra water *allows* the expansion to occur by removing a constraint, it does not *make* it happen by providing a driving force.

It is also worth remarking that the use of this term is fundamentally incorrect, having nothing whatsoever to do with the chemical phenomenon of hygroscopicity, which is the tendency of a substance to absorb water from the atmosphere; a term such as **hygrogenic expansion** would be better. Nevertheless, it is unlikely that the dental usage could be changed in the foreseeable future.

●5.3   Expansion measurement

As indicated above, the expansion or contraction actually observed depends on the confinement of the sample, and in fact on the method of recording the dimensional change as well.[5] For the same product and similar mixing conditions a range of results may be demonstrated (Fig. 5.6). If the total (apparent) volume of the mixture is monitored, the initial contraction will be observed, followed by some expansion. In the more usual circumstances, where the top of the mixture is open in a trough-shaped 'extensometer' (Fig. 5.7), and thus may subside under gravity, the initial contraction cannot be seen as insufficient force is generated to move the end-wall, the position of which is monitored by a dial gauge or other instrument. However, this contraction cannot have any bearing on the dimensional changes experienced in practice, except that the amount of water used for mixing will play a part in determining when crystal growth has proceeded far enough for the crystals to impinge on one another: the more water, the later that expansion will occur and therefore the less that can then occur (because there is less hemihydrate remaining to convert). Hence, it is important that the correct, specified, water : powder ratio be used, whatever other controls are employed.

The components of the dimensional changes occurring during the setting process are summarized in Fig. 5.8. Note that the relative values depend on several conditions and so cannot be specified with any precision.

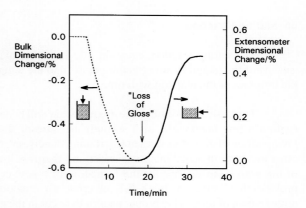

Fig. 5.6 Effect of test method on the setting dimensional change observed. Left, broken line, measuring 'total' volumetric change; right, solid line, using a V-trough 'extensometer'. Notice that the scales have been adjusted so that the final parts of each curve overlap.

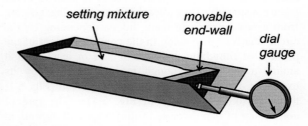

Fig. 5.7 A V-trough 'extensometer' for observing the expansion of setting gypsum products and similar materials.

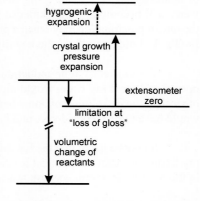

Fig. 5.8   Schematic summary of dimensional changes in setting gypsum products. (NB: not to scale.)

A matter of some importance to arise from these observations is that the setting expansion of a model material in an impression will be constrained unevenly. In the bucco-lingual direction, the impression tray walls may provide considerable constraint, while in the mesio-distal direction, along the arch, there may be substantially less, but perhaps not zero constraint. In the (tooth-)axial direction (antero-posterior) there will be little restriction to expansion because the mould is open that way. We therefore expect to find distortions in the model tooth shape because the plaster or stone was not able to expand isotropically. This would also be affected by the thickness and stiffness of the tray wall, the thickness of the impression material between the tooth and the tray wall, and on the modulus of elasticity of the impression material.

§6. Setting Reaction

The setting reaction, equation 1.1, is exothermic (Fig. 6.1), as might be anticipated from the thermal aspects of the manufacture of hemihydrate, and the temperature of the setting mixture can be used to monitor the progress of the reaction with time.

Typically, even though the mixing water will become saturated with respect to hemihydrate very quickly (within a few hundredths of a second), because of the large surface area of the powder and the stirring, precipitation of dihydrate does not start immediately, even though it is thermodynamically required. As in many other instances in dental materials' reactions, there are **kinetic** limitations. In this case, it is the non-availability of the **activation energy** for the formation of a new crystal, that is, for the **nucleation** of that crystal, that provides that limitation. It is a difficult process to create a new crystal, and it may take some time for the chance arrangement of ions in the proper pattern to occur in the normal course of diffusion, which is a random process. However, the activation energy for the addition of ions to the surface of an existing crystal is very small. Once precipitation has been initiated crystal growth is limited only by the diffusion rate (kinetics, again) of ions to the surface through the solution. But as the crystals grow, so heat is evolved. Experimentally, the greatest rate of heat evolution, as shown by the greatest rate of temperature rise, and therefore of crystal growth, is seen to occur some considerable time after the start of mixing, ranging from a few seconds to many hours, depending on the circumstances. The delay in the commencement of the process of dihydrate precipitation, or at least until it attains a rate great enough to be noticeable by its exotherm, is known as the **induction period** (Fig. 6.1; this is also seen clearly in Fig. 5.4). The rate of conversion of hemihydrate to dihydrate will depend, at least in part, on the total surface area of existing dihydrate, and thus on the number of crystals and their size, and this in turn depends on the number of nucleation sites. At a certain point, however, the rate of dissolution of hemihydrate, replenishing the stock in solution as it were, will become limiting (kinetics yet again), so that the rate of precipitation may reach a plateau value before finally declining rapidly as the hemihydrate is used up.

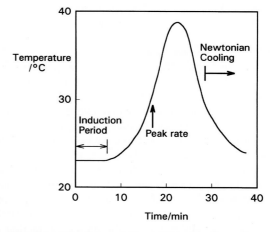

Fig. 6.1 The form of the reaction exotherm for a setting gypsum product.

The existence of an induction period longer than the time needed to prepare and use the mixture, the operational **working time**, would mean that a reaction allowance (§2.4) would not be required as no change other than the dissolution of some hemihydrate would occur. If the setting process is accelerated (§7.1), this would have to be reconsidered, and in all practical dental contexts this must be the case.

It may be noted that the loss of gloss does not correspond to the end of the induction period as the latter marks only the start of appreciable crystal growth, not a full contact network, and loss of gloss is not detectable in the reaction isotherm because it is a mechanical, not a chemical event.

It may also be noted that the peak temperature corresponds only to a balance being struck between the rate of heat generation and the rate of heat loss to the surroundings. Since the latter depends on the exact circumstances (mass of material, surface area, conductivity of mould, draughts, *etc.*) it provides no information about the chemistry of the setting process. While the rate of reaction falls off rapidly after the peak temperature has been passed, it does not go to completion for many hours. This is thought to be due to the last remnants of hemihydrate being extensively occluded by the precipitated dihydrate, making the path for the diffusion of ions the very narrow intercrystalline spaces. These spaces tend to close up as, of course, the conditions for precipitation of dihydrate are correct there. Changes in strength can be observed for at least 24 h after mixing – the so-called **ageing** of the set plaster or stone. This is despite the fact that there is usually ample water remaining. The presence of any anhydrite (Fig. 1.1) in plaster, because its manufacture is not a very precisely controlled process (Fig. 3.1), will also affect the progress and rate of reaction, according to form and amount.

●6.1   Nucleation

A crystallization nucleus may be any site on existing solid that favours or allows the exact arrangement of the constituents of a crystal. The first few ions to precipitate must go into the correct configuration or crystal growth cannot occur. These nuclei can be either random sites on other types of solid material, or small crystals of the same compound. This kind of nucleation is called **heterogeneous**; a solid substrate is involved as it is more favourable in terms of the activation energy for the process (see 11§2.5). The alternative to this is **homogeneous** nucleation, where the crystal nucleus is formed spontaneously in the solution. This requires a large assembly of ions in an arrangement which could occur randomly only with very low probability. For this process the induction period is necessarily long, and may explain the slow initiation of setting with unmodified very pure hemihydrate – it may take up to 24 h to commence. It is obvious that hemihydrate itself cannot provide nucleation sites, because the very large quantity present in the slurry when it is first mixed would otherwise appear to require the immediate initiation of crystallization. This does not happen. However, and yet again for kinetic reasons, the industrial production of plaster does not necessarily convert all gypsum to hemihydrate, and the small residue will be more than enough to promote crystallization. The decomposition reaction is reversible and, for the required dehydration to occur, time is required for either diffusion of water vapour from deep within lumps, in the case of plaster production, or for the growth of hemihydrate crystals in the case of artificial stone manufacture. Thus, the approach to completion of the process requires longer heating times, which costs more, and therefore – on the basis of 'diminishing returns' – purity may be sacrificed for price.

Even so, the presence of dihydrate in plaster powders is an unreliable feature, and samples of commercial products may show great variation in induction period. To avoid the difficulties that this inconsistency would produce, some products incorporate a small quantity of dihydrate, finely powdered, in order to have a more nearly constant amount of nucleation, and therefore predictable setting behaviour. This is why scrupulously clean mixing bowls and spatulas must be used to avoid undesired extra nucleation and an inappropriate increase in the setting rate. Such additions may be made deliberately (by the producer, not the user) especially when modifiers are used to control expansion, and particularly when these retard the setting. In a similar manner, the amount of spatulation used with the relatively fast-setting dental products can affect the *number* of nuclei by breaking up crystals as they are forming: crystal fragments are very effective in this way. **Over-spatulation** is therefore to be avoided. Recall the **seeding** of supersaturated solutions with a small crystal of the substance to be crystallized and the dramatic effect this can have to appreciate the significance of this. As a result of these circumstances and the need for reliably consistent behaviour, it may be that no appreciable induction period is discernible for some products and the temperature rise indicating precipitation commences more or less immediately.

Mechanical mixing machines may also be used for both speed and ease of mixing as well as operating under reduced pressure to minimize the presence of bubbles. It is to be expected that the vigorous stirring given by such devices will tend to fragment crystals even more than for hand-mixing, and this may shorten the setting time noticeably. Over-mixing is again to be avoided.

●6.2   Thermal expansion

It will be noticed that the setting expansion is occurring at about the same time as the temperature is rising. They are not directly causally related. That is, the majority of the expansion is not primarily *thermal* but rather is due to crystal growth pressure (§5.1) and necessarily can only start after a certain amount of reaction has already occurred and the growing crystals have become large enough to come into contact with other dihydrate, the walls of the container or other solids. This can be seen by the fact that the temperature starts to rise (because the reaction is exothermic) before the expansion is noticed.

Even so, the temperature rise on setting of gypsum products is substantial. This prompts the question of whether this affects the dimensional accuracy of the cast. The thermal expansion coefficient for plaster and stone is about 12 MK^{-1},[3] so that cooling from, say, a peak of 40 °C to 25 °C the shrinkage is expected to be about 0.02%. This can be considered as entirely negligible in comparison with the usual scale of the setting expansion (0.2 - 0.6%). Note that this is entirely unaffected by the porosity (or, equivalently, the water : powder ratio) of the material (*cf.* 18§4.6).

●6.3   Exotherm
The setting exotherm is not a problem in any sense for the material itself, but in the case of impression plaster some care is required, especially since its setting rate needs to be increased so that it sets in a rather short time (§7). Even though ordinarily it may be used in relatively thin sections, the common use of an impression tray of thermally-insulating material means that cooling will not occur as quickly as might be expected. In addition, an exotherm is by definition a rise above ambient, and in the case of soft tissue (an open mouth temperature of, say, 35 °C) this may then become important with respect to the temperature at which inflammation occurs, ~42 °C. The temperature rise may be unexpected, and worrying for the patient, if not explained. It is also of some concern in surgical contexts, where tissue damage is to be minimized.[6]

§7. Modifiers

The setting reaction was said earlier to depend on the relative solubilities of the hemihydrate and dihydrate, but more detail is required to understand the effect of modifiers.

●7.1   Potassium sulphate
Potassium sulphate is a very effective accelerator for the setting process. Adding it results in the formation of the distinct crystalline double salt known as **syngenite**: $K_2Ca(SO_4)_2 \cdot H_2O$. The solubility of this compound, in g/L, is apparently higher than that of dihydrate, but not if expressed in terms of molarity with respect to the calcium it contains:

	solubility @ 20 °C	
$CaSO_4 \cdot \frac{1}{2}H_2O$	8.8 g/L	60.6 mmol/L
$CaSO_4 \cdot 2H_2O$	2.1 g/L	12.2 mmol/L
$K_2Ca(SO_4)_2 \cdot H_2O$	2.5 g/L	8.6 mmol/L

The solubility of ionic compounds is actually determined by the ionic product for the constituent species which are at equilibrium in the presence of the solid substance. This is the **solubility product**, K_{sp}, which expresses the fact that an independent increase in the concentration of one component by the addition of another salt can lead to crystallization at lower concentrations of the other. So we may write for dihydrate:

$$K_{sp} = \frac{[Ca^{2+}][SO_4^{2-}][H_2O]^2}{[CaSO_4 \cdot 2H_2O]}$$

$$= [Ca^{2+}][SO_4^{2-}] \approx 2.6 \times 10^{-5} \quad \text{at } 20\,°C \tag{7.1}$$

(that is, taking the conventional activity of both the solid being formed and the water to be unity); and for syngenite:

$$K_{sp} = [K^+]^2[Ca^{2+}][SO_4^{2-}] \approx 1 \times 10^{-13} \tag{7.2}$$

In this way the solubility of any substance depends on the concentrations of the common ions from other solutes already present. Thus, in the presence of calcium and sulphate ions, the addition of any potassium ions to exceed the required ionic product for syngenite leads to the preferential crystallization of that salt, even if the formation constant for dihydrate were not attained. Potassium chloride would therefore be effective as a source of potassium ions to achieve this, but if potassium sulphate is used, the sulphate concentration is also increased and the effect is magnified. Notice that the effect of these ions is expressed as the square of their concentrations, so that the system is really quite sensitive to that salt, which might account for its well-documented effectiveness as an accelerator.

---

[3] 1 MK^{-1} is the same as 1×10^{-6} K^{-1}; say "per megakelvin".

However, this solubility consideration on its own is insufficient. There are two distinct requirements which must be met: firstly, the accelerator must itself cause the crystallization of a solid, and then that solid must facilitate the precipitation of the dihydrate. Since acceleration does occur in this system, syngenite itself seems to have a low activation energy for nucleation. It also seems that its formation does not have a significant induction period of its own. Nevertheless, it is not yet clear whether syngenite nucleates homogeneously, or heterogeneously on hemihydrate. But the presence of another solid material will not, of itself, accelerate setting unless it is also effective as a base for heterogeneous nucleation. That is, the syngenite must act as a **template** for the formation of dihydrate crystals while, as mentioned above, dihydrate does not nucleate on hemihydrate.

●7.2 Sodium sulphate

In the same way we may consider the effect of sodium sulphate. However, there is no stable sodium analogue of syngenite, and any acceleration of the setting reaction will be due essentially to the increase in sulphate concentration. But, when the sulphate concentration is increased in this way, the dissolution of the hemihydrate will be inhibited as its solubility product will be attained by the dissolution of less hemihydrate. The net result, therefore, depends on the balance between these two effects. A further mechanism for inhibition exists because of the limit of solubility of sodium sulphate itself (about 230 g/L at 20 °C). If during the reaction, which is of course *consuming* water, the solubility product for sodium sulphate is exceeded, sodium sulphate decahydrate will precipitate (below 32·4 °C, above which the anhydrous salt is stable) and may physically block dissolution of hemihydrate. This obviously depends on how much mixing water is present. So there is an optimum concentration (Fig. 7.1) for the accelerating effect of sodium sulphate, but somewhat dependent on the water : powder ratio used.[7]

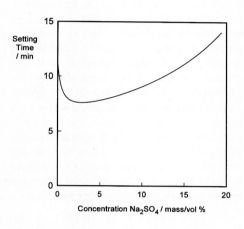

Fig. 7.1 Effect of added sodium sulphate on setting time.

●7.3 Salt effect

Strictly speaking, the solubility product is calculated not in terms of concentration but from the activity of the ions concerned. In dilute solution the activity or 'apparent concentration' may be close to the actual concentration. In concentrated solutions the effects of the ions on each other and the effect of other non-reacting ions influence the activity coefficient and reduce the activity, so the rate of reaction to produce dihydrate will be reduced. This is known as the **salt effect**. The dissolution of hemihydrate will also be affected so that the actual interrelationships operating in the system are quite complicated. For example, the overall setting rate goes through a maximum with the addition of sodium chloride (Fig. 7.2). The precipitation of crystals of sodium chloride can also occur with low mixing water and high concentrations of salt, in the same way as with sodium sulphate.

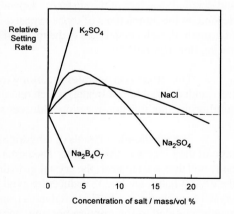

Fig. 7.2 Effect of various added salts on setting rate (schematic).

●7.4 Borax

Borax (sodium tetraborate decahydrate, $Na_2B_4O_7 \cdot 10H_2O$), on the other hand, has a rather more direct effect as an inhibitor. The solubility product for calcium borate is low and all concentrations of borax result in the precipitation of a layer of calcium borate[4] on the hemihydrate particles, effectively reducing their rate of dissolution, but also on any pre-existing dihydrate particles so that they are rendered ineffective as crystallization nuclei.[8] Thus the setting reaction may be retarded. This effect is a purely kinetic limitation (although more complex factors may also be operating[9]). Many other substances are also effective in this respect: many biological organic materials such as proteins (*e.g.* gelatine, saliva, blood) and polysaccharides (agar, alginate) interfere with crystal growth, possibly by adsorbed layers preventing either dissolution of hemihydrate or by

---

[4] Sometimes said to be colemanite, $CaB_3O_4(OH)_3 \cdot H_2O$, although the evidence is obscure. However, the exact chemistry is of no concern here, it is just that a precipitate forms as a coating.

occluding the dihydrate growth areas. This provides one reason for rinsing off such contaminants from impressions, to permit the proper setting of the dental stone when the model is poured. Other retarders include $Fe_2(SO_4)_3$, $Al_2(SO_4)_3$ (despite the extra sulphate), acetates and citrates.

●7.5   Crystal habit

All of the substances mentioned above affect not only reaction rate but also the form or **crystal habit** of the dihydrate produced (compare the effect of solution conditions on the form of the hemihydrate crystals in artificial stone, §3). Crystals do not grow by accretion of ions all over their surface in a smoothly random fashion (this would lead to nearly spherical and very porous 'crystals'!) but generally by the growth of one layer at a time on each crystal face to form **terraces**, the actual active sites being very few in number and generally restricted to **kinks** in the **ledges** at the edge of these terraces. This arises because the activation energy for the start of a new layer

Fig. 7.3 Three terraces to show some types of crystal growth site. A: hole in a terrace, B: corner at a kink in a ledge, C: edge or ledge site, D: surface, in order of decreasing energy of capture.

is rather higher than that required to continue an existing layer (Fig. 7.3).[10] Thus, to take a crystal of NaCl as an example, the relative energy benefit for adding an ion at some of those sites are:

   B : 0.874   C : 0.181   D : 0.066

Site type A can be expected to be even more energetically favourable, although kinetically it may be of lower likelihood that it gets filled. In addition, the energies of the various naturally-forming crystal faces, or **habit planes**, are different, so that the crystal grows in some directions faster than in others. These different energies arise because the combination of ions or other entities exposed by a cut through a crystal – the 'composition' and detailed structure of the surface – depends on where and at what angle the cut is made. These energies also depend on the interactions of those exposed species with all the species present in the solution. Thus, the rate of growth of each crystal face is the result of all of these factors operating. This has a bearing on the observed setting dimensional changes.

This effect also operates according to the presence of excess ions from reacting additives (such as of K^+ from added potassium sulphate) which remain in the solution in order to satisfy the solution *vs.* solid solubility equilibrium, not just non-reacting additives such as sodium chloride.

The needle-like crystals of dihydrate formed in a pure system may be described by their **axial ratio**, the ratio of their length to their width. The larger this is the more rapidly the growing crystal may reach some other body and thus start to exert the crystal growth pressure and cause expansion. Many other ions in solution affect the crystal habit, tending to reduce the axial ratio (it is naturally so large in the first place that any change in crystal face energies is likely to reduce it), and therefore reduce the setting expansion. Potassium sulphate, for example, added to the mixture not only accelerates the setting, it also reduces the expansion. However, the acceleration that would result from an addition giving sufficient reduction in expansion is too great to be workable, and a retarder is necessary to balance the effects. **Anti-expansion (AE) solution** is therefore a mixture of potassium sulphate (4%) and borax (0.4%) to attain just that balance whilst reducing expansion appreciably, typically to about 0.05%. Both additives individually cause a reduction in expansion, but neither is satisfactory on its own because of the effects on rate. It is probably not possible to eliminate expansion altogether, since some crystal contact, and therefore growth pressure, must occur, no matter what.

As described in §3.1, dental stones are made by crystallizing hemihydrate from solution. Thus, as distinct from plaster, the purity of the product is much higher and little or no contamination with dihydrate can be present. The same induction period problem as described in §6 would therefore occur. It would seem that, to get the prompt setting shown by most dental stone products, manufacturers add the components of anti-expansion solution to the powder, thus ensuring suitable setting behaviour with low expansion. It is therefore quite unnecessary – indeed it would be wrong – to use AE solution with such materials as the appropriate balance has already been attained.

The involvement of a contribution from an activation energy in a process implies the temperature dependence of rate. This is true for the setting of gypsum products also, and an increase in the setting rate (noticeable primarily as a *decrease* in the induction period) can be obtained by using mixing water at higher temperatures, up to about 30 °C or so. However, the energy terms themselves, as well as ionic mobilities and solubilities, also vary with temperature and overall only a slight increase in rate is observed on further raising the temperature to about 50 °C. Thereafter, the rate decreases steadily up to ~100 °C when it goes to zero, which has to do with the relative instability of dihydrate (§11.4).

●7.6   Mixing water purity

It is relevant here to remark that the use of tap-water for mixing may be detrimental to gypsum product properties. Such water is far from pure, and may contain calcium, magnesium, sulphate, phosphate, silicate, nitrate, chloride, chlorine, fluoride, aluminium, iron, copper, lead (old piping, but also from soldered copper piping joints) and possibly many other species, depending on the water source, treatment, and the type of plumbing. All of these species might be expected to have some effect on the activation energies of nucleation and growth. Certainly, it is observed that setting rates may be altered. Expansion will necessarily be affected also. It is then essential to avoid such complications by the use of distilled, reverse-osmosis or deionized water only for mixing gypsum products or making up anti-expansion solution. Bottled drinking water is also not acceptable because of the deliberate mineral content. It must be remembered that a primary requirement for a model or impression is dimensional accuracy, but a second consideration is consistency from time to time. One needs to rely on the reproducibility of all of the processes of dentistry to attain the highest quality results.

§8. Porosity

Because excess water is always required to produce a usable slurry, this excess must be left at the end of reaction. The set gypsum product will therefore have both solid and liquid phases present. In addition, the setting expansion will have drawn either air or more water (if this was provided for hygroscopic expansion) into the structure (Fig. 8.1). However, neither air nor water has any strength. The relative strength of the set mass therefore depends strongly, and roughly inversely, on the volume fraction of pores, φ_p, (*i.e.* non-solid material). In fact, the strength becomes zero at a value of φ_p much less than unity (Fig. 8.2). This is the **critical porosity**, when the structure is incapable of supporting itself, either because the skeleton that is left is too weak to carry its own weight or the structural elements as they grow do not make contact to create such a skeleton.

This behaviour is observed in many diverse systems (Fig. 8.3), and it is no less applicable in other dental contexts such as cements, investments, amalgam and porcelain, irrespective of their variety of chemistry and structure.[11] The excess volume fraction

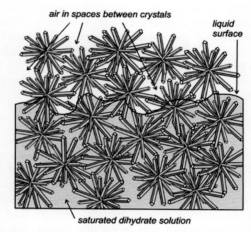

Fig. 8.1 The spaces between dihydrate crystals may be filled by air or liquid – neither has any strength.

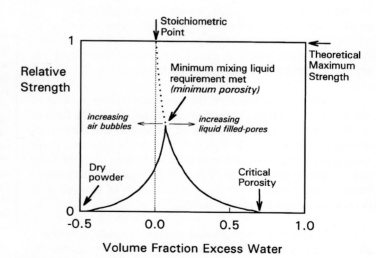

Fig. 8.2 Strength of plaster and stone as a function of water added. The abscissa is calibrated according to formula F of §4. Thus, reading from the left, increasing reaction gives more conversion and better binding with decreased volume of air-filled voids. After the cusp, at the point where the mixing liquid requirement is met, excess water leads to increasing liquid-filled porosity and decreasing strength. (Negative excess = insufficient)

of fluid is the key factor in determining the strength of this type of material as it represents a phase of zero strength in the final structure. The effect is the same whether it is water or, after evaporation, air, or a mixture of both, permeating the whole body.

It is worth noting that this behaviour is in marked contrast to that which might otherwise be expected on a 'remaining solid cross-section' prediction, *i.e.* the straight line in Fig. 8.3:

$$\sigma = \frac{A_s}{A_p + A_s} \cdot \sigma_0$$

$$= \frac{V_s}{V_p + V_s} \cdot \sigma_0 = \varphi_s \cdot \sigma_0 \qquad (8.1)$$

where A_p and A_s are the effective cross-sectional areas of porosity and solid respectively. Porosity is so detrimental because of the stress-concentration it causes on the remaining solid but also because in the Griffith sense (1§7.1) the probability of a critical flaw being found in such a stress-concentration rises rapidly, compounding the effect.

●8.1   Bubbles
A source of porosity that is no less detrimental to strength is that of the air bubbles introduced during mixing. If the mixing is done by hand, and with at least the minimum mixing water present, air bubbles are readily incorporated. They will, of course, simply add to the effect of the excess water and, perhaps, air in the final structure discussed above and generally reduce the strength. However, the size of the bubbles thus formed will in general be much greater than the typical space between dihydrate crystals (Fig. 8.4) and so, by reference to the Griffith criterion (Fig. 1§7.2), lowers the strength more rapidly than would otherwise be expected. This is the reason why mixing under (even partial) vacuum and casting with vibration (to allow bubbles to escape, due to shear thinning, 4§7.9) are the preferred and recommended techniques.

Similarly, if less than the minimum mixing volume of water is used, a complementary amount of air, also in the form of discrete bubbles, must necessarily be incorporated at the start of mixing, primarily due to dilatancy. Although this does not prevent such mixes being made and used (or tested), they are very pasty and are then more prone to further air incorporation as trapped bubbles, and hence a very rapid deterioration of strength with decreasing mixing water is observed.

In fact, both strength and surface indentation hardness show a very pronounced peak or cusp in their plot against the volume fraction of excess water added, indicating the great sensitivity of the system in this respect (Fig. 8.2). It follows then that deviation from recommended mixing ratios is to be strongly discouraged, as it is usual for these to correspond very closely to the minimum mixing volume, now identified from a graph such as Fig. 8.2 as corresponding precisely to that point. Notice that the theoretical strength of solid crystalline gypsum is not attainable. We have here one of the many essential compromises of dental materials.

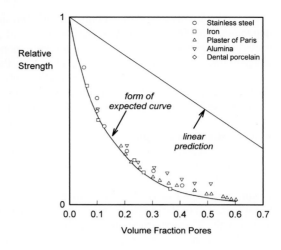

Fig. 8.3 The effect of porosity on the strength of several materials. Note that the linear prediction fails. (See also Fig. 14§8.1)

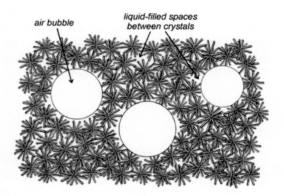

Fig. 8.4 There are two kinds of porosity: small scale distributed porosity between dihydrate crystals and larger, discrete air bubbles

Since the water : powder ratio approaches infinity for very small amounts of hemihydrate, it is unhelpful to use this ratio in practice to assess the critical porosity. Conversion must be made through the equation for excess water (§4) or a direct calculation made. Indeed, while it is a very convenient means of specifying what the mixing proportions are in the clinic or laboratory, water : powder ratio has no fundamental chemical or physical significance and should be avoided when behaviour is being studied, *i.e* scientifically.

It is therefore helpful to distinguish between **intrinsic** and **extraneous flaws**. Intrinsic flaws are those that are present as an unavoidable consequence of the nature of the material, in this case the distributed porosity between the crystals that arises because of the need for a minimum mixing volume greater than required for stoichiometric reaction. Bubbles introduced during mixing or pouring, as well as scratches on the surface, are extraneous – due to faulty technique or inattention.

It is the need to avoid incorporating air bubbles that underlies the instruction to sift powder onto the surface of the measured out water. This means that the effects of capillarity (10§2.9) drive the liquid front in one direction through the powder as it 'soaks in', displacing the air. Otherwise there is a great risk of enfolding a bubble, containing dry powder, which is then very difficult to break down and eliminate. Similar problems are encountered with other dental products, especially alginate impression material (7§9).

Note that the strength of plasters is generally reported to be lower than that of artificial stones, whose strength is lower in turn than that of die stones. This can be seen to be due primarily to the effects of hemihydrate particle size and the corresponding minimum mixing volume requirements of the three types of product (Fig. 8.5). The volume fraction of solid decreases with increasing water : powder ratio (Fig. 8.6). Chemically, the reaction products are identical. It is essentially the residual pore volume which controls the strength, if dihydrate crystal shape effects are held constant (which of course they may not be in practice because of differing additives between products).

In a porous body, whatever the cause of the porosity, the indentation hardness (1§8) depends at least in part on the collapse of the structure, compacting it. In a ceramic system such as a gypsum product, there will be negligible plastic deformation of the structural material itself, the crystals, and their brittle fracture will be the principal mechanism of collapse. This behaviour may be termed **quasi-plastic**, macroscopically it appears to be plastic, but microscopically it is not. This is to be contrasted with the displacement sideways and upwards at the sides of the indenter which occurs with the resulting plastic deformation of metals and polymers – a true flow phenomenon.

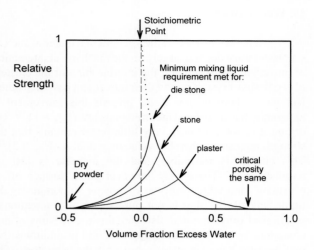

Fig. 8.5 Note that the principal effect of varying the type of hemihydrate product is to vary the minimum mixing volume and thus the strength attainable.

Fig. 8.6 Increasing the water : powder ratio results in a more 'dilute' structure, with greater spaces between crystals.

As a result of this difference in mode of indentation, a key distinction can be made between hardness and other strength tests: the hardness value obtained is not sensitive to flaws in the Griffith sense, no matter what is happening on a small scale, locally. Thus, a single bubble that might be a critical flaw in a flexure strength test would represent only a small fraction of the total porosity that is obliterated by consolidation; its effect on the depth of penetration, and thus the hardness value determined, will be similarly small. Accordingly, a sharp cusp at the minimum mixing liquid point will not be observed, and the hardness continues to increase with reduction in mixing volume, diverging only slowly from the ideal curve. A gradual roll-over to a decreasing curve at some substantially lower mixing volume is therefore expected, even though in practical terms this may not be attainable. However, as bubbles here (and other flaws in other materials) will be randomly distributed, what exactly occurs under an indenter is variable: increasing scatter will be observed. Furthermore, this scatter will increase the smaller is the indentation (*i.e.* lower the test load), as a smaller volume of material is consolidated and less averaging of effect occurs.

§9. Wet or Dry?

It is easy to see that the strength of a material such as set plaster must depend largely on the mechanical interaction of one crystal with another; the intermeshing of the clusters of needle-like crystals inhibit their movement past each other. But while there is any water present this movement is apparently lubricated and the strength is low (Fig. 9.1).[12] On drying it is not until only very little water remains that the strength increases, when the value may double (Fig. 9.2).[13] This change is reversible, and the increase is lost on remoistening. There may however be a further effect operating, one which may be of greater importance: the liquid left after setting is not pure water, it is a saturated solution of dihydrate and also contains the remains of any contaminants of the original gypsum, and of additives that may have been used in manufacture or in the laboratory. The evaporation of this solution will first of all lead to some crystal growth, but then as this approaches completion, and the liquid is separated into small droplets in corners and at contacts, it will leave a residue of solid dihydrate and these other substances cementing together the dihydrate crystals at their points of contact, much of which would redissolve on rewetting. Thus, in Fig. 5.5, every remnant of liquid at contact points will leave some solid that increases the mechanical interference to relative movement even if bonded by no more than van der Waals and ionic forces (10§3).

Being composed of a well-crystallized ionic substance, set plaster and stone are very brittle materials. That is, they show no plastic deformation on fracture and can withstand very little strain before breaking. Care must therefore be exercised in their handling, but there is a definite advantage to this in the case of impression plaster: should the impression break, or need to be broken in order to remove it from the patient's mouth because of an undercut, the pieces can be reassembled (and held with sticky wax) for the model to be poured, without loss of accuracy.

Some laboratory techniques, such as when wax is to be formed on a model, require the model to be wet. But because the solubility of dihydrate is appreciable, significant amounts can dissolve if the set gypsum product is exposed to fresh water. This will result in the loss of material from the surface, usually the most important aspect of a cast, making it rough, friable and, of course, changing the dimensions of the piece. Obviously, the more frequently this is done, the worse the effect. Consequently, washing or soaking models in fresh water should be minimized, and preferably avoided altogether. Similarly, when the base of a model is ground to a convenient shape on a model trimmer, which uses running water to remove the debris and keep the grinding wheel from clogging (20§2.5), if the model has dried at all in the meantime it becomes similarly vulnerable. A saturated solution of calcium sulphate dihydrate,

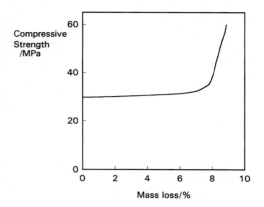

Fig. 9.1 Effect of drying on strength. The material was mixed with a water : powder ratio of 0·30, equivalent to about 9% excess water by mass.

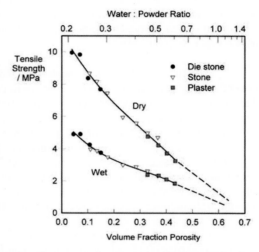

Fig. 9.2 Variation of gypsum product strength wet and dry. Notice the correspondence of water : powder ratio and volume fraction pores (which is the same as volume fraction excess water).

Fig. 9.3 Models should always be wetted with slurry water to avoid dissolution.

Fig. 9.4 "Slurry water" can be collected as the waste stream from a model trimmer.

commonly known as **slurry water**, or sometimes **SDS**, should always be used to soak the model first (Fig. 9.3). This may be prepared by soaking pieces of waste plaster or stone, or by taking the some of the waste stream from a model trimmer (while grinding a waste piece of set material) (Fig. 9.4), but anyway always from a known, non-biologically contaminated source – no patient contact.

In passing, we might notice that a gas flame is commonly used to warm the wax being used to make a pattern in a gypsum model, and render it glossy by superficial melting. Were it not for the model being wet, it would be very easy for the edges of an inlay preparation, for example, to be heated enough for the gypsum to dehydrate (Fig. 1.1). Such flames have temperatures over 600 °C, and an edge would heat very quickly. This would make this critical region very weak and friable and therefore easily damaged, if not already destroyed, thus spoiling the mould for the pattern, apart from the problems caused by the wax soaking into the gypsum. Clearly, the cast needs to be maintained wet to prevent this, and drying by repeated warming carefully avoided.

§10. Storage

Because of the large quantities of hemihydrate products used as a matter of course, the appropriate storage conditions for stock and working supplies need to be understood. Less obvious is the need for care with plaster and stone models: stability is not guaranteed.

●10.1 Hemihydrate

The reactivity of hemihydrate with water suggests that there may be a problem in storing powders in unsealed containers, as is commonly the case for plaster and stone, which are used in large quantities in dental laboratories. However, a similar problem arises with respect to nucleation as in ordinary mixing: it is not simply a case of a desiccant action such as shown by phosphorus pentoxide. As with most materials in contact with the atmosphere, even at moderate humidities, hemihydrate adsorbs something approaching a monolayer of water, but reaction cannot occur because this water cannot hold anything in solution. At contact points, and in the pores of plaster particles, surface tension effects may lead to small accumulations of what could be considered liquid water, which could then permit reaction. Such accumulation may be encouraged by the presence of (true!) hygroscopic salt additives so that some commercial products may be more sensitive than others, but in any case nucleation is still required. However, this does occur quite readily, and rapid reaction commences for plaster of Paris after only about 15 h exposure to 100% RH, or 4 days at 85% RH (Fig. 10.1). Other 'pure' hemihydrate products behave similarly, but gypsum-bonded investment (17§3) is much more sensitive, presumably because of additives.[14][15]

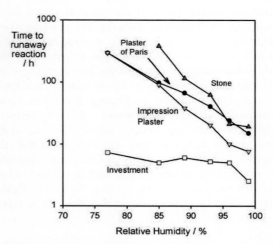

Fig. 10.1 Effect of humidity on safe exposure time of hemihydrate products – some experimental results at 25 °C. That is, how long before nucleation of dihydrate occurs.

Once the process has started, the hemihydrate is steadily consumed, with a corresponding steady decrease in the strength of the set gypsum which might be made from it, until the material would be incapable of setting at all on mixing. Meanwhile, the presence of more and more dihydrate crystals reduces the setting time, because they themselves would act as nuclei for crystallization when the slurry was mixed, thus rendering the material unworkable. Care should therefore be taken to expose only that material required for a day or two's work at a time if the humidity is or could be high, to avoid wastage and spoilt models, or – less obviously – models more easily damaged because they are weaker. Necessarily, other properties such as setting expansion will be altered, and with investment materials as well this would be disastrous.

It also follows that if a 'working' container is refilled from a bulk package, whether or not kept closed, new material should not simply be added on top each time. The older material will have accumulated time of exposure to the atmosphere, and could have reacted far enough (indeed, must eventually) that it will nucleate rapid reaction when it contaminates a portion. Ideally, such a container will be emptied completely, cleaned and thoroughly dried (!) before being refilled. Compare the effect of mixing bowl contamination (§6.1).

●**10.2   Dihydrate**

As can be deduced from the nature of the reaction (1.1), the formation of dihydrate is in fact an equilibrium, and so depends on the effective 'concentration' of water, in other words, its activity or vapour pressure (whether as liquid or gas), as well as temperature.   In fact, calcium sulphate dihydrate is thermodynamically unstable with respect to hemihydrate at less than 100% RH above about 40 °C or so (Fig. 10.2).[16]   This is the fundamental basis of the manufacture of plaster of Paris.   However, the rate of decomposition is very low and not commercially-useful at lower than about 120 °C, hence the need for moderate heating (§3.1).

It can also be seen that the manufacture of dental stone (§3.1) depends on the exact same instability. Providing that the boiling point of the water is raised far enough, whether by dissolved salts or by pressure (*i.e.* in an autoclave) – or both, all that is required is that the stability line is crossed for the dehydration to proceed. Obviously, recrystallization is only possible in the presence of liquid. Notice that the activity of liquid water is not appreciably affected by pressure, so heating a gypsum-liquid water mixture under pressure above ~100 °C results in the dihydrate becoming unstable, especially since the dissolved calcium sulphate ions have solvation shells that reduce the activity, *i.e.* the 'availability', of the water. Including high concentrations of soluble salts such as calcium chloride also lowers the water activity, moving the system further away from stability.   Even without that, the rate is very high at 140 °C.[17]

It then follows that plaster and stone models are not stable above about 40 °C under ordinary circumstances, since a water-saturated atmosphere is not a normal environment. Yet, it is perfectly feasible for the temperature in model stores to exceed that in many countries, and especially at low humidity, and particularly if a building's air-conditioning (if any) is turned off overnight or at weekends. The surface of such models becomes powdery, accuracy being gradually lost, and ultimately they collapse. Obviously, drying models by leaving them on a radiator, in front of a hot-air duct or blower, or in an oven, is not sensible; forced-air ventilation at ordinary temperatures is quite effective enough. If models are intended to be kept for extended periods, storage conditions should be considered and modified as necessary: cool and not desiccating. This also means that experimental work that requires dry specimens should not involve elevated temperatures.

There is a more serious problem. Given the emphasis now on infection control, the need to sterilize models, which will have been poured in impressions that can only be disinfected, is of interest. Autoclaving has been considered as one possible approach. However, typical procedures involve 5 min at 134 °C or 15 min at 121 °C in saturated steam (which can also be with a subsequent drying stage), conditions which are clearly similar to those used for making dental stones. Indeed, substantial conversion occurs even at

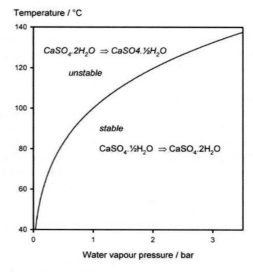

Fig. 10.2   Stability curve for calcium sulphate dihydrate. The equilibrium line closely follows the saturation vapour pressure curve for water.

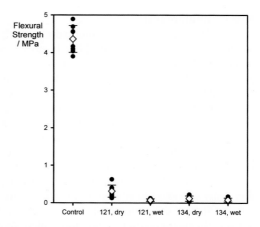

Fig. 10.3   Effect of autoclaving on the flexural strength of a dental stone. Temperature indicated in degrees Celsius; control at 25 °C. Mean (diamond) and 1 standard deviation error bars with data points.

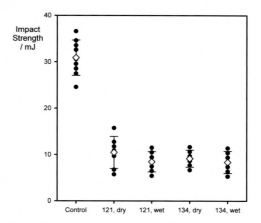

Fig. 10.4   Effect of autoclaving on the impact 'strength' of a dental stone.

such short exposure times, with loss of compressive strength and indentation hardness.[18] In addition, there is considerable expansion which may be attributed to the crystal growth pressure arising from the formation of crystalline hemihydrate. Compressive strength is not in fact a very sensitive measure of damage (1§6.3), and the effect on flexural strength (Fig. 10.3) and energy to fracture under impact (Fig. 10.4) is much more marked. (In both cases, the effect of the drying is similar to that for untreated material, §9.) In all of this, it should be borne in mind that the outcome depends on the thermal history – time and temperature – as well as that of the water status of the system.

§11. Setting Time

In many cases a knowledge of the **setting time** of a dental material is of direct use and interest, for example, with impression materials and gypsum products. There is, however, a major difficulty in defining setting time because no material exhibits a sharp end to the reaction (or corresponding physical change); setting reactions may (and frequently do) continue for many hours. "Setting time" is therefore not a physically-definable quantity in any absolute sense. Instead, we must rely on some intermediate stage or condition where a certain desired or intended operation can be carried out with little risk of significant damage to structure or dimensions, and thus create a functional or **operational** definition of practical use. For example, an impression material must be sufficiently elastic so as not to distort too much on removal, silver amalgam must be strong enough to bite on, plaster strong enough to handle. In development studies too, changes in setting behaviour might be monitored to ensure that a proposed change in formulation does not have undesirable side-effects.

●11.1 Gillmore needle

The **Gillmore needle** is frequently used as a quick and easy way of testing setting time, particularly for gypsum products. As this device applies a fixed stress on a small circular area of the flattened surface of the setting material, using a flat-ended cylindrical indenter, it may be viewed as another kind of hardness test (1§8). There are two such indenters that have been defined for dentistry, the one giving 16 times the stress of the other, for so-called 'final' and 'initial' set determination. If the material has sufficient strength to withstand the applied stress, leaving no appreciable permanent indentation, it is said to have set according to that test. However, the arbitrariness of such tests must be recognized (see §11.3). We have to begin by asking what is meant by 'set'.

In the context, the word 'set' has a prejudicial, implicit and absolute mechanical interpretation: solidification has occurred. This is weakened by talking about 'initial' set – what can that mean? Likewise, 'final' set seems to be redundant. The only way out is to assume that 'set' is a material property which varies continuously during the setting process. This semantic confusion is unhelpful. Unlike a proper indentation hardness test, a Gillmore needle can only indicate when the penetration-resistance 'strength' (as defined by the conditions of the test) has exceeded a certain value. It can give no information whatsoever on whether a reaction is complete or how far it has to go. Supposing, then, that indentation hardness can be determined for a material, we might have curves such as those in Fig. 11.1. Curve A then represents a material for which a 'final setting time' can be determined using the Gillmore needle. Curve B, on the other hand, represents a material which appears never to set according to the same criterion simply because its strength is insufficient ever to withstand the test stress, even though it may have more or less completed reaction well before A, and by any objective measure of reaction progress be considered to have set earlier. However, it is clearly possible for an 'initial' setting time to be recorded for A longer than for B. These problems apply no matter what class of material is being considered. In other words, the Gillmore needle test is naïve (and others like it), confusing material properties with the underlying process. Extreme caution is therefore necessary when interpreting such data. A similar test, the Vicat needle, is based on the idea of full penetration rather than a superficial indentation, but the same kinds of remarks apply: it is arbitrary, and uninterpretable in any sense of a material property.

Clearly, what is lacking is a proper definition of the information that is required. Is it, for example, a particular proportion of the ultimate strength? There could be no

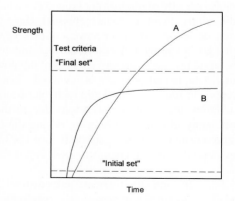

Fig. 11.1 The form of Gillmore needle test results.

fundamental reason for choosing such a number. Conversely, is it the degree of completion of the reaction that is of concern? It may well be for some materials, but finding the proper means of determining that could also be very difficult, even if the number could be chosen. The point is, the decision must be taken based on the true purpose of the test and the context. It must be accepted that setting time (of whatever type) is an arbitrary concept that has yet to be soundly defined in any context (despite its importance for handling and usage instructions and standardization testing). A more meaningful operational definition for actual dentistry might be the minimum strength that must be attained to permit handling, with caution, safely. (It has to be noted, however, that there is no suggestion to the effect that the Gillmore needle 'final set' represents such a strength, nor could there be.) This would appear to be at least relevant in practice, although it remains difficult to identify what value it should be for any class of material. Ultimately, it comes down to a trade-off between the risk of damage (which might go unseen but have far-reaching consequences) and the economic benefit to the dentist or technician of shorter time before the next step can be taken; another compromise.

●**11.2   Setting rate**

There is a related concern in terms of the perception of the development of strength. This may be expressed as **working time**, that is, how long manipulation may continue without (overt) detrimental effects. Thus, we know that strength depends on the powder : liquid ratio, and the approach to final strength is asymptotic as reaction progresses, so we might have a set of curves as in Fig. 11.2 for a given product under otherwise similar conditions. This kind of behaviour might be anticipated in many systems. However, it is commonly said, for a variety of products, such as zinc oxide-eugenol cement (9§2) and others, that the rate of setting can be reduced by adding more liquid (in other words, to buy time). This can be seen to be a faulty conclusion. Obviously, adding liquid reduces the strength that will be attained (and thus the adequacy of the end result),

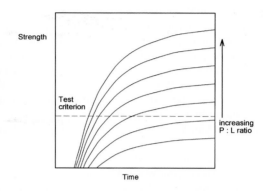

Fig. 11.2 Possible form of development of strength as powder : liquid ratio (P : L) is varied.

and so the rate at which strength varies with time must also be reduced. But there can be no effect whatsoever on the kinetics of the chemistry at the crucial powder-liquid interface as the chemical environment has not changed in any way. To imply that the rate of the reaction here has been changed is simply nonsense. This confusion is widespread.

Factors that affect the chemical kinetics include temperature, pH, and other species in solution which might affect surface energies, diffusion rates, or activation energies. Changing the amount of mixing liquid does not alter these factors in any immediate sense because the activities of the reactants in such systems are unaffected by mixing proportions – it is not a concentration-dependent matter as it would be in a single solution, for example. That said, it is quite common to adjust the temperature up or down to achieve some control of reaction rate, or to prevent temperature excursions that would interfere in this way.

A contrast can be drawn between the strictly kinetically-defined rate of a reaction at an interface (or in solution) and the bulk conversion rate, by which is meant the rate of formation of the reaction product in, say, moles per litre of mixture per second (which, of course, itself varies with time as, for example, the length of the diffusion path for reactants increases). Here, we are more concerned with the total surface area at which reaction may occur, so that a finer powder allows a greater rate of conversion. Plainly, reducing the amount of powder in a given volume reduces the surface area in proportion, and likewise the bulk conversion rate. Similarly, as reaction proceeds, the surface area of the powder decreases. In this special sense, the overall "concentration" of reactants and products is involved. But, to achieve a given strength in a more powder-dilute mixture, replacing zero-strength liquid with load-bearing structure, the reaction has to convert more of the reactants. In addition, there are complications due to the mixing process: continued stirring brings reactants together, offsetting diffusion limitations, raising the rate of conversion (but possibly also breaking down developing structure and so reducing strength – see 15§7.2). So whilst the time to a given strength will thus vary (Fig. 11.2), evidently the actual chemistry itself is not affected.

In summary, there are three distinct aspects to this to be considered:
*   chemical kinetics: reactivity, temperature, diffusion rates ...
*   mass conversion: surface area, stirring, diffusion path length ...
*   property development: structure, volume fractions

Whilst these aspects are perhaps not totally independent of each other, the underlying facts must be recognized and distinguished. Hence, if structural differences are expected due, for example, to changing powder : liquid ratio, which alters the amounts of matrix and core in cements (essentially by diluting the elements on which strength depends) (Chap. 9), or volume fraction of porosity in gypsum products, it is meaningless to speak of 'setting rate' being controlled by varying the liquid proportion, at least, not without defining precisely what that phrase is supposed to mean. Such a proposal also fails to notice that all other properties are thereby affected, and presumably for the worse as a departure from recommended or optimized proportions is involved.

●11.3   Related concepts

In §6, reference was made to working time in operational terms – how long is actually required to get a job done effectively, while above (§11.2) the reference is to a property-based criterion. These are clearly very different in principle, and care must therefore be taken in discussing the term. Thus, if it is believed that one can detect the point at which damage will ensue from further manipulation, the tendency will be to continue to shape and adjust, in what may be considered a materials version of Parkinson's Law[19] – filling the time (apparently) available. This is a dubious approach as our sensitivity in this sense is never very great. A similar distinction was pointed out at the beginning of this section regarding setting time: the practical, operational sense is quite distinct from that associated with any underlying chemical or physical process. Likewise, mixing time (Chap. 15) might be labelled according to what is necessary as opposed to what is perceived to be appropriate.

There is a similar problem with the terms "initial" and "final" set commonly used in connection with a wide range of dental materials. These are always arbitrarily defined, whether operationally or instrumentally. The words themselves have an implied absolute sense that tends to lend a false confidence in the nature of the process of setting. No system has a clear, sharply two-stage process with definite endpoints. Whether chemical or physical processes are involved, the kinetics generally imply gradual, asymptotic approaches to the equilibrium (or quasi-equilibrium) state (except perhaps for first-order processes such as melting and freezing). Indeed, many reactions in real systems do not complete, ever. In addition, there is no dental system where a process depends on the completion (a very strong, absolute term) of a prior process before it can start. All reactions are 'local': if conditions permit, it will happen, irrespective of the state of the system elsewhere; if a path exists (in the sense that the activation energy required is not prohibitive), it will be taken – this is the **thermodynamic imperative**. The rate of the process is a separate matter (the **kinetic reality**). The use of such timing terms may be a convenience, and it would certainly be pedantic to insist on replacing them with a wordy but meaningful phrase, but their absolute nature should not be interpreted as implying that the underlying process is also absolute in either start or finish.

In each case, it has to be decided what is meant, that is, what is the purpose of the definition. If personal judgement is to be taken out of such important matters, to realize the best attainable performance in service, it is essential to comprehend and respect the chemistry and physics of the processes involved.

●11.4   Effect of temperature

Given that the setting of gypsum products depends on dissolution, solubility, diffusion, and crystallization – all activated processes – it is no surprise that the setting time is dependent on temperature, but the ordinary expectation would be of an exponentially increasing rate: the Arrhenius rule-of-thumb is to double the rate for a 10 K rise in temperature. The actual behaviour is rather more complicated (Fig. 11.3).[20] Indeed, there is an appreciable variation over the range of temperatures that might be encountered in a dental laboratory, in the expected direction, and also sensitivity therefore to the temperature of the mixing water. Nevertheless, there is a clear albeit broad minimum in the setting time curve, after which the rate falls rapidly. Since the setting process is exothermic (§6), starting at 'ordinary' temperatures, the setting is therefore auto-accelerating up to a point, and then at least partially self-limiting. Even so, it is not possible to say much more

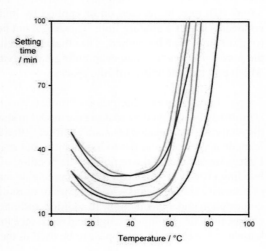

Fig. 11.3 Variation of Gillmore needle setting time with mixing water (and ambient) temperature for a variety of plaster products, all at 2 : 1 water : powder ratio. For comparability of such data, within and between curves, it is assumed that crystal habit and nucleation density are constant for the final structures to be identical (see 11§2).

than that the behaviour observed will depend on the mass of material, its shape, and the nature of the container (thermal mass, conductivity) as this affects the thermal behaviour.

The shape of the curve, however, can be connected with the thermal stability of the dihydrate (§10.2). This falls as the temperature rises above about 40 °C, so that the driving force for the formation of dihydrate reduces. This is reflected in the rise in the induction period that occurs above that point.[21] Notice that the activity of the water in a hemihydrate-dihydrate slurry is close to but just below unity, *i.e.* an effective water vapour pressure of not quite 1 bar, due to the effect of solutes (*i.e.* mostly calcium sulphate). So, the conditions for stability are met all the way up to ~98 °C, when setting in fact never occurs – clearly, at this point, the stability line has just been crossed.

In practical terms, it is enough to be aware that the time taken for a gypsum material to set (however this is defined, see above), is dependent on an additional factor that may be overlooked. However, there may also be some unexpected effects as dihydrate crystal habit is also affected, from very fine particles through the ordinary spherulitic form (Fig. 5.1) to 'sheaf'-like bundles over the range 0 ~ 46 °C. This must affect expansion substantially as it affects loss of gloss, crystal growth pressure, strength and hardness, and therefore both the dimensional accuracy of work and the robustness of the model. Thus, a consistent set of working conditions is to be preferred, and (as ever) following instructions.

§12. Hardeners

The surface of set gypsum products is necessarily porous and somewhat rough (§8), and the individual crystals are necessarily brittle and fragile. Accordingly, contact with any hard object can abrade the surface quite easily. For dimensionally-critical applications, as for die stone, where the use of metal instruments to prepare wax patterns is normal, the margins of preparations for inlays and crowns can become damaged, with obvious problems for the eventual fit of the cast device. Likewise, the forming of a platinum foil matrix for powder-method porcelain work (25§4.7) can also result in accidental damage.

●12.1 Surface treatment

Whilst die stone products have been manufactured to provide the densest possible surface, as indicated above (§3.1, Fig. 8.5), they are not immune to such damage. Accordingly, there are products sold which are claimed to "harden" stone surfaces. There appear to be two general types. Commonly, a solution of a polymer such as poly(methyl methacrylate), PMMA, in a solvent such as 2-butanone (methylethyl ketone, MEK; b.p. ~79.6 °C), may be painted on or applied by immersion, after which evaporation of the solvent (in a fume cupboard) leaves a binding film. This might not be sufficient to seal the surface completely, and a second coat (or more) may then be applied. The other approach is to use a cyanoacrylate (10§6.2) which similarly soaks into the porosity but rather than evaporating has polymerization initiated by the unavoidable adsorbed water film, which polymer therefore fills the spaces.

In both cases, penetration of molten wax into the porosity may be limited or prevented (although this is more easily achieved by soaking the model in slurry water [§9]). While it would appear that the binding effect may improve the abrasion resistance a little,[22] there is apparently little or no evidence to suggest that the material is made appreciably harder to indentation, but may be made worse.[23] The reasons for this may be two-fold. Firstly, there can be little or no specific bonding of the polymer to the dihydrate, only perhaps a little hydrogen bonding (10§4), so there can be no appreciable matrix constraint (6§2.2). Secondly, polymers in solution are of necessity simple chains – no branches or cross-links (3§6.2) that would raise the strength and prevent plastic deformation under indentation. In addition, ordinary cyanoacrylates are monofunctional and can only form short chains, thus also limiting strength and resistance to deformation. Cross-linking difunctional cyanoacrylates are known[24] but these do not appear to be used in the context.

This second problem may be addressed by the use of a solution of the components of an epoxy resin in acetone (dimethyl ketone, b.p. 56 °C),[25] allowing the otherwise very viscous components to be mixed and readily taken up by the porous gypsum (see 10§2.9). When the solvent has evaporated (apparently it prevents premature reaction), setting would proceed spontaneously (27§5), but accelerated by heating (37 °C, which avoids decomposition of the dihydrate – §10.2). This increases the material hardness, abrasion resistance and strength, with the added advantage that the gypsum does not have to be completely dry. Unfortunately, this procedure cannot be recommended for general use. Many of the amines used in such resins are known to be carcinogenic: all contact with unset resin and its components should be assiduously avoided (even in domestic contexts). Worse, such amines can migrate through ordinary plastic or rubber gloves, and especially so in a solvent such as acetone. The protocol required for safe handling would be onerous and perhaps quite impractical. (Equally, great caution is also required for handling epoxy die materials; 27§5.)

While all such treatments create a film of polymer on the surface, for one coat or so it may represent an insignificant dimensional change – that is, not appreciably affecting the fit of the device. The use of multiple coats, however, in an attempt to improve the mechanical properties of the surface, may compromise that fit by building up a sufficiently thick layer (because it can no longer soak in), even with so-called 'air-thinning'– the application of a jet of compressed air in an attempt to blow away excess.

●12.2   Mixing liquid

A special mixing liquid, 'gypsum hardener', is sold for use with die stone using an apparently undocumented approach, that is, with colloidal silica (~15 mass%), said to improve the fluidity of the mix and the strength when set. The effects appears to be purely mechanical in that there is no apparent mechanism capable of affecting nucleation or crystal habit. Colloidal silica itself can be stable enough (at pH > 7), but the suspension is freezing-sensitive so that in storage or transit if the temperature experienced is low enough the silica separates and the liquid becomes useless. Accordingly, some 4 mass% ethanol is included, but although the actual freezing point depression is then only to –1.5 °C (Fig. 12.1),[26] the colloid appears to be stabilized somewhat.

The special liquid affects the fluidity markedly (Fig. 12.2). Changing the water : powder ratio for the die stone had very little effect, yet using the special liquid at a typical ratio gave a dramatic increase in the slump diameter. This cannot be accounted for by the ethanol, as pure ethanol-water mixtures also had very little effect (inset, Fig 12.2), and the viscosity actually increases in mixtures (Fig. 12.3), due to the formation of ethanol clusters.[27] The explanation may be that colloidal silica reduces the friction between tumbling particles (Fig. 2.1), as if by a layer of bearing balls, so to speak. Note that what this means in practice is that the mixing ratio can be reduced substantially before the fluidity falls to near the original value with water.

The special mixing liquid also has an effect on strength (Fig. 12.4). This is attributable to the ability to use a lower mixing ratio because of the improved fluidity. In the wet condition it is expected that the silica can add little to the strength because no bonding is possible (cf. Fig. 17§3.3), even so some increase is seen at low mixing ratios. This may be due to the silica occupying spaces between crystals and carrying some of the load (the material has a

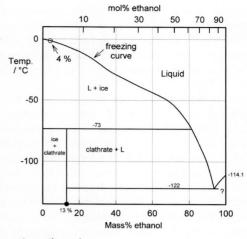

water-ethanol

Fig. 12.1 Phase diagram for water-ethanol showing the freezing curve. (For explanation see Chap. 8.) (The higher-temperature portion is at Fig. 10§9.10.)

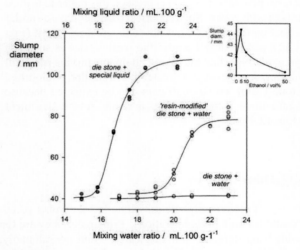

Fig. 12.2 Variation in fluidity of dental stone mixes with deionized water or colloidal silica 'special liquid' using the slump test from a 40 mm-diameter, 55 mm-high cylinder (see 4§7.8), and for resin-modified stone. The upper abscissa scale refers to the volume of special mixing liquid used, the lower scale to the water used or the water content of the special liquid. Inset: effect of ethanol alone in the mixing liquid.

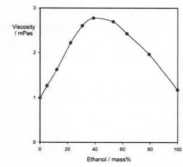

Fig. 12.3   Viscosity of ethanol-water solutions at 20 °C.

greater volume fraction of solid). However, if this material becomes too voluminous, it must detract: the proportion of binding matrix is diminishing and the silica behaves as a zero-strength dispersion. Indeed, the presence of the ethanol is also detrimental in the sense that this is in effect extra porosity – unreactable liquid. These two components mean that the decrease in strength with increasing mixing ratio is at a greater rate than for water alone, meaning that using a lower mixing ratio is essential for any benefit.

It can be seen that the actual water available for reaction can fall below the stoichiometric value (18.6 g/100 mL). The higher fluidity presumably means fewer bubbles because the minimum mixing volume (§3.2) does not need such a large dilatancy or dilution allowance. However, there is still a limit and it can be seen that insufficient water for reaction causes a drop in strength (Fig. 12.4). Note that these results were obtained with a dual asymmetric centrifugal mixing machine: with hand-mixing the lowest ratios are impossible to prepare usefully.

That the effect of a non-reacting component reduces strength we may compare the properties of a so-called resin-modified die stone, in which a proportion of a resin powder (*cf.* 5§2.2) is present; this material is to be mixed with water. Such spherical particles have a similar effect on the fluidity, but not to as great an extent (Fig. 12.2), presumably because the particle size is so much greater. (Note that the lowest attainable mixing ratio still gives a slump diameter greater than a similar, unmodified stone.) But, no strength gain can be expected because again there is no bonding, a rather severe detriment (Fig. 12.4).

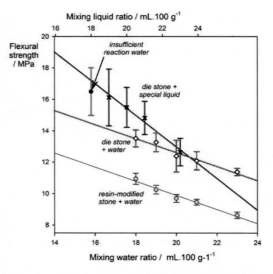

Fig. 12.4 Variation of the 3-point bend strength of die stone with deionized water or colloidal silica 'special liquid', and for resin-modified stone. Scales as for Fig. 12.2.

§13. Other Uses

Calcium sulphate dihydrate has a number of other applications in a clinical context (besides plaster 'casts' for limb immobilization). Principally, they are for space-filling and healing promotion[28] in connection with bony defects, both surgically[29] and periodontally, as well as for extracted tooth-socket packing and endodontic repair.[30] These uses are based on a number of properties:
- because of its solubility, it is fairly rapidly resorbed, on a timescale of a few weeks, depending on the scale of the implanted material;
- it can be used as pre-cast shapes or as 'set' granular material, as hemihydrate in a paste, or a mixture of hemihydrate and dihydrate;
- it is highly biocompatible, provoking no adverse reactions;
- the resorbability means that it can be used as a scaffold into which bone growth may occur;
- the released calcium ions may encourage osteoblastic activity for that bone growth;
- it lends itself to use as a delivery vehicle for pharmacologically-active substances;
- combined with various other substances the mechanical properties and resorption behaviour can be modified;
- it is cheap (at least, the raw material is).

Of course, factors such as purity (§3.3) and sterility are important to all of the above, and this requires both more strict quality control and special handling. Indeed, it may be that direct synthesis is the preferred route for such critical use.[31] Some care is required with setting materials as blood and other biological fluids will interfere, although calcium sulphate is described as haemostatic.[32]

References

[1]   Cole WF & Lancucki CJ. A refinement of the crystal structure of gypsum, $CaSO_4(H_2O)_2$. Acta Crystallographica B 24: 1968, 1982.

[2]   Xu K, Cao PG, Heath JR. Graphene visualizes the first water ad layers on mica at ambient conditions. Science 329(5996): 1188 - 1191, 2010.

[3]   https://www.bgs.ac.uk/downloads/start.cfm?id=1359

[4]   Docking AR, Chong MD & Donnison JA. The hygroscopic setting expansion of dental casting investments. Part 2. Austral Dent J 52: 160 - 166, 1948.

[5]   Franz G. Das Dimensions-verhalten zahnärztlicher Hartgipse. Hanser, München, 1978.

[6]   Nissan J, Gross M, Ormianer Z, Barnea E, Assif D. Heat transfer of impression plasters to an implant-bone interface. Implant Dent 15(1): 83 - 88, 2006.

[7]   Jarvis RG & Earnshaw R. The effect of alginate impressions on the surface of cast gypsum. II. The role of sodium sulphate in incompatibility. Austral Dent J 26(1): 12 - 17, 1981.

[8]   Buchanan AS & Worner HK. A study of the action of borax in retarding the setting of plaster of Paris. J Soc Chem Indust 65: 23 - 26, 1946.

[9]   Jiang WG, Pan HH, Tao JH, Xu XR & Tang RK. Dual roles of borax in kinetics of calcium sulfate dihydrate formation Langmuir 23: 5070 - 5076, 2007.

[10]   Guy AG. Introduction to Materials Science. New York, McGraw-Hill, 1972.

[11]   Scala E. Composite materials for combined functions. Hayden, New Jersey, 1973.

[12]   Earnshaw R & Smith DC. The tensile and compressive strength of plaster and stone. Austral Dent J 11: 415 - 422, 1966.

[13]   Craig RG. Restorative Dental Materials. 6th ed. Mosby, St. Louis, 1980.

[14]   Torrance A & Darvell BW. Effect of humidity on calcium sulphate hemihydrate. Austral Dent J 35: 230 - 235, 1990.

[15]   Chan TKC & Darvell BW. Effect of storage conditions on calcium sulphate hemihydrate-containing products. Dent Mater 17: 134 - 141, 2001.

[16]   Anon. Drying Plaster Casts. IG502/rev. 7-00 United States Gypsum Company, Chicago. 2000.

[17]   Combe EC, Smith DC. Studies on the autoclave dehydration of gypsum. J Appl Chem Biotechnol 21: 283 - 284, 1971.

[18]   Whyte MP & Brockhurst PJ. The effect of steam sterilization on the properties of set dental gypsum models. Austral Dent J 41(2): 128 - 133, 1996.

[19]   Parkinson CN. Parkinson's Law: The Pursuit of Progress. John Murray, London, 1958.

[20]   Worner HK. The effect of temperature on the rate of setting of plaster of Paris. J Dent Res 23: 305 - 308, 1944

[21]   Ridge MJ. Effect of temperature on the structure of set gypsum plaster. Nature 182: 1224 - 1225, 1958.

[22]   He LH, van Vuuren LJ, Planitz N & Swain MV. A micro-mechanical evaluation of the effects of die hardener on die stone. Dent Mater J 29(4):433 - 437, 2010.

[23]   Harris PE, Hoyer S, Lindquist TJ & Stanford CM. Alterations of surface hardness with gypsum die hardeners. J Prosthet Dent 92(1): 35 - 38, 2004.

[24]   https://www.threebond.co.jp/en/technical/technicalnews/pdf/tech34.pdf

[25]   Sanad MEE, Combe EC & Grant AA. Hardening of model and die materials by an epoxy resin. J Dent 8: 158 - 162, 1980.

[26]   Potts AD & Davidson DW. Ethanol hydrate. J Phys Chem 69 (3): 996 - 1000, 1965.

[27]   Wakisaka A & Ohki T. Phase separation of water–alcohol binary mixtures induced by the microheterogeneity. Faraday Discuss 129: 231 - 245, 2005.

[28]   Thomas MV & Puleo DA. Calcium sulfate: properties and clinical applications. J Biomed Mater Res B Appl Biomater 88B: 597 - 610, 2009.

[29]   Maeda S, Bramane CM, Taga R, Garcia RB, Gomes de Moraes I & Bernardineli N. J Appl Oral Sci 15(5): 416 - 419, 2007.

[30]   Zou L, Liu J, Yin S-H, Tan J, Wang F-M, Li W & Xue J. Effect of placement of calcium sulphate when used for the repair of furcation perforations on the seal produced by a resin-based material. Int Endod J 40(2), 100 -105, 2007.

[31]   http://www.wipo.int/patentscope/search/en/detail.jsf?docId=WO2001005706

[32]   Scarano A, Artese L, Piattelli A, Carinci F, Mancino C & Iezzi G. Hemostasis control in endodontic surgery: a comparative study of calcium sulfate versus gauzes and versus ferric sulfate. J Endod 38(1): 20 - 23, 2012.

Chapter 3   Polymers

Polymeric systems form the basis of many classes of product and application in dentistry. Broadly, they appear in such contexts as impressions (e.g. silicone rubber), restorations (filled resin), prostheses (denture base acrylic), equipment (mixing bowls) and tools (polishing disks). No matter what the application or the chemistry of the polymer, the same general principles are involved. These explain the properties, handling, behaviour in service and the limitations of all types and products.

*Polymer properties are primarily dependent on the molecular structure, the arrangements of the atoms and groups within the giant molecules. However, their size means that they do not crystallize readily (if at all); their flexibility leads to the possibility of all kinds of arrangements in space. The behaviour of these **random coils** and rotations about bonds are therefore of central interest.*

*Polymer properties are strongly temperature dependent. This can be traced to the behaviour of portions of the polymer molecule known as **chain segments**. A major shift in mechanical and physical properties is observed in passing through the **glass transition temperature**, which is a principal means of characterizing polymers in general.*

The nature of polymer molecules leads to a strong time-dependence of their strain response to stress challenges, especially the glassy-rubbery shift in properties, and sensitivity to the presence of plasticizers and cross-links in the structure. Compromises are usually necessary because it is not possible to optimize all aspects simultaneously.

Materials Science for Dentistry
https://doi.org/10.1016/B978-0-08-101035-8.50003-1

Polymers, both synthetic and natural, form the basis of very many dental materials, and these exhibit an equally wide range of properties. This variation in properties is intimately bound up in the structure and behaviour of the molecules. But far from being a mere catalogue of variations, polymer physics and chemistry is explicable, and predictable, in systematic terms of chain structure, side-groups, and cross-linking. This can be taken to the stage of the design and creation of a polymer having particular properties to suit a specific purpose, and indeed this has already been done in many contexts. To understand these polymer properties and design principles in dentistry, or any other applications, it is necessary to deal first with the basics.

§1. Basic Structure

A **polymer** was originally understood to be the result of **polymerization**, a process whereby covalent bonding between two or more identical molecules yielded a larger molecule with the same **empirical formula**, that of the **monomer**. The usual example of this is the cyclic molecule paraldehyde, $(CH_3CHO)_3$, formed from acetaldehyde, CH_3CHO (Fig.1.1). However, with the identification of many kinds of giant molecule, the usage of these terms was considerably extended. Thus, an **addition** polymer molecule is a simple multiple of the monomer, as in paraldehyde, whereas a **condensation** polymerization proceeds with the elimination of a small molecule, often water, for each monomer molecule reacted. The empirical formula of a condensation polymer therefore differs from that of the monomer; that of an addition polymer is identical to it. As a result of this emphasis on indefinitely large polymers, the smaller molecules, up to say 6 or 10 units, are now preferably called **oligomers**.

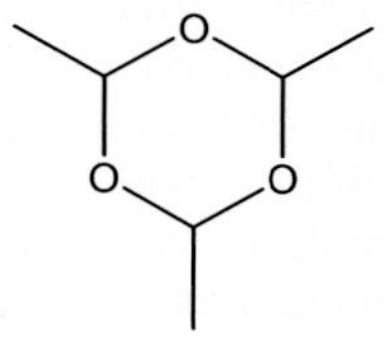

Fig. 1.1 The structure of paraldehyde, one possible polymer of acetaldehyde, CH_3CHO. This is an addition polymer.

The condensation reaction may be illustrated by the example of acetic anhydride, $(CH_3CO)_2O$, which is formed from two molecules of acetic acid, CH_3COOH, with the elimination of water (Fig. 1.2). A similar reaction, that of esterification, can also be used (as in the reaction of acetic acid and ethanol to form ethyl acetate), again with the elimination of water. But for polymers to be formed it is necessary that, after each unit has been reacted, the chain continues to have a reactive site for the link to the next unit. Thus, in the case of polyesters, a mixture of diacid and diol may be used, or even an acid-alcohol (hydroxyacid) (see 27§4), so that each end of the chain can grow independently.

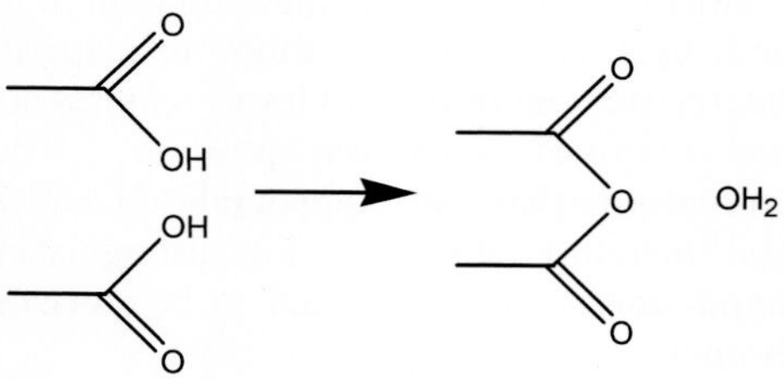

Fig. 1.2 The condensation of acetic acid to acetic anhydride eliminates a water molecule.

In forming both addition and condensation polymers, more than one monomer may be reacting: this is known as **copolymerization**, *i.e.* generating a **copolymer**, with the implication that the sequence is random, but depending on relative concentrations and reactivity. This kind of process leads to a slight difficulty of description, for what is truly a repeating unit in a simple addition polymer, and identifiable exactly in empirical formula with the starting material, does not exist as such in either random copolymers or condensation polymers, and especially not in condensation copolymers. The general term **mer** can thus be used to indicate a structural unit in a polymer, without prejudice as to chemical formula or indeed to its uniqueness in the structure, as long as it is referable as being derived directly from one or other of the starting compounds. A **block copolymer**, in which alternating sequences of identical mers of one kind or another occur, can also be made for special purposes, a kind of structure further requiring care in description.

It should be noted that, without further qualification, a polymer is commonly considered to be just a simple chain of units, one after the other, with no branches, loops or other complication. Indeed, such a **primary structure** is described as a straight-chain or **linear polymer** (although this has nothing to do with the way the molecule is arranged in space – as will be seen (§2), it is unlikely to be physically straight). In addition, the word polymer on its own implies nothing about composition: many biological materials are extremely complex polymers, proteins in particular (even if linear) (27§3), but also RNA and DNA, which are linear. To be clear about the fact that only one monomer is involved, the word **homopolymer** may be used, although this is usually explicit when the monomer is mentioned directly or indirectly. Other primary structures are possible, indeed are common. By introducing reactants with a higher number of functional groups or because of side reactions, branches may be created in the chains, and tree-like molecules or random three-dimensional networks may be made. In each case, the mechanical properties are modified, and this becomes the basis of design for applications. In biological systems in particular, regular sheets and networks may be found. A number of these larger systems are of relevance to dentistry, and will be dealt with later in the appropriate sections.

§2. Configuration and Conformation

●2.1   Conformation

Single σ-bonds between atoms permit the relative rotation of those atoms. Thus, a methyl group CH_3- may rotate with respect to the rest of the molecule to which it is attached. In addition, where two bonds to the same atom have an angle between them which is not 180°, such as the tetrahedral angle of the sp³-hybridized atomic orbitals of carbon, such a rotation changes the relative position in space of the atoms and groups either side of that rotating bond. In larger, more complex molecules where several such rotatable, angled links exist, the molecule may adopt a wide range of shapes or layouts in space. That is to say, it may exhibit various **conformations**. Notice that interconversion between conformations depends only on bond rotation. There is then, in principle, a simple path between any two conformations without breaking any bonds (assuming no other form of steric hindrance) simply by rotation about one or more bonds.

●2.2   Configuration

When an sp³-hybridized carbon atom has four different substituent groups attached to it, there are just two distinct ways of doing this, which are the mirror image of each other (Fig. 2.1). The central atoms are then said to be **asymmetric**. The two forms cannot be interconverted at ordinary temperatures by mere rotation of atoms or groups, just as left- and right-hand gloves cannot be superimposed by mere rotation. Interconversion of the two forms requires at least one bond to be broken and reformed in a new position. This situation gives rise to **stereoisomerism**: two compounds of identical composition and bonding, and therefore structure, are distinguishable by their **chirality** or **handedness**. They are said to be **stereoisomers** or **configurational isomers**.

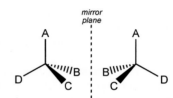

Fig. 2.1   The four substituents of asymmetric carbon atoms can be arranged in mirror-image forms, non-superimposable by rotation.

When the monomer has four distinguishable substituents on a carbon atom that will form part of the backbone chain of a polymer, or such asymmetric

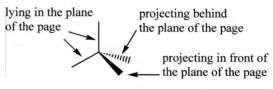

To assist in the interpretation of three-dimensional molecular structures, there are various conventions for drawing the stereochemistry of molecular bonds. This is one such:

lying in the plane of the page

projecting behind the plane of the page

projecting in front of the plane of the page

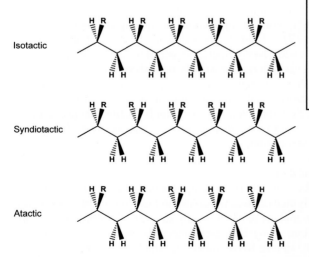

Fig. 2.2   Some possible configurations of polymers with asymmetric carbons in the chain. Note that the conformation has been adjusted to show the patterns clearly.

carbon atoms arise as a result of the polymerization reaction itself, as occurs in poly(methyl methacrylate) (Chap. 5), several possibilities arise for the **configuration** of the side chains or substituents of that polymer. Hence, polymers with asymmetric carbons are distinguishable by configuration, and are permanently locked into their particular condition, allowing discrimination between the various kinds.

●2.3   Tacticity

While conformation relates to internal movements of molecules, and configuration the way in which a chain is built at any asymmetric carbon, **tacticity** refers to the ordering of the configuration *sequence* of a polymer. Thus, if the relative positions of the substituents are always the same, from asymmetric centre to asymmetric centre, along the length of the chain, the polymer is said to be **isotactic** (Fig. 2.2), whereas an *alternating* configuration is called **syndiotactic**. The occurrence of these two arrangements commonly depends on complex steric constraints and they require specialized reaction conditions and methods to be realized. However, for many systems, such as with methyl methacrylate as monomer, no such conditions apply, and the polymer produced is **atactic**: the substituents are randomly arranged - there is no regularity along the chain. This condition of **atacticity** has important consequences for the structure of the solid. Because the former two regular types have a repeating structure, not just chemically but also in configuration, they can readily be formed into regular arrays, given that this will require some rotation about successive chain bonds to achieve the correct orientation at each carbon atom. Such polymers therefore can often crystallize quite readily, whereas atactic polymers simply can *not* crystallize. They are amorphous, resulting in materials which are less brittle than their crystalline counterparts. Crystallinity in polymers is explored further below (§4).

Polymer chains tend to lend the property of plasticity to the material (and hence the common term for manufactured polymers is 'plastic') because the chains – or parts of those chains – can be rearranged readily from the usual random coils (Fig. 2.3) to extended chains under moderate forces (Fig. 2.4). It should be noted that this kind of rearrangement for chain extension[1] does in fact depend on rotation occurring at each bond in the chain. It is the relative ease, or even the feasibility, of this rotation that is a highly-significant factor controlling both the physical and the mechanical properties of the bulk polymer. We proceed to examine the energy changes for changes of conformation by rotation, and then the geometrical implications.

●2.4   Ethane

If we consider an ethane molecule, C_2H_6, it

Fig. 2.3 A computer-generated image of a polymer random coil (polyethylene, C_{1000}). For clarity, the two chain ends have been marked with larger, darker atoms.

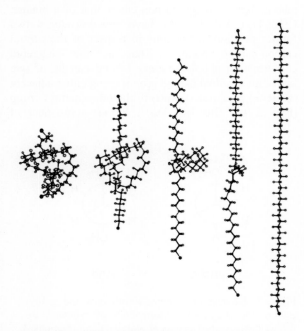

Fig. 2.4 Polymer chain extension from a random coil to a completely straight molecule requires continuous conformation change (polyethylene fragment, C_{50}). [This model was generated by 'pulling' on the ends and annealing out any bond length or angle strain.]

can be seen that because of the rotation that can occur at the carbon-carbon σ-bond, the two methyl groups can adopt any relative rotational position with respect to each other. Two of these positions are particular distinct: these are labelled **staggered** and **eclipsed** (Fig. 2.5), depending on whether the hydrogen atoms are superimposed or not when viewed along the C-C bond axis, 'end on'.

These two positions correspond to the minima and maxima respectively in the plot of energy *versus*

---

[1]   Care must be taken to distinguish between chain growth (*i.e.* adding mers) and arranging the chain to be laid out so that its measured overall length (that is, in a straight line) is increasing – which process is called chain extension.

dihedral (rotational) angle (Fig. 2.6). Energy is required to rotate the one methyl group with respect to the other because of the small but important mutual steric interference of the hydrogen atoms. The energy difference between the two positions is about 12 kJ/mol. Because there are three equivalent positions for each hydrogen, the energy variation is cyclic, with three peaks and three troughs, and each portion of the curve is equivalent to the others (Fig. 2.6). The important point here is that the peaks represent an energy barrier to the free rotation about the central bond, the eclipsed form representing the less favourable (*i.e.* higher energy) position.[1] Rotation is therefore an **activated process**, requiring an **activation energy**. It follows that it is a temperature-dependent process. We can describe the three 'detent' positions of the energy minima as **rotational isomers** or **rotamers**, even though in this case they are not distinguishable.

●**2.5   Butane**
       To continue the building of the model, we may take a slightly longer carbon chain, that of n-butane: CH_3-CH_2-CH_2-CH_3. We have now introduced two further position distinctions in respect of the central carbon-carbon bond. That is, the staggered conformation about the central bond can be of two kinds: the terminal methyl groups being either adjacent (**gauche**) or opposed (**anti**) to the other methyl group (Fig. 2.7). This leads to a further component of

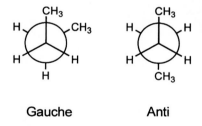

Gauche   **Anti**

Fig. 2.7 Because of its extra substituents, the methyl groups, the staggered conformations of n-butane are further distinguished as *gauche* and *anti* conformations.

variation in the energy with rotation about the central carbon-carbon bond (Fig. 2.8) because the bulky methyl groups have greater mutual steric interference in what may be termed the fully-eclipsed position, but also because the interference between methyl and hydrogen is greater then hydrogen-hydrogen, making the rotation much more difficult. This general result applies in principle to rotation about any bond in polyethylene, where the methyl groups of butane can be considered as replaced by effectively infinite -CH_2- chains, but the peaks and troughs of the energy curve become slightly more exaggerated.

       Extension of the idea can be made to chains

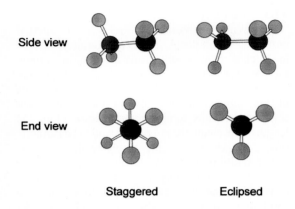

Fig. 2.5 The two extreme conformations of the ethane molecule, permitted by rotation of the C-C bond.

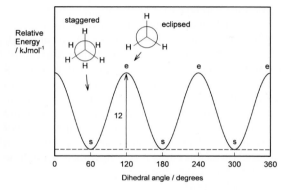

Fig. 2.6 The cyclic variation in energy of ethane as a function of the angle of rotation of the C-C bond. (e: eclipsed, s: staggered)

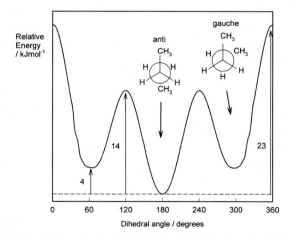

Fig. 2.8 The energy of n-butane as a function of the angle of rotation of the central C-C bond showing the extra periodicity. The highest energy state is with the two methyl groups fully eclipsed. The occupancy of the three energy wells (distribution of rotamers) would be about 14, 72 and 14%, respectively, at 25 °C, on a time-average – this is a dynamic system.

with larger substituents than hydrogen. The hindrance is not just steric in the sense of the substituents themselves physically clashing, but the interaction of the bonds themselves is involved, as must be evident from the geometry of ethane. Ball-and-stick models do not adequately represent the true situation, and must not be taken literally. The effect has been considered as the mutual repulsion of the electron clouds of each molecular orbital, although this is a simplistic view, [2][3] a more definite stabilization process being involved. The ultimate effect is, nevertheless, extremely important. Polypropylene, for example, which can be 'derived' from the polyethylene structure by replacing a hydrogen on every second main chain carbon by a methyl group (*i.e.* arranged as are the groups R- in Fig. 2.2), is somewhat stiffer than polyethylene, primarily for reasons of rotational steric hindrance.

As the size of the side groups increases, so the height of the energy barrier to rotation increases. For n-butane, complete rotation requires about 20 kJ/mol. In poly(methyl methacrylate) the side groups are methyl and methyl carboxyl, and rotation would be expected to be much inhibited in comparison with polyethylene, an observation supported by the relative flexibility of the two bulk polymers (*cf.* Fig. 7§2.2). It should be evident, therefore, that the work required to extend a polymer chain depends to a large extent on the energy required to rotate each chain link.

•**2.6   Bond geometry**

It remains to be demonstrated that chain extension does indeed depend on bond rotation - that there can be no extension unless rotation does occur. This should be clear from Fig. 2.9. The angle between the bonds to a carbon atom when it has sp^3-hybridization, the tetrahedral angle, φ, is given by:

$$\varphi = 2\sin^{-1}(\sqrt{2/3}) = 109.47...°$$

then, $a = -\cos\varphi = 1/3$

$b = \sin\varphi = 2\sqrt{2}/3$

hence $d_1 = 1 + 2a = 1.666... = 5/3$

and $d_2 = \sqrt{4b^2 + d_1^2} = 2.516... = \sqrt{19/3}$

hence $d_1 : d_2 :: 1 : 1.51$ (2.1)

To put this in scale, the difference in the distance between C_1 and C_4 in its two extreme positions, $d_2 - d_1$, is just about 0.85 of the C-C bond length. Thus, a tensile force acting across C_1-C_4' must result in greater separation, and this

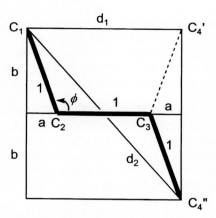

Fig. 2.9 The geometry of the n-butane molecule, and therefore of any four carbon fragment of a polymer chain, for the "eclipsed" C_4 and "anti" C_4'' conformations.

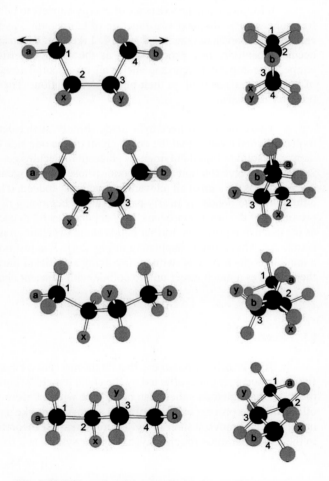

Fig. 2.10 Effect of applying tension to the axial hydrogen atoms of butane in the "eclipsed" conformation (top). Rotation occurs leading to the "anti" conformation (bottom). Left: lateral view; right: slightly oblique view along chain from the C_4 end. The path taken is the one of least energy, and takes into account the steric interference effects of the hydrogen atoms.

must involve relative rotation of the intervening atoms, with their substituents, about all three bonds. It is therefore plain that any deformation of a polymer that results in a change in dimensions does so by doing work against the steric hindrance to rotation. It is also worth noting at this point that once the fully extended conformation has been attained any further deformation must be because of bond-angle and bond-length strain. In both senses, C-C bonds are very stiff in comparison with the work required for bond rotation. One would therefore expect a dramatic increase in elastic modulus once this point has been reached.

The fact of the central bond rotation can also be demonstrated with a molecular model (Fig. 2.10). Applying tension to the axial hydrogens (a, b) of butane in the eclipsed conformation, the relative positions of the hydrogens on C_2 and C_3 (x, y) can be observed to change. Thus hydrogen atoms x and y start out on the same side of the C_2-C_3 bond (they too are "eclipsed", *cf.* Fig. 2.5), but end up on opposite sides, a full 180° apart, their positions having rotated in opposite directions relative to the tension axis. This can only have happened by rotation of the C-C bonds.

Notice also that the H_a-C_1 and H_b-C_4 bonds do not remain aligned. This is also due to rotation of the methyl groups about the C_1-C_2 and C_3-C_4 bonds, driven by steric interference with the hydrogen atoms on C_2 and C_4. Thus, for longer chains, such as illustrated in Fig. 2.4, rotation at any C-C bond has implications for the adjacent groups. The overall energy of the chain depends on all such effects.

●2.7 Activated process

It follows from the existence of energy barriers to bond rotation that the rate of crossing of those barriers from one state to the next is a function of temperature, as temperature affects the vibrational and rotational energy of all bonds, as concluded at §2.4. Likewise, because deformations of the kind shown in Fig. 2.10 must occur sequentially in a process such as the chain extension of Fig. 2.4, they take time. Such time- and temperature-dependent rearrangements are an essential aspect of the behaviour of not just single molecules but of bulk materials as well, in their plastic deformation. These then are flow processes, and the characterizing property is viscosity.

Two other points need to be made. Firstly, such calculations as are required for Fig. 2.8 are for isolated molecules, and assume that the entire chain on the one side of the swivelling bond moves freely. This cannot happen in solid or liquid polymers as adjacent molecules get in the way. Secondly, the extension shown in Fig. 2.10 must involve rotation at three bonds (those to hydrogens **a** and **b** are free), and the net activation energy is determined as the sum of all relevant interactions. Indeed, if hydrogens **a** and **b** are replaced by extended chains, although those bonds are nearly parallel at the beginning of the sequence (there is some strain because of the proximity of the other hydrogens, they are at about 6° to each other), they become increasingly out of parallel as the extension proceeds, further bonds along the chain must also rotate to achieve even this relatively simple action (*cf.* what must be happening in Fig. 2.4). It is just possible to conceive of swinging carbons 2 and 3 as a unit about the **a-b** 'axis' without involving significant movements in adjacent parts of such a chain, but that then involves a much larger unit to collide with adjacent chains and appreciable bond strain energy.

§3. Viscosity

The viscosity of polymers, that for the moment we may simply describe as the ease with which liquid-like flow may occur, is much affected by the chain length.[4] This is largely due to the fact that the longer the chain the more entanglements there can be. There will therefore be more resistance to the chains sliding past each other bodily (a process called **reptation** – movement like a snake), and so to gross flow since this will require many chains to move in such a way. There is a simple theoretical prediction that viscosity, η, is just proportional to the molecular mass for any given polymer system at a fixed temperature, *i.e.*

$$\eta \propto M \qquad (3.1)$$

Experiment, however, shows clearly that the actual effects are more complicated, and an equation of the form

$$\eta \propto M^\beta \qquad (3.2)$$

is required to account for the data, with the exponent β never having the exact value 1. It usually takes a value somewhere between 1 and 2 for low **degrees of polymerization**, *i.e.* chains less than about 1000 units (mers) long. This effect can be explained by observing that the chain *ends* are much *less* entanglable than the rest. The terminal sections of the polymer chain are restrained at only one end, where they are connected to what may be termed the middle section. The ease with which a chain can be entangled therefore depends on the length of the 'middle', not the whole length of the chain; in other words the contribution of the ends to the overall viscosity is much *less* than would otherwise be calculated. The effect of increasing the degree of polymerization is consequently proportionally rather *greater* than expected, because the ends are effectively of fixed length, and we may consider that it is the 'middle' portion that grows, hence the value of $\beta > 1$.

●3.1   Entanglement

On the other hand, at degrees of polymerization greater than about 1000, the value of β is nearly always about 3.5 (Fig. 3.1). Without attempting to explain the precise value, this abrupt change indicates a **critical entanglement chain length** which is independent of the chemistry of the polymer (Fig. 3.1). Crudely, chains may now be tied in knots in such a way that slippage (reptation) caused by tension cannot unravel them. In effect, these entanglements behave like cross-links (§6.2). This critical chain length also corresponds to a critical point in the strength of such polymeric materials. Now, after a certain amount of deformation, the applied load will be borne by the primary bonds of the polymer chain, which bonds are much stronger than the van der Waals forces otherwise holding the chains together. There is no slippage possible that will allow the dissipation of that stress (*i.e.* no relaxation). Hence, those primary bonds break rather than the chains becoming disentangled. A noodle analogy works here: short noodles separate readily, long ones must be broken when pulled because they become entangled.

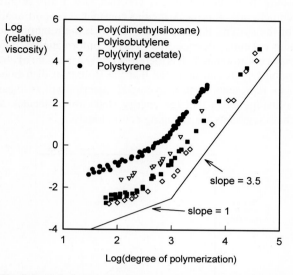

Fig. 3.1 Relative variation of polymer melt viscosity with chain length (N) in mers. Note the critical value for N where the slope changes abruptly.

●3.2   Chain segments

For flow to occur, a force must be applied so as to generate a shear stress in the material (Chap. 4). Shear means the sliding of one layer or molecule past another, but we cannot expect that polymer chains will be aligned in the direction of shear and permit this conveniently. In fact, the opposite is true: the majority of polymers are amorphous – random tangled coils – and thus generally will lie across the direction of flow. In order, then, to obtain the deformation, portions of chains must be moved in a piecemeal and incremental manner. One loop being moved permits another to move into its place, which in turn permits some other rearrangement. We are therefore concerned with a unit of action in a relatively small section of the chain, a so-called **chain segment**. These are not well-defined as to length. The apparent size depends on a number of factors, not least just what the local conditions are. In some places the entanglements will be severe, and the effective chain segments small. Elsewhere, they may be large. A chain segment is also not a fixed region of the chain: the 'ends' will move as adjacent chains move and as the chain itself undergoes random thermal motions. A chain segment is therefore not an isolatable, discrete entity, but rather it is a vague, indeterminate, fleetingly-existent statistical object. We may imagine a mean chain segment length because the actual effective values over the volume of interest at any instant show some dispersion, though what the mean value or that dispersion might be is difficult to say. However, these values characterize the polymer's behaviour under the prevailing conditions.

●3.3   Temperature-dependence

Primarily, then, it is the ability of a portion of a polymer chain to move into an adjacent space which is of concern. This movement is similar to the diffusion of smaller entities, atoms and small molecules, except for one very important difference: the movements of the ends of the moving section are inhibited since they are bonded to a larger section of chain, or possibly just locked at an entanglement. Thus, attention is focused on the **diffusivity** of the chain segments. Because we are concerned with a diffusion process, the viscosity of polymers is, not surprisingly, temperature-dependent. As was described above (§2), changes in chain conformation (necessary for motion) require an activation energy, E_a, because bond rotation is necessary. The availability of this energy is determined by the Boltzmann distribution, and is shown by a Maxwell-type distribution for the velocities – and thus the kinetic energies – of the moving segments (Fig. 3.2), and an **Arrhenius equation** may therefore be written for the diffusivity, D (m²/s):

$$D = A.e^{-E_a/RT} \qquad (3.3)$$

where R is the molar gas constant and T the absolute temperature. A is the Arrhenius constant of proportionality for the system; this is the rate at which the rotational and adjacent molecule energy barriers for conformational change may be surmounted. It is therefore really a measure of chain mobility. (It can be noted that the Boltzman distribution also represents a time-averaged description, as clearly collisions mean that the energy states of molecules and chain segments are changing continuously.)

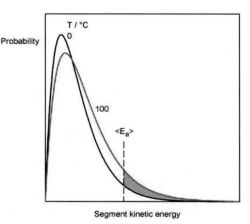

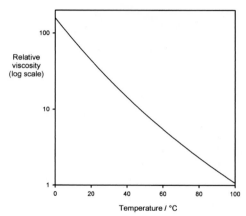

Fig. 3.3 Effect of temperature on the viscosity of polymers (above T_g)

Fig. 3.2 Form of the effect of temperature on the distribution of chain segment kinetic energy according to the Maxwell molecular velocity distribution (the area under the two curves is the same. $\langle E_a \rangle$ represents the velocity corresponding to a particular (kinetic) activation energy, the shaded area to the right between the curves represents the extra proportion exceeding that energy at the higher temperature.

In addition, there is also a component of viscosity which may be viewed as due to the effect of molecular friction, and this is proportional to absolute temperature:

$$\eta \propto T \qquad (3.4)$$

This is because raising the temperature increases the agitation of the molecules and thus the space they try to occupy, increasing the interference between them. An analogy for this might be a crowd of people moving in opposite directions along a path: the more often and the further out members of this

crowd throw out their elbows or their arms, the more those moving in the opposite direction are impeded – the rate of flow will decrease because the effective viscosity has increased. The random movements of chain segments have the same effect on adjacent polymer chains.

Combining the two relationships we obtain an expression for the overall temperature dependence of the viscosity:

$$\eta \propto T.e^{E_a/RT} \qquad (3.5)$$

The exponential term varies very rapidly with temperature, and this dominates the relation, which then explains the extreme sensitivity of polymer viscosity to temperature that is usually observed. It was noted earlier (Fig. 2.8) that the rotational energy barriers are of the order of 20 kJ/mol for small molecules with only hydrogen as side groups, taking butane as a guide. For melts of the more common polymers a figure of about 45 kJ/mol is more typical. The relation 3.5 shows that over the range 0 ~ 100°C the viscosity may change by a factor of 100 or more (Fig. 3.3). This is a very large effect indeed. It has important consequences for the

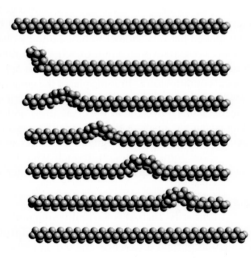

Fig. 3.4 Reptation consists of the movement of chain segments by the diffusion of kinks along the chain, aided or hampered by applied forces and steric hindrance from adjacent molecules.

behaviour of dental polymers, such as poly(methyl methacrylate). If the temperature is raised, appreciable plastic deformation – flow – may occur under even very small stresses. In particular, stresses that are 'frozen in' during the manufacture of a denture base, for example, may be released by heating due to overzealous polishing (frictional heat) or immersion in hot water.

To return to reptation, it can now be seen that a whole molecule cannot simply be slid out from the mass in one smooth, continuous movement (this is where the spaghetti type of model breaks down). It requires successive incremental movements as chain segments diffuse back and forth and sideways, even if the net effect is in one direction (Fig. 3.4). Random thermal motions in a chain mean that a kink ordinarily will diffuse (one-dimensionally) backwards and forwards along the chain in a random walk. However, if diffusion in one direction tends to reduce the energy of the system (or, equivalently, the stress) it will be favoured: it will become **stress-directed**. Thus gross plastic flow under some general applied stress is itself controlled by the diffusive nature of the chain segments. Such flow is not organized or concerted, nor indeed is it cooperative. Each molecule is affected by both its own chain segment behaviour and those of neighbours. In other words, the response to a pull is not instantaneous; one has to wait until it just so happens that a chain segment movement occurs in the right place in the right way for an increment of movement to occur (which must lower the local stress), and then repeated many times. Fig. 2.4 can only represent a single, isolated and idealized molecule's behaviour; the reality of a bulk polymer is much more complicated.

§4. Glass Transition

If a solid polymer is cold enough, it will be stiff, hard and brittle. A piece may be broken with little plastic deformation, and the fracture surfaces will show the kind of pattern seen with broken silica-based glass: a smooth, wavy, reflective form known as a **conchoidal fracture** (meaning a sea shell-like appearance). Indeed, this behaviour is characteristic of glasses in general, and can properly be called **glassy**. It depends on the amorphous structure of the material. However, if such a polymer is heated, we would of course expect some variation in mechanical properties with the change in temperature, but there comes a point where there is a very rapid and marked change from being glassy and brittle to distinctly ductile or rubbery. Note that the polymer has not melted; the melting temperature - if it can be reached without the polymer decomposing - will be rather higher. The body will still be solid, as we ordinarily understand the term, but the elastic modulus will have fallen considerably, and on fracture there will be much plastic strain: the fracture surface will now be rough and irregular. The temperature at which this change occurs is called the **glass transition temperature**, and is usually labelled T_g.

The glass transition temperature is the principal characterizing property of polymers in general. Although it may be defined in terms of the brittle to ductile change in behaviour, there are also a number of other effects. There is, for example, a more or less sudden alteration in the rate of change of specific volume (the inverse of density) with temperature (Fig. 4.1). That is to say, there is a discontinuity in the value of the coefficient of volumetric thermal expansion, and thus in linear expansion (Fig. 4.2) (see 17§1.1). These effects are reversible and reproducible.

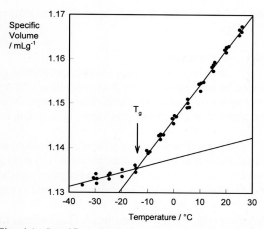

Fig. 4.1 Specific volume *vs.* temperature for a typical polymer at around its T_g.

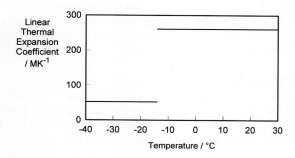

Fig. 4.2 The variation of the coefficient of linear expansion with temperature for the same polymer to which Fig. 4.1 refers., *i.e.* the slopes of those lines.

That a fundamental change in some aspect of the structure of the material is occurring at this point is shown by the variation in specific heat with temperature. Such a plot also shows a discontinuity, although perhaps not as sharply (Fig. 4.3). The size of the jump is a measure of the energy required for the process occurring at the transition. It can be likened to the specific heat of melting or boiling: energy is required for the change to a new state.

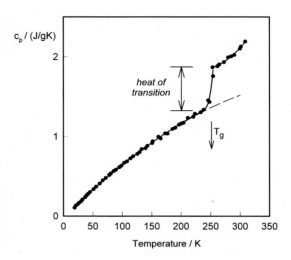

If polymers are solidified quickly from the melt, without any further mechanical or thermal treatment, their chains are typically in random, entangled coils. We may describe them as 'raw' polymers. This condition is the basis of describing their structure as amorphous. However, we may notice three things about such a material:

Fig. 4.3 Typical variation of specific heat vs. temperature for a polymer going through its T_g.

(1)   by chance, sections of a number of chains may be more or less aligned in a crystal-like manner;

(2)   because the packing in a random heap of coiled chains is far from perfect, there will be a large number of substantial 'holes' of dimensions similar to the chain thickness scattered randomly throughout the structure; and

(3)   thermal energy causes agitation of the polymer chains and substituents through bond stretching and bending.

The effects of these features need to be examined to understand better the bulk mechanical properties of polymers.

●4.1   Crystallinity

In order to achieve the required properties, primarily toughness, most polymers used in dentistry should indeed be amorphous rather than crystalline. To understand the circumstances in which this condition is met, it is easier to consider the question for the opposite position: crystallinity. It is therefore the *lack* of crystallinity and absence of the factors predisposing to it that are of concern now.

It was emphasized in §2 that only isotactic and syndiotactic polymers are capable of crystallizing. This is true, but it is not a sufficient condition. Crystallinity is also a relative phenomenon in the context of the giant molecules of a polymer. Even in a polymer such as polyethylene, where there are no side groups to interfere, and the chain is unbranched, complete crystallization cannot occur. There are four major reasons for this.

Polydispersity: Firstly, the molecules have a large range of sizes, because chain growth is random. Hence, some 'unmatched' ends would always be present in a random selection of adjacent molecules and thus have to be left over as irregularly coiled material even if the bulk were crystalline.

Periodicity: Secondly, the molecules are periodic: polymers are by definition comprised of repeat units. This property allows the matching of a section of one chain with an equivalent piece on another. This matching may occur anywhere along the two chains, without regard to the size of the unmatched end pieces. The energy difference between two such matches will be negligibly small and provide no thermodynamic drive for change. In other words, adjusting the position of one chain with respect to another, along its length, will not result in a lower energy state on average.

Reversibility: Thirdly, the chains in simple polymers may be reversible, in the sense that there is no way of distinguishing the direction of travel along it, so that a chain section may align with another section of the same chain that has folded back on itself. In fact, after a double fold, even chains with a 'directionality' will be capable of such matches. This kind of process will leave loops which could not be made ultimately to disappear. They would therefore remain in disordered regions.

Probability: Fourthly, in purely statistical terms, there is only a very small probability of obtaining a substantial amount of crystallization directly from the melt at moderate cooling rates. The chains will be highly entangled and the viscosity very high. In fact, the molecules are so big that they can have sections in several different 'crystalline' regions at one time. Nevertheless, there will be regions where, by chance, chains will align sufficiently well to give a semblance of crystallinity, a greater degree of regularity or order than elsewhere; these regions may be termed **crystallites**. The size of such regions, that is, the lengths and numbers of chains involved, will depend on a variety of factors, including the size and configuration of substituents. Thus, we can talk about crystallinity in relative terms, ranging from a fully aligned, regular array, all the way down to totally amorphous.

Fig. 4.4 The fringed micelle model of polymer crystallinity.

Putting all of these ideas together we arrive at the so-called **fringed micelle** model of polymer crystallinity (Fig. 4.4). Following a chain, it may wander from crystallite to crystallite, through amorphous regions, sometimes folding back on itself. The boundaries of crystallites are poorly defined as quite where the fringe of disordered chain begins is hard to say. The degree of crystallinity in this sense thus depends very much on the type of polymer, as well as the time and temperature at which solidification is allowed to occur.

One of the implications of this structure concerns the optical appearance of the bulk polymer. Since the more ordered regions are better packed, they have a higher density than the surrounding amorphous, more poorly-packed material. Accordingly, the refractive index is also higher in these more crystalline regions than elsewhere. This means that the path of a ray of light is far from straight in traversing many such regions, *i.e.* it is scattered (24§5.11). This scattering results in lower transparency. Thus, bulk-polymerized poly(methyl methacrylate) is a so completely transparent polymer that it is used for aircraft cockpit windows, due to its being completely amorphous. Polyethylene, however, which can crystallize relatively easily because of its simple chain, is merely translucent except in thin films, and even then can be seen to introduce distortions on a small scale to images viewed through it.

> It should be noted that so-called 'crystal glass', as used in drinking vessels and chandeliers, is not crystalline in any way to any extent. The term has no relation to crystallinity in polymers, even though the silica-based glass is a polymer, albeit inorganic (25§1.3). It refers to the crystal-like facets which are commonly cut in or on it to reflect and refract light – aided by the fact that its refractive index is high because of its lead oxide (PbO) content (typically 35% by mass). Again, this is an indication that care should be taken with common-usage words in technical contexts.

Even so, no matter how well such a polymer crystallizes, the presence of the amorphous inter-crystallite regions is inevitable. This is one of the reasons why high polymers that crystallize are not completely brittle and still show a clear glass transition: the volume fraction of glass cannot be zero.

●4.2 Free volume

In the amorphous regions of randomly coiled, intertwined and tangled chains, much as appeared in the molten polymer, there will be large numbers of relatively large interstitial spaces, as occur in a pile of noodles. The packing is poor, and the overall density must be low. The difference between the bulk volume of a piece and the actual volume of the molecules in it is known as the **free volume**. In amorphous polymers it is particularly large, but it is still expected to be temperature dependent - this is the nature of thermal expansion. Importantly, the rate at which the value of the free volume increases goes up substantially as the temperature is raised through the T_g (see Fig. 4.1).

●4.3 Thermal motion

The free volume is of great importance because it allows room for portions of the surrounding adjacent polymer chains to move over relatively large distances. In a crystal, where 'contact' between contiguous chains is present, there is very little room for movement except for bond vibrations. But in amorphous regions, larger sections of chain can move because the holes available are larger. These larger sections of the polymer molecule are the **chain segments** referred to above as the means of diffusion in polymers. Again, these are not related in any way to the individual bonds or atoms of the chain, nor to the notional building units, the individual mers.

The chain segment is a concept which refers to the length of chain which may swing or writhe, more or less unhindered, between end-points determined by entanglement or a branch point in the chain. In the noodle analogy, if all the contact points between noodles were fixed or glued, there would still remain loops and sections which could be moved. In a polymer, the contact points are only temporary and may slide or change. Branch points are permanent, but, if the surroundings permit, even they (with their attached chains) may move bodily. However, thermal agitation of the polymer will result in greater motions of the portions of the chain between those pinned points than of other sections; the chain segments therefore become the dominant structural unit of the system. All consideration of the mechanical behaviour becomes centred on the behaviour of these entities. Indeed, as was said above (§3.3), the diffusivity itself is really a measure of chain segment mobility.

●4.4 Statistical nature

The glass transition is a very complex phenomenon which is still not fully understood. However, the principal feature of interest is the substantial and relatively abrupt change in the diffusivity of chain segments. The response to applied stresses is therefore changed, and hence all mechanical properties are affected. In a sense, this increase in diffusivity on heating can be viewed as a kind of partial melting, whereby the van der Waals forces which hold adjacent (contacting, not bonded) sections of different chains together are overcome. This is therefore an activated process, and the energy required is thermal. The effect of this process is an increase in the free volume into which chain segments may 'diffuse', and thus the size and number of spaces available for this to occur.

Diffusivity is also affected by the relative ease of rotation about bonds in the polymer chain, in other words the flexibility of the backbone (because all conformational change requires bond rotation), but more strongly by the regularity with which the polymer chains can pack. That is, the better the packing the shorter the range over which the van der Waals forces must operate, and at more points. However, the fact that an activation energy is required for the change also has one extremely important side-effect: the T_g is not a sharply defined temperature, the transition in properties occurs gradually over a small range. This arises because the availability of the energy for a movement, such as overcoming a steric barrier, is a probabilistic matter: the longer one waits the greater the chance of the event having occurred. Also, steric barriers will not be all identical, some more severe, some weaker, so that change is progressive. Thus, measured values of T_g depend very much on the method used, and considerable variations in tabulated data are common. As with any test result, the meaning depends on the context, and the value to be used must be chosen to be from a method appropriate to the context.

●4.5 Side chains

Large side-chains inhibit crystallization by blocking the alignment and close approach of chains. The longer and more flexible they are, the lower the energy required for their own rearrangement and this in turn affects the rearrangement of the chain segments bearing them: if they can be moved out of the way or rearranged easily, the main chain segment motion is relatively free.[5] This can be seen very clearly in the data for two series of acrylic polymers (Fig. 4.5). As the ester side chain lengthens from methyl to n-octyl the glass transition temperature falls steadily and substantially in both series. It is also very clear that the steric interference to rotation of chain segments introduced by the methyl group in the methacrylic series (as compared with the other, non-substituted series) causes a marked increase in T_g, as would be expected from the argument of §2.

Observing such relationships, it thus starts to become possible to design a polymer with specific properties by appropriate choices of side chains.

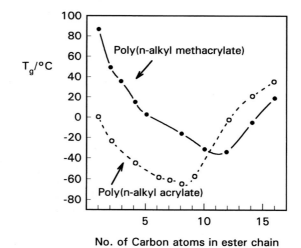

Fig. 4.5 The variation of T_g with side-chain length for the two series of polymers poly(n-alkyl methacrylate) and poly(n-alkyl acrylate).

Thus, the poly(n-alkyl acrylate)s would all be too soft for use in the mouth as denture base materials, but (as we already know) poly(methyl methacrylate) is suitable. For comparison, it is worth noting that the T_g of natural rubber (polyisoprene, see 27§1) is about −75°C and that of a typical silicone rubber is about −120°C; both of these materials are well-known for their rubbery properties at room temperature. The process of side-chain lengthening can be taken too far, however, and eventually at a critical length they can be sufficiently long to start to 'crystallize' themselves (Fig. 4.6), in effect forming their own micelle, reversing the trend to lower T_g (Fig. 4.5). This works because the tiny regions of 'crystallized' side chains are held together better than are the main chains one with another, and the overall free volume is reduced. Effectively, these regions behave as cross-links, inhibiting backbone chain segment movements. This is essentially because the side chains start to behave as polymer chains themselves, almost leading an independent existence.

Fig. 4.6 Side-chains can become long enough to 'crystallize' themselves in their own micelle.

●4.6 Supercooling

Glasses, whether of the familiar silica-based varieties (which are themselves polymers), organic polymers, or any other system (such as the exotic metallic glasses,[6][7] or toffee), can be considered to be **supercooled** liquids, by which it is meant that the random, disordered structure of the liquid persists at a temperature below the freezing point. In each case this arises essentially because during the freezing process insufficient time is allowed for diffusion to occur to form the crystalline structure. For metals it might require cooling rates of the order of 10^6 K/s or more for crystallization to be avoided and the glass to form.[2] For less easily diffusible systems, such as polymers, even very slow cooling may not be sufficient to permit crystallization.

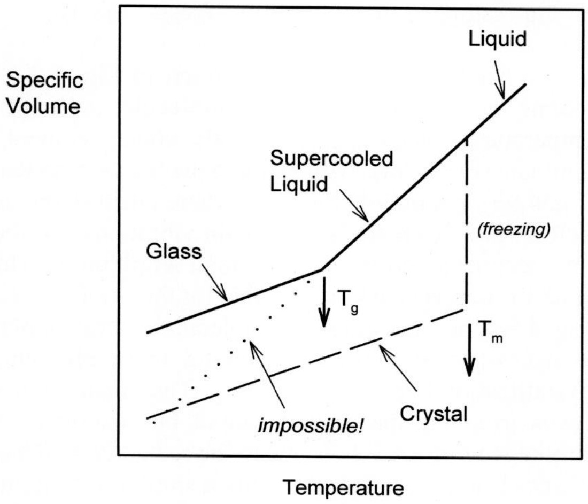

The freezing point itself, of course, is associated with a latent heat for the abrupt change of state as well as the structural change of crystallization. The existence of a **latent heat** has one very important connotation: freezing occurs at constant temperature. In other words, the addition or subtraction of heat at the freezing point is expected to result only in a change in the *amounts* of the solid and liquid, but not in their temperature.

Liquids are characterized by, amongst other things, the complete lack of long-range order, and indeed the short range order may be very limited. Compared with a crystalline solid, the energy content of a liquid (at the same temperature) is rather greater, principally due to the difference in entropy, a measure of disorder; this is the latent heat. In crystallizing, the molecules of a liquid (usually) tend to pack more closely together (water is an obvious exception), but in any case in a much more ordered manner. This normally leads to an abrupt discontinuity in the plot of specific volume *vs.* temperature at the freezing point, T_m (Fig. 4.7). Note that specific volume is the sum of the free volume and the true molecular volume (which hardly changes with temperature as the thermal expansion coefficient of covalent bonds is low). Hence, the changes shown in Fig. 4.7 are almost entirely due to change in free volume.

When supercooling occurs, there is no change in the slope of the plot as that point (T_m) is reached because there is no change in structure – the randomness of the arrangement of the liquid above the true melting point is maintained. But in polymers, further cooling is required before the **glass point** is reached and chain segment motions cease to dominate the properties (*cf.* Fig. 4.1), still preserving a random structure. Nevertheless, the T_g is not as sharply defined a temperature as the T_m of simple

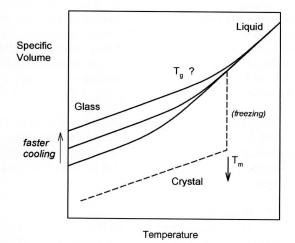

Fig. 4.7 A comparison of the volume effects observed with materials that crystallize and those that form a glass. The 'impossible' line is inaccessible because of molecular contact in the amorphous condition. The slopes of the lines are the respective thermal expansion coefficients (*cf.* Fig. 4.1, 4.2).

Fig. 4.8 The value of the glass transition temperature is not precisely defined, and it is dependent on cooling rate.

---

[2] However, special alloying can now allow this to be done with much greater ease.

molecules because (as explained above, §4.4) the onset of the characterizing chain segment motions requires an activation energy. As a change of state is not involved, the behaviour of the material is probabilistic. Thus, the likelihood of a chain segment becoming free to move increases with temperature, but also the likelihood of it having been free in some time interval increases with the length of that interval. What this means is that the higher the temperature, the shorter the time that it would take on average for any given chain segment to move. The temperature of the glass transition therefore depends on time-scale(cooling rate) (Fig. 4.8), and is not a fixed point, quite unlike the freezing (or melting) point for simpler substances. (The behaviour on heating may be slightly different, with some overshoot, also dependent on rate of change of temperature,[8] in a kind of hysteresis (7§8.4).)

Another view of this is given in Fig. 4.9. The free volume associated with each molecule is large at high temperatures when the large-scale chain segment motions dominate (Fig. 4.9a). This volume decreases with temperature (Fig. 4.9b), but only to the point where chain segment motions more or less cease (only side chain vibrations and the like are now occurring and these have small amplitude). This is now about the densest packing possible for the amorphous condition (Fig. 4.9c), because extensive molecular rearrangements would be necessary to make the packing more efficient, i.e. for crystallization to occur (Fig. 4.9d). The random pile of chains leaves irregular spaces that cannot be occupied. The only shrinkage now possible is due to the reduction in the amplitude of side chain movements, and this is similar in magnitude to the effect that is seen in the crystalline material, i.e. the thermal expansion coefficients are similar.

Because the **relaxation time** is short, diffusivity being high, a liquid above the melting point very rapidly approaches equilibrium. However, on cooling below the T_g diffusivity falls rapidly, and the approach to equilibrium becomes very slow if not non-existent at a sufficiently low temperature. Such polymers are then clearly in a non-equilibrium state, and it is impossible to avoid this: crystallization cannot occur. However, while the rate of heating or cooling has an effect on the observed value for the T_g, time at a temperature sufficiently close to the current T_g will allow relaxation to a denser state and therefore a lower value for the T_g (Fig. 4.8).

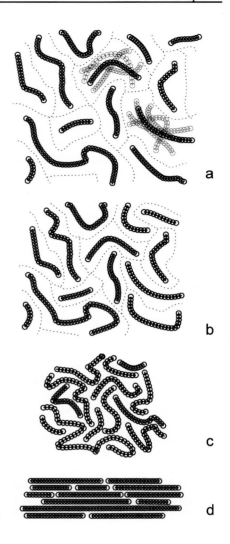

Fig. 4.9   (a) Because of thermal motion molecules occupy a volume that shrinks on cooling (b). However, there is a minimum volume that they can occupy in an amorphous condition (c), which is less efficient packing than in a crystal (d).

The melting points of amorphous polymers are also subject to the same kind of uncertainty, for more or less the same molecular and statistical reasons. There is no change in structure, just a rapidly increasing molecular mobility, but the randomness of entanglements and the processes which undo them simply prevent a single temperature being identified with 'melting', and indeed the apparent melting temperature will depend on heating rate.

●4.7   **Drawing**
The application of a tensile load across any two points of a polymer chain, which ordinarily will be in a random coil, must cause a general straightening of the chain (Fig. 2.4). Much of the energy of deformation goes into this process *via* segment rotation, which is very much easier in comparison with the straining of bonds, but there also tends to be slippage of one chain past another. This kind of deformation is called **drawing**. But, in the course of this, the molecules tend to become more aligned with each other than they were in their random coils, and the overall regularity of the structure is increased. In effect, the material becomes more crystalline (Fig. 4.10).[9]

As a result of this it becomes much stiffer in tension, *i.e.* the modulus of elasticity increases, because there is then less capacity remaining for extension *via* chain segment rotation (Fig. 4.11). This arises because the deformation can then only be of the C-C bond length and deformation of the tetrahedral angle. For example, fully crystalline polyethylene (*i.e.* fully aligned chains) is about 80 times stiffer along the chain axis (E = 235 GPa, see Fig. 1§14.1) than the bulk amorphous material.

More drawing would take the process of alignment and increase in tensile modulus of elasticity further. The original **isotropic** structure (that is, uniform in all directions, lacking directionality) becomes highly **anisotropic** as the chains become **oriented**. The increase in modulus is also anisotropic as it is in the drawing direction only that the material's capacity for further chain deformation by segment rotation is reduced. In directions perpendicular to the drawing axis, deformation would represent moving in the sense of an approach to the original random coil (although this could never be recovered by deformation) and is therefore relatively unhindered. If the tensile deformation is not very small, relative chain movement (slippage) will mean that when the load is removed, although there will be some elastic recovery, the deformed material will preserve to some extent these new properties. It is then said to be **strain-hardened** (this must be carefully distinguished from the strain-hardening of metals, which has a fundamentally different mechanism, 11§6).

●4.8 Toughness

Toughness in polymers, their ability to absorb energy before fracture, is clearly related to the difficulty with which chain alignment may occur during the drawing process. Work is done overcoming steric hindrances both within and between chains, that is, during both chain segment rotation and slippage. The ease of these two mechanical processes depends on the availability of the activation energy, that is, where in relation to the glass temperature the test temperature lies. The toughness of a polymer can therefore be predicted to be highly dependent on both temperature and time, the latter in the sense of the *rate* at which the drawing process is attempted. In other words, instead of the atomic and grain boundary motions referred to in Fig. 1§9.1, which motions apply to metals and other substances, a new process, chain segment diffusion, must be written into the diagram for polymers. Its position in that diagram, however, will depend on many factors.

●4.9 Properties *vs.* temperature

The glass transition in fact marks only one of several stages in the sequence of changes which have been observed and documented for the thermo-mechanical properties of polymers. A set of conventional descriptive terms has been developed which are broadly applicable to all polymers, regardless of chemistry, and these terms therefore can convey much about their behaviour,

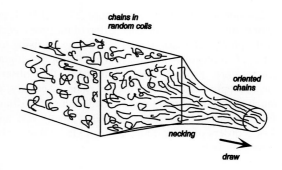

Fig. 4.10 Drawing a polymer changes it structure, chains becoming well-aligned in the process.

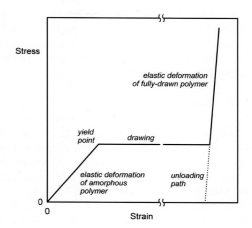

Fig. 4.11 Idealized stress-strain plot for an amorphous polymer. It is assumed here that there is no strain hardening during drawing, that there is no chain slippage after drawing is complete, and that the specimen does not break. Unloading assumes no recoiling of chains.

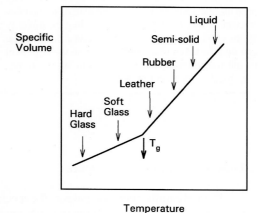

Fig. 4.12 The correlation of the mechanical properties of a polymer with temperature relative to the T_g.

independently of actual temperature (Fig. 4.12). It is informative in this sense to plot elastic modulus against temperature, and note the correlation of the gross mechanical properties at each point with the principal features of the plot (Fig. 4.13). Each step down with rising temperature corresponds to the 'unfreezing' of a particular type of molecular motion. This latest or highest-temperature process then dominates the observed macroscopic behaviour of the polymer. Actually, it is much more convenient to plot log(modulus of elasticity), as is done here, because the range is typically very large, 4 or 5 orders of magnitude between glassy and rubbery behaviour being commonplace.

It is noteworthy in such diagrams that neither the glass transition temperature nor any other feature is specified for particular chain lengths or molecular weights, that is, the behaviour is independent of degree of polymerization above some sufficiently large number (*cf.* §3.1). This is due to the fact that the rotation of an individual chain segment cannot be influenced by the chain more than a few bonds away; it is in effect invisible to it. It is therefore of no consequence whether that unseen chain is 10 or 10,000 mers long.

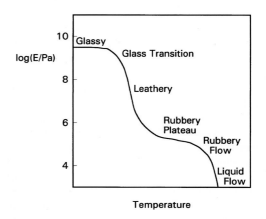

Fig. 4.13 Typical variation in polymer modulus of elasticity *vs.* temperature, correlated with observed mechanical properties.

● **4.10 Thermal expansion**
The dominance of chain-segment effects in controlling polymer properties may perhaps be judged by the similarity of values across a wide range of straight-chain polymers. Thus, in the rubbery region, *i.e.* $T > T_g$, the volumetric coefficient β is in the region of 700 MK^{-1} ($\alpha \sim 250$ MK^{-1}), while in the glassy region, $T < T_g$, $\beta \sim 200 - 300$ MK^{-1} ($\alpha \sim 70 - 100$ MK^{-1}). These figures may be compared with the range of values for many low molecular weight organic liquids where $\beta \sim 500 - 1500$ MK^{-1}, and with other types of material to put this in context (Fig. 4.14). In other words, polymer chain segment motions are close to liquid-like in their capacity for large scale movements.[3]

> The graph axis and table column label convention used here might need explanation. Essentially, only pure numbers can be plotted or tabulated, so the label is made to be dimensionless by dividing the quantity by the units, thus "Length/m". Likewise, a logarithm can only be taken of a pure number, so the argument of the transformation is expressed in similar terms, *e.g.* log(E/Pa), as in Fig. 4.13.

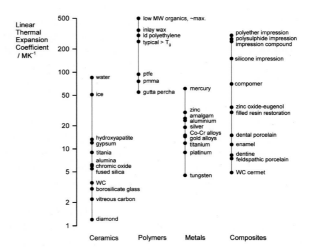

Fig. 4.14 Comparison of values of the linear thermal expansion coefficient of some dental examples of the four major classes of material. All at ~25 °C, except ice (0° C). For liquids, the equivalent value for isotropic expansion is given.[3]

§5. Time-Temperature Equivalence

There are further consequences of the special structure and molecular behaviour of polymeric materials. Those relating to the strength of the material are particularly important for the selection of a polymer in the context of its service conditions. The point has already been made that both time and temperature are significant, and their influence extends to all aspects of the mechanical properties of polymers. This therefore applies to any

---
[3] $\beta \approx 1 - (1 - \alpha)^3$.

materials containing them. For example, the apparent stiffness, strength and toughness of polyethylene vary very clearly with the strain rate (Fig. 5.1): the greater the strain rate, the more rigid the material appears (*i.e.* the higher is Young's Modulus), as shown by the slope of the first part of the curve. Also, there is a decrease in the amount of strain that can be sustained before fracture, as well as an increase in the value of the yield point. All of this is traceable to the time required for molecular rearrangements to occur, otherwise known as the **relaxation time**, a concept previously introduced (Fig. 1§9.1), only now chain segment rearrangements are included and indeed they dominate the behaviour of polymers from about the T_g upwards in temperature. Thus, low strain rate is equivalent to long duration tests, *i.e.* putting a specimen under load for long periods, simply because it allows time for chain segment diffusion.

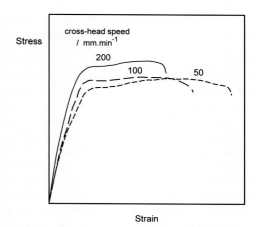

Fig. 5.1 Variation in the loading behaviour of a sample of polyethylene at different deformation rates.

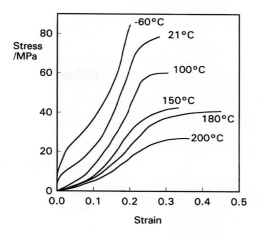

Fig. 5.2 The effect of temperature on the stress-strain curve for one variety of nylon.

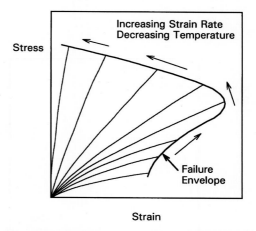

Fig. 5.3 For polymers, the ends of all possible stress-strain plots define the failure envelope; this may be mapped in either temperature or strain rate.

Similarly, the higher the temperature the lower the modulus, the greater the ductility, and the lower the yield point (Fig. 5.2). Notice the reciprocal relationship between the two variables: that because there is less energy available for molecular rearrangement as the temperature is lowered, the relaxation time required becomes longer. *Decreasing* the temperature is equivalent to *increasing* the strain rate, that is, reducing the time available for any activated process, such as chain segment diffusion, to occur.

For many polymers it is possible to take advantage of this **strain rate - temperature equivalence** and draw a family of stress-strain curves, the ends of which (representing fracture) form a characteristically-shaped enclosing **failure envelope** (Fig. 5.3).[10] The path that any particular tensile test takes depends on both temperature and strain rate. (Notice that the stress-strain plot is not a straight line, even initially, and the remarks on non-Hookean systems, 1§4 and Fig. 1§13.1, must apply.)

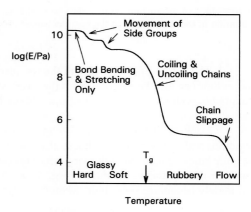

Fig. 5.4 The polymer modulus *vs.* temperature plot may also be correlated with the dominant molecular processes. Notice that other processes associated with lower temperatures have not stopped, it is only that their effect has become swamped by the larger scale of the just enabled process.

The elastic modulus that is observed experimentally can be correlated with the relaxation time associated with different molecular processes (Fig. 5.4). Thus, at low temperatures (or high strain rate) the stiffness of primary valence bonds dominates the behaviour, when there is either no time for other movements to occur or insufficient energy to activate those processes. At temperatures greater than the T_g chain segment movement becomes possible and deformation is associated with coiling and uncoiling of chains – writhing movements arising by virtue of chain segment diffusion. However, molecular entanglements prevent or at least limit longer range motions, *i.e.* slippage. Thus the region about the T_g is associated with rubbery behaviour. Ultimately, when the temperature is high enough for gross plastic flow is observed, the chains slipping past each other give a very low modulus. This effects arise because either time is being allowed for slow processes, or the energy is available for them to occur at a high enough rate for their influence to be detectable.

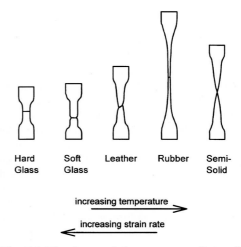

Fig. 5.5 The characteristic appearance of a tensile test specimen at the point of failure in the various regions of behaviour.

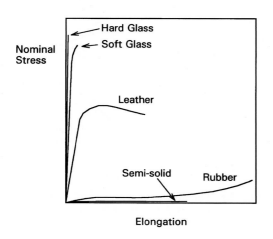

Fig. 5.6 The variation in tensile test curve shape with temperature or strain rate corresponds to characteristic variation in the appearance of the test piece after failure.

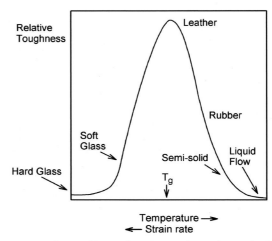

Fig. 5.7 The toughness of a polymer also varies markedly with temperature and is correlated with the molecular processes.

The nature of the fracture of the test piece also reflects these properties (Fig. 5.5), and this variation in behaviour can also be inferred from the stress-strain diagram (Fig. 5.6). Notice that elongation at failure and toughness both show maximum values, although these do not coincide.

Looking at Figs 5.3 and 5.6 again, it is apparent that they reflect more than the variation in elastic modulus and strength: the toughness also varies considerably. Thus, the total work to failure of the brittle glass is quite small, but as chain drawing becomes more dominant so the plastic work increases, and this reaches a peak in the 'leathery' condition (Fig. 5.7). After that, the resistance to chain slippage decreases rapidly, ultimately becoming very slight when the polymer has melted.

§6. Plasticizers and Cross-linking

Since relative chain segment motions are now seen to be the dominant issue in determining polymer properties, it should be clear that any other factor which can aid or inhibit the motion of one chain or chain segment past another is also relevant as a factor to be considered in understanding the mechanical properties of, or indeed for designing, a polymer.

● **6.1 Plasticizers**

An important method of facilitating chain segment movement, thereby lowering the elastic modulus or making drawing easier, is to incorporate what may be termed 'internal lubricants', more properly called **plasticizers**. These are substances with small, mobile molecules which can be dissolved in the solid polymer, but which do not form any strong bonds with the polymer. In effect, plasticizers provide more space for chain segment rotation, firstly by holding the chains apart and reducing the van der Waals forces operating between them, and secondly by being very much more mobile than the polymer chains themselves, unconstrained in both orientation and position because they are small, free, molecules. This freedom allows them to be pushed out of the way by moving chain segments much more easily than they would another chain. It is as if the free volume of the polymer were increased. For polymers that absorb appreciable amounts of water (typically, polymers with polar groups) these effects can even be seen with changes in the relative humidity of the test environment – another reason for being careful with the conditions of testing.

Nylon shows just such a significant decrease in both tensile strength and elastic modulus, depending greatly on the water content (Fig. 6.1). This effect can be anticipated for any polymer which can absorb water, and polar polymers such as acrylics are most affected. In fact, the absorption of water is a principal reason for nylon not being acceptable for use in the mouth (although the injection moulding method needed to shape devices also has severe problems in the dental context).

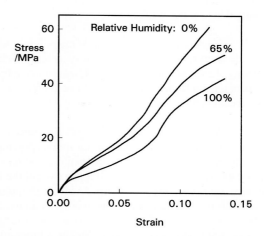

Fig. 6.1 The water vapour pressure prevailing during storage of a test specimen may affect results, as with those illustrated here for a type of nylon.

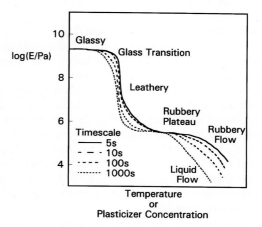

Fig. 6.2 Variation in plasticizer content is equivalent to a variation in temperature in its effect on polymer modulus.

If the polymer is exposed to the plasticizer (*i.e.* after it has been formed into the required shape), the sorption results in swelling, as interactions between polymer chains are weakened. In some cases, of course, we may obtain complete dissolution, as with PMMA in chloroform (the basis of a very useful 'plastic cement' for joining acrylic sheets together). Otherwise, the change in dimensions might be of concern, as with water in a denture base or a resin-based restorative material.

Side-chains (Fig. 4.5) also act as plasticizers. They are necessarily more mobile than a chain segment of the same size because they are fixed at only one end. They are thus readily displaced by a moving chain segment. However, the fact that they are attached to the main chain means that they are not as effective in this respect as a free molecule(although, of course, they cannot then leach out).

In terms of their effect on mechanical properties, the addition of plasticizers can be seen to be directly equivalent to increasing temperature because the glass transition temperature is *lowered* - remember the effective increase in free volume.[11] Information on plasticization can therefore be incorporated into one figure to show its interrelationship with the effects of strain rate and temperature. These three factors are very important in the design and application of polymers as well as the interpretation of test results (Fig. 6.2). The different curves here are identified by the duration of the test used to evaluate the modulus, that is, each is for a different strain rate. Slight differences in the shape of each curve now start to be apparent, but the validity of the general equivalence is not weakened.

Obviously, the toughness of the polymer will also be affected by the presence of plasticizing molecules or side chains, in much the same way as temperature operates. However, in the case of side-chains, the effect of long chains, as shown in Fig. 4.5, would also have to be taken into account. A further factor is that the thermal expansion coefficient would tend to be increased, again because the interactions between the backbone chains are weakened.

A related effect is that of the rate of cooling, in conjunction with the idea that an amorphous polymer is a random structure. Some regions will settle into a compact arrangement more easily than others on cooling, so overall the transition to a glassy condition cannot be sharp. In effect, the actual value of T_g varies from place to place in the mass. Secondly, cooling faster allows less time for relaxation into a compact arrangement, so the glassy state will occur sooner (higher temperature), in a less dense condition. The situation is more like that indicated in Fig. 4.8 than that of Fig. 4.7. This also means that **annealing** a rapidly-cooled polymer at an appropriate temperature will change the properties by permitting relaxation by chain segment diffusion, primarily by increasing the density.

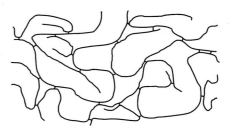

Fig. 6.3 Polymer chains may be also be cross-linked to inhibit chain slippage by direct connections.

●6.2   Branching and cross-linking

So far we have only considered straight chain polymers, although these may have had bulky side groups. But the arguments based on ease of movement of chain segments and whole chains apply with even more force when the chains are branched (Fig. 6.3), or even cross-linked (Fig. 6.4). A **branch** consists of a polymer chain attached at one end covalently to some point along the length of another chain, while a **cross-link** is a direct connection between two chains at points somewhere along their length (*i.e.* not at the end of either chain). Thus the **node** of a branch has a **connectivity** of 3 – three chains arise from it, while the node of a cross-link has a connectivity of 4 – four chains arise from it.

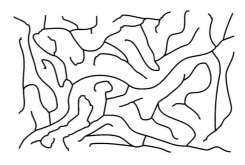

Fig. 6.4 Polymer chains may be branched, which would   inhibit chain slippage by increasing entanglement.

Clearly, branching and cross-linking will severely inhibit the motion of chains past each other, as well as the more local effects of rotation of segments, because the 'entanglements' are fixed within chains, and cannot ever be unravelled by pulling on one end. The strength of cross-linked polymeric materials may be increased, the primary covalent bonds tending to carry more of the load, but in particular the elastic modulus may be greatly increased. Sufficient cross-linking also prevents true fluid-like flow, and a descent from the rubbery plateau may not occur on heating, at least, not before decomposition.

Increasing molecular weight (*i.e.* degree of polymerization) also increases entanglements (§3.1), and thus this also has an effect in solid polymers, tending to push the onset of flow to higher temperatures, although the T_g may not be much affected because chain-segment diffusion is not much limited, unlike cross-linking and branching. The general effects are summarized in Fig. 6.5.[12] Note that cross-links also raise the modulus of elasticity on the rubbery plateau.

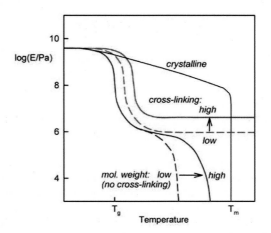

Fig. 6.5 General effects of raising molecular weight and increasing cross-linking. For comparison, a fully-crystalline low molecular weight polymer (oligomer) collapses at the melting temperature.

Highly cross-linked polymers may however be very brittle, exhibiting little or no plastic deformation even at low strain rates, as the glass transition temperature is raised very effectively by this means. Then, drawing of the polymer becomes impossible, strain hardening therefore does not occur, and even softening to the extent that fluid flow occurs simply cannot be achieved by raising the temperature without decomposition first occurring.

The creation of cross-links would also be expected to contribute to polymerization shrinkage (see 5§2.4), as chain segment motions are prevented, and also to a decreased thermal expansion coefficient as the constraints on chain separations are increased. As was indicated in §4.3, the thermal expansion of polymers is largely due to the behaviour of highly-mobile chain segments. If these are reduced in number by introducing further nodes, the scope for large-scale movements is reduced. Thus, what may be considered as the most highly cross-linked polymer, diamond, has a thermal expansion coefficient of 1.2 MK^{-1}.

●6.3 Summary
Thus, to summarize all the effects: increasing temperature is (effectively) equivalent to
(1)   increasing plasticizer
(2)   decreasing strain rate
(3)   decreasing degree of polymerization (mol. wt.)
(4)   decreasing branching and cross-linking, and indeed
(5)   lowering the T_g by changing the nature of the polymer itself by means of different side chains or less sterically-hindered rotations.

There are two main points to emerge from this. Firstly, polymer design can permit the properties to be matched to the application. This has, of course, certain fundamental limitations: both the upper limit to working temperature and strength are set by the strength of the primary bonds of the polymer chain. In addition, these designed properties hold only so long as the service conditions do not vary too much. Secondly, the results obtained in the mechanical testing of polymers depend intimately on the test conditions: time(or strain rate), temperature, and intentional or unintentional plasticizers, much more so than for almost any other class of material. Unless the testing is done under conditions that mimic those of service very closely, the results will be at best misleading, at the worst quite useless. The history of polymer development has depended on the recognition of these two points, and no less in dentistry. Adjustments in basic chemistry, product formulation, and processing have always been made with a view to getting improved performance under actual service conditions. All materials' behaviour must be understood in that light.

References

[1]   Eliel EL. Stereochemistry of Carbon Compounds. McGraw-Hill, New York, 1962.

[2]   Weinhold F. Chemistry: a new twist on molecular shape. Nature 411 (6837): 539 - 541, 2001.

[3]   Pophristic V & Goodman L. Hyperconjugation not steric repulsion leads to the staggered structure of ethane. Nature 411(6837): 565 - 568, 2001.

[4]   Schultz J. Polymer Materials Science. Prentice-Hall, New Jersey, 1974.

[5]   Saunders KJ. Organic Polymer Chemistry. Chapman & Hall, London, 1973.

[6]   Demetriou MD, Launey ME, Garrett G, Schramm JP, Hofmann DC, Johnson WL & Ritchie RO. A damage-tolerant glass. Nature Materials 10: 123–128, 2011.

[7]   http://en.wikipedia.org/wiki/Amorphous_metal, http://en.wikipedia.org/wiki/Liquidmetal

[8]   Badrinarayanan P, Zheng W, Li QX & Simon SL. The glass transition temperature versus the fictive temperature. J Non-Cryst Solids 53: 2603 - 2612, 2007.

[9]   Vincent JFV. Structural Biomaterials. Macmillan, London, 1982.

[10]   Scala E. Composite Materials for Combined Functions. Hayden, New Jersey, 1973.

[11]   Tobolsky AV & Mark HF. Polymer Science and Materials. Wiley, New York, 1971.

[12]   Aklonis JJ. Mechanical properties of polymers. J Chem Educ, 58 (11): 892 - 897, 1981

Chapter 4   Rheology

*An important behaviour in nearly the whole of dentistry is that of **flow**. This subject addresses dynamic aspects of the deformation of materials, picking up themes that were introduced in Chapter 1, which was otherwise more to do with static tests. However, no sharp division exists between the two areas. While 'static' tests tend to be concerned with the final outcome, flow focuses on deformation while it is happening. Rheological considerations are involved in many types of material, including impression materials, restorative materials and cements, waxes and the casting process. It is also highly important in the context of gypsum product slurries.*

*Simple analogue mechanical **models** again provide a useful means of classifying the types of process. By dealing with the model elements in this way, more complicated systems can be dealt with simply in terms of additive effects. These elements are the Hookean spring and static friction block as before, but now adding the Newtonian dashpot.*

*A simple **classification** of flow behaviour shows the implications for the handling of real materials. For proper control of mixing, impression taking, model pouring and so on, the flow properties must be taken into account.*

*An example of detrimental flow behaviour is illustrated by the **compression set** test, applicable to flexible impression materials which suffer deformation during removal from the mouth. The permanent deformation that results is the major source of dimensional inaccuracy.*

Fillers *are very widely used to modify the properties of materials, usually by increasing the viscosity of the medium in which they are suspended. However, the amounts that can be used are limited, and this is another source of compromise that must be accommodated in the design of materials.*

Some examples of flow systems of importance are discussed, including settling, waxes, and through tubes.

Materials Science for Dentistry
https://doi.org/10.1016/B978-0-08-101035-8.50004-3

In practically every aspect of dentistry the flow of a material is vitally important to the outcome of the procedure: taking impressions, pouring models, making wax patterns, investing, casting, cementing, fissure sealing and placing restorations are just some of these aspects. Such procedures are essential because only very rarely will the material for a procedure or a device not be required to flow or deform plastically at some stage, and no practical substitute methods for the same purposes can presently even be imagined which do not involve flow at one stage or another. In view of the likelihood that flow processes will remain part of dentistry, some appreciation of these phenomena should be gained.

§1. First Thoughts

Rheology is concerned with the time-dependent deformation of bodies under the influence of applied stresses, both the magnitude and rate, whether the bodies be solid, liquid or gaseous. That is to say, the plastic deformation or flow of the material is the focus. The discussion of the mechanical properties of solids (Chap. 1) centred largely on the stress limits below which elastic behaviour is observed or outright failure does not occur. Nevertheless, plastic deformation was of importance in considering toughness, and it is central to the outcome of an indentation hardness test. However, it is usually considered as an essentially instantaneous response to the stress in those contexts. It was only when we considered creep (1§11) that the *rate* of deformation as such assumed any importance. Yet it should be clear from the idea of a **relaxation time** (1§9) being applied to all processes at the atomic or molecular levels, and especially from the time-dependent behaviour of polymers (3§5), that it may be difficult, even dangerous, but certainly misleading, to ignore the rate at which these events occur. Thus, making a simple distinction between the assumed instantaneous recovery of elastic deformation and the *permanency* of plastic deformation in solids, ignores the possibility of other behaviour. In the case of fluid flow, superficially it would seem that only permanent plastic deformation is possible, but a great many solids and liquids exhibit time-dependent *recovery* of deformation, a kind of slow elasticity, and this behaviour must be included too in a general description of rheology.

'Description' here is the important word, for we will not attempt to explain rheological properties in terms of detailed atomic or molecular processes but rather resort only to bulk, macroscopic accounts of the responses to applied stresses, except for broad indications of mechanisms in one or two special cases later. Nor is this necessary. In other words, we can ignore atomic and molecular level structure, and treat the materials as continua. For the time being, this **phenomenological** approach will be entirely sufficient to understand the terminology, the observations, and the practical implications of the various kinds of behaviour in the applications of the materials exhibiting those properties. Certain structural aspects can then be introduced to provide physical explanations of events where these are relevant to the formulation or handling of a product. In practice, failure to recognize these structural causes may lead to faulty technique and wasted work.

§2. Elasticity

Elasticity is ordinarily thought of as an instantaneous response to an applied stress, and does not appear to be a time-dependent property to our general perception. It was shown, however, that even truly elastic deformation takes time (1§9). However, flow is about permanent deformation, and elasticity is by definition completely reversible. Nevertheless, there are two reasons why it must be included in any discussion of rheology. Firstly, all materials exhibit some elastic behaviour. Even undeniable liquids such as water have a compressibility, that is, a finite bulk elastic modulus (1§2.6) as demonstrated by hydrostatic testing. The conduction of sound (an elastic tension-compression wave) through any medium is a consequence of this. Thus, if we are to attempt a general description of deformation, elasticity is necessarily included. Secondly, it will be shown that the idea of instantaneous pure elasticity is but one extreme of the whole range of possible rheological behaviour, and that it is logically and mathematically part of the overall descriptive equation. We cannot ignore it in this context and it is, therefore, dealt with here.

As before, we draw attention to the fact that the term elasticity only implies a return to zero strain at zero stress, without regard to the linearity of the deformation (1§2), the distinction between Hookean and non-Hookean systems being essentially irrelevant and unnecessary (if we were to be excessively pedantic, no material is Hookean, see 10§3.6). However, to simplify the description we will now only refer to Hookean deformation as the ideal behaviour, although this must be understood to be without prejudice to the possibility of other elastic responses. Furthermore, we must make the same restriction as before (1§2), but now in respect of all processes, that the geometry is unchanged by the deformation – an impossibility except in particular special cases, but a basic working approximation.

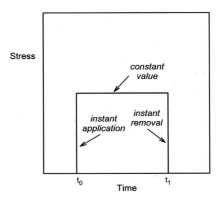

Fig. 2.1 A rectangular stress pulse.

Since now we are interested in behaviour with respect to time, we will plot responses to changes in applied stress and introduce the notion of the **stress pulse** (Fig. 2.1). That is, a constant stress is to be applied instantaneously to a body for a fixed period of time and then removed completely, again instantaneously. We may then plot the corresponding strain against time (Fig. 2.2). Ideal solids stressed in their elastic region, that is below the yield point if there is one (1§3), always return exactly to their original dimensions upon release of the stress. But, while the constant stress is being applied there is no change in the strain whatsoever. This is obviously implicit in the relevant equation (1§2.5) as time does not appear there as a variable.

The above result applies for any kind of stress, but because flow requires shear, in the present context it will be implicit that shear stress (τ), shear strain (γ), and shear modulus of elasticity (G) are meant exclusively. Unfortunately (for the novice), it is conventional to represent the ideal elastic body by a simple coiled spring loaded in tension for its mechanical analogue (*cf.* 1§5). It must be remembered that this is only a conventional model representation, and one which has been adopted simply because of the inconvenience of drawing a corresponding shear model.

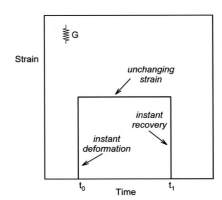

Fig. 2.2 The strain response of a Hookean body to a rectangular stress pulse. Inset: the mechanical analogue, a spiral spring.

We may then write the shear deformation equation for the Hookean body as follows:

$$\gamma(t) = \frac{\tau}{G} \qquad (2.1)$$

The important addition to this equation is the notation for time, t, at which the observation is made: read "$\gamma(t)$" as "shear strain at time t". This serves as a reminder that the stress which has been applied is explicitly a function of time. Nevertheless, time does not appear on the right hand side of the equation in this case. The units are as before: stress in pascals (Pa), as is the modulus of elasticity, while the shear strain is dimensionless.

§3. Newtonian Flow

It is basic to the idea of flow that there is shear. No relative movement of one part of a continuous unbroken body with respect to another adjacent part can occur without shear. All descriptions of flow thus depend on shear images. The shearing that occurs in fluids as they flow is perhaps best visualized as between two parallel plates (Fig. 3.1), extending the idea of Fig. 1§2.8. It is a requirement now that the material of the plates is wetted by the fluid so that there is a layer effectively stationary on the surface of each, the **boundary layer**. In the absence of an applied force, nothing happens. The system is static and at equilibrium.

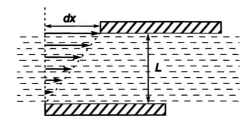

Fig. 2.3 The geometry of the displacement of a fluid between parallel plates. The upper plate is moving from left to right with respect to the lower plate. The size of the arrows indicate the relative velocity of the corresponding fluid layer with respect to the lower plate.

If now a force is applied to the upper plate, acting in its plane (and that of the page in Fig. 3.1), that plate is expected to move with respect to the lower plate (which we will hold stationary for reference). In so doing, the boundary layer to the top plate is carried along, and this in turn tends to cause the next layer to move in the same sense by a frictional effect, and so on from layer to layer to the opposite boundary layer and plate. Thus, the relative motion of one layer exerts a shear force on the next layer. This establishes the **shear gradient** across the thickness of the fluid. There is complete uniformity across that thickness in the force acting *between layers* and the relative velocity of one layer with respect to the next. This is the **principle of uniformity** again (1§2.1).

The friction between layers is generated by the **transfer of momentum**. In a fluid, all molecules have kinetic energy by virtue of their thermal velocities: they diffuse. These velocities have, of course, a wide range of values, but more importantly the directions of travel are random. Thus, in a non-flowing system, some molecules will move from one region to another and other molecules in the opposite direction, on average in equal numbers and on average with identical total momentum (just opposite sign). The net effect is no change. However, in a system undergoing shear, it is clear that one layer is moving faster in the shear direction than an adjacent layer (Fig. 3.2). All molecules in that faster layer thus have a little more momentum in that direction than those in the slower layer. Thus, if some molecules are exchanged between those layers, by diffusion, the net effect is to reduce the total momentum of the faster layer, and to increase it in the slower layer – in the direction of shear. The faster layer therefore suffers friction, slowing it down, whereas the slower layer tends to be speeded up. Ultimately, the momentum is transferred to the moving plates (in a system such as that of Fig. 3.1), and this therefore represents work done by the external system. The same principle applies for polymers, whose macromolecules clearly cannot be exchanged in this way. Instead, the effect is served by chain segments (loops of backbone) and side groups. In addition, momentum can be transferred simply through collisions.

The representation of a single force is, of course, a simplification. In fact, there exists across the fluid a turning moment because, by Newton's Third Law, there must be an equal and opposite force acting on the lower plate when this is stationary (Fig. 3.3). It is conventional to ignore that moment, but its existence should not be forgotten. Now, because any fluid flow, no matter what the geometry, is dependent on shear between adjacent layers, any general remarks applicable to the simple system of parallel plates can be extended to the more complicated systems of actual applications. In fact, all practical **rheometers**, devices for observing and measuring rheological behaviour, have more complicated geometry because of the practical difficulty of building a machine on the basis of Fig. 3.1. This does not detract from the usefulness of the model.

In all the discussion in this chapter, all accelerations are ignored. That is, we simplify the consideration by treating all components – and especially the fluid of interest – as massless and therefore inertialess. This includes gravity, and thus the self-weight of all components.

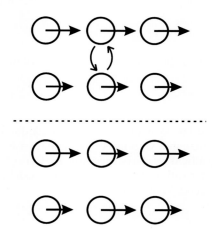

Fig. 3.1 Two layers suffering shear (top). The lengths of the vector arrows represent the momentum of the particles. Exchange of particles between layers (or momentum transfer by collision) results in the net momentum of the upper layer being reduced, that of the lower layer being increased (bottom).

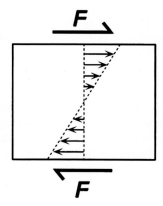

Fig. 3.2 Magnitude and direction of the shearing forces acting in the system of Fig. 3.1, viewed as a couple and referred to the centre line.

●**3.1   Velocity gradient**

Because the material between the plates is a fluid, it therefore responds continuously to the applied shear stress. Thus, in order to maintain that stress across it, and therefore maintain the shear gradient, the upper plate must be kept in motion. It might reasonably be expected that, ignoring accelerations, a constant rate of displacement of the upper plate would require the application of a constant force, when no other aspect of the system is changing. There is now an explicit involvement of time in describing behaviour, since the equation for the displacement in the x-direction of the upper plate in unit time is its velocity v (with respect to the lower plate):

$$v = \frac{x_t - x_0}{t} = \frac{dx}{dt} \qquad (3.1)$$

From the definition of shear strain (equation 1§2.20), for the fluid thickness L, the increment of shear strain developed for the increment dx in the displacement of the upper plate is given by:

$$d\gamma = \frac{dx}{L} \qquad (3.2)$$

But this has occurred in increment of time dt, hence the *rate* of change of shear strain is given by:

$$\frac{d\gamma}{dt} = \frac{dx}{L}\cdot\frac{1}{dt} = \frac{v}{L} = \dot{\gamma} \qquad (3.3)$$

The last symbol, $\dot{\gamma}$, (say "gamma dot") is a common shorthand[1] for this quantity (the dot means the first differential with respect to time, that is, the rate of change). This **shear strain rate** is also called **velocity gradient**, as can be seen from the form of v/L (see Fig. 3.1). Then, following the style of calculation of an elastic deformation (1§2), the shear stress can be related to the shear strain rate by a constant of proportionality, a sort of modulus of flow:

$$\tau = \eta\dot{\gamma} \qquad (3.4)$$

Thus, in our idealized system, for a constant applied shear stress, the rate of change of shear strain is constant. On removal of the stress the accumulated strain remains fixed at its last value; there is no recovery of strain because there is no elastic aspect to such a system. That is, there is no driving force to return the system to its original state – no stored energy. We might suspect this because all the work was being done against 'friction' between layers, and the energy put in was therefore dissipated, not stored. This was the basis of Joule's experiments on the mechanical equivalent of heat (*c*. 1847) in which water was stirred – stirring causes flow, and flow depends on shear. However, the temperature rise must change viscosity, because diffusion is an activated process, and care must be taken experimentally to control temperature. In some circumstances, the frictional heat of mixing will also affect reaction rates.

● **3.2   Viscosity**

The modulus η is called the **coefficient of viscosity** (or, just the **viscosity**[2]) of the fluid. It is defined as a constant of proportionality, just as E and G (1§2.1, 1§2.5) are constants of proportionality in Hookean solids:

$$\eta = \frac{\tau}{\dot{\gamma}} \qquad (3.5)$$

Equation 3.4 is **Newton's Law of Viscosity**,[3] and a fluid with this type of response (its behaviour) is described as **Newtonian**, or said to be showing **pure flow**. (The reciprocal of this quantity may be called the **fluidity**.) The shear deformation equation, that is, the amount of strain produced in a given time at a constant stress, for the Newtonian body is therefore:

$$\gamma(t) = \frac{\tau}{\eta}\cdot t \qquad (3.6)$$

Time is now explicitly present on the right hand side. This is therefore time-dependent behaviour. Stress is still in pascals, so the viscosity – the resistance to flow of a fluid – is expressed in pascal-seconds (Pa.s). The important aspect of the system now therefore is that the fluid suffers continuous deformation whilst a stress is applied (Fig. 3.4). Thus,

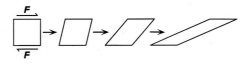

Fig. 3.3 A volume element of a fluid under shear undergoes continuous deformation.

---

[1]   The notation was introduced by Isaac Newton (1642 – 1727).

[2]   Also called the **dynamic viscosity**, to distinguish it from the **kinematic viscosity** = η/ρ (ρ = density) – which not a viscosity.

[3]   Sometimes called Newton's Law of Friction.

applying the stress pulse of Fig. 2.1 to a Newtonian fluid, the behaviour illustrated in Fig. 3.5 is observed: steadily increasing deformation whilst the stress is applied, no subsequent change when it is removed. The greater the viscosity (*i.e.* the value of η), the shallower the slope of the plot of strain against time.

The mechanical analogue for a Newtonian fluid body is the so-called "dashpot". This is represented as a cylindrical vessel containing a loosely fitting disc or piston immersed in a (Newtonian!) fluid such as an oil (Fig. 3.6). When a force is applied to the piston the fluid is forced through the narrow annular gap. Since the gap is small, a small portion of the circumference approximates the appearance of Fig. 3.1, where the upper plate is a portion of the piston and the lower plate the wall of the container. This can also be seen from the cross-sectional view of Fig. 3.7. It is this real mechanical equivalent that leads to the use of the model and the symbol. Such devices are in fact used in shock-absorbers for motor vehicles and energy-absorbing buffers of various kinds, "dampers". We might also note in passing that the flow in the gap may be described as **laminar**, that is, like the sliding deformation of a pack of cards. (It is also assumed that there is no turbulence outside the annular gap, which is in practice means a very low displacement rate.)

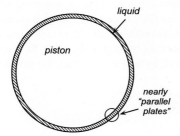

Fig. 3.6 The dash-pot model approximates a parallel plate system.

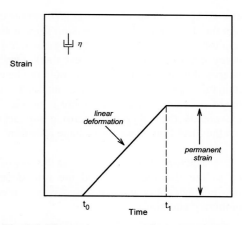

Fig. 3.4 The strain response of a Newtonian body under the stress pulse of Fig. 2.1. Inset: the mechanical analogue, the conventional drawing of an oil dashpot.

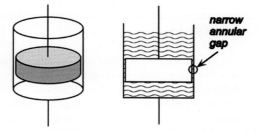

Fig. 3.5 The design of a dash-pot – the model for a Newtonian viscous body.

Fig. 3.7 A pack of playing cards when sheared shows "laminar flow".

§4. The Maxwell Body

Having developed the ideas of the two extreme forms of deformation, ideal elasticity and ideal flow, we may proceed to consider the intermediate conditions which are more realistic for most real materials. The easiest way to do this is by continuing the model-building approach of 1§5, using the two ideal mechanical analogues as elements, and examining the behaviour of these models.

The simplest combination of the spring and dashpot is one of each in series (Fig. 4.1). A material showing the corresponding kind of behaviour is known as a **Maxwell body**. When the instantaneous stress of the pulse of Fig. 2.1 is applied to this system, the Hookean element responds by extending proportionally to the stress, instantaneously. The deformation due to the viscosity of the dashpot then proceeds linearly with time (since the stress is constant). The removal of the stress at the end of the pulse results in the instantaneous recovery of the elastic strain but, of course, there is no recovery of the viscous strain. The body has therefore suffered a permanent deformation. We can see this directly from the dashpot model - there is no force acting to push the plunger back towards the starting position (there is no energy stored elastically to achieve this). Obviously, any material showing such behaviour could not be expected to retain its original dimensions under any load - even its own weight - for any time. Since there is no yield point (or, equivalently, the yield point is zero), it is in fact little better than the ideal fluid if we want shape retention.

The total strain at any time is the *sum* of the strains due to the two elements, and the time-dependent and time-independent components are easily written down by combining equations 2.1 and 3.6:

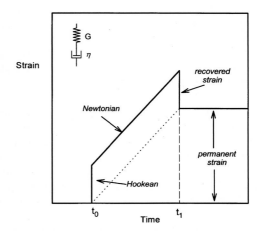

$$\gamma(t) = \frac{\tau}{G} + \frac{\tau}{\eta}.t$$

$$= \tau\left(\frac{1}{G} + \frac{t}{\eta}\right)$$

(4.1)

Fig. 4.1 The strain response of a Maxwell body to the stress pulse of Fig. 2.1. Inset: the mechanical analogue.

The two strains are therefore algebraically and physically independent of each other, and simply additive. There is no interaction.

On reflection, one can see that such a model must better represent the behaviour of real fluids such as water. We have already noted that water has an elastic modulus (as seen through the existence of its bulk modulus), as indeed must all real liquids. However, the value of that elastic modulus is so high that the deformation produced by the stresses necessary to make water flow is too small to be observed when flow is permitted. Water is therefore a good approximation to a Newtonian fluid because the elastic component in its Maxwell body representation is too stiff to cause any ordinarily noticeable effect. We start to see here how it is the relative magnitude of the two moduli, G and η, that influences the observed behaviour. If η is made large, the graph of Fig. 4.1 approaches that of Fig. 2.2. On the other hand, as G is made large, the graph looks more and more like Fig. 3.5.

§5. The Kelvin-Voigt Body

The alternative arrangement of the two basic elements, the spring and dashpot, is in parallel (Fig. 5.1). This represents the **Kelvin-Voigt** body. In this model the extension of the elastic component, the spring, is controlled by the flow in the dashpot. Plainly, no strain can occur instantaneously, since the dashpot requires the elapse of time to move (Fig. 3.5), which is easily seen to be the case even if the value of G were zero. Thus, at t_0 the spring is carrying no stress whatsoever. If it were, then it would have extended. There must also be an overall limit to the strain that could occur for a given applied stress if sufficient time were to be allowed. This upper limit will be given by the modulus of elasticity of the spring element exactly as in equation 2.1:

$$\gamma(t=\infty) = \frac{\tau}{G}$$

(5.1)

that is, the same result as would be obtained with η = 0. Hence, the stress on the spring alone is zero at t = 0 (since no extension has occurred), and must

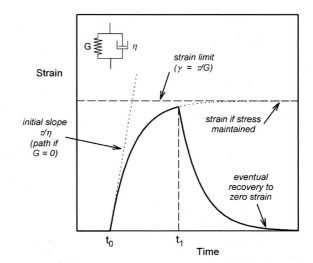

Fig. 5.1 The strain response of a Kelvin-Voigt body to the stress pulse of Fig. 2.1. Inset: the mechanical analogue.

be τ at $t = \infty$ (when it is fully extended). Conversely, the stress on the dashpot element is τ at $t = 0$ and zero at $t = \infty$. In other words, during the course of the overall fixed stress application, that stress is transferred (in the model) progressively from the dashpot to the spring. Now, the *rate* of strain depends only on the stress on the viscous element, τ_v (see equation 3.4), and this stress decreases according to how much strain *remains* to be accumulated as a proportion of the total strain possible:

$$\frac{d\gamma}{dt} = \tau_v(t) \propto 1 - \frac{\gamma(t)}{\gamma(t=\infty)} \tag{5.2}$$

(recall Newton's Law of Cooling: the rate of loss of temperature is proportional to the temperature excess, 11§2.7). The integration of that differential equation yields the following exponential equation as a result:

$$\gamma(t) = \frac{\tau}{G}\left(1 - e^{-t.G/\eta}\right) \tag{5.3}$$

It can be seen that when $t = \infty$ this reduces to equation 5.1 ($e^{-\infty} = 0$), and when $t = 0$

$$\frac{d\gamma}{dt} = \frac{\tau}{\eta} \tag{5.4}$$

which is one of the defining conditions. It should be apparent from equation 3.6 that this ought to be the case.

The quantity $t.G/\eta$, the exponent in equation 5.3, must be a dimensionless quantity (a pure number), and the ratio θ, defined by

$$\theta = \frac{\eta}{G} \tag{5.5}$$

can therefore be seen to have the dimension of time (Pa.s/Pa). It is usually known in the rheological field as the **retardation time** of the system, but it has exactly the same sense as the **relaxation time** of 1§9; we may use the terms interchangeably to describe the process of approaching equilibrium.

If then we make the substitution in equation 5.3 we get

$$\gamma(t) = \frac{\tau}{G}\left(1 - e^{-t/\theta}\right) \tag{5.6}$$

which shows this idea more clearly. In other contexts, such an exponential term will have a **time constant**, with units such as s^{-1}, and this would correspond to θ^{-1}, but the sense of the equation remains the same.

●5.1   Reversibility

The important point about the Kelvin-Voigt system is that the entire deformation is recoverable on removing the stress - it is an *elastic* process. The driving force for this **relaxation** comes from the stress stored elastically in the spring element - in an ideal system there is no dissipation of energy. It is therefore an *internal* stress, having no external manifestation except the changes in dimensions.

The rate of strain **recovery** is controlled by the viscous aspect again, the "dashpot", and the behaviour of the system can be seen to be entirely symmetrical. The same retardation time (or time constant) applies to this reverse process, and the graph for recovery has exactly the same shape (just inverted) as that for the original deformation. As time is required for the approach to the equilibrium position, whether on initial stressing or during recovery to zero strain, the dimensional changes are said to be **damped** or **retarded**. A material exhibiting such behaviour would have to be allowed a certain period to recover after exposing it to any stress if the original dimensions were important, and this will assume considerable significance in the context of impression materials and waxes in particular.

In a sense, Kelvin-Voigt deformation might be thought not to be relevant to flow as such, simply because it is an elastic, recoverable deformation. However, if we take the microscopic view that the movement of an entity from one location to another is what we are concerned about, that process is indeed flow. In any case, if it is only 'real' flow that we are concerned about, we still need to be able to identify other components of deformation, as observed in practice, in order to discount them.

In that light, we can see that the Kelvin-Voigt model must be a better representation of the behaviour of real solids in the sense discussed in 1§9. That is to say, if we attempt to measure the elastic modulus of a body by applying a stress, it requires a finite time for the deformation (which we need to measure to calculate the strain) to go to completion. For many solids, that time is so short – of the order of $1 \sim 100$ μs (Fig. 1§9.1) – that ordinarily we do not notice the effect. However, when we deal with polymers, where the controlling factor is the movement of chain segments, the relevant time-scale may be far longer (1 ~ 1000 s, or more). Thus, there is no real or clear distinction to be made between so-called static mechanical tests and the rheology of Kelvin-Voigt bodies; it is a matter of the relevant observational time-scale that determines our view of the processes involved. Thus, there can be no such thing as a perfectly Hookean body as in Fig. 2.2 because the response cannot be instantaneous, even though for many purposes it is entirely satisfactory to assume that it is a good working description of actual behaviour.

§6. Viscoelasticity

Many materials, no less in dentistry, show more complicated behaviour than those so far described. We may continue the model-building to consider a series combination of the Maxwell and the Kelvin-Voigt systems (Fig. 6.1). Here we have a combination of an instantaneous elastic deformation, a time-dependent reversible deformation, and irreversible flow, all superimposed. This class of behaviour is called **viscoelastic** (also known as the **Burgers** model), and the plot of strain against time for the application of the pulsed stress of Fig. 2.1 clearly shows all of these components.

The instantaneous elastic deformation is followed by a stage of decreasing rate of strain as the Kelvin-Voigt component is extended. The Newtonian or pure viscous deformation continues for as long as the stress is applied. On removal of this stress, the Hookean elastic strain is instantaneously recovered, followed by the negative exponential recovery of the Kelvin-Voigt strain. Of course, the Newtonian deformation cannot be recovered, and this is the cause of the permanent strain.

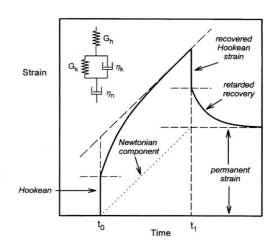

Fig. 6.1 The strain response of a viscoelastic body to the stress pulse of Fig. 2.1. Inset: the mechanical analogue.

Again, the total strain at any time after the application of the stress can be calculated as simply the sum of the strains from the various components. That is, the system remains purely additive in terms of its individual components. We may write this out algebraically:

$$\gamma(t) = \frac{\tau}{G_h} + \frac{\tau}{G_k}\left(1 - e^{-t/\theta}\right) + \frac{\tau}{\eta_n}.t$$
$$= \tau\left[\frac{1}{G_h} + \frac{1}{G_k}\left(1 - e^{-t/\theta}\right) + \frac{t}{\eta_n}\right] \tag{6.1}$$

where now it is necessary to distinguish between the shear moduli of the Hookean and Kelvin-Voigt components by appropriate subscripts, since they clearly can take different values. Also, θ here is given by $\theta = \eta_k/G_k$, following equation 5.5, for the Kelvin-Voigt component. Notice, too, that the order of the elements in the model is of no concern, as the sum of equation 6.1 is unaffected by the order of the terms – in other words, both may **commute**.

If we continue the argument at the end of each of the previous two sections, that is, that the Maxwell body is a better approximation to the behaviour of real liquids and the Kelvin-Voigt body the same in respect of real solids, we can see that the viscoelastic body is a generalization that can accommodate both extremes. The comments about the relative magnitudes of the moduli and the time-scale of observation still apply, as these determine which aspect of behaviour is dominant.

●6.1   Impression materials

This kind of mixed behaviour is shown in a very obvious fashion by all set polymeric dental impression materials, albeit to various degrees. Thus, the pure (Hookean) elasticity is a basic requirement if the mould is to approximate the dimensions of the original faithfully; the fact that they are polymers means that the retarded elasticity is noticeable; and various reactions ensure that creep or viscous flow causes enough permanent deformation to reduce the accuracy enough to be important. It can be inferred from this that, if applied stresses are of short duration, the purely viscous deformation will be small. But, equally, some time is required for the full recovery of the reversible strain and the attainment of the maximum accuracy after removal. Hence the recommendation must be for the removal of flexible impressions from the mouth in a single 'snap' action (to minimize irrecoverable flow) and for a small delay before pouring the model (to allow for recovery of the retarded deformation – the delay varies between types of material).

It should therefore be apparent that the most important problematic aspect of an impression material is the non-recoverable, viscous deformation, as it is this that ultimately limits the dimensional accuracy. Put another way, the design intention of an ideal rubbery impression material is that when set
•   the elastic modulus is low,
•   the retardation time is zero, and
•   the viscosity is infinite.
What we are prepared to work with, or are obliged to work around, by the fundamental limitations of the chemistry and physics of the chosen materials themselves, is quite another matter (*cf.* 1§15).

●6.2 Compliance

An alternative view sometimes taken in describing the deformation properties of a body is to consider not the elastic moduli, which measure stiffness in terms of the stress required for unit strain, but their inverse, the strain observed for unit stress. This kind of measure is referred to as a **compliance** (J) of the material. Thus, in the present context, the (shear) **elastic compliance**, J_0, is given by

$$J_0 = G_h^{-1} \tag{6.2}$$

(*cf.* flexibility, equation 1§2.4c). The (maximum) **retarded elastic compliance**, J_r, is therefore

$$J_r = G_k^{-1} \tag{6.3}$$

The **creep compliance**, J_c, is similar except that it refers to unit time as well as unit stress:

$$J_c = \eta^{-1} \tag{6.4}$$

As defined here, all three are scale-free (and time-independent) material properties.

Unfortunately, the usage of these terms in some sources is a little inconsistent, sometimes referring to the total strain observed, or even just the total deformation, regardless of stress or time. Thus, as shown in the example of Fig. 6.2, what is labelled "creep compliance" is a test result that is a direct function of the time taken. Care is thus necessary in reading where these terms are employed to ensure that the definition being used is understood. However, it is usually possible to estimate the magnitude of each compliance term graphically (Fig. 6.2). Since the creep compliance component of the behaviour gives a straight line plot against time, the value of the associated coefficient of viscosity can be determined so long as sufficient time is allowed for the retarded compliance to be fully expressed (but, again, on the condition that the geometry is not appreciably changed in the process), when scaling by the time taken is easy.

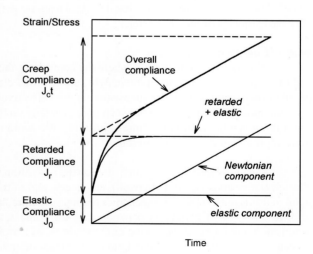

Fig. 6.2 The components of the deformation of a viscoelastic body can be expressed as compliances. Notice that here "creep compliance" is not a material property but depends on the duration of the test, *i.e.* the total observed.

Equation 6.1 can therefore be written in a simpler form, expressing the total deformation as the sum of the deformations due to the individual compliances, again stressing the simple additivity of those components:

$$\gamma(t) = \tau [J_0 + J_r(1 - e^{-t/\theta}) + J_c t] \tag{6.5}$$

•6.3   Complex materials

The above analysis is valid for homogeneous, single phase materials in which only one retarded stress relaxation process can occur. However, from Fig. 1§9.1 and the description of polymers in Chap. 3, this is plainly not going to be appropriate to many common materials: there may be several relaxation processes possible. If the structure is a gel, or a composite material such as an emulsion of one type of material in another, or even a mixture of several different materials, each will make its own contribution to the overall behaviour. Notice that any combination of Hookean elements only, in series, parallel or any more complicated arrangement, will behave as if it were one, single equivalent spring element; the individual elastic compliance effects cannot be resolvable. Similarly, the combined effect of a number of Newtonian compliances would be that of an equivalent single element. However, the retarded elastic compliance attributed to each structural aspect will be identified with a separate Kelvin-Voigt component in the model, each having a distinct retardation time. The model of such a system has the appearance shown in Fig. 6.3, in this case

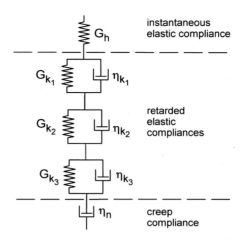

Fig. 6.3   The mechanical analogue of a multi-component viscoelastic material.

showing only three retarded compliances; there may be more. Thus, the equation describing this behaviour has a summation of the effects of the retarded compliances in the place of the single term of equation 6.1:

$$\gamma(t) = \tau\left[\frac{1}{G_h} + \sum_{i=1}^{n}\frac{1}{G_{k_i}}\left(1 - e^{-t/\theta_i}\right) + \frac{t}{\eta_n}\right] \qquad (6.6)$$

It is important to notice that since the summation of the retarded compliance effects is not algebraically equivalent to a single term of that kind (since a sum of exponentials is not itself an exponential[4]), the observed plots of strain against time may be complicated curves. Sometimes one can clearly see a rather rapid recovery of a large portion of the strain (but obviously not instantaneous), followed by a slower stage. In practical terms, this kind of behaviour also means that if several tests are performed, the curves themselves cannot be averaged, only the estimates of the parameters.

•6.4   Compliance spectrum

With polymeric materials, where the controlling factors are chain segment length (which is a variable) and the activation energy for a movement, the retarded elastic compliance will have a spread or continuous distribution of values; n in the summation is then effectively very large. In fact, taken to a logical limit, this equation indicates a continuum between pure elasticity on the one hand, when there is no retardation, and pure viscosity, when the retardation time is infinite. We could therefore speak of the **spectrum** of the retardation times represented in a given material.

The deviations from ideality, such as the relaxation time of an elastic solid (1§9; Kelvin-Voigt) and the elasticity of viscous liquids (Maxwell), are clearly part of this general model of the rheology of materials. The point of this is not to make the mechanics of materials intractably complicated, but to allow for the manifestation of different kinds of behaviour depending on the context, and primarily that of time-scale. One chooses the description most appropriate to the circumstances, as simple as need be, but without losing sight of the other aspects. Thus, ordinarily, one can ignore the relaxation time and creep of ceramics (at ordinary temperatures), just as one can usually ignore the elasticity of water.

§7. Classification of Flow Behaviour

So far we have considered materials being tested with a fixed stress in order to investigate their strain behaviour as a function of time. The strain rate of a viscous body is determined by the definition of the coefficient of viscosity, hence the velocity gradient is fixed. It is found, though, that the measured value of the

---

[4]   This means, for example, that experimental data curves of this kind cannot be meaningfully averaged for a 'mean' curve.

viscosity of many materials varies as a function of the velocity gradient, *i.e.* the shear strain rate. In other words, the observed viscosity is dependent on how fast the material is deformed. Several types of such behaviour can be identified (Fig. 7.1).

●7.1   Newtonian flow

The reference behaviour is **Newtonian**, that is for an ideal liquid. The plot of shear stress against shear strain rate is a straight line, *i.e.* one of constant slope (Fig. 7.2). In other words, the viscosity is constant at all shear rates (equation 3.4). Typical such materials of low viscosity are water, mineral oils, and other solvents which have small molecules; they are homogeneous. Their flow behaviour is therefore well-described by the one coefficient.

●7.2   Viscous plastic flow

Even so, it is perhaps unreasonable to expect such behaviour to persist at extreme values of shear strain rate. In the sense that elastic solids may show a breakdown to plastic flow at high enough strain (1§3), or simply fracture, so viscous fluids may show failure, a total breakdown, at some high enough strain rate. Even liquids must have a tensile strength (although it must be measured hydrostatically). This is called **viscous plastic** behaviour and can be modelled by adding a static friction element (1§5) to the dashpot in the model (Fig. 7.3). This friction element is imagined to slip at some critical stress which corresponds to the **shear strength** of the fluid. No examples in dentistry of this kind of breakdown are known for certain, shear rates being, in general, rather modest. The only possibility that springs to mind is in the context of ultrasonic scaling instruments, the oscillation frequencies of which are perhaps high enough to enter this region.

●7.3   Plastic flow

Previously, the stress-strain plastic behaviour of solids was modelled using just a friction element (Fig. 1§5.1b). There, no notice was taken of time; the flow behaviour itself was not considered. We can now improve the model of ideal **plastic** flow by adding a Newtonian dashpot behind the friction element (Fig. 7.4): for flow to occur the stress must exceed a certain value, the **yield point**, but after that the slope of the plot of stress against strain rate is constant. This type of system is known as a **Bingham body**. We can illustrate this by using equation 3.6, with τ replaced by the 'excess' shear stress, $(\tau - \tau_{yield})$, and only a positive value is allowed. The definition of the effective coefficient of viscosity must then also be in terms of that excess shear stress:

$$\eta = \frac{\tau - \tau_{yield}}{\dot{\gamma}} \qquad (7.1)$$

Real materials, however, generally tend to show a falling slope with increasing shear rate (Fig. 7.5) and may

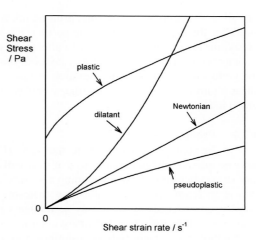

Fig. 7.1 Some of the principal types of flow behaviour. (It is the *shapes* of the curves that are important, not the relative scales of slopes, *etc.*)

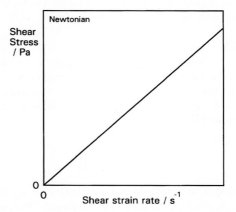

Fig. 7.2 The slope of the plot of shear stress against shear strain rate, the flow curve, for a Newtonian body is constant.

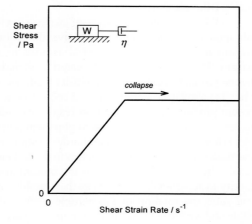

Fig. 7.3 The flow curve for an ideal viscous plastic body. Inset: the mechanical analogue.

be called **viscoplastic**. Chocolate, printing ink and some similarly thick pastes show this type of shear-thinning behaviour (see §7.9). If we were to attempt to describe such a non-ideal material it is clearly necessary to specify the yield stress as well as the variation of viscosity with shear strain rate, a job that is perhaps best done by the graph itself.

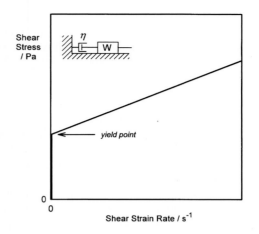

Fig. 7.4 The flow curve of an ideal plastic material. Inset: the mechanical analogue.

●**7.4   Pseudoplastic flow**

In contrast, **pseudoplastic** materials (Fig. 7.6) have no measurable yield point (so that flow will occur at any stress) but they show a declining viscosity as shear rate is increased (thus distinguishing themselves from Newtonian materials by what may be termed a negative deviation). Into this class fall many common materials such as molten polymers, polymer solutions, adhesives and suspensions. In fact, waxes are in this category, as they show a marked stress-dependent viscosity – but no yield point (16§3.1). Both plastic and pseudoplastic behaviour influence the ease with which fluid surfaces may be levelled (that is, for example, the brush marks in paint tending to disappear) and the spreading of these materials over other surfaces, as in painting itself.

●**7.5   Tangent viscosity**

We mentioned non-Hookean behaviour in 1§13 and the need to define the 'local' modulus of elasticity. Evidently, a similar approach is necessary here for non-Newtonian flow: the local or 'tangent' viscosity is defined by

$$\eta_t = \frac{d\tau}{d\dot\gamma} \qquad (7.2)$$

(this can be compared with equation 3.5); in words, this is the slope of the plot at any given point. Thus, we are able to discuss variation in observed viscosity with shear rate or strain rate. This allows us to make sense of plots such as that of Fig. 7.4 or 7.5, where an average or secant viscosity (*cf.* Fig. 1§13.1) would not be very meaningful.

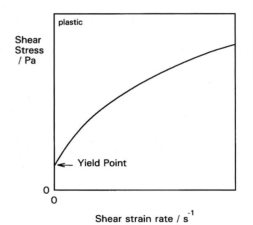

Fig. 7.5 The flow curve of a typical real plastic material, one showing viscoplasticity; the viscosity falls with increasing shear strain rate.

●**7.6   Dilatant flow**

When the material shows an increase in viscosity with increasing strain rate, a positive deviation from Newtonian behaviour, the material is said to exhibit **dilatant** flow (Fig. 7.7).[1] Such materials include some silicone products, pastes with high filler loadings – below or not far above the gauging liquid requirement (as for gypsum products, 2§2, filled resin restorative materials, and ceramic slurries such as dental porcelain powders in water) (but see §7.7). It becomes harder to stir or extrude these materials the faster one attempts to do it. This arises because the suspending medium is not present in enough quantity to separate the particles as they tumble. At high shear rates the rotation of the particles causes them to collide and wedge against each other, or they may already be in contact. If they are to pass each other they must move apart, effectively causing an increase in the overall volume and increasing friction. This jamming cannot be

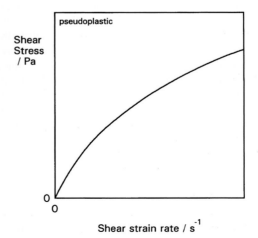

Fig. 7.6 The flow curve of a typical pseudoplastic material. There is no yield point.

avoided, given the small amount of liquid present.

Here, and in 2§2, dilatancy is described in terms of the rotation of irregular particles. This is indeed the common reason for the effect, but it is not essential to have non-spherical particles for it to occur. Even spherical particles, which tend to sit in the recesses of the layer below, need to be lifted up and over neighbours for flow to occur. Taken over many sites, this effect leads to a measurable increase in volume under stirring and therefore dilatant behaviour. Having a mixture of sizes of particle cannot change this. The effect is clearly seen in denture base acrylic, which consists of a spherical particle powder mixed with the monomer liquid (Fig. 5§2.2).

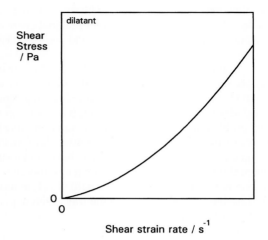

Fig. 7.7 The flow curve of a typical dilatant material; the viscosity rises steadily with increase in shear strain rate.

The key issue here is that the proportion of solid is sufficiently high that particle-particle interactions occur. If the slurry is sufficiently dilute there will be no dilatancy. The boundary between these two conditions occurs at the **random loose-packed limit**, which for uniform spherical particles has the value $\varphi_f \approx 0.56$, that is, the minimum particle concentration which can allow a 'rigid', just stable, three-dimensional, continuous network path of contact throughout the assemblage.[2] Of course, that boundary value will depend on particle shape and size distribution, but this is a **critical point** in all such systems, a point of discontinuity in behaviour.[3] If the fluid matrix phase has no yield point, then neither will the mixture below that value, dilatancy must then occur above it. But now it is clear that there is a structure associated with the particle network, and this must be broken for flow to occur. Such systems then tend to have a yield point, and the material must be plastic even if the matrix fluid is not.

●7.7 Plastic dilatant flow

Thus, slurries of some materials such as porcelain, gypsum products and investment powders, as well as freshly mixed cements and amalgam, also very obviously tend to show a yield point – a minimum shear stress which must be applied before flow will occur. We may better term this **plastic dilatant** behaviour (Fig. 7.8). This is evident from what is termed the **slumping** behaviour, or, rather, the lack of it. We can deduce this from some common observations: the fact that sand castles do stand without collapsing into a puddle of wet sand; the fact that porcelain powder can be built up into the form of the restoration without collapse; the fact that investment and plaster mixes stay on the spatula without running away. The self-weight of the mixture under local gravity represents a stress in the material, and this depends on the density and height of the heap. If this stress exceeds the yield point, then the mass **slumps** until this criterion is not quite met at the bottom of the heap (where the stress is maximum). Notice that this is the condition after slumping flow has actually stopped, so it is not just a case of a high viscosity slowing down the flow. 'Plastic dilatant' is therefore a better description of behaviour than merely dilatant when there is any evidence at all of a yield point.

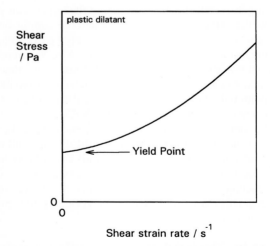

Fig. 7.8 The flow curve of typical plastic dilatant material. A distinct yield point is present.

●7.8 Slump test

Some dental materials showing this behaviour have been tested in the following manner: a cylindrical mould like a casting ring, open at top and bottom and resting on a slab, is filled with the slurry, which is levelled, then the cylinder withdrawn (vertically) (Fig. 7.9). The diameter of the mass when the movement is complete is called the 'slump diameter'. The stress on the bottom surface due to the self-weight of the material decreases to the point of equilibrium with the yield point in shear.

It is not certain quite what the significance of this is, except as a crude measure of the yield point of the mixture. However, the complicated and varying geometry precludes detailed analysis. If slump diameter presents any problem in practice for powder-liquid mixtures, it is easily overcome by vibration. Of course, plastic materials in general, not just dilatant ones, may have circumstances when slumping behaviour is apparent or important: some impression material products may deliberately exploit this to limit the loss of paste from the tray when upper arch impressions are attempted.

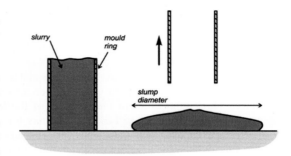

Fig. 7.9 The 'slump' test: the mould ring is filled, lifted away, and the final diameter of the puddle measured.

●7.9   Shear thinning

Both plastic and pseudoplastic materials get easier to stir as the stirring rate is increased: this is the extremely important property of **shear thinning**, and in many such materials it is commonly due to continued structural change. That is to say that the shearing process alters the interactions of molecules, by changing their alignment, orientation, folding and coiling, or even their bonding, in such a way that their mutual interference is reduced. Obviously, the reverse process occurs fast enough that the effect is apparently microscopically reversible, on a timescale short enough to be ordinarily unnoticeable (*cf.* 1§9). Such a process depends on the activation energy for the structural change being supplied by the work of shearing.

A similar appearance may be given by dilatant materials under certain conditions, and the term can be applied, but a careful distinction must be made. For example, a sand pile will slump further on being vibrated, and the pouring of concrete or dental gypsum products and investment slurries is often accompanied by vibration to enable flow. This vibration temporarily and locally increases the shear strain rate. This is enough to reduce the apparent viscosity (or for the stress locally to exceed the yield point) and allow flow to occur. Even knocking the side of the container is often enough to cause a noticeable effect. However, it should be noticed that the vibration does not itself cause bulk flow – that comes from the overall applied shear stress. What it does is provide the activation energy to overcome the local friction of particle on particle, allowing them to jump past each other. Even slight relative movements of particles otherwise jammed against each other, or short duration larger overall shear forces, permit the larger scale movements to occur. A macroscopic parallel can be found in the effect of vibration on frictional sliding between large bodies (1§5.5), and in the liquefying of certain soils under the shaking of earthquakes. What is important is that the amount of liquid present is just below that of the random loose-packed limit: too much liquid and the yield point will vanish, too little liquid and useful flow will not occur even under vibration. This threshold behaviour can be compared with the mixing requirements (2§2.4), noting that such vibration can mean that the dilatancy allowance (and thus the overall mixing ratio) could be reduced – except that normal mixing by stirring (whether by hand or machine) requires the full allowance if bubbles are to be avoided. Dual-centrifugation mixers appear to permit that mixing-ratio reduction.

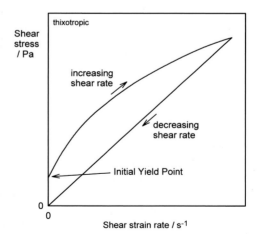

Fig. 7.10 The flow curve for a typical thixotropic material. The characteristic features are the yield point, decreasing viscosity with increase in shear strain rate, and the failure to return along the same path when the shear strain rate is then decreased. (Constant temperature is assumed; heating effects due to stirring are not implicated.)

●7.10   Thixotropic flow

There is one class of behaviour which is widely and frequently misunderstood: **thixotropy** (Fig. 7.10). As with plastic materials there is a very definite yield point, below which there is no flow. Above the yield point, again as with plastic materials, increasing shear rate decreases the viscosity, but this is due now to a physical breakdown of some aspect of the structure of the material.

For example, some mineral powders interact strongly with the liquid vehicle and with each other, effectively to form a macroscopic three-dimensional network structure (Fig. 7.11). Such a structure, which may be due to oriented particles, coiled fibres, emulsions or aggregated fillers, may be reformed on removal of the shearing forces. This structure is more or less fragile, and its state of assembly controls the flow of the material. Also, some substances that form gels, that is, a molecular network held together by hydrogen bonds for example, can be broken down by stirring. When the material is allowed to rest the gel network structure can reform, causing the viscosity to rise again. Notice that reforming the structure takes time because it requires diffusion at the molecular level, or the migration of particles under small forces, leading to an ordinary thermodynamic energy state minimization. Such reversible structure alteration is the key to true thixotropic behaviour.

A crucial distinction for thixotropic behaviour as opposed to simply plastic flow (Figs 7.4, 7.5) is the fact that microscopic reversibility of the stress-strain rate plot does not apply. Thus, in Fig. 7.10, on steadily increasing the shear rate the (tangent) viscosity falls steadily, but on reversing the direction of change of shear strain rate the original path is not retraced. This may be so extreme that no yield point is detectable immediately afterwards. Were this kind of test cycle to be applied to a plastic, pseudoplastic, dilatant or even plastic dilatant material then exact retracing of the path of the graph would be expected – these are microscopically reversible processes because no time-dependent change of structure is involved (Fig. 7.12), at least, not on a timescale that we can perceive. We may therefore describe thixotropy as shear-thinning with **hysteresis**.

●7.11   Time-scale

We must, of course, consider the time-scale on which the structure breaks down and reforms. In water, which has structure due to hydrogen bonding, this occurs extremely rapidly and no detectable change in structure follows stirring, however fast, and it cannot be considered as thixotropic. The speed of reformation is due to the fact that no diffusion is necessary to obtain the position necessary for the bond to form – the nearest neighbour is enough. In a gel, however, where the network-forming molecules are very large and polymeric, the diffusion of chain segments is the limiting factor. This is much slower than for a water molecule, but as hydrogen bond cross-links form, the diffusion rate falls quickly. For structures involving large entities such as oriented solid particles, the rate of reformation could be even slower.

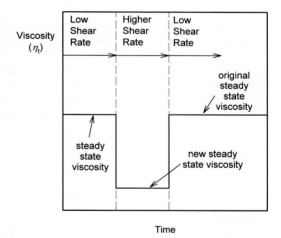

Fig. 7.11 Example of a material conferring thixotropy in an aqueous suspension, fumed silica or 'aerosil'. Hydrogen bonding between particles cause the reversible formation of a fragile network.

Essentially, what is being established in such systems is an equilibrium between two processes:

$$\text{Structured} \underset{k_2}{\overset{k_1}{\rightleftharpoons}} \text{Unstructured} \qquad\qquad (7.3)$$

Rate constant k_1 depends on the shear strain rate, k_2 depends on diffusion and the chemistry of the structure-forming reaction. The equilibrium thus depends on the ratio of the two: k_1/k_2. The faster the stirring, the further to the right that the equilibrium will shift. It is emphasised that complete reversibility is implied. There must be no history effect, *i.e.,* properties and changes of properties that depend on the previous sequence, beyond that of the equilibration timescale itself.

However, equilibrium is not established instantaneously, even this takes time. Thus, another view of thixotropic behaviour to help understand this aspect is to consider the viscosity as a function of time at constant shear rate. At a high shear rate (Fig. 7.13) an initial very high viscosity reduces steadily to some constant value. If the viscosity is now followed when the shear rate is changed to a lower value (Fig. 7.14), it will be seen to increase steadily again to a new constant value. It is this time-dependency of the apparent viscosity that characterizes thixotropic materials.

Such a process of change in the viscosity is not just a one-step event: successive changes in shear strain rate produce successive commensurate changes in viscosity. Thus, a sequence of increasing shear strain rates will generate a sequence of progressively lower viscosities (Fig. 7.15). Reversing the sequence allows the viscosity to increase in similar fashion (Fig. 7.16).

This kind of behaviour is typical of certain foodstuffs which are based on starch or gum thickeners; tomato ketchup is the most commonly cited example. After being allowed to stand for some time, very little or no flow occurs on inverting the bottle. However, after vigorous shaking (when the yield point will have been exceeded and structural breakdown has occurred), the material will flow very readily.

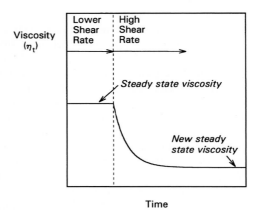

Fig. 7.13 Decrease in viscosity with time of shearing due to breakdown of structure in a thixotropic material.

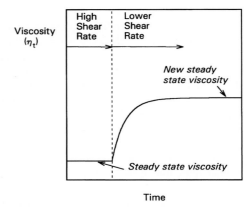

Fig. 7.14 Recovery of viscosity with time due to rebuilding of structure in a thixotropic material at low shear rate.

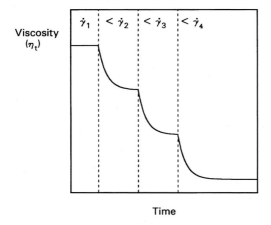

Fig. 7.15 Successive reduction of the viscosity of a thixotropic material in a sequence of increasing strain rates.

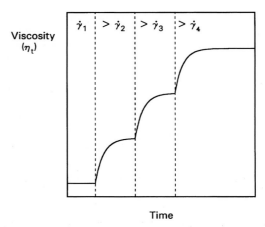

Fig. 7.16 Successive increase of the viscosity of a thixotropic material in a sequence of decreasing strain rates.

It can now be seen that the curve shown in Fig. 7.10 is a simplification. Obviously, the amount of structural breakdown that occurs depends on the total shearing that has occurred in a given time. Fig. 7.10 represents the output of a 'scan' from zero to some value of shear strain rate, $\dot{\gamma}$. If the scan were very slow, equilibration would be more or less continuously achieved, as suggested by Fig. 7.15. Thus the faster the scan, the less time for breakdown, and the curve would tend to be elevated and flatter. Similarly, the return curve would change. For extremely slow returns, the structure would reform, and a yield point would be apparent. The faster the return, the more it would represent the behaviour of a system whose structure was more broken down than that value of $\dot{\gamma}$ would appear to represent. Such a graph as Fig. 7.10 is therefore dependent on the scan rate, $\ddot{\gamma}$, the rate of change of shear strain rate, with respect to the relaxation time of each of the processes occurring internally. Great complexity is therefore possible. It is enough to summarize the behaviour as being dependant on the *recent* shear history as, of course, no irreversible changes are occurring in an ideal thixotropic material.

> There is another class of rheological behaviour which might be encountered: **rheopexy**. Although no examples are currently known in dentistry it is possible that they might occur because of the increasing complexity of dental product formulation. **Rheopectic** behaviour is essentially the inverse of thixotropic behaviour: the viscosity increases with time of shear, and with magnitude of shear strain rate. It is due to the *development* of structure on shearing, which breaks down relatively slowly by some diffusive process when the shearing is reduced or stopped. The behaviour is illustrated by interchanging the high and low shear rate labels in Figs 7.13 and 7.14, and by exchanging the "<" and ">" symbols in Figs 7.15 and 7.16. In addition, the circuit direction of Fig. 7.10 is reversed.

These effects should be distinguished from those of temperature (*e.g.* 3§3.3). Obviously, at high shearing rates appreciable heating could occur, and this would add a complication to the behaviour. However, in practice, assuming no chemical reactions, such effects should also be completely reversible.

●7.12   Dental relevance of thixotropy

A number of materials have been advertised in dentistry as being thixotropic when in fact they merely exhibit plastic or pseudoplastic behaviour. Such simple shear thinning has no relation to the structural changes which are the basis of thixotropy. For example, a topical fluoride "gel" may (incorrectly) be described as thixotropic (10§7.3). The job that is to be done is that of keeping the material in the application tray when this is inverted for placing it on the lower arch. Nevertheless, the material must flow well enough to coat all tooth surfaces. Such products are usually supplied in a soft squeeze bottle, and are to be dispensed by squirting through a narrow nozzle. Now, if the material were really thixotropic, the structural breakdown that would ensue from the very high shear rates to which it will have been subjected while being squirted out would ensure that the viscosity was then low, and that there would be a negligible yield point. The material would flow out of the tray under its own weight, away from the teeth, assuming of course that it could actually be got there in an inverted tray in the first place.

Likewise, a thixotropic impression material would be of little benefit. The mixing process would break down the structure (especially with a Kenics-type mixer [15§7.2]), so that it becomes a runny, Newtonian fluid with consequential handling problems. Similarly, a filled resin or adhesive described as "anti-slumping" or having "shape memory" is clearly not thixotropic but rather would have both a clear yield point and appreciable elasticity, neither behaviour being of particular benefit in use, but both of which would be lost on manipulation. There is a contradiction between what is desired and what is claimed.

Thixotropy as such is therefore of no value to dentistry, whether for fluoride preparations, bleaching agents, impression materials, filled resin restorative materials – or anything else, whether or not it is actually achieved in any product. The structural breakdown would defeat the object, unless that structure reformed rapidly. It may be concluded that what is supposedly needed, and probably what is being mislabelled as thixotropy, is simple plastic flow (§7.3). Such a material would flow under sufficient stress, but ideally not under its own weight: it should not slump. It would therefore stay where it was put, and not (for example) run out of an inverted tray. One can suppose that the word 'plastic' is, firstly, not understood and, secondly, not considered exotic or esoteric enough for a product to be thought 'advanced' and therefore a successful sale. Such abuses devalue the scientific terminology, demean attempts at being rigorous, and insult the practitioner. Even so, it is clear that shear-thinning could be an advantage in aiding the flow of such materials when placed under stress (Fig. 7.5). Taking this to the limit, a near zero-viscosity Bingham body (zero slope in Fig. 7.4) might be considered ideal, a kind of instantaneous pressure melting – like ice under a skate blade (*cf.* Fig. 1§5.1 d; see Fig. 8§2.1).

We may take this a step further. A yield point requires a minimum stress for flow to occur. However, the locations where the material is most needed is commonly where the driving force is least, *e.g.*, proximally (for treatment), line angles and corners of prepared teeth (for impressions). Wetting (10§1) cannot (in general) provide enough line stress to overcome the yield point, so spontaneous spreading is not going to happen. We therefore have another compromise, trading off the handling convenience of plasticity against the complete coverage of the target surfaces. Plastic materials, therefore, are not optimal in such circumstances. Pre-injection may help to overcome such difficulties as well as displacing air and possibly tissue fluids, if the flow path is appropriate.

Curiously, at least some toothpaste is said to be thixotropic.[4] While the general consequences of this in practice are unclear, it is implied to be an undesirable characteristic in that the material extruded from the tube should stay well-shaped on the brush rather than sinking in,[5] presumably purely from the effect of its appearance on the user's perception rather than its function as a dentifrice. It would arise from the tendency of certain gums and clays in such formulations to form structure through gelation (*cf.* 7§8) and hydrogen bonding (Fig. 7.11).

Oddly, there is some evidence that waxes may show some slight thixotropic behaviour (16§3.1).

●7.13   **Other models**
Although a variety of simple rheological models have been used in the above account, the list is not exhaustive. Depending on the nature of the components, additional types can be identified, for example **plastic-viscoelastic**, in which a collapse occurs at some high-enough shear strain rate, but where a retarded compliance was also involved. However, it should not be thought that it is necessarily possible to assign a given material to such an exact, perfect classification, or indeed to use a simple analogue representation; elaborate schemes are possible, but mechanical ingenuity does not imply clarity. Each aspect described here is idealized for the sake of simplification, and no more should be read into these examples than the elementary concepts they embody.

●7.14   **Phenomenology**
It can be seen from the above descriptions of the various kinds of rheological behaviour that the treatment is purely **phenomenological**. That is to say, a macroscopic overview is taken, as in a purely observational experiment. The important issue is the ability to describe, essentially graphically, the way in which various types of fluid behave when sheared. The equations are descriptive of the shapes of the plots, and not at all descriptive of the actual molecular processes which cause that behaviour. In effect, we treat the bodies as structureless continua. That is, neither the actual physics of momentum transfer, nor the chemical kinetics of molecular conformations and associations, nor the mechanics of bond forces and strengths are addressed in any way (although some suggestions of mechanism have been made). Whether or not it is possible to explain any of this behaviour in principle or in practice is, however, irrelevant here. What we are concerned about is the way in which these behaviours affect dental processes of one kind or another. The ability to control those processes depends only on an understanding of their nature at that phenomenological level. An example of just this kind is given in the next section.

§8. The Compression Set Test

The dimensional accuracy of a flexible impression material is of primary importance to its success as a mould: ideally, it should be perfectly elastic. Unfortunately, for a variety of reasons, perfection in this sense is unattainable, there being varying degrees of Newtonian-like flow and retarded elasticity. But a material may be *sufficiently* good for it to be serviceable in one context or another, despite the existence of the imperfections. The study of the rheology of impression materials has usually been not through the examination of flow curves, but through the use of the **compression set test**.

●8.1   **Test procedure**
In this test, a cylindrical specimen (Fig. 8.1) is

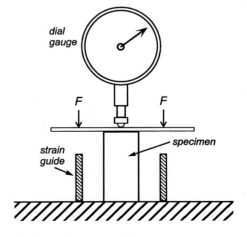

Fig. 8.1 The apparatus for a compression set test.

subjected to a strain pulse (Fig. 8.2). No regard is given to the stress that is required to achieve this strain (it is not measured), but the strain is held at the predetermined value, C_i, for a short time, then released. What happens next depends on the material. For an ideal elastic material there will, of course, be an immediate return to zero strain, as in Fig. 2.2. However, for typical real impression materials the elastic recovery is only partial and followed by a retarded recovery of only part of the total strain (dotted line). The recovery of the specimen is monitored, and the acceptability of the material is gauged by the accuracy of its return to its original dimension. The discrepancy between the original length, L_0, and the final length (at some predetermined time), L_t, is called the compression set, C_t:

$$C_t = \frac{L_0 - L_t}{L_0} \qquad (8.1)$$

Its definition therefore is simply that of a strain observed under specified conditions.

The response of the material when the stress maintaining the predetermined strain is removed depends on the flow properties of the material, and therefore on the *duration* for which that strain was held. The exception, of course, would be the ideal elastic material, whose behaviour is represented entirely in Figs 2.1 and 2.2.

●8.2 Newtonian

No Newtonian fluid material (*i.e.* assuming an infinite elastic modulus) could be tested in this way, even if such a specimen could be made, as an infinite stress would be needed for the required strain to be achieved instantaneously. However, flow will ensure that the intended strain is always attained at the end of the test:

$$C_t = C_i \qquad (8.2)$$

There will, of course, be no recovery. This is not a very meaningful test but it represents an illustrative extreme case.

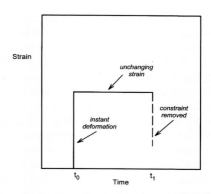

Fig. 8.2 A strain pulse, as applied in a compression set test. Note that the strain is only defined during the "holding" period.

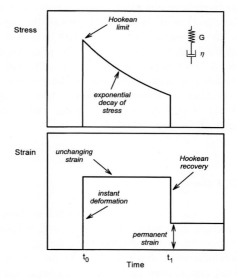

Fig. 8.3 The variation of stress and strain in a Maxwell body during and after a compression set test.

●8.3 Maxwell

Turning to a Maxwell body, the peak stress observed would correspond to that expected for its Hookean element alone (Fig. 8.3), but that stress will decay exponentially as deformation of the Newtonian element occurs. Release of the constraint will see the immediate recovery of the elastic deformation that corresponds to the Hookean element's deformation from the stress existing at the moment of release. Such a material then exhibits a compression set, the amount of which increases with the duration of C_i, but which does not change any further after release of the constraint.

●8.4 Kelvin-Voigt

A Kelvin-Voigt body again would suffer an infinite stress to achieve C_i instantly, and this stress would then vanish as in a Newtonian system. It is, therefore, not possible to treat such a material in this fashion either. However, it can be deduced what the recovery curve would be like with the constraint released, for the situation would be exactly equivalent to that of Fig. 5.1 when the stress was removed: a negative exponential return to the exact starting dimension.

●8.5 Viscoelastic

A viscoelastic material, and therefore one rather more realistic as a model for the impression materials in which we are interested, can now be considered. The initial stress will be that determined by the Hookean element, and this stress will decay according to both the viscous and retarded elastic components, acting simultaneously (Fig. 8.4), while the constraint is maintained. On release of the constraint there will be immediate recovery of the Hookean elastic deformation corresponding to the stress acting at the moment of release. The retarded elastic compliance that has been accumulated during the predetermined strain period will also be recovered entirely, but at a rate depending on its retardation time. The Newtonian flow is, of course, irrecoverable as before, and this again constitutes the basis of the compression set, but it is not observable until the retarded compliance has been recovered. Again, it is necessary to wait for a sufficient time for this to have occurred before a model is poured if the full accuracy of the material is to be realized in practice.

•8.6   Stress set

It should be noted that the underlying mechanism of compression set is operating not only under that kind of stress, but entirely similar effects can be observed in tension, flexure, shear and torsion as well. Experimentally, these other types of behaviour are rather more difficult to measure, but this does not diminish the importance of the phenomena (even if they are all – including compression set – ultimately due to only the tension and shear behaviour of the material; see Chap. 1). Removing an impression from a tooth (Fig. 7§11.1) will involve complicated deformations with components of all types: distortion will result.

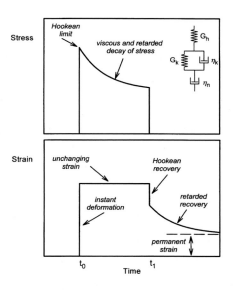

Fig. 8.4 The variation in stress and strain of a viscoelastic body during a compression set test.

§9. Fillers

The effect of filler particles on flow is very important in practical terms as they may be added to materials for a variety of reasons. In filled resin restorative materials, for example, the filler is largely present to increase the elastic modulus, strength and abrasion resistance of the material, as well as to reduce the polymerization shrinkage (6§2.3). In alginates, they are added to modify the stiffness of the set material. In other applications they can reduce the amount of an expensive component, modify the optical properties of colour and translucency, or even modify the electrical properties to make the material more or less conductive. Metallic fillers would also increase the thermal and electrical conductivity, ceramic ones would reduce these values. In dentistry, fillers are often added for their primary mechanical effect, that of modifying the viscosity of the medium in which they are suspended.

•9.1   Shear velocity

Extending the idea of Fig. 3.1 by introducing rigid filler particles at intervals over the depth of the fluid medium, it can be seen that the shear stress τ is transmitted to a fluid element without any deformation of the filler (Fig. 9.1). This shear stress is then applied across the depth of the fluid element, L_m (subscript m for 'medium'), so that for the viscosity of that medium, η_m, the resulting displacement in unit time is dx_m (equation 3.3). However, over the depth of the mixture, L_o (subscript o for 'overall'), the total displacement in unit time is only the sum of the fluid element displacements, dx_o:

$$\frac{dx_o}{L_o} < \frac{dx_m}{L_m} \qquad (9.1)$$

since the filler shows no shear. Thus, from equation 3.5, since the overall shear strain rate is now smaller, the viscosity of the mixture

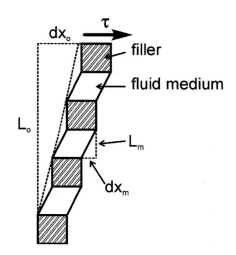

Fig. 9.1 The effect of including filler particles in a fluid medium on the shear velocity gradient.

must be higher. Effectively, the viscosity is expected to be higher by the factor $1/\varphi_m$, where φ_m is the volume fraction of the fluid medium, since this is the same as the proportion of the fluid element depths in the overall depth: L_m/L_o. A more formal calculation of this is follows. However, it is worth noting that by a "rigid" filler we mean that its elastic deformation is negligible, not that its elastic modulus is infinite. More to the point, it is just elastic – no flow.

In this system, the shear stress acting on each element is the same (this situation is known as **isostress**), so that by definition

$$\tau_o = \tau_m = \tau_f \tag{9.2}$$

(subscript f for 'filler'). Since the total shear velocity is the sum of the shear velocities of the two components, and the velocity depends on the length of the component over which the shear is occurring (equation 3.3):

$$v = \dot{\gamma}L \tag{9.3}$$

so we can write:

$$v_o = \dot{\gamma}_o L_o = \dot{\gamma}_m L_m + \dot{\gamma}_f L_f \tag{9.4}$$

Thus, the overall shear strain rate would be given by:

$$\dot{\gamma}_o = \dot{\gamma}_m \varphi_m + \dot{\gamma}_f \varphi_f \tag{9.5}$$

that is, weighted in proportion to the volume fractions, except that the shear strain rate in the solid filler must be zero and that term can therefore be dropped. We then substitute for shear strain rate from equation 3.5:

$$\frac{\tau_o}{\eta_o} = \frac{\tau_m \varphi_m}{\eta_m} \tag{9.6}$$

Therefore, because from equation 9.2 the shear stresses are equal, we have

$$\frac{1}{\eta_o} = \frac{\varphi_m}{\eta_m} \tag{9.7}$$

or, rearranging,

$$\frac{\eta_o}{\eta_m} = \frac{1}{\varphi_m} = \frac{1}{1 - \varphi_f} \tag{9.8}$$

In other words, on this very simple model, the viscosity of the mixture would be expected to rise slowly at first, and only quickly at high filler loadings.

Experimentally, however, it is found that viscosity rises much more rapidly with the volume fraction of filler, φ_f (Fig. 9.2) than that equation suggests. This imposes a practical limit when the viscosity is so high as to make handling (dispensing and shaping or moulding) difficult. But there is clearly also a physical upper limit to the system when the filler particles are close packed (*i.e.*, when the arrangement is most efficient for filling space), which is at $\varphi_f = \pi/3\sqrt{2} \approx 0.74$ for spherical particles. More reasonable, however, is the limit set by the random packing of uniform spheres, $\varphi_f \approx 0.64$, more precisely, the **maximally random jammed** state, when there are no loose particles anywhere, no 'rattlers'.[6] Nevertheless, something might be expected to happen when the random loose-packed limit is reached at $\varphi_f \approx 0.56$, when a continuous, three-

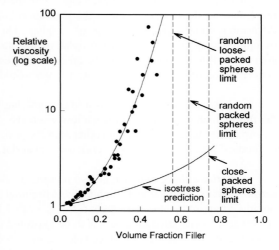

Fig. 9.2 The effect of amount of filler on the relative viscosity of a medium, for uniform spherical filler particles.

dimensional contact network can first be formed, that is, a switch to a different kind of behaviour. In fact, this is where the transition to dilatant behaviour occurs (§7.6). Note that it does not require that all particles are restrained by contacts in this way, only that enough are to establish the network structure on a large enough scale that stress is transmitted across the body.

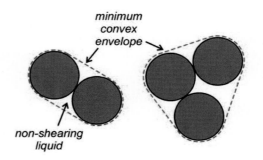

However, particle-particle interactions must occur randomly and transiently during shear. Collisions in this way would tend to make the effective volume fraction φ_f rather larger because in a cluster in contact, *i.e.* that moves as one object for the moment, there is liquid trapped in the interstices which cannot move at all (Fig. 9.3). The liquid affected in this way is more or less defined and enclosed by the **minimum convex envelope**, the three-dimensional

Fig. 9.3 When particles make contact, they become in effect a larger particle, and liquid inside the envelope ceases to be shearable, for the time being.

surface that wraps the assemblage without indentation. The number, frequency and size of such clusters will increase rapidly with φ_f until there is continuity across the entire volume. Thus, it does not require the overall system to be dilatant for extra work to be done to get particles to move past each other, merely that a temporary contact chain is established which creates in effect a large but rather porous particle, in which is held some unshearable liquid. On this basis, the rapid rise in viscosity shown in Fig. 9.2 is not unexpected.

Although many attempts have been made to fit a general equation to the rheological behaviour of (dilute!) suspensions of solids in liquids (and even of emulsions of immiscible liquids), no such general theory is yet available. However, the reasons for the observed dramatic results can be approached from a consideration of the molecular-level effects.

●9.2   Boundary layer effects

Each particle must be wetted by the liquid. If it were not it would not stay suspended but be ejected, like wax from water or metal oxides from mercury. This means that each particle has a shell of interacting liquid molecules around it (Fig. 9.4); this is the expected stationary boundary layer mentioned above (§3). This shell may effectively be very much more than one molecule thick, depending on the strength of the molecular interactions, but in any case this represents liquid that is no longer *fluid* in the same way as the bulk. Although it is still labile (*i.e.* not permanently bonded) and capable of some movement and exchanges with the bulk, it is much less so than 'free' liquid – the activation energy for diffusion is much higher – its viscosity is higher. The volume of liquid tied up in this way is proportional to the total surface

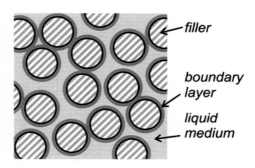

Fig. 9.4 The particles of a disperse phase have a boundary layer of less mobile molecules.

area of the filler. Consequently, the amount of "free liquid" decreases much faster than simply the increased concentration of filler would suggest. Eq. 9.6 would then require φ_f to be increased by some factor.

Taking all of these aspects into account, high filler contents lead to very stiff materials indeed, even if the fluid which suspends them has a low viscosity. This effect is used to advantage in the production of, for example, the range of silicone and other impression materials with different viscosities (loosely called "weights" or "bodies"), from a thin, runny, syringeable material for 'wash' techniques to a very stiff paste (called a **putty**).

Particle shape and size is also important. A departure from sphericity must increase the ratio of the surface area to volume (since a sphere has a minimal surface), and therefore this controls directly the amount of liquid tied up in the boundary layer (*cf.* Fig. 17§3.7 for the effect of subdivision of particles on surface area). However, an even stronger effect is to be expected on reducing the particle size at constant volume fraction. Since the area of a sphere's surface is given by $A = 4\pi r^2$, and its volume by $V = 4\pi r^3/3$, we have:

$$A = \frac{3V}{r}$$

(9.9)

that is, halving the particle size doubles the surface area, and thus doubles the volume of the boundary layer material. Then if the particles are irregular or rough, the effective boundary layer can be much increased, and be similar to the situation described above for collisions (Fig. 9.3). That is, liquid in the roughness (or even connected porosity) – inside the minimum convex envelope that contains the particle, ceases to be fluid in the sense of being shearable (Fig. 9.5). Furthermore, rough surfaces will not slide on contact, but will tend rather to interdigitate and lock, and so will increase the lifetime of assemblages such as in Fig. 9.3. However, there is an extra effect for shape due to the effects of tumbling (Fig. 2§2.1), even if the material overall is not dilatant, particles must collide if close enough, and those with high **aspect ratio**, *i.e.* length to width, will collide more readily. Likewise, the relative size distribution of the powder particles is important to viscosity through the same surface area mechanism, but also because very high packing densities can be achieved with an appropriate choice of the sizes and proportions of particles.

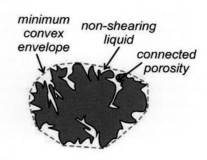

Fig. 9.5 Rough particles and connected porosity trap liquid, effectively increasing the volume fraction of the filler.

Some filler systems, such as the fumed silicas (Fig. 7.11) and carbon blacks (as used in commercial tyre rubbers, 7§3.4), and particularly at the nanometric scale for many materials, readily form **agglomerates**, clusters of particles that are more or less stable because of specific bonding, but which enclose a large volume. These then have more complicated behaviour because the effective volume fractions of matrix and filler can be very different from the actual, 'true' values. Particles in the form of rods or platelets, and especially whiskers and fibres, which change both specific surface area and physical collisions during shear – if not actual entanglements, also tend to raise viscosity above what might otherwise be expected.

The failure to find a general theory for suspensions, mentioned above, is now seen not to be so surprising. It would have to take into account the specific interactions of the medium with the disperse phase, that is, molecular orientation and ordering according to polarity, hydrogen bonding, solvation of solutes and so on (*cf.* the effects of hydrogen bonding in sugar syrups, 7§8.2). Again, we are obliged to fall back on the phenomenological approach in these matters (§7.13).

These effects of fillers are in addition to the other types of rheological behaviour described above. Generally speaking, the rates of all flow phenomena are decreased, whether due to 'Newtonian' or retarded elasticity, and this has given one more dimension to the possibilities of design in a material to achieve the desired properties, for example, in impression materials. On the other hand, these effects are present whether or not they are desirable, and some compromise is inevitable if fillers are required for reasons other than flow control. For example, increasing the filler proportion in a resin-based restorative material increases the viscosity so that it becomes more and more difficult to pack the cavity properly. There must be a practical limit to the amount that can be used therefore, when the material becomes simply unusable, well before the densest packing limit is reached (Fig. 9.2).

●**9.3   Settling**

In the above we have ignored gravitational settling. That is, because a filler will usually have a density higher than the liquid in which it is suspended, there will be a force acting on it to cause it to fall to the bottom:

$$F = Vg(\rho - \rho_1) \tag{9.10}$$

where V is the volume of the particle, ρ is the density of the filler, ρ_1 is the density of the liquid, and g is the local acceleration due to gravity (see box, 1§10.1). This means that over time, even for viscous media such as the resins of direct restorative materials and elastomeric impression materials, some separation of the components may occur. This accounts for the appearance of a clear oily portion when some products are dispensed from tubes or syringes. Not only is this portion not going to behave as expected (because of the lack of filler), but also the remainder will be that much more concentrated, even to the point of being unusable. Either way, problems can be expected. This is one reason why stocks of materials of these kinds should not be too large so that they do not sit for very long periods, slowly settling. It also means that filler-containing fluids should, if possible, be stirred before use to ensure uniformity, for example, paint (but not thixotropic types!). It is certainly the reason for the instruction on many bottles of suspensions, medicinal or otherwise: "Shake the Bottle". In addition, if there is more than one solid component, and these have different densities, they will settle at different rates and cause more complicated changes. Of course, if a dispersed component is less dense than the medium, it will tend to migrate to the upper surface – inverted settling. (See also the end of §10.)

Settling will even occur in systems where the minimum loose-packed condition (§7.6) is exceeded. That is, even if a continuous-contact structure exists in the mixture, shear-thinning (§7.9) can overcome the frictional constraints. So long as a path to a denser packing exists, even if only by moving one particle at a time, then because this leads to a lower-energy, more stable state, it will happen. Of course, vibration in transit is to be expected, so even products freshly-delivered may require stirring before use. This is the reason for the disclaimer on cereal boxes: "full when packed", and affects dental products such as alginate powders (15§4.1).

We may note in passing that if the density difference is negative, the dispersed material will be buoyant and therefore rise. Either way. if the medium is plastic, the force has to create a stress to exceed the yield point to move at all.

●9.4   Powders

While we have been discussing systems that implicitly have always involved a liquid matrix, we should not forget that 'fluid' includes gases. That is to say, that the flow of powders (typically, for dentistry, in air) is also of concern, whether this is the dispensing of powders such as of cements, plaster or acrylic for proportioning mixtures, or the delivery of the abrasive in grit-blasting (20§4.1).

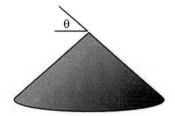

For dry powders, there is a natural tendency to settle under gravity into a random loose-packed structure ($\varphi_f \approx 0.6$), although some vibration may allow the bulk density to increase further (*cf.* tap density, 2§2.2; alginate powder,15§4.1) and thus approach the random-close packed state. This is the reason for the common kind of labelling of foodstuff containers: "full when packed, contents may settle". However, there is a limit to the amount of settlement that can occur, as can be shown by the slump behaviour (§7.8) in a test exactly as in Fig. 7.9: when the cylinder is removed (slowly, to avoid dynamic effects), a conical pile will be produced whose slope, called the **angle of repose** (θ) (Fig. 9.6), is controlled by friction between the particles, as might be expected. An approximation to this behaviour is given by

Fig. 9.6 Angle of repose for a spontaneously formed pile of dry, granular material.

$$\tan \theta = \mu \qquad\qquad (9.11)$$

where μ is the coefficient of friction (1§5.5) between layers of the powder; θ then has a maximum of 45°, in the absence of other effects (this also assumes that the lowest layer does not slip on the substrate). This demonstrates that there is a yield point for such a system: clearly, the gravitational force – the weight – of the granular material is now just balanced by that friction. (However, the onset of powder flow also depends on the initial packing density, that is, vibrational history.[7]) The effects can be observed in the flowing sand of an hour-glass or egg-timer. Of course, if the powder is just damp or slightly wet, contacts will be held together by surface tension (Fig. 25§2.2) and the behaviour considerably modified – essential for powder-method porcelain (and sand castles) but otherwise often just a nuisance, like damp table salt in a salt shaker.

As above (§7.9), vibrational shear thinning is effective and can cause such heaps to collapse and spread. This is turned to advantage when dispensing small amounts of powder, such as dental stone or cements, for weighing: tapping the spatula or container allows fluid-like flow that enables fine metering, avoiding the random addition and subtraction otherwise used (indeed, both mechanical and electric vibrating spatulas are sold). Vibration may also be needed if transferring a powder to another container using a funnel – intermittent jamming may occur at or above the critical volume fraction, $\varphi_f \approx 0.56$, when dilatancy sets in, and certainly at $\varphi_f \approx 0.64$ when no flow can occur. So marked is this latter limit that it is a critical value even for bubbles![8]

§10. Stokes' Law

There are several different kinds of test which may be used to determine viscosity, each suited to particular ranges of viscosity and strain rate. Waxes (Chap. 16) pose some particular problems because of their very high viscosities when solid, and most methods are impractical. Stokes' Law can be used to design an approximate but practical test for this case.[9]

Picking up the idea of a particle sinking in a fluid from equation 9.10, at equilibrium the resistance due to the viscosity η of an infinite sea of liquid acting on a small sphere, radius r, at its terminal velocity u (steady-state free fall) must, from Newton's Third Law of Motion, be equal to the gravitational force acting on it (equation 9.7) (Fig. 10.1):

$$6\pi\eta u r = (4/3)\pi r^3 g(\rho - \rho_1) \qquad (10.1)$$

frictional   gravitational
retardation   acceleration
force   force

where the buoyancy effect is again taken into account. This can be rearranged to allow the viscosity to be calculated from observation of the free-fall velocity:

$$\eta = \frac{2gr^2(\rho - \rho_1)}{9u} \qquad (10.2)$$

The 'infinite sea of liquid' is not, of course, realistically approximated in most experimental circumstances. In practice, a container must be used, and then serious interferences may arise from the proximity of the walls and base of the container, both acting to retard the fall of the ball. (If we take this to extremes, we can see that the equator of the ball starts to look like the plunger in the Newtonian dash-pot model, Fig. 3.6.) The wall effect is corrected to a reasonable approximation by the factor λ:

$$\lambda = 1 + \frac{2\cdot4r}{R} \qquad (10.3)$$

where R is the radius of the container. The end effect is similarly corrected by the factor μ:

$$\mu = 1 + \frac{3\cdot3r}{h} \qquad (10.4)$$

where h is the total height of the liquid. The latter correction applies only to the middle third of the liquid column, and this is where the velocity measurement should be made. The velocity u in equation 10.2 is then replaced by:

$$u_\infty = u\lambda\mu \qquad (10.5)$$

Unfortunately, ρ cannot be made large enough for a useful velocity to be obtained in waxes, but it can be made to look high by fixing the ball on a narrow rod and pushing with an external force, relying on the fact that the wax does not flow well enough to close in behind the ball as it passes (Fig. 10.2). This effect, however, means that the conditions under which equation 10.1 was derived are not quite satisfied, so only an apparent viscosity η_a is obtained. This loading is done in practice by adding weights to the rod (Fig. 10.3). So, the *apparent* density of the sphere is calculated as if it contained the total mass, M, of the load assembly. This then means that usually the buoyancy correction becomes small enough to

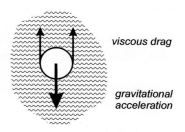

Fig. 10.1 A falling ball is subject to acceleration under gravity and frictional retardation by the medium. There is therefore a 'terminal velocity' for steady-state fall, as for rain drops and parachutists.

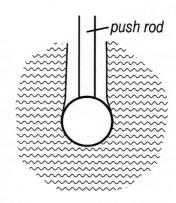

Fig. 10.2 With wax as the medium, it does not close in behind the ball as it passes to contact the push rod, avoiding spurious friction.

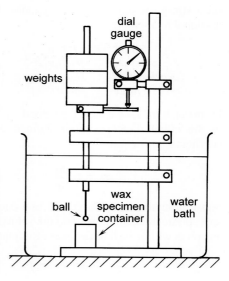

Fig. 10.3 The experimental set-up for determining the viscosity of wax using Stokes' Law.

ignore, and instead of equation 10.1 we can write:

$$6\pi\eta_a ur = gM \qquad (10.6)$$

so that, making the end effect corrections again, we have

$$\eta_a = \frac{gM}{6\pi r u \lambda \mu} \qquad (10.7)$$

A further point may be made about settling based on equation 10.1: the geometry of the particles also matters. Thus, if again there is more than one filler component, even if of the same density, they will settle at different rates if the particles have different sizes or shapes. The left hand side contains an explicit geometrical factor for the sphere, in this case, $6\pi r$. As sky-divers and parachutists in free-fall and hawks know, changing aspect ratio and cross-section has a dramatic effect on free-fall velocity.

§11. Poiseuille's Law

There are a number of circumstances in dentistry where fluid materials need to be delivered through tubes, most notably liquids for injection using hypodermic needles. Other examples include low-viscosity 'wash' impression materials and certain kinds of cement that are delivered through nozzles from a syringe-like mixing capsule. It is worth noting the factors that control such activities.

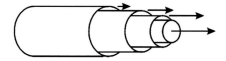

Fig. 11.1 Flow in a tube may be treated as occurring by the sliding of cylindrical layers.

In Fig. 3.1, we considered the movement of parallel plane layers of a Newtonian fluid suffering friction, generating a velocity gradient. In a circular tube, radius R, we may similarly consider thin concentric tubular layers, annuli sliding one within another, like an extending telescope (Fig. 11.1).[10] At the tube wall, we still have a stationary boundary layer. However, now the velocity gradient is non-linear, and depends on the radial position of the point of interest. This is seen by considering a force balance as at equation 10.1, *i.e.* under conditions of steady flow – Newton's Third Law of Motion again. Thus, in Fig. 11.2, the small disc element of the flowing fluid is accelerated by the pressure difference across it (dP) acting over its circular area with radius r, while the shear stress τ is acting over the cylindrical area πr^2.dx which provides the frictional retardation:

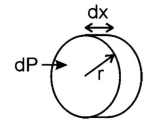

Fig. 11.2 The force balance is determined for a small disc-like element moving coaxially in the tube.

$$\begin{array}{cc} dP.\pi r^2 = \tau.2\pi r.dx \\ \textit{pressure} \qquad \textit{frictional} \\ \textit{acceleration} \qquad \textit{retardation} \\ \textit{force} \qquad \textit{force} \end{array} \qquad (11.1)$$

From equation 3.4 we have the velocity gradient in the radial direction du/dr defined by the coefficient of friction:

$$\tau = \eta\dot\gamma = \eta.\frac{du}{dr} \qquad (11.2)$$

Substituting this in equation 11.1, and doing the appropriate integration, we find the velocity u at any point within the tube distance r from the axis:

$$u = -\frac{1}{4\eta}.\frac{\Delta P}{L}.(R^2 - r^2) \qquad (11.3)$$

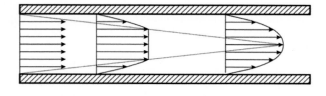

Fig. 11.3 Conceptual development of the parabolic flow profile in a tube: successive layers are held back by friction with the next outer layer. (This is also the pattern for the development of the full flow profile at the entry end of a tube – see box.)

Here, $\Delta P/L$ is the pressure gradient over the entire tube length L and R is the tube radius (the negative sign is because the pressure gradient is negative). It can be seen that the radial velocity profile is therefore parabolic, the accelerating force decreasing as r^2 but the area being sheared decreases only as r. Thus, in Fig. 11.3, initially all fluid can be considered to have the same velocity across the lumen area. However, the layer next to the wall is brought to rest to form the boundary layer, which then impedes the next layer, and so on.

A further integration can be performed to determine the volumetric discharge rate, dV/dt:

$$\frac{dV}{dt} = \frac{\pi R^4 \Delta P}{8\eta L}$$ (11.4)

This is known as the **Poiseuille Law**, and thus flow of this type is said to be **Poiseuille flow**. From this some important results are obtained. Firstly, as might be expected, for a given flow rate, the pressure required increases in proportion to the length of the tube and the viscosity of the fluid. Thus, for hypodermic needles, more effort would be required with a long as opposed to a short needle. However, much more significant is the dependence on lumen size: halving the diameter requires sixteen times the pressure or, equivalently, would cause the task to take sixteen times as long. The consequences of using a syringe or nozzle device (Fig. 11.4) to deliver impression materials, cements or filled resins should follow clearly from this since their viscosities are typically very high.

In practical situations, the pressure is generated by manual force applied to a plunger of various diameters, according to the size of the syringe. Thus the force to be applied needs to be controlled according to that size. In addition, there will be some friction of the plunger with the wall offsetting that, meaning that the actual external force applied to the plunger must be increased. For high viscosity materials (such as freshly-mixed glass ionomer cement in a capsule, or long path-length devices (impression material mixing nozzles, 15§7.2), lever-operated devices with a high mechanical advantage are essential (Fig. 11.5). For hypodermic and other needles, the pressure may be sufficient to detach a 'Luer slip' needle connection, necessitating the use of the 'Luer lock' kind.

Under certain circumstances there will be a back-pressure, for example acting at the tip of a needle as the injected fluid builds up, dependent on its ability to infiltrate the tissue.[11] This pressure is to be subtracted from that exerted by the user, hence the use of the term ΔP, the pressure difference, in the calculation. However, as far as the operator is concerned, a greater force for injection is required. Thus, a small diameter syringe would be an advantage as pressure is force × area, that is, area of the plunger. In such contexts, lever devices can also be used (Fig. 11.6).

Several assumptions are made in this section: that the fluid is incompressible (K_{water} : 2.2 GPa – which is good enough at ordinary pressures); that the flow rate is low enough for it to be streamline flow, *i.e.* non-turbulent (see 15§7.1); that the fluid is Newtonian; that the tube is long in proportion to its diameter so that the parabolic flow profile can be fully developed. The latter two are not usual in dentistry, and the last hardly ever. Nevertheless, the approximations are good enough for the present purposes to establish the main points.

Fig. 11.4 Viscous materials require high forces for delivery through narrow tubes.

Fig. 11.5 Long levers may be necessary to extrude some materials from capsules. This device has a mechanical advantage of ~3 : 1.

Fig. 11.6 Periodontal injection may require very high pressure. Devices such as this can generate pressure in excess of 80 bar (8 MPa), although whether this is good for even dense tissue is another matter altogether.

References

[1]   Reynolds O. On the dilatancy of media composed of rigid particles in contact. *Phil Mag* ser. V: 20 (127): 469 - 552, Pl. X, 1885.

[2]   Jaeger HM & Nagel SR. Physics of the granular state. *Science* 255:1523 - 1531, 1992.

[3]   Onoda GY & Liniger EG. Random loose packings of uniform spheres and the dilatancy onset. *Phys Rev Lett* 64: 2727 - 2730, 1990.

[4]   Ardakani HA, Mitsoulis E & Hatzikiriakos SG. Thixotropic flow of toothpaste through extrusion dies. J Non-Newtonian Fluid Mech 166: 1262 - 11271, 2011.

[5]   http://www.brookfieldengineering.com/education/applications/texture-toothpaste-firmness.asp

[6]   Torquato S, Truskett TM & Debenedetti PG. Is random close packing of spheres well defined? Phys Rev Lett 84(10): 2064 - 2067, 2000.

[7]   Gravish N & Goldman DI. Effect of volume fraction on granular avalanche dynamics. Phys Rev 90: 032202, 2014.

[8]   Lespiat R, Cohen-Addad S & Höhler R. Jamming and flow of random-close-packed spherical bubbles: An analogy with granular materials. Phys Rev Lett 106: 148302, 2011.

[9]   Darvell BW & Wong NB. Viscosity of dental waxes by Stokes' Law. *Dent Mater* 5: 176-180, 1989.

[10]   Massey BS. Mechanics of fluids. 5th ed. Van Nostrand-Reinhold, Wokingham, 1983.

[11]   Pashley EL, Nelson R, Pashley DH. Pressures created by dental injections. J Dent Res 60 (10): 1742 - 1748, 1981.

Chapter 5   Acrylic

Poly(methyl methacrylate) – so-called 'acrylic resin' – is usually the material of choice for full denture bases and 'gumwork' for removable devices. It has also been the chemical model for many other material developments in dentistry, such as restorative materials. The properties, behaviour and handling of poly(methyl methacrylate) likewise form a basis for understanding those other materials.

*A **polymerization** reaction is performed as part of the normal dental procedures in the laboratory, and the proper control of this is important for controlling the properties of the product. It is essential to understand both the **heat-cured** and **cold-cured** types, whose properties and limitations differ because of their processing differences.*

*Since the methyl methacrylate monomer is a very reactive compound, **storage** is a problem. Its chemistry has a bearing on a number of issues.*

*The **mechanical properties** are the ultimate concern in determining suitability to task, and the relationship of these to processing and service conditions are clearly important. This topic is dependent on the special behaviour of polymers. The properties of the plain product are not ideal, however, and various modifications to the chemistry are possible to obtain better behaviour.*

*Dental acrylic resin represents a case study in polymer science of wide interest in itself. Even so, there are many points which are relevant in other dental materials contexts, and a grounding in this type of product as a **model** is therefore essential to the understanding of the more advanced and complicated materials which are, in a very direct sense, derived from it.*

Materials Science for Dentistry
https://doi.org/10.1016/B978-0-08-101035-8.50005-5

Polymers based on methyl methacrylate, broadly termed 'acrylic resins', are used in very large quantities for full denture base fabrication and for adding 'gum work' to cast metal frameworks. Indeed, despite much effort to find alternatives with better properties, it remains essentially the only polymer system in current use for these applications. In modified forms, such polymers may be used as direct restoratives, denture repair materials, soft liners and denture teeth. Whilst plain poly(methyl methacrylate), PMMA, is a typical polymer in many mechanical and physical respects, it can be considered the archetype for a very large number of related products as many aspects of its chemistry are relevant in those as well. Indeed, its importance in dentistry is such that it warrants detailed treatment.[1]

§1. Polymerization Reactions

●1.1 Initiation

Acrylic resins are prepared by a **free-radical addition polymerization chain reaction**, where the opening of one double bond results in the formation of another free radical which can attack and join at another double bond, resulting in yet another free radical. The mechanism is straightforward. The vinyl group of methyl methacrylate is susceptible to attack by a free radical, resulting in the opening of the π-bond, the formation of a new σ-bond to one carbon, and the creation of a lone (that is, unpaired) electron on the central carbon atom (Fig. 1.1); this constitutes the **initiation reaction** in the sense that the polymerization chain has been started. The attack is selective on the more-exposed carbon atom, as opposed to that leading to a terminal position for the free radical electron. This is driven by the steric hindrance of the methylcarboxyl and methyl groups – it is simply easier to get at. It is of course possible that some attacks may take the less-favoured route, and indeed a small proportion do, since collisions are random and not 'intentional'[1]. However, the resulting radical is also extremely unstable and suffers a rapid reaction, probably by hydrogen abstraction (§1.4), so that it is not a significant species in the overall chain reaction.

●1.2 Propagation

The new free radical is equally capable of attacking another double bond in exactly the same way, and the resulting radical another bond, and so on. This process of repeated reaction of the same type is termed **chain propagation** (Fig. 1.2). It can be seen that because of the bulk of the parts of the molecule surrounding the new unpaired electron, steric hindrance effects for attack on the next double bond are now even greater, and it may be supposed that as a consequence nearly all attacks result in the methyl methacrylate **residues**, the monomer units, being linked by methylene bridges, -CH$_2$-. Polymer chains carrying an active free radical in this way are termed **growing** or **live chains**.

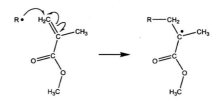

Fig. 1.1 Initiation of polymerization of methyl methacrylate by the attack of a free radical on the double bond. (Single-barbed curly arrows indicate the shift of single electrons.)

A **free radical** is chemical species carrying an **unpaired electron** where otherwise it would be expected to part of an ordinary covalent bond. Because it is equivalent to a broken bond, such a species is necessarily high-energy and typically very reactive when the electron remains in a single, localized orbital. The term comes from the outdated concept of chemical 'units' such as methyl, CH$_3$-, as building blocks referred to as radicals. Hence, when not attached, i.e. not bonded, they are 'free'. The implication of being free other than by being ionized is that the bond has been broken and the molecule simply separated, with one electron going to each part, so that each is electrically neutral. However, while the term free radical is commonplace and accepted, the focus is really on the unpaired electron as the chemically-relevant entity. The 'radical' part is generally of secondary interest.

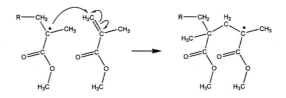

Fig. 1.2 Chain propagation repeats the process. Note the methylene -CH$_2$- bridge between units.

---

1 It must always been borne in mind that chemical reactions are not purposeful or directed from outside; what is desired or desirable has no bearing. Accordingly, any discussion of chemistry has be viewed in the context of the process being stochastic or probabilistic, the result of an accident in time and space – the collision – that happens to lead to a reaction pathway or mechanism that ultimate leads the species involved, and thus the whole system, to a lower energy state.

•1.3 Termination

Chain termination can occur at any time. It is not a direct function of the chain length already created (which cannot affect the reactivity of the radical nor be 'seen' in any way by incoming monomer or radical – the "equal reactivity" hypothesis[2][3]), but it depends on the concentration of free radicals in the system. Since the reaction may be written:

$$R_i \bullet + R_j \bullet \Rightarrow R_i\text{-}R_j \qquad (1.1)$$

we can then see that the rate depends on $[\Sigma R_i]^2$, that is, the square of the total free radical concentration.

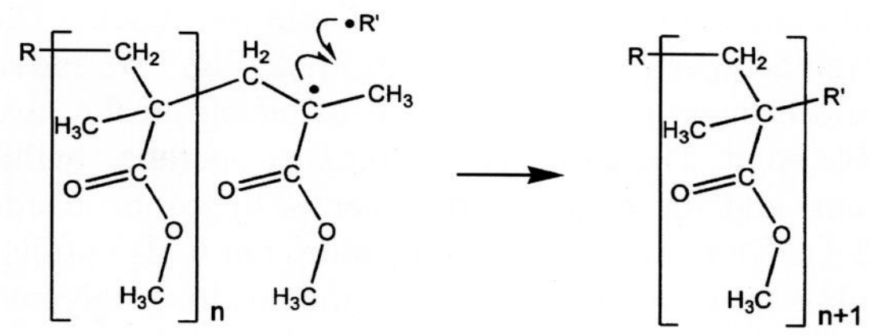

This step, which accounts in large part for the self-limitation of the reaction, is the simple **mutual annihilation** of two free radicals which happen to come close enough together and collide (Fig. 1.3). That is to say, the unpaired electrons as such, in reacting to form the covalent bond, simply cease to exist. This could as easily occur by reaction with another live chain (of any length) as with another initiator radical. There is a third possibility.

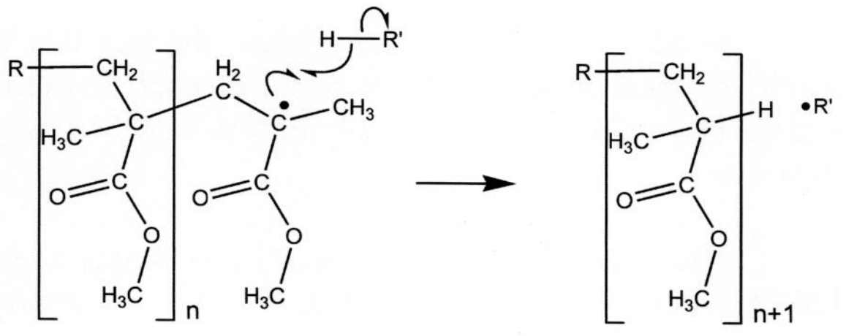

Fig. 1.3 Chain termination results from the mutual reaction of two free radicals. Notice that R' can be any free radical, including a live polymer chain.

Fig. 1.4 Chain transfer reactions terminate one chain and simultaneously start another, which may be a branch on an existing chain.

•1.4 Chain transfer

As indicated above, free radicals tend to be high energy, reactive species, and one of the reactions that may occur in systems such as this is **hydrogen abstraction**. This is the simple transfer of a hydrogen atom, from nearly anywhere at all, to the attacking radical. Of course, this leaves an unpaired electron residing on the attacked species, which may then itself go on to attack methyl methacrylate double bonds, initiating a new live chain. The important thing about this kind of reaction is that the hydrogen may have been removed not just from a monomer molecule but from any of the possible sites on existing polymer, such as the esteric methyl group. In this case, a **branch** will have been introduced into the polymer chain, the active site then continuing to propagate just as before to create a new polymer chain. Since the growth of one chain is stopped, and another begun, this process is called **chain transfer** (Fig. 1.4). This random reaction also tends to limit the average chain length that may be obtained in the final product, at least, in the sense of the length of backbone between branch points, although the molecule itself can only get bigger. It will also be realized that, given the random-walk conformation expected for such polymer chains (Fig. 3§2.3), the transfer could even be with another point on the same molecule.

•1.5 Tacticity

When, as is usual for the steric reasons mentioned above (§1.1), the result of attack on the monomer is an unpaired electron residing on the central carbon atom, there are two possible outcomes to the next step, depending on which 'side' of that atom the single electron happens to be lying at the time that it attacks the next double bond (Fig. 1.5). Bear in mind that the free radical part, which has sp^3-hybridized orbitals in the usual tetrahedral arrangement (Fig. 3§2.1), will naturally be 'flipping' or **inverting**, back and forth between the two forms (as well as rotating about the bond to the carboxyl group) at a very great rate (Fig. 1.6). The **configuration** is easily invertible to its mirror image because the lone electron is mobile. So, even if there is a preference initially (which is unlikely), there will soon be a scrambling and

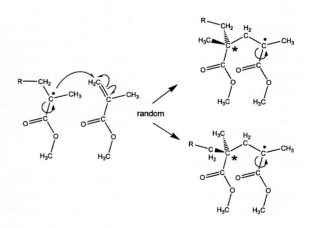

Fig. 1.5 The free radical formed from the methyl methacrylate double bond is asymmetric, so therefore is the environment of that carbon atom (starred) after further reaction.

equal numbers of both forms available for attack. The free radical species is asymmetric (3§2.2; Fig. 1.6), and so therefore must also be the result of attack on the next double bond. There is no steric guidance operating in this system, and the resulting polymer is therefore atactic (3§2.3). This prevents crystallization, but it also affects the glass transition temperature of the resulting polymer. For anhydrous PMMA, the T_g has the following values: isotactic, ~60 °C; atactic, ~110 °C, syndiotactic, ~130 °C. This indicates the subtleties of the interplay between chain packing (and thus free volume) and ease of chain segment rotation, bearing in mind that the molecule is polar and so has appreciable intermolecular interactions.

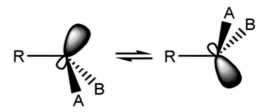

Fig. 1.6 Inversion of an sp³ free radical occurs readily, with the the single electron oscillating freely between the two lobes of the orbital.

In addition to the tacticity issue, the fact that the site of attack can vary, as indicated in §1.1, albeit heavily biassed, means that there is a further source of irregularity in the chain. Chain transfer reactions can also lead to branching and other irregularity. PMMA as used in dentistry therefore does not – and cannot – crystallize.

The question that has not been addressed is, how to start the process of polymerization? The identity of that first radical has not been stated. What is apparently required is a reaction **initiator**.

§2. Heat-cured Acrylic

As is commonly the case in many such practical polymerization systems, the reaction for the resins used for dentures is generally initiated by benzoyl peroxide[2] (Fig. 2.1) (often abbreviated 'BPO'). The central O-O bond is weakened by the electron-withdrawing power of the adjacent oxygen and carbonyl groups. Thus, this compound readily dissociates simply on heating to produce a pair of benzoate (or, benzoyloxy) radicals. These are themselves not very stable, and will break down further. Even at room temperature, solutions of this compound slowly liberate carbon dioxide and produce phenyl radicals in the process. Either of these radicals – benzoate or phenyl – may attack any compounds that are available, extracting hydrogen atoms and creating a new radical in the process. It will be seen later that the phenyl radical will not be very effective in such reactions, due to

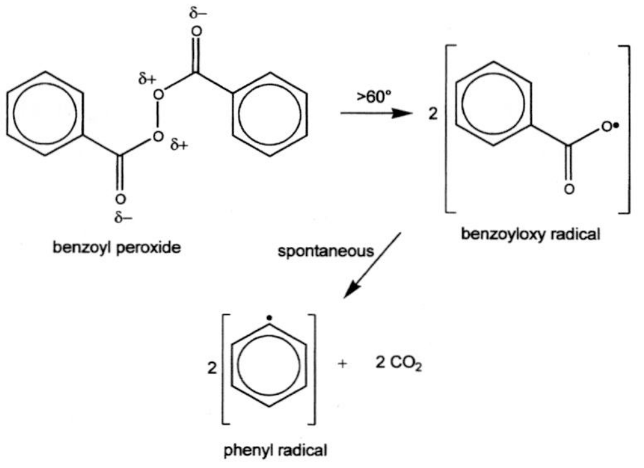

Fig. 2.1 The dissociation of benzoyl peroxide.

delocalization. However, the most important reaction, and the one that proceeds most readily in the present context, is the attack on one carbon atom of the acrylic double bond, which results in an unpaired electron on the second carbon atom. This, in turn, is more likely to attack another double bond although, as we have seen, chain transfer and termination reactions can also proceed. Although benzoyl peroxide does dissociate at room temperature, the rate is too low to be of practical use and heating must be used to generate radicals at a sufficiently high rate that the polymerization occurs within a suitably short time. This then is the basis of the normal dental laboratory process for preparing acrylic devices – and why such products are commonly referred to as **heat-cured** acrylic.

●2.1   Manufacture

This reaction is also the basis of the industrial manufacture of denture acrylic powders, a process often called **emulsion polymerization**, but more accurately **suspension polymerization**. A suspension of methyl methacrylate is prepared and maintained in water containing some surfactant (10§8) by vigorous stirring to

---

[2] This compound is sometimes incorrectly described as a catalyst. Since it is broken down, incorporated into the polymer, and cannot be recovered from the final mixture – even in principle, it cannot be so described. A catalyst does not initiate reaction, but merely increases the rate of a reaction while being unchanged chemically at the end of the process.

obtain very small globules. A small amount of benzoyl peroxide in solution in, say, ethanol is then added,[3] the temperature raised slowly, with a reflux condenser fitted to the reaction vessel to prevent loss of the monomer. If the stirring is maintained properly and the concentration of monomer is not too high, the product which is filtered off, washed and then dried is a fine powder consisting of nearly spherical and nearly uniformly-sized polymer beads[4] (Fig. 2.2).

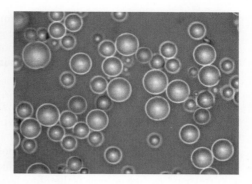

Fig. 2.2 Beads of PMMA.

●2.2   Laboratory technique

It is the same polymerization reaction that is used, in a slightly modified form, in fabricating a denture base. The supplied PMMA powder has had added to it a small amount of benzoyl peroxide. In this form it is stable more or less indefinitely, and such powders do not, therefore, spoil or 'go off'. When the monomer is added to the powder to make the dough, the benzoyl peroxide dissolves slowly, but because the dissociation of benzoyl peroxide is so slow at room temperature, very little polymerization occurs during the normal preparatory stages. (However, the dough cannot be kept too long – it will eventually harden. The dissociation rate of benzoyl peroxide in solution at room temperature is not zero.) The changes that are observed in the dough result solely from the physical processes of the monomer dissolving in the resin beads and the resin dissolving, more quickly, in the monomer. Initially, the polymer beads are merely wet, giving the appearance of 'wet sand'. As polymer dissolves in the monomer, so the viscosity of the liquid rises and the mixture becomes sticky. As the concentration of dissolved polymer continues to rise so this stickiness is lost. Meanwhile, some monomer dissolves in the polymer, and although these particles are plasticized by that, the presence in the fully-cured resin of spherical structures identifiable as the original beads shows that they are not destroyed by the manipulation of the dough. The mutual solution of the two components assists in obtaining a good bond between them by entanglement, in addition to the covalent bonds that would be obtained by chain transfers.

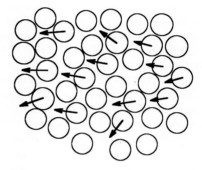

Fig. 2.3 Even spherical particles give dilatant behaviour. For some particles to move forward, others must move apart. This causes a general increase in total volume occupied on stirring or deformation.

●2.3   Mixing volume

It is usually said that when making up an acrylic dough enough monomer should be present to completely 'wet' the polymer powder. A little thought will show that this is quite incorrect: only a very little monomer is in fact required actually to wet the surface of even a fine powder, if wetting means to create a continuous film of a liquid on a surface (*cf.* 2§2.5). What must be meant, and what is clearly required in this context to minimize the porosity in the final product, is that the free space in the bulk powder be completely filled by liquid: that is, the gauging volume must be supplied so that the bulk powder is **saturated** (2§2.5). However, systems such as this are still dilatant (4§7.6), despite the spherical particles (Fig. 2.3). This arises because to move any single particle past another it must "rise up" slightly. This means that adjacent particles must on average be moved further apart than their equilibrium, 'settled' positions. Some dilution may also be necessary to obtain

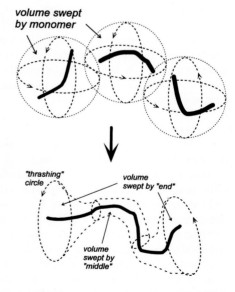

Fig. 2.4 Thermal motions cause a molecule to occupy more space than its actual volume. Polymerization reduces the scope of such movements.

---

[3] But this appears to be quite unnecessary – see §2.7.

[4] There has been no production for some time of denture polymer powders by milling bulk polymerized resin, which method would have resulted in highly irregular and rough particles. The gauging liquid requirement would have been much higher – see §2.3.

sufficient plasticity in the dough. The minimum mixing volume is necessarily a little larger than the gauging volume (2§2.4). But this begs the question of why powder and liquid are used in these systems. There are two main reasons.

●2.4   Shrinkage

Firstly, **polymerization shrinkage** is reduced. Each single monomer molecule has an effective volume, due to its **thermal motion**, larger than its actual size (Fig. 2.4). However, a monomer residue – a **mer** – at a polymer chain end is less free to move because of the chain to which it is attached; its corresponding free volume is therefore less than that of the equivalent unreacted monomer molecule. Thus, for each new bond formed (*i.e.*, each mer added to the chain), there is a reduction in the total free volume in the mixture. In the case of PMMA, the total change in volume which occurs with

> **Thermal motions** are of several kinds. Bond length and bond angle stretching produces the ordinarily-observable thermal expansion in many materials. Bond rotation of small side groups might not have very large effects but bond rotation in polymers also involves chain segments and the movements are correspondingly larger in size and effect. The end of a chain-like molecule would behave somewhat like an angry snake held by the head, thrashing at high speed. Increase in these swept volumes with temperature, given the usually weak intermolecular forces acting in polymers, is the cause of most of the thermal expansion of polymers.

complete polymerization from pure monomer amounts to some 21% of the original volume (22.5 mL/mol). Such a shrinkage would be very difficult to handle in practical denture base moulding. Hence, the more polymerization that can be done before moulding, the better. However, the commonest industrial method for fabrication with polymers, injection moulding (where the material injected is all polymer, no monomer), has so far quite failed to produce a viable denture under dental laboratory conditions, the construction of a strong enough mould with sufficient accuracy being a major problem. The practical compromise used in dentistry is to mix polymer and monomer, then polymerize the monomer. The ratio of the two is limited by the minimum mixing volume requirement, so the best that can be achieved in practice is a net volumetric shrinkage of about 6 or 7%. Denture fabrication techniques must take this into account.

●2.5   Heat dissipation

Secondly, by reducing the number of covalent σ-bonds to be made, the amount of heat to be dissipated is reduced in proportion. The C-C σ-bond has an energy of about 350 kJ/mol, while the π-bond is at about 270 kJ/mol; thus some 80 kJ/mol would be expected to be evolved as heat on polymerization (in fact, the actual figure is about 56 kJ/mol because other factors are involved as well, but this is still substantial). Thus, by using a proportion of polymer in the mixture, the risk of overheating and the amount of thermal contraction to be experienced are both reduced as well.

Other reasons which could be cited, such as the ease of handling a plastic mass as opposed to a mobile, volatile liquid; avoidance of mould leakage problems; absence of the need to handle benzoyl peroxide (a strong oxidizing agent) in bulk or high concentration, are all important but are of less practical significance than the main two. All such considerations, to a greater or lesser extent, apply to cold-cure materials as well, including direct restorative (filling) materials.

●2.6   Heating

The polymerization reaction is **activated** by immersing the packed flask in water and raising its temperature. As indicated above, the reaction is strongly exothermic, and because the thermal conductivity of acrylic material is low (as is that of the gypsum mould) there comes a point in the heating of the dough when more heat is being produced by the reaction than can be conducted away (Fig. 2.5)[4]. Since the decomposition of the peroxide is an activated process, the rate has an Arrhenius-like dependency on temperature, and the ensuing rise in temperature accelerates the reaction further. The temperature of the mixture therefore rises very quickly. Notice that initially the flask and the dough are at successively lower temperatures than the external water bath, limited by the rate of conduction of heat inwards. But at about 70 °C the reaction rate becomes so high that the reaction 'runs away' and the

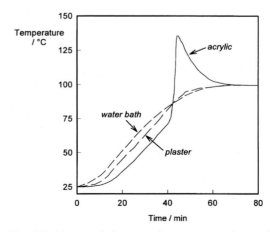

Fig. 2.5 Nature of changes of temperature observed when a thick denture base is processed by raising the water bath temperature continuously to 100 °C.

internal temperature may reach 130 °C or more if the material is thick. This is undesirable for two reasons:
- Monomer has a boiling point only just over 100 °C and any remaining unreacted at this point would boil, generating porosity which would remain in the finished denture. This would both spoil the appearance and reduce the strength.
- The stresses induced in the now-solid polymer as it cooled again might result in severe distortion of the denture.

Accordingly, several steps are taken to avoid these problems. The first is the controlled heating of the flask (Fig. 2.6). As the heating rate of the flask is reduced so the onset of the strongly exothermic 'run away' stage is delayed, allowing more reaction heat to be dissipated so that the peak temperature is reduced. By choosing a suitably low heating rate, the peak temperature can be controlled to just about 100 °C or less, although this depends on the thickness of the section (the thicker, the hotter). Alternatively, heating to only, say, 70 °C will limit the peak temperature, although this has other problems, as we shall see.

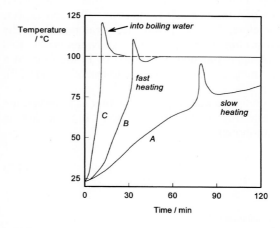

Fig. 2.6 Form of temperature changes observed for various rates of heating of thick denture base at ambient pressure. A: gradually to 70 °C, B: directly to 100 °C, C: flask placed in water at 100 °C.

There is a simple solution to this first problem: pressure. The boiling point of a liquid is, by definition, the temperature at which its vapour pressure is equal to the local ambient pressure: raising the pressure raises the boiling point. This behaviour is described approximately by a form of the **Clausius-Clapeyron equation**:

$$\ln\left(\frac{P_1}{P_2}\right) = \frac{H_{vap}}{R}\left(\frac{1}{T_2} - \frac{1}{T_1}\right) \tag{2.1}$$

where T_1 is the (absolute temperature) boiling point at normal atmospheric pressure, P_1, and T_2 the boiling point at the pressure of interest, P_2. H_{vap} is the heat of vaporization (37.5 kJ/mol), and R the molar gas constant. This is plotted in Fig. 2.7.

Since a denture base mould is normally overfilled somewhat, when this is clamped for processing the internal pressure can be elevated substantially, given that the flow of the very viscous dough through the very small space between the mould halves will be slow. The pressure increases further on heating due to the thermal expansion of the dough, which is over six times greater than that of the mould (Fig. 3§4.14; 18§4.7). Consequently, if the clamping is done properly, even the temperature spike from typically rapid heating will not cause porosity, even in thick sections (Fig. 2.7).[5]

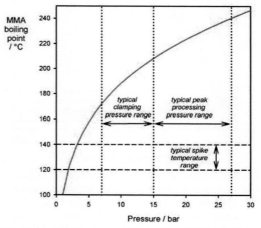

Fig. 2.7 The effect of pressure on the boiling point of methyl methacrylate (MMA).

The clamping also ensures that the denture base remains constrained by the mould, minimizing distortion. Some stress relief occurs by viscous flow while the material remains above the glass transition temperature.

●2.7 Residual monomer

It is important to recognize that polymerization is not complete once the exotherm has passed. It has long been known that there is always residual monomer present in processed acrylic.[6] There are two factors operating here. Firstly, as in any simple chemical reaction between two species, the concentration of each is important. Thus, the concentration of free radicals [R•] and the concentration of monomer [M] need to be considered, the rate of reaction being expected to depend on the product of the two: [R•][M]. But, of course, the value of [M] is falling steadily as reaction occurs, and [R•] also falls because of termination reactions (§1.3). However, the viscosity of the medium is rising as the degree of polymerization increases and simultaneously the plasticizing effect of monomer is decreasing as [M] falls. Hence, the diffusion of monomer becomes the rate-limiting step as this is essential for reaction to occur: the live polymer ends are more or less fixed in location because they have very limited diffusibility. As diffusion is an activated process, it is temperature-dependent (3§3.3), and it takes time for any such reaction to approach completion. We can expect, therefore, at a given temperature, that the concentration of monomer falls asymptotically; to a first approximation this would be in the form of a declining exponential (similar to that of Fig. 11§2.9).

Free-radical polymerization is, however, not a one-way, irreversible process. It is in fact an equilibrium:

$$MMA \underset{k_2}{\overset{k_1}{\rightleftharpoons}} PMMA \qquad (2.2)$$

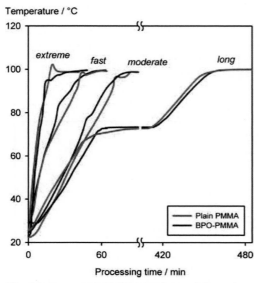

Fig. 2.9 Temperature changes observed for various heating schemes (under pressure), with and without benzoyl peroxide: start at ambient for 'long' and 'moderate'; at 60 °C and 100 °C for 'fast' and 'extreme' respectively; held at 72 °C until 6.5 h in the 'long' cycle.

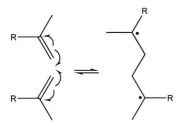

Fig. 2.10 Spontaneous formation of a dimer diradical from methyl methacrylate. This species can initiate polymerization.

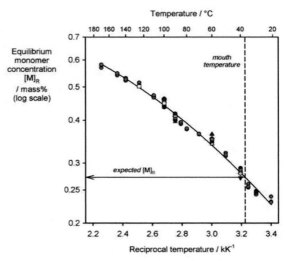

Fig. 2.8 The equilibrium between methyl methacrylate (MMA) and its polymer (PMMA), pressurized to prevent the monomer boiling. Experimental data for the pure system are marked by the filled circles, commercial dental products by the other symbols. There is no difference. The expected value at 37 °C is indicated.

and the equilibrium concentration of monomer is temperature-dependent because the rate constants, k, are for activated processes (Fig. 2.8).[7] One of the consequences of this is that complete polymerization, meaning zero residual monomer, is simply not possible. But since MMA is irritating to the mucosa, and it also affects the mechanical properties of the polymer, there is a need to minimize the amount remaining. Another implication is that if equilibration occurs at high temperature, say 100 °C, the concentration of residual monomer $[M]_R$ will be higher than at mouth temperature (which is expected to be around 0.27%, at equilibrium). If the cooling after processing is fast in comparison with the equilibration time, as ordinarily it will be, "excess" monomer must then be present as the system has been quenched in the high-temperature condition. It will take time for this excess to evaporate, or react, so delay in delivery might be seen as beneficial from this point of view.

The fact that is an equilibrium has a further curious consequence: it will occur anyway in the absence of any initiator: the use of benzoyl peroxide is, strictly speaking, quite unnecessary for processing a denture base, and indeed perfectly acceptable results can be obtained without it, with barely any distinction between the two conditions (Fig. 2.09). It would appear to be an early misunderstanding of the nature of such polymerization that led to the assumption only that it was necessary. That is to say, free radicals do not need to be supplied for polymerization to occur, they will be generated spontaneously. One possibility is the dimer diradical formed from two monomer molecules in another simple equilibrium (Fig. 2.10)[8], evidence for which is found in the form of a cyclobutane.[9][10] (This, incidentally, is why inhibitors are required [§4].) In addition, oxygen (which would be hard to exclude without special effort being made) can itself initiate the polymerization, at least above 100 °C.[11]

The kinetic considerations above are clearly affected by the temperature-dependent end-point. When all of these factors are combined (and leaving the complicated effect of initiator out of consideration) a relatively simple relationship emerges (Fig. 2.11). In effect, there is a time and temperature equivalence: short time at high temperature gives the same outcome as a longer time at a lower temperature, which is intuitively reasonable. However, the end point, as described above, rises slowly with temperature. A section at constant time then has the appearance of the solid line in Fig. 2.12.[12] Accurate determination of actual residual monomer concentrations follows this line except for two aspects: values are lower than predicted, and the decrease continues for a very long time.

These two observations may be accounted for noticing that the processing mould, being gypsum, is porous, and considerable losses of monomer may occur by diffusion out of the polymer. Since the rate of formation of monomer by decomposition of polymer is relatively low, so the actual concentration falls below the equilibrium value. Indeed, this will be true in service at mouth temperature. Even though some small amount of monomer is generated spontaneously all the time, because it is an equilibrium, the diffusive escape into the surroundings in the mouth upsets this, leading to lower values than otherwise expected.

We can now consider what processing time is actually required. It needs only that the minimum groove in Fig. 2.11 be mapped, and this is done in Fig. 2.13. It can be seen that processing at 70 °C, as is sometimes advocated, would require over 300 h to achieve the same effect as 24 h at 100 °C, which is in itself rather longer than is normally used. Given the diffusion losses, however, a nominal overnight cure of ~16 h might be a practical alternative. This can be compared with a common kind of schedule (Fig. 2.14).

A further point to emerge from consideration of these plots is that because the residual monomer is a thermodynamic consequence, the amount is not controlled by the initial mixing ratio, as is often suggested, but rather depends only on sufficient processing time at a suitable temperature – it is the combination that matters. No differences are to be expected between products (Fig. 2.8), nor between classes of product (light-cure, microwave, or cold cure), if time at temperature is adequate .

Figure 2.13 suggests that processing at yet higher temperatures might be advantageous. However, since there is a compromise between processing speed and dimensional accuracy, and thermal contraction becomes increasingly problematic as the temperature is raised, this may result in some practical difficulties. However, as a process it is apparently feasible in principle, using a 'pressure cooker' or autoclave.[13][14] Even so, the times then proposed (10 or 20 min) are seen to be quite inadequate to achieve monomer minimization.

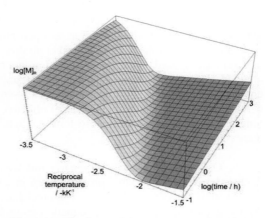

Fig. 2.11 The form of the response surface for the concentration of monomer [M] in the processing of PMMA. The reciprocal temperature axis is reversed to put high temperature to the right.

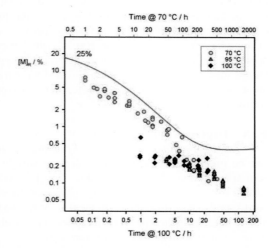

Fig. 2.12 Effect of processing time at three temperatures on actual residual monomer concentration (literature values, many sources) compared with the theoretical curve (solid line) for an initial monomer content of 25 mass%.

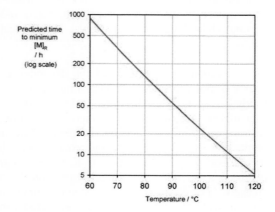

Fig. 2.13 Time to reach the theoretical minimum $[M]_R$ as a function of processing temperature.

In any case, it is extremely wasteful to try to attain 100 °C literally by boiling water in an open processing bath: a great deal of energy is required to do this. Should a thermostat be used, the set-point would in fact never be attained. Looking again at Fig. 2.13, it is apparent that a compromise may lie in using a thermostatically-controlled bath at 95 °C without significant loss of effectiveness, so long as the processing time were long enough, that is, increased to compensate.

On the other hand it is clearly not feasible for denture bases to be prepared properly even in a boiling-water bath if this is to be done at any appreciable altitude. Applying the Clausius-Clapeyron equation again, this time to the boiling point of water (Fig. 2.15), it is apparent that many populated places have this difficulty: even 95 °C could not be attained at altitudes greater than 1500 m. This might be circumvented by using a pressure vessel (as indicated above), but this has risks. A dry-heat oven system may provide a solution, but noting the thermal expansion problem identified above, this perhaps should be at no higher temperature than 100 °C, which would be another compromise. The decomposition of the plaster mould may also be a problem at elevated temperatures, and especially in an autoclave at the temperatures used for sterilization (2§10.2).

● **2.8   Degree of polymerization**
The strength and toughness of polymers in general can be greatly affected by the molecular weight[15] or, equivalently, the chain length (Fig. 2.16),[16] where a certain critical length is necessary. This can be seen to correspond to the critical entanglement point shown in Fig. 3§3.1, suggesting that the work of reptation in the absence of such entanglements is actually very low.

A low degree of polymerization can occur in two ways: too many live chains and insufficient reaction. The latter is not really a problem in practice in that in excess of 99% conversion is easily achieved (Fig. 2.12), although residual monomer acts as a plasticizer, facilitating chain drawing and this lowers both strength and stiffness. The main reason is that if too much initiator is used, the average chain length must be low. This is dealt with below (§3.3).

Another consequence of the dependence of the degree of polymerization on temperature is found in the rigidity of sections of differing thickness. Because the heat of reaction cannot be conducted away from thick sections so efficiently, locally the temperature may be higher than in resin elsewhere and remain higher than the mould for longer, resulting in a higher degree of polymerization and concomitant mechanical property changes. The converse is actually more important: thin sections will be weaker than would be expected from thick section measurements, for current typical processing regimes.

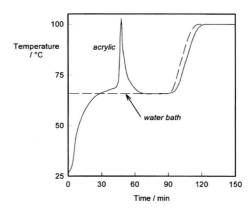

Fig. 2.14 A typical heating cycle intended to minimize overheating and maximize degree of polymerization. The low temperature stage is quite unnecessary, and the high temperature stage far too short.

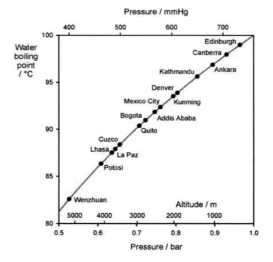

Fig. 2.15 Boiling point of water at altitude. Many places could not process a denture base well enough by boiling for a typical time of a few hours.

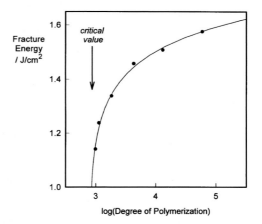

Fig. 2.16 Variation in fracture energy (toughness) of a typical polymer as a function of molecular chain length (*cf.* the critical chain length in Fig. 3§3.1).

●2.9   Trommsdorff effect

The 'run-away' reaction noted earlier (§2.6), whose effect is seen very clearly in Figs 2.5, 2.6 and 2.14, remains to be explained. Certainly, as the temperature rises so the rate of generation of free-radicals would be expected to increase, and so the rate of polymerization. However, the increase in the rate of heat-generation is far greater than is accounted for by this.

The rate of polymerization certainly depends on the concentration of free radicals, and can be written as proportional to the product of this and that of the monomer: $[R\bullet][M]$. At the same time, the rate of termination (§1.3) is ordinarily proportional to $[R\bullet]^2$. But, the diffusivity of the live chain end decreases as the chain length increases and also as the viscosity of the surroundings increases (although, as noted earlier, its reactivity does not change). That is, the rate of mutual annihilation of live chains drops rapidly when they can no longer diffuse to collide, and as noted at §2.7 the diffusion of monomer becomes the rate-limiting step. However, at this point there is still so much monomer present that it is the severe curtailment of the mutual termination reaction of live chains, while new radicals continue to form, that leads to an overall increase in the rate of polymerization. As seen earlier (3§3.1), the increase in the rate of change of viscosity with molecular weight can be quite abrupt, and in a polymerizing system the time of the change is referred to as the **gel point**, hence this **auto-acceleration**[17] is also called the **gel effect**, but commonly it is named after one of its discoverers as the **Trommsdorff effect**.[18] Quantitatively, if the termination rate falls by a factor of 4, the overall polymerization rate doubles. In addition, the average chain length is increased markedly because active chains now live longer.

In some systems, and methyl methacrylate is one, the effect is particularly dramatic when the quantities are large. Slight variations in viscosity in industrial contexts can lead to marked variation in temperature from place to place. Bulk polymerizations may catch fire, or even explode, they can get so hot, but boiling is often seen even on a scale of just a few millilitres. The larger the quantity, the greater the care required to dissipate heat. This is certainly a reason for the use of a mixture of powder and liquid for denture bases (§2.5).

Subsequently, the rate of polymerization drops dramatically when the mobility of the monomer itself becomes limited. This generally occurs a little after the material has become glassy (that is, the glass transition temperature has reached the current ambient temperature of the mass), even though substantial amounts of monomer remain. This effect is known as **auto-deceleration**.

●2.10   Equilibria

The essential equilibrium between MMA and PMMA was stressed at equation 2.2, and it is of value to explore some ramifications. The equilibrium constant K is defined in the conventional way as the ratio of the concentrations of reaction product to reactant, but also inversely as the ratio of the rate constants, roughly:

$$K = \frac{k_1}{k_2} = \frac{[PMMA]}{[MMA]} \tag{2.3}$$

However, the two ks have different temperature dependencies because the activation energies for the forward and reverse processes are different (the diffusion and collision requirements for MMA clearly do not apply to the polymer), hence the value of K must also be temperature-dependant. In fact, the 'unzipping' process is seen to become relatively more important as the temperature rises (hence Fig. 2.8), even though the rate of polymerization is increasing rapidly (Fig. 2.11). The part played by bulk conversion rates can be seen by rearranging equation 2.3:

$$k_1[MMA] = k_2[PMMA] \tag{2.4}$$

emphasizing the fact that if the equilibrium is disturbed by there being a high concentration of monomer, the overall conversion rate to restore that equilibrium is high (**Le Châtelier's principle**), but the rate slows dramatically as equilibrium is approached (curve of Fig. 2.11 – note the logarithmic axis scales). Even so, one must be careful to keep distinct the quite general ideas of the **thermodynamic imperative** – where the system would end up if given the opportunity, and the **kinetic reality** that the rates of processes limit progress towards equilibrium, not only in the sense that it takes time to equilibrate, but also that equilibrium may never be achieved. There is therefore a fundamental need for compromise in the practical context of many systems.

§3. Cold-cured Resins

Although the heat-cured resin provides a generally acceptable product, the process may not be appropriate in all circumstances. Such a cycle of heat and pressure could not be applied in the mouth for a direct restoration, and denture-base repair in this way would damage the original device (by stress relief on heating) without resorting to an extremely elaborate process more expensive and time-consuming than preparing a new denture. An effective method of processing at low temperature is required, but this is only partially successful.

The method depends on the reaction of a tertiary amine with benzoyl peroxide to generate free radicals at ordinary temperatures, hence the commonly-used term **cold-cure acrylic**. This amine is present already dissolved in the liquid component of these products. Liquid ("monomer") sold for cold-cure products must therefore not be used for heat-cure acrylic, and *vice versa*. However, no important chemical differences are expected between the two product powders.

●3.1   Toxicity

As has already been pointed out, benzoyl peroxide does not dissociate at a useful rate at room temperature. However, the reaction can be promoted without heating. Tertiary amines (Fig. 3.1) are active in this respect. The choice available for dental use seems to be somewhat restricted. The simplest tertiary amine, trimethylamine, is too volatile for use (b.p. ~3 °C), while dimethyl aniline is a severe poison. All tertiary amines are toxic and care in handling them and formulations containing them is necessary. N,N-dimethyl-paratoluidine combines the required chemical properties with perhaps a lesser degree of overt toxicity, although commercial expedience (price!) is likely to be dominant. Nevertheless, the presence of such substances in 'cold-cure' monomer liquid is reason enough for all contact with the skin to be avoided. Not only will the degreasing effect of the monomer increase the risk of infections and dermatitis that occurs with all such organic solvents, heat-cure monomer included, but the amines will be detrimental through skin absorption. The problem here, as with many of the special chemistries now being used for dental materials of many kinds, is that the desired high reactivity of some components is not selective. These substances will also react readily with living tissue. Considerable care and attention is required to avoid exposure during one's working life.

●3.2   Free radical generation

The setting reaction of cold-cure acrylic proceeds by the amine first forming a salt with the benzoyl peroxide; this disproportionates readily at room temperature to generate a single benzoate (benzoyloxy) radical (Fig. 3.2). This is then available to initiate a polymerization chain as before with the heat-cured materials. The amine is properly called an **activator** in these systems, as its only role is to make available the polymerization initiator – the free radical.

As one would expect, the polymers produced by these methods are weaker and less rigid than heat-cured resins because the degree of polymerization is always much lower. There is also a considerable proportion of unreacted monomer dissolved in the polymer, and the plasticizing effect of this may be considerable.

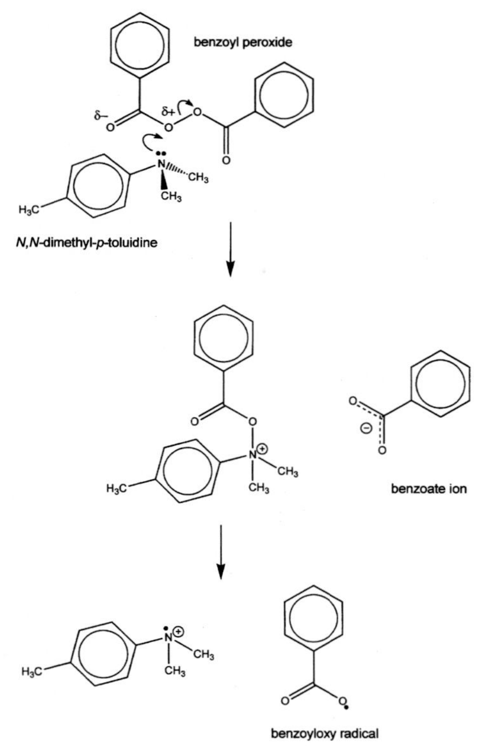

Fig. 3.1 Some tertiary amines which could serve as cold-cure activators.

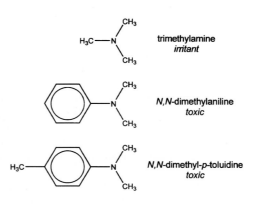

Fig. 3.2 The activation of benzoyl peroxide by a tertiary amine.

●3.3   Chain length

As has been stressed already (§2.7), the molecular weight or chain length of a polymer controls the mechanical properties. Thus the number of actively growing chains in a reaction mixture would seem to need to be minimized to ensure as high a chain length as possible. But chain termination and chain transfer reactions tend to limit the chain length. In any case, the polymerization would then proceed so slowly as to be of little use, leaving large quantities of monomer unreacted. As it is, it is difficult to approach 100% reaction of monomer even under the most favourable conditions (§2.7). Conversely, increasing the amount of initiator, and so the number of free radicals originally created, would increase the initial number of active chains but reduce their mean length at the end of reaction. Also, the reaction rate and heat evolved would be so great as to seriously damage the prosthesis and certainly produce much porosity. A compromise has to be found between overall reaction rate and chain length in order to develop adequate properties in a suitable time.

The chain length issue can be illustrated as follows. Suppose that there are 100 monomer molecules, and that terminations and side reactions are ignored. Clearly for one initiating free radical only one chain could form, 100 units long. But with two initiating species, the average chain length must already only be 50 units. With 4, the average is 25, and so on: it is a reciprocal relationship (Fig. 3.3). Furthermore, as the concentration of active free radicals increases, whether initiating species or live chain ends, the rate of mutual annihilation must also increase, so that there would be a disproportionately large number of very short chains, even though joining two chains together in this way increases chain length. Such reactions reduce the free radical concentration, so that the remaining radicals successively produce relatively longer chains. Even so, the number of small chains adversely affects the mechanical properties irreversibly.

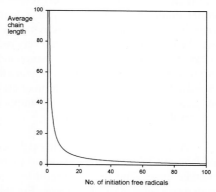

Fig. 3.3 Effect of number of initiating free radicals on the average chain length of the resultant polymer chains.

Because of the need to chemically-activate the dissociation of the benzoyl peroxide in cold-cure acrylics, the onset of the temperature rise due to the exothermic polymerization reaction, and thus that of the Trommsdorff effect (§2.9), is faster than with heat-cured resins (Fig. 3.4). The temperature does not rise as high on setting as with the heat-cure process (at least, not if the mass and thickness of the resin is limited). In addition, there is no extended high-temperature curing stage for completing the reaction. Cold-cure resins thus always have much unreacted monomer and are therefore weaker than heat-cured material because of both the plasticizing effect and shorter average chain length. This difference between the cold-cured matrix and the emulsion-polymerized beads means that such acrylic is not as clear as heat-cured material: there is a sufficient difference in the refractive index to cause scattering (3§4.1). (Indeed, even heat-cured acrylic is often not completely clear and transparent because there remains a sufficient difference between the matrix and bead materials, having been made under different conditions.)

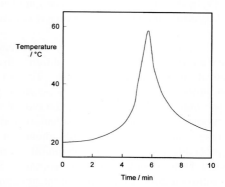

Fig. 3.4 A typical plot of temperature against time for a cold-cure acrylic resin.

Even though denture-base repair is done under pressure (typically 2 bar gauge, that is, over atmospheric, 3 bar absolute) to limit the porosity, there is a tendency for some to remain. But recognizing that there is a problem with extent of reaction, this is commonly done at temperatures around 45 ~ 50 °C. Obviously, if the temperature were any higher, there is the risk of the exotherm exceeding the boiling point of the monomer (Fig. 2.6) in thick sections, but clearly a very long time would still be required for equilibration (Fig. 2.13). In addition, higher temperatures would allow the denture base to warp through the relaxation of frozen stress.

●3.4   Combined approach

'Cold-cure' chemistry provides an additional approach to dealing with the problem of the 'heat spike' of normal heat-cured acrylic. Some products have included a small amount of amine activator in the monomer liquid so that the polymerization starts on mixing. This is intended to dissipate some of the heat of reaction before the water-bath heating process is started, although it clearly relies on the termination rate being high enough that this polymerization is only partial. However, some issues need to be taken into account. Firstly, early polymerization will increase the rubbery nature of the dough at the packing stage, possibly making it more difficult and prone to error. Secondly, the presence of the amine in the monomer liquid increases the toxic hazard of handling the dough. This does not affect the time required for residual monomer minimization.

§4. Inhibitors

Free radicals from benzoyl peroxide are used to initiate methyl methacrylate polymerization, but oxygen can provide such radicals, especially if certain transition metal impurities are present. Ultra-violet light may also open the double bond and so provide a double free radical (see 24§6.2). Background ionizing radiation (such as cosmic rays and natural radioactivity; see 26§5.4) generates many ions and radicals that can be active in this sense. Molecular oxygen is in fact a very reactive material as it has two unpaired electrons. It is a free diradical and can react with many organic compounds, including monomer, to produce peroxides which readily break down to generate more radicals; these are then active in initiating polymer chains (*cf.* 6§6). However, as has been seen, the spontaneous formation of the dimer diradical (§2.7) means that even in the purest, best-protected storage, methyl methacrylate (and by implication, any such vinyl monomer) must inevitably polymerize sooner or later if there is no intervention.

Any of these processes, although occurring at low rates, would be sufficient to make storage of methyl methacrylate monomer impossible for any useful length of time without it setting into a solid mass of polymer. For this reason **stabilizers** or **inhibitors** must be added to monomer (Fig. 4.1). The one most commonly used is hydroquinone, although any one of a variety of chemically-similar compounds may be used. As it is, methyl methacrylate must always be stored in the dark, or at least in UV-absorbing glass (dark brown is common).

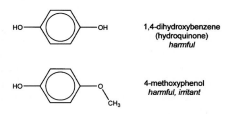

Fig. 4.1 Some possible inhibitors of free radical polymerization.

Polymerization inhibitors work because they have a very high probability of reaction with any free radicals with which they collide. To be effective inhibitors, this reaction must necessarily proceed much more readily than that of the radical with the monomer itself. The inhibitor radical so produced, of course, must not react with monomer at an appreciable rate. In any case, an inhibitor can only be useful if it is effective at low concentration. Since by definition it does not take part in the polymerization process, and must behave as a plasticizer (a small, dissolved molecule) when finally the monomer is processed, high concentrations would not be acceptable.

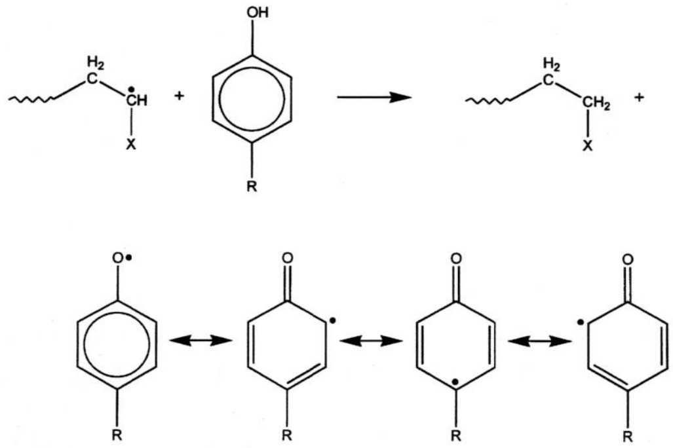

Fig. 4.2 Reaction of a polymerization inhibitor with a free radical, showing possible resonance structures.

At room temperature, a methacrylate free radical may undergo many collisions before it achieves the right orientation and activation energy to react with another methacrylate double bond. Hydroquinone and other phenols can very readily lose the phenolic hydrogen to a free radical (Fig. 4.2). If this radical happens to be a polymer chain which has just started to grow, the chain is terminated. The free radical produced by this action does not itself initiate the formation of a new polymer chain because the unpaired electron is delocalized. That is, resonance structures of the kind shown in Fig. 4.2 'spread' the electron over several atoms, making the effective 'concentration' lower. This situation exists because it represents a lower energy state and is therefore more stable.

Despite the name 'inhibitor', such substances must be seen simply as **competitors**. They have no effect on the activation energy or the probability of a favourable collision, and so they cannot and do not exert any effect on the rate of the undesired reaction, they simply compete for the active species, the free radical. They can only work by preferential reaction, that preference being by a very large factor because, by necessity, they should only be present at low concentration.

As a consequence of the lower energy of this free radical, it is incapable of opening the acrylate double bond and so initiating polymerization itself. Even so, it can eventually react with another radical of its own or another species, so removing both from the system entirely (see Termination, §1.3).

Naturally, if a sample of monomer contains inhibitors, these must consume benzoate radicals as they are formed in the normal course of the activated breakdown of benzoyl peroxide until the inhibitor itself is used up. Then the benzoate radicals can initiate polymerization chains. The same applies to the live radical of growing chains, should any benzoate radicals escape and actually initiate a chain reaction.

●4.1   Cold-cure liquid

Curiously, aromatic tertiary amines, as used for activating benzoyl peroxide in cold-cure acrylics, are themselves inhibitors because they can act as free radical sinks (Fig. 4.3). These are stabilized in a similar way to the phenolic inhibitors, by resonance and delocalization. Obviously, there is a conflict between the production of radicals and their consumption by the same compound. However, the reaction rate for the former process is much higher than the latter, and the polymerization is initiated normally. Even so, the presence of these amines in cold-cure monomer would enhance the stability of the liquid and may permit the omission of a phenolic inhibitor. Certainly, too much amine added to the mixture will prevent the polymerization.

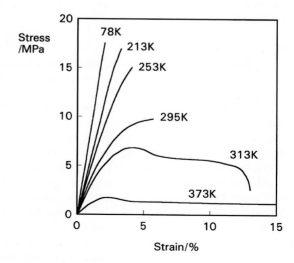

Fig. 4.3 Aromatic amines can also function as polymerization inhibitors as well as activators.

§5. Cracks & Crazes

Poly(methyl methacrylate) has a glass transition temperature (Fig. 5.1) of about 72 °C as detected, say, by dilatometry.[19] This marks the region of transition between glassy and rubbery behaviour. But there is still considerable variation in mechanical properties with changes in temperature apart from this (Fig. 5.2), as has been discussed already (3§5). The glass transition is where the temperature sensitivity of such changes is very high. The temperature has to be as low as 78 K (– 195 °C) before completely brittle behaviour is observed, *i.e.* for the polymer to behave as a hard glass. As the temperature is raised, so more and more plastic deformation is observed until at 40 °C it is really quite ductile. Notice, though, that at 100 °C the behaviour is quite different, enormous extension occurring before failure.

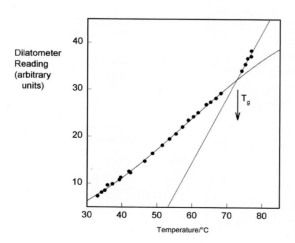

Fig. 5.1 The thermal expansion of PMMA. The T_g here is about 72°C.

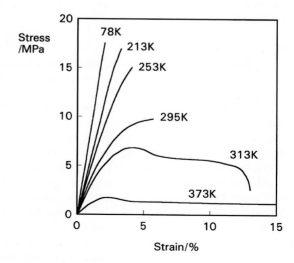

Fig. 5.2 The effect of temperature on the stress-strain behaviour of PMMA. The plot for 373 K extends to 180%.

Despite the considerable viscoelastic component of this behaviour at room or body temperature, PMMA still follows the Griffith criterion (1§7) in respect of cracks at a high enough strain rate. That is, for a crack of given size (Fig. 5.3), there is a stress at which the crack will grow to failure spontaneously (Fig. 5.4).[9] The solid line is the expected relationship of the stress to cause crack growth,

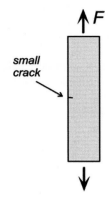

Fig. 5.3 A tensile test to check the applicability of the Griffith criterion to PMMA.

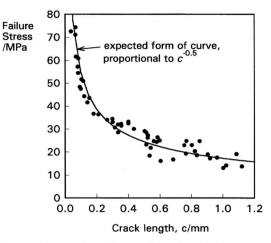

Fig. 5.4 The results of the test shown in Fig. 5.3.

being proportional to the reciprocal of the square root of crack length. The experimental data fits quite well. The cracking of polymers in general, however, is not quite so straightforward as would appear from this.

●5.1 Crazes

Cracking in polymers is in fact normally preceded by **crazing**. A **craze** consists of a region of polymer which has undergone localized drawing, where the polymer chains are aligned and drawn out from the adjacent unaltered material into microscopic fibrils. Crazes usually originate at a free surface, from where they grow in two directions: they propagate outwards and downwards, but always transverse to the principal, axial, tensile stress (Fig. 5.5).[20] This behaviour makes it easy to identify the direction of the tensile stress vector that caused the crazing at any point, even if that stress has been removed – crazes will not undo themselves. This arrangement can be seen to follow necessarily from their structure. Clearly, they cannot form in response to a local compressive stress.

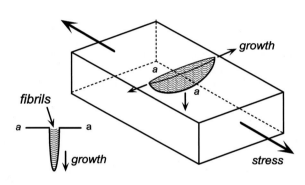

Fig. 5.5 The geometry of a polymer craze.

●5.2 Generation

But, to go back a step to before the craze has formed, we can consider the process in more detail. Plastic deformation in polymers depends on chain drawing (Fig. 3§4.10). As the drawing of polymer chains proceeds under the influence of the applied (axial) tensile stress, the trend would be for the polymer to occupy a larger volume (elongation with constant cross-section – ignoring Poisson contraction for the moment). But this increase in volume would be opposed by the intermolecular forces. If the piece is small enough in cross-section, it can be seen from Fig. 3§4.10 that the elongation must in fact be accompanied by large lateral contraction to preserve the volume (assuming that the volume occupied by the polymer is more or less constant). However, the surrounding of unchanged polymer in larger sections tends to prevent this bulk lateral contraction. Thus, lateral tensile stresses are generated in the drawing region, that

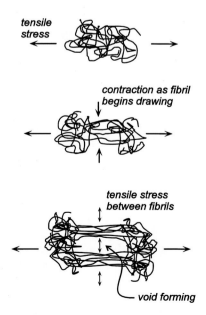

Fig. 5.6 The formation of a void in the process of chain drawing is a spontaneous result of weak interchain forces.

is, acting in the plane perpendicular to the draw direction. These stresses tend therefore to separate the now more-aligned polymer chains (Fig. 5.6). Overall, then, fibril formation is at constant volume for the solid polymer. It is the void volume that increases during crazing.

The weakness of the intermolecular bonding, that is, in the directions normal to the drawing axis, is easily seen in a common material: polypropylene 'string'. This string is in fact a long strip of very thin film that has been produced by melt-spinning – the drawing process is performed on the hot polymer to achieve an extremely high level of orientation in the chains. This means that the strength and stiffness in tension along its length are both high, as is appropriate. But if the crumpled sheet is flattened out, it will be found that it can readily be separated into narrow strips, which can with care be subdivided indefinitely (Fig. 5.7). In effect, a length of such string is one monster craze!

Alternatively, consider the outermost polymer chain in the neck (Fig. 3§4.10): this cannot be straight, but the applied axial tensile force is tending to straighten it, therefore its central region is being pulled away from the midline of the neck. Since typically it is only weak, van der Waals forces acting between chains to hold them together, these generated lateral tensile stresses may become large enough readily to pull the chains apart from their neighbours.

As this process continues, separate bundles of polymer chains – called **fibrils** – form. Each fibril grows as if drawn out of the surface of the craze – the interface between the unchanged and the aligned material – at both ends (Fig. 5.8). Such a process is usually initiated by a flaw in the surface of the piece but, as the fibrils form and separate, the stresses act more on adjacent regions and the drawing process and fibril formation continues in an increasingly large volume. This zone has a characteristic half lens-like shape (Fig. 5.5) and this is the **craze**. It is important to recognize that a craze is *not* a crack because it contains solid material connected to both sides of the zone, albeit in a finely fibrillar form.

●5.3 Cracking

During the drawing process, each fibril is carrying a share of the applied load. As the crazing continues, the distance between the walls of the craze increases, which means that the fibrils continue to be drawn out of those walls until they can be drawn no more. If then a fibril fractures, the load is thrown more on adjacent fibrils, which may in turn fracture if their strength has been exceeded (Fig. 5.9). Remember that the drawing process is time-dependent – *i.e.* strain rate-sensitive – and with a sudden application of an extra load a fibril will not draw out further material because that material will behave in a brittle, not ductile, manner. There would then be a rapid failure of many fibrils, and the craze would be converted to a

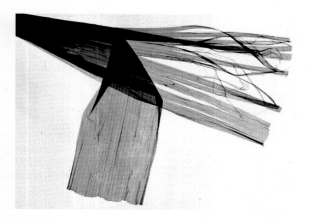

Fig. 5.7 Highly-drawn polypropylene film can easily be separated into fine strands because interchain forces are weak.

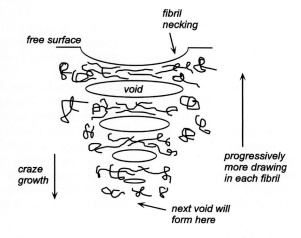

Fig. 5.8 As drawing proceeds, the craze grows deeper into the body by successively opening new voids.

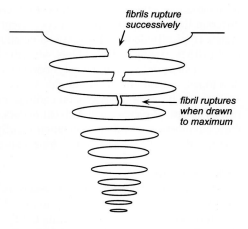

Fig. 5.9 Cracking begins as fibrils begin to rupture, having been drawn to their maximum extent for the strain rate prevailing.

true crack. A real crack in a polymer showing ductile behaviour will therefore only be propagated through a craze which has already been formed.

Craze breakdown can therefore be a gradual process, but even so, when it has proceeded sufficiently far the point may be reached when catastrophic failure occurs. Increasing actual stress will increase the local strain rate, causing more glassy behaviour to emerge, *i.e.* a brittle failure may follow when a critical crack size is attained and crazing is no longer possible.

Notice that the stress required to draw the polymer and thus generate crazes is less than the ultimate strength of the material. Here it is again necessary to notice the distinction between yield point and the stress at collapse.

> Crazing in polymers must be carefully distinguished from the phenomenon of the same name in brittle materials such as porcelain and ceramic ware glazes where there is formed a network of true cracks, without any bridging material, although macroscopically they may appear very similar.

●5.4   Optical properties

Because of the separation of the fibrils, there is a considerable amount of void space in the craze regions, and this accounts for a number of features. The first is the altered refractive index because the material is much less dense – the voids created between the fibrils would be filled by air (if the free surface of the polymer piece were in air). Crazes are thus very readily detectable by eye (even if of minute size) if the lighting is at the right angle, because of the refraction and scattering which can be caused. Visibility is also enhanced by the diffraction occurring in the array of very fine parallel fibres, which cross from one wall to the other and which therefore resemble a diffraction grating. Spectral ('rainbow') colours may then be seen.

Crazes need not be so large as to be visible individually by eye. Many polymers craze very readily, such as on bending a piece of polystyrene (not foam). The region that has suffered the crazing becomes visible as it turns white – **stress whitening** – because the many small crazes scatter incident light. Such an effect is used to advantage in what are called "self-hinges", particularly in objects such as polypropylene boxes that are moulded in one piece, lid and base. On first bending, the whole hinge region is crazed, becoming much less stiff as a result, and the stress whitening can be seen as marking out the affected region. Of course, repeated flexing will accumulate further damage and eventually cause fracture; such hinges do not last indefinitely. Nevertheless, it is a good illustration of the mechanical effects of such localized drawing, and an interesting application of deliberate damage in reducing the costs of fabrication and assembly.

●5.5   Processing stresses

Because of the unavoidable inhomogeneity of the heating of a denture base during processing, there will always be miscellaneous residual stresses after cooling due to the variation of polymerization shrinkage from point to point. In addition, as the material cannot be considered homogeneous in terms of chain length, stresses will also appear on cooling from variation in thermal contraction. Furthermore, PMMA has a markedly different linear thermal expansion coefficient (*c.* 70 MK^{-1}) compared with metals (typically about 30 MK^{-1}) and ceramics such as denture porcelain teeth (10 MK^{-1}) (Fig. 3§4.14). Thus, when the acrylic carries metallic inserts such as clasps or, more obviously, porcelain teeth, large stresses may be developed around the insert at the surface. Crazes may then appear spontaneously in these regions on cooling, but even if not, they may still readily appear if contact with solvent occurs. They may be recognized in these circumstances by their orientation, chiefly in a radial pattern around the insert, as the tensile stress acts in a circle, a so-called **hoop stress** (Fig. 5.10).

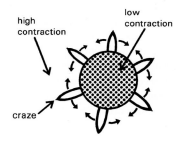

Fig. 5.10 Hoop stresses give radial crazing, and cracks if the stress is high enough.

●5.6   Solvents

A matter of considerable interest is the facilitation of crazing by solvents applied to the outer surface of the polymer. There are two aspects to this. Firstly, any solvent will tend to dissolve in the solid polymer and therefore act as a plasticizer, lowering the T_g. This then acts to lubricate the relative movement of the polymer chains, facilitating their alignment during the drawing process, lowering the work that needs to be done in this sense and therefore making crazing easier. Secondly, since the formation of the fibrils means the creation of a great deal of new surface area (many times that of the two 'faces' of the craze), work also needs to be done to make that new surface (1§7). The surface energy of a body depends on the medium with which it is in contact

(10§1) – the so-called interfacial energy is what we are really concerned with, and this is necessarily much lower for a solvent which wets the polymer than against air. The energy required for the formation of the craze fibrils will be much lower if solvent is available to be drawn into the void as it is being formed. The presence of a solvent thus facilitates the direct formation of a craze.

Thus, should there be for any reason stresses present in the surface of a polymer body, and the polymer is capable of being drawn (if it is not too highly cross-linked, for example), the application of a solvent (even a very weak one such as ethanol) may promote almost instantaneous crazing. This situation arises for PMMA in dentistry. On the other hand, if the polymer is sufficiently cross-linked, chain drawing is inhibited and no crazes develop. The material is then stronger, and when failure occurs it is with the immediate formation of true cracks. This behaviour can also emerge in some polymers as **environmental stress crazing** (or even cracking) where a body is under stress and some component of the environment (such as water) facilitates the slow development of crazes when that service stress would not otherwise do so. This effect is due to the nucleation of the craze being made easier.

Crazes in dentures should be avoided if at all possible as they are sufficient flaws in the surface to initiate cracking and gross failure or, at the very least, discolouration as debris accumulates in the voids between fibrils. Unfortunately, they are very readily triggered by methyl methacrylate which, because it is chemically nearly indistinguishable from its polymer, has a very nearly zero interfacial energy with it. The work of formation of the craze is therefore very small indeed. The repair of a denture base is thus a risky proposition if it has metallic or ceramic inserts or if it was badly processed.

This kind of behaviour has another implication of much wider concern for dental materials (in fact, as a matter of principle for any context). The mechanical testing of polymeric materials must be undertaken in an environment resembling that of actual service conditions: that is, in this case, wetted by a medium approximating human saliva, and at the correct temperature, in the present case. Misleading results of little or no value to understanding the actual application are otherwise obtained. (Indeed the remark also applies to all other classes of material.)

§6. Modifications

That the glass transition temperature is concerned with a real change in the mobility of polymer chains, and thus the internal space available for their movement (3§4), is illustrated by the water absorption curve for PMMA (Fig. 6.1). The equilibrium water absorption changes only very slowly up to the glass transition, but above that point there is a marked increase. But, as remarked previously, even the small amount absorbed at mouth temperature is enough to alter the mechanical properties, emphasizing again the need for realistic test conditions. Equilibration itself, however, may take many days (about 3 weeks for an ordinary acrylic denture[21]), depending amongst other things on the square of the thickness of the piece immersed, because it depends on the diffusion of the water through the solid polymer, so that time of testing becomes an important variable.[7]

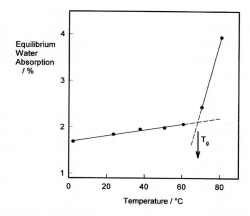

Fig. 6.1 Water absorption against temperature for PMMA. The T_g here is seen to be about 68 °C, lowered slightly by the effect of the water itself (*cf*. Fig. 5.1).

It should also be borne in mind that absorption of any solvent must also result in swelling. Whilst it is true that there is increased free volume in an amorphous polymer compared with the (often theoretical) crystalline condition, this does not mean that solvent molecules can just fit into vacant places. Diffusion into amorphous polymers is certainly easier, but it still requires chain-segment diffusion to allow space to occur (it is an activated process still). But on average, when solvent is present the interchain forces are weakened, allowing both more solvent in and greater separation of the chains: the bulk volume must increase, and roughly directly in line with the volume of solvent absorbed. In other words, there are dimensional changes in the body that must be taken into account.

●**6.1   Softening**

The absorption of water by a resin can be deliberately increased by using a monomer with hydrophilic substituents. For example, the use of hydroxyethyl methacrylate (Fig. 6.2) (commonly abbreviated as 'HEMA') gives a polymer with pronounced water absorption, up to 20% by mass, compared with the roughly 2% of normal PMMA. This makes the material very soft and flexible, a property enhanced by the use of the ethyl as opposed to the methyl ester, the longer chain giving an extra pronounced drop in the glass transition temperature (3§4). While not very strong, these materials are useful as soft liners for denture bases as well as for items such as contact lenses.

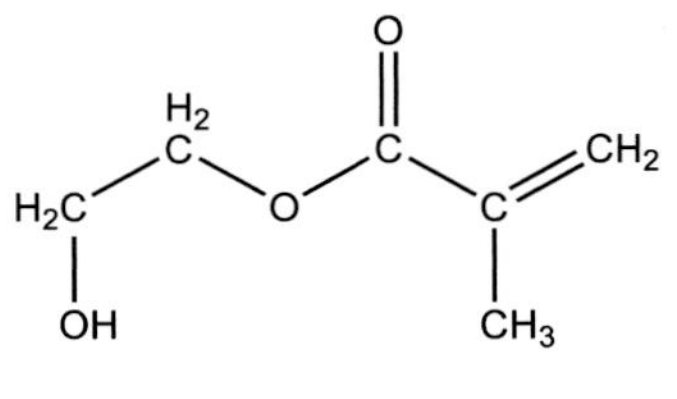

Ethanol is also used as a plasticizer for poly(ethyl methacrylate) soft lining materials but this, of course, is volatile as well as soluble in saliva and so readily leaches out, giving a relatively short lifetime for the original properties. There is also a tendency now to avoid the uncritical use of ethanol in pharmaceuticals and cosmetics, and the prolonged exposure of mucosa to an ethanol-plasticized soft liner may soon be deemed inappropriate.

Other plasticizers that may be used are intended for longer term use and accordingly are not so volatile. Dibutyl phthalate (Fig. 6.3) used to be one of the commonest, but in principle any long chain ester is usable. These compounds, because they are not reacted with the polymer, are still slowly leachable. This property also raises concerns over long-term exposure to substances whose pharmacology and toxicology are not fully understood. Phthalates generally, however, now are believed to be a distinct hazard in several respects and are deprecated. Substances such as acetyl tri-alkyl citrates (Fig. 6.4) appear to be much more benign[22] and less soluble when R is reasonably large (butyl: 5 mg/L), when the boiling point is also high (butyl: ~330 °C). The products of hydrolysis are also thought to be benign.

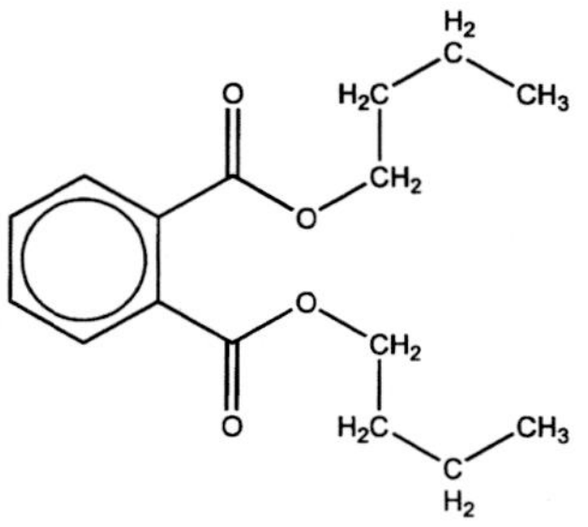

The use of reactive monomers such as octyl methacrylate in small proportions is also effective, and of course such additives are not then leachable. The alternative approach of incorporating more flexible links in the polymer chain, by omitting side groups, is sometimes used to advantage in the form of copolymers of methacrylates with unsubstituted vinyl monomers.

●**6.2   Hardening**

The opposite requirement of producing a resin with greater rigidity, as opposed to flexibility, can also be satisfied by designing the polymer chain appropriately. For example, this can be done by the incorporation of di- or tri-functional monomers (Fig. 6.5) which lead to branching of the polymer chain. Thus, every so often another, independent chain can grow out from the original backbone chain, behaving quite independently and normally. This results in greater entanglement and more difficult relative motion of chains under stress. Carried further, a higher concentration of branch points leads effectively to cross-linking between chains, especially if the ends of branches can react together to close loops or the second functional group is attacked by a growing chain (which is more probable than being attacked by an initiator radical). This results in all polymer chains becoming part of a single, giant three-dimensional network. The behaviour of the polymer will therefore depend on the number of branch points introduced, which will affect the length of the **internodes**.

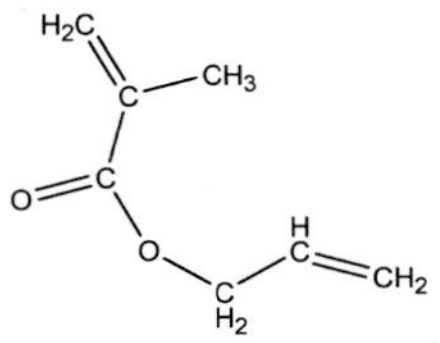

Such changes modify the mechanical properties still further, making the susceptibility to the formation of crazes less. This is the result of the much greater difficulty of the drawing process to form fibrils of aligned chains when covalent cross-links must be broken (high energy σ-bonds) to allow relative chain movement.

hydroxyethyl methacrylate

Fig. 6.2 An example of an hydrophilic monomer for use in PMMA – hydroxyethyl methacrylate. The hydroxyl group is sufficiently polar (and forms hydrogen-bonds) for this to be a strong effect.

dibutyl phthalate

Fig. 6.3 The flexible side chains allow dibutyl phthalate to work as a good plasticizer in many polymer systems.

allyl methacrylate

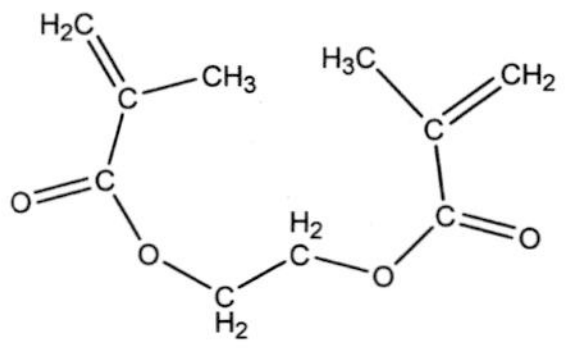

glycol dimethacrylate

Fig. 6.4 Two acrylic cross-linking agents; each has two vinyl groups for independent polymerization.

However, this can be taken too far and extreme amounts of cross-linking lead to extreme brittleness. Even so, because of its dependence on the breaking of covalent bonds, the ultimate strength of such materials is little improved at temperatures much below the glass transition.

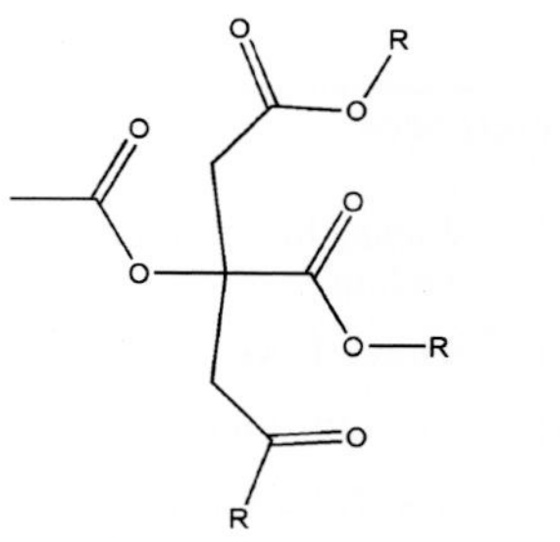

●6.3   Water

Because denture bases are commonly prepared in a water bath over an extended period, it can be expected that the PMMA is then at least partially saturated with water, saturation that will go to completion when in service in the mouth, even for products that are cold-cured (§3), microwave-processed or light-cured. The swelling that occurs (typically of the order of +1%, linear) depends on diffusion through the solid and, depending on the material thickness, can take a few weeks to come to equilibrium. Accordingly, as the outer layers suffer greater absorption initially, stresses will be generated, acting in various directions (*cf.* 6§2.2). These will reduce as the process goes to completion. However, should the denture base now be allowed to dry, the reverse will occur: the outer layers

Fig. 6.5 Acetyl tri-alkyl citrates may be safer alternatives as plasticizers in denture base acrylic. R may be ethyl, butyl, hexyl, and so on.

will tend to shrink over the still-swollen core. While these stresses may not be enough to cause crazing, because the material is viscoelastic there will be a tendency to stress-relaxation through some chain drawing, that is to say, distortion (warping) will occur. Since such processes are time-dependent, essentially irreversible, and thus cumulative, repeated drying-wetting cycles may lead to sufficient distortion as to cause problems of lack of fit.

It is, therefore, important that once a denture base has been soaked it is necessary to keep it so: during finishing in the laboratory, prior to delivery, and in service – with the appropriate instructions to the user.

If the mould is incorrectly sealed (7§13), or is not closed properly in the clamp, then the water of the bath will be in good contact with the dough. At the normal processing temperature, which is well above the glass transition, the then high diffusibility of water means that saturation is approached rapidly, especially at the surface. However, this amount is then well above the equilibrium value at room temperature. Accordingly, there is excess present when the device is cooled after processing, and now because escape by diffusion is far too slow, a **phase separation** occurs (8§4.1): fine droplets of liquid water dispersed in the PMMA matrix. The difference in refractive index is enough to cause the PMMA to be very cloudy, and essentially opaque. This is most obvious when the acrylic is clear, but it might also alter the appearance of the usual pink opaque material. If it is marked enough at the surface, then grinding or polishing may expose it as an irreducibly, albeit finely, rough surface, and this may have implications for hygiene and discolouration.

References

[1]   Combe EC. Acrylic Dental Polymers in: Williams D. Concise Encyclopedia of Medical and Dental Materials. Pergamon, Oxford, 1990.

[2]   Flory PJ. Molecular size distribution in linear condensation ppolymers. J Amer Chem Soc 58: 1877 - 1885, 1936.

[3]   Flory PJ. Principles of Polymer Chemistry. Wiley, New York, 1953.

[4]   Tuckfield WJ, Worner HK & Guerin BD. Acrylic resins in dentistry, part II. Austral Dent J 47: 172 - 175, 1943.

[5]   Yau WFE, Cheng YY, Clark RKF & Chow TW. Pressure and temperature changes in heat-cured acrylic resin during processing Dent Mater 18: 622 - 629, 2002.

[6]   Smith DC. The acrylic denture base: some effects of residual monomer and peroxide. Brit Dent J 106: 331 - 336, 1959.

[7]   Lung CYK & Darvell BW. Methyl methacrylate monomer–polymer equilibrium in solid polymer. Dent Mater 23 (1): 88 - 94, 2007.

[8]   Stickler M & Gunther Meyerhoff G. The spontaneous thermal polymerization of methyl methacrylate: 5*. Experimental study and computer simulation of the high conversion reaction at 130°C. Polymer 22: 928 - 933, 1981

[9]   Lingnau J, Stickler M & Meyerhof G. The spontaneous polymerization of methyl methacrylate – IV. Formation of cyclic dimers and linear trimers. Eur Polym J 16: 785 - 791, 1980.

[10]   Lingnau J & Meyerhoff G. The spontaneous polymerization of methyl methacrylate: 6. Polymerization in solution: participation of transfer agents in the initiation reaction. Polymer 24: 1473 - 1477, 1983.

[11]   Bamford CH & Morris PR. The oxidative polymerization of methyl methacrylate. Die Makromolekulare Chemie 87: 73-89, 1965.

[12]   Lung CYK & Darvell BW. Minimization of the inevitable residual monomer in denture base acrylic. Dent Mater 21 (12): 1119 - 1128, 2005.

[13]   Xia CM, She CX & He WZ. Rapid-processing procedure for heat polymerization of polymethyl methacrylate in a pressure cooker with automatic controls. J Pros Dent 76: 445 - 447, 1996.

[14]   Durkan R, Ozel MB, Bagis B & Usanmaz A. In vitro, comparison of autoclave polymerization on the transverse strength of denture base resins. Dent Mater J 27: 640–642, 2008.

[15]   Mark FF. Future trends for improvement of cohesive and adhesive strength of polymers. in: Weiss P. Adhesion and Cohesion. Elsevier, Amsterdam, 1962.

[16]   Causton B. Acrylic and BIS-GMA polymers. in: O'Brien WJ & Ryge G. An Outline of Dental Materials and their Selection. Philadelphia, Saunders, 1978.

[17]   O'Shaughnessy B & Yu J. Autoacceleration in free radical polymerization. Phys Rev Lett 73 (12): 1723 - 1726, 1994.

[18]   Trommsdorff E, Köhle H & Lagally P. Zur polymerisation des methacrylsäuremethylesters. Makromol Chem 1 (3): 169 - 198, 1948. DOI: 10.1002/macp.1948.020010301

[19]   Braden M. Polymeric prosthetic materials. in: von Fraunhofer JA (ed) Scientific Aspects of Dental Materials. Butterworths, London, 1975.

[20]   Schultz J. Polymer Materials Science. Prentice-Hall, New Jersey, 1974.

[21]   Braden M. The absorption of water by acrylic resins and other materials. J Pros Dent 14: 307 - 316, 1964.

[22]   Johnson W. Final report on the safety assessment of acetyl triethyl citrate, acetyl tributyl citrate, acetyl trihexyl citrate, and acetyl trioctyl citrate. Int J Toxicol 21(suppl 2): 1 - 17, 2002.

Chapter 6   Resin Restorative Materials

In efforts to achieve an 'invisible' repair of teeth a great deal of work has been done on polymer resin-based restorative materials. While there have been many types of polymer explored and marketed for this purpose, they have all been derived from the simple vinyl, free-radical polymerized methyl methacrylate model, whose basic properties therefore underlie the behaviour of these materials.

*There is, however, a range of deficiencies of chemical, physical and mechanical kinds that have encouraged the development of many modifications. The nature of those **deficiencies** in the context of restorative materials is now set out, and some of the attempts that have been made to overcome them are described.*

*The mechanical and physical problems are largely addressed through the incorporation of **fillers** to create a **composite** structure. The effects of fillers are of such a general nature that they account for the nearly-universal occurrence of composite structures in dental materials. The principles and behaviour exemplified by the filled resin materials can therefore be applied to understanding the properties and the handling of all of those other products.*

*The **chemistry** of the polymer can also be modified to reduce the disadvantageous effects otherwise obtained with PMMA. That is, by the choice of polymer backbone structure and density of cross-linking, the mechanical properties can be adjusted. This is the issue of polymer design, a matter that has implications in several dental materials areas (such as impression, tray and suture materials), and will become increasingly important in the future. Choice of product may depend on an understanding of the benefits versus the cost of such differences.*

*Although the use of fillers overcomes or reduces many problems, one deleterious effect arises directly: difficulty with mixing. To avoid this, **command set** materials are now commonplace, employing light to initiate the setting reactions. Correct handling and use depends on an understanding of these aspects.*

Resin-based restorative materials are a significant part of modern dentistry. Both existing products and the likelihood of future developments place considerable emphasis on comprehension of the general principles set out in this Chapter for proper selection, handling and use.

Materials Science for Dentistry
https://doi.org/10.1016/B978-0-08-101035-8.50006-7

Although silver amalgam has served as a convenient and durable direct restorative material for over 150 years, the fact that it does not look like tooth material is often now considered a major disadvantage. There are also concerns (essentially unfounded) that the mercury in it is a source of generalized toxic effects. However, environmental concerns over waste disposal are a cause to be taken seriously. Considerable effort has therefore been expended on the design and development of tooth-like materials, for restorations that are undetectable to a casual inspection, that is mimicry of the appearance of tooth tissue is the goal . One of the main ideas in this endeavour is that of the polymer resin-based restorative (the other is that of the ion-leachable glass: silicate cement, now obsolete, and glass ionomer cement, 9§8). Again, it is found that there is no ideal solution of this kind to the problem, and a number of compromises are involved.

§1. Direct Acrylic Resin Restorative Materials

The direct fabrication of a polymer resin restoration *in situ* is feasible because the polymerization of methyl methacrylate can be made to proceed at low temperatures with the aid of a suitable activator (5§3). Such 'cold-cure' resins extend the applications of acrylic polymers away from denture bases and related areas. They are usable in the mouth because the temperature rises involved during reaction are low enough not to pose an overt hazard to a vital tooth (due to the small volume of a typical restoration, heat is lost quickly), and the setting reaction rate can be made high enough that chair-time is kept within reasonable limits. In fact, so-called light-cured materials are even more convenient. But now, to examine the successes and failures of this line of development, a semi-historical account will be given.

●1.1   Unfilled acrylic

The earliest materials used were plain and simple, unfilled, poly(methyl methacrylate). This was in effect just cold-cure acrylic resin, but with colouring more appropriate to tooth tissue. The technique consisted of moistening a small brush with monomer, and then using that to transfer a small amount of the PMMA powder to the cavity. The polymerization reaction would already have started in that first increment when the next portion was added, and so on until the cavity was filled. Benzoyl peroxide could still be used as the initiator, being a relatively cheap source of free radicals. However, the amine **activators** ordinarily used (5§3.2) are aromatic, and have conjugated systems of π-bonds which absorb ultra-violet light (24§6.2) to create high energy states. Such activated molecules or side-groups to the polymer chain can react with oxygen, which would be dissolved in the resin from the air, or with other aromatic groups, or even with small organic molecules derived from foodstuffs which could diffuse into (*i.e.* dissolve in) the polymer, to form larger, more extended conjugated systems. The creation of these larger systems – **colour centres** or **chromophores** – pushes the absorption of light into the visible region, starting with the blue, with the consequence that the resins become yellowish (24§6.2).

The colour stability of such a resin is more critical for restorations than for denture base materials because of the general paleness and translucency required to match tooth tissue. Since some ultra-violet light illumination is unavoidable (from sunlight, Fig. 24§4.2, as well as fluorescent lighting, 24§4.9, and quartz-halogen lights, 24§4.3), especially in the anterior regions of the mouth, such restorations would quickly become obvious and unacceptable. One possibility is the use of ultra-violet absorbing molecules, intended to play no part in the polymerization reaction, much as such absorbers are used in commercially-produced bulk polymers (then called UV-stabilizers) or as are used in sun-screen products for our skin. However, these were found to be of limited value, and it was therefore necessary to find alternative activators.

●1.2   Activators

There are a variety of chemical systems which have now been found to be workable as activators in the dental context, more or less overcoming the colour stability problem. For example, tributyl borane and the sodium salt of *p*-toluene sulphinic acid have been used as these do not form such coloured compounds so readily. Alkyl boranes (Fig. 1.1) readily form adducts with electron donors – such as benzoyl peroxide. These polar compounds are unstable and so they then rapidly decompose, generating two radicals which may be active in the initiation of polymer chains. The resulting boron ester compound continues to be active until the tri-benzoate ester has been formed. The essence of this sort of reaction is only that of encouraging the dissociation of the benzoyl peroxide. However, it may be noted that the boron tribenzoate is a small, polar, organic molecule: it is therefore (a) a plasticizer, (b) going to encourage water absorption, and (c) potentially leachable. As in many areas of materials design, there is a compromise: the mechanical properties of such a material will not be quite as otherwise expected. Even so, there are several serious drawbacks to using such a resin system as PMMA as a restorative material.

●1.3　Polymerization shrinkage

This is of the order of 21 % by volume for polymer from pure methyl methacrylate. It was partly overcome by using beads of polymer to the extent of about three-quarters of the initial mixture, as in denture fabrication or the use of cold-cure materials elsewhere. However, the 5 or 6 % shrinkage which remains demanded that special techniques be used if its effects were to be minimized. Minimized, perhaps, but not eliminated, since the polymerization must continue for some time after the initial 'setting', that is, when the material is apparently hardened. The shrinkage correspondingly proceeds at about the same rate. Such shrinkage led to marginal leakage, discolouration, sensitivity, and secondary caries.

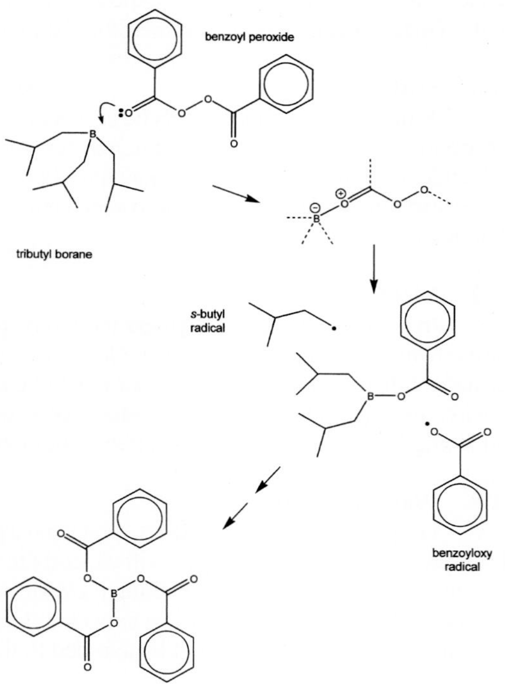

●1.4　Thermal expansion

The coefficient of **thermal expansion** of polymers in general is very large, about 70 MK^{-1} for PMMA, and very much larger than that of enamel (11 MK^{-1}) or dentine (8 MK^{-1}). Consequently, even the modest excursions of temperature normally experienced in the mouth will produce marked relative dimensional changes of the restoration in the tooth. This might not be so bad if it were reversible.

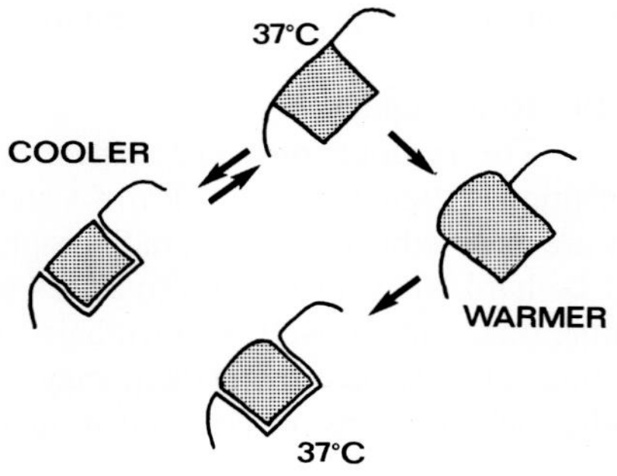

Shrinkage on cooling will open a gap at the periphery and permit the percolation of oral fluids, with the possible consequence of a brief period of pain and some discolouration at the margin. Rewarming to normal temperatures would be expected to close the gap again. On the other hand, expansion due to higher temperatures than normal would result in a rise in the pressure in the material, if – as is intended to be the case – there is contact between the filling and the cavity walls. This pressure will tend to be relieved by plastic flow or creep (1§11), an extrusion of the resin beyond the original margin (Fig. 1.2). This permanent deformation may only accumulate slowly, as the viscosity of such a polymer even a few degrees above mouth temperature would be very high, but it would be enough to ensure that on cooling it would leave a progressively larger and persistent peripheral gap. Subsequent cooling would therefore expose a larger than expected gap, and clearly could not undo the damage already done. The dimensional change of cooling (by itself) is recoverable for a simple, one-surface restoration; that of heating will not be entirely recoverable. In more complicated cavity designs, involving two or more surfaces, or a narrow isthmus, other forms of distortion will occur, as stress relaxation must proceed for both heated and cooled restorations. Distortion or extrusion would also expose edges to extra wear. A key requirement for an ideal restorative material is thus the matching of its thermal expansion coefficient to that of enamel.

●1.5　Strength

The **strength** of PMMA, a polymer, is relatively low compared with enamel, a ceramic composite, and a relatively large bulk of resin would be required to withstand masticatory forces in the absence of supporting tooth tissue. However, the abrasion resistance is very much lower than that of tooth and wear occurs readily, due to the abrasiveness of foodstuffs, opposing dentition and toothpaste. The consequent relief of the restoration

Fig. 1.1 The reactions of the tributyl borane activation system.

Fig. 1.2 The effects of thermal cycling on a viscoelastic material. Cooling shrinkage may be reversible, but expansion stresses may cause flow and irrecoverable dimensional change.

surface leaves the enamel margins exposed and thus prone to chipping and as enhanced abraders of an opposing tooth. **Differential wear** should therefore be avoided.

●1.6   Stiffness

Young's Modulus for acrylic resins is very much lower than that of enamel (Fig. 1§14.1) and so the deformation on loading is greater. Such differences in strain also cause shear stresses to appear at the interface, and while a simple retention form may have been used for such restorations, *i.e.* an undercut design, such shear would also tend to damage any micro-mechanical key. Elastic modulus matching is thus a further requirement of an ideal material.

●1.7   Yield

In addition, because polymers tend to be **plastic** (4§7.3) – as is implied by the use of that term as their common name – and with a low yield point, ordinary masticatory stresses applied in service may cause permanent distortion and extrusion, slowly but cumulatively, resulting in deleterious changes to the peripheral fit with consequent leakage and therefore pathology. This is interrelated with the issues of thermal expansion and strength, discussed above, for effects after extrusion.

●1.8   Water absorption

For polymers such as PMMA, which have polar substituents, water absorption has two important effects:
(1) it produces a swelling which will exacerbate the extrusion mentioned above due to heating, and
(2) since water is a plasticizer, making the resin softer (lower elastic modulus and lower viscosity), it therefore increases the deformation to be expected during such episodes of stress. The strength is also reduced.
Exposure to water plainly cannot be avoided in the mouth.

●1.9   Appearance

This is obviously of keen interest if indeed the tooth material is to be mimicked and the restoration be invisible to casual inspection. Well-prepared polymers such as PMMA are optically clear, and indeed this polymer is used for spectacle lenses and aircraft windscreens, and some modified types as contact lenses. Teeth obviously are relatively opaque, scatter light, and are somewhat coloured. The latter can be dealt with by adding dyes, but the other properties remain to be addressed (see 24§5.11).

●1.10   Radio-opacity

The **radiographic** appearance may also be a problem. Being composed of light elements, X-ray absorption is slight (26§3) and PMMA and other organic polymers are essentially **radiolucent**. This means that they are impossible to detect in radiographs, and so distinguishing restoration from pathological effects (caries) may be hard. This problem also arises in a more life-threatening manner in the context of inhaled PMMA denture fragments, as well as with inhaled, swallowed or otherwise internalized plastic toys and parts in children: locating, let alone recovering them, may be very invasive, and although magnetic resonance imaging (MRI) can handle such problems, its cost puts it out of the reach of routine work.

●1.11   Mixing

This is a general problem affecting any product where reaction commences immediately on contact between reactants as separate components. Restorative materials must set in a reasonably short time to be practical. This implies high reactivity at room or mouth temperature and therefore little time for mixing and placement. Questions of the homogeneity of the mixture (15§7), and thus of the uniformity of properties if not the completeness of setting, of the accuracy of proportioning to attain the correct outcome, as well as of the incorporation of air bubbles, must all be considered. The fact that small amounts are used does not make this any easier.

●1.12   Reaction exotherm

Polymerization and cross-linking reactions are exothermic, *i.e.* they raise the temperature of the material above that of its surroundings, aided by the poor thermal conductivity of polymers. This may have two consequences. Heating and thus irritation and inflammation of underlying pulp or odontoblastic processes might lead to post-operative pain or discomfort,[1] and secondly, thermal expansion of the restoration (§1.4).

●1.13   Composites

All of these points explain the lack of success of a plain acrylic restoration. Clearly, modifications to the polymer itself (size and polarity of substituents, for example) could only have limited effects, and only in some areas because of the very nature of organic polymers. However, the addition of a stiff, inert, radiopaque and optically-suitable (24§5) filler to any such resin is instrumental in producing much larger and more useful changes in some properties, and goes a considerable way in ameliorating those inherent difficulties. Such materials are said to have a **composite structure**, a mixture of (at least) two materials with differing properties of one kind or another. The properties of the composite are intermediate in many senses to those of the components, and effective compromises are possible, utilizing one or more aspects of each component to advantage when individually they would be unusable or unserviceable, as indeed we have now found plain resins to be in a particular context. Nevertheless, we do find that there is compromise, as one gain is traded against another loss.

It is, however, appropriate here to draw attention to the fact that most materials in dentistry are composite in structure, and that the term 'composite' is not in any way restricted in application to these particular resin-based restorative products. Thus, all cements are composite, consisting of a **core** of unreacted powder in a **matrix** of reaction products (9§4); amalgam is a composite, unreacted original alloy is embedded in a metallic matrix of reaction products (14§5); all flexible impression materials (except some agar products) contain a finely particulate filler to modify viscosity and stiffness (7§3); carbide-containing alloys such as steel and cobalt-chromium as well as cermets (21§3), casting investments, porcelain, and so on, all include more than one discrete and recognizable substance; skin is composite; teeth are composite. We could, without stretching the point too far, describe PMMA prepared from powder and liquid as composite. The polymer created from the liquid, as a matrix for the commercially prepared powder, must have (slightly) different mechanical properties because the polymerization conditions are different (5§2). Similarly, dental waxes, since they consist of a mixture of crystalline core and amorphous matrix (16§1), are also composite, even if the compositions of the two regions are similar. A trivial case would be that of dental plaster: a dihydrate matrix with a core of air or aqueous solution. It is still a clearly two-part structure. Since the overall behaviour depends on the structures and behaviours of the components, one approach to such systems is to consider the combination, whether by class of component (Fig. 1.3) or by structure (Table 1.1). However, some care is required in applying simple schemes such as these: such classifications are not absolute or comprehensive – there are other combinations and variations possible.

It is, therefore, not actually wrong in any way to refer to filled-resin restorative materials as 'composites', but it is certainly wrong to use the term as if it applied only in that very limited connection: such a usage is quite incomprehensible outside of dentistry, where the design and study of composites such as boron fibres in metal alloy matrices is central to much 'high performance materials' research. Within dentistry the usage devalues the scientific meaning of the term and draws attention away from the fact that so many dental other materials are of this kind of structure, and that their behaviour and properties all can be understood in precisely the same terms no matter what the composition or application. Whilst it is true that the phrase **'filled-resin restorative material'** is a little long, it is at least the minimum acceptable, accurately descriptive term. The short-hand form 'filled resin' will suffice here because the immediate context is taken to be understood, but the mechanical properties must be discussed in terms of the general theory of composite

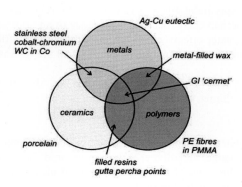

Fig. 1.3 Some kinds and dental examples of composite, according to classes of components. Like-with-like lie in the main regions. (PE = polyethylene)

	Table 1.1	*A classification of composites relevant to dentistry.* (Core and matrix cannot be distinguished in bicontinuous systems.)	

Core *Matrix*	**Crystalline**	**Amorphous**
Crystalline	metal alloys, inc. carbides dental silver amalgam	?
Amorphous	dental porcelains filled impression materials zinc phosphate cement 'fringed micelle' polymers wet gypsum materials enamel dentine	glass ionomer cement glass-filled resins soft tissue

structures. It will be understood, therefore, that much of what is said here also applies in one form or another to all of those other materials, in dentistry or outside, whether technical or commonplace, natural or artificial, food or fabric.

§2. The Effects of Fillers

The effects of the rather large polymerization shrinkage of PMMA can be reduced by adding a proportion of already polymerized material (5§2.4). The overall shrinkage that is then observed is in direct proportion to the volume fraction of the polymer newly produced (or, of the mole fraction of monomer added, reckoning the added polymer in terms of mers). However, even relying on this, as did the early direct-filling acrylic materials (§1), the shrinkage is inconveniently large. Similarly, the amount of heat released during the polymerization reaction is reduced in proportion, as the number of bonds to be formed is reduced. This kind of effect clearly also applies no matter what the 'filler' is, PMMA or other material, but such properties as the water absorption and thermal expansion are quite unaffected if it is PMMA that is used. However, if a non-absorbent filler were to be used, the overall water absorption would be similarly *proportionately* reduced (although the absorption by the polymer itself must remain unaffected). This is an interesting lead, and the effects of such fillers on properties such as thermal expansion, strength and modulus need to be explored, for it is not at all obvious that similarly simple relationships apply. We return to this topic below.

●2.1   Thermal expansion

If the filler is chosen to have a coefficient of thermal expansion lower than that of the matrix, experimentally it is found that the relationship with proportion used is not that of a straight line, as would perhaps be expected on the basis of an arithmetic mean effect (Fig. 2.1). The observed behaviour indicates a greater effectiveness for the addition of the filler than its volume fraction would suggest. The explanation for this can be found in the mechanical interaction of the two components.

The defining equation of the coefficient of volumetric thermal expansion, β, is:

$$\Delta T . \beta = \Delta V / V \qquad (2.1)$$

where ΔT is the temperature interval over which the change in volume, ΔV, is measured (see also 17§1.1), while that of the bulk modulus of elasticity (1§2.6) was given by (equation 1§2.23)

$$K = \sigma / (\Delta V / V) \qquad (2.2)$$

$$i.e., \quad \sigma = K . (\Delta V / V) \qquad (2.3)$$

so, substituting from 2.1, we have

$$\sigma = \beta \Delta T K \qquad (2.4)$$

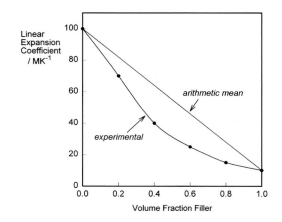

Fig. 2.1 Variation of the coefficient of thermal expansion with volume fraction of filler.

This is the stress that would be obtained in a particle trying to change dimension against a perfect constraint, *i.e.* embedded in a rigid, non-expanding environment when there is no actual volume change at all. That is, a particle in this situation would be squeezed by the matrix – hence the use of the bulk modulus. However, we must allow for the fact that the matrix will also be changing dimension to some extent, so that the effective coefficient of expansion to be used in that equation will be given by the difference $(\beta_c - \beta_i)$, where β_c is the *overall* expansion coefficient, and $i = 1, 2, ...$ identifies which of the n components is of interest. Thus, the stress in any component of the filled resin system (and this is true for any particulate composite material) is therefore given by:

$$\sigma_i = (\beta_c - \beta_i) \Delta T K_i \qquad (2.5)$$

That is, the (hydrostatic) pressure locally depends simply on the difference between the local and the overall expansion coefficients. However, if the body is under no externally applied constraint or load, the overall mean hydrostatic pressure must be zero (for if it were not, the dimensions must change until equilibrium were achieved). Thus, summing over all i components:

$$\sum \sigma_i \phi_i = 0 \qquad (2.6)$$

because tension and compression are taken to have opposite signs (ϕ is volume fraction, as before, 2§2). Once again, this is Newton's Third Law of Motion being invoked: if there were to be a non-zero resultant force, something would have to move until equilibrium were established. Hence, from 2.5 and 2.6, and in the simplest case of one filler in a continuous matrix (n = 2), we can write:

$$(\beta_c - \beta_1)\Delta T K_1 \phi_1 + (\beta_c - \beta_2)\Delta T K_2 \phi_2 = 0 \quad (2.7)$$

which rearranges to show that

$$\beta_c = \frac{\beta_1 K_1 \phi_1 + \beta_2 K_2 \phi_2}{K_1 \phi_1 + K_2 \phi_2} \qquad (2.8)$$

which is a kind of weighted mean, indeed, an example of a **rule of mixtures**.

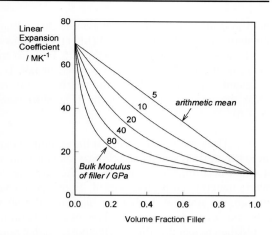

Fig. 2.2 Expected thermal expansion behaviour according to equation 2.8 for fillers of varying bulk elastic modulus. The thermal expansion coefficients of the matrix and filler are 70 and 10 MK^{-1} respectively. $K_{matrix} = 5$ GPa.

What this analysis shows is that the overall thermal expansion depends not only on that of the components, and their volume fractions, but also rather critically on their bulk moduli, their compressibilities (Fig. 2.2). We may also notice from equation 2.5 that as one of β_1 and β_2 in a two component system must be greater, and the other lower, than β_c, the stresses in the two components must ordinarily be in opposite senses, there being only one temperature at which the stresses are both zero. The presence of such stresses has implications for the interface between the components: these must be under stress too. The importance of this we return to below.

Ceramics, which are usually stiff, hard and strong, and therefore apparently suitable as strengthening fillers, typically have coefficients of expansion about one tenth of those of typical polymer resins (Fig. 3§4.14). Indeed, at least one is known and used in dentistry with a *negative* coefficient over the temperature range of interest.

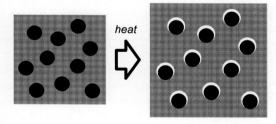

Fig. 2.3 The thermal expansion of a non-bonded matrix is unconstrained by a low expansion filler.

If, as is obviously intended to be the case with filled-resin restorative materials, the filler has a lower expansion coefficient than the resin matrix, raising the temperature could not develop a stress across the interface if there were no bond between matrix and filler. The matrix would expand unhindered (Fig. 2.3) as there can be no restraint, and the previous equation would be inapplicable. The filler particles would be loose within the matrix and just rattle around as the holes get relatively bigger. There would thus be no effect whatsoever on thermal expansion on heating due to the filler (Fig. 2.1 and 2.2 would therefore then only apply for cooling). Indeed, the filler could be absent altogether (equivalent to having pores or bubbles in the resin) and not be noticed as missing at all from this point of view.

●2.2   Matrix constraint

If the matrix of such a system has expanded, in order to return to it contact with the filler it must be hydrostatically compressed (Fig. 2.4). Thus, when there is bonding, and the temperature is raised, the filler particles would be put in tension (as must be the interface between the two itself). To balance the stresses, the matrix must then be in compression (Fig. 2.5, top). This 'holding back' of the expansion of the matrix, or **matrix constraint**, due to interfacial bonding is thus absolutely crucial to the effectiveness of a filler for this purpose of

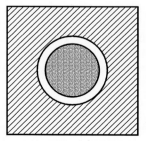

Fig. 2.4 If an insert is smaller than the hole, the surround must be compressed hydrostatically to fit it, or the insert stretched – or both. Thus, letting go, if the two are bonded together, the insert will be in tension – and the surround must remain somewhat compressed.

modifying thermal expansion behaviour. On cooling such a material, *i.e.* with β(matrix) > β(filler), the stress at the interface becomes compressive (Fig. 2.5, bottom) and the nature of the interface, whether bonded or not, is quite irrelevant. The stresses within the matrix are now tensile while the core is in compression (Fig. 2.6) (see also 18§4.6).

There are limits to this. If the cooling is extreme these stresses may result in cracking within the matrix, that is when according to equation 2.5 the strength of the matrix material has been exceeded. Similarly, on heating enough, the stress at the bonded interface of matrix and filler may exceed its strength. The resultant bond failure would then render the filler ineffective from that point on. Of course, outright failure is not the only possibility with polymeric matrices, and stress relaxation through flow processes might complicate the picture (see box).

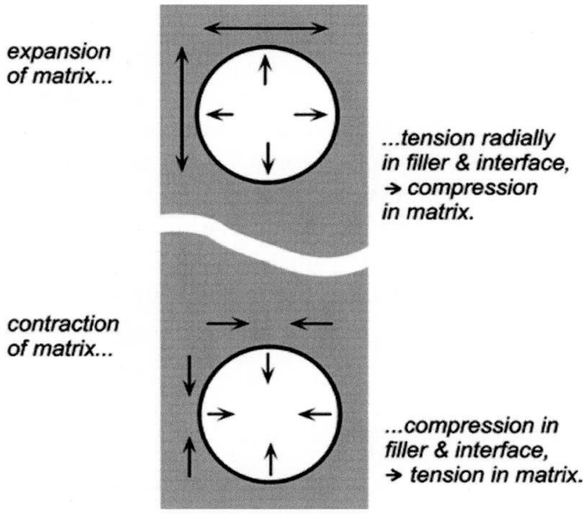

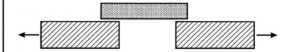

Fig. 2.5 Relative expansion of bonded matrix is constrained by the filler.

> In fact, if the flow effects dominated, the overall volumetric thermal expansion coefficient would approach the arithmetic mean (*i.e.* weighted by volume fraction):
>
> $$\beta_c = \beta_1\varphi_1 + \beta_2\varphi_2$$
>
> This equation (another rule of mixtures) would apply to a system such as sand in water. The line representing this equation is shown in Fig. 2.2 for comparison, but it is emphasised that it is not the proper equation for a bonded, particulate-filled composite except if the bulk moduli happened to be identical, when equation 2.8 reduces to the form here.

Fig. 2.6 If an insert is larger than the hole, the surround must first be stretched to accommodate it. Thus, letting go, the insert will be under compression – and the surround must remain somewhat stretched and therefore in tension.

If visualizing how an expanding matrix produces compression in itself when bonded to the filler causes any difficulties, consider the entirely equivalent situation of a contracting filler (Fig. 2.7, bottom). Clearly, the interface is in tension again, but the squeeze exerted on the matrix is now perhaps more obvious. Again, if the filler were to expand it would put the matrix into tension as it was obliged to stretch over the now larger insert (Fig. 2.7, top). This latter situation is the case in the rusting of reinforcing steel in concrete: the expansion bursts the concrete. It is also the situation in impression plaster when this contains starch (27§2.4). The point is that these systems are symmetrical: if one case can be worked out, the others follow automatically by reversing the signs of all the stresses.

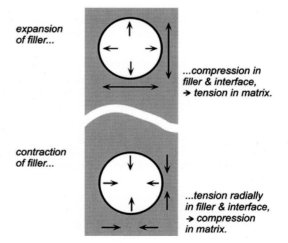

Fig. 2.7 Relative expansion of the filler throws the matrix into tension.

Similar considerations apply in mechanical testing machines. If the specimen in Fig. 1§3.6 is in tension, then the side columns, supporting the two fixed crossheads, must be in compression. The point is that all stresses occur in pairs if static equilibrium is to be observed – Newton's Third Law of Motion.

●2.3 Polymerization shrinkage

In a filled material, polymerization shrinkage leads to a similar condition as does cooling (Fig. 2.5, bottom). While the matrix is still relatively plastic and not cross-linked enough, viscous flow would permit the relief of the stresses that this shrinkage causes. However, at some stage, the viscosity will become sufficiently high that this cannot happen fast enough (§4.5). Hydrostatic tensile stresses then begin to develop within the matrix (§4.6), increasing (asymptotically) as the reaction proceeds, with cross-linking and viscosity rising, until the reaction and the flow come effectively to a halt. The system is then described as **stress frozen**. Under these conditions, and assuming that the rate of reaction has been uniform over the whole volume of the material, the system looks as though it had been cooled – the volume of the matrix would, at equilibrium, have been decreased, but the presence of the filler has prevented this. The matrix is therefore in tension. The temperature of the fully-set material must then be raised by an amount sufficient to reduce these tensile stresses to zero before further heating causes the interface stresses to change from tension to compression.

The polymerization reaction exotherm is also relevant. Thus, when the cross-linking has proceeded to a stage where viscous flow is much impeded, the temperature may be near its maximum. This may be only a few degrees above 37°C (although it could be much higher), but this effect would contribute to the residual tensile stress in the matrix on cooling to 'normal' mouth temperature, adding to that stress produced by the polymerization shrinkage. Clearly, whether there is a net hydrostatic tension within the matrix when it is fully set and cooled to its working temperature depends on the balance between the rate of relaxation through viscous flow and the rate of cross-linking. Unfortunately, since it is a dynamic system, detailed calculations are not feasible. However, it seems likely that there does remain some hydrostatic tension in the matrix of the fully set material. The main evidence for this is that polymerization shrinkage as a bulk phenomenon affecting the fit of restorations is still serious enough to need to be taken into account with modern materials. Hence the use of incremental 'wedge' techniques and similar procedures to minimize the stress placed on the bonded interface between restoration and tooth.

●2.4   Water sorption

The oral environment is wet – internally and externally. To understand polymer behaviour in service in such circumstances, we must therefore consider the effect of water sorption. As already described (5§6), PMMA and similar polymers with polar groups will absorb appreciable amounts of water, one or two percent being typical. This water has two principal effects, which apply in the present systems as much as anywhere. There is the plasticizing effect: the lowering of modulus and increase in flow rates due to the increase in effective free volume for chain segment motions. There is also a volumetric expansion as the absorbed molecules do not just fit into available space, but allow the polymer chains to move further apart as well. The resulting expansion would also be expected to operate as does thermal expansion in terms of the stresses generated (Fig. 2.5, top). This absorption, and thus the expansion, will occur after setting is largely complete (because 'setting' for clinical purposes takes only a matter of a minute or two) because it is only then that exposure to water or saliva will occur. But it is still expected to take several or many days even to approach equilibrium because the diffusion of water through resin is a slow process.[2]

How much such expansion actually compensates for polymerization shrinkage depends very much on the chemistry of the resin – primarily the relative hydrophobicity or hydrophilicity,[3] as well as other factors, but the existence of the effect illustrates the complexity of the analysis of such systems in service. For example, if the absorption expansion exceeds the polymerization shrinkage, then the restoration would fit, but would suffer extrusion as in Fig. 1.2, and the tooth will be distorted, even without warming. Also to be taken into account is the fact that the 'open mouth' tooth temperature during clinical work may be several degrees below that of the normal situation. Thus warming of the material after completion of the clinical work should also be taken into account.

A non-absorbent filler is expected to reduce the overall water absorption of the composite material in direct proportion to its volume fraction, *i.e.* the absorption, W_c, is in direct proportion to the volume fraction of the resin (φ_m) doing the absorbing, given the absorption, W_m, of that matrix resin itself:

$$W_c = W_m . \varphi_m \qquad (2.9)$$

Unfortunately, if the interface between resin and filler is not well-bonded, or is subject to hydrolytic breakdown (§2.15), this region can itself hold water, increasing the total absorption above that of this simple expectation.

●2.5   Elastic modulus

The effect of fillers on the modulus of elasticity is of very great importance in the general field of composite materials. We shall return to it in particular in the context of impression materials (7§3), but since nearly all materials used in dentistry are composite, this is perhaps the single most important aspect.

Given the (differing) mechanical properties of two homogeneous and isotropic components (Fig. 2.8), the problem is one of how to calculate the mechanical properties of a composite of the two. The typical case in dentistry is of a particulate filler in a continuous matrix, but since this is too difficult to handle theoretically we shall use a simpler system, the so-called 'sandwich' model, consisting of alternating layers of the two components. Having done that, there is clearly now a directionality to the structure that must be taken into account (Fig. 2.9). More than that, the sandwich composite is fairly obviously going to be anisotropic, that is, generally have different properties in the two load directions.

If the elements of the structure are in parallel and aligned with the load axis, the strain of each element is by definition identical since we are assuming that the end faces remain plane and normal to the load axis. (We imagine that the loading is applied over the entire upper and lower faces of the block by a perfectly rigid and flat anvil.) This then is known as the **isostrain** model (sometimes also called the **Voigt** model – not to be confused with the Kelvin-Voigt model of retarded elasticity, 4§5 – elements in parallel is the common idea). However, the resulting *stresses* in the two components must be different because the material with the lower modulus requires a lower stress to be extended by the same amount as the stiffer component.

For the alternative orientation, where the layers are normal to the load axis, and therefore connected in series along the load axis, the *stress* acting on each must be identical. If the elastic moduli of the layers differ,

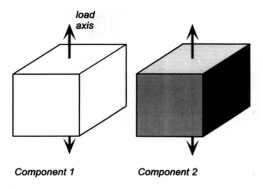

Fig. 2.8 Two components from which a composite material is to be built.

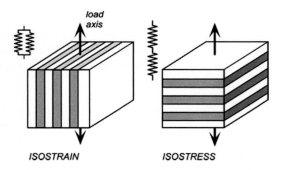

Fig. 2.9 The isostrain or 'equal strain' (left) and isostress or 'equal stress' (right) sandwich models used to establish the upper and lower bounds respectively of the modulus of a particulate composite loaded in tension.

necessarily they suffer differing strains. This **isostress** model (also known as the **Reuss**[1] model) thus represents the opposite extreme of the orientation of the structure with respect to the load axis.[4][5]2

●2.6 Rules of mixtures

The overall modulus of elasticity of the composite in the direction of loading can be calculated in the two cases using the relevant so-called **rules of mixtures**.[6] These are no more than the arithmetic or harmonic mean of the individual components' elastic moduli, weighted according to their volume fractions. The derivations of these are quite straightforward.

Thus, for the **isostrain** model, we have, by definition

$$\varepsilon_c = \varepsilon_1 = \varepsilon_2 \qquad (2.10)$$

where ε_c is the overall strain on the composite body. But the overall stress on the composite σ_c must be distributed according to the volume fractions of the two components (which can be seen to be the same as the *area* fractions α_f in a cross section, or indeed the *length* fractions λ_f in the stacking direction, $\varphi_f = \alpha_f = \lambda_f$):

$$\sigma_c = \varphi_1\sigma_1 + \varphi_2\sigma_2 \qquad (2.11)$$

This is the principle of uniformity again (1§2.1). But now we can replace the stress according to Hooke's Law (equation 1§2.4 a):

$$E_c\varepsilon_c = \varphi_1 E_1\varepsilon_1 + \varphi_2 E_2\varepsilon_2 \qquad (2.12)$$

Therefore, because the strains are identical (equation 2.10),

---

[1] Pronounced "Royce".

[2] Paul (1960) was apparently the first to apply the Voigt- and Reuss-type models to this problem, but without citing them.

$$E_c = \varphi_1 E_1 + \varphi_2 E_2 \tag{2.13}$$

i.e., the overall modulus of elasticity for the **isostrain** case is given by the volume fraction-weighted **arithmetic mean** of the elastic moduli of the components (since $\varphi_1 + \varphi_2 = 1$).[7]

For the **isostress** system we have now, again by definition,

$$\sigma_c = \sigma_1 = \sigma_2 \tag{2.14}$$

The overall strain is therefore given by

$$\varepsilon_c = \varphi_1 \varepsilon_1 + \varphi_2 \varepsilon_2 \tag{2.15}$$

Again, we make the substitution from Hooke's Law, this time for the strain:

$$\frac{\sigma_c}{E_c} = \frac{\sigma_1 \varphi_1}{E_1} + \frac{\sigma_2 \varphi_2}{E_2} \tag{2.16}$$

Therefore, because the stresses are identical (equation 2.14), we have

$$\frac{1}{E_c} = \frac{\varphi_1}{E_1} + \frac{\varphi_2}{E_2} \tag{2.17}$$

Solving this for E_c gives:

$$E_c = \left(\frac{\varphi_1}{E_1} + \frac{\varphi_2}{E_2} \right)^{-1} \tag{2.18}$$

i.e., the overall modulus of elasticity for the **isostress** case is given by the volume fraction-weighted **harmonic mean** of the elastic moduli of the components. This is the same as the volume fraction-weighted arithmetic mean of the compliances, $J = 1/E$ (see equation 1§2.4 c):[8]

$$J_c = \varphi_1 J_1 + \varphi_2 J_2 \tag{2.19}$$

This can be understood by noticing that the deformation of each layer is assumed to be completely independent of that of the other layers, *i.e.*, no shear stresses at the interfaces due to Poisson deformations, that is to say that $v_1/E_1 = v_2/E_2$ must hold if the lateral contraction in the two kinds of layer is to be the same. Actually, the same assumption of independence is also made for the isostrain case, equation 2.13, which is equivalent to saying that $v_1 = v_2$ if the lateral contraction in the plane of the layers is to be the same in the two component materials.[3] Clearly, both cannot be true simultaneously unless $E_1 = E_2$, when the system is effectively reduced nearly[4] to a mechanically-homogeneous material and the "composite" properties are therefore unaffected according to these equations.

It is not possible to make a structure any more or any less stiff than these simple equations imply (importantly, assuming both no holes and that the layers are properly bonded in the isostress case) and thus the graphs of these functions are the absolute **upper** and **lower bounds** to the achievable modulus of elasticity of

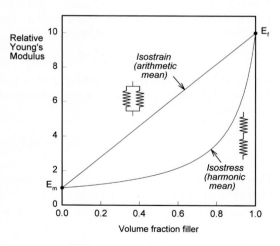

Fig. 2.10 The Voigt (upper) and Reuss (lower) bounds on modulus derived from the models of Fig. 5, assuming that the ratio of elastic moduli for filler (E_f) and matrix (E_m) is 10:1.

---

[3] The effect of these assumptions is actually rather slight and may be ignored given the wider context of estimated bounds.

[4] "Nearly" because the bulk and shear moduli could still differ, equation 1§2.25.

any composite, including those with particulate fillers (Fig. 2.10). These are rather wide limits, and as such are not particularly useful for understanding or predicting behaviour, but they do form the basis of the development of more precise but much more complex descriptions of composite behaviour. It has to be said that there is, as yet, no complete theory, but some of the underlying principles can be identified.[5]

In passing, we may note that the elastic modulus or compliance averaging can be applied to composites of more than two phases to obtain the upper and lower bounds by the obvious extension of equations 2.13 and 2.17, as we can for all such rules of mixtures. It may also be noted that the layers are not required to be of uniform thickness (spacing) or of any particular number – the results are identical however the structures are built. In particular, the outcome is **scale-free**.

The hydrostatic tension arising within the matrix due to thermal contraction has already been mentioned. If there is a perfect, non-slipping bond at the matrix-filler interface, there is a similar condition of matrix constraint by the filler during loading. That is, during the application of a uniaxial load, say in tension, a hydrostatic tension is developed because the Poisson contraction (1§2.2) otherwise expected is limited by the filler. In other words, we need to go beyond the assumption of independence of deformations that was relied upon in the derivation of the isostress and isostrain bounds.

●2.7   Apparent modulus of elasticity

Using the relationships of Hooke's Law and the Poisson Ratio, the strain in the matrix of a particulate-filled composite in the load axis (*x*) direction can be derived. We shall assume, for simplicity, that there is no deformation of the filler (*i.e.* it has an infinite modulus of elasticity). Thus, we start with the basic results

$$\varepsilon_x = \frac{\sigma_x}{E_m} \qquad (2.20)$$

$$\varepsilon_y = \varepsilon_z = -\nu\varepsilon_x \qquad (2.21)$$

where σ_x is the tensile stress (for example) in the *x*-direction and the subscript m refers to the matrix. We now take constraint to mean that the Poisson contraction is opposed. This is equivalent to transverse tensile stresses just large enough to achieve this being superimposed:

$$\sigma_y = \sigma_z = E_m \cdot \varepsilon_y \qquad (2.22)$$

But if both of these stresses were to be applied on their own, they would of course produce a Poisson contraction in the *x*-axis direction (Fig. 2.11):

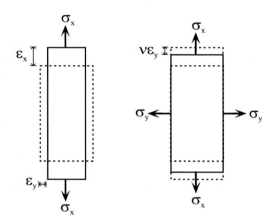

Fig. 2.11 The application of a tensile stress σ_x results in Poisson contraction ε_y (left). This strain can be cancelled out by the application of a lateral stress σ_y, which reduces the axial strain by the amount $\nu\varepsilon_y$ (right). The z-direction effects are similar.

$$\Delta_x = -\nu(\varepsilon_y + \varepsilon_z) = -\nu \cdot \frac{(\sigma_y + \sigma_z)}{E_m} \qquad (2.23)$$

Thus, when σ_x, σ_y and σ_z are applied simultaneously, the net axial strain would be given by the sum of the right hand sides of equations 2.20 and 2.23:

$$\varepsilon_x = \frac{1}{E_m}\left[\sigma_x - \nu(\sigma_y + \sigma_z)\right] \qquad (2.24)$$

This equation represents what is known as the **generalized Hooke's Law**, taking into account a full **triaxial stress** state. But now, thinking about the behaviour internally at local level, under the assumed condition of complete **matrix constraint**, the matrix loading is required to be hydrostatic, so that:

$$\sigma_x = \sigma_y = \sigma_z \qquad (2.25)$$

and since $\nu \approx 0.3$, typically, from equation 2.24 we have

---
[5] See also 29§4.1

$$\varepsilon_x \approx \frac{1}{3} \cdot \frac{\sigma_x}{E_m} \qquad \text{or} \qquad \frac{\sigma_x}{\varepsilon_x} \approx 3E_m \qquad (2.26)$$

In other words, for a very stiff filler and a good interfacial bond, we expect the *apparent* matrix elastic modulus to be some three times greater than its actual value (*i.e.* measured by itself).

This is a very crude analysis, and it is intended to do no more than illustrate the nature and importance of the effects of matrix constraint. The condition of perfect constraint is, however, very severe and has never actually been observed in this type of material (due to stress relaxation from reptation and chain-segment diffusion (3§3)). Nevertheless, the analysis shows the direction of the effect to be anticipated, if not its precise magnitude, *i.e.* that the matrix will behave as if it were substantially more stiff than is assumed in the isostress layer model.

●2.8   Refined upper and lower bounds

The best theoretical treatment of the problem (so far) for particulate composites takes into account the interaction of bulk and shear moduli and leads to expressions for the upper and lower bounds of much greater complexity;[9] again, this is given only to illustrate the point. Thus, for the lower bound is obtained:

$$\frac{1}{E_1} = \cfrac{1}{3\left[G_m + \cfrac{\varphi_f}{\cfrac{1}{G_m - G_f} + \cfrac{6(K_m + 2G_m)\varphi_m}{5(3K_m + 4G_m)G_m}}\right]} + \cfrac{1}{9\left[K_m + \cfrac{\varphi_f}{\cfrac{1}{K_f - K_m} + \cfrac{3\varphi_m}{3K_m + 4G_m}}\right]} \qquad (2.27)$$

where G and K are the shear and bulk moduli given by equations 1§2.22 and 1§2.23 respectively. (It can be seen that this takes the general form of equation 1§2.25.) The upper bound is obtained by exchanging the subscripts for filler (f) and matrix (m) everywhere. This result again depends crucially on the assumption of bonding at the interface.

The graphs of these so-called **Hashin-Shtrikman** bounds calculated for nominal values for resin and filler in dental restorative materials of this type are shown in Fig. 2.12. Although the calculation is clearly only a moderate improvement over that of the simple models in this case (because the component moduli are so different), the new bounds are seen to lie between the two extremes previously calculated, indicating that the mechanical behaviour of particulate composite systems is expected to be some kind of average of the isostress and isostrain models – which intuitively seems about right (see also Fig. 21§3.3, 29§4). This has been borne out in practice by observations in a great number of systems outside dentistry. Thus, the effect of constraint on the matrix by the filler is seen to be of central

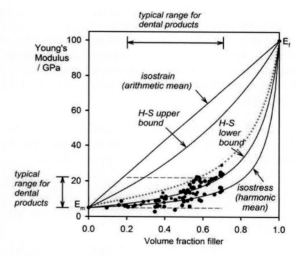

Fig. 2.12 The refined best estimates of the upper and lower (Hashin-Shtrikman, H-S) bounds for the modulus of a typical particulate composite added for assumed values of $E_m = 5$ GPa and $E_f = 100$ GPa. The ranges for dental filled resin restorative materials are also indicated, as well as values for a variety of products. (The dotted line is the Paul mean, 29§4.1.)

importance to the mechanical properties of materials in that it produces hydrostatic loading. However, in the absence of bonding not only will there be no constraint, but the material will behave as if full of holes, and be both significantly less stiff and significantly less strong than the unfilled material (*cf.* Fig. 2§8.3; see also 9§4).

Unfortunately, as is obvious from Fig. 2.12, actual dental filled-resin restorative materials[10][11][12][13][14] do not in fact perform very well. This may be due to a combination of factors: inadequate bonding of matrix and filler (see §2.11), and the presence of defects such as bubbles in the matrix. (The variability of the actual values of E_m and E_f will not affect this conclusion.) A further comparison to put these result in perspective is that the modulus of elasticity of dental enamel is roughly in the region of 70 ~ 100 GPa. That is, about 4 ~ 5 times stiffer than the restoration. This mismatch must jeopardize the tooth as it is not then well supported by the restoration and differential movement or shear stress at the interface can be expected on loading. For comparison, the minimum modulus of elasticity of dentine is around 10 GPa, which is still stiffer than most products.

However, it may be noted that the values for the real dental products in Fig. 2.12 lie close to the lower, Reuss limit, which implies that the system is more nearly in the isostress condition. This is equivalent to saying that there is more strain in the matrix than expected. The cause of this may be found in the relaxation mechanisms available to the polymer: chain segment diffusion, reptation (3§3), and even chain rupture (7§2.2, 7§3, 7§11.6). That is, under load, the molecules of the matrix are rearranged somewhat, not merely loaded *in situ*; the local geometry changes and a purely elastic model is no longer appropriate. (This may be simply stated by saying that it is *not* an **affine deformation**, 7§2.4). Taking into account fully this viscoelasticity (4§6) is obviously a much harder proposition than even equation 2.27, but it may be sufficient to say simply that, for this class of materials, the Reuss, isostress model is the best working approximation. What this implies is that physically it is simply not possible to do appreciably better, no matter how stiff the filler, unless the matrix is changed fundamentally.

●2.9 Stress transfer

Similar arguments can be developed in respect of the shear modulus. The shear modulus of the composite is always larger than that of the matrix on its own – assuming bonding at the interface, as can be deduced from Fig. 2.13. It can be seen that:

$$\frac{\Delta s_o}{L_o} < \frac{\Delta s_m}{L_m} \qquad (2.28)$$

That bonding is a necessary condition is easily seen from the need to transmit the stress from one element to the next for the deformation to be spread over the whole thickness of the sample. An alternative view, therefore, of the role of interfacial bonding in any composite structure is that it is important for **stress transfer**, whether this be in shear or tension. In other words, this is the necessary condition for load to be carried by the reinforcing filler.

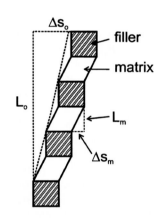

Fig. 2.13 When the modulus of the filler is very high in comparison with that of the matrix, under shear (as with any mode of loading) the strain is essentially confined to the matrix.

We may also apply the methods of §2.6 to the shear case (Fig. 2.14). Thus, for the **isostrain** model we have, by definition, for shear strain γ, shear stress τ and shear modulus G:

$$\gamma_c = \gamma_1 = \gamma_2 \qquad (2.29)$$

$$\tau_c = \varphi_1\tau_1 + \varphi_2\tau_2 \qquad (2.30)$$

$$G_c\gamma_c = \varphi_1 G_1\gamma_1 + \varphi_2 G_2\gamma_2 \qquad (2.31)$$

Therefore, $$G_c = \varphi_1 G_1 + \varphi_2 G_2 \qquad (2.32)$$

i.e., the overall shear modulus of elasticity for the **isostrain** case is again given by the weighted **arithmetic mean** of the shear elastic moduli of the components. (Although there is a clear orientation to the layers, the isostrain condition ensures that the behaviour is isotropic – the same result across or along the layer direction because the value of φ_f is unaffected.)

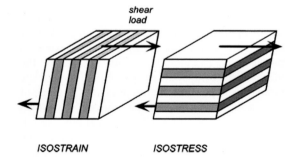

Fig. 2.14 The isostrain or 'equal strain' (left) and isostress or 'equal stress' sandwich models used to establish the upper and lower bounds respectively of the modulus of a particulate composite loaded in shear.

Likewise, the for the **isostress** system we have instead, by definition again,

$$\tau_c = \tau_1 = \tau_2 \qquad (2.33)$$

$$\gamma_c = \varphi_1 \gamma_1 + \varphi_2 \gamma_2 \qquad (2.34)$$

$$\frac{\tau_c}{G_c} = \frac{\tau_1 \varphi_1}{G_1} + \frac{\tau_2 \varphi_2}{G_2} \qquad (2.35)$$

Therefore,

$$\frac{1}{G_c} = \frac{\varphi_1}{G_1} + \frac{\varphi_2}{G_2} \qquad (2.36)$$

That is, for the **isostress** case it is again the weighted **harmonic mean**.

●2.10 Strength

The same principle of hydrostatic constraint that increases the apparent modulus of a composite also applies to its strength. Since the strength of a material can be interpreted as the stress required to achieve the limiting value of the strain that the material can bear, that is, the extension of a bond before it breaks (*i.e.* the strain at failure is actually the material property in which we are primarily interested; *cf.* 10§3.6), the higher the effective matrix modulus the greater is the stress required to achieve that strain and hence to obtain failure. In other words, the strength of the matrix would be increased in proportion to its effective modulus (equation 2.26).

We can see the nature (if not the magnitude) of the effect by considering the Voigt isostrain model of Fig. 2.9. The strain of the body subject to a tensile stress is limited by the stiffer phase (say, white, here). But this is the same strain as occurs in the grey phase, which therefore requires that a much greater overall stress be applied before it can reach the strain at which it fails. The composite is therefore stronger than the grey phase alone (assuming that the 'white' phase itself does not fail, but then ceramics are usually stronger than polymers (Fig. 1§14.2). This can only arise because the deformation of the less-stiff phase is constrained. Nevertheless, this is a simplistic approach, and the accurate prediction of strength values of real composite materials is even more complicated than that of elastic moduli.

To re-emphasise the importance of the bonding of matrix to filler, consider what happens when a tensile stress is applied: the filler holes elongate (Fig. 2.15). Even if the Poisson contraction provides some lateral constraint, it is clearly much less effective since the load is carried by the reduced cross-section of the matrix. The holes now act as crack nucleators in the usual way.

●2.11 Coupling agents

The need for bonding the filler particles to the matrix has already been emphasized, and the more bonding between the two that can be made the greater the constraint of the matrix, the higher the net modulus, and the higher the strength. By 'more bonding' is meant the areal density, the number of covalent bonds per unit area. Note that covalent bond strengths or bond energies do not vary over a very great range for the atoms of possible interest, while other bond types are much weaker (see 10§3).

While it is necessary to use special **coupling agents** to obtain covalent bonds between filler and matrix, the problem, in effect, is one of ensuring that there is total coverage of the filler surface, a complete monolayer, for the maximum matrix constraint to be achieved. It follows then that increased constraint can also be obtained by increasing the surface area of the filler for a given volume – the

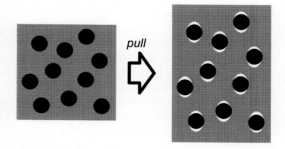

Fig. 2.15 Under tension, unbonded filler provides negligible constraint to deformation and therefore little if any improvement in stiffness. However, strength is reduced.

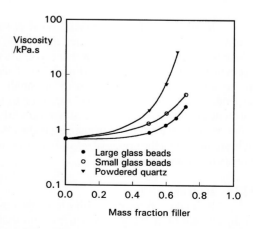

Fig. 2.16 The effect of filler specific surface area on the viscosity (log scale) of the uncured resin matrix.

specific surface area. This means changing the shape or reducing the particle size, or both.[15] It must not be overlooked, however, that increases in filler volume fraction, or of surface area, increase the viscosity of the material before curing (Fig. 2.16, Fig. 4§9.2), at the time of placement (4§9). If this is to be a practical proposition, there is a limit to how much can be achieved, and this limit is well below the ultimate packing density limit.

●2.12   **Prestress**

In the use of dental filled resins that are polymerized *in situ* (*i.e.*, as are direct filling restorative materials), the position may be further complicated by the residual stresses due to polymerization shrinkage. As stated above, these stresses are hydrostatic tensile in the matrix. Any external tensile load will then be in the sense of tending to increase those tensile stresses, and this must result in reaching the strain for failure earlier. Thus, the inevitably **prestressed** nature of the resin composite matrix can make it appreciably weaker in tension than otherwise would be expected (see 23§2.8), although the strength is still better than that of the resin alone by virtue of matrix constraint (§2.10). This assumes that any testing is done under simulated oral conditions. The results obtained in such tests are expected to be strongly temperature-dependent, especially if the curing of the resin was at 37 °C and testing follows at 23 °C when the greater thermal expansion coefficient of the matrix must result in differential shrinkage that increases the tensile stress in that phase. Such differential dimensional changes must also result in shear stress at the interface, so it can be expected that the shear strength will be similarly affected (and therefore "compressive strength"). The role of absorbed water in this system, in the sense of the expansion that it causes (§2.4), has not been fully elucidated. However, if the absorption is such as to neutralize the polymerization shrinkage, or even to place the matrix in compression, the material will fail in tension at a higher stress than otherwise would be the case (assuming no other effects are operating).

●2.13   **Toughness**

The energy required for crack propagation, the toughness, is also increased by the presence of the filler if this has a higher strength than the matrix. As a growing crack meets a filler particle, it must travel around the particle if it is to continue to propagate. This increases the length of the crack path, the **tortuosity**, and therefore the total surface area. As the energy required for propagation is in proportion to the total area of new surface created, the toughness of the composite is thereby increased (*cf.* 25§3). This effect is enhanced by the matrix constraint raising the overall stress at which the strain in the matrix is sufficient to initiate cracking (§2.10). Thus, even if failure is brittle, the work to failure is increased as both the stress and strain at failure are increased, increasing the area under the curve (Fig. 1§4.2).

In addition, if the interfacial bond is weaker than the matrix, as is often the case, the crack tip may be blunted and so reduce the stress concentration at that point, the **Cook-Gordon** mechanism of crack-stopping[16] (Fig. 2.17). There is in front of an advancing crack a zone with a small tensile stress perpendicular to the principal load axis (due to Poisson strain). This tends to tear a weak interfacial bond slightly before or just as the crack tip reaches it. This increase in the effective crack tip radius requires that the principal tensile stress be increased before propagation can again occur.

●2.14   **Internal work**

It should be remembered that toughness is also dependent on the amount of drawing of polymer chains that occurs during crack growth (3§4.7). This is due to the work required for chain segment rotations against the steric hindrance to that rotation

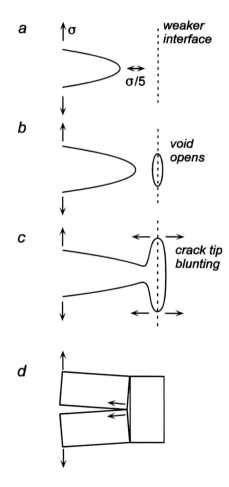

Fig. 2.17   A crack approaching a weak interface (a) may open a tension crack at that interface (b) or just be blunted when it reaches that point (c). The reduction in stress concentration reduces the rate of propagation. The interfacial crack tends to be opened by Poisson tension (d).

within the chain itself as well as simply pushing neighbouring chains and groups out of the way. This plastic deformation both consumes energy (some of which appears as heat) and causes crack tip blunting (Fig. 2.18).

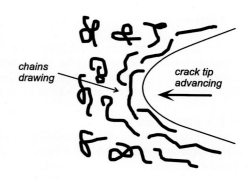

Consequently, those factors that diminish this behaviour must reduce toughness. Thus, cross-linking inhibits drawing and therefore makes the polymer more brittle. On the other hand, the presence of plasticizers can increase toughness. If any mechanisms exist for doing **internal work**, they will increase the toughness of the material. However, the outcome of a given set of conditions is not easy to predict: reference must be made to the failure envelope (3§5) for the polymer to determine the variation in toughness (area under the curve) with the variables of interest.

Fig. 2.18 Chain drawing blunts the crack tip and consumes energy, thus increasing the toughness of the material.

The addition of a filler to a poly(methyl methacrylate) resin system thus offers considerable improvement in mechanical and physical properties, as compared with the unfilled polymer, assuming that a satisfactory interfacial bond can be achieved. But part of the problem lies in the chemical nature of the matrix resin itself: its elastic modulus and strength are low, water absorption and polymerization shrinkage high, each of which limit the improvements that can be obtained with a filler. The selection of a resin system with a less flexible polymer chain can therefore be expected to increase the modulus by limiting chain movement during deformation. Extensive cross-linking has the same effect, bearing in mind that if this is carried too far the matrix will be brittle, not tough.

The optical properties of fillers are of considerable importance because upon these the tooth-like appearance of the restoration is critically dependent. Aspects of this issue are dealt with in 24§5.

●2.15 Water

Water has further roles affecting the strength of filled-resin restorative materials. The plasticizing effect, facilitating chain drawing, has already been mentioned (3§6.1), but there is a longer-term problem: **hydrolysis**.[17] It is certainly the case that 'wet is weaker', in part due to the reduction of surface energy and thus the energy cost of cracking (1§7, 10§1.2). But the crucial factor is the strength of the coupling of matrix to filler (§2.11). Unfortunately, these bonds are subject to attack (*cf.* 25§8), and this will be assisted by stress. Thus, in service, where there is a wet environment, and stress cycling is part of the usual effects of mastication on restorations, aided by abrasion (Fig. 20§4.1), progressive breakdown of the matrix-filler interface will occur, leading to failure of the restoration itself. In fact, the glass of the filler itself may be subject to hydrolysis,[18] which clearly contributes to the breakdown of the material as a whole. This process affects the observed water absorption of such materials (§2.4).

●2.16 Reaction exotherm

It can be noted that the net heat of formation of a C-C bond (kJ/mol) from a vinyl group is much the same, whatever the starting materials, so that the reaction exotherm remains significant for all present system chemistries. Obviously, the volume fraction of filler reduces the total amount of heat produced per unit volume in strict proportion, although it will be similarly affected by the degree of conversion attained (§5.13). However, the rate at which the heat is generated depends on many chemical and physical factors which affect the rate of reaction and cannot be summarized in any simple fashion. Accordingly, the mechanical effects will be more complicated (§2.3), although the biological significance of the lesser temperature rise will be more directly important. In addition, the larger surface : volume ratio for a small restoration, in comparison with a denture base, also limits the temperature excursion. Nevertheless, it has appreciable effects on the subsequent thermal shrinkage.

§3. Filler Loading

It is now apparent that, despite some fundamental limitations, the use of a filler is extremely important in making resin-based materials functional as tooth filling materials and luting agents. Many if not most of the requirements (§1) have already been seen to be addressed. Some of the considerations relating to the choice of filler, in particular particle size, need to be exercised.

●3.1   Differential wear

The early filled resin materials used a fairly coarse filler powder, up to about 50 μm in diameter. High volume fractions were achieved, without making the unset material too viscous to use (4§9.2), and both high elastic modulus and strength were obtained. However, this design had two serious flaws:

(1)   polishing was extremely difficult, if not impossible, even using high-speed diamond instruments (Chap. 20), essentially because of the great difference in hardness between the two components;

(2)   that same difference in hardness led to **differential wear** in service, the soft resin being relieved by foodstuff and toothbrushing abrasion. The surface became rougher with time, and therefore readily discoloured by embedded débris.

Further, the exposed edges of the filler particles then made them easier to dislodge, and rapid erosion of the restoration followed (Fig. 20§4.1). Even though some finer powders were later introduced, at about 5 ~ 10 μm, the problem was only slightly abated. These products are frequently referred to as 'conventional', although the so-called convention has long ceased to be current (*cf.* 'conventional' amalgam alloys, Chap. 14). Of course, wear works both ways: exposed filler (Fig. 20§4.1) can act as an abrasive on opposing dentition and restorations. Reductions in the size of the filler particles were therefore made, although the improvements proved to be relatively minor.

●3.2   Micro-fill

The response to these problems was the introduction of the so-called 'micro-fill' products. These used extremely fine filler particles, a silica 'flour' called 'aerosil' or fumed silica. This material has a typical particle size of the order of 0.05 μm, one tenth of the wavelength of yellow light, and less than one hundredth of the size of previous 'fine' particles, hence these are sometimes referred to as 'micro-fine'. It therefore has no visible effect on the optical properties compared with unfilled resin. However, while this did allow products that were easy to polish (because the filler particles were much smaller than the size of any abrasive particle), the very large surface area of such powders caused unusably high viscosities when present in even a modest volume fraction (4§9.2). The only answer to this, of course, was to use substantially lower volume fractions of filler, the practical limit being about 30 %, and this had the expected effects on elastic modulus (lower), strength (lower), and wear rate (higher). This example illustrates very well the type of compromise which crops up so often in dentistry (as it does in all materials applications).

Since then there have been various other attempts at resolving the conflict between these opposing effects of filler loading. Thus, using the greater power of machines, very high volume fractions of fumed silica can be incorporated into a resin which is then polymerized. This mass is then broken up and reduced to a fine powder, which then is mixed – as if it were a filler itself – with resin containing only a moderate amount of fumed silica. This then is the product delivered to the dentist. This means that the lower viscosity of the uncured mixture allows ready handling; but when polymerized polishing is straightforward since all of the actual filler is very fine. The surface of the restoration thus presents only a varying filler loading from point to point, and the overall wear experienced is less than that of the simple 'micro-fill' materials, while the strength and modulus of elasticity are somewhat higher. However, the wear rate remains unacceptably high. 'Hybrid' materials have been created using both 'fine' and 'micro' fillers in the same matrix, and carefully graded particle size distributions used to minimize the gaps between particles supposedly by allowing them to fit together more efficiently. It should be remembered that the total filler surface area is increased by such means for a given volume fraction, and therefore so must be the viscosity. It follows that the volume fraction overall must be reduced. From time to time, minor variations on these themes have appeared. None of these designs has overcome what should be seen as the inevitable and irresolvable compromise between the demands of viscosity and service performance. The expected life of a filled resin restoration is realistically only a very few years. Regular replacement is necessary, and this means progressive loss of tooth tissue at each treatment. The consequences are obvious.

Whether or not filled resin materials have a place in dentistry is not the issue now, but a recognition of certain fundamental limitations in the nature of the material itself is required. Firstly, the carbon-carbon bond is the weak link which determines ultimately the achievable strength of any such material. Secondly, the viscosity of the uncured material controls the ease with which it may be placed in the cavity, and therefore the amount of filler which may in practice be incorporated. Allowing for the demands on the design of the resin matrix itself, cross-linking, water absorption, shrinkage, and so on, the likelihood of substantial improvement in the performance of the system as it stands must be very small. There must be a limit because the component materials involved themselves have absolute limits (see Fig. 1§14.1).

●3.3   Mass *vs.* volume fraction

Most often, advertising literature is the only source of information that a practising dentist will see. There is, understandably perhaps, a tendency on the part of the advertiser to be somewhat selective in what is communicated, and this is always in the best possible light. It is worth noticing, therefore, that the amount of filler in a resin restorative material is usually (if not always) quoted as a percentage by mass, even if this is not explicitly stated. Remembering that for close-packed spheres their *volume* fraction is about 74 % (4§9), the viscosity of a material described as having the 'typical' "70 ~ 75 %" filler would be impossibly high, especially since such particles will not be spherical and therefore pack badly (as well as exhibiting plastic dilatant behaviour (2§2.3, 4§7.7)). A material with a claimed figure of "80 %" would be essentially solid, not a viscous paste, if this were by volume.

We can easily test the consequences of such filler loadings. The overall density, ρ_c, of a mixture is given by the appropriate rule of mixtures: the arithmetic mean of the volume fraction-weighted densities:

$$\rho_c = \phi_f \rho_f + \phi_m \rho_m \tag{3.1}$$

Then, since

$$\phi_f + \phi_m = 1$$
$$\rho_c = \phi_f \rho_f + (1 - \phi_f) \rho_m$$
$$= \phi_f (\rho_f - \rho_m) + \rho_m \tag{3.2}$$

Hence, the corresponding volume fraction of filler is given by:

$$\phi_f = \frac{\rho_c - \rho_m}{\rho_f - \rho_m} \tag{3.3}$$

Here, ρ_c is unknown for the moment. But mass divided by the density gives the specific volume. Therefore, mass fraction, μ, divided by density gives the actual volume of that component in unit mass of the mixture. The total (relative) volume of the object is simply the sum of the component relative volumes:

$$\frac{1}{\rho_c} = \frac{\mu_f}{\rho_f} + \frac{\mu_m}{\rho_m} \tag{3.4}$$

since $\mu_f + \mu_m = 1$. This can be read as saying that the harmonic mean of the densities, weighted by the mass fractions, also yields an expression for the overall density (yet another rule of mixtures). Then, substitution for ρ_c in equation 3.3, by taking the reciprocal from equation 3.4, yields the required value:

$$\phi_f = \frac{\mu_f \rho_m}{\rho_f - \mu_f (\rho_f - \rho_m)} \tag{3.5}$$

Since the density of many organic polymer resins is in the region of 1.1 g/mL, while that of quartz, a common and cheap filler, is 2.66 g/mL, use of equation 3.5 shows that for a *mass* fraction of filler of 0.75, the *volume* fraction would be about 0.554. This gives a substantially different impression of the filler loading. If this transformation is applied to the data of Fig. 2.16 the effect can be clearly seen (Fig. 3.1). (If, as is common, there are bubbles in the matrix resin of the as-supplied product, effectively lowering its density, the effect is even greater.) It follows that the appropriate way to study

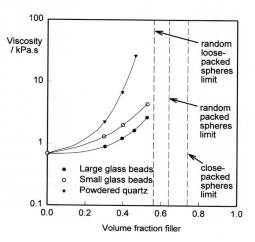

Fig. 3.1 The data of Fig. 2.16 recalculated in terms of volume fraction filler, showing the changed impression of the system that this allows.

the properties of any composite material is by volume fraction, using mass fraction may be a convenience but essentially quite unhelpful. Care is therefore also required in reading the literature.

So that such restorations are detectable in radiographs, there is a tendency for radio-opaque fillers to be used in these products. This requires the use of heavy elements, such as barium, in the glass of the filler powder. These then confer a higher density on the glass, and this will make the above calculation of true filler loading even less favourable, although it would appear to increase the amount of filler present if it is the percentage by mass which is stated. It helps to view the overall effect of this confusion of mass and volume fractions (Fig. 3.2) when it becomes obvious that the 'misdirection' is greatest at volume fractions in the middle of the range, at the low end of the normal dental formulations, where it matters most. 'Misdirection' here means the numerical value of the difference $\mu_f - \varphi_f$. Obviously, such advertised figures must be viewed with circumspection: *caveat emptor*.[6]

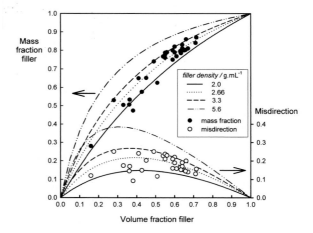

Fig. 3.2 Dependence of mass fraction of a filler on its volume fraction and density (left ordinate); data for a variety of products and experimental materials as reported in the literature. Also shown (right ordinate) the 'misdirection' – the numerical excess of the mass fraction over volume fraction. The filler densities shown range from an arbitrary lower limit, through quartz and strontium-glass, to that of zirconia (which, incidentally, cannot be bonded to the matrix).

●3.4   More product types

Filled-resin materials are not the easiest to handle. They are sticky and viscous (of course), require considerable care to place properly, and do not suit all circumstances equally well. Personal preferences for handling characteristics also play a part. Consequently, a variety of products are available described variously as **condensable** or **packable**, and **flowable**. 'Condensable' essentially means that the unset material is capable of being used with a condensing point, in the manner of dental silver amalgam, and so implies that it has a non-negligible yield point (even though no "condensation", in the sense of making it more dense or compacted, is actually occurring; "packable" is preferred). Conversely, 'flowable' means that the unset material has a relatively low viscosity and negligible yield point so that it will flow easily under gravity: it is runny.

Notwithstanding the use of elaborate or novel filler chemistry, particle shape or size distributions, there are only two main factors of relevance: filler loading and resin formulation. If the volume fraction of filler is high enough, the material will become plastic dilatant (4§7.7), thus exhibiting a clear yield point and requiring some packing effort to mould it to the cavity. Indeed, some products have included glass fibres (5 ~ 10 μm × 50 ~ 200 μm) to achieve this effect (see 29§4.5). On the other hand, if the filler loading is low enough, possibly with adjustments to the types and proportions of the various monomers in the resin, the material can be made arbitrarily fluid.

Such adjustments to rheological properties are associated with the concomitant mechanical property changes that can be attributed to volume fraction of filler (Fig. 2.12) and to set resin structure (§4.3, §4.5, 3§6.2, 5§6.2). The problems seem to be that the flowable materials must inevitably be weaker and less stiff than usual, and the packable materials be less able to conform to the details of a cavity, especially line angles and corners, and to run increased risks of extra porosity being incorporated (*cf.* silver amalgam, 14§8.1). Again, it is a matter of the essential compromise between properties and handling. Product selection must be made very carefully to take into account all relevant factors, including the circumstances of use, if the highest quality of outcome is to be attained – given the unavoidable limitations of filled resins in the first place (§3.2).

---

[6] "Let the buyer beware."

§4. Resin Chemistry

●4.1   Innovation

The design of a polymer for dental use of any kind is severely inhibited by just one factor – cost. An examination of dental materials in general reveals a startling fact: hardly anything has been designed deliberately from scratch with its dental application as the motivating target. In most cases, a material long used in another application has been spotted as offering some useful property, and adapted for dentistry by relatively minor changes. Development has then proceeded by modification of that starting product. (The major exceptions to this are the polycarboxylate and glass ionomer cements.) The trouble is that dentistry is, in industrial chemical terms, relatively a very small market (example: polysulphide rubber impression material had just one throwaway line in the definitive textbook on that class of polymers). More aggressive marketing techniques have expanded the scope, but there remains relatively little incentive for truly innovative research into new materials for the field. These observations underline the generalization that very little is so special to dentistry that it cannot be understood by reference to other, broader, applications. (The corollary of this is that a knowledge of dental materials provides a thorough basis for understanding most everyday materials.) This dependence of dentistry on other fields of development is well-illustrated by the subject of the present chapter.

●4.2   Goals

Given that the limitations of filled PMMA arise essentially because of the properties of the PMMA itself, the desirable aspects of a resin system for use in restorative materials may be readily identified:
- a large monomer molecule to reduce polymerization shrinkage;
- reduced polarity to reduce water absorption and plasticization;
- sterically-hindered monomer to increase the elastic modulus of the resin;
- more than one functional group per unit for extensive cross-linking;
- non-toxic[7];
- cheap.

●4.3   Bisphenol A

The most convenient and cheapest starting point for the majority of resins used in filled restorative systems in dentistry is bisphenolacetone, commonly called **bisphenol A** in dentistry (Fig. 4.1), one member of a large class of similar compounds. This is cheap because, industrially, it is prepared very easily from phenol and acetone, both of which are industrial chemicals available in bulk at low cost. It is, in fact, used commercially for various resin applications. This molecule can be

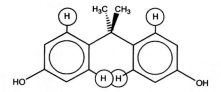

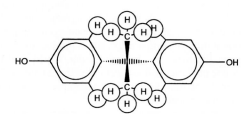

Fig. 4.1 The structure of 'bisphenol A', 2,2-bis(4-hydroxyphenyl)propane, showing aspects of the steric hindrance to rotation caused by the phenyl hydrogens at positions 2 and 6 on each benzene ring.

incorporated into polymer chains in a number of ways, but its usefulness lies in its structure.

Rotation of the benzene rings about the central bonds to the propane residue is impossible at ordinary temperatures (Fig. 4.2). The steric hindrance produced by the overlapping hydrogen atoms is so great that only very high temperature

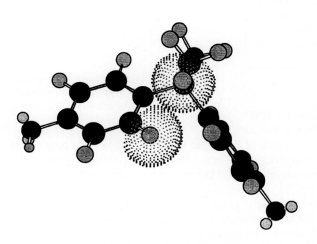

Fig. 4.2 A diagram of a model of bisphenol A showing the interference at the van der Waals radius of the phenyl and methyl hydrogens on rotating the phenyl group.

---

[7] See box "Toxicity", below.

could permit enough energy to be available for that rotation (Fig. 4.3). However, a C-C σ-bond has an energy of about 350 kJ/mol. In other words, decomposition of the molecule would occur first! Thus, any polymer chain incorporating such a unit can be expected to be very much more rigid than an equivalent length section of a methacrylate polymer.

Advantage has been taken of the rigidity of this structure in many filled-resin restorative materials by incorporating bisphenol A in a molecule, terminated at each end by the methacrylate group: bisphenol A glycidyl methacrylate, usually just called **bis-GMA**[8] or "Bowen's resin" after its inventor[19] (Fig. 4.4). This monomer can, of course, be polymerized by any of the usual initiator-activator systems such as benzoyl peroxide-alkyl borane. Having two vinyl functional groups, the cured resin would be very highly cross-linked, and so be very rigid. It may be easier to understand the final resin structure by viewing it as poly(methyl methacrylate) which has been cross-linked with a bisphenol A-based molecule. In addition, the size of the bis-GMA molecule is such as to give a shrinkage of only ~8 % when polymerized alone (no filler). All of the points listed above in §4.2 can be seen to be satisfied by this material, a remarkable achievement, although the presence of the two hydroxyl groups raises the water sorption unnecessarily as well as increasing its viscosity through intermolecular hydrogen bonding.

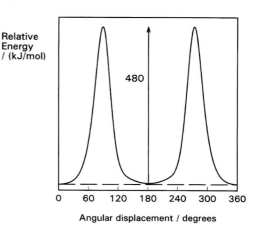

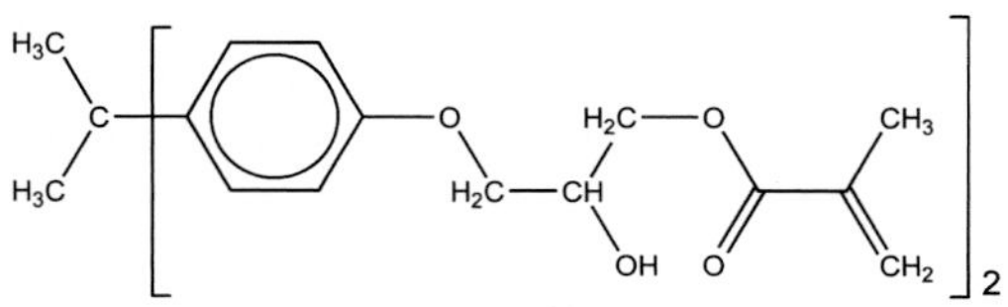

Fig. 4.3 The energy of bisphenol A as a function of the angle of rotation of one phenyl group (for scale, *cf.* Fig. 3§2.8).

Fig. 4.4 The "bis-GMA" molecule, used as a common basis for the matrix of filled-resin restorative materials.

Toxicity
There has been much debate about the oestrogen-mimicking and endocrine-disruptive properties of bisphenol A (further abbreviated to "BPA"), because of its widespread use in consumer products and packaging, leading to bans in certain applications.[20] Note that although bisphenol A is not an ingredient as such of any dental product, there may be some remnant as an impurity, as well as the possibility of being a degradation product of the matrix of filled resin, that may very slowly leach out. Whilst a definitive statement is not possible, it does seem that the amount of exposure from dental materials is very small in comparison with other sources of this and related substances.[21][22]

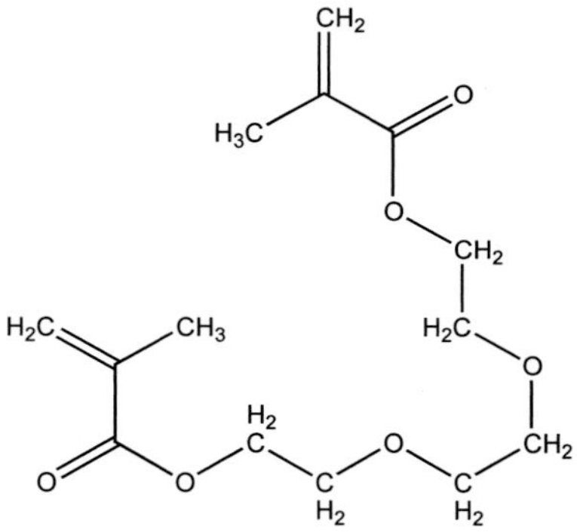

Fig. 4.5 TEGDMA – triethylene glycol dimethacrylate – a very flexible monomer used to reduce the viscosity of matrix resins.

●**4.4   Dilution**
Difficulties arise, however, in that the monomer bis-GMA is itself a very viscous substance. Firstly, it has a high molecular weight (~513 Da), it has a large rigid section, and is capable of hydrogen bonding (10§4) to its neighbours because of the presence of the hydroxy group and carbonyl oxygen. Then, when mixed with the very high filler loadings required from the other considerations, discussed earlier (§2), it becomes an unusable, nearly solid mass.

Methyl methacrylate or other small molecule with a vinyl group, could be added as a thinner or diluent for the uncured paste. However, the volatility of MMA is potentially a problem in that its loss would affect properties adversely. The vinyl group is necessary to ensure that any diluent is also a polymerizable molecule and therefore is itself incorporated in the final polymer and so not leachable. But a singly-functional monomer would reduce the cross-linking density of the final matrix. Hence, what is required is a non-volatile (*i.e.* large), flexible, bifunctional monomer.

---

[8]   An erratically-formed abbreviation: **bis**phenol A **g**lycidyl **m**ethacrylate

Triethyleneglycol dimethacrylate (TEGDMA, sometimes TEDMA) is commonly used for this purpose. It is an extremely flexible molecule because the ether links in the chain present little steric hindrance to rotation (Fig. 4.5). The use of such a molecule unfortunately reduces the stiffness of the matrix polymer somewhat, because there is now a lower concentration of the rigid bisphenol A moiety. Nevertheless, this reduction in stiffness may in fact be an advantage in preventing completely brittle failure (*i.e.* soft rather than hard glass), that is to say that the polymer (and therefore the composite material) is tougher. A further consequence is that the hydrophilicity of the resin is also increased (*cf.* polyether impression materials, 7§5) and so the water-absorption is a little higher as a result. Again, it is a compromise between conflicting demands that must be balanced.

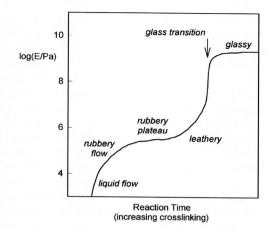

Fig. 4.6 Schematic variation in modulus of elasticity during setting of filled-resin restorative materials.

●4.5   Property development

It is worth noting at this point that the process of setting in filled-resin restorative materials, involving as it does the extensive cross-linking of the structure, is necessarily accompanied by a steady increase in the glass transition temperature (3§4). As a result, all of the characteristic kinds of behaviour, from viscous flow through rubbery, leathery, glassy, may be observed in sequence during this process (which is envisaged as occurring at more or less a constant temperature). It is as if a graph such as Fig. 3§4.13 were read from right to left (Fig. 4.6) as the reaction proceeds.

The general changes in mechanical properties can also be illustrated as in Fig. 4.7. Initially, the resin matrix is more or less simply viscous, perhaps with some retarded compliance (4§6.2) that causes some recovery after deformation, although this may be difficult to see because these materials are very sticky and adhere to instruments readily. As cross-linking proceeds, the viscosity rises steadily, but because this process involves the creation of a three-dimensional network, the rubbery behaviour rapidly becomes very marked and the true (*i.e.* Hookean) elasticity rises – it gets stiffer.[23] However, because chain segment movements in response to applied stresses involve diffusion through a viscous environment, the retarded compliance component increases as well.

Further cross-linking continues to cause both the viscosity and stiffness to go on rising, but the retarded compliance goes through a maximum as the average distance between cross-links shortens, eventually becoming negligible (although probably not zero). The increasing viscosity also has an effect on the reaction rate: the diffusion of reactive species is significantly impeded, and it becomes more and more difficult for them to come together and react (Fig. 4.8). (It is unlikely, however, that the Trommsdorff effect, 5§2.9, occurs in these systems, or has an appreciable effect, because their starting viscosity is already very high.) In the present case this **rate limiting step** is the diffusion of monomer reactive groups to free radical-bearing polymer chain ends. The rate of increase of

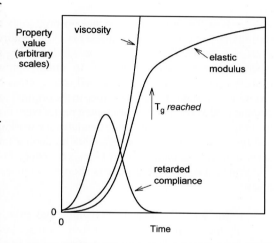

Fig. 4.7 Schematic variation in mechanical properties during the setting of filled-resin restorative materials. (The retardation time would also be expected to go through a maximum.)

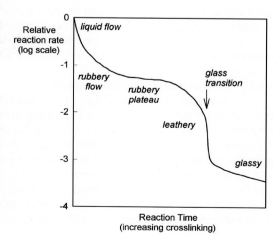

Fig. 4.8 Schematic variation in reaction rate during setting of filled-resin restorative materials showing the corresponding type of mechanical behaviour at each stage.

elastic modulus therefore then decreases. In fact, as the glass transition temperature is rising because of the cross-linking, there comes a point when it reaches and then exceeds the actual temperature of the material. The diffusion rate of reactants then must fall rather abruptly on passing through this point, even though the reaction is not completed. It is then, *i.e.* when the glass transition temperature of the setting material is equal to the actual temperature of the object ($T_g = T$), that the material may be considered to have set effectively, in practical terms. Afterwards the value of the elastic modulus rises only very slowly, asymptotically to the eventual limiting value. Completion of the reaction is very much diffusion-limited. In fact, the **degree of conversion** (the proportion of double bonds that are reacted) may never get very high because of this viscous limitation (although there may be some topological or stereochemical difficulties in connecting all functional groups when connection is at random).

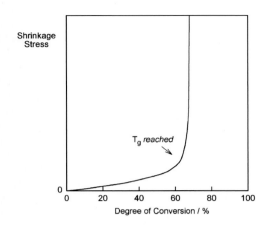

Fig. 4.9   Schematic variation in polymerization shrinkage stress during the setting of filled-resin restorative materials showing the effect of the T_g being raised to ambient. (If the stress exceeds the strength of the weakest part of the system, then a rupture must occur.)

It is the progressive change in properties during setting that imposes a major clinical handling requirement. Thus, it is extremely important to avoid disturbing a resin when it has reacted sufficiently to attain appreciable rubbery properties, and certainly well before cross-linking has taken it to the **gel-point**, the state of the object-spanning, "infinite" molecule. Since in the original 'two-paste' products (see below) the polymerization reaction commenced as soon as the component pastes were in contact with each other, the reaction rate had to be designed to compromise between the time available for working, that is before appreciable rubbery behaviour developed, and the time required for an effective hardening (and this also applies to impression materials, cements and several other systems). This problem is of reduced significance in the one-paste, light-cured materials since the duration of the vulnerable stage is shorter and placement occurs before setting is initiated, but it has not gone away.

In 3§6.2, the ideas of branching and cross-linking were introduced in addition to chain growth, and certainly these aspects of polymer formation are key aspects of the resins discussed here. We might note that when long, flexible, difunctional monomers are used, and the network becomes as complex as this, other kinds of reaction may occur, that is, loop formation or **cyclization**.[24] This may involve a free radical attacking a vinyl group attached to the same chain or the mutual reaction of two free radicals on the same chain or monomer. Such **intramolecular** reactions may result in closed loops with or without included branch points, and although not contributing to the extension of the network they must still affect the properties of the bulk material because they represent further constraints on chain segment movement. Indeed, if the loop wraps around a network **internode** or another loop, separation by drawing is not possible and chain rupture must occur if the load on the chains is high enough. Chain flexibility is important: TEGDMA (Fig. 4.5) forms primary loops (that is, within the same monomer molecule), at some three times the rate that bis-GMA (Fig. 4.4) does.[25]

●4.6   Polymerization shrinkage stress

As a direct indication of the effect of reaching the T_g by the cross-linking reaction, the development of polymerization shrinkage stress may be examined (Fig. 4.9). The experimental set up is of the resin being placed between walls, representing those of the cavity; these walls are wetted by the resin. As polymerization proceeds, the shrinkage generates a stress transmitted to those walls, and this may have detrimental effects.[26] It is possible, through spectrophotometric means, to monitor concentration of double bonds during this process so that the extent of reaction can be directly mapped. It is found that whilst the elastic properties are developing, and the viscosity is rising, so the stress rises slowly. But at the point of **vitrification**, when the T_g has reached the temperature of the resin body, the stress rises rapidly, and very little conversion of vinyl groups to cross-links occurs after that point. Essentially, vitrification has reduced substantially a path for stress relaxation (*cf.* §2.3).

The consequences of this stress can be examined further. Thus, in Fig. 4.10, if a bar of material is allowed to shrink unconstrained from length L_1 to L_2, such that the stress in state S_2 is zero, we have the following linear shrinkage coefficient:

$$\alpha = \frac{L_1 - L_2}{L_1} = \frac{\Delta L}{L_1} \qquad (4.1)$$

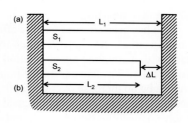

To return it to the state S_1 requires a stress to be applied to generate the equivalent strain in reverse:

$$\varepsilon = \frac{L_1 - L_2}{L_2} = \frac{\Delta L}{L_2} = \alpha \frac{L_1}{L_2} \qquad (4.2)$$

– which is close to α when the shrinkage is small. Then,

$$\sigma = E.\alpha \qquad (4.3)$$

Fig. 4.10   Shrinkage generates a stress acting on the walls of the cavity, that required to return S_2 to the length of S_1.

This is then the stress applied to the wall of the cavity, *i.e.* at the interface (assuming a completely rigid wall material). If this exceeds the strength of the union, then a marginal gap opens. If this failure is not immediate, then plastic deformation – yield – is expected to occur with time, assuming that no other changes are occurring, and that the failure is slow and progressive. Shrinkage, of course, is not unidirectional, and the stresses are more complicated, depending on how much constraint there is (Fig. 4.11): the stresses on the walls will not be uniform, and also involve appreciable shear. The consequences can range from simply detaching the resin material from the cavity wall to cracking the attached enamel if the adhesion is good enough, especially in the presence of pre-existing flaws.

    The values are such that this would happen as a matter of course. Accordingly, various techniques have been explored for reducing the shrinkage stress, in particular incremental build-up (Fig. 4.12). By using small portions, cured in turn, a large proportion of the shrinkage can be compensated, although this is only really practicable with light-cured materials (§5) (which for restorative purposes are now used essentially exclusively used).

Fig. 4.11   Shrinkage is not unidirectional, and stresses and distortions are complicated, depending on the constraints present.

    The assumption of a rigid cavity is severe. In reality, the finite modulus of elasticity of enamel allows the walls to flex appreciably, and to an extent dependent on the depth and width of the cavity. The stress generated is, accordingly, less than might be expected.[27] Even so, enamel fracture can occur.[28] Of course, the forces of mastication may produce increases in stress when the resultant flexure of the wall is outward.

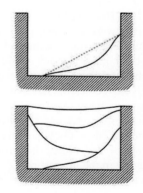

    In dentistry, the effect of constraint has been described in terms of the so-called C-factor ("configuration factor"), defined as the ratio of the number of bonded to unbonded walls of an idealized cuboidal restoration. This is, however, a rather crude measure that ranges from 1 to 5 in unit steps, and is clearly not linear (*cf.* water : powder ratio, 2§4). **Shrinkage constraint** (or, generalized to **dimensional constraint**, to include expansion) is better considered in terms of the fraction (or percentage) of the total surface that is constrained, A_b/A, but even then that fails to take into account the shape of the piece. The greater the total surface area in relation to the volume, A/V, the greater the constraint that a given proportion represents (this is an exact parallel of matrix constraint, but on a macroscopic scale). A long narrow restoration is clearly more constrained than one of the same volume that is more equiaxed but with the same number of bonded walls, and even more so for a re-entrant shape such as a typical molar fissure outline. In addition, the flexure of the walls modifies the outcome. Such considerations should be involved in deciding on the increments to be used (Fig. 4.12).

Fig. 4.12   Incremental build-up can compensate for much of the shrinkage and reduce the residual stresses substantially.

●4.7   Filler bonding

It has been stressed that the effectiveness of the filler in such resin-based materials depends on bonding the two together. It should be noted that the critical step is done at the manufacturing stage, when the filler is treated with a coupling agent solution so that molecules carrying a polymerizable vinyl group are covalently attached (10§5.1). The number of potential links to the resin matrix polymer network is thereby fixed. Thus, there can be no increase over this in the bonding of resin to filler at the filler particle surface during the setting process. Of course, it could be argued that the bonding is becoming more effective in the sense that there may be more connections through to the network and therefore more effective matrix constraint results, but if the filler were not treated nothing would happen and the filler would not have the desired effect.

●4.8   Two-paste problems

Originally, then, filled-resin restorative materials were presented as two pastes, one containing initiator with an inhibitor, the other the activator, after the pattern of cold-cure acrylic. Shelf-life was rather limited, although usefully extended by refrigeration, but the biggest difficulty was mixing the two pastes together in an homogeneous mass that would set uniformly well (15§7). The very high viscosity (4§9) meant that the incorporation of large air bubbles was inevitable, with the expected effects on modulus of elasticity and strength (Griffith criterion, 1§7), but with polishing or wear these were exposed and contributed to discolouration. In fact, the manufacturing difficulties are such that many small bubbles already exist in the paste of most such products as presented (including light-cured materials). The second result of the mixing difficulty was the practically unavoidable unevenness of the mixing itself: this necessarily meant that the degree and timing of polymerization was variable throughout the mass with, again, undesirable effects on strength and elastic modulus. It was therefore desirable to eliminate the stage of mixing such materials for use in the clinic. This was partly the motivation for developing 'command set', light-cured materials (§5).

Despite this, two-paste products continue in use in one important context: for **luting** crowns, inlays and some kinds of bridge. Described as 'resin cements' (9§1.5), such materials are used under opaque devices and therefore could not be light-cured. The chemistry is then essentially that of the materials described in §1.1, §1.2. The need for proper mixing is possibly helped by the requirement that such materials have sufficiently low viscosity that they can be reduced to a thin film beneath the device easily (9§9) and so have lower filler content than otherwise similar 'restorative' products. Nevertheless, particular attention must still be paid to proper mixing (15§7).

●4.9   Alternative chemistry – monomer

Bis-GMA (Fig. 4.4) is commonly referred to as if it were a single, pure substance. In fact, in manufacture, the addition of the acrylic acid may proceed through attack on either of the two epoxide carbon atoms of the precursor, bisphenol A diglycidyl ether, or "BADGE"), leading therefore to two isomeric forms, not quite randomly, although the factors involved are complex. This means that commercial products contain a mixture of the two forms, *a* and *b* (the iso- form, Fig. 4.13), say, roughly in the proportions 2 : 1.[29] However, since the two sides of the bis-epoxide of bisphenol A will react independently, there are three kinds of molecule possible: *a-a*, *a-b* and *b-b* (*b-a* is, of course, indistinguishable from *a-b*). Since the overall probability of *a* is ~2/3, that of *a-a* is $(2/3)^2$, and so on, leading to the proportions of the three kinds of molecule being ~ 4 : 4 : 1, respectively. This has little bearing on the outcome, in fact, as the

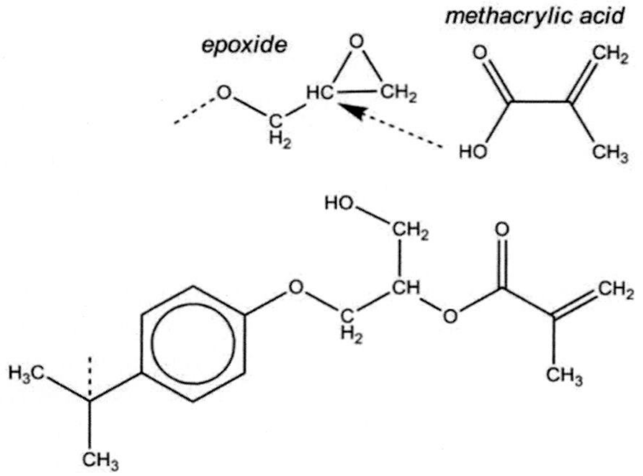

Fig. 4.13 Formation of iso-'bis-GMA'. The 'internal' attack point on the epoxide is arrowed.

reactivity of the methacrylate vinyl groups is unaffected, and the set resin would be amorphous anyway, although perhaps with a slightly lower T_g because of the little side chain (but offset perhaps by slightly easier hydrogen bonding). However, it does point out that simplistic assumptions (which occur often in the field) pose some risks. To this complication may be added the likelihood of oligomers of the epoxide forming during manufacture (27§5). There are also many other analogues of bisphenol A,[30] any one of which conceivably is usable in preparing a dental resin.

The hydrophilicity of bis-GMA, due to the hydroxyl group on the glycidyl group (Fig. 4.4, 4.13) is a concern because it raises both the viscosity of the unset material because of hydrogen bonding, and the water sorption of the set material (§2.4). An alternative monomer which eliminates that aspect is so-called **bis-EMA**, or ethoxylated bisphenol A glycol dimethacrylate (Fig. 4.14). It can be see that this produces etheric chain segments similar to those of TEGDMA (Fig. 4.5), which further lower the viscosity, but which are still also associated with some water sorption effect (*cf.* 7§5.3).

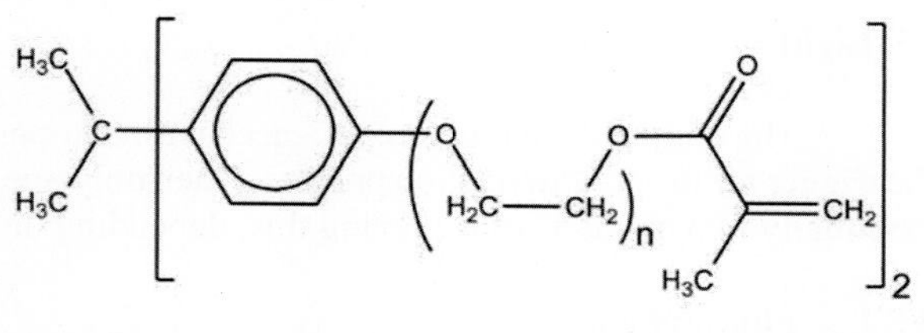

Urethane-based monomers are also widely used (see §7.5), but of several kinds, including isomers. They all appear to be dimethacrylates, but all appear to be given the abbreviation **UDMA**.[31] Clearly, some care is required in reading the literature. A typical example is shown in Fig. 4.15.

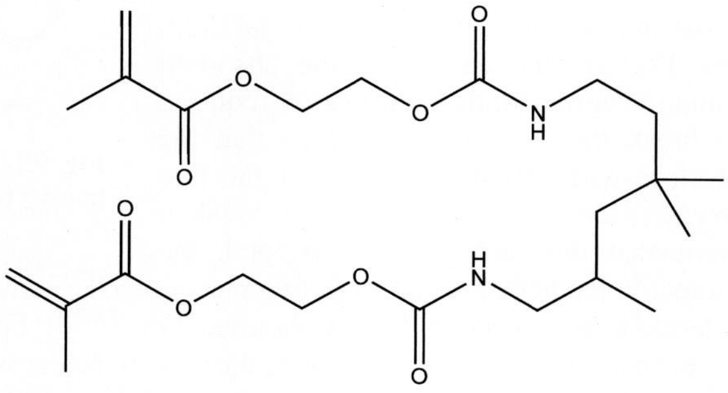

Reasons for urethanes such as this to be used include much increased flexibility in comparison with bis-GMA but not as much as TEGDMA. Such factors are important for the effects they have on the rate of polymerization[32] as the diffusivity of the live chain end controls the relative rates of chain extension and of termination by mutual annihilation (Fig. 4.16). In addition, it controls the degree of conversion – the proportion of vinyl groups that are reacted (4.17). This is correlated with the cross-linking density, and thus is important for the development of the mechanical properties of the set resin, although the details of the chemistry of the chain are also significant as they affect chain segment mobility.

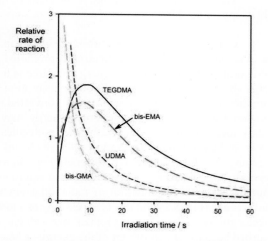

Fig. 4.14 Bis-EMA: ethoxylated bisphenol A glycol dimethacrylate. Typically, n = 1 ~ 3, each side, average 2 overall.

Fig. 4.15 Diurethane dimethacrylate (one isomer – the central alkyl chain has several possibilities).

Fig. 4.16 Relative rates of reaction of light-cured pure monomers.

Of course, in practice, various mixtures of these and other monomers are used in an attempt to optimize the outcome. Again, however, it is clear that there is a trade-off between them. There is an essential compromise involved because the various factors act in different directions.

These examples lead us into the topic of light-cured resins.

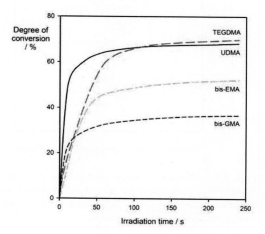

Fig. 4.17 Variation in degree of conversion for light-cured pure monomers.

§5. Light-Cured Materials

The mixing difficulties experienced with two-paste systems are eliminated completely by using light as the trigger for the polymerization process, when only one, ready-for-use paste is required. Chemically, there are essentially two methods of achieving this, depending on whether light energy is used directly or indirectly.

●5.1   Photolytic initiation

This is, as its name implies, the use of light to break a bond directly and create free radicals that can start the polymerization chain reaction. Compounds such as benzoin and its ethers (Fig. 5.1) (and similar substances) readily absorb ultra-violet light. But because the energy of the absorbed radiation corresponds to the energy required to break the central bond, given that the electron-withdrawing tendency of the two oxygen atoms is sufficient to weaken somewhat the carbon-carbon σ-bond, the

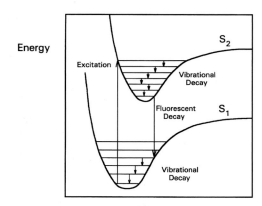

benzoin methyl ether

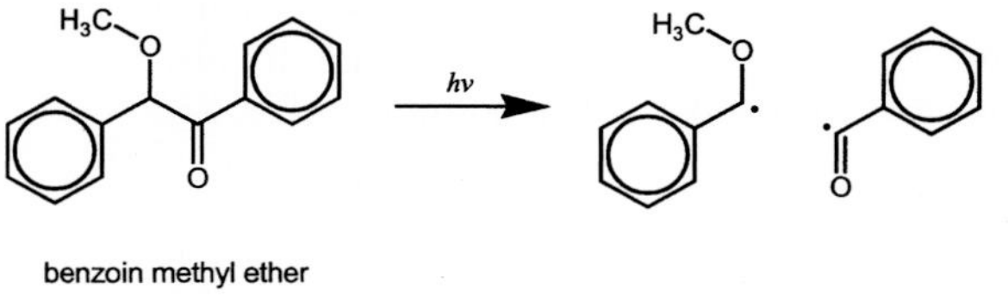

Fig. 5.1 The photolysis of benzoin methyl ether by ultra-violet light to give two free radicals (λ_{max} ~247 nm).

molecule tends then to fall into two pieces, each a radical. This is called **homolytic scission**. (This may be referred to as a **Norrish Type I** photoreaction.[33][34]) For the present example and its kind, there are a number of resonant structures that delocalize the unpaired electron (*cf.* 5§4), but these radicals are nevertheless energetic enough each to initiate the polymerization chain directly without using benzoyl peroxide.

●5.2   Photosensitizers

There is an alternative route to the generation of free radicals, that is, to use a **photosensitizer**. Here, light energy is used indirectly, and a number of steps and conditions are needed for it to operate.

Consider a molecule of a substance which, in the **ground state**, is capable of absorbing radiation, such that in so doing a non-bonding electron is promoted to an **excited state** (*cf.* 24§6). The emphasis here is on *non*-bonding electrons since the energy required is lower than for bonding electrons. There is usually excess energy absorbed because the condition for absorption is that the energy must be at least as great as the minimum promotion energy. The excess energy is converted into vibrational energy, but this vibrational energy may be quickly lost through collisions with other molecules, to appear as heat (as indeed the molecule which did the absorbing was then itself 'hot'); this energy is said to be **thermalized**. The excited electron may now return to its ground state with the emission of radiation known as **fluorescence** (Fig. 5.2), but this is necessarily at a longer wavelength (*i.e.* lower energy) than that which was absorbed because of loss of part of the energy as heat.[35] The time scale for this emission process is of the order of 10 ns, compared with about 0.1 ns for the loss of the excess vibrational energy. To put these figures into perspective, compare the typical bond vibration period of about 0.1 ps (= 0.0001 ns). These electronic events are thus very fast. Consequently, there is usually insufficient time for a collision to occur which could involve any kind of chemical interaction with another molecule, no matter how much energy was associated with the original absorption because diffusion is so much slower. Free radicals, in particular, are therefore not produced.

Fig. 5.2   Potential energy curves showing the absorption of radiation in the ground singlet state S_1 to create an excited singlet S_2, spontaneous vibrational decay, and fluorescent re-emission.

●5.3   Singlet states

Now, electrons are usually paired in their molecular orbitals and, according to the **Pauli exclusion principle**, their spins must be **anti-parallel**, that is, in opposite directions. This is known as a **singlet state** (S). When radiation is absorbed to promote an electron, the spin of the excited electron is conserved – the same as it was in the ground state (because the simultaneous change of both electronic energy and spin is forbidden according to the quantum number rules). The result is called an **excited singlet state** (S_2 in Fig. 5.2). This conserved spin permits the ready return of the electron to its previous state (S_1) through a fluorescent emission since it can, of course, re-enter the original orbital without breaking the Pauli exclusion principle.

●5.4　Triplet state

There are, of course, many electronic energy states available to complex organic molecules. Some of these correspond to the condition where the one excited electron has its spin *parallel* to that of the corresponding unpaired electron remaining in the ground state. Such an electronically-excited molecule is said to be in a **triplet state** (T) (Fig. 5.3).

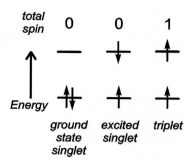

Fig. 5.3　Electronic state of a molecule for various spin conditions. The arrows are symbolic for the 'up' ($+\frac{1}{2}$) and 'down' ($-\frac{1}{2}$) spin states.

Occasionally, compounds have the energies of the excited singlet and triplet states so similar that there is some overlap of the energy curves (Fig. 5.4). Now it becomes possible that, during the decay of the excess vibrational energy of the excited singlet state following radiation absorption, at the point where the potential energy curves cross, the electron may, by chance, have its spin flipped by a collision (without a change of energy), and thus pass into the triplet state. In so doing, of course, the spins of the two electrons are now parallel, and the return to the ground singlet state becomes **'forbidden'**. This process of entering the triplet state is known as an **intersystem crossing** (Fig. 5.4). Such an event is relatively improbable, and the usual course will be for fluorescent decay of the excited singlet. But, because the return to the ground state from the triplet state involves the forbidden reversal of spin at the same time, an increasing proportion of excited electrons becomes trapped in the triplet state. We obtain what is called a **population inversion**, where the normal distribution of molecules over the possible energy states (*i.e.* lowest energy state most occupied, *etc.*) is changed dramatically such that the excited states dominate.

●5.5　Phosphorescence

However, such excited triplet molecules do not remain like this forever. Due to the complicated electric fields involved, collisions are sometimes capable of activating the decay to the ground state by causing the excited electron to change spin again, back to antiparallel. This requires that the vibrational state in S_1 closely matches that of T for the molecule at that moment. A transition of this kind will then be followed by the emission of radiation, and the process is called **phosphorescence**.[36] The radiation emitted is again at longer wavelength than that originally absorbed. This singlet-triplet transition is also 'forbidden' to occur spontaneously. It may take very much longer to occur than does fluorescence, perhaps thousands of times longer, occurring anything from 10 μs to many minutes after the original absorption. Hence, the number of molecules in this triplet state tends to build up with continued irradiation.

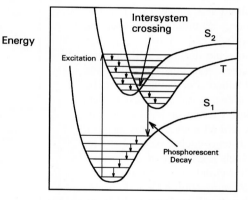

Fig. 5.4 Potential energy curves for a phosphorescent system. Vibrational decay of the excited singlet leads to an intersystem crossing, $S_2 \rightarrow T$. The transition from triplet state T to S_1 is 'forbidden' so the number of molecules in that state T increases.

●5.6　Radiationless transfer

It is now a case of a dynamic equilibrium between the populations of molecules with electrons in the three states, but favouring larger numbers in the triplet state if the irradiance is increased. In effect, the irradiation pumps the molecules up to a higher energy state in which they become trapped, and from which they take a long time to escape. Now, if the colliding molecule that causes the triplet excited molecule to return to the ground state is itself capable of absorbing the energy, rather than activating phosphorescent reradiation, that return to the ground state will be **radiationless**. In other words, the collision transfers the energy from the one molecule to the other. As the one returns to the unexcited condition (to be recycled), the other is excited to a high energy state. The fate of the transferred energy is now our concern.

That the triplet state has a long lifetime is essential because diffusion is required for the collision with the transfer 'acceptor' species to occur, the rate of reaction being proportional to the product of the concentrations of the two. It should also be borne in mind that for reaction to occur between the reactants, in any kind of system, the correct parts of the two molecules must collide in the right relative orientation. Thus, there may be a number of collisions between the reactants that are ineffective. In this case, the survival time of the triplet state must be great enough to accommodate this kind of inefficiency.

●5.7 Benzophenone

Benzoyl peroxide does not have a useful absorption in the ultra-violet or the visible region (it is colourless) that would permit its direct photolysis to free radicals, hence the use of benzoin methyl ether as described above. This is despite the fact that the central bond of benzoyl peroxide can be broken by moderate heat alone, in other words, it is in the thermal vibrational range (5§2). The energy required can however be transferred from another molecule during a collision, no matter how that energy was obtained. For example, it can be the energy of a trapped triplet-state excited molecule, such as described above. Thus, using a substance such as benzophenone, which does have a convenient ultra-violet absorption (λ_{max} ~ 250 nm) (but which does not photolyse), irradiation will, in the first instance, produce an excited singlet. Ordinarily, this will decay rapidly with a fluorescent emission. But, should the hoped-for but very rare collision with a molecule of benzoyl peroxide occur, the energy may be transferred. The central bond must break, so that two radicals are formed (Fig. 5.5).

●5.8 Solvent cage

This looks promising. Unfortunately, the radicals have a short lifetime and are therefore extremely unlikely to be useful in initiating polymerization. This is because these radicals have retained their antiparallel spins (total spin is conserved in the reaction, change is forbidden) so that decay through mutual collision is easy. The recombination is aided by the fact that the two will be held together by the surrounding molecules for a relatively long time compared with the recombination decay time. This effect is known as the **solvent cage** (Fig. 5.6). Thermal molecular motions occur on much longer timescales than

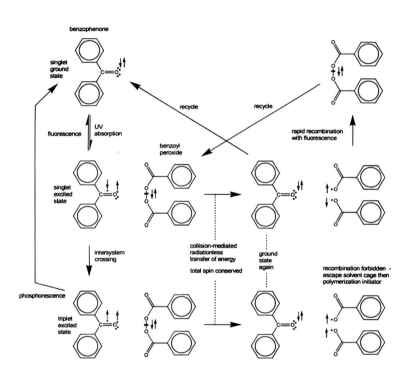

Fig. 5.5 The photosensitized dissociation of benzoyl peroxide. The antiparallel spin pair of radicals are both unlikely to form and likely to recombine rapidly – if they form at all. The parallel spin pair are the effective polymerization initiators.

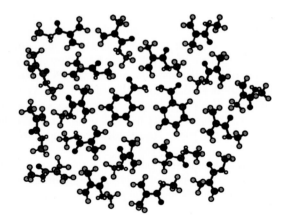

Fig. 5.6 The solvent cage for a pair of newly formed benzoyloxy radicals (with the rings) in methyl methacrylate: they must diffuse out to avoid recombining if they have antiparallel spins. (It will of course be denser than in this diagram.)

intramolecular vibrations. The probability of polymerization initiation is much further reduced by the extremely short lifetime of the benzophenone excited singlet itself, *i.e.* fluorescence occurs before it has a chance to collide with a peroxide molecule. The collision transfer reaction is therefore unlikely in the first place.

Figure 5.6 has been drawn for methyl methacrylate as the solvent. However, in bis-GMA and other systems with much larger and more rigid molecules, the viscosity of the surroundings is accordingly higher, and thus the diffusivity of the newly-formed radicals that much lower because the activation energy required for diffusion will be higher (equation 3§3.3). The requirement for a long lifetime to allow escape from the solvent cage is then all the greater. As polymerization and cross-linking proceed, so the requirement becomes more severe.

> In diagrams such as Fig. 5.2, increasing molecular temperature is represented by the increasing amplitude of the vibrational energy levels. If the temperature is high enough, the bond breaks (the separation goes to 'infinity') (10§3.8). In the thermal processing of acrylic (5§2), the benzoyl peroxide dissociates in just this way, resulting in a pair of anti-parallel spin free radicals. These are a equilibrium processes since this thermal energy is Boltzmann-distributed (*cf.* Fig. 3§3.2). Whilst such radicals are hot enough (*i.e.* maintained so by the surroundings), recombination is not favoured even though the spins do not forbid it (or if it occurs separation soon occurs again). Thus, thermally-generated radicals usually have a life-time long enough to escape the solvent cage and initiate polymerization.

However, benzophenone does exhibit an intersystem crossing, and on irradiation excited-triplet state molecules are formed. Because these have a long lifetime, they have a much higher probability of colliding with and transferring their energy to benzoyl peroxide. But (and most significantly), because spin is conserved, the electrons of the two benzoyloxy free radicals formed now have parallel spins. Recombination is therefore forbidden. These radicals now have plenty of time to escape the solvent cage, and therefore have a good chance of being effective as polymerization initiators. The benzophenone itself, in any case, whether through fluorescent or phosphorescent emission, or through collision, is returned to the singlet ground state and is available for re-excitation by further absorption. It is therefore an **energy-transfer agent**.[9] A photosensitizer only works this way if the intersystem crossing can occur, giving a high energy species with a sufficiently long lifetime. There are many substances that can operate in this fashion, with various absorption spectra, efficiencies of triplet formation, and cost, but the essential mechanism is the same for all of them (Fig. 5.7). The advantage of using this process (as opposed to photolytic initiation) may be that the lack of delocalization of the unpaired electron on the peroxide radical makes polymerization initiation more efficient.

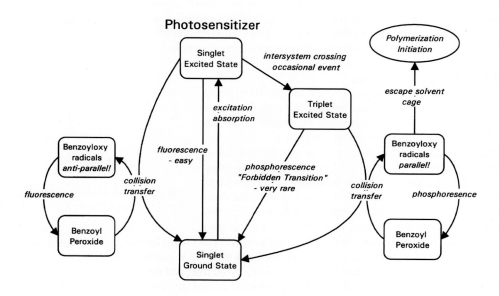

Fig. 5.7 Flow chart for the processes occurring on irradiating a photosensitized polymerization system.

---

[9] Note that the benzophenone on its own does nothing, and therefore is not an initiator of anything.

●5.9   UV problems

However, the use of ultra-violet light in this context has some serious attendant problems: it is injurious to soft tissue ('sunburnt' gums have been reported) and especially to the eyes, so that the user needs to employ eye protection as a matter of course and to exercise considerable care generally. Accidental exposure of the patients' eyes to the light is also of concern. The lamps used as the source of the light are expensive, and have the unfortunate characteristic of a steadily declining efficiency with age, that is, reducing output intensity, which makes consistent curing difficult if not impossible. Compensation would entail a steadily increasing exposure time being used, but with little accuracy if reliance were placed on guesswork. A radiometer ("intensity" meter) would add substantially to the cost. Ultra-violet light is strongly absorbed by tooth tissue, making the resin accessible for exposure effectively only from its free surface. In addition, the strong scattering by the filler and the absorption due to other components of the resin causes a rapid decrease in intensity with depth; **depth of cure** thus depends on intensity and time. Consequently, the polymerization essentially proceeds from the surface down, making it difficult to cure fully deep restorations without very long exposure times. Polymerization shrinkage under these conditions tends to cause contraction away from the walls and floor of the cavity, towards the more rigid surface layer, jeopardizing fit. A method of reducing the effect of this is to use a few small increments, say 1 or 2 mm at a time, and cure each before adding the next (Fig. 4.12). This, indeed, was the principle of the method employed with self-curing unfilled acrylic resins to minimize shrinkage effects, only with smaller increments.

●5.10   Visible light

Camphorquinone (Fig. 5.8) **(CQ)**, which is a diketone, provides an alternative to the use of ultra-violet light. The energy for excitation to the singlet state is lowered so much that the corresponding radiation comes into the visible range. Blue light is thus the activator for such 'light cured' products, and this has several advantages:

(1)   it is very much less damaging to tissue than ultra-violet;
(2)   it is available from simple tungsten filament (quartz-halogen) lights, rather than a mercury discharge lamp, such lamps being relatively quite stable in their output intensity for long periods of time; subsequently, also available from light-emitting diodes (LEDs), which have further advantages of size, low heat generation, and lifetime;

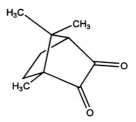

Fig. 5.8 One structure of camphorquinone (the molecule is chiral, having a mirror image form; the absorption spectra are the same).

(3)   it is better transmitted by both tooth tissue and by the resin and filler components so that a full cure can be effected rather more easily, so-called **transillumination**, *i.e.* through the tooth tissue, being said to be a feasible adjunct to the normal top (free surface) illumination (although the attenuation is then substantial and the time required much greater).

●5.11   Diketone – amine systems

The design of the one-paste 'light-cured' filled resin material has now come some way in this account, but there remains a problem: benzoyl peroxide is itself unstable. Although normally it requires heating to get it to decompose at a useful rate, as for denture base acrylic (5§2.6), this is not to say that at room temperature nothing will happen (5§2.7). Indeed, this is why denture base dough cannot be left indefinitely before packing. Thus, consistent good refrigeration would be required, but even then a limited shelf-life would be expected (we return to this point below). It follows that the benzoyl peroxide must be replaced

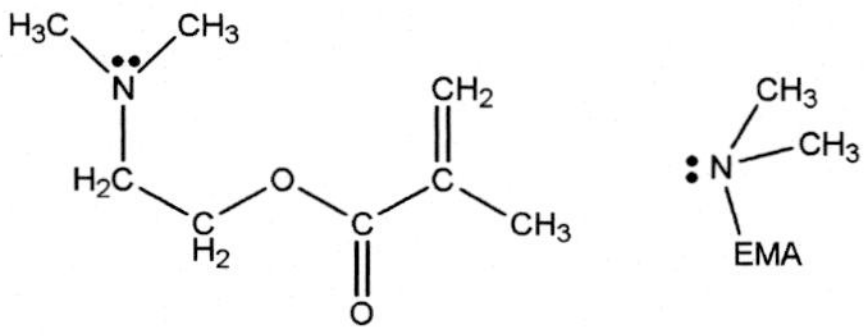

Fig. 5.9 Dimethylaminoethylmethacrylate, and the abbreviated structure used in Fig. 5.10.

with another source of free radicals. In fact, there are a number of systems known and in use, but one will suffice as an example – that of a tertiary amine. As pointed out in §1.1, aromatic tertiary amines are unacceptable because of the discolouration problem they cause. So to avoid this, and have a molecular weight high enough that the toxic amine is not volatile, and indeed polymerizable so that any remaining is not leachable, a compound such as dimethylaminoethyl methacrylate, which lacks an extended π-bonded system, can be used (Fig. 5.9). The use of a such a polymerizable initiator also means that another branch point is created in the network as the amine carries the radical which reacts with monomer, *i.e.* starts a chain, while the vinyl group becomes incorporated in a separate polymerization chain. (Di- and tri-functional such amines may also be used, increasing the potential for cross-linking.)

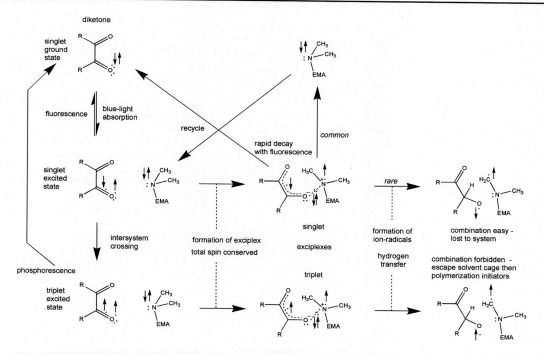

Fig. 5.10 The diketone-amine photosensitized free-radical polymerization system. The parallel spin pair of species (bottom right) are again the effective polymerization initiators. Notice that resonance structures tend to stabilize the exciplexes by delocalizing the unpaired electron on the diketone moiety.

The reaction scheme is very similar in overall plan to that of the benzoyl peroxide system, but the intermediate steps are somewhat different (Fig. 5.10).[37] The excited-state diketone reacts with the tertiary amine to form what is called an **exciplex** – an "excited-state complex". The triplet state version has a longer lifetime (it is in effect trapped in this state), and thus a greater chance of undergoing an internal hydrogen transfer and conversion to a pair of free radicals. These radicals cannot combine because they preserve the parallel spin condition, as before. These are then the active polymerization initiation species (although the diketone-derived species can rearrange and delocalize the unpaired electron, making it very much less effective in this regard). If the singlet-state exciplex lasts long enough (*i.e.* the expected rapid fluorescent decay does not occur first), free radicals may also be formed in a similar fashion, but these can combine and be removed from the initiation system (although still polymerizable). Otherwise, the singlet exciplex normally rapidly decays, recycling the components. This type of polymerization initiation system is also known as a **redox** system because the transfer of an electron between molecules is involved in the crucial step (exciplex formation), not simply a homolytic bond cleavage. The amine is therefore an electron donor, *i.e.* a reducing agent, and not in itself an initiator. Likewise, the diketone is not the polymerization initiator, despite commonly being labelled as 'photoinitiator'. (Furthermore, this is *not* a Norrish Type II photoreaction,[38] as is sometimes asserted – the mechanism is different and a 1,4-diradical is not formed.)

It can be seen that a key step in the process is the hydrogen abstraction from the amine. In fact, any suitable species can provide it, if the transfer is easy enough. Components such as TEGDMA (Fig. 4.5), with the electron-withdrawing oxygens in a favourable position, have easily abstractable hydrogens and polymerization in such systems is then easy, and of course such an effect leads to branching, which can be seen to be advantageous, offsetting to some extent the extra flexibility of the polyether chain. (Curiously, this kind of reaction does mean that the amine is then not strictly necessary for the polymerization to proceed at a useful rate.)

> The amine is sometimes described as a "synergist". This is chemically incorrect: it is required for the generation of the free radicals by reaction with the diketone excited triplet but has no polymerization activity on its own, nor indeed has CQ, so there is no 'synergy'.

It can be seen then that in the absence of light that can be absorbed by the diketone, it might be expected that there could be no reaction for this type of product and shelf life should be much longer if not indefinite. However, it can be expected that diradical formation will occur (5§2.7) which must lead to spontaneous polymerization. Even so, since background ionizing radiation is still present (potassium, carbon-14, radon, cosmic rays, X-ray sets! [26§5.4]), refrigeration (which device provides some shielding as well, at least to X-rays) is still sensible to reduce the rate of reaction for those free radicals that are formed accidentally. Indeed, so-called 'inhibitors' (5§4 and box there) must still be used in this type of product and these lead to a short period of irradiation over which no discernible setting may occur: that is, a **threshold** exposure will first need to be exceeded when the inhibitor is consumed. Nevertheless, any inhibitor will gradually be consumed as it does its job so that even under refrigeration these materials have a limited storage life, that is, for which there is no deleterious partial reaction and corresponding development of rubbery properties that would affect placement. A typical inhibitor is 2-methylhydroquinone (Fig. 5.11).

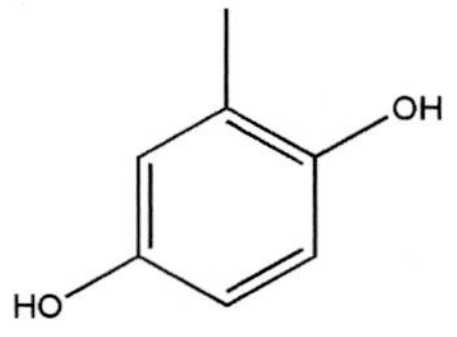

Fig. 5.11 Typical free-radical sink for use as an inhibitor in light-cured filled resins, 2-methyl hydroquinone.

Since the reactions that occur spontaneously are activated processes, the Arrhenius equation (3§3.3) controls the behaviour. As a very rough rule of thumb, this leads to the expectation that such a rate may double for a rise in temperature of 10 K. Taking 'room temperature' as around 24 °C, and the normal lowest refrigerator temperature (at the bottom) to be around 4 °C, the best that can be expected is about a 4-fold extension of shelf life. The temperature further up in the cabinet may be appreciably higher. Conversely, the temperature during warehousing and transport may be substantially higher. Proper attention to storage and stock control is clearly important. Likewise, inhibitors are essential.

●5.12 Hazards

The obvious problem with light-cured materials is the need to store them in the dark to avoid premature reactions, hence the opaque, usually black, syringes in which they are normally packed. Exposure of the unset material to sunlight or light from chair-side lights, in particular, because of the UV content of their light, must be very carefully avoided. There is, however, a much more serious concern associated with the curing light.

The eye is not very sensitive to the wavelengths of interest now (Fig. 24§4.8), so a suitable curing light source will not be bright enough to elicit the usual avoidance reaction on the part of the observer. A special filter in the lamp housing prevents other (longer) wavelengths getting through (especially for quartz-halogen lights) – essentially to prevent avoidable overheating of both the material itself and the tooth being irradiated (§5.17), but this limits the perceived brightness. Even so, while the perceived brightness of the light may still indeed be high, yet the actual intensity of the blue light will necessarily be very much higher and eye damage can and does occur on extended or repeated exposure.[39] The problem is that irreversible photochemical damage is done.[40][41] The relative risk of damage is quantified by the **blue-light hazard function**[42] (Fig. 5.12), which curve can be compared with the emission spectra of common lights (see below). While one can look away from such sources, this is not condusive to maintaining correct exposure of the filled resin.[43]

It is therefore very important to use an effective filter screen or protective eyewear (which are necessarily orange or brown), both in size and absorption, to view the work in progress in order to minimize such exposure (Fig. 5.13) whilst ensuring correct irradiation. Auxiliary staff should also be advised not to look at the light, while the dentist should take care to avoid exposure,[44] with special attention paid to avoiding the patient's eyes. There would

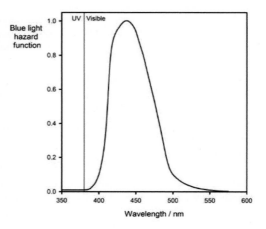

Fig. 5.12 The blue-light hazard function.

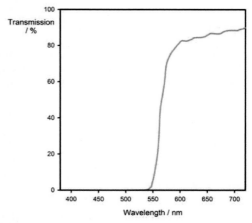

Fig. 5.13 Desirable kind of transmission spectrum for blue-blocking shields and eyewear, as in an effective dental product for use with light-curing.

be no harm in all wearing such protection. Even so, the mere fact that a shield is orange is no guarantee that it is effective.[45] There is no need to obtain transmission spectrum: if you can see the blue light through it, it is not working as it should.

In addition, if a quartz-halogen (QH) or plasma-arc light is used as a source, the wavelengths emitted do extend into the ultra-violet,[46] in the case of QH because the filament temperature has to be very high to obtain sufficient intensity in the blue region (Figs 24§4.3, 24§4.5). Unless a second special filter is used to remove that ultra-violet from the light transmitted into the light-pipe, there remains the risk of 'sun-burnt' finger tips and additional eye damage for the operator. The original claim that because it was 'visible' light it was necessarily harmless is not valid, as is now recognized.[47] The UV component is unlikely to be a problem for the patient because the total exposure is very limited, however, it is not known whether there are any effects on soft tissue (including the dentist's fingers) from the very high exposure to blue light (but see §5.18). Even so, patients with a history of photobiological reaction, or using photosensitizing medicaments, should not be exposed to any such irradiation.

●5.13 Incompleteness of setting

It should be recognized that the use of radiation to start the setting process is just that – a start. The mistake should not be made of thinking that setting reactions stop when the light is turned off. The changes of properties that occur in the early stages, during irradiation, are indeed very striking, and it is proper to say that some setting has occurred. However, the live polymer, the chains with free radicals at their ends that have not been terminated, must continue to survive and react for some time. This is despite the fact that the rate declines rapidly as the concentration of double bonds available for attack decreases and the viscosity of the matrix rises, inhibiting diffusion. The mechanical properties will therefore continue to change measurably with time, perhaps for many hours.

The term **command set** which is often applied to 'light-cured' materials is therefore not completely accurate and is in fact somewhat misleading. It follows that the **degree of conversion**, a measure of how far the setting reaction polymerization has gone (proportion of double bonds remaining) is not proportional to the illumination **exposure** (irradiance × time). It is a complex kinetic balance between the rate of initiation, rate of propagation and the rate of termination whilst the light is on, and then between rate of propagation and rate of termination when it is off. It does not follow that the reaction can ever be complete in the sense that 100 % conversion of double bonds to polymer chain links is never actually achieved (Fig. 4.9). In fact, estimates of only 50 ~ 75 % conversion under normal circumstances are typical. The only real difference between this type of product and the chemically-activated sort is that the initiation in the latter depends on diffusion to get the reaction started, otherwise the processes are similar.

Nevertheless, the progress of setting in relation to exposure has a particular kind of form, the **response surface** (Fig. 5.14). As mentioned above (§5.11), the inhibitor present must first be consumed before any network formation can occur. After that, a logarithmic setting ramp sees the steady development of that network and the values of the properties that depend on it until vitrification causes this to come to a halt on the plateau that corresponds to the maximum possible reaction. The transition between these two surfaces, the 'shoulder' as it were, represents the locus of the minimum acceptable exposure, although because of the failure of reciprocity this is not a

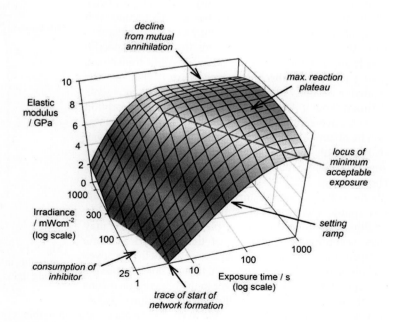

Fig. 5.14 Typical form of the response surface for the setting of a light-cured filled resin. Products will vary in values and sensitivities, but the various features must all be present.

locus of constant exposure (its slope on the base plane will vary between products). After that, continued exposure can do no harm (other than the effects of overheating). The slight slope that this plateau has toward higher irradiance at constant time may be due to heating, when the value of the T_g is driven up (§4.5). However, there is a maximum irradiance that can be used before mutual annihilation of radicals begins to dominate, and network formation is then less than maximal.[48]

•5.14 Curing lights

At first sight exposure might be thought to be a simple matter, if not in terms of degree of conversion, then at least in terms of the amount of polymerization initiation that occurs. However, there are several aspects to this that need consideration. We need to cover the properties of the photosensitizer, the light source itself and the material.

A photosensitizer works by absorbing energy, that is a quantum from the incoming radiation, which absorption creates an excited state by promoting an electron to a higher energy orbital. As we might imagine, the efficiency of absorption is affected by the concentration of the absorbing species, and this is expressed by the **absorbance** A through the **Beer-Lambert Law**[10]:

$$A = \log\left[\frac{I_0}{I}\right] = \varepsilon c L \qquad (5.1)$$

where the ratio of the incident (I_0) and exit (I) light intensities is affected by the path length L, molar concentration c, and the so-called molar extinction coefficient ε for the substance concerned. ε is a measure of the efficiency with which the quantum can be absorbed or, more to the point, the probability that the particular electronic transition will occur when the interaction with the quantum occurs. This efficiency varies considerably with the wavelength of the radiation for a specific transition, and there may be many possible transitions, giving rise to an **absorption spectrum**. This may be plotted in a variety of ways (always check the *y*-axis scale definition) but it illustrates the effectiveness of absorption of each wavelength. (Bear in mind that not all absorptions will be effective for the transition that is required, and that there may also be competition from overlapping bands.) An absorption spectrum for camphorquinone (Fig. 5.8) is shown in Fig. 5.15.[49] Although this is a commonly-used photosensitizer, others may be employed and the absorption spectrum thus can vary accordingly. In addition, the concentration of photosensitizer used may vary according to the shade of the material, as some compensation for the ineffectual absorption of light by dyes and pigments. However, there is a limit to this because the absorption by the photosensitizer in the blue region means that the material takes on a yellowish colouration (24§3.4)

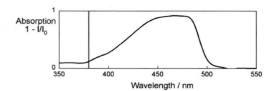

Fig. 5.15   Part of the absorption spectrum of a solution of camphorquinone, strongest absorption occurring between about 430 and 480 nm.

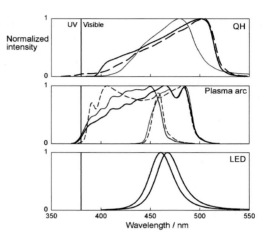

Fig. 5.16 Comparisons of the emission spectra of a some examples of each main kind of visible-light curing light: QH – quartz-halogen, high-intensity plasma arc (both after the long-wavelength filtration used remove heat and ineffective radiation), and LED – light-emitting diode.

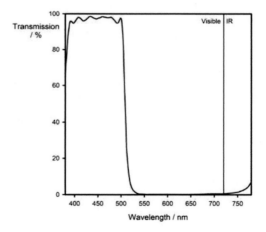

Fig. 5.17 Typical band-pass spectrum of a dichroic (thin film, interference) filter as might be used in some curing lights. Note the detail in the blue region.

and too much would be undesirable in what is meant to be an invisible repair of the tooth if there were to be any left over, although this would fade with time if continued reaction were possible.

---

[10]   The log term on the left is also known as **optical density**, see 26§5.2.

The second factor is the light emitted by the curing light. This will show a variation in intensity with wavelength, according to the physics of the light-generation process (which we need not worry about here) (Fig. 5.16). However, sources such as quartz-halogen (QH) (24§4.7) and plasma-arc lights also generate a great deal of heat, as well as wavelengths in both the red to green range that will not be effective and in the ultraviolet that are not needed. Any absorptions in these regions also end up as heat. A heat filter is also essential, and usually some kind of **band-pass filter** to limit the range of wavelengths delivered to that most appropriate. Filters, unfortunately, do not show very sharp cut-offs and perfect transmission where desired, although **dichroic filters** come very close (Fig. 5.17).[50] Thus, depending on the design, they too will show a more or less complex absorption spectrum (or transmission spectrum, depending on one's point of view), and some undesirable wavelengths will be represented in the light. The output spectrum from a curing light device is therefore dependent on the nature of the light-source proper and the filtration that is applied to it. For light-emitting diodes (LEDs), which have a tightly-defined emission process, there may be very little heat or unwanted light and no filter may be necessary. From the output spectra of these three kinds of device illustrated in Fig. 5.16, it is immediately apparent that there is considerable variation possible, even within one type.

However, it is not just the shape of the spectrum that varies. According to the device, the total power emitted also varies widely. LEDs are intrinsically relatively low-power output devices (although rapid improvements are being made, especially by being used in groups), while plasma-arcs are capable of very high power, and QH lights are somewhere in between (Fig. 5.18). Lasers are capable of much higher power, and may yet be used for curing,[51] but would need a beam-spreading device to allow adequate coverage of the target area.

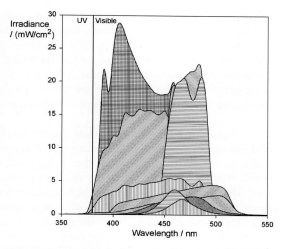

Fig. 5.18 Comparisons of the irradiance spectra for the curing lights of Fig. 5.16 (compare the curve shapes).

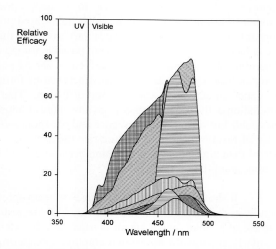

Fig. 5.19 Calculated efficacy spectra for the same lights as Fig. 5.18, assuming the use of camphorquinone with the absorption spectrum of Fig. 5.15. (The shading indicates the corresponding curves in the two figures.)

Variation between restorative material products according to the concentration of photosensitizer is a factor as mentioned above, but, for a given product, the light delivered can effectively only be used according to the absorption spectrum. Combining this information we can examine the relative efficacy spectra by calculating the rate of delivery of quanta (moles/s) (see box, 24§6.1), and multiplying this by the absorption efficiency, for a given concentration and path length, wavelength by wavelength. This gives a picture such as is shown in Fig. 5.19, where it can be seen, for example, that much of the short-wavelength output of the plasma arc lights may be wasted, as is the long-wavelength component of the QH output.

Then, assuming that a quantum absorbed is used effectively by leading to the production of the appropriate singlet excited state, the area beneath each curve is a measure of the overall effectiveness of the light-product combination, and the irradiation time required for a sufficiently-cured material varies

---

Since both the emission (X) of a light source and the absorption (Y) of the photosensitizer vary with wavelength – they have **spectra** – the net effect, here termed the **relative efficacy** (RE), is the product of the two values at each wavelength (on appropriate linear scales) as if they were normalized probabilities:

$$RE(\lambda) = X(\lambda).Y(\lambda)$$

This is a **spectral convolution**. The same process is used for calculating the effect of filters and other absorption processes (such as by deliberate colouring agents in the resin).

accordingly. This assumption may not be good. In the first place, it is clear from Fig. 5.4 that the initial excited state, S_2, must have an energy greater than that of the crossing-point, else vibrational decay to reach it cannot occur, and so no triplet formation. Secondly, there may be more than one excitation process (that is, giving different excited states, in different molecular orbitals), and only the one may lead to the intersystem crossing of interest. Absorptions of these other kinds can only lead to fluorescence and cannot contribute to the free-radical generation process. Hence, the relative efficacies of Fig. 5.19 can only be upper bounds. It does not appear to be known what the effective absorption spectrum of camphorquinone is in this sense.

However, there are other, kinetic factors which affect the efficiency of the process: including rate of decay of the singlet state, rate of formation of the triplet state, rate of formation of the exciplex, efficiency of chain initiation and so on. These in turn will depend on the material with respect to such matters as the concentrations of photosensitizer and amine, the viscosity – as it affects the diffusion of reacting species, the choice of substances for photosensitizer, amine, inhibitor and monomer – as these affect reactivity and diffusion. In other words, there is a lot more scope for variation between products – which all means that what is appropriate for one system is likely to be inappropriate for another. Despite the ease with which light-cured products appear to be used, great care is necessary to ensure that they are in fact treated properly, as it is not possible for the user to determine that setting has been driven as far as possible. Temperature is also a factor which affects setting through its influence on activation energies and diffusion.

> It is sometimes said that the 'ideal' curing light spectrum should match the shape of the absorber (*i.e.* CQ), in the mistaken belief that this is efficient. However, if uniform efficacy, according to the spectral convolution, is to be attained, then the illuminant must have a reciprocal "U" shape to the absorption spectrum. This is not possible. The best that is achievable is for a flat illuminant spectrum over the range where there is effective absorption, and none outside that range. However, in terms of the most effective use being made of the light delivered, what is required is a single wavelength corresponding to the peak of the absorption spectrum. A laser might achieve this.

This is affected by the reaction exotherm and heat from the irradiation delivered – whether from unwanted or unused wavelengths or from the energy dissipated through the inevitable but wasteful fluorescence and phosphorescence processes. Low-intensity irradiation will give lower rates of reaction, less temperature rise (as heat dissipation is allowed to occur) and possibly longer reaction chains than high-intensity irradiation because of the lower concentration of radicals produced, and thus a lower mutual-annihilation rate (5§3.3).

It should also be noted because of beam spreading, the light intensity reaching the resin material falls rapidly as the light-guide is moved away from the restoration. Further, if access is difficult, such as on the distal surface of a molar, and a mirror is used, greatly extended irradiation times must be expected to be necessary. Similarly, if **transillumination** is employed, scattering and absorption within the tooth will reduce the available intensity dramatically. It should go without saying that the cleanliness of all optics in these systems is important (especially the exit window of the light guide) to maintain output, as is the avoidance of damaged filters to prevent improper exposure.

●5.15 Depth of cure
On top of the effect of shade, absorption by the photosensitizer itself means that the intensity of irradiation falls as it passes through the material, and most rapidly at the wavelengths most efficiently absorbed, according to the Beer-Lambert Law (equation 5.1). Thus the efficacy of curing must fall at successively deeper positions in the irradiated material, implying that longer and longer irradiation time is necessary to achieve the same effect. Notice the trade-off in that increasing photosensitizer concentration exaggerates this effect. But absorption is not the only factor: scattering by the filler particles (and bubbles!) (24§5.11) means that the intensity falls with depth more rapidly than otherwise might be expected, and so effects due to particle size and distribution, as well as volume fraction of filler, will add to the variability between products.

The clinical importance of this is that, for a given exposure (intensity × time) at the surface, there is a maximum practical thickness for which the material can be expected to be properly cured at the bottom as well. Attention should therefore be paid to the manufacturer's instructions in this regard. However, it is clearly impossible to determine (without destroying the restoration) whether the exposure was adequate, especially given the difficulty of assessing the thickness of the increment. It is therefore suggested that the most conservative or cautious approach is appropriate: erring on the side of thinner increments and longer irradiation times is preferable to ensure that the material is properly cured, even at the bottom.

The practical testing-laboratory importance of this is that test specimens cannot be prepared to have uniform properties across their thickness. There must be variation in actual exposure in the irradiation direction, and therefore variation in curing rate if not in final completeness of reaction. Measured property values then reflect the behaviour of this non-uniform outcome; interpretation may be difficult.

Thus, the effectiveness of light curing will depend on:

- product   - manufacturer's choice of chemistry
- shade   - darker shades may require extended irradiation
- type of light   - spectrum and total output
- age of light   - envelope darkening, deterioration
- light filtration - spectral range matching that of photosensitizer
- matrix   - reflection occurs, reducing irradiance (24§5.15)
- cap   - losses due to mechanical protective devices
- sheath   - losses due to the use of an infection control barrier
- distance   - the light emitted from the tip is not collimated
- angulation   - reflection losses increase with angle (24§5.3)
- irradiation time - completeness of cure
- thickness   - Lambert Law reduction in irradiance
- usage   - distance, angulation, access, cleanliness
- age   - change in inhibitor concentration

The possibility of transillumination (§5.10) is superficially encouraging. However, it should be noted that substantial scattering occurs, as well as some absorption, so that the intensity reaching the material to be cured may be very much lower (say, 1/100th), than otherwise might be the case. The required irradiation time is consequently that much longer, and may therefore become impractical.

●5.16 Units

The usual manner of discussing the curing of filled resins now is in terms of the **radiant exposure** (improperly, **energy density**), in the units of J/cm^2, representing the total energy delivered to the free surface.[52] However, it should be apparent that the curing process – specifically the absorption by the photosensitizer of a single photon – is a quantum process, not one that is directly dependent on energy as such. As will be apparent also, once absorption has occurred, any excess energy is lost as heat through vibrational decay, the energy of the required triplet state being rather lower than that initially absorbed, so that the same outcome is attained irrespective of the photon's energy. Of course, the energy of the light photons varies with wavelength (Fig. 5.20). The energy

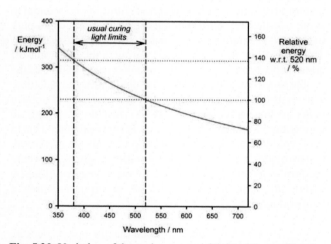

Fig. 5.20 Variation of the molar energy of light with wavelength.

at 380 nm is some 37 % greater than at 520 nm, the extremes of the visible range used for curing. One could invert this and say that 37 % more photons need to be delivered at 520 nm than at 380 nm for the same radiant exposure.

Complicating this is the fact that wavelengths are not in fact equally efficacious in terms of their absorption by the photosensitizer (*e.g.* Fig. 5.15). Thus 1 J/cm^2 at 400 nm, say, cannot be as effective as at 480 nm. Indeed, it would require some eight times as many photons for CQ – that is $8.4 \times 1.2 = 10$ times as much energy (absorption ratio × energy ratio) at those exact wavelengths – to achieve the same effect (assuming that all wavelengths absorbed by the photosensitizer are equally effective in creating the required triplet state, and that is by no means certain[11]).

Under certain weak additional assumptions – use of a camphorquinone-based material with an LED light

---

[11] No work appears to have been done on this aspect.

spectrum such as indicated in Fig 5.15, calibrated against a photometer that truly measures spectral energy across the relevant wavelength range – a curing exposure in terms of energy delivered may be an adequate approximation. However, the fact that from 16 to 35 J/cm² has been considered necessary for 'adequate' curing[53] undermines the concept of a single universal figure, even if the proper usage of a curing light is taught and observed as it should be.[ibid.] Indeed, given the variety of curing lights available, a single figure for a single shade of one product seems infeasible. The manufacturer's instructions prevail for maximum increment thickness and irradiation time for a specified illuminant, according to shade. Ensuring a known light output is another matter altogether.

●5.17 Shade

One point to notice is that in light-cured materials the diketone is consumed in the effective initiation process (Fig. 5.10), so it might not be altogether accurate to describe it as (just) a photosensitizer in the same sense as benzophenone (§5.7). Since the photosensitizer absorbs in the blue region it is actually a bright yellow substance, and imparts a yellowish colouration to the material it is used in. This has two immediate consequences. Firstly, this yellow colouration (Fig. 5.15) means that a colour match can not be expected with fresh paste (24§5.11), but this also fades during the course of irradiation – the reaction product is colourless since there is only one keto- group. It is of course intended that selection be made for the shade it will be after curing. Secondly, the penetration of the light improves slightly at the same time, and therefore slightly improves the depth of cure from that theoretically expected from the uncured absorbance. There are some other factors operating as well (24§5.11).

●5.18 Heat

At several points above the reaction exotherm has been mentioned, but in general the temperature rise that this causes in adjacent living tissue is insufficient to cause immediate concern in that respect. It has also been stressed that for visible light-curing materials, filtration is essential to avoid overheating, ensuring that the irradiation occurs only over a wavelength range relevant to the photosensitizer. Thus, the filter in a (non-LED) curing light has a cut-off around 500 nm or so (Fig. 5.17), which is in the middle of the visible spectrum, and is not just to exclude infra-red radiation. This is because any absorption of visible light (i.e. by a material that is coloured, 24§3.4) results in a rise in temperature as the absorbed energy is **thermalized**. This in itself could be hazardous to the pulp (20§7). That is to say that no matter how efficient heat filtration might be, or where the heat is generated (a light centimetres away or close-by LED), all light is energy and all absorption causes a temperature rise.

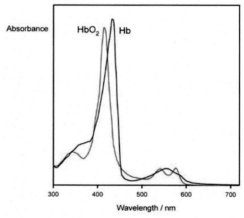

Fig. 5.21 Absorption spectra of haemoglobin (Hb) and oxyhaemoglobin (HbO₂). Such so-called **Soret peaks** are also characteristic of many haeme-containing enzymes (cytochromes).

However, it should be noted that the pulp is dark in colour, primarily due to the blood present, and so will absorb strongly in the blue region (Fig. 5.21). Even plasma has an appreciable absorption in the relevant region (Fig. 5.22). Indeed, tooth tissue itself is sufficiently coloured to cause appreciable absorption of the irradiating light, and so experience a temperature rise. In addition, the excess energy of the initial S₂ excited singlet state appears as heat through vibrational decay. As a result, the temperature rise of the pulp just on irradiation may be sufficient to cause inflammation, given that this starts at only 42 °C. Rises of 8 ~ 12 K have been reported, with a further 2 or 3 K on top due to the reaction exotherm. Although there may be some benefit from the cooling that necessarily occurs from an extended period of open mouth while the preparatory work is done, in addition to the cooling that occurs during cutting under a water stream, the risk is clearly there and it may account for some reports of post-operative sensitivity

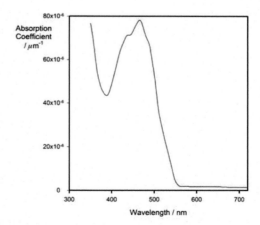

Fig. 5.22 Spectral absorption coefficient for blood plasma.

otherwise unexplained. Certainly, the emission from modern curing lights can be sufficiently great that the irradiated site gets noticeably hot in a few seconds. Exposure of the skin, say, on the back of one's hand, in contact or at a few millimetres distance commonly results in an avoidance reaction. A check like this can be salutary as regards what may be done inadvertently to the patient who will probably have been anaesthetized and so not respond in such a way. Indeed, burns to lips have been reported in practice[54] with only a modest output power device – temperatures over 90 °C were easily obtained experimentally, and gingival burns may appear to be simply a mouth ulcer.

The problem might be dealt with by the following approaches, given that temperature rise depends on rate of heat deposition and the thermal diffusivity, the balance between input and dissipation:
- not using high-irradiance curing lights
- ensuring that all filters are in place and undamaged
- using a light-tip just large enough for the job
- ensure accurate placement and aim of the tip
- using an intermittent irradiation protocol
- providing additional cooling with an air stream between increments
- shielding all soft tissue

Clearly, there is a need to ensure adequate irradiation – which means a sufficiently long total exposure, $\Sigma(I \times t)$, and this should not be compromised in the course of keeping the tooth cool. It then follows that the problems with transillumination (§5.15) are that much more serious.

A further problem arises from such heating: the consequent thermal shrinkage of the now-solid restoration. This is superimposed on polymerization shrinkage, and thus exacerbates the stress acting on bonded walls (§4.6). However, the thermal history is unlikely to be uniform across the whole restoration, unless it is rather small. Firstly, the low thermal conductivity of such materials means that the reaction exotherm (§2.16) will cause the central portion to be appreciably warmer than the periphery. This temperature variation is auto-accelerating because of the general Arrhenius-dependance on reaction rates and diffusion – this is positive feedback (polymerization of many resins in large bulk can cause them to decompose and ignite). This in turn would lead to spatial variation in extent of reaction and T_g and thus of both polymerization and thermal shrinkage. Added to this is the likelihood of uneven irradiation, whether from the non-uniformity of the intensity across the face of the light-tip[55] or carelessness and inattention[56], and the consequential variation in both the reaction it initiates and the heating it causes directly. It can be seen, then, that there is potential for a complicated frozen-stress field, therefore complexity in the stress on the bond to the cavity wall, and also variation in the response to water absorption, but primarily there will be variation in the mechanical and other properties of the material.[57]

•5.19 Alternative chemistry – photoinitiation

In the reaction scheme of Fig. 5.10 the essential feature is a visible-light chromophore, that is, one which generates a useful singlet state. There are many such: the groups R in Fig. 5.10 can vary widely. Two that have been used in dental resins are shown in Fig. 5.23, and whose absorption spectra are shown in Fig. 5.24. Lucirin TPO is in fact a photolytic initiator (§5.1), primarily because sterically it is a very crowded molecule (Fig. 5.25) and thus does not need the use of an amine. Ivocerin also generates free radicals directly by homolytic scission.[58] Crucially, for these last two, which are also are therefore also of Norrish Type I (§5.1), the absorptions are usefully in the visible range, in comparison with that of benzoin methyl ether (Fig. 5.1).

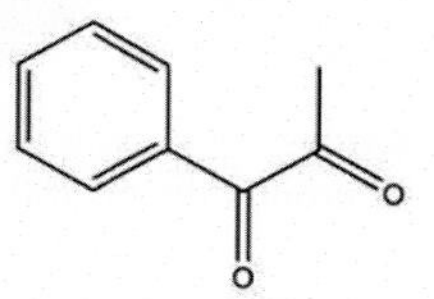

The selection of a photosensitizer or photoinitiator can depend on a number of factors, for example quantum

1-phenyl-1,2-propanedione (PPD)

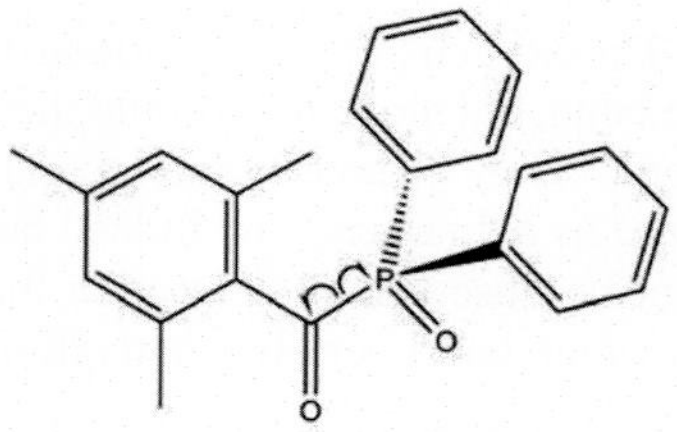

Lucirin TPO (LTPO)

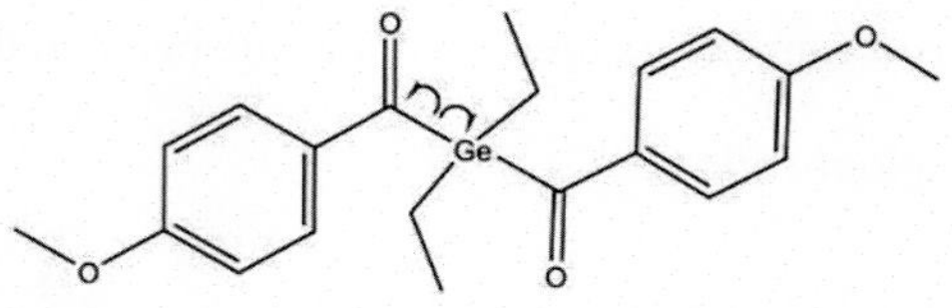

Ivocerin

Fig. 5.23 Alternative photosensitizers. (Lucirin TPO is an old name for Irgacure TPO, both of which are trademarks.) LTPO and Ivocerin suffer homolytic scission at the indicated P-C or Ge-C bonds on irradiation.

efficiency, solubility, toxicity, and price. However, it is obvious that there are important constraints arising from the absorption spectra. Thus PPD will not be as effective as CQ for the usual blue light, while Lucirin TPO will be even worse. This requires that the irradiation moves to shorter wavelengths, and corresponding violet-emitting LEDs are available and used in some curing lights. It is then apparent that the blue-light hazard can be increased (the eye-sensitivity and thus the avoidance reaction are also much reduced), as well as the heating effects in soft tissue. Where such LEDs are included in the same unit as blue-emitters, by way of making such curing lights "universal", many complications may ensue in terms of equivalence (and thus required exposure, especially at depth), and uniformity of the emission pattern. It remains true that such systems represent a compromise in that optimization cannot obtain for any product combination.

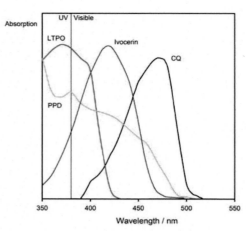

Fig. 5.24 Absorption spectra for alternative photosensitizers compared with that of camphorquinone (CQ) (not to scale). (Key: see Fig. 5.22).

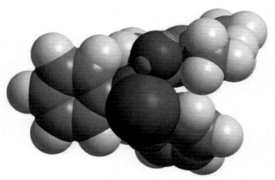

Fig. 5.25 Space-filling model of Lucirin TPO to show the crowding around the central P-C bond.

The terminology in this area can be tricky. All free-radical polymerization is initiated by a free radical: that is the chain initiator. The absorption of light to start a sequence of reactions is, broadly, photoinitiation, but unless the substance doing the absorbing generates the chain-initiating radicals directly it cannot be termed a photoinitiator. Rather, it is a photosensitizer as it is the change in a second molecule that it causes that is the important one generating the free radical.

§6. Oxygen Interference

The oxygen of the air is primarily in the form of the diatomic molecule O_2. This may be thought to have a double bond, but molecular orbital theory indicates that there are two electrons in antibonding orbitals, one on each atom. They ordinarily have parallel spins (oxygen is **paramagnetic**). Molecular oxygen can thus be considered as a diradical, and indeed behaves in this fashion. In the present context of polymerization this is important because it may react with the free radicals at the ends of propagating chains, a **radical coupling** reaction which has a very low activation energy in the absence of steric inhibition:[59]

reaction with live chain : $R\cdot \; + \; \cdot OO\cdot \; \Rightarrow ROO\cdot$ (6.1)

chain transfer : $ROO\cdot \; + \; R'H \Rightarrow ROOH \; + \; R'\cdot$ (6.2)

(Atmospheric oxygen itself, despite being in the triplet state, cannot abstract hydrogen in this way.[60])

multiplication : $ROOH \Rightarrow RO\cdot \; + \; \cdot OH$ (6.3)

chain transfer : $RO\cdot \; + \; R''H \Rightarrow ROH \; + \; R''\cdot$
                                     $HO\cdot \; + \; R''H \Rightarrow H_2O \; + \; R''\cdot$ (6.4)

and from then on many more reactions are possible (feeding $R'\cdot$ and $R''\cdot$ back into reaction 6.1, for example), including mutual reaction of any combination of radicals (termination) – the parallels with the processes of 5§1 will be apparent – and the generation of formaldehyde following polymerization of the peroxide (Fig. 6.1) [61][62] (see 30§3). Of course, the radicals that are produced by these reactions can react with more oxygen, and the cycle continues. In particular, the multiplying effect of reaction 6.3 is important, the **homolytic scission** of the hydroperoxide group, but then the rate of mutual annihilation of radicals further down the cascade of reactions also rises very rapidly. (It should be noted that both radicals produced by the initiator [Fig. 5.10] can be involved as any of the "R" species in those reactions, providing effects very early in the process.) The net outcome is the production of very many short chains and the effect is readily observable: the surface of the resin that is exposed to the air does not cure hard as does the bulk but remains sticky in what is known as the **oxygen inhibition layer** (sometimes referred to as the 'OIL'). It may be some 10 ~ 50 μm thick, depending on the product and conditions. The behaviour is not restricted to light-cured materials: any free-radical polymerization system must be affected, so two-paste systems are also included (see below). Indeed, it is frequently seen in many other resin systems outside of dentistry, such as those used in fibreglass fabrications.

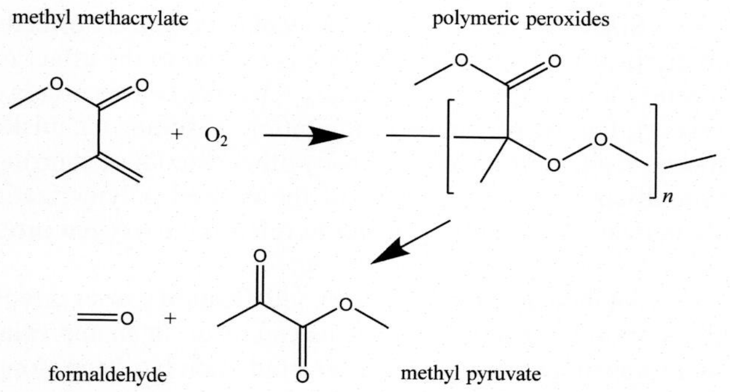

Fig. 6.1   Polymerization in the presence of oxygen can generate formaldehyde.

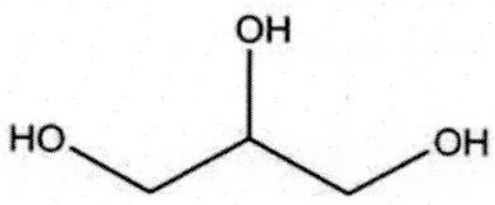

Fig. 6.2   Glycerol, propane-1,2,3-triol.

A matrix strip is a sufficient barrier to prevent the formation his happening on some restoration surfaces, but obviously on more complicated contours it will not be possible to obtain the necessary contact. To avoid having to remove this uncured or, at least, incompletely cured layer (it is not known what proportion of vinyl groups remain), barrier-layer products are sold which appear to be (or based on) merely glycerol (glycerine) (Fig. 6.2), which is a colourless viscous liquid (~1.4 Pa.s; *cf.* water: 1 mPa.s at 20 °C). These appear to operate by reducing substantially the diffusion rate of oxygen ($D_{O2} \sim 0.176$ cm^2/s in air at 25 °C), but also the solubility of oxygen in it is rather low (Fig. 6.3),[63] in comparison with water, for example. These materials would seem to be useful only if no reduction or finishing of the restoration's surface were necessary or feasible. However, glycerol is transparent in the visible range (it is colourless), so will not affect curing adversely. Indeed, if excess is used and the tip of the curing light fully immersed (no bubbles), the otherwise inevitable reflection at each air interface is mostly eliminated and the irradiation is increased (24§5.15). Its solubility in water means it can be washed away afterwards easily; it is non-toxic.

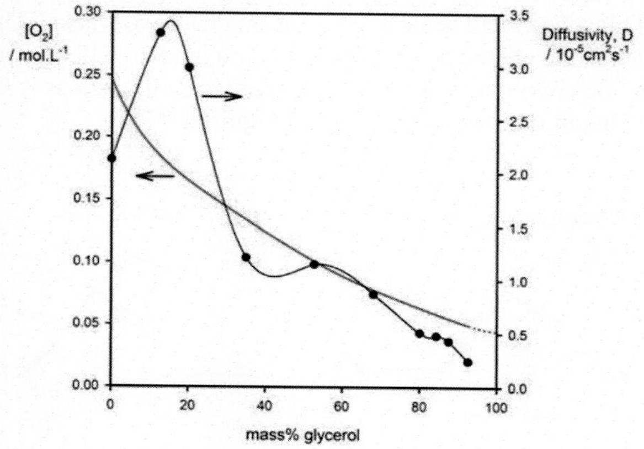

Fig. 6.4   Saturation concentration against air (left), and diffusivity (right) of oxygen in water-glycerol mixtures at 25 °C.

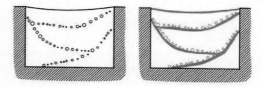

Fig. 6.3 Incremental build-up can trap bubbles, but also all air-exposed surfaces must suffer oxygen interference.

However, a barrier layer cannot be used in the incremental technique – it is not possible to clean up each surface before the next layer is added. Thus, it can be expected that the material of each inhibition layer except the last is necessarily incorporated into the body of the restoration.

In addition, it is likely that air bubbles are trapped between increments, and these of course must also suffer inhibition, resulting in a situation similar to that indicated in Fig. 6.4. It would perhaps not be practical to scrape away the inhibited layer each time, but this would in any case increase the trapping of bubbles on a rough surface (*cf.* Fig. 10§2.16). The net outcome expected is of a reduced strength between increments, compared with the bulk, thus 'bulk cure' materials will not suffer such internal effects. In any case, the lower surface will probably trap some bubbles, especially if placed on a liner, and thus may then also have an inhibited layer.

Since the reactions with molecular oxygen are effectively preventing the generation of a long-lived live polymerization chains, the outcome is similar to the effect of so-called inhibitor (§5.11, 5§4): delay until the concentration is reduced sufficiently – by maybe two orders of magnitude – that the competition is negligible. However, the use of the word 'inhibition' is somewhat misleading: while the desired outcome is delayed, if it happens at all, there are clearly many other reactions occurring, indeed are caused by the oxygen. It is not a case of modification of the kinetics of the desired polymerization. As the oxygen interferes with the required polymerization, it might be better to refer to the **oxygen interference layer**.

As indicated earlier (§4.5), vitrification causes a rapid and substantial drop in reactant diffusion rates and therefore in reaction rate. This can result in an appreciable concentration of free-radical, live chain ends being trapped, persisting long after irradiation has been stopped and the material is apparently fully set.[64] Of course, these radicals contribute to so-called **post-cure** reaction, when the degree of conversion rises slightly, but equally some may not be able to react at all in this way. Such radicals then are subject to reactions 6.1 ~ 6.4 and so on with oxygen that subsequently dissolves in and diffuses through the matrix.[65] This causes some network degradation through chain transfer processes. In two-paste systems, the mixing process ensures that exposure to air occurs whilst free radicals are being generated, indeed bubbles would be mixed in. It would seem likely that the overall setting is therefore compromised to some extent. not just the surface.

●6.1   Bleaching agents

Most bleaching agents for use in dentistry (principally hydrogen peroxide or "carbamide peroxide") generate a variety of oxygen species: molecular O_2, •OH radicals, HO_2^- and others (30§1.3). As these materials are expected to diffuse into enamel and dentine, and as reaction takes time, such species may be present for some considerable period after the treatment. Clearly, they would interfere with a free-radical polymerization as does atmospheric oxygen,[66] as indicated above. Accordingly, it has been recommended that resin-based restoration of bleached teeth be delayed by as much as two weeks. The polymerization inhibition would mean that the resin tags in the etched tooth would not cure, and the effect would be one of reduced bonding strength. This perhaps can be circumvented by using an antioxidant (such as is shown in Fig. 6.5)[67], although an appropriate time for diffusion into the enamel is also required.[68]

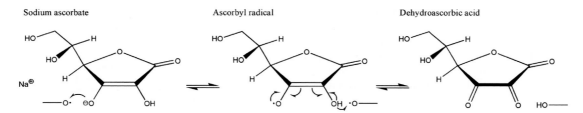

Fig. 6.5 Schematic of the electron transfer reactions of ascorbate as an antioxidant. This reaction is accelerated at pH > 6, and reaction with atmospheric oxygen is also rapid. A solution of the sodium salt has pH ~7.

§7. Temporary Restoration Resins

In a variety of contexts it may be necessary to provide a temporary 'interim' or 'provisional' restoration – for reasons including appearance, function, speech, protection of vital teeth and occlusion, and comfort – while the intended permanent device is fabricated. Thus, for the most part, the same requirements in terms of structure and mechanical behaviour, chemical response to oral exposure when set can be expected to pertain, although in broad terms the requirements for strength and accuracy of fit and contour may be less severe as long-term wear is not anticipated. The most significant difference, however, may be found in the need for rapid and convenient handling, yet perhaps at less cost than the materials intended to be permanent.

As was pointed out in 3§4, 3§6 and 5§6, there are many ways of controlling polymer system properties, and as has been stressed earlier here, there are a variety of other options. Thus, there are many products from many manufacturers with a great many varieties of formulation. It is not possible, therefore, to give a single summarizing description except that the majority of products are filled resins – composite materials with all that that entails. Instead, some examples of the variety of products that may be encountered are given with some observations of distinguishing characteristics.

●7.1   PMMA

Broadly, this is just unfilled cold-cure acrylic (5§3), with all of the problems described above (§1). Although functional in the context, the exotherm is marked, which poses a marked risk to vital pulps when cured *in situ*, and the shrinkage is relatively large. The unreacted monomer is distasteful.

●7.2   PEMA

Poly(ethyl methacrylate) materials are very similar to PMMA-based products, but with a lower T_g (Fig. 3§4.5) and so have a lower modulus of elasticity. At the same time, the additional free volume means greater water sorption and, reportedly, more discolouration (although this may have more to do with the use of amine initiators; *cf.* §1.1, 5§3). The larger molecule means that on a mass basis the heat evolved is less than for PMMA, but also the diffusivity is less so the reaction rate tends to be lower, which in turn means that the exotherm peak is lower. Even so, care and cooling is required. Poly(butyl methacrylate), which has been used in various combinations, continues all of these trends.[12] The monomers remain unpleasant in smell and taste. Plasticizers such as dibutyl phthalate (Fig. 5§6.3) may also be used (but see 5§6.1). The problems outlined in §1 remain.

●7.3   Bis-acryl

This is a very broad class of filled materials based on more or less the same chemistry as the filled resins intended for 'permanent' restorations: a bis-phenol A-derived difunctional acrylate monomer is usual. This may, indeed, be bis-GMA itself, or any one of a variety of similar molecules where the trade-off between reactivity and stiffness when set is managed through modifying the linking groups (*i.e.* instead of the glycidyl moiety), mixtures of monomers (including polyurethane methacrylates), plasticizers, and polymerization initiation systems. The filler is commonly silanized for bonding to the matrix. The name itself, 'bis-acryl' (which is not a standard chemical term), seems to be an attempt to distinguish these materials from ordinary filled restorative materials, but as can be seen there is no fundamental distinction to be made. Similarly low exotherms, low shrinkage, high stiffness, low water sorption, and so on, are found. The viscosity of the unset material tends to be kept low, whether by the use of diluent monomers or less filler, or both, so that the product can be mixed automatically (15§7), *i.e.*, these are two-paste systems, and this then limits appreciably the strength and stiffness attainable – although this may be acceptable in the context. Alternatively, light-cured and dual-cure products are also available. This then continues the spectrum of possibilities to the ordinary micro-fill type of restorative resin product (6§3.2), which is sometimes used for temporary restorations (and which perhaps raises, again, the question of the appropriateness of this kind of material for permanent restorations).

●7.4   Epimine

The chemistry of this type of material is rather different from the above, being a **cationic chain polymerization**. The monomer is very similar to that of bis-GMA (Fig. 7.1), but the reactive, polymerizable group is a strained, 3-membered ring called an **epimine**. The lone pair on the nitrogen naturally will tend to bond with a positively charged species (much like the formation of quaternary ammonium compounds), and this species is provided by an aromatic sulphonate, supposed sometimes to be methyl toluene-4-sulfonate

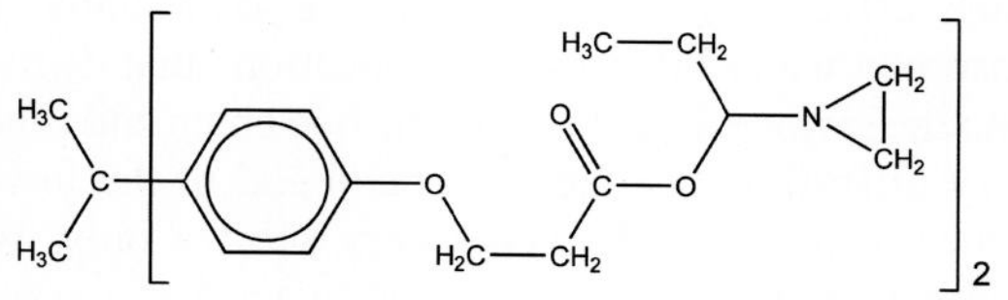

Fig. 7.1   The bis-phenol A-based epimine resin monomer.

(Fig. 7.2) (but the use of this exact compound is doubted).[69] In such compounds, the strong electron affinity of the sulphonate causes the electron density on the terminal methyl to decrease, leaving it with a net small positive charge. But, when this bonding occurs, the ring springs open as this lowers the energy of the system, and in so doing creates a new – and fully charged – cation, which of course can then react with another epimine group in a chain reaction (Fig. 7.3). Since this reaction can occur at both ends of the monomer, a highly cross-linked polymer results.

The material has been found to be filled with a spherical-particle polyamide powder (sometimes reported as nylon), but it was later described by the manufacturer as being PMMA-filled. In either case (and whether or

---

[12]   Certain products are very frequently described in the literature as being based on "vinyl ethyl methacrylate" (which is a badly constructed chemical name); this is quite incorrect. One is poly(isobutyl methacrylate), another poly(ethyl methacrylate), according to their respective manufacturers; the error for the latter at least apparently due to an advertising copywriter's ignorance. Were these materials based on the implied type of bifunctional monomer (3-butenyl 2-methacrylate, or the like – which are not known to be used in dentistry), they would be highly cross-linked. Because such errors are propagated freely and uncritically in the literature and advertising, primary sources should be checked wherever possible, but even so chemical sense is paramount.

not the formulation was changed), such a filler would not result in the hardness, stiffness or strength associated with ceramic-filled resins as the mechanical properties would be similar to those of the matrix (and bonding could not be expected). The main effect would therefore be to reduce the size of the exotherm.

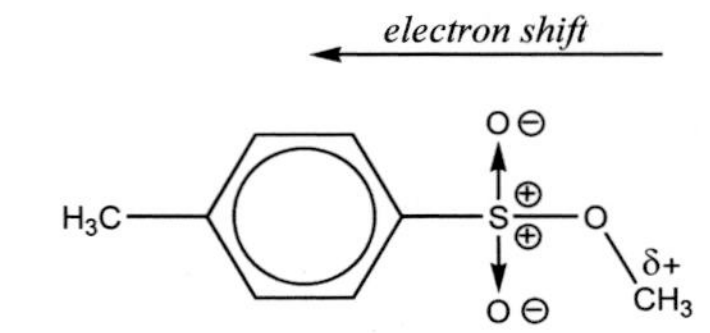

Again, on a monomer mass-basis, the exotherm for this reaction is low in comparison with plain PMMA, although there is still a risk of tissue damage *in situ*, and the material is less water-absorptive. The mechanical properties are complicated by the rigidity of the bis-phenol A moiety (which would tend to limit the degree of conversion), with the flexibility of the chain to the epimine group and the ethyl side-chain, as well as the bulk of the unreacted epimine rings, all of which would lower the glass transition temperature of the set material (3§4.5). Indeed, this has been reported to be around mouth temperature, and this may account for reports that in function the material is noticeably rubbery (in comparison with a typical filled resin). Because the monomer is difunctional, the proportion of completely unreacted monomer (*i.e.*, unbound) is low, and its size means that it would not readily diffuse out of the set material.

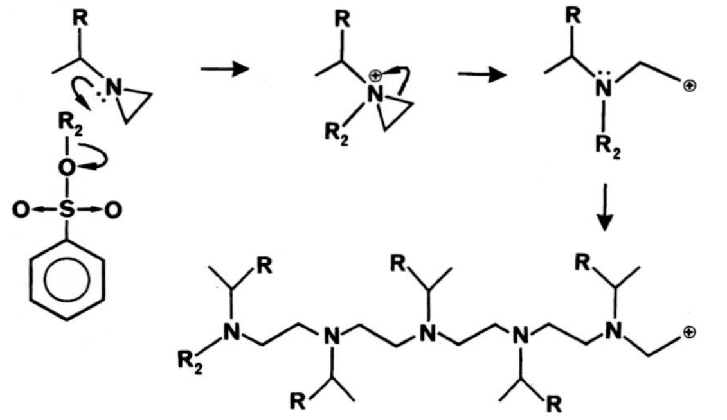

The material is sensitive to the presence of water, presumably due to the high reactivity of the **carbocation** ($>N-CH_2CH_2^+$) of the opened ring, which can be expected to suffer rapid **nucleophilic attack** (*i.e.*, positive charge-seeking) and the attachment of an hydroxyl group (Fig. 7.4). Thus, the live chain end would not survive long on exposure to oral conditions. Furthermore, despite the expectation that some unreacted epimine rings remain in the set material, repairs or additions cannot be made (*i.e.*, to stick effectively). This might be attributable to absorbed water killing any carbocations that diffuse into the surface, but of course the live chain end is not very diffusible (see the right hand end of the lower structure in Fig. 7.3), being very short – only two carbons long – and typically tied to a rather massive and relatively rigid network immediately behind it through the nitrogen.

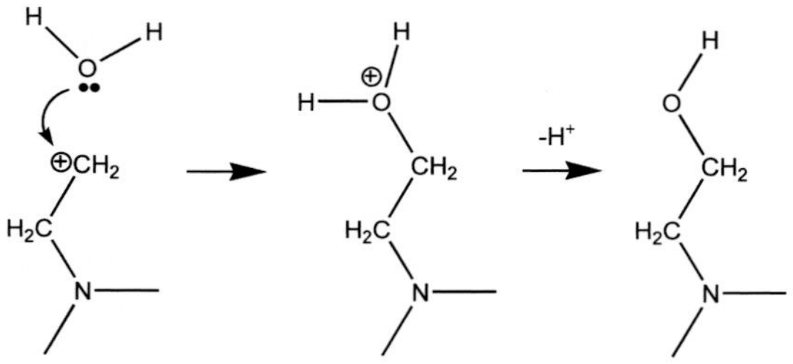

Fig. 7.2 Possible initiator for epimine cationic polymerization: the charge on the right-hand methyl group is due to the electron-withdrawing power of the sulphonate.

Fig. 7.3 The polymerization mechanism for epimine resin.

Fig. 7.4 Rapid reaction of a carbocation with a nucleophile to form an oxonium ion, which then promptly deprotonates.

Unfortunately, benzene sulphonates are known to be contact sensitizers, and adverse reactions were sometimes encountered (*cf.* 7§5.4). Whether this was a contributory factor is unknown, but the only known such product has been discontinued.

●7.5   Urethanes

Ethyl carbamate, or urethane (Fig. 7.5A), is the notional parent compound of a very large series of polymerizable substances derived from isocyanates, R–N=C=O (see also 10§6.3). In principle, and ignoring the chemistry that leads to the formation of such derivatives, the urethane unit can be extended by modifying the ester part, typically by attaching an acrylate, thus making it a polymerizable species (Fig. 7B). It can also be doubled-up, so to speak, by reaction with an isocyanate at the amine to form an imidodicarbonate (IDC) (Fig. 7C), so that combining this with the first modification leads to a bifunctional, free radical-polymerizable monomer that could be subject to all the usual processing possibilities. Furthermore, if R_1 is methacrylate (MA), say, then R_2 can lead through other chain types both to extend the length of the whole molecule and to modify the properties of the resulting polymer. The possibilities are clearly very numerous, and there is no point in a catalogue of types. Indeed, many have been used for filled-resin matrices, whether neat or mixed with other monomers to adjust the behaviour of the set material, instead of – or as well as – the bis-GMA-based types.

What can be seen immediately is that such a polymer would be both more hydrophilic and subject to inter-chain hydrogen bonding involving the secondary amine, >N-H (*cf.* polypeptides, 27§3), although the chain itself would have little steric hindrance to segment rotation. Its irregularity would contribute to the maintenance of amorphousness (3§4.1). Unfortunately, the hydrophilicity and susceptibility to hydrolysis limit the applications and service life of such polymers, although there is some variability.

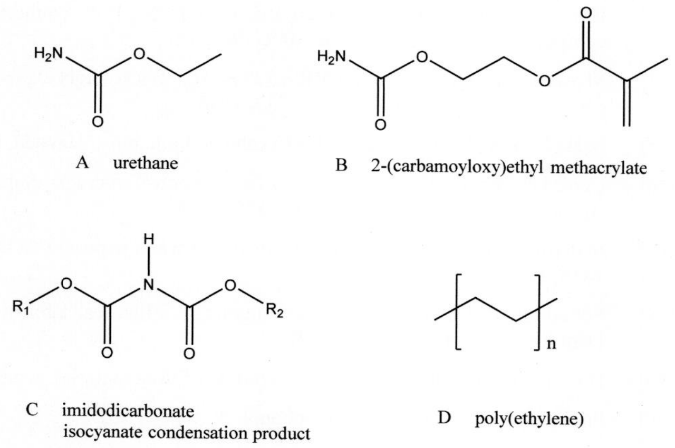

Even so, in the context of temporary materials, where service life is less of a problem, one formulation has used an inserted segment of hydrogenated polybutadiene (HPB) in a more or less symmetrical molecule thus:

A   urethane

B   2-(carbamoyloxy)ethyl methacrylate

C   imidodicarbonate isocyanate condensation product

D   poly(ethylene)

Fig. 7.5 Some aspects of urethane chemistry.

MA–IDC–R–IDC–HPB–..., where R is a branched alkane (and therefore sterically-hindered for rotation). Now polybutadiene (27§6.1) itself is a very good rubber, but by hydrogenating it – that is, removing all the remaining double bonds, 'saturating' it – would convert it in effect to polyethylene (Fig. 7.5D), albeit probably with some side groups and branches (Fig. 27§6.1). This would then cease to be a rubbery material in itself, but the likelihood is that some limited phase separation occurs (27§6.3) to form micelles (Fig. 3§4.4). If the cross-linking is extensive enough in the polymerized material, then this would lead to a glassy matrix with the micelles as a dispersion of weak, plastically-deformable particles, but bonded to the matrix. The effect of this would be to provide a means of blunting crack tips (Fig. 2.18) and thereby raise the toughness of the material. However, to describe such a system as 'rubberized' is an error. Indeed, the matrix could be viewed as being more rubbery: highly cross-linked but with very long flexible internodes, portions of which may also be micellar.

References

[1]   CG Plant, DW Jones, BW Darvell. The heat evolved and temperatures attained during setting of restorative materials. Brit Dent J 137(6): 233 - 238, 1974.

[2]   Alrahlah A, Silikas N & Watts DC. Hygroscopic expansion kinetics of dental resin-composites. Dent Mater 30 (2): 143 - 148, 2014.

[3]   A Versluis, D Tantbirojn, MS Lee, LS Tuc R DeLong. Can hygroscopic expansion compensate polymerization shrinkage? Part I. Deformation of restored teeth. Dent Mater 27(2): 126 - 133, 2011.

[4]   Paul B. Prediction of elastic constants of multiphase materials. Trans TMS-AIME 218: 36 - 41, 1960.

[5]   Pabst W & Gregorová E. Critical Assessment 18: elastic and thermal properties of porous materials – rigorous bounds and cross-property relations. Mater Sci Technol 31 (15): 1801-1808, 2015.

[6]   Vincent JFV. Structural Biomaterials. Macmillan, London, 1982.

[7]   Voigt W. Lehrbuch der Kristallphysik, pp 410 - 420. Taubner, Leipzig, 1928

[8]   Reuss A. Berechnung der Fliessgrenze von Mischkristallen auf Grund der Plastizitätsbedingung für Einkristalle. Z Angew Math Mech 9(1): 49 - 58, 1929.

[9]   Hashin Z & Shtrikman S. On some variational principles in elasticity and their application to the theory of two phase materials. Office of Naval Research, Contract 551(42) Technical Report No. 1, 1961.

[10]   Braem M, Lambrechts P, Van Doren V & Vanherle G. The impact of composite structure on its elastic response. J Dent Res 65(5): 648 - 653, 1986.

[11]   Braem M, Davidson CL, Vanherle G, Van Doren V & Lambrechts P. The relationship between test methodology and elastic behavior of composites. J Dent Res 66(5): 1036 - 1039, 1987.

[12]   Braem M, Van Doren VE, Lambrechts P & Vanherle G. Determination of Young's modulus of dental composites: a phenomenological model. J Mater Sci 22: 2037 - 2042, 1987.

[13]   Braem M, Finger W, Van Doren VE, Lambrechts P & Vanherle G. Mechanical properties and filler fraction of dental composites. Dent Mater 5: 346 - 349, 1989.

[14]   Miyazaki M, Oshida Y, Moore BK & Onose H. Effect of light exposure on fracture toughness and flexural strength of light-cured composites. Dent Mater 12: 328 - 332, 1996.

[15]   Scala E. Composite Materials for Combined Functions. Hayden, New Jersey, 1973.

[16]   Cook J & Gordon JE. A mechanism for the control of crack propagation in all-brittle systems. Proc Roy Soc London. Series A, Math Phys Sci 282 (1391): 508 - 520, 1964.

[17]   Söderholm K-JM & Roberts MJ. Influence of water exposure on the tensile strength of composites. J Dent Res 69(12): 1812 - 1816, 1990.

[18]   Söderholm K-JM, Mukherjee R & Longmate J. Filler leachability of composites stored in distilled water or artificial saliva. J Dent Res 75(9): 1692 - 1699, 1996.

[19]   Bowen RL. Synthesis of a silica-resin direct filling material: progress report. J Dent Res 37: 90 [M13], 1958.

[20]   https://en.wikipedia.org/wiki/Bisphenol_A

[21]   https://www.bda.org/dentists/policy-campaigns/public-health-science/fact-files/Documents/bisphenol_a_in_dental_materials.pdf

[22]   http://www.bisphenol-a.org/human/dental.html

[23]   Braem M, Lambrechts P, Vanherle G & Davidson CL. Stiffness increase during the setting of dental composite resins. J Dent Res 66(12): 1713 - 1716, 1987.

[24]   Elliot JE, Lovell LG & Bowman CN. Primary cyclization in the polymerization of bis-GMA and TEGDMA: a modeling approach to understanding the cure of dental resins. Dent Mater 17: 221 - 229, 2001.

[25]   Lovell LG, Berchtold KA, Elliott JE, Lu H & Bowman CN. Understanding the kinetics and network formation of dimethacrylate dental resins. Polym Adv Technol 12: 335 - 345, 2001.

[26]   Ferracane JL & Hilton TJ. Polymerization stress – Is it clinically meaningful? Dent Mater 32: 1–10, 2016.

[27]   Wang ZZ & Chiang MYM. Correlation between polymerization shrinkage stress and C-factor depends upon cavity compliance. Denta Mater 32: 343 - 352, 2106.

[28]   Tabata T, Shimada Y, Sadr A, Tagami J & Sumi Y. Assessment of enamel cracks at adhesive cavosurface margin using threedimensional swept-source optical coherence tomography. J Dent 61: 28 - 32, 2017.

[29]   Vankerckhoven H, Lambrechts P, Van Beylen M & Vanherle G. Characterization of Composite Resins by NMR and TEM. J Dent Res 60(12): 1957-1965, 1981.

[30]   en.wikipedia.org/wiki/Bisphenol

[31]   Polydorou O, König A, Hellwig E & Kümmerer K. Uthethane [sic] dimethacrylate: a molecule that may cause confusion in dental research. J Biomed Mater Res B: Apply Biomater 91B (1): 1 - 4, 2009.

[32]   Sideridou I, Tserki V & Papanastasiou G. Effect of chemical structure on degree of conversion in light-cured dimethacrylate-based dental resins. Biomater 23: 1819 - 1829, 2002.

[33]   http://goldbook.iupac.org/html/N/N04219.html

[34]   Braslavsky SE. Glossary of terms used in photochemistry 3rd ed (IUPAC recommendations 2006) Pure Appl Chem 79 (3): 293 - 465, 2007.

[35]   Barrow GM. Physical Chemistry, 4th ed. McGraw-Hill, New York, 1961.

[36]   Lewis GN & Kasha M. Phosphorescence and the triplet state. J Amer Chem Soc 66: 2100 - 2116, 1944.

[37]   Andrzejewska E, Lindén L-Å & Rabek JF. The role of oxygen in camphorquinone-initiated photopolymerization. Macromol Chem Phys 199: 441 - 449, 1998.

[38]   http://goldbook.iupac.org/html/N/N04218.html

[39]   Labrie D, Moe J, Price RBT, Young ME & Felix CM. Evaluation of ocular hazards from 4 types of curing lights. J Can Dent Assoc 76: b116, 2010.

[40]   Ham WT, Mueller HS & Sliney DH. Retinal sensitivity to damage by short-wavelength light. Nature 260 (5547): 153 - 155, 1976.

[41]   Sliney DH. Ocular hazards of light. Int Lighting in Controlled Environments Workshop. Tibbitts TW (ed). NASA-CP-95-3309 183 - 189, 1994. www.controlledenvironments.org/Light1994Conf/4_2_Sliney/Sliney%20Text.htm

[42]   Ziegelberger G. ICNIRP guidelines on limits of exposure to incoherent visible and infrared radiation. Health Physics 105 (1): 74 - 96, 2013.

[43]   Federlin M & Price R. Improving Light-Curing Instruction in Dental School. J Dent Educ 77 (6): 764 - 772, 2013.

[44]   S.E. Kopperud SE, Rukke HV, Kopperud HM & Bruzell EM. Light curing procedures – performance, knowledge level and safety awareness among dentists. J Dent 58: 67 - 73, 2017.

[45]   Soares CJ, Rodrigues MdeP, Vilela1 ABF, Rizo ERC, Ferreira LBF, Giannini M & Price RB. Evaluation of eye protection filters used with broad-spectrum and conventional LED curing lights. Braz Dent J 28 (1): 9 - 15, 2017.

[46]   Lee SY, Chiu CH, Boghoran A & Greener EH. Radiometric and spectrophotometric comparison of power outputs of five visible light curing units. J Dent 21: 373 - 377, 1993.

[47]   Chadwick RG, Traynor N, Moseley H & Gibbs N. Blue light curing units – a dermatological hazard? Brit Dent J 176: 17 - 21, 1993.

[48]   L Musanje & BW Darvell. Polymerization of resin composite restorative materials - exposure reciprocity. Dent Mater 19 (6): 531 - 541, 2003.

[49]   Cook WD. Spectral distributions of dental photopolymerization sources. J Dent Res 61: 1436 - 1438, 1982.

[50]   http://opticalfiltershop.com/shop/color-filter/color-filters-blue-508nm-2/

[51]   Ro JH, Son SA, Park JK, Geon GR, Ko CC & Kwon YH. Effect of two lasers on the polymerization of composite resins: single vs combination. Lasers Med Sci 30 (5): 1497 - 1503, 2015.

[52]   Price RB, Ferracane JL & Shortall AC. Light-curing units: A review of what we need to know. J Dent Res 94: 1179 - 1186, 2015.

[53]   Mutluay MM, Rueggeberg FA & Price RB. Effect of using proper light-curing techniques on energy delivered to a Class 1 restoration. Quint Int 45: 549 - 556, 2014.

[54]   Spranley TJ, Winkler M, Dagate J, Oncale D & Strother E. Curing light burns. Gen Dent e210 - e214, 2012.

[55]   Michaud P-L, Price RBT, Labrie D, Rueggeberg FA & Sullivan B. Localised irradiance distribution found in dental light curing units. J Dent 42: 129 - 139, 2014.

[56]   Price RB, Strassler, Price HL, Seth S & Lee CJ. The effectiveness of using a patient simulator to teach light-curing skills. J Amer Dent Assoc 145: 32 - 43, 2014.

[57]   Price RBT, Labrie D, Rueggeberg FA, Sullivana B, Kostylev I & Fahey J. Correlation between the beam profile from a curing light and the microhardness of four resins. Dent Mater 30: 1345 - 1357, 2014.

[58]   Moszner N, lFischer UK, Ganster B, Liska R & Rheinberger V. Benzoyl germanium derivatives as novel visible light photoinitiators for dental materials. Dent Mater 24 (7): 901 - 907, 2008.

[59]   Ingraham LL & Meyer DL. Biochemistry of Dioxygen. Plenum, New York, 1985.

[60]   Borden WT, Hoffmann R, Stuyver T & Chen B. Dioxygen: What makes this triplet diradical kinetically persistent? J Amer Chem Soc 139 (26): 9010 - 9018, 2017.

[61]   Rawe A. Principles of Polymer Chemistry. p. 255. Springer, New York, 1995

[62]   Øysæd H, Ruyter IE & Sjøvik Kleven IJ. Release of formaldehyde from dental composites. J Dent Res 67: 1289 - 1294, 1988

[63]   Jordan J, Ackerman E & Berger RL. Polarographic diffusion coefficients of oxygen defined by activity gradients in viscous media. J Amer Chem Soc 78 (13): 2979 - 2893, 1956.

[64]   Leprince JG, Lamblin G, Devaux J, Dewaele M, Mestdagh M, Palin WM, Gallez B & Leloup G. Irradiation Modes' Impact on Radical Entrapment in Photoactive Resins. J Dent Res 89: 1494 - 1498, 2010.

[65]   Leprince J et al. Kinetic study of free radicals trapped in dental resins stored in different environments. Acta Biomater 5: 2518 - 2524, 2009.

[66]   Titley KC, Torneck CD, Ruse ND. The effect of carbamide-peroxide gel on the shear bond strength of a microfil resin to bovine enamel. J Dent Res 71: 20 - 24, 1992.

[67]   Gökçe B, Çömlekoglu ME, Özpinar B, Türkün M & Kaya AD. Effect of antioxidant treatment on bond strength of a luting resin to bleached enamel. J Dent 36: 780 - 785, 2008.

[68]   Lai SCN et al. Reversal of compromised bonding in bleached enamel. J Dent Res 81: 477 - 481, 2002.

[69]   Braden M, Causton B & Clarke RL. An ethylene imine derivative as a temporary crown and bridge material. J Dent Res 50(3): 536 - 541, 1971.

Chapter 7   Flexible Impression Materials

The problem of designing a mould material for making dental models is formidable. It includes severe requirements for flow, setting in a reasonable time and perfect elasticity as well as absence of dimensional changes on setting and after. No perfect solution has been found, although many attempts have been made. This chapter describes a range of common types of dental impression material and how they meet – or fail to meet – those demands. The good as well as the bad points of each must be taken into account in the selection of the type of product and its manipulation for the task in hand.

*All flexible mould materials are polymeric and must be **cross-linked** to be rubber-like, that is, **three-dimensional random networks**, and they are markedly non-Hookean in their deformation behaviour; this can be traced to features of their structure.*

*Fillers are again often important in order to obtain adequate viscosity in the unset material, and appropriate stiffness when it is set. However, the large strains that these materials must endure may result in a structural breakdown called **strain-softening**. The properties are also dependent on the nature of the filler.*

*Polysulphide, polyether and silicone impression materials all give **covalently-bonded** networks when set, although each has its own particular dimensional stability problems arising from their chemistry. Two **polysaccharide** materials, agar and alginate, are strongly dependent on hydrogen bonding for network formation; alginate relies on a chelation mechanism as well. These two types are sensitive to water gain and loss.*

*The many sources of **permanent deformation** after exposure to a stress mean that perfection is unattainable in practice. Again, compromise is necessary to balance the various beneficial and detrimental properties.*

Flexible impression materials play a crucial role in the fabrication of many dental devices, from full dentures to inlays, from orthodontic appliances to implanted prostheses. A knowledge of these structures and the chemistry is essential in order to make an intelligent selection.

Materials Science for Dentistry
https://doi.org/10.1016/B978-0-08-101035-8.50007-9

Since many of the restorative devices that are used in dental treatment need to be fabricated outside of the mouth, ranging from inlays to full dentures, there are obvious problems in getting them to fit if reliance is placed on trial and error methods. This used to be done when full dentures were carved from materials such as ivory, but it was quickly realized that a model of the mouth made the task much easier. The problem then is one of making the model. For the model to be made, leaving aside the question of from what, a mould is required. The earliest mould material was in fact wax, soft enough that the shape of the teeth could be impressed into it, hence the term impression, even though very little pressure is used with most modern materials.

§1. Basic Requirements

The most important characteristic for an impression material is fairly obvious: it must accurately reproduce the entire surface upon which the device to be made will fit. All angles and distances must be preserved. This means that both small scale detail and larger scale dimensional accuracy must be attained, and particularly over the full size of the region being reproduced, in other words: no distortion. This requirement is more critical for larger devices. For example, an error of 10% when a 1 mm distance is considered may not seem very much, but clearly over 100 mm such an error would be laughable - more than the width of a tooth.

The second required characteristic is that of elasticity, in conjunction with a low elastic modulus and large elastic range (1§2.1). The reasons for this combination of properties are entirely due to the usual morphology of oral structures. Human teeth are not smoothly conical, but 'waisted'. The cervix of the tooth is narrower in at least one direction than the crown, that is, there are 'undercuts'. Thus, in order to remove an impression from an undercut it must be deformed, stretched, to allow the crown of the tooth to be slid out. This must only be a temporary deformation, so that the original dimensions are recovered perfectly, but with the least convenient force being used to avoid pain for, or injury to, the patient. The principal material requirement in this respect is its permanent deformation after stress, as shown, for example, by the compression set test (4§8). The implication here is that the impression should be removed in one 'snap', a quick operation that does not allow time for flow.

It will be found that these requirements cannot be met fully in practical materials, and to some extent are mutually incompatible. We have to find a workable compromise between these properties. The elastic modulus and flexibility requirements can be dropped for certain special applications, in particular for whole mouth edentulous impressions (assuming no undercuts in the bone-mucosa structure), when other properties assume dominance. Wax- and gypsum-based impression materials, used in these special cases, will not be dealt with in this chapter.

There are other considerations. Whatever process is used to create the elastic mould, time is an important factor: patient tolerance is limited, and a busy practice is said not to be able to afford the economic implications of disproportionately lengthy procedures. Whether this last item is a legitimate consideration needs to be judged in terms of the alternative: what is the most favourable course from the point of view of patient care? The toxicity of materials is of concern where soft-tissue contact occurs, and excessive heat must be avoided. Once a mould has been made, its useful life (duration, number of models that can be made) constrains its handling severely. Even colour has been found to be important: inspection by the clinician to determine the adequacy or otherwise of fine detail reproduction is limited by the opacity and contrast of the material. Given that a compromise is involved anyway, the cost of the material must be balanced against the benefit - the clinical outcome, which usually means asking whether the device being made ultimately fits well enough to be able to serve its clinical function indefinitely, assuming no other factors are operating.

§2. Deformation

The flexibility requirement arises essentially only because oral structures are undercut: a rigid impression could not be removed without cutting or breaking it. This can prove to be extremely difficult and distressing. A patient's mouth will be fairly filled by the impression material and tray (which is often metal) and access for cutting tools is very limited and hazardous. Yet even if the material is deformable, the elastic modulus must not be so high that the load required to get the impression off the tooth exceeds that required to extract the tooth - or come anywhere near that value. It is obvious that an impression ought to be perfectly elastic to be withdrawn over such undercuts and then recover exactly the original shape. Equally obviously, it will not matter if the deformation is not Hookean, that is, proportional to stress, so the proportional limit of these materials is of no practical concern. In fact, the deviation from ideal behaviour in this sense can be very great without the slightest influence on the acceptability of the material. Conversely, the yield point is very important: no plastic deformation is, in principle, acceptable in the solid material.

It will be recalled that polymers are, as a general rule, characterized by their glass transition temperature, above which they exhibit rubbery behaviour (Fig. 3§4.13). To summarize what constitutes such a rubbery or **elastomeric** material, we may cite:
- free rotation of bonds,
- amorphous structure (*i.e.* not crystallize readily),
- cross-linking of chains to prevent relative slippage, and
- lack of interaction between chains.

Because of these structural factors, the mechanical behaviour of elastomers is characterized by two properties. Firstly, **isotropic deformation**; the lack of long-range structural regularity means that all directions in the mass are equivalent (at least, until appreciable deformation and therefore chain alignment has occurred). Secondly, deformation is expected to be at **constant volume**[1] (assuming constant temperature), which means that the Poisson ratio would be exactly 0.5. This freedom of chain segments to move means that the free volume (and thus average distance between chains) is, ideally, not affected.

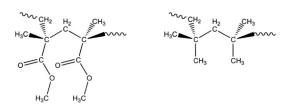

Fig. 2.1 The structural formulae for (left) poly(methyl methacrylate) and (right) poly(isobutylene).

Thus, polyisobutylene (PiB)[2] (Fig. 2.1), for example, is a much better elastomer than poly(methyl methacrylate) (above its T_g ~72 °C, and thus in a rubbery state; Fig. 5§5.1) because, despite the apparently bulky methyl groups, chain segment rotation is easier by some 6 kJ/mol at the maximum, but also the average energy is about half that of the the PMMA (Fig. 2.2), which is why PMMA needs to be hotter to be rubbery at all. In contrast, PiB has a T_g around −70 °C. Such rotation activation energy calculations, however, refer only to the gas phase – an isolated molecule. In the solid state, chain interactions must also be considered (3§2.7). Thus collision between a rotating segment and its neighbours makes the activation energy for that rotation higher. Broadly, the larger the side group, the more difficult it becomes to get past such an obstacle. Thus, the

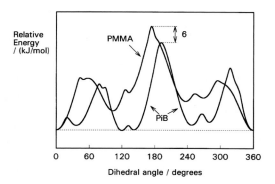

Fig. 2.2 Energy *vs.* dihedral angle at the -CH$_2$- link in the backbone of poly(isobutylene) (PiB) and poly(methyl methacrylate) (PMMA).

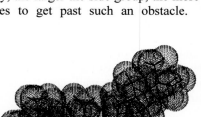

Fig. 2.3 Fragment of a space-filling molecular model of poly(isobutylene).

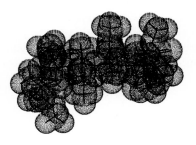

Fig. 2.4 Fragment of a space-filling molecular model of poly(methyl methacrylate). Notice the side-chain 'knobs' that would interact with adjacent chains.

---

[1] However, this critically depends on structure and allowing chain segments to move freely. Some unusual polymers have been designed which expand in volume on being pulled and offer some special properties. These are known as 'auxetic' materials. See ref. [1].

'smoother' chain of PiB (Fig. 2.3) also permits chain segment rotation and reptation more easily than does the more 'knobbly' PMMA (Fig. 2.4) because steric interaction between chains is much less. Furthermore, being non-polar, the bonding between chains is only of the very weak van der Waals type (10§3), while the dipole on the carbonyl group of the PMMA strengthens inter- and intra-chain bonding.

The effect of interchain **steric hindrance** means that plasticizers (3§6) also work to make elastomers softer, whether these are present deliberately (§5), as by-products (§4, §6.1), or through accidental or procedural exposure to solvents, including water.

In contrast to the above structurally-based explanation of an elastomer, we may also consider a more operational view, even though the structural reasoning is the same. Hence, we may define an elastomer as a **three-dimensional network polymer** which has been designed to have both good elasticity (in terms of its perfection) and have a large elastic range (1§2.1). Implicit in this is the assumption of being within an intended working range of temperature. Thus, $T > T_g$ is required for it to be in the rubbery region (Fig. 3§4.13), while clearly oxidation or decomposition sets an absolute upper limit to the working temperature range (it is clear that if the polymer is sufficiently cross-linked to be a good rubber, it will not have a meaningful melting or 'rubbery flow' point). We should also bear in mind that there will be strain-rate effects on the rubbery plateau as well (Fig. 3§6.2).

●2.1 Equations

Using a slightly different notation to that used in 1§2, we can express the constancy of volume in a simple equation. Thus, under uniaxial strain (in the x-axis), in tension or compression,

$$\begin{aligned} V &= x_0 y_0 z_0 \\ &= x_1 y_1 z_1 \\ &= (\alpha x_0)(\beta y_0)(\beta z_0) \end{aligned} \tag{2.1}$$

where the subscript 0 refers to the original dimensions, and subscript 1 the dimensions after deformation; α and β are therefore the factors by which the original dimensions are changed, *e.g.*

$$\alpha = \frac{x_1}{x_0} \tag{2.2}$$

This α is termed the **extension ratio**,[2] and it is more usual to refer to that than strain in the context of elastomers because these materials usually have low elastic moduli and therefore show large values of strain at low stresses. Extension ratio and strain, ε are, however, simply related:

$$\alpha = 1 + \varepsilon = 1 + \left[\frac{x_1 - x_0}{x_0}\right] = 1 + \left[\frac{x_1}{x_0} - 1\right] \tag{2.3}$$

If α is measured, the lateral extension ratio, β, is easily found from equation 2.1:

$$\beta = \frac{1}{\sqrt{\alpha}} \tag{2.4}$$

as clearly we must have $\alpha\beta^2 = 1$. The mobility of chain segments is sufficient to allow the rapid relief of lateral stresses, limited only by the relaxation time for the diffusion of chain segments. The Poisson ratio is therefore taken to be exactly one half, at least at low values of strain. Amongst the consequences of this is the impossibility of calculating the true modulus of elasticity or the bulk modulus from Young's Modulus and the conventional formulae (equations 1§2.10, 1§2.23) since the divisor would become zero. This underlines the fact that there are always approximations in such calculations, *i.e.* dropping non-negligible higher powers of strain is not appropriate in these circumstances.

An equation relating the axial stress (σ_x) to the extension ratio in elastomers has been derived:

$$\sigma_x = vkT(\alpha - \alpha^{-2}) \tag{2.5}$$

The constant of proportionality involves the absolute temperature, T, the Boltzmann constant, k, and v the number of **effective chain segments** per unit volume of the polymer. A segment can be considered as that length

---

[2] Extension ratio is also commonly given the symbol λ.

of polymer chain whose ends are more or less immobile (on the time scale of the observation) because of entanglement, branches or cross-links (3§3, 3§6.2). This equation has been shown to hold over a wide range of compression and extension (Fig. 2.5), although deviations start to appear at very high extensions. This illustrates the importance of the effect of polymer structure on the properties, and further permits the deliberate design of polymers with specific values for those properties. However, it is obvious that the stress-strain curve is nowhere linear. For this class of materials an approach to proportional extension may only be seen at very small stresses. We can see that they are not Hookean materials because the deformation *internally* is a rearrangement of chain segments and not the straining of primary interatomic bonds. The restoring force is therefore due to conformational changes, making the system more ordered. Return is therefore due to **entropy elasticity**. The mechanism of deformation is therefore fundamentally

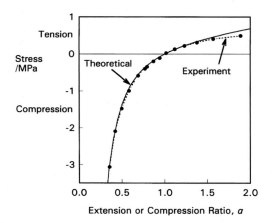

Fig. 2.5 Theoretical and experimental stress-strain curves for an elastomer. Note the strongly non-Hookean behaviour.

different from that of most other types of material. Indeed, this behaviour may be termed **hyperelasticity**.

●2.2 Failure

During such deformation, the rearrangement of the polymers chains is such as to tend to align them. This is equivalent to drawing (3§4.7) except that macroscopic plastic strain – flow – is not required. Indeed, this is deliberately meant not to happen by the provision of cross-linking, and enhanced by the inevitable entanglements present.

These cross-links and entanglements are the defining feature of elastomers, preventing macroscopic chain drawing, because they are the source of the desired elasticity. However, once a section of chain has become straightened, the cross-link and entanglement points, as well as the length of chain between, must from now on carry proportionately more load than adjacent chains. That chain is taut, while the adjacent chains are carrying relatively little load (Fig. 2.6). The energy of deformation is now stored in bond length and angle strain – **energy elasticity**. The (tangent) modulus of elasticity now tends to start to rise again (*cf.* Fig. 3§4.11). But if the load on any one bond in the chain exceeds its strength that bond must break. Such broken chains can then no longer contribute to the stiffness of the body in the region of the break. In other words, applying a sufficient stress to the body must result in some chains being broken even if no obvious plastic deformation or failure occurs, and this causes a reduction in elastic modulus. This is **strain-softening**.

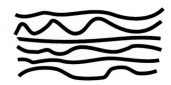

Fig. 2.6 Polymer chains will reach full extension at different stages in the drawing process.

There are further consequences. The rupture of a polymer molecule commonly results in the formation of a pair of free radicals[3] (**homolytic scission**) (Fig. 2.7). These radicals are active and high-energy entities which may take part in a number of further reactions. One of these is chain transfer (see 5§1.4) through **free-radical attack**, a kind of bond exchange. Ordinarily, this would not be a problem, but when the polymer is under stress further damage can be done. Thus, if the bond that the free radical attacks is under load, the rupture is easier (because of the stored energy) and leads to **stress relaxation** as the newly-formed radical-bearing chain end is pulled away (Fig. 2.8). This then throws the load onto adjacent chains. As they in turn become taut through localized drawing so they will become subject to attack by the new radical. The process can then be repeated indefinitely, or at least until another free radical is encountered and they mutually annihilate (*cf.* 5§1.3). In this way the nuclei of cracks will have been generated, and these will

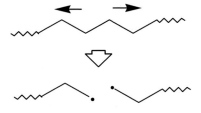

Fig. 2.7 Polymer chain rupture generates a pair of free radicals.

---

[3] Radicals are created in preference to ions (*i.e.* a carbocation-carbanion pair) because this is the lowest energy outcome for a homopolar bond – the principle of least action.

grow as the damage accumulates, in effect being **sub-critical cracks** in the Griffith sense (1§7). When they have grown to a critical size, then crack growth will initiate leading to spontaneous collapse.

Even if the attacked chain is not under stress, **entropic recoil** means that the new chain end will tend to move away, diffusively, from its original position. Thus, the extra freedom that the chain end has, in the context of thermally-activated diffusion, means that the number of random coil conformations available has increased. An extended chain is relatively improbable, so recoil is expected, and would enhance the effect of stress relaxation.

There is one further aspect of this, which applies to many polymers under stress, not just elastomers: **autocatalytic damage**. When there is oxygen present (which in dental contexts cannot be avoided) this may react with the scission free radical. Reactions of the same type as occur with oxygen inhibition then can occur (6§6), the key feature being the multiplication of the numbers of radicals when the hydroperoxide formed itself undergoes homolytic scission spontaneously (reaction 6§6.3). This is perhaps a fairly slow process under most ordinary conditions, but it does account for the inability of many polymers to sustain tensile loads for extended periods and beyond the effects of simple diffusive creep, which is clearly not a significant process in many cases (Fig. 2.9).

As a further consequence of this kind of chemistry, we can note that PMMA and other free-radical polymerization systems are essentially equilibria (5§2.7). The "unzipping" process generates free radicals spontaneously. Clearly, these also permit stress-relaxation in the same way: even a small stress, below the strength of the bond, would result in chain-end separation, even if entropic recoil did not achieve this. Such a mechanism may be part of the creep process in filled-resin restorative materials (6§1.4), as well as the formation of crazes in PMMA that is left under stress for extended periods, such as bent sheet material or around overtightened screws.

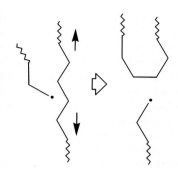

Fig. 2.8 Free radicals can attack other chains, leading to stress relaxation if such a chain is under tension but in any case reducing constraint and allowing entropic recoil.

Fig. 2.9 Tension cracking in a polymer tube cap.

●2.3 Cause of elasticity

It should perhaps be emphasised why elastomers are indeed rubbery, what gives them their elastic behaviour. Given that the natural state of a non-crystallizing polymer chain is an amorphous, random coil (Fig. 3§2.3), which represents the most probable state, this is maximum entropy – disorder. Entropy is, of course, energy. Work has to be done to straighten such chains (Fig. 3§2.4), raising the energy of the system. Initially, this may be as strain in primary bonds, but then 'frictionally' in rearranging the molecules – which raises the temperature. That is, to get a chain segment from one position to another the activation energy has to be provided to get over the energy barrier, but once 'over the hill' that energy is thermalized, converted to heat. Since crystallization as such is not available as a path for lowering the energy of the system, the stress is maintained by the sideways thermal motions of chain segments, but the greater ordering of the chains as they become drawn and aligned means that they are relatively more crystalline, therefore of lower degree of disorder, which is lower entropy. That energy must then also be released as heat, raising the temperature further. Assuming that there are no other mechanisms operating to reduce that stress (such as chain ruptures), it will be maintained indefinitely at a constant temperature, indeed, the force generated will be a measure of temperature. If the constraint is removed, randomization again occurs through the rapid diffusion of chain segments to restore random coils everywhere. This is **entropic recoil**. It is clear that the speed of this is much lower than for bond length or angle strain, and indeed is a retarded recovery (4§5) as it is necessarily diffusive and therefore time-dependant. Note that the effect of a small sideways force requires a large resultant axial force to maintain the length: any displacement sideways must shorten the distance between the ends if the chain is not stretchy (Fig. 2.10). Again, this is **entropy elasticity**. Of course, some measure of energy elasticity will occur when chains are fully extended (§2.2).

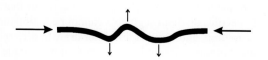

Fig. 2.10 Small lateral thermal displacements generate large axial forces in a polymer chain.

An experiment may be done to illustrate this (Fig. 2.11). Take a thick elastic band, and keeping it slack touch it to your lip to gauge the temperature. Stretch it quickly as far and as fast as you can, and again touch it to your lip: it is now warm from the work done and thermalized entropy. Keeping it stretched, wave it about in the air for a few seconds to cool again. Touch it to your lip to confirm this, then let it relax quickly and again touch to your lip: it is now cold because energy was required to re-establish the high entropy of the random coils. An analogy is melting – a disordered state created from an ordered one with the input of the latent heat of melting. The (relatively) ordered condition of the polymer chains when drawn requires the

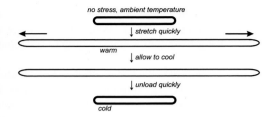

Fig. 2.11 The rubber band experiment: the thermal changes on stretching and recovery are easily detectable and reveal changes in chain ordering.

"latent heat" of randomization; this is taken from the material itself and the surroundings, hence the cooling. Equivalently, the warming on stretching is like the evolution of the latent heat of crystallization.

A second experiment might be to stretch such a band in the same way (but not with your fingers!) and immerse it while stretched in liquid nitrogen (77 K, – 196 °C), converting it to a hard glass (for natural rubber, $T_g \sim -65$ °C; 27§1). It is now rigid, and retains its length when let go as there is no chain segment diffusion. There is no restoring force. Allow it to warm up and recovery becomes rapid when the temperature of the rubber rises above its T_g. A similar effect can be seen when a drawn polymer filament, such as nylon fishing line, is heated: it will shrink when the activation energy for diffusion is available and entropic recoil occurs, even though it is not an elastomer.

●**2.4 Affine deformation**

In developing the ideas of elastic deformation (1§2), and their application to elastomers above, it is implicit that we are treating the bodies as being **continua**, that is, structureless and homogeneous: no atoms, molecules or other identifiable entities on any scale. Essentially, this is a macroscopic approximation, a purely geometrical approach to an idealized behaviour. The implication of this is that all points within the body must remain in the same spatial arrangement and relationship to one another, microscopically and reversibly, except for distances and angles. Literally, it is as if simple geometrical transformations were applied, scaling and shear, and all combinations, as mimics of the assumed behaviour under axial and shearing forces (Fig. 2.12). Such an

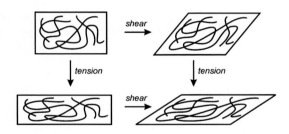

Fig. 2.12 Affine deformation: simple geometric distortion (assumed constant volume).

exact process is called an **affine deformation**. (Translations and rotations are also affine, but trivial here.) An equivalent is the distortion of objects in a graphics program obtained by manipulating the 'handles', as indeed was used to create Fig. 2.12. Affine transformations such as this preserve collinearity (*i.e.*, all points lying on a line remain on a line) and ratios of distances (*e.g.*, the midpoint of a line segment remains the midpoint). Similar (but more elaborate) arguments can be created for more complicated distortions (see below). The implication is that for a homogeneous continuum the deformation is also microscopically affine, and therefore completely reversible. What we may now do is apply what should be an affine transformation to a body as a whole and then examine it at a small scale – that of the molecules and atoms – to determine whether the transformation is everywhere uniform.

Unfortunately, in polymeric systems, and especially in elastomers, where the strains may be very large, the uncoiling and slippage of chains mean that locally the spatial relationships between atoms on adjacent molecules or chain segments cannot be maintained. Such chains vary in length and orientation, and in pattern of folding, and so must unfold or draw in different ways. There is necessarily differential movement between these atoms, with no homogeneity of magnitude or direction in the displacement vectors. At this level, then, polymer deformation cannot be affine: the topology cannot be preserved. Accordingly, in cross-linked systems, for convenience, the deformation with respect to the nodes of the network (the cross-link or branch points) might be treated as affine, but only as a first approximation. This means that the spatial relationships of those nodes are treated as if they alone were subject to the simple geometrical deformations, leaving the internode chains to find their own way, as it were, but still assuming that with respect to those nodes the system is microscopically and completely reversible. However, if the chains surrounding a node do not deform affinely, then the forces acting on the nodes, whether through bonds or molecular contacts, cannot be isotropic or even show a smooth consistent gradient over the whole piece: the nodes must themselves show relative movement beyond the affine prediction. In other words, in any such polymer there will be permanent deformation after any strain that causes chains or chain segments to slide past each other to any extent, or even a side group to be rearranged.

A simple demonstration of the failure of the affine assumption can be made with a thick elastic ("rubber") band, as above. Grasp a portion of the loop between fingers and thumbs so that it can be stretched to near its maximum extension ratio, and compare the colour with that of the unstretched portion alongside: it will have become slightly whiter. This arises because the random cross-linking means that the alignment of chain segments on drawing is not uniform. This leads to variation in the density of the rubber, and therefore its refractive index – a micellar structure (Fig. 3§4.4) will have been created, although all the micelles will be aligned in this case. Thus, extra light scattering occurs, producing a colour change similar to that seen with crazes (5§5.4) – this is **stress whitening**. Clearly, the lack of homogeneity implies a non-affine deformation even at small strains.

It is therefore the case that, for a simplistic approach, we may use the affine assumption cautiously as a useful approximation, but if we wish to consider local molecular effects it will break down. There are macroscopic phenomena due to the cumulative effects of the departure from an assumed ideal deformation, and there are consequences for dentistry arising from this. These issues will be further addressed in §11. It may also be noted that these considerations apply to any polymer system, not just elastomers, and to filled resins in particular (6§2.8). In both kinds of system, the filler particles may be taken as the reference points in the deformation, but there are further issues to consider when a filler is present. In fact, there are many other kinds of deformation, such as "keystone", bending (Fig. 23§2.12) and torsion (Fig. 1§2.09), in which collinearity and ratios of distances would not be preserved (these are described as non-affine transformations), but where local relationships would still be maintained in the homogeneous model. This does not reduce the force of the argument above, but indeed strengthens it as making it more general: no deformation of a polymer can preserve local chain topology.

§3. Effects of Fillers

As pointed out earlier (4§9), fillers are often used to modify the properties of a wide range of materials. They are useful in rubbers generally;[3] for example, car tyres are heavily filled, and most dental impression materials contain fillers to a greater or lesser extent. Essentially, the problem is that while we require a low modulus in an impression material, that of pure elastomers is usually too low: an impression that distorted by sagging under its own weight would not be useful. We return to this point later (§12.9). Thus, we may view the design of an impression material in terms of the two factors: increasing the degree of cross-linking to attain good rubbery behaviour without the T_g being raised too high, and then adding in a filler to raise the value of the elastic modulus to suit the conditions of use.

To examine the mechanical effects of fillers in elastomers we can consider a small cylindrical element containing just one filler particle, and then apply an axial tensile stress (Fig. 3.1). The overall extension of that cylinder is defined by the positions of the end faces relative to the start, but if the filler particle has itself a higher modulus of elasticity than the surrounding rubber it will not have been extended to the same amount as the adjacent material. In fact, typical fillers are ceramic or similar very high elastic modulus materials, and their extension will be entirely negligible. Hence, in order that the end faces of the cylinder remain plane, the column of material above and below the filler particle must be extended by a proportionally greater amount than that of the surrounding cylinder (disregarding for the moment that the two sections of elastomer are connected and that the notional boundary between them is in shear).

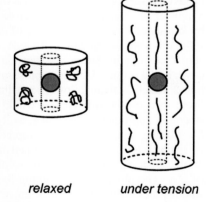

relaxed　　　　under tension

Fig. 3.1 The extension of a volume element of a filled elastomer must take into account the failure of the filler to deform.

These extensions are easily calculated (Fig. 3.2). Taking the length between filler particle centres as the reference unit, the column with the filler particle is shorter by the filler particle diameter. The displacement, ΔL, is the same for both. It is then a matter of writing down the relative dimensional changes, *i.e.* the strains of the two sections. The strain in the filler column is expressed as a function of the cube root of the volume fraction of filler, φ_f, the value of which will be known from the formulation of the material. Thus,

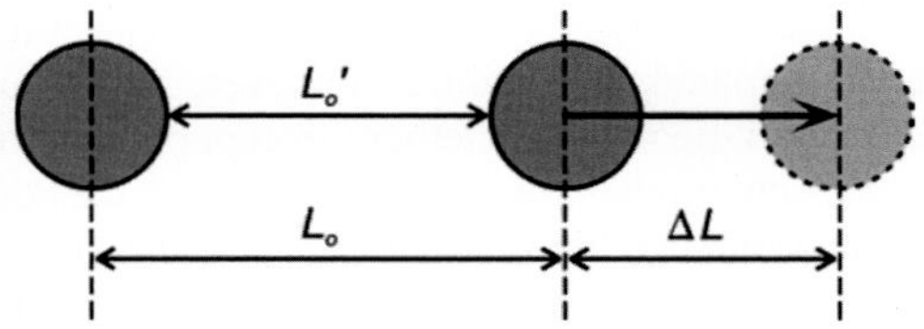

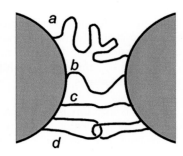

Fig. 3.2 Calculation of strain in filled elastomers.

$$\frac{\Delta L}{L_0} < \frac{\Delta L}{L_0{'}} = \frac{\Delta L}{L_0 - \varphi_f^{1/3}} \qquad (3.1)$$

It is apparent that at high filler contents the strain locally may become quite large, *i.e.* when filler particles are very close together, even though the overall strain is small.

●3.1 Mullins effect

One of the conditions essential for the proper functioning of a filler in any system, and no less in impression materials, is that there is bonding between the matrix and the filler (6§2). If we take a close look at the possible arrangements of polymer chains between filler particles (Fig. 3.3), it becomes clear that the effects of overall deformation are not uniform for each one (*cf.* Fig. 2.6). Since the bonds will be made at random, the lengths of chain between such link points will also be random: $a > b > c$. Therefore, the amount of extension that each chain can withstand before being fully straightened (by bond rotation, 3§2) will vary, according to their lengths, entanglements (*d*) and cross-linking. Even at overall low strain some chain segments might be expected to break because the load that each now carries exceeds its strength, or, for some of the bonds to the filler to fail. Clearly then, these broken bonds or chains reduce the stiffness of the material at strains *up* to that which produced the breaks. By being converted to 'free ends', such sections of chain can contribute little or nothing to the resistance to deformation.

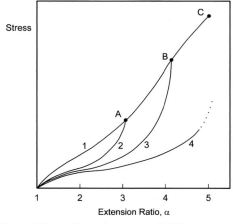

Fig. 3.3 Polymer chains linking filler particles will be of random length and entanglement and so have differing capacities for extension.

Thus, in the stress-extension curve 1 (Fig. 3.4), from the origin through ABC..., we observe unremarkable behaviour; superficially it is just as is shown in Fig. 2.5. But a number of chains or matrix-filler bonds must have been broken successively in the process of reaching point C. So, if the strain was taken only as far as A, and then returned to zero, the *unloading* stress *vs.* extension curve follows line 2 down to the origin. However, on reapplying the load, it is line 2 which is then followed, not line 1, back up to A. From then on the line ABC... would again be followed, smoothly continuous with line 1, as it would have done if the loading had not been interrupted. If, however, the

Fig. 3.4 The cyclic deformation of a filled elastomer reveals the strain-softening of the Mullins effect. Path 1 is followed on initial loading, path 2 on unloading from A; path 2 is then followed on reloading, path 3 on unloading from B; and so on.

extension were instead to be stopped at B, when the load relaxed to zero, line 3 would be followed back to the origin. Loading for the third time would follow line 3 from the origin, back up to B, and from there following BC... . The point is that straining beyond the previous maximum strain results in further bond breakages, whether they are within a chain segment or to a filler particle, making the material less and less rigid. Ultimately, a curve such as number 4 will be produced. It should be noted that it does not require every loading-unloading step of this example to be followed: if the interruption at point A were omitted, line 3 would still be the unloading curve from point B. This strain-dependent softening is called the **Mullins effect**. It can be seen to be similar to the strain-softening described in §2.2, but it is exaggerated by the presence of the filler.

It is now also apparent that similar effects will operate in filled-resin restorative materials, where bonded filler and a high degree of cross-linking is necessary (Chap. 6). Any stress which causes a polymer chain to break results in the material being less stiff afterwards below that stress. In other words, the effective number of cross-links has been reduced. Stress relaxation due to flow is thereby facilitated. This will be a contributing mechanism to long-term deterioration.

●3.2 Removing impressions

The Mullins effect is a second strong reason (see 4§6.1) for the removal of impressions in flexible materials from the mouth in one snap, and in the correct direction of least resistance. This is because each attempt involves some deformation, and each such attempt may result in some further softening of the structure in one region or another. This recommendation is not because the Mullins effect can be avoided, it cannot. However, in the course of several attempts, levering first one way then another, the total strain applied may easily exceed that which was strictly necessary. This makes the risk of distortion of critical parts of the mould under the forces of casting the model that much greater. Such parts are the margins of cavity preparations and the interproximal sections, which are very thin. This effect is in addition to the time-dependent viscous flow which will also be occurring during each of these extra periods of stress, the complications of the permanent deformations called 'compression', 'tension' and 'shear set' (4§8).

Strain softening has another more subtle effect. Ordinarily, one might measure a secant modulus (Fig. 1§13.1) for a filled rubber by noting the deformation of a specimen under a fixed load. It might be concluded from that result that the material was stiff enough to perform well in handling after taking the impression. However, those regions which have suffered large strains during removal, such as were in the interproximal spaces, will be significantly softer than such a simple measurement would imply, and especially at low loads, so that appreciable deformation in critical places may occur during the pouring of the model. Again, it is a matter of the design of the test reflecting the actual service conditions.

It can be seen that the ultimate tensile strength of a filled elastomer is not at all affected by the operation of the Mullins effect: all the bonds that are broken in succession would be broken in the same sequence anyway (if we assume that the direction of loading is constant). However, after stress cycling to any level below failure, the area beneath the curve for a subsequent cycle is reduced. Hence, the energy required for failure decreases according to the extent of prior loading; this is equivalent to a reduction in the tear strength of the material. In other words, the work of fracture has already been partially supplied in the earlier stress cycles; there is no violation of physical law.

●3.3 Importance of shear

The rheological properties of impression materials are important to their dental use - even when they are set and supposedly not capable of flowing. This arises from, on the one hand, the deficiencies of the structure (in the chemical sense) permitting flow, and on the other hand the internal stress conditions which are generated during any kind of loading except pure hydrostatic. It is shear that causes flow (4§3), and where there is both tension and compression acting at a point (even if it is only the one arising from the other as an 'effective' stress due to Poisson strain; 1§2.2, 6§2.7), there is a resultant shear stress. To this is added the local shear due to the mismatch of strain in adjacent columns of material (Fig. 3.1). This shear then causes deformation of the shape of a polymer chain (Fig. 3.5).

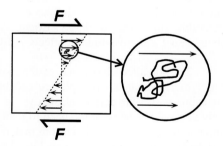

Fig. 3.5 Shearing forces acting on a polymer molecule during flow deform its random coil conformation.

It then follows that since the polymer chains in their sheared condition must relax, and because this depends on chain-segment diffusion, the process takes time. Thus the Kelvin-Voigt or retarded component of elastomer rheological behaviour arises naturally from the structure. If any portion of the system is not constrained by cross-links, entanglements or bonding to filler, true flow must occur to give permanent deformation as well (§2.4). This points out the underlying necessity of viewing these processes at the molecular level if the clinical behaviour and handling requirements are to be understood.

•3.4 Filler factors

The amount of filler has, as might be expected, a direct effect on the modulus of elasticity of the composite structure that is the filled impression material (6§2). In this respect the properties are directly controllable to give the desired values (bearing in mind the compromises which may be required due to viscosity effects; 4§9, 6§3). There are, however, additional effects operating which have to be taken into account: particle shape, surface activity, specific surface area, and degree and strength of agglomeration. Fig. 3.6 shows the effect of increasing the volume fraction of two different fillers on the modulus of a rubber. It is too simplistic just to require 'bonding' between matrix and filler, the strength of that bonding (which need not be restricted to covalent bonds) and the areal density of the bonds, determine how much the matrix is constrained. In fact, for the very large quantities of filler involved for impression materials, it is probably not economic to treat them with special coupling agents. Instead, reliance may be placed on mechanical key or natural reactivity. This would be enhanced by irregularity or roughness, as well as agglomeration.

In addition, the specific surface area of the filler, or the particle size, is also a means of control effective with some materials. As with the effects on viscosity, mediated through a stationary boundary layer (Fig. 4§9.4), the more material is held in this layer the less is 'available', as it were, to deform. The nature of the filler-matrix interaction is also dependent on the detailed nature of the polymer, as indicated by the differences between two types of rubber for the same fillers (Fig. 3.7); the one material is practically unaffected. It should be apparent from Fig. 3.4 that if a filler is to be effective in increasing the modulus of an elastomer, and therefore any matrix material, it must be bonded to that matrix. Were it not, the Mullins effect would not be observable to such a large extent, although chain entanglement and cross-linking effects would remain.

The so-called elastic impression materials are based on polymers which in principle are capable of very large extensions with near perfect elasticity. In practice, this is rarely achieved because of the need to realize curing at low temperatures and in relatively viscous materials (which inhibit reactant diffusion), when the completeness of the reaction must therefore be relatively low. The demands of the clinical context require that any chemical reaction for cross-linking be very rapid, that is, using highly-reactive substances. But reaction can only occur after diffusion has brought the reactants together, and this becomes more difficult as cross-linking proceeds. In addition, high filler contents are often necessary to obtain sufficiently stiff materials as end-products. The principal types currently available are based on polysulphide, polyether and silicone rubbers.

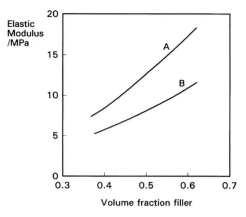

Fig. 3.6 The effect of volume fraction of two kinds of carbon black filler on the modulus of a rubber. The surface activities of the two are believed to differ.

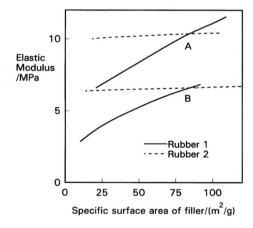

Fig. 3.7 Variation of effect of specific surface area of two different carbon black fillers in two different rubbers. The volume fraction is held constant.

§4. Polysulphide

The polysulphide rubber impression materials, sometimes very loosely called 'rubber base', or 'thiokol' or 'mercaptan' rubber, are based on a single bulk commercial liquid polymer product labelled "Thiokol LP-2". In order to understand some aspects of the structure and behaviour of the dental product, it is necessary to review the manufacture of the main ingredient, although it is stressed that this is not the chemistry of the setting process, which is described in the following section.

●4.1 Manufacture

The industrial manufacture of LP-2 involves first reacting bis(2-chloroethyl)formal:

$$Cl - C_2H_4 - O - CH_2 - O - C_2H_4 - Cl \qquad (4.1)$$

which has reactive chlorine end-groups, with sodium polysulphide. This reaction eliminates NaCl, replacing pairs of chlorine atoms with an -SS- group, creating a high polymer. This on its own would necessarily be composed of all straight chain molecules, and so would be quite useless for making a rubber because it could not form a network structure. Branch-points are therefore introduced by including in the reaction mixture 2 mole percent of 1,2,3-trichloropropane, a cheap and readily available trifunctional reactant, the chlorine reacting in the same way as for compound 4.1. The result of this first reaction is therefore a polysulphide rubber, but since it is now already nearly completely polymerized it is unusable at this stage. Hydrolysis with a strong base breaks this structure down again, cleaving the polymer at the disulphide linkages, and leaving the new terminal sulphhydryl groups, -SH (also known as thiol or mercaptan). This may seem a strangely roundabout means of obtaining the material, but it can be relied upon as being an economically valuable route.

The golden yellow, somewhat viscous liquid thus obtained is the polymerizable product which, with its moderately high number average molecular weight of about 4000, is best described as *pre*polymer. It is this material that forms the basis of the product supplied to dentistry. Contrary to many descriptions, the most common type of molecule in this liquid is the *monomeric* dithiol version of the dichloride precursor above:

$$HS - C_2H_4 - O - CH_2 - O - C_2H_4 - SH \qquad (4.2)$$

with dimers, trimers and so on in diminishing proportion (Fig. 4.1).[4] However, the average degree of polymerization is about 24. There are, of course, a variety of other kinds of molecule which have incorporated the trichloropropane moiety and which are therefore branched, to give a variety of Y-, H- and more complicated shapes which by being polyfunctional maintain the ability to create a network when the polymer is (re)formed.[4] The viscosity and the average molecular weight arise from the presence of the relatively smaller numbers of very large molecules (Fig. 4.1). The presence of low molecular weight (and therefore volatile) thiol impurities give this class of material their characteristic and unpleasant smell.

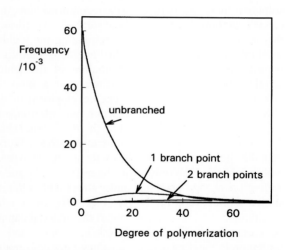

Fig. 4.1 Distribution diagram for polysulphide prepolymer (Thiokol LP-2). Plotted is the frequency of occurrence of molecules as a function of total chain length.

●4.2 Setting reaction

Now, to convert the prepolymer into a rubber, by reaction between the thiol groups (the 'setting' that occurs when taking an impression), an oxidizing agent is required to remove the hydrogen. The most commonly used is lead dioxide (N.B. *not* peroxide), the very dark brown substance which gives some polysulphide impression material products their characteristic colour. (A number of other oxidants have been used, but appear to have been less successful; §4.3.)

The initial step is the oxidation of two thiol groups to form a disulphide bridge (Fig. 4.2), with the elimination of water: this is therefore a condensation reaction. Notice that the reaction

---

[4] Notice that the commonly asserted distinction between 'terminal' and 'pendant'-SH groups on branched fragments is meaningless.

$$Pb^{IV} + 2e^- \Rightarrow Pb^{++} \qquad (4.3)$$

is a reduction of the lead, corresponding to the oxidation of the thiol groups.

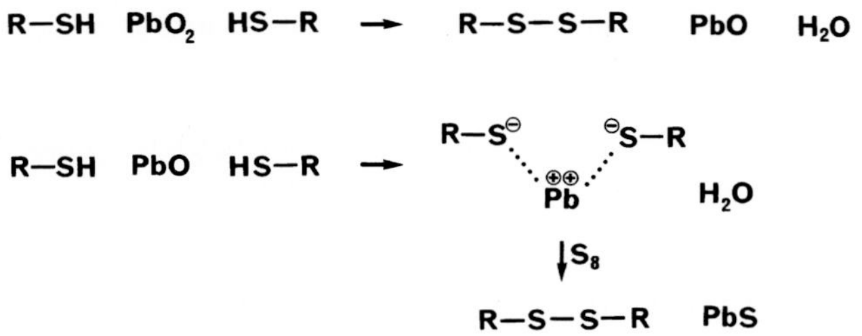

However, there is a tendency for the lead monoxide by-product formed here to react with further thiol groups itself, because it is still sufficiently reactive as a base with respect to the acidity of the thiol hydrogen. However, since this is only a salt-forming reaction, again eliminating water, it forms an ionic bridging complex instead of a covalent bond. (Essentially, Pb^{++} is not a strong enough oxidizing agent to abstract the electrons from the sulphide terminals to create a covalent S-S bond and be reduced to metallic Pb in the process.) These ionic coordination bonds are very weak and labile, exchanging with one another readily, so that under even moderate stresses considerable flow would be displayed by the cured material. More importantly, the presence of the sulphide terminals, -S⁻, is detrimental to the existing covalent links because they can attack and exchange with them (Fig. 4.3).[5] These processes are the exact parallels of the stress relaxation due to free radical action seen in other polymers (§2.2). A small amount of elemental sulphur is therefore included in the reaction mixture; this can accept those electrons and the sulphide ions formed then react with the lead ions, precipitating the highly-insoluble PbS (Fig. 4.2). This addition, therefore, permits the more complete formation of disulphide covalent bonds and the establishment of a properly effective network (Fig. 4.4). The development of the elastic properties is then much better, but unfortunately the -S-S- bond is still weaker than the -C-C- bond (Fig. 4.5), leading to preferential rupture under stress. Worse than this, however, is the tendency for sulphides to react with sulphur to form polymeric chains (Fig. 4.5), and to undergo exchange reactions under stress with adjacent similar bonds very easily (*cf.* Fig. 2.8), even with disulphide. As a result, the flow of the material cannot be eliminated. There is, therefore, a basic and unavoidable limitation to the dimensional accuracy after deformation that may be attained by impressions made with this type of rubber.

In view of the manufacturing route, which has to be carefully distinguished from the setting reaction, the latter is conveniently described as *re*forming the polymer. However, it should be recognized that this process still results in reaction shrinkage because the free volume associated with the SH-carrying ends of the molecules is reduced when the -SS- bond is formed (*cf.* Fig. 5§2.4). In fact, all such 'connecting up' reactions in impression materials result in this effect, and this is always a factor in limiting their accuracy. Strictly, this setting reaction is not a polymerization, then, because only a proportion of the initial material is actually monomer, even if it is the most common species (Fig. 4.1).

The PbS eliminated by reaction with the sulphur is black, accounting for the darkening of PbO_2-based materials on setting. The second by-product of the setting reaction, the water eliminated in both oxidation steps, is volatile and so can evaporate from the set mass, leading to an appreciable (and clinically significant) shrinkage

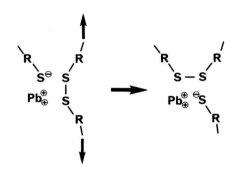

Fig. 4.2 The chemistry of the setting reactions of polysulphide rubber.

Fig. 4.3 The stress-relaxation mechanism of polysulphide rubbers, the cause of the majority of the flow observed under stress.

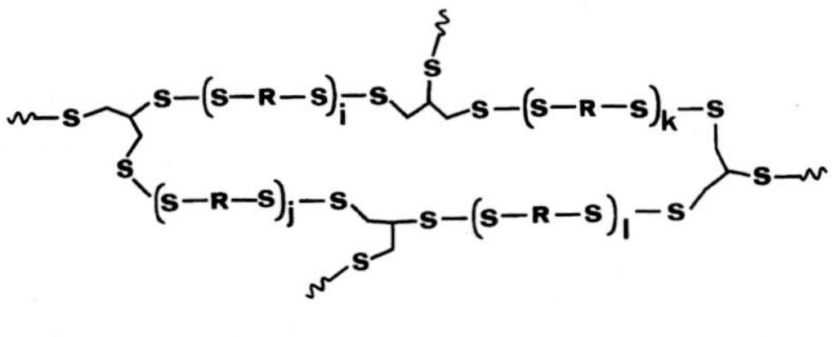

Fig. 4.4 Possible fragment of the polysulphide network.

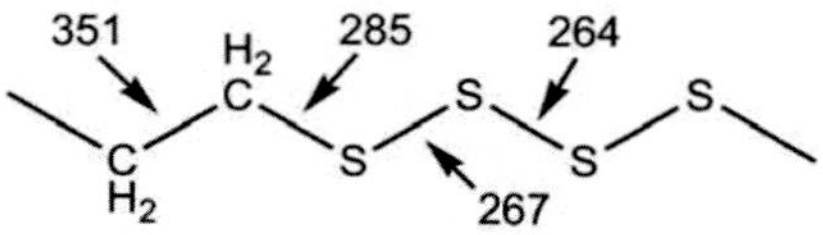

Fig. 4.5 Bond energies (kJ/mol) for bonds relevant to polysulphide cross-links.

on standing. Since this is superimposed on the reaction shrinkage that occurs due to the reaction continuing after removal from the mouth, polysulphide impressions are not considered to be very stable. This evaporation is unavoidable, suggesting that the pouring of the model should not be delayed more than, say, 30 min.

The water eliminated in the setting reaction is, of course, a plasticizer for the network, although its contribution might be expected to be relatively small, such that the modulus of elasticity would change with time as it is lost. The detrimental effect of the shrinkage, however, would override such considerations. The point is that the behaviour of a system depends on all components present, whether deliberate or consequential, and whether or not desirable.

Polysulphide impression materials are supplied in two separate tubes. One tube contains the prepolymer mixed with an inert filler such as a silica powder (to increase the viscosity of the fresh mix and the modulus of elasticity of the set rubber), the other containing a paste of the PbO_2 and sulphur in an inert oil as a carrier or **vehicle**. This oil must behave as a plasticizer in the set rubber, a further source of softening that has to be compensated by raising the filler content. Notice that the lead dioxide is *not* a catalyst as it is frequently described, especially by the manufacturers in labelling the tube. Any material which takes part in a reaction and is not recoverable (even in principle) unchanged at the end is not a catalyst but a *reactant*, which is the best label for PbO_2 now. There are, in fact, only very few true catalysts in use in dental materials.

●4.3 Other setting agents

The oxidation of thiols can also be effected merely by the oxygen of the air, particularly in thin films where diffusion throughout the bulk occurs most rapidly. This accounts for the appearance of set or partially set material around the neck of a partly-used tube of the 'base' component, and even of a tube 'going off' during prolonged storage. Many other setting agents are possible. Zinc oxide, for example, is a strong enough base to work in the same manner as PbO, *i.e.* as a salt-former. This is true for many metal oxides, but is also unhelpful since a covalently-bonded network is not formed. The formation of a salt bridge alone is enough to effect 'setting', but such a rubber would be quite unsatisfactory for dental impression use because of its poor dimensional stability under stress. Some dental products use compounds called hydroperoxides, ROOH, as the oxidizer (although these are also volatile and reduce dimensional stability), and still others are used in industrial applications, but the principle of the setting reaction is always the same.

●4.4 Need for filler

The polymer chain of the polysulphide rubber itself is expected to be very flexible as there is little steric interference to bond rotation to be found; the only substituent is hydrogen and that is spaced out by the sulphur and oxygen atoms in the chain. This implies that chain segment rotation in the polymer will be relatively unhindered, so that the pure rubber itself is very soft. Filler is therefore necessary to control the distortion that would occur in the impression under its own weight (sagging) and especially under that of a gypsum slurry. This can also better suit it for the various impression taking techniques that may be used. It is worth stressing that this product (in common with other rubbery materials) is only a rubber because of the cross-linking present to inhibit chain slippage.

●4.5 Property development

It is also worth noting that because of the mixture of a variety of prepolymer molecules, straight chain, Y- and H-shaped fragments, and the more complicated molecules, cross-linking reactions will be occurring from the very start of mixing. There is no distinction between the chemical rates of reaction for 'chain extension' and 'cross-linking' because, quite simply, there is no distinction between the reactivities of the thiol groups on the various fragments. All that is happening is that network fragments grow, gradually joining together to make larger and larger molecules, until one single 'infinite' molecule, spanning the entire piece, is formed.

Appreciable viscoelastic properties are present at the moment of insertion of the loaded tray since the reaction begins immediately on starting mixing, and it is only the fact that the viscous flow dominates that allows the impression to be taken. Nevertheless, this aspect of the behaviour (which is also common to silicone and polyether materials) means that mixing should be done promptly to the manufacturer's instructions, and that delay in loading and inserting the tray must be minimal to avoid sufficient elasticity being present to distort the mould on removal. In other words, a graph similar to Fig. 6§4.7 applies in all such systems as well, although the final material is rubbery rather than glassy (the T_g does not rise above – or even approach – the working temperature).

§5. Polyether

•5.1 Manufacture

Polyether impression material is manufactured as a **copolymer** of ethylene oxide and tetrahydrofuran to create a prepolymer chain with frequent oxygen atoms (Fig. 5.1). Those two substances are very cheaply available industrial feedstocks. By copolymer is meant that the two reactants are present simultaneously in the reaction mixture and their order of incorporation in the chain is random, depending only on their concentrations and relative reactivities. The need for a copolymer arises because a **homopolymer** (with all repeat units the same and therefore with a completely regular repeating pattern) of either starting material would readily crystallize and increase the modulus of elasticity of the set material excessively. This would also make the starting paste as supplied to the dentist very viscous or even solid, even though the prepolymer at this stage is neither cross-linked nor branched. As with the polysulphides, these copolymer chains are expected to be extremely flexible as the energy barriers to chain segment rotation will be very low because of the absence of side-groups and the presence of frequent oxygen 'spacers'.

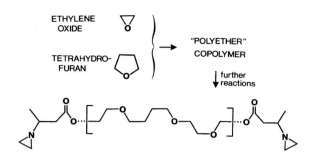

Fig. 5.1 Basic starting materials (monomers) used in the manufacture of polyether prepolymer and the typical resulting chain structure with reactive end-groups added *via* an ester.

•5.2 Setting reaction

Further reactions are necessary in manufacture to add reactive groups to the ends of the prepolymer chains. The group chosen is **ethylene imine** (also called **aziridine** or **azacyclopropane**), linked through the nitrogen to the main chain – identical chemistry, in fact, to that of the epimine temporary resin (6§7). Again, this group is reactive for the setting reaction because three-membered rings are highly-strained (the ring angle is about 60°, while the tetrahedral angle is about 109°) and the activation energy for ring opening is low. The nitrogen atom can readily donate its lone pair of electrons to a positively charged species to form an ammonium-like salt. Therefore, the setting reaction **initiator**, an alkyl benzene sulphonate, is chosen to be effective in such a reaction by virtue of the small positive charge on the terminal methyl group (Fig. 5.2). Thus, on forming the quaternary ammonium salt, the 3-membered ring opens spontaneously, exactly as described before (Fig. 6§7.3). The newly-created carbocation behaves in much the same way as the original **electrophilic** sulphonate, reacting just as easily if not more strongly with another imine nitrogen. The sequence then repeats itself, forming a so-called **daisy-chain**[5] type of structure, that is, the head (-N<) of one unit is connected to the tail (-C_2H_4-) of the next, and so on.

Because both ends of the prepolymer molecule carry reactive groups, the effect of these reactions is to produce a very highly cross-linked structure (Fig. 5.3). From the

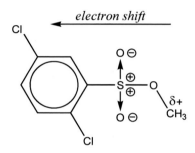

Fig. 5.2 The initiator molecule for polyether impression material depends on the electron-withdrawing power of the sulphonate group. This can be enhanced by, for example, chlorines (which are themselves electron withdrawing) on the benzene ring.

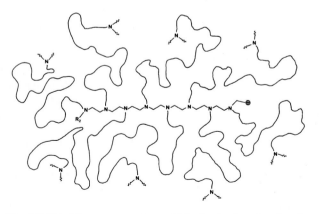

Fig. 5.3 Possible structure of cross-linked polyether showing the 'daisy-chain' arrangement. The same type of structure is expected to be found at the nitrogen centre shown at the other end of each former prepolymer chain.

---

[5] Named after a children's game using small flowers, linking them together in a repetitive manner.

amount of initiator included in the reactor paste, groups of up to a maximum of about eight chains might be expected to be linked in this way. The setting is thus characterized by a rapid increase in viscosity and the equally rapid development of elastic properties (again, no separation of these trends is possible). The degree of cross-linking accounts for the high modulus of elasticity observed in commercial materials, despite the flexibility of the polymer chain itself, which packs well (lacking bulky side-groups), so there is little free volume for chain segment movements. The cross-linking reaction may be viewed as an ionically-initiated addition polymerization, *alias* **cationic chain polymerization**.

These materials tend to be rather stiff and limited in application to small impressions, of only a few teeth at most, in order to avoid the force required for removal being too high (§1). Even so, a plasticizer is used, whether added by the manufacturer or in the clinic as a 'thinner'. This is also a polyether, to be compatible with the polymer although, of course, it must be of shorter chains and lacking reactive end groups. A range of viscosities is now offered, but without affecting either the setting rate or final stiffness.

•5.3 Dimensional stability

In contrast to polysulphide materials, however, the presence of the highly polar sulphonate salts and the numerous tertiary amines (all of which may be readily and strongly solvated by water), not to mention the etheric oxygen atoms and ester groups in the polymer chain itself, means that the solid polyether can easily and rapidly absorb large amounts of water, in fact up to about 15% by mass. This necessarily produces a large volumetric expansion (Fig. 5.4).[6] But the behaviour is rather more complicated because the plasticizer used (*i.e.* the thinner) is itself a low molecular mass polyether and therefore water-soluble, and thus may be leached out. In dry air the (unsoaked!) material is quite stable, but storage in high humidity may be expected to cause problems through absorption.

Even so, the hydrophilicity of this material is supposed to be an advantage on damp surfaces.

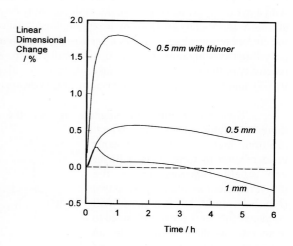

Fig. 5.4 Dimensional changes of set polyether material in water showing both swelling and dissolution.

However, perhaps because of the reactivity of carbocations with absorbed water (Fig. 6§7.4), the impression surface is sometimes found to be spoilt if the site is too wet. (There would also be boundary layer separation problem, 23§10.2.) In addition, it is not possible to add more material in a two-stage technique, were that desirable: it does not stick to itself very well, for the same reasons.

•5.4 Hazard

The initiator used for polyether (methyl 2,5-dichlorobenzene sulphonate) is a known sensitizing agent, *i.e.* of adverse physiological action in contact with living tissue.[7] It should therefore never come into contact with the skin. Not only does this mean that care should be taken with the initiator-containing component of the product, but also that the material should not be mixed with bare hands. In fact, this kind of warning applies to many modern dental materials: they frequently contain highly reactive ingredients, which means essentially that they are certain to be toxic. Contact with them could be detrimental to dentist and auxiliary staff and, if improperly mixed, to the patient. Even then, severe allergic patient responses have been observed on occasions, presumably due to sensitization on an earlier treatment exposure.

§6. Silicones

●6.1 Condensation silicones

Silicon does not form stable long chains with itself as does carbon, although in terms of bonding geometry it is very similar. But stable polymers based on silicon can be made by interpolating an oxygen atom between each pair of silicon atoms to make compounds called **polysiloxanes** (Fig. 6.1), commonly called silicones.[6] However, hydrogen directly bonded to silicon, Si-H, as in the **silanes** (the compounds equivalent to the carbon-based alkanes (16§1)) is extremely reactive and it is necessary to substitute an alkyl group to obtain unreactivity (Fig. 6.2), the Si-C bond being almost as inert as a C-C bond in a paraffin. If such a polymer chain is terminated at each end with a hydroxyl group (Fig. 6.2), the molecule may be called a **silanol**, and more specifically an α,ω-dihydroxypolydimethylsiloxane,[7] although the current terminology refers to these simply as **hydroxy-terminated polydimethylsiloxanes**. These hydroxy groups are the key to the useful reactivity of these materials as it makes them acidic, albeit weakly (Fig. 9§3.3), and thus subject to condensation and hydrolysis reactions.

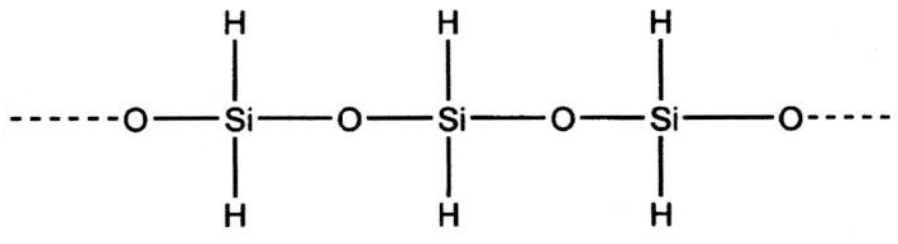

Fig. 6.1 The structure of a polysiloxane.

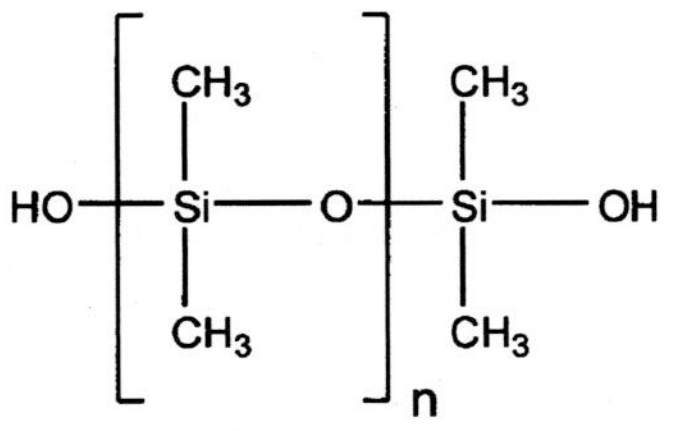

Fig. 6.2 The structure of the principal chain of condensation silicone materials, an α,ω-dihydroxypolydimethylsiloxane. The hydroxyl hydrogens are acidic.

The manufacture of silanols is *via* the hydrolysis of dimethyldichlorosilane, in which reaction there is a spontaneous condensation to form various linear and cyclic species since the dihydroxydichlorosilane is not stable (Fig. 6.3). Indeed, this material will condense further if water is allowed to evaporate, as there seems to be an essential equilibrium (Fig. 6.4) and

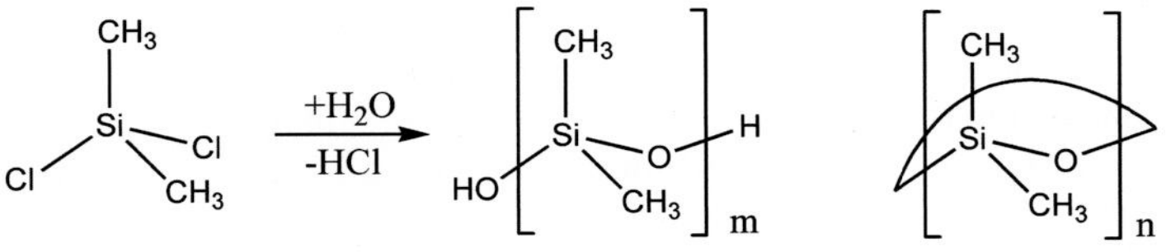

Fig. 6.3 Primary manufacturing route for silanols.

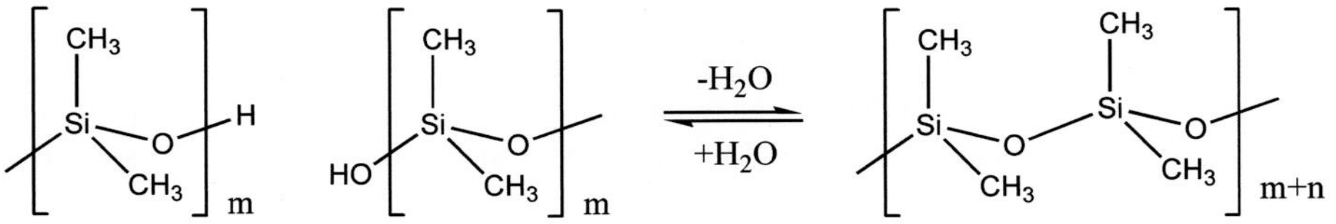

Fig. 6.4 Condensation equilibrium of silanols. This reaction can occur intramolecularly to create cyclic species.

this is how a sufficiently high molecular weight is obtained, by driving this reaction to the right. This is, even so, a rather slow process, and the activation energy reasonably high. Thus, the basic material is an equilibrium mixture of (potentially polymerizable) chains and unpolymerizable cyclic species. If the dental material is left exposed to the air, if dry enough, for long periods, that equilibrium will shift to the right, the average chain length increase, and viscosity rise, as water is lost. Note that the atmospheric moisture is not the cause of this condensation, as has been claimed, but rather it would tend to inhibit or reverse it.

A similar condensation can also occur with alkyl silicates, with the equivalent elimination of the corresponding alcohol:

$$-\text{Si-O-C}_2\text{H}_5 \; + \; -\text{Si-OH} \; \Rightarrow \; -\text{Si-O-Si}- \; + \; \text{C}_2\text{H}_5\text{OH} \qquad (6.1)$$

---

[6]   Note the spelling: the element is silic**on**, the polymer silic**ōne**.

[7]   α,ω- means at the beginning and end of the chain (say "alpha-omega") – first and last letters of the Greek alphabet.

This is a much more one-way process, as alcohol is not basic enough to attack the Si-O-Si linkage. Tetraethylorthosilicate or TEOS, $Si(OC_2H_5)_4$, is the compound ordinarily used, and by offering four such functional groups can lead to cross-linking. However, its rate of reaction is too slow to be of use to dentistry, but it has been found that the inclusion of a tin salt accelerates the process. This salt must be miscible with (*i.e.* soluble in) the hydrophobic prepolymer, and tin octanoate,[8] $Sn(C_7H_{15}COO)_2$, is commonly used, where the long paraffin tail achieves the compatibility with the silicone. Even so, octanoic acid is a weak carboxylic acid, and in the presence of water, the tin salt partially hydrolyses in the presence of the trace of water normally found in the prepolymer by virtue of the chain condensation equilibrium (Fig. 6.4) to an active tin hydroxide compound:

$$RCOO\text{-}Sn\text{-}OOCR + H_2O \rightleftharpoons RCOO\text{-}Sn\text{-}OH + RCOOH \qquad (6.2)$$

It is this species that then reacts with the TEOS cross-linking agent, eliminating ethanol:

$$RCOO\text{-}Sn\text{-}OH + Si(OC_2H_5)_4 \rightarrow RCOO\text{-}Sn\text{-}O\text{-}Si(OC_2H_5)_3 + C_2H_5OH \qquad (6.3)$$

This tin species, by virtue of the slight positive charge it carries, then can attack and exchange a silanol hydroxy group (Fig. 6.5). It will be noticed that in this process the -Sn-OH species is regenerated, so that the overall reaction remains exactly as shown at reaction 6.1. Thus, it would appear that this active tin hydroxide may properly be considered to be the **catalyst** for the reaction, and this means that the original tin octanoate should be called a **catalyst precursor**.

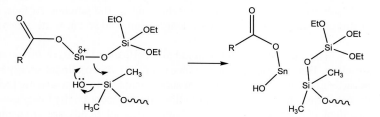

Fig. 6.5 Silanol exchange reaction with the active tin intermediate. ["Et" means the ethyl group: C_2H_5-]

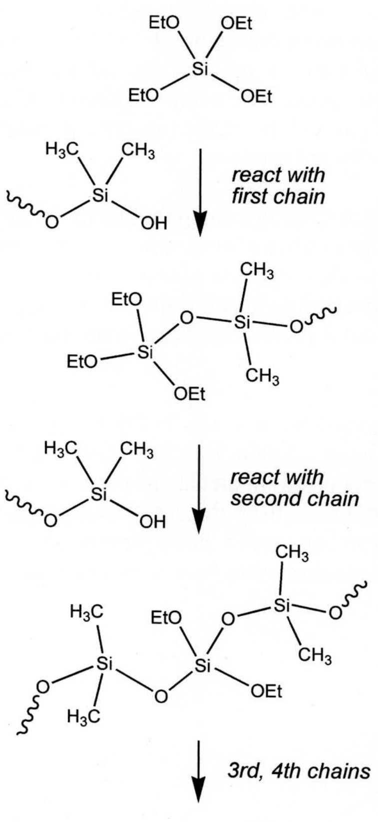

This condensation process is sequential (Fig. 6.6). The first such reaction at a given silicone chain end does no more than add a new polyfunctional group to the end of the chain. The second reaction effectively only extends the chain by joining two ends together. Of course, if both chains are already attached to branched molecules at their far ends, this step would constitute a cross-linking. However, the third step is where a branch point is now introduced, a node where three chains are attached. The fourth reaction then increases the connectivity of the node from 3 to 4, enhancing the cross-linking if any of the attached chains are themselves branched.

The active tin hydroxide can also condense with a silanol, in a reaction parallel to that of 6.3, with the elimination of water:

$$RCOO\text{-}Sn\text{-}OH + HO\text{-}[\underline{Si}]_x \rightarrow RCOO\text{-}Sn\text{-}O\text{-}[\underline{Si}]_x + H_2O \qquad (6.4)$$

where $[\underline{Si}]_x$ stands for the polysiloxane chain. This can then react with another such silanol (Fig. 6.7), exactly parallel to the reaction at Fig. 6.5, such that only chain lengthening is obtained and the cross-linking process is not fully efficient. However, the water generated at reaction 6.4 may serve to enhance the generation of the active species (reaction 6.2), and so provide some acceleration of the setting process.

The condensation reactions that lead to setting may also occur **intramolecularly** – there can be no reactivity difference to prevent this and the desirability of cross-linking only is irrelevant.

Fig. 6.6 Chain extension occurs at the second condensation, branching can only occur at the third.

[8] Sometimes also, incorrectly, called 'octoate'.

Thus, it can be expected that a proportion of the bonds that are formed only generate cyclic species, loops, and (on a statistical basis) typically these will be rather small. They therefore cannot contribute to setting as such, although should a loop form around another chain an *effective* cross-link would be present.

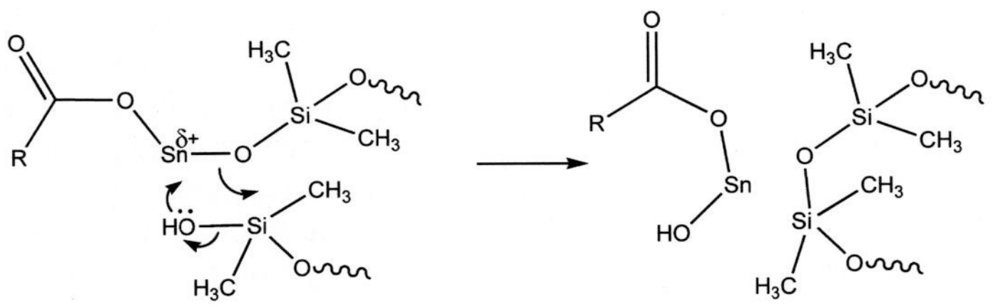

Silicone rubbers in general are intrinsically hydrophobic because the 'surface' of a chain is dominated by the paraffinic methyl groups. Such poor

Fig. 6.7 Silanol chain condensation is also facilitated by the active tin intermediate. This can also be intramolecular. Notice that the tin entity is attached to the end of a polysiloxane chain here (on the top right), as opposed to the TEOS in Fig. 6.5. Otherwise, the reaction is identical.

wetting of impression materials by water is a disadvantage in that impressions of saliva or blood globules may also be taken unless the site is carefully dried. These globules, of course, may obscure essential detail. In addition, the quality of the model poured in such a mould may be compromised. This has been overcome in part, not by changing the polymer design, but by adding non-ionic surfactants to the mixture (10§8.4).[8] However, such substances are prone to be leached out of the silicone surface, and especially if exposure to water is prolonged, such as for washing or disinfection, and so the wetting behaviour may deteriorate.

The use of the polyfunctional alkyl silicate (TEOS) results in an extensive three-dimensional network structure. But, as with polyisobutylene (Fig. 2.1), the rotational energy barrier for the chain segments is not large because of the uniformity of the two substituents. The extra spacing provided by the oxygen atom links reduces the activation energy required markedly, giving exceptionally low glass-transition temperatures ($< -120\,°C$) and excellent elastomeric behaviour. Notice that all bonds are covalent, so there is none of the lability that compromises the performance of polysulphides, although there is still some appreciable compression set. This arises from the fact that the stress-induced bond exchange can still occur, particularly in the presence of water or more acidic materials, mediated by the hydrolysis shown in Fig. 6.4 (leftward reaction).

The alcohol eliminated in the setting reaction is, of course, a plasticizer, being a small mobile molecule. We can also note that the cyclic silicone species, with four-membered rings predominating, which are present in the commercial precursor material (Fig. 6.3), cannot take part in network formation except if hydrolysis converts them to chains (although the equilibrium of Fig. 6.4 includes the possibility of reforming these). Thus, there will be proportion of small, unconnected and mobile silicone molecules present in the set material, and these therefore add to the plasticizing effect.

Necessarily there will be some appreciable shrinkage with this system because the TEOS has a large associated free volume (four mobile side chains), but there is a further issue. Again, this is a condensation reaction, and a principal drawback is the volatility of the alcohol, usually ethanol, eliminated in the process. This leads inevitably to some shrinkage with time as it is readily lost by evaporation from the surface (any water present will also contribute, Fig. 6.4, although this is likely to be small). Again, as a result, fairly prompt pouring of models is recommended, within about half an hour. An alternative reaction scheme has been developed which overcomes this disadvantage, although with an increase in cost.

●6.2 Addition silicones

If, instead of hydroxyl, the polysiloxane chain is terminated at each end with a vinyl group, that is, **vinyl-terminated polydimethylsiloxanes**, these may be made to react with a silane (that is, with reactive Si-H hydrogen) to effect the cross-linking (Fig. 6.8). The net effect is to transfer the silane hydrogen and its silicon to separate carbon atoms of the vinyl group, and the process is called **hydrosilylation**. Being an **addition**[9] reaction, there is no by-product,

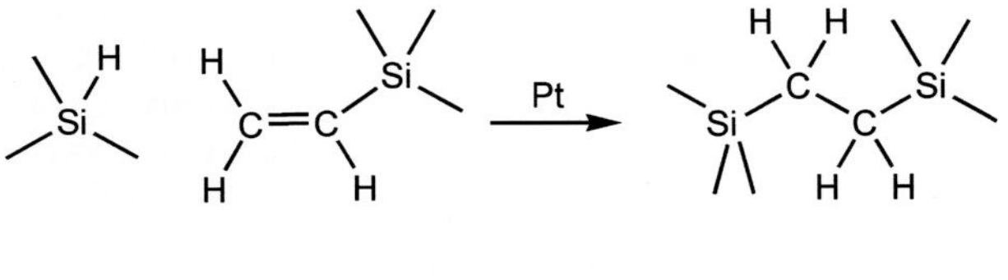

Fig. 6.8 Addition of a silane across a vinyl group.

---

[9] NB: not "additional".

and such materials − informally called **addition silicones** − achieve possibly the best dimensional accuracy and stability of all available dental impression materials despite there being, of course, still some setting reaction shrinkage. Clearly, the silane must have at least two Si-H hydrogens (on different Si atoms), and one or both of the reactants in Fig. 6.8, the divinyl polysiloxane and the silane, must be branched (or at least trifunctional) if network formation by cross-linking is to occur, and not just chain extension, as explained in Fig. 6.6.

> Terminology is again in need of being considered carefully in this context. Such materials in dentistry are sometimes called **polyvinyl siloxane** or **PVS**. It can be seen that this is incorrect as it is only the the terminal groups that are vinyl, and these do not polymerize as in acrylic and other free-radical resin systems. The correct short-form label must be **vinyl polysiloxane** and the abbreviation therefore **VPS**.

The setting reaction itself is effected by the presence of platinum atoms, Pt^0, as catalyst (the superscript zero emphasises the point that the metal atom is in the zero-valent state). Pt^0 is not unreactive, and can form π-bonded complexes, coordination compounds, with vinyl groups, as indicated in Fig. 6.9. Here, the short divinyl siloxane is used as a convenient vehicle in what is ordinarily known as **Karstedt's catalyst** (despite it not being the actual catalytic entity; *cf.* the discussion above on a catalyst precursor).[9][10] The vinyl groups are relatively labile, and easily exchanged with those of the longer-chain prepolymer molecules. The catalyst is necessary since there is no reaction between vinyl and silane at ordinary temperatures.

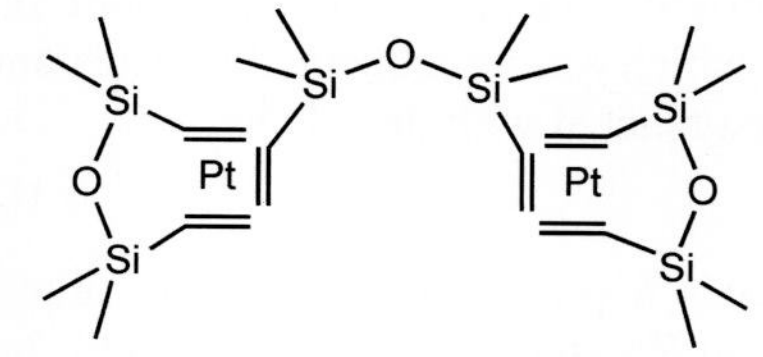

Fig. 6.9 The nature of "Karstedt's catalyst", as used in addition silicone impression materials.

The first step (A, Fig. 6.10) on mixing a silane-containing component with vinyl-terminated material in the presence of the catalyst is that the silane splits and adds to the Pt atom. This makes the hydrogen atom available for transfer to an adjacent 'solvating' vinyl group, which is now bonded by the other carbon atom to the metal atom (step B). This leaves room for another vinyl group to coordinate with the Pt (C), which makes it easy for the silane silicon to transfer to the adjacent carbon atom and so allow the now-joined components to leave the metal centre (D). This in turn allows a silane to add again (step A), and the reaction continues in an example of true catalysis: the Pt is not used up in any way.

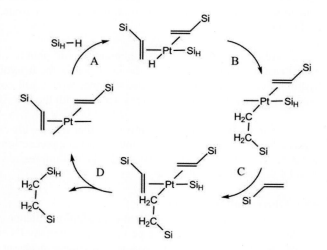

The dental two-paste product therefore consists of, one the one hand, a mixture of silane and vinyl-terminated prepolymer, and on the other, a mixture of the vinyl-terminated prepolymer and Karstedt's catalyst (whose own vinyl-terminated siloxane "solvent" will be reacted in the same way with the silane).

Fig. 6.10 An outline of the setting reaction of vinyl siloxane impression materials. This is an example of true catalysis. The Pt atom is shown with only one 'solvating' vinyl group for clarity: others may come and go ("spare" coordination bonds indicated). The silicon atom of the silane is labelled with a subscript H simply to facilitate identifying it in the successive reaction products.

One difficulty with this kind of chemistry is that the silane may react with any water present (whether as a contaminant of the 'base' component or from the environment), eliminating hydrogen, due to the polarity of the Si-H bond :

$$—Si^+-H^- + H_2O \Rightarrow —Si-O-H + H_2\uparrow \qquad (6.5)$$

Not only does this reduce the effectiveness of the cross-linking, it also produces gas bubbles in the rubber, spoiling the surface of the impression, and bubbles at the surface of the model poured against it. The formation of such bubbles at the interface may be due to the relatively low solubility of hydrogen in water at ordinary temperatures and a high diffusion rate in the polymer (because of its necessarily large free volume and total lack of bonding interaction): it can escape from the silicone but not dissolve in the water fast enough. Sub-surface bubbles would form when the partial pressure of the hydrogen could not be contained by the low stiffness of the rubber near the surface: a phase separation occurs similar to divers' decompression sickness or the fizzing of a rapidly opened bottle of carbonated drink. Hydrogen-absorbing substances (such as very finely-divided palladium, **palladium black**) may be added to the material to reduce the effect of this undesirable side-reaction, but reduction of water as a contaminant is probably the more general solution. However, before such formulations were made, it was recommended to delay the pouring of model for an hour or so, or even to treat the impression with hot water, to ensure complete reaction and escape of the hydrogen.

In older, now discontinued products, a slightly different formulation was used where the 'catalyst' was a bright red solution in isopropanol of chloroplatinic acid, H_2PtCl_6, known as **Speier's catalyst**,[11] a few drops of which was mixed with the paste containing the mixture of silane and vinyl siloxane. It is now known that this first reacted with the silane:

$$H_2PtCl_6 + 4 \equiv Si\text{-}H \Rightarrow Pt^0 + 6\,HCl \tag{6.6}$$

that is, to produce neutral (*i.e.* metallic) platinum atoms, which then formed essentially the same kind of structure as in Karstedt's catalyst (Fig. 6.9), by coordinating with the vinyl siloxane (and therefore the actual setting reaction was identical, Fig. 6.10), but also hydrochloric acid. (Other reactions may also have generated hydrogen, adding to the bubble problem above.[12]) It was rather difficult to ensure good mixing of components with such disparate viscosities, but the main problem was the acidity. This promoted bond exchange, as above, causing the stress-set behaviour to be much poorer than otherwise would be the case, but also led to depolymerization, especially in the presence of water (as in atmospheric humidity), so that the long-term stability was compromised. The catalyst solution is also very toxic, allergenic and corrosive, and any skin contact problematic.

Addition silicones are also inherently hydrophobic, and so still require the inclusion of surfactants for model casting. An alternative is the inclusion in the network of polyether chains (Fig. 5.1) whose intrinsic hydrophilicity, while diluted by the silicone, would still contribute to the wettability of the material. The expected cost of this is higher water sorption, with the risk of phase-separation of micro-domains (8§4.1).

It must be pointed out, however, that even with current addition silicone materials there remains appreciable viscoelastic behaviour and that all due precautions to avoid excessive permanent deformation must be taken. Even so, this is the most stable class of impression material and the model pouring may be delayed, even by several days, with little effect.

●6.3 Gloves and inhibition
There is an important problem with addition-silicone materials: that is, when they are mixed in the hand, as must be done with the putty-type very high viscosity ('very heavy body') versions, and using latex rubber gloves, they may fail to set. Gloves are, of course, necessary for such tasks because it is not advisable to let highly-reactive chemicals contact the skin (silanes will react with any hydroxyl-containing substance). However, in this case polyethylene or similar non-reactive types must be used. This may in part be related to the chemistry of latex rubber (27§1) which has many double bonds remaining. Since the platinum catalyst specifically reacts with the vinyl group of the polysiloxane, it seems likely that the very high density of double bonds in the latex rubber, which are chemically similar, binds too much catalyst for the setting reaction to proceed normally.[13] However, a greater effect is that substances used to promote the cross-linking of the rubber in manufacture, sulphur-containing accelerators (27§1.3), react with the platinum catalyst, poisoning it, because of their sulphur content.[14]

The problem is not restricted to the use of putty-type materials. The residues of such accelerators may be transferred to any surface that is handled or touched: impression trays, teeth and soft tissue, and the wires used for root post impressions. The effect here is that the surface of the impression in contact with the contamination remains unset, obviously interfering with its accuracy but also being rather messy. The solution is to use a non-reactive glove for all procedures where addition-silicone rubbers are to be used, and obviously care has to be taken to avoid contaminating any instruments (such as the mixing spatula), equipment and other surfaces that might be used subsequently. In fact, washing alone may not remove the contamination. Thus, even if polyethylene gloves are used over latex gloves, the external surface must not itself be contaminated by being handled with or touched by the latex glove. Glove powder may be a means of spreading the contamination. Similar effects are seen with rubber dam,[15] and indeed can be expected with any cured-latex product.

The reaction of the catalyst with the vinyl group indicates that exposure of any addition-silicone impression material, no matter how it is mixed, to any source of double bonds would interfere with its setting, spoiling the surface. Thus, exposure to eugenol-containing materials, or surfaces contaminated by it, would be suspect. Likewise, residual double bonds in filled resins, whether temporary (6§7) or permanent, or even resin-modified glass ionomer cement,[16] would also be expected to be reactive in this sense. This appears to be a general problem[17] (it should be noted that there are no sulphur-containing reactive substances or groups in these materials). Residual unreacted material can also contaminate the surfaces from which temporary restorations have been removed and interfere with subsequent impressions. Decontamination with hydrogen peroxide appears to be effective in this circumstance. This would be due to reactions similar to those that occur in the oxygen-inhibited layer of such resins, eliminating the vinyl groups, and acting in the inverse sense to that of residual peroxide interfering with the setting of vinyl-based resins (6§6).

Note that 'vinyl' gloves are made from poly(vinyl chloride), and have no remaining double bonds. On the other hand, 'nitrile' ('NBR') gloves have many remaining double bonds (27§6.2) and so can be expected to be as reactive as natural latex. In addition, ordinary nitrile rubber is cross-linked with sulphur (from thiocarbamates, for example) and so will be similar to natural latex rubber (27§1.3). While nitrile medical examination gloves that are accelerator- and sulphur-free are available, the double bonds remain. These may be reduced (but not eliminated altogether) in hydrogenated nitrile rubber ('HNBR'), but such gloves would be appreciably more expensive, and are not known to be available.

It is sometimes said that astringents such as ferric and aluminium sulphate, *i.e.* as used for haemostasis in retraction cord, may also inhibit the setting reaction because of the sulphur in the sulphate; this is chemically infeasible, and no effect has been found.[18][19]

§7. Polysaccharides

A quite different class of impression material is represented by the so-called hydrocolloids, which are often classified as reversible (agar) and irreversible (alginate) types. None of these names is particularly useful, and two are not very accurate. Thus, it is difficult to see how substances which form proper solutions, even if macromolecular, can properly be called colloids. The high degree of solvation which will become apparent is associated with these substances is such that it is not really sensible to attempt to identify a phase boundary: colloids are essentially two-phase systems (see 8§2).[20] (There are, of course, polymers which disperse in water and which forms because the molecules form globular rather than extended structures. Nevertheless, the distinctions are difficult to make in any absolute sense.) Secondly, calcium alginate yarn was made at one time on the premise that it was 'insoluble'. However, when used to make fabric for bathing costumes, which promptly dissolved in the sea, the error was appreciated (if not by the wearer). This nevertheless proved to be a beneficial property, as surgical sutures for internal use, and swabs and gauze which could be accidentally left inside the patient, would dissolve in body fluids and so vanish harmlessly if made from such material. Chemical 'irreversibility' is very often a matter of degree, and careful thought is required to make sure of the value of the statement in any given context.

> A **sol** is a suspension or dispersion of particles larger than ions and small molecules, generally taken (arbitrarily) to lie in the range 1 nm to 1 μm in diameter. They are sufficiently large that they cannot be considered to be in true solution, but they are maintained in suspension by the thermal impacts of the surrounding solvent molecules, as is shown by Brownian motion. That is, they do not settle out under the effect of normal gravity. Typically, particulate sols scatter light. Their solution-like behaviour – no structure – means that they do not have a yield point if the suspension medium is purely viscous.

Both agar and alginate are extracts of marine algae and belong to the enormous and complex chemical group called the **polysaccharides**.[21] These, as the name suggests, are formed by the linking of sugar molecules into long chains. Industrial synthesis is not attempted and the links are formed enzymically by the algae as part of their normal metabolic processes, the polymers being structural elements of the plant cells. Sugars (Fig. 7.1) are themselves identified as polyols, that

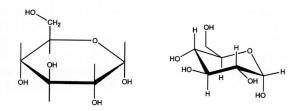

Fig. 7.1 Conventional and conformational structural diagrams of a typical hexose sugar (α-D-glucose).

is, having several hydroxyl groups. It is these which give them their characteristic properties, and it is condensation with the elimination of water at pairs of these groups that produces the polymers. The various monosaccharides are distinguished in part by the particular orientation of each hydroxyl group in turn around the ring, and a complex nomenclature has been developed to identify these structures, although this need not be a concern now. Notice that the actual conformation of the sugar molecule is not a flat ring (as it is commonly drawn) but a puckered shape usually in the so-called 'chair' conformation (Fig. 7.1, right). In forming the polymer chains, the number of possible linkages between rings (*i.e.* which carbon is connected to which) is large, the number of possible combinations of sugar residues is astronomical, as indeed is the range of properties. But as the choices of agar and alginate have been made essentially through commercial convenience (and the availability of the algae), there is no need here to discuss this area.

However, the most important aspect of these materials is the very large number of hydroxyl groups that they possess. All of these can form hydrogen bonds with water, creating a remarkably tightly-bound **solvation shell**. This bound water is thought not to freeze above $-40°C$ (that is, convert to the ice structure), and it is incapable of acting as a solvent for anything else. These hydroxyl groups can also form hydrogen bonds between themselves, both within and between chains, binding chains strongly to each other. To illustrate just how strong, it has been shown that the strength of *dry* cellulose is entirely due to hydrogen bonds between chains.

§8. Agar

Agar-agar is the Malay name applied to an edible marine red alga, still used as a foodstuff in several countries. This name, or simply agar, as it is now more commonly called, is also given to the extracted polysaccharide, which is essentially a polymer of mainly two different sugars in an alternating sequence three or four hundred units long.[10] This polymer is known chemically as **agarose** (Fig. 8.1). Notice that even though one of the sugar groups has undergone an internal condensation to form a bridge across the ring – which stiffens the chain slightly by preventing the 'chair' from flipping back and forth, there are still plenty of hydroxyl groups sticking out of the chain to form hydrogen bonds. Consequently, these materials can form viscous solutions at very low concentrations, and even set to form a good **gel** (see box) at concentrations as low as 1·5 mass percent. The question is, how is this achieved?

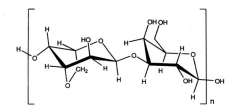

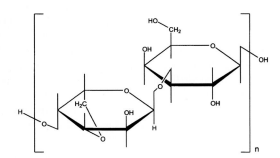

Fig. 8.1 Conformational and conventional structural diagrams for the basic repeating unit of agarose.

●8.1 Water

It is necessary to consider the structure of water itself to begin to answer this.[22] Water is a very special substance, as is discussed elsewhere (10§4). On the basis of comparison with similar molecules, water might be expected to freeze at $\sim-90°C$ and boil at $\sim-80°C$, but it obviously does not follow the expected pattern: hydrogen bonding is believed to explain its properties. It is calculated, for example, that in (liquid) water at $0°C$ an average of about 90% of hydroxyl groups are involved in hydrogen bonds, and this proportion falls to only just below 80% at $100°C$. This has a startling implication: water has short range order. In other words, the commonest liquid does not have the completely random structure that we ordinarily think of as characterizing liquids, even when boiling. This structure is closely related to that of ice, with a hexagonal lattice (analogous to the structure of tridymite, Fig. 17§2.12, except that the oxygen of water is where the silicon would be, and the hydrogen atoms lie slightly to one side of where the silica oxygen would be). However, in this liquid structure, the spacing between neighbouring oxygen atoms is significantly greater. This structure persists in transient clusters all the way up to the boiling point, with a typical lifetime of about 10^{-8} s (this time-scale is short

---

[10]   Natural products are notoriously variable in composition, by species, by location, and by season, so that published accounts vary accordingly. Further specification is therefore of dubious value, and as no finer detail is necessary for an understanding of the properties in the dental context, it will not be attempted.

enough that water appears properly fluid – Newtonian – under most ordinary circumstances). The size of a cluster at around $20 \sim 25\,^{\circ}$C is estimated to lie between 200 and 400 molecules or, assuming that the cluster is roughly globular, a diameter of about 7 to 9 molecules.

●8.2 Polysaccharide template

The next point to notice is that the spacing between the hydroxyl oxygens on polysaccharides often matches the spacing in the ice-like lattice of liquid water. It is considered that this is sufficient to act as a template upon which a cluster may become more permanently centred. For example, noting the agar concentration given above as forming a good gel, it can be easily calculated that each sugar residue is responsible for essentially immobilizing at least 550 water molecules on average, or a shell out to a *radius* of about 8 or 9 molecules from the chain axis. The unexpectedly high viscosity of mono- and di-saccharide sugar solutions (syrups) can be attributed to similar effects. The polymer-associated water lattice can be viewed as structural water: it is not really *liquid* water anymore.

> A **gel** is a kind of composite material in which there are two continuous, interpenetrating phases, one of which is a liquid. The second 'phase' may consist only of a molecular network (in which case calling it a phase is a bit dubious, much as is the boundary to distinguish between solutions and colloids) or be built up of a **micellar** structure of aggregated polymer chains. The essential characteristic of this part is nevertheless the presence of a three-dimensional network with definite bonding establishing the spatial relationships of the component molecules or the segments between nodes. These materials are typically rather brittle, even at modest strain rates, and should be distinguished from plastic solids. Even so, the structure means that there is a yield point. Solutions that are merely viscous are not gels. Cross-linked polymers which have swollen by absorption of large amounts of solvent may be also called gels; these are effectively two-phase systems. In addition, polymerizing systems with sufficient cross-linking are said to pass through a gel stage where the material is soft, weak and brittle. The 'infinite' molecule at this stage encloses a substantial proportion of unconnected material.

●8.3 Junction zones

Other structures are possible in polysaccharide gels, their frequency and importance depending on concentration and the details of the structure of the polysaccharide itself. Thus, association between chains may lead to partially crystalline zones or **micelles** appearing in which sections align to allow hydrogen bonds between them (remembering our relaxed view of what constitutes crystallinity in polymers, 3§4.1). This can be done in several ways. In unsubstituted polysaccharides such as agar, double helices may form what are termed **junction zones** (Fig. 8.2) (reminiscent of the DNA structure) and these helices may themselves assemble into bundles (without twisting) called **superjunctions** (Fig. 8.2).[23] Obviously, if a portion of chain is already fixed by a such a structure forming, preventing rotation, a helix cannot form. However, the formation of coordinated bundles through hydrogen bonding is not excluded, and these 'cross-linked' micellar regions would continue to grow with time, given that all such processes are diffusion-limited. Such structures would be expected to greatly increase both the modulus of elasticity and the strength of the gel. Indeed, effective cross-links of this kind are essential for the formation of a three-dimensional network if there is to be a gel at all. Furthermore, as the network grows, and the number and density of chain associations increases, so the strength and stiffness of the gel would be expected to increase.

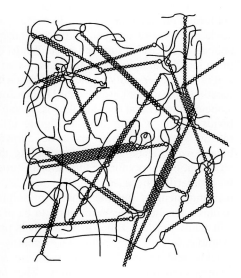

Fig. 8.2 Some structures due to chain association found in polysaccharide gels. Single chains are in the sol state (the gel is in the process of forming), double chains are 'junction zones', multiple junction zones are 'superjunctions' (these are drawn as straight for clarity – they need not be, and probably are not).

The micellar structure of a gel of this kind is shown by its optical properties: it is slightly **turbid**, that is, light is scattered. It is inherently a two-phase structure, with water in the cells created by the network. This water has a refractive index different to that of the micelle material, which is of the order of 200 nm or more in thickness, and so refraction occurs (24§5.11). Pure agar gels are therefore normally slightly 'cloudy' (and much more so than the solution), and blue-tinged when viewed from the side, because of the scattering. This is made very obvious by the fact that the addition of solutes to the water (such as a sugar) to raise its refractive index to match that of the micelle material can make such gels completely transparent. The formation of the gel is therefore a kind of **phase-separation** process (8§4.1), and results in two interpenetrating or **bicontinuous** networks: the polymer bundles and an aqueous solution of everything else.

It will be noticed that the formation of junction zones and superjunctions markedly decreases the amount of thermal motion that can be exhibited by the chain segments involved. In other words, as the chains become more ordered there is a decrease in the free volume of the system (*cf.* 3§4.2). This means that there is dimensional change on setting similar to that observed in polymerizing materials (although, of course, the setting of agar is *not* a polymerization process). We may term this **ordering shrinkage** (*cf.* the decrease in specific volume at T_m for ordinary solids on crystallizing – see Fig. 3§4.7). There is therefore a limit to the dimensional accuracy that may be attained with such materials quite apart from such issues as stress set (§11).

Dental products are typically formulated to give around 15 mass percent of agar overall. Some products also contain a substantial quantity of a filler. The filler is present because the elastic modulus otherwise would not be high enough for the needs of impressions. It is also necessary to increase the viscosity of the unset material (which would simply be too runny to stay in the tray). In some products borax (sodium tetraborate) is added, although the reason for this is not very clear. Consequently, if it is present, potassium sulphate has then also to be included to avoid inhibition of setting, and therefore a poor surface, for the gypsum model material (2§7.1) that will be cast into it. In the absence of the potassium salt, the two materials are said to be **incompatible**, a term broadly used whenever some kind of reaction occurs between two products brought into contact.

●8.4 Reversibility

Agar gel materials, whether used for dental impressions, microbiological culture media, or simply food, rely for their usefulness on the reversibility of the gelation process, that is, the solution- or sol-gel transition is **thermally labile**. In other words, on warming the structure may be broken up to form a **sol**, a colloidal solution, only to reform the gel on cooling. This may be repeated indefinitely. However, there is considerable **hysteresis** even for moderate rates of temperature change: breakdown to a solution occurs at a higher temperature than does gel formation on cooling (Fig. 8.3), at least, for ordinary rates of temperature change. This certainly reflects the steric problems of unravelling the secondary structure helices, but there is a further aspect. Whilst the lifetime of an individual hydrogen bond is very short, say 10^{-11} s, it is the cooperativity of the many bonds involved in junction zone formation and in the lattice structure of the associated water that prevents the loss of any one or two bonds having a more than negligible effect on the overall structure. Time must be allowed for the breakdown to occur. Similarly, on cooling, there will be an activation energy for the formation of structure. This is primarily because of the extensive spatial rearrangement of polymer chains in solution that is required for the proper orientation to be obtained that will allow junction zones to form. It is possible, for example, to chill some solutions of gelling agents too quickly and to too low a temperature so that the solutions remain liquid indefinitely. They may be said to be **supercooled** (*cf.* 11§2.1). Warming then causes **gelation** as the activation energy becomes available.

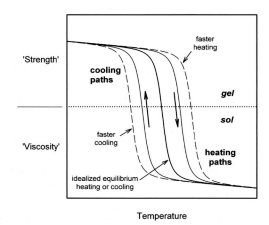

Fig. 8.3 The sol-gel transition for polysaccharides such as agar is sensitive to the rate of change of temperature.

Dental agar materials have to be heated to temperatures approaching 100 °C for melting (in a convenient time) but, in view of the fact that they will be placed in contact with living tissue a certain amount of pre-cooling – so-called **tempering** – is required before use. This may be to around 50 °C, so that by the time the tray has been loaded, sufficient further cooling will have occurred to avoid trauma. Bear in mind that inflammation is triggered by raising tissue temperature to as little as 42 °C. Special thermostatted water baths are therefore required to prepare agar impression material for use. Gelation is usually encouraged by using water-cooled trays, a further set of specialized equipment.

A complication to the practical effects of the phenomenon of hysteresis is due to the low thermal conductivity of the material. It takes time for a mass to be heated through to a given temperature, and likewise to cool again everywhere below the necessary temperature for gelation. (Convection cannot occur until it has melted, and even then the high viscosity limits this.) Thus partial melting or partial solidification will be seen during these processes. The latter is especially risky for the accuracy of an impression, and cooling must be thorough to be certain of no ill effects, bearing in mind that this will not be a visible condition and that cooling will be slowest adjacent to the tissue surfaces of primary interest.

The reversibility (*i.e.* for re-use) of agar is of little practical advantage in the clinic, given that it must inevitably be contaminated by patient contact. This would not be a problem for laboratory model duplication agar materials if it can be ensured that the original model is first decontaminated. This may not be possible: transfer from impression to model to duplicator of an infectious agent remains problematic.

The substantial temperature changes, and thus volumetric shrinkage, involved in the process of using agar as an impression material set limits on the accuracy that may be attained (in addition to problems with loss and gain of water, §10). Thus, given the complexity of the handling process and the availability of more convenient materials, the usage of agar is declining.

§9. Alginates

The alginates are based on a different chemistry, although there are many aspects in common with agar. Alginic acid, derived from an extract from a brown marine alga, is insoluble in water. Each sugar residue has a carboxyl group attached, and it may be presumed that the chain-chain interactions are stronger than even water can overcome (*cf.* the dimerization of acetic acid, Fig. 10§4.1). The alkali metal and ammonium salts of this acid, however, are quite soluble, the mutual repulsion of the carboxylate anions being sufficient encouragement for dissolution. On the other hand, polyvalent metal ions can form chelation complexes with alginate in much the same way as occurs in polycarboxylate cements (9§6, 9§8). It is this structural element that can provide the necessary cross-links between chains to form a three-dimensional network and thus an elastic gel. Monovalent ions simply do not have enough charge to form stable complexes.

Alginic acid is basically a copolymer of two closely-related acid sugar residues (Fig. 9.1) which we may, for convenience, label M and G.[6] They differ only in the relative position of the carboxylate group. It is thought that, within the polymer chain, they occur in three kinds of region, which may be represented by -MMMMM-, -GGGGG-, and -MGMGMGM-. These regions are found in each chain in a random sequence, the proportions of which depend on the source. This kind of structure very effectively prevents large scale crystallization of the polymer.

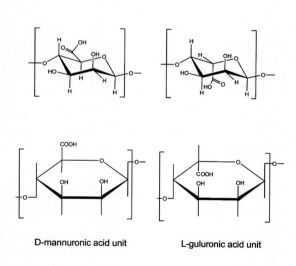

D-mannuronic acid unit   L-guluronic acid unit

Fig. 9.1 The two principal acid sugar residues found in alginic acid.

The -GGGGG- sequences may adopt a conformation in which the carboxylate groups are ideally placed to coordinate in pairs with, say, a calcium ion, or any other di- or trivalent cation. A corresponding pair from an adjacent chain completes the complex, which has the 'egg-box' structure (Fig. 9.2). Such chelation is extremely favourable energetically, and is not broken down in plain water even at 100°C.

The gel structure forms extremely rapidly when suitable ions are present, and for dental use it is necessary to modify the chemistry somewhat to avoid the premature – nearly instantaneous – setting that occurs when, say, a solution of $CaCl_2$ is mixed with the alginate solution. The remaining hydroxyl and carboxyl groups can be expected to be involved in immobilizing water as does agar, as well as in further chain-chain interactions, and therefore contribute to the gel structure.

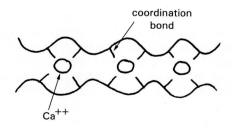

Fig. 9.2 Schematic diagram of the 'eggbox' structure of alginate gels.

●9.1 Setting reaction

Alginate impression materials usually contain the sodium or potassium salt of alginic acid, sometimes other monovalent cation salts (such as of ammonium), and these are, as explained above, quite soluble, dissolving readily on mixing the powder with water to give alkaline solutions (weak acid, strong base). To control the rate at which the cross-linking cations are made available, a sparingly and slowly soluble salt is required. Using powdered calcium sulphate dihydrate, $CaSO_4.2H_2O$, as the **source** is a very convenient and cheap means of meeting this demand, calcium ions then being the cross-linking agent:

$$Ca^{++} + 2Alginate^- \Rightarrow Ca(Alginate)_2 \tag{9.1}$$

Even so, setting would be prompt and the material unusable.

But to delay the onset of gelation, a **sink** for these metal ions is needed, a means of consuming them as fast as they are formed, at least for a time long enough for mixing, and then loading and seating the impression tray. This function is commonly fulfilled by phosphate ions. The calcium phosphates are fairly insoluble at neutral or high pH, and by using di- or trisodium orthophosphate (Na_2HPO_4, Na_3PO_4), for example, calcium ions can be removed from solution almost as fast as they dissolve from the sulphate:

$$3Ca^{++} + 2PO_4^{3-} \Rightarrow Ca_3(PO_4)_2 \tag{9.2}$$

(Notice that it is essential that the cation in the sink must itself be monovalent to avoid being a cross-linking agent for the alginate.)

However, as soon as the phosphate ions are used up, the calcium ions supplied from the still-dissolving source become available to the alginate and setting commences quite suddenly (Fig. 9.3).[24] It is possible to design several variants on this theme, using different compounds as sources and sinks of calcium, or other divalent metal ions, but the principles are always the same. Indeed other sinks that have been used at one time or another include Na_2CO_3, and alternative cross-linking agents have included $PbSO_4$, Pb_2SiO_4 (understandably discontinued on concerns about toxicity) and $BaSO_4$. Although various brands of alginate impression materials are known to use, or have used, such differing ingredients, the setting reaction version used here as an illustration is only one example of the many which are feasible.

These 'sink' compounds are commonly called **retarders**. However, we must be careful to distinguish their action from the kind of activity shown by borax in hemihydrate slurries (2§7.4), for example, which merely slow down the reaction, *i.e.* reduce its rate. Here, it is a **competing reaction** that leads to the delay in the onset of the actual setting reaction, what we may term a **kinetic competition**.

As with agar, the formation of the cross-links between chains in alginate materials causes a reduction in thermal motion and free volume, *i.e.* there is an overall shrinkage on setting due to ordering. Again, this sets a limit to the achievable dimensional accuracy of impressions in these materials (Fig. 9.4). What is curious about this shrinkage, and this is true for all flexible impression materials, is that it is not a uniform scaling of the mould space. Rather, the magnitude of the effect, say bucco-lingually, is roughly proportional to the thickness of the material between the point of interest and the wall of

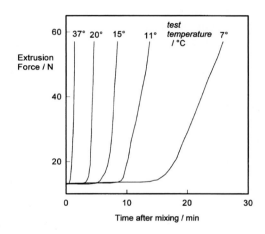

Fig. 9.3 Some force-time curves for the extrusion (as if from a syringe) at a constant rate of setting alginate impression material at various temperatures. As the temperature falls, so the delay to onset of cross-linking is increased.

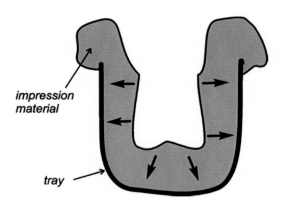

Fig. 9.4 Reaction or ordering shrinkage leads to dimensional errors in the impression.

the tray. This means that cervical regions will be relatively more enlarged than at the coronal bulbosity. On top of this, shrinkage in the other two directions is constrained by the material being adhered to the tray wall, exacerbating the shrinkage towards the tray wall. It is therefore a complex distortion that results, and no subsequent processing of model, wax pattern or device can compensate for this properly.

When the primary source of the cross-linking ions is used up, that is not necessarily the end of the story. The salt formed by reaction with the sink need only be less soluble than the source to prevent cross-linking occurring in the early stages. However, no salt is totally insoluble. Solubility is a matter of equilibrium between the solid salt and its solution. Thus, when the reaction with the alginate has reduced the concentration of the cross-linking ions sufficiently, then the calcium phosphate (in this example) may then start to dissolve and act as a source of cross-linking calcium ions itself. In this way, the setting reaction may continue for a very long time, well beyond the time at which the impression would be removed from the mouth. Since this continued reaction will contribute to dimensional changes, it provides one reason for the pouring of the model promptly, after making allowance for the retarded recovery of the deformation occurring on removal from the mouth. The overall reaction scheme is summarized in Fig. 9.5.

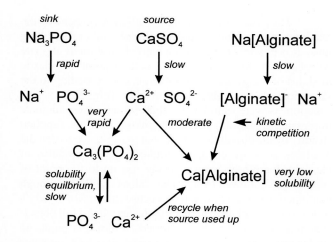

Fig. 9.5 Reaction scheme for a typical alginate impression material.

One of the consequences of the reactions occurring in the course of setting is that the pH varies with time (Fig. 9.6). Thus, for example, Na_3PO_4 has a very alkaline reaction on dissolving in water (strong base, weak acid), and the initial pH is therefore high. As it is calcium phosphate that is being precipitated, effectively what is left in solution is sodium sulphate (strong base, strong acid), so the pH approaches neutral. However, in the third stage, when the calcium phosphate redissolves, it is again more or less sodium phosphate that is left in solution, when the pH is again very high. Of course, the actual pH observed depends on the exact ingredients used, their concentrations, and their relative rates of dissolution and reaction as well as their diffusion rates through an increasingly viscous medium. Nevertheless, a consistent pattern is found for alginate impression materials in general (Fig. 9.6).

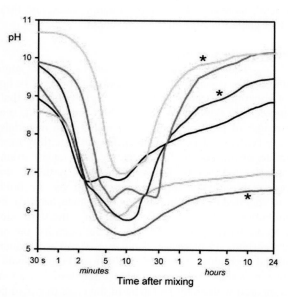

Fig. 9.6 Variation in pH with time for a variety of alginate impression materials setting at 25 °C. Curves marked with a star are for products with pH indicators intended to identify adequate setting.

The pH variation can be so marked that a number of products, so-called **chromatic alginates**, have incorporated a pH indicator dye, the change of colour being used to judge when the material is sufficiently set to be safe to remove from the mouth. Clearly, the rate of pH change needs to be rather high near the desired end-point, and the indicator value close to the minimum attained, bearing in mind that most such indicators are only good to about 1 pH unit (meaning that the colour change is always gradual over such a range), some caution would be required in deciding when the change has occurred sufficiently. It would appear that nothing special has been done to the chemistry of such materials to enable this effect, there being no clear distinguishing features in the pH plot. It is also clear from Fig. 9.6 that, as indicated above, the setting reaction does indeed continue steadily for a long time, not approaching equilibrium even at 24 hours in some cases.

There is one possible advantage associated with the temperature-sensitivity of the setting rate, and in contrast to agar. Exposed material will (commonly) be cooler than that against tissue and so set more slowly. Thus, once this portion has reached the correct condition it would seem likely that the rest of the bulk has also. However, subjective judgements (a "thumbnail test") cannot be considered reliable in such contexts. The manufacturer's instructions should be respected.

The structure and strength of alginates depend on the relative concentrations of polysaccharide and cross-linking ions. It is possible to perform a kind of titration of alginate with calcium, for example: at a certain point the elastic modulus of the gel ceases to increase, corresponding to the saturation of available chelation sites (Fig. 9.7). This situation is very similar to that in a setting alginate impression material. The cross-linking ions are supplied steadily by the dissolution of the source substance so that the available sites are occupied gradually. The modulus of elasticity therefore shows a steady rise up to the point of 'saturation', followed by a plateau value (Fig. 9.8), although continued reaction is expected to cause this to rise slowly a little more. The plateau perhaps indicates when a full object-spanning network has been formed. From this point, no further cross-link junctions can form because chain segments are no longer mobile enough to reach over large distances. Mechanically, the network structure has been defined. The continued reaction shown by Fig. 9.6 must therefore be more 'local', increasing the size of junction zones, and titrating other chelation sites on the same chain.

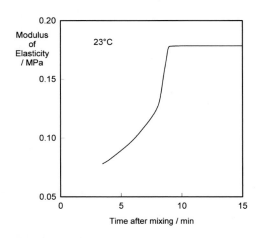

Fig. 9.8 The variation of stiffness of an alginate gel during a titration with calcium ions. [10]

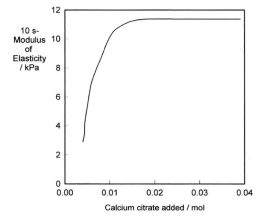

Fig. 9.7 Variation in modulus of elasticity with time of a setting dental alginate product. (This material is much stiffer than the gel of Fig. 9.7 because of the presence of a filler.)

●9.2 Gel brittleness

The gel structures of both agar and alginate immobilize water so well that it seems no longer to be liquid, as mentioned above. It is so tightly bound that no appreciable flow can occur (on short time-scales), and the strength of junction zones is such that both materials are very brittle, that is, showing little or no ductility even on moderate timescales. Unfortunately, little tensile strain can be tolerated before failure and, once initiated, cracks propagate quickly (this is not helped by air bubbles, incorporated by the hand-mixing of alginate, acting as crack initiators). More mobile water in otherwise similar polysaccharide gel systems can confer ductility, but not in the systems of interest in dentistry. In fact, such ductility would be highly undesirable. Even so, both alginate and agar materials show appreciable viscoelasticity, although compared with many members of the class this is in fact a relatively minor effect. Nevertheless, it is of a magnitude sufficient to be of concern in dentistry, and great care is necessary to avoid excessive permanent distortion. The performance of both alginate and agar is appreciably poorer than that of a number of other impression materials, and they are generally not suitable for precise work.

●9.3 Hazard

The viscosity of the freshly mixed alginate and the elastic modulus of the set material are both increased to suitable values by the incorporation of a filler. This has commonly been **diatomaceous earth**. This is an ancient marine sediment composed largely of the skeletons of certain algae called diatoms. These skeletons are essentially pure silica. As such, they are capable of inducing the slow debilitating lung disease **silicosis**, which is more usually associated with miners. It is therefore important that care is taken to avoid inhaling the dust, especially when opening the container after shaking it to 'fluff up' the powder. Some recent products are reportedly 'dust free', having been formulated to reduce this risk. This probably means the addition of a small amount of a substance such as glycerol (Fig. 6§6.2) which, while not allowing reaction, makes the powder particles 'wet' enough to cling together through better hydrogen bonding. The problem is that such a substance, called a **humectant**, is **hygroscopic** (in the true, chemical sense) – it absorbs water from the atmosphere. Such water is then available as a medium for the setting reaction, so that unless care is taken to keep containers tightly sealed against 'atmospheric moisture' the material will "go off" and fail to set properly. At the very least, setting time, strength and stress-set behaviour (§11; 4§8) will be adversely affected, although it may be difficult or impossible for the dentist to notice these effects, at least in the early stages.

§10. Water Gain and Loss

Water-based gels such as the above systems are very sensitive to loss and gain of water, depending on the environmental conditions. Thus both agar and alginate are moderately stable at 100% relative humidity, but they will dry out rapidly if exposed to dryer air. The cause of the gain of water observed if these materials are wetted is less obvious, but it is still important as causing an unwanted dimensional change (Fig. 10.1).[25] Both materials contain soluble salts for one reason or another, and these raise substantially the osmolarity of the contained water in comparison with that of tap water or even saliva. If allowed time, water would therefore be steadily absorbed from an external source by **osmosis**. This process is sometimes referred to as **imbibition**. It leads to a generalized expansion which reduces the mould space and thus the size of the model (the opposite effect of that in Fig. 9.4). Thus, although it is important to rinse away blood and saliva to avoid inhibiting the setting of the model material (by the proteins) and thereby spoiling the surface, this must be done quickly and the surface water must be carefully shaken off to remove free liquid before putting the impression into a container for sending to the laboratory. Damp gauze or paper towel is commonly used to raise the humidity in that container to prevent drying. Needless to say, it should not act as a source of water for imbibition.

There are other sources of water. With the emphasis now placed on infection control in dentistry, impressions must be disinfected before being sent to the laboratory for pouring up the model. This involves immersion in aqueous liquids for various lengths of time, during which at least some absorption must occur. Extended soaking should not be used. In addition, the dental stone slurry that is poured is also a source of water which might be expected to lead to expansion, at

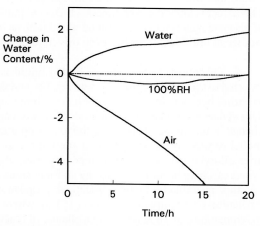

Fig. 10.1 Water gain and loss by agar gels under various conditions of storage.

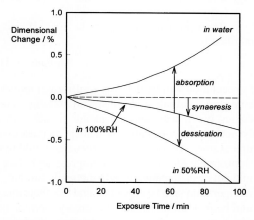

Fig. 10.2 Dimensional instability in freshly formed polysaccharide gels.

least while the slurry was still fluid. However, the hydrostatic pressure due to the 'head' of this would tend to oppose the expansion, but to an extent varying by depth – in other words, there would be distortion, not just scaling. The magnitude of this effect does not appear to have been studied.

In principle, if the alginate were to be immersed in a solution with precisely the same osmotic potential (*i.e.* water activity) as the aqueous phase in that gel, then there should be no net movement of water in or out of the impression material and thus no dimensional change from that source. However, while it can safely be assumed that the osmotic potential differs between products such that a single recommendation could not be made, it is also likely that the value changes with time as the setting reactions proceed: the concentrations of the relevant species are changing with time, and their solvations also differ. Even so, there is an overriding behaviour that makes any such attempt futile.

•10.1 Synaeresis

In both agar and alginate the gel structure results from the coordination of lengths of the polysaccharide chains in junction zones to form the three-dimensional network. As has been said, this results in a 'setting shrinkage' for these materials. However, since the greater the coordination the lower is the energy of the system, there is always a tendency for this chain association to increase – this is the thermodynamic driving force. After the initial stage when cross-linking nodes are first established, any further coordination results in a shortening of the uncoordinated sections as the lengths of the junction zones grow. This is then necessarily a slow process because of the activation energy requirements and the viscosity of the system, but it does cause a net *decrease* in the enclosed volume of each cell of the network, raising the pressure on the enclosed aqueous phase. As a result, free water (or, rather, the solution of the non-reacted materials) is gradually extruded from the surface. This effect is known as **synaeresis**. It can be quickly demonstrated in a simulation: Squeeze a 1 cm or so cube of alginate material between finger and thumb as hard as possible for a few seconds. Liquid will appear on the surface at the sides. This has been driven to percolate through the gel by the raised hydrostatic pressure.

Of course, because the extruded liquid is no longer bound, being external to the gel structure, it has no rigidity and the impression can be seen to shrink bodily (Fig. 10.2); by 24 h the extent may be around 1% or more. This synaeresis process can <u>not</u> be stopped, nor can it be ameliorated by any change in technique. It is by far the most important reason for pouring the model as quickly as possible after the impression has been taken (that is, after allowing for viscoelastic recovery, say, about 15 min). Certainly, synaeresis may be slowed down by cooling, since it is an activated process, but this is unlikely to be a practical proposition. Thus, assuming an Arrhenius-type temperature effect (3§3.3), where fairly typically a change in temperature of 10 K results in a change of reaction rate by a factor of two, cooling by 20 K might therefore reduce the rate by a factor of about four. Overnight, say 16 h, in a refrigerator at 4 °C would then be equivalent to about 4 h at 24 °C. Ordinarily, 4 h would not be an acceptable delay before pouring a model, and therefore refrigerated storage overnight should not be used as it cannot prevent the deleterious changes.

> The effects of synaeresis are sometimes confused with evaporation. If evaporation occurs at a high enough rate then the extruded liquid will not be seen. Of course, the effect of evaporation simply adds to the shrinkage (Fig. 10.1).

There is a further aspect to this. If one writes the coordination of water with the polysaccharide symbolically:

$$\text{Alginate} + \text{H}_2\text{O} \rightleftharpoons \text{Alginate·H}_2\text{O} \tag{10.1}$$

the expectation will be that this equilibrium moves to the right at low temperature. That is, the solvation shell becomes more stable and thicker on cooling. The free volume associated with the water molecules thereby becomes much less, and shrinkage ensues. In addition, the tendency for hydrolysis and breakdown of polysaccharide coordination structures increases with temperature; in other words, they are more stable at lower temperatures. Although these structures would tend to break down again on rewarming (as presumably would be appropriate for a refrigerated impression before pouring the model), the reversal of the stabilization process would not be instantaneous but kinetically limited. Thus, even without continued 'setting', shrinkage as a result of refrigeration is to be expected, and this therefore compounds the problem of overnight storage. These effects would be in addition to that of ordinary thermal expansion and contraction (§12.3).

Structurally, the cause of synaeresis is straightforward. At the point of removal of the impression from the mouth, when it has been judged to be (well-enough) set, the reactions have not, of course, been completed. In fact, as reactants are used up, and diffusion becomes more difficult in the gel, the reaction simply becomes slower and slower but still continues for many hours. Since in alginates the principal setting or gel-forming mechanism is the 'zipping up' of pairs of chains, this process also continues, but coordination of chains – or pairs of chains – through hydrogen bonding also occurs (as in agar). Each step in that process applies a small force to the chains (as it is an energy minimization process as well), and this force gradually draws the chains

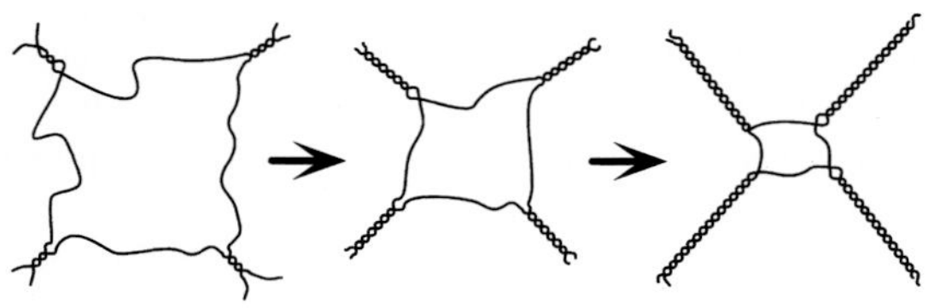

Fig. 10.3 The growth of junction zones reduces the enclosed volume of gel cells, causing synaeresis. Here it is a two-dimensional model of what is a three-dimensional structure, but the effect is the same. Note that helix formation is not essential, only coordination between chains in micellar structures, whether purely hydrogen-bonded or chelation-derived.

together. This enables a further link to be added, and so the process continues. However, at the same time, the 'cells' that are formed within the gel, defined by the enclosing uncoordinated alginate chains, must be getting smaller (Fig. 10.3). The liquid that is contained within these cells thus gets squeezed out, eventually to appear on the surface. There is therefore a small positive pressure acting on the liquid in the gel, which is therefore driven to percolate through the material. It can be seen, therefore, synaeresis is a further sign of continued reaction, as calcium ions are slowly being consumed (Fig. 9.6). It is also obvious from this alone that refrigeration is bound to be ineffective. Again, 'set' from an operative clinical point of view does not, and cannot mean that reactions are complete.

Synaeresis also affects agar gels but perhaps to a lesser extent because the forces of zipping up junction zones are less (remember that these gels are heat-labile). However, agar is at least as sensitive to loss by evaporation and gain of water on contact as alginate.

§11. Stress Set

In §1, flexibility was introduced as a basic requirement for an impression material to be functional in the presence of undercuts (Fig. 11.1). This means that after deformation it is desirable that there should be a return, more or less promptly, to zero strain. The failure to do so is revealed in a 'compression set' test (4§8); the same considerations of course apply to the related ideas of 'tension set' and 'shear set'. The main effect is of the ordinary viscous component of the deformation, the creep compliance (4§6.2) – the longer the material stressed, the greater the permanent deformation. Unfortunately, no material is perfect in this respect, they are all viscoelastic, but there are six additional principal reasons for this that exacerbate the difficulties.

●11.1 Continued reaction

As was shown specifically in the case of alginate, setting reactions can continue for long periods. Indeed, this must be true of all flexible impression materials since reaction depends on diffusion, and this becomes more and more difficult as the cross-linking raises the viscosity and increases the steric hindrance to the movements of small molecules.

It should be recognized that most materials will be never be near being completely reacted, and especially on timescales similar to those of normal dental use, *i.e.* at removal of the impression, when it is a matter of balancing the sufficiency of the progress of setting against the boredom or tolerance of the patient or the pressure of the appointment diary – although the influence of such factors must be firmly resisted so that instructions are followed and a repeated procedure is avoided. There will thus always be, in practical situations, scope for such further reaction during the time of deformation, even if the strained state is of short duration. Such reaction makes new bonds that oppose the elastic recovery. It does, of course, give one more reason for reducing the number of tries taken to remove the impression. Hence, a simple continuation of the setting process, but in the deformed condition, ensures that the new shape is to some extent taken up.

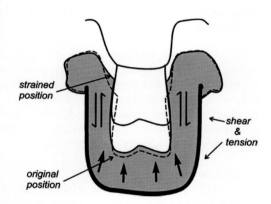

Fig. 11.1 The presence of an undercut in an impression means that various portions of the impression material will be subjected to tension and shear, not just compression.

●11.2 Stress-aided reaction

This is a chemically similar process to that of continued reaction (§11.1). Diffusion of polymer chain segments in the viscous pastes of unset impression materials is slow, and it is inhibited altogether once appreciable cross-linking has occurred. Consequently, there will be functional groups which are incapable of reacting by reason only of their separation being too great. Should those functional groups now be brought sufficiently close by the rearrangements of chain segments under the influence of an applied stress (because deformation is not microscopically affine, §2.4), then they may well react (time and activation energy permitting). The newly-formed bond, however, now prevents the return of those chain segments to their original positions when the external stress is released, as above. Repeated over many sites, such new bonds will cause noticeable permanent deformation of the body. This applies even to the hydrogen bonding in agar and alginate.

●11.3 **Bond exchange**

This has been discussed in the context of polysulphide rubbers (§4), but it can operate to some degree in all practical flexible impression materials used in dentistry. In polysulphide, the -S-S- bond is of relatively low energy and so may be made to exchange under stress. Similarly exchanges occur, but more readily, with ionic bridges or coordination bonds, again as in polysulphide but also in alginate, where the structure is dependent on ionic cross-links with a cation, commonly calcium. It is, of course, even easier with the more highly-labile hydrogen bonds of agar and alginate. Should local stresses exceed the strength of a bond it will break, but the important step is then that there will be relative motion of the two polymer chain segments in response to the applied stress (entropic recoil, §2.2). A new bond can then be formed at a different site, between the old chains and new ones that have now been brought close. On removal of the external stress, the local stresses tending to reverse the process cannot, in general, be large enough to break the bond, so a permanent strain has thereby been introduced.

●11.4 **Chain segment conformation**

The fourth point is a *statistical* one. For a given set of polymer chains, with a given set of cross-links, there are many ways of arranging the lengths of polymer chain between those cross-links. Even if one allows for loops and entanglements of a random nature, there will still be many ways of rearranging the chain segments with respect to one another. The structure formed on setting such a rubbery material will be only one such, random, arrangement. On deformation by an external stress, all chain segments will be rearranged to some extent even if the actual cross-links, loops and entanglements in a topological sense are not affected (that is, if there is no bond exchange or stress-aided reaction). For the original, dimensions to be exactly recovered on removing the stress requires an exact retracing of the path taken by every single chain segment and side chain. This is kinetically impossible, given thermal agitation and the fact that movements of this kind require an activation energy: each step is itself a probabilistic process, depending on local conditions and the time allowed. Some folds in chains may have been undone in such a way that there is no possible driving force to return them the way they came. In fact, any one such segment failing to return to its original position must prevent another from doing so by occupying its space, and so on. This is the reason for the failure of the assumption of affine deformation (§2.4).

Ultimately, on removing the stress, the actual arrangements of chains cannot possibly be the same as before, and some permanent deformation must result. Thus, even if a rubber is perfect in the sense of absence of bond exchange and stress-aided reactions, it cannot possibly be perfect in this respect. The effect could be reduced by increasing the number of cross-links, thereby reducing the lengths of chains between those points. But this would also increase the modulus of elasticity, rendering the rubber too stiff to be of use as an impression material and, ultimately, glassy. Again, it is a matter of compromise to obtain a product that is adequate for the job in hand, since ideality is unattainable.

●11.5 **Loose ends**

Fifthly, there is the issue of the failure of **chain-end rethreading**, which may be seen as an extension of the statistical argument above. During any deformation, at least some chain segments will be subject to tensile stresses along their length. If such chain segments are near the end of a chain, *i.e.* a free terminal, then that tension will tend to act so as to slide the chain out of its initial surrounding tangle of other chains – it will be drawn out (reptation). However, when the external stress is removed, the restoring force on the chain segment in question will not be able to rethread the chain into its former environment, even assuming that nothing else has changed, in particular because the chain is not a rigid entity. It would, in preference, buckle and fold or coil up; this is the energetically-simple and statistically most-likely outcome. The chain end is, in any case, a dynamically moving object because of thermal agitation. Similar arguments apply to loops that are not entangled. These are very strong reasons for the failure of the affine deformation assumption (§2.4).

●11.6 **Other factors**

The redistribution of portions of polymer chain therefore contributes to overall permanent deformation. Chain rupture (§2.2), enhanced by the Mullins effect (§3.1), and the related entanglement-generated intra-chain bond breakage, causes changes in elastic modulus, *i.e.* strain softening. This leads therefore to a reduced driving force for elastic recovery, whether true or retarded elasticity. This is therefore the sixth effect operating. However, each such rupture generates two free ends, each of which must then behave as described above (§11.5). In other words, strain softening of this kind must be associated with permanent deformation, and increasing the volume fraction of filler must increase the magnitude of the effect. Note that the Mullins effect is not itself plastic deformation, it merely makes it easier for this to happen.

●11.7 Consequences

The clinical implications of all of the above are clear. Primarily, the duration of the application of any stress must be kept to a minimum, that is, remove the impression in one fairly rapid direct motion, usually described as a "snap" removal. This can be summarized as the need to minimize the creep compliance (4§6.2), but also the time-dependent (kinetically-dependent) reaction factors mentioned above. It must be recognized that these effects are cumulative: repeated attempts and rocking motions all count as contributing to the permanent deformation. The time-dependence of the flow occurring during the application of a stress is a function of the diffusivity of the various components of the structure and reactants. One would expect this to be controlled by how far the working temperature exceeds the T_g. However, bond exchange reactivity appears to be dominant: polysulphide shows a far greater creep rate than does silicone, for example.

Secondly, the design or selection of the impression tray must allow a sufficient thickness of material where any stress is to be experienced on removal – which means everywhere – that the *strain* resulting is also kept to a minimum (Fig. 11.2). This latter condition cannot, of course, be met in the embrasure (and especially not in the interdental space); the results here are seen all too clearly in any impression.

Nevertheless, it should be remembered that distortions cannot ordinarily be seen by visual inspection, only gross defects could be picked up this way. Reliance must be placed on correct technique, not trial and error, as the discrepancies will not become apparent until after the device has been fabricated and fitting is attempted.

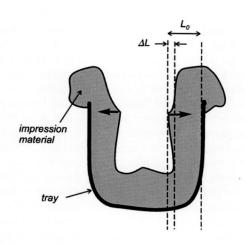

Fig. 11.2 To minimize the effects of stress set on the dimensional accuracy of an impression, there must be as much material as possible (L_0) between the greatest dimension of the tooth and the tray wall so that the deflection ΔL and thus the strain $\Delta L/L_0$ is minimized.

§12. Other Accuracy Issues

The accuracy deficiencies of flexible impression materials because of their departure from truly elastic behaviour have been stressed, and indeed a large part of Chapter 4 was leading up this. However, there are several other factors that need to be considered in the context of their use in practice.

●12.1 Retarded compliance

Since the three-dimensional networks of flexible polymeric impression materials are viscoelastic (4§6), there cannot be instantaneous recovery of part of the deformation, irrespective of whether the material is otherwise ideally Hookean (Fig. 1§2.5) or non-Hookean (Fig. 2.5). In other words, there is a retarded component to the compliance (Fig. 4§6.2). This is due to the nature of the materials: polymeric. It is the issue of chain segment conformation again (§11.4) that explains this. Not only do the chain segments take time to diffuse in response to the applied stress, they take time to diffuse back again when the stress is removed (see Fig. 4§8.4). Hence some allowance must be made for this when pouring the model. Sufficient time must elapse for that recovery to be essentially complete. In the case of the rubbery materials, polysulphide, silicone, polyether, only a very few minutes are generally necessary. However, for alginate and agar the time-constant for the process is much longer (probably because the internal restoring forces are much smaller), and typically a period of 15 minutes or so is recommended as the minimum time before pouring. Remember, though, that further delay is not recommended because of the effects of continued reaction.

●12.2 Setting shrinkage

As has been mentioned, the process of joining up ends in the formation of polymer chains and cross-links reduces the associated free volume and thus show what may generally be called reaction or setting shrinkage (5§2.4). All flexible impression materials show this effect, although there is some variation between them (Table 12.1), depending in part on the number of bonds being formed. The point of this is that it tends to make the model oversize on *external* dimensions (Fig. 9.4), although cavities, diastemata, embrasures and similar *internal* dimensions will be undersized as a result. The use of two-stage impression techniques, such as putty-wash, must therefore offer considerable improvement in this regard, as the actual shrinkage of the thin wash layer will be proportionately negligible. Similar reaction shrinkage effects are also present in agar and alginate materials: the reduction in free volume occurs whether there is coordination or hydrogen bonding. However, two-stage impression are not attempted with these as primary impression materials, and the distortions (§9.1) remain uncompensated.

As an aside, it might be noted that two-stage impressions are made in a variety of other ways for essentially the same reason: reduction of errors due to the imperfection of the first step. Thus, zinc oxide-eugenol (9§2), plaster or alginate may be used with impressions made in 'impression compound' (16§5); wax may be used to effect an adjustment to a partial denture. Indeed, the use of "special trays" may be seen as a similar means of achieving higher accuracy, and thus the employment of impression compound may be viewed as just a version of that technique. Another approach involves the simultaneous use of two viscosities of material: syringing the low viscosity material around the teeth while loading the tray with the higher viscosity material. This is supposed to reduce the pressures involved, and thus deformations, and avoid cutting relief channels in a first-stage impression. Whatever approach is taken, it is recognition of the factors operating, the properties of the materials used, and an intelligent control of the processes that will lead to a high-quality outcome.[26]

Table 12.1 *Setting shrinkage of impression materials at 24 h.*

	%
Polysulphide	0.7
Condensation silicone	0.5
Polyether	0.1
Addition silicone	0.04

Some care with terminology is also appropriate here. Since polymerization is strictly the process of joining units to form new chains, the more general terms of reaction or setting shrinkage are preferable. Thus, while polyether involves a true polymerization reaction, only part of the polysulphide setting process can be described as such, and the setting of silicone and polysaccharide materials involves only cross-linking.

●12.3 Thermal expansion

Elastomeric impression materials are also subject, of course, to thermal expansion and contraction. Typical values for the linear coefficient lie in the range $150 \sim 300$ MK^{-1} (Fig. 3§4.14), rather more than is expected for other types of polymer because of the necessary condition that the glass transition temperature for the elastomers must be well below room temperature (see Figs 3§4.1, 3§4.13), *i.e.* an elastomer is rubbery (and therefore more liquid-like), not glassy. However, this is affected by the filler content in the normal way (6§2.1). For example, for typical addition silicone products, the value steadily decreases in progressing from 'light', through 'medium' and 'heavy' body to 'putty' materials, reflecting the steadily increasing volume fraction of filler. What this means in practice is that on removing an elastomeric impression, and cooling from say 35 to 20 °C, the impression material must shrink between about 0.2 and 0.3%. This gives the same effect as setting shrinkage (Fig. 9.4) in terms of the dimensional errors introduced and of course these two will be cumulative – no compensation is involved. While the material is warming, whether from the heat of reaction or from contact with the mouth, and it is still fluid, there will be no adverse effects. However, as soon as appreciable elasticity develops, distortion must develop as the expansion will be constrained.

For comparison, the (equivalent linear) thermal expansion coefficient for water is about 100 MK^{-1} over the same temperature range. Therefore, since this is the major component of alginate and agar impression materials, this figure will dominate the thermal expansion behaviour of these materials, giving similar values overall to the elastomers.

The immediate implication of this, and of reaction shrinkage in general, is that the thickness of material in the tray should be minimized. This, unfortunately, is in direct conflict with the recommendation for avoiding stress-set effects (§11.7). Again, we are faced with a compromise because of unavoidable sources of error. The main means of avoiding problems are: (a) follow instructions accurately; (b) be consistent in the procedures and processes used. This way, any technique that allows compensation for scaling errors of these kinds at the point of final fabrication (say, mould expansion for casting, Chap. 17) will be effective once calibrated. Needless to say, distortions arising from any of these sources of error cannot be compensated, as was discussed above (§9.1).

It is also worth nothing that the tray material also has an appreciable thermal expansion coefficient. Thus, if the tray were to be at the same temperature as the impression material when that was set, with the same value of the coefficient, cooling would compensate for the impression material's shrinkage by reducing the enclosed volume (*cf.* 18§4.6). However, it is likely that tray will not achieve the same temperature (with probably an appreciable gradient from tooth surface to tray wall), and coefficient-matching is not feasible (especially, metal trays will be very different – Fig. 3§4.14). The net effect is complicated. In any case, setting shrinkage cannot be compensated thereby. It should be borne in mind, however, that the width of the arch will decrease because of tray cooling when a full arch impression is taken if the tray warms at all.

●12.4 Tray retention

A tray is used for impressions in order to contain the initial fluid material as well as to provide mechanical support to the impression during subsequently handling, and model pouring in particular, to avoid distortion of the mould (*e.g.* due to the hydrostatic pressure of the dense hemihydrate slurry). However, this presupposes that the impression material is attached to the tray itself in order to resist distortion during removal from the mouth, when large forces are often involved. This attachment may be done by using perforated trays or 'tray adhesive', but it is important to ensure that this retention is achieved everywhere. Gross detachment may be obvious, but a partial failure will result in significant distortion (Fig. 12.1) that may not be readily detectable by eye. It therefore follows that the retentive features of the tray must be small and closely-spaced, or the adhesive applied over the entire surface – localized, macroscopic features or spots of adhesive are unsatisfactory.

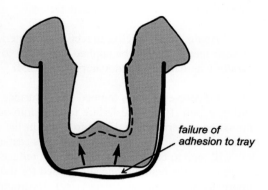

Fig. 12.1 Failure to achieve perfect retention of the impression in the tray will result in distortion.

*failure of
adhesion to tray*

●12.5 Mixing

It may seem obvious that proper mixing for any material is important, but the effects of failing to achieve it with impression materials may not be so plain. For hand-mixed two-paste as well as (formerly) paste and liquid systems, as has been used for several elastomeric products, strong contrasting colours are used in order to readily detect unmixed or poorly-mixed streaks by eye (but see 15§7.2). The material in such streaks will not cross-link properly, if they set at all, and so not be as stiff as adjacent material. However, more to the point, these regions will also show much more flow under any shearing forces arising during removal. This must exaggerate stress set distortion, and in a very irregular manner. The presence of any visible streak on the surface certainly indicates incorrect mixing, but also a great risk of an unacceptable impression.

The uniformity of dispersal of the filler is another critical point. If there are streaks of material with lower filler content there will be shear between these and adjacent more highly filled volumes. There will be relative strain effects, similar to the situation of Fig. 3.1, but on a macroscopic scale. This will contribute to distortion. Thus, it is always essential to achieve thorough mixing of components in the preparation of impression materials to eliminate variation in filler content which may arise through differences in manufacture, settling or other effects of storage, or even actual differences between components' particles themselves. Examining sections of set materials is a very good way of checking how well mixing was done.

This need for, and the recognized difficulty of attaining by hand, thoroughly uniform mixtures is the motivation for the design of disposable self-mixing nozzles (15§7.2). Such devices offer consistent mixing at a speed impossible to match by hand. This presumably offsets the cost of the device and material wastage that it necessarily causes. On the same basis as the effects of unintentional filler loading inhomogeneity, distortion might be expected across the interface of multilayer impressions taken with materials of different filler contents.

●12.6 Bubbles and voids

The instructions for the use of any impression material include the use of a mixing technique that minimizes the incorporation of air bubbles, and for the tray to be loaded in such a way that no voids (*i.e.* air spaces) are included. Both aspects require careful technique and attention to detail. The reason for this is again the avoidance of distortion, but in two separate ways.

Firstly, air bubbles and larger spaces are easily compressible during the seating of the tray, and that compression will only slowly be released because of the viscosity of the material. That release may not be completed in the time before the material sets. However, when the impression is removed, part of the constraint on the compressed air is removed, and so the impression may be distorted on the surface overlying the air space. This may take the form of a visible blister, or a more subtle and undetectable bulge or distortion. Either way, it could ruin the impression. This effect is more likely to occur with putty-wash systems where the escape of the low viscosity material through narrow spaces is difficult.

Secondly, there can be no stress transfer across an airspace. This means that shear and tensile stresses generated during removal from the mouth will result in much greater local displacements. These mean larger stress set effects, and thus greater distortion in the impression. In fact, the deformation of the airspace during the flow of the material as it is seated may lead to more crack-like spaces, with sharp (narrow radius of curvature) edges. These would act as stress-concentrators during removal, and may result in some tearing internally, again contributing to distortion, again undetectably.

Again, it is instructive to section set materials to see how well the aims of avoidance of bubbles and voids has been met. This can be applied from time to time as a **quality control** check on the mixing skill of the operator where hand-mixed materials are used.

Of course, bubbles trapped on the surface of dental tissues under an impression material, due to lack of wetting (Chap. 10) or the failure to flow into surface detail, are also problematic. The use of hydrophilic impression materials is unlikely to assist in the first context (10§9.3), but the technique of syringing some material into critical sites before the tray is seated (even with the putty-wash method) is valuable because the flow will tend to carry away bubbles and reduce the risk of boundary layer separation (23§10.2). Likewise, ensuring that the hemihydrate slurry used for the model wets the impression and displaces all bubbles is also important (with vibration as necessary). Hydrophilic impression materials are partially beneficial, but it cannot be assumed that their use will compensate for poor technique or carelessness.

●12.7 Evaporation

It is worth remembering that where there are volatile components in a material, evaporation must lead to shrinkage. This is most obvious with agar and alginate materials, but as has already been mentioned it is also an issue with polysulphide because water is the by-product of the condensation reaction; and with condensation silicones where an alcohol is produced. The result will be similar to reaction shrinkage in effect (§12.2, Fig. 9.4): concave spaces becoming oversize, convex impressions becoming too small.

●12.8 Soft tissue

There is a debate in the area of prosthetic dentistry as to whether and when a so-called mucodisplacive or mucostatic impression technique is most appropriate, indeed whether the latter is even possible. However, without entering that particular argument, it should be noted that soft tissues, and the mucosa in particular, are viscoelastic. That is to say, there is a time-dependent, retarded deformation under the forces exerted by the act of taking an impression (if there is flow of the impression material there must be a pressure gradient), although it must be presumed that the Newtonian flow is negligible. This also applies to the recovery of the displacements occasioned by the wearing of a denture. What this last point means is that if a mucostatic impression is attempted for a patient who is already wearing a denture, then the tissues must first recover from that wearing. When such an impression is planned, recommendations not to wear the denture for some hours beforehand may be made.

However, there is another implication which may not be so readily recognized. The mere act of taking a mucodisplacive impression (whether intended or not), results in a change in that tissue that will require a similar time to recover. A soon-repeated impression will therefore be different from the first, and the third from the second, noting that tissue displacements will then tend to be at least partially cumulative under such repeated and relatively lengthy loading. Care must be taken that one is not caught unawares by such behaviour when accuracy of contour is required, as is assumed to be the case.

●12.9 Hydrostatic distortion

Brief reference was made earlier to the effects of fillers in respect of self-weight distortion (§3). Ordinarily, much of this is prevented by the impression tray giving support. However, the hydrostatic head of the dense hemihydrate slurry when the model is poured would have even greater effects. This would be of importance in several respects. Clearly, material that was overhanging the impression tray would be most vulnerable (even if this was not supposed to happen, it probably often does), but if the slurry were poured unevenly into the tooth spaces, the thin material representing the embrasure – thinning to nothing at the contact point – must be displaced. In fact, it is probably impossible to avoid such distortion, one can only minimize it by paying attention and taking sufficient care. Thirdly, were there to be a subsurface bubble the surface would be depressed slightly, creating a slight bulge on the model. This might make removing a wax inlay or crown pattern difficult (causing it to distort), or affect the clasp or rest of a removable device.

●12.10 Constraint distortion

Setting shrinkage, thermal contraction and evaporative loss lead to dimensional changes, but they also lead to distortion for a separate reason: tray constraint. That is, the mechanical or adhesive attachment of the impression material to the tray provides a lateral dimensional constraint no matter what process is involved. Parts more distant from that surface will shrink towards it, but while there is also shrinkage in the transverse direction on that surface, this cannot occur adjacent to the tray wall. The dimensional change is necessarily anisotropic, and certainly not homogeneous.

●12.11 Diffusional inhomogeneity

Any effects that depend on the absorptive gain or evaporative loss of water, alcohol, and so on, will not – cannot – occur uniformly: diffusion within the material is much slower than its gain or loss from the surface, equilibration will not be achieved unless all such changes are complete. Such expansion or shrinkage that occurs is thus more rapid in thin sections and those with large exposed surface areas, leading to complicated stresses and strange distortions. Obviously, no losses can occur at the interface with the tray itself. These are not simple scaling errors that might be compensated by the casting investment, for example. However, if the time for such changes is kept short, only a thin layer of material may be affected. While the relative dimensional change for the process may be large, the absolute amount is limited by the thinness of the affected zone. In effect, then, the effects may be insignificant under a regime of 'best practice'.

●12.12 Tray distortion

Some properties of the tray itself are also critical to the outcome. That is, the design of the tray (wall shape and thickness) and the material used. Under forces from the dentist's fingers, the tray may distort. This may take the form of side-wall spreading (especially if a putty-type of material is present), and torsion and warpage of the channel – such C-channel devices are not in general stiff in torsion. Metal is thus preferable to plastics, and thick walls rather than thin, while lack of direct contact with teeth and low force are required. Thus, errors both locally (size and shape of a tooth preparation) and remotely (across the arch) may ensue when the distorting forces are lost on removal of the impression from the mouth and elastic recovery of the shape of the tray occurs and the impression material is thereby deformed. Warping of this kind (mostly tooth axis rotation, especially distally, but also shape and dimensions) will not be detectable until the final device is tried in, and may be especially problematic for precision bridges and implant superstructures.

§13. Mould Seal

Although not an impression material, the fact that this type of material is commonly an aqueous solution of sodium alginate makes it convenient to consider it here.

When PMMA denture bases are made in a gypsum mould, it is important to reduce the exposure of the reacting polymer to the water present in the mould material (which would make the denture base material somewhat opaque) and to prevent the monomer from soaking into the mould surface before polymerization, which would bond the mould to the denture base, making recovery and clean-up difficult. Pure tin foil used to be used for this purpose, its ductility permitting it to be burnished smoothly onto the surface of the mould. It was, of course, impermeable. However, this is a time-consuming and painstaking process not without its own difficulties. Thus, the foil thickness might affect the fit by changing the mould volume, the margins adjacent to teeth are critical, and some problems can arise in removing the foil from the PMMA after processing (this sometimes used to be done by rubbing with mercury! 14§1.2, 28§6).

The application of a film of sodium alginate solution to the mould by brushing achieves a similar effect rather more rapidly. As above (§9), the alginate reacts with calcium ions from the gypsum mould, immediately precipitating a membrane-like film of calcium alginate, thus blocking the porous surface and preventing monomer ingress, thereby making the removal of the gypsum less problematic. The membrane is only semi-permeable, reducing (but not eliminating) the effect of water on the acrylic (or indeed diffusion of monomer, 5§2.7). The stability of the coordination structure of the calcium alginate at 100°C is therefore crucial to this application.

A similar effect was obtained with a concentrated aqueous solution of sodium silicate (so-called **water glass**) (see Fig. 9§7.1) or even sodium oleate or stearate (*i.e.* soap), since the corresponding calcium salts have low solubility. The soap reaction may be familiar as the effect seen when washing in 'hard' water, when the dissolved calcium produces a scum. A borax solution has also been used for a similar effect where the precipitated calcium borate tends to block the porosity (2§7.4). All of these reactions are distinct from the barrier films formed by the evaporation of solvent (2§12.1, 9§1.2, 9§11).

References

[1]   Burke M. A stretch of the imagination. New Scientist No. 2085, p. 36, 7 June 1997.

[2]   Kunal K, Paluch M, Roland CM, Puskas JE, Chen Y & Sokolov AP. Polyisobutylene: a most unusual polymer. J Polym Sci: Part B: Polym Phys 46: 1390–1399, 2008.

[3]   Schultz J. Polymer Materials Science. Prentice-Hall, New Jersey, 1974.

[4]   Darvell BW. Aspects of the chemistry of polysulphide impression material. Austral Dent J 32: 357-367, 1987.

[5]   Tobolsky AV. Structure and Properties of Polymers. Wiley, New York, 1960.

[6]   Braden M, Causton B & Clarke RL. A polyether impression rubber. J Dent Res 51 (4) : 889 - 899, 1972.

[7]   Nally FF & Storrs J. Hypersensitivity to a dental impression material. Brit Dent J 134: 244, 1973.

[8]   Braden M. Elastomers for Dental Use. in: Williams DF. Concise Encyclopedia of Medical and Dental Materials. Pergamon, Oxford, 1990.

[9]   Faglioni F, Blanco M, Goddard WA & Saunders D. Heterogeneous inhibition of homogeneous reactions: Karstedt catalyzed hydrosilylation. J Phys Chem B 106: 1714-1721, 2002.

[10]   Stein J, Lewis LN, Gao Y & Scott RA. In situ determination of the active catalyst in hydrosilylation reactions using highly reactive Pt(0) catalyst precursors. J Amer Chem Soc 121: 3693-3703, 1999.

[11]   Speier JL, Webster JA & Barnes GH. The addition of silicon hydrides to olefinic double bonds. Part II. The use of Group VIII metal catalysts. J Amer Chem Soc 79 (4): 974—979, 1957.

[12]   Lewis LN. On the mechanism of metal colloid catalyzed hydrosilylation: proposed explanations for electronic effects and oxygen cocatalysis. J Amer Chem Soc 112: 5998-6004, 1990.

[13]   Darvell BW. Problems with addition-cured silicone putty. Brit Dent J 161 (5) : 160, 1986.

[14]   Causton BE, Burke FJT, Wilson NHF. Implications of the presence of dithiocarbamate in latex gloves. Dent Mater 9 (3) 209-213, 1993

[15]   Noonan JE, Goldfogel MH & Lambert RL. Inhibited set of the surface of addition silicones in contact with rubber dam. Oper Dent 2: 46-48, 1986.

[16]   Moon MG, Jarrett TA, Morlen RA & Fallo GJ. The effect of various base/core materials on the setting of a polyvinyl siloxane impression material. J Pros Dent 76: 608 - 612, 1996.

[17]   Al-Sowygyh ZH. The effect of various interim fixed prosthodontic materials on the polymerization of elastomeric impression materials. J Pros Dent 112: 176 - 181, 2014.

[18]   de Camargo LM, Chee WWL & Donovan TE. Inhibition of polymerization of polyvinyl siloxanes by medicaments used on gingival retraction cords. J Pros Dent 70: 114 - 117, 1993.

[19]   Machado CEP & Guedes CG. Effects of sulfur-based hemostatic agents and gingival retraction cords handled with latex gloves on the polymerization of polyvinyl siloxane impression materials. J Appl Oral Sci 19: 628 - 633, 2011.

[20]   Glasstone S & Lewis D. Elements of Physical Chemistry, 2nd ed. Macmillan, London, 1960.

[21]   Percival E & McDowell RH. Chemistry and Enzymology of Marine Algal Polysaccharides. Academic, London, 1974.

[22]   Coultate TP. Food - The Chemistry of its Components, 2nd ed. Royal Society of Chemistry, London, 1989.

[23]   Vincent JFV. Structural Biomaterials. Macmillan, London, 1982.

[24]   Braden M. Impression materials. in: von Fraunhofer JA. Scientific Aspects of Dental Materials. Butterworths, London, 1975.

[25]   Swartz ML, Norman RD, Gilmore HW & Philips RW. Studies on syneresis and imbibition in reversible hydrocolloid. J Dent Res 36: 472 - 478, 1957.

[26]   Wassell RW, Barker D & Walls AWG. Crowns and other extra-coronal restorations: Impression materials and technique. Brit Dent J 192: 679 - 690, 2002.

Chapter 8   Composition and Phase Diagrams

Pure materials are very rarely used in dentistry because they hardly ever offer the properties required. Mixtures of two or more substances are therefore generally encountered. It is necessary therefore to be able to relate properties to composition in terms of the mixtures. This would lead to difficulties of data management if only the enumeration of all relevant properties for all possible mixtures of all relevant substances – at all relevant temperatures – were attempted. Means must be found to reduce the scale of the problem and facilitate exploration and comprehension of the effects of the mixtures.

*The first concern is to make a **map** of the range of compositions available for a given set of ingredients for direct visual appraisal of a system. This can be done on an elementary geometrical basis on a line for two ingredients and in a triangle for three. The meanings of terms such as **state**, **phase** and **component** must be established.*

*Whenever a reaction is involved, it is necessary to enquire what the equilibrium condition is towards which the system tends to move, that is, we enquire about its **thermodynamics**. This is also true when we deal with systems such as alloys, where chemical reactions in the normal sense do not occur, and for changes of state as in melting and freezing.*

*There are two separate ways of describing the conditions which hold at equilibrium. The first is the thermodynamic, in the sense of no further reduction of overall energy being possible, reference being made to the **chemical potential** of each component. The second is a direct consequence of that thermodynamic statement, that is, the actual compositions of each phase and the environmental variables such temperature and pressure which are needed to make a complete observational **physical description**. This leads directly to the **Phase Rule** and the ability to read and interpret **constitutional diagrams**, to make predictions of behaviour which would arise from changes of composition, conditions or processing, and to understand the properties and reactions of systems as diverse as dental cements, amalgam, cermets and casting alloys.*

Materials Science for Dentistry
https://doi.org/10.1016/B978-0-08-101035-8.50008-0

It is only rarely that pure materials, a single element or compound, will be of interest in practical applications for engineering and related purposes such as in dentistry, where only pure gold (19§1.1, 28§4), platinum (25§4.7, 25§6.3), silver (or nearly so) (11§5.1, 28§9), tin (7§13) and titanium (28§1) have specific uses in that form. Much more often than not, the desired properties cannot be found in a pure substance, and mixtures of one kind or another must be used. This leads to the question of organizing the information on the properties of all investigated mixtures. Certainly, it is feasible merely to tabulate data for many mixtures, but the sheer volume of data needed for practical work would become overwhelming. It would be extremely difficult, if not impossible, to digest such a mass of data. More importantly, it is usually impossible to extrapolate or interpolate such data as it stands to find values for unlisted compositions or conditions, and generally obtain an appreciation of what is going on in a system. We are led inevitably to seeking graphical methods of understanding and communicating composition-dependent data, and in particular to graphical methods for handling the thermodynamic data of the equilibria and equilibrium descriptions of multicomponent, multiphase systems. The concise manner in which such data is then presented makes phase diagrams indispensable tools in such studies and, as will be shown in the next chapter, capable of providing new insights otherwise inaccessible.

> In discussing 'pure' substances we immediately run into a difficulty. The practical examples given here are of 'relatively pure' or 'high purity' materials. That is, they will always contain contaminants of one kind or another. Reading the labels of 'ultra-high purity' analytical chemicals is salutary. Obtaining absolutely pure substances is a matter of very great difficulty and expense, if it is at all possible, and often restricted to just a small number of atoms of a single isotope. This has, however, no effect on the theoretical aspects of composition and phase diagrams, and in practice all that is implied is that (probably) all other substances or phases mentioned are likewise contaminated. This may affect some physical properties, and even phase field boundaries, but the effects are (hoped to be) slight because the contamination is low. Examples of such contaminants will be mentioned from time to time (*e.g.* for amalgam alloy [14§2.4] and titanium [28§1]).

§1. Plotting Composition

In displaying data graphically in terms of the composition of a mixture, whether expressed by mass, volume or molecular proportions, the key problem is representing that composition. It may appear to be self-evident, but there are implications that require some consideration and it is worth being systematic. Obviously, a pure substance can only be represented as a single point (Fig. 1.1) as it is defined to have no possible variation. Also, any property can only have a single value under given conditions and so needs no graphical representation.

●1.1   Two components

If we now consider two different substances, we have to be able to represent all possible mixtures. This is done on a **composition line** (Fig. 1.2). Between two points representing the pure substances (as in Fig. 1.1) a line is drawn and divided, the divisions being taken as representing the proportions of the two amounts of substance, measured in any convenient units, such as mass, volume, or molarity, for example. Thus, the amount of A is determined by the relative distance *towards* the 100% A point, that is, measured *from* the 100% B point, where by definition there is 0% A. The amount of B is determined in a complementary fashion (Fig. 1.3) since necessarily

$$[A] + [B] = 1 \qquad\qquad (1.1)$$

A ☉

Fig. 1.1 The diagram required to represent the composition of a pure substance is no more than a single point.

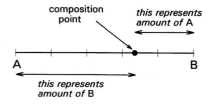

Fig. 1.2 The lengths of the line segments cut off by the composition point represent the proportions of the components in the mixture – measured from the *opposite* end.

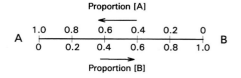

Fig. 1.3 For a mixture of two substances a composition line can be drawn to represent all possible mixtures.

The composition point is therefore a kind of weighted arithmetic mean (*cf.* equation 6§2.13), where the 'weights' of the components could be hung at their respective end points and the system would balance about the composition point as a fulcrum. (This is the **barycentric calculus**.[1]) It now becomes feasible to represent variation in some property, such as strength or density, as ordinates raised on this composition line as abscissa (Fig. 1.4). This then is an extension of a familiar kind of graph (*e.g.* Fig. 2§8.3) and has already been used to express the behaviour of composite structures (*e.g.* Fig. 6§2.10).

●1.2   Three components

Depicting the composition of a mixture of three components is only slightly more difficult. The sum of the perpendicular distances of a point within an equilateral triangle to the three sides is a constant. That this is so is readily shown by **Viviani's Theorem**.[2]

ABC is an equilateral triangle where h is the height, and s is the side length. P is any point inside the triangle, and a, b, c are the (perpendicular) distances of point P from the sides. Then, since

$$\Delta ABC \;=\; \Delta ABP + \Delta ACP + \Delta BCP \qquad (1.2)$$

The areas are:

$$\frac{sh}{2} \;=\; \frac{sa}{2} + \frac{sb}{2} + \frac{sc}{2} \qquad (1.3)$$

Therefore, dividing through by $s/2$:

$$h \;=\; a+b+c \qquad (1.4)$$

If, then, the height of the triangle is taken as unity or 100%, and the vertices of the triangle represent the pure components, the distances from the opposite sides may represent the proportions of these components in any mixture (Figs 1.6, 1.7). A grid may be drawn in such a triangle to facilitate reading the composition represented by a point. From this it can be seen that each side of the

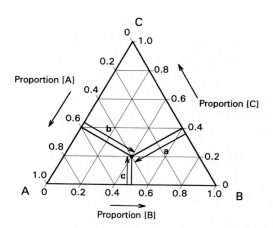

Fig. 1.7 ... for example, [A] = 40%, [B] = 40%, and [C] = 20%.

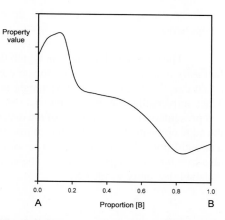

Fig. 1.4 Property variation with composition can be expressed simply with a graph raised on a composition line.

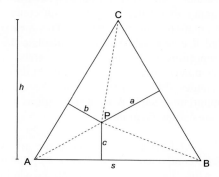

Fig. 1.5 The sum of the perpendicular distances of a point from the sides of an equilateral triangle is constant.

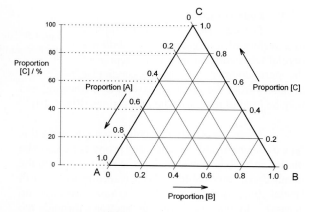

Fig. 1.6 The overall composition of a mixture of three substances can be mapped onto a composition triangle ...

triangle is in fact a composition line for the components represented at its ends. In other words, the edges of the triangle represent all compositions in which the amount of the third component is zero. The relationship of the proportion scale on the triangle sides to the height may be seen from the extra scale shown on the left in Fig. 1.6.

Similarly, 'height' scales could be drawn for the other two components.

The equilateral triangle on which the above is based is in fact only a special case, chosen for its convenience. It will be noted that the actual size of the triangle is obviously irrelevant. Thus, universally, the proportion of a component in the mixture is represented as the *proportion* of the distance from any point on the edge opposite the apex for that component, to that apex, along a straight line. This is therefore true for any size triangle. It is also true for any shape of triangle: the sum of the proportions is constant, *i.e.* unity (Fig. 1.8):

$$\frac{a}{A} + \frac{b}{B} + \frac{c}{C} = 1 \tag{1.5}$$

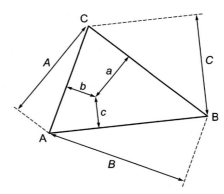

Fig. 1.8 A composition triangle can be any shape or size because the key calibration property is the proportion of the height above the base in the altitude of the triangle in the direction of the relevant apex.

This, too, has a weighting analogy in the sense of the 'barycentric calculus', as if suspending weights from the triangle apices to arrange that it lies horizontal. Of course, orientation is of no concern. This means that it is always possible to calculate the proportions of components represented at the apices at any point in an arbitrary triangle (Fig. 1.9). This becomes important when looking at sub-regions in a composition diagram.

Clearly, if any property of a three-component mixture were to be plotted as a function of composition over the whole range, the ordinate would be perpendicular to the plane of the triangle to give a three-dimensional graph like a prism (see for examples Figs 14§3.7 and 19§1.10).

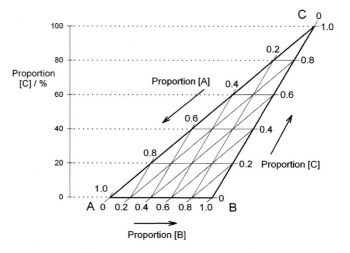

Fig. 1.9 The calibration and interpretation of a composition triangle is unaffected by either scaling or distortion of the triangle.

●1.3   Four components

The same principles can be applied to a system of four components: within a regular tetrahedron whose apices represent the pure components, the perpendicular distances from the faces again sum to a constant, and may be used to represent the proportions (Fig. 1.10).[3]

There are clearly difficulties with drawing this type of diagram, and evidently any property needs to be plotted in an axis orthogonal, *i.e.* at right angles, to the tetrahedron, impossible without resorting to a fourth spatial dimension (to which, of course, we have no access). In this case, various sections through the tetrahedron have to be used to suit each region of interest, plotting a triangular prism as before.

Mixtures of more than four components present similar but more formidable difficulties, although the formal geometry for such systems is known and straightforward (just hard to visualize).[4] Nevertheless, we have established that for the simpler systems compositions may be plotted according to some elementary geometrical rules, and this will facilitate the

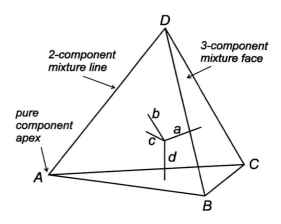

Fig. 1.10 For four substances, a tetrahedral plot is required just to plot composition.

later discussion of a number of systems of immediate relevance to dentistry.

Referring again to Fig. 1.6, it was pointed out that the edges of the triangle are composition lines of the type shown in Fig. 1.2, that is, for mixtures of two components. Similarly, the apices are really as in Fig. 1.1. Likewise, each face of the tetrahedron of Fig. 1.10 is a triangle as in Fig. 1.6, and each edge as in Fig. 1.2, and so on.

●1.4   Special cases

These diagrams also have other properties of interest. Again in Fig. 1.6, a line from an apex, say A, to cut anywhere on line BC represents a series of mixtures of constant **ratio** of B : C for varying A (Fig. 1.11). Such a line is called an **isopleth**, *i.e.* this is the locus of all points satisfying B : C = constant. Likewise, any line *parallel* to BC represents constant A and varying ratio B : C. Similar arguments apply to sections through tetrahedral composition diagrams.

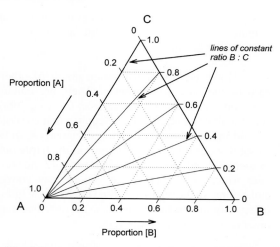

Fig. 1.11 In a three-component diagram, lines from an apex to the opposite side are lines of constant proportion, here of B:C.

Furthermore, a line drawn between *any* two points in a 3- or more component diagram is itself a composition line for the mixture of the mixtures represented by the end points. That is, one can discuss the outcome of mixing two complicated liquids, say, milk and sea water, in the same terms.

So far, all that these diagrams have achieved is a means of plotting an overall composition for a mixture, or of a property's value as a function of overall composition. There is no indication yet as to the condition of the mixture, its structure, or the compounds present. Further information is required, but it is a strength of such a diagram that this extra information can be plotted in it simply, a mere map.

§2. States and Phases

There are three **states of matter** encountered at ordinary temperatures and pressures: solid, liquid and gaseous. Gases tend to expand to fill completely the available space and have no bounding surface except by virtue of being adjacent to another type of phase. This means that two identifiably different gas phases cannot exist adjacent to each other – all gases are freely miscible (in the absence of reaction). Liquids take the shape of their container under gravity and show a sharp boundary at the free surface, *i.e.* adjacent to a gaseous phase. Sharp boundaries between immiscible liquids can also be seen, for example mercury, oil and water together. Solids have a very definite shape with sharp boundaries. Evidently, such boundaries can be maintained between solids of different kinds. Liquids and solids are known as **condensed** states, because the component atoms or molecules are closely approximated.

Judging from common experience, all three states can exist together for a given material. For example, water as the solid ice, the liquid itself and its vapour may be in contact, although the ice may be melting, or the liquid evaporating or freezing, depending on conditions. Chemically, the three are indistinguishable, but clear sharp boundaries exist between them. These boundaries represent the transition between the differing structures of the regions in the different states, even though (for a pure substance) the composition is the same in all three. No difficulty of identification occurs for changes in subdivision: raindrops are still liquid, snowflakes are still solid. Nor is there a problem of scale: the states maintain their identities down to the molecular level.

●2.1   Environmental variables

In general, for a single pure substance, we may examine the states present as a function of temperature and pressure (Fig. 2.1), when the relative amounts of material in the various states are stable for fixed conditions. Each point in such a graph, which is called a ***P-T* diagram**, represents the description of the system at **equilibrium**. The effect of any change in these **environmental variables** can be readily mapped. For example, for any given temperature, the condensation of the vapour to liquid, or its deposition as solid, as the pressure is increased can be noted.

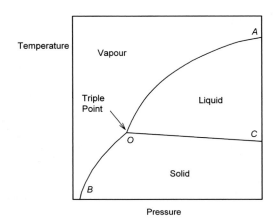

However, it is not a question simply of either one phase or the other everywhere. There are conditions where more than one phase can coexist at equilibrium. Thus, the line *OA* represents the equilibrium coexistence of liquid and vapour, *OB* of solid and vapour, *OC* of liquid and solid, while the point *O* – known as the **triple point** – represents the only conditions of temperature and pressure under which solid, liquid and vapour may coexist at equilibrium. In the regions marked out by those lines, but not actually on the lines, there can be only one state of matter represented at equilibrium.

Fig. 2.1 Water provides an example of a pressure-temperature diagram for a pure substance, a map of the equilibrium existence limits for each phase and their coexistence conditions. (Notice the negative slope of line OC: ice melts under increased pressure because of the reduction of volume that then occurs.)

●2.2 Kinetic limitation
This brings out an important point which is relevant in many circumstances in dental materials, especially where there is some chemical or physical process in operation: what may be observed may not be (indeed, usually is not) an equilibrium condition. That is to say, by equilibrium is meant the lowest energy state available to the system, the thermodynamic end-point of the process. Thus, in the 'common experience' example above, of ice, water and vapour, observation of coexistence does not necessarily imply equilibrium. Rather, kinetic limitations of diffusion, heat flow, or the availability of the activation energy for a process may limit the approach to equilibrium, or even prevent it altogether. This should always be borne in mind. Thermodynamic considerations provide the driving force for a process – the imperative, kinetics controls the rate of approach to the end-point – reality. Nevertheless, we shall nearly always approach a mixture problem first from the point of view of equilibrium, and then consider the kinetic aspects, for in order to understand the process (whether it be a setting reaction or some other behaviour) we must first understand where that process is going, then the limitations. The motivation now in this chapter is therefore the depiction of equilibrium systems: time is not a factor.

●2.3 Definition of a phase
We need now to extend these ideas to systems containing more than one substance. To reiterate: gases mix perfectly and no boundary between any pair may exist, but between liquids such boundaries readily occur for many combinations such as oil and water. Again, subdivision and scale have no effect; there are evidently two distinct regions which are liquid but chemically identifiable. Equally, we may have two solids, such as the core and matrix of a filled resin restoration, in intimate contact, yet chemically and structurally distinct. Similarly, gas-liquid, liquid-solid, gas-solid and gas-liquid-solid systems may be envisaged where in addition to the state differences there are chemical differences. There is yet one more possibility: water ice, for example, goes through a series of different crystal structures (15, at the last count) as the pressure is greatly increased and the temperature is lowered. They are chemically identical, and certainly in the same state, they may yet be distinguished on the basis of structure alone. From that point of view, ice, liquid water and water vapour are distinguished by structure alone. We can summarize the state, chemical and structure distinction of each type of bounded region by referring to it as a **phase** of the system. All regions of the same description belong to the same phase, irrespective of shape, size, subdivision or separation.

We may define a phase as

any homogeneous and physically-distinct part of a system which is separated from other parts of the system by definite boundary surfaces.

bearing in mind that 'physically-distinct' means by virtue of state or other structural description. Note also that the existence of definite boundaries implies that, in principle at least, a mixture of phases are mechanically-separable without affecting the existence or stability of those phases (if the environmental conditions are preserved). Fig. 2.1 may therefore be called a **phase equilibrium** or **constitutional diagram**. The definite boundary surface is an important requirement. In contrast, for example, composition gradients within a single phase are perfectly reasonable. Consider a beaker of water into which some sodium chloride crystals were put, but without stirring. After a while the salt will have dissolved, but the concentration at the bottom will be very high while that at the top will be low. Despite variation in refractive index, density and other properties from top to bottom, no definite boundary exists along those gradients. If one were to contrive to arrange a layer of pure water on top of a concentrated salt solution, a boundary may appear to be present, but in reality it must have a substantial thickness, and diffusion on standing will blur and eventually eliminate it. Simple stirring will show immediately the complete miscibility of the two regions, and thereby demonstrate that the apparent boundary was not a phase boundary. Indeed, such a system will not have been at equilibrium until it had been well stirred.

It is the definition of a phase which allows us to recognize when we are dealing with composite structures (6§1.13), the implications of which are so far-reaching. It is by study of the conditions that control the number of phases, which phases are present, or the reactions between them, that the properties and handling of many materials can be understood.

●2.4   Calcium sulphate – water

As another example we may consider the system H_2O - $CaSO_4$ in a **T-X diagram**, that is constitution as a function of temperature and composition (Fig. 2.2). We can identify six distinct phases between 0° and about 250°C: liquid (*i.e.* calcium sulphate solution), dihydrate, hemihydrate, hexagonal and orthorhombic anhydrite, and vapour. We would, of course, find a further phase if the temperature were lowered below the freezing point of the solution (*i.e.* to give ice). The pressure here is taken to be fixed at one atmosphere (and this will always be assumed unless stated otherwise). Each bounded region of the diagram is called a **phase field**, and the dividing lines are **phase field boundaries**. Invisible, because of the scale of the drawing, are the fields corresponding to each of the pure solids. The reason for this is that they are stoichiometric compounds and so have no *range* of composition. Even so, they occupy their own (rather narrow) phase fields. The only field to contain just one phase and to have an appreciable width in this diagram is that of the solution of calcium sulphate in water (shown enlarged in Fig. 2.2). The right hand boundary of this field corresponds to the just-saturated solution (with respect to dihydrate).

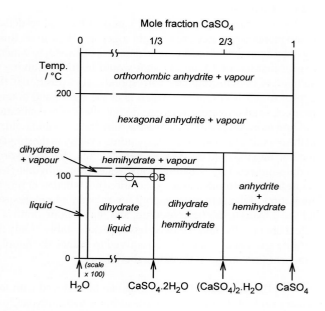

Fig. 2.2 A hypothetical* equilibrium phase diagram for the system H_2O - $CaSO_4$. Note the increased scale to the left of the break in order to show the very narrow field corresponding to the solution of calcium sulphate in water.
(* See 2§11.4 regarding the stability of the dihydrate.)

●2.5   Definition of a component

Fig. 2.2 has been constructed on a composition line corresponding to H_2O - $CaSO_4$, and these two substances quite clearly are in themselves quite sufficient for describing all possible mixtures between those limiting compositions. There is no need to consider separately the oxygen, hydrogen, sulphur and calcium because, within the substances chosen to generate the mixtures, the ratios of the elements are fixed. It is certainly possible to draw diagrams for, say, the Ca - O or H - S systems, but they are quite irrelevant to the system in hand.

We therefore consider the H_2O and $CaSO_4$ as **components** of the system. We are, however, not restricted to these substances in writing descriptions or labelling the composition line. It is perfectly feasible to use, for example, $(CaSO_4)_2.H_2O$ - H_2O or $(CaSO_4)_2.H_2O$ - $CaSO_4$ as all compositions in the range H_2O - $CaSO_4$ may be generated *algebraically* from these other components. In these cases, the composition lines might appear to be somewhat awkward, having negative values in some places. But the principle stands, and it is sometimes of great value, particularly when looking at sub-regions of a larger diagram. An example of negative values on a composition line is seen in Fig. 2§8.2, where it was simply more convenient to do so. This illustrates the thermodynamic point that it is not the substances themselves but their number which is of importance.

A **component** of a system is therefore defined as

> **one of a minimal set of independent chemical constituents by means of which the composition of every possible phase of that system can be expressed.**

Notice the phrase "a minimal set"; there is not necessarily a unique such set. Thus in Fig. 2.2 only two components are required to define (algebraically) the composition of any phase. Anything more than that would be unnecessary or redundant, that is to say, not independent. That said, it is more usual to use the 'terminal' components to express composition. However, one further point needs to be made: an **analytical** composition statement says nothing about the chemical condition of a component. For example, in solution calcium sulphate is ionized, solvated, and forms a number of chemical species that do not appear in any way in a diagram such as Fig. 2.2.

●2.6   Reactions

Diagrams such as Fig 2.2 can also be used to deduce the likely progress of the reaction of the two components when they are brought together. We can do this in the form of a thought experiment: for example by imagining that we add minute amounts of $CaSO_4$ to water at around room temperature. We would naturally expect that the first kind of event would be the dissolving of that solid in the liquid to give, very simply, a solution of calcium sulphate, corresponding to a position in the 'liquid' field. Adding more solid, we eventually expect to reach saturation (which is at the right hand boundary of the liquid field). The next addition must produce solid dihydrate since we do not allow supersaturation – which would be a non-equilibrium condition. Adding more solid still, we cross the dihydrate + liquid field, gradually using up the water until there is none left at the overall composition corresponding exactly to the dihydrate. Adding a little more $CaSO_4$ then takes us over the line into the field dihydrate + hemihydrate. In other words, anhydrite is not stable in contact with dihydrate and it must react by 'stealing' some water from dihydrate to create hemihydrate. Notice that these statements relate to the thermodynamic expectation, not whether or how fast such a process will actually occur – the kinetic limitation of the activation energy for any and all processes is not addressed in such a diagram. Finally, when all the dihydrate has been used up, and only hemihydrate is present, all further additions of anhydrite have no effect chemically, and the two phases coexist stably. Only the proportions are now affected, and obviously an infinite amount of anhydrite is required to reduce the hemihydrate concentration to zero; there are sometimes practical limitations in thought experiments.

The direction of the addition can be reversed without affecting the logic of the process, *i.e.* addition of water to anhydrite, in respect of the ultimate thermodynamic endpoint. However, we sometimes need to enquire about the sequence of reactions when the components are brought together in bulk, which is the way things are usually done in practice. It seems to be a general rule that the first reaction product phase formed is the one (isothermally) closest to the fastest-diffusing phase or component in the diagram. Thus, putting water and anhydrite together we expect dihydrate to form first, irrespective of the actual proportions. In other words, even if there is insufficient water to correspond to hemihydrate alone at equilibrium, dihydrate will still form first, reacting later (if possible) with more anhydrite to generate hemihydrate.

This kind of behaviour is expressed by **Ostwald's Rule of Stages**,[5] or of **Successive Transformation**, which notes that, over time, a thermodynamically unstable phase tends to undergo a sequence of reactions which form in turn progressively more stable phases. This may be because the formation of each **metastable** phase lowers the total free energy of the system faster than would the immediate formation of the most stable phase or phases. Alternatively, the formation of a more soluble, less stable phase is kinetically favoured because the more soluble phase has a lower interfacial energy (10§1.3) with the more-diffusible phase than a more stable, less soluble one. In essence, a solid-liquid system preferentially forms the phase with the fastest precipitation rate under the prevailing conditions.[1]

This kind of consideration is relevant in several areas of dentistry in addition to the gypsum product context illustrated above. Thus, the setting of zinc phosphate cement (9§5) and of dental amalgam (14§1) are both examples. Ostwald's Rule may also apply to systems that have undergone abrupt cooling, where intermediate phases may appear to have been by-passed on considering the endpoint, but which are involved in the transformation reactions.

---

[1]   It will be seen that the outcome depends on the balance of thermodynamic and kinetic factors: it is a guide, not a universal "law".

§3. The Phase Rule

The fundamental expression of equilibrium in a system lies in the **Phase Rule**. Through its use, we can ensure that we are indeed considering an equilibrium system, and how much scope there is for changing any aspect while maintaining exactly the same phase description or constitution (*i.e.* which phases are present, not their proportions). The development of the phase rule requires the consideration of some algebraic identities: firstly regarding what is known as the **chemical potential** of each component of the system in each phase, and then some simple accounting for the total amount of each component present. We begin by considering the **partition** of substances between phases.

●3.1 Partition law

A common and easily-visualized example to illustrate the nature of chemical potential and partitioning is that of iodine in a mixture of tetrachloromethane (carbon tetrachloride) and water. On adding a small quantity of iodine to such a mixture of solvents, at a given temperature, some will dissolve in each layer, giving at equilibrium (speeded up by giving it a good shake) concentrations c_1 and c_2, say, respectively (Fig. 3.1). If now a little more iodine is added, and again equilibrated, the concentration in each layer will have increased, but in such a way as to keep the *ratio c_1:c_2* the same as before. This may be repeated over a wide range of total iodine added, and the ratio will be maintained constant. Alternatively, some more water could be added to the mixture, which would therefore dilute the aqueous iodine solution. Iodine will then move (diffuse) from the tetrachloromethane phase to the aqueous phase until the ratio of concentrations is again back to the same value. This is an illustration of the so-called **Partition Law** which describes the distribution of the solute between the two phases.

●3.2 Chemical potential

Now, the chemical potential μ for the solute, the iodine in the present example, in each phase is represented by an equation of the form

$$\mu = \mu^0 + RT.\log_e a \qquad (3.1)$$

where R is the molar gas constant, T is the absolute temperature, a is the **activity** of the solute iodine and μ^0 is the chemical potential in the so-called standard state (the reference value under defined conditions) for the solute in the phase being considered. But, for equilibrium to exist, μ_1 and μ_2 for the iodine in the two phases must be identical everywhere, *i.e.* no driving force for change:

$$\mu_1 - \mu_2 = 0 = \mu^0_1 - \mu^0_2 + RT(\log_e a_1 - \log_e a_2) \qquad (3.2)$$

Therefore, since the two standard state chemical potentials μ^0_1, μ^0_2 are each by definition constant, so a_1/a_2 must also be constant at a given temperature, since:

$$\log_e a_1 - \log_e a_2 = \log_e(a_1/a_2) \qquad (3.3)$$

There can also be a third phase present, in this example it is in the space above the liquid containing the vapours of the two solvents. The iodine too has an appreciable vapour pressure at ordinary temperatures. So at equilibrium an amount of iodine will have evaporated from the solutions into the closed space so as to give the corresponding equilibrium partial pressure. Should the amount of iodine in the solvent system be increased, then automatically the vapour pressure of the iodine will be increased in proportion so as to maintain the constancy of the

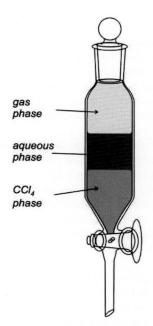

gas phase
aqueous phase
CCl_4 phase

Fig. 3.1 Equilibration must attain identical chemical potentials for all mixture components in all phases. This is iodine in water and tetrachloromethane.

In ideal systems or, for practical purposes, systems which do not depart appreciably from ideality, for activity we can read 'mole fraction', which for dilute solutions is nearly the same as concentration. In systems where there is a strong chemical interaction between components, their activity is decreased; where there is a chemical incompatibility, the activity is raised. In such cases, concentration is no guide to chemical potential.

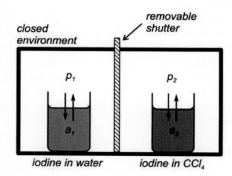

closed environment   removable shutter

p_1 p_2

a_1 a_2

iodine in water   iodine in CCl_4

Fig. 3.2 An experiment to show that a_1/a_2 must attain the same value as if the solutions were in contact.

ratio a_1/a_3, where the subscript 3 refers to the gas phase, and also – of course – of a_2/a_3 as well.

There does not need to be direct communication between all phases, so long as a path exists (and the activation energy for the diffusion), equilibration will occur. Thus, in Fig. 3.2, while the shutter is in place, the equilibrium vapour phase concentration of iodine corresponding to each solution activity will be attained eventually. But when the shutter is removed, if p_1 and p_2 are not equal, iodine will evaporate from one solution and dissolve in the other until the chemical potential is the same everywhere and the activity relations above are satisfied, *i.e.* $p_1 = p_2$, as it must. The initial mixed vapour will be wrong for both solutions: p too high for one (thus driving dissolution) and too low for the other (driving evaporation).

It can be seen from the equation 3.1 that the actual value of the chemical potential depends on the temperature of the system. Thus, if the temperature of the mixture is changed, then the individual concentrations in all three phases will also change, and therefore the activity ratios also change. In other words, the partitioning of a solute between phases ordinarily depends on temperature, but at a fixed temperature the ratios of activities are constant irrespective of concentration or, most importantly, the actual amount of any phase. One drop of tetrachloromethane in a litre of water would give the same result as one drop of water in a litre of tetrachloromethane, assuming that both phases are present.

In fact, we can discuss the dissolving of a small amount of the tetrachloromethane in the water, as well as the simultaneous dissolving of a small amount of water in the tetrachloromethane, in exactly the same terms. The activity ratio for water in the two phases must be constant, and the activity ratio for the tetrachloromethane must similarly be constant (although with a different value to that for the water), no matter what the actual amounts of the two substances are, again assuming that both phases are present. Clearly, we can go on to consider the vapour pressure of each solvent in the space over the liquids in precisely the same way, and the dissolved oxygen and nitrogen in the two liquids. The chemical potential of a given component must be everywhere the same in the defined system – at equilibrium. This statement can be made in turn for all components, although (of course) the actual value will vary between components.

These principles also apply to the partition at equilibrium of the components of metal alloys when there is more than one phase present. Thus, for example, silver and copper are only slightly soluble in each other at ordinary temperatures, and the situation is very similar to the case of tetrachloromethane and water. No matter what the relative proportions of the Ag- and Cu-rich phases, the activity ratio for the Ag is fixed, and that for the Cu is fixed, at a given temperature, *i.e.* the chemical potential μ[Ag] is the same in each, and μ[Cu] is the same in each. Now, if to this system we add a small amount of gold, which dissolves in both phases, it will distribute itself between the two phases according to the Partition Law such that its chemical potential, μ[Au], in the two phases is identical. Again, these values of the chemical potentials change with temperature, but the corresponding activity ratios are fixed for any given temperature, and therefore the identity of the chemical potential for any one component can be guaranteed at all points in a system at equilibrium.

The requirement for the identity of chemical potential for a component in each phase is maintained even if reactions occur in one or more of those phases. Thus, if the iodide ion is present in the aqueous phase of the model system (Figs 3.1, 3.2), the total amount of iodine as I_2 molecules over all three phases must be reduced as the reaction to form the tri-iodide ion 'consumes' some. This reaction, too, involves an equilibrium (as indeed do most reactions, no matter how far to the right they lie). Thus, while the reacted iodine is effectively put into another "compartment", the activities of the elemental iodine in the three phases must remain in the correct ratios if equilibrium is attained whilst simultaneous satisfying the equilibrium constant for the distribution of iodine across species in the aqueous phase (Fig. 3.3).

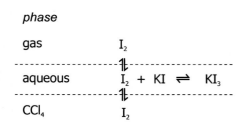

Fig. 3.3   Simultaneous equilibrium will be thermodynamically-required, even if side-reactions occur in addition to simple dissolution.

●3.3   Equilibrium

At equilibrium there can be no change with time in the amounts of each phase present, nor any driving force for this to happen. If the temperature is high enough for diffusion to occur, the equilibrium will be dynamic, with equal amounts of any component entering and leaving a given phase in any time interval; we say that the **net flux**[2] across a phase boundary is zero. A simple example would be of water and its vapour in a sealed tube at a fixed, uniform temperature: molecules escape from the liquid continually, but just as often condense into it, leaving the amount (and therefore the pressure of) the vapour constant. Indeed, this kind of system provides the definition of **vapour pressure**. Alternatively, we may also say that the chemical potential, μ, of the water is the same in both phases, or indeed that the activity of the water is the same in both phases.

Another view of activity or chemical potential is to describe it as the **escaping tendency** of a species: when this takes the same value on either side of a boundary, there must be balance in the numbers of particles crossing that boundary in each direction. Such a statement is therefore both a necessary condition and a sufficient definition of equilibrium in this thermodynamic sense. Again, as noted above, if a component is present in three phases, its chemical potential must be the same in each. Such a statement must therefore also be true for each component of a system (although, again, the chemical potentials of each component will not, in general, be equal).

> The phrase "in general" is commonly used in discussing the way systems behave, and it is worth recognizing its implications. The interpretation is that for the great majority of circumstances or combinations of conditions, the statement is true. However, it is possible that some special combinations can be found where the statement is not true. Thus, "in general" means "usually", but not "always". Thus, here it is conceivable that the chemical potentials of each component are identical, but it would be a rather unusual circumstance. Likewise, in Fig. 2.1, one could state that, in general, only one phase is present at equilibrium, although particular precise combinations of temperature and pressure lead to two phases being stable, and one special combination results in three. This does not diminish the generality.

Thus, extending this idea to each of C components, distributed over P phases,[3] we can write out the C sets of equalities expressing the principle that (for a given fixed temperature, pressure, *etc.*):

**at equilibrium, for each component separately,
the chemical potential has the same value in every phase.**

If this were not so, that component would tend to migrate from one phase to another until equilibrium was obtained. This implies that the free energy of the whole system is being minimized, and so this represents thermodynamic predetermination of phase compositions. Thus, we need to ask how many equalities need to be established in order that we can know for certain the chemical potential of every one of C components in every one of P phases, that is, that equilibrium exists. We answer this by explicitly writing out these equalities:

	Phases				Number of Independent Equations
	a	b	...	P	
Components					
1	μ_{1a} =	μ_{1b} =	... =	μ_{1P}	P - 1
2	μ_{2a} =	μ_{2b} =	... =	μ_{2P}	P - 1
$\vdots$	$\vdots$	$\vdots$		$\vdots$	$\vdots$
C	μ_{Ca} =	μ_{Cb} =	... =	μ_{CP}	P - 1
				Total:	C(P - 1)

$$(3.4)$$

There are P - 1 *independent* equalities in each line because it is unnecessary to write down that the first and last are equal (*i.e.* if $a=b$ and $b=c$, then $a=c$ automatically[4] [6]). There are thus C(P - 1) independent equations for chemical potential for the whole system. Since these expressions were written down directly from consideration of the condition of equilibrium and are sufficient to express that condition, these are the only thermodynamic descriptors necessary for the system. Thus, in the experiment of Fig. 3.1, the chemical potential of the iodine is the same in each of the three phases at equilibrium, as is true of the water (albeit with a different value), and the carbon tetrachloride, and the nitrogen of the air, the oxygen, and so on, for all substances present. But, in each case, we only need to know of the truth of two equalities to guarantee the third. That is, there is no point in writing $\mu_{1P} = \mu_{1a}$, for example – this is implicit.

---

[2]   **Flux** here refers to the amount of flow or transport of substance per unit time.

[3]   Some care needs to be taken with symbols, although context should make the distinction clear between pressure and number of phases (P), between a component and number of components (C), and so on.

[4]   Transitivity equivalence: Common Notion 1. Things which equal the same thing also equal one another.[6]

●**3.4   Mass balance**
A second approach to describing the equilibrium condition is to determine the mass balance since it follows that the concentrations of species in each phase will settle at definite values, as indicated by the discussion in §3.2. Concentrations can be expressed in various ways, but it is commonly most convenient to work in the chemically-meaningful terms of **mole fraction**. The sum of the mole fractions for all the components present in a given phase is by definition unity (*cf.* volume fraction, 2§2). We can, therefore, writing Φ_{ij} for the mole fraction of the *i*th component in the *j*th phase, tabulate the values phase by phase:

	Components						Number of Independent Mole Fractions
	1	2	...	C			
Phases							
a	Φ_{1a} +	Φ_{2a} +	... +	Φ_{Ca}	=	1	C - 1
b	Φ_{1b} +	Φ_{2b} +	... +	Φ_{Cb}	=	1	C - 1
:	:	:		:		:	:
P	Φ_{1P} +	Φ_{2P} +	... +	Φ_{CP}	=	1	C - 1

$$\text{Total:} \quad P(C - 1) \tag{3.5}$$

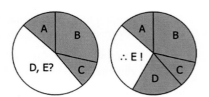

Fig. 3.4 For each phase, the composition is not fully specified until the penultimate value is given, then the last one is known automatically.

Since, for a given component, the value of one mole fraction can be deduced when C - 1 are known (Fig. 3.4), because the total must be 1, there are $P(C - 1)$ independently variable concentrations in the whole system. These are physical descriptors of the system. What this means in principle is that by experiment we could determine the composition of each phase when equilibrium had been obtained (repeated observations would show constancy of values). We may then phrase the question as follows when trying to determine whether another example of the system is at equilibrium: How many concentrations need to be measured to know absolutely that we have the same condition as before? The answer is $P(C - 1)$.

●**3.5   Environmental variables**
It is also apparent that the environment of a chemical system (including its temperature and pressure, gravitational, magnetic and electric field conditions, for example) is part of the physical description of that system, and indeed clearly is capable of affecting the thermodynamic description. The number of independently selectable **environmental variables**, E, required to fix the total physical description of the system can therefore be added to the number of mass balance descriptors. Therefore, the system is then necessarily fully *physically* described by exactly $P(C - 1) + E$ variables.

●**3.6   Degrees of freedom**
The question then is, how sufficient are the thermodynamic descriptors in determining the physical description of the system? Do the two approaches always give the same outcome unambiguously? Comparing the numbers of the physical and thermodynamic descriptors we find the following:

$$\underset{\textit{physical}}{[P(C - 1) + E]} \quad - \quad \underset{\textit{thermodynamic}}{[C(P - 1)]} \quad = \quad C - P + E \tag{3.6}$$

This means that the physical descriptors supply more pieces of information than do the thermodynamic ones, $C - P + E$ more; that is, this information is necessary for a properly-described equilibrium. In other words, the values of this number of variables must be *predetermined* or chosen in some sense by the observer in order for it to be possible that the thermodynamics yields an unambiguous and full description of every aspect of the system. This equation thus provides the definition of the number of **degrees of freedom**, F, which apply to the system under consideration:

$$F = C - P + E \tag{3.7}$$

This, then, is known as the **phase rule**. Thus, when the values of this number of variables are fixed, the description of the system at equilibrium is completely defined; that is, its **equilibrium phase description** or **constitution**.

Notice that in effect some choices have already been made: the number of components in the system, and the number of environmental variables we shall permit as capable of being varied (if you like, what axes are going to be created for our constitutional map of the system). Even so, by electing not to consider variation in gravity, magnetic or electric fields we have in effect chosen and fixed the values to what is current or normal – but they are unlikely to be zero, or fixed (and indeed on the surface of the Earth they are neither).[7][8][9][10] These are implicit assumptions on the basis that ordinarily variation in such factors is of no consequence or irrelevant to the circumstances of interest, especially if we have no choice or possibility of control. What remains then, if F > 0, is to choose some values for the remaining unknowns. These may be the concentration of one or more components in one or more phases, or the value of one or more environmental variables, or maybe a combination of these.

●3.7   Water

The implications of the phase rule are best illustrated by example, and we may take the *P-T* diagram for water to start with (Fig. 3.5). Here there is only one component, so C = 1. Then, if we determine (or choose) that only vapour, say, is present, *i.e.* there is one (named[5]) phase *p*, we have P = 1. Thus F = 1 – 1 + 2 = 2, so that of the items left unconsidered, the environmental variables, both temperature and pressure must be fixed to define the system fully. The system is then said to be **bivariant**. Clearly, as judged from this diagram (Fig. 3.5), there are temperatures and pressures where the phase description "only vapour" is not true; but since then this would change the phase description, it is beyond the range of validity of the statement as it stands. We may apply this analysis to the phase fields "only solid" and "only liquid" with the same outcome. Nevertheless, we are essentially only trying to identify a point in a two-dimensional graph. For this we need only the ordered pair of coordinates of the point (*P*, *T*) – such a graph only ever has a maximum of 2 degrees of freedom, two pieces of information to locate the point. The phase rule cannot have P = 0, of course, which leaves two things remaining to be specified after identifying *p*: the pressure and temperature.

If we choose to consider a condition where any two (named) phases *p* and *r* are required to be present, then we have F = 1 – 2 + 2 = 1. Thus, *either* temperature *or* pressure can be predetermined to define which point on the selected phase field boundary represents the now **univariant** system (Fig. 3.6). Indeed, it is impossible to choose both freely as either one value determines the other. Notice, though, (amplifying the point made for one phase in the footnote) that *which* two phases are to be considered must be stated, not just their number. Without that particular information there is an ambiguity: we do not know which boundary to look at, which means that the system is incompletely specified – the condition of two phases in contact occurs in three places. If we specify the name of just one phase, there are two places that meet the condition. We are obliged to name the second phase, using up two degrees of freedom (" – P"), to have the system

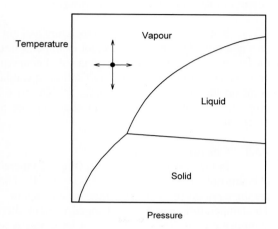

Fig. 3.5 Temperature and pressure may be adjusted independently (that is, the values chosen) without affecting the constitution of the system: it has two degrees of freedom when just one phase is present.

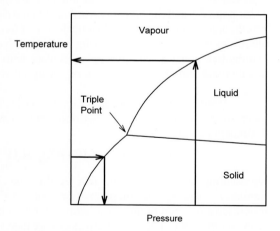

Fig. 3.6 When two phases are present in this system, either temperature or pressure may be chosen, but not both. If three phases are present, there is no choice.

fully specified by constitution. However, if three phases are to be present at equilibrium, F = 1 – 3 + 2 = 0, there is now no choice in any respect since there is only one point, the triple point, where that statement can be true.

---

[5] It should be noted that simply saying how many phases is not a sufficient specification. For F = 2, at a temperature just below the triple point in Fig. 3.5 there are three possibilities: vapour, solid and liquid, depending on the pressure. The count is right, but the system is not specified. Strictly, the algebra holds (as the thermodynamics requires) but the state of the system is not fully defined.

The system is then said to be **invariant**. We have no choice whatsoever but to accept that this description holds at only one temperature and one pressure.

In dentistry, as in much of everyday experience, we are concerned only with working at atmospheric pressure, which ordinarily varies over only a small range. Certainly, if we are concerned with the boiling points of liquids or gas expansion, we need to take it into account very carefully (*e.g.* Figs 5§2.7, 5§2.15). However, melting points, solid and liquid state chemical reactions, and metal solid solubilities are little affected by pressure over normal ranges, even at high altitude. This environmental variable can thus frequently be safely ignored, as indeed we ignore fields (gravitational, magnetic, *etc.*) which have no effect on the structures with which we are concerned in their normal range of variation, and therefore E = 1 for most purposes in the present general context. In effect, we have predetermined the values of these other variables, thus using up the corresponding degrees of freedom in advance, as it were.

Notice though that it is temperature and not heat which is counted in E. That is to say that at any point on a phase field boundary where, for example, a liquid-solid equilibrium applies such as on line *OC* in Fig. 2.1, the latent heat of the phase change is not of concern. The proportions of phases here lie anywhere in the range between an infinitesimal amount of solid to a similarly vanishingly small amount of liquid without moving the point on the equilibrium diagram. Within those limits, heat can be put in and taken out, changing these proportions continuously, without changing the phase description. This applies at any such boundary or invariant point.

●3.8   Calcium sulphate

In the system H_2O - $CaSO_4$ (Fig. 2.2) there are two components and the pressure is arbitrarily fixed at one atmosphere, so that F = 2 – P + 1 = 3 – P. Then, for example, at about 100 °C at a point such as A, there are three phases present: dihydrate, liquid, vapour. In other words, there are no remaining degrees of freedom. The temperature may not be changed without altering the constitution of the system. Notice that the overall composition can lie anywhere in a large range without affecting this outcome. This is because only the *proportions* of the *phases* will change in this kind of circumstance on changing the overall composition, with which the phase rule is not at all concerned.

Point B on the other hand might seem to be a little ambiguous. However, it must be taken to lie on one side or other of the single-phase field (that is, the vertical line is considered to have some width, however small) representing pure dihydrate. On the left there are then three phases in contact: dihydrate, liquid and vapour (just as for point A), and the point is invariant. On the right, there are only two phases in equilibrium: dihydrate and hemihydrate. There must then be one degree of freedom here so that there is a choice of temperature where this phase description is valid. It helps to notice that nowhere in this diagram (at atmospheric pressure) can hemihydrate exist at equilibrium in contact with liquid. There is a similar point for the co-existence of hemihydrate, dihydrate and vapour on the dihydrate field 'line'.

Reading such a diagram as Fig. 2.2 vertically, *i.e.* at a given overall composition, the region where a temperature-dependent reaction such as:

$$2CaSO_4 + H_2O(g) \rightleftharpoons (CaSO_4)_2 . H_2O \qquad (3.8)$$

representing an equilibrium, can readily be identified. Similarly, reading the diagram from side to side, the progressive reaction of anhydrite with water may be mapped at any temperature. Diagrams such as this are then very useful for exploring the behaviour of a system, for doing 'thought experiments' otherwise impractical, or for selecting compositions and conditions to achieve chosen objectives.

●3.9   Isothermal tie-lines

One further point may be made: if a horizontal (*i.e.* isothermal) line is drawn in any two-phase field of a composition-temperature diagram such as Fig. 2.2, the compositions of the phases present at any point along that line are represented by the overall compositions corresponding to

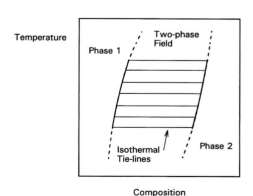

Fig. 3.7 Isothermal tie-lines drawn between the boundaries of a two-phase field. The compositions of the two phases in equilibrium are indicated by the ends of the tie-line at any temperature.

the ends of the line. Such a line, which must not cross any boundaries, and therefore has its ends lying at the boundaries of single phase fields, is known as an **isothermal tie-line** (Fig. 3.7). Such a line, since it represents a mixture of two phases, can itself be treated as a composition line as in Fig. 1.2, and can be calibrated accordingly with a 0 - 100% scale for each of the two phases concerned (Fig. 3.8). That is, treating those two phases (at their defined compositions in terms of the components of the overall system) as the components of the range of mixtures represented by the tie-line (Fig. 3.9).

●**3.10 Lever rule**

        In other words, at any overall composition (*i.e.* in terms of the components for the whole diagram) the relative proportions of the two *phases* present are in direct proportion to the distances of the point of interest from the ends of the line. That is, Figs 1.2 and 1.3 apply anywhere such a line can be drawn. This is known as the **lever rule**. Bear in mind that the measurement of the distance for a given phase is from the *other* end of the line, *i.e.* from 0% of the phase in question. The phases present at either end of a tie-line are known as **conjugate phases**.

        If a phase field boundary for the two-phase region is not exactly vertical, it implies that the composition of the phase concerned varies with temperature. The composition can be read off by dropping a vertical to the system composition line for each temperature of interest from the point where the tie-line meets the phase field boundary. The overall composition of the mixture can lie anywhere between those extremes without limitation. Thus in Fig. 3.8, phase 1 has the composition indicated by *A*, phase 2 the composition corresponding to *B*, while the overall composition of the mixture may be given by any point in between such as *C* (Fig. 3.9).

        As far as the phase rule is concerned in such a case, we have C = 2, P = 2, and E = 1, so F = 1. So, for a chosen temperature (using up that one available degree of freedom), the compositions of both phases are fixed exactly, but *not* the overall composition. The ratio of the amounts of the two phases is not of thermodynamic concern. This echoes what was said about the amounts of the two solvents in the iodine example at the beginning of this section. The result is to be contrasted with the effect of choosing P = 1 for C = 2. Now F = 2, and both the composition of the phase (which must be the same as the overall composition) and the temperature need to be selected to fix the system's description completely (Fig. 3.10).

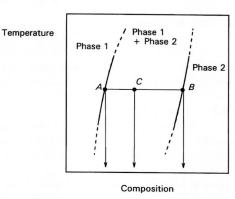

Fig. 3.8 An isothermal tie-line can be treated as a composition line. The compositions of the conjugate phases are fixed at a given temperature, no matter what the overall composition while both phases coexist at equilibrium.

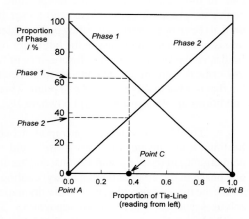

Fig. 3.9 Calibration of isothermal tie-lines as composition lines (refer to Fig. 3.8), according to the lever rule.

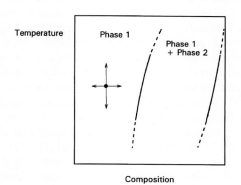

Fig. 3.10 Both the composition of the phase and temperature may be varied freely in a one-phase field.

§4. Critical Points

While the phase diagrams of wholly-liquid systems are generally of little interest in dentistry, there being few relevant examples, some cases can provide insights that are applicable to solid systems which are of concern. They are perhaps more easily visualized and grasped because they are liquids and thus in a more familiar form.

●4.1   Two liquids

It is commonly the case that two liquids are not entirely **miscible**. There are several possibilities for the phase diagram of such systems, but one type is of interest as a model for alloy systems to be considered later, that is a system showing an **upper consolute** or **critical solution point**. An example is water-*iso*butanol (Fig. 4.1). As before, an isothermal tie-line in the two-phase field indicates the compositions of the two liquids at equilibrium, and the intersection with the ordinate raised at the overall composition point divides the tie-line into segments whose lengths represent the proportions of the two phases, exactly as before (Fig. 3.6). However, it can be seen that, in this kind of case, as the temperature is raised those conjugate phases become closer and closer in composition, until at the critical temperature they become identical as the tie-line is of zero length. Because the boundary of the two-phase field is viewed as representing the presence of two phases, then this critical point might be considered somewhat anomalous, but that is what makes this a *critical* point, where the transition conditions are only just met, and nowhere else.

This kind of system has an important behaviour. Whilst heating from inside the two-phase field results in the convergence of the two compositions, and the dissolution of the one phase in the other at the boundary, cooling from somewhere above the two-phase region has a more dramatic effect. When the solvus is reached there is a **phase separation**. That is, throughout the body of liquid microscopic droplets of one phase will suddenly appear as the remainder becomes a second phase. Typically, an **emulsion** will form, a **suspension** of the one **disperse phase** in the other **continuous phase**. Optically, the liquid becomes cloudy, as there is usually a marked difference in the refractive indices of the two phases (24§5.11) – the term **cloud point** is therefore commonly used to describe the temperature at which this occurs for a given mixture. Normally, such emulsions are not stable and the droplets will steadily coalesce (due to droplet collisions or Ostwald ripening), leading to two distinct layers (as in Fig. 3.1), aided by the general density difference between the phases. This is also an energy-minimization process as the total interfacial area is decreased at each droplet merger or growth (10§1, 10§8.1). Similar behaviour is seen with some disinfectant solutions added to water, and also with various aniseed spirits – the ouzo effect[11], but in dentistry cloudy PMMA is the result of just such a process (5§6.3).

The disperse phase that forms at the cloud point is the one with the lesser volume (*cf.* the random loose-packed spheres limit, ~0.56, in Fig. 4§9.2). However, near the composition for the critical solution point, this separation is not quite so obvious in terms of which phase can contain the other. Also, because the compositions of the conjugate phases are very close, there is little refractive index difference between them. In addition, the normal

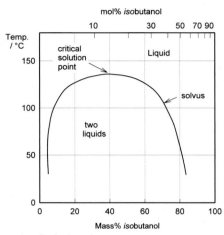

water-*iso*butanol

Fig. 4.1   An example of a two-liquid binary system showing an upper consolute point. (The diagram is from a study of sealed tubes in which the liquids are not allowed to boil.) *Iso*butanol is 2-methyl 1-propanol).

The existence of phenomena such as critical opalescence underlines the dynamic, stochastic nature of all such systems: locally, fluxes and concentrations must be varying with time, and therefore are not absolutely defined – they have a range of values statistically, even if centred on some average, system-representative values. Indeed, for such reasons, even the boundary between equilibrium solid and liquid, say, is uncertain in that molecules will be arriving and leaving continuously.

Phase separation as in Fig. 4.1 may seem puzzling. However, this in this case it can be viewed as the limit of the clustering reported for ethanol in water: molecular-scale self-association. The alkane part of the ethanol slightly favours similar neighbours, and allows more hydrogen-bonding in the water. This tendency is increased in propanol-water mixtures, and is taken to the limit of forming a distinct boundary with the butanols. The extreme case would be like oil in water (10§8.1). It is, of course, the trade-off of relative energies underlying the thermodynamics.

thermal fluctuations and local composition variation due to random diffusion effects leads to a kind of not-quite-decided condition just above the critical solution temperature. The liquid takes on a bluish, hazy appearance known as the **critical opalescence**. The scale of the fluctuations obviously is large enough to interfere with light, but neither phase can be said to be disperse. On further cooling, this ambiguity is resolved, and phase separation occurs. Bear in mind that phase separation is a diffusion-dependent process. Despite this seeming sharp distinction, it should be borne in mind that even in the one-phase region the alcohol molecules form clusters (2§12.2).[12] On cooling these must grow in size until they become detectable optically (at the point of critical opalescence), even though they remain labile and ephemeral as objects. One's view of a solution as a perfectly homogeneous, random mixture at the molecular scale must, in this kind of case, be modified.

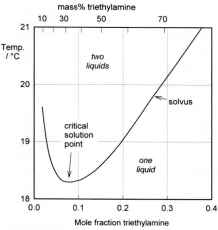

water-triethylamine

Fig. 4.2 An example of a two-liquid binary system showing a lower consolute point.

There is also the possibility of a **lower consolute point** (Fig. 4.2).[13] That is, on heating a system of fully-miscible liquids, a phase separation occurs. All effects are therefore reversed. Although there is no apparent occurrence of such a liquid system in dentistry at present, it does provide a useful model for the behaviour of certain polymers (27§6.3) and does occur in potentially biologically-relevant contexts.[14]

More importantly, phase separations can occur when water is absorbed by certain polymer systems. Normally, the water would be randomly distributed in the free volume (3§4.2), but if, acting as a plasticizer, it facilitates the relaxation of the matrix polymer chain segments, effectively forming better micelles (3§4.1), a lower energy state is obtained with the water as a separate phase (at least part of it). This would be driven further by the presence of solutes that partition into the aqueous phase, when osmotic effects become important. The usual laws of diffusion would not then apply. By implication, raising the temperature of a water-containing polymer may provide the activation energy for the relaxation, but this would not be expected to be reversible. The reverse situation can occur in denture base acrylic (5§6.3).

●4.2   Ternary liquid system

A similar critical point occurs in ternary liquid systems where complete miscibility does not occur, and can be extended to solid solutions, and in particular dental gold alloys (19§1.6). One example[6] is acetic acid-chloroform-water (at 1 atmosphere pressure) (Fig. 4.3).[15] This diagram may be read in a very similar way to Fig. 4.1, even though it is an isothermal section. Tie-lines connect the compositions of the two-phases that are in equilibrium across the two-phase field. However, because of the particular chemistry of each such system, in general these tie-lines are not parallel to each other in the plane of the section. Even so, the compositions of the conjugate phases converge to another special limiting case where the tie-line is of zero length and the two compositions are identical. This is known as **the isothermal critical point** or **plait point**. It is necessarily unique at any temperature.

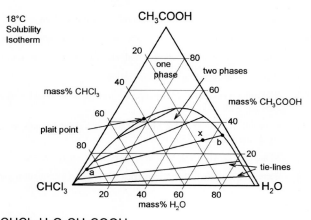

CHCl$_3$-H$_2$O-CH$_3$COOH

Fig. 4.3 18 °C isothermal section of the system acetic acid-chloroform-water showing the mutual solubility limit and experimentally determined tie-lines.

---

[6] This is a so-called Type 1 system, where only one pair of the components shows immiscibility. There are other types, as might be imagined (2 pairs, 3 pairs, and some variations), but these are not relevant here.

Such tie-lines in ternary systems cannot be calculated easily because there is no simple theory to explain the partitioning of the components – it requires a detailed thermodynamic calculation. Thus the compositions can in practice only be determined from analyses of the two phases present for a series of compositions across the two-phase field. Applying the phase rule (§3.6) to the two-phase field of such a system at a fixed temperature and pressure gives:

$$F = C - P + E = 3 - 2 + 0 = 1 \qquad (4.1)$$

That is, the composition of only one phase may be fixed; the composition of the other is predetermined by the tie-line through that point. Unfortunately, there is no easy way to predict where that line will lie.

An experiment to generate such a diagram might go like this: Mix in a separating funnel a suitable measured combination of the three components such that there are two layers at equilibrium (much like Fig. 3.1), for example corresponding to point x in Fig. 4.3. A portion of each layer can then be analysed, allowing points a and b to be plotted. The tie-line then represents all proportions of the two liquid phases, remembering that the thermodynamics of equilibrium does not involve the amounts of the two. Progressively draining off the lower phase in Fig. 3.1 does not alter the equilibrium at all. The process can be repeated until the entire two-phase field is mapped. This is efficient because only one overall composition per tie-line is required, and a short series running along near the 40% $CHCl_3$ line, from the $CHCl_3 - H_2O$ edge, would be sufficient. In fact, it is not even necessary to measure the quantities of the three components accurately, x does not have to be known for any such point, so long as two phases are formed that can be analysed separately.

References

[1]   Möbius AF. "On the composition of line segments and a new way of founding the barycentric calculus that arises from it." J f reine und angew Math 28: 1 - 9, 1844. Translated by D. H. Delphenich : http://www.neo-classical-physics.info/uploads/3/4/3/6/34363841/moebius_-_barycentric_calculus.pdf

[2]   http://mathworld.wolfram.com/VivianisTheorem.html

[3]   Abboud E. On Viviani's Theorem and its Extensions. http://arxiv.org/PS_cache/arxiv/pdf/0903/0903.0753v3.pdf

[4]   http://en.wikipedia.org/wiki/Simplex

[5]   Ostwald W. Lehrbuch der allgemeinen Chemie. 2 (ii) : 444. Engelmann, Leipzig, 1893.

[6]   Euclid. The Elements. Book 1. Alexandria, Egypt, c. 300 BCE.

[7]   http://en.wikipedia.org/wiki/Gravity_of_Earth

[8]   http://en.wikipedia.org/wiki/Earth%27s_magnetic_field

[9]   http://science.nasa.gov/science-news/science-at-nasa/2003/29dec_magneticfield/

[10]   http://en.wikipedia.org/wiki/Natural_electric_field_of_the_Earth

[11]   http://en.wikipedia.org/wiki/Ouzo_effect

[12]   Wakisaka A & Ohki T. Phase separation of water–alcohol binary mixtures induced by the microheterogeneity. Faraday Discuss 129: 231 - 245, 2005.

[13]   Counsell JF, Everett DH & Munn RJ. Recent redeterminations of the phase diagram of the system: triethylamine + water Pure Appl Chem 2: 335 - 338, 1961.

[14]   Nakayama D, Mok YB, Noh MW, Park JS, Kang SY & Lee Y. Phys Chem Chem Phys 16: 5319--5325, 2014.

[15]   Perry RH, Chilton CH & Kirkpatrick SD (eds). Chemical Engineers Handbook 2nd ed. New York, McGraw-Hill, 1941.

Chapter 9   Cements and Liners

*Cements in dentistry are **composite** materials based on mixtures of a solid and a liquid which, typically through an acid-base reaction, produce a solid matrix that binds the mass together. There are two types of task to which these materials are applied: to fill a hole in a tooth, that is, as a restorative material, and to retain an indirectly-fabricated device in place, such as an orthodontic bracket or a crown. In both cases the fixation may be either temporary or permanent, but it may also be fulfilled by other types of material such as filled resins (for both) and amalgam (as a restorative material). The appropriate choice depends on the mechanical properties, the appearance, the handling characteristics, and the behaviour in service. All of these depend entirely on the chemistry of the cement: composition, setting reactions, degradation mechanisms and structure. A detailed understanding of each is therefore necessary to make those decisions.*

*Zinc oxide-eugenol is an old, but still used material, although it has serious limitations. Zinc phosphate, too, has been in use for a long time. Its setting is dealt with in detail as a case study in the use of constitutional diagrams to understand the process. This stresses that the explanation of clinical observations and handling instructions can be obtained in no other way. Zinc polycarboxylate is also based on reaction with zinc oxide, but it offers the possibility of true adhesion as opposed to merely mechanical **retention**.*

***Ion-leachable glasses** form a group of materials of increasing importance with the development of glass ionomer cements. These have the additional advantage of tooth-like appearance, but of this group only glass ionomer is adhesive.*

*The most recent class to be introduced is that of the **hydraulic silicates**, and these appear to have particular application in endodontics, being well-tolerated and promoting healing.*

*An important additional property in the context of retention applications is **film thickness**, a purely geometrical consideration that affects the design of crowns and inlays. Cavity liners and varnishes are also discussed briefly.*

*The cements, as a group, illustrate very well the **compromises** that must be made in the design and use of dental materials. There are serious deficiencies in all types, and developments will no doubt continue. The benefits can only be judged by relating structures to behaviour.*

Materials Science for Dentistry
https://doi.org/10.1016/B978-0-08-101035-8.50009-2

Cements in dentistry have three general roles to play. The first is in the sense of a gap-filling mortar to join two solid objects together, such as an inlay or crown onto a tooth; in this context they are frequently referred to as **luting agents**. This sense is borrowed from the alchemists who were concerned to make air-tight seals in their apparatus by filling the gaps with a special paste or **lute** (taken from the Latin, *lutum*, mud – which is appropriately illustrative). Indeed, this idea of sealing is still relevant in dentistry. The second role, overlapping somewhat with the first, is that of a retentive or adhesive material for applying devices such as orthodontic brackets and buttons to tooth surfaces, or simply holding something in place when the geometry is not intrinsically stable. The third job is of a direct restorative material, filling large spaces with a material which is intended to take masticatory loads and abrasion. The success or failure of a cement in any of these contexts depends on the chemistry of the setting process, on the structure of the resulting material, and therefore on the way that it is mixed and handled.

§1. General Principles

The basic principle behind all cements is that of a fluid, semi-fluid or plastic material which undergoes some kind of change to yield a rigid solid and which therefore provides mechanical retention for one part on another. This change of condition may be achieved in several ways.[1]

●1.1 Change of state
The first is the simplest: the purely physical change of state involved in a material freezing, or at least being cooled to a glassy or highly viscous condition. Such a material is illustrated by **sticky wax**; the change is entirely reversible and no reaction with the substrate occurs. Retention of one part on another is due to the interlocking of the hardened wax with the roughness of both surfaces, and also due to the wetting of the substrate allowing van der Waals forces to operate (Chap. 10). But although sticky wax has great utility in the laboratory, this system or its like cannot be used effectively in the mouth because of the trade-off between two critical aspects.

Materials in general rapidly become weaker as their melting or softening points, T_m, are approached. The working or service temperature, T, must be sufficiently far below that temperature for the material to be useful, typically $T/T_m < \sim 0.9$ (temperatures in kelvin), assuming that the strength is high enough in the first place. Waxes are not strong (Chap. 16). Secondly, oral tissues cannot tolerate large temperature rises without severe damage; tissue temperatures of 45 °C are quite sufficient to cause a severe response, indeed, burns begin at around 44 °C, while just 42 °C is enough to cause inflammation (and, systemically, brain damage). This combination of factors effectively prevents any material from being used in this context, as a suitable strength with a low enough melting point could not be obtained. For comparison, one could note that one extra-oral application in dentistry where such a process is successful is soldering (Chap. 22); the temperatures here are much higher, and intrinsically stronger materials are involved – metals. However, even this is a second-class joint, not being as good as the metals being joined.

●1.2 Solvents
Solvent evaporation is the second means of effecting solidification. Materials relying on this process are usually called glues or adhesives. Solutions of polymers find application in a number of industrial and household products, the most obvious being water in gelatine (partially hydrolysed collagen) or starch glues, as used to be used for carpentry and bookbinding. Polystyrene and other 'plastic' so-called cements rely on the evaporation of volatile ketones, for example, while 'rubber cements' utilize petroleum fractions. Clearly the solvent must be matched to the solute for good solubility, and high concentrations of the solute are necessary to assist in gap filling.

However, there is no application in dentistry for the purposes of 'cementing' devices or restorations for this type of system. This arises from the need for there to be an appreciable rate of evaporation of the solvent for the strength of the cement to be high enough, early enough. There are time limitations in processes such as this in the clinic. Such evaporation cannot happen under impermeable metallic crowns or brackets, for example. Secondly, these materials are inherently very weak, relying for their effect on large surface areas of close contact, much as does sticky wax. Alternatively, when used on plastics, the solvent is meant to partially dissolve the substrate so that some chain mingling can occur, the most intimate 'contact' possible (*cf.* the repair of acrylic denture bases). But this means both a loss of surface precision and the application of a substance that will dissolve in the substrate polymer. Such solvents act as plasticizers, and therefore might permit the undesirable deformation of the object. They certainly cause swelling, and then shrinkage again when they finally escape (not necessarily in an accurately reversible manner). There is also the problem of toxicity. Many organic solvents are known to cause liver damage, and certainly must interfere with cell membranes (which are lipid) because they are solvents for fat-like substances. It is therefore inappropriate to expose the patient to them, or permit any to inhale the vapour unnecessarily. Thirdly, aqueous solvents are unusable in an aqueous environment as the cement would dissolve in the oral fluids. Even so, a major problem with these adhesives is the large shrinkage, or even porosity, which inevitably results from the evaporation of the solvent.

The only examples of these solvent-based systems in dentistry are found in the tray adhesives for impression materials, cavity varnishes and some dentine sealants, where strength is not a major factor. Some types of cavity liner essentially consist of calcium hydroxide or zinc oxide suspended in a volatile vehicle. These powders, having no appreciable cohesive strength on their own, may be bound together by a small amount of resin (*i.e.* a polymer) left behind when the solvent in which it was dissolved evaporates. Solvents such as ethanol or chloroform may be used in these circumstances. Generally, however, they are not satisfactory materials, failing to achieve what is intended with any reliability or longevity.

●1.3 Hydraulic reactions
Chemical reaction with water provides a third class of cementing materials, as typified by gypsum-based products and Portland cement. In these, water is a reactant. In general, such systems suffer problems of solubility in an aqueous environment, and rate of development of strength (very slow for Portland cement), while manipulative factors have generally excluded these and similar systems from dental use. Gypsum products, of course, are used in other contexts, but a variation on the theme of Portland cement has a specialized use in dentistry: **hydraulic silicate cement** ("HSC") (§12). These types of system must be distinguished from those in which water is the medium for other reactions, whether or not it may be involved in a structural sense (*cf.* alginate and agar, Chap. 7).

●1.4 Acid-base reactions
Thus, because of the difficulties generally encountered, we find that in practice we are mostly concerned with what are called **cementitious** materials, powder-liquid systems in which there are direct chemical reactions between compounds other than water itself, although it is true that water still plays a vital part in all dental systems. What we are left with, then, is a class of materials which are in fact based on the reaction between an acid and a base to produce some form of salt and, of course, water. These we may term the 'true' dental cements, which is not to dismiss the HSC type.

●1.5 Polymerization
Sometimes referred to as **resin cements**, filled resins of the general kind dealt with in Chap. 6 are also used for luting crowns and inlays (6§4.7), that is, non-aqueous materials that harden by polymerization or covalent cross-linking *in situ* by various means. While they indeed have uses which overlap those of cements proper in the dental sense, there are no extra details to be discussed here. Unfortunately, there is also a proliferation of products which try to combine resin-like chemistry with a cement-like system, in a vain attempt to get the best of both worlds, but combinations which confuse the issue of classification. We shall return to these products later.

§2. Zinc Oxide - Eugenol

The oldest type of dental cement still in use is that produced by reacting zinc oxide and eugenol. Eugenol[1] is the trivial name for a phenol (Fig. 2.1) whose special feature is the methoxy group at the 2-position, *i.e.* adjacent to the phenolic hydroxy group. Phenolic hydrogen is weakly acidic (Fig. 2.2) and may be removed to form a phenoxide ion. The strength of the acidity is about the same as that of the hydrogen in the bicarbonate ion, and so salts can be formed in an analogous manner:

$$ZnO + 2(H\text{-}eug) \Rightarrow Zn^{2+}(eug)_2^- + H_2O \qquad (2.1)$$

This is the setting reaction.[2]

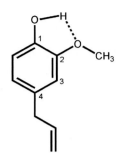

Fig. 2.1 Eugenol, 4-allyl-2-methoxyphenol.

Many multivalent metal ions can form complexes where they are coordinated to between 4 and 6 centres of high electron density. This coordination represents a state of often considerably lower energy than the dissociated ions and so may be very stable. Biological examples are the iron in haeme and the magnesium in chlorophyll. The environment of the metal ion is such as to shield as much as possible the strong positive charge it carries. This effect has already been seen for the calcium ions in alginates (7§9). Similarly, zinc ions react with the eugenolate ions formed by the acid-base reaction to form such a complex (Fig. 2.3). The molecule of water produced in the neutralization is also coordinated, lowering the energy even more. The essential point is the formation of the type of compound called a **chelate** (from the Greek word for a crab claw) by components which have the electron-dense oxygen atoms in such a position as to form a ring readily. Five-membered rings, as formed here, are generally the most stable types of chelate.

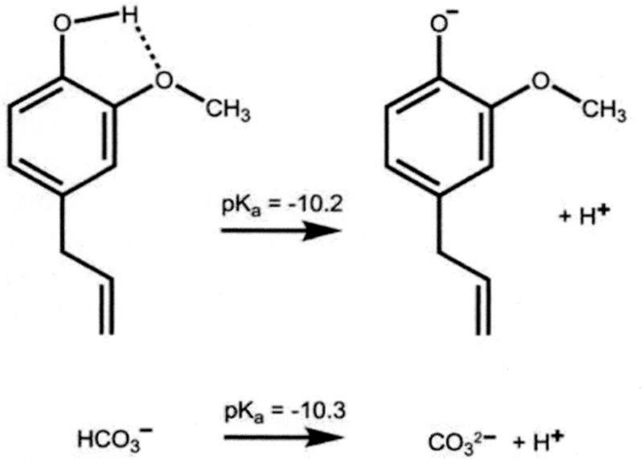

It is important to notice in that in this kind of cement the intermolecular bonds (*i.e.* between the complexes) are weak, being of the van der Waals type only. This sets a limit to the attainable strength of the material as a whole.

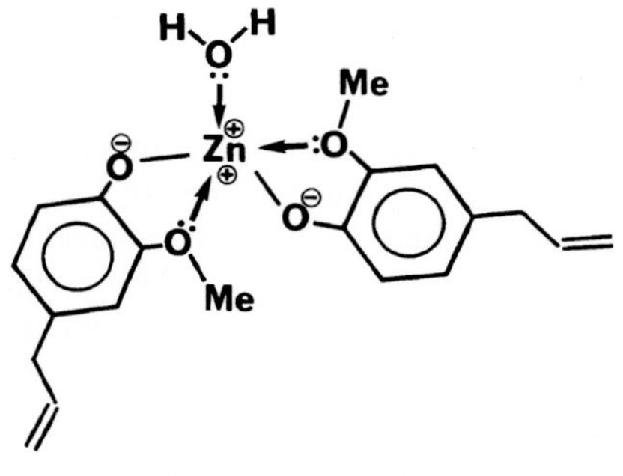

Fig. 2.2 The phenolic hydrogen is of comparable acidity to that of the bicarbonate ion.

●2.1   Acceleration of setting

Depending as it does on the reaction of zinc *ions* with the eugenol, the setting process does not proceed at an appreciable rate in the complete absence of water – it is a **solution-mediated** process. Zinc oxide does dissolve slightly and slowly in water, but it then readily forms a tetrahedral hydrate which is unreactive. The hydrolysis of this to produce hydrated zinc ions is only slight, and proceeds too slowly for a clinically useful cement. Hence, any additive to the system to increase the availability of the zinc ions would be expected to

Fig. 2.3 Probable structure of the zinc eugenolate chelate. In the crystal, stacks of these structures are thought to form, sandwiching the coordinated water molecule.

improve the rate of setting. Quite clearly, water is going to be an accelerator as it permits the reaction in the first place. But the addition of hydrogen ions in the form of an acid would greatly increase the concentration of zinc ions by dissolving the oxide and hydrolysing any undissociated hydroxide. Zinc salts act in a similar way, both initially by raising the concentration of zinc ions, and subsequently because their consumption is accompanied by the liberation of hydrogen ions. Thus the pH will drop and the dissolution of zinc oxide increases. Commercial products take advantage of these effects in their formulations, small additions of water and zinc acetate or acetic acid being made. Even so, the setting rate is markedly affected by the ambient humidity at the

---

[1]   The principal constituent of oil of cloves. The clove is the dried flower bud of the tropical tree *Caryophyllus aromaticus*, a native of the Moluccas, Indonesia, now more widely grown.

point of mixing, and care must be taken in the storage and use of such materials to avoid excessive variation (Fig. 2.4).[3] On the other hand, water present at the surface of dentine will encourage setting and there is no point in attempting to dry this other than to remove evident overt liquid.

●2.2   Dissociation

Unfortunately, the chelate formed with eugenol is not particularly stable and the reaction is slowly reversed in the presence of excess water:

$$Zn^{2+}(eug)_2^- + 2H_2O \Rightarrow Zn(OH)_2 + 2(H\text{-}eug) \quad (2.2)$$

Thus this is, in fact, an equilibrium, lying to the left here because of the excess zinc oxide. However, since eugenol is slightly soluble in water and volatile, it will thus diffuse away from the site and escape from the system. This steady hydrolysis limits the life and applicability of the material as a cement. Indeed, even as a liner – where one might imagine it

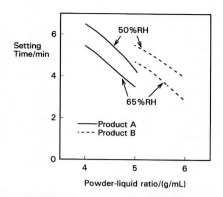

Fig. 2.4 The effect of powder : liquid ratio on the setting time (at 23 °C) of two brands of zinc oxide eugenol cement for two values of ambient relative humidity.

to be isolated from its surroundings – it may not be very long lived. There are reports that it has disappeared from beneath amalgam restorations, presumably dissolved in the fluid present, which is percolating because of a lack of seal (14§6.2). In any case, the eugenol would tend to diffuse away through the dentine, which is permeable, even in the absence of leakage as such, albeit at a lower rate.

Furthermore, it is not a particularly strong or abrasion resistant cement. Attempts have been made to improve the material by using various additives such as resins which may dissolve in the eugenol and provide some further binding when set, as well as modifying the rheological characteristics when freshly mixed. None has achieved any prominence.

Zinc oxide-eugenol cement (or 'ZoE') has been frequently and euphemistically described as obtundent ("dulling or blunting sensibility"), or even "palliative", yet apart from the lack of pain there seems to be no evidence of any other beneficial effects under such material: there is no reparative dentine, and only inflammatory responses are found, mostly ascribed to the eugenol. It would appear that the pulp is, at best, damaged by these materials.[4] (ZoE is toxic in contact with bone and soft tissue, as used formerly in 'periodontal packs', and also to nerves.) In addition, eugenol is a known sensitizer, and may provoke a strong physiological response in some persons. Even so, 'clove oil' and cloves themselves remain a common and popular folk-remedy for toothache, precisely because it is obtundent.[5][6]

●2.3   Discolouration

The presence of the phenolic hydrogen in eugenol and similar substances has an untoward effect on resin restorative materials placed on top: it inhibits polymerization because it is a free radical sink (5§4).[7] It will thus prevent proper curing at the bottom of the restoration. Furthermore, eugenol and other substances that may be included in the cement's liquid will dissolve in the restorative resin, discolouring it in the process to a rather dark yellow (this colour may be due to the reactions with free radicals creating extended conjugated systems [24§6.2]). This will not occur immediately but over some hours, but it will very effectively destroy any colour match that may have been achieved. There will, of course, be the other consequences of exposure to a plasticizer: softening and swelling – and therefore distortion, then eventual shrinkage as it is slowly lost by diffusion, if not reacted.

Eugenol is itself subject to attack by oxygen, and suffers reactions akin to those shown in 6§6 (*cf.* Fig. 5§4.2) probably aided by exposure to UV light. These reactions are responsible for the gradual darkening of the liquid in storage, despite the use of brown glass bottles. As will be noticed, eugenol has a vinyl group which would be expected to be polymerizable, and although the competition with 'inhibitory' aspects of its chemistry complicate matters, there is some evidence that such a reaction may occur slowly in ZoE to yield a complex polymer network very much like lignin (the principal binding polymer of wood).[8] Indeed, lignin is formed from coniferyl alcohol and similar molecules, to which eugenol is closely related. While it is not clear whether discoloured eugenol is appreciably affected in terms of its reactivity with zinc oxide, if the polymerization reaction could be promoted it might lead to a cement which is both stronger and less soluble, even though this would not affect the initial irritancy or the effect on resins.

●2.4 Variations

In view of the general weakness of zinc-oxide-eugenol cement, many attempts have been made to improve on it. These attempts include the use of inert fillers, polymeric binders dissolved in the eugenol (as mentioned above), and a wide variety of other formulations. In addition, alternative chelating agents have also been much investigated, for there are many compounds meeting the basic structural requirements for chelation (Fig. 2.5) and indeed these have given rise to various commercial products. In each case, a more or less acidic hydrogen can be replaced by the metal ion.

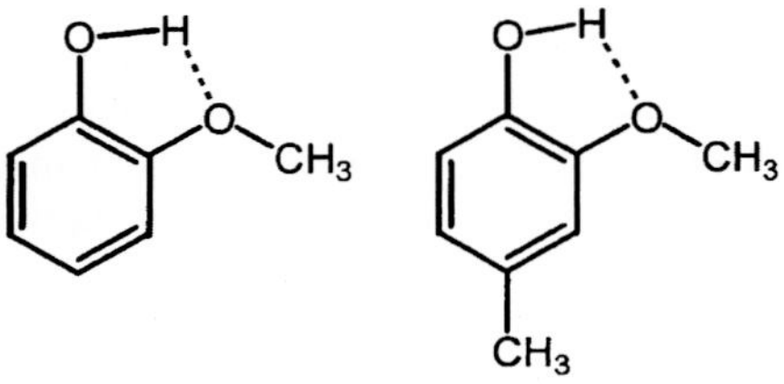

Fig. 2.5 Two alternative reactants for a zinc oxide cement: 2-methoxyphenol (left) and 2-methoxy-4-methyl-phenol. Note the acidic phenolic hydrogen.

The most successful of these alternatives seems to be 2-ethoxybenzoic acid or 'EBA' (Fig. 2.6). Being a carboxylic acid, this is much more acidic than eugenol, and therefore much more reactive in salt formation than the phenol; this also leads to fewer problems with control of the setting rate. However, the more strongly polar benzoate is more readily solvated by water and the solubility of the set cement accordingly much greater. The six-membered ring is perhaps a little less stable, although probably offset by the increased polarity of the carboxylate. Even so, with the addition of inert filler (such as alumina), a proportion of eugenol, and other materials in small quantities, some slight net improvement of properties over plain zinc oxide-eugenol can be obtained. The principal advantage, however, seems to be a reduced tendency to cause the burning sensation that eugenol does on its own.

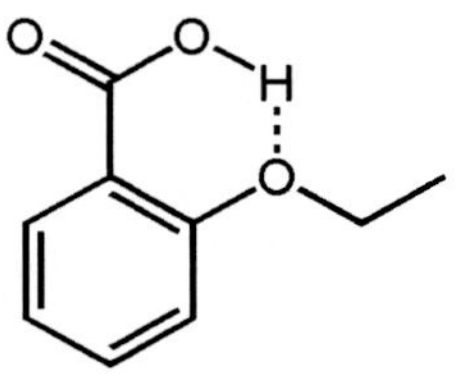

Fig. 2.6 2-ethoxybenzoic acid, "EBA".

In endodontic applications, cements need to be radio-opaque, and substances such as calcium tungstate ($CaWO_4$); bismuth oxide (Bi_2O_3), carbonate or nitrate (which have less definite compositions); or barium sulphate ($BaSO_4$) are often used, even metallic silver.

●2.5 Presentation

Originally sold just as powder and liquid, ZoE cement has also been presented as a two-paste system. Here, the zinc oxide and accelerator will be suspended in an inert vehicle, an oily or resinous mixture that plays no part in any reaction. In addition, the eugenol is made into a paste with an inert filler. The main advantage of this approach perhaps being apparent ease of dispensing and mixing. Certainly, no great improvements in mechanical properties are expected: the inert filler is unbonded and the vehicle for the ZnO can only weaken the system. The use of the label "catalyst" for the eugenol component is also incorrect for these materials (*cf.* 7§4.2), and "accelerator" is similarly unhelpfully wrong.

Zinc oxide-eugenol is also used as an impression material in prosthetic dentistry, when it is called a 'paste' rather than cement, indeed, it is often presented in the two-paste form. There is no fundamental difference between this and the cement: the setting reaction is the same. Some adjustments may be made to the proportions of the reactants and inert components to adjust the rheological properties of the unset material, and the pastes may be coloured to assist in monitoring their mixing (15§7). Its use frequently results in a burning sensation; the relatively short duration of exposure limits the toxic effects of the eugenol (see above).

§3. Background Chemistry

●3.1 Zinc oxide

Zinc oxide is used in a remarkable number of dental products, mostly cements,[9] for reasons of reactivity, radio-opacity, whiteness, and low cost, but also – given those other benefits – low toxicity. Zinc is an essential element in human physiology as a component of many enzymes, but is apparently tolerated at levels well above those needed to maintain health. This may account for the widespread use of zinc oxide and the absence of overt problems in biological contexts. A little background chemistry is helpful in understanding some of behaviour in certain applications.

Zinc oxide exists in the so-called **wurtzite** crystal structure (Fig. 3.1) which may be viewed as a network of tetrahedra formed by the oxygen atoms in a structure reminiscent of that of diamond.[10] It can be seen that of the possible tetrahedrally-coordinated sites in the structure, between groups of oxide ions, only half should be occupied by zinc ions in order to satisfy exactly the stoichiometry of the composition ZnO. If zinc oxide is heated, a slight decomposition occurs:

$$2ZnO \rightleftharpoons 2Zn + O_2 \qquad (3.1)$$

which effectively removes some oxygen atoms from the lattice. The excess zinc atoms (not ions, notice) then dissolve in the remaining solid oxide and occupy some of the vacant holes (*i.e.* **interstitial** sites). The extent of this depends on temperature, and up to 0.03 atom percent of Zn in excess of the stoichiometric ratio may be present. This results in a slightly coloured material, ranging from pale yellow to deeper shades. As can be deduced from the equilibrium above, Zn vapour will increase the colour and oxygen gas will bleach it.

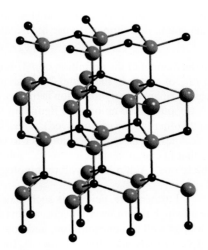

Fig. 3.1 The crystal structure of zinc oxide. Note the type of site which can accommodate an extra zinc atom.

    The important point in the present context is that the presence of the excess Zn atoms reduces the reactivity of the oxide with respect to acid-base reactions. In the case of the reaction with eugenol this leads to excessively long setting times, especially as the reaction is sluggish to start with, and care has to be taken with the preparation of the oxide for this reason. Reactive zinc oxide can be made by decomposing by heat the carbonate or the hydroxide at low temperatures. On the other hand, heating to high temperatures can be used deliberately to reduce reactivity. For these reasons, zinc oxide supplied with one type of product should not be used with any other product, or even a different brand of the same type of product. The reactivity is likely to be wrong.

●3.2 Oxide selection
    The choice of zinc oxide for so many uses may seem a little odd, until one arrives at its near-inevitability by a process of elimination. Taking out from the list of elements those that are firstly chemically-inappropriate as not being metals, those that are extremely reactive, radioactive or otherwise grossly toxic, then those that are strongly coloured, finally those that are simply too expensive (bearing in mind that these are frequently overlapping categories, not mutually-exclusive labels), we are left with eight possibilities (Fig. 3.2). Of these, Mg, Al, Ca and Ti are not very radio-opaque with respect to tooth tissue (Chap. 27), and Sn can be toxic, simply

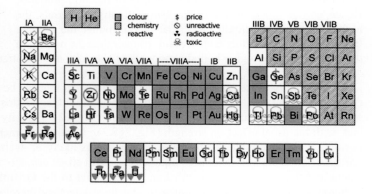

Fig. 3.2 The choice of metals for dental cement and related uses is limited by a number of factors, all of which need to be considered.

because it is a heavy metal – as are Sr and Ba, when in a soluble form. That leaves Zn (and even this can be toxic in sufficient quantity[11]). This is not to say that in the right context, with strong enough justification or special chemical context, others might not be used (*cf.* §10), but on balance it turns out to be, at the least, a cheap and convenient material.

●3.3 Acidity
    Reference is often made to the relative weakness or strength of an acid (or base) (*e.g.* §2, 7§9.1). It is well to remember that this phrasing refers to the ease or completeness of ionization:

$$HA + H_2O \rightleftharpoons A^- + H_3O^+ \qquad (3.2)$$

Given that this dissociation is an equilibrium, we write:

$$K_a = [A^-][H_3O^+]/[HA] \qquad (3.3)$$

where K_a, the equilibrium constant for reaction 3.2, is specifically labelled the **acidity constant**. Since the range of possible values is very large, and the values themselves typically rather small, it is simpler to use the negative logarithm:

$$pK_a = -\log_{10}(K_a) \qquad (3.4)$$

– in the manner of the definition of pH. Indeed, by this formulation the pK_a of water is the same, 7. It should be noted that this is measure of how easily the anion is made available for reaction, and the strength of the anion as a **ligand** in some contexts (although several other factors are involved). Thus, talking about the strength or weakness of an acid does not refer to concentration, despite the fact that in common speech a 'weak' solution, for example, means a dilute one. Again, care with terminology is required.

Furthermore, acidity as such is not a direct measure of the rate at which reactions will occur, as this depends on many kinetic factors as well as concentration, and does not necessarily indicate relative corrosiveness towards metals, for example.

§4. Cements as Composite Materials

It is most important to realize that with the zinc oxide-eugenol system, as indeed with all other dental cement systems discussed here, the set material is essentially a *composite* structure. Always much more of the solid component (and not just zinc oxide) is used than is required for complete reaction with the liquid, and so it persists as an appreciable volume fraction of the set material as a stiff and relatively strong core in the matrix of reaction products. Hence, as was found for the filled-resin restorative materials, the higher the powder : liquid ratio (that is, the higher the volume fraction of core, when this is stiffer and stronger than the matrix, in the set material) the higher both the modulus of elasticity and the strength of the set cement are expected to be.

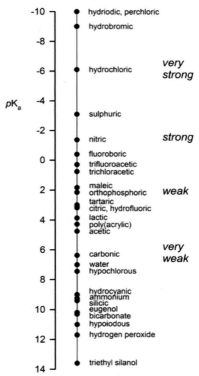

Fig. 3.3 Acidity of some acids of relevance to dentistry, and the general terms that may be used to describe them. (For polybasic acids, only the value for the first ionization is given.)

Naturally, there is a limit to the amount of powder that can be used because the viscosity of the mixture rises rapidly (4§9) and would easily become unworkable at high ratios. This is apart from the more critical limit set by the minimum mixing volume of liquid required (2§2). Porosity will rise rapidly if this requirement is not met, with the usual effect on strength (Fig. 2§8.3). It must be remembered, however, that in general mixing powders with viscous liquids to make thick pastes is likely of itself to include air bubbles regardless of the presence of the 'correct' amount of liquid. As a result all set cements are inevitably porous to a greater or lesser extent and thus have their strength limited according to the Griffith criterion (1§7). Mixing and handling techniques must endeavour to minimize this intrinsic defect and therefore its detrimental effects on the reliability, longevity and efficacy of the treatment of which it is a part.

It is worth emphasizing here, however, that the minimum mixing volume for cements is *required* to be less than the stoichiometric amount for complete reaction if there is to be a core remaining. This is in contrast to the situation for gypsum products and investments, where the minimum mixing volume is *required* to be at least that necessary for complete reaction. Given that such mixtures are plastic dilatant (4§7.7), it would be expected that at the completion of the procedure the surface must still be glossy if the requirements have been met. This provides a simple visual check of the adequacy of the mixing ratio.

It must be assumed that there will be changes in total volume between reactants and reaction products, and since in general there is a reduction in the mobility of the liquid components this change must be a net shrinkage (by a reduction of free volume), although contact between reactant core and solid reaction products means that a loss-of-gloss event can be anticipated. In addition, crystal growth pressure effects may be present when crystallization is involved. However, since the scale of use is small, the absolute value of the dimensional change is probably insignificant in most contexts.

Porosity not only affects strength, it affects modulus of elasticity as well, and this applies to all classes of materials, not just cements: impression materials, acrylic, filled resins, amalgam, investments, porcelain, and so on – noting that some such porosity is normal and essentially unavoidable. As might be deduced from a consideration of the rules of mixtures (6§2), the behaviour is complicated.[12] It can be approximated for closed porosity, which pores are assumed to have $E_p = 0$ (and therefore both bulk and shear moduli are also zero) and volume fraction ϕ_p, by

$$E / E_0 = 1 - (1+A)\phi_p + A\phi_p^2 \qquad (4.1)$$

where A is a function of the Poisson ratio, ν (Fig. 4.1).[13] (The effective Poisson ratio also changes.) It is therefore clear that a porous body must be treated as a composite with the porosity as a phase. Thus, wherever the true value of a non-porous material elastic modulus is of interest and this is to be determined with a macroscopic experiment, any porosity present must be taken into account, and *vice versa*. In other words, where stresses or strains are to be found in an experimental or service context, the effect of the porosity must be recognized, quite apart from any flaw-related strength considerations, as it is clear that there is considerable sensitivity. However, to the extent that such porosity is normally present, it is the properties of the actual material under such service conditions that is of primary importance, irrespective of theoretical 'perfect' values. Values from test methods that do not account for this composite behaviour may be misleading.

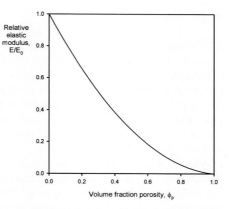

Fig. 4.1 Variation in relative stiffness with porosity (closed bubbles). $\nu = 0.35$

Machine mixing of encapsulated cement products may result in higher porosity,[14] with air being 'beaten into' the mixture, and this may even be exacerbated by the lower powder : liquid ratios used in order that this kind of mixing procedure works at all (15§3). The lower ratio must mean lower strength anyway, but the increased porosity would make it worse still.

§5. Zinc Phosphate

Another cement system based on zinc oxide that has been used in dentistry for a very long time is that which relies on the reaction with orthophosphoric acid.[2] The overall reaction resulting from simply mixing zinc oxide and aqueous phosphoric acid is essentially:

$$3ZnO + 2H_3PO_4 + H_2O \Rightarrow Zn_3(PO_4)_2 \cdot 4H_2O$$
$$2 \times 98 + 1 \times 18 \equiv 91.6 \text{ mass\% acid} \qquad (5.1)$$

This, however, is ordinarily a very fast reaction, and since it is highly exothermic ($H^+ + OH^-$ releases 57 kJ/mol) the temperature rises greatly. When the mixture cools this results in a friable mass of crystals which is of no use as a cement, even were it possible to get it into place before setting. The problems are therefore threefold:
- to reduce the rate of setting,
- dissipate the heat (the total amount of which is, of course, entirely unaffected by the rate of setting), and
- improve the strength in order to obtain a practical cement.

●5.1   Oxide modification

As mentioned above (§3), the reactivity of the zinc oxide can be altered by the method of preparation, and for phosphate cements this means firing at a very high temperature for some time. This may have several effects. Firstly, the excess Zn^0 atoms (§3) affect reactivity. Secondly, sintering will occur to reduce surface area, which will involve both particle roughness and agglomeration, as well as remove some kinds of crystallinity defect, again reducing reactivity. Magnesium oxide is added as this also reduces the reactivity, and indeed other oxides such as those of aluminium and silicon may be added for the same reason. In addition, the presence of the aluminium may suppress crystallization of the reaction product (§5.15), and other precipitation may occur to coat the oxide particles with a diffusion barrier (*cf.* 2§7.4). Unfortunately, little seems to be known for certain in this area, the products as sold are empirically derived – *i.e.* from trial and error.

---

[2] Orthophosphoric acid, H_3PO_4, is so named to distinguish it from metaphosphoric (HPO_3), pyrophosphoric ($H_4P_2O_7$) and hypophosphoric ($H_4P_2O_6$) acids. In dentistry, however, it is usually just called phosphoric acid; the two terms will be used interchangeably here.

●5.2 Liquid

The liquid might present another possibility for controlling reaction rate but this is, in fact, insignificant in practice. Orthophosphoric acid is a relatively weak acid (although much stronger than eugenol), and the ions present in solutions containing it depend on the pH, such that at pH 0 only very small amounts of $H_2PO_4^-$ may be present (Fig. 5.1). Similarly, because pH falls steadily with increasing concentration through the range of dental interest (Fig. 5.2), the only ionic phosphate species effectively present is $H_2PO_4^-$ and then in low concentration. It is only at very low overall phosphate concentrations or much higher pH that other ions appear. Therefore the ionic constitution of the acid liquid is essentially irrelevant to the setting reaction of zinc phosphate cement[3] and its rate. This is contrary to the statements sometimes made that the reaction products depend directly on the proportions of the ions present, a misunderstanding of the nature of equilibria.

The chemistry can nevertheless be understood by examination of the P_2O_5 - H_2O - ZnO constitutional diagram. This turns out to show the power of such an approach in dealing with a system as complicated as this one, where elementary acid-base considerations fail completely to explain both the empirically-derived clinical handling procedure and known properties.[15] In that sense, this system may be treated as a case study to put into a practical context a number of the points made in Chap. 8, and for that reason the explanation will be detailed and stepwise through the entire process.

●5.3 Diagram layout

To study the chemistry of this system, we start by identifying the known stoichiometric compounds and plotting these on a ternary composition diagram (Fig. 5.3). Since we are only dealing with whole-number ratios of the components (even though the empirical formulae may be reduced to the lowest numerical terms by taking out common factors), various **isopleths** can be drawn through those compound points to indicate the relationships: *e.g.*, the sequence from zinc metaphosphate, $Zn(PO_3)_2$, towards the water apex, adding two, then two more, moles of water to form the primary phosphate, then its dihydrate. Similarly, a line drawn between any of the water-free salt points (on the ZnO-P_2O_5 edge) and the H_2O apex are lines of constant Zn-P ratio, and the corresponding hydrates therefore lie on such lines.

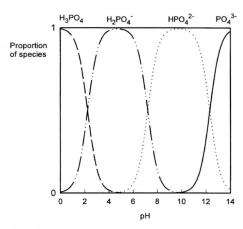

Fig. 5.1 The variation in the proportions of the various dissociated species of phosphoric acid as a function of pH.

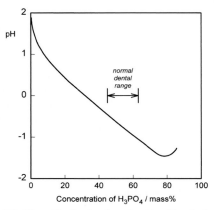

Fig. 5.2 The pH of orthophosphoric acid solutions.

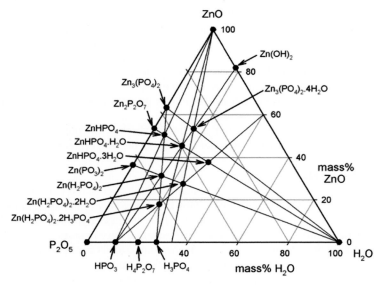

Fig. 5.3 The stoichiometric compounds of the ZnO - P_2O_5 - H_2O system, and some isopleths indicating certain fixed ratios, irrespective of chemistry.

---

[3] This used to be called, and sometimes is still sold as, zinc oxyphosphate cement; there is no distinction and the name has no chemical validity.

This kind of plot is a purely numerical exercise: there is no necessarily implied chemistry. However, the line between the points corresponding to the compositions of H_3PO_4 and $Zn_3(PO_4)_2$ does, for example, represent the anhydrous reaction products of the neutralization of orthophosphoric acid itself by ZnO, including the compound $Zn(H_2PO_4)_2 \cdot 2H_3PO_4$, which has **acid of crystallization**.

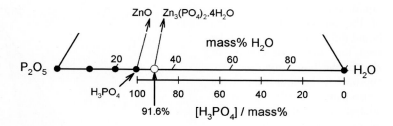

Fig. 5.4 Recalibration of part of the P_2O_5 - H_2O edge as a composition line for H_3PO_4 - H_2O. The calculation of equation 5.1 is thus equivalent to the intersection with the relevant isopleth, the required acid concentration indicated by the open circle.

One matter of immediate interest is the orthophosphoric acid point, H_3PO_4. The edge of the composition triangle between this and the water apex can be recalibrated as the concentration (mass%) of this acid in aqueous solution (Fig. 5.4). This is an example of the application of the lever rule (8§3.10). The formation of H_3PO_4 can be written thus:

$$P_2O_5 \;+\; 3H_2O \;\Rightarrow\; 2H_3PO_4 \tag{5.2}$$

It can then be seen that the isopleth which runs through the point for $Zn_3(PO_4)_2 \cdot 4H_2O$ identifies the acid concentration required for complete reaction, as in equation 5.1.

The isothermal section of a partial equilibrium constitutional diagram at 25 °C is shown in Fig. 5.5, with several special points labelled. The detail of the area of the diagram on the left-hand side is omitted as it is of no relevance to the dental system and can safely be ignored. In addition, zinc hydroxide has not been reported to form under normal dental conditions, so can be omitted from consideration. As indicated above, the effective mixing liquid of the dental product is H_3PO_4 solution, so that composition line is used for clarity, rather than P_2O_5 - H_2O. To simplify the description, phase labels as in Table 5.1 will be used (see also the box).

Table 5.1 Phase labels for zinc phosphates relevant to the dental cement system.

α	$Zn_3(PO_4)_2 \cdot 4H_2O$
$β_1$	$ZnHPO_4 \cdot H_2O$
$β_2$	$ZnHPO_4 \cdot 3H_2O$
γ	$Zn(H_2PO_4)_2 \cdot 2H_2O$
δ	$Zn(H_2PO_4)_2 \cdot H_3PO_4$

The compounds of relevance are marked by the solid points with the phase labels. The **solidus**, the boundary between the regions of all solid and those of solid + liquid, is represented by the irregular line drawn between the points ZnO - α - $β_2$ - $β_1$ - γ - δ - H_3PO_4. The **liquidus**, the boundary between the regions of all liquid and those of solid + liquid, is the line drawn between the points H_2O - **P1** - **P2** - **P3** -...- H_3PO_4. The points **P1** - **P3** do not correspond to stoichiometric compounds, but are critical solution compositions at 25 °C. These will be explained below.

The extremes of the reported commercial range of orthophosphoric acid concentrations for the dental cement, approximately 45 ~ 63 mass% as H_3PO_4, are marked on the acid composition line (Fig. 5.5).

By convention, solid crystalline phases are given lower case Greek letter labels as a shorthand notation for ease of labelling and for writing out reactions symbolically. They are usually in a simple continuous sequence: α, β, γ, δ ... , but not necessarily – historical usage and other factors may make it convenient to employ some out of sequence. They are essentially arbitrary (and especially what is chosen to start the list) and have no intrinsic meaning. Subscripts and other distinguishing marks may be used, likewise for convenience, but often to show similarity in some respect. Thus here $β_1$ and $β_2$ are both 'secondary' phosphate (as it happens), differing in their amounts of water of crystallization. Effectively, such a label only has meaning for the particular phase diagram in question, and unless that is referenced directly it cannot be interpreted: usage can vary by author and over time, and indeed according to the way the diagram is drawn. (See also 12§4.1)

Since the dental cement is prepared by mixing zinc oxide with just one such acid liquid, isopleths can be drawn from the acid line to the ZnO apex (constant P_2O_5 : H_2O ratio = constant H_3PO_4 : H_2O ratio) to represent all such possible mixtures. The dashed lines then represent the limits to the overall compositions which may be obtained with those commercial liquids, that is, at all possible powder : liquid ratios.

The reported upper limit of about 10 mass% Zn (*i.e.* calculated as metal) in those same commercial liquids corresponds to the region of the hatched ellipse in the liquid field in Fig. 5.5. It will be seen that the incorporation of zinc in the liquid represents an approach to the liquidus, with the consequent benefit of partial neutralization of the acid and therefore the loss of that portion of the heat of reaction prior to clinical use. In addition, the reduction of the concentration of 'free' acid means that its reactivity is also reduced, so the initial rate of evolution of heat is also lower. The more zinc or other metal dissolved in the liquid, the closer to the liquidus does its composition lie, and the more sensitive to loss of water it will become, when solid might appear. Cloudy liquids, or ones containing crystals, are spoilt and should be discarded, as should the remains of liquid from a pack whose powder has been used up. (This, of course, applies to any powder-liquid cement system, not just zinc phosphate.)

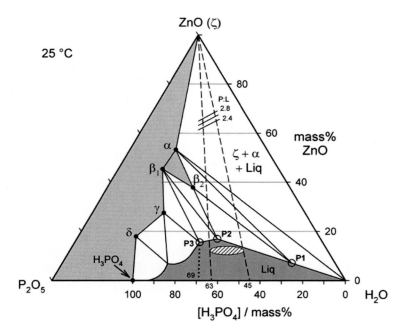

Fig. 5.5 Partial constitutional diagram for the system P_2O_5 - H_2O - ZnO. The all-solid region is indicated by the grey area on the left; the liquid field by the darker grey region. The region of the compositions of commercial liquids is indicated by the hatched ellipse in the liquid field. The recommended powder : liquid (P:L) ratio range 2.4 ~ 2.8 g/mL is indicated by those constant-proportion lines.

Commonly, the liquid is said to be 'buffered' by the zinc, and sometimes other metals such as aluminium that are present in the liquid as supplied. Given the remarks above about the constitution of the liquid, and the reduction in the total heat of neutralization that this dissolved metal represents, such a description is seen to be of little value and plays no part in understanding the system.

•5.4   Setting reactions - Stage 1

To consider the setting reactions of the dental cement, we conduct a thought experiment in the form of imagining the consequences of a very slow addition of ZnO to the chosen acid composition, effectively performing a titration of the acid with a base. This is realistic in the sense that it is the ZnO that dissolves in the acid, and not the other way around. Thus, if we take the liquid composition as, say, 45 mass% H_3PO_4, the first step is to draw the **mixing line** from that point on the acid composition line to the ZnO apex (Fig. 5.5). This is then a composition line itself, representing all possible mixtures of the acid liquid and the powder. We can also refer to Ostwald's Rule of Stages (8§2.6) to describe the sequential formation of successively more stable phases, noting that α-$Zn_3(PO_4)_2$·$4H_2O$ is not stable in contact with the original liquid.

It is worth stressing that such a mixing line represents the only (overall) compositions accessible when mixing a given liquid with ZnO – it is only the proportions of these two that can change. Hence, in doing the thought experiment we must stay rigidly on that line, but we must refer to the landmarks on the phase diagram map to understand what is happening. Of course, should water be allowed to evaporate from the liquid during mixing, or condensation on a cooled slab be incorporated, the consequences can be ascertained in the obvious manner (see §5.13).

The first increments of the ZnO powder to be added to the acid liquid only dissolve, since this part of the mixing line is in the Liquid phase field. Whether or not zinc is complexed by phosphate in solution is irrelevant and the ionic constitution of the liquid does not require consideration in the reaction scheme. It is, however, quite improper to say that "primary phosphate" exists in solution.[4] This simple dissolution continues

---

[4]   So-called primary phosphate, $Zn(H_2PO_4)_2$, cannot play a part in the reaction in the usual range of acid concentrations as is frequently stated elsewhere, neither as an identifiable solution species nor as a solid.

until the liquidus is reached, somewhere between points **P1** and **P2**. Most of this part of the addition is, of course, represented by the zinc content of the liquid as supplied commercially. However, this cannot be taken too far as solid would appear on cooling, or even on slight evaporation of water from the liquid.

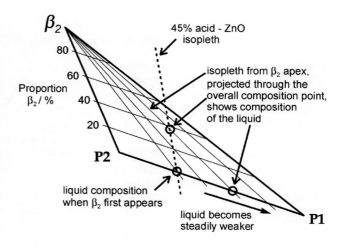

Fig. 5.6 Stage 2: The β_2 - Liquid field; ZnO reacts with liquid and precipitates β_2.

P1 and **P2** represent the compositions of two liquids. But since they are both aqueous solutions in the same system they are, of course, miscible. This means that all points lying between them on the **P1** - **P2** composition line represent only single liquids – not two phases – which may be viewed as the result of mixing different proportions of the two solutions represented by **P1** and **P2**. (This also applies in equivalent terms to the lines between **P1** and H_2O, and between **P2** and **P3**.)

●5.5   Stage 2

Continuing along the acid–ZnO mixing line, crossing the liquidus takes us into the Liquid + β_2-phase field (Fig. 5.6). The precipitation of "secondary" phosphate trihydrate therefore commences. We can be sure that the phase formed first is indeed β_2 because we see from the diagram that it is stable in contact with the liquid – while that liquid continues to exist. Conversely, there is no phase field with the description ζ + Liq; the ZnO is therefore unstable with respect to that liquid, hence it dissolves. The reaction may be written as:

$$\zeta \ + \ Liq \ \Rightarrow \ \beta_2$$
$$ZnO \ + \ H_3PO_4 \ + \ 2H_2O \ \Rightarrow \ ZnHPO_4 \cdot 3H_2O \tag{5.3}$$

As further ZnO dissolves, so more β_2 is precipitated, but now the acid concentration of the liquid declines as a result. This may seem a little odd, but note that the molar ratio of $H_3PO_4 : H_2O$ is 1:7 at about 44% mass% acid, so that removing these components in the ratio 1:2 (according to reaction 5.3) must result in acid dilution. This can be seen graphically by drawing a line from the β_2 apex through the overall composition point (on the dashed mixing line), extended to cut the liquidus between **P1** and **P2**. This intersection point moves towards **P1** as the amount of ZnO increases until, of course, it coincides with **P1**. The liquid then corresponds to about 20 ~ 25 mass% acid, while the overall composition lies on the β_2 - **P1** line. In addition, as this phase field is crossed, the proportion of liquid in the mixture is decreasing from 100 mass% at the liquidus to about 42 mass% at the β_2 - **P1** line, which is therefore being treated as a composition line itself.

●5.6   Stage 3

When the β_2 - **P1** line is crossed (Fig. 5.7) by the addition of more ZnO, the first step must be the production of the "tertiary" zinc phosphate tetrahydrate:

$$\zeta \ + \ Liq \ \Rightarrow \ \alpha$$
$$3ZnO \ + \ 2H_3PO_4 \ + \ H_2O \ \Rightarrow \ Zn_3(PO_4)_2 \cdot 4H_2O \tag{5.4}$$

(*i.e.* precisely the same as the overall reaction 5.1 – it is only now that it actually happens directly). The acid of the liquid must therefore be made more dilute because the effective concentration of the acid required (91.6% acid, reaction 5.1) is greater than that of the **P1** liquid itself, *i.e.* about 22% acid. The new liquid would therefore correspond to a composition on the H_2O side of point **P1** on the liquidus. But clearly β_2 is now unstable in contact with that more dilute liquid, since it does not appear in a phase field adjacent to such a composition.

The only stable solid phase for contact with liquids of compositions lying between **P1** and the H_2O apex is α. Hence, after such further addition of powder, some of the β_2 already precipitated must decompose, enough to cause the liquid composition to return to that corresponding to **P1**. A disproportionation reaction therefore occurs:

$$\beta_2 \Rightarrow \alpha + \text{Liq}$$
$$3\text{ZnHPO}_4\cdot3\text{H}_2\text{O} \Rightarrow \text{Zn}_3(\text{PO}_4)_2\cdot4\text{H}_2\text{O} + \text{H}_3\text{PO}_4 + 5\text{H}_2\text{O} \qquad (5.5)$$

The 'liberated' liquid is thus essentially 52% 'free' acid, and indeed this replenishes the diluted acid liquid, bringing its composition back to that composition corresponding to **P1**.

An alternative view of the changes occurring as the $\alpha + \beta_2 +$ Liquid phase field is crossed, still following the dashed mixing line from 45% acid, is to write the overall reaction as:

$$\zeta + \beta_2 \Rightarrow \alpha + \text{Liq} \qquad (5.6)$$

It must be emphasized that the small amount of liquid produced overall in this process must correspond exactly to **P1**, whose composition does not change during this stage of the process. The fact that some liquid is produced is shown by the mixing line (dashed in Fig. 5.7) being at a slight angle to the α-β_2 side of the composition triangle. This is clearly true for any such isopleth passing through this field since they must by definition pivot on the ZnO apex.

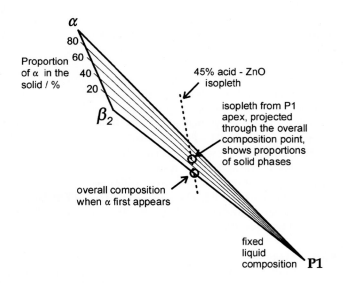

Fig. 5.7 Stage 3: The α - β_2 - Liquid field; β_2 reacts and α is precipitated.

●5.7 Stage 4

Once the β_2-phase has been used up, the overall composition enters the $\alpha +$ Liquid phase field (in the triangle marked by α - **P1** - H_2O) (Fig. 5.8). Continuing to follow the mixing line, direct precipitation of the α-phase now continues according to reaction 5.4. Now, since the molar ratio of $H_3PO_4 : H_2O$ consumed in this process is 2:1, the liquid is yet again diluted, the composition moving quickly towards the H_2O apex, following the liquidus. In fact, the liquidus cannot quite meet that apex because ZnO and "tertiary" zinc orthophosphate are not completely insoluble, but the distinction is too fine to be drawn in this diagram.

Nevertheless, once the α-H_2O boundary is crossed, there is no further reaction. Any subsequent addition of ZnO remains as unreacted core in the reaction product matrix, which is of course what is required for the composite structure to have the necessary mechanical properties.

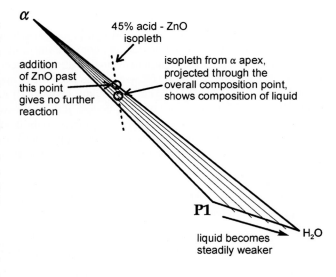

Fig. 5.8 Stage 4: The α - Liquid field; α is precipitated by direct reaction with ZnO, as the liquid becomes dilute.

The overall setting process is therefore first the precipitation of β_2, then α with the concomitant conversion of β_2, followed by the formation of α alone. This is summarized in Fig. 5.9.

●5.8 Kinetic limitation

The above scheme is based essentially on the assumption in the thought experiment of the attainment of equilibrium after each infinitesimal increment of ZnO. However, the disproportionation reaction 5.5 proceeds only slowly, even in contact with quite dilute acid (composition to the right of **P1**), and it is this that constitutes the basis of the final setting reaction of the clinical cement: the slow release of acid delays the completion of the formation of α-phase. In other words, the formation of β_2 represents a reservoir for a proportion of the acid that is only slowly made available for final reaction. Without this kinetic limitation, the cement would set far too quickly to be of any clinical use. This mechanism also accounts for the long-continuing low pH measured on the surface of apparently set cement, because the remaining liquid is continually replenished with free acid. Completion of the reaction would be indicated by a tendency for the pH to rise again from its intermediate, dynamic quasi-equilibrium value when the $\beta_2 \Rightarrow \alpha$ and $\zeta \Rightarrow \alpha$ reactions are proceeding at equivalent rates.

●5.9 Stronger acid

To return to the thought experiment: if the acid liquid is sufficiently concentrated initially, that is, greater than about 58 mass% H_3PO_4, β_1 will be the first solid precipitated during the addition of powder:

$$\zeta + \text{Liq} \Rightarrow \beta_1$$
$$\text{ZnO} + H_3PO_4 \Rightarrow \text{ZnHPO}_4 \cdot H_2O \quad (5.7)$$

We can see this by choosing as acid a concentration corresponding to about the upper limit of commercial liquids, 63 mass% H_3PO_4. We therefore follow the left-hand dashed mixture line towards the ZnO apex in Fig. 5.5. This isopleth goes initially into the field β_1 + Liquid (Fig. 5.10). As this phase field is crossed, so more β_1 is precipitated, following the same pattern as described for reaction 5.3. Thus the liquid composition moves towards **P2** along the line **P3** - **P2**, and the amount of liquid declines from 100 mass% to about 80 mass%.

However, if now a little more ZnO is added, this would take the overall composition into the $\beta_1 + \beta_2$ + **P2** field (Fig. 5.11). The immediate reaction would then be precisely that of 5.3, but this means that the liquid has become more dilute with respect to the acid. Now it is the β_1 that is unstable, so in effect it undergoes an hydration reaction:

$$\beta_1 + \text{Liq} \Rightarrow \beta_2$$
$$\text{ZnHPO}_4 \cdot H_2O + 2H_2O \Rightarrow \text{ZnHPO}_4 \cdot 3H_2O \quad (5.8)$$

This reaction obviously consumes water, increasing the concentration of the liquid back to **P2** again.

There seems to be no evidence that this is other than a minor detour in the reaction path and there

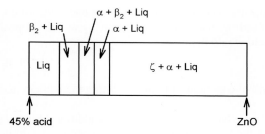

Fig. 5.9 Summary of the 45% acid, gradual addition of ZnO thought-experiment at 25 °C, *i.e.*, reading from left to right.

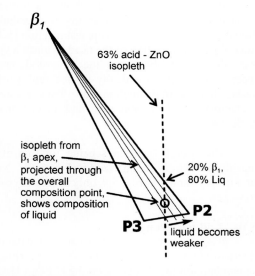

Fig. 5.10 The β_1 - Liquid field; with stronger acid, ZnO first reacts with liquid and precipitates β_1.

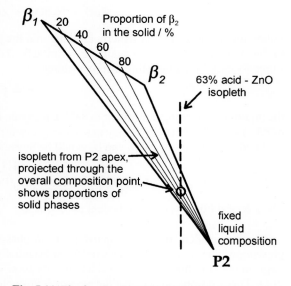

Fig. 5.11 The β_1 - β_2 - Liquid field; β_1 disappears as β_2 is formed.

appears to be no obvious detriment due to its presence. It can be taken simply as part of the ordinary retardation of setting that makes the cement a practical proposition. The overall reaction while crossing the $\beta_1 + \beta_2 + $ **P2** field would therefore be:

$$\zeta + \beta_1 + Liq \Rightarrow \beta_2 \qquad (5.9)$$

with the amount of liquid declining to about 55 mass%.

However, because of this additional kinetic limitation, it is likely that the β_1-phase persists long enough for the liquid, in the meantime, to become so dilute by reaction with further added ZnO that β_2 would not be stable and a direct conversion of β_1 to α then occurs:

$$\zeta + \beta_1 + Liq \Rightarrow \alpha$$
$$ZnO + 2ZnHPO_4 \cdot H_2O + 2H_2O \Rightarrow Zn_3(PO_4)_2 \cdot 4H_2O \qquad (5.10)$$

Bear in mind that all of the above reactions are mediated through the liquid phase, that is, dissolution followed by reprecipitation. This last reaction (5.10) is therefore an illustration of the need to make a very clear distinction between the thermodynamic expectation and the kinetically-limited reality of a process: this kind of distinction is so often a feature of the chemistry of dental materials and, indeed, many practical systems in other areas.

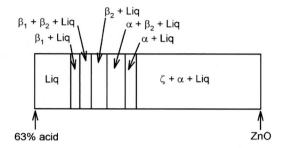

Fig. 5.12 Summary of the 63% acid, gradual addition of ZnO thought-experiment at 25 °C, *i.e.*, reading from left to right.

The overall reaction scheme for this more concentrated acid is summarized in Fig. 5.12. Note that it is very similar to Fig. 5.9, only adding the extra two stages.

●5.10 Acid too strong

It is known from direct experiment that there is an acid concentration above which the setting time begins to increase sharply, namely ~69 mass% H_3PO_4 (Fig. 5.13), *i.e.* a little above the commercial product upper limit. The isopleth joining this point on the H_3PO_4 - H_2O composition line and the ZnO apex (part shown dotted in Fig. 5.5) passes through point **P3**, one of the apices of the β_1 + Liquid phase field (Fig. 5.14). The critical issue is that to the left of this point primary phosphate dihydrate, γ-phase, would be the phase first precipitated on crossing the liquidus:

$$\zeta + Liq \Rightarrow \gamma$$
$$ZnO + H_3PO_4 \Rightarrow Zn(H_2PO_4)_2 \cdot 2H_2O \qquad (5.11)$$

Following the now familiar pattern, crossing into the $\beta_1 + \gamma + $ Liq field, the first reaction with further addition of ZnO would be reaction 5.7 again, *i.e.* forming β_1. This must make the liquid more dilute, leaving the initially formed γ unstable; it therefore disproportionates:

$$\gamma \Rightarrow \beta_1 + Liq \qquad (5.12)$$

the liquid formed being equivalent to 84% acid, again maintaining the liquid phase constant at the composition of **P3**. Yet again, if there is a delay in achieving this for kinetic reasons, the concentration of the liquid may have declined so much that even β_1 and β_2 would be unstable, so direct conversion to α would occur:

$$\gamma \Rightarrow \alpha + Liq$$
$$3Zn(H_2PO_4)_2 \cdot 2H_2O \Rightarrow Zn_3(PO_4)_2 \cdot 4H_2O + 4H_3PO_4 + 2H_2O \qquad (5.13)$$

The released liquid is then equivalent to 92% acid.

It is evident from a comparison of the phase diagram with setting-time data[16] (Fig. 5.13) that the presence of this compound, γ, does indeed substantially delay the setting reaction by tying up a proportionately large amount of phosphate which can then only be released to the system very slowly for further reaction with ZnO. This is despite the fact that the γ-phase is unstable with respect to acid more dilute than 69%. The stronger the original acid, the more γ is produced in the initial mixing stages, and the greater the delay.

It is considered clinically impractical to allow these longer setting times, even though the final strength of the cement continues to rise with increasing initial acid concentration (Fig. 5.13): the risk of damaging the bond between the device and the tooth is too great and, of course, out of the control of the dentist if the patient has left the chair.

●5.11 Temperature

The rate of setting of zinc phosphate cement increases with temperature (as would also be true of most other systems), and two specific aspects of the clinical mixing procedure are designed to minimize the effects. Firstly, a cooled mixing slab may be used; secondly, the mixing is done in small increments. This increase in rate can also be explained by a further aspect of the system, in addition, that is, to the general kinetic effect expected on raising temperature. However, the consequences are more serious than merely reduced working time.

Figure 5.5 is the 25 °C isotherm of the constitutional diagram, one section through the triangular prism that would be the full diagram. At higher temperatures the Liquid field expands, the liquidus moving generally in the direction of the ZnO apex (Fig. 5.15) (the stoichiometric compound composition points of course do not move). This means that the solubility of each solid phase increases with temperature, which is a common type of behaviour. Further, point **P1** moves slowly back towards **P2** as the temperature is raised, thus reducing the amount of β_2 that can form initially, *i.e.* α will form directly at progressively higher acid concentrations as the temperature is raised, so that no retardation of setting occurs through the creation of a phosphate reservoir in β_2-phase solid. (Indeed, **P2** and **P3** also move.)

When the ZnO powder is added to the liquid, the precipitated phases (whatever they are) will tend to coat the oxide and slow down further reaction: a diffusion barrier limitation to rate. But if much oxide is added at once, the very large surface area presented to the liquid will mean a very large amount of reaction in a short time; the temperature will thus rise rapidly, increasing the solubility of all possible reaction products.

Now, when the mixture cools, given that the acid is now very dilute, only α-phase is expected to precipitate directly from solution to give large, well-formed crystals which are very weak and brittle. Not only does β_2 not form, but the matrix is clearly crystalline and the setting time unhelpfully short. Hence, the addition of only a small increment of ZnO can be seen as adding only a small surface area for reaction so that the heat released can be dissipated. Subsequent increments repeat the process, with rapid formation of a protective shell of β_1 or β_2 on

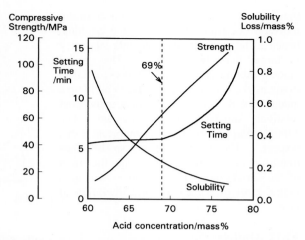

Fig. 5.13 Variation in some properties of zinc phosphate cement with the water content of the liquid. ('Solubility Loss' means from a given surface area in a standardized time.)

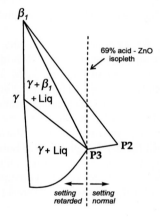

Fig. 5.14 Acid stronger than 69% results in the formation of γ phase initially and retarded setting.

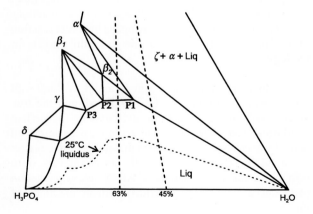

Fig. 5.15 Nature of the shift of the liquidus at raised temperatures. Note that the stoichiometric points corresponding to the various solid phases do not (cannot!) move, but that the key liquid points **P1**, **P2** and **P3** do move. The reaction paths are therefore changed.

the ZnO, at least until the acid concentration moves to the right of **P1** along the liquidus; hence the last one or two increments can safely be larger than the first few. The special clinical mixing procedure is thus essential to dissipate the heat of reaction and to build up a quantity of β_2 which can subsequently react relatively slowly to develop the set. The ambient temperature clearly is important in this respect. The particle size also becomes important to the setting rate through the effect of the surface area: finer powders will give higher rates of conversion and liberation of heat.

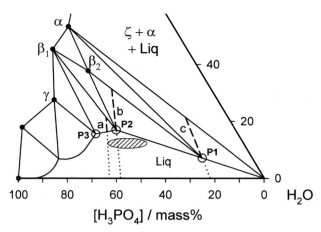

Fig. 5.16 Mixture paths for the assumption of delayed reaction of intermediate phases β_1 and β_2.

●5.12 Outline

Although the setting reactions are set out above in some considerable detail, as said before this is more by way of a case study and a demonstration of several key issues in a practical context. In particular, it was based on an equilibrium process, with only a brief mention of the kinetic limitation at §5.8. We can go a little further with an extra assumption.

We suppose that a solid once formed does not (for the moment) react any further. Then, again starting with the normal upper limit 63% acid, the path with solid production (of β_1) is as already described in Fig. 5.10, that is line segment **a** in Fig. 5.16. The liquid then reaches composition **P2**. Since the β_1 does not now react, the mixture line isopleth for further addition of ZnO has in effect been moved to pass through **P2**, so that the line segment **b** (Fig. 5.16) is the path to be followed. While β_2 is precipitated, as in Fig. 5.6, the liquid composition again shifts until it reaches **P1**. As the β_2 is not reacting further, the new mixture line corresponds to line segment **c**, producing α directly, during the course of which the liquid becomes steadily weaker, as before (Fig. 5.8) until the α + Liq phase field boundary is reached, when any added ZnO simply remains. We now have a mixture of $\beta_1 + \beta_2 + \alpha + \zeta +$ Liq. Of course, β_1 and β_2 are unstable in contact with that liquid, and so must now be permitted to react slowly, consuming some more ZnO to produce α, finishing off the setting process. The reaction for β_2 is shown above (reaction 5.5), while that of β_1 is, similarly:

$$\beta_1 \Rightarrow \alpha$$
$$3ZnHPO_4 \cdot H_2O + H_2O \Rightarrow Zn_3(PO_4)_2 \cdot 4H_2O + H_3PO_4 \tag{5.14}$$

consuming some water, and releasing phosphoric acid. Thus these β phases represent a delaying reservoir of phosphate for the continuation and completion of setting. Again, it must be remembered that all of these reactions are solution-mediated: nothing happens by direct reaction of solid phases as the activation energy for diffusion is far too high.

In actuality, of course, the β_1 and β_2 will not wait for the addition of ZnO to be completed, and the true process will lie between these extremes of the perfect equilibrium sequence and the fully-limited case described here. Furthermore, it can be seen that it is a simple matter to extend the scheme to include the formation of γ, or to start with weaker acid than 63%. All the relevant reaction paths can be summarized as in the flow diagram scheme of Fig. 5.17.

We have found several things in the course of this. Firstly, a phase diagram is a powerful tool for understanding reactions and clinical observations where no other approach is feasible. This approach has provided

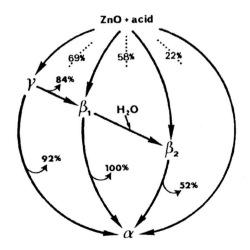

Fig. 5.17 Summary of the principal reaction paths for dental zinc phosphate cement. Where reactions result in the release of acid (curly arrows), the nominal concentration (mass percentage) of the released acid is shown.

detailed explanations of the observed behaviour and of the handling instructions derived pragmatically from experience rather than theory. Secondly, the tools for obtaining that understanding were no more than the composition lines and three-component mixture diagrams, applied repeatedly, but using as components the various solids and liquid compositions as appropriate. This could be done despite the apparent complexity of the overall diagram: breaking it down into small pieces and small steps is the way.

●5.13   Water content

An examination of the stoichiometry of the reaction is revealing (reaction 5.1). Since the final reaction product is the tetrahydrate, and three-quarters of that water comes from the neutralization reaction itself, the liquid should have no more than about 8 mass% of water if there is to be no excess at the end of the reaction. This concentration of acid would be so viscous as to be clinically unusable, as well as causing the setting time to be excessively long (the δ-phase would also be formed). Any excess water from the liquid, therefore, is left over as the insoluble α-phase, zinc triorthophosphate tetrahydrate, is precipitated. This water remains as liquid distributed in the set mass, much as is the excess water in gypsum products (2§8). In terms of the strength of the material, such liquid will behave as voids (*i.e.* a modulus of elasticity and strength of zero) and reduce the strength (Fig. 2§8.3).[17] This accounts for the shape of the strength curve of Fig. 5.13. The optimum mixing ratio has been shown to be in the region 2.6 ~ 2.8 g/mL[18], indicated by the "P:L" lines in Fig. 5.5 (where correction for the density of the acid liquid leads to their slopes). From the position of these lines, it is clear indeed that a substantial amount of liquid is left in the set cement, but also that there is no benefit in pushing toward higher powder : liquid ratios (higher final ZnO content) because of the limitations set by the minimum mixing liquid requirement (2§2.4).

The improvement in strength obtained with high powder : liquid ratios can be seen not merely as the improvement to be expected from high filler loadings in a particulate composite material, but also as aided by the reduction in effective pore volume caused by (slightly) smaller amounts of left-over liquid. In addition to the loss of strength, the porous nature of the set mass so produced effectively increases the rate of dissolution. Zinc phosphate is not entirely insoluble. The standard tests of the amount of material lost during a fixed time of extraction with water show higher values for higher water content (Fig. 5.13).

There are therefore two additional reasons for not allowing water to contaminate or evaporate from the liquid, even while it is on the slab. The liquid is sufficiently **hygroscopic** (in the proper chemical sense!) as to absorb moisture from the air if the humidity is high enough, but a chilled slab will become wet from condensation if below the dew-point[5] for its environment, and so contaminate and dilute the mixture.

●5.14   Mixing left-overs

Despite the retardation of setting that arises from the special chemistry of the system, the rate of reaction is still far too fast to be useful with most zinc oxide, and the effects described in §3 must be utilized. The oxide is fired at temperatures up to 1400°C in order to reduce its reactivity sufficiently. Other substances may be included, such as MgO, to further modify the powder. It is thus necessary to recognize that the powders of zinc oxide-eugenol and zinc phosphate cements (and, indeed, of zinc polycarboxylate or any other product) are not interchangeable. In fact, it is ill-advised to mix nominally-similar powders or liquids from different brands of the same type of product. Each manufacturer can be expected to have achieved a balance of properties in a different way, whatever class of product is being considered, not just cements. If there is any surplus when one component of a two-part product is used up, it is best to discard the excess, especially if that component is liable to deteriorate, as is all too easy with cement liquids. It is not usual to be able to buy one component part at a time to make saving the excess an economic proposition. It is curious how much wastage of this kind occurs throughout dentistry.

It might be noted that the incorporation of a proportion of MgO in the cement powder will not change the outcome substantially. The ionic radii of Mg^{2+} and Zn^{2+} are 65 and 74 pm respectively, radius ratio about 1 : 1.14. The Mg^{2+} ion may substitute for Zn^{2+} in all the relevant compounds, forming continuous solid solutions – no new phases appear to be formed, although the liquidus and **P1**, **P2**, **P3** may suffer some slight adjustment.

●5.15   Cement structure

The final matrix material of zinc phosphate cement has been called a gel, but true gels cannot be formed

---

[5] **Relative Humidity** is defined as the ratio of the water vapour pressure actually present to the pressure of the saturated vapour at the same temperature; it is usually expressed as a percentage. **Dew-point** is the temperature at which the water vapour already present in the air would be saturated and thus would begin to condense, *i.e.* dew begins to form.

in this system (see box, 7§8), and the cement depends for its strength on the microcrystalline, nearly amorphous mass which bonds the unreacted excess zinc oxide particles together. The very large number of crystallites means an effectively very small grain size and a less brittle material. In addition, the presence of aluminium and other metals which are dissolved in the liquid or included in the powder may help to prevent good crystallization or promote the creation of amorphous precipitates. However, on exposed surfaces of the cement in the presence of water, the normal oral environment, recrystallization tends to occur to develop the thermodynamically-expected larger and more perfect crystals of α-phase. It is supposed that the presence of these larger brittle crystals would reduce the strength of the mass in the long term, and this may be related to the cement's relatively poor service life.

Given the inevitable excess liquid in the set material (§5.13), we again have the need to recognize the presence of **intrinsic** and **extraneous flaws** in the set cement (2§8.1), as bubbles resulting from the mixing process affect the strength of this brittle material on top of the effects of powder : liquid ratio.

•5.16 Other systems
Since the reaction in zinc phosphate cements may be envisaged in part as being between an acid phosphate and zinc oxide (reactions 5.6 and 5.9), some cement products have been formulated so as to mix a powder with water only. The soluble acid phosphate included (either of calcium or magnesium) provided the anions required for the reaction. Difficulties such as the solubility of the set cement, the hygroscopic nature of the powder, and large setting contraction have precluded their adoption as practical systems although the idea remains attractive as a means of avoiding one source of variability in the final material. Obviously, unless the water required for the minimum mixing volume (2§2.4) can be fully-reacted as water of crystallization, it must remain as liquid and so act as voids in the set cement, limiting the attainable strength.

There are many other metal oxide - acid systems, *i.e.* acid-base reactions, that can be regarded as cement-forming, bearing in mind that the aim is to produce a solid composite structure from a paste (Fig. 5.18). However, only copper oxide has been used in clinical preparations. The idea underlying their use was that copper ions are bacteriostatic (which means generally toxic – being a heavy metal) and so they were recommended for severe lesions and deciduous teeth. However, no real advantages have emerged and their use in dentistry has declined.

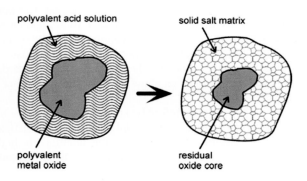

Fig. 5.18 General structural change on setting for a metal oxide - acid cement.

•5.17 Etching
It should be emphasized that the retentive effect of cements such as zinc oxide-eugenol and zinc phosphate is entirely mechanical. That is, they rely on the interlocking of the roughness of the two bodies to be joined with that of the set cement. The cement, of course, must be sufficiently fluid at the start to flow into that roughness and achieve intimate contact with all available surfaces. Equally, the two bodies to be joined must be rough. A prepared tooth surface is usually fairly rough from the effects of the tools used, but a cast metal surface may need to be grit-blasted (20§4) to ensure that it is rough enough. Orthodontic brackets may have special patterns formed in the cementing surface for this reason.

There is a further contribution made by the liquid of the cement where this is acid enough to dissolve tooth mineral, hydroxyapatite – which is true of all dental cements. Some differential etching almost certainly must occur, allowing a better interlocking (see 10§6), and acid attack may also assist by removing some of the more loosely-held debris that would otherwise weaken the cementation. Some dental products are supplied with what are called **conditioners**, acid solutions that are meant or claimed to act in this cleansing manner only. However, it is obvious that if such a solution is acid then some tooth material etching must occur. For example, citric acid solutions have been used. This etching is likely to be beneficial in the sense that the mechanical interlocking or **key** is improved. Nevertheless, it should not be thought that because the *intention* is only to clean that cleaning is all that happens. Failure to recognize that the chemistry of such solutions unavoidably entails etching of enamel and dentine could result in problems.

Sometimes the interlocking roughness is described as **micromechanical key** to distinguish it from the 'gross' mechanical retention form used for amalgam and other direct restorative materials (undercut cavity walls). The important issue is slightly different in the two cases. Gross retention is simply a matter of preventing the restoration falling out. Key on the smaller scale of roughness is the only means of obtaining resistance to shear in devices such as inlays and crowns where no gross undercuts can be employed. In the absence of specific chemical bonding, in neither case can there be significant resistance to tensile forces, such as would be experienced on changes of temperature or deflections on loading of either tooth or device, unless there were an appreciable amount of 're-entrant' roughness, *i.e.* microscopic undercuts, that the cement could penetrate.

§6. Zinc Polycarboxylate

A cement system that takes advantage of the chelation effect described for zinc oxide-eugenol, but which replaces the van der Waals bonding between molecules in the solid by much stronger covalent bonds, is that between zinc oxide and a polycarboxylic acid. This is done by using an acid whose functional groups are substituents on a polymer chain. Poly(acrylic acid) is used in aqueous solution to react with a zinc oxide powder, which has been modified by the addition of magnesium oxide and high temperature sintering to control the reactivity in a similar manner to that used for zinc phosphate. The setting reaction is an ordinary acid-base neutralization. This system[6][19] is one of the few materials which has been deliberately designed for a purpose in dentistry.

●6.1   Setting

A solution of poly(acrylic acid) is a rather stronger acid than eugenol and so also forms stronger chelating bonds than does that phenol. While tetrahedral coordination is commonly expected for the zinc ions,[20] water is probably involved to create an octahedral complex (Fig. 6.1, 6.2[7]) as this lowers the energy of the structure. Clearly, the carboxylate groups involved can be on the same or different polymer chains. When the latter case occurs, or they are enough apart, this in effect cross-links the polymeric acid to create a three-dimensional, tangled, network. However, these ionic bonds are not as strong as covalent cross-links, as in some impression materials, and so some time-dependent stress relaxation is therefore to be expected as these bonds interchange (as seen in the polysulphides, for example).

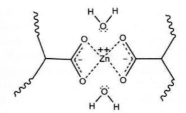

Fig. 6.1 A possible chelation structure binding chains together in zinc polycarboxylate cement, also showing coordinating water.

The cross-linking to form the polymeric matrix in this type of cement results in the setting material progressing continuously from being a high-viscosity, probably dilatant, plastic paste (Fig. 4§7.7) when the powder-liquid ratio is high enough, to showing more and more rubbery behaviour and retarded deformation, before it becomes effectively solid. This is similar to the sequence of changes observed for filled resin restorative materials (Figs 6§4.6, 6§4.7), and it has some important implications. Reaction starts as soon as powder and liquid come together, so mixing must be prompt and may benefit from using a cooled slab (again avoiding any condensation that would dilute the mixture).

Fig. 6.2 Model of the kind of structure shown in Fig. 6.1, with the coordinating atoms shown in 'space-filling' style to show the compact structure around the central metal ion.

The effective working time is over as soon as rubbery characteristics develop, which is indicated by a 'stringy' appearance. This 'stringy' condition implies that the reaction has proceeded too far for clinical use, as the rubbery aspect of the behaviour would compromise the flow necessary for good cementation of devices. Thus, prompt placement and seating of the cemented device is important. This property of steadily

---

[6] Invented in 1968 by DC Smith.

[7] Images of molecular models are have been 'relaxed' to a minimum energy state; the structure is natural, not guessed.

increasing viscosity does create handling difficulties. Oddly enough, they seem to have been overcome to some extent by adding some tartaric acid to the liquid, which may have an effect similar to that in glass ionomer cements (discussed below, §8.3): the working time is extended without affecting strength.

The initial viscosity of the mixed system is high, having a large amount of powder in an already viscous liquid. This inhibits diffusion and, given the ready formation of ionic coordination bonds and the use of an atactic polymer, the material is incapable of crystallizing. Thus, the final cement structure consists of residual oxide particles embedded in an amorphous mass of polysalt as the matrix. The acid liquid is composed of no more than about 40 mass% of polyacid, but the water from the liquid, as well that produced by the neutralization reaction, is so tightly bound to the extremely polar acrylate polymer that it is effectively immobilized as a solvation shell. This is exactly parallel to the situation in alginates (7§9) – the water is structural. It therefore allows the solid cement matrix to be considered (properly, this time) as a gel. As regards the liquid, it might be noticed that the change in polymer from a poly(ester), *i.e.* PMMA, to a poly(acid) results in the change from a totally insoluble but slightly water-absorbent polymer to a very soluble one; solvation due to the increased polar nature of the carboxylate group, also now ionizable, is the reason.

●6.2   Bonding

Zinc polycarboxylate was the first cement to be deliberately designed for a dental application, and one particular design feature leads to perhaps its most important property: it is adhesive to enamel. The tooth enamel presents many calcium ions at the surface of the hydroxyapatite crystals and, while these are held in place by ionic forces due to adjacent phosphate groups, there is still scope for some coordination with external ions (Fig. 6.3), only now the coordinating oxygen atoms for a square-based pyramid with the calcium ion at the apex (ignoring the rest of the surface). It is probable, for example, that even in zinc oxide-eugenol, some attachment of eugenol to the surface occurs. However, whether or not this bond is strong, stress cannot be transferred effectively to the bulk of the cement because the eugenol-eugenol bonding is weak. But with the link between acid residues being through the covalently-bonded polymer chain, such stress transfer becomes effective and, multiplied over the many billions of such sites per square millimetre, results in a properly adhesive cement. Zinc oxide-eugenol, zinc phosphate, and many other cements have only

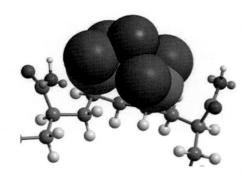

Fig. 6.3   Model of the coordination of a polycarboxylic acid fragment with a calcium ion (top, centre) in a solid surface, with the coordinating atoms shown in 'space-filling' style to show compact structure. (Note: the adjacent surface ions are not shown.)

ever relied on a **mechanical key**, the interlocking and resistance to shear provided by the intimate contact of the cement with the roughness of the substrate. Polycarboxylate cement thus laid a foundation for 'adhesive' dentistry. However, it can still be expected that some etching of tooth tissue occurs before the material is set so that some "self-etch" key will occur regardless, contributing to the strength of the union.

The chelation-bonding adhesive effect is not restricted to enamel surfaces but extends to any solid containing polyvalent metal ions. Thus, stainless steel and cobalt-chromium alloys, coated inevitably with Cr_2O_3 (13§5, 19§2, 21§2), have such a surface and may be cemented effectively, even to the extent of bonding stainless steel orthodontic brackets onto smooth enamel surfaces without (deliberate) etching. Gold alloys (which have no oxide film), porcelain (mostly monovalent metals), and platinum (again, no oxide) will not bond, but tin-plated platinum, subsequently oxidized, as used in foil matrix porcelain crowns (25§6.3), will bond through the tin ions present. There is also a disadvantage to this chemistry: the cement sticks to stainless steel instruments. Care must be taken to clean well the mixing spatula and other tools before setting has occurred. A solution of sodium bicarbonate has been recommended for removing set cement. The large excess of sodium ions displaces the chelated ions, both within the cement and at any interface, permitting it to dissolve (*cf.* the disruption of calcium alginate gel in sea-water, 7§7).

§7. Ion-leachable Glasses

The ion-leachable glass cements are chemically similar in principle to the foregoing types inasmuch as their setting reaction is one of acid-base neutralization. The differences lie in the chemical presentation of the metal oxides in the powder and the structure of the matrix in the set cement.[21]

The principal phenomenon upon which this class of cements depends is the hydrolysis of silica-based glasses. Such substances have a very wide range of possible compositions (Fig. 7.1), even if the metal oxide is limited to just that of sodium. Some of those based on anhydrous formulations are used in window and other glasses. Despite the need for such products to be durable, even these undergo some etching as do glass roofs, for example, on exposure to acidic rain in industrialized areas. Similarly, formulations which dissolve readily in water can be found. These represent unusable extremes for dentistry; water solubility is particularly undesirable in the mouth.

Naturally, compositions are not limited to using sodium as the cation. Potassium, calcium and zinc have been used for modifying properties without imparting any colour (see 24§6), and lead is used to increase the refractive index for wine glasses and the like ("lead crystal"; see Box in 3§4.1). Indeed, beryllium silicate was at one time used as a flux (see below), with the further effect of lowering the refractive index of the glass. However, this kind of addition is perhaps not appropriate in the dental context because of its toxicity (28§7): for example, lead is known to be dissolved from lead crystal glass left in contact with wine, which is acidic, for long periods. Thus, although the total amount would be small in typical uses of dental cements, and the dissolution occurring over years rather than days, it is unlikely to be considered acceptable.

●7.1   Basic chemistry

Silicates have a complex aqueous chemistry which depends on the formation of a variety of polymeric anions under a wide range of conditions; this arises because of the ability to share oxygen atoms in the silicate ions (Fig. 7.2) in reactions of the kind represented by:

$$2SiO_4^{4-} + 2H^+ \Rightarrow Si_2O_7^{6-} + H_2O \qquad (7.1)$$

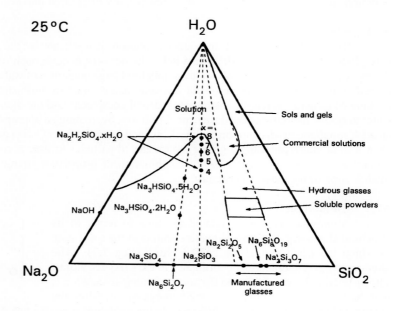

Fig. 7.1 The system Na_2O-SiO_2-H_2O, showing the ranges of solubility and various stoichiometric silicates.

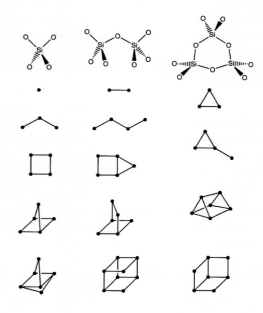

Fig. 7.2 The structures of some of the polymeric silicate species that have been identified in aqueous solution. The occurrence of each depends on pH. (The top row indicates the structures corresponding to the symbolic forms shown in the row below.) Note that because each non-bridging oxygen has a negative charge, these are all polyvalent anions, and each may be variously protonated, depending on pH – *i.e.*, to have -OH groups. Such species are weak acids.

Such a scheme can be seen to be extended easily to lead to the formation of very large species in a kind of polymerization. It leads ultimately to an extensive, irregular three-dimensional network under acid conditions.

●7.2   Silica gel

By adjusting the pH of an aqueous solution of a silicate by the addition of acid, the silica can be precipitated as a substance called **silica gel**. The network is quite open (there is little tendency to crystallize during such a rapid reaction) and can reversibly bind large quantities of water in its molecular-scale pores. This, indeed, is the basis of the desiccant of the same name found in those small packets included in the original packing of cameras, scientific instruments, dental equipment and the like. Moderate heating can remove the water to enable the reuse of the desiccant. Similarly, by treating an appropriate solid silicate glass powder with an acid, simultaneous dissolution of the silicate and precipitation of silica gel can occur. The silica gel can then form a solid matrix, binding together the unreacted particles to form a composite structure with cement-like properties. However, to get a useful combination of properties, strength, setting time, and so on, further modification of the glass is necessary.

●7.3   Aluminium

Aluminium is closely related to silica in terms of its structural chemistry. It forms aluminate anions in the solid state in which the basic unit is the $[AlO_4]$ group (Fig. 7.3). The aluminium ion is tetrahedrally coordinated to the 4 oxygen atoms. The distinction between this and the corresponding silicate ion is that the latter has a nominal associated charge of -4, while the aluminate has -5, reflecting the difference in the oxidation states of the two central atoms. In glasses, aluminium may also replace silicon one for one so long as an appropriate adjustment is made in the number of positive counter ions present to maintain overall electrical neutrality. A range of glasses containing aluminium can also be made with similar hydrolysis and dissolution properties as

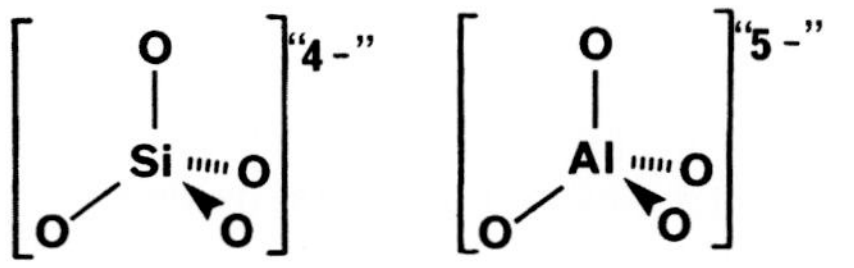

Fig. 7.3 The notional charge on the silicate and aluminate units of glasses. That is, for every 'Al' included in the network, there is an associated negative charge that must be balanced by a cation, such as Na^+,

the purely silica-based substances. The metal oxides which were included in dental silicate cement glasses included those of calcium and sodium, plus zinc in small quantities. The metal cations are also dissolved from the surface of such glasses at appreciable rates, thus aiding the disruption of the glass structure and allowing it to be hydrolysed more readily.

The fusion of the silicon, aluminium and other oxides to produce a glass would ordinarily require very high temperatures (see, for example, Fig. 25§2.3) but, in common with many other systems, the deliberate addition of 'impurities' lowers the melting point substantially. In glass technology, such substances dissolved in the melt are known as **fluxes** because they facilitate flow.[8] One of the best species for this effect in such glasses is the fluoride ion. Sodium or calcium fluoride, or **cryolite** (Na_3AlF_6), may be added for this purpose. The reactivity depression that this causes in the final glass powder is offset by the addition of a small quantity of a phosphate such as $Ca_3(PO_4)_2$ to the melt. The amount of fluoride flux added does not dissolve completely, however, and a large proportion is found in a second phase in the form of small (solid) droplets dispersed in the final glass. These act as light-scattering centres (24§5) and contribute to the final translucent, but not transparent, opal-like appearance.

●7.4   Setting reactions

Phosphoric acid was again used for the liquid in these now all-but obsolete **silicate cements**, although the concentration and additives were different from those used with zinc phosphate cement. The reason for this choice may be based on a combination of factors. Phosphoric acid is a relatively weak acid, so that dissolution of the glass would be relatively slow. However, the particular chemistry of the aluminium ions released may dominate the resultant properties. It would tend to give a stronger product due to the stronger ionic interactions between the tribasic phosphoric acid and a trivalent cation. There would also be better water binding as well because of the high charge and opportunities for hydrogen bonding.

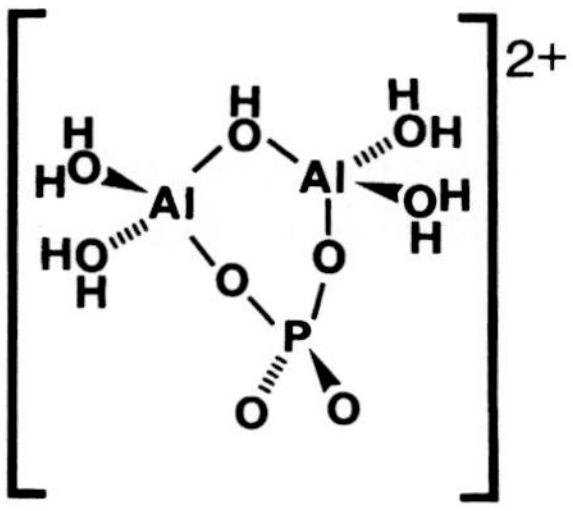

Fig. 7.4   Aluminium phosphate complex bridge structure found in solution.

---

[8]   Such fluxes must be carefully distinguished from their soldering namesakes (22§2): the latter are washed away when their oxide-removal job is done.

Aluminium ions also readily form polymeric species in solution, with polyvalent anions forming bridges between the similarly highly-charged aluminium cations (Fig. 7.4); phosphate thus performs this role well. These bridged structures (Fig. 7.5) extend as the concentration of aluminium rises during the hydrolysis and dissolution of the glass, and eventually result in the coprecipitation of highly hydrated aluminium phosphate and silica gels. There may even be some covalent bonding between these two making, in effect, a single network. These inter-penetrating insoluble gels thus form the matrix of the cement in which are embedded the unreacted glass particles, which again must be in excess to form the strong core of the composite structure. There are also small crystallites of the only very slightly soluble calcium fluoride formed at the same time. The reactions may be summarized as follows:

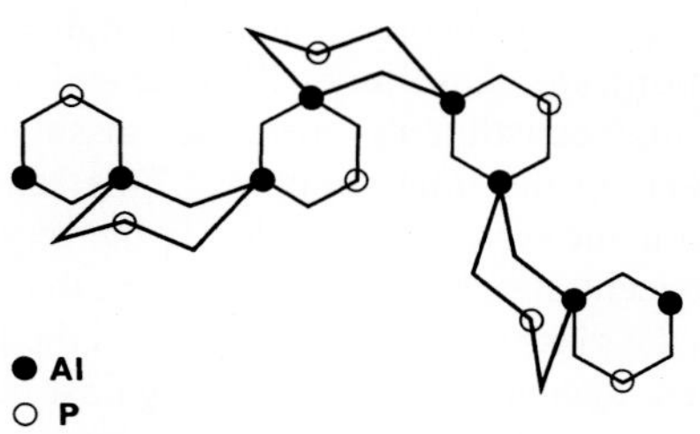

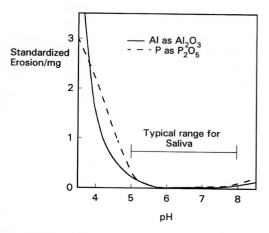

● Al
○ P

Fig. 7.5 Possible structure of a fragment of the aluminium phosphate gel polymer. The rings probably adopt the 'chair' conformation, as in cyclohexane (see Fig. 17§2.14).

$$glass + H^+ \Rightarrow Al^{3+} + Ca^{2+} + Na^+ + SiO_4^{4-} + F^-$$

$$nSiO_4^{4-} + 4nH^+ \Rightarrow (SiO_2)_n \cdot 2nH_2O \quad (silica\ gel)$$

$$xAl^{3+} + xPO_4^{3-} \Rightarrow (AlPO_4)_x \quad\quad (phosphate\ gel) \quad\quad\quad (7.2)$$

$$Ca^{2+} + 2F^- \Rightarrow CaF_2$$

$$Na^+ + 2H^+ + PO_4^{3-} \Rightarrow NaH_2PO_4$$

(No attempt has been made to balance these reactions, as the glass is not a stoichiometric compound and the exact chemistry of most of the reaction products is uncertain. For example, silica gel is a very indefinite entity. They must instead be treated as purely symbolic tokens rather than rigorous statements.) These reactions lead to the general kind of structure shown in Fig. 7.6.

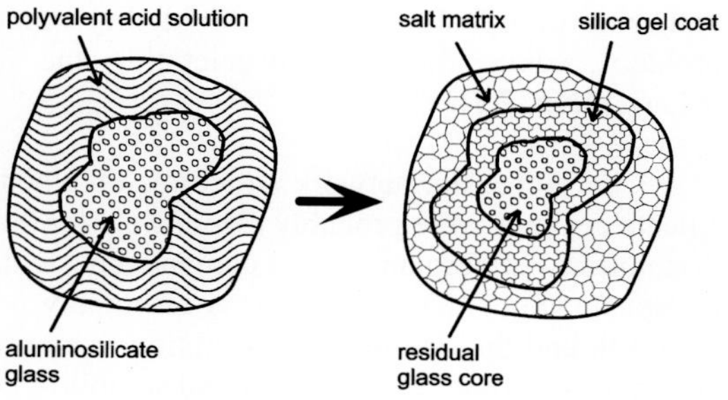

Setting is accompanied by shrinkage. Given that fact and that these materials are non-adhesive, there was therefore a tendency to marginal leakage. A further problem is that a certain proportion of the reaction products are very soluble, in particular the sodium phosphate (Fig. 7.7), although this dissolution is nearly complete very quickly (Fig. 7.8).

polyvalent acid solution   salt matrix   silica gel coat

aluminosilicate glass   residual glass core

Fig. 7.6 General structural change on setting for a silicate glass - acid cement.

●7.5   Clinical use

Because of the translucency of both the core and the matrix (in stark contrast to all materials based on the extremely opaque zinc oxide) silicate cements were not used as cements proper, for holding something else in place, but as direct restorative materials. With careful colouring (transition metal oxides, 24§6), they could be made to resemble tooth tissue extremely well; their principal use was therefore in the anterior part of the mouth.

A major difficulty with the use of silicate cements was associated with the gel nature of the matrix, since it depends for its stability on the water of hydration associated with the polar silica and aluminium phosphate 'polymers'. If this is allowed to dry out the structure is destroyed (because it depended on the solvation shell), and the surface then shrinks and cracks, allowing staining and mechanical degradation. Conversely, if excess water

Standardized Erosion/mg

— Al as Al_2O_3
- - - P as P_2O_5

Typical range for Saliva

pH

Fig. 7.7 Leaching from silicate cements of aluminium and phosphorus, expressed as their oxides, as a function of pH.

is allowed to be absorbed by the gel before it has formed fully (*i.e.* by osmosis into a highly concentrated solution), some swelling will occur and some of the gel-forming constituents may be dissolved out, thus weakening the final structure. This sensitivity to the water in the system was evident in the very narrow range of concentrations of phosphoric acid that could be used with the glass powder. Accordingly, the freshly-placed cement needed to be protected against both water loss and gain. However, in 'mouth-breathers' sufficient dehydration can occur, even when the material is fully set, to begin the degradation of the surface in normal service.

Setting time as well as strength is affected by the acid composition, but in a way that is not open to simple explanation. Note that this type of cement system does not have the simple stoichiometry and limited range of compounds formed in cements even as complex as zinc phosphate. Some compromise between the hydration of the gels and the completeness of the aluminium phosphate structure seems to be involved. Either way, small changes in water content could effect great changes in properties. The addition of some Zn and Al to the acid liquid resulted in some strength improvement (as indeed has been done with zinc phosphate cements). Solubility remained a problem, acidic foodstuffs accelerating the loss of material (Fig. 7.7).

Even so, the network structures obtained in the silicate cements were probably responsible for the better dissolution performance in comparison with zinc phosphate (Fig. 7.8). The latter is essentially a simple ionic salt and thus subject to dissolution in an ordinary, continuous fashion, even if the actual solubility is low.

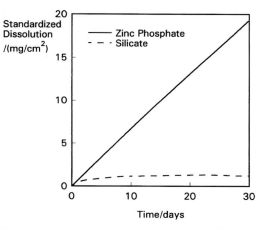

Fig. 7.8 Comparative dissolution rates of silicate and zinc phosphate cements.

The term **solubility** is frequently used in the context of cements. It should be noted that this is not solubility in the sense of a saturated solution (*e.g.* 2§1.2, 2§7.1, 12§3.2), but rather the quantity that is lost from a designated specimen under particular conditions of medium, temperature, time, *etc.*; that is to say, it is a **standardized dissolution**. In this sense, it is not a property, but a characterization under entirely arbitrary circumstances. The manner of expression also varies: see Figs 5.13, 7.7 and 7.8 for examples. Care is therefore required in reading and interpreting such values and reports: comparability is not to be taken for granted. Even less is relevance to the oral environment and ordinary service to be assumed.

●7.6   Fluoride

The presence of fluoride in the substances released during the hydrolysis process has generally been thought to be beneficial in that it could provide a steady 'topical' treatment of the enamel, especially in the crevices, so inhibiting recurrent caries. The effect was inferred from the clinical observation of old silicate restorations which showed such lesions relatively infrequently. This 'observation' has had a great effect on the design of many products. Plain zinc polycarboxylate cement was soon followed by a version containing stannous fluoride, SnF_2, and even amalgam alloys and filled resins have been sold with the substance added. Similarly, fluoride-containing varnishes (§11) are sold for topical application. While the latter three types of product do have serious deficiencies, it is curious to note that the supposedly beneficial effect of leached fluoride has never been tested experimentally. This is not to dispute the efficacy of topical fluoride treatments, but to query whether the amounts involved in a leaching system could possibly be adequate for such a result. (See also §8.12.)

§8. Glass Ionomer Cement

The **glass ionomer cements**, now also known as **glass polyalkenoate cements**, are a development of an idea similar to that of the silicate cements.[6][22][23][24] Thus, an alumino-silicate glass containing calcium and fluoride is reacted with an acid to produce a gel of hydrated silica, although the glass is of a substantially different composition in these materials compared with that for silicate cements. However, and most importantly, the liquid used is based on the design idea of polycarboxylate cement: a solution of mostly poly(acrylic acid) which permits the formation of an ionically cross-linked polymer network exactly as in the zinc polycarboxylate cements (Fig. 8.1), but now interpenetrating that of the silica gel. Aluminium and calcium ions form the cross-links just as zinc operates in polycarboxylate cements. The metal ion is octahedrally coordinated with carboxyl groups, water, and other electron-donor groups that are sterically able to fit around the cation (Fig. 8.2).

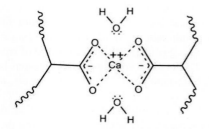

Fig. 8.1 An octahedrally coordinated calcium ion in a polycarboxylic acid, as may occur in glass ionomer and polycarboxylate cements. In the former, Al^{3+} may be complexed in the same way.

•8.1   Glass composition

The Al_2O_3 - CaF_2 - SiO_2 system (Fig. 8.3) is the basis of the glass ionomer cements. Certain compositions in that range have the necessary combination of properties. Thus, there must be sufficient alumina present for the glass to become basic enough for it to be susceptible to acid attack. However, too much results in the separation of Al_2O_3 as a separate phase ('corundum') and this makes the glass increasingly opaque. Similarly, the presence of the calcium fluoride is essential for a good cement-forming material, but too much again makes the material opaque. A certain amount is required to give the 'opal' appearance, due to the presence of a CaF_2-rich separated phase, which assists in creating the tooth-like optical properties of the final set cement (see 24§5). The CaF_2 also lowers the fusion temperature of the glass. Other glass modifiers are also used, such as $AlPO_4$ and Na_3AlF_6 ('cryolite'), to adjust the setting and other properties. In addition, deliberate inclusion of other dispersed phases in the glass, such as ZrO_2, can increase the final strength. Thus even the glass itself typically has a composite structure.

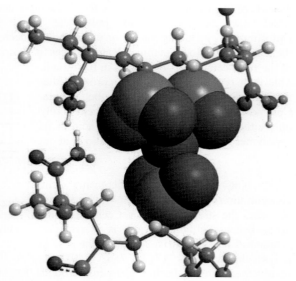

Fig. 8.2 Various coordination structures may be present. It is possible to fit three carboxyl groups around one metal ion, but this results in the oxygens forming a distorted octahedron (nearly a triangular prism).

•8.2   Acid composition

A solution of poly(acrylic acid) is certainly an effective cement-forming medium, but it suffers from the drawback that on standing it forms a gel. This is also promoted by cooling. Hence, such liquids (including those for zinc polycarboxylate cements) should never be stored in a refrigerator. The reason for this gelation is the establishment of hydrogen bond cross-links between chains, the number of which increase steadily with time, as the polymer chain segments gradually

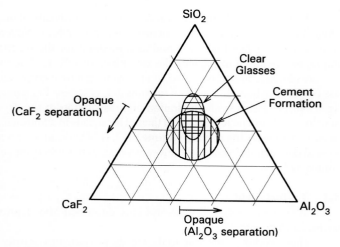

Fig. 8.3 The region of the Al_2O_3-CaF_2-SiO_2 system that forms cements with polycarboxylic acids. The clear glasses are unsatisfactory for tooth mimicry.

diffuse in the viscous solution. Essentially, this means that highly-ordered micellar structures similar to those of polysaccharides (Fig. 7§8.2) would form. It is a kind of crystallization. While this could be reversed by stirring or shaking (making this an example of true thixotropy, 4§7.10 ~ 12), or by warming (such hydrogen-bonded gels are heat-labile, as in agar, 7§8.4), it is a practical problem.

Inhibition of this crystallization and cross-linking process can be achieved by making the polymer chain less regular (*cf.* 3§2.3). This may be done by using copolymers of a variety of unsaturated acids, such as itaconic, maleic and tricarballylic acids, as well as acrylic acid itself (Fig. 8.4). These units also differ in their acid strength and so affect the reactivity of the polyacid with the glass, and therefore the setting rate. In addition, the varying side chain lengths affect the ease of formation and stability of the chelation structures which may be formed with polyvalent metal ions. Unfortunately, including maleic or itaconic acids decreases the bond strength to tooth mineral, and reduces the resistance of the final cement to acid attack in the mouth. This then represents another compromise that must be found to balance these behaviours. Commercial products, however, do seem to rely only either homopolymer of acrylic acid or its copolymer with maleic acid in the ratio of about 2:1.

Not surprisingly, increasing the concentration of the polyacid in the mixing liquid increases the strength of the set cement. This arises because although water is a structural element of the gel matrix, similar to the case in polysaccharide gels (7§7, 7§8, 7§9), its

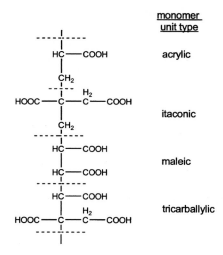

Fig. 8.4 Some of the possible units of copolymeric polycarboxylic acids. Such copolymers inhibit crystallization.

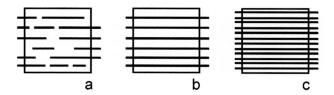

Fig. 8.5 The strength of the cement depends on stress transfer between volume elements (a) and is improved by increasing polymer molecular weight (b) and concentration (c).

involvement can only be through hydrogen bonds. Replacing water by polyacid means increasing the proportion of covalent bonds, *i.e.* in the polymer chain, and therefore better distribution of stresses through the structure. In addition, increasing the molecular weight (*i.e.*, degree of polymerization) has a similar effect (Fig. 8.5). This is through the increased chelation cross-linking and chain entanglements. Increasing concentration and polymer size also shortens the setting time (bearing in mind what this means, 2§11).

Unfortunately, both factors also increase the viscosity of the mixing liquid, thereby making mixing more difficult and increasing the risk of incorporating air bubbles. Nevertheless, there is a minimum amount of water required for the silica hydration and as a medium for the reactions to proceed. There is thus an essential compromise required to balance the two kinds of effect. The mixing problems arising from the increased viscosity of concentrated polyacid can be overcome by incorporating that as a dried powder mixed in with the glass powder – which, so long as it remains dry, is stable. Plain pure water can then be used as the mixing liquid (Fig. 15§6.1). Because the hydration and dissolution of the dry polyacid would take a short time, the full viscosity would not be reached until then. This means in addition that a higher powder : liquid ratio is attainable, that is, a lower **working mixing volume** (2§2.4). This kind of cement can therefore be made at a higher effective polyacid concentration and is therefore stronger as described above and as a result of the higher volume fraction of unreacted glass.

●8.3   Tartaric acid
Whilst the polyacid-glass systems described above provide good cements, there are two serious drawbacks in practice:
1.   short working time, due to the steady increase in viscosity from the beginning of mixing, and
2.   long setting time, meaning that the approach to the final strength is very slow.
Adjustment of glass composition is inadequate as a means of overcoming these limitations, even though setting rate and strength are affected, sometimes considerably, by such changes. Part of the slow-setting problem may

be due to the relatively weak acidity of the polyacid itself. But certainly the chelation cross-linking reactions would commence immediately that any di- or tri-valent metal ions were available, in much the same way that alginate would start setting in the absence of retarder (7§9).

The answer to these two problems appears to be the incorporation of some 5 ~ 10 mass% tartaric acid, $(HO.CH.COOH)_2$, in the mixing liquid. Whilst this is a somewhat stronger acid than the usual polyacids, the net effect is to 'sharpen' the set of the cement (Fig. 8.6). [25] That is, there is a delay before the viscosity of the mix starts to rise appreciably, and then that rise is rather quicker than without the tartaric acid present. This is explained as follows.

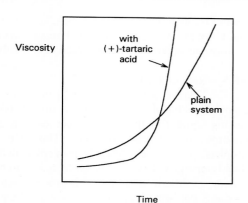

Fig. 8.6 Variation of viscosity of a setting glass ionomer cement with and without tartaric acid in the mixing liquid.

The tartrate ions **sequester** the first metal ions released, that is, chelating them strongly, and thereby preventing the direct cross-linking of the polyacid chains. This reaction first forms a complex of only one metal ion per tartrate ion: Ca-Tartrate (chemical equilibrium requires this form to predominate). When more than this equivalent amount of metal ions has been released from the glass, a second ion can be bound to the tartrate, Ca_2-Tartrate, tartaric acid being a di-functional acid. In the Ca-Tartrate complex, the ion can also be bound to the polyacid acid by coordinating with one or two $-CO_2^-$ groups. However, steric hindrance would prevent those groups coming from *two* chains, so a cross-link cannot be formed. However, when the Ca_2-Tartrate complex is formed, the same thing can happen at each Ca^{++}, effectively forming a bridge between the two chains: -polyacid-Ca-tartrate-Ca-polyacid-. The formation of the di-Ca complex seems to be delayed while the tartrate is 'titrated' by the ions coming into solution. Thus, the formation of bridges is also delayed, hence the longer working time without an increase of viscosity. Further metal ions must then go immediately into forming bridges, hence the rapid increase in viscosity once that point is reached. This may also be the explanation of the effect in zinc polycarboxylate cements (§6). Tartaric acid may therefore be viewed as a kind of **retarder**, in a use of the term similar to that in the context of alginate impression material.

●8.4   Stereochemistry

There is a subtlety to this process. Tartaric acid has two chiral centres, the two ends being asymmetric (Fig. 3§2.1).[26] This means that there are three distinct configurations – **stereoisomers** – possible (Fig. 8.7). Either of the optically-active stereoisomers, (+)- or (−)-tartaric acid, or mixtures of them [including the so-called racemic acid, (±), which is a mixture of equal parts of the (+) and (−) forms], have the desired effects on the setting of the cement. The *meso* form does not work. This can perhaps be explained by noting that the chelation of the first metal ion by tartrate is between the

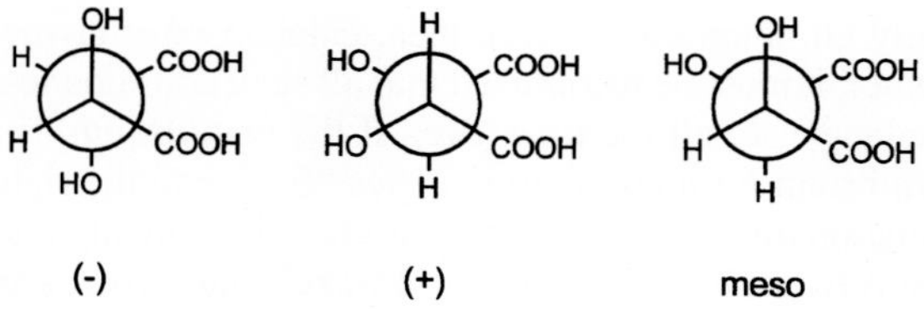

Fig. 8.7 The stereochemistry of the tartaric acids. The differences between the forms can be appreciated by rotation of the nearer axis carbon so that the carboxyl groups are mutually eclipsed. The (+) and (−) forms are then seen to be mirror-images of each other; the *meso* form is distinct. NB: they cannot be interconverted by rotation.

carboxyl on one asymmetric carbon (say, C_1) and the hydroxyl on the other, C_2. This creates a six-membered ring, which is stable. The second metal ion can then be bound by the carboxyl on C_2 and the hydroxyl on C_1, without strain – but only in either of the optically active configurations. The *meso* form can react in this way to give the first chelate, Ca-Tartrate, but then the C_1-COO⁻ and C_2-OH are too far apart to form a ring with the metal ion; too much strain would be required. The second metal ion, essential for bridge-forming, cannot be added. Thus, the only way in which a bridge can be formed to link polyacid chains is by either a (+) or (−) form. Stereochemical effects are therefore seen to be of great importance in such reactions. Of course, once the tartaric acid has been used up in this way, ordinary coordination cross-links directly between chains may form, given that chain segments may diffuse to obtain the correct relative positions.

The progression of the mechanical and rheological properties during the setting process follow those discussed for zinc polycarboxylate cement (§6), that is, with the same kinds of change as observed in filled resin restorative materials (Figs 6§4.6, 6§4.7). Consequently, the handling precautions are the same: keep it cool, mix and place promptly. The only difference might be that the powder-liquid ratio is rather higher for restorative purposes as opposed to cementation, and the rheological properties vary accordingly.

●8.5   Role of aluminium

There is a further aspect of the setting process of considerable importance in the context of glass ionomer's clinical use. As was mentioned above, there are two metal ions which can be released by the acid mixing liquid from the glass powder: calcium and aluminium. There is an important difference in the rates of reactions involving the two ions. The calcium ions, whether in the glass proper or in the phase-separated droplets, do not require hydrolysis of covalent bonds and are therefore liberated readily by the acid. Being only doubly-charged, these will be quite mobile in the matrix. The calcium is thus responsible for the initial cross-linking, as described above. This process yields an initially-set material which is carvable – the strength and hardness not being high enough to resist sharp instruments. On the other hand, aluminium ions are released much more slowly than are calcium ions because hydrolysis of aluminate from the covalently-bonded glass matrix is first required. These Al^{3+} ions also diffuse much more slowly because they are much more strongly hydrated, their higher charge leads to a larger hydration shell. But it is these ions which are responsible for the development of the final set strength, again because of their higher charge.

The strength of the coordination bond is related to the **cation field strength**, $\mathbb{F}$:

$$\mathbb{F} \propto \frac{z}{r^2} \tag{8.1}$$

where z is the charge and r the ionic radius (nm), that is, applying the inverse square law at the ionic radius. For Ca^{2+} we therefore have $\mathbb{F} \propto 2/0.114^2 = 154$, and for Al^{3+}, $\mathbb{F} \propto 3/0.053^2 = 1070$. In other words, it is ~7 times greater for aluminium. The chelates that form with aluminium ions are therefore rather stronger and more stable than those with calcium, and replacement of the one by the other may also occur because it leads to a more stable (lower energy) coordination. The second and final stage of setting, in which the service strength is developed, is due essentially to the aluminium cross-linking becoming dominant.

Care must be taken in speaking of "stages" in this way: it is to be understood that there is no sharp boundary between the two. Calcium will continue to be leached as the glass is hydrolysed, and some aluminium must be available early on. The proportion of the one to the other changes gradually from nearly all calcium to nearly all aluminium. Even then, calcium must always contribute to the linking between chains. In a like manner, it must be recognized that all such coordination bonding systems are equilibria: exchange must occur spontaneously all the time, even if the equilibrium lies far in the favour of the complex. On this view, the equilibrium for aluminium ions lies "further to the right" than for calcium. Even so, there is a major kinetic limitation involved. Calcium ions are only ionically associated with the glassy polymeric matrix and so can be dissolved fairly readily by an ion-exchange process with hydrogen ions from the liquid. The aluminium, however, is covalently part of that polymeric glass structure and so can only be released by the relatively slower hydrolysis of those bonds. Thus calcium ions are available promptly, but at a lower rate as the attack front penetrates the glass particle and diffusion comes to dominate. On the other hand, aluminium ions must arise gradually but increasingly (at first) as the area exposed to hydrolysis increases; effectively, the leaching of calcium makes the outer coating of the particles permeable.

The fact that some aluminium ions must be released early on may be contributory to the setting behaviour in the absence of tartaric acid. Again, the cation field strength is critical: any such ions are likely to cross-link the polyacid fairly promptly. It can be expected, though, that these would also react with tartrate at a high rate and strongly, in the same sense that Ca^{2+} would, so that the effectiveness of this additive is at least in part due to this action preventing aluminium ion cross-linking.

●8.6   Setting reactions

The setting reactions can therefore now be outlined in the following way, using the symbolic approach used above for silicate cements rather than exact chemical statements, which again cannot be made for just the same reasons. Thus, the initial process is hydrolysis of the glass, releasing calcium ions and silicate:

$$glass + H^+ \Rightarrow Ca^{2+} + SiO_4^{4-} \tag{8.2}$$

The silicate promptly forms silica gel, and the calcium ions are then in due course (after the effects of the tartrate retarder, that is) chelated by the polyacid ('PA'), reaching the stage known as the initial set:

$$nSiO_4^{4-} + 4nH^+ \Rightarrow (SiO_2)_n \cdot 2nH_2O \qquad (silica\ gel)$$
$$Ca^{2+} + 2PA^- \Rightarrow Ca(PA)_2 \qquad\qquad (initial\ set)$$

(8.3)

Hydrolysis continues, breaking down the glass further and releasing aluminium ions and more silicate. The silicate continues the formation of silica gel, but now the aluminium can be chelated itself, even displacing calcium because of its higher charge and stronger bonding. This constitutes the final set condition:

$$glass + H^+ \Rightarrow Al^{3+} + SiO_4^{4-}$$
$$Al^{3+} + 2PA^- \Rightarrow Al(PA)_2^+$$
$$Ca(PA)_2 + Al^{3+} \Rightarrow Al(PA)_2^+ + Ca^{2+} \quad \Big\} \quad (final\ set)$$

(8.4)

The charge is left on the aluminium complex to show that the strength of the bond must be greater, and that there is capacity for binding water more strongly than would calcium, enhancing the strength of the gel. Of course, the displaced calcium ions may still be chelated. They are smaller, and so more mobile and diffuse more easily than the hydrated aluminium throughout the matrix until a binding site is found. The overall setting reaction, taking into account the retarding effect of the tartrate, is set out in Fig. 8.8.

●8.7 Adhesiveness

Glass ionomer cements are adhesive in the same way as zinc polycarboxylate (§6.2) and lend themselves to types of restoration with minimal preparation of the tooth, for example cervical erosion lesions, and as a fissure sealant. As a cement, of course, they will be similar in their behaviour to the zinc polycarboxylate materials, and likewise they would be expected to have a contribution from the mechanical key arising from the self-etch effect. Nevertheless, attention must be paid to the soundness of the surface to be bonded and the avoidance of contamination as with all such work.

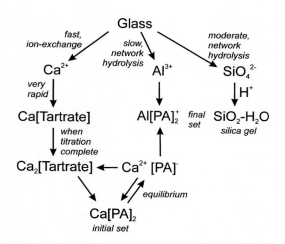

Fig. 8.8 Overall setting reaction scheme for glass ionomer cement on exposure to polycarboxylic acid solution. Early-released Al^{3+} would be expected to contribute to the titration of the tartrate.

●8.8 Cermet cements

The relatively poor strength (compared with, say, amalgam) of such gel-matrix materials has led to the investigation of some possible means of improving the system so that load-bearing occlusal restorations might become feasible. For no apparent reason except that a portion of the core is stronger and less brittle (which may improve the abrasion resistance of the material) mixtures with metal powders have been investigated. In its simplest form, a mixture with ordinary dental silver amalgam alloy powder has been proposed (see also 21§3.4).

A more elaborate product has been prepared and marketed containing silver powder, and referred to as a **cermet cement**. This is manufactured by mixing the glass and metal powders, compressing the mixture into pellets under high pressure, firing at 800°C or so, then regrinding to a fine powder. It is claimed that this results in the metal being bonded to the glass. However, it is not clear what chemical basis there might be for this as the metals involved are oxidation-resistant (see 13§4) and interactions such as occur between polycarboxylate-based cements and materials such as stainless steel (§6.2), or in porcelain bonded to metal (25§6), seem not to be possible. In the set cement there is still a composite structure, consisting in this case of a dispersion of metal in the (nearly) ceramic matrix, *i.e.* two major disperse phases are present. However, after reaction, any metal particles which are released from the glass matrix, or simply on the surface of a glass fragment, would still seem not to be bonded to the polyelectrolyte matrix for the same reason as above: no adherent oxide coat or recognized mechanism for interaction. This seems to violate the requirement for a good interfacial bond in composite materials of all kinds if strength is to be improved (6§2). Whether or not the claims for improvements are justified (there is no strength benefit, Fig. 8.10, nor in fracture toughness[27]), the material is no longer tooth-like in appearance, and the return to metallic-appearance occlusal restorations is a curious reversal of the claims that amalgam is 'unaesthetic' (see 24§4.14). However, there is no change in the chemistry of setting to result from this modification. Gold, platinum and palladium have also been tried; the analysis is the same.

Stainless steel, in contrast, as indicated above, would appear to offer the benefits of a bond to the matrix, and thus matrix constraint,[28] an observation that also applies to polycarboxylate cement.[29] However, despite patents, neither seems to have been commercialized. Recognizing the importance of the interface, oxidized silver amalgam-like alloy has also been explored, although the oxides formed could not be bound well to the underlying metal.[30] It may be possible to overcome the bonding problem in the case of silver and some other metals which form strong, hydrolysis-resistant bonds to thiols, R-SH, to form compounds like R-S-Ag. This is chemically-similar to the sulphide tarnish that forms on silver (13§6.6, 15§1). If the R- group is functionalized to be reactive with the matrix material (such as with a vinyl group), then the required stress-transfer and matrix constraint might be obtained. In any case, the appearance remains an issue for glass ionomer cement.

There is a further problem which does not seem to be generally recognized, that any such 'additions' must be as replacement for reactive glass; the mixing requirements must continue to be met (see §8.13).

●8.9 Hybrid products

The setting time of glass ionomer cements has been considered by some to be a problem, in comparison with the so-called 'command set' of light-activated curing filled-resin restorative materials (6§5). In an attempt to remedy this supposed deficiency, a variety of products have been offered with 'hybrid' formulations, essentially including the chemistry of the above-mentioned resin materials. These products range from the addition to the polycarboxylic acid chain of functional groups carrying vinyl double-bonds (*cf.* vinyl siloxane silicones, 7§6.2), through the use of a mixture of polymerizable monomers (typical hydroxyethyl methacrylate) and polycarboxylic acid so that each reacts in its own way, all the way to using glass ionomer cement glass powder in an otherwise conventional resin matrix. The latter simply cannot be called 'glass ionomer' in any sense as there is no reaction involving the glass.

There is at present no point in detailing such products. It is enough to say that, so far at least, none has shown any substantive improvements over either glass ionomer cement or ordinary filled resin materials. In fact, they are always poor compromises between the two, the strength, for example, lying between the two extremes.[31] What seems to have happened is that the design of these materials has been based on convenience to the dentist and not sound materials science principles.[32] The question is whether the trade-off of service properties against a minute or two of saved time is an appropriate course to take.

The nature of this compromise can be understood from the network competition involved.[33] That is, the creation of a network progressively reduces the diffusivity of all molecules and chain segments, such that if the free-radical polymerization network is over-represented, the glass ionomer network cannot form properly (Fig. 8.9), while if the glass ionomer network forms first, forms first, the polymer network will be prevented from forming (Fig. 8.10). There is direct

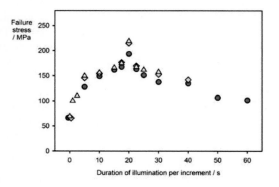

Fig. 8.9 Effect of irradiation on failure stress (in compression) of a "resin-modified" GI cement: competition between the networks leads to a clear maximum when both can form, although neither can be fully-developed.

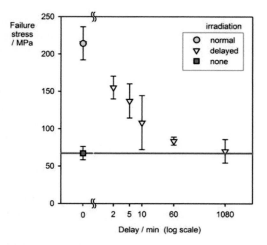

Fig. 8.10 Delaying the irradiation of a "resin-modified" GI cement limits the formation of the polymer network by allowing the GI reaction to proceed, with ultimately a weaker structure.

inhibition of the one process by the network from the other. Neither can be fully formed if the other proceeds at all, but then a network can form only to the extent that its components are present, that is, to the maximum represented by its mole fraction of the reactable systems. This in turn means that on their own, neither system will approach the strength of the non-hybrid equivalent, and that together some intermediate value will be obtained, in some rule-of-mixtures sense of less than 100% reaction of both. The actual strength therefore depends on irradiation time and irradiance for light-curing systems beyond what might occur for 'pure' filled-resin materials.

●8.10 Water status: shrinkage and expansion

One area affected by the modification of these cements by including polymerizable systems is setting shrinkage. As might be expected, the immobilization of polymer chain segments, metal ions and other species such as tartrate leads to a loss of free volume, which means an overall volumetric shrinkage. In fact, for ordinary glass ionomer cements this seems to be rather small and so not a clinical problem, but the picture is complicated by the absorption of water from the environment which ordinarily results in a slight expansion. The water activity in the set cement is approximately 0.8 of that of pure water, which means that if the relative humidity is greater than about 80% the cement will gain water, below that value and it will lose it and so shrink. Premature exposure to water leads to leaching of ions, swelling and weakening; loss leads to shrinkage and cracking. All this leads to three recommendations.

First, the setting cement should be protected from exposure to water – primarily meaning saliva – until it is set. Since the strength increases steadily for at least two months, that is a difficult point to judge. Conventionally, and entirely arbitrarily, 24 h may be taken as the time required for adequate setting. A barrier layer must be used to provide protection for this time, after which final finishing, if necessary, can be done. There are several nominally-suitable barrier materials, solvent-based or polymerizing varnishes that are sufficiently durable and impervious to water, although none is yet ideal. Certainly, petroleum jelly and the like will not be adequate here. The most effective may be something like a light-cured bonding resin.

Secondly, the set cement should be protected from loss of water. This means that the use of these materials in mouth-breathers, and especially in anterior positions, is likely to fail. A further implication is for the protection of such restorations when other dental work is done if, for example, they are isolated outside of a rubber dam such that they may suffer desiccation. Here, the use of petroleum jelly is probably the most effective solution.

Thirdly, glass ionomer cements should not be used to cement into place ceramic crowns. The swelling associated with the slight absorption of water is enough to generate a stress sufficient to fracture the crown, given that the stress will be tensile (a hoop stress) and the inner, rough surface will be sensitive to crack initiation in the Griffith sense. This problem is made even worse in the 'resin-modified' materials where the need for a hydrophilic monomer or pre-polymer, in order that they be soluble in the cement liquid, leads to greater water absorption.

Since it is the hydrous gel structure of the cement matrix that is concerned in these changes, it might be thought that reducing its volume fraction by the (partial) substitution of an unaffected polymer network would reduce the sensitivity to these changes in proportion. Unfortunately, the shrinkage associated with the polymerization of the monomer will still be present, and this cannot be compensated. In other words, while some of the undesirable aspects of glass ionomer cement may be overcome by these 'hybrid' designs, this is at the cost of a further problem that has yet to be overcome.

●8.11 Radio-opacity

It is of benefit to be able to identify the presence of foreign materials in oral radiographs. Unfortunately, glass ionomer cements are rather radiolucent, being composed of light elements. To improve their visibility in this sense, some products have substituted strontium for the calcium. The chemistry of the setting reaction is essentially unaffected, since strontium is very similar to calcium in many respects in its chemical behaviour, but because its atomic number, Z, is 38 as opposed to calcium's 20, it is substantially more absorptive of X-radiation. The 'cermet' type (§8.8), will of course be much more radio-opaque (for Ag, Z = 47; Au, Z = 79).

●8.12 Fluoride

The fluoride ions due to the glass-forming flux again are slowly leachable and are thought by some[34] to be similarly beneficial as for silicate and polycarboxylate cements (§7.6). Such an effect could really only be beneficial on the cut surface of the cavity facing the cement (assuming that the interface leaks and thus the tooth tissue is prone to acid attack – which is not supposed to happen because of the adhesion of GI to tooth tissue), not on the adjacent free surface because saliva flow will carry away the majority of any leaching ions, and the effect is short-lived anyway.

Because of the relatively short duration for appreciable leaching of fluoride to be detected in laboratory experiments, it has been proposed that glass ionomer restorations be "recharged" by treating them with a topical fluoride preparation. This concept is deeply flawed. Firstly, unless the restoration were very leaky (and thus in need of replacement anyway), no fluoride would reach the cement surface where it would be most effective as a 'reservoir' – but were it to do so it would treat the facing tooth surface as well and anyway far more effectively than subsequent leaching possibly could. In other words, only the free surface would be treated and the fluoride then escaping subsequently would be ineffectual and wasted. Secondly, only the very surface of the cement would be treated, and a negligible amount of fluoride could possibly diffuse to the tooth-cement interface through the solid cement matrix (and take a very long time to achieve it). Thirdly, the mere act of applying the fluoride solution would treat the adjacent tooth free surface anyway – there is no reason to avoid doing so, and there is no practical way that the treatment could be restricted to the GI surface. Fourthly, topical fluoride solutions, if of the highly acid 'APF' type (10§7.3), are likely to etch and roughen the cement surface. This damage cannot be justified on the hope that the escaping fluoride somehow knows that it is supposed to migrate to the adjacent tooth instead of being washed away: diffusion cannot be directed by intent. The effective salivary concentration must be very low, and the topical treatment of the whole mouth is bound to be far more effective than any possible local effect. It is a pity that so much effort has been wasted on such futile research. Indeed, it has been shown that for cement more than about 1 month old there is little or no uptake anyway.[35]

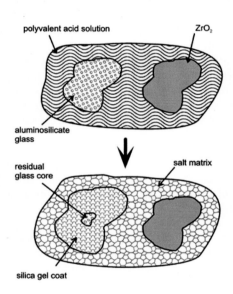

Fig. 8.11 Possible outcome of mixed core GI reaction: the volume of fraction of residual glass is smaller, that of the other core unchanged: the total residual volume fraction of core is unaffected.

●8.13 Other fillers

Other attempts (cf. §8.8) at improving the strength of GI have been made, one of which is to use zirconia (ZrO_2) powder as a separate filler,[36] that is, not added to the glass melt. The same requirements as before apply: bonding to achieve matrix constraint and thus improvements in strength and stiffness. As indicated above, this is expected to be good on such metal oxide surfaces. However, it should be noted that the additional filler can only *replace* the glass, not be in addition to it, as the minimum mixing volume – which is determined by the physical conditions (2§2.4) – is unchanged. (The reaction allowance, which is a small proportion of the total mixing liquid, may be reduced slightly as the reduced surface area of GI glass will mean a correspondingly reduced rate of reaction and so a lesser extent of reaction at the time of use. We can ignore that here.) The total volume fraction of the two fillers is therefore also not changed. Thus, the amount of liquid required is much the same, and so the extent of reaction when fully set is also the same, as this depends primarily on the total number of carboxyl groups present. Accordingly, the volume of the glass powder which will then have reacted at completion is unchanged. Since the total volume fraction of filler is fixed, the volume fraction remaining at the completion of reaction is unchanged (although the glass particles will, on average, be correspondingly smaller then, assuming no reaction with the zirconia) (Fig. 8.11). There can be no expectation, therefore, of any improvement in mechanical properties as the volume fraction and bonding of the filler overall are no different. This is, indeed, what is found experimentally (Fig. 8.12). The same argument applies

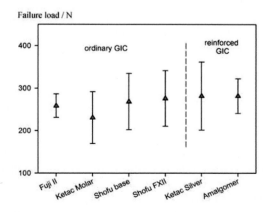

Fig. 8.12 Comparison of the failure load under ball-on-disc ("Hertzian") testing of 'reinforced' and ordinary GI cements. No reinforcement is detectable.

to all cermet systems (§8.8) – the minimum mixing volume remains a critical concern, and the metal must replace glass.

Of course, if any dissolution of the zirconia occurs, coordination with the polycarboxylic acid can be expected in the usual way, with little change in the conclusions, but only on a cation-equivalent replacement basis for the glass. This is also assuming that the amount of GI glass present is at least enough for the liquid to react fully. If there were not, there would necessarily be a weaker material as there would left-over acid groups and thus less cross-linking per unit volume. We may further note the cation field strength for Zr^{4+}: $\mathbb{F} \propto 4/0.086^2$ = 540 (equation 8.1), about half that for aluminium. The point here is that the consequences of the presence of the alternative filler need to be explored properly, and that those consequences are inevitable. The minimum mixing volume remains a crucial constraint.

Curiously, the use of a metal oxide in this way results in an opaque white material, similar to zinc polycarboxylate in appearance. Nothing would seem to have been gained in making such a substitution, although increased radio-opacity would follow (for Zr, Z = 40).

One of the difficulties in evaluating materials which differ in such respects arises from the usual proportioning by mass per liquid drop volume. Obviously the volume fractions depend on the densities (6§3.3), and proper comparisons can only be made on that basis. The density of monoclinic ZrO_2, for example, is 5.6 g/mL, that of silver, 10.5 g/mL. The ratio distortion resulting from such figures is very large in comparison with that of filled resins.

As with cermets, similar considerations also apply for the attempts to use glass and carbon fibres (the latter being black) in terms of mixing liquid requirements, but the greater length of fibres large enough to have any beneficial effect, even assuming bonding, makes mixing and placement even harder. With porous or absorbent fillers, the mixing liquid requirement is further complicated by the fact that extra liquid is required to fill those spaces, meaning that more glass reaction must occur and that the final volume fraction of core is less.

§9. Film Thickness

For the purposes of cementing crowns and inlays, the most important properties are those of the rheology, as it is these that control the ease with which excess cement may be extruded, and the film-thickness or minimum thickness of cement that can be obtained under normal restoration seating pressures. Matters such as strength and solubility are irrelevant unless the device can be put in place properly.

Film-thickness is understood to mean the thickness finally obtained for a portion of the cement when it is squashed between flat plates under a defined load (Fig. 9.1). This is relevant because the thickness of the cement layer controls both the accuracy of seating and the strength of the bond. The strength of a wide range of types of bonding system (in the sense of 'retention') depends inversely on the thickness of the layer of glue, cement or adhesive (*cf.* the Griffith crack theory, 1§7). In addition, the thicker the cement-line, the exposed surface at the margin of the restoration,

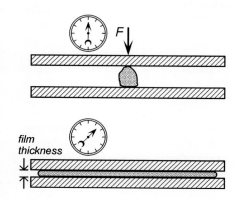

Fig. 9.1 Film thickness is defined as the thickness of the cement after squashing between flat plates for a predetermined time.

the greater the rate of dissolution of the cement. The rheology may be controlled to a certain extent by the formulation of the liquid and the powder : liquid ratio used. Indeed, different mixes may be specified by the manufacturers for different applications. The limiting factor in any case is the particle size of the powder, after allowing for the partial dissolution that may occur during the initial stages of the setting reaction.

Clearly, the minimum separation of tooth and restoration corresponds to the diameter of the largest particle present (as shown schematically in Figs. 9.2, 9.3, 9.5), assuming that the rate of flow at normal clinically applied pressures and times is sufficient for full extrusion of excess and that the particle does not crush.

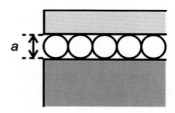

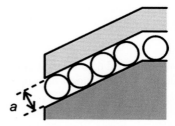

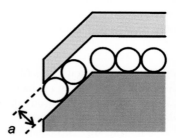

Fig. 9.2 The effect of cement film thickness depends on the geometry of the device being cemented and the maximum particle size in the mixture.

●9.1   Geometry

The geometry of the mating surfaces is an important factor (Fig. 9.2).[37] Only in the case of a flat plate on a flat surface will the cement-thickness correspond to the measured standard film-thickness, which is determined in just this way. As an example, for a crown, as the angle of the preparation gets steeper so the effect of the cement becomes more pronounced. With a perfectly perpendicular-sided preparation no cement can be present, if the fit is accurate.

The effect may be calculated from the geometry and the measured film-thickness (Fig. 9.3). Thus, for a crown of perfect dimensions, for a given cement the vertical discrepancy, x, is proportional to the film-thickness a and the reciprocal of the sine of the angle of the preparation, φ:

$$x = \frac{a}{\sin \varphi} \qquad (9.1)$$

(This is assuming that the angle is uniform all around the crown preparation.) This equation gives the results shown in Fig. 9.4. The discrepancy for small slopes increases very rapidly indeed. The situation is identical for inlays (Fig. 9.5). This has clear implications for the preparation design for crowns, or of the cavity for inlays. In particular, it can be seen that an 'ideal' casting would not have exactly the dimensions required to fit the preparation, but would have an allowance made for the cement film. Requirements for the accuracy of impressions, models, patterns and investments must be viewed in this light.

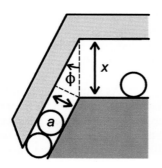

Fig. 9.3   The detail of the geometry of film thickness effects.

It should be noted that sometimes the preparation for a crown or inlay is specified in terms of the **taper angle** (sometime called the **convergence angle**), θ, such that

$$\varphi = \frac{\theta}{2} \qquad (9.2)$$

(see Fig. 9.5), where the intrinsic asymmetry of the preparation is allowed for. More to the point, it is not actually feasible to use a predetermined angle-bisecting path and prepare the angle φ accordingly, as if it were independently and accurately controllable.

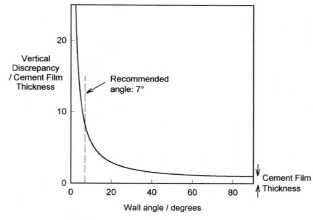

Fig. 9.4 The vertical discrepancy of a seated crown or inlay calculated as a function of wall angle φ.

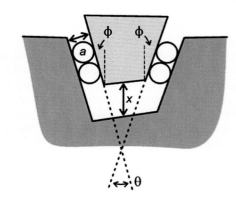

Fig. 9.5 The behaviour of inlays is similar to that of crowns. Note the definition of taper angle, θ.

●**9.2   Seating force**

The practical film thickness will depend not just on particle size but also on the viscosity η of the mix. For the situation shown at the top of Fig. 9.2, *i.e.* for parallel plates where the excess material can be extruded at the edges, the relationship between the rate of change of separation *a* under an applied force *F* is given by:

$$\frac{da}{dt} = \frac{2Fa^3}{3\pi\eta r^4} \tag{9.3}$$

where *r* is the radius of the circular plates.[38] (This is on the assumption that the behaviour of the fluid between the plates is Newtonian which, of course, cement mixes most definitely will not be.) In other words, the force required must increase as $1/a^3$, *a* decreasing, if a constant rate of settling into place is to be obtained. This would rapidly become an extremely large force. Conversely, for a given applied force (in practice this is usually applied by the dentist's manual effort, or by the patient biting), the rate of seating will decline very quickly.

Thus, although equation 9.3 refers to a situation slightly different from the usual designs of objects that need to be cemented (but compare orthodontic brackets and buttons), it can be seen that the extrusion of excess cement will proceed more and more slowly as the crown or inlay is seated further and the gap through which the excess is to be extruded becomes smaller. In fact, more realistic designs would be more difficult than expected on the above model. It can also be seen that further resistance would be felt, and the rate of closure decline more rapidly, as the cement sets and the viscosity rises. It may require very high forces actually to realize the theoretical film thickness at the margin. In view of this behaviour, some venting by way of a small hole in the occlusal surface of crowns has been suggested as a way of easing the difficulty. The vent is later filled with a polymeric material, or even with pure gold foil, which can be cold-welded (28§4) into gold alloy crowns.

Even so, continuous flow down to a gap size corresponding to a single particle cannot be relied on if the mixture at the point of use has a sufficiently high volume fraction of solids for it to be dilatant under the high shear strain rates that will necessarily occur as it is being extruded through such small gaps. It is possible then for the powder particles to jam up and lock, preventing any further movement (*cf.* the flow of dry powders, 4§9.4). Getting the mixing ratio correct is critical: maximizing strength, avoiding porosity, permitting sufficient flow.

●**9.3   Burnishing**

The fact that there is a practical limitation to the narrowness of the gap that will be left after seating a crown or inlay (whether this is due to viscosity or particle size effects) means that the cement would ordinarily be exposed to saliva, and thus risk degradation and dissolution. The rate at which this occurs can be minimized by **burnishing**[9] the margins of metallic castings. That is, applying pressure with a smooth, bluntly-rounded tool to deform the metal plastically (this obviously does not apply to ceramic or ceramic-veneered restorations), getting a better approximation of the margin to the tooth and therefore making the gap narrower. Burnishing is easiest (and most visible) with the softer, gold-based alloys, but the feather-edges of harder alloys may still be deformed under pressure to close the gap somewhat.

§10. Calcium Hydroxide Liners

Reference was made earlier (§1) to cavity liners that have little intrinsic strength because they are conveyed to the site suspended in a vehicle that evaporates. It has been believed that the supply of calcium ions encourages the formation of reparative dentine (which Zn ions do not, being of a heavy metal and toxic, notwithstanding the physiological role of Zn as a trace element). Calcium hydroxide is therefore thought to be a better lining material for deep cavities, in contact with living tissue. In fact, there is little objective evidence for this assertion. If anything, one would cite the very high pH (around 10) of fluid adjacent to an excess of $Ca(OH)_2$ as a toxic challenge to living tissue. Current thinking suggests that it is more likely that the high pH, and the **macerating** effect that this has on living organisms, serves more to prevent inflammation, pain and caries by sterilizing the site, bacteria being implicated in each case.

The need for some strength in lining materials (whatever their rationale) to resist displacement as stronger cements and restorative materials are placed on top, has led to some *setting* lining materials being developed.

---

[9] Strictly, burnishing means making metal bright by rubbing with a smooth metal tool. In dentistry, it commonly refers to the **swaging** of the margins of metal castings for crowns and inlays.

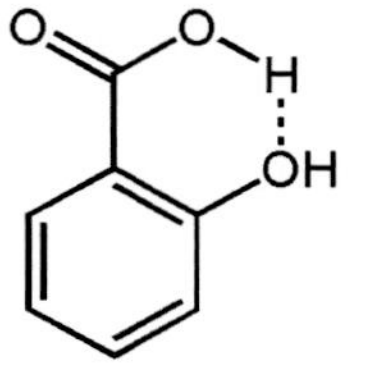

This means that essentially all of these materials are cements in the sense of §1.4. As an example, one of these is based on the acid-base reaction with 2-hydroxybenzoic acid (Fig. 10.1), otherwise known as salicylic acid, a substance closely related in structure to 2-ethoxybenzoic acid (Fig. 2.6). The formation of a chelated calcium complex is readily inferred, on the pattern of ZoE and EBA cements, leading to a solidified matrix embedding excess unreacted calcium hydroxide as core; again, a composite material.

Fig. 10.1 Salicylic acid, 2-hydroxybenzoic acid.

A drawback with using this reactant is that setting is too rapid. It has been found that using di-functional compounds, such as shown in Fig. 10.2, where esters have been formed thus reducing the overt acidity of the reactant, the chelation reaction still proceeds but more slowly, giving an adequate working time. There is also the benefit, presumably, of better bonding between chelation centres because of the 'cross-link' represented by the glycol-derived bridge. As indicated at §2.4, many similar substances may also be used for such reactions.

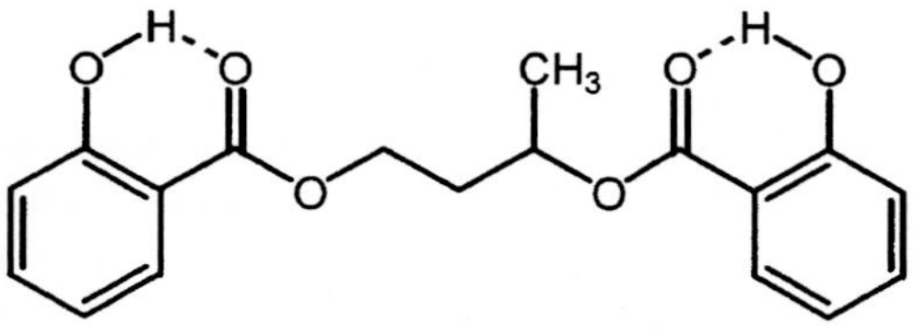

Fig. 10.2 An alternative reactant for setting calcium hydroxide liners: butan-1,3-diol disalicylate.

It is curious to note that salicylic acid has been shown to be the active moiety of aspirin, which is acetylsalicylic acid, and which is hydrolysed after ingestion. Aspirin is used because it is somewhat less irritant to the stomach than the plain acid. (In fact, salicylic acid is the principal ingredient of many of the tissue-corrosive wart- and corn-removal preparations which are available.) However, although it might be expected that the analgesic, anti-inflammatory and antipyretic properties would also be realized through, in effect, topical application on a compromised pulp, there is at present no evidence that it works in this way. Unless its irritancy supervenes, the effects of such a lining may prove a curious and useful contrast to those of eugenol (§2).

It is to be expected that the polymerization of composite resin restorative materials placed in contact with this type of liner will be inhibited by the phenolic hydrogen on any such compounds, as with eugenol. Whether discolouration also occurs has not yet been determined. However, a bigger problem may be the hydrolytic instability of the reaction products in the same way that ZoE is unstable (§2.2),[39] although the sterilization effected by the very high pH of the $Ca(OH)_2$ may provide some justification, there can be no mechanical support for a base or restoration placed on top. Even so, the mechanical strength of such materials when set must be compromised by the use of various compounds described as "plasticizers", for example: paraffin oil, N-ethyl toluene sulphonamide and polypropylene glycol (cf. §2.5). It is evident that the use of the word plasticizer can only apply to the unset mixture, where the rheological properties affect its handling and placement. Obviously, in the set material, such fluid, non-reacting substances can only act to weaken the mass, and if a phase separation occurs, as seems likely, this weakening would be exacerbated by the presence of numerous Griffith flaws. It would appear that the material properties are subordinate to the user's perception and convenience.

§11. Cavity Varnish

It is sometimes deemed necessary to protect dentine exposed in a prepared cavity from the irritant effects of acids or other constituents of filling materials, for example, zinc phosphate or, formerly, silicate cements. When a cavity varnish is used it is intended to place a thin, impermeable, insoluble layer over the prepared surface for that purpose. Consisting of a natural resin or other polymeric material dissolved in a volatile organic solvent, such as polystyrene in acetone, no chemical reaction is involved, merely the evaporation of the solvent. Were the layer continuous only one application would be necessary, but in attempting to keep it thin it is difficult to ensure that continuity, especially with the relatively crude delivery on a cotton wool pledget which is frequently employed. Small bubbles may be formed which burst as they dry, leaving pin holes. Two or three such layers in succession must be used to have any confidence in the film's integrity. However, each application must disturb the previous layers since no chemical reaction is involved and more solvent is supplied. What is likely, therefore, is a thick, irregular and defective coating.

Were the cavity wall sufficiently rough to provide a mechanical key, such a varnish might be stable. But the fact that the tooth tissue is damp and permeable to water, and the varnish itself lacking any chemical bond to that tissue, simple mechanical challenges such as flexure under biting forces are likely to result in the detachment of the varnish. It therefore fails in its primary function.

Before the advent of etching and bonding techniques for resin restorative materials, such varnishes were sometimes advocated for use with them: the monomer present would have had the tendency to soften or redissolve the varnish, and with any mechanical interference during packing the film would probably have been damaged.

More common now, though, is the use of varnish beneath amalgam restorations, supposedly to reduce leakage. As is shown later (14§6.2), amalgam must always leak when freshly placed, and the presence of varnish on the cavity wall can have no influence on this. Even so, if post-operative sensitivity is attributable to leakage, then the liquid (saliva, foodstuffs) might be prevented from reaching the dentine by a layer of varnish. But, again, it can do nothing to prevent the corrosion of the amalgam that follows leakage. It may be deduced therefore that, in general, whatever the restoration or cement, cavity varnishes could not possibly prevent leakage, only access of that leaked material to sensitive tissue.

Perhaps recognizing that leakage is inevitable, some varnishes incorporate fluoride in an attempt to reduce or avoid recurrent caries. However, fluoride that is to be effective as a local topical treatment for enamel must be soluble in or react with water to release that fluoride. It cannot therefore be soluble in such solvents as acetone, alcohol or chloroform which are used as the solvent for the varnish. Hence, if it is present it is in the form of fine particles, and their dissolution can only risk increasing the porosity of the varnish (if they are 'available', unvarnished, at the surface in the first place). Such designs seem to be based on wishful thinking rather than scientific principles.

In broad terms now, cavity varnishes are seen as a failed treatment. Because an idea is good in principle does not mean that an embodiment of it will actually be successful. Common sense requires that the idea be tested explicitly. Nevertheless, if we accept that there is merit in sealing the cavity surface, the logical products to use are the dentine bonding agents and similar materials now available. Whilst intended to provide a bonded interlayer between tooth and restoration (usually filled resin, but some are also advocated for amalgam), such materials are intended – on a properly prepared tooth surface – to provide both chemical bonding and sound mechanical key to that surface. The use of several layers is still appropriate, but the polymerization of each in turn prevents disturbance. Whether this is in fact a functional treatment remains to be seen.

§12. Hydraulic Silicate Cement

The remaining class of so-called **hydraulic** cement-forming reactions was introduced to dentistry in the form commonly referred to as "**MTA**", which is a trade name.[10] The expansion of this label, "mineral trioxide aggregate", in fact has no chemically-meaningful sense at all and can be ignored as useless. **Hydraulic Silicate Cement (HSC)** is an appropriate generic label.[40] In broad terms, the original brand of HSC consisted primarily of calcium silicates, with a variety of other phases present, depending on additions made to modify properties. The similarity to ordinary Portland cement is not accidental, being both the inspiration and preliminary trial material. However, Portland cement itself is unworkable in the dental context, and the modifications that have been made have been said to address concerns of particle size, setting rate, solubility and, possibly, toxicity (by the reduction of heavy metal content). It is used particularly for endodontic applications, where its strongly alkaline nature is considered to be beneficial, as for calcium hydroxide (§10); it is believed to be well-tolerated biologically and shows little setting shrinkage.

●12.1 Manufacture

Manufacture involves the firing at 1400 ~ 1500 °C of a finely-ground mixture of limestone (calcium carbonate) and clay silicate minerals such that the main reaction is the formation of tri- and di-calcium silicates, after the decomposition of the limestone:

$$CaCO_3 \Rightarrow CaO + CO_2\uparrow \tag{12.1}$$

---

[10] The wide use of this as a genericized trademark is both a case of "trademark dilution" and a misapprehension of chemical sense.

which occurs readily from about 800 °C, and the CaO (lime) then reacts:

$$CaO + MO.SiO_2.xH_2O \Rightarrow 3CaO.SiO_2, 2CaO.SiO_2 + MO + xH_2O\uparrow \tag{12.2}$$
$$\mathbf{C_3S} \qquad\qquad \mathbf{C_2S}$$

This firing process or **calcination** is sometimes known as **clinkering**, and the output as **clinker**. As before, no attempt is made here at stoichiometric equations because the natural starting materials are of variable composition, the intimacy of the ground materials cannot be perfect, and all such reactions depend on diffusion and take time. Notice that a notation is used similar to that employed for other ceramics (25§1.5), but in this field (*i.e.* Portland cement) a further shorthand form called **Cement Chemist Notation**[41] is common: **C** for CaO, **A** for Al_2O_3, **S** for SiO_2, **H** for H_2O ... ; these labels will be introduced as we proceed. The clay minerals also contain a variety of other metals in minor quantities, such as sodium, potassium and magnesium (commonly expressed as oxides, "MO"), and these will be present in the fired material as, for example, sulphates. Silicate minerals typically also contain a proportion of alumina, and this reacts on firing to form tri-calcium aluminate:

$$CaO + Al_2O_3 \Rightarrow 3CaO.Al_2O_3 \tag{12.3}$$
$$\mathbf{C} \qquad \mathbf{A} \qquad\quad \mathbf{C_3A}$$

The proportions are such that some calcium oxide, CaO, may remain unreacted in the final mixture. The clinker is then ground to a very fine powder for use. It is at this grinding stage that calcium sulphate, typically in the form of dihydrate (gypsum), may be incorporated as a means of controlling the rate of the setting reaction. However, if it is added before clinkering, then anhydrite and hemihydrate (2§1) will be present.

●12.2 Setting reactions

Despite a long history, much research effort, and being the single largest volume production industrial material we have, the setting of Portland cement remains in many respects obscure, with much controversy and uncertainty, despite its economic importance.[42] The setting of HSC, as with Portland cement, is also an extremely complicated process that depends on the exact proportions of phases present, their purity, and the temperature of the mixture (which of course varies during setting as the reaction is exothermic). In addition, there is a sequence of reaction products formed according to the kinetics of the various processes, and following an **Ostwald Succession** (8§2.6), such that some products are transient, but also affected by the physical barriers to diffusion created by reaction product layers on particle surfaces, much as calcium sulphate setting may be inhibited by borates and other materials (2§7.4). Dissolution, recrystallization, and inter-reaction of products also occurs. Thus, the situation will be similar to that described for zinc phosphate (§5.12): a trade-off between the thermodynamic expectation and the kinetically-limited reality. In that light, then, and following the style used for GI above (§8.6), we may outline the setting process, approximately according to the relative reactivities of the components.

Any excess calcium oxide reacts almost immediately:

$$CaO + H_2O \Rightarrow Ca(OH)_2 \tag{12.4}$$
$$\mathbf{C} \qquad \mathbf{H} \qquad\quad \mathbf{CH_2}$$

In the context **CH₂** is sometimes called **portlandite**. Tri-calcium aluminate also has a very fast reaction, and ordinarily would form a succession of hydrates:

$$\mathbf{C_3A} \Rightarrow \mathbf{C_2AH_8} \Rightarrow \mathbf{C_4AH_{19}} \Rightarrow \mathbf{C_3AH_6} \tag{12.5}$$

which provide diffusion barriers as described above. However, if sulphate ($\bar{\mathbf{S}}$, for SO_3) is present, competing reactions dominate:

$$\mathbf{C_3A} + \mathbf{\bar{S}} + \mathbf{H} \Rightarrow \mathbf{C_6A\bar{S}_3H_{32}} \Rightarrow \mathbf{C_4A\bar{S}H_{12}} \tag{12.6}$$

which effectively prevents further reaction with water until all sulphate has been consumed ($C_6A\bar{S}_3H_{32}$ is also known as **Ettringite**, which goes on to form a **monosulphate**). Such reactions may account for the early setting of the material. Diffusion of water through this barrier layer, and the consequent swelling due to the now kinetically-limited reactions (12.3), eventually bursts it, and the simple hydration reactions (*i.e.* 12.3) resume at a more normal rate. Meanwhile, the majority constituent of the powder, **C₃S** (also called **Alite**), and the similar **C₂S** (also called **Belite**), react with water (**C₂S** much more slowly) to provide the principal hardening mechanism:

$$\mathbf{C_3S, C_2S} + \mathbf{H} \Rightarrow \mathbf{C\text{-}S\text{-}H} + \mathbf{CH_2} \tag{12.7}$$

C-S-H is unfortunately not (or capable of being) well-characterized; it may vary over a wide range of compositions and is largely amorphous. However, the important thing to notice is that CH_2 is left over (the $C : S$ ratio of $C\text{-}S\text{-}H$ is ultimately less than $2 : 1$) and it remains as a distinct crystalline phase in the set material. Reaction of calcium sulphate hemihydrate and anhydrite to dihydrate (2§1) is not relevant, if it occurs at all.

The general description of the setting is thus the formation of more-or-less crystalline hydrated calcium aluminates and sulphate aluminates in an amorphous hydrated calcium silicate matrix. Excess calcium hydroxide is also present as a dispersion of crystals, and is responsible for the very high continued pH of the material, ~10. The reaction scheme is outlined in Fig. 12.1. Even so, it seems likely that unreacted material will persist for an extended time because of the diffusion barrier created by at least some of the silicate reaction products; only about 60 % completion has occurred at 1 week,[43] with reaction continuing for many months. The clinical implications of this are unknown as yet.

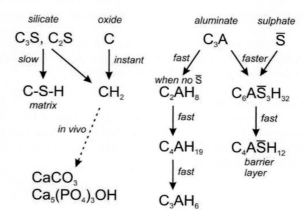

Fig. 12.1 Main setting reaction scheme for an hydraulic silicate cement in Cement Chemist Notation. Each solid arrow represents reaction with water, H. Note the *in vivo* further reactions with components of tissue fluids.

●12.3 Other constituents

The raw materials for an HSC, especially the silicates, frequently contain substantial quantities of iron oxide and this remains in the clinker (some 2 ~ 5% by mass, commonly), mostly in the form of calcium aluminoferrite, C_4AF ($F = Fe_2O_3$). This makes both the powder and the set material grey, which is considered undesirable in some contexts. Accordingly, a low-Fe version (< 0.5 %), 'white MTA' has also been produced. C_4AF also hydrates, in a manner very similar to that of C_3A but more slowly, thereby producing similar reaction products but with iron incorporated in solid solution. There are, therefore, some differences in the overall rates of the various setting reactions between the white and grey types, but no general effect on strength. Overall, these differences can be ignored as being minor variations.

The elements of HSC are mostly low Z, and so the radio-opacity of the material is less than is considered desirable, despite the presence of some iron. Accordingly, bismuth oxide, Bi_2O_3 may be incorporated to about 2 at% overall to rectify this (for Bi, Z = 83). This addition plays no appreciable part in the setting process, and so simply acts as a radio-opaque filler.[44] In fact, as with other non-reacting fillers (§8.8, 8.13), such particles would act as flaws in the Griffith sense, but here the small proportion in a non-loading bearing context probably means that this is unimportant.

●12.4 Further reactions

Portland cement is sensitive to loss of water during the early stages of setting, and care must often be taken to prevent or reduce this loss. HSC, on the other hand, is for use in circumstances where dehydration is very unlikely to occur, in marked distinction to GI and the older silicate (§7.4) cements. Accordingly, no special procedures are required. However, it should be noted that (apparent) 'setting' takes about 2 or 3 h, because of the slowness of the hydration and other reactions.

However, HSC is noteworthy for the excess calcium hydroxide, which must remain reactive and accessible by diffusion through the matrix, as well as microbicidal because of the high pH, which causes the hydrolysis of proteins, cell membranes and so on. In Portland cement, reaction occurs in the long term with atmospheric carbon dioxide to reform calcium carbonate, as indeed it did with the older lime mortar. This reaction is therefore expected to occur in HSC as it is used in contact with physiological fluids where the concentration of CO_2 (and carbonate) is appreciable. Likewise, physiological phosphate is also present in the cement's working environment, and calcium phosphates (principally hydroxyapatite-like) would be expected to form. Thus, in the long term, one would expect to see changes in hardness and strength due to such reactions, but also a continued and improving seal against tooth tissue as precipitation would be expected to occur where calcium ions diffuse to the surface, that is to say, all over. Similar reactions would also be expected to occur for calcium hydroxide materials (§10) if these are sufficiently exposed to tissue fluid.

Curiously, evidence of exposure to moisture before packing has been detected, as well as the presence of carbonate, in commercial material as-supplied.[45] The implications of the latter are unknown, although no adverse reactions are expected. Clearly, the carbonate should not be left over from the firing (reaction 12.1), so must be absorbed from the atmosphere. Prior reaction with water suggests that considerable batch variation might be expected in the absence of adequate quality control.

The high pH of the set material also means that the adjacent protein matrix of the dentine will be hydrolysed. The implication of this is that the strength of the root is adversely affected at a rate greater than that of the usual breakdown of a non-vital system.

●12.5 Variation

The success of the Portland cement-type of HSC, but yet the difficulties caused by the uncertainties of composition and particle size, has led to the introduction of materials based on synthetic (and therefore purer) tricalcium silicate (C_3S), possibly with a variety of other ingredients. The setting reaction is just as reaction 12.6 and the left hand side of Fig. 12.1.

References

[1]   Wilson AD. Dental Cements - General. in von Fraunhofer JA. Scientific Aspects of Dental Materials. Butterworths, London, 1975.

[2]   Douglas WH. The metal oxide/eugenol cements. I The chelating power of the eugenol type molecule. J Dent Res 57: 800 - 804, 1978.

[3]   Batchelor RF & Wilson AD. Zinc oxide-eugenol cements. I: The effect of atmospheric conditions on rheological properties. J Dent Res 48: 883 - 887, 1969.

[4]   Ingle JI & Bakland LK. Endodontics. 5th ed. Vol. 1. BC Decker, Hamilton, 2002. pp. 125 - 126.

[5]   Park C-K et al. Eugenol inhibits sodium currents in dental afferent neurons. J Dent Res 85(10): 900 - 904, 2006.

[6]   Hume WR. In vitro studies on the local pharmacodynamics, pharmacology and toxicology of eugenol and zinc oxide-eugenol. Int Endod J 21(2): 130 - 134, 1988.

[7]   Fujisawa S & Kadoma Y. Effect of phenolic compounds on the polymerization of methyl methacrylate. Dent Mater 8: 324 - 326, 1992.

[8]   Weinberg JE, Rabinowitz JL, Zanger M & Gennaro AR. ^{14}C-eugenol: I. Synthesis, polymerization, and use. J Dent Res 51: 1055-1061, 1972.

[9]   Wilson AD. Zinc Oxide Dental Cements. in von Fraunhofer JA. Scientific Aspects of Dental Materials. Butterworths, London, 1975.

[10]   Wells AF. Structural Inorganic Chemistry, 3rd ed. Oxford UP, 1967.

[11]   Doherty K, Connor M & Cruickshank R. Zinc-containing denture adhesive: a potential source of excess zinc resulting in copper deficiency myelopathy. Brit Dent J 210: 523-525 (2011).

[12]   Pabst W & Gregorová E. Critical Assessment 18: elastic and thermal properties of porous materials – rigorous bounds and cross-property relations. Mater Sci Technol 31 (15): 1801-1808, 2015.

[13]   Grenoble DE & Katz JL. The pressure dependence of the elastic constants of dental amalgam. J Biomed Mater Res 5: 489 - 502, 1971.

[14]   Nomoto R & McCabe JF. Effect of mixing methods on the compressive strength of glass ionomer cements. J Dent 29: 205 - 210, 2001.

[15]   Darvell BW. Aspects of the chemistry of zinc phosphate cements. Austral Dent J 29: 242 - 244, 1984.

[16]   Worner HK & Docking AR. Dental materials in the tropics. Austral Dent J 3: 215, 1958.

[17]   Fleming GJP, Shelton RM, Landini G & Marquis PM. The influence of mixing ratio on the toughening mechanisms of a hand-mixed zinc phosphate cement. Dent Mater 17: 14 - 20, 2001.

[18]   Fleming GJP, Marquis PM & Shortall ACC. The influence of clinically induced variability on the distribution of compressive fracture strengths of hand-mixed zinc phosphate dental cement. Dent Mater 15: 87 - 97, 1999.

[19]   Smith DC. A new dental cement. Brit Dent J 125: 381, 1968.

[20]   Cotton FA & Wilkinson G. Advanced Inorganic Chemistry, 5th ed. Wiley, New York, 1988.

[21] Wilson AD. Dental Cements Based on Ion-leachable Glasses. in von Fraunhofer JA. Scientific Aspects of Dental Materials. Butterworths, London, 1975.

[22] Wilson AD & McLean JW. Glass-Ionomer Cement. Quintessence, Chicago, 1988.

[23] Katsuyama S et al. (eds). Glass Ionomer Dental Cement - The Materials and their Clinical Use. Ishiyaku EuroAmerica, St. Louis, 1993.

[24] Wilson AD. Discovery! A hard decade's work in the invention of glass ionomer cement. J Dent Res 75 (10): 1723 - 1727, 1997.

[25] Nicholson JW, Brookman PJ, Lacy OM & Wilson AD. Fourier transform infrared spectroscopic study of the role of tartaric acid in glass-ionomer dental cements. J Dent Res 67: 1451 - 1454, 1988.

[26] Eliel EL. Stereochemistry of carbon compounds. McGraw-Hill, New York, 1962.

[27] Lloyd CH & Adamson M. The development of fracture toughness and fracture strength in posterior restorative materials. Dent Mater 3: 225 - 231, 1987.

[28] Kerby RE & Bleiholder RF. Physical properties of stainless-steel and silver-reinforced glass-ionomer cements. J Dent Res 70 (10): 1358 - 1361, 1991.

[29] Brown D & Combe EC. Effects of stainless steel fillers on the properties of polycarboxylate cement. J Dent Res 52(2): 388, 1973

[30] Sarkar NK Metal-matrix interface in reinforced glass ionomers. Dent Mater 15: 421 - 425, 1999

[31] Wang Y & Darvell BW. Failure behavior of glass ionomer cement under Hertzian indentation. Dent Mater 24: 1223-1229, 2008

[32] Musanje L, Shu M & Darvell BW. Water sorption and mechanical behaviour of cosmetic direct restorative materials in artificial saliva. Dent Mater 17: 394 - 401, 2001.

[33] Yelamanchili A & Darvell BW. Network competition in a resin-modified glass-ionomer cement. Dent Mater 24: 1065 - 1069, 2008.

[34] Mickenautsch S, G Mount G & Yengopal V. Therapeutic effect of glass-ionomers: an overview of evidence. Austral Dent J 56: 10 - 15, 2011.

[35] Czarnecka B & Nicholson JW. Maturation affects fluoride uptake by glass ionomer dental cements. Dent Mater 28: e1 - e5, 2012.

[36] Gu YW, Yap AUJ, Cheang P & Khor KA. Zirconia-glass ionomer cement–a potential substitute for Miracle Mix. Scripta Materialia 52: 113-116, 2005.

[37] Marxkors R & Meiners H. Taschenbuch der zahnärztlichen Werkstoffkunde. Hanser, München, 1978.

[38] Stefan MJ. Versuche über die scheinbare adhäsion. Sitzungberichte der kaiserlichin Akademie Wissenschafter Mass Naturwisseclasse 69: 713 - 735, 1874.

[39] Prosser HJ, Grossman DM & Wilson AD. The effect of composition on the erosion properties of calcium hydroxide cements. J Dent Res 61: 1431 - 1435, 1982.

[40] Darvell BW & Wu RCT. "MTA"—An Hydraulic Silicate Cement: Review update and setting reaction. Dent Mater 27: 407 - 422, 2011.

[41] http://en.wikipedia.org/wiki/Cement_chemist_notation

[42] Hewlett PC (ed). Lea's chemistry of cement and concrete. 4th ed. Arnold, London, 1998

[43] Coleman NJ, Li Q. The impact of zirconium oxide radiopacifier on the early hydration behaviour of white Portland cement. Mater Sci Eng C 33: 427 - 433, 2013.

[44] Li Q & Coleman NJ. Early hydration of white Portland cement in the presence of bismuth oxide. Adv Appl Ceram 112 (4): 207-212, 2013.

[45] Coleman NJ & Li Q. The hydration of ProRoot MTA. Dent Mater J 34(4): 458 - 465, 2015.

Chapter 10   Surfaces

Relationships across and between surfaces are a large part of the interaction of materials with their environment, as contrasted with the more 'internal' mechanical properties upon which we have so far concentrated. Many dental materials function in a specialized environment, the mouth, being a biological as opposed to a purely physical or chemical context. Even so, if these interactions are to be understood, the physics and chemistry of surfaces must first be explored. It is the purpose of this chapter to introduce the fundamental concepts and illustrate their relevance to dental applications.

*The major surface interaction of interest is the **wetting** of a solid by a liquid. The importance of surface energy or, equivalently, surface tension is stressed, as well as the relevance to cracking in solids.*

*It is the energetic considerations of wetting that drive the movement of liquids over surfaces, in particular the phenomenon of capillarity. When this driving force is combined with the limitation to flow provided by viscosity, we obtain a measure to describe the **penetration** of fluids into spaces. This is of great importance to the proper functioning of topical treatments such as fluoride products as well as fissure sealing and etching for bonding purposes.*

*True **adhesive dentistry** depends on the attainment of chemical bonding across the interface, and the nature of this is discussed. Hydrogen bonding is both a benefit and a nuisance, depending on the context, but control of the outcome again depends on understanding the process. However, there are many difficulties in achieving an adhesive bond, and some of the chemical approaches that have been tried, and their deficiencies, are also discussed.*

Surface energies are involved in many other areas: the setting expansion of gypsum products, the investing of wax casting patterns, oxides on amalgam and casting alloys, as well as soldering and porcelain. It is the pervasiveness of these effects and their influence on the success of dental procedures of so many kinds that demands an understanding of the basic principles.

Materials Science for Dentistry
https://doi.org/10.1016/B978-0-08-101035-8.50010-9

It has been suggested that dentistry is very largely concerned with interfaces: the adhesion of plaque, the retention of full dentures, the cementation of devices, the application of fissure sealants and fluoride treatments, and so on; all depend on the properties and behaviour of tissue and other surfaces. Thus, in order to understand the effects and processes of dental procedures and treatments, their successes and failures, it is necessary to consider surface chemistry and physics. This will then guide the selection of designs and treatments in many contexts.

§1. Wetting

Whilst restorative dentistry is needed (which will be so long as tooth tissue continues to be lost through caries or trauma), the repair of developmental defects is attempted, or even cosmetic changes are desired by the patient, the retention of the restoration is of central interest. Hence, the search for direct restorative or cementation materials that are truly adhesive in dentistry is based on ideas such as the desirability of restorations being permanent, or at least as long-lived as possible, and orthodontic appliances being firmly held for as long as is required. This begs the question of what actually constitutes adhesion. In the circumstances, it is probably better to drop the use of the word adhesion, at least for the time being, and instead to discuss the nature of the bond between two substances. The use of the term 'specific adhesion' has been used to emphasize the difference between purely mechanical and purely chemical effects. But apart from the gross mechanical retention afforded by dovetail joints and similar undercuts in restorations, as exemplified by amalgam or a partial denture framework utilizing clasps, it is not really that clear-cut a distinction in practical dentistry. The reason for this is that to obtain mechanical interlocking of the roughness of tooth tissue on a microscopic scale, whether or not etched by, say, a resin-based restorative material, the wetting of the one by the other is an essential prerequisite.

●1.1   Contact angle

The fundamental effect of interest is the degree or ease of wetting, and this is conveniently studied through measurements of the **contact angle** exhibited by the fluid material on the substrate (Fig. 1.1). This is defined as the angle included between a plane tangent to the surface of the liquid and a plane tangent to the surface of the solid, both being at a point on their line of contact, measured through the liquid. It can be seen that liquid drops may exhibit a variety of extents of spreading on a surface (see, for example, water on wax or alcohol on glass). The greater the tendency of the liquid to wet and spread, the smaller the contact angle, until it vanishes when perfect spreading is observed. It is now necessary to explain how this comes about, and what the physical driving forces are that underlie this behaviour.

It has already been said that in order to fracture a solid material work is done in creating the two new surfaces (1§7): there is a characteristic amount of energy required per unit of area of new surface, and this is called the surface energy, γ (J/m^2). This is no less true for liquids. A raindrop in free fall always tends to assume a perfectly spherical shape, as this is the one unconstrained surface that minimizes the surface area, and hence the total surface energy, for a given volume. Any system will, of course, tend towards a state of minimum total energy: this is a thermodynamic requirement. Because the water is fluid, and the droplet can therefore change shape, a mechanism exists for this minimization to occur. Thus, work has to be done on the raindrop by the friction of the air in falling to disturb it from its spherical shape. Likewise, droplets of one liquid in another, such as an emulsion (8§4.1), tend to become spherical – viscosity and time permitting.

●1.2   Surface tension

One physical interpretation of this tendency of liquids to change their shape spontaneously after a disturbance is the concept of surface tension (Fig. 1.2) where the apparent surface skin of the liquid is attributed to the difference in the number of molecules mutually attracted by molecules at different positions. Simplistically, this mutual attraction of molecules explains

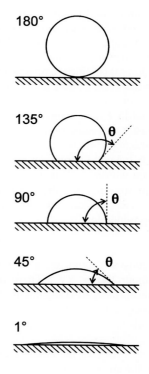

Fig. 1.1 Liquids may exhibit a variety of contact angles on solid substrates. (Note that the drawings do not represent constant drop volume.)

the tendency to minimize surface area because this also minimizes the number of 'unbalanced' molecules at the surface.

Surface tension, also given the symbol γ, is expressed as a force per unit length (N/m) (that is, a **line stress**), but since a joule is defined as 1 N.m (force × distance acted over), surface energy and surface tension can be seen to be entirely equivalent: $J/m^2 = N.m/m^2$. It is just the conceptual approach that differs, one's point of view, so to speak. (Note that the surface energy here is understood to be *specific* rather than *total.*) Hence, the work done in extending a surface, counting both sides of the film (Fig. 1.3), is identical to the work done creating the same area by splitting a column of liquid (Fig. 1.4), again counting both new surfaces:

$$E_{new\ area} = 2A\gamma = 2Ld\gamma = Fd$$
$$(F = 2L\gamma) \tag{1.1}$$

Surface tension may be measured directly as the force required to extend a film (Fig. 1.3) but more usually by a method equivalent to that shown in Fig. 1.5, *i.e.* by pulling on an object attached to a liquid surface and measuring the force required.

●1.3 Interfacial energy

So far we have discussed the fluid as if it were in vacuum, unaffected by any gas molecules above its surface. Indeed, when we consider the surface energy of a solid, we take it for granted that we are working in air at 'normal' temperature and pressure. Solids usually have entirely negligible vapour pressures (iodine is a notable exception, 0.31 mm Hg at 25 °C), so the composition of the air is unaffected. This is not, in general, the case with liquids. They will be more or less volatile, having appreciable vapour pressures, so that the 'air' must have some proportion of the vapour present. But the attraction from molecules in any vapour from the liquid tends to balance the forces acting on surface molecules of the liquid from within it. This is why we must refer to such systems at equilibrium, *i.e.* with saturated vapour, if any measurements of surface tension are to be meaningfully accurate and reproducible.

In other words, if γ_s is the surface tension of the solid against a vacuum, and γ_{sv} the value equilibrated against some vapour, then $\gamma_s > \gamma_{sv}$. There are in fact adsorbed molecules on that surface, although not necessarily a full layer, and the extra energy for the surface against a vacuum can be explained as the work required to remove those adsorbed molecules.

In fact, the mere existence of a liquid surface depends on the attractive forces operating between the dense liquid and its much less dense vapour, and indeed any other gas molecules present, this force being less in total magnitude than between the molecules of the liquid. The implications of this may perhaps be seen when a liquid substance is raised above the critical temperature (at the critical pressure): the otherwise clear boundary between liquid and vapour simply ceases to exist as the densities of the two phases become indistinguishable.

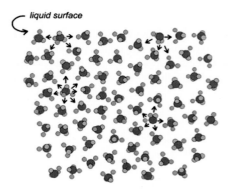

Fig. 1.2   Surface tension may be viewed as arising from unbalanced attractive forces on molecules at the surface in comparison with those in the bulk.

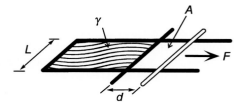

Fig. 1.3 Film extension (as for a soap film on a wire frame) ...

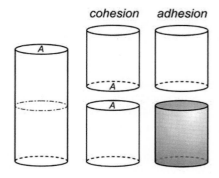

Fig. 1.4 ... and column splitting are equivalent in that both create new surface – which requires work to be done. The separation of adherent materials is similar.

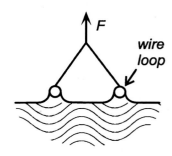

Fig. 1.5 One means of measuring surface tension directly.

Because there is no surface, there is then no surface tension. It has in fact diminished steadily to zero as the temperature is raised.

Similarly, any other liquid in contact with the first will have its own cohesive forces acting, and there will also be some interaction between the molecules of the two liquids. Since the second liquid will usually be very much denser (more compact) than the vapour of the first (*i.e.* the molecules more numerous per unit volume), even quite weak interaction between the two liquids leads to a generally lower **interfacial** energy, as it is now called, than either of the individual liquids' surface energies when in contact with their vapours.

●1.4   Energy minimization

So it is with a liquid on a solid: very often there will be a lower interfacial energy between the liquid and the solid than between the liquid and its vapour. There is then a tendency for the liquid to spread on the solid – which is energetically favourable. However, this spread necessarily implies a change in the surface area of the liquid against the vapour – which needs to be allowed for because the total liquid-vapour surface energy there is changed. There is also the decrease in the area of the solid against the vapour of the liquid to take into account.

We shall consider the simple case of a liquid drop on a plane, uniform[1], homogeneous[2], rigid, isotropic[3] solid surface, but the results are extendable (with some effort) to other conditions. We know first that the shape of the liquid-vapour interface must be a spherical cap (Fig. 1.6) because, for any given liquid-solid interfacial area, this is the shape that minimizes that surface. We derive this as follows: a sphere is the closed surface of minimum area. The two parts cut off by an intersecting plane must also each be of a minimum area for the fixed intersection line since if either were not it implies that their sum, the original spherical surface, was not at a minimum in the first place. It follows then that the liquid-solid interface is a circular disc, since any section through a sphere gives a circle.

A sense of the nature of the variations in the interfacial areas as a drop spreads can be gained from Fig. 1.7. In order to understand this and its consequences, an equation for the total energy, E_{TOT}, of the system must be written. This takes the form of the sum of the product of the surface area A of an interface with its specific surface energy, γ, for each of the interfaces present:

$$E_{TOT} = A_{sv}\gamma_{sv} + A_{sl}\gamma_{sl} + A_{lv}\gamma_{lv} \qquad (1.2)$$

where the subscripts s, l and v refer to solid, liquid and vapour respectively and the pairs identify the interface in question. It is this sum that must be minimized for thermodynamic equilibrium.

By the application of a little geometry, it can be shown that the minimum in equation 1.2 is obtained for the spherical cap whose contact angle θ is obtained from

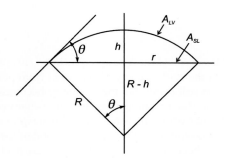

Fig. 1.6 Geometry of a drop on a plane surface.

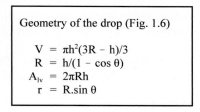

Geometry of the drop (Fig. 1.6)
$V = \pi h^2(3R - h)/3$
$R = h/(1 - \cos\theta)$
$A_{lv} = 2\pi Rh$
$r = R.\sin\theta$

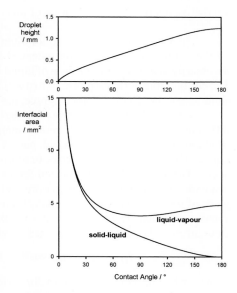

Fig. 1.7 Variation in droplet height (top) and the liquid interfacial areas (bottom) as a function of contact angle for a unit volume droplet (1 mm³) (see Fig. 1.6). The minimum for the liquid-vapour plot occurs at $\theta = 90°$. The maximum for both interfacial areas corresponds to a monomolecular layer; for water that would be about 2 m² (2§2.5). The solid-vapour curve is the complement of that for solid-liquid with respect to the original solid area (which is arbitrary).

---

[1] *e.g.*, if crystalline, only one crystal plane showing.

[2] single phase

[3] *e.g.*, not drawn polymer or edge-on layer-structure crystal

$$\cos\theta = \frac{\gamma_{sv} - \gamma_{sl}}{\gamma_{lv}} \tag{1.3}$$

This is called the **Young equation**. This result is on the assumption that the solid surface is ideal in the sense that the surface energy is constant from point to point. Even if it were not, the thermodynamic minimization must still occur, but the geometry becomes much more complicated and our ability to calculate a solution is somewhat impaired. Gravity, too, is ignored in equation 1.3, but for small drops the flattening that this causes, due to hydrostatic head, can safely be ignored. In addition, it is assumed that viscosity does not play a role, and that dissolution of the substrate – and of the liquid in the substrate – can be ignored. That is, all kinetic aspects have been allowed to equilibrate or are negligible. Despite all this, the general principle of the minimization of the total energy still stands as the driving force for the adjustments in shape that are observed when any liquid drop is placed on any surface. Systems of relevance to dentistry include fissure sealants into crevices, mercury on silver amalgam alloy, investment slurry on wax, and so on.

It is possible to deduce from equation 1.3 what conditions are required to promote wetting or non-wetting, depending on the application and the intention. Low surface tension liquids clearly will wet more effectively, and particularly if they are on high energy solid surfaces, aided by a strong interaction between the liquid and the solid to lower that interfacial energy. Conversely, low surface energy solids (such as polytetrafluoroethylene, 'ptfe' or 'Teflon') will be difficult to wet, and high surface tension liquids, such as mercury and other molten metals, do not wet most surfaces, and in particular not ceramics.

●1.5   Force balance

Given that we found an equivalence between surface energy and surface tension above, there is an alternative approach to the solution of this problem based on the surface tensions which are operating.

In any given surface (at equilibrium), the surface tension acting on one side of an arbitrary line drawn on that surface must be exactly balanced by the identical tension on the other side – the line stresses are necessarily identical. But if a drop of liquid is placed on a surface a line is established that disrupts that balance. The **contact line** or **three-phase line** (the 'TPL') is the boundary line along which solid, liquid and vapour meet simultaneously. Consider, for example, the view from the solid: the solid-vapour interface no longer exists on one side of that line, and it is replaced there by a solid-liquid interface whose interfacial tension is in general different from that of the solid-vapour interface. Across that contact line is therefore an imbalance. Equally, from the point of view of the liquid, part of the liquid-vapour interface has been replaced by that of the solid-liquid. There is again an imbalance of tensions across the contact line, but this time generally at some arbitrary angle. Because of these unbalanced forces acting on the line, the line must move, and it will move until the resultant of all three forces, $F_=$, is precisely zero in the plane of the solid surface at the contact line. This force-vector approach (Fig. 1.8) leads to the following straightforward equation being written down directly:

$$F_= = \gamma_{sl} - \gamma_{sv} + \gamma_{lv}.\cos\theta = 0 \tag{1.4}$$

each term having a magnitude and a direction. This can be seen to be only a rearranged version of equation 1.3, and thus equally well expresses the thermodynamic result at equilibrium.[1]

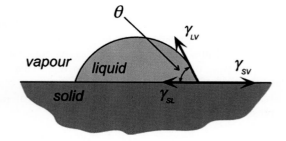

Fig. 1.8 The force vectors which may be considered to act at the contact line of a drop on a solid surface.

It should be clear from this that contact angle is not, in fact, just a property of the liquid itself at all, but a *geometrical* consequence of the particular values that the three interfacial energies have for a given system (and these, of course, vary with temperature). In other words, it is not an independent, intensive physical constant but a behaviour dependent on all local circumstances that influence any element of the system. Again, it is a property of the combination of the particular solid, liquid and gas phases – composition, constitution and structure, as appropriate.

In Fig. 1.8 it might be noticed that there is an additional force component, normal to the surface, of $\gamma_{lv}.\sin\theta$, that is left apparently unbalanced. To see why the complementary downward force does not appear in equation 1.4, consider the situation of a drop of a liquid A floating on immiscible liquid B (Fig. 1.9). Here there are again three interfaces to consider, but the force balance is complete because of the deformation of the surface of liquid B. Liquid A has two spherical section surfaces of different radii of curvature. In other words, in Fig. 1.8, the solid surface cannot deform (measurably), so the downward balancing force of $-\gamma_{lv}.\sin\theta$ has no bearing on the calculation: the three-phase line can only slide in the plane of the solid surface. The normal forces still exist, but are accounted for by equilibrium elastic deformation of the solid (but which is usually far too small to be of any importance). (The 'vapour' in Fig. 1.9 must, of course, be saturated with respected to both liquids for thermodynamic equilibrium to be attained.)

Consideration of equation 1.4 will show that there are circumstances when no real value of θ can cause the sum to be zero. Thus, if

$$(\gamma_{sl} + \gamma_{lv}) < \gamma_{sv} \qquad (1.5)$$

then $F_=$ is negative (limiting $\cos\theta = +1$), which means a net tension pulling the contact line over the unwetted solid surface. This corresponds to **perfect wetting** and is accompanied by spontaneous spreading, the limit presumably being a monomolecular film. On the other hand, if

$$\gamma_{sl} > (\gamma_{sv} + \gamma_{lv}) \qquad (1.6)$$

then $F_=$ is positive (limiting $\cos\theta = -1$), and the contact line would be driven to retreat across any already wetted surface until the contact area, the solid-liquid interface, vanishes.

These conclusions can also be seen in the equivalent equation 1.3 where quite clearly it is very easily possible to construct ratios whose values lie outside of the range $[-1, +1]$. This is not an indication of the failure of the theory, but rather an illustration of the fact that the intermediate condition of **partial wetting**, $-1 \le \cos\theta \le +1$, is a rather special occurrence (Fig. 1.10). This observation reinforces the remark above that contact angle is not in any way a property of the liquid but a characteristic of the system as a whole.

A further point may be made here. Sometimes the word 'wetting' is interpreted as only applying to complete wetting, *i.e.*, if and only if $\cos\theta = 0°$. However, noting that the interaction of liquid with solid that leads to any adhesion (Fig. 1.4), that is, lowering the energy of the system, means that some adsorption has occurred: this is wetting. Thus it is only for $\theta = 180°$ that there is no wetting. Care must be taken to distinguish between the dynamic, physical process of the spreading of a liquid on a solid and the thermodynamic driving force for this that derives from a chemical interaction across the interface. "Partial" wetting is still wetting.

●1.6 Critical surface tension

There are considerable difficulties with measuring individual solid interfacial tensions, and it is certainly easier to measure contact angles. But, in the region of perhaps most interest, θ approaching $0°$, the determination of angles is difficult and an indirect approach is sometimes of value. For a given substrate, θ is measured for a series of homologous liquids (the aliphatic alcohols for example, or the paraffins themselves). The surface tensions of the liquids are independently measurable by other means (Fig. 1.5), and $\cos\theta$ is then plotted against γ_{lv} (Fig. 1.11).

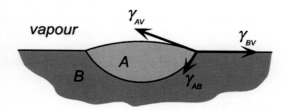

Fig. 1.9 The force vectors which may be considered to act at the contact line of a drop floating on another (immiscible) liquid (gravity-free).

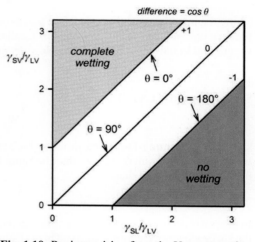

Fig. 1.10 Regions arising from the Young equation. By scaling the γ_{SV} and γ_{SL} values, and plotting the contours of constant difference, the intermediate partial wetting condition is seen to be a special occurrence.

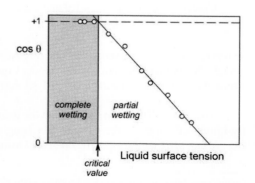

Fig. 1.11 Plot to determine critical surface tension.

This is known as a **Zisman plot**. The plotted points tend to fall on a straight line and an extrapolation can be made to estimate the **critical surface tension**, the value above which full wetting will not occur, and below which spontaneous spreading will be observed – for that substrate. Necessarily, this approach has to be used with caution because the result only applies to a similar liquid, one whose chemical interaction with the substrate is similar to that of the others of the test series.

●1.7   Work of adhesion

The necessity of counting *both* new surfaces in crack formation, film extension and so on can be demonstrated using the above results. The **reversible work of adhesion**, W_a, is the work required to separate the liquid from the solid (Fig. 1.4), and is given by:

$$W_a = \gamma_{sv} + \gamma_{lv} - \gamma_{sl} \tag{1.7}$$

(using the interchangeability of surface tension and surface energy). This equation is to be read as saying that for each unit of area of solid-liquid interface that is separated (*i.e.*, removed), one unit of area of each of solid-vapour and liquid-vapour is created. Now, from equation 1.4,

$$\gamma_{sv} - \gamma_{sl} = \gamma_{lv}.\cos\theta \tag{1.8}$$

Substitution from 1.8 in 1.7 gives:

$$W_a = \gamma_{lv}(1 + \cos\theta) \tag{1.9}$$

known as the **Young-Dupré equation,** which applies to any system where chemical changes do not occur on wetting. But if we imagine now that the same liquid is substituted for the solid substrate, we have of course $\theta = 0°$, $\cos\theta = +1$, so that

$$W_c = 2\gamma_{lv} \tag{1.10}$$

In other words, the work done on joining together – or creating – two identical liquid surfaces is precisely twice the surface energy (ignoring sign). This is usually then referred to as **cohesion**, Fig. 1.4, hence the change of subscript to 'c'. The result necessarily applies by extension to solids also, and serves to illustrate the reason for existence of the work of fracture previously discussed.

The difference between the work of adhesion of a liquid on a solid and the work of cohesion of that liquid

$$W_a - W_c = \gamma_{lv}(1 + \cos\theta) - 2\gamma_{lv} \tag{1.11}$$

is often given the (quite misleading and erroneous) name **spreading coefficient** (or **parameter**), the equation reducing to:

$$S_{eq} = \gamma_{lv}(\cos\theta - 1) \tag{1.12}$$

where the subscript 'eq' is specifically to indicate the equilibrium condition, that is when droplet spreading (or retraction) is complete. S_{eq} is, of course, not a coefficient, as it has the same dimensions as γ_{lv} (neither is it properly a parameter, *i.e.* a controlling dimensionless value in a mathematical expression). Strangely, as the term in parenthesis has a maximum value of zero (for complete wetting), minimum value −2, S_{eq} might be better viewed as a measure of the thermodynamic driving force (*i.e.* change of energy) for *non*-spreading or, in the surface tension sense, the line stress tending to drive the TPL *back* from a fully-spread, complete wetting state.[4]

Comparison of equations 1.9 and 1.12 then shows that

$$W_a = -S_{eq} \tag{1.13}$$

which is reasonable in the sense that one refers to the work of forming the solid-liquid interface, the other to its removal. S_{eq} is therefore not an independent measure of anything, just a complementary view of the same wetting phenomenon (indeed, equation 1.12 may also be found referred to as the Young-Dupré equation).

The above discussion depends on the existence of an equilibrium TPL. However this does not apply to fully-immersed systems – there is no contact angle (they are always fully wetted unless there are bubbles – §9.2). Here, we can see that the (specific) **work of immersion**, is given straightforwardly just by:

---

[4] Confusingly, this is sometimes defined with the opposite sign, $S >= 0$. This would make sense if the sign of the energy change on joining bodies is defined to be negative.

$$W_{imm} = \gamma_{sv} - \gamma_{sl} \qquad (1.14)$$

where the solid is imagined first to have been equilibrated against the vapour. If this condition is not allowed, as is normal, say for adding a dry powder to a liquid (although adsorbed water is likely always to be present, 2§2.5), then the vapour subscript of γ_{sv} needs to be very carefully considered and perhaps another value defined and used. The emphasis on contact angles in connection with fully-immersed systems such as implants is therefore seen to be misplaced.

●1.8 Boundary layer

Reference was made to the existence of a boundary layer on a filler and the effect that this has on viscosity (4§9.2). It is now possible to see that since wetting implies a lowered interfacial energy, the molecules forming that layer are by definition in a more stable position than those further away: they are adsorbed to the filler. They do not exchange with the bulk of the liquid so readily, and are more organized because they are constrained to be a two-dimensional entity, following the surface of the other phase. Indeed, depending on the strength of the binding to the substrate, the boundary layer may be described as a two-dimensional liquid or even crystal-like (indeed, for water it is probably ice-like for at least two layers[2]). The ordering this represents may therefore have an influence on the next nearest neighbours, and so on, so that the effective thickness of the boundary layer may be several molecules thick, depending on the strength of interactions between molecules. The hydrogen bonding (§4) of water is clearly important in this context, and comparison may be made with the template effect of polysaccharides (7§8.2). The variation in behaviour between fillers as to their effect on viscosity can therefore be seen to depend in part on the chemical nature of the surface as well as surface area and matrix chemistry.

●1.9 Roughness

The calculations above involving surface energy implicitly assume that the respective surfaces are smooth – perfect geometrical planes and spherical caps. In such cases the specific surface energy is appropriate because the actual area of the interface enclosed by the relevant boundaries is exactly the area of the ideal delineated shape. For liquids, this is a perfectly acceptable assumption and – while flow can occur – inevitable thermodynamically to minimize area. However, real solid surfaces are rough, having a true surface area perhaps as much as two or three times the apparent size. Clearly, it is the true area of interface that must be used in the calculation of energy in equation 1.2, for both solid-liquid and solid-vapour. It is obvious that doubling the area inside a perimeter is numerically the same as doubling the specific surface energy (*i.e.*, the work of formation):

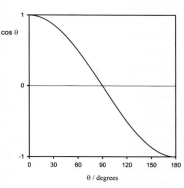

Fig. 1.12 Graph of the cosine function.

$$2A_{sv} \times \gamma_{sv} = A_{sv} \times 2\gamma_{sv} \qquad (1.15)$$

Thus, if the areal roughness factor, r, is given by:

$$r = \frac{\text{actual area}}{\text{delineated area}} \qquad (1.16)$$

then equation 1.2 becomes

$$E_{TOT} = rA_{sv}\gamma_{sv} + rA_{sl}\gamma_{sl} + A_{lv}\gamma_{lv} \qquad (1.17)$$

such that equation 1.3 becomes:

$$\cos\theta_r = \frac{r(\gamma_{sv} - \gamma_{sl})}{\gamma_{lv}} = r\cos\theta \qquad (1.18)$$

where θ_r is sometimes called the **Wenzel** angle. That is, the right hand side of the Young equation (1.3) is multiplied by the roughness factor. In other words, since the cosine function has the form shown in Fig. 1.12, the effect of roughness is to magnify the effect of the gain or loss of energy occurring in replacing the sv interface with the sl: if this is favourable (positive), it becomes more so; if not (negative), it is made worse[3] (noting again the meaning of

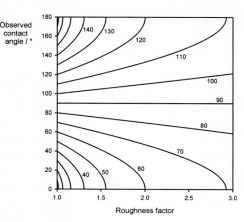

Fig. 1.13 Effect of roughness factor on observed contact angle. The contours are labelled with the value of the contact angle on a smooth surface. Thus, for roughness factor $r = 1$, observed angle = smooth surface angle.

values >1 or <-1, Fig. 1.10). The effect is illustrated in Fig. 1.13. For a 'perfect', flat surface, contact angle of 90° (when $r = 1$), there is no change, but if there is any better wetting than this on a smooth surface, a rough surface improves it, and this can be quite dramatic for even a modest value of roughness factor of 1.5, where complete wetting (and thus spreading) is observed for a 'smooth' contact angle of about 47°. Conversely, droplets would roll cleanly off a surface with $r = 2$ for $\theta = 120°$. It is a combination of microscopic texture and the waxy surface of many leaves that allows raindrops to be shed from them so easily (the 'lotus effect'), and likewise from the treated fabric of raincoats. We can therefore amplify earlier remarks by saying that contact angle is a geometrical consequence of the particular liquid, solid and solid surface texture, and clearly not a material property in any sense.

●1.10 Levelling

One manifestation of the spontaneous tendency to minimize the surface area of a liquid is **self-levelling**. That is, a rough or wavy surface will become smooth, even in the absence of gravity effects. This is seen very clearly in paint, where brushmarks tend to disappear while the paint is still fresh enough to be just viscous (and not appreciably elastic), and in flamed wax (16§2.6). The same thing drives the smoothing of a fractured glass surface at a high enough temperature (so-called 'flame-polishing' – really flow when above the T_g; *cf.* 25§4.6), and can be seen to a lesser extent in freshly mixed dental cements, filled resins, and so on, depending on whether there is a yield point, time allowed, and the radius of curvature.

●1.11 Experimental considerations

It may seem that the experimental determination of contact angle is a straightforward matter, but a number of factors must be borne in mind for meaningful results. Firstly, it has been emphasized that the equilibrium state that is the goal has to be with respect to the vapour of the liquid. More specifically, it has to be the saturated vapour if the thermodynamic conditions are to be met. Failure to achieve this not only affects the interfacial energies concerned, but might lead to appreciable evaporation of the liquid. If this liquid is itself a solution, then its concentration must change, and with it the interfacial energies. Secondly, any dissolution of the substrate, or components of it, changes both liquid and solid, including the roughness of the latter (§1.9). Even slightly soluble substances could have significant effects. Thirdly, the liquid should not dissolve in the substrate, for example, water in polymers. Testing polymeric materials as used in the mouth implies that they are saturated with water. Each of these can give time-varying behaviour. In addition, the presence of surface-active substances in the solution, such as glycoproteins in saliva or other body fluids, means that the true substrate surface is coated (almost instantly) and obscured such that what is tested is the chemisorbed (multi-molecular) layer instead. This is a quite different matter since this will tend to be chemically similar for a given solution no matter what substrate is underlying. Indeed, the adsorption of other active contaminants from the air means that the testing of truly clean surfaces is a serious challenge, while testing with liquids other than water is problematic in view of the near-universal presence of an adsorbed water layer (2§2.5).

Physically, observations of contact angles may be affected by electrostatic charges, especially where the substrate is not an electrical conductor. Such charges, which reside on the surface (as required by Coulomb repulsion) must – by that same repulsion – tend to increase the area of those surfaces, driving spreading. This is effectively the same as lowering the surface tension. Such charges can arise in a variety of ways, most commonly through the **triboelectric effect**, but also through chemical processes.[4] Such effects must be avoided for accurate measurements.

Another field in which interfacial energy is of considerable importance is, as mentioned above, that of fracture. In particular, the Griffith criterion for cracking (1§7) depends on that work of formation. Thus, in any experiment in which crack formation and growth is involved, the environment of the new surface is critical. Effectively, as was discussed there, the strength of a brittle material depends on whether or not the test object is immersed, and in what kind of fluid. Obviously, for materials seeing service in the mouth, saliva is the principal medium of interest, but it is conceivable that foodstuffs and drinks might provide a sufficient change of conditions to permit cracking under otherwise survivable stress. For other types of material, variation in humidity might be important in affecting the value of the critical stress. It should go without saying that any attempt to measure a work to failure, or the toughness of a material (including fracture toughness, K_{Ic} [29§5.2]), should also be under a realistic service environment in all respects (including temperature) if meaningful values and comparisons are to be obtained (equation 1§7.2). Indeed, any process in practice that depends intrinsically on the formation of new surface, such as tearing, machining, cutting and abrasion, will also depend on the interfacial energy (see also 20§2.7), and thus the environment.

§2. Capillarity

Surface tension has another important manifestation which has many implications in dentistry: that of **capillary rise** or **capillarity** (Fig. 2.1). The same forces that cause the boundary of the liquid to move on the solid when this is horizontal, adjusting the position of the contact line, may also be observed to operate on vertical surfaces. The lens-like surface of water in a glass tube (such as a burette) – the **meniscus**, and the curved edge around the liquid in a cup should be familiar phenomena. It is this curvature that causes capillary rise. However, to demonstrate this, another system must first be examined.

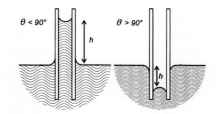

Fig. 2.1 Capillary rise and depression.

●2.1   Droplets and bubbles

Consider a spherical droplet of radius r (Fig. 2.2). For any line segment drawn on a liquid surface, the total force acting on it is given by the product of its length and the line stress due to the surface tension of the liquid. Thus, if the line is a meridian (*i.e.*, a full circle from pole to pole) on a droplet, length $2\pi r$, the total force F acting on it is given by

$$F = \gamma.2\pi r \qquad (2.1)$$

(We can drop the subscripts 'lv' when this interface is understood.) This generated force is acting in the sense of minimizing the area overall, and thus can be seen to be squeezing the contents of the droplet. It must therefore act over the entire cross-section (area = πr^2), so the increase in pressure, ΔP, inside the drop with respect to the outside is given by:

$$\Delta P = \frac{F}{\pi r^2} = \frac{2\pi r\gamma}{\pi r^2} = \frac{2\gamma}{r} \qquad (2.2)$$

which is known as the **Laplace equation**.

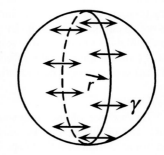

Fig. 2.2 The forces acting on a meridian of a liquid drop.

For a droplet, the (excess) pressure is positive (compressive) in the liquid, and this is taken to define the curvature, C = 1/r, of the drop's surface to be positive (Fig. 2.3, left). This point is important because it establishes the sense of the pressure difference in any system. Thus, in a bubble of gas in a liquid, which can be represented by the same diagram, Fig. 2.2, the pressure in the gas of the bubble is higher than the pressure in the liquid by the amount given by equation 2.2. From the point of view of the liquid, the curvature of the interface is negative (curving away from the observer), so the pressure in the liquid is relatively negative *with respect to the gas*. We could change the viewpoint, and just as accurately say that from the point of view of the gas the surface curvature is positive (wrapping around the observer), and that therefore the pressure in the gas is higher than that in the liquid (Fig. 2.3, right). In brief, the pressure is relatively higher on the concave side, towards the centre of curvature, because the energy minimization requires that interfacial area to be minimized: whether bubble or droplet, it is 'trying' to shrink.

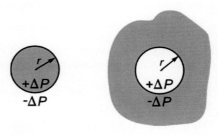

Fig. 2.3 Pressure differences across droplet (left) and bubble (right) surfaces.

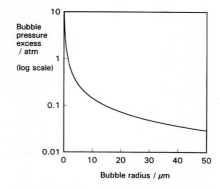

Fig. 2.4 Dependence of the Laplace pressure difference in a bubble on its size. Water at 20°C.

This effect is not insignificant: very small bubbles will have very high pressures (Fig. 2.4). Put simply, it is merely the curvature of the boundary that generates the pressure difference across the interface. In other words, it is of no consequence what lies either side of the boundary, the physical effect and its sign is the same. It is only the magnitude of the effect that depends on the surface energy of that interface.

In passing, we can observe that a soap bubble has two boundary surfaces, an inner and an outer, either side of the liquid soap solution. There are therefore two contributions to the excess pressure inside the bubble, which is very nearly double that of the single surface value because the two radii are very similar, since the soap film is thin. Of course, the mere fact that a bubble has to be blown (Fig. 2.5) shows that a pressure difference is involved.

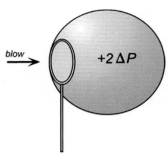

Fig. 2.5 Blowing a soap bubble requires that the pressure inside is greater than outside, but it also has two surfaces.

In deriving equation 1.3 we noted that a spherical surface has minimum area and therefore minimum energy. We can obtain a further view of what this means by considering the curvature. If the curvature of the surface of an unconstrained droplet or bubble varied from point to point across the surface, that is, not spherical, then the pressure just beneath the surface would accordingly vary. This is not an equilibrium condition. Pressure differences within a fluid mean that flow must occur in response, in order to eliminate all gradients (we are ignoring the effect of gravity here). Thus, the curvatures become everywhere identical and the surface is spherical.

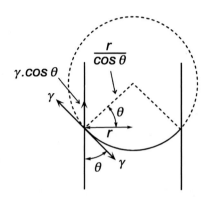

Fig. 2.6 A meniscus in a capillary.

•2.2 Capillary

The surface of the liquid-vapour interface in a circular capillary, the meniscus, can be viewed as sliced from a bubble (dotted in Fig. 2.6), a spherical cap. Now the tension forces in the liquid-vapour interface at the circular boundary are not balanced by forces in the same kind of surface, but rather by those at the contact line with the solid wall of the capillary tube. The total force acting at the meniscus in the direction of the axis of the tube is given by:

$$F = 2\pi r \gamma . \cos\theta \qquad (2.3)$$

Dividing by the cross-sectional area of the tube, πr^2, the pressure difference across the meniscus is therefore given by:

$$\Delta P = \frac{2\gamma}{r}.\cos\theta \qquad (2.4)$$

This is known as the **capillary pressure**. This equation reduces to equation 2.2 when $\theta = 0°$, *i.e.*, when the meniscus is hemispherical, and to zero when $\theta = 90°$.

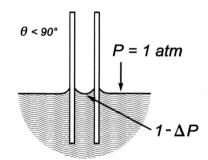

Fig. 2.7 The hydrostatically disallowed situation of a curved meniscus in capillary level with the outside liquid.

If the capillary is standing vertically in a large container of liquid with $\theta < 90°$, *i.e.* as measured on the material of the capillary, and supposing that the meniscus in the tube is level with the flat liquid surface outside (Fig. 2.7), it can be seen that there is an imbalance of pressures: immediately beneath the meniscus pressure must be low because of equation 2.4, but equally the pressure immediately beneath the surface of the liquid outside the tube is atmospheric because the curvature there is zero. Hydrostatically, this condition is obviously disallowed, and the pressure difference forces the liquid up into the capillary until the pressure in the capillary at the same level as the outside surface is also atmospheric. This means that there is then a column of liquid of height h great enough to balance the reduction in pressure due to surface tension (Fig. 2.1, left):

$$\Delta P = -h\rho g \qquad (2.5)$$

where ρ is the density of the liquid and g is the local gravity (the height in this sense is a negative depth, measured from the level of the outside liquid, to match the negative curvature of the meniscus). This situation may be compared with that in a closed tube (Fig. 2.8).

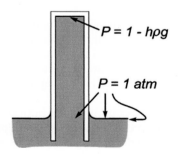

Fig. 2.8 At the top of the column of liquid in a closed tube, the weight of the liquid causes a pressure reduction.

The reduction in pressure at the top of the tube is precisely that due to the weight of the liquid column.

At equilibrium, the pressures calculated from equations 2.4 and 2.5 must be numerically equal:

$$h\rho g = \frac{2\gamma}{r}.\cos\theta \qquad (2.6)$$

so that the pressure difference inside the capillary at the level of the outside liquid is precisely zero (as required by hydrostatics).[5]

The same situation applies in a Torricellian **barometer**, mercury in a closed-end glass tube (Fig. 2.9). Since there is a vacuum in the space above the mercury (ignoring the mercury's vapour pressure, Fig. 28§6.1), the pressure just inside the surface of the liquid must also be zero because the height of the column balances atmospheric pressure. That is, it is the pressure of the atmosphere driving the liquid up the tube, not the vacuum 'dragging' it up. The same effect would be obtained if the tube were

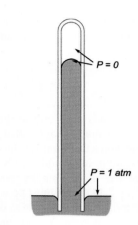

Fig. 2.9 In a mercury-in-glass barometer the height of the column balances atmospheric pressure.

open at the top and it reached into space, when the pressure in it, above the liquid, must fall to (very near!) zero. This therefore demonstrates that it is the pressure difference between points just in the liquid beneath the two surfaces (*i.e.* inside and outside the tube) which drives the capillary climb in an open tube, such as on the left in Fig. 2.1.

On the other hand, if $\theta > 90°$, there will be a depression of the liquid surface in an open capillary because of the positive curvature of the meniscus (Fig. 2.1, right). The height, h, would then have a negative value. The pressure just beneath the meniscus must then be higher than atmospheric by equation 2.5, and is of course identical to the pressure at the same depth outside the capillary, which is the hydrostatic condition. For the barometer, the actual height of the column is in fact slightly less than would ordinarily be expected simply because of this depression effect due to the positive curvature meniscus, which opposes atmospheric pressure slightly. Thus a slight correction is necessary to allow for the effect of the diameter of the tube (which is deliberately quite large to keep the effect small).

These phenomena can also be viewed as being driven by the imbalance in forces at the contact line, as was done above in the lead up to equation 1.4, with the weight of the liquid column 'hanging' on the contact line. Naturally enough, at $\theta = 90°$ exactly there will be neither capillary rise nor depression and the meniscus will be quite flat, curvature zero. (As an aside: for certain of the polymers now used for cheap chemical volumetric ware such as burettes and pipettes, instead of glass – which has $\theta < 90°$ for aqueous solutions – the meniscus is almost completely flat, $\theta \approx 90°$, making reading somewhat easier.)

Clearly, with low contact angles the equalization of pressures draws more liquid from the mass, but if this reservoir is not present, a hydrostatic negative pressure is developed within the liquid and does not disappear. Thus, if we consider a liquid droplet placed beneath a supported microscope slide cover slip such as in Fig. 2.10, the tendency to spread at the TPL boundary on each glass surface (to increase the area of the wetted interface) the drop is placed under a negative pressure. The area of the interface with the air is minimized simultaneously, accounting for the strange shape of that surface.

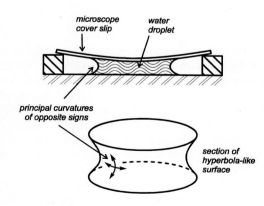

Fig. 2.10 The reduction in pressure in a drop can bend a cover-slip.

It may be noticed that the curvature of the liquid-air surface is negative in the radial planes, while the curvature in the other sense (*e.g.* the horizontal plane at the 'waist') is positive; this is an **anticlastic** surface. Without derivation, we can note that for all situations equation 2.2 may be generalized to take account of both curvatures (which are measured in orthogonal directions, at any point on the surface, all orientations having the same net result):

$$\Delta P = \gamma \cdot \left(\frac{1}{r_1} + \frac{1}{r_2} \right) \qquad (2.7)$$

When $r_1 = r_2$ this is of course the same as equation 2.2. Equation 2.7 is read as saying that, at equilibrium, the sum of curvatures on a free liquid surface is everywhere constant if the pressure in the liquid is to be uniform. If the sum of curvatures is positive, then the pressure is greater inside as a result (Fig. 2.11) (as it must be with a **synclastic** surface, *i.e.* both curvatures positive). On the other hand, if the sum is negative, the pressure is correspondingly negative. This is the situation shown in Fig. 2.10. Here the negative curvature outweighs the small positive curvature so that the sum is indeed negative. The term in parenthesis in equation 2.7 may be called the **total curvature** of the surface.[5] It follows that if the total curvature varies from place to place on a (gravity-free) liquid surface then the system is not at equilibrium: there must be a pressure gradient, and flow must follow.

Note that in the kind of system shown in Fig. 2.11 (a non-wetting droplet squeezed between two plates), the definition of the Young contact angle (equation 1.3) does *not* apply because the geometry of the system is constrained. Here, what appears to be the contact angle is not a function of properties of the solid and liquid but a geometrical consequence of the system. Thus the included angle at the TPL in this kind of case is *not* the Young contact angle. This can be seen in Fig. 2.11 where the "contact angle" at the TPL is clearly not that of mercury on glass (~130° [6] [[6]]). To emphasise the point, as the plates are squeezed together, the disk becomes thinner, the radius of curvature must increase, so the 'wall' of the sphere section gets flatter, and the apparent contact angle approaches 90°.

This kind of wetting effect (Fig. 2.10) also accounts for the close adhesion of wet microscope slides, and partly for the retention of full dentures (§9.10). Notice that the strength of such a bond in shear is negligible because the 'adhesive' is a fluid (those microscope slides slip about quite freely), but in tension it can be appreciable. Capillary forces, therefore, lead to the retention of fluid in fine crevices and, conversely, when saliva is removed from these sites, drive the flow into them of fissure sealants, topical fluoride preparations, resin bonding agents, cements and similar materials.

Such forces also account for decreased expansion of setting gypsum products (2§5), the coherence of wetted porcelain powders (25§2.1), and many other such phenomena. Thus, in the case of plaster and the like, the reduction in volume of the liquid as setting proceeds causes the loss of gloss. But this means that the liquid surface is now locally curved. This in turn means that the pressure inside falls, so opposing crystal growth pressures. This applies in all systems where such a phenomenon occurs (Fig. 2.12; 14§6.3, 17§6.1).

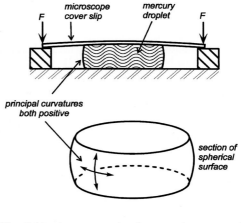

Fig. 2.11 A mercury droplet must be squeezed between the slip and the slide to deform it from a full sphere (doubly-truncated), although the surface remains 'spherical', with the two curvatures identical in value. The pressure due to the curvature is exerted on the glass.

It should be noted that the use of the 'radius of curvature' in equation 2.7 for the situation of a drop between two surfaces does not, in general, mean that a normal cross-section of the liquid surface is an arc of a circle. This is true in only a few special cases. It especially is not true if the Young contact angle (equation 1.3) is non-zero. Serious errors in calculations result from making that assumption. Curvature is a local property of a line on a surface representing the rate of change of slope at that point in a given direction.

Hydrostatic pressure within a drop means that at the (gravitational) bottom the total curvature is greater (more positive) than at the top. This results in distortions from the ideal shapes of such drops, and although the geometry of the drop is greatly complicated by this, the physical principles are not altered. We ignore gravity effects completely here only to simplify the picture.

---

[5] Differential geometers do not like this physics usage of the term, which they define as either half the sum (= mean curvature, H), or as the product of the two reciprocals (= Gaussian curvature, K), depending on context. The point is here, as always, is to check definitions and symbols to ensure that the correct formulation is being used, no matter the context. 'Total' here is more logical.

[6] That this is not 180° is attributable to van der Waals forces: §3.3, §3.4. Small droplets will cling to many surfaces (even in the absence of electrostatic effects due to charging).

●2.3   Capillary flow

Of course, the above remarks are based on equilibrium conditions and, since these would be achieved only through the flow of the fluid, its viscosity, η, would be expected to be a controlling factor in the rate of attainment of equilibrium. Dentistry is concerned about such behaviour because it controls the effectiveness of such treatments as the use of fissure sealants, infiltration of lesions,[7] topical fluoride, etchants, bonding agents and, to a lesser extent, impression materials and cements.

Flow in tubes was dealt with in 4§11, where the force applied was external. Poiseuille's Law can again be used to deal with the situation where the force arises from wetting, *i.e.* as at equation 2.3, but now where the length of the tube filled varies with time, applying the same force-balance method. The rate of penetration of a fluid into a horizontal capillary (Fig. 2.13) can thus be calculated. The capillary is considered to be placed horizontally so as to eliminate gravitational effects, and penetration must continue indefinitely. We then have, from equation 4§11.4:

$$\frac{dx}{dt} = \frac{\Delta P\, r^2}{8\eta x} \qquad (2.8)$$

where ΔP, the driving force, is given by equation 2.4 and x is the distance travelled by the meniscus of a Newtonian liquid into the capillary of radius r.[8] This may be integrated with respect to time to give:

$$x^2 = \frac{r\gamma.\cos\theta}{2\eta}.t \qquad (2.9)$$

which means that $x \propto \sqrt{t}$ (Fig. 2.14).[7]

We can understand the process in the following terms. The wetting of the tube wall creates a meniscus that generates a force, given by equation 2.3, acting to draw the liquid into the tube. This force creates a shear stress at the interface between the liquid and the tube that essentially depends on the area of the interface, $2\pi rx$. The wetting force is constant, and so therefore is the pressure difference across the ends, but the length is increasing, so the pressure gradient declines as the interfacial area increases with time. Therefore the shear strain rate decreases as does the shear stress, *i.e.* the meniscus velocity, or equivalently the bulk or volumetric flow velocity, decreases with time.

●2.4   Penetrativity

If we separate the 'external' variables and the physical constants of the system we obtain the following relationship:[9][10]

$$\frac{x^2}{rt} = \frac{\gamma.\cos\theta}{2\eta} \qquad (2.10)$$

The expression on the right-hand side has been called the **penetration coefficient** for the system (also, **coefficient of penetrance**), but **penetrativity** is the preferable term. It has the units of m/s. Notice that this is

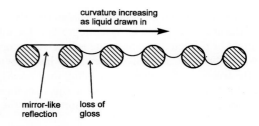

Fig. 2.12 For a setting gypsum product, loss of gloss indicates the generation of a negative pressure inside the mass due to the curvature of the liquid surface.

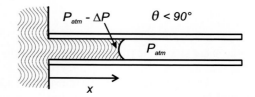

Fig. 2.13 Flow into a horizontal capillary, such as is used for blood collection, is driven by the advance of the contact line, which lowers the pressure behind the meniscus with respect to that in the fluid at the entrance, thereby causing flow.

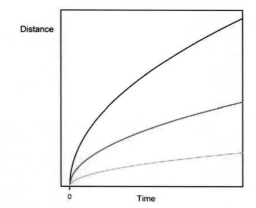

Fig. 2.14 Shape of the penetration curve for a wetting liquid into a uniform open capillary for penetrativity (§2.4) values in the ratio 100 : 25 : 4.

---

[7] This is an approximation as clearly any such flow must be 'undeveloped' (see 4§11). Nevertheless, experimentally it is a fair fit. In addition, the initial supposedly infinite velocity, when $x = 0$, which is physically impossible, because mass and acceleration are ignored, is overlooked for convenience. Likewise, pinning (§9.6) is ignored. Experimentally, measurements would best be taken when the flow has settled down somewhat, *i.e.* the flow approaches the 'developed' condition.

defined for the whole system, *i.e.* including the substrate and gas phase, and not just for the test liquid because of the presence of cos θ, which represents the interaction of solid and liquid under the given atmosphere. As might be expected, the flow rate is inversely proportional to the viscosity. Clinically, the behaviour of interest would be how fast such penetration into crevices and capillaries might occur, as this would control the effectiveness of treatments, or rather the time which must be allowed for adequate completion of the process. In practice, the slope of a plot of the square of the distance travelled against time would be measured and, scaled by the capillary radius, this is proportional to the penetrativity.

It is worth noting that the numerator of the right-hand side of equation 2.10, γ.cos θ, represents the thermodynamic driving force (the scaled line-stress) for the system, while the denominator represents the kinetic limitation (the viscosity). That is, this system is a good example of the trade-off so often encountered in dental materials: the compromise between the goal and the path, between ideality and reality, or between expectation and practical outcome. We may compare this with a rearrangement of equation 4§3.5:

$$\dot{\gamma} = \frac{\tau}{\eta} \tag{2.11}$$

– the shear strain rate (the flow) is given by the ratio of shear stress to viscosity: drive over retardation.

On the assumption that a transparent capillary material could be found such that the same contact angles would be formed with the test materials as would occur with, say, enamel, then direct comparisons of the penetrativity can be made without first having to obtain surface energy and viscosity data. That assumption may not be capable of realization, but some clinical correlation remains possible with, for example, glass capillaries. However, such a test system is capable of yielding much more information if the material, such as a self-cured fissure sealant, is in the process of setting (although this is of diminished interest with the current almost exclusive use of light-cured materials). However, it is still important to know how long to wait before the use of the curing light to allow adequate penetration.

●2.5 Working time
A self-curing material undergoing a process of cross-linking will show an increase in viscosity with time. However, as the network becomes extensive, the capacity for flow will be lost as the material is now rubber-like (see Fig. 6§4.7). Penetration ceases. Whether or not the setting reaction is anywhere near completed, the existence of the network is sufficient to stop flow under the small forces which are operating.

On this basis it will be seen that, for such materials, the penetrativity will be insufficient as a figure of merit for comparison of brands, for the 'working time' may be so short as to negate an apparent advantage, or short enough to override a seeming disadvantage (Fig. 2.15). Equally, the effect of delay in placement can also be judged, as the working time must be measured from the start of mixing (as with all materials). A vertical displacement of the plotted curve, so that its intersection with the abscissa indicates the actual time of placement, will indicate this directly because, of course, the working time is unaffected by placement if there is no change in temperature.

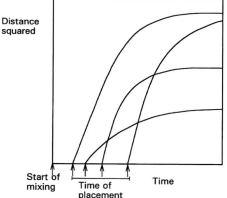

Fig. 2.15 Penetration plots to show working time effects for materials such as self-cure fissure sealants having a variety of setting behaviours.

The implication of this is that the use of 'command-set' materials, *i.e.* light-cured resins, will avoid this problem. Nevertheless, it is still essential that time be allowed for penetration to occur.

●2.6 Bubbles
It is worth mentioning again the extreme sensitivity of polymer viscosity to changes in temperature (3§3) as well as molecular weight. It is obvious that much variation in either of these can drastically change the behaviour of the material. However, it is not enough merely for the penetrativity to be high and setting time to be long. Unlike the experimental system described above, the spaces into which sealants and bonding agents

are intended to go will be blind, *i.e.* closed at one end (§2.10). There will be nowhere for the air to go. For large fissures it is possible to place the fluid material at one end and allow the flow to fill the space gradually. This is not possible with small scale irregularities, and in particular with the pores of etched enamel. The air will be compressed because of the pressure difference across the meniscus which is driving the penetration. As a result the air tends to dissolve in the resin, but in any case the penetration rate will be decreased, and the extent may be substantially reduced. Sufficient time must therefore be allowed for air dissolution in the process of penetration of etched surfaces.[11][12] Clearly, there must only be air in these places, as water

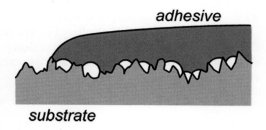

Fig. 2.16 Air may be trapped in roughness under an applied material.

could not be displaced or dissolved. Careless placement of sealants on rough surfaces may easily trap bubbles (Fig. 2.16). This may be difficult to avoid. It reduces the total area of contact, but also detracts from the mechanical keying effect which resists shear forces. Such bubbles are also flaws in the Griffith sense (1§7).

Notice that the roughness effects described in §1.9 will apply here, so that if the resin is inherently going to wet the substrate, a rough etched surface, or even one that has simply been prepared with a rotary cutting instrument, will allow better wetting and spreading. The point is that it must be allowed to do so spontaneously, rather than simply covering the surface and thereby trapping bubbles.

●**2.7   Contamination**
It should be clear that flow driven by wetting is an important aspect of many dental systems. Because of this, the effects of contamination should not be forgotten. Contaminants of interest are nearly all of the kind that would reduce wetting. Thus, lubricating oil from a handpiece is potentially a serious problem,[13] although the usual copious water spray may help to reduce the severity of this. Not so with the air used for drying surfaces prior to the application of bonding agents, cements or fissure sealants. This is one reason why oil-free compressors are (or should be) used to supply air for dentistry: a fine oil mist is always produced by oil-based compressors that cannot adequately be removed by filtration. Similarly, traces of wax or petroleum jelly in the wrong place will diminish if not destroy the intended effect. Even finger grease, which cannot be entirely removed by normal washing (nor is it desirable to try), is enough to spoil wetting. As a matter of principle, no surface which has been prepared for bonding, whether by etching (such as Maryland bridges or enamel), grit-blasting, or degreasing with detergent or solvent, should be touched by anything else prior to the bonding agent.

It should be noted that silicone oils and greases, which are commonly found in so-called barrier creams for hands and certain types of release agent used in dental technology, are particularly difficult to remove and should be avoided carefully. Of course, it is these very properties that make silicones suitable for those applications where wetting is not desired. Contamination from silicone impression materials (which always leave oily residues) is also to be avoided in critical contexts, as would be the inert oil used as the vehicle for the lead dioxide in polysulphide rubber. There are many sources of contamination. Care and forethought are therefore required to avoiding spoiling work.

●**2.8   Penetration by plastic materials**
There are two issues to be addressed in considering the delivery of a material to a site: getting it there and keeping it there. These are conflicting demands and compromise again is inevitable. It is apparent that wetting and viscosity affect delivery, but often a product is supplied in a viscous form to prevent it running away – out of a tray or off the tooth. Clearly, in this case, penetration is necessarily slow and it may be so slow that it is ineffective. Indeed, some products are plastic – they show a distinct yield point (4§7.3). It should now be clear that unless the wetting force (equation 2.3) is enough to exceed the yield point for the cross-section, nothing will happen. There will be no penetration. It is difficult if not impossible to force such materials into crevices and other small spaces when no seal can be made to allow the pressure to be raised. Thus, it is inevitable that topical fluoride products, for example, with a yield point (4§7.12) will not reach the target in pits and fissures without special effort, if at all. Likewise, impression materials will not record the necessary detail of embrasures, gingival margins and cavity line angles. Accordingly, it is usual first to apply light-bodied impression materials directly to the sites of principal interest with a syringe, relying on the pressure required to cause flow to drive the materials into the important features, before seating the filled tray. Flow requires a pressure gradient: fast enough flow requires a great enough excess pressure over the yield point.

●2.9   Penetration of porous bodies

If instead of a single capillary (§2.3) a porous body is considered, it is found that the penetration of a (Newtonian) liquid has exactly the same relationship, that is, distance travelled depends on the square root of time elapsed. The progress of such an experiment may be followed by weighing so that we may write, after correcting for the liquid density:

$$\frac{V}{A} = S.t^{1/2} \tag{2.12}$$

where V is the volume of liquid absorbed, A the cross-sectional area of the body being tested, and S is thereby defined as the **sorptivity**, the tendency to absorb liquid by capillarity (units m.s$^{-\frac{1}{2}}$).[14] Obviously, the porosity has to be continuously connected for this to work. Equally, the same general influences of interfacial energy, contact angle and viscosity must be involved as providing the driving force and the retardation for the process. What is not known is the scale factor. While a bundle of capillaries could be treated simply by a natural extension of the ideas of §2.3, normally the channels in a porous body are convoluted, irregular in cross-sectional shape and area, merging and diverging (typically) in a random manner. Thus, we need to consider the average behaviour assuming only that the body is similar in its porosity from place to place. The penetrativity is a property of the solid-liquid combination (as is contact angle itself, §1.5), and thus remains the same. All that is left is the effective scale factor representing the average porosity, r_{eff}, so we can write (either measuring the distance travelled directly or calculating it – making sure that evaporation does not affect the result):

$$x^2 = \frac{r_{eff}\,\gamma.\cos\theta}{2\eta}t \tag{2.13}$$

as at equation 2.9. Thus, from the right-hand side of equation 2.12, the sorptivity is then given by:

$$S = \sqrt{\frac{r_{eff}\,\gamma.\cos\theta}{2\eta}} \tag{2.14}$$

so that if the penetrativity were known, the porosity can be characterized, but in any case the behaviour of the liquid in that body has been characterized by S. It can therefore be seen that coarse porosity leads to rapid absorption, as does improving the wetting ability of the liquid on that solid.

This behaviour is seen in the absorption of the solvent in thin-layer and paper chromatography. Contexts of dental interest include the initial absorption of water by a gypsum product powder (2§8.1), monomer in acrylic powder, glass into an alumina body (25§9.3), bonding resin into etched tooth tissue, including exposed collagen, fluids into endodontic paper points (27§2.2), water into a filter paper used to dry porcelain powder during build-up (25§2.1), and soaking a dental stone model. The latter three are not critical, perhaps, but there is an aspect of some importance for others.

Because the porosity in such bodies is generally random, and has a range or spectrum of characterizing values on the local scale, however these are defined, r_{eff} is in fact an overall average derived experimentally. Locally, however, penetration may proceed at varying rates, depending entirely on local conditions – including variation in surface energy. The absorption front, the 'position' of all the individual menisci in the absorption direction, may be far from planar (Fig. 2.17). The absorption front becomes irregular, and because the tortuosity of the path of wetting in the porosity may cause the path to turn back on itself, it may then enclose bubbles which cannot be swept out. That is, because the tendency is for the front at any moment to advance in a direction normal to itself, and because this direction is subject to randomization as the path taken locally, at the level of the single channel, deviates randomly (in a kind of "drunkard's walk"), it may meet itself, closing off regions of a random range of sizes and shapes. In other words, the volume fraction of liquid absorbed will not attain the volume fraction of the porosity. In the case of investments and gypsum products, while soaking the powder like this is beneficial in removing a large volume of air, it cannot be complete and vacuum mixing remains necessary (18§1.2). The embedding of porous bodies in resin for microscopy, likewise, cannot in general be perfect, and the same applies with infiltrated ceramics (25§9.3), where remaining flaws might be critical.

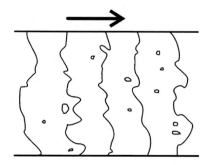

Fig. 2.17 The absorption front for a liquid in a porous body may be very irregular, and even cut off regions leaving trapped air.

In a system such as dried enamel, rehydration will occur by such a process. Accordingly, some air may remain trapped. However, in this case, the Laplace bubble pressure helps to drive the dissolution of that air. **Henry's Law** says that the solubility of a non-reacting gas is proportional to the partial pressure, and the bubble pressure is necessarily greater than that of the surroundings. This establishes a concentration gradient between the air dissolved adjacent to the bubble and the air outside, and this drives the diffusion of the dissolved air out of the imbibed saliva. As the air dissolves so the bubble shrinks, and this raises the bubble pressure further, accelerating the process. After a while, the enamel will be completely bubble-free. This effect would have to be allowed for in any quantitative work on the mineralization of enamel, where imbibition of sections with special dense solutions ('Thoulet's solutions') and fluorescent dyes is involved. Such dissolution may be of some value in contexts such as air bubbles trapped in denture base acrylic, although this must be distinguished from porosity generated by boiling monomer (5§2.6, 25§4.4).

It will be noticed that there is a parallel between the superficial roughness described at §1.9 and the internal character of a porous body. Indeed, any section will be rough in that sense. Accordingly, we can predict that the wetting internally will be affected in a similar manner so that if the smooth surface value of θ is < 90°, penetration will be enhanced, and *vice versa*.

We can extend these ideas to cover various other circumstances. For example, the penetration of a liquid between two rough surfaces. If these surfaces are placed close together it will be equivalent to a thin slice through a porous body, and the path that the liquid takes will again be convoluted and may pinch off bubbles. Some examples might be soldering (22§1.2), fissure sealing, and even casting metals into an investment mould in very thin sections.

•2.10 Penetration of closed capillaries

While the penetrativity (§2.4) and the sorptivity (§2.9) may be a useful means of characterizing liquids and solids in open systems, in dentistry at least there are other types. Very often, closed capillaries or porosity is relevant, and this leads to another complication: rising pressure in the closed space now also opposes the flow.

Boyle's Law, for a fixed mass of (ideal) gas, says

$$PV = constant \qquad (2.15)$$

at fixed temperature. So, in Fig. 2.18, the original gas trapped in the full length L of the closed capillary has a pressure given by:[8]

$$P_t = \frac{P_{atm}\pi r^2 L}{\pi r^2 (L-x)} = \frac{P_{atm}L}{(L-x)} \qquad (2.16)$$

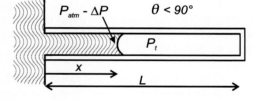

Fig. 2.18 Flow into a closed horizontal capillary is opposed by the compression of the included gas.

Thus the 'excess' pressure generated over the ambient pressure is then:

$$P_{atm}\frac{L}{(L-x)} - P_{atm} = P_{atm}\left(\frac{L}{(L-x)}-1\right) = P_{atm}\frac{x}{(L-x)} \qquad (2.17)$$

and this must be subtracted from the surface tension-generated driving pressure, ΔP, in equation 2.8 because it is opposing the flow:

$$\frac{dx}{dt} = \left(\Delta P - P_{atm}\frac{x}{(L-x)}\right)\frac{r^2}{8\eta x} \qquad (2.18)$$

Obviously, flow stops when the two pressures are equal:

$$\Delta P = P_{atm}\frac{x}{(L-x)} \qquad (2.19)$$

Unfortunately, the integral solution of equation 2.18 is

> Compression of a gas raises its temperature, and this should be taken into account in a full analysis. However, because the volumes in the dental context are very small and the compression not rapid, very little heat is generated, and this may be rapidly dissipated through the relatively large surface area of a narrow capillary. Equilibration is therefore fast. The approximation of a constant temperature is therefore satisfactory for this discussion.

---

[8] The volume of the meniscus sphere segment has been ignored for simplicity.

complicated and cannot be written explicitly for x as a function of t. However, it can be handled numerically, and an example of the form of the plot is shown in Fig. 2.19. From this it is clear that the trapped gas is a serious impediment to the penetration, as might be expected. But, of course, cylindrical capillaries would be unusual in practice, although dentinal tubules might provide a reasonable approximation. More common might be a tapered tube (Fig. 2.20), as is implied by the shape of the resin tags formed in etched enamel. The effect here is even more marked (and the integral even more complicated), as the excess pressure as in equation 2.17 then depends on $[L/(L-x)]^3$, and the flow comes to halt substantially sooner (Fig. 2.19). For effective penetration, one has to rely on the dissolution of the trapped gas (by Henry's Law, §2.9), which would take a relatively very long time.

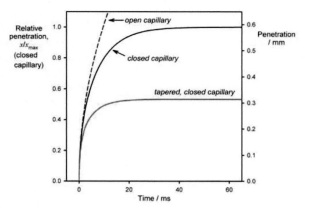

Fig. 2.19 Flow into a closed horizontal capillary is opposed by the compression of the included gas. If that capillary is also uniformly tapered to a point, the reduction in penetration is even more marked. (r = 1 μm, L = 1 mm, η = 1 mPas, γ = 72 mNm, θ = 0°)

Equating ΔP from equations 2.4 and 2.19, and rearranging, we get:

$$\gamma.\cos\theta = P_{atm}\frac{x}{(L-x)}\frac{r}{2} \qquad (2.20)$$

This is the actual driving force, expressed as the line stress (N/m), for the spreading of the liquid on that particular substrate; this is then a meaningful emergent property of the system. This might allow the direct use of tooth tissue substrates (with a drilled hole), where penetrativity could not be measured (§2.4).

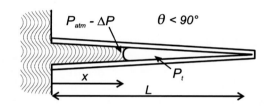

Fig. 2.20 Flow into a conically-tapered, closed horizontal capillary.

The conceptual treatment can be extended to cover closed porous bodies (§2.9), although it is much harder to deal with the flow rates, using the effective pore size. In any case, the net result is that despite good wetting, one cannot expect porous structures to imbibe applied fluids very well unless they are vented. The trapping of gas may be accidental or unavoidable, but the limitations this represents to the outcome need to be recognized. This is why care must be taken to add powder to liquid in many dental materials so that air is not trapped. In the case of gypsum materials, a short wait for such soaking-in to occur is also recommended.

Further extension of these ideas can be made for casting metals (18§3.3 *et seq.*) where contact angles are unfavourable – no wetting, casting pressure needs to be generated, gas needs to escape, and cooling is occurring. Indeed, they are applicable to any system which involves the moulding of a fluid or semi-solid material into a constrained space which entails gas being trapped: filled resins (Chap. 6), impression materials (Chap. 7), cements (Chap. 9), and so on.

§3. Aspects of Chemical Bonding

So far only vague references to the attractive forces between molecules have been made. But to understand how bonding in the macroscopic sense may be achieved or improved, it is necessary to consider the mechanism of bonding at the atomic or molecular level.

The strength of chemical bonds is measured by the energy required to break them (Fig. 3.1). The **covalent** bonds that hold molecules together are very strong, typically about ten times more energy being required to break them than the **electrostatic** forces holding ionic materials together (Fig 3.2). But there is a range of bonding from the weak to very weak collectively referred to as due to **van der Waals forces** (Fig. 3.3). These are the forces that hold many condensed, that is solid or liquid, materials together. Liquid air and paraffins are good examples. There is no chemical reaction between the molecules (that is, no formal bonds are formed), but their mutual attraction holds the liquid together, preventing it flying apart or boiling. The strength of the bonds is measured in these simple cases by the boiling points of the components. When enough energy is provided, the molecules escape. The origins of surface energy or surface tension in terms of these forces is obvious, but the forces themselves need some explanation.

●3.1 Ions

The inverse square law measures the force of attraction between two (assumed point-like) ions:

$$F = \frac{q_1 q_2}{d^2} \qquad (3.1)$$

where d is their separation and q the charge (Fig. 3.2). When the charges are dissimilar in sign, the force is negative, meaning tension or attraction. This is the basis of the **electrostatic** bonds of ionic crystals.

●3.2 Dipoles

The same law applies to molecules which are polar through what are known as **Keesom** forces. That is, due to differences in the electronegativities of the constituent atoms, the electrons are not shared evenly but slightly displaced, revealing some of the nuclear charge on some atoms, and placing a net negative charge on others. If the overall effect is that the centre of the negative charge does not coincide in space with the centre of the positive charge, there is said to be a **permanent dipole** (Fig. 3.3). Dipoles interact in the same way as ions (that is, electrostatically), although more weakly because the effective charges are less, but also because each charge of one dipole interacts with each charge of the other, and two of the interactions are attractive while the other two are repulsive. It obviously depends on the relative orientation of the two dipoles whether the net effect is attractive or repulsive, but in liquids at least molecules and side groups are capable of rotation and reorientation to achieve a state of minimum energy, which means that if possible there will be a net attractive force. It does not require that the molecule has an overall dipole from such effects. Polar side-groups or substituents, such as hydroxyl or carboxyl, will act in the same way on their own account to strengthen intermolecular forces.

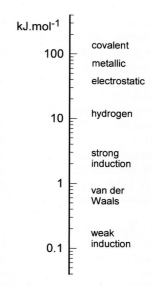

Fig. 3.1 The scale of chemical bond energies. Note that each type has a substantial range of values, and that these ranges overlap considerably.

Fig. 3.2 Ionic solids are held together by electrostatic molecular interaction.

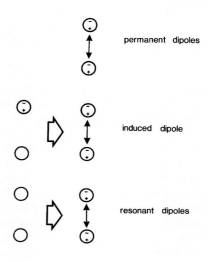

Fig. 3.3 Van der Waals types of molecular interaction.

●3.3   Induced dipoles

The electron 'cloud' around molecules and atoms is not rigid. Under the influence of the electrostatic field of a nearby ion or dipole it is capable of distortion. That is, the electron density distribution may become altered. The relative amount of distortion under a given influence is referred to as the **polarizability** of the molecule or atom. The effect is to create a new or **induced dipole** from the originally non-polar molecule. The forces between the inducing ion or dipole and the induced dipole, known as **Debye** forces, are about as strong as permanent dipole interaction forces, but because the induced dipole is induced, the closest poles are necessarily of unlike charge and attraction is always the net result.

●3.4   Resonant dipoles

Electrons can (simplistically) be viewed as orbiting the nucleus, and so at any instant there is a dipole. Over a period of time the average charge separation is of course zero, so that there is no net dipole, but the instantaneous dipole can be seen to oscillate. While a dipole exists it can induce another dipole in an adjacent molecule, which is itself oscillating. If these oscillations are in step, and the tendency will be for them to become so because this reduces the energy of the system, the attractive forces of adjacent dipoles will be operating, although there is no permanent dipole or ion being present. These dipoles are said to be **resonant** and the attraction is in fact stronger than with the other dipole interactions. These interactions are known as **dispersion** or **London forces**.

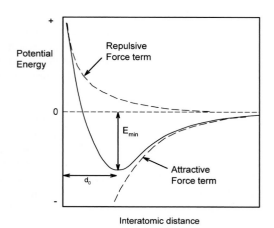

Fig. 3.4 The variation with distance of the potential energy of a system of two atoms due to attractive and repulsive forces between them. [NB: Schematic only – not at all to scale.]

●3.5   Minimum energy

The forces themselves do not determine stability, but the resultant energy of the system. The energy of attraction (E_a) depends on how close the two molecules or atoms are and the magnitudes of the charges on the ions or of the dipoles. Clearly, however, there is some limit to how close atoms may get, as strong repulsive forces may be expected to develop as the electron clouds begin to overlap due to the mutual repulsion of like charges (Fig. 3.4). The energy of the system will tend to be lowered by closer approach due to the attraction forces, but also tend to be increased as electron cloud overlap increases. The net effect is the sum of the two energy changes. These two energy terms depend on different powers of the separation d, understood to be the internuclear distance:

$$E_a \propto d^{-1} \tag{3.2}$$

$$E_r \propto d^{-n}, \quad n \sim 5 - 12 \tag{3.3}$$

where E_r is the corresponding repulsion energy (the value of n depends on the atoms involved and their chemical state, that is, the kind of bonding involved). The resultant energy curve goes through a pronounced minimum at an 'equilibrium' separation d_0.

Because of thermal vibration, the actual separation varies over a small range. As energy is supplied (*i.e.* the temperature is raised) so the oscillations increase in magnitude until complete separation occurs (d very large). E_{min} (Fig. 3.4) can thus be viewed as the activation energy for the dissociation, as well as the stabilization energy for the association. Note also that the atoms must approach one another very closely for the attractive forces to be effective, only one or two atom diameters. This kind of treatment can be given for all atom-atom interactions of the van der Waals and electrostatic types, but similar behaviour is seen for covalent and metallic bonds as well. Ion-ion interactions can thus be seen as one extreme of a continuous range of strengths of such interactions. But there is another type of association as strong as those between ions which is of great importance biologically, in general chemistry as well as in many dental materials, *i.e.* the hydrogen bond (§4).

●3.6   Origin of strength

While considering bonding in this way, we can also identify the origin of the strength of materials. Solid and liquid materials are held together by the forces acting between atoms and molecules and it is evident that these forces must be overcome if rupture of the material is to occur. Obviously, this has happened when a material has been heated enough to supply the energy corresponding to the bond energy. Thus, the melting point and heat of fusion of a substance are related to the depth of the energy well, E_{min}. To consider mechanical strength requires an alternative viewpoint. We may consider a diatomic molecule as a model system. The bond between the two atoms may be considered to act as a kind of spring, and we can plot the force acting when that spring has been compressed or stretched (Fig. 3.5). This curve in fact represents the slope of the potential energy plot for the bond (*i.e.* as in Fig. 3.4):

$$F = \frac{dE_p}{dd} \qquad (3.4)$$

If there is no applied force (and ignoring molecular vibration), the separation of the two atoms is represented by its equilibrium value, d_0. It can be seen that the force rapidly becomes very negative when the bond is compressed, but on stretching it goes through a clear maximum. Thus, starting at d_0 and pulling, the separation

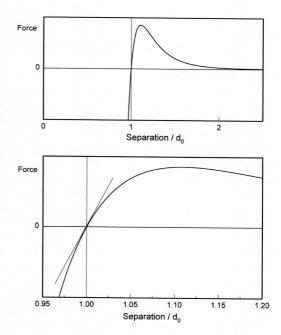

Fig. 3.5   Force acting in a diatomic molecule as a function of separation. The lower graph is an expanded view of part of the upper.

will increase steadily as the force is increased, but as soon as the separation reaches the value corresponding to the peak the bond must break as now it has become an unstable system (the slope is zero at that point) and the force cannot be sustained (*cf.* the load-control test of 1§3.1).

Now we can see that the work of bond rupture, the area under the curve, is equivalent to the work to failure described earlier (1§4), at least for brittle materials. This follows since that area is the integral from d_0 to the maximum (Fig. 3.5), and the force curve was obtained as the differential (slope) of the potential energy curve. It can also be noted that the tangent 'spring constant' or stiffness of the bond, K (1§2; *i.e.*, underlying the modulus of elasticity), is given in turn by the slope of that line:

$$K = \frac{dF}{dd} = \frac{d^2 E_p}{dd^2} \qquad (3.5)$$

Several other things may be noticed in the force-separation plot. Firstly, Hooke's Law (1§2) can be seen to be an approximation that holds good only over a very small range. This corresponds to the observation that when the strain is large, deviations from linearity in the stress-strain curve can be more easily detectable (this does not apply in polymers at temperatures above the T_g because other mechanisms are operating, 3§4). Collapse of the bond corresponds to the point at which the stiffness is zero, that is, at the peak. Secondly, the maximum relative separation, in terms of separation/d_0 (which is defined in the same way as extension ratio, 7§2.1), is actually quite small, but it is this strain that is the controlling, limiting factor for the 'strength' of a material. Thirdly, it is clear that bond failure cannot occur in compression: if the compressive force were large, one would expect shear to occur – if not exactly aligned in the direction of force application, one atom would slip to one side (*cf.* 1§6.3), and thermal agitation ensures that this always happens. Fourthly, the general tendency for strengths to decrease as temperature increases can be understood by noting that thermal energy is equivalent to the work done mechanically in stretching a bond (*e.g.* Fig. 14§7.2), and that the force required to rupture an already stretched bond must then be that much less. Fifthly, the modulus of elasticity of a material is also expected to decline steadily with temperature.

Since real materials on a macroscopic scale consist of very many molecules (or in a metal, atoms), it is clear that many bonds must be broken to fracture a real object. A simple calculation would consist of counting how many bonds were involved (of the order of 10^{13}/mm^2), and multiplying the force accordingly – assuming that they were all loaded uniformly. The reason that this is quite inadequate was explored in 1§7, where structural defects are seen to play a crucial role in real materials. Nevertheless, the arguments here can be seen to set an absolute upper limit to the strength attainable under any circumstances in the Griffith sense (equation 1§7.3), and to show how materials can show spring-like deformation.

●3.7 Thermal expansion

Whilst considering the chemical bond, we can usefully consider one aspect – thermal expansion. In a graph such as Fig. 3.5, for simplicity the atoms may be treated as if static, but of course, as explained in §3.5, they will be oscillating between the limits represented by the curve, according to the vibrational energy. As will be appreciated, the vibrational states of a bond are discrete or quantized (Figs. 6§5.2, 6§5.4), that is, only particular values of energy are possible, although there are very many such levels. Near the bottom of the energy well, the curve is very close to a simple parabola in shape, so the vibrating bond may be treated as a simple harmonic oscillator, to a first approximation. That is, it behaves like a Hookean spring (Fig. 1§2.1), with no loss of energy due to damping. The velocity is then zero at the extremes (all potential energy – hence the curve of Fig. 3.4) and at a maximum at the midpoint of the vibration (all kinetic energy) – very similar to the behaviour of a perfect pendulum (1§10.1). The mid-point then represents the ordinary, low temperature bond length, d_0 (§3.5). At progressively higher vibrational energy levels, the shape of the curve departs more and more from the simple parabolic. The midpoint therefore tends to move away from the low-energy d_0 value, increasing steadily (Fig. 3.6), and so the average bond length increases – which means that the atoms are, on average, moving further apart. Of course, not all bonds of a given kind have the same energy at a particular temperature, but rather the occupancy of the various vibrational energy levels follows a Boltzmann-type distribution (see Fig. 3§3.2). Even so, as the temperature is raised, so the mean of that distribution moves to higher values, and therefore so must the mean bond length, averaged over all such bonds in the body. In other words, the body is seen to exhibit thermal expansion.

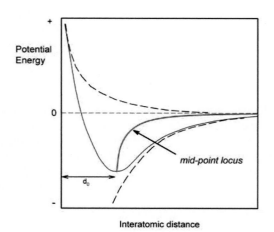

Fig. 3.6 The mid-point of the potential energy curve represents the average separation at a given energy. The locus of these point therefore represents the thermal expansion curve for the system. [NB: Schematic only.]

The same argument applies to all kinds of bond, whether covalent, ionic, metallic or van der Waals. In the latter case, the forces required are low and the bonds 'soft', so that greater thermal expansion coefficients are observed for materials such as small molecule liquids, waxes and organic polymers (Fig. 3§4.14). Giant molecules, fully covalently bonded, the best example being diamond, have low coefficients because van der Waals bonding is not involved. Other materials, with a mixture of bonding types and stiffnesses, show intermediate values. The differences in bond character affect the depth and shape of the potential energy curve. Broadly, the deeper it is, the lower the thermal expansion tends to be because the mid-point locus (Fig. 3.6) is near-vertical at ordinary temperatures.

The ideas of the energy to break a bond and thermal dissociation are therefore seen to be complementary, similar to the complementarity of the load and strain views of the mechanical breakage of a bond (§3.6). It is also clear that any bond must break at a high enough temperature. In addition, the escape of a molecule from a surface (whether liquid – evaporation, or solid – sublimation), can be seen as simply the realization of this breakage event for van der Waals bonding.

●3.8 Bond lability

In discussing matters such as bond length and strength it is easy to fall into the trap of thinking in terms of static systems. But, as made clear in §3.7, bonds are not of fixed length: atoms oscillate about a mean position. That is, all bonds are dynamic, and not just in simple stretching but in bending as well. Even this is simplistic. While at given temperature we can state what the average vibrational energy of a bond is, in fact over all such bonds there will be a distribution of values of the Boltzman kind (*cf.* 3§3.3), as indicated above – some will be low but, more importantly, some will be high. Thus, because of this high-energy tail to the distribution, there is a chance that any bond may break spontaneously at any temperature. However, at a low enough temperature this chance may be entirely negligible and materials and substances will appear to be quite stable. Of course, a broken bond may reform promptly, and nothing will appear to have occurred. We can see immediately, then, that bond-breaking is an activated process. Vapour pressure – molecules overcoming van der Waals forces – has a clear basis in this probabilistic view: the higher the temperature, the higher the rate of bond breaking and the higher the number of molecules in the gas phase at equilibrium:

$$[X]_L \leftrightarrow [X]_V \qquad (3.6)$$

Equally, diffusion and creep (1§11) depend on bonds being broken, at least for a short time, for the movement to occur; these are activated processes (equation 3§3.3). Likewise, bond exchange in impression materials (7§11.3), whether these are covalent or hydrogen bonds, will occur at a rate dependent on the temperature. Of course, in such cases nothing seems to happen except when a load is applied and distortion results, but we talk about such bonds as being more or less **labile** – subject to being broken.

In this sense, all reactions and processes where bonds are to be broken can be seen to be activated and thus show a rate dependant on temperature. Thus the strength of a material is itself temperature-dependent as the thermal energy fluctuation is superimposed on the energy of the mechanically-stretched bond.

It should be noted that there is considerable confusion in some places over what is meant by the "strength" of a chemical bond, confusing at least three distinct aspects. The first is the chemically-meaningful bond energy – the work to failure, E_{min} in Fig. 3.4 (*alias* the work of formation); the second is the load at which collapse occurs, the peak value in Fig. 3.5 – the force required to break it; while the third is the stiffness, the spring constant (sometimes called the force constant), as shown by the tangent to the force-separation curve. These are parallel to the ideas of toughness (1§4), failure stress (1§3.2), and elastic modulus (1§2.1), respectively, which are indeed quite distinct even if in common speech they may be confused or conflated. (We can also draw a parallel with the confusion of impact 'strength', 1§10.3.) While there may be a correlation between these values – a large value in one tending often to be associated with a large value in another, this does not mean that they are interchangeable. Primarily, this is because of the complexity of bonding: the nature of the molecular orbitals involved, that is, their three-dimensional shape; their polarizability (and so their dependence on their intra- and inter-molecular environment); as well as the mere fact that bonds are not independent – deformation changes the intramolecular environment, which changes the molecular shape, at least locally. In fact, in some contexts **bond length** (d_0) is taken as an inverse measure of bond strength, without specifying what is meant by strength, while **bond order** (essentially measuring how many pairs of electrons are involved) is an altogether separate idea, although the two are often inversely correlated. Accordingly, great care is required to determine which aspect is in fact under discussion when the term "bond strength" is used.

§4. The Hydrogen Bond

When hydrogen is attached (covalently) to strongly electronegative atoms, oxygen and (to only a slightly lesser extent) nitrogen being the ones of prime importance in the present context, it tends to acquire a small positive charge. That is, a local molecular dipole is produced. The same electronegative atoms acquire a negative charge by virtue of their electronegativity, their 'electron-withdrawing' power. When lone-pairs are available on others of these kinds of atom, a bond of the form:

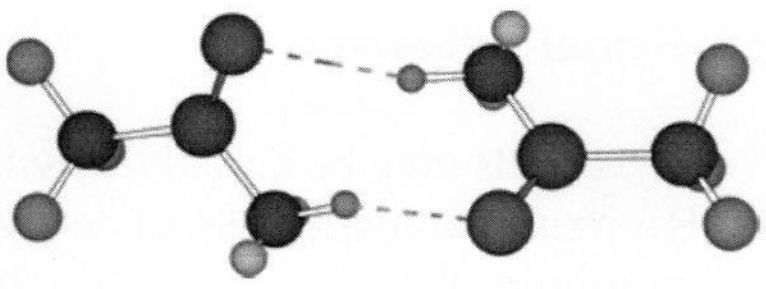

Fig. 4.1 Hydrogen bonding in acetic acid dimer.

$$\overset{\delta-}{>O:} \ldots \overset{\delta+}{H} - \overset{\delta-}{O} -$$

is possible.[15] These **hydrogen bonds** are responsible, for example, for the dimerization of acetic acid (Fig. 4.1), the extraordinarily high boiling point and heat of vaporization of water (Figs 4.2, 4.3), the network structure of polysaccharide gels (7§7, 7§8), the complex tertiary structure of proteins (27§3), the viscosity of poly(acrylic acid) (9§8.2), as well as the strength of dry cellulose (7§7, 27§2), and so on. The structure of ice (which is tridymite-like for the normal type, Ice I_h;[16] – see Fig. 17§2.12) is all due to hydrogen bonds.[17]

Hydrogen bonds have energies of about 4 ~ 40 kJ/mol, depending on the circumstances, and so are not quite as strong as typical covalent bonds (see Fig. 3.1).[18] However, the frequent appearance of hydroxyl groups and oxygen or nitrogen means that hydrogen bonds can occur in very many circumstances, often with several per molecule. Note that it is the total energy of the association which determines the stability of the system.

Notice also that hydrogen bonds are specific, between particular atoms, rather than general as with van der Waals forces. This confers a spatial stability and directional (or steric) properties to any structure involving them. This, of course, is crucial in the case of the secondary and tertiary structure of proteins.

As a result of this ready formation of hydrogen bonds, water will adsorb strongly on very many polar surfaces containing appropriate groups, as well as dissolve in polymers such as poly(methyl methacrylate). Indeed, the polymers called nylon all contain many amine and carbonyl groups which are so readily solvated by this mechanism that they are unusable as denture base materials (Fig. 3§6.1), the plasticizing effect of the water so absorbed being excessive. A further point is that under normal circumstances polar surfaces and other high energy surfaces tend to be covered by a monolayer of water molecules, and the higher the water vapour pressure in the air, the more complete will this layer be. Obviously, we are not ordinarily aware of this, but potentially it represents a source of interference with the otherwise expected surface chemistry and physics.

It is apparent from previous remarks that any interface between a solid and a liquid with a contact angle less than $180°$ has a lower energy than the 'clean' solid surface. Clearly then, any material which is adsorbed does so because of the consequent lowering of the interfacial energy, *i.e.* due to the **heat of association**. Hydrogen bond formation will make a major contribution to such energy lowering, when it is possible.

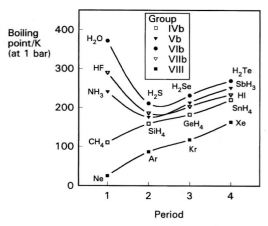

Fig. 4.2 The boiling points of the hydrides. The effect of hydrogen bonding in water, hydrogen fluoride and ammonia is clear.

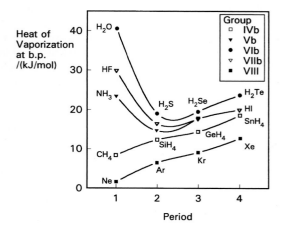

Fig. 4.3 The latent heats of vaporization of the hydrides. The hydrogen bond contributes about 30 kJ/mol in the case of water.

§5. Dental Adhesives

Solids may be broadly classified by their surface energies:
- High surface energy, associated typically with high melting point, hard, inorganic materials, *i.e.* where strong intermolecular forces operate, *e.g.* tooth enamel, porcelain, metals.
- Low surface energy: usually associated with low melting, soft, organic materials, *i.e.* with weak intermolecular forces, *e.g.* protein, polymers (such as acrylics).

The adsorption of proteins into the polar surface of hydroxyapatite is thus energetically favourable and liable to be very stable (*cf.* §1.11).[19] Such a film, commonly known as **acquired pellicle** when on enamel, and which may be very thick in molecular terms, will prevent other larger molecules which are designed to interact with the hydroxyapatite surface from doing so. Even so, for displacement to occur by small molecules, a further reduction of the interfacial energy must be associated with wetting by the incoming compound (*i.e.* a smaller contact angle). Such displacement does not occur in practice.

Consequently, for bonding metallic or ceramic devices to enamel, a major consideration is the cleanliness of the surfaces. Grit blasting (20§4.1) metallic items will remove much contamination, whereas the firing process is likely to burn off any organic contaminant on a porcelain restoration. It is then important to avoid subsequent contamination of those critical surfaces by contact with fingers or from other sources. Tooth surfaces must be cleaned thoroughly ('prophylaxis') to rid tooth surfaces of protein films, and subsequent contact with saliva scrupulously avoided. The contrast with the contamination discussed in §2.7 is that adsorbed protein may not reduce the wetting obtained with hydrophilic materials because they are so polar. However, they will prevent any intended mechanical key from being so effective because the protein layer will be mechanically weak, hydrolyse in the long term, and prevent altogether any intended covalent or coordination bonds to structural components from forming.

Polymeric 'adhesives' relying on hydrogen bonds may then bond sufficiently well to enamel thus cleaned and dried. However, due to the water absorption of such polymers, water soon becomes available at the interface by diffusion through the solid if not through the enamel itself (which, of course, is not impermeable), even if perfectly excluded initially (which is unlikely). Consequently, bonds between such surfaces can never be entirely satisfactory in the long term because the water will compete with the polymer. Other methods must be used for a strong bond. Various special **coupling agents** have been proposed but many have failed to perform satisfactorily clinically, if indeed any improvement could be detected otherwise.

●5.1   Silanes

Alkoxy groups on silicon are readily hydrolysed by water to the corresponding hydroxysilane (Fig. 5.1), which in turn can condense (water being eliminated) with hydroxy groups on other materials (*cf.* condensation silicone impression materials' setting reaction, 7§6.1). If the hydroxy group is on the surface of a solid, as for example ionically-bound hydroxyl in the surface of tooth enamel hydroxyapatite, the net effect is the 'grafting' of the silane onto the surface. If the silane also carries other reactive functional groups, such as vinyl or methacrylate, further reactions can then attach polymer chains at these sites, thus effecting a chemical bond of the bulk polymeric restoration or fissure sealant through to the tooth. Theoretically, the use of a methoxysilane, for example, can obtain the bond along with the elimination of adsorbed surface water molecules. Again, though, clinical results have been disappointing, possibly because the attachment site density (number of bonds per unit area) has been too low. Clearly, since the weakest link in the chain is the bonding of the silane oxygen to the substrate, dissolution of that substrate will free the link, as well as hydrolysis of the link itself (*cf.* Fig. 25§8.1). As was pointed out, water is available at such interfaces, and the exact layout of the hydroxyapatite crystal surface, ion by ion, cannot be expected to be maintained indefinitely.

●5.2   Coordination bonds

Chelation of metal ions by carboxylic acids has been mentioned in the contexts of alginate impression materials, and polycarboxylate and glass ionomer cements (Fig. 9§8.1). Because these are multiple coordination bonds, great stability may be achieved. These effects are also apparent on ions residing in the surface of a solid, and moderately good bonds with polyacrylates to enamel have been achieved. Indeed, this was the rationale behind the invention and development of those cements. Metal ions in oxide films such as are found on stainless steel and cobalt-chromium alloys are also susceptible to this action, leading to the usefulness of polycarboxylate cements for orthodontic bracket bonding, and the need for care in cleaning up when using stainless steel instruments with such cements.

§6. Enamel & Dentine Etching and Bonding

As the surface area of the substrate is important for providing a sufficient number of bonding sites, and while mechanical retention still has to be relied on for many systems, etching is used as a surface pre-treatment. This was first found to be useful on tooth enamel. The action of the acid solutions used is twofold:
(1)   by removing contaminants from surfaces inaccessible to prophylaxis, the surface energy may be raised;
(2)   the etching rates of different sites in the enamel structure may be greatly different, resulting in deep etch pits or pores being created; this is **differential etching**.
If the bonding material can flow into these pits before setting, many 'tags' may key the material to the surface and so resist shearing, although the resistance to tension normal to the plane of the substrate may not be improved very much in the absence of a more specifically chemical bonding. This approach is used for filled-resin restorations and fissure sealants in particular. However, the presence of the particulate filler in the restorative materials will inhibit the effect by blocking the pores and increasing the viscosity. Usually, then, an unfilled 'primer' resin would be used first, on top of which is placed the filled material.

●6.1   Dentine

Dentine presents rather different problems, however. It comprises a large proportion of low surface energy, hydrolysable, protein. Although sophisticated methods can actually graft polymeric materials onto the protein molecule using, for example, trialkyl boranes (which are thought to generate free radicals on proteins) (*cf.* 6§1), the poor mechanical strength and biological instability of protein does not permit good permanent bonds to be achieved. Chelating acids offer some chance of a bond but the exposed proportion of hydroxyapatite is very low and, since etching will remove this material and not the protein, the position can only deteriorate with this procedure. It is also quite impossible to dry the surface of dentine without damage to the pulp as the latter provides a good continuous supply to the surface film via the tubules.

Dentine bonding thus depends on several factors combined to achieve the best results, although whether they are yet – or can be – satisfactory in the long term remains to be determined.[20] The factors are:
- mechanical key,
- interpenetration of structures, and
- specific bonding.

Mechanical key probably remains dominant, but it depends on penetration of dentinal tubules rather than a differentially etched structure as in enamel. This is problematic in that all that is needed here is the clearance from the tubules of the smear layer and debris plugs due to the preparation of the cavity by rotating instruments. Going further means the removal of structurally-important hydroxyapatite. Thus, the acidic agents used for this are described (erroneously) as 'conditioners' rather than etchants, but it must be recognized that the chemistry of the system takes no notice of the dentist's intentions and etching necessarily does occur. What matters is controlling the extent of that etching; calling it a different name does not change the action (see 9§5.17).

Secondly, by a proper design of polymer in terms of viscosity and hydrophilicity, the exposed protein matrix resulting from the etching of the dentine that does occur may be infiltrated by the resin to create a composite structure called the '**hybrid layer**' which is structurally continuous with the underlying dentine through the protein fibrils and appears to be important to achieving a strong interface.[21] This may be valuable in providing a gradual transition in mechanical properties, specifically modulus of elasticity, from the unfilled resin through to the unaltered dentine. However, the stability of the protein has yet to be proved in these circumstances.

The third factor, **specific bonding** refers to covalent bonds between the molecules of the adhesive and some structural molecule (such as a protein) of the substrate. Such bonding is, of course, important in the same sense that a composite's matrix must be bonded to its core to be effective (6§2.9): stress transfer. There are many proprietary versions of attempts to achieve this, and it is as yet unclear which is to be preferred. Even so, it appears that there is no firm evidence that these bonds are formed, at least, not in significant numbers. Despite this uncertainty, non-specific bonding, in the sense of wetting, perhaps aided by hydrogen bonds, can go some way towards achieving the necessary condition. This failure to attain covalent bonds does not, however, affect the importance of the principles which must be followed for bonding between tooth and resin to be effective, *i.e.* to have sufficient strength to be functional. In any case, since the substrate molecules to which the bonding must be made are themselves rather labile and subject to hydrolysis, the long term prospects for a reliable bond are poor.[22] Indeed, enzymes called matrix metalloproteinases (MMPs) are present in dentine and odontoblasts and appear to be active in degrading the attachment to dentine, especially if this has been treated with acids or acidic bonding materials. Water in any case cannot be excluded from any site since it can diffuse through the materials involved; it does not require a gap to be present.

The chemistry of adhesives on the market changes rapidly, and in particular the presentation in terms of the numbers of preparation steps and their intentions. Given that at each stage there is a compromise, and that pressure of time is commonly overriding (sustaining the use of 'self-etching' adhesives[23]), here as elsewhere attention to detail is essential for the best outcome.[24] Even so, the longevity of such systems is likely to be affected by water sorption and hydrolysis.[25] Some of the principal systems are used as illustrations here.

●6.2   Cyanoacrylate

In addition to the simple methacrylates and their numerous bis-GMA and other variants, as used in filled restorative resins, fissure sealants based on other reaction systems have been proposed. The cyanoacrylates, which have attained considerable fame (or maybe notoriety) as instant 'super-glues' on the domestic market, are characterized by their polymerization being initiated by water (Fig. 6.1).[26] The cyano- group is so very electrophilic that some net positive charge is apparent in the terminal vinyl carbon, which thus becomes susceptible to attack by as weak a base as the hydroxyl ion or even water. The carbanion so formed can go on to attack other vinyl groups at the same site, thus effecting polymerization. This is therefore an **anionic addition polymerization chain reaction**. The parallel with the methyl methacrylate polymerization (5§1) should be clear.

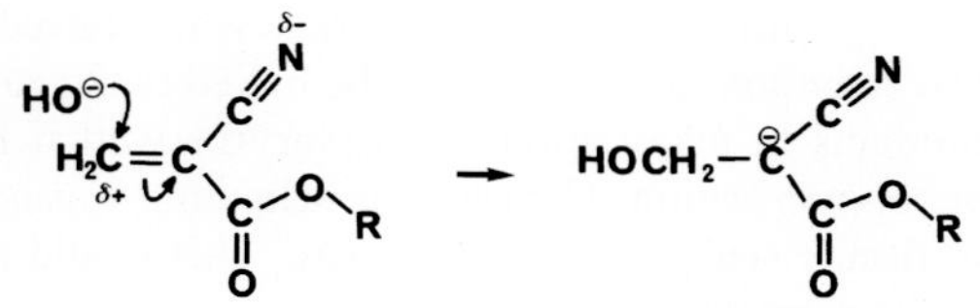

Fig. 6.1 Initiation of polymerization in cyanoacrylates. (R = methyl, ethyl, propyl, butyl, ... *etc.*) Water itself is also nucleophilic enough to initiate polymerization.

The reaction with adsorbed surface water molecules is, of course, both the polymerization initiation step and advantageous for bonding in that it removes a barrier to polymer-substrate interaction. Since proteins have hydroxyl and carboxyl side groups (some five amino acids have -OH groups, two have a second carboxyl group, and even one with -SH), bonding to and initiation by them of cyanoacrylate polymers is possible. Such a reaction can be seen to have **grafted** the polymer onto the substrate molecule by forming a covalent bond to it, which is necessarily quite strong, bearing in mind that the strength of the union of the two parts depends on the areal density of such bonds. Hence, some true bonding to the protein matrix of enamel and dentine, as well as fingers, is possible. The rate of chain termination is rather high so that only thin films polymerize successfully, limiting clinical application, although their low viscosity allows good penetration.

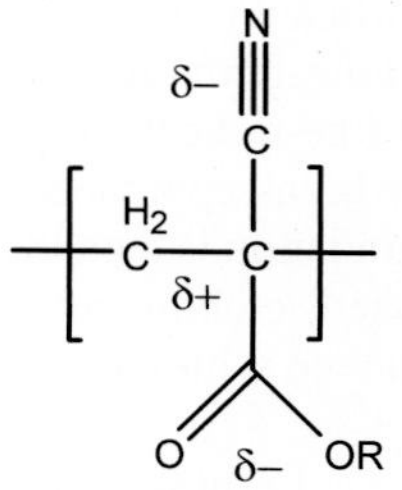

Fig. 6.2 The electron withdrawing power of the substituents leaves cyanoacrylate polymers open to nucleophilic attack.

Hydrolysis of the polymer is also a problem, especially with the lower alkyls (R = methyl, ethyl). The strong electron-withdrawing power of the cyano and carboxylate groups on the same chain carbon atom leaves this with a positive charge, weakening the bonds of the backbone chain and thus leaving them vulnerable to attack by hydroxide, so-called **nucleophilic attack** (Fig. 6.2). It is only with the butyl ester that the hydrolysis rate is low enough for it to be use effectively as a medical tissue adhesive (presumably by providing a sufficiently polarizable source of electron density). The tensile strength and abrasion resistance of these materials is also poor, although some improvement might be expected with cross-linked systems or the addition of a filler.[27] So, despite the clinical successes observed with, for example, six-monthly reapplications, usage has waned. However, as an adhesive for dry service it has its uses, and variants are used industrially for vibration-proofing nuts and bolts, for example.

●6.3 Polyurethane

Polyurethanes (6§4.9, 6§7.5) have also been much discussed in the context of fissure sealants. Isocyanates react with alcohols to produce a so-called urethane (Fig. 6.3). Clearly, with di-isocyanates and polyhydric alcohols, cross-linking with polymerization can occur, leading to a solid material. If water reacts with the isocyanate group, instant decarboxylation occurs to yield an amine and CO_2 (*i.e.* when R = H in Fig. 6.3). This reaction is used in the manufacture of polyurethane foam. The amine this forms can, however, then react with further isocyanate to form a substituted urea; again a cross-link or chain growth in the polymerization is possible.

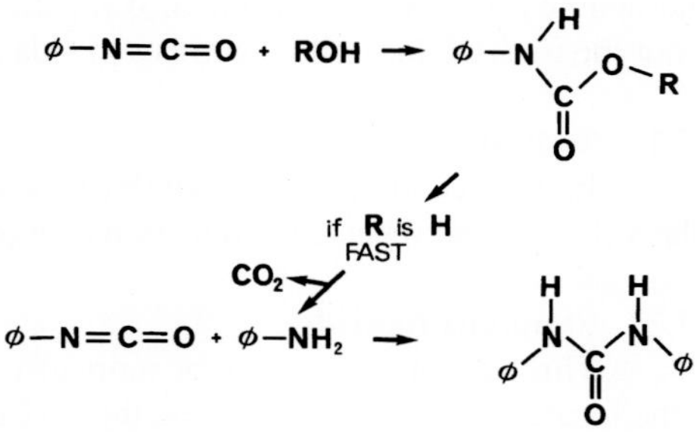

Fig. 6.3 Reactions of the urethane group. The CO_2 elimination occurs if pH > ~7.

An advantage of this system is therefore the elimination of surface adsorbed water, permitting the closer interaction of the polymer with tooth enamel. If it is only adsorbed water that is present, the small amount of gas produced probably just dissolves in the polymer. If there were larger quantities of water present either the urethane would be used up so that no little or no polymerization reaction is possible, or the CO_2 produced would appear as bubbles and compromise the strength of the interface. It is presumed that the reactive hydroxyl group could be on biological macromolecules, when some specific bonding is in theory possible, again grafting the polymer onto that substrate.

Unfortunately, the materials which have been tested in this class proved unsuitable clinically, with very low retention rates. This may be linked to the ease with which urethanes are hydrolysed: commercial polymer products in many applications survive well if kept dry, but even high humidity can initiate a catastrophic depolymerization. Clearly, this behaviour was not taken into account in connection with the service conditions of fissure sealants: wet and warm. This would also apply to any specific bonds formed to proteins and other tissue elements.

●6.4   **Bis-GMA**

Pit and fissure sealant systems based on bis-GMA-methacrylate polymers (6§4.3) seem to be the most effective currently. Certainly, vinyl-type polymers are not capable of being hydrolysed – there is only a carbon-carbon backbone. But whether or not the methacrylate ester (see Fig. 6§4.4) linkage is hydrolysed at a clinically-significant rate is not known, although such a reaction is feasible. It would be a possible contributory factor to the long-term breakdown of both filled-resin restorations and sealants. However, it should also be remembered that free-radical polymerizations represent an equilibrium between monomer and polymer (5§2.7) so that bonds may break spontaneously and the chain ends be subject to entropic recoil and stress relaxation (7§2.2); slow degradation is then inevitable, especially if any leaching occurs. Nevertheless, no matter what type of material is used, etching remains essential for mechanical retention as covalent chemical bonds have not yet been achieved with the desired stability.

Light-activation offers most working time, but other systems based on chemical activation are successful. The use of glass ionomer cements as fissure sealants has been investigated, and indeed is showing much promise, and this may well achieve prominence in time, largely due to their good adhesion to enamel. The advantage may be that if ion mobility at the tooth surface causes a bond to be broken, it is capable of reforming, unlike typical coupling agents. One drawback is the poor penetration obtained because being a paste initially, with a high filler loading, the viscosity is very high. They therefore require some small sacrifice of tooth material to ensure retention.

§7. Fluoride

The wetting of enamel surfaces is no less important for topical fluoride applications to be successful than for bonding purposes. This is, in fact, of little difficulty as all systems are aqueous and water readily adsorbs to enamel, making the ion-exchange mechanisms required straightforward. The general aim of fluoridation is the replacement of a proportion of the hydroxy groups in hydroxyapatite by fluoride, on the grounds that fluorapatite is less soluble than hydroxyapatite, so reducing caries susceptibility. However, there is some doubt about the truth of this, with a number of side reactions occurring, such as the formation of calcium fluoride.

●7.1   **Sodium fluoride**

In a concentration of about 2% this was extensively used early on, but the ion-exchange 'activity' is relatively low and repeated applications are essential. The solution is at about pH 8.

●7.2   **Stannous fluoride**

This substance is said to be more effective, requiring fewer applications. The effect may be partly due to the formation of some tin salts on the surface, but it is mostly due to the low pH (~2!) of the solution, which gives it its unpleasant taste. This is due to the extensive and immediate hydrolysis of the salt:

$$3SnF_2 + 4H_2O \Rightarrow [Sn_3(OH)_4]^{++} + 4H^+ + 6F^- \tag{7.1}$$

Some dissolution of hydroxyapatite must follow in such an acid treatment. Further reactions in the solution result in the precipitation of various stannous oxide hydrates (which are of complicated and variable composition), leaving essentially a dilute solution of hydrofluoric acid. Oxidation of Sn^{II} to Sn^{IV} also occurs readily in contact with the oxygen of the air and in the light, so that the overall stability of the system is poor. Decreased activity in the fluoride-treatment sense is said to follow this. The solution should therefore be made freshly before use. So-called 'stabilized' solutions have been sold, but their composition and relative effectiveness is not presently known.

●7.3   APF

As indicated above, low pH seems to be advantageous for topical fluoride treatments, but it also necessarily causes dissolution of the enamel. The tendency to dissolve can be opposed by using a solution with a high concentration of phosphate to prevent this ion being removed from the surface. This cannot be done simply with stannous fluoride solutions because of the insolubility of tin phosphates. However, so-called 'Acidulated Phosphate Fluoride' (APF) materials incorporate fluoride ion in a solution of orthophosphoric acid. Such a solution may have a pH ~3. Commonly, these solutions are presented as a very viscous solution for easier manipulation. These are usually described as gels and to be thixotropic. However, they are not gels in the chemical sense, having no stable, three-dimensional network structure (see 7§8), but are simply solutions of substances such as sodium carboxymethylcellulose which in low concentration produce very high viscosity (27§2.3). Neither are they thixotropic, but are probably only plastic, showing a definite yield point (see 4§7.3).

The need for wetting by and a high penetrativity (§2.4) for these materials should be apparent, but it is a curious contradiction, apparently not recognized, that the treatment that is meant to deliver fluoride ions to caries-susceptible sites (which perhaps are the least accessible) is attempted with a material that has very poor penetration.

The rationale for the formulation of APF was apparently that the phosphate present would suppress the dissolution of the enamel, despite the low pH. Strangely, the original data show the opposite effect.[28] Indeed, such phosphate is associated with very strange solubility behaviour.[29]

●7.4   Surface energy effects

The incorporation of fluoride ions into the enamel surface results in a change in surface chemistry. Primarily, this exchanges hydrogen bonds of the type S-O-H···O= for S-F⁻···H-O- (where S is the solid substrate). Thus there will be strong interaction with water, as opposed to, say, carboxyl groups. This might make subsequent treatments with 'adhesive' materials more difficult or less successful. Certainly, anything that relies on the presence of hydroxide ions in the surface for coupling or other reactions must be compromised. Indeed, it is recognized that care has to be taken to avoid using fluoride-containing abrasive pastes prior to such procedures. Even so, reduction in surface energy has been cited as one reason for lower caries incidence by reducing the adhesion of plaque, although it does not necessarily follow that the surface energy falls. What is certain is that the nature of interactions will change.

This effect has to be carefully distinguished from that of the reduced solubility of tooth tissue as a result of fluoride treatment. It is believed that a common reaction is the formation of CaF_2 under normal conditions, and possibly other low-solubility substances. Fluorapatite, where the hydroxyl ion is substituted by fluoride ion does not form very readily except possibly at the very surface, or in the very long term through solution-mediated recrystallization. Even so, the net effect is believed to be an increased resistance to acid dissolution. Since this is not a matter directly of a surface energy effect (because of the primary facts of the relevant chemical equilibria), it cannot be claimed that dental caries – or even deliberate etching for restoration retention – is affected by other than a change in solubility or rate of dissolution.

This is not to say that there is no contribution at all from surface energy. Small crystals are more soluble than large ones, essentially because the larger **specific surface area** of small particles (area per unit mass) provides a greater driving force for dissolution, *i.e.* the lowering of the energy of the system on dissolution is greater. This is the underlying reason for the recrystallization of finely divided excess solid in a saturated solution to a few large crystals, given enough time. Related issues are discussed in connection with metal nucleation (11§2.2), grain growth (11§4.1), corrosion (13§6.3) and electrolytic polishing (20§5).

§8. Surfactants

Control of wetting is important in a number of areas, specifically when it is initially or ordinarily poor and needs to be much improved for a process to be effective. Examples are the wetting of wax patterns on being invested, and the pouring of hemihydrate slurries for casting models in elastomeric impressions.

●8.1   Oil and water

Water molecules interact so strongly with each other through the formation of hydrogen bonds that if a non-polar molecule, such as a paraffin, is introduced into a body of water, the energy of the system is raised. This is because some hydrogen bonds are prevented from forming because the paraffin molecule is in the way. A second paraffin molecule would have the same effect. But if the two molecules were side by side there would be less disruption of hydrogen bonds than if they were separate. In effect, the introduction of a non-interacting molecule into water creates a small cavity. That cavity has a surface, so the total surface area of the water is increased, therefore the total surface energy is raised. Putting the two non-polar molecules together lowers the surface area of the cavity required to contain them. Even better, if the paraffin molecules were moved outside the body of water, the interface with the water is now minimized, therefore the energy of the system is minimized. This is what accounts for the spontaneous separation of oil and water.

●8.2   Amphiphiles

A molecule which is paraffin-like at one end and polar at the other presents an interesting problem in energy minimization. Clearly, the polar part would lower the energy of the system by being solvated, but the hydrocarbon part would be better separated from the water. The solution to this is to have the molecules arranged in **micelles**, near-spherical groups where all the polar parts are outside, interacting strongly with the water, while the non-polar parts are inside, gathered together (Fig. 8.1). Such molecules are called **amphiphilic** – hydrophilic at one end, hydrophobic at the other.[30] Such micelles can take a variety of forms depending on concentration, but are typically spherical at modest values. Note that their formation is so energetically-favourable that the true solubility of such substances is ordinarily very low indeed. In other words, the concentration of fully-separated or isolated single molecules of an amphiphile in water is very small. Spherical micelles typically contain ~100 molecules, but depending on the compound and the conditions it may vary from a few to thousands.

Fig. 8.1   A micelle of amphiphilic l,

If then a paraffin molecule is introduced, it would still raise the energy of the system if it were placed so as to be surrounded by water. But if it were placed inside a micelle, surrounded by the paraffinic tails of the amphiphilic molecules, no such disruption of water-water interactions occurs, except for that due to the small increase in the volume of the micelle (Fig. 8.2). It is therefore energetically favourable for the oil molecule to be surrounded by a coat of the amphiphile, not because of any significant interaction between the non-polar molecules but because disruption of polar interactions is avoided. Increasing the amount of oil in the system leads to small droplets, coated by the amphiphile, dispersed in the water. This action therefore stabilizes oil-in-water **emulsions**, and the action is said to be one of **solubilization** of the oil. However, this does not form a true solution, even after much agitation, and the result is probably best described as a **colloidal emulsion**. So long as there is sufficient amphiphile present, subdivision of the droplets increases the amphiphile-oil interfacial area, and thus lowers the overall energy.

Fig. 8.2   The non-polar oil is in the lowest energy position surrounded by the amphiphile micelle.

●8.3   Detergents

Such amphiphilic molecules are the soaps and other detergents. Thus, sodium stearate is a typical soap: $CH_3(CH_2)_{16}COO^-Na^+$ – the long paraffin tail is a very clear feature. There are several other types of detergent, mostly designed to be more "soluble" in water than soaps, and which in one way or another are designed to improve on the efficiency of the interactions. Such substances are therefore effective in **degreasing** surfaces prior to some treatment which depends on a clean substrate – such as bonding – when the use of organic solvents would be inappropriate.

Fig. 8.3   Dirt is better suspended when the adsorbed surfactant presents a hydrophilic surface to the water.

There is one consequence of using such molecules which has wider implications. Any molecule which finds itself at the surface of the water will also be able to contribute to the energy minimization by sticking its tail out of the water – out of the surface.[31] Thus, the surface of an aqueous solution of a detergent tends to be a monolayer of the detergent molecules, ionic heads in the water – solvated, non-polar tails pointing out (Fig. 8.3). The water is thus covered by a low-energy layer. In other words, the surface tension of the liquid body is lowered. This means that on many surfaces the contact angle is reduced, even to zero. Detergent solutions therefore spread well. This wetting also applies to dirt, particulate contamination that may well be oily or otherwise poorly wetted by water. Consequently, such dirt is readily suspended in the aqueous phase and removed (Fig. 8.4). The detergents are broadly described as surface active agents or **surfactants**. Note that the properties of the water itself have not been changed by the presence of the detergent.

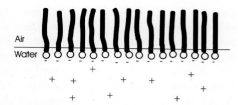

Fig. 8.4 The surface of the water carries a monolayer of the detergent molecules.

The lack of wetting and spreading of water and simple salt solutions on non-polar surfaces leads to the problem of air bubbles being trapped when plaster slurries are cast against hydrophobic impression materials (such as silicone rubbers, 7§6) or investment against wax casting patterns. In the latter case, a surfactant solution (commonly called a **debubblizer**) is normally applied to the surface of the impression or pattern, and then dried in order to leave a film which will be wetted by the slurry (Fig. 8.5). That is, the wax or impression material itself is no longer 'seen' by the water.

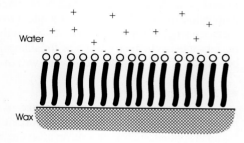

Fig. 8.5 Wax patterns and hydrophobic impression materials can be wetted when they are first coated with a surfactant.

● **8.4 Non-ionic detergents**
The basic type of general surfactants is ionic, that is, the molecule dissociates in solution into two ions. the charged amphiphilic part and, typically, a small ion. Soap (§8.3) is an **anionic** detergent; there are also **cationic** detergents. Increasingly now, **non-ionic detergents** are being used. These may be selected because they do not foam as much, are less toxic, or are not as strong in their action, so permitting a wider range of applications, particularly in biological contexts. A typical example group are the *n*-alkyl polyglycol ethers, $CH_3(CH_2)_{n-1}(OCH_2CH_2)_mOH$. It can be seen that such a molecule remains amphiphilic: there are still the paraffinic and water-solvatable regions (*cf.* the behaviour of polyether impression materials, 7§5). Clearly, with this type of chemistry the balance between the hydrophilic and hydrophobic parts may be tuned by adjusting chain length, so enabling control of the properties.

§9. Wetting in Practice

Although a number of situations have been discussed above, there are some further aspects that need to be explored in order to assess the importance of wetting in dentistry. It is necessary to give some thought to the practicalities, that is, the actual situations in which a liquid is in contact with a solid.

● **9.1 Types of context**
Wetting situations for dentistry can be grouped into just four classes:
1. mixing, as of the liquid and powder components of cements, acrylic, plaster and silver amalgam;
2. penetration of small spaces, as in fissure sealants, bonding agents, solder and impression materials;
3. immersion;
4. prevention.
The first class depends much on mechanical intervention (Chap. 15), but while the second depends on the spontaneous movement of the three-phase line (TPL), and therefore in this sense depends on contact angle through the penetrativity (§2.4), there is generally required to be an excess of the liquid available so as to fill the spaces. Immersion speaks for itself, but it must be recognized that contact angle is then irrelevant.

Thus, there is no instance in dentistry where a small, metered droplet is placed on a surface with the specific intention that it spread spontaneously to cover that surface (that is, other than for experimental purposes). In all practical instances, an excess of the liquid is provided, mechanical intervention is common to spread it (*i.e.* with a implement of some kind, such as a spatula or brush) and, if necessary, excess is subsequently removed.

Coating like this is in effect immersion. Examples include treating a wax pattern with surfactant solution, veneering porcelain-on-metal, silane treatment, and simply cleaning an object with a detergent solution, such as hands or hand instruments. Even if a small drop is used, such as of an adhesive or primer, spreading is secured by the approximation of the parts or the use of an implement, not through the action of unbalanced forces at the TPL. Contact angle as such, therefore, is again quite irrelevant as a concept or measure of goodness of wetting in such contexts, and the emphasis it is given misplaced. All that is required, in fact, is that the work of adhesion, W_r (§1.7) be non-zero for there to be an interaction between the two components (even if this can be expressed in terms of θ; equation 1.9). Of course, it might be argued that in some situations the larger is W_r the better, such as when there is a change of state of the liquid to solid and retention is important. However, the key here is that a liquid with any value of contact angle θ < 180° on a given substrate can fully wet that surface when it is present in excess: molecular-level contact is obtained. In other words, wetting occurs of any body in any liquid on immersion irrespective of contact angle, provided only that it is < 180°, and it is not a function of substrate surface energy or liquid surface tension – only the strength of the interaction is affected, and that has no bearing on subsequent chemistry (such as for silanes). This means that, further to the discussion of 2§2.5, we have to be very careful in what we mean when we talk about wetting.

•9.2   Bubbles

Even so, there is a circumstance in the context of immersion where contact angle is important. When excess liquid is applied, such as casting an investment slurry around a pattern, or instruments are placed in an ultrasonic bath, air bubbles may be stick to the surface. The analysis of this situation (Fig. 9.1), is exactly the same as for the droplet (Fig. 1.8) except that liquid and vapour are interchanged. It is necessary to bear in mind that contact angle is measured through the liquid: the Young equation (1.3) still holds, which can be seen from a consideration of the force vectors – there is symmetry. This means that for θ > 0, such bubbles are stable, and unless the air dissolves (which would take a relatively long time), mechanical means would be required to dislodge them. Remembering that the contact angle will be large on low-energy surfaces such as oily or greasy contaminants, bubbles will tend to stick in such areas, hence the problems with investments on wax patterns (§8.3).

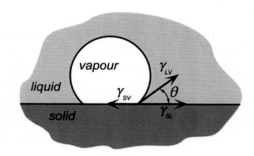

Fig. 9.1 Force vectors acting at the three-phase line of a bubble on an immersed solid surface.

One can understand the system through a small thought experiment. Consider such a bubble at equilibrium (Fig. 9.1) in, say, pure water. Now lower the surface tension of the liquid, *i.e.* the value of γ_{lv}, such as by adding an alcohol or detergent, which then diffuses to a uniform concentration, so that the forces now acting tend to move the TPL under the bubble, reducing the contact area, A_{sv}. This process continues with each addition. When the critical value is reached (Fig. 1.10), θ = 0 – and the bubble detaches from the surface. This can be seen quite clearly to occur in a physical experiment of this type: adhering bubbles spontaneously separate from the surface.

•9.3   Oil

A similar analysis applies when a surface is coated with oil or grease (Fig. 9.2), with the exception now that one has to recognize the existence of two contact angles, through the oil (θ_o) and through the water (θ_w) (depending on one's point of view), but these are, of course, complementary angles: $\theta_o + \theta_w = 180°$, and their cosines have equal magnitudes but opposite signs: $\cos(180 - \theta) = -\cos(\theta)$ (Fig. 1.12), both indicating the net force balance – there is no contradiction. Now, if the same kind of thought experiment

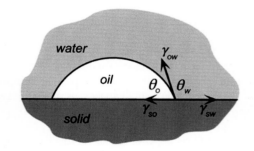

Fig. 9.2 Force vectors acting at the three-phase line of an oil droplet on an immersed solid surface.

as in §9.2 is conducted, lowering the interfacial tension γ_{ow} by the addition of a detergent (with an effect as in Fig. 8.5), so that θ_w decreases, then the TPL again tends to move under the drop, with the contact area A_{so} decreasing, ultimately leading to the release of the droplet from the surface. Simple observation shows that this occurs, although it might not be very rapid because it involves the flow of the oil or grease, whose viscosity will play its part. Of course, raising the temperature must lower that viscosity and speed up the process, regardless of the changes that must occur in all interfacial energies: hot water is better for washing the dishes.

A similar situation arises when taking impressions of wet tooth surfaces: the TPL involves water (or saliva), tooth enamel and the impression material. Broadly, the system will look like that of Fig. 9.1 (with the impression material in the place of the bubble) because the wetting by the saliva is better. It is generally thought that this is problematic, especially for silicones (7§6) so that consequentially 'hydrophilic' materials have been produced. However, it is clear that the wetting must be substantially better than by water for spreading to occur (even if the spreading force resulting from the lack of equilibrium was enough, given the high viscosity of the material). That is, on a wet surface the water or saliva is still favoured in terms of contact angle (Fig. 9.3). While the aqueous liquid may be displaced, it is clear that this must be a mechanical action rather than a physical wetting and spreading process. Indeed, detail would not be reproduced when the impression material may ride over liquid in depressions (boundary later separation may be facilitated by that liquid – see 23§10.2). Thus, any suggestion that 'hydrophilic' silicone materials can displace water because of that property is in error. Nevertheless, the pouring of the model (*i.e.* dental stone slurry) in that material may be better. Such considerations apply to all impression materials, but for aqueous systems such as alginate and agar (7§8, 7§9) it is clear that they cannot be much better than saliva, if at all. The consequence then is that the contact angle is around 90°, and this must represent the limit for any attempt to improve the hydrophilicity. That is, no better than on a dry hydrophilic surface, if the behaviour shown in Fig. 9.3 (top) is indicative. The point here is that hydrophilicity is irrelevant: affinity for the substrate surface is what matters. Only if this

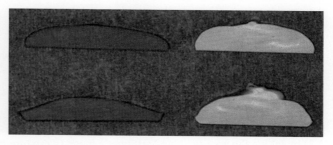

Fig. 9.3 Diametral sections of blobs of 'hydrophilic' silicone impression materials extruded onto (top) a dry hydrophilic surface, and (bottom) the same surface with a film of water present. The difference in contact angle is plain. Photo: Tom Smethurst

is higher than that of water can active water displacement possibly occur. Active spreading still remains a problem if the viscosity is high, and impossible if the material has an appreciable yield point (*i.e.* plastic, 4§7.3).

●9.4   **Other geometries**

As has been stressed in §1, contact angle is not a material property in any sense whatsoever but arises from the geometrical requirements dictated by the energy minimization (equation 1.2), or equivalently the force balance (equation 1.4). It is generally defined in the context of an isolated drop on a plane substrate, which despite its common use for teaching and experimental purposes is a special case – very few practical examples are encountered. The case of a meniscus in a circular capillary was found to satisfy the exact same condition of the Young contact angle (§2.2). In these two special cases the liquid surface is a spherical cap. There are other cases where the gravity-free solution can be easily derived and shown to have the same expression (equation 1.3) for a simple surface shape, for

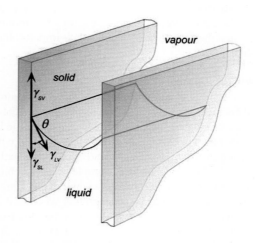

Fig. 9.4 Force vectors acting at the three-phase line of a liquid between infinite parallel plates.

example the "semi-infinite" filling between infinite parallel plates (Fig. 9.4). Here the constancy of curvature arises from the circular section of the liquid surface in the plane normal to the plates and the zero curvature in the direction along the meniscus; it is cylindrical.

The constancy of curvature is a strong requirement: as explained above: if it is not, the Laplace pressure differences must cause flow to occur until equilibrium is reached (bearing in mind that we have, for convenience, ignored gravity). We found this for a non-wetting drop pressed between two plates (Fig. 2.11). Thus, in the case, for example, of a droplet on a rod (Fig. 9.5), that droplet might (simplistically) be assumed to have a spherical surface. But this would then mean that the TPL does not show a fixed value of θ along its length according to the Young equation, varying from a maximum at point **a** to a minimum at point **b**, for the circumstances drawn here. This, then, is a contradiction: it cannot be (if the TPL is free to move), and so the equilibrium surface cannot be exactly spherical. Thus, in general (see box, 8§3.3), sphericity of the droplet surface is not required. There are many complex surfaces of constant curvature. There is a further difficulty.

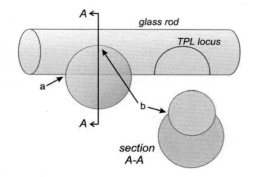

Fig. 9.5 A droplet on a rod: the contact angles at **a** and **b** clearly would be different if the droplet had spherical segment surface.

●9.5 Non-homogeneity of surface energy

We have been assuming a number of conditions (§1.4) to keep the discussion simple, although we have noted briefly the effect of contamination (§2.7). However, most of the materials used in dentistry are composite, *i.e.*, of more than one phase, or polycrystalline, such as metals. Clearly, the surface energy can be expected to vary between phases in any system, but even the mosaic of crystal sections presented at metal surfaces (Fig. 1§2.12) have different surface energies because of the differing arrangements of atoms, as shown for example by reactivity variation (Fig. 12§2.5) and the variety of crystal growth habits (2§7.5). Indeed, even grain boundaries are different enough to matter here, being higher energy because they are disordered.

In principle, it is easy to deal with heterogeneity of this kind in calculating the energy balance (equation 1.2) by taking the area-weighted arithmetic mean (a mixture rule) and inserting this, much as a modified value was used for a rough surface (§1.9). Unfortunately, this ignores the fact that the force balance (§1.5) must also be satisfied at every point along the TPL. While these views are equivalent for homogeneous systems, it clearly is a bad assumption for polyphase materials.[32] Essentially, if the contact angles differ but are fixed on each phase, there is a discontinuity in the liquid surface at the boundary, which is impossible. The shape of the surface in the vicinity of the boundary thus becomes very complex to maintain the constancy of the Laplace pressure; contact angle must vary smoothly and continuously around the TPL. Sphericity, again, is impossible. There is yet another difficulty.

●9.6 Pinning

In discussing roughness (§1.9) we ignored the local variation of contact angle in exactly analogous fashion: the total interfacial areas were calculated for the energy balance, but the local requirement for force balance was ignored. Obviously, local contact angle varies on a rough surface (Fig. 2.14), and again the actual liquid surface shape near the TPL must be very complex. The problem that emerges may be seen from Fig. 9.6. A very small displacement of the TPL causes the contact angle to vary depending whether it is an "uphill" (θ_u) or "downhill" (θ_d) location. Assuming that the true value lies between them, the one is being driven forwards, the other backwards.

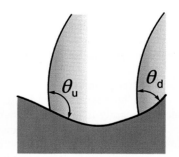

Fig. 9.6 A droplet on a rough surface would show locally-varying contact angle over microscopic displacements if the droplet has a perfectly spherical surface.

Depending on how steep the surface is locally, and especially if it is a sharp roughness, this can result in a considerable force being required to cause the TPL to advance or retreat, according to net energy balance demands. The roughness can be at a very small scale indeed, and not necessarily visible, for this to operate. Effectively, these surface tension-derived forces may not be great enough to overcome the effective activation energy barrier for movement, and the TPL gets stuck: it is said to be **pinned**. The same behaviour occurs on polyphase surfaces, even if atomically smooth. Contamination, such as by a minute amount of grease, or microscopic particles, have the same effect. It can readily be seen by the highly asymmetric shape of raindrops on windows, and their jerky, irregular movement. However, on an oiled surface, which is molecularly smooth (§1.10), the droplets are more nearly spherical and move steadily. In fact, if it were not for pinning, a droplet on a plane surface would immediately start to roll off if that surface were tilted at all.

This behaviour is quite general, whether due to roughness, heterogeneity or contamination, so general that there are many reports of "advancing" and "retreating" angles (Fig. 9.7), a behaviour known as **hysteresis**.[33]

Unfortunately, it is often not recognized that such angles are artefacts and do not represent a clear thermodynamic property of the system.[34] However, the behaviour of a pinned TPL is such that it appears to have a (line) shear strength, with the same dimensions and therefore units as surface tension (N/m) – sometimes (badly) described as 'friction'. If the force acting is great enough, whether through the effects of gravity, intervention through some instrument or tool, a blast of air, or other agent, then it will move, otherwise not. This is therefore a **metastable** state, unaffected by small (enough) perturbations. Thus, the mere fact of a favourable contact angle under ideal conditions cannot be taken to imply that wetting (*i.e.* spreading) will occur spontaneously. The remarks at §1.11 must then be extended to take into account these other interferences. It is this behaviour that for all $\theta > 0°$ causes water droplets to cling to even vertical 'smooth' surfaces, and even to leave behind a thin film, against which, of course, the contact angle is zero. Again, were the activation energy for movement zero, which is the same as saying that the shear strength of the TPL is zero, *i.e.* no pinning, all droplets of all liquids would be shed immediately. (One can measure the **roll-off** or **slide angle** of the droplet on the substrate for such systems as a measure of hydrophobicity which takes into account pinning. Effectively, it is a measure of the component of the weight of the droplet in the plane.) Likewise, one may observe that an evaporating droplet may

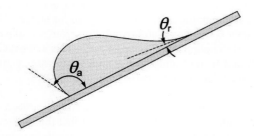

Fig. 9.7 Contact angles observed on a pinned drop: a "advancing", r "retreating". They may be grossly different, as shown here, and observed commonly for raindrops on window panes.

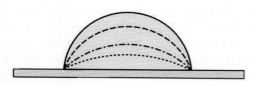

Fig. 9.8 A drying droplet may not show the expected retreating TPL but keep a nearly constant outline for some time, because of pinning.

not see the contact area shrink (Fig. 9.8) until the force vector imbalance becomes too great, when the TPL may jump substantially. In practical terms, for dentistry, steps must be taken to intervene to obtain the desired outcome in many systems, whether this be the removal of drops (such as by a towel or absorbent paper, where by implication wetting and capillarity are favourable, 27§2.1), or the full wetting of a surface by spreading.

By symmetry, similar remarks apply for bubbles in liquids (§9.2). Pinning can be seen clearly enough in carbonated drinks in a glass. It complicates the pouring of models and casting investments, and is another reason why the vibration for shear-thinning (4§7.9) is beneficial, along with the use of 'debubblizer' (§8.3): providing the activation energy for TPL movement and thus the dislodging of bubbles. Likewise, oil droplets may need some encouragement to move, even with detergent.

●9.7   Drying for bonding

For resin-bonding of restorations (§6), especially to dentine, there is considerable debate as to how wet or dry the substrate should be, and what other conditions are required. Part of the debate concerns the wetting of the prepared tooth surface by the bonding agents, and their penetration into the etched surface and demineralized collage. As has been discussed above (§2.3, §2.4), capillarity and penetration depend in part on the surface tension of the applied material and the contact angle exhibited on the substrate.

One means of achieving better wetting is to include in the applied mixture substances of low surface tension, such as ethanol or acetone, where a small proportion can mean a large drop in the surface tension of the mixture (Fig. 9.9).[35][36] Thus, water absorbed or desorbed from the substrate does not interfere substantially with the treatment, at least, while such substances are present. The lower viscosity caused by the presence of such diluents also assists the spreading and penetration of the bonding agent, although,

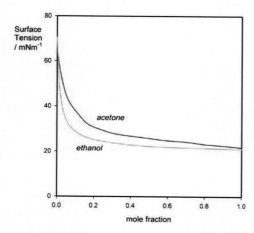

Fig. 9.9 Small proportions of water-miscible solvents such as acetone and ethanol reduce the surface tension of the mixture with water substantially, giving better wetting and spreading.

of course they should not remain and steps are taken to evaporate then, usually just by an air stream. The problem then is that any water in solution from the substrate tends to be left behind, incorporated as droplets in the resin, its boiling point being higher, and therefore its vapour pressure lower (see Clausius-Clapeyron equation, 5§2.1).

In this context, it has sometimes been claimed that acetone is a "drying" agent, and that this is because it forms an azeotropic mixture with water, in essence claiming that the evaporation of the acetone 'assists' or 'encourages' the evaporation of the water. Neither of these claims is true.

An **azeotrope** is a "constant-boiling" mixture, meaning that the composition of the vapour is the same as that of the liquid. This means that there is a minimum, called the **azeotropic point**, in the **dew curve** (where liquid would first appear on cooling the mixed vapours), and this is where it coincides with the **bubble curve** (where bubbles would first appear on heating, assuming nucleation), which is also at a minimum (Fig. 9.10). The behaviour can be traced using an isothermal tie-line in the liquid-vapour field on the left-hand side of the azeotropic point: the composition of the vapour is richer in ethanol than the liquid. On the other hand, if the initial mixture has more than this proportion of ethanol, then in complementary fashion the vapour is of a lower proportion of ethanol than the original liquid.

However, the phase diagrams are, of course, maps of closed systems at equilibrium. In a typical distillation process, which is 'open', the vapour is condensed and then effectively repeatedly redistilled, meaning that from the left, the ethanol concentration in the successively condensed liquid would rise, but only to approach, not exceed, the azeotropic composition. (A stepwise sequence similar to that shown in Fig. 12§1.8.) Conversely, from the right, it would fall. Thus, in this sense, then, a small quantity of water would be boiled off from an excess of ethanol because it is an azeotropic system. Distillation of such a high-ethanol liquid would, in an equilibrium system, approach the azeotropic point from the right.

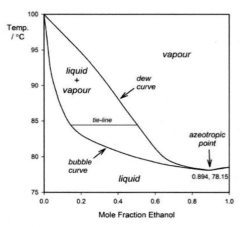

Fig. 9.10 Partial phase diagram for ethanol-water at 1 bar (see Fig. 2§12.1 for the lower-temperature portion). (Above the azeotropic point the two curves are too close together to resolve properly in this diagram, but there is a gap. The behaviour is parallel to that on the left-hand, low-ethanol side.)

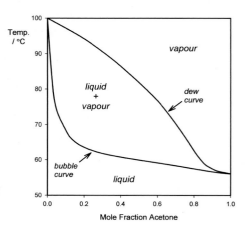

Fig. 9.11 Phase diagram for acetone-water at 1 bar.

Even so, this only applies at the temperatures attained during distillation at atmospheric pressure (1 bar). As the total vapour pressure is lowered, that is, for equilibration occurring at lower temperatures, the azeotropic point moves rightward and has vanished by 66 mbar, when the b.p. for ethanol is around 21 °C and that for water is about 38 °C. At open-mouth temperature, therefore, it cannot be expected that even if a large amount of initially pure ethanol were used (*i.e.* no water), effective drying of water adsorbed on the substrate could not occur: no preferential 'encouraged' evaporation of water then occurs, even with a jet of air to keep the sum of the vapour pressures low in the vicinity of the liquid surface (which of course cools the system even more, 30§5). Such behaviour is also affected by the other substances present.

For acetone-water, however, there is no azeotrope at 1 bar (Fig. 9.11), hence the 'not true' remark above, although it is known that at higher pressures (> ~1.4 bar) such behaviour does occur. Thus, the equilibrium vapour is always richer in acetone, and water gets left behind preferentially. In fact, boiling behaviour is altogether irrelevant in considering the effects of solvents on water in the present context of evaporation at, say, 20 ~ 37 °C; azeotropes are of no value here. Indeed, the equilibrium phase diagrams are of no help at all (remembering that they are thermodynamic statements). What are important are the relevant rates of evaporation.

Unfortunately, it is not a just a simple matter of comparing vapour pressure curves calculated from the Clausius-Clapeyron equation, it requires a bit more elaboration because in such systems, where there is marked hydrogen bonding, 'ideal' behaviour (*i.e.* following **Raoult's Law** for the sum of vapour pressures) does not occur. The activities of the components in the liquid vary in non-independent fashion, not strictly proportional to concentration, hence their vapour pressures are not proportional either. Thus the **partition** (8§3.1) between liquid and vapour phases needs to be examined, but also the kinetics of escape need to be considered since it is not a closed system in practice. This partition is portrayed in so-called **x-y** vapour-liquid equilibrium diagrams. Those for acetone-water and ethanol-water at 35 °C are shown in Fig. 9.12 (they do not vary much with modest changes of temperature). Here it is clear that the vapour pressures of the acetone and ethanol are both much greater than would be expected, which suggests that they would be

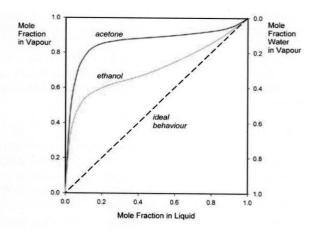

Fig. 9.12 Vapour-liquid equilibrium diagrams for acetone-water and ethanol-water at 35 °C. Note the complementary values for the mole fraction of water in the vapour, as read from the right-hand ordinate. For an azeotrope, the curve crosses the ideal line; here, neither does, so water cannot be 'driven off'.

readily lost from an open system, leaving the water concentrated in the liquid. Indeed, it is known that the lower boiling component, in general, has the greater escaping tendency, *i.e.* evaporates faster. This can be seen clearly for ethanol-water (Fig. 9.13).[37][38]

It is clear from that figure that the ethanol is lost steadily until it has disappeared completely before 14 s have elapsed, well before the water itself had evaporated. Water, however, was initially absorbed from the air before its evaporation commenced. In fact, it is a common observation that if a film on, say, a glass surface of some such pure and initially water-free solvent is allowed to evaporate, tiny droplets of water remain; there is no "drying" effect, but on the contrary, water is deposited. The same can be expected to occur for acetone, only more strongly. This simple demonstration is enough to negate the wishful thinking. In fact, water absorption in the high humidity of the mouth can be expected to occur far more rapidly as well, compounding the problem. Thus, it is not a meaningful treatment to wash a surface to be bonded with acetone (or any similar solvent) and let it dry naturally. In fact, it is very difficult to keep such solvents water-free:

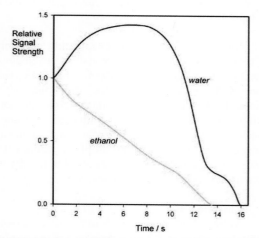

Fig. 9.13 An evaporation experiment for a 50 μL drop of 99% ethanol-water in air at 50 %RH, 25 °C and 1 bar, using an "IR microscope".

mere exposure of the contents of the bottle to the air is enough for water to be absorbed (at any non-zero humidity). In any case, absolutely dry ethanol and acetone are expensive, but clearly quite pointless in the context.

However, the avidity of such hydrophilic solvents for water is such that if one is placed in excess on a wet substrate (such as dentine), water will migrate – partition – from the substrate into the bulk of the solvent, a diffusive process.[9] If then, before it is able to evaporate, it is removed in bulk, as liquid, by blotting or being displaced by a blast of air (that is, not allowing much evaporation to occur), then that water will have been at least partially removed. This, then, is a dilution mechanism of relative dehydration. Indeed, 99% ethanol is sold

---

[9] A similar process is used in histology for dehydrating tissue, using 'ascending' concentrations of ethanol, from 30 to 100%, with the bulk simply decanted after each step. The sequence is used to avoid osmotic distortion of the tissue, but it takes time, 15 min ~ 24 h per step, because the partitioning is a diffusive process.

for just this purpose as a dental product. Of course, if the hydrophilic solvent is present in a prepolymer resin mixture, water will be absorbed in the same way, but without the means of removal, as mentioned above.

It should also be recognized that such solvents will also migrate by diffusion into the substrate dentine. Apart from any potential biological effects, this would mean that complete evaporation is not immediately possible. Remnant solvent diffusing back out could be involved at the interface with the bonding agent and also the resin.

●9.8   Marangoni effects
Solvents such as acetone and ethanol are sometimes said to be "water chasers" and to "displace" fluids in dentinal tubules and the water suspending the collagen fibrils exposed by acid-etching. This is a misinterpretation at best. The primary question would be one of to where such fluid could be displaced, ignoring miscibility: there is no space for this to occur, and quite obviously such water does not get 'ejected' from the system, like oil (§9.3). As indicated above, the water would be dissolved in the solvent, at the same time as the solvent dissolves in the water, and the mixture then occupies the whole space. Bulk removal would leave a remnant of the mixture in the collagen, tubules and the etched topography, from where evaporation would result in the same outcome as above: loss of the more volatile component first, then the water, more slowly.

This kind of remark may arise from observations of what are called **Marangoni effects**, which can indeed produce strange behaviour, due to surface tension gradients in imperfectly mixed systems. Noticing the large surface tension difference between water and solvent (Fig. 9.9), it can be seen that bringing the two liquids together will result in a large mismatch at the junction surface. Accordingly, there will be dramatic movements of the liquids because of the imbalanced forces, the water tending to be pulled back by its greater surface tension. The values change continuously as they each diffuse into the other (which occurs quickly), with corresponding continued movements, until equilibrium is attained at uniform composition. However, this can only occur in an open system with a free surface containing the liquid junction. Nothing like this can occur in a closed tubule or etch pit: the water has nowhere to go.

Of course, if solvent is evaporating unevenly from the surface of a mixture with water, then the imbalanced forces will again cause visible movements, but tending to concentrate the water locally rather than displacing it from the site of concern. This is the reason for the phenomenon of "tears of wine".[39] Alcohol solutions wet glass better than water – the contact angle is smaller – so they tend to climb higher up the side of the glass. However, the rapid change of concentration by the preferential evaporation of the alcohol from the thin film (Fig. 9.12) leave a liquid with higher surface tension, that therefore 'rounds up' into tears which then fall back down into the bulk liquid. Similar dramatic movements can be observed on a powder-sprinkled water surface, or one with scattered oil droplets: adding a drop of detergent or an alcohol, with the concomitant low surface tension, causes the particles or oil to rapidly withdraw from that point, as the surface area of the high surface tension pure water is minimized.

●9.9   Cassie's Law
At §9.5, mention was made of the effect of non-homogeneity of surface energy. We can expand on this by extending equation 1.2 for a drop on a plane composite surface with solid phases s1 and s2:

$$E_{TOT} = A_{s1v}\phi_1\gamma_{s1v} + A_{s1l}\phi_1\gamma_{s1l} + A_{s2v}\phi_2\gamma_{s2v} + A_{s2l}\phi_2\gamma_{s2l} + A_{lv}\gamma_{lv} \qquad (9.1)$$

where ϕ is the area fraction of the phase ($\Sigma\phi_i = 1$), and which after the usual manipulation reduces to:

$$\cos\theta_c = \phi_1 \frac{(\gamma_{s1v} - \gamma_{s1l})}{\gamma_{lv}} + \phi_2 \frac{(\gamma_{s2v} - \gamma_{s2l})}{\gamma_{lv}} \qquad (9.2)$$
$$= \phi_1 \cos\theta_1 + \phi_2 \cos\theta_2$$

This is known as **Cassie's Law**,[40] which is a type of mixture rule (*cf.* 6§2.1, 6§2.6). (Obviously, this does not depend on the shape, size or distribution of the second phase in the first.) It has the obvious extension for more than two phases, and for the roughness(es), if any, of the individual phases as at equation 1.18. However, this can only be an approximation because, as noted above, the liquid-vapour surface cannot be exactly spherical everywhere, or – equivalently – the TPL cannot be circular, for a continuously varying value of contact angle. Nevertheless, it appears to be a workable approximation, assuming that the scale of the structure, the grain size of the various solid phases, is small in comparison with the size of the droplet (which is the usual condition for all such mixture rules).

If the second phase is in fact vapour (*i.e.* the droplet lies on a porous or highly irregular surface; Fig. 2.14), where the contact angle is taken to be $180°$,[10]

$$\cos\theta_c = \phi_1\cos\theta_1 + \phi_2(-1)$$
$$= \phi_1\cos\theta_1 + (1-\phi_1)(-1) \quad (9.3)$$
$$= \phi_1(\cos\theta_1 + 1) - 1$$

What this means is that if the effective (solid) surface area of the porous body is small, the apparent contact angle can be driven to near $180°$ even if θ_1 corresponds to reasonable wetting (Fig. 9.14).[41] Thus, going further than the conclusions of §1.9, highly-sculptured surfaces can exhibit very high apparent (*i.e.* viewed macroscopically) contact angles[42] because the geometry prevents spreading, despite apparently adequate 'flat surface' contact angles. This seemingly paradoxical behaviour – superhydrophobicity – can be seen in many contexts: droplets of water may not soak into paper towels, or will just sit on many other fabrics, even without special treatment, and in particular mixing liquids on powders may not immediately soak in (Fig. 9.15) (similar effects can be seen with raindrops in the dust). However, such systems may only be metastable, because once spreading starts (whether by some diffusive process or pinning being overcome by an external force), the wetted system does not undo – it is not a reversible process.

Now, a comparison between the 'Wenzel' behaviour on a rough surface (§1.9) (which involves wetting all of the enclosed surface) and the superhydrophobic Cassie Law effect shows that the critical value for the contact angle, when the energy minimization for the two equations (1.14, 9.3) gives the same apparent contact angle, is given by:

$$\cos\theta_{crit} = -\frac{1-\phi_1}{r-\phi_1} \quad (9.4)$$

In other words, the superhydrophobic behaviour is shown when the flat surface value, θ, gives a Cassie value greater than this:

$$\cos\theta_c < \cos\theta_{crit} \; ; \; \theta_c > \theta_{crit} \quad (9.5)$$

This is shown in Fig. 9.15, where for various values of the roughness factor, r, the critical value is plotted on top of the contours of Fig. 9.13.

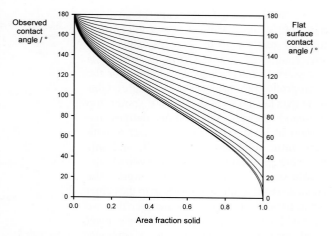

Fig. 9.14 The apparent contact angle of a liquid on a porous or highly-sculptured surface. First choose the line for the contact angle that the liquid gives on a flat surface (right-hand ordinate scale), then read off the observed contact angle for the given area fraction being wetted. Note that even well-wetting systems appear able to be made 'superhydrophobic'.

Fig. 9.15 Droplets may rest on the rough surface of a powder even if wetting is expected. Left: water on dental stone; right: water on alginate impression powder.

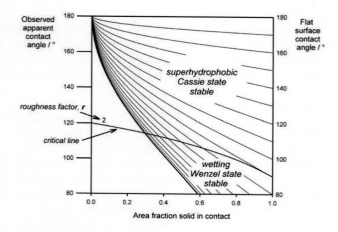

Fig. 9.16 The upper half of the graph of Fig. 9.14 with the critical angle of equation 9.4 plotted for roughness factor 2. Above the critical line, the drop rests on the roughness; below it and the whole surface inside the contact circle TPL is wetted.

---

[10] This is an approximation as the work of separation of a gas is not zero in the sense of Fig. 1.4; there must still be intermolecular van der Waals forces to overcome. Nevertheless, as vapour densities are typically very low in comparison with solids and liquids, at ordinary temperatures and pressures, the approximation is good enough for the present purposes.

We can try a thought experiment to see how this works. Consider a droplet being lowered gently onto the rough surface with, say, roughness factor 2 (Fig. 9.16). It is necessary to trace along the contour for the appropriate "flat surface" contact angle (right-hand axis labels) from the (0, 180°) point (top left-hand corner), which line defines the system in hand. The initial contact is with very little solid, so φ_1 is small, and this therefore corresponds to the superhydrophobic Cassie state. As the droplet is lowered, the contact area fraction will increase to the actual value for the substrate. (Remember that we start with a spherical droplet, and any contact is of zero area. Distortion to a spherical segment, lens-like shape is required to obtain increasing contact area – work has to be done in departing from the spherical shape, §1.1. Gravity provides the driving force for this.) If this point now lies above the critical line, it

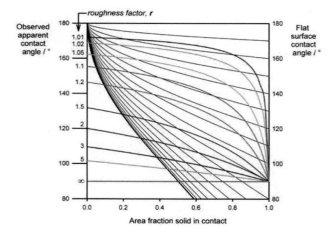

Fig. 9.17 Critical line variation with roughness factor. Very small contact areas on even low-roughness surfaces can allow metastable 'superhydrophobic' behaviour to be observed.

will be stable in that condition. If it is not, and it has crossed to the other side, the liquid is obliged to wet the whole of the roughness inside the contact circle (assuming that bubbles can escape).

The effect of varying roughness can now be seen in Fig. 9.17: identify the flat surface contact angle to choose the correct line, determine the area in contact, and examine that point with respect to the critical line for the relevant roughness.

However, if there is any pinning – which generally seems inescapable – a metastable superhydrophobic state may easily be attained, even to the extent of falling below the $r = \infty$ line, where thermodynamically it is disallowed without that restraint on spreading. In this context, the effect of gravity matters, as the weight of large drops may force the contact area to be large enough to cross the critical line, while very small drops may survive, at least for a while. (An extreme example of this is shown by mercury porosimetry: the liquid can be pumped into any connected porosity in any material against the effects of surface tension by a sufficiently large external pressure.)

In dentistry, such behaviour is frequently seen when powder-liquid systems have to be mixed. The first contact may result in a powder-covered droplet that does not get absorbed by the mass, or the liquid simply sitting on the powder (Fig. 9.15), until some mechanical intervention disturbs the metastable state of the system or reaction changes its description. Such effects are possibly part of the process described in §2.6.

●9.10 Denture base retention

A subject in which wetting behaviour is of some interest is that of the retention of acrylic (PMMA) full denture bases. It is, of course, true that were there to be no wetting, no force would be needed to be applied to separate the denture from saliva and there would be no retention (Fig. 2.11). Acrylic does, however, wet with water, although this is not particularly good, and some efforts have been made to modify the material. Firstly, however, anything done to the polymer itself to improve wetting would increase its water sorption and therefore it would be plasticized and less stiff. Secondly, coatings may be applied which wet better. However, the wetting with saliva is much better than with water. This is due to the proteins and mucopolysaccharides present which adsorb to the acrylic rapidly and fairly strongly (multiple hydrogen bonds but through a covalent backbone, as with polycarboxylic acids: 9§6, 9§8). But, in so doing,

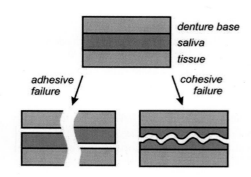

Fig. 9.18 Neither interfacial separation (adhesive failure), nor rupture of the liquid film (cohesive failure) can occur.

they create a surface which is more wettable: a condensed, multimolecular film of hydrophilic substances that then wets much better than the underlying acrylic, indeed the contact angle may be zero. Such a film is laid down very rapidly on contact with saliva, and on any coating, so that what is presented to the liquid saliva is always much the same.[43] (Of course, it then becomes possible for a **biofilm** to develop: a microbiological ecosystem, but this remains perfectly wettable.) Thus, interfacial failure by a simple separation of denture and saliva, *i.e.* adhesive failure, does not occur (Fig. 9.18). This strength is therefore quite adequate and its insufficiency can be discounted as a factor of any importance. Likewise, the tensile strength of water (of the order of 70 MPa[44]) is sufficient for cohesive failure not to occur, although any bubbles present may give the appearance of much lower apparent strength.

Disregarding mechanical devices such as clasps, screws and clips, or even magnets, there are only two mechanisms available for retention as such of a plain acrylic denture. The first is gravity, where the weight of the lower denture may hold it in place, but clearly this is of no help for an upper denture. The second is surface tension, depending on just the same pressure reduction as occurs between two plates (Fig. 2.10), providing that there is indeed a meniscus. A fully-immersed system cannot provide any retentive force (Fig. 9.19), only resistance to movement through the viscosity of the medium at low velocity. At high enough separation rate Stokes' Law-dependent drag becomes noticeable. If we consider only the effect of the viscosity of the liquid between such plates (and assume that they remain parallel and aligned), the **Stefan** equation[45] shows that the velocity depends on the cube of the separation, h, for a given force:

$$-\frac{dh}{dt} = \frac{8Fh^3}{3\pi\eta R^4} \tag{9.6}$$

where R is the radius of the circular plates. After integration and rearrangement, we get:

$$\frac{1}{h} = \sqrt{\frac{1}{h_0^2} + \frac{16Ft}{3\pi\eta R^4}} \tag{9.7}$$

This gives plots as in Fig. 9.20. From this simplified[11] model we can deduce several things. Firstly, while breakaway when it occurs is extremely abrupt – the system collapses – the initial separation is very slow, meaning that there is an effective retention simply because the gap is narrow. An unsupported upper denture can therefore be expected to appear to remain stably in place for some appreciable time. Secondly, the better the fit, *i.e.* the smaller the initial gap, the longer it will take before breakaway occurs. Not only does this mean that the better the denture is made the better it will be retained, but the consequence of biting, which 'seats' the denture, squeezing out saliva, is to ensure that the gap is returned to a minimum. Thirdly, raising the viscosity of the medium in the gap can increase the time required for breakaway

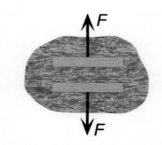

Fig. 9.19 No forces are acting on a fully-immersed system to oppose separation except those due to flow into the gap (ignoring Stoke's Law drag; *cf.* equation 4§10.1).

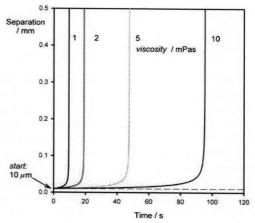

Fig. 9.20 The separation of parallel plates with a filled gap is quite abrupt, and depends on the viscosity of the fluid. The effect of increasing the initial separation gap can be seen by moving the time '0' to the right to correspond to the chosen value. (calculated for R = 30 mm, F = 0.5 N) Viscosity of water: 1 mPas.

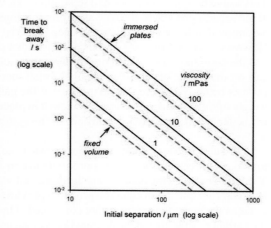

Fig. 9.21 Breakaway time varies as the inverse square of the initial separation and the viscosity of the liquid. (Other conditions as for Fig. 9.20.)

---

11   Again, the inertia of the fluid is ignored – it is treated as massless – and therefore the acceleration is also ignored, for simplicity.

substantially. This is the basis of the action of many denture 'fixatives' or 'adhesives', given that they wet (adhere to) both mucosa and acrylic. The effect of gap and viscosity on breakaway time for immersed plates is illustrated in Fig. 9.21 (solid lines). However, it is not often the case that one's mouth is completely full of liquid.

Thus, if the plates are not completely immersed, but simply have the gap filled by a fixed volume of liquid, then of course as separation occurs air is admitted to the space. Since the viscosity of air is about 0.02 mPas at ordinary temperatures, negligibly low compared with that of water, 1 mPas, we can expect the retardation of separation to be diminished. The equivalent equation to 9.7 for this case (which ignores the air) is:

$$\frac{1}{h} = \sqrt[4]{\frac{1}{h_0^2} + \frac{128\pi F t}{3V^2 \eta}} \qquad (9.8)$$

where V is the initial liquid volume between the plates, filling the space. This means that breakaway occurs about twice as quickly and more abruptly (Fig. 9.22; Fig. 9.21 – dashed lines). Otherwise, the breakaway behaviour is similar to that in the immersed plates case (Fig. 9.20). This kind of behaviour is the cause of what is commonly referred to as 'suction', and is an everyday kind of observation when trying to separate wetted surfaces. However, this 'suction' must be seen as an **emergent** effect, a reaction to the applied force because of the difficulty of the liquid flow – there is nothing going on otherwise in an undisturbed system.

What has been ignored in that last scenario is the effect of surface tension. That is, as the liquid surface is drawn into the edge of the plates, the meniscus curvature generates a lowering of the pressure (Figs. 2.4, 2.10). This would be appreciable at small separations. However,

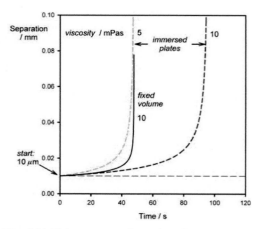

Fig. 9.22 If the gap contains only a fixed volume of liquid, the separation occurs earlier and more abruptly. (Other conditions as for Fig. 9.20.)

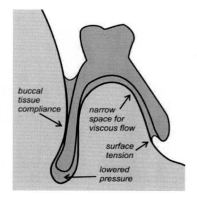

Fig. 9.23 The main factors promoting denture retention.

at the separations where breakaway is prompt (Figs. 9.21, 9.22), the contribution would be negligible. This is assuming that there is no relatively deep reservoir of saliva in contact with the margins – which is unlikely. However, there are two topographical factors that promote retention: the enclosure of the ridge by the flanges, and the compliant seal occasioned by the soft tissue on the buccal surface (Fig. 9.23). The first means that the gap on the flanks of the ridge changes much more slowly than the displacement so the viscous retardation persists better, while the second means that the path for fluid ingress is possibly more limited. Of course, the weight of the lower denture tends to retain it in place, whilst the upper suffers that permanent force. Essentially, the better the fit, the better the physics of the system provides retention.[46]

Such a system is not static, but is instead subject to intermittent forces, both tending to seat and tending to remove. It is therefore more appropriate to consider the response of the system to stress pulses (Fig. 4§2.1). The liquid between the denture and the tissue, saliva, is close to Newtonian in behaviour at low shear strain rates, and therefore the Stefan equation probably provides a suitable working model over small displacements, at least as far as the main dependencies are concerned. Once breakaway is approached, collapse is prompt. Of course, the variation in the size of the denture between wearers is relatively small, so the principal controlling factor then becomes saliva viscosity.

Unfortunately, fit cannot be retained for very long, no matter how well the denture is made, because of the biological changes in soft and hard tissue that occur. Accordingly, materials described as 'denture fixatives' of 'adhesives' are sold. These are intended to provide a highly viscous, sticky layer between the soft tissue and the denture base that also adheres to both. A common type of formulation includes a proportion of zinc oxide as a supposedly benign filler (9§3.2) in a mineral oil or aqueous carrier. The difficulty arises when reapplication several times a day is required because the denture fit is so poor that retention is even then inadequate, noting that the lost material is being swallowed. While such ingestion in small amounts (*i.e.* used as directed) has no apparent ill consequences, over a long period excessive consumption may result in a zinc-induced copper deficiency which has been associated with a number of problems (30§5).[47] Alternative formulations including substances such as carboxymethylcellulose (27§2.3) and copolymeric polycarboxylates (*cf.* Fig. 9§8.4) may be used that offer no such risks.

●9.11 Separating agents

Other than the direct blocking of porosity, preventing mechanical interlock, that mould sealants are intended to achieve (7§13), there is a group of materials whose purpose is to prevent wetting and adhesion. There are several aspects to this. Mechanically, if a coating remains liquid then no rigid interconnection can develop unless undercuts are severe. Physically, if a surface offers a large contact angle with the material to be overlain, then even on a moderately rough surface a full conformity will not be achieved (Fig. 2.14). Chemically, the complete avoidance of covalent and chelation (9§6.2) bonding, and the minimization of other interactions (Fig. 3.1), means that the work of adhesion (§1.7) is minimized. In particular, reaction with the moulded material must be avoided. If the mould is disposable, effects on that side are of lesser importance providing that dimensions are preserved accurately. Accordingly, thin coatings such as mineral or silicone oil, petroleum jelly, waxes (16§1.2) (perhaps applied in an evaporating solvent) and polytetrafluoroethylene (PTFE) may be used for many combinations of materials. Care must be taken in every case to ensure that contamination of important surfaces by these materials does not occur – silicones are especially problematic as they spread easily, are difficult to remove, and interfere with bonding even in very small amounts that are essentially undetectable by ordinary means.[48] There are ways of removing silicones, but these are slow and hazardous,[49] ill-suited to a dental context. While the reaction of Fig. 7§6.4 can be reversed by strong base or strong acid, this is not a practical proposition.

References

[1]   Young T. An essay on the cohesion of fluids. Phil Trans Roy Soc (Lond) I 95: 65 - 87, 1805.

[2]   Xu K, Cao PG, Heath JR. Graphene visualizes the first water adlayers on mica at ambient conditions. Science 329(5996): 1188 - 1191, 2010.

[3]   Wenzel RN. Resistance of solid surfaces to wetting by water. Indust Eng Chem 28(8): 988 - 994, 1936.

[4]   Choi DW, Lee HR, Im DJ, Kang IS, Lim GB, Kim DS & Kanga KH. Spontaneous electrical charging of droplets by conventional pipetting. Sci Rep 3: 2037, 2013. https://www.ncbi.nlm.nih.gov/pmc/articles/PMC3687225/

[5]   Roura P. Contact angle in thick capillaries: a derivation based on energy balance. Eur J Phys 28: L27–L32, 2007.

[6]   Smithwick RW. Contact-angle studies of microscopic mercury droplets on glass. J Coll Interface Sci 123: 482 - 485, 1988.

[7]   Meyer-Lueckel H & Paris S. Improved resin infiltration of natural caries lesions. J Dent Res 87(12): 1112-1116, 2008.

[8]   Washburn EW. The dynamics of dapillary flow. Phys Rev 17, 273 - 283, 1921.

[9]   Fan PL, Seluk LW & O'Brien WJ. Penetrativity of sealants: I. J Dent Res 54(2): 262 - 264, 1975.

[10]   Retief DH & Mallory WP. Evaluation of two pit and fissure sealants: an in vitro study. Paed Dent 3(1): 12 - 16, 1981.

[11]   Asmussen E. Penetration of restorative resins into acid-etched enamel. I Viscosity... Acta Odont Scand 35: 175 - 182, 1977.

[12]   Asmussen E. Penetration of restorative resins into acid-etched enamel. II Dissolution of entrapped air... Acta Odont Scand 35: 183 - 189, 1977.

[13]   Pong ASM, Dyson JE & Darvell BW. Discharge of lubricant from air turbine handpieces. Brit Dent J 198(10): 637 - 640, 2005.

[14]   Philip JR. The theory of infiltration: 4. Sorptivity and algebraic infiltration equations. Soil Sci 84: 257 - 264, 1957.

[15]   Latimer WM & Rodebush WH. Polarity and ionization from the standpoint of the Lewis theory of valence. J Am Chem Soc 42: 1419 - 1433, 1920.

[16]   Tribello GA, Slater B, Zwijnenburg, MA & Bell, RG. Isomorphism between ice and silica. Phys Chem Chem Phys, 12(30): 8597-8606, 2010.

[17]   Pauling L. The structure and entropy of ice and of other crystals with some randomness of atomic arrangement. J Amer Chem Soc 57(12): 2680-2684, 1935.

[18]   Chaplin M. Water's hydrogen bond strength. https://arxiv.org/ftp/arxiv/papers/0706/0706.1355.pdf

[19]   Lendenmann U, Grogan J & Oppenheim FG. Saliva and dental pellicle - a review. Adv Dent Res 14: 22 - 28, 2000.

[20]   De Munck J et al. A critical review of the durability of adhesion to tooth tissue methods and results. J Dent Res 84: 118 - 132, 2005.

[21]   Perdigão J, Lopes M, Geraldeli S, Lopes GC, & García-Godoy F. Effect of sodium hypochlorite gel on dentin bonding. Dent Mat 16 (6): 311 - 323 (2000)

[22]   Hashimoto M et al. In vivo degradation of resin-dentin bonds in humans over 1 - 3 years. J Dent Res 79(6): 1385 - 1391, 2000.

[23]   Moszner N, Salz U & Zimmerman J. Chemical aspects of self-etching enamel-dentin adhesives: A systematic review. Dent Mater 21: 895 - 910, 2005.

[24]   Peumans M, Kanumilli P, De Munck J, Van Landuyt K, Lambrechts P & Van Meerbeek B. Clinical effectiveness of contemporary adhesives: A systematic review of current clinical trials. Dent Mater 21: 864 - 881, 2005.

[25]   Ferracane JL. Hygroscopic and hydrolytic effects in dental polymer networks. Dent Mater 22: 211 - 222, 2006.

[26]   https://www.threebond.co.jp/en/technical/technicalnews/pdf/tech34.pdf

[27]   US Patent 3940362 A: Cross-linked cyanoacrylate adhesive compositions. 1972

[28]   Brudevold F, Savory A, Gardner DE, Spinelli M & Speirs R. A study of acidulated fluoride solutions.1. in vitro effects on enamel. Arch Oral Biol 8: 167 - 177, 1963.

[29]   Liu Q, Chen ZF, Pan HB & Darvell BW. The effect of excess phosphate on the solubility of hydroxyapatite. Ceramics Int 40: 2751 - 2761, 2014.

[30]   Hargreaves T. Surfactants: the ubiquitous amphiphiles. Chem Brit 39(7): 38 - 41, 2003.

[31]   Smith LO & Cristol SJ. Organic Chemistry. Reinhold, New York, 1966

[32]   Murray MD & Darvell BW. A protocol for contact angle measurement. J Phys D: Appl Phys 23: 1150 - 1155, 1990.

[33]   Rodríguez-Valverde MA, Montes Ruiz-Cabello FJ, Gea-Jódar PM, Kamusewitz H, Cabrerizo-Vílchez MA. A new model to estimate the Young contact angle from contact angle hysteresis measurements. Colloids and Surfaces A: Physicochemical and Engineering Aspects 365: 21-27, 2010.

[34]   Darvell BW, Murray MD & Ladizesky NH. Contact angles: a note. J Dent 15 : 82 - 84, 1987.

[35]   Vazquez G, Alvarez E & Navaza JM Surface tension of alcohol + water from 20 to 50 °C J Chem Eng Data 40: 611 - 614, 1995.

[36]   Enders S, Kahl H & Winkelmann J. Surface tension of the ternary system water + acetone + toluene. J Chem Eng Data: 52, 1072 - 1079, 2007

[37]   Innocenzi P et al. Evaporation of ethanol and ethanol-water mixtures studied by time-resolved infrared spectroscopy. J Phys Chem A 112(29): 6512 – 6516, 2008.

[38]   Sefiane K, Tadrist L & Douglas M. Experimental study of evaporating water-ethanol mixture sessile drop: influence of concentration. Int J Heat Mass Transfer 46(23): 4527 - 4534, 2003.

[39]   Fournier JB & Cazabat AM. Tears of Wine. Europhys Lett 20(6): 517 - 522, 1992.

[40]   Cassie ABD & Baxter S. Wettability of porous surfaces. Trans Faraday Soc 40: 546 - 561, 1944.

[41]   Hosono E, Fujihara S, Honma I & Zhou HS. Superhydrophobic perpendicular nanopin film by the bottom-up process. J Am Chem Soc 127(39): 13458 – 13459, 2005.

[42]   Movafaghi S, Leszczak V, Wang W, Sorkin JA, Dasi LP, Popat KC & Kota AK. Hemocompatibility of superhemophobic titania surfaces. Adv Healthcare Mater 1600717, 2017. DOI: 10.1002/adhm.201600717

[43]   Murray MD. Investigation into the wettability of poly(methylmethacrylate) in vivo. J Dent 14: 29 - 33, 1986.

[44]   Williams PR, Williams PM, Brown SWJ & Temperley HNV. On the tensile strength of water under pulsed dynamic stressing. Proc Roy Soc: Math Phys Eng Sci 455(1989): 3311 - 3323, 1999.

[45]   Stefan MJ. Versuche über die scheinbare adhäsion. Sitzungberichte der kaiserlichin Akademie Wissenschafter Mass Naturwisseclasse 69: 713 - 735, 1874.

[46]   Darvell BW & Clark RKF. The physical mechanisms of complete denture retention. Brit Dent J 189 (5): 248 - 252, 2000.

[47]   Bodnar DC, Muntianu LAS, Bucur M-B, Ionescu I & Burlibaşa L. Theoretical and practical aspects regarding zinc excess-induced copper deficiency pathology following the use of denture adhesive creams. Metalurgia Int 17(3): 159 - 162, 2012.

[48]   Petrie EM. Addressing silicone contamination issues. Metal Finishing 111 (4): 27 - 29, 2013.

[49]   http://www.dowcorning.com/DataFiles/090277018417382f.pdf
          http://www.dowcorning.com/DataFiles/09027701840cf487.pdf

Chapter 11   Metals I : Structure

Metals have an extraordinarily important place in dentistry, as in many technologies, because of their special combinations of properties. These properties have implications for handling and design which cannot be ignored: they control absolutely the applications and what can be achieved. In view of the relationships between structure and properties, it is the purpose now to describe the fundamentals of metal structure in preparation for more detailed consideration of specific types of dental product.

*After first identifying what it is that is a metal, the formation of solid in the **freezing** process is explored as it is the results of this process that underlie and explain much of subsequent structure, properties and behaviour.*

*Metals are typically **crystalline**, and the relevant varieties of crystal structure are explained, the relationships between them, and some of their features, especially the existence of close-packed planes which are very important in determining mechanical properties. Normally, metal objects are composed of many crystals, and this **grain structure** is the second important feature which must be understood to explain mechanical and other behaviour. The concept of solid solutions is introduced.*

*Whether it is deliberate, as in fabrication processes, or unrequired, as in exceeding the strength of the metal, the deformation of metals is due to **slip** between planes of atoms. The mechanism of **dislocation** movement, and some of the essential features of this process are briefly described. The number of dislocations and their ease of movement depends on the structure and both the thermal and deformation history of the piece. Controlling these aspects is critical to success in obtaining the desired blend of properties, although again it is a matter of compromise.*

In view of the fact that all discussion of metal properties ultimately refers back to structure, a thorough grounding in these aspects is essential to the understanding of dental applications and service behaviour.

Materials Science for Dentistry
https://doi.org/10.1016/B978-0-08-101035-8.50011-0

Metallic alloys, and a few pure metals, are used in many dental procedures. From the stainless steel of hand instruments, through cast gold devices and amalgam to titanium implants, the special properties of metals (such as their strength and ease of fabrication in complex shapes) have been advantageous and will continue to be so. Indeed, they permit dental procedures not otherwise possible. However, in addition to the wide range of properties shown by the metallic elements themselves, by the means of suitable treatment and alloying an enormous range of combinations of properties are possible. This range is so broad that it allows the design of alloys – as opposed to mere passive acceptance of what is already known – for specific purposes with definite values of the necessary properties. How this comes about can be understood from a study of the structure of metals from the atomic scale up to that of the most characteristic entity of metals, the grain, and then how the assemblage of grains behaves in the bulk metal, whether wrought or cast.

The scope is so enormous that new alloys continue to be invented and added to the already huge catalogue. Even in dentistry this process continues, especially for casting alloys, but also for orthodontic and implant use. However, it is not necessary to go into great design detail to understand the dentally-important systems, but it is important to understand basic structural matters and behaviour in order to make effective product selection, as well as to handle metallic materials and devices – including fabrication and finishing – to the best advantage of both dentist and patient.

§1. What is a Metal?

●1.1   Delocalization

Metals are essentially distinguished from other types of substance (in the condensed states, solid or liquid) by their special type of bonding. This is known simply as **metallic bonding** although, as will be seen, it also occurs in substances not usually considered to be metallic. Metallic bonding means that the valence electrons (that is, the 'outermost' electrons that are normally involved in chemical reactions) are not confined to specific atom-atom bonds or molecular orbitals but are each spread over all possible sites, that is to say that they are completely **delocalized** over the entire object, as well as being mobile. There are no underlying directional bonds in true metals.

Fig. 1.1 The resonance view of bonding in benzene.

Delocalized electrons will be familiar in the context of conjugated π-bonds in organic chemistry. Thus, in benzene (Fig. 1.1) three double bonds can be viewed as resonating between two arrangements, and the structure overall may be considered as a hybrid of the two. The theoretical picture is of overlap of all of the atomic orbitals to give one molecular orbital (Fig. 1.2). As more and more rings are added to the structure, as

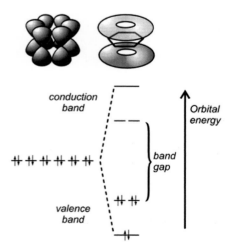

Fig. 1.2 The π-bond molecular orbitals of benzene delocalize the electrons and split the energy levels. The lower set are bonding, the upper anti-bonding.

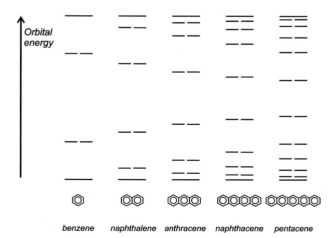

benzene   naphthalene   anthracene   naphthacene   pentacene

Fig. 1.3 Increasing the size of ring systems increases the number of available energy levels and thereby decreases their spacing. The band gap also decreases, making conduction require less energy.

in the series naphthalene to pentacene (Fig. 1.3), so the delocalization is greater and the number of energy levels available for the electrons increases, becoming closer together in the process.[1] This means that the energy required to promote an electron to the conduction band, where it is free to move across the system becomes lower, *i.e.*, electrical conductivity rises.

● **1.2   Graphite**

In the ultimate planar carbon ring system of graphite (Fig. 1.4) the number of possibilities of 'double bond layout' becomes astronomical. The electrons are then spread over the whole planar network, with very many energy levels available, very close together. Promotion of an electron to a higher level is thus very easy (*i.e.* requires little extra energy, thermal energies are more than enough), and this leads to the good electrical conductivity of graphite: it is some 1000 times greater in the plane of the rings (~300 kS/m) than across them (~0.3 kS/m).

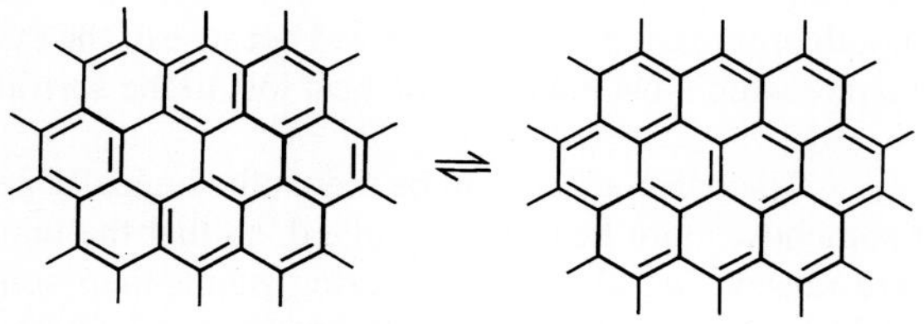

The delocalization in graphite is, however, restricted to one plane. Indeed, because of this special structure, graphite is described quite legitimately as a 'two-dimensional' metal. (This becomes very clear if the single sheets of graphene are considered, where there is no band gap.) This is in marked contrast to the structure of diamond: the completely covalent, localized and directional bonding gives no trace of metallic properties (the band gap is substantial: 5.5 eV = ~96 kJ/mol).

Fig. 1.4 Two of the many formal valence structures of a single layer of graphite (or graphene): π-bonding electrons are delocalized over the entire plane. There is no band gap, and conduction in the plane is metal-like.

In graphite, the bonding between layers is relatively weak, supposedly being due to van der Waals forces only, allowing the layers to slide past each other comparatively easily. This is said to confer its familiar dry lubricating properties (as well as its usefulness in ordinary pencils: its name comes from the Greek γραφειν [graphein], to write). However, ease of sliding depends on intercalated water and gas molecules from the air (that is, absorbed between the layers). Graphite is not a lubricant in a vacuum, when the intercalated molecules are desorbed.

The electrical conductivity of graphite is important in dentistry when it is used as the conductive coating on an elastomeric impression prior to making an electroplated metal shell die in copper or silver. The graphite is applied as a colloidal suspension in water ('AquaDAG'[1]). The same material is also used to provide an electrical leakage path (when it has dried, of course) for metallic specimens to be examined by scanning electron microscopy. Non-metallic specimens may be coated with a very thin film of evaporated carbon (which is graphitic) for the same reason.

● **1.3   Metals**

In the true metals themselves, as ordinarily understood, the electron delocalization is complete and in three dimensions, making the structure stronger and electrically conducting in all directions. The energy levels are even closer together than in graphite, and this band of energy levels confers the light-reflecting properties so characteristic of metals, the 'metallic appearance' only partially seen in graphite. High thermal conductivity also follows from this structure, which has been called an electron 'sea' or 'gas', in which the metal atoms are said to be embedded.

Iodine is not, of course, a metal in any chemical sense, although the element has a slight metallic lustre. This can be viewed as the last trace of metallic character due to its low ionization energy giving some mobility to the outer electrons, which also allows some slight electrical conductivity.

§2. Freezing

Pure metals, as with other pure substances, have very definite sharp melting points. That is, at the melting point only, solid and liquid will be in equilibrium, *i.e.*, according to the Phase Rule (8§3),

$$F = 1 - 2 + 1 = 0 \qquad\qquad (2.1)$$

---
[1]   Trade name: Aqueous Deflocculated Acheson Graphite

at atmospheric pressure; *i.e.* it is an invariant point, meaning that there is only one way in which the equilibrium coexistence of solid and liquid can be achieved.

If we allow the liquid metal to cool, plotting the temperature against time to obtain a 'cooling curve', initially we will merely observe a declining exponential curve (Fig. 2.1), as is ideally expected for Newtonian cooling (rate proportional to temperature excess – see §2.6, §2.7). But at the freezing point an arrest in this smooth progression will be observed because of the evolution of the (latent) heat of fusion. This heat is evolved at a rate which just balances the heat lost to the surroundings.

The balance here can be seen to be logically necessary. If somehow more heat were evolved, so that the temperature were raised, *i.e.*, above the freezing point, then some metal would be obliged to remelt, absorbing that heat so that the temperature fell again to the freezing point. Likewise, if the temperature were to fall below the freezing point while molten metal was still present, some would be obliged to freeze, evolving heat to return the temperature to the freezing point and thereby restore the balance.

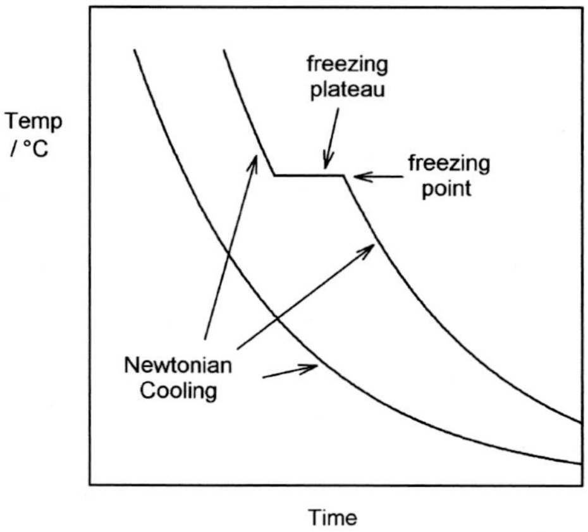

A typical value for the specific heat of metals is about 25 J/mol.K, while the latent heat of fusion may be about 10,000 J/mol. There is then indeed much heat to dissipate during freezing (and this is why liquid metal burns are very damaging). At the end of the freezing plateau, when there is no more liquid present, Newtonian cooling again resumes (Fig. 2.1).

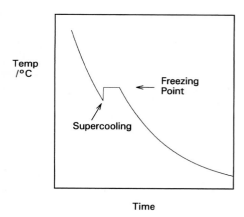

Fig. 2.1 A cooling curve for a pure metal, showing the freezing plateau, compared with a simple Newtonian cooling curve.

•2.1   Supercooling
Frequently, however, freezing does not commence immediately the 'melting' point is reached: there may be some noticeable overshoot or **supercooling** (Fig. 2.2). This means that the temperature continues to decrease smoothly until crystal formation starts, when the temperature rises rapidly back to the melting point, a phenomenon called **recalescence**, after which the normal freezing plateau will be observed. Usually, only very little solid is required to form for this return to occur. Thus, taking the above typical figures for the specific and latent heats, some 400 g of metal would be returned to the freezing point after supercooling by just 1 K when 1 g of metal has solidified. Obviously, the temperature cannot exceed the freezing point because heat was lost during the supercooling itself, and if it did all the now-solid metal would remelt. The total heat content after recalescence must be less than that when the cooling first reached the freezing point.

Fig. 2.2 Supercooling may lead to temperatures below the freezing point being observed before solid forms.

An extreme example of this is very pure gallium which, if left undisturbed as it cools and in a non-nucleating sealed container such as of PTFE, can sit indefinitely in the liquid state around normal room temperature (say, ~23 °C), which is well below the melting point of about 29.8 °C. However, if this is disturbed such as by shaking, crystallization commences immediately, when the temperature rises promptly to the freezing point. Some explanation of this phenomenon is required.

•2.2   Energetics of crystallization
Crystalline solids are distinguished from liquids by the long-range ordering of the atoms (or molecules, remembering that there are no multi-atom molecules in either solid or liquid metals) into regular arrays. It is the change in the degree of ordering that accounts in part for the latent heat of a change of state: the local change in entropy (the thermodynamic measure of disorder) results in a large free energy change for the system.

Now, in creating a solid phase in liquid, there is necessarily the formation of a definite boundary between the two (see the definition of a phase, 8§2.3). As this boundary will be associated with an interfacial energy, γ_{sl} (10§1), work must be done to create it. The work of formation of that surface, W_s, is therefore proportional to the surface area or the square of the radius. To keep it simple, and illustrate the underlying principles, we can treat the particle of solid as spherical. We then have:

$$W_s = 4\pi r^2 \gamma_{sl} \qquad (2.2)$$

On the other hand, the work associated with the phase change, W_v (due to the loss of entropy), is proportional to the number of atoms involved, which is therefore proportional to the volume of the solid:

$$W_v = \frac{4}{3}\pi r^3 . \Delta H . \left(1 - \frac{T}{T_m} \right) \qquad (2.3)$$

where ΔH is the latent heat per unit volume of the solid formed, T is the actual (absolute) temperature, and T_m the melting point. The net work of formation, W_f, of the crystal of a definite size, r, is then the difference of these two terms:

$$W_f = W_s - W_v \qquad (2.4)$$

since the one is a gain, the other a loss. If this total work of formation is plotted against r (Fig. 2.3) we can see that it is not a monotonic function: it goes through a maximum and then changes sign. Thus, initially there is an increase in free energy as the crystal comes into existence and begins to grow; this is therefore unfavourable.

However, if a certain minimum radius (r_0) is just exceeded, the slope of the net energy curve becomes negative so that there is then a tendency for the grain of solid to grow spontaneously. This critical point is found by differentiating equation 2.4 with respect to r and setting the slope to zero. We then have:

$$r_0 = \frac{2\gamma_{sl}}{\Delta H \cdot \left(1 - \frac{T}{T_m} \right)} \qquad (2.5)$$

From this we can see two things of interest. Firstly, when $T > T_m$, this **critical radius**, r_0, will be negative – which means that it does not exist and no crystals are formed (which is only reasonable, above the melting point). Secondly, the critical radius depends on the amount of supercooling, $T_m - T$: the lower the temperature the steeper the W_v curve becomes and the smaller the value of critical radius (Fig. 2.4). In fact, exactly at the melting point, the critical radius can be seen to be infinite! Hence some considerable supercooling is actually required to initiate solidification. Strictly speaking, this means that while the melting point is very well defined and realizable, the freezing point – in the absence of any solid – is not. However, once solid has formed, the equilibrium temperature is automatically obtained.

> It is important to remember that all calculations of a thermodynamic kind involving temperature must use the absolute temperature scale, *i.e.* be in kelvin (K), even if it is more convenient to talk or plot in terms of degrees Celsius.

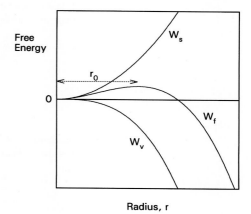

Fig. 2.3 The sum of the surface and volume energies goes through a maximum at the critical radius for the crystal nucleus.

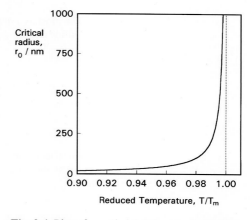

Fig. 2.4 Plot of equation 2.5 for a typical metal.

●2.3   Crystal nucleus

A statistical or probabilistic view of the process is helpful. Because of the rapid and continuous thermal motion of atoms in a melt, the spatial arrangement characteristic of the crystalline solid will occur spontaneously over short distances in the liquid randomly at any temperature, but be just as readily broken up or lost by further motion. The lower is the temperature, the larger such transient structures will tend to occur more readily and tend to persist longer (Fig. 2.5). Such a cluster of atoms constitutes the **crystal nucleus**; it may or may not grow into a crystal. Above the melting point the thermal energy of the atoms is so great that any such cluster must fly

Fig. 2.5 Homogeneous nucleation depends on the random formation of crystal-like arrangements of atoms.

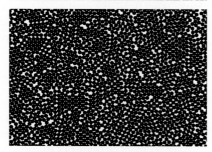

a: In the liquid very little structure is found, and what is there is transient and on a very small scale.

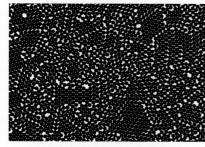

b: Nucleation at the freezing point involves the formation of small ordered clusters.

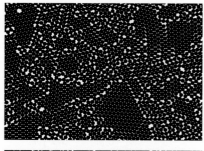

c: The clusters grow into recognizably crystalline regions.

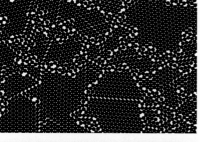

d: These incipient crystals grow steadily now that they are large enough to be stable.

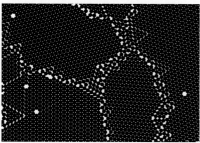

e: Crystal growth continues. Some defects that were present may disappear as equilibration of the structure can occur most easily near the melting point.

f: The metal is now solidified into well-formed crystals – with a few vacancies.

apart after only a very short time. Below the melting point, if the initial cluster survives just long enough for the accretion of enough atoms to cause it to exceed the critical radius, then growth will occur as the net energy change on freezing is then favourable.

The critical radius thus corresponds to the size of crystal nucleus at which growth and melting have an equal chance. The process could go either way at this point. For any particular cluster, the processes of addition and subtraction of atoms are random and continuous. The greater the supercooling the more centres of growth there will be throughout the liquid (as the probability of their forming is increased). With only slight supercooling, but allowing plenty of time for centres to be created, only a few will grow. The difference then is one of either many or few crystals. If many, they must be small because there is a fixed mass of metal to freeze; conversely, few nuclei yield large crystal grains. The difference between a fine- and a coarse-grained crystal structure is thus essentially dependent on the rate of cooling or, more accurately, the rate of abstraction of heat.

The fact that the formation of crystallization nuclei seems to require an input of energy needs further consideration. That is, in approaching the peak of the W_f curve in Fig. 2.3 from the left, the energy of the nucleus is greater the larger it is, and therefore more unstable with respect to the proper liquid state. It may therefore look as though a rising temperature would favour larger nuclei. But since such a rise in temperature can only increase the critical radius (equation 2.5), this is seen to be a false conclusion. It is the randomness of the formation and breakdown of the nuclei that leads to the outcome. This random, probabilistic nature of nucleation may perhaps be better understood by considering the local fluctuations in energy that may occur, even for a closed system of defined total energy.

Consider a bucket of balls that are repeatedly thrown at a ramp (Fig. 2.6). The velocity with which the bucket is moved can be tightly controlled so that the total energy of the assemblage of balls is fixed. Yet it is obvious that they will not all reach the same height on the ramp for any one throw. Because of internal collisions in the group, some will rise higher than others, some not reach so far. On average, one might imagine a spread of distributions of maximum height attained on the ramp by all balls over time, much as the thermal energy of the atoms in a melt will have a range of values (Fig. 2.7). Naturally, the highest position attained by any one ball will also vary from throw to throw. Thus, on occasion, one might observe a ball go over the peak of the ramp, despite the fact that the average energy per ball is quite insufficient for this event to occur. If a ball does go over, then it must continue on down the other side, to lower and lower energy.

Fig. 2.7   Although average energy is well defined, random collisions result in a spread of individual energies.

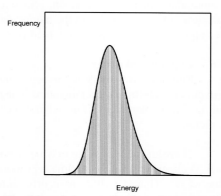

Fig. 2.8   A large collection of atoms as in a liquid show a distribution of thermal energy states with a small, but important tail at high values.

The physical counterpart of that model in the present context is as follows. In the highly-agitated liquid metal, at any given instant, there will be a random collection of nuclei of a range of sizes (Fig. 2.5a). These regions of crystal-like structure have formed by accident, and have a very short lifetime because collisions from other atoms are likely to break them apart again. The probability of finding a given size of nucleus decreases with increasing size, but it remains possible, below the melting point, for one (or more) nuclei to reach or exceed the critical radius. That is all that is required: once that condition is met, that nucleus will then grow steadily, assuming that heat continues to leave the system. The lower the temperature the greater the number of nuclei, the greater the average size of nucleus, and the greater the probability at any moment of exceeding the critical radius somewhere. Thus, crystallization is time-dependent in that the repeated 'sampling' of possible states ('snapshots') increases the probability of finding a greater-than-critical nucleus as the time interval of observation increases. It is cooling-rate dependent in that the lower the temperature the lower the critical radius and the greater the probability of finding many nuclei that can grow spontaneously.

•2.4   Homogeneous nucleation

Crystallization throughout, and within, the melt in the manner described above is termed **homogeneous nucleation**. The crystal nuclei are of the same material as the melt and formed from it. It is important to recognize that the nuclei of such crystals are indeed themselves crystalline, and are not formed on anything else. A liquid melt is like a violently agitated bucket of marbles: there is no long range order or structure within it. It is by chance only that locally, and only fleetingly, a few atoms may achieve an arrangement like that in the solid, crystalline material.

Thus, since this nucleation is an essentially random process (§2.3), that is, amongst other things, the locations of the nuclei are randomly distributed, it follows that the sizes of grains also have a random distribution, albeit affected by temperature gradients across the metal. This arises because a grain can only grow until it contacts an adjacent grain (which has itself been growing), when there is no more liquid available to freeze onto either side of that interface. Further, since the 'concentration' of nuclei also depends on cooling rate (§2.3), the mean size is greater at lower cooling rates: the total amount of metal is divided over a smaller number of grains, so their size must be greater. There is a further element of randomness: crystallographic orientation. As is detailed below (§3), crystal structures are directional in that there are layers: they cannot be structurally isotropic. Hence, as there is nothing in the melt to constrain the orientation of the layers of a nucleus, the crystallographic axes of the grain may assume any orientation in the solid. This leads to the kind of structure discussed earlier (1§2.7), and so to the effects on overall properties of polycrystalline materials that differ from single crystal behaviour.

•2.5   Heterogeneous nucleation

As with crystallization from solutions (*cf.* 2§6.1), **heterogeneous nucleation** is the process of starting the growth of crystals from the melt on, and aided by, an already solid but foreign surface. That foreign surface has, of course, a specific interfacial energy for contact with the melt so that adding atoms to it, although different, does not appreciably change either the total surface area of that contact or the net surface energy. The value of W_v therefore dominates in equation 2.4, and the critical radius may be negligibly small. In other words, its value can be rapidly exceeded ($r_0 < 1$ atom!) so that solidification commences immediately. The other aspect of this is the template. Most surfaces are rough at the atomic scale, with steps in the faces of crystals, defects of various kinds (see Fig. 2§7.3) that provide ready niches for further atoms to settle into. Once such an atom has been added there still remains a step which can accommodate another, and so on. So long as the imposed spacing is sufficiently close to the natural spacing of the metal freezing out, the crystal will grow readily.

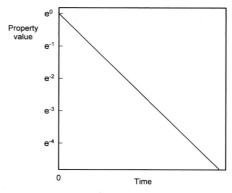

Fig. 2.9   When the rate of decline of a quantity is proportional to its current value, it gives a straight line on a semi-log plot, slope -k.

•2.6   Exponential decay

Reference was made at the beginning of the section to the exponential form of the temperature-time plot for Newtonian cooling. The explanation of this will now be set out to complete the picture.

There are many systems in which the rate that a quantity, say B, declines with respect to time is proportional to the current value of B. This is written as a differential equation in the following way:

$$-\frac{dB}{dt} = kB \qquad (2.6)$$

where k is the constant of proportionality and the leading negative sign indicates that it is a decreasing function. The question to be answered is: how does the value of B actually vary with time? The answer is given by integrating this expression, and since it is so important this will be set out rather than just stating the result. Rearranging to separate the variables:

$$-\frac{dB}{B} = k.dt \qquad (2.7)$$

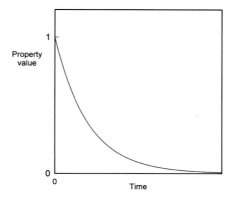

Fig. 2.10   The familiar shape of an exponential decay process plot.

so that:

$$-\int_{B_0}^{B} \frac{1}{B}.dB = k.\int_{0}^{t} dt \qquad (2.8)$$

From standard tables this is:

$$\ln B_0 - \ln B = kt \qquad (2.9)$$

or

$$\ln B = \ln B_0 - kt \qquad (2.10)$$

which gives a straight line plot (Fig. 2.8), or:

$$B = B_0 e^{-kt} \qquad (2.11)$$

which is the exponential decay function (Fig. 2.9).

The simplest example, perhaps, is that of radioactive decay where B is the number of unstable nuclei in the sample. It is also the form for a so-called first-order chemical reaction where B is the concentration of the reacting species. If B is substituted by ΔB, meaning the difference $B - B_\infty$, where B_∞ is the value that is eventually attained, we have the form for Newtonian Cooling (Fig. 2.1, §2.7); B then of course is temperature. Similarly, the deformation of a Kelvin-Voigt body (4§5) is obtained from the same analysis where B is replaced by the strain in the system. Lambert's Law is derived likewise, only instead of time as the 'elapsed' variable we now have distance traversed. This applies to X-radiation in matter (26§3), visible light through filters (26§5.2) and similar systems.

●2.7 Cooling curves

The determination of a cooling curve is the simplest form of thermal analysis. The typical basic form of the experiment is shown in Fig. 2.10. A sample of the material is heated to some appropriate starting temperature, and then the temperature is monitored as it cools in an undisturbed manner. Heat (Q) is lost to the surroundings at temperature T_R (across the tube wall, for example, in Fig. 2.10) according to the temperature difference ($\Delta T = T - T_R$, sometimes called the **temperature excess**), the area of the interface, A, and the efficiency with which that heat is transferred across the interface, what is called the "exterior conductivity", H (*i.e.* in $W/K/m^2$). So, we can write:

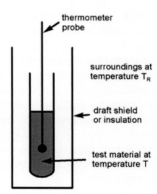

Fig. 2.11   Basic set-up for a cooling curve experiment.

$$-\frac{dQ}{dt} = AH \cdot \Delta T \qquad (2.12)$$

The effect on the temperature of the sample that this loss of heat has depends on the specific heat, c_p,[2] and the mass of the material:

$$Q = \Delta T c_p M \qquad (2.13)$$

However, we are usually not in a position to know A, H or c_p (which may vary with temperature), and M is arbitrary, so we can substitute for Q from equation 2.13 into the left-hand side of equation 2.12 and write:

$$-\frac{dT}{dt} \propto \Delta T \qquad (2.14)$$

(since M and T_R are constant, and we assume now that so is c_p). This is now in a form similar to equation 2.6, and is effectively the definition of **Newtonian Cooling**. Hence, assuming that there are no other processes occurring at any temperature to absorb or evolve heat, the plot of T *vs.* t is an exponential decay (Figs 2.1, 2.7).

The value of such an experiment comes from the fact that changes of phase such as solidification are associated with changes in the energy of the system – latent heat – due to the change in ordering (entropy). Even if only the value of c_p changed, the slope of the cooling curve would change more or less abruptly by some small amount, giving a kink which would be easily detectable. Thus the appearance of deviations from the expected smooth curve indicates the temperatures at which thermodynamically important events occur. These can then be related to structural and other aspects of the test system using, for example, metallography and X-ray diffraction (12§2).

---

[2] The subscript 'p' means at constant pressure.

§3. Crystal Structure

Having introduced the matter of the crystallization of metals, it is now necessary to consider the nature of crystals, and in particular the kinds of structures that metals exhibit. These patterns are fundamental to the mechanical behaviour of metals and alloys, and so to the design of alloys and the control of properties to suit the intended purpose.

●3.1   Space lattice

Disorder represents high energy, therefore energy minimization requires long range order. Order means that the arrangement pattern at one locality can be recognized at many localities; indeed, that it must occur throughout the crystal. If a structure is of just such a repeated pattern, there can be no region that does not fit that pattern in a perfect crystal. A crystal can thus be considered as being built of a three-dimensional series of identical pattern units, laid side by side, row upon row, layer upon layer, in what is called three-dimensional **translational periodicity**.[3] The pattern unit consists of a set of points in space, each point having the property that its surroundings are identical to those of every other point. In other words, for all points, each set of neighbouring points has the same set of relative coordinates, *i.e.* referred to the point of interest as the origin each time. Such an array of points is called a **space lattice** (geometrically, there are only 14 such **Bravais lattices** possible).

The (minimum) set of points used to define the lattice is to some extent arbitrary, there frequently being several ways to choose that set. But by joining up with lines certain points, convenient box shapes can be identified, the array of lattice points within which are identical one to another. Such a box, a generalized parallelepiped, is called a **unit cell**. These unit cells are then the building blocks from which the crystal may be considered to be built (Fig. 3.1). The key point is that all the corners of such a cell are equivalent, otherwise they could not be stacked to reproduce the structure – a corner must be common to eight cells.

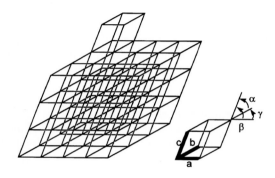

Fig. 3.1 Crystal structures are built up of many unit cells stacked in regular arrays.

●3.2   Unit cells

Unit cells are described in two separate ways. Firstly, in terms of the geometry, by the lengths of the edges: *a, b, c*, and the angles between those edges: α, β, γ (Fig. 3.1). These six descriptive parameters are called the **lattice constants**. The edges are taken to be parallel to the **axes** of the crystal system. These axes are the equivalent of the Cartesian or rectangular coordinate system axes of ordinary three-dimensional graphs, except that the angles between them are not restricted to 90°.

Secondly, the positions of the actual atoms in the cell are plotted according to the coordinates on those axes, the actual distances from the origin. This position pattern leads to a simple classification of types. For example, placing atoms at only the corners achieves what is described as the **primitive cell**; adding an atom at the centre of the cell as well as the corners gives

Table 3.1 *The conditions for the 7 types of unit cell.*

System	Cell edges	Cell angles
Cubic	$a = b = c$	$\alpha = \beta = \gamma = 90°$
Tetragonal	$a = b \neq c$	$\alpha = \beta = \gamma = 90°$
Orthorhombic	$a \neq b \neq c$	$\alpha = \beta = \gamma = 90°$
Monoclinic	$a \neq b \neq c$	$\alpha = \gamma = 90°, \beta \neq 90°$
Triclinic	$a \neq b \neq c$	$\alpha \neq \beta \neq \gamma \neq 90°$
Hexagonal	$a = b \neq c$	$\alpha = \beta = 90°, \gamma = 120°$
Rhombohedral	$a = b = c$	$\alpha = \beta = \gamma \neq 90°$

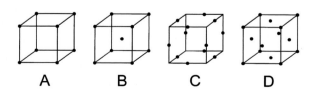

Fig. 3.2 Atom counting in unit cells. (A) Primitive, (B) body-centred and (D) face-centred cells of the cubic system. (C) is hypothetical.

---

[3]   Quasicrystals apart; they lack translational periodicity. see http://en.wikipedia.org/wiki/Quasicrystal

the **body-centred cell**; or alternatively adding atoms in the centre of each face as well as the corners for the **face-centred cell** (Fig. 3.2). An arrangement such as these can only be considered as a unit cell if translation along one or another crystal axis direction by an amount equal to the length of the corresponding edge results in exact superimposition of all atoms. Notice that although Fig. 3.2 is drawn as if the unit cell were a cube, $a = b = c$ and $\alpha = \beta = \gamma = 90°$, other shapes are possible (Table 3.1), although just two systems are very common for metals, the cubic and the hexagonal.

●3.3 Atom counting

It is sometimes necessary to count how many atoms are 'in' a unit cell. The following straightforward rules apply:

If the centre of an atom lies

(i)   anywhere **within** the unit cell, that atom belongs to that cell only; count 1.

(ii)   anywhere on a **face**, it belongs to the 2 cells that share that face, simultaneously and equally, and therefore is counted as half an atom for the unit cell.

(iii)   anywhere on an **edge**, it belongs to the 4 cells that share that edge; count one-quarter.

(iv)   at a **corner**, it belongs to the 8 cells that share that corner; it is therefore counted as one-eighth.

We can therefore write:

$$N_{cell} = N_{inside} + \frac{N_{face}}{2} + \frac{N_{edge}}{4} + \frac{N_{corner}}{8} \qquad (3.1)$$

Thus, in Fig. 3.2, unit cells A, B, C and D contain 1, 2, 3 and 4 atoms respectively. (Note that Fig. 3.2C is hypothetical only for the sake of the illustration here of the counting method; it is not a real crystal structure as it stands.)

●3.4 Common structures

There is thus scope for an immense variety of crystal structures, but fortunately for us now the majority of metals and their alloys tend to have one of only three structures (Fig. 3.3):

body-centred cubic   (b.c.c.)
face-centred cubic   (f.c.c.)
hexagonal close-packed (h.c.p.)

These structures are illustrated in Figs 3.4 to 3.14, drawn in different ways to emphasize first the unit cell idea, second to show how the atoms fill space in each structure, and then to show the set of nearest neighbours. (It is a good idea to build or handle models of these unit cells and layers to gain a better three-dimensional appreciation of the structures.)

Unit cells are usually drawn in an expanded form better to show the spatial relationships of the atoms (Figs 3.4 - 3.6). In the case of b.c.c. (Fig. 3.4) and f.c.c. (Fig. 3.5) this is quite straightforward. However, for reasons that are not really clear, the h.c.p. structure is usually illustrated by a diagram such as Fig. 3.7,

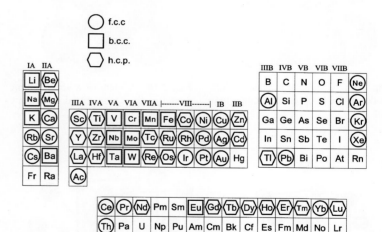

Fig. 3.3 The elements whose normal low temperature (c. 20 °C) crystal structure is one of the common three. There is no very obvious pattern, illustrating the subtlety of the factors involved and the closeness of the energies of all three. All those not shown are more complex (or, in a few cases, unknown).

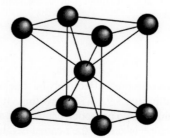

Fig. 3.4 The arrangement of atoms in the body-centred cubic unit cell. This contains just 2 atoms.

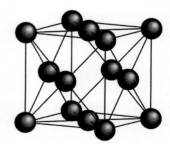

Fig. 3.5 The arrangement of atoms in the face-centred cubic unit cell. This contains 4 atoms.

which is actually equivalent to three unit cells. It is, however, misleading in the sense that it must be viewed as either the three cells having different orientations, which is against the principle of defining a unit cell, or as two whole cells and two half cells. Referring to Fig. 3.8, which shows four unit cells as clearly derived from Fig. 3.6 by the copying and translation process that defines lattice generation from the unit cell, it can be seen that the neat hexagonal array of atoms on the top and bottom layers of Fig. 3.7 is obtained by cutting off two corners.

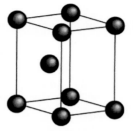

Fig. 3.6 The arrangement of atoms in the hexagonal close-packed unit cell. This too contains just 2 atoms.

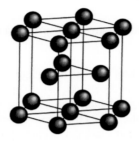

Fig. 3.7 The h.c.p. arrangement as it is more usually shown – which is misleading – equivalent to 3 unit cells.

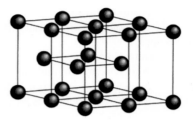

Fig. 3.8 A better view of multiple h.c.p. unit cells. Here all four have the same orientation.

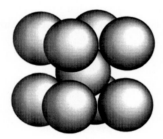

Fig. 3.9 A view of a space-filling model of the body-centred cubic unit cell.

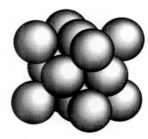

Fig. 3.10 A view of a space-filling model of the face-centred cubic unit cell.

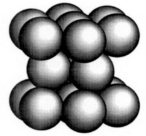

Fig. 3.11 A view of a space-filling model of the hexagonal close-packed structure, corresponding to Fig. 3.7

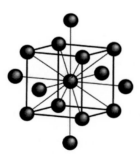

Fig. 3.12 The b.c.c. structure, showing the nearest neighbours are of two types at different distances.

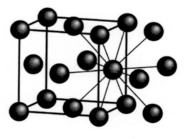

Fig. 3.13 The f.c.c. structure, showing the nearest neighbours all at the same distance.

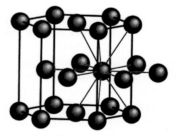

Fig. 3.14 The h.c.p. structure, showing the nearest neighbours, also all at the same distance.

Of course, in the solid the nearest atoms must effectively be touching, the distance between atomic nuclei then being the sum of the two atomic radii; Figs 3.9 - 3.11 show this. It is now clear that in the b.c.c. structure the atoms occupying the corners of the unit cell do not touch each other, as do those in the f.c.c. and h.c.p. cases, but only the atom at the centre. This is emphasized by the nearest neighbours being of two kinds for the b.c.c. structure, eight 'corner' atoms and six 'body-centre' atoms slightly further away (Fig. 3.12). This is in marked contrast to the f.c.c. and h.c.p. arrangements where there are 12 equidistant neighbours, all touching (Figs 3.13, 3.14). It is worth drawing attention again to the fact that every atom in these structures has an identical environment, each lattice position being, by definition, equivalent to the others. In other words, the origin of a unit cell could be taken to be what is shown as the 'body' or 'face' atoms in Figs 3.4 ~ 3.6, redrawing the lattice, and coming to exactly the same views.

Of the three structures we are discussing, the f.c.c. and h.c.p. can be seen from Fig. 3.3 to be the most common. However, hexagonal close-packing is in fact very closely related to face-centred cubic, as will now be shown.

●3.5   Close-packing

A single layer of identical atoms, or uniformly-sized spheres in general, may be packed efficiently in a layer in what should be a familiar pattern (Fig. 3.15); this is called a **close-packed** plane. A second similar layer is now put on top of this (Fig. 3.16), such that each of its atoms is resting in the hollow formed by and touching the three atoms below (*i.e.* in the minimum energy, most stable arrangement). There are apparently two possibilities for the position of that second layer, but they are not distinguishable; that is, only one *relative* position can be found. In fact, the two possibilities are superimposable after only a rotation of one by 30°. All displacements of the second layer to similarly 'settled' positions result in an identical-looking arrangement.

Close examination of this structure will show two sorts of interstice or 'hole' between the layers. One type is surrounded by just four atoms, the other by precisely six atoms (Fig. 3.16). It will be seen that this is quite unavoidable, and persists no matter how the one close-packed layer is adjusted with respect to the other, so long as the minimum energy requirement is met of allowing the second layer to settle into the recesses of the first. For these two types of hole, the centres of the atoms surrounding a hole lie at the apices of a geometrical (Platonic) solid; these solids are the tetrahedron and octahedron respectively (Fig. 3.17), and the 'holes' are named tetrahedral or octahedral accordingly. This also becomes important later in the context of steel (Chap. 21). The tetrahedral holes can also be seen to be oriented alternately "base up" and "apex up" with respect to our viewing direction, but there is only one type of octahedral hole.

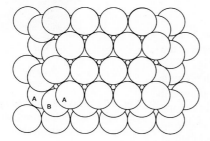

Fig. 3.18 Third layer – first possibility: each atom lies immediately over one in the first layer in an A-B-A pattern.

To continue building: if we wish to place a third layer of atoms on this stack, we find that there are two possible relative settled positions, but these are now geometrically distinguishable. The first possibility is for each atom of the third layer to lie directly

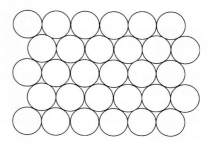

Fig. 3.15 One layer of close-packed spheres.

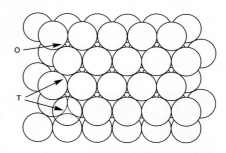

Fig. 3.16 The second layer of close-packed spheres creates two kinds of hole between the layers. Sites of octahedral (O) and tetrahedral (T) holes are indicated.

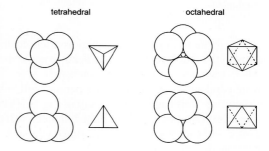

Fig. 3.17 The atom arrangements around tetrahedral and octahedral holes in close packed structures.

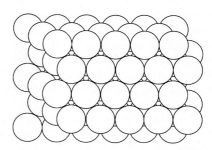

Fig. 3.19 Third layer – second possibility: each atom lies immediately over an octahedral hole between the first two layers.

over a "base up" tetrahedral hole between the first and second layers, and thus to lie directly over an atom in the first layer (Fig. 3.18) . The second possibility is that each atom lies directly over an octahedral hole between the first and second layers (Fig. 3.19). This is the way that greengrocers stack oranges and the like for display; it is sometimes called "cannon-ball stacking". The **packing density** is the same in both cases, the maximum attainable for hard spheres of one diameter.

When the fourth layer is added, following the respective patterns, in the first case each atom lies immediately over one in the *second* layer, while in the other case it returns to the relative position of the *first* layer (Fig. 3.20). The two structures are therefore distinguished by the repeat pattern of the layering when this is continued in the same manner. The first may be represented by A,B,A,B... , while the second is of the form A,B,C,A,B,C... . The first type is in fact **hexagonal close-packing** (h.c.p.) (Fig. 3.21) while the second is the **face-centred cubic** (f.c.c.) structure (Fig. 3.22). The similarity of these two structures is emphasized by the fact that face-centred cubic is, in some texts, referred to as **cubic close-packing** (c.c.p.).

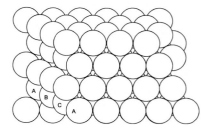

Fig. 3.20   The fourth layer in the face-centred cubic structure lies directly over the atoms of the first layer.

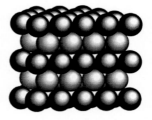

Fig. 3.21 A side view of the layers of h.c.p. The atoms of every second layer are in alignment.

Fig. 3.22 A view of the layers of the f.c.c structure. The atoms of every third layer on the cube diagonal are in alignment.

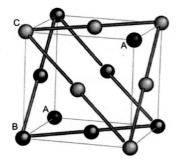

Fig. 3.23 A view of the f.c.c. structure showing the relative orientation of two of the close-packed planes.

Note in Fig. 3.16 that there is a tetrahedral hole above every atom in the first layer and also beneath every atom in the second layer. Also, each atom creates an octahedral hole. Thus, there are 4 octahedral and 8 tetrahedral holes per unit cell in both f.c.c. and h.c.p. structures. This can be seen easily in Fig. 3.10: each of the 12 edges represents one quarter of an octahedral hole, with one whole one at the centre; and each corner atom is resting on three others thereby creating 8 tetrahedral holes inside the cell.

While the best way to appreciate these ideas is to handle models of the structures, a key point is that in the f.c.c. arrangement the close-packed planes lie perpendicular to the cube diagonals (Fig. 3.22). There are two such planes which lie through the unit cell, each passing through three non-adjacent corners (Fig. 3.23). The diagonal directions (and there are four altogether in the cube) are the directions in which the layers may be seen to be stacked in the A,B,C,A,B,C... pattern. Thus the octahedron that can be drawn to mark the octahedral hole at the centre of the f.c.c. unit cell (Fig. 3.24) can be seen to have its faces lying in and defined by those diagonal close-packed planes.

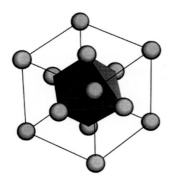

Fig. 3.24 Another view of the f.c.c unit cell with an octahedron drawn in to emphasis the octahedral hole at the centre of the cube.

The difference between h.c.p. and f.c.c. may also be seen by reference to Figs 3.25 and 3.26. These show the 12 nearest neighbours to any atom in the two structures. Note in particular the orientation of the top triangle of atoms in relation to that at the base.

Comparison of Fig. 3.2C with the f.c.c. structure as shown in Fig. 3.23 will reveal that if one atom were to be added at the centre of the cube in the former diagram, the two structures would be identical, even if defined with different origins (the unit cell atom count would then be 4). This makes the important point that a unit cell can be defined in any arbitrary manner so long as it remains the 'repeat' structure, *i.e.* with all corners identical. However, the cell is normally chosen specifically to be the simplest possible that communicates the essential symmetry of the structure. For this reason Figs 3.5 (which is the same as Fig. 3.2D) and 3.10 are in the preferred style for presenting the f.c.c. structure. For the h.c.p. structure the version shown in Fig. 3.7 may be useful for visualization purposes, even if not a unit cell, as explained above. In addition, it is preferred by convention that the unit cell be defined with an atom at each corner, rather than the spaces of Fig. 3.2C.

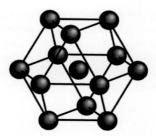 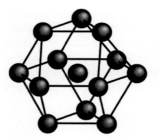

Fig. 3.25 The relationship of the 12 nearest neighbours to the central atom in the f.c.c. structure. (*cf.* Fig. 3.13)

Fig. 3.26 The relationship of the 12 nearest neighbours to the central atom in the h.c.p. structure. (*cf.* Fig. 3.14)

●3.6 Coordination number

These comments re-emphasize the important fact that every point in a space lattice has *identical* surroundings; bearing in mind the atoms present in adjacent unit cells, this can be seen by study of Figs 3.4 -3.14. Another way of expressing this is to say that the unit cell can be defined starting with *any* atom. For example, the atom at the centre of Fig. 3.2B is entirely equivalent to a corner atom. Arising from this is the idea of the **coordination number** (C.N.) of the atoms; that is, the number of *other* atoms each one is touching. For the body-centred cubic arrangement the C.N. is 8, while for h.c.p. and f.c.c. it is 12.

●3.7 Solid solution

It is evident, though, that for these contacts to occur in a completely regular pattern, all atoms must have the same diameter – the condition that was stated at the beginning of the packing discussion. Thus, the inclusion of an undersize or, especially, an oversize atom must lead to a distortion of the packing pattern, the amount of which distortion can be seen to be dependent on the relative sizes of the atoms concerned. However, it is observed in practice that in many cases the distortion of the lattice may be tolerated (*i.e.* the structure is stable), at least over some range of compositions. Such a mixed-atom crystalline solid is known as a **random substitutional solid solution** (Fig. 3.27), by analogy with liquid solutions in which the solute is dispersed randomly in the solvent. That is to say that *substitution* of one kind of atom for another may be done *randomly* (this is important), without causing a fundamental change in the crystal structure. The factors that are conducive to this behaviour are that the radius ratio of the two atoms is typically not more than about 1:1.14, that the metals are chemically similar, and in the pure state they have the same type of crystal structure (Table 3.2). The extent of the solid solution, that is the compositional range, varies with extent to which these conditions are met.

In particular, when the range of stability extends from one pure metal to the other, that is, over all compositions of two component metals, from 100% of one to 100% of the other, it is termed a **continuous solid solution**. Obviously, in this case the

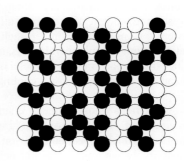

Fig. 3.27 Example of a possible crystal layer in a random substitutional solid solution.

Table 3.2 *Factors for solid solubility – the **Hume-Rothery Rules***

Radius ratio <~ 1 : 1.14
Chemical similarity: electronegativity difference < ± 0.4 e.u.
Lack of tendency for intermetallic compounds: valence the same
Crystal structure of pure constituents similar (continuity)

Also temperature-dependent: higher T/T_m is conducive

crystal structure type must be identical for the two metals. Some examples of this are Ag-Au (radius ratio 1.01) and Ag-Pd (1.05), both being of dental interest (these are dealt with in 12§1). Notice that for a structure to be considered as unchanged, it is only necessary to note the relevant equalities and inequalities as shown in Table 3.1. The actual dimensions of the unit cell are not critical to this assessment. Obviously, if gold atoms are to be substituted into a crystal of silver, it must get bigger overall because the gold atom is bigger than that of silver. Unit cell dimensions are very sensitive to this effect, and crystal compositions can be accurately calibrated using X-ray diffraction to measure the average separation of the various planes of atoms.

What happens to the structures when the continuous solid solubility conditions are not met is dealt with in Chap. 12. However, it is sufficient to note here that if the lattice strain is too great, determined only by whether or not there exists an alternative structure for that composition with an overall lower energy, the tendency must be for a rearrangement to occur, driven by the thermodynamic stabilization of the system. Even so, meeting the conditions of Table 3.2 is no guarantee that continuous solid solution will be observed: Ag-Cu (12§3.1) is a good example. In this case, lattice strain is not so dominant.

§4. Grain Structure

Crystals that are formed in the melt by homogeneous nucleation may tend to grow in all directions, but naturally none can grow any further when its surface meets that of another crystal. The interface is called the **grain boundary** (Fig. 4.1), as this defines the extent of each metal crystal or 'grain'. The nucleation of a freezing metal is a random process (§2.3); that is, amongst other things, the locations of the nuclei are randomly distributed. It follows that the sizes of grains also have a random distribution since their separations, and hence the distances over which they may grow in any direction, are similarly random. In the absence of any constraint, the crystallographic orientation (that is, in three dimensions) is randomly and uniformly distributed. Such conditions mean that there is no inherent **directionality** (structural anisotropy). Likewise, since all crystals can be expected to grow at similar rates, and also all crystallographically-similar faces, for a given crystal the amount of growth in all directions tends to be of similar extent. Consequently, under such conditions, grains tend to have random convex, polyhedral forms, and with a range of sizes. Further, since the 'concentration' of nuclei depends on cooling rate (§2.3), the mean size is greater at lower cooling rates.

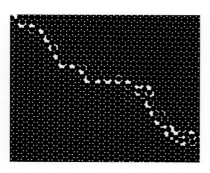

Fig. 4.1 The grain boundary between crystals of differing lattice orientations.

Fig. 4.2 Expanded polystyrene provides a convenient macroscopic model for the grain structure of metals.

Grains whose diameter is similar in every direction are known as **equiaxed**. The faces of such grains bear no relation whatsoever to the facets which appear naturally on crystals that have been allowed to grow freely without contacting neighbours, and none to the lattice planes, because their orientation and separation with respect to their neighbours is random. A model of metal grains that emphasizes the constraint aspect of their shape, as being due to collision between adjacent growing grains, is that of a soap bubble foam. Expanded polystyrene also has this kind of structure (Fig. 4.2), and with a little care individual 'grains' can be picked out of a fracture surface.

The situation is only slightly different with heterogeneous nucleation and where the crystals are growing from the walls of the container. In this case we may have a more columnar form at the surface, but this is most marked with very large castings that are cooled slowly. This therefore is uncommon in dental castings.

●4.1 Grain growth

The balance between the surface and volume energies was discussed in §2, but we might easily have framed the discussion in terms of the 'escaping tendency' of atoms from the surface of the crystal nucleus, an idea similar to that of the 'vapour pressure' of a substance over its solid or liquid. Being a process that requires an activation energy, such escape or 'evaporation' does not stop abruptly at some point as the temperature is lowered, but simply becomes slower and slower until at some stage, and at some time-scale, it may be considered to have become negligible. The same applies within a solid metal in respect of escape from one grain to another. There is expected to be a continuous exchange of atoms by diffusion across grain interfaces.

However, because the surface energy of small grains is proportionately greater than that of large, and by virtue of the greater curvature, the vapour pressure or escaping tendency is similarly greater. But transferring atoms from one grain to another reduces the surface area of the smaller grain (A_1) more than it increases the area of the larger one (A_2) (Fig. 4.2). Thus as the total grain area (A_T) decreases so the total free energy must decrease in such a process since the volume of the two grains is by definition constant. There is a thermodynamic driving force for the transfer of atoms from one grain to the other: one grain grows as the other shrinks (Fig. 4.3). There comes a point, of course, for the radius of the smaller grain to fall below the critical value for that temperature and it therefore 'evaporates' spontaneously. Viewed on a larger scale, fewer grains for a given mass of metal must mean a lower total grain boundary area, which is therefore a lower energy state.

A comparison may be made with Fig. 8§3.2: if the equilibrium vapour pressures over the two grains are different, mass transport from the one with the higher value to the other must occur.

Diffusion rates are of course strongly temperature-dependent, and such **grain growth** occurs readily only at higher temperatures (high, that is, in relation to T_m: the measure is T/T_m). This is one common result of annealing: average grain sizes are generally increased (while the number of grains decreases – there is a fixed mass of metal). This is known as **coarsening** of the grain structure.

A familiar example of grain growth of this kind can be found in the case of ice-cream. After long storage it will be found that it has become gritty in texture due to the presence of rather large ice crystals instead of the uniform dispersion of very small crystals which are undetectable to the tongue. This process of coarsening can be slowed to insignificance by storage at a very low temperature, but domestic refrigerators are not cold enough to achieve this. This accounts for the storage time recommendations for frozen foods in general: it has to do with damage to structure or texture by ice crystal growth rather than bacterial spoilage.

Grain growth can have a marked detrimental effect on yield strength: the empirical **Hall-Petch** relation[2][3] (Fig. 4.4) implies that the basic strength of the material, σ_0, is increased by an amount which is proportional to the reciprocal of the square root of the average grain diameter, d:

$$\sigma = \sigma_0 + k/\sqrt{d} \qquad (4.1)$$

The reasoning supporting this experimental observation is complex (and indeed has been shown to be an

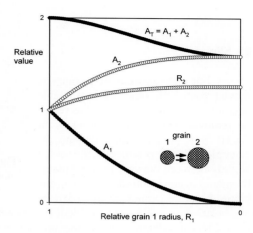

Fig. 4.3 Variation of the relative surface areas of two spherical grains, and the total area, A_T, when one grows at the expense of the other. R_2 is the radius of grain 2.

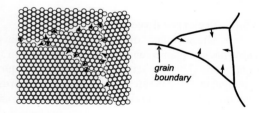

Fig. 4.4 Grain growth depends on the transfer of atoms from one grain to another at a greater rate than they transfer in the opposite direction.

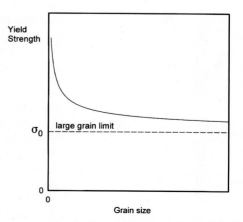

Fig. 4.5 Form of the variation of the strength of metals with average grain size known as the Hall-Petch relation.

oversimplification[4]), but on its own the curve shape illustrates clearly enough the main point: larger grains deform more easily. However, to understand its implications, we need to explore yield (1§3) at the atomic level. (There is obviously an upper limit to the strength that can be achieved, because of the finite strength of chemical bonding, 10§3.6, which is not taken into account here, but in addition the relationship breaks down for very fine grain metals, *i.e.* nanocrystalline, d < ~1 μm.[5])

§5. Slip

For yield to occur atoms must be moved past each other to new stable positions from which spontaneous return on removal of the stress cannot occur (1§3). Now the atoms in one plane, for example in the h.c.p. structure, are settled slightly between those of the next plane, the planes of atoms interdigitate. The one plane therefore cannot be moved from

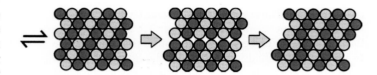

Fig. 5.1 Homogeneous slip would mean that all atoms of a plane (and all others attached to it) are required to move simultaneously.

one position to another of equivalent energy without being lifted up bodily to avoid the interference of one atom on another: the layers cannot slide past each other without first increasing their separation. But for close-packed planes the normal displacement of one plane with respect to the other, to enable that movement laterally, is quite small (Fig. 5.1). If, however, that kind of movement were attempted in a direction other than along a close-packed plane, the interference would be much greater, and the normal displacement required correspondingly larger (Fig. 5.2).

●5.1   Preferred directions

These other directions are so unfavourable that slip along close-packed planes tends to occur no matter in what direction the loading is applied. This is facilitated by the many equivalent planes in the three principal metal structures (Fig. 5.3); in other words, their high symmetry will guarantee that a favourable slip direction can be easily found. Thus there are **preferred** directions even when they are not parallel to the load axis. For a hypothetical single crystal of a pure metal taken along a random load axis, the test piece elongates because slip occurs in these preferred planes, rotation of each segment with respect to the load axis occurring simultaneously (Fig. 5.4).

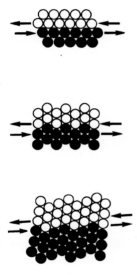

Fig. 5.2 Slip along directions other than those of close-packed planes is much more difficult.

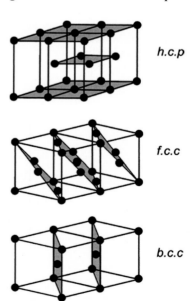

Fig. 5.3 Some of the close-packed planes in the principal metal crystal structures (in the case of b.c.c. some clos<u>est</u> packed planes are shown).

Real samples of metals, however, are usually polycrystalline, that is, they have many grains, each of which has a random orientation of its crystal axes with respect to those of its neighbours (Fig. 1§2.12). There will therefore always be many grains that will allow some slip easily, thereby transferring stress to those with perhaps less favourable orientations. Yield thus becomes a gradual process macroscopically, but can be a stepwise and intermittent process at the scale of the grain. Ultimately, all grains in a polycrystalline metal develop slip as in Fig. 5.4, with microscopic ledges becoming visible on the outside of every such grain. These features are known as **Lüders lines**, and account for the roughening seen in the neck of Fig. 1§3.3. In some systems they can be large enough to be visible to the naked eye.

In any case, it is now easy to see why solid solution alloys provide strength advantages over pure metals: the distortion of the lattice introduced by random substitution of a second element (Fig. 5.5) increases the energy required for slip. So effective is this solid solution strengthening mechanism that even amounts of alloying elements which ordinarily would be considered merely as impurities, say, less than 0.1%, can produce marked changes in properties. Indeed, extremely pure metals are typically very soft in comparison with commercial 'pure' grades for this very reason.

Slip is important for the working properties of metals: **ductility** (the capability of being drawn into fine wires), **malleability** (the capability of being hammered or rolled into thin sheets), and the ability to be **burnished** (edges shaped by pressure and surfaces smoothed). Indeed, there cannot be plastic deformation of metals without slip. This remark is related to the statement that all flow depends on shear (4§3). Hence the more close-packed planes there are in a structure, and the closer the packing, the greater the ease of these macroscopic deformations.

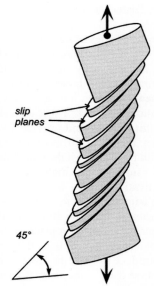

Fig. 5.4 Slip occurs along close-packed planes even if not aligned with the expected direction of slip. In this case shear should occur at 45°.

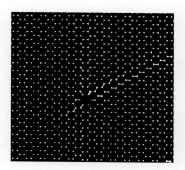

Fig. 5.5 Substitutional solid solution atoms introduce distortion into the crystal lattice. Here, just one atom distorts the packing many atoms away.

Some examples of pure metals used in dentistry are silver points for endodontic obturation (28§9), gold foil as used for direct gold restorations (28§4), tin foil as was formerly commonly used to line denture base moulds as a water barrier (7§13), and platinum foil as used as a matrix for porcelain jacket crowns (25§4.7). In each case it is the ductility and malleability, the softness if you will, of the metal that is important to allow it to conform to the shape desired using just manual techniques.

●5.2 Inadequacy of model

However, the view represented by Fig. 5.1 is far too simplistic. The energy required to simultaneously lift an entire plane of atoms to enable it move over another is simply too great (Fig. 5.6) and the entire system would then be extremely unstable. Thus **homogeneous slip** just cannot happen.

A consideration of the energetics of such a system shows how to understand this. On loading, energy will be stored elastically in compressing each column of atoms. However, it is a thermodynamic necessity that if a path exists for lowering the energy of a system, it will be taken. Buckling offers an easy path to relieve the elastic strain. That is, the regularity of the crystal would be destroyed by the creation of damage – imperfections caused by sideways movements. Buckling is therefore an event that would occur before the energy necessary for homogeneous slip could be delivered and stored in the system.

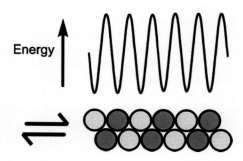

Fig. 5.6 Although the displacement is small, the total energy required to lift an entire plane of atoms enough to slip is too great. It cannot occur.

The explanation for the observed ductility of metals arises, oddly enough, from the imperfections of the crystals. That is to say that while the underlying cause of the phenomenon of ductile metals – slip – is not in doubt, the mechanism that permits this needs elaboration.

§6. Dislocations

•6.1   Types of defect

Crystals are rarely perfect in the regularity of their lattices. Quite often holes known as **vacancies** are left in a crystal lattice (Fig. 6.1). These are called **point defects**, and provide a very easy means of diffusion of atoms within the lattice if the temperature is sufficiently high that the activation energy is available for an adjacent atom to move into the vacancy. Impurity atoms at lattice sites may also be considered to be point defects in that they may introduce strain in the lattice, but of course there is the alternative view that these are simply in solid solution (§3.7); it is a question of whether they are intentional or not.

Defects of structure also arise from larger scale **stacking faults**, where the rows of atoms get out of alignment (Fig. 6.2). Thus, it has been found by X-ray diffraction, for example, that what appear superficially to be single crystals are in fact three-dimensional **mosaics** of **crystallites** (*i.e.* small crystals) which can be distinguished by slight differences in the orientations of their crystal axes (Fig. 6.3). The boundaries between such regions are thus known as **low-angle boundaries**. Importantly, there is a definite continuity of structure, or **coherence**, across these boundaries (Fig. 6.4), the layers of the crystal being traceable across from one region to another. This coherence is totally lacking between grains, which are therefore identified by large discrepancies in axis orientation (see Fig. 4.1). Grain boundaries are therefore described as **incoherent**.

The angle between the crystal axes of two crystallites is due to the **defect structure**, which may be typified by part of a layer of atoms occasionally being totally missing (Fig. 6.5). Such **edge defects** or **dislocations** are described as **linear**. A second important type of linear defect is the **screw dislocation** (Fig. 6.6), so called because crystal growth on the top surface may be envisaged as proceeding in a spiral around the **screw axis**. One circuit brings the locus back to a level one layer above the starting point. It is the commonest form of defect in many crystalline substances, and in fact provides the main crystal growth mechanism as there is always an edge at which to add new atoms, whereas the filling of surface holes and corners must terminate (*cf.* Fig. 2§7.3).

Stacking faults are **line defects**, in contrast to the point defects of vacancies. By analogy, grain boundaries may be described as **surface defects**, as the imperfections extend two-dimensionally over the entire grain itself. There are also **plane defects** in which the stacking of whole planes of atoms does not follow the

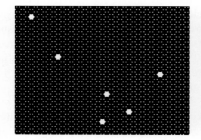

Fig. 6.1 Vacancies are where atoms are missing from a crystal lattice.

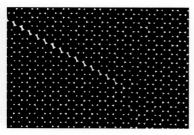

Fig. 6.2 Stacking faults can cause rows of atoms to get out of alignment.

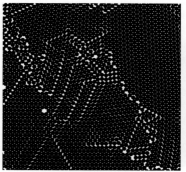

Fig. 6.3 Real crystals are composed of many small regions of high regularity with only slightly differing crystal axis orientations. Compare the mismatch of the grain on the top left where there is a distinct grain boundary.

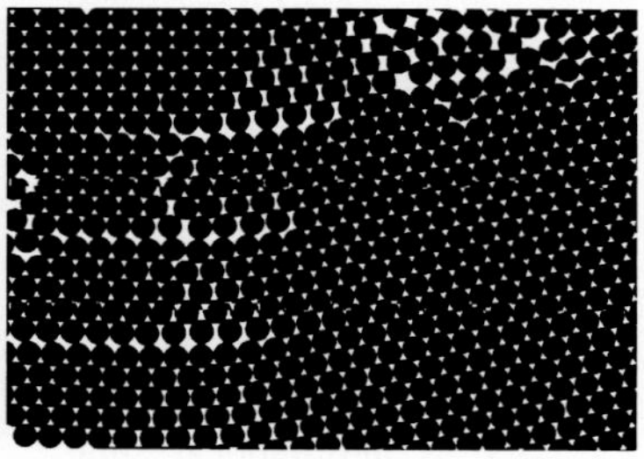

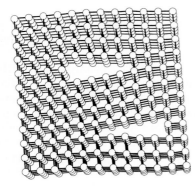

Fig. 6.4 Coherence means that there is continuity of structure between one crystallite and the next, even if there is distortion.

Fig. 6.5 Stacking faults include incomplete layers...

pattern implied by the long-range regularity of a crystal, even though locally (within the plane) there may be no problem. A special kind of such a defect is **twinning** (28§1.2).

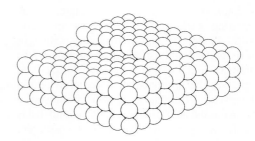

Volume defects are also possible: odd clusters of atoms, voids or 'bubbles' larger than a single atom, second phases that do not belong, and **domains** between which some crystallographic aspect such as the orientation of an axis varies (19§1.4)

Looking again at a coherent, low-angle boundary between crystallites it can be seen that it represents a series of stacking faults (Fig. 6.7), in the region of which the lattice is distorted. There is therefore a **strain field** associated with such defects, extending several atom diameters out in all directions, and of course this is associated with a higher energy – elastic energy is stored in the lattice. (Note the symbol ⊥, which represents the end of the extra row or layer that constitutes the dislocation.)

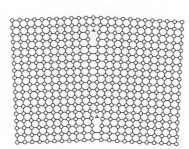

Fig. 6.6   ... and malalignment to give a screw dislocation.

Fig. 6.7   A coherent or low-angle boundary between crystallites.

It is worth noting that when lines are drawn linking atoms in diagrams such as Figs 6.5 and 6.7, as well as diagrams such as in Figs 3.4 - 3.8 and so on, they do not represent bonds. Metallic bonding is not directional, and occurs no matter how metal atoms are arranged. Thus, there is still bonding at boundaries, whether coherent or incoherent. If there were not, the piece would separate into individual grains. The lines are only meant as a guide to show spatial arrangement, just as they are used in unit cell and lattice diagrams.

●6.2   Inhomogeneous slip

Dislocations are exceptionally important because they provide the means for **inhomogeneous slip**, which is the actual mechanism for yield[6] (*i.e.* plastic deformation) in any metal or metal alloy (note: they have nothing whatsoever to do with yield in amorphous polymers). At the dislocation just one atom at a time may be dislodged into a new position, so that the actual energy required for each step is very small. This is an activated process, requiring energy to be supplied to pass an energy barrier, but one which leads to a new and lower energy state which relieves strain. Thereby, consecutive rows of atoms can be moved into new positions, and ultimately whole planes. In so doing the site of the dislocation is translated from one place to another, and the locus of its movement defines the slip plane. The dislocation's position at any moment therefore defines the boundary between the slipped and unslipped regions. We can now see that the buckling instability referred to in §5.2 would generate dislocations, enabling slip at lower stresses than otherwise might be expected, even if the crystal were initially absolutely perfect (which is probably impossible).

●6.3   Dislocation pile-up

Dislocations are necessarily sites of high strain energy, because of the local distortion of the crystal lattice away from its ideal regularity. Atom movements across a dislocation result in the movement of the dislocation in the opposite direction, for the nature of a dislocation is such that the rearrangement of only a few atoms cannot correct the defect. Thus the dislocation moves an atomic diameter at a time as slip progresses across a crystal. Slip is therefore discontinuous, stepwise in fact, at the atomic level also. Eventually the dislocation will come to a grain boundary, but (typically) the orientation and type of dislocation is such that it cannot be 'lost', too many atom rearrangements would be required, and its movement is arrested at the grain boundary. (It must be remembered that dislocations have length and extend great distances, in terms of numbers of atoms, in the grain, and usually not in straight lines.)

Continued strain generates new dislocations, and these behave in the same way as the pre-existing crystal defects. But these other dislocations moving in the same direction, and particularly along the same slip plane, similarly cannot be annihilated or joined together to make one larger dislocation because the strained zone around the defects themselves prevents the two from approaching each other closely (an even higher energy state is implied). They can be thought of as repelling each other like similar magnetic poles. In fact, if the grain boundary is thought of as being highly disordered, and therefore strained, it represents a massive dislocation, repelling the approaching dislocation within the crystal. (There is one special circumstance when mutual annihilation can occur, and that is when a pair of dislocations represent incomplete layers in a complementary sense: ⊤ and ⊥. Obviously, these must lie in the exact same glide plane.)

As dislocations **'pile-up'** at grain boundaries, the opposing plane of atoms is of course moving steadily in the opposite direction. This eventually leads to the opening up of microscopic cracks in the grain boundaries themselves (Fig. 6.8), cracks which will eventually lead to failure (Griffith criterion! 1§7[2]). (Dislocation pile-up is where the dislocations form a queue, as it were, not literally superimposed or stacked one on top of another.) This is the consequence of slip (Fig. 5.4) in a polycrystalline metal (Fig. 1§2.12). Such micro-cracks can be considered to be **volume defects**.

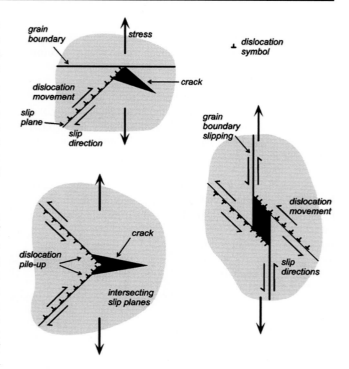

Fig. 6.8 Under stress, dislocations move to grain boundaries and pile up, but the slip thereby generated opens up microscopic cracks which lead to failure.

To put this in perspective, in a heavily strained metal dislocation densities of up to about 10^{12} cm^{-2} may be found in a random cross-section. For an atom diameter of, say, 250 pm, there will be around 1.6×10^{15} atoms/cm^{-2}, or about 1600 atoms/dislocation, roughly equivalent to a square 40 atoms on a side.

●6.4   Cold work

Dislocations are generated by the deformation of a crystal as well as being present initially, and the more deformation the more dislocations are generated, each of which can move as before. However, as dislocations pile up there is less and less room for new ones, and the capacity for plastic deformation becomes less. It requires more and more work to create and move dislocations, and this is referred to as **strain-hardening**. When no more yield is possible, brittle failure follows. This is a very important consequence of **cold-work**, which tends therefore to convert ductile metals, with much capacity for plastic deformation, into hard, brittle ones. In other words, the toughness of the metal has been reduced. This is readily seen in the repeated bending of wire: after a certain point, *i.e.* a certain amount of cold-work, no more permanent deformation can be accommodated without fracture. Metals that have been affected in this way are usually described as **work-hardened**.

●6.5   Cumulative damage

Notice that once a dislocation has been generated, it cannot be undone by reversing the stress. They may move in the opposite direction, to pile up at the opposite side of the grain, but their numbers can in general only increase. Similarly, once a crack has been formed at a grain boundary, it cannot be closed up by reversing the stress. In other words, the accumulation of dislocations and crack growth – structural damage – are irreversible processes unless some other treatment is applied.

●6.6   Annealing

After deformation, there are many dislocations distributed throughout all grains (Fig. 6.9a).[7] The number of these dislocations can be reduced by heating at a high enough temperature for diffusion to occur. Since it is diffusion that permits the changes, and this is an activated process (there are potential barriers to be crossed), time at a given temperature is required. Such a treatment is called **annealing**. The changes involved in this process are as follows.

On annealing, the initial stage is known as **recovery**, where only the major sites of internal strain will have been affected, at around 0.2 T_m and 0.3 T_m (temperatures in kelvin). In this process, atom movements are the least possible to achieve the relaxation, and whilst they are thermally-activated, diffusion distances are very short. This generates an extensive mosaic structure of crystallites within each grain (Fig. 6.9b) but has no effect on incoherent grain boundaries. More or less continuous with this (there is no definite boundary between the two conditions), the greatly-distorted crystallites start to reorganize in the process of **recrystallization** to create identifiable but small, new grains – grains because their boundaries are not now coherent (Fig. 6.9c). However, a minimum **recrystallization temperature**, T_r, is required for this to occur spontaneously and promptly: typically T_r is between about 0.35 T_m and 0.5 T_m, depending on the metal (Fig. 6.10). This stage coincides with a great softening of the metal as the strain-hardening is entirely undone, a process which is complete when all material has been recrystallized (Fig. 6.9d). The dislocation density is typically reduced to around 10^7 cm^{-2}. The next stage, that of grain growth (Fig. 6.9e), is slower as has already been discussed (§4.1), when a further steady decrease in yield point is observed, according to the Hall-Petch relation.

Although this all sounds fairly definite, it should be recognized that recrystallization temperature is not a fundamental thermodynamic property. Since it is primarily a consequence of diffusion, and therefore thermally-activated, the process is associated with a relaxation time (1§9): the rate depends on temperature. It is also affected by stress, whether 'residual', internal stress because of the cold work done, or externally applied. The effects may be seen in 'bead bath' sterilization of endodontic files or the over-heating of wax knives in a flame.

Fig. 6.9 The effects of annealing on cold-worked metals.

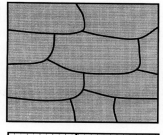

a. After cold-work, all grains have many dislocations.

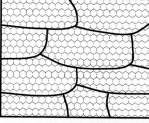

b. On heating, the most strained sites recover to generate a marked mosaic of crystallites.

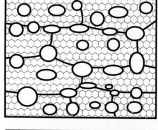

c. Further heating above the recrystallization temperature sees the nucleation of new grains (shown as white ellipses).

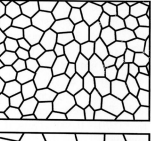

d. When all metal has recrystallized, the grain size is small.

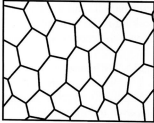

e. Continued heating causes grain growth at the expense of the smaller ones, which disappear.

During recovery both micro- and macroscopic scale residual stresses (introduced by the deformations of grains by cold-work) are reduced – they cannot spontaneously vanish when below the yield point, *i.e.* without diffusion – but the strength and therefore the hardness remain high. During recrystallization, as new grains form, residual stress is eliminated, but strength and hardness decrease rapidly. The mechanical changes can be summarized in a diagram such as Fig. 6.11. The yield point and ductility vary in a complementary manner. What this means in practice is that the mechanical properties of metals are dependent on and sensitive to the history of the piece, both in terms of deformation and of time and temperature.

The meaning of 'cold work' (§6.4) is brought out by a particular pair of examples of procedures: burnishing tin foil as a separator for denture base acrylic, and the swaging of a stainless steel denture base. As discussed in 7§13, pure tin foil can be formed by burnishing to create a well-adapted coating on the gypsum mould. There is no limitation to this in terms of how much plastic deformation is permissible. Similarly, a thin stainless steel sheet can be swaged to fit the model surface by hammering or pressing (28§3). However, in this case, to complete the job, the piece must be annealed to remove the effects of cold-working before continuing so as to avoid cracking. The difference in these two cases is the relative or **reduced temperature**, $T_R = T/T_m$ (temperatures in kelvin). Thus, for tin (m.p. ~232 °C, 505 K) working at room temperature (say, 300 K), $T_R \sim 0.6$, well above its recrystallization temperature. Thus, tin cannot work-harden. Atomic diffusion occurs at a rate high enough that not only is **recovery** spontaneous, but the metal is said to undergo **dynamic recrystallization**. In fact, for such a metal this kind of treatment is called **hot work**. Of course, this does not mean that grain boundary damage does not accumulate (§6.3), but the ductility of the metal is unaffected, allowing the burnishing to proceed with consistent ease.

In marked contrast to this is the situation for stainless steel. The solidus temperature (12§1.2) for Fe-18Cr-8Ni (a common stainless steel; 21§2) is about 1475 °C, so 300 K corresponds to $T_R \sim 0.17$, too cool even for recovery, which occurs at around 200 °C ($T_R > 0.27$). Deformation at room temperature or thereabouts is therefore properly called **cold**

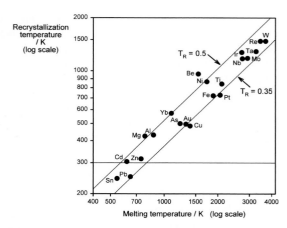

Fig. 6.10 Relationship of recrystallization temperature to melting temperature for a range of metals.

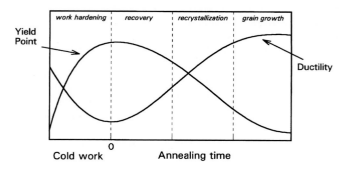

Fig. 6.11 Schematic diagram of the changes occurring on annealing a cold-worked metal. (Not to scale.)

work. To achieve recrystallization, annealing for a few minutes at around 950 - 1050 °C has been recommended ($T_R = 0.70 \sim 0.76$). Thus, if the shape required (*i.e.* the denture base model) is such as to require substantial deformation, and the risk of cracking is present because of work-hardening, such heating before continuing working is the only means of achieving the target (although it should be noted that there are other issues to be taken into account, such as weld-decay, 21§2.5, and oxidation of the surface).

●6.7   Diffusion

Vacancies are very important defects for another reason. As mentioned above, in an otherwise regular and undisturbed array of atoms in a lattice there may be from time to time an atom missing from one of the sites supposed to be occupied. Evidently, if diffusion within a metal crystal relied on the activation energy required to squeeze between close-packed layers, very little would be achieved: the gaps are simply too small. But the presence of vacancies permits the (relatively) ready transport of atoms through solid. Of the order of 1 in 10,000 atom vibrations leads to that atom jumping into a neighbouring vacancy at 'normal' temperatures. This means that exchanges of position may occur at about 10^8 Hz. Since the process depends on an activation energy, higher temperatures give higher rates. Annealing can therefore reduce the number of vacancies by allowing them to diffuse to grain boundaries where they will vanish.

Thus the mechanical behaviour of metals is closely dependent on their grain structure and the defects within the grains. As will be seen later, controlling mechanical properties depends very much on controlling these structural aspects. But before this matter is treated it will be necessary to consider a further layer of complexity in that structure: the polyphase condition which arises in alloys when the continuous solid solution condition is not met (Chap. 12).

●6.8   Strength

As was explained in 1§3.2, there are various definitions of strength which are used according to need. However, unless one is careful, there can be confusion over the effects of work-hardening and annealing on 'strength'. Thus, in Fig. 6.11, it is the yield point which is graphed. If yield point is the strength of interest then it is fair to say that the strength has been increased – the material has become harder (see 1§8).

However, the true breaking strength cannot be changed in such a process. For example, if an annealed specimen is tested, it may show yield at a certain (true) stress, say σ_1 (Fig. 6.12), and then plastic deformation would occur, passing along the line to B, then through C and D, ultimately to failure. If instead the loading were terminated at B, then plastic strain corresponding to A_1-A_2 would have been accumulated so that if the whole process were started again, it would be as if A_2 were the origin for the new stress-strain curve. Yield would now occur at some higher stress σ_2, and the plastic deformation observed would correspond only to the segment BCD, but the ultimate breaking stress would be the same. This is because the same cumulative work-hardening would have been obtained. The same can be said successively for stopping the test at C, and restarting, effectively at A_3, when the yield stress would be higher at σ_3, and then likewise for stopping at D.

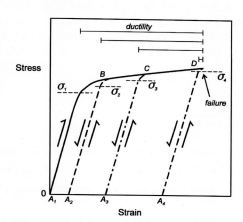

Fig. 6.12 Cold work uses up the capacity for plastic deformation without affecting the ultimate strength (arrows show possible multiple loading paths).

The difference would be seen if the specimen to be tested had already received cold work equivalent to the plastic deformation up to B. There would be no knowledge of the previous history, the yield strength would be higher, and the ductility would be less than the fully-annealed material. We can now interpret ductility in a metal that is *capable* of work-hardening as the plastic deformation capacity remaining, the **ductile range**. Obviously, a specimen giving a curve such as A_4 - D - failure would be called brittle because it showed very little ductility. However, this can be turned around to say that its plastic deformation capacity had been used up by the prior cold work. Thus, yield strength and ductility are affected by cold work, but not the ultimate tensile strength. One view of this behaviour is that brittle failure occurs when crack tip blunting by localized plastic deformation is no longer possible, because no further dislocation generation or movement can occur.

A further point may be made by reference to Fig. 6.12: the modulus of elasticity is unaffected by the history. The slopes of the line segments from A_1, A_2 ... to their respect yield points are all the same. Since the stiffness of a material depends on the averaging of the force per unit displacement (as in Fig. 10§3.5, the tangent slope) across all the atoms of a section, and unless the crystallographic structure is changed in the course of the plastic deformation, this relationship cannot change. Thus, work-hardening and annealing only affect yield strength and ductile range, but not stiffness. However, caution is required in making a 'manual' assessment of the stiffness of a wire, for example. It is easy to be misled by the little effort required to permanently deform a soft wire into thinking (in everyday speech) of it as "not stiff" – this is to confuse yield strength with modulus of elasticity; they are, of course, not the same.

It may have been noticed that grain boundaries are very imperfect structurally, and there would be the expectation therefore that they are weak, acting as crack nuclei in the Griffith sense in the ordinary manner (slightly larger average interatomic distance than in crystals). However, in many metals the stress for plastic deformation (slip) is very much lower than the Griffith critical stress, and yield dominates (Fig. 6.13). Even so, as the yield point is raised by cold work, the value may be raised above that required for intergranular cracking. Thus, at this point brittle fracture will ensue, following the intergranular path of least work. Of course, no further plastic deformation means that no further cold work can be done, and the metal becomes unworkable. Likewise, should

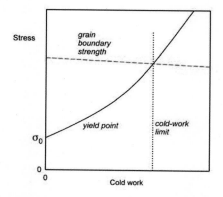

Fig. 6.13 Failure takes the path of least work: the process with the lower initiating stress must dominate.

intergranular cracking become the least-energy outcome, because of the accumulation of dislocations and micro-cracks (Fig. 6.8), that will be the path taken; the effect is the same.

It may therefore be wondered how it is that polycrystalline metals in fact hold together at all. However, we may note that while a grain boundary is disordered (similar to random close-packing, *see* Fig. 4§9.2, but very thin), there is still necessarily atom-to-atom contact. This means that metallic bonding (§1) must be continuous, even if apparently a little more 'dilute'. Clear demonstrations of this are given by the electrical and thermal conductivity of polycrystalline metals, which depend on metallic bonding: these are not appreciably different from single-crystal values. It can be seen that, by extension, the same applies to polyphase metallic alloys.

References

[1]   Smith LO & Cristol SJ. Organic Chemistry. Reinhold, New York, 1966.

[2]   Petch NJ. Ductile fracture of polycrystalline α-iron. Phil Mag, Ser 8, 1(2): 186 - 190, 1956.

[3]   Armstrong R, Codd I, Douthwaite RM & Petch NJ. The plastic deformation of polycrystalline aggregates. Phil Mag 7: 45 - 58, 1962

[4]   Dunstan DJ & Bushby AJ. Grain size dependence of the strength of metals: The Hall–Petch effect does not scale as the inverse square root of grain size. Int J Plasticity 53: 56 - 65, 2014.

[5]   Schiøtz J, Di Tolla FD & Jacobsen KW. Softening of nanocrystalline metals at very small grain sizes. Nature. 391: 561 - 563, 1998.

[6]   Guy AG. Introduction to Materials Science. McGraw-Hill, New York, 1972.

[7]   Humphreys FJ. University of Manchester Institute of Science & Technology, 1994.

Chapter 12   Metals II : Constitution

In dentistry, metal alloys are widely used because pure metallic elements are very limited in terms of strength and stiffness. Alloying allows a much wider range of properties for two reasons: one, the types and detail of crystal structure can be modified over very wide ranges; two, metals containing more than one type of structure can be created. **Constitutional diagrams** *can be read to interpret the structures in alloys as a preparation for the study of dental alloys. In particular, the expected effects of variation in temperature and composition can be determined directly.*

Only under special circumstances do alloys exhibit sharp melting points. Ordinarily, solidification leads to non-equilibrium structures in which the components are **segregated**. *This is nearly always a bad thing in that very weak regions may be present, corrosion is enhanced, and the structure is prone to change with time or heating.*

When complete solid solutions of one metal in another are not possible, more than one phase may be present at equilibrium. In the simplest case, this leads to a **eutectic** *type of system in which a minimum melting temperature is observed. The freezing behaviour, and the types of structure to be found, are characteristic.*

Although metal phases often have very wide composition ranges, metal alloy systems produce **intermetallic compounds** *in which the composition is more narrowly defined. Despite this, even the more complicated constitutional diagrams can all be read in the same way, allowing descriptions of the effects of changing overall composition and the reactions that result from heating and cooling, whether these reactions are solid-liquid or completely solid state.*

The scope of dental alloys is very broad and ranges from amalgam fillings to gold for inlays, from cobalt-chromium for bridges to stainless steel for appliances. Discussion of the behaviour of these and other systems requires reference to phase diagrams. Understanding dental applications is therefore dependent on a grasp of the fundamentals.

Materials Science for Dentistry
https://doi.org/10.1016/B978-0-08-101035-8.50012-2

The properties and behaviour of metals have been described in the previous chapter in terms of atomic- and grain-scale structure. There are however effects on a further level, those directly due to the presence of alloying elements. Apart from the crystal disorder introduced by their presence in solid solution, the melting and freezing of alloys differ from these processes in pure systems, and compositional and structural effects follow. Since alloys are far more varied and useful than pure metals in general, we must consider these effects of alloying in order to understand the properties of the alloys. In so doing, some of the rules for reading phase diagrams will be elucidated. These rules apply in all systems, not just for metals.

§1. Segregation

●1.1 Solution

Gases, it will be recalled, mix freely with one another in all proportions (that is, assuming no chemical reactions). The mixture will be thoroughly homogeneous, each kind of gas molecule being distributed randomly throughout the space available. Liquids, too, may exhibit complete miscibility in all proportions, such as with ethanol and water, but equally may show two layers or phases. An example of this latter type of case would be chloroform and water. In such combinations each liquid is apparently totally insoluble in the other. More usually, each liquid shows some degree of solubility in the other. Each such phase may then be referred to as a solution, which must of course be saturated with respect to the other component if the other solution phase is in contact with it at equilibrium. A phase diagram can be used to show the variation of solubility with temperature, and examples are readily found in physical chemistry textbooks. Similarly, a temperature-composition phase (**T-X**) diagram can portray a system consisting of a solid and a liquid in which it is soluble (sugar and water) for example. The liquid consists of an entirely homogeneous random dispersion of sugar molecules in the water, but the solid (in this case, at least) contains no water.

These ideas concerning solution are relevant also in the solid state: within the limitations of chemical similarity, lack of reaction and radius ratio (11§3.7), atoms of one metal may be mixed with those of another, one metal dissolved in another, with the implications of randomness of lattice site occupied and homogeneity. This is **solid solution**. The complete range between totally insoluble, through limited mutual solubility, up to complete solid solubility in all proportions may be observed.

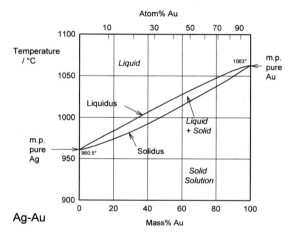

Fig. 1.1 Phase diagram for the system Ag-Au.

●1.2 Continuous solid solution

Two examples of complete solid solubility were introduced in 11§3.7, namely Ag-Au and Ag-Pd. Their phase diagrams are shown in Figs 1.1 and 1.2. Two points are of immediate interest. Firstly, it can be seen that, below the melting point of Ag, there is a complete absence of structural change as the one element is dissolved in the other, that is changing composition from one extreme to the other (depending on the point of view, or from which side of the diagram one works). Indeed, there is only one all-solid phase field, and this only contains one phase. Such a system is known as a **continuous solid solution**. Secondly, there is now a new kind of region to consider, one that lies between (all) Liquid and (all) Solid. This region corresponds to the presence of

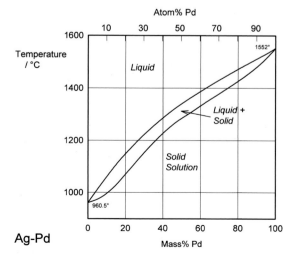

Fig. 1.2 Phase diagram for the system Ag-Pd.

both states simultaneously, at equilibrium. The upper and lower boundaries of this region are known as the **liquidus** and **solidus** respectively (Fig. 1.1) (and are taken by convention themselves to represent the presence of two phases). The key point is that there is a temperature difference between these two lines at any composition except for the pure metals, when the lines coincide. The complications of alloy metallurgy arise from precisely this kind of feature in a phase diagram, and the effects of changing the temperature of a mixture through such a phase field is what we shall now discuss.

> The constitutional diagrams for binary systems exhibit a pair of features known as **point reactions**, one at each end. At these two points the reaction on heating is of the kind
>
> $$\alpha \rightarrow Liq$$
>
> that is, the solidus and liquidus coincide at this point. However, since both points correspond to the melting of the respective pure component, they are known more specifically as **component reactions**.

Note the characteristic pattern of phase field boundaries at the left- and right-hand boundaries, the terminals. There is a K-shape, a sideways 'V', < or >, formed by the liquidus and solidus meeting at the point where they touch the vertical boundary. In addition, note the occupancy of or number of phases in the phase fields around this point: 1,2,1 – liquid, solid + liquid, solid. This is a universal pattern.

●1.3 Partition

To understand the implications of this kind of pattern, and one of the ways phase diagrams may be interpreted, we proceed with another thought experiment. So, starting at low temperature somewhere in the all-solid phase field, and raising the temperature, no structural change is to be observed so long as we remain in that solid solution region. But at some point the solidus will be reached, and melting will commence. We now have two phases in contact; both are solutions, but in different states. These states are distinguished by the presence or absence of long range order. In other words, they are solid and liquid. However, the chemical potentials of each of the components will have changed in going from a solid to a liquid of the same composition, and changed (in general) by different amounts. However, for a system to be in equilibrium it is required that the chemical potential of any element is the same in all phases which are then present (8§3).

Thus, for a given component in the just-melting system, the chemical potential will be different in the two kinds of environment. It will then adjust its concentration in the two phases, by diffusion one way or the other across the boundary, until the potential is equilibrated (and there is no further driving force for change). In other words, that component will **partition** itself between the two phases until that thermodynamic requirement is met (*cf.* the effect of temperature discussed in 8§3.2). This must occur simultaneously for all components present. Consequently, the compositions of the solid and liquid in contact at any temperature at equilibrium will differ. Clearly, if this is the case, one phase must become richer in one component and poorer in the other, and *vice versa*, since the overall composition is fixed. As with all processes this kinetic activity depends on the availability of the activation energy for the process, and the time for it to occur. We shall for the time being assume perfect equilibrium at all stages.

A stylized diagram of the continuous solid solution type for just such a system is shown in Fig. 1.3. If we consider a typical isothermal tie-line XZ, its ends are, by definition, at the solidus and liquidus. Consequently, at an overall composition corresponding to X, and at that tie-line's temperature, there would be solid having the corresponding composition (in this example, 20% B) in equilibrium with an infinitesimal amount of liquid (lever rule, 8§3.10). Similarly, at an overall composition corresponding to Z, liquid with the corresponding composition (*i.e.* 60% B) is in equilibrium with an infinitesimal amount of solid of composition corresponding to X. To reiterate the point: since it is an isothermal tie-line, and both points represent solid-liquid equilibrium, and are in the same phase field, both solid and liquid must have the same individual compositions at the two points. In other words, at both X and Z, the solid has the composition X and the liquid has the composition Z. This must also be true for any other overall composition point in between, such as at Y. This

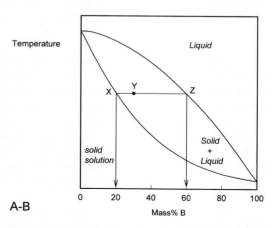

Fig. 1.3 Compositions of the conjugate phases for a continuous solid solution system in the two-phase field.

principle of the interpretation of isothermal tie-lines within a two-phase field is perfectly general for two-component systems. Indeed, in a sense, this is the embodiment of the thermodynamic demands of equilibrium.

●1.4   Heating

We now proceed to raise very slowly the temperature of the system with the overall composition corresponding to Y in Fig. 1.3, *i.e.* about 31% B (Fig. 1.4). When the solidus is reached, at T_1, the first infinitesimal amount of liquid, of composition corresponding to L (*i.e.* richer in B than is solid Y) has appeared. The solid composition is hardly affected because the amount of liquid is so small. It is still just about 100% solid that is present.

Heating a bit further, to T_2, and allowing the system to come to equilibrium, we now have a solid corresponding to S_2 (approximately 21% B) coexisting stably with a liquid of composition corresponding to point L_2 (approximately 61% B) (Fig. 1.5). In other words, S_2 has a larger proportion of A than the overall composition Y; while L_2 is richer in B than Y, it is in fact richer in A than L_1, the first liquid to have formed. The amounts of liquid and solid are in the mass proportions $\mu_L : \mu_S$, according to the corresponding line segment lengths of the tie-line. What has happened is that in going to temperature T_2 more atoms of B have left the solid surface than of A, leaving that surface richer in A. These A atoms then have diffused (slowly) into the solid so that it is of uniform composition throughout. The system continues this slow dynamic adjustment until the equilibrium situation is attained.

This process continues on further heating to a temperature such as T_3 (Fig. 1.6), making the solid (S_3) richer and richer in A while the liquid (L_3) does the same (but not necessarily at the same rate, it depends on the slopes of the liquidus and solidus at any given temperature). Of course, as melting is occurring, there is proportionately more liquid and the ratio $\mu_L : \mu_S$ is now larger.

These changes occur progressively until the liquidus is reached (Fig. 1.7). At this point, T_4, only an infinitesimal amount of solid remains, with composition corresponding to S_4, while the liquid L_4 is now just the same as the original overall composition Y. This is reasonable when (very nearly) all original solid has melted. Beyond that temperature there is no further phase or composition change except for the disappearance of that theoretical tiny amount of solid.

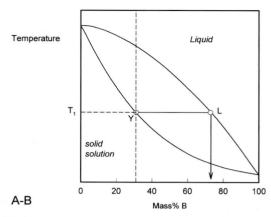

Fig. 1.4   The first melting liquid L is B-rich.

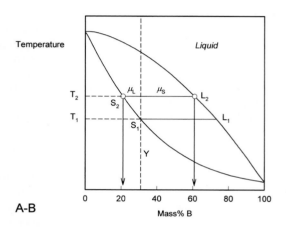

Fig. 1.5   Subsequent melting leaves the solid richer in A than the initial composition.

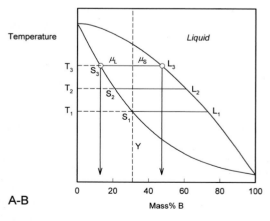

Fig. 1.6   The drift in composition of both liquid and equilibrium solid continues with melting.

This kind of analysis is on the assumption that complete equilibrium can be maintained at every stage, which means allowing time for diffusion in the solid, such diffusion being the rate-limiting step. The overall composition is of course guaranteed to be constant by mass balance. For such a system it is only at pure A and pure B that the liquid can have the same composition as the solid at equilibrium.

●1.5 Cooling

The converse process can be traced just as simply by taking the figures in the reverse order. Thus, taking the overall composition Y again, and starting in the melt, the first solid to appear has the composition corresponding to S_4. As cooling progresses the solid becomes richer in B: S_4 - S_3 - S_2 - S_1 (which last is identical to the overall composition), while the liquid also does: L_4 - L_3 - L_2 - L_1. The proportion of liquid to solid, μ_L : μ_S gradually declining to zero at the solidus temperature of Y.

●1.6 Segregation

Now the accuracy of all of this depends ultimately on the equilibrium of *all* of the solid with the liquid at every step of the sequence, whether in heating or cooling. But because during cooling successively-formed solid will be in layers on earlier already-formed grains, so equilibration tends to be inhibited because the diffusion rate in the solid state is less than in the liquid state by a factor of about 10^{-4}. So the assumption made in §1.3 is now abandoned as unrealistic. In effect, those atoms which do not belong in the solid at the current temperature are trapped because the rate of advancement of the solidification front is faster than they can diffuse. Consequently, in practice the composition of the solid changes steadily from the nucleus of each crystal out to the grain boundary, becoming steadily richer in the lower melting component. If this is occurring the *effective* overall composition of the solidification system will drift towards pure B, as A is made unavailable by being locked-up or sequestered inside the earlier-freezing solid. In this way the last solid to form may have a composition closely approaching pure B.

The process can be imagined as occurring in a stepwise cooling sequence. In Fig. 1.8, the first solid to form on cooling from Y to T_1 has the composition corresponding to S_1. If we now jump to T_2, the solid which forms corresponds to S_2, which leaves the liquid with the composition of L_2. Since equilibration is not occurring it is now that liquid L_2 which on cooling to T_3 starts to freeze, so that the next solid to form corresponds to S_3, leaving liquid L_3. It is then this liquid that is cooled to T_4, producing solid S_4, leaving L_4 – and so on. It can be seen that this ultimately leads to the final tiny amount of liquid to freeze being essentially pure B. This is the case no matter how small the cooling steps are. In other words, even continuous smooth cooling results in a non-equilibrium, layered or gradient structure (Fig. 1.9).

This is the process of **segregation**: the variation in composition from point to point in the solid when homogeneity was expected. The structure is now said to be **cored**. It usually requires extensive **annealing** at a temperature high enough for diffusion to occur readily (and yet below the solidus or any other transition) for an equilibrium structure to be formed eventually; this is **homogenization**. It is a seeming paradox that, up to a point, the *slower* the cooling rate the further from

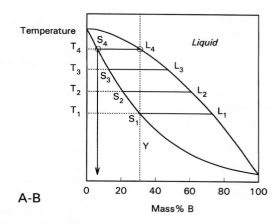

Fig. 1.7   The drift continues until the liquid reaches the overall composition.

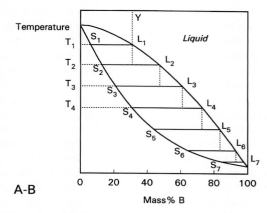

Fig. 1.8 On cooling under non-equilibrium conditions, the liquid composition becomes progressively richer in the low-melting component, and the solid formed at any moment similarly changes in composition.

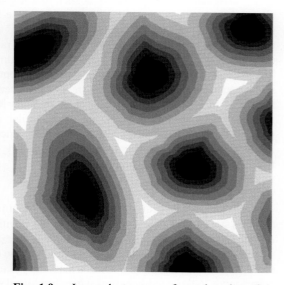

Fig. 1.9   Layered structure of metal grains after stepwise non-equilibrium cooling of a continuous solid solution alloy. Black: core of high-melting component-rich alloy; white: final alloy close to pure low-melting component.

equilibrium the phase description and phase compositions may be. But the faster is the rate of cooling the less material will actually freeze out at a particular temperature; there is a kinetic limitation to local equilibration. This is in effect supercooling, but instead of the temperature rising of necessity back up to the melting point as with a pure metal and establishing equilibrium, the composition of the then solidifying metal will have the composition corresponding to that indicated by the tie-line at *that* temperature (but only because the rate of loss of heat is so great). The last metal to solidify will have a composition closer to the overall composition, and the cast ingot would require relatively little annealing to redistribute the components (by diffusion) to achieve an equilibrium structure.

Fast cooling like this is referred to as **quenching**, the intention being to try to prevent the segregation referred to above, or at least to reduce it markedly, by not permitting time for the required diffusion to occur. Liquid is solidified at close to its overall composition. In addition, the effective supercooling produces a large number of crystallization nuclei, and so the average crystal (grain) size will be small. This makes the distances over which diffusion will be required for subsequent equilibration correspondingly small, and the required annealing time short, even if the segregation is marked.

It is worth stressing that the outcome of the solidification of a melt is dependent on two equilibration systems: liquid with fresh solid, and that fresh solid with underlying material. In effect, the diffusion rate of atoms in the melt is so high that under most ordinary circumstances it can be considered as nearly homogeneous, no matter what the composition is of the solid currently forming. Thus, that solid can be treated as if formed in equilibrium with the current liquid composition, as dictated by the tie-line: exchange of atoms between liquid and solid surface is rapid. Exchange between any pair of layers in the solid, including the outermost, depends on solid state diffusion, via vacancies, which may be ten thousand times slower. However, this view does suggest that if the cooling rate is fast enough, equilibration at the liquid interface cannot occur and the composition of the solid that forms is closer to that of the liquid, which therefore does not show as much drift in composition over the whole process. However, extremely high cooling rates simply cannot be attained if the thickness of the metal is too great (more than a few tenths of a millimetre) because thermal conductivity then becomes the limiting factor. Nevertheless, such cooling rates may be achieved (or approached, at least) for the layer of metal in contact with a cool wall when the melt is poured into a mould, *i.e.* in a casting. The constitution and grain size of the surface may then differ from the bulk, which is obliged to cool rather more slowly.

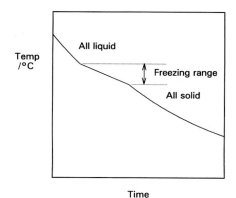

Fig. 1.10 Cooling curve for a system giving a single phase solid solution at low temperature.

The nature of segregation is such that impurities in the alloy, elements that were not intended to be there but that are present in small quantities, and that have low solid solubility in the main phase(s), will tend to remain in the melt until the very end of solidification as the partitioning is given time to occur. This means that these impurities tend to be concentrated at the grain boundaries, with various potential consequences. They may lead to weaker-than-expected grain boundaries and thus lower strength overall by allowing intergranular cracking to proceed more readily. The composition difference may also allow intergranular corrosion to occur more readily, because of the dissimilar phases now present (13§4.10).

●**1.7   Cooling curves**
The cooling curve (11§2.7) for a continuous solid solution system should appear as in Fig. 1.10, *i.e.* with a **freezing** or **solidification range**, because the effective freezing point is varying continuously, just as the composition of the equilibrium solid varies continuously. Note that the freezing range corresponds to those temperatures where liquid and solid are co-existing. So for a series of alloys in an A-B, **binary**, system (Fig. 1.3), a series of cooling curves (Fig. 1.11) will be obtained. From data such as these the phase diagram itself (Fig. 1.3) would be constructed, plotting the temperatures where discontinuous

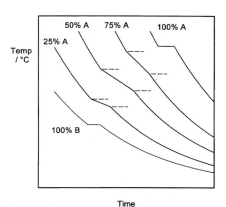

Fig. 1.11 A set of cooling curves for a variety of compositions in a binary complete solid solution system. (*cf.* Fig. 1.3)

changes in the slope of the cooling curve are found against composition, although usually this approach must be supplemented by metallography – direct observation of grain structure. The difficulty is, of course, that the lower limit of the freezing range can be very uncertain because of the difficulties arising from segregation. However, as mentioned above, for moderate rates of cooling (*i.e.* not too slow) a better approach to the expected condition can be obtained and cooling curves are often a practical means of obtaining such data in a preliminary way.

One consequence of segregation that is worth noting relates to the experimentally-observed freezing range. While the liquidus will be determined quite accurately, the solidus temperature will not correspond to the end of the freezing range. The last liquid is necessarily of lower freezing point than the solidus for the equilibrium constitution at the given overall composition. The slower the cooling, the worse this might be. Thus, depending on the alloy system, such values from a cooling curve experiment need to be interpreted carefully.

§2. Metallography

One of the approaches which may be taken in determining the actual constitution of a sample of a metal is simply to look at it: **metallographic examination**. In practice this requires an optical microscope, but this is really only because grain sizes ordinarily are in the range of one to a few hundred micrometres.

The technique in principle is very straightforward. A specimen of the metal, which may be anything from a very thin foil, through wires, to large objects such as a piece of a casting, is first usually embedded in a block of resin (such as cold-cure acrylic) to facilitate handling. One face is then ground away flat to expose a section of the specimen (Fig. 2.1), using a very coarse grinder or perhaps a saw followed by abrasive paper of decreasing coarseness. Then, using a succession of abrasives such as alumina and diamond dust, the cut surface is polished so that no scratches remain visible (< ~0.2 μm), taking care to keep it flat. Sometimes this is sufficient because the phases present have sufficiently different colours

Fig. 2.1 Some titanium bone screws embedded in resin and ground down to their midline ready for polishing.

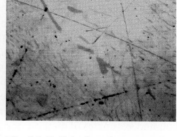

Fig. 2.2 Polished section of Ag-Cu-Sn alloy showing slightly darker colour of one phase (ε-Cu$_3$Sn).

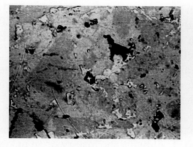

Fig. 2.3 Polished section of dental amalgam, slightly out of focus, to show the effect of differential removal of the γ$_2$ Hg-Sn phase to give relief to a surface.

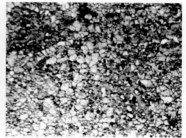

Fig. 2.4 Polished section of dental amalgam etched to show the grains of all phases present.

that they can be distinguished at this stage (Fig. 2.2). Or, if the hardnesses of the phases differ sufficiently, one or more may be removed more quickly than the surroundings, giving relief effects that show up clearly under the microscope (which must, of course, use reflected light rather than transmitted as with biological sections) (Fig. 2.3).

Normally, however, these effects are too weak to give clear images, and etching is used. Metals vary in the rate in which they dissolve in acids, while some develop oxide coats under oxidizing conditions, some dissolve in alkali, and some are extremely resistant. Sometimes a treatment such as this produces a variety of colours, sometimes it simply leaves holes, sometimes it makes the surface rough and therefore dark because it scatters light. Taking advantage of these variations in reactivity, treatment with carefully chosen solutions, at particular temperatures, or carefully timed, can yield striking pictures of the distribution of all of the phases present (Fig. 2.4). These can have different colours or shades (from the deposition of reaction products – tarnished), and clearly defined boundaries, even between grains of the same phase because crystal lattice orientation differs (and therefore reactivity) (Fig. 2.5). Notice how scratches are differentially-enhanced according to

Fig. 2.5 Polished and etched section of β-Ag-Sn alloy showing grain orientation effects on etching rate. Scratches etch and show very clearly on low-rate surfaces because they expose other crystal planes.

Fig. 2.6 Dental amalgam alloy particle polished and etched to show grain boundaries in the γ-Ag-Sn phase.

the crystal planes they expose. Finally, grain boundaries, because they are disordered and therefore of higher energy than the adjacent crystalline region, tend to react more readily, often leaving a minute groove between grains, which feature is then clearly visible under the microscope (Fig. 2.6).

Of course, none of this on its own identifies a phase or its composition. Series of alloys must be prepared, covering the possible range of compositions at small intervals. These may then be held at a series of known temperatures for long periods to allow the structure to equilibrate, followed by rapid quenching to room temperature to preserve the structure. Metallographic examination then permits the phases to be counted, estimates of their proportions made, and characteristics such as colour or etching behaviour to be noted. Correlation of this data with cooling curve information then permits the building of the constitutional diagram.

In addition, X-ray diffraction is used to determine crystal structure. Unknowns can frequently be identified simply by referring to an index of diffraction data (Fig. 2.7) noting the positions and relative intensities of reflection – pattern matching.

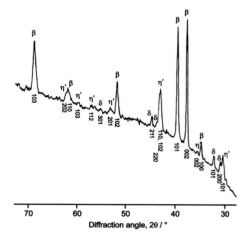

Fig. 2.7 Variation of reflection intensity with reflection angle for X-rays: a diffraction pattern from an amalgam alloy ingot surface showing the phase and crystal plane (Miller indices) assignments.

Even direct analysis of individual grains in the polished section can be made with techniques such as the electron microprobe, on a scanning electron microscope, which uses characteristic X-rays (see 26§1) to identify and quantitate the elements present with fair accuracy.

If then the variation of thermal history (time and temperature) is added, the course and speed of equilibration processes can be mapped, allowing the history of an unknown specimen to be deduced.

Many systems have now been mapped, some in more detail than others, but many remain to be studied. The number of binary alloys possible is rather limited, but ternaries (three components) are clearly much more numerous. Assume 70 alloying elements, and we have nearly 5,000 binaries to consider; there are then well in excess of 300,000 ternary possibilities, each of which may be considered at a minimum of 50 or so compositions. Modern alloys may contain six or a dozen components, and contaminants are frequently of interest at this level of complexity. There is yet plenty of scope for metallographic study.

§3. Eutectics

When either the radius ratio of the two atoms is outside of the stable range (11§3.7) or chemical considerations prevent it, solid solubility may not be observed over the whole range of composition. Thus, if substitution of, say, a larger atom into a pure metal crystal lattice is made one atom at a time, randomly throughout the crystal, there may come a point where the structure is no longer stable, the accumulating distortion being too much for the lattice to contain. One way in which stabilization may then occur is through the separation of a second solid solution phase. That is to say that the strain energy is too great to be offset by the structure energy, and an alternative arrangement of the atoms in two separate structures gives an overall lower energy. There is therefore a third condition for the stability of a solid solution: the *lack* of an alternative pair of structures (*i.e.* for the two phases) that would be more stable. This is the thermodynamic imperative: if a path exists, it will be taken (kinetics permitting).

●3.1 Ag-Cu

A system of dental interest that demonstrates this is that of Ag-Cu (Fig. 3.1). Here the radius ratio Ag : Cu is about $143/128 = 1·12$ and so represents a borderline case for the stability from that point of view. The constitutional diagram shows only limited solubility of Cu in Ag (so-called α-phase; see §4.1) and of Ag in Cu (β-phase), even at high temperatures. But what is most significant is the lowering of the melting point of both metals by the presence of the other, as shown by the solidus near to both pure metal boundaries, and the corresponding lowering of the liquidus temperature in each case. The idea of impurities lowering the melting point of substances may be familiar; in organic chemistry it is used as a simple check of the purity of a compound.

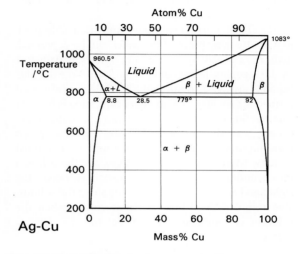

Fig. 3.1 Phase diagram for the system Ag-Cu.

The liquidus in the vicinity of a pure metal boundary may be taken as representing the plot of the limit of solubility of the solid solution phase in the liquid phase as a function of temperature. Since the solubility of β-phase (Ag in Cu) falls as more Ag is added to it, and the solubility of α-phase (Cu in Ag) also falls as more Cu is added to that, there is obviously a point where these two separate plots cross. We cannot go to any lower temperature on either plot without lying beneath the solubility limit of the other plot. Below the temperature of this crossing point the solubility of *both* α and β phases is exceeded, and both must crystallize out simultaneously. There can be no gap between the solidus and the liquidus at the precise overall composition corresponding to the cross-over point, and the cooling curve of a melt of that exact composition will show a simple arrest, indistinguishable from that of a pure metal (Fig. 3.2). Since at this **eutectic point** ('easily melting') there are both α and β solid phases in equilibrium with liquid, application of the phase rule (8§3.6) gives:

$$F = 2 - 3 + 1 = 0$$

In other words, this is an invariant point. We can write the equilibrium there as follows:

$$Liq \Leftrightarrow α + β \qquad (3.1)$$

or conversely we can write the **eutectic reaction** on heating:

$$α + β \Rightarrow Liq \qquad (3.2)$$

or cooling

$$Liq \Rightarrow α + β \qquad (3.3)$$

In constitutional diagrams, the eutectic-type equilibrium system always has the appearance shown in Fig. 3.3, the key feature being the 'V' resting on a horizontal

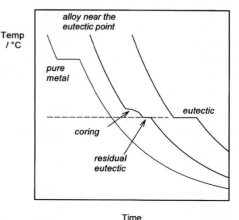

Fig. 3.2 Cooling curves from a binary system showing a eutectic point.

phase-field boundary.[1] This pattern can be universally recognized and understood, even in more complicated diagrams, because the behaviour of the system in its vicinity is always the same.

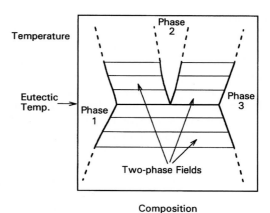

Fig. 3.3 The general layout of a phase diagram in the region of a eutectic-like point.

●3.2 H₂O - KI

There are some very important issues in the subject of eutectics and it may be of value to examine another system to illustrate these. Thus, we may consider the simple system of water - potassium iodide. This has the advantage that the two components do not exhibit any solid solubility in each other, thus simplifying the process and the discussion.

Thus, starting with pure water and adding KI to it the freezing point of the water will be depressed in the normal way by the presence of the salt, depicted by the **freezing-point curve** (Fig. 3.4). Note that the liquid is always a solution of KI, while the only solid formed is pure ice. The immediate question is how far can that line be extended downward and to the right? Clearly, as the temperature falls for any given concentration of KI, so the amount of ice must increase (draw a tie-line to see this), and the solution must simultaneously become more concentrated – the right hand end of the tie-line moves along the solidus. One might deduce that the limit comes when the solution is saturated with respect to KI. This is correct.

We may consider the system from the other side (Fig. 3.5). We can pick a temperature, such as 20 °C, and determine how much KI dissolves at equilibrium, repeating this for a variety of temperatures to draw the **solubility curve**, to the right of which is a two-phase field: KI + Liquid. This can of course also be done by choosing a concentration and cooling the hot solution until crystals first form (or use the cooling curve method). The question now is how far can that line be extended down and to the left? One might deduce here that the limit comes when the water freezes. This, too, is correct.

It will now be seen that the two limits must in fact be one and the same point. Logically, if the first approach leads us to identify a temperature and composition corresponding to the solidification reaction:

$$\text{Liq} \;\Rightarrow\; \text{Ice} + \text{KI} \qquad (3.4)$$

and the second line of reasoning similarly brings us to a point where

$$\text{Liq} \;\Rightarrow\; \text{KI} + \text{Ice} \qquad (3.5)$$

occurs, they must indeed be identical. We therefore combine the two diagrams, Figs 3.4, 3.5, to obtain the full picture (Fig. 3.6). The last tie-line that we can draw in the Ice + Liquid field, at the temperature

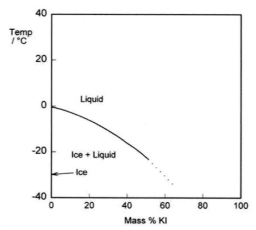

Fig. 3.4 Portion of the H₂O - KI system showing the freezing point depression of the water. Note that the pure ice phase field is infinitesimally narrow along the left hand edge of the diagram below 0 °C since KI does not form a solid hydrate.

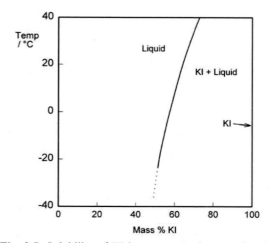

Fig. 3.5 Solubility of KI in water. Again, note that the pure KI phase field is infinitesimally narrow on the right hand border because water does not dissolve in KI.

where solid KI is about to appear because the solution is saturated, must be continuous with the last tie-line in the KI + Liquid field when the water is about to freeze. This line is at the eutectic temperature, and there is only a single, two-phase field, Ice + KI, beneath it.

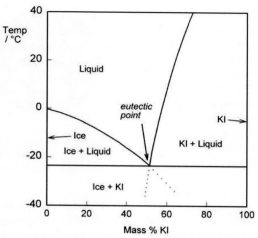

Fig. 3.6 Phase diagram for the binary system H$_2$O - KI showing the eutectic point.

It can now be seen that the dotted line extensions of the two solubility lines (ice in KI solution, so to speak, and KI in water) are inaccessible. This is because cooling a mixture either side of the eutectic composition must reach the liquidus corresponding to the other solid first. (Unless, of course, some way of inhibiting the nucleation of that solid could be found, such as using an ultra-clean and containerless apparatus. However, it would not then be an equilibrium system.)

Returning now to the Ag-Cu system the parallel is seen clearly. Nevertheless, there is an important difference: both α and β phases are not pure substances, but are themselves solid solutions whose equilibrium compositions vary somewhat with temperature. This is, however, only a slight complication. It does not affect the argument just given, or the consequence of the simultaneous solidification of the two phases in the same way that Ice and KI solidified simultaneously at that eutectic's temperature.

We may make two important points arising from this development. In Fig. 3.6, in the field labelled KI + Liquid, the liquid is necessarily saturated with KI – if it were not, more would dissolve until equilibrium was reached. Likewise, the Ice + Liquid field is effectively "saturated" with respect to the ice – no more can melt. Hence, the liquid of the 'Liquid' field is unsaturated everywhere. This applies in all systems: in fields lying between liquidus and solidus the liquid is saturated, while above the liquidus it is unsaturated. Indeed, this is a meaning of the liquidus, liquid solution solubility, as the solvus represents solid solubility (see below, §3.5). The Phase Rule can therefore be interpreted from this perspective as well. At the eutectic point, the liquid is saturated with respect to both solutes simultaneously, so there is no degree of freedom remaining; on the liquidus only one saturation is specified; in the liquid-only field there is none, and so here there are two degrees of freedom (for a binary system).

●3.3   Microstructure
Solid which has crystallized from a eutectic composition melt frequently has a characteristic structure consisting of many very closely spaced, very thin plates ('lamellae') of alternating phase, or similarly closely spaced thin rods of the one phase in a matrix of the other, perhaps a mixture of the two forms. This arrangement arises from the physical fact that atoms need only diffuse very short distances in the liquid to achieve the right local concentrations for the structure of each phase, that is, the minimum path to achieve **phase-separation**. However, in distinction to the two-liquid case discussed earlier (8§4.1), since it is two solids that are formed, surface tension-driven formation of more or less spherical droplets of one phase in the other does not occur, and the initial 'minimum path' form is preserved. It must be stressed that a eutectic solid structure is *two* phase, and distinct boundaries are observable between these phases (see the definition of a phase, 8§2.3), although the grain size may be very small indeed, perhaps even undetectable optically.

If we consider the cooling of a melt of, say, Ag-70% Cu (by mass),[1] we find from the phase diagram (Fig. 3.1) that the first solid to appear does so at about 950 °C. This solid has the composition indicated by the isothermal tie-line at the boundary of the β-phase, and is about Ag-93% Cu. Continued cooling leads to the liquid becoming richer in Ag as more Cu-rich solid separates, even though the Ag-content of that β-phase solid then in equilibrium also rises slowly. The last tie-line possible in the β + Liquid phase field corresponds at the Cu-rich end to β-phase with 92% Cu, and at the Ag-rich end to the eutectic point itself, which means that this is the composition of the liquid at that moment. However, it can be seen that this tie-line now extends through the eutectic point and, quite suddenly, extends across not to the liquid field but to that of the α-solid solution

---
[1]   This convention for specifying the composition of an alloy arises directly from the basis of the equations of 8§3.4: one proportion is redundant and is therefore not stated, its value is implicit.

(8·8% Cu); this identifies which other phase will appear when the eutectic-composition liquid remaining does freeze. So, when that **eutectic temperature** has been reached, the next solid to form is a mixture of α and β, in fact the eutectic α + β intimate mixture.

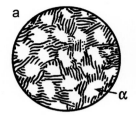 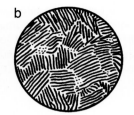 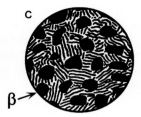

Fig. 3.7 Sketches of the metallographic appearance of structures possible in a binary eutectic system, as if etched. a: α-core in eutectic matrix (*e.g.* Ag-15% Cu); b: eutectic without coring; c: β-core in eutectic matrix (*e.g.* Ag- 60% Cu).

The cooling curve for such a condition will have the appearance shown in Fig. 3.2 (centre); that is, a combination of the freezing range expected for a solid solution system plus the eutectic plateau. The structure of a solid formed in this way is described as having **primary** crystals of *one* of the solid solution phases (*i.e.* the first to form) in a **matrix** of eutectic. This is also known as a **cored** structure (Fig. 3.7).

●3.4   Mechanical properties

Mechanically, eutectics tend to be hard, strong and tough, which arises from their special two-phase structure. The fact that the density of grain boundary (area per unit volume) is very high and there are only short distances to travel between adjacent boundaries, means that dislocations pile up readily, but also that there is room for only very few. Thus, plastic deformation is very limited and high stresses are needed. Eutectics tend therefore to work-harden very rapidly. Secondly, when cracks do form they cannot travel very far

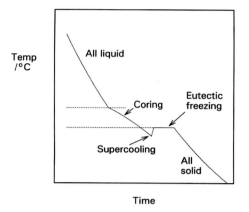

Fig. 3.8 Cooling curve for a eutectic binary system showing supercooling, the composition being near the eutectic point.

without meeting a grain boundary, which must usually force a deviation in its path so that the total surface area created is larger than expected, absorbing more energy, which means such alloys are tougher.

It is sometimes the case that proper supercooling of eutectic material can occur (Fig. 3.8) even in the presence of a primary core because of the particular nucleation requirements, but once started crystallization leads to the evolution of heat, which must bring the temperature back up to that of this invariant point. The remainder of the freezing and cooling process is then straightforward.

●3.5   Solid solubility variation

From an examination of the left- and right-hand compositional boundaries of the α + β phase field of Fig. 3.1, it is evident that the solid solubility of both Ag in Cu, and Cu in Ag, varies with temperature. Greater solubility is observed in each case at higher temperatures, with a maximum corresponding to the eutectic freezing point. These boundaries are referred to as **solvus** lines or **solvi** (solid solubility limits). Their importance can be judged from an example: when an alloy with, say, 5 mass% Cu at 800 °C (which at equilibrium consists only of α-phase) is cooled below about 660 °C, the solubility limit of Cu in Ag is exceeded. So, just as with a salt from an aqueous solution, the excess solute is expected to precipitate from the solid solution. In the present example this would be β-phase with about 96% Cu. However, unless the rate of cooling is very slow there will be no time for this diffusion-controlled solid state reaction

$$\alpha \;\Rightarrow\; \alpha + \beta \tag{3.6}$$

and the high-temperature structure will be preserved. This is again a kinetic limitation of a thermodynamic requirement. Notice that the composition of the α-phase changes slightly in the course of this, but since the structure is unchanged it is correct to refer to it as the same phase. Of course, if such a high-temperature structure were quenched (*i.e.* cooled rapidly), subsequent reheating (but remaining in the two-phase field) could make the diffusion activation energy available and cause the precipitation to occur. There are two possibilities here. If we desire the retention of the high-temperature structure (say, because the mechanical properties are useful), then reheating – deliberately or accidentally – will spoil the material. Alternatively, quenching and reheating may allow fine control of the composition of the phases and the extent of reaction, and therefore fine control of the resulting properties. Again, though, further heating must spoil that carefully obtained outcome. There are a number of instances in dentistry where this thinking is applicable.

We can note that the curvature of the solvi in diagrams such as Fig. 3.1 is the expression of the temperature-effect remark in Table 11§3.2. Higher temperature allows more atoms of disparate sizes to be accommodated, as the thermal expansion lessens the constraint and thus the distortion of the lattice.

●3.6 Heat treatment

Similarly, the compositions of the phases in a eutectic mixture as first formed are unstable with regard to both α and β phases at lower temperatures. Continuous adjustment is necessary during cooling, transferring atoms from one phase to the other, to correct the compositions in order to maintain strict equilibrium at all times. This is inefficient and difficult to manage, but **annealing** will achieve this. The higher the temperature at which this is done the faster it will be completed. Annealing to allow the approach to equilibrium to occur is described as **ageing**, as if the alloy were cooled much more slowly.

But note that this can only ever achieve (or approach) equilibrium for the temperature at which the annealing was done. If the composition boundaries of the phase field of interest are not perfectly vertical, *e.g.* the solvi are similar to those in the Ag-Cu system, subsequent cooling may be equivalent to quenching, preserving the higher temperature structure. This then must imply a non-equilibrium structure, which in turn implies locked-in strain and a higher energy state. This in turn may affect surface properties and the chemistry of the alloy, *e.g.* in its corrosion behaviour. From this is will be seen that given the essentially unavoidable conflicts between thermodynamic demands and kinetic limitations, some form of compromise is inevitable.

Fig. 3.9 summarizes the transformations which may be obtained on ageing in a generalized **eutectoid** (*i.e.* eutectic-like) system.[2] The principles and behaviour are identical whether β in this diagram is a liquid or a solid phase – the same 'V on a line' pattern is present. In addition, the equivalent but reversed reactions can be identified for equilibration at a *higher* second temperature (*i.e.* reverse the arrows). Thus structures which may have been created on cooling but which were not desired can be removed. Of course, quenching is then necessary to prevent this being undone again on cooling. The commonest form of this is **solution heat-treatment**, creating a single-phase solid solution, that is, dissolving all precipitates.

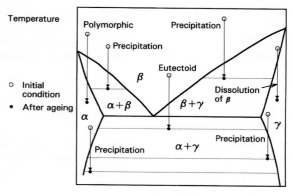

Fig. 3.9 The various heat treatment possibilities for change in a binary, generalized eutectoid system. The tie-lines indicate the compositions of the conjugate phases present after equilibration at the lower temperature.

We may also note a further pattern. The tie-line at the eutectic temperature necessarily extends between the single phase fields corresponding to the two conjugate phases. At these points there is another characteristic pattern, ⊢ or ⊣, perhaps better described as a Y, with the 'leg' horizontal, although the two arms can lie at a variety of angles to the leg – which being an isothermal tie-line must be exactly horizontal. Above and below the leg are two-phase fields; between the arms is a single phase field. This, too, is a universal pattern.

It should be borne in mind that these transformations may be kinetically-limited. Thus, during the eutectoid reaction:

$$\beta \rightarrow \alpha + \gamma \qquad (3.7)$$

there will be a period in which all three phases are present, a clearly non-equilibrium condition unless at exactly the eutectoid (or eutectic) temperature. Diffusive processes are time-dependent, not instantaneous. A heat-treatment in either direction can be continued to completion, or interrupted to leave intermediate conditions, according to need. We can also note that this kind of reaction is a phase separation in the same sense as 8§4.1

except that here we have crystalline solid solutions.

It will be noticed that a **precipitation** reaction also involves entry into a two-phase field, *e.g.*

$$\beta \Rightarrow \beta + \gamma \quad (3.8)$$

(as before, reaction 3.6). This means that the composition of the 'parent' phase must change to be in equilibrium with the precipitate – there is a tie-line whose ends are necessarily not at the overall composition (Fig. 3.9). Thus, the time taken for

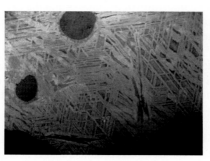

Fig. 3.10 Polished and etched section of a meteorite showing the Widmanstätten pattern resulting from a phase separation, in this case very slowly so that the precipitate grain size is large (this image has a field width of about 250 mm).

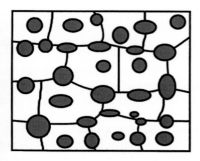

Fig. 3.11 A polymorphic transformation involves nucleation and grain growth of a second phase in the matrix of the parent phase, which is by definition not thermodynamically-stable at that temperature.

equilibration depends on large-scale and long-distance diffusion. Such reactions often give rise to a characteristic metallographic appearance known as the Widmanstätten[2] pattern (Fig. 3.10).

In contrast, a **polymorphic** transformation involves only a change in crystal lattice: the same atoms are rearranged locally only, large diffusion paths are not involved. Recrystallization thus occurs, nucleated randomly, each new grain growing as the parent matrix is consumed (Fig. 3.10). This is similar to the process described at Fig. 11§6.9 c, except that there it was crystals of the same phase that were growing.

We are therefore now able to predict and describe the structural and compositional changes expected for the heat treatment of any binary alloy, given the constitutional diagram.

§4. Phase Diagram Conventions and Layout

Constitutional diagrams are tools to facilitate the handling of phase data, maps of the layout of systems. Certain conventions are used and must be appreciated to read these maps easily.

●4.1 Letter sequence

As mentioned in the box in 9§5.3, the Greek letters conventionally used to label solid phases in metal (and other) constitutional diagrams have no particular significance in themselves. They are merely labels, and have no general structural or compositional connotations which can be inferred simply from that label. When such a diagram is drawn for a binary system, it is now (usually) conventional for the composition axis to be labelled in alphabetical order of the symbols from the left. The first phase to appear at low ('normal' or 'room') temperature on the left is then given the label α, the next is β, and so on, working to the right and taking into account phases which appear at higher temperatures. Sometimes the existence of a new phase may be discovered some time after a diagram has been published. Then the next, or another, letter may be used for it, but generally the entire diagram is not then relabelled, to avoid confusion. Thus, in the Cu-Sn system

The composition lines of constitutional diagrams can be expressed in terms of mass or atomic proportions. These are calculable, one from the other, as follows, where a is the atomic proportion, m the mass proportion, and M the molar mass (g/mol) (or atomic weight), since $m = a.M$ if m is in grams and a is expressed in moles.

<u>Mass proportion</u>

$$m_2 = \frac{a_2 M_2}{a_1 M_1 + a_2 M_2}$$

which is a simple weighted arithmetic mean.

<u>Atomic proportion</u>

$$a_2 = \frac{m_2 / M_2}{m_1 / M_1 + m_2 / M_2} = \frac{m_2 M_1}{m_1 M_2 + m_2 M_1}$$

which is a weighted harmonic mean.

M can also be expressed as molecular weight, and a in moles, as necessary.

---

[2] "Vid-man-state-tun". The Austrian, A.J. Widmanstätten (1754 - 1849), indeed discovered the pattern in polished and etched sections of some meteorites.

(Fig. 5.1, later), the ζ-phase is apparently out of (Greek) alphabetical order, but this has no effect on our ability to read the diagram.

●4.2   Layout

For metallic systems, the convention now is to refer to a system with the element symbols in alphabetical order (*e.g.* Au-Cu), even though this does not correspond with the English name, and the first named therefore usually goes on the left of the diagram. Again, for historical reasons, this may not always be the case or convenient. In ternary and higher alloy systems it is clearly a lot less easy to establish any formal labelling system, and when the various binary systems from the same elements have already been labelled, it can become problematic to relabel for that application. Thus, in the dental amalgam system (Chap. 14) where the three principal phases happen each to have been called γ in their respective binary diagrams, the distinctions have been made conventionally in dentistry by adding subscript numbers, *e.g.* $γ_2$. Overall, we find that historical usage tends to prevail.

What this means is that, to avoid confusion, the label should always be used with the relevant alloy system clearly specified, and in the context of an identifiable diagram, just in case there is more than one version around, which unfortunately could occur for a considerable while yet. In any case, if reference is made to old publications in the normal course of research, such variations must be expected.

●4.3   Phase field sequence

Necessarily, any field at the left- or right-hand extremes of any binary T-X diagram (and so on equivalently for multicomponent systems) can only contain a single phase. This follows because a pure component can only exist in a single phase at equilibrium except when it is undergoing a phase transition such as melting:

$$α \quad ⇔ \quad Liq \qquad\qquad (4.1)$$

but such an event is on a phase field boundary and not *in* a field. Were it otherwise, one phase would have a lower energy than the other and spontaneous transformation of one phase must occur to minimize the energy of the system. If then a second component were to be added to the system in very small amount, the expectation would be the formation of a solid solution – even if that were 1 atom in 10^{10} and might be better considered as contamination. Only when sufficient of the second component has been added to exceed its solubility in the first, at that temperature, would a second phase appear. Thus, no matter how low the solubilities might be, there is still a single phase field at each edge of the diagram -- the **terminal phases**.

If a pure component is **dimorphic** or **polymorphic**, that is, has more than one stable crystal form, depending on temperature, then there will be more than one terminal solid solution phase. However, the temperature ranges of stability will not overlap and only simple equilibria will exist between them:

$$α \quad ⇔ \quad β \qquad\qquad (4.2)$$

A good example of this is iron (see left margin of Fig. 21§1.1).

Following from the phase rule, the intervals (in terms of composition) between single phase fields can only ever contain two phases in binary systems:

$$F = 2 - 2 + 1 = 1$$

(*i.e.* we can choose either the temperature or the composition of one phase to determine the description of the system). These phases must, of course, be those represented on either side of the two-phase field. In other words, neither two two-phase fields nor two one-phase fields can ever be found adjacent horizontally. Thus, in Fig. 4.1, for increasing amounts of element B, the sequence must be of the type: α / α+β / β / β+γ / ... and so on. This is only true for isothermal changes in composition, but it always means that the 'occupancy' of an

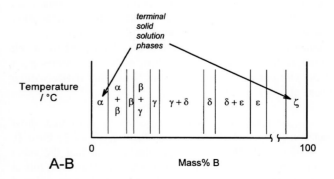

Fig. 4.1 Alternation of singly and doubly occupied phase fields in any horizontal traverse of a binary phase diagram. (The widths of the phase fields as shown have no significance.)

unlabelled phase-field can be worked out. So strong is this rule that commonly, to avoid clutter, constitutional diagrams are deliberately left partially unlabelled because the required information is still easily accessible, with never an ambiguity.

•4.4   Electron compounds

The composition limits of some phases are very narrow. Examples relevant to dentistry are found in the Ag-Hg (Fig. 14§1.2) and Cu-Sn systems (Fig. 5.1). Often they correspond closely to simple integer ratios, as in formulae such as Cu_3Sn and Cu_6Sn_5. Such observations suggest that fairly definite relationships are involved. Put another way, the solubility range of either one of the components is extraordinarily small in such phases. Phases showing these characters are known as **electron compounds**,[3] not because such substances are isolatable as individual molecules, but because the attainment of definite ratios of valence electrons to atoms seems to underlie many such phases. It is useful to explore the idea briefly since it crops up in the dental literature.

Electron compounds are based on the concept of the electron : atom ratio (e/A) of the constituents. This is sometimes called electron concentration, and is defined as follows. If elements X and Y have i and j valence electrons respectively, then in an alloy X_mY_n we have:

$$\frac{e}{A} = \frac{im + jn}{m + n} \tag{4.3}$$

The number of valence electrons is generally determined from the element's position in the periodic table (*i.e.* its Group). (Group 8, the noble gases, are assigned a value of 0.) For example, in Cu_3Sn, Cu (Group 1) and Sn (Group 4) give

$$\frac{e}{A} = \frac{1 \times 3 + 4 \times 1}{3 + 1} = \frac{7}{4} \tag{4.4}$$

By inspection of a wide range of alloy systems it appears that there are three e/A ratios commonly represented, and they tend to have particular crystal structures associated with them. Thus:

$$\frac{21}{14} \text{ or } \frac{3}{2} \text{ phases are often b.c.c. or h.c.p.}$$

$$\frac{21}{13} \text{ phases are, unfortunately, particularly complex, but}$$

$$\frac{21}{12} \text{ or } \frac{7}{4} \text{ phases are usually h.c.p.}$$

Other factors such as radius ratio evidently modify these 'ideal' ratios as thermodynamically a total energy statement is required. So, for a series of such alloy systems (Fig. 4.2), we see only a broad agreement with this theory, some systems showing marked departure from expectations – see Ag-Hg, for example. In addition, temperature affects even the appearance of a given phase. Nevertheless, the idea of electron compounds has sometimes been found useful for a general classification of metal alloy phases, although the exceptions and deviations limit its application. Care has to be taken to avoid imputing too much rigour to the 'rules' of these ratios, as has sometimes been done with not very helpful results. No further notice will be taken of them.

To return to labelling: in alloy systems that show electron compound effects in a tidy fashion (Fig. 4.2), the α, β, ... sequence is effectively predetermined, so that the same labels tend to be applied to similar phases. However, this is coincidental – it obviously depends on which element is placed at the left of the diagram. In fact, in a number of systems historical usage means that the electron compound labels are not used. Overall, they are best ignored.

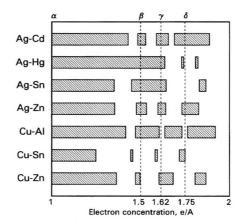

Fig. 4.2 The ranges of composition of 'electron compound' phases in some binary systems at normal temperatures.

The narrow range and apparently definite (*i.e.* small integer) ratios of electron compounds has led to the term **intermetallic compound** also being used for them, more or less interchangeably. However, even when the ratios observed in actual phase diagrams do not correspond very well to the theoretical values (*cf.* Fig. 4.2), small integers are used to express the composition and the term intermetallic compound is still used. It should be recognized that this is a convenience and a shorthand only. There is no 'chemical' compound,[3] and the composition ranges can sometimes be quite broad. Such intermetallic compounds also commonly show random substitutional solid solutions with a variety of other metals, extending their compositional range.

§5. The Cu-Sn System

Constitutional diagrams incorporate much information. They are useful not just for simple statements of structure, but for predicting reactions and accounting for properties. For example, it is necessary to be able to read the diagrams for the requisite information in order to understand the handling and behaviour of dental casting alloys (Chap. 19) and silver amalgam (Chap. 14). Part of this process of reading such a diagram has already been given in the context of the Ag-Cu system (§3).

We may conveniently employ the Cu-Sn system (Fig. 5.1)[4] as a case study to illustrate the various possibilities and the reading rules. Copper-tin alloys are called **bronzes** and have been known from antiquity, although as such are not important in dentistry. However, the Cu-Sn system is of interest because of its

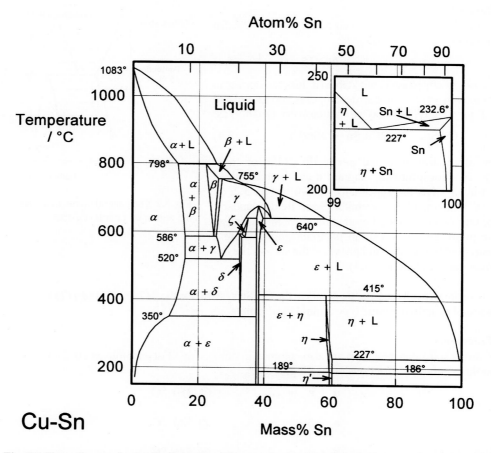

Fig. 5.1 Phase diagram for the Cu-Sn system. The occupancy of unlabelled two-phase fields can be found from the horizontally-adjacent single-phase fields, *e.g.* β + γ. The inset shows the detail for 99 - 100% Sn.

---

[3] Group I and II elements can form true ionic, stoichiometric compounds known as Zintl phases with some main group III - VII elements, where the electronegativities are sufficiently different. None is known to be relevant to dentistry.

involvement in dental silver amalgam. It is (unfortunately) one of the more complicated. Nevertheless, the rules for reading constitutional diagrams are few and consistent, no matter how complex the diagram.

To start with, we can notice the very narrow composition limits of the ε (Cu_3Sn) and η (Cu_6Sn_5) phases, which will reappear with some importance in the context of silver amalgam (14§1). We may also notice the 1/2/1 occupancy sequence of the phase fields. Thus, for example at 500 °C, the sequence from the left is: α / $\alpha+\delta$ / δ / $\delta+\varepsilon$ / ε / $\varepsilon+L$ / L. The labels for the few fields that remain unlabelled can therefore be worked out very simply, and even though there is duplication in one instance. The solidus can be traced as being the lower or leftmost borders of the following fields: $\alpha + L$, $\beta + L$, $\gamma + L$, $\varepsilon + L$, $\eta + L$, although some of these have strange shapes. Lastly, the solid solubility of Cu in Sn is so small that this field and the corresponding Sn + L field do not show in the diagram, so this detail is shown in the inset enlarged portion instead.

Reading the diagram vertically, *i.e.* at a constant composition and just changing temperature, various kinds of transformation may be found. Thus, at the solidus at about 10% Sn we find:

$$\alpha \;\Leftrightarrow\; \text{Liq} \tag{5.1}$$

corresponding to the freezing or melting of the α-phase with the variations in composition discussed earlier (§1). Far on the right is the barely visible eutectic point (see the inset, Fig. 5.1) where the reaction

$$\eta \;+\; (Sn) \;\Leftrightarrow\; \text{Liq} \qquad @\ 227\ °C \tag{5.2}$$

is expected.

●5.1   Peritectics

However, at 415 °C between about 38 and 92% Sn there is a reaction of a new kind:

$$\eta \;\Leftrightarrow\; \varepsilon + \text{Liq} \qquad @\ 415\ °C \tag{5.3}$$

In other words, the η-phase is not stable (it does not appear in the diagram) above 415 °C and so decomposes into Liquid and ε-phase. Equivalently, this may be written (very roughly) as:

$$Cu_6Sn_5 \;\Rightarrow\; 2Cu_3Sn + 3Sn(L) \tag{5.4}$$

This type of reaction is called **peritectic** ('like' or 'near melting'). In a constitutional diagram it can be recognized because it always has the appearance shown in Fig. 5.2: an inverted 'V' underneath a horizontal phase field boundary. The tip of the 'Λ' in this case lies at about 59% Sn. This is a **peritectic point**, and it is also an invariant point. Note a similar pattern of boundaries and phases at the ends of the tie-line as found for the eutectic (§3.3).

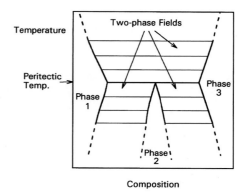

Fig. 5.2 The general layout of a phase diagram in the region of a peritectic-like point.

There are two other peritectic points in the Cu-Sn system, for the β (798 °C) and γ (755 °C) phases. The reactions are of exactly the same type around these points.

●5.2   Eutectoids

Entirely solid state reactions are also common in this system. For example, at 20% Sn, we can trace the following three **eutectoid** reactions on heating:

$$\alpha + \varepsilon \;\Rightarrow\; \delta \qquad @\ 350\ °C \tag{5.5}$$

$$\alpha + \delta \;\Rightarrow\; \gamma \qquad @\ 520\ °C \tag{5.6}$$

$$\alpha + \gamma \;\Rightarrow\; \beta \qquad @\ 586\ °C \tag{5.7}$$

followed by the peritectic melting of the β-phase:

$$\beta \;\Rightarrow\; \alpha + \text{Liq} \qquad @\ 798\ °C \tag{5.8}$$

●5.3   **Peritectoids**

There are also two peritectoid reactions in the Cu-Sn system:

$$\delta \;\Rightarrow\; \gamma + \zeta \qquad @ \; 593 \; °C \qquad\qquad (5.9)$$

$$\zeta \;\Rightarrow\; \gamma + \varepsilon \qquad @ \; 640 \; °C \qquad\qquad (5.10)$$

As with eutectoids, these involve only solid phases. The field pattern is the same as that of Fig. 5.2.

●5.4   **Other reactions**

Apart from the gross phase compositional and structural changes illustrated by solid state reactions such as these, more subtle transformations can occur. Thus, the reaction

$$\eta' \;\leftrightarrow\; \eta \qquad\qquad\qquad (5.11)$$

is seen to occur at 189 °C below about 60% Sn, and 186 °C above. The same Greek letter is retained because the composition and structure are very similar in the two cases (indicated by the prime), just a minor shift in symmetry being involved (they are both essentially hexagonal phases; η' is **ordered** -- see 19§1.3), but enough for the transition to be detectable and an equilibrium possible.[5] This is therefore a **crystallographic transformation** similar to that of the formation of an allotrope: structurally distinct, compositionally unchanged, but the change from disordered to ordered means that this is a **diffusive** process.

On heating an alloy with 50% Sn, after the peritectic melting of η at 415 °C, we find the following reaction:

$$\varepsilon + Liq \;\Rightarrow\; \gamma \qquad @ \; 640 \; °C \qquad\qquad (5.12)$$

In other words, the ε-phase has become unstable and decomposed to yield a new solid phase (actually consuming some liquid in the process); the reaction is known as **metatectic**. Note that this does not involve a change in the composition of the liquid (the right hand end of the tie-line is fixed), just its amount.

A further crystallographic transformation can be seen at the composition Cu_3Sn, related to 5.12:

$$\varepsilon \;\Rightarrow\; \gamma \qquad @ \; 680 \; °C \qquad\qquad (5.13)$$

ε-phase (h.c.p.) transforms to γ, which is orthorhombic. This is a **congruent transformation** as there is no change of composition, and it is also **non-diffusive** as only a distortion of the structure is involved, with no other rearrangement of the atoms.

●5.5   **Cooling**

The reactions described above are in terms of on heating, with the implicit corollary that on cooling the same equilibria would apply. This is true, in principle, but again there is a kinetic limitation. As with segregation (§1.6), equilibrium depends on solid-state diffusion, and this is slow. Thus, a peritectic reaction requires the precipitation of a second phase on top of another as that first phase is being consumed (the reverse of reaction 5.3). Clearly, in this the newly forming η phase would block further reaction with ε, so the cooling liquid must now take a different path as its composition drifts, along the lines indicated in §1.4. Some very complicated outcomes can be expected, depending on alloy and cooling rate.[6] The implication of this is that obtaining the equilibrium constitution requires rapid cooling and careful annealing. This might not be feasible industrially, but certainly will not be the case or even possible for dental castings (Chap. 18, 19).

References

[1]   Guy AG. Introduction to Materials Science. McGraw-Hill, New York, 1972.

[2]   Lyman T (ed). Metals Handbook, Vol. 8. Metallography, Structures and Phase Diagrams. Amer. Soc. for Metals, Ohio, 1973.

[3]   Wood WA. The Study of Metal Structures and their Mechanical Properties. Pergamon, New York, 1971.

[4]   Brandes EA & Brook GB (eds). Smithells Metals Reference Book. 7th ed. Oxford, Butterworth-Heinemann, 1992.

[5]   Larsson AK, Stenberg L & Lidin S The superstructure of domain-twinned η'-Cu₆Sn₅ *Acta Cryst* B50: 636 - 643, 1994.

[6]   Castanho MAP *et al.* Steady and unsteady state peritectic solidification. Mater Sci Technol 31 (1): 105 - 114, 2015.

Chapter 13   Corrosion

Corrosion is the chemical reaction of a metal with components of its environment. Since in the dental or more general biomaterials context metals may be exposed to wet, warm, salty, acidic and oxygenated conditions, the possibility of such reactions must be considered. This chapter sets out the types of corrosion mechanism, methods of control, and factors influencing the outcome. This is to enable the correct decisions to be taken in choosing alloys for specific applications, and the recognition of risk factors, to achieve the most favourable long-term solution in treatment.

*Corrosion is an **electrochemical** process, and the electrode processes operating in spontaneous and driven corrosion systems are explained. The identification of what types of reaction and where they are occurring are key issues.*

*There are a number of methods of **protection** against corrosion which are in common use, whether by providing a physical barrier or deliberate control of the corrosion reaction, but unfortunately very few possibilities are appropriate or feasible in the oral environment. In addition, corrosion once started tends to be self-perpetuating.*

*Two approaches are feasible: the use of either inert or passive metals or alloys. **Passivity** is obtained through an unreactive oxide coating, but this is not a guaranteed cure. Corrosion can still occur under strongly acidic or alkaline conditions, and then can continue in a much more severe fashion. Even so, passive metals are of increasing importance in dentistry, especially titanium.*

__Stress corrosion__ is a risk whenever a metal object is stressed under potentially corrosive conditions as the stress increases the driving force for the reaction. This applies whether the stress is continuous or intermittent.

*Electrochemical processes are also used for deliberately **etching** or polishing metals as well as for plating. These techniques allow a close control that would otherwise be difficult to achieve, and in some cases permit a process that would not be practical another way.*

A major factor in the design and selection of alloys for use in dentistry is the corrosion resistance. It is only by being aware of the factors involved, and understanding the mechanisms and processes operating, that the correct choices can be made for effective long term treatment. There are also implications for the tools and instruments used in dentistry, where sterilization offers more serious challenges.

Materials Science for Dentistry
https://doi.org/10.1016/B978-0-08-101035-8.50013-4

There are a number of applications for metallic materials in dentistry, both within the mouth and in the numerous instruments, tools and equipment associated with clinical and laboratory work. The prime demands on a metallic structure are usually those of strength and rigidity, but it seems self-evident that in addition there must be a lack of chemical reaction with the substances found in the working environment. It is plain that not all metals are as unreactive as gold or platinum, and the chemistry of a metal or alloy must therefore be taken into account when designing for a particular application. The more aggressive the environment, the more serious the problem. This aspect of metal chemistry is usually referred to as the corrosion properties; corrosion resistance or corrosion rate are the relevant concerns.[1] In particular, the oral environment presents a corrosion challenge to metallic devices: it is warm, wet, acid and salty. While these conditions are physiologically benign and normal (and not obviously challenging), for many metals they represent substantial problems, especially in the context of the many years of exposure that are expected. To understand, therefore, the design of alloys for dental use, and the limitations in application or handling that arise from their corrosion behaviour, the principles must first be established.

The corrosion of metallic objects has a number of possible consequences in the dental or biomedical context. Firstly, the metals most often encountered as structural materials are those from the transition periods of the table of the elements, groups VIA - IIB. Their oxides and salts are typically strongly coloured (24§6.4). Thus, their corrosion products will tend noticeably to discolour the metal itself, if they adhere, or the surrounding tissue or other materials. Secondly, the fact that metal is being removed from the surface of the object by the corrosion reactions means that its roughness may increase. This would spoil the appearance if the object was originally highly polished but also, in the mouth, it would be more retentive of plaque, in itself undesirable. The loss of material may go further, perhaps intergranularly or in pitting, and reduce the mechanical strength of the object, causing failure. Lastly, except for a few that are required in very small amounts for special physiological or biochemical reasons (the so-called 'trace' elements) the 'heavy' metals are mostly toxic; iron is the obvious exception to this. Corrosion products therefore may pose a threat of local or systemic effect on the organism. These problems are not mutually exclusive, and various combinations usually occur. Whatever the combination, from the point of view of the task being performed by the metal object, corrosion of any kind is usually undesirable. It must therefore be avoided or controlled to be very limited.

§1. Basic Considerations

●1.1   One electrode
Whenever a metal is in contact with an aqueous solution such as saliva or blood (Fig. 1.1) there is a spontaneous tendency for metal ions to go into solution, leaving electrons behind. This reaction may be written:

$$M \underset{\text{REDUCTION}}{\overset{\text{OXIDATION}}{\rightleftharpoons}} M^{n+} + ne^- \qquad (1.1)$$

Notice that the reaction moving to the right involves the removal of electrons, the **oxidation** of metal atoms to positive ions (cations), while the converse reaction is a **reduction** of the cations to metal, *i.e.*, the addition of electrons.

Because a charge separation is involved in the process of oxidation (*i.e.* work is being done), it is appropriate to define the tendency of the reaction to move to the right by a voltage or potential

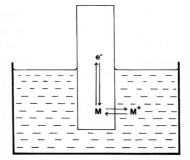

Fig. 1.1 Any metal in contact with an electrolyte spontaneously establishes an equilibrium with its own ions.

difference between the solution and the metal immersed in it – this is the **Volta potential**.[1] This might also be termed the **escaping tendency** for the ion. There is, however, no way to measure this voltage directly because any other contact with the solution would involve a similar reaction, operating in the opposite direction, obscuring the value of interest. Even so, it can be readily appreciated that one factor of relevance in establishing the equilibrium must be the effective concentration or activity of the metal ion in the solution, and so the electrode potential is dependent on the composition of the solution. The potential is also dependent on temperature (equation 8§3.1). A metal in contact with a solution in this manner is termed an **electrode**.

---

[1] Note that the term 'corrosive' refers to the ability of, say, acids to attack metals. It does *not* mean low corrosion resistance.

•1.2   Galvanic couple

If two metals of differing **oxidation potential** are put into contact, again a spontaneous potential difference is present – the Volta potential – as a result of a shift in electrons from one to the other (Fig. 1.2). (This voltage is not measurable as such because it would require other metals to form a circuit, and all the differences must cancel out.) Another way of saying this is that the metals have different **electron affinity**.

If two such different metals are immersed in the same solution (without touching), the equilibria set up will be largely independent of each other. But if the metals, say Cu and Zn, are in contact (Fig. 1.3), allowing electrons to move freely between the two, and the escaping tendencies of their ions are different, only one reaction can move to the right. Essentially this is because the higher electron 'concentration' resulting from metal with the higher tendency to dissolve unbalances the equilibrium (1.1) from the point of view of the second metal, forcing reduction. Thus, should a supply of the other metal's ions be already available from the solution, the reaction in respect of that metal will move to the left, and metal atoms will be deposited on that electrode.

The electrode associated with the metal ion reduction process is known as the **cathode**, and that associated with the oxidation process is the **anode**. This pair of definitions represent the single most important distinction to remember for, once having identified an electrode *process*, all else follows.

Thus, in the course of this **spontaneous** reaction, summarized as:

$$Cu^{++} + Zn(s) \Rightarrow Cu(s) + Zn^{++} \qquad (1.2)$$

(where the 's' refers to the solid state), electrons must be transferred from one electrode to the other. It does not matter what path is taken so long as the electrical connection is made, and this may easily be outside of the solution or **electrolyte** (Fig. 1.4). In comparison with the arrangement of Fig. 1.3 there is no change in any aspect of this system electrochemically, assuming that there is no electrical resistance in the external circuit.

If such a resistance is incorporated, however (Fig. 1.5), the transfer of electrons from one side to the other is delayed, creating a backlog as it were. The net *surplus* of electrons at their source (the anode) leads to this electrode carrying a negative charge, while the *deficit* at the cathode leads to it carrying a positive charge. The voltage measured now across the terminals, if the resistance is very large, corresponds to the **potential difference** between the electrodes. Such a system of dissimilar metals with an electrical connection, both immersed in an electrolyte, is known as a **corrosion cell** or **galvanic couple**.[2] The only requirement for the relevant reactions to tend to go is that the electrode potentials of the two metals, under the prevailing conditions, are different.

Note that it is incorrect to say that the potential difference measures the *rate* of reaction. The rate also depends on circuit resistance and concentrations, as well as other factors.

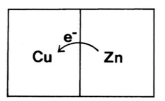

Fig. 1.2 Two metals simply in contact with each other generate a potential difference.

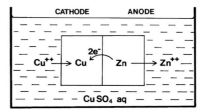

Fig. 1.3 Two metals in contact with each other and an electrolyte form a spontaneous electrochemical cell.

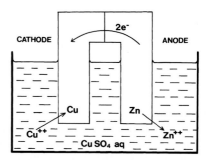

Fig. 1.4 Rearrangement of the cell of Fig. 1.3 to create an external circuit.

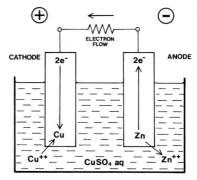

Fig. 1.5 The external circuit of Fig. 1.4 is replaced by an external resistance, over which a potential difference can be measured.

---

[2] After Luigi Galvani (1737 - 1798) who experimented with the electricity generated by such pairs of metals.

•1.3   Chemical potential

If we write a hypothetical equilibrium constant equation for the anode reaction (1.1):

$$K = \frac{\{M^{n+}\}\,\{e^-\}^n}{\{M\}} \qquad (1.3)$$

not only does K depend on the metal ion activity in the solution (as already discussed) and on the 'electron activity' or potential at that point, there is also a term for the activity of the solid metal (8§3.2). This is normally by convention taken to be unity, which can be interpreted very simply as meaning that the activity of the solid is independent of its bulk – remembering that thermodynamics does not consider the quantity of material as such. However, this assumes that the metal is in a **standard state**, usually understood to be a perfect, perfectly pure crystal. The actual activity is affected by lattice defects, strain, roughness, temperature, impurities and deliberate alloying, all of which affect the energy of the system.

If the atoms of the metal of interest are not in pure solid but dissolved in another metal, its **chemical potential** is evidently altered (*cf.* 8§3.2), and so will be its electrode potential too, and in a concentration-dependent manner (albeit not necessarily in a linear fashion). So the electrode potential of a single-phase alloy will reflect the elements present as well as their proportions. The electrodes of the system shown in Fig. 1.5 could just as well be of Cu-Sn and Sn-Zn alloys[3] as of the pure metals. The potentials will be different and the rate of reaction may be different, but the net effect will be the same. Even single-phase alloys from the same two metals but with different compositions, for example α and β Ag-Cu (Fig. 12§3.1), will show this kind of electrochemical behaviour because each component may have different potentials, both electrical and chemical.

•1.4   Driving forward

The rate of the electrode reactions in Fig. 1.5 depends on the diffusion of Cu ions to the cathode, the diffusion of Zn ions away from the anode (to permit the reaction 1.1 to move to the right), and transfer of electrons from anode to cathode. Stirring would obviously help the first two, but the rate could also be increased if the electrons could be delivered to the cathode at a higher rate. So if some kind of electron 'pump' were available to charge up the copper electrode with more electrons so that it acquired a lower positive charge or even a net negative charge (Fig. 1.6) when immersed in the electrolyte, the same reactions would still go but at a greater rate (Fig. 1.7).

Notice that the charges on the electrodes are now of opposite sign to those in Fig. 1.5, although electrochemically there is no change in the description. In other words, the charge on an electrode is no guide to the electrode process occurring at its surface. This illustrates the fundamental point stressed above on the nature of electrochemical cells, however they are formed: it is the electrode *reaction* that is to be taken into account, and this gives the name of the particular electrode being considered. The net charge at any point is quite irrelevant to this, although it must account for external factors.

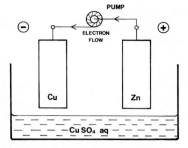

Fig. 1.6 An hypothetical electron pump is added to the external circuit to change the signs of the spontaneous charges; the circuit is not yet completed by immersion in the electrolyte.

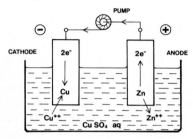

Fig. 1.7 Immersion of the electrodes of Fig. 1.6 results in the same reactions as the spontaneous cell of Fig. 1.5 at a greater rate.

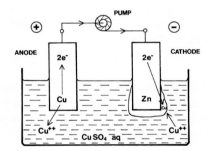

Fig. 1.8 An hypothetical electron pump is added to the external circuit of Fig. 1.5 to oppose the spontaneous potential.

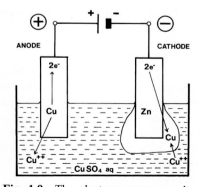

Fig. 1.9 The electron pump may be a battery of spontaneous electrochemical cells.

---

[3]   The electrode potential of Sn lies between those of Cu and Zn in most cases.

•1.5   Driving back

The same electron pump might be reversed so as to increase the net charges on the electrodes in the same sense as found in Fig. 1.5 (Fig. 1.8). At some point the electrode potential spontaneously generated would be exactly balanced and no reactions would occur, there being no net driving force. When that potential is exceeded, however, the reactions will tend to be driven in the opposite directions, Cu dissolving at the (new) anode and Cu would also be deposited at the (new) cathode (Fig. 1.8). This then is an electrolytic or **plating cell**. The 'pump', of course, is some form of battery (which strictly speaking is itself be made up of a series of spontaneous electrochemical cells) or other electrical voltage source (Fig. 1.9).

Each terminal on a dry cell, battery or other voltage source is labelled with the sign of the charge at that point, and this is therefore the same as that of the net charge residing on the attached electrode itself (Fig. 1.5). Thus the 'positive' terminal of a dry cell is the cathode. If you need to work out what is going on in any system, merely determine the *electrode reactions*, and all else follows.

§2. Polarization

Corrosion cells have been discussed as though the rate of reaction were independent of time, but this is an over-simplification in most cases. In a spontaneous cell such as that of Fig. 1.3, if the reaction is allowed to continue, the concentration of metal ions around the dissolving anode will rise. Consideration of the equilibrium that generates the potential (equation 1.3) shows that as this occurs there will be less tendency for the anode metal to dissolve, therefore a lower rate of dissolution, gradually reducing to zero. This is described as the cell becoming **polarized** due to the generation of a **back e.m.f.** (electromotive force). The back e.m.f. due to these concentration effects can eventually equal the original cell potential, resulting in an equilibrium with no further dissolution occurring. Such a closed system would be self-limiting. However, there are two major ways in which this limit may be broken, both of which are highly pertinent to dentistry.

•2.1   Limitations

The first arises because most of the metals used have rather insoluble hydroxides. This means that they react readily with water:

$$M^{n+} + nH_2O \rightleftharpoons M(OH)_n + nH^+ \tag{2.1}$$

This has two effects. Primarily, the concentration of metal ions is kept down to correspond with the solubility of the hydroxide. Secondly, hydrogen ions are generated, lowering the pH, thereby increasing the solubility of the hydroxide somewhat, but also increasing the potential for dissolution. The fate of the hydrogen ions will be discussed in a moment.

The second process occurs when the system is open (as opposed to the closed systems illustrated so far), which means that as fast as metal is dissolved, the ions are carried away by diffusion or bulk flow in the electrolyte so that the ion concentration cannot build up. Corrosion in the mouth frequently occurs under such circumstances because of the flow of saliva and foodstuffs. The corrosion product ions are maintained at a low concentration in the vicinity of the corrosion site, and thus the corrosion rate remains high, very similar to the initial rate.

•2.2   Depolarization

In a closed system the build-up of metal ions in the vicinity of an anode results in an approach to equilibrium and the cessation of dissolution, limited by diffusion. In addition, the rate of deposition of metal ions at the cathode may exceed the capacity of diffusion to replace them. **Depolarization** may be then effected by stirring, which redistributes the solutes in the electrolyte. The potential measured across a cell such as is shown in Fig. 1.5 will be strongly dependent on such effects, and the experimental determination of cell e.m.f.s is made more difficult as the *exact* conditions at each electrode become difficult to control. Convection, due to changes in density resulting from the dissolving metal or temperature differences, may also cause variation in potentials over time.

§3. Cathode Processes

So far it has been assumed that certain suitable metal ions already exist in the electrolyte for the cathodic reduction reaction. This obviously will not always be the case (and especially not so in the mouth) but there may be other reactions possible to serve as **electron sinks**, such as:

$$Cl_2 + 2e^- \rightleftharpoons 2Cl^- \tag{3.1}$$

which may be relevant in swimming pools and chlorinated drinking water;

$$2H^+ + 2e^- \rightleftharpoons H_2(g) \tag{3.2}$$

which tends to be difficult unless on specially-prepared catalytic surfaces; and

$$H_2O + \tfrac{1}{2}O_2 + 2e^- \rightleftharpoons 2OH^- \tag{3.3a}$$

or, equivalently,

$$H^+ + \tfrac{1}{2}O_2 + 2e^- \rightleftharpoons OH^- \tag{3.3b}$$

If going to the right, these are all reduction reactions, consuming electrons. The second example may be familiar as one half of the electrolysis of water, the complementary oxidation (of hydroxide to form oxygen) being the reverse of the third example.

The third example, as written, is one of the most important cathodic reactions in any dental context (and indeed in many others) because of the nearly universal presence of the very reactive gas oxygen dissolved in the electrolyte, be it saliva, blood, sterilization liquid or whatever. We can generally assume, therefore, that the cathode reaction in the oral environment is the reduction of oxygen. Note that this does not in any way affect the arguments above in terms of electrochemical cells, but merely facilitates their occurrence in a wider range of contexts. It can be seen that hydrogen ions are effectively consumed in this process and so must diffuse from the anode, so decreasing the tendency for the pH to fall there and, conversely, rise at the cathode. These pH changes provide one very simple means of detecting such electrochemical reactions and so identifying directly which electrode is which. Either by adding an appropriate pH indicator dye to the system, or by using a 'pH electrode' (itself a special type of electrochemical half-cell), the zones of altered pH may be visualized directly or mapped.

An alternative view of the meaning of reaction 3.3b concerns the driving force for moving to the right. It is apparent that, first, acidic conditions promote reaction. The hydrogen ions (or, hydroxonium, H_3O^+, if you prefer) of acids are thus seen to be directly involved in corrosion. Secondly, an increase in dissolved oxygen will also drive the move to the right, so agitation, stirring or flow to promote aeration will promote corrosion.

§4. Corrosion Protection

Having seen how corrosion may occur, we now consider how it may be avoided. In other words, how are materials selection and design affected?

●4.1 Noble metals

The obvious metals to use if corrosion is to be avoided are those which do not oxidize under the conditions to be encountered during the fabrication or service of the device. Thus, the so-called **platinum group** elements of ruthenium, rhodium, palladium, osmium, iridium and platinum (the second and third triads of group VIII) are usually found in nature in the 'native' state, *i.e.* as metals, indicating their resistance to oxidation. Gold, of course, normally is as well, while silver is found as metal occasionally. These metals are therefore often described as **'noble'** (although this is not a strictly chemically interpretable

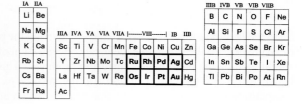

Fig. 4.1 The 'noble' metals.

label and is generally deprecated) (Fig. 4.1), and they do find many uses where the resistance to oxidation is important, in chemistry and electronics, for example. Jewellery is perhaps the most obvious use for gold, and platinum is used for mass standards. These properties put the metals in great demand, for various reasons, and so the price tends to be very high. Gold, being relatively plentiful, continues to be used in restorative dentistry and silver for endodontic points, but in general price is a major factor in determining the choice to be made. Coupled with the problems of strength that arise with such metals, because of their structure (Chap. 12), other metals and alloys must be considered. Consequently, dealing with corrosion is an important aspect of selection.

Even so, the resistance to corrosion of even the 'noble' metals cannot be taken for granted. Thus, pure silver may corrode slowly but steadily in the presence of chloride ions, as is well-known to shipwreck treasure hunters: silver coins may be destroyed by this process in the sea. Indeed, silver endodontic points commonly suffer from the same reaction whilst in place in a tooth (28§9).[2][3] Similarly, Ag-Pd alloys, which are single-phase, solid solutions (Fig. 12§1.2) may react in the same way, as do Ag-Pd-rich phases in more complicated alloys.

●4.2   Tin plating

Iron is used in many circumstances in the form of steel: food cans, car bodies, bridges, *etc.* because of its cheapness, strength and ease of fabrication, but it is readily oxidized. In all of those and similar applications protection is required from the corrosive environments in which it is used or which it contains. Food cans, for example, are made of steel coated with a thin, impermeable layer of tin (Fig. 4.2 A). The relatively unreactive tin thus keeps the surface film of

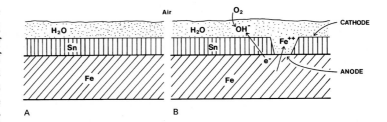

Fig. 4.2 (A) Tin plate protection of steel when the plating is intact. (B) When the tin plating is damaged, or has naturally occurring pores, the galvanic couple present exaggerates the corrosion of the steel.

moisture, always present in moderate and especially in high humidity, and the oxygen of the air separated from the reactive iron. Ordinarily there is no perceptible reaction. But should the thin plating become damaged by a scratch or perforated in any way to expose the iron, immediately a galvanic couple is set up and the iron, having the higher electrode potential, begins to corrode (*i.e.* is the anode) (Fig. 4.2 B):

$$Fe \rightleftharpoons Fe^{2+} + 2e^- \tag{4.1}$$

The cathode reaction is then the reduction of oxygen as set out above. Ferrous ions, however, are readily oxidized to ferric by dissolved oxygen:

$$2Fe^{2+} + \frac{1}{2}O_2 + H_2O \rightleftharpoons 2Fe^{3+} + 2OH^- \tag{4.2}$$

Further, ferric ions may be hydrolysed readily:

$$2Fe^{3+} + 6H_2O \rightleftharpoons 2Fe(OH)_3(s) + 6H^+ \tag{4.3}$$

and are therefore effectively removed from the system. The net reaction can thus be written:

$$2Fe + \frac{1}{2}O_2 + 5H_2O \rightleftharpoons 2Fe(OH)_3 + 4H^+ + 4e^- \tag{4.4}$$

Even in the absence of oxygen, hydrolysis occurs:

$$Fe^{2+} + 2H_2O \rightleftharpoons Fe(OH)_2 + 2H^+ \tag{4.5}$$

So as the precipitation of the hydroxide is occurring the pH is dropping, making the potential for the dissolution of Fe greater. At the same time the local concentration of oxygen is decreasing, and replenishment by diffusion is inhibited by the precipitated corrosion products. This adds a further component to the potential difference between the free surface of the Sn and the surface of the Fe in the scratch, because there is no longer the competition with the oxygen reduction process which otherwise might tend to occur at the anode. Notice also that the pH at the (cathodic) Sn surface is rising because of the production of hydroxyl ions, ensuring that any Fe^{3+} which does diffuse out is precipitated, contributing to the isolation of the active corrosion site.

●4.3 Chromium plating

Chromium plating of steel relies on a similar mechanism: the chromium itself is protected by its oxide film (§5), but wherever this plating is interrupted, whether by a scratch or an unavoidable bolt hole for fixing, corrosion of the underlying metal occurs. Such corrosion is not helped by the fact that layers of copper and nickel are required on the steel first if the chromium is to adhere, these layers also forming galvanic couples. The protection is at best a temporary decoration.

●4.4 Zinc plating

An alternative method of protecting steel is by zinc-plating, so-called **galvanized** steel. Because zinc is a reactive metal, the corrosion reaction results only in attack on the plating, so that the structural integrity of the steel is preserved (Fig. 4.3A). If in this case the plating is damaged so as to expose the iron (Fig. 4.3B), it remains the zinc which is attacked because it has the greater electrode potential, the normal anodic oxidation reaction occurring.

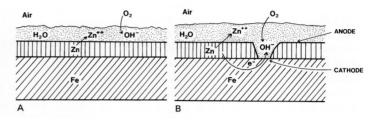

Fig. 4.3 (A) Zinc plating on steel corrodes slowly when the plating is intact. (B) When the plating is damaged, the zinc corrodes more rapidly, still protecting the steel.

But because the cathodic reaction is the reduction of oxygen, the underlying iron is completely unaffected. Thus, although the corrosion of the zinc may be going at a *higher* rate than otherwise would be the case, the iron remains protected while zinc is present. This then is known as **sacrificial plating**, and the iron is **cathodically protected**.

●4.5 Sacrificial anodes

A related method of protection is employed on oil and gas pipelines, where the use of galvanized steel would be impractical because the plating would last a far shorter time than the intended life of the pipeline. Here, large ingots of magnesium are buried at intervals in the adjacent soil and, for best effect, connected electrically to the pipe by a cable. Such **sacrificial anodes** must be renewed regularly, but this is still cheaper than replacing the pipeline. An alternative method is to use scrap steel as the sacrificial anode, but to impose sufficient voltage across the two that it is indeed the scrap that dissolves in preference to the pipe (*cf.* Fig. 1.9). Both kinds of anode must be repeated at intervals, as the (electrical) resistance of the pipeline steel means that the effect diminishes with distance. The approach, however, using an magnesium billet, has been shown to be effective in protecting air-turbine handpiece bearings (which are usually stainless steel, 21§2) from corrosion during autoclaving.[4]

●4.6 Other barriers

Some metals, especially steels, can be treated with phosphate (PO_4^{3-}), chromate (CrO_4^{2-}) and similar salts to create a barrier of a monolayer of ions strongly adsorbed to the surface, or even conversion to the salt, effectively preventing contact with the environment (see §4.12). Such a coating is necessarily very thin and has little resistance to abrasion. Whilst it may be used as a pretreatment of a metal surface before painting (to overcome some of the deficiencies of paint), and so may be used in large dental surgery equipment such as furniture frames, it has no practical biological-context use.

In any case, no such protection for metallic devices for use in the mouth or in more intimate physiological contexts is workable. Barriers (as with Sn, paint or lacquer) are difficult to apply to ensure complete coverage, are subject to wear and prone to damage, all of which negates their effect. Sacrificial systems (as with plating or a separate anode) are inappropriate, if for no other reason than that the taste of the dissolving metal would make it unacceptable.

●4.7 Gold plating

As a case in point, consider the gold-plated screw posts which have been (and still are) sold for reinforcing restorations. The gold is certainly corrosion-resistant, and if the plating is intact (no pinholes), the post will survive all ordinary corrosive media. However, if the plating is damaged, as will almost unavoidably occur when it is placed because gold is so soft, the brass or steel beneath will become exposed and corrode but at a faster than usual rate, driven by the extra boost given by being part of the galvanic couple with the gold. On the other hand, if such a post were to be used with amalgam, as is expected, the gold will be dissolved by the mercury in the unset mixture and corrosion is unavoidable, given that amalgam does not wet steel and leakage at the interface must occur. In addition, for plated steel pins that are intended to be sheared off from the mandrel when in place, the fracture surface is necessarily high-energy and unplated. The gold plating in all such cases is worse than useless – it is positively detrimental by driving corrosion. Likewise, the suggestion[5] that silver amalgam restorations could be coated using gold powder (28§4) is untenable. Gold plating is useful in unchallenging situations, such as printed circuit board contacts, but there is no application in dentistry where it is a practical proposition. Demand for such products is unjustifiable.

The mistake here is in thinking that because the gold (or any other such plateable metal) is corrosion resistant on its own, it is therefore inert. It is not: it contributes to the behaviour of the electrochemical system because it is a conductor and because it has an electrode potential of its own. It is crucial to the success of any such barrier that it remains intact and impervious, truly isolating the object of the protection from the environment that presents the problem.

●4.8 Oxygen concentration cells

Local changes in the oxygen concentration are seen to ensue from some electrode reactions, but the prior existence of a difference in oxygen concentrations in systems

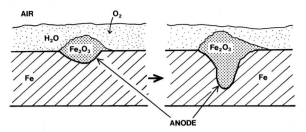

Fig. 4.4 Formation of an active pit beneath oxide through an oxygen concentration cell.

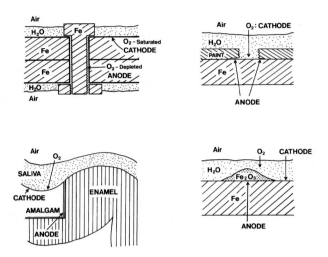

Fig. 4.5 Some examples of the kinds of sites where oxygen concentration cells may form.

otherwise devoid of galvanic couples can still lead to the formation of a corrosion cell. The driving force is simply the difference in activity or chemical potential at two electrically-connected sites. If, for example, under a speck of oxide resulting from direct reaction of, say, iron with the oxygen of the air, there is then some oxidation of ferrous to ferric ions, the oxygen concentration is thereby lowered locally. The tendency for more Fe to dissolve beneath that oxide is then greater as oxygen is reduced to OH⁻ on the free surface. This leads to the formation of a pit directly beneath the oxide. As it does so the pH is dropping in the pit, *i.e.* at the anode, because of hydrolysis of the iron ions (reactions 4.3 and 4.5), and the corrosion cell is said to be **activated** (Fig. 4.4). This then is a self-sustaining system with, curiously enough, the maximum rate of metal dissolution occurring at the site most remote from the atmospheric oxygen. Because the supply of oxygen to a cathode in an oxygen concentration cell may be diffusion limited, such systems are also subject to polarization effects (§2).

It need not only be the presence of oxide that initiates this process. The crevices between bolts and screws and the parts they are holding, the metal beneath a coat of paint *adjacent* to a scratch (since this will not in general be chemically bonded to or wet the metal), the marginal crevice between an inlay or amalgam restoration and the tooth, or merely beneath dirt or, in the dental context, plaque; these are all sites prone to **oxygen concentration cell** or **differential aeration** attack (Fig. 4.5). **Pitting** or **crevice corrosion** are other descriptive terms often used for such processes.

Under plaque the oxygen concentration will fall because of consumption by aerobic micro-organisms, as well as the pH falling due to fermentation processes (there will be an extra driving force due to chelation of metal ions by lactate, reducing their concentration, *i.e.* driving equation 1.1 to the right, §6.5). In the fabrication of dental items such as castings or amalgams, if failure through this kind of corrosion is to be avoided it is crucial to avoid pores communicating with the surface and fine crevices between different parts, both sites where an oxygen deficit can form. The development of cracks at the roots of clasps provides such corrosion opportunities, as does the crevice between the acrylic parts and the metal mesh of a partial denture, where no seal is possible.

●4.9   Dissimilar metals

The other principal method of avoiding corrosion should be apparent: dissimilar metals or alloys should not be allowed to come into contact. This means that, amongst others, bridge clasps should not rest on amalgam restorations, nor should amalgams oppose gold alloy crowns or inlays (Fig. 4.6). Likewise, a metallic crown should not have an electrical connection to silver endodontic points in the same tooth, or metallic crowns be placed over a dissimilar-alloy cast root post. On contact such a circuit would be closed and the external circuit, passing as it does partially through oral tissues *via* the pulp in each tooth, may result in a corrosion current large enough to cause a significant shock, detectable at least by the pulp. This is the basis of what is called 'dental galvanism'. Such shocks can also be experienced with some kinds of cutlery when they accidentally contact a metallic restoration. In fact, even if it is not felt, such corrosion will lead to failure of the treatment.

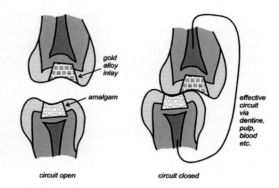

Fig. 4.6 An electrical circuit may be formed in the mouth when dissimilar metallic restorations are brought into occlusal contact, perhaps resulting in a shock.

Aluminium temporary crowns are particularly prone to this problem if allowed to contact another metallic restorative material because aluminium is such a reactive and electropositive metal. For example, placing such a temporary crown over the remains of an old amalgam, or even an amalgam core which has been prepared for the permanent crown (which should not be done anyway, see §5.2), can cause severe pain. The amalgam core is a major risk anyway under any metallic crown because there will always be some leakage through the cement, enough to provide the electrical path necessary for corrosion. Solder joints, *e.g.* silver solder on stainless steel

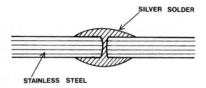

Fig. 4.7 Silver solder on stainless steel is an inevitable corrosion risk.

(Fig. 4.7, 22§3.3), are also inevitable corrosion risks because the solder necessarily has a different composition to the parts it is joining. To a certain extent it is a matter of balancing the limited lifetime of temporary devices against the convenience of fabrication but, for 'permanent' work, even low rates of reaction may be problematic in the long term.

●4.10   Multiphase alloys

It follows from this that multiphase alloys are also a potential problem. As was pointed out in §1.3, the electrode potentials of single phase alloys must vary with composition. Similarly, the two or more (metallic) phases present in a multiphase alloy will, in general, show different electrode potentials, the most electropositive under the prevailing conditions tending therefore to dissolve first.

Bear in mind that it is not so much the 'standard' potential that is relevant here, which is measured against the (arbitrary) hydrogen electrode, but the potential *difference* between two electrodes under the service conditions that drives the reaction: if this is large enough the reaction will go. It is therefore a general principle of alloy design for corrosive conditions (the mouth and biological implant sites included) that single phase alloys are to be preferred, irrespective of strength or other considerations. Dental silver amalgam, of course, is a notable exception (14§9), but this arises for historical reasons. One can presume that it would not be considered an acceptable alloy design were it proposed today. Even so, its continued use underlines the fact that in the context of corrosion, as well as elsewhere in biomedical materials, compromise may be necessary.

●4.11   Sterilization

The corrosion challenge can be very severe when sterilization of instruments and so on is necessary. When this involves a wet process, either at high temperature and pressure in steam in an autoclave or by low temperature 'chemical' exposure, the risk is increased. Inhibitors may then be used in the process to good effect. This approach is particularly necessary for carbon steel items such as burs, but stainless steel (see below) is still at risk.

In the case of autoclaving it is important to use distilled water to avoid chloride-enhanced corrosion. It is even important to clean the instruments well before processing to avoid introducing electrolyte contaminants into the autoclave. The heating elements and pressure chamber of the autoclave itself are at risk of damage and consequent failure through this means. The problems of dissimilar metals should also be noted here. While it may not be possible to avoid contact altogether, at least different types of metal should be kept from touching to reduce the risk of galvanic effects. Bear in mind also that equipment such as air-turbine handpieces may contain bearings and other parts which, because of the mechanical properties required, as with burs, will probably be made from alloys that are corrosion prone. In such cases, the use of a coating of oil, as through the normal pre-autoclaving lubrication routine, is beneficial in isolating the metal from the environment.

●4.12 Corrosion inhibitors

There are a number of circumstances where corrosion-resistant metals cannot be used, and then corrosion inhibitors may be effective in preventing or reducing the rate of reaction. There are many types, the choice of which depends on the metal to be protected and the service conditions. Broadly, they operate in one of a few distinct ways.

Oxygen scavengers: where oxygen is involved in the cathode process (§3), the use of anti-oxidants such as ascorbic acid may be effective, simply by consuming that oxygen (Fig. 6§6.5).

pH control: also called cathodic inhibition. Since low pH generally encourages corrosion reactions, alkaline substances may be used to raise the pH to greater than, say, 8. Unfortunately, this is not feasible with aluminium alloys as these then corrode:

$$2\,Al + 2\,OH^- + 10\,H_2O \;\rightarrow\; 2\,[Al(OH)_4(H_2O)_2]^- + 3\,H_2 \tag{4.6}$$

– which includes solutions of common laboratory and other 'decontamination' detergents.

Passivation layer: also called anodic inhibition. Isolation of the metal can be attained by a reaction to create an adherent, insoluble coating (see §5). This has been done with phosphates, chromates (very toxic, carcinogenic) and by inducing γ-Fe_2O_3 to form on steel by substances such as nitrites (*e.g.* of sodium or dicyclohexylammonium – toxic, probably carcinogenic). Only this latter approach might be of general utility in dentistry, and then only in contexts such as shipping or extended dry storage (vapour-phase inhibition, VPI) or autoclaving. Passivating layers need to be complete otherwise local anodes are formed. Such inhibitors that are alkaline (including nitrites) also cannot be used on aluminium or its alloys or similarly reactive metals.

Otherwise, simply providing a strong diffusion barrier to water, the medium through which most corrosion occurs, by oil or grease, allows corrosion-prone devices to be stored for extended periods. Obviously, this is of little applicability in the context, and clean-up would add to the burden.

●4.13 Dealloying

In some alloy systems, even though they may be single-phase, by virtue of a strong difference in the reactivity of the components towards some aspect of the corrosive environment, differential reaction will occur. Thus, the more reactive element is removed, leaving a porous surface which is the more or less pure second element (in a binary alloy). This behaviour is commonly seen for brass (Cu-Zn), especially in marine use, when the bright pink but matte surface of copper makes it very obvious. But of course there is now a galvanic effect from the (cathodic) copper coating against the more electropositive alloy beneath, with automatically active pitting. In this case it is called **dezincification**, but similar **dealloying** occurs with Cu-Pd dental alloys (Fig. 19§1.19),[6] where the Pd forms the coating, and also occurs with the aluminium bronzes (Al-Cu) which have some use in dentistry and in which Cu again is enriched in the surface (although corrosion in these latter alloys can be dominated by galvanic effects from Fe which may be present[7]). Such variation in behaviour means that corrosion testing under relevant conditions is an important screening process for alloys, and especially where these are to go into the mouth or implanted, when toxicity is of great concern even if the mechanical implications can be tolerated.

§5. Passive Metals

Corrosion can be seen to be dependent on the metal concerned coming into contact with an electrolyte as a vehicle for ion transport and a source of reactants, no matter what other conditions are necessary or present. This, of course, may even be the adsorbed film of moisture. In really dry conditions very little reaction is possible. It is to be noted, then, that one of the most general approaches to protection is to apply a barrier layer, whether an inert metal, paint, or even grease, the purpose being to prevent metal and water coming into contact. These approaches we have already seen to be unserviceable in the dental (oral) context. Certain metals, however, achieve this effect directly because of their special chemistry.

●5.1 Passivation

The process which affords the protection is called **passivation**, and this occurs in magnesium, aluminium, and the transition metals of Groups IVA: titanium, zirconium and hafnium; VA: vanadium, niobium and tantalum; and VIA: chromium, molybdenum, and tungsten (Fig. 5.1). It consists of the ready formation of an oxide layer on contact with air, since all of these metals are reactive towards oxygen. But, in marked contrast to metals such as iron, the oxides that are formed are all:

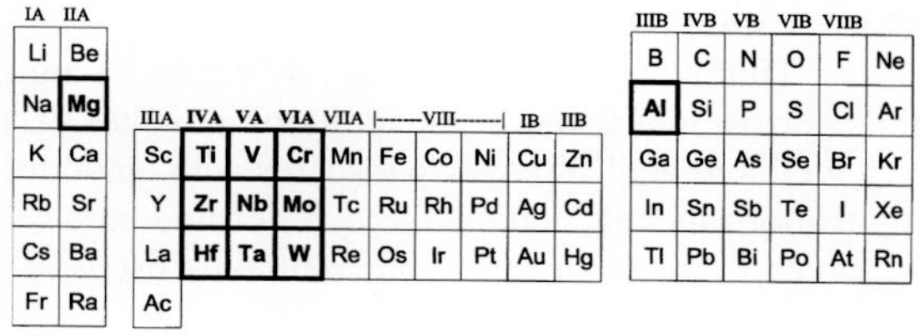

Fig. 5.1 The metals that can be passivated.

• coherent – closely adherent to the substrate metal which acts as a template for oxide growth,
• isovolumetric with the metal – no swelling occurs which would cause disruption,
• continuous – covering the whole surface, and
• impermeable to water and oxygen at ordinary temperatures.

The oxide film thus limits the reaction that would otherwise spontaneously consume the metal. Such oxide films are usually so thin that they do not impair the characteristic metallic appearance of these elements, and they form so quickly that to all intents and purposes they are always present, even on freshly-cut metal (unless under an oxygen-free atmosphere). Objects fabricated from such metals therefore are functionally corrosion resistant, but it must be recognized that it is a resistance of a fundamentally different kind to that observed with metals such as gold and platinum.

●5.2 Aluminium

Aluminium can thus be used for many domestic and industrial applications without the need for paint for reasons of protection, and is commonly seen in such objects as window frames and storm shutters as well as drinks cans. The oxide, Al_2O_3, is nevertheless soluble in both acid and alkaline media, albeit slowly, and care must be taken to avoid or severely limit such exposure. The film can, however, be made much thicker and more robust by a process known as **anodizing**, which is just as its name suggests. The workpiece is made the anode in an electrolytic cell, with a carefully-designed solution composition, and current passed. The oxide film is permeable enough under these conditions that more metal is oxidized and added to the protective layer. Such thick coatings, being ceramic, are very hard (alumina is a common dental abrasive) but very brittle, and they cannot survive bending without rather obvious cracks being produced or, in extreme cases, flakes of oxide becoming detached.

Thinner, naturally-formed films are barely noticeable, but the effect of their presence can be demonstrated readily. A droplet of mercury can be placed on an aluminium surface with no apparent effect for a long period. However, if the aluminium is then abraded through the mercury, such as with a sharp instrument, so that the mercury can wet and amalgamate with the aluminium, the resulting metal surface cannot sustain an adherent oxide. The aluminium then steadily oxidizes simply by contact with the oxygen of the air, white powdery alumina seeming to grow from the abrasion as one watches, revealing the previously hidden reactivity. This effect can be seen in those predosed amalgam capsules that have the mercury contained by an aluminium diaphragm which is to be burst by pressure on the cap: after mixing, inspection of the diaphragm will reveal the feathery oxide growing from the fracture surfaces and where any abrasion has occurred. The same effect can occur with an aluminium crown placed over a fresh amalgam core. Should any scratch occur to cause close contact, which is hard to avoid, the crown will be consumed, with the evolution of much heat. Incidentally, this the reason why mercury in any form, including mercury-in-glass thermometers, is banned from carriage on aircraft – whose bodies are largely built from aluminium alloys: any spillage would be disastrous. You will recognize how difficult it is to retrieve spilt mercury (28§6.1), and the smallest droplet would be enough to cause a serious problem.

●5.3   Chromium

Chromium is a rather harder metal than aluminium, but it is nevertheless readily oxidized to form a similarly thin coat of Cr_2O_3. This substance is green in bulk, although its colour does not affect the appearance of the metal when it is that thin. Formerly used extensively for decorative bright metal work, most conspicuously on cars, it has found further and much wider application because its corrosion resistance can be conferred on its alloys, in common with several of the metals discussed in this section. Thus, if Cr is present to at least 12 or 13 mass percent, the oxide which is still inevitably formed will be continuous and impermeable. A long series of so-called 'stainless' steels has been developed on this basis (21§2). Similarly, the high-temperature service alloys known as **stellites**, adopted and now familiar in dentistry as the cobalt-chromium casting alloys (19§2), are similarly corrosion resistant.

●5.4   Titanium

Titanium (28§1) is extensively used for surgical implants because of its strength, but primarily because of its corrosion resistance. In addition, it evokes very little adverse reaction from living tissue. It is said to be very well tolerated, a property it shares with tantalum for the very same reason: its adherent oxide film. Dental applications include certain alloys for orthodontic wires, which have corrosion resistance as a requirement and for which stainless steel is most commonly used for the same reason, except that the steel is much cheaper.

●5.5   Risks

It is important to recognize that corrosion-resistant metals and their alloys are just that: *resistant*, not corrosion *proof*. It might otherwise be assumed that corrosion cannot occur, when all that is really implied is that corrosion does not occur under most 'ordinary' circumstances; that is to say, near neutral pH and low concentrations of salts. 'Stainless' steel is therefore effectively an advertising lie: it is not true. Certainly, it survives much longer than ordinary steels, which rust very readily. It is for this reason that it is valuable in marine applications, for example. But under any circumstances where the oxide film may be dissolved, by acids or alkalis (all except that of magnesium are more or less **amphoteric**), the usual 'self-healing' behaviour of the protective film is lost, and uncontrolled corrosion occurs. Thus, certain cleaning agents such as hypochlorite bleaches should not be used on cobalt-chromium frameworks (or stainless steel for that matter) as they attack the oxide. (See also 21§2.6.)

Oxide coats of this kind are ceramic. The thermal expansion coefficient will therefore be substantially different from that of the underlying metal so that for large temperature falls the compressive stress in the oxide may lead to it being broken off (*cf.* 20§7, 25§5.4) while large rises would cause cracking. While they may appear flexible or plastic when very thin (29§2), thicker layers will be brittle and can crack on flexing and this may lead to further damage to the film that forms on the now exposed metal when the flexure is reversed. Furthermore, at high temperature oxygen diffusion through the film can become appreciable and thickening will occur, which then affects the mechanical behaviour further.

However, there is an aspect of the behaviour of corrosion-resistant alloys which makes any corrosion in them unusually dangerous. The tendency is for corrosive attack, once started, to continue only at the point where it first occurred. This results in an activated pit forming, the acidic (and low oxygen) conditions associated with the anode being maintained in the confined space. Such pits corrode very rapidly, even though they may amount to no more than a pinhole. However, even that pinhole may be enough to cause the failure of a Co-Cr clasp if it occurs near its root, and stainless steel dental instruments may break in service if such pitting goes undetected near the working point. In fact, corrosion resistant alloys corrode *more* rapidly, not less, once corrosion has been allowed to commence. A somewhat more cautious attitude towards the treatment that they receive is therefore warranted: they are not immune to abuse.

§6. Other Factors

●6.1   Stress corrosion

The 'escaping tendency' of a metal ion is a function of its energy state in the surface of the electrode, and any factor which modifies this in relation to the solution in contact with it will have an effect on overall corrosion rates. Hence the presence of strain, due to cold-work (dislocations) or because the piece is under load, will increase the electrode potential and it will therefore tend to corrode at a greater rate.

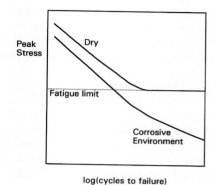

Fig. 6.1   Corrosive conditions reduce the fatigue tolerance of steels and many other metals.

The effect is readily demonstrated by bending a test anode and observing the change in the potential difference. Such **stress corrosion** is evident in rivet and nail heads, which will have been formed by cold working, but it is also relevant to dental bridge clasps which may have been adjusted after casting to obtain a better fit. This adjustment must leave some residual strain, because it is cold work, and corrosion of the root of such a clasp is liable to follow. This takes the form of a crack developing which soon becomes an activated corrosion cell, and rapidly becoming self-sustaining as it converts to an oxygen concentration cell, corrosion products diffusing out. This **stress-corrosion cracking** results in greatly decreased service life even though the stress is static and well below that which would cause immediate failure.

If cyclic loading is present (as with a repeatedly flexed clasp, which must occur during chewing to a greater or lesser extent), the effect is more marked. The number of cycles leading to failure may be substantially reduced in comparison with a non-corroding test piece by what is now termed **corrosion fatigue**. Even steels which show a fatigue limit in air (*i.e.* do not fail below a certain stress no matter how many cycles; 1§12) will fail quite quickly in a corrosive environment (Fig. 6.1).[8] This example provides further evidence of the need for laboratory testing to be performed under conditions which mimic those found in service if the results are to be meaningful, notwithstanding the difficulties of doing so.

It should be noted that the dislocations of cold-worked metal themselves indicate the presence of local stresses. The strain in the crystal associated with stacking defects implies an equivalent stress, and therefore a higher energy state. Such locations are therefore subject to preferential dissolution, and may therefore be the initiator of an active anode site.

●6.2   Roughness

Pursuing the idea of surface energy affecting corrosion, we turn to roughness. Noting that a rough surface consists of an irregular series of hills and valleys, the radius of curvature, r, of the surface clearly varies from place to place. The specific surface energy in part varies as does r^{-1} (10§2.2), so points have very high energy, and are thus liable to corrode first. But this does not mean a rough surface will automatically become smooth in a form of electrolytic self-polishing (20§5): the deposition of insoluble corrosion products will tend to fill the valleys, and in so doing inhibit the diffusion of oxygen to the deeper parts, thereby setting up oxygen concentration cells which will cause the valleys to deepen, sometimes to convert to active pits. The roughness may actually be enhanced by such a process. Differential polishing, such as with a two-surface amalgam where only one surface can be easily polished, could set up a cell with the polished surface as the cathode, actually encouraging the corrosion of the rougher part. Thus, while it is important to have smooth surfaces to reduce corrosion, a uniform finish is preferable. Special effort must therefore be made for those less-accessible regions.

However, even if a highly-polished surface has been produced it may still show a high corrosion potential. The act of polishing causes deformation of the surface material and is therefore a form of cold-work (see Fig. 20§2.3). As discussed above (§6.1), such strained material may initially show a high rate of corrosion, perhaps only for a few seconds but, as the strained metal is dissolved, the potential (and therefore the rate) falls. Very carefully annealed electrodes are necessary for accurate potential measurements, free of extraneous effects.

●6.3   Surface energy

As has been pointed out elsewhere (Chap. 10), surface energy is in part controlled by the environment since this is a function of interactions across the interface. Hence any factor which changes that interfacial energy may affect the corrosion tendency of a metal. Examples might be polarizable molecules which adsorb onto the surface, or ions in solution. Some thought will, however, show that many aspects of corrosion have surface energy aspects or interpretations – as for roughness, above. The point here is that these are not mutually exclusive ideas, but the most relevant can be invoked as necessary for any given situation.

●6.4   Temperature

The temperature of a system affects two aspects of its corrosion: rate and outcome. The rate of a chemical reaction is of course usually affected by temperature. If it is an activated process, the activation energy is more available at higher temperatures, and the Arrhenius equation shows that, in many cases at least, the rate increases very rapidly indeed. In particular, the rate of diffusion – an activated process – increases, allowing reactants to come together, and reaction products to get away, thus preventing or limiting polarization (§2). Secondly, the outcome is affected because the position of a chemical equilibrium depends on the relative values of the rate constants for the forward and backward reactions, or the free energies of the reactants and products. It is not always possible to predict the change in outcome for a change of temperature, but in general higher temperatures mean faster corrosion. Thus, corrosion may proceed at entirely negligible rates at room temperature but be disastrous under the conditions found in autoclaves or simply in boiling water.

●6.5   Chelation and complexation

The local conditions at the anode are important, as explained above, in the polarization of the corrosion cell. Apart from precipitation (§2.1) and diffusion (§6.4), there is another means of reducing the concentration of metal ions: the formation of other species in solution. Thus, in the manner of metal ions being bound by electron donors in cements (Figs 9§2.3, 9§6.1, 9§7.5, 9§8.1, 9§8.2) and alginates (Fig. 7§9.2), small molecules or ions may form chelates or other coordination complexes with metal ions, remaining in solution but reducing the concentration of the free metal ion itself and thus shifting the equilibrium, permitting corrosion to continue more readily. Such species as hydroxyl do this, but citrate, oxalate and tartrate ions (*cf.* 9§8.3) are much more powerful – *i.e.* form more stable complexes. Of particular interest in dentistry is lactate, which is present in dental plaque (Fig. 6.2).

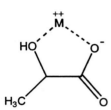

Fig. 6.2 Lactate may chelate metal ions and thus influence a corrosion process.

●6.6   Tarnish

If the rate of corrosion is sufficiently high the corrosion products may not, even if insoluble, remain in the vicinity of the anode. They may be sufficiently fragile or precipitated in such a way as to be readily washed or abraded away. At low rates, however, the corrosion products may adhere tightly to the surface, perhaps excluding further reaction at that site and avoiding oxygen concentration cell corrosion. The resulting discolouration of the metal is usually called **tarnish**. This usage gives the impression that it is somehow fundamentally different from 'corrosion'. This is not so. They are chemically similar processes, the mechanical state or disposition of the corrosion product is all that changes. Even direct reaction of a metal with sulphide produced by oral bacteria from protein and sulphate:

$$-SH \Rightarrow H_2S \tag{6.1}$$

$$SO_4^= \Rightarrow S^= \tag{6.2}$$

is corrosion because metal is being removed from the structure and it is, ultimately, the electrochemical relationships which determine whether the reaction goes. Whether it is discolouring metal sulphide or oxide films or powdery, easily-removed, corrosion products that are formed depends on diffusion, solubility, flow and bulk of solution available, as well as reaction rate, topography and the surface chemistry of substrate and product. It is therefore not satisfactory to assume that tarnish can be ignored.

●6.7   Medium

In discussing the corrosion of dental devices in the mouth, it is implicit that saliva is the medium. Likewise, for implants, tissue fluid or plasma are relevant. These fluids are generally of modest concentration, and benign pH: blood is at about pH 7.4, while saliva varies from about pH 8 when under strong stimulation to as low as pH 5 when in the 'resting' state. Even so, the chloride and other salt content makes them appreciably active in the corrosion sense such that they cannot be ignored.

For intraoral devices, however, there are further challenges. The effects of bacteria have just been mentioned as they affect the device environment under plaque and through their metabolism. In addition, foodstuffs and drinks may be strongly acidic and very salty, they may contain high concentrations of chelating acids such as tartrate and citrate, and they certainly can contain appreciable concentrations of sulphur-containing proteins (egg yolk, cabbage family). To this can be added the effect of some kinds of mouthwash (zinc chloride solution, potassium alum), as well as some fluorides in toothpaste.

Often overlooked are iatrogenic effects. Topical fluoride treatments, for example, may be at very low pH ("APF gel") and high fluoride content. These are known to corrode silver amalgam, dissolve glass ionomer cement, and cause the embrittlement of titanium devices (28§1.5, 28§2.4). Care also needs to be taken with materials such as zinc phosphate cement, which remains acidic for a long period, as well as etchants used on teeth such as phosphoric and citric acids.

There are, of course, other corrosion contexts in dentistry. Autoclaves have been mentioned, but in the laboratory water baths for processing denture bases, pressure vessels for rapid 'cold cure' repair, much wet equipment for handling and processing gypsum products and investments, as well as the spatulas and other instruments used with them. Stainless steel is used almost exclusively now for dental instruments and auxiliary devices such as matrix band clamps, but the reason for the choice – and the remaining vulnerability – should not be lost sight of. Even so, for some applications, such as the bearings of air-turbine handpieces, carbon steel (21§1) is still necessary; these corrode unless protected (although non-corroding ceramic bearings are now available).

§7. Deliberate Corrosion

Corrosion as a chemical process is not necessarily always a bad thing. Electrochemical processes chemically indistinguishable from corrosion are used deliberately, under carefully-controlled conditions, to achieve certain specific effects. Electropolishing, alluded to above, is dealt with in 20§5. It uses a set-up similar to that indicated in Fig. 1.9 except that the workpiece is made the anode, to dissolve unwanted roughness, rather than the cathode, which is plated.

Etching for metallographic purposes is usually done without the benefit of an externally applied voltage. However, the effects can sometimes be assisted by the imposition of a further potential, in addition that is to the potentials due to the galvanic couples that must be present (because alloys are frequently not single phase systems).

●7.1   Bonding alloys

Differential etching is also deliberately used in dentistry with alloys intended for bonding, as in the so-called Maryland bridge. The point is here that obtaining a chemical bond between a casting alloy and a cement of any kind is difficult, and even the polycarboxylate types may not provide enough strength or durability. While resin 'adhesives' may be stronger in themselves, there is a similar difficulty in obtaining a covalent bond to metal or tooth substance. As is used for resin-based direct restorations, etching of the tooth enamel may be employed, not to achieve a chemical bond but to provide a mechanical key. On the other face of the cementing resin a similar effect may be used, *i.e.* between metal and resin. Ordinarily, etching the casting alloy will not produce a deep enough pattern of etch pits for such a key to be effective because they normally have a single (metallic) phase. The risk is of a second, etchable phase being so electropositive as to readily corrode under oral conditions and leave holes which would become active pits and cause the failure of the device. However, by an appropriate choice of composition, a *two-phase* alloy can be designed which can be deeply etched using an electrolytic cell such as shown in Fig. 1.9, and essentially the same as that used for electropolishing. The main requirements are that the phase which is to be dissolved has (a) the more electropositive potential, but not by so much that galvanic corrosion becomes a problem in service, and (b) that it is interconnected – a **continuous phase**, so that a three-dimensional network of pores can exist into which the resin may flow and so achieve a satisfactory key. Such a network can arise from the contact of dendritic crystals of the primary phase (*cf.* Fig. 14§3.1), or equally the matrix or secondary material. By using an electrolytic process the etching can be driven at a high rate and quite selectively.

●**7.2   Die plating**

Plating as a decorative or protective layer in dentistry is not a practical proposition, as has been indicated above, but there is one application for it that is of some importance. A die is a model of a single tooth in or on which a restoration is to be prepared. For this purpose the die needs to be hard and abrasion resistant; artificial stone (gypsum) may not be considered robust enough. In that case, the impression may first be electroplated. Copper plating uses an acid solution containing copper sulphate with a copper anode; silver plating uses an alkaline electrolyte containing silver cyanide with a silver anode. In each case the impression must first be made conductive by coating with a fine copper or silver powder as appropriate (or graphite can also be used). The key point is that the anode is dissolved as the plating proceeds, the electrochemical process being essentially that indicated in Fig. 1.9. The actual chemical and electrical conditions – concentration, pH, additives, current, temperature – are all important in controlling the quality of the plated metal, but the details are beyond the present scope.

●**7.3   Porcelain bonding**

As described in more detail in 25§6.3, in order to obtain a bond between a porcelain coating and a platinum matrix, it is necessary first to create a metal oxide layer. Platinum does not oxidise appropriately, but an electroplated tin layer on that platinum can be made to, thus obtaining suitable conditions for bonding.

References

[1]   https://en.wikipedia.org/wiki/Volta_potential

[2]   Seltzer S, Green DB, Weiner N & Derenzis FA. A scanning electron microscope examination of silver cones removed from endodontically treated teeth. Oral Surg Oral Med Oral Path 33: 589 – 605, 1972.

[3]   Margelos J, Eliades G & Palaghias G. Corrosion pattern of silver points in vivo. J Endod 17: 282 – 287, 1991.

[4]   Wei M, Dyson JE & BW Darvell BW. Factors affecting dental air-turbine handpiece bearing failure. Oper Dent 37 (4): E1 - E12, 2012.

[5]   Ghassan IN, Taylor RR. Powdered gold as a veneer over silver amalgam restorations: A preliminary report. J Pros Dent 20 (6): 561-563, 1968.

[6]   Sarkar NK, Berzins DW, Prasad A. Dealloying and electroformation in high-Pd dental alloys. Dent Mat 16 (5) 374 - 379, 2000.

[7]   Tibballs JE & Erimescu R. Corrosion of dental aluminium bronze in neutral saline and saline lactic acid. Dent Mater 22(9) 793 2006

[8]   Hanks RW. Materials Engineering Science: An Introduction. Harcourt, Brace & World, New York, 1970.

Chapter 14   Silver Amalgam

Dental silver amalgam has been an extremely effective restorative material for a very long time and, despite challenges from cosmetic dentistry and doubts about safety (it is safe), it is likely to continue to be used. While it is capable of good service, it is also subject to abuse. Correct handling will give good results but its behaviour, and the compromises that are involved, need to be understood and respected for high quality work.

*In the setting **reactions** of dental amalgam a series of intermetallic compounds are formed as a solid matrix embedding unreacted alloy. These reactions can be traced through the relevant constitutional (phase) diagrams, although equilibrium is probably never reached. The mechanical and corrosion properties, which involve several compromises, are directly dependent on this polyphase structure. However, the path of these reactions can only be understood in terms of the constitution of the alloy, which varies according to the method and conditions of manufacture.*

*The mixing **proportions** of alloy to mercury are critical in that excess alloy is necessary for strength, yet the safety margin is small. Recent alloy compositions, often including large amounts of copper, can result in some improved properties, but the safety margin for mixing is narrowed.*

*The setting process is associated with **dimensional changes**. These must be controlled if excessive leakage or pain and cracked teeth are to be avoided. Again, this is very much a function of composition and the condition of the alloy.*

*Amalgam is a polyphase alloy and consequently is prone to **corrosion** by galvanic effects. However, this corrosion is beneficial as it normally leads to a seal being formed at the margin. More corrosion resistant formulations (high copper) may not achieve this, and involve other compromises.*

Although dental amalgam is a relatively forgiving material, as a long term restorative it is jeopardized by lack of attention to the critical factors. By understanding the reactions, and recognizing the limitations of this material, sufficient control can be exercised for successful dental use.

Materials Science for Dentistry
https://doi.org/10.1016/B978-0-08-101035-8.50014-6

An amalgam generally is an alloy of a metal with mercury, but in dentistry the term has adopted a more specific interpretation: the result of mixing a silver-based alloy, in the form of small particles, with mercury into a paste which then sets hard. It has been the dominant direct restorative material for many years, and it owes this popularity to a combination of useful properties: relative cheapness, easy preparation in the clinic, simple placement procedures, rapid setting, high strength. However, the apparently forgiving nature of the material is, as with many other materials in dentistry, deceptive. It must be handled with considerable care if its full potential as a very long-lived restoration is to be realized. This care can best be applied with a good understanding of the structure and reactions of the material.

The first mention of such a material being used in dentistry appears to be that in a Chinese *Materia Medica* of the 7th century AD. However, its use in the West seems to date from the beginning of the 19th century. Although the convenience alone of the setting reaction appeared to be the reason for its adoption, no attempt at optimizing properties was really made until about 1896.

There is a second amalgam associated with dentistry, copper amalgam (28§6.5). This is not now known to be in use, and the term "amalgam" is commonly understood in dentistry to refer now only to silver amalgam. In this chapter, 'amalgam' will be used to refer only to the silver-based material, irrespective of any copper present.

§1. Basic Setting Reactions

Most modern dental amalgam alloys are based on or derived from the intermetallic compound Ag_3Sn, the γ-phase of the Ag-Sn system (Fig. 1.1).[1] In the early stages of development, when only Ag-Sn alloys were considered, this overall composition gave what seemed to be a balance between setting expansion (high Ag) and contraction (high Sn) as well as good mixing and handling properties, with good strength. It can be seen from the Ag-Sn constitutional diagram that the γ-phase has very narrow compositional limits, as befits the label 'intermetallic compound'. Compared with that composition, an excess of Ag will lead to the appearance of the β-phase ($\sim Ag_5Sn$ at its high-Sn phase field boundary), while excess Sn merely gives the solid solution of Ag in Sn, which is very nearly pure Sn.

●1.1   Ag-Hg

The initial reaction product of Ag with Hg is the so-called γ_1-phase, approximately Ag_4Hg_5[1] (Fig. 1.2), as can be deduced by reading the diagram from the right at the level of room temperature. We can apply the thought-experiment approach of considering pure mercury to which is added a small portion of silver – effectively what happens as the alloy reacts when amalgam is mixed (*cf.* the zinc phosphate analysis, 9§5). This is an example where **Ostwald's Rule of Stages** (8§2.6) applies quite nicely because of the presence of the highly-diffusible, liquid mercury. We can be sure that γ-Ag-Hg is formed first at normal temperatures because we see from the diagram that it is stable in

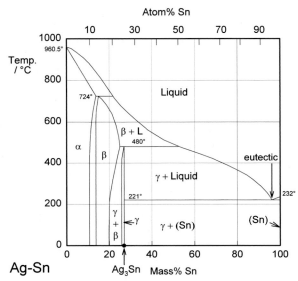

Fig. 1.1 The Ag-Sn system equilibrium phase diagram. The δ-phase, nearly pure tin, has too narrow a terminal phase field to show.

Dental silver amalgam is based on the system Ag-Hg-Sn, and is discussed in terms of and by reference to the three binary systems: Ag-Sn, Ag-Hg, Hg-Sn. Since the phase-labelling in the corresponding diagrams was done independently in each, the phase labels themselves are repeated in the three binary systems, *i.e.* starting from α. To avoid ambiguity, and where there is overlap, in dentistry the phases of the Ag-Hg system are subscripted "1", and those of the Hg-Sn system are subscripted "2", while the labels for the Ag-Sn system are left unsubscripted, as are the Cu-Sn phases discussed later because the labels are not re-used.

---

[1]   This is often quoted as being Ag_2Hg_3 or Ag_3Hg_4; the composition given appears to be more accurate in the dental context, notwithstanding theoretical intermetallic compound compositions.

contact with liquid mercury – while that continues to exist. Equivalently, there is no phase field with the description β + Liq at temperatures below about 127 °C; β is therefore unstable with respect to liquid Hg, hence any silver added must react to give γ immediately. The second Ag-Hg phase, β_1 (approximately Ag_5Hg_4), is expected to form in the presence of excess silver, that is when the Hg has been used up. But since this would be the result of a solid state reaction, it would consequently be expected to appear only very slowly at normal temperatures. We shall return to a consideration of this phase later.

●1.2   Hg-Sn

There is only one reaction product of Sn with Hg at normal temperatures, namely the γ-Hg-Sn phase (Fig. 1.3), which is given the label γ_2 in the context of dental amalgam. This may have a wide range of compositions if only Hg and Sn are present, but in dental amalgam this phase seems to be best described as approximately $HgSn_{7.6}$. This phase has been the focus of much attention from the point of view of strength, creep and corrosion. We will return to these points below.

●1.3   Alloy + Hg

The reaction products of Hg with pure γ-phase Ag-Sn alloy are just the same as with the pure metals, so that the basic setting reaction of such a mixture may be given as[2]:

$$Ag_3Sn \overset{Hg}{\Rightarrow} Ag_4Hg_5 + HgSn_{7.6}$$
$$\text{or:} \quad \gamma \overset{Hg}{\Rightarrow} \gamma_1 + \gamma_2 \tag{1.1}$$

It is worth emphasizing that, contrary to common statements, there is no evidence that Hg can dissolve *in* Ag_3Sn to any measurable extent at the beginning or during the reaction. The decomposition of γ-phase Ag_3Sn is the only consequence of its contact with mercury. Amongst the evidence for this is the failure to detect any transformed zone in the surface of particles of alloy in etched polished sections.[2] Notice that it does not matter whether such a composition is possible (such as by using high- temperature preparation techniques), only that it does not happen under normal conditions.

The conversion of the originally liquid Hg phase to a solid matrix, in which is embedded the remaining unreacted (or **residual**) alloy particles (assuming that there is present *insufficient* Hg for complete reaction of the alloy), constitutes the production of the composite-structured restorative material called dental amalgam, and is responsible for the development of its strength (Fig. 1.4).[3][4]

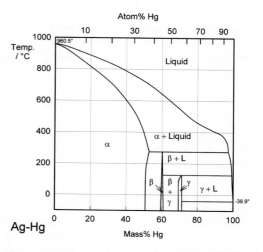

Fig. 1.2 The Ag-Hg system equilibrium phase diagram.

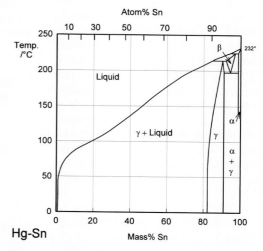

Fig. 1.3 The Hg-Sn equilibrium phase diagram.

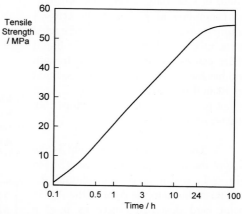

Fig. 1.4 The development of strength with time in a typical setting amalgam (time on logarithmic scale).

---

[2] Notice that there is no need, indeed it is quite wrong to put the residual alloy on the right-hand side of this reaction as is commonly done: it is not formed in the process, it is merely that some remains unreacted, which is understood.

The reaction of β-phase Ag_5Sn is essentially the same as for γ-phase but it occurs at a very much greater rate, although necessarily with different proportions of products:

$$\beta \overset{Hg}{\Rightarrow} \gamma_1 + \gamma_2 \qquad (1.2)$$

If there were excess Sn in the original alloy this would, of course, merely produce more γ_2-phase.

These reactions can be summarized in the (non-equilibrium) constitutional diagram shown in Fig. 1.5.[5] This is to be read by first drawing a line from the Hg apex to cut the Ag-Sn composition line at a point corresponding to the alloy of interest; that is, the line of constant Ag-Sn ratio, which must always be the case for mixing a given alloy with mercury. This Ag-Sn edge of the triangle, of course, corresponds to an isothermal section in Fig. 1.1 at about room temperature. Then, moving along this reaction line **isopleth** from the alloy composition point on the Ag-Sn edge, the consequences of slowly adding Hg to the mixture (or *vice versa*) may be deduced, again doing a thought experiment in the manner used for the discussion of zinc phosphate cement (9§5).

It can thus be seen that only mixtures with the phase description $\gamma + \gamma_1 + \gamma_2$ can be formed from using Ag_3Sn, at least until all that γ has been consumed. There is then only a very narrow field corresponding to $\gamma_1 + \gamma_2$ to be crossed before liquid mercury remains unreacted in the mixture. The presence of a liquid phase causes a dramatic loss in strength and hardness. The effect of composition on Brinell hardness can be judged from Fig. 1.6, where the isopleth is drawn in, and on strength from Fig. 8.5.[6]

The only remaining point is the fate of the nearly pure tin phase, (Sn) in Fig. 1.1, which is present should the composition lie to the right of the Ag_3Sn point, and designated here as β_3 – it simply forms γ_2 very promptly.

$$\beta_3 \overset{Hg}{\Rightarrow} \gamma_2 \qquad (1.3)$$

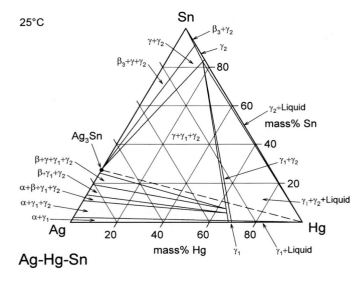

Fig. 1.5 An isothermal section through the constitutional diagram for the Ag-Hg-Sn system as approached from the reaction of an Ag-Sn alloy with Hg, corresponding to the reaction stage that would be considered 'set' in amalgam. This is not an equilibrium diagram.

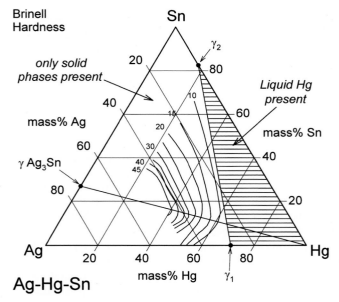

Fig. 1.6 Variation in Brinell indentation hardness (numbered contours) with composition in the Ag-Hg-Sn system corresponding to the reaction stage shown in Fig. 1.5. The line between the points labelled γ_1 and γ_2 represents compositions of stoichiometric complete reaction; it is the solidus.

Tin is a **polymorphic** element, it has two allotropes. Grey or α-tin has a diamond-like lattice and is non-metallic, stable below 13.2 °C. White or β-tin has a body-centred tetragonal lattice and is metallic – the form normally encountered. For this reason the polymorphism is commonly ignored in phase diagrams involving tin, for example in Fig. 1.3 the α-phase is β-Sn! To try to avoid confusion, where it is necessary to give it a phase label now in the dental amalgam context the label is β_3.

●1.4   Cu

The addition of copper to the basic Ag-Sn alloy was found to improve the strength of the resulting amalgam, but too much (greater than ~6 mass%) led to excessive expansion on setting, which risked splitting the tooth or at least pain for the patient. Such alloys were accordingly deemed as not acceptable according to many national standards specifications on these products.   The acceptable compositions are now referred to as low-copper alloys, for reasons that will become clearer shortly. As now the Ag-Cu and Cu-Sn systems make a contribution to the net constitution of the alloy, these two phase diagrams (Figs 12§3.1, 12§5.1) should be reviewed.

The   room   temperature isothermal diagram of the Ag-Cu-Sn system is given in Fig. 1.7, with the detail of the dentally relevant area shown enlarged in Fig. 1.8. In this latter figure the compositions corresponding to formerly 'standardized' and 'acceptable' alloys are indicated by the dotted lines. Notice that *eight* possible phase descriptions may apply to such an alloy at equilibrium. The Cu will be present in part in solid solution in the β and γ Ag-Sn phases. But, depending on its quantity and the amount of Sn in the alloy, either or both of the intermetallic compounds Cu_3Sn (ε-phase) and Cu_6Sn_5 (η') may be present.[7] The solid solubility of Cu in β_3-Sn is very small (see box).

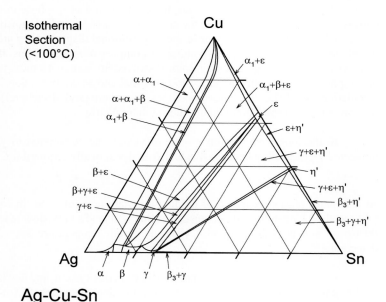

Fig. 1.7 Possible structure of the equilibrium phase diagram for the Ag-Cu-Sn system.

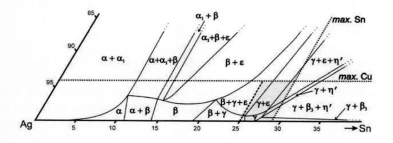

Fig. 1.8 Detail of the region of Fig. 1.7 of interest in dental amalgam alloys based on Ag_3Sn. The dotted lines correspond to the former specified maximum Cu and Sn contents, the dashed line to the practical maximum Ag, so that the allowable composition lay in that enclosed region.

Some Cu can be seen to be in solid solution in the γ-phase (about 1 %). However, the solubility of Cu in the γ_1-phase is appreciably lower (less than a third of the value), and so a Cu-Sn phase must form during the setting process, probably η'.

●1.5   Slow reactions

Additional reactions to those already outlined are, of course, possible when such alloys are mixed with Hg. The reaction rate of β- or γ-phase Ag-Sn with Hg is sufficiently fast that the γ_2-phase is always formed initially, but Cu_3Sn (ε-phase) and γ_2 cannot exist together for long and the relatively slow reaction

$$\varepsilon \; + \; \gamma_2 \;\; \Rightarrow \;\; \eta' \; + \; (Hg) \tag{1.4}$$

must occur. This is at a relatively low rate because it is a solid state reaction. The rate is even slower than might be expected because these two phases will be distributed as small grains throughout the amalgam without necessarily touching, and all diffusion of metal atoms must be through other phases or grain boundaries. The Hg liberated from the γ_2-phase partially reacts further with γ-phase alloy remaining (according to reaction 1.1), and partially is incorporated in the η'-phase in solid solution.

Because of this last behaviour, the composition of η′-phase in the matrix of a set amalgam does not correspond to Cu_6Sn_5, but because the crystal structures are indistinguishable, and there is no boundary formed between them either physically or in the constitutional diagram, they are one and the same phase. Such extra dissolved elements are frequent features of complex mixtures. Thus, it is known that both the γ_1 and the β_1 phases of a dental amalgam contain some dissolved Sn, somewhat more in the latter phase than the former. This has implications for structure, electrode potential and so on, but it does also mean that detailed stoichiometry calculations are difficult. All compositions given here (and elsewhere!) are necessarily approximations, used for convenience, more symbolically than stoichiometric.

In the long term (*i.e.* on a scale of years), the conversion of γ_1 to β_1 may occur in the presence of unreacted γ-phase alloy because this is thermodynamically required (*cf.* Ostwald's Rule of Stages, 8§2.6) :

$$\gamma + \gamma_1 \;\Rightarrow\; \beta_1 \tag{1.5}$$

As this latter phase is capable of holding a certain amount of Sn in solid solution, this may come from the γ-phase also. Under normal circumstances this is indeed a very slow reaction, because it depends on solid-state diffusion, and periods of several years at mouth temperature are required for the product phase to be detectable.[8][9]

Only one other reaction is of significance in this system, and that is in the presence of excess Hg (in terms of the reaction with Ag-Sn phases), namely the decomposition of Cu-Sn phases to give the phase Cu_7Hg_6 (β_2-phase) (see 28§6.5):

$$\varepsilon, \eta' \;\overset{Hg}{\Rightarrow}\; \beta_2 + \gamma_2 \tag{1.6}$$

Ordinarily, however, β_2 should not be detectable. If it is found it can be taken as an indicator of excess mercury in the original mixture (see §4.2). Naturally, γ-phase Ag_3Sn will not then be detectable.[3]

•1.6   Importance of γ_2

It will have been noticed that the γ_2-phase in a set amalgam may be reduced in quantity, or even eliminated, by the presence of ε-phase Cu_3Sn (reaction 1.4). This is a potentially valuable reaction for several reasons. With the exception of the very small amounts of zinc present in the alloy in solid solution (§2.4), tin is the most electropositive element present. In the γ_2-phase it also has a high activity, making this the most electropositive phase, and hence the most easily corroded. Since the corrosion properties of a polyphase alloy system depend on the potential difference between the most and the least electropositive phase, the elimination of the γ_2 would be ordinarily expected to be a distinct improvement. The γ_2-phase is also extremely weak and soft, deforming readily, and the strength of amalgam is generally thought to be limited by its presence. Thirdly, because it is so weak it is thought to contribute to the static creep of amalgam; this is the continuing deformation which is observed under loads well below the conventional yield point (1§11).

•1.7   Added Cu

As has already been said, more than ~6 % Cu may not be included in the so-called 'conventionally' formulated alloy because excessive expansion on setting results. However, if the extra Cu is incorporated in the form of a second alloy powder, mixed with the first (such a mixture of two powders is called an **admixed** alloy), the difficulty may be avoided. One such second alloy which has been found to be particularly effective is the eutectic of the Ag-Cu system; this corresponds to the composition 3Ag.2Cu (Fig. 12§3.1). Recall that this is a two phase alloy (α-Ag + α_1-Cu solid solutions) and not a compound or single phase solid solution (hence the special way of writing its composition). It has the characteristic finely-lamellar

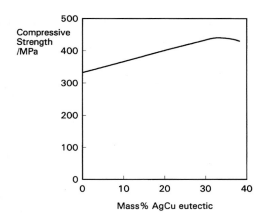

Fig. 1.9 Effect of added Ag-Cu eutectic particles on the strength of a conventionally-formulated, Ag_3Sn-based, amalgam.

---

[3]   The detection of phases may be done metallographically, on polished sections, or by X-ray diffraction as each crystal structure produces a characteristic diffraction pattern (12§2).

structure (Fig. 12§3.7 b). The relevant reactions may be written together as:

$$\alpha + \alpha_1 + \gamma_2 \Rightarrow \eta' + \gamma_1 \qquad (1.7)$$

Again, because this is a solid state reaction, it does not proceed very rapidly, but it may be essentially complete within a few days, but can take a lot longer.[10] The elimination of, or at least great reduction in the quantity of, the γ_2-phase results in improved properties as indicated above, particularly in the strength in compression (Fig. 1.9). There will also be some direct reaction of the kind

$$\alpha \overset{Hg}{\Rightarrow} \gamma_1 \qquad (1.8)$$

This alloy system was not, however, designed on the principle of γ_2-phase elimination. It was originally thought that, in the usual manner of composite structures, the incorporation of a very strong, stiff and hard alloy in the form of fine particles would impart these properties to the resulting amalgam.[11] It was not for some time that the true reason for at least the major part of the improvement in strength was discerned, although the reaction with mercury is thermodynamically required. However, all alloy compositions in which there is substantially more than the previous limit of 6 % Cu are now called **high-copper**, and thus also the amalgam made from them.[4]

●1.8   Summary

The reactions of dental silver amalgam may be summarized as in Fig. 1.10. Unfortunately, there are now many variants on the market, having various modified compositions, some with extra elements such as palladium (Pd) and indium (In). In all cases it is a matter of affecting the ratios of reaction products, reaction rates, and detailed constitution and phase compositions. However, the basic phase description and setting reactions will be much the same. Nevertheless, the discussion above has treated the system as if it were more or less at equilibrium in both the initial alloy and the resulting amalgam. That this is far from the case will be shown. But first, in order that this be understood in context, it is necessary to indicate briefly alloy powder manufacturing methods to show how variation can come about.

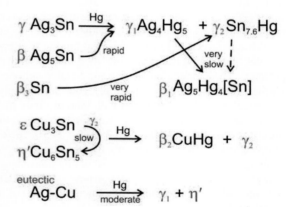

Fig. 1.10 Summary of the principal reactions of dental silver amalgam. The reaction of mercury with β-Ag-Sn is considerably faster than that with γ-Ag-Sn.

§2. Alloy Powder Manufacture

The manufacture of an alloy for dental amalgam essentially consists of the melting together of the component elements to form a single phase, a liquid solution – the melt. This can then be cooled to form the solid which is then **comminuted**, that is, reduced to a powder. It is at the cooling stage that several possibilities arise for the constitution of the alloy, as detailed in §3. The oldest and most obvious comminution method is dealt with first.

●2.1   Lathe-cut

Pouring the melt into a cylindrical iron mould and allowing it to cool unassisted in air gives a solid **ingot**; typically this may be about 300 - 400 mm long by 50 mm diameter. This is then mounted on a lathe and **turned** to produce a very coarse powder. Amalgam alloys are fairly brittle, and the chips fragment readily as they stream from the cutting tool (see Fig. 20§3.2). The stub, where the ingot was gripped in the chuck, and a thin column of alloy that is not readily turned, can be remelted in a subsequent batch. The coarse powder is then **ball-milled** to reduce it to finer particles. This consists of tumbling the powder with large steel or ceramic balls in a large rotating pot, relying on the alloy's brittleness for each particle to be crushed. Simply sieving, or now more conveniently **air-elutriation** (whereby a whirling column of air sorts powders by lifting the particles to different heights according to their mass), then yields a powder of the desired particle size range. These particles are

---

[4] It is stressed that this type of material has nothing to do with 'copper amalgam' (28§6.5).

necessarily irregular in shape (Fig. 2.1). Of importance is the fact that they are heavily cold-worked.

When dental silver amalgam was first invented, it was the practice to use a **file** to create the powder from a solid piece of metal, such as a silver dollar coin. The term **filings** was and may still be applied to the powder prepared by turning and ball-milling.

●2.2 Spherical

Clearly, the above process is time-consuming and requires a great input of energy at the turning and ball-milling stages, much of which is wasted as heat. Spraying from the melt, or **atomization**, avoids both of these stages altogether. The molten alloy is squirted through a small jet into a non-oxidizing atmosphere, whereupon the stream breaks up into fine droplets which, aided by very high surface tension and low viscosity (compare Table 18§2.1), rapidly assume a spherical shape (Fig. 2.2). The very high surface area causes very rapid cooling to solidification, with no time for an equilibrium structure to be developed, but the resulting powder then only needs sorting to give the desired product. These materials are therefore called spherical particle amalgam alloys, or simply (but confusingly) 'spherical alloys' (when mixed with mercury they give 'spherical amalgams'!). By way of contrast, the mechanically-comminuted types are called 'conventional', a term which embraces the compositions near Ag_3Sn, but which is rapidly becoming nonsensical as both composition and presentation now vary so widely in commonly-used commercial products. 'Lathe-cut' is a preferable term.

Fig. 2.1 Some lathe-cut amalgam alloy particles. ×600

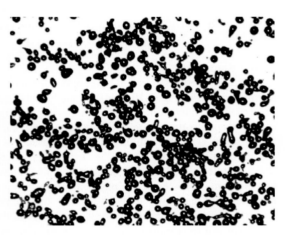

Fig. 2.2 Some spherical amalgam alloy particles. ×45

The Ag-Cu eutectic particles added to certain high-copper amalgam alloys is invariably produced by spraying from the melt. This is the only practical way of producing such powders in bulk because the eutectic is so very tough (characteristic of eutectic alloys); machining such as lathe-turning is virtually impossible.

Even faster cooling is possible. By squirting the melt directly onto a high thermal conductivity surface, cooling rates measured in *millions* of degrees per second can be attained. This may be done on a rapidly rotating aluminium drum, for example. Some (non-dental) alloy systems produce metallic glasses when they are treated in this way. With amalgam alloys, however, merely a very fine-grained and highly-strained structure is formed in the ribbon of metal, and this fragments readily to a powder. The crystal structure is also very imperfect, and this causes certain problems, discussed below. No commercial exploitation of this is known at present, although as a method it has been investigated extensively.

●2.3 Blending

The manufacturing techniques described above result in powders that have different internal structures and different particle shapes. The phase constitution of the alloy is of great importance to the behaviour on setting, as will be detailed in the next section, and this depends on the composition to a large extent. It may be that a single alloy with a given composition has some undesirable feature. In addition, the particle shape affects the 'feel' of the material when it is being packed; this is dealt with in §4.1. Sometimes the desired behaviour and properties cannot be obtained by preparing a powder from just one composition melt using just one method. Sometimes it is only a matter of creating a product which is patentable, perhaps avoiding a rival's patent. In each case, however, blends of alloys may be used. That is, powders of different composition, thermal history and particle shape may be mixed to create the commercial product. There are indeed many variations possible and that have been sold. Whilst these blends may introduce differences from the descriptions in this chapter in various ways, they do not seriously affect the principles described here. They are neither good nor bad in any general sense, and such products must be judged on properties such as strength, expansion and cost. Such alloy products may be described as **admixed**, with the implication of the **admixture** being in lesser amount than the 'basis' alloy powder (Fig. 2.3).

•2.4   Contaminants

Whatever the technique subsequently used to create the powder, all processes must first melt the alloy components. Tin and copper, however, oxidize in air, and molten silver-rich alloys may dissolve large quantities of oxygen from the air, encouraging the oxidation of the Cu and Sn. The composition changes that then occur make such oxidation undesirable. Even if the metals are melted under an inert gas blanket, such as argon, it is difficult to exclude air completely, and even the original silver ingots may contain oxygen from their previous melting. Sulphur is also frequently present as an impurity in many metals. Such contaminants as these may also change the properties of the resulting

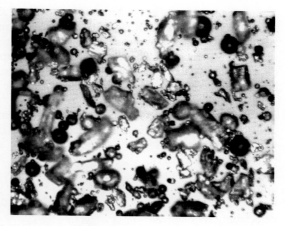

Fig. 2.3 Some admixed amalgam alloy particles. ×135

alloys. However, both oxygen and sulphur can be readily **scavenged** from the melt by the addition of a small quantity of a more electropositive metal, and zinc is convenient: it is both cheap and reactive. Typically, about 2 % by mass may be added to the melt, although it is expected that much of this will be oxidized or volatilized and therefore lost from the system, reducing the amount left in the final alloy, perhaps to very low levels. This remaining zinc will be present in solution in the γ-phase, and end up in the γ₁-phase after reaction.[12] These **deoxidized** alloys are typically less brittle than their oxygen-containing counterparts, and this increased ductility is associated with a more easily polished and potentially tougher amalgam that has been reported to perform better in service.[13] Remaining zinc also has marked effects on other mechanical behaviour and properties.[14]

A variety of other contaminants may be present, generally at low concentration, such as Sb, Cd, In, and Pb,[15] according to the nature of the source ores for the principal components and the purity (price!) of those component metals as bought. Generally, these will be present in solid solution in the alloy, and thus affect mechanical properties. Their fate in a set amalgam is unreported.

§3. Cooling Rate and Constitution

In order to understand a number of aspects of the setting reaction process and outcome it is necessary to examine in some detail the constitution of the alloy. As previously discussed (12§1), the rate of cooling from the melt has profound effects on the phase description of the resulting solid, which may be very far removed from that expected at equilibrium; this applies equally to amalgam alloys. The manufacturing process is, however, not concerned with creating a thermodynamically-ideal equilibrium system, but rather a product that behaves in a useful fashion. It is therefore useful to illustrate the segregation and phase transformations that occur using a combination of metallographic images and phase diagrams.

•3.1   Solidification

Using a typical low-copper alloy composition (< 6 % Cu), where the equilibrium condition would be expected to consist only of γ + ε, (Fig. 1.8), moderately fast cooling results in very clear coring.

The first phase to form is β-Ag-Sn, in the form of dendrites (branching, tree- or fern leaf-like crystals), embedded in a matrix γ-phase which, of course, must have frozen later (Fig. 3.1). This outcome can be seen to be required by tracing the cooling of Ag₃Sn alloy in Fig. 1.1 where the β + Liquid field is the first to be encountered from the melt:

$$\text{Liq} \;\Rightarrow\; \beta_1 + \text{Liq} \tag{3.1}$$

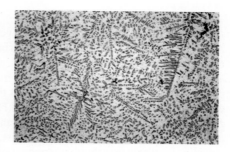

Fig. 3.1   Ag-26 %Sn-5 %Cu : Cooled quickly after casting; polished section etched to show β-Ag-Sn dendrites as the fern-like structures surrounded by a matrix of γ-Ag-Sn. ×150

Appropriate etching of a polished section of such an alloy specimen reveals a fine dispersion of eutectic (Fig. 3.2), the material last to freeze (see Fig. 1.1). This material consists of a matrix of β_3 (which is nearly pure Sn) containing an extremely fine dispersion of a small quantity of γ-phase, too small to be detectable at the present scale but which we know from the phase diagram must be present.

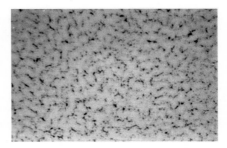

Fig. 3.2 The same alloy as in Fig. 3.1, but etched to show just the finely distributed eutectic particles. ×550

Somewhat slower cooling results in more prominent β-phase dendrites (Fig. 3.3) and coarser eutectic (Fig. 3.4), although the Cu-Sn phases (if present) remain relatively fine-grained and well dispersed and not yet visually detectable (although X-ray diffraction would show these phases to be present (Fig. 12§2.7)). Much slower cooling produces very large dendrites (Figs 3.5, 3.6) which are expected to show a gradation of Ag-content according to the variation of the composition of the liquid as solidification proceeds. This has shown up in the etched section by variation in the colouration (*viz.* low Ag ≡ dark staining). Note that the dendrites are still all β-phase structure; it is because that phase field is broad (Fig. 1.1) that there is a wide composition range possible. Surrounding each dendrite is a layer of γ-phase (Ag$_3$Sn) which has not been etched; embedded in this are grains of ε-phase as well as eutectic.

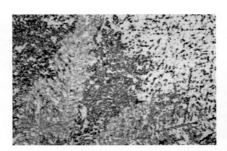

Fig. 3.3 Same alloy as Fig. 3.1, treated in similarly, except lower rate of cooling so that the dendrites are larger, representing a larger volume fraction of the metal. ×150

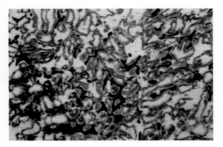

Fig. 3.4 The same specimen of alloy as in Fig. 3.3. The voids (dark spaces) indicate where the eutectic was before etching. ×550

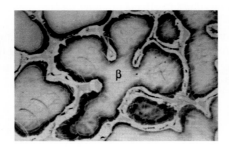

Fig. 3.5 Again, the same alloy as in Fig. 3.1, treated similarly except rate of cooling very low so that the dendrites are very large. The Sn-rich boundary of the β-dendrites is stained dark – the more Sn, the darker. Surrounding γ-phase layer is pale, with slightly darker islands of ε-Cu$_3$Sn. The eutectic was removed by the etching (black). (*cf.* Fig. 12§1.9) ×150

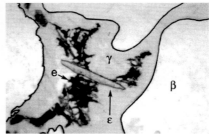

Fig. 3.6 The same specimen of alloy as in Fig. 3.5. This shows more clearly the grains of ε-Cu$_3$Sn embedded in the γ-Ag-Sn, and the voids (dark spaces) where the eutectic (e) was before etching. ×550

For a full appreciation of the range of possible phase descriptions, reference must be made to the full Ag-Cu-Sn ternary diagram (Fig. 3.7), and in particular to the **melting grooves** on the liquidus surface. This kind of diagram can be imagined as a part of a hilly landscape, where the stream beds are the exact topographical equivalent of the melting grooves. The main melting groove is the locus of the eutectic point for a series of pseudobinary (vertical) sections through the prism-shaped body. This "minimum melting" path starts at the eutectic for the Ag-Cu system (Fig. 12§3.1), and then goes through the points labelled *B*, *C*, *D*, *F*, *G*, *H* – where *H* is the eutectic point of the Ag-Sn system (Fig. 1.1). The 'tributaries' correspond to other phase changes. In a binary alloy with a eutectic point the last liquid to be present when it is cooling and segregating is indeed the eutectic. Similarly, in a three-component system the liquid has a composition which follows the liquidus surface (as in 12§1.5), but now it is for a line corresponding to the path of steepest descent on the 'hillside'. It does this until its composition corresponds to that of a melting groove, and then the liquid composition follows that path – because this is now the path of steepest descent. The liquid composition therefore follows the track that a ball would take if placed on the surface and allow to roll.

In a diagram such as Fig. 3.7, the faces of the prism correspond to two-component T-X diagrams, while the base is a three-component mixture triangle. Thus, the visible faces of Fig. 3.7 are, from the left, the two-component diagrams of Fig. 1.1 and Fig. 12§5.1, the hidden face is Fig. 12§3.1, and the base is Fig. 1.7.

A partial map of the melting grooves for the dentally-relevant part of the Ag-Cu-Sn system is shown in Fig. 3.8. The phase label that is shown in a region is that of the solid to form from a liquid composition lying within that area, that is, with the melting grooves as boundaries. If the liquid composition lies exactly on a boundary (*i.e.* is actually in the melting groove), the two solid phases shown either side of the line at that point will form simultaneously – just as in a binary eutectic. Once a liquid composition and its temperature correspond to lying in a groove, the liquid composition must follow that track all the way to the final eutectic (*H*).

Also shown in Fig. 3.8 are a few of the kind of tracks expected for the appropriate isothermal tie-lines (see Fig. 8§4.3) which pass through the (low-copper) composition range area of dental interest (dotted) – bearing in mind that these are projected onto the plane. Briefly, these tie-lines connect the solid forming at any instant with the liquid from which it is forming – the conjugate phases, in the same way as do tie-lines in a two-component diagram (Fig. 8§3.7). Thus, given a composition for the alloy melt, as solidification and segregation is occurring we can see how the composition of the liquid varies: following first the tie-line track until a melting groove is reached, and then following that groove path.

Following those rules, studying these diagrams shows the following possible sequence of reactions for a dental amalgam alloy on cooling from the melt at a rate too fast for equilibration of the solid

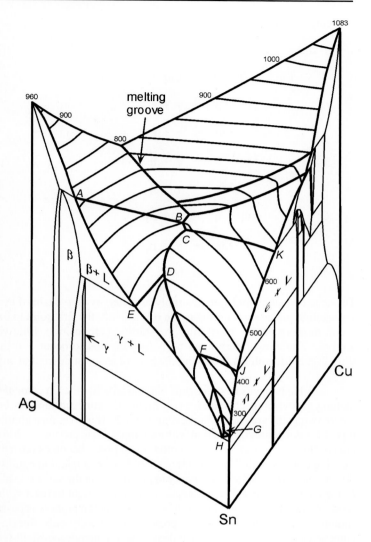

Fig. 3.7 The liquidus surface of the Ag-Cu-Sn system, where height above the base represents temperature. The isothermal contours on that surface are shown with the 'melting grooves'. The points labelled A, B ... correspond to those in Fig. 3.8.

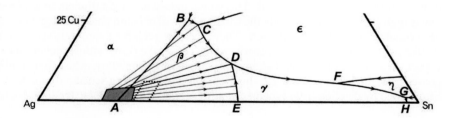

Fig. 3.8 A partial map of the melting grooves of Fig. 3.7, also showing some possible tie-lines relevant to the cooling of a melt of a conventional amalgam alloy formulation (dotted rhomboid - see Fig. 1.8). The tie-lines connect the β-solid solution phase field (shaded) to the melting groove, and so are not all at the same temperature. This is a plan view of the part of the liquidus surface shown in Fig. 3.7. The phase labels show the solid being formed for a composition in that field; for a point on a melting-groove boundary both phases indicated by the adjacent labels form together. The letters A, B ... correspond to the labels in Fig. 3.7.

to occur, *i.e.* normal rates of cooling:

$$
\begin{array}{l}
\text{Liq} \;\Rightarrow\; \beta \\
\qquad\qquad \nwarrow\;\searrow \\
\text{Liq} \;\Rightarrow\; \beta + \varepsilon \qquad \text{Liq} + \beta \;\Rightarrow\; \gamma \\
\qquad\qquad \searrow\;\swarrow \\
D:\quad \text{Liq} + \beta \;\Rightarrow\; \gamma + \varepsilon \\
\qquad\qquad\qquad\downarrow \\
\text{Liq} \;\Rightarrow\; \gamma + \varepsilon \\
\qquad\qquad\qquad\downarrow \\
F:\quad \text{Liq} + \varepsilon \;\Rightarrow\; \gamma + \eta \\
\qquad\qquad\qquad\downarrow \\
\text{Liq} \;\Rightarrow\; \gamma + \eta \\
\qquad\qquad\qquad\downarrow \\
G:\quad \text{Liq} + \eta \;\Rightarrow\; \gamma + \beta_3 \\
\qquad\qquad\qquad\downarrow \\
\text{Liq} \;\Rightarrow\; \gamma + \beta_3 \\
\qquad\qquad\qquad\downarrow \\
\qquad\qquad\qquad H
\end{array}
\qquad (3.2)
$$

> In this account, reference is made to the high-temperature phase η-Cu_6Sn_5, as this must be the first to form. It will transform to η' on cooling, as indicated by Fig. 12§5.1, but as this is without change of composition, the constitution of the final alloy is only affected by that small label change.

The scheme can be understood simply as a step by step chain of events representing the various stages of the segregation process, according to the rules. Thus, the first solid to form is always β, and the composition of the liquid moves steadily away from the starting composition, following a tie line, until it reaches the melting groove. The temperature must be falling for this to occur and therefore the track on the liquidus surface (Fig. 3.7) is 'downhill', following the path of steepest descent at any point.

What happens next depends only on whether the Cu content of the initial melt was high enough that the melting groove is reached above (*i.e.* higher Cu concentration) or below the junction point D; this is the point where two paths are shown in the diagram 3.2. If above, then simultaneous freezing of $\beta + \varepsilon$ occurs; if below, then there is a peritectic reaction with existing solid to produce γ. Either way, because these two melting grooves are confluent, the next step is common: the β phase becomes unstable, and reacts with the liquid while $\gamma + \varepsilon$ is freezing out. Downhill from D there should, in principle if at equilibrium, be no β present (although solid state diffusion is too slow for this to occur in practice), but the same two phases continue to freeze. Then, below point F, the ε becomes unstable and again in principle ought to react with liquid to form $\gamma + \eta$ until that ε is in turn used up, but the same precipitation continues. When point G is passed the η just produced is now unstable and should react with the liquid now present. Eventually only freezing of $\gamma + \beta_3$ is occurring, at least until the eutectic point, H, is reached when there is simultaneous, constant-temperature freezing of the same two phases as a eutectic mixture.

Thus all five possible phases plus the eutectic mixture may, and usually do, result from an alloy whose nominal composition corresponds to just one phase, *i.e.* γ-Ag-Sn with Cu in solid solution. Mostly, the complexity can be seen to arise from the failure to attain equilibration where a reaction of the form

$$
\text{Liq} + \varphi \;\Rightarrow\; \cdots \qquad (3.3)
$$

(φ is an arbitrary phase) is expected. This would indeed occur if sufficient time were allowed, but if the cooling rate is high enough, as is normally the case, crystals of existing phases may get coated with new solid and only be accessible by solid-state diffusion. They therefore get left behind, so to speak, as the reactions then become ineffectually slow. It is just this process which causes the composition gradient seen in Fig. 3.5. We therefore expect, as a matter of course, that an dental silver amalgam alloy, when first solidified, contains all possible phases from the list in scheme 3.2. In effect, the actual cooling scheme for an Ag_3Sn alloy (*i.e.* no Cu) would be obtained by deleting from scheme 3.2 all reactions of type 3.2, *i.e.* the lines labelled D, F and G.

The usual process, then, may be illustrated as in Fig. 3.9. Taking as an example a composition near Ag-Cu6-Sn25 (top left corner of the formulation range rhomboid, Fig. 3.8) so that the lowest tie-line reaches the melting groove C-D, the first solid to form is β. The liquid composition now tracks the liquidus as segregation occurs (12§1.6) and β continues to form, although steadily decreasing in Ag-content. When the temperature has fallen to correspond with that of the melting groove, the situation is now that of a **quasi-eutectic**, where the solidification of two phases occurs simultaneously:

$$\text{Liq} \;\Rightarrow\; \varphi_1 + \varphi_2 \qquad (3.4)$$

in this case, $\beta + \varepsilon$. After this point, the liquid composition path is that of the melting groove.

If an ingot is poured for lathe-turning, the initial cooling at the mould walls can be rather rapid although necessarily the first phase to freeze is β. However, the remainder will cool relatively slowly, and a structure very far removed from equilibrium will be produced, one which shows the effects of the sequential changes given above. This would be a problem in itself, as the constitution of the powder would be complex and uncontrolled from batch to batch, but the reactivity to mercury would be so high as to render the alloy unusable. Further, the method of comminution does much cold-work on the metal, leaving a highly-strained structure that would be even more highly reactive, making setting far too fast.

●3.2   Heat treatment

The first difficulty is to overcome the coring and segregation in the solid ingot. This is done by

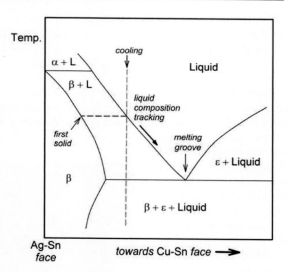

Fig. 3.9 Sketch of a quasi-eutectic pseudo-binary section of Fig. 3.7, taken along a tie-line leading to the melting groove between C and D, to illustrate the solidification process for one alloy composition example.

annealing it for **homogenization**. By raising the temperature the diffusion rates of atoms in the solid are raised greatly, and a better approach to an equilibrium constitution is thereby attained. This annealing must clearly be done at a temperature below that of the peritectic decomposition of the γ-phase, which is 480 °C in the pure Ag-Sn system (Fig. 1.1) but somewhat lower in the presence of the dissolved Cu (in Fig. 3.7, point D is at a lower temperature than E), perhaps 465 °C. Other additives, such as Zn, will lower this even further. Temperatures in the range 350 to 450 °C ($T/T_m \sim 0.84 - 0.98$) for as much as a day or as little as a few hours have been used. With such treatment there is an extensive reorganization and redistribution of the elements present, but as explained before (12§3.6) this is towards equilibrium at the annealing temperature, very far removed from the temperature at which the alloy will be used. On cooling, then, the alloy would still be unstable and over-reactive, but it has yet to be reduced to a powder. When this has been done the strain introduced by the comminution process must be removed. To a certain extent this wastes the annealing that was done on the ingot, and a further high-temperature treatment is necessary.

For this reason the homogenization may be performed after turning rather than before, with similar effect but in addition allowing relief of cold-work. In fact, the defects present after cold-work may enhance the rate of homogenization by encouraging diffusion. Nevertheless, it is still a high-temperature structure that has been produced and, with time, even at room temperature, some continued diffusion occurs, leading to a steady decline in reactivity (*i.e.* setting rate) over some months. This is unacceptable since the commercial product would vary from batch to batch, depending on how long it was in storage before sale. Accordingly, the alloy powder is *artificially* **aged** by heating to about 100 °C ($T/T_m \sim 0.5$) for a few hours. This accelerates the stabilization, and commercial alloy powders then show no detectable change in setting rate with time.

Despite annealing, in commercial alloys all combinations of phases may still be found. This is largely due to the large distances over which diffusion has to occur after segregation, the difficulty of this in the solid state, and the fact that the relatively small further gains in stability and properties would not justify the expense of the heating for the longer times necessary. This then can give a wide variety of phase descriptions in the set amalgam because there may be many grains of various phases within particles of alloy, inaccessible to the mercury and reaction products. In addition, due to the varying rates of reactions, the phase description may change with time in the long term. It is thus impractical to rely on the overall composition of alloys in attempting to understand the behaviour of amalgams prepared from them. Rather, actual tests must be conducted to determine clinical suitability, as thermal history, mechanical treatment and particle size are expected to contribute to the overall pattern of behaviour and to control actual property values.

The effect of annealing a phase with a composition which can be at equilibrium only at some higher temperature has been discussed previously (12§3.6); this can readily be observed in the case of β-Ag-Sn.

Reference to Fig. 1.1 will show that a portion of the solidus between the β and β + Liq phase fields lies over the β + γ field. In other words, the reaction

$$\beta \;\Rightarrow\; \beta + \gamma \qquad\qquad (3.5)$$

is expected to occur at low temperatures. During the annealing process crystallites of the relatively Sn-rich γ-phase precipitate from the β, which therefore remains as a matrix phase, but with a higher proportion of Ag, yielding the characteristic Widmanstätten structure (Fig. 12§3.10). This can be clearly seen in Fig. 3.10, where the β-phase has been etched in a sample of a commercial alloy, otherwise consisting of γ-phase.

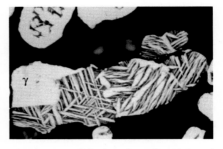

Fig. 3.10 Polished section of particles of commercial lathe-cut amalgam alloy, etched to show Widmanstätten structure: γ-phase precipitated in matrix of β-phase. Remainder of alloy mostly γ-phase. ×500

Fig. 3.11 Polished and etched section of amalgam from a similar alloy as in Fig. 3.10. Remains of Widmanstätten structure (arrowed) still visible in residual γ-phase (grey). γ₂-phase removed (black). ×500

The β-phase is very reactive towards mercury and will be rapidly quickly consumed after mixing. The effect of this is detectable in the set amalgam where the unreacted crystallites of precipitated γ-phase remain in the characteristic pattern (Fig. 3.11); this pattern survives because of the large quantity of γ present, and the fact that a near minimum amount of mercury is used.

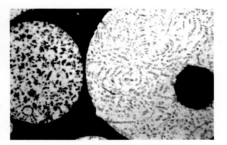

Fig. 3.12 Polished and etched section of spherical amalgam alloy powder showing β-dendrites. ×500

The alternative commercial method of producing alloy powder, atomization, gives cooling rates many times faster than occur in cast ingots; the β-phase dendrites are accordingly less well developed (Fig. 3.12, *cf.* Fig. 3.1). Even so, the structure is important as, again, the β-phase reacts preferentially with Hg, and the pattern may be seen to persist in the set amalgam (Fig. 3.13). Hence, the reaction product γ₁ may be seen to pervade the residual γ-phase and the outline of the original alloy particles may be only slightly reduced in diameter compared with the original alloy particle.

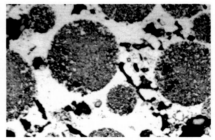

Fig. 3.13 Amalgam from alloy shown in Fig. 3.12, sectioned, polished and etched to show residual γ-phase (grey) and γ₂-crystal sites (black). ×500

●3.3 'Preamalgamated' alloys

A minor constitutional point may be raised here. Some alloys have been available described as **preamalgamated**,[16] by which was meant that an attempt had been made to 'plate' the surface of each particle with mercury, on the naïve assumption that it will dissolve in the alloy and so make subsequent mixing easier. As has already been stated, this solid-solution assumption is false. The manufacturing process consists of stirring the alloy powder in an aqueous solution of mercuric chloride when a displacement reaction occurs:

$$HgCl_2 + Cu \;\Rightarrow\; Hg + CuCl_2 \qquad\qquad (3.6)$$

Even though a very small amount of Hg is deposited (of the order of 2 or 3 mass percent) it is all found as γ₁-phase, as identified by X-ray diffraction, and distributed in minute grains over the alloy surface. The reaction is, of course, the normal one (reaction 1.1). There is no detectable change in the crystal lattice constants for the γ-phase, which would be expected if solid solution occurred, even after heating, which converts all the γ₁ to β₁-phase (reaction 1.5).[2] Similar effects are found for electroplated material.[17] The supposed easier mixing appears to be an unsupported conjecture, although it is possible that the tiny specks of γ₁-phase interrupt the film of oxides that subsequently forms on the alloy surface on exposure to the air, forming a slightly easier path for the mercury to follow when starting to wet the alloy. The advantage would seem slight if it does exist. On the other hand, a greater amount of mercury will be present after mixing since the γ_1 would appear as part of the solid alloy powder. This has implications for the mixing ratio (§4.2).

§4. Mixing Proportions

Set dental amalgam is a composite material consisting of a core of unreacted alloy embedded in a matrix of reaction products. It depends for its strength on the much stronger original alloy in the same way that other composite structures rely on the strength and hardness of the core. Again, the target is to get as high a volume fraction of core as possible, which means in effect the least possible mercury in the finally placed restoration. Precisely the same principles of gauging, dilatancy, dilution and reaction allowances were described for gypsum products (2§2) apply here and a **minimum mixing volume** is required to minimize porosity. Immediately after mixing, when only little reaction has occurred, we can treat the mixture as a simple slurry and understand viscosity effects just as in many other systems.

●4.1   Packing

The process of packing an amalgam mixture into a prepared cavity is commonly known as **condensation**, although a better term might be **consolidation** or compaction. The irregular particles of lathe-cut alloy inhibit plastic deformation of the whole mass through friction and interference. This gives a faintly crunchy 'feel' to the mix under the packing instrument and its ability to be compacted with moderate to high forces without being displaced. Packing an amalgam consists of the squeezing together of alloy particles, expressing some of the still-liquid mercury in the process, thus raising the 'core' alloy content. The more mercury that is squeezed out, the better.

Spherical particle powders, however, although having an appreciably lower gauging volume and a lower dilatancy allowance because of their particle shape, show very low resistance to penetration by the packing instrument, merely moving aside as pressure is applied. This kind of behaviour demands a different technique, with lower forces and wider packing instruments, but because of that it often fails to allow satisfactory moulding of the mixture to the detail of the cavity. The tendency for the spherical particle powders to give faster setting amalgams, because of the amount of β-phase, complicate their use due to this handling difficulty. The initially-perceived advantages of spherical powders (ease of manufacture, low mixing mercury) were somewhat diminished in importance by these observations. In a compromise between the feel of lathe-cut powders and the low Hg requirements of the spherical type, mixtures of the two have been sold.

●4.2   Mercury equivalence

That the remaining alloy is important has been stressed for reasons of the mechanical properties (§8.2). To ascertain the adequacy of meeting this criterion, the amount present under any given conditions needs to be calculated. If we consider the basic setting reaction:

$$\gamma \;\overset{\text{Hg}}{\Rightarrow}\; \gamma_1 + \gamma_2 \tag{4.1}$$

and we take its stoichiometry, for convenience of calculation only, as being approximately

$$8Ag_3Sn + 31Hg = 6Ag_4Hg_5 + Sn_8Hg \tag{4.2}$$

the amount of mercury required for complete reaction is thus found from the atomic weights of the elements concerned to be about $1 \cdot 76$ g/g alloy. This corresponds to a mass fraction, $\mu[\text{Hg}]$, of $0 \cdot 637$ $[1 \cdot 76/(1 + 1 \cdot 76)]$ or, as it is more usually quoted, $63 \cdot 7$ % by mass. For a lathe-cut alloy powder the gauging Hg might typically correspond to about 40 % by mass. As mixing amounts of Hg of about 50 or 60 % by mass are commonly needed, there is clearly a considerable amount extra required to allow for dilatancy and dilution, not to say reaction, to obtain a suitably plastic mass. The coarser the alloy powder (if of irregular particles) the larger the amount of mercury required. However, there would seem to be little risk of stoichiometric excess, given that some mercury is always expressed during packing, and assuming that there is no delay in the completion of this process. With spherical particle alloys this risk is further reduced because the dilatancy requirement is much lower (but not zero – see Fig. 5§2.2).

The situation changes, however, with high copper content alloys. In order to assess the effect of composition more easily than working out the alloy constitution and then all possible reactions, a shortcut can be taken by calculating only what may be termed the **molar mercury equivalent**, E(Hg), of the alloy, expressed in moles/mole. Thus, 1 mole of Ag is taken as the equivalent of 5/4 mole Hg, because that is the molar ratio in the γ_1-phase, irrespective of the phase in which the Ag is initially found in the original alloy. By multiplying the mass fraction for that Ag in the original alloy by that molar ratio and the atomic mass number ratio, Hg/Ag = 200·6/107·9, we obtain the mass equivalent of the mercury for that Ag. Thus, applying this procedure to the Sn and Cu as well as the Ag, we find the following:

$$E(Hg) = \left[\frac{\mu[Ag]}{107 \cdot 9} \cdot \frac{5}{4} + \left(\frac{\mu[Sn]}{118 \cdot 7} - \frac{\mu[Cu]}{63 \cdot 5} \cdot \frac{5}{6} \right) \cdot \frac{1}{7 \cdot 6} \right] \times 200 \cdot 6 - \mu[Hg] \qquad (4.3)$$

where the correction to the Sn term according to the Cu content is because of the formation of η'-phase, which uses up Sn and prevents it tying up some Hg. This internal bracketed quantity clearly is only valid if non-negative, otherwise it is set to zero. Of course, any Hg already present (see §3.1) must be discounted in this calculation, hence the last term. This equation simplifies to:

$$E(Hg) = 2 \cdot 32 \mu[Ag] + (0 \cdot 22 \mu[Sn] - 0 \cdot 35 \mu[Cu])_{>0} - \mu[Hg] \qquad (4.4)$$

In other words, this is the number of moles of mercury required to fully react with the unreacted Ag and Sn in 1 mole of the alloy in question. In effect, this is equivalent to a titration of the alloy with mercury, where the end-point is the disappearance of all Ag-Sn phases. For Ag_3Sn itself, this works out at 1·78 g/g, or 64 % by mass of the mixture, slightly more than was found in the cruder calculation above. Even so, uncertainties in the composition of product phases, because of the presence of other dissolved metals, mean that there is some similar error in any calculated equivalent, say ±1 %. We have ignored the formation of CuHg because, if it is going to form, the implied absence of the strengthening core of γ-phase means an unsatisfactory amalgam anyway. Hence, we may calculate the molar mercury equivalent for any alloy, and from this the mass-percentage of mercury in the just fully-reacted mixture with alloy, E(%), to enable comparison with the manufacturer's recommended proportion. Some analyses and their corresponding mercury equivalents are given in Table 4.1. These calculations are the same as determining where on the mixture line in Fig. 1.5 or 1.6 the final amalgam composition would be plotted for a Ag-Sn alloy, allowing also for the presence of the Cu.

Table 4.1.   *Mercury requirements for four representative commercial alloy compositions.*

Ag	Sn	Cu	Hg	E(Hg)	E(%)	mix Hg%	
69·9	26·0	3·9	-	1·67	62·5	58·0	safe
59·2	27·9	12·8	-	1·39	58·2	43·5	safe
42·8	29·9	25·5	1·6	0·98	49·4	53·5	unsafe!
39·9	30·2	29·9	-	0·93	48·1	46·0	safe?

E(Hg)   is the molar equivalent of mercury for 1 mol alloy
E(%)   is the mixture mass percentage of mercury corresponding to E(Hg).
mix Hg%   is the amount recommended by the manufacturer for that product.

From this it can readily be seen that there is a great danger of the total amount of mercury left in the mixture at the end of packing, the so-called **residual mercury**, exceeding the reaction equivalent, such that there is in fact **excess mercury** present (which will be liquid):

residual mercury - alloy-equivalent mercury = excess mercury $\qquad (4.5)$

It is then clear that any mercury initially present (§3.1) actually makes the situation worse.

An amalgam with excess mercury like this is necessarily rather weak and unserviceable (§8.2). Whether such amalgams present a toxicity hazard, possibly by having free Hg available for an extended or indefinite period, remains to be determined, but obviously the use of high-copper alloys demands greater caution and care in use as regards proper packing than do 'conventional' low-copper alloys.[18][19]

Incidentally, we refer to 'residual mercury' because this figure is generally intended to be less than the proportion present at mixing. Packing is intended to remove some, and the reference is to the amount not expressed. Obviously, this process is operator-dependant, which is where the risk of excess mercury being present arises. The term had greater significance when amalgam was hand-mixed and a much larger proportion of mercury was required for effective mixing, much of which extra mercury was then expressed before packing (15§1.1). We may note the contrasting (and confusing) connotation for the phrase 'residual alloy': that which remains unreacted when the amalgam has set.

In this context, the common process of packing 'through' the mercury-rich layer with each successive increment is clearly beneficial in reducing the porosity that might be trapped between layers. However, it is important to ensure that the packing is efficient so that the expressed, mercury-rich band rises (without leaving pockets behind), eventually to emerge above the intended final surface before carving is started: overfilling is essential. If this is not done, the concentration in the topmost portion of the restoration will possibly (indeed, very likely) exceed the mercury equivalent of the alloy present there, and so be weak and unserviceable. The greatest risk will be in the regions hardest to pack – the margins – which are at the greatest risk of fracture.

●4.3   Setting rate

We can now assess the meaning and implications of "slow" and "fast" setting amalgam products. Whilst it is apparent that reaction rate can be affected by both composition and history (thermal and mechanical), it appears that in pre-dosed capsule products a simple variation of the amount of mercury corresponds to the rate of setting labels applied. For example, ranging from 46 to 49 % for fast to slow, respectively. Plainly, the mercury equivalent is not changed, nor indeed is the actual rate of the reaction in any chemical sense, but the residual mercury must be increased, if for no other reason than that a longer time will be taken by the operator, by choice. The risk of an unsafe mixture in the above sense is therefore increased for so-called "slow-setting" amalgam. Similar arguments apply for the recommendations regarding mixing bulk alloy powder.

The discussion in 2§11.2 is relevant here, but we can extend it with a consideration of the perception of working time. It would seem that, in a subjective manner, the force required to condense the amalgam is what matters. In other words, the pressure under the condensing point that the material can withstand without being deformed or rearranged. Obviously, this depends on the diameter of the tip, but the rheological properties as the amount of liquid changes are the key. As the volume fraction of solid increases, so does the interference between those particles, eventually 'locking up' as the material becomes dilatant. From this point, deformation can only generate cracks, and attempts to reclose these will only produce voids in the mass. It can then be seen that the use of "wet" techniques, where excess extruded mercury is allowed to lie on the surface, trying to consolidate the amalgam through it, is beneficial to the extent that any cracks will then fill with mercury rather than air. However, this cannot be taken too far, as the mercury equivalent may be exceeded locally and the upper portions of the restoration may contain no residual alloy.

●4.4   Tin solubility

Although it is clear from Fig. 1.6 that the strength of amalgam is affected by alloy composition (as well as mercury content – see also Fig. 8.5), what has been generally overlooked is that the solubility of tin in mercury is appreciable, around 0.8 mass%, 1.3 at% (Fig. 4.1), while that of silver is low (0.04 mass%, 0.08 at%) (Fig. 1.2).[20] What this means is that when the amalgam is packed and the 'mercury rich' layer is created and then scraped away, a much greater proportion of tin is removed than of silver, disturbing the ratio in the set material. Clearly, the more mercury present in the original mixture, the more tin is lost. That is, the 'wetter' the technique the more the final composition is affected. This may be beneficial in that the amount of the weaker γ_2-phase is reduced, but it does imply that there is an extra source of variation in experimental results when the mixing ratio is altered, even for the same alloy. At (open) mouth temperature the effect would be increased.

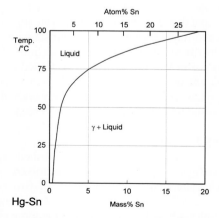

Fig. 4.1   Detail of Fig. 1.3, showing the appreciable solubility of tin in mercury at ordinary temperatures.

§5. Constitution of Set Amalgam

The γ_1-phase is the principal matrix phase of dental amalgam. Moderately strong, it accounts for the bulk of the reaction products, as can be judged by the first term in equation 4.4 above. However, it is not the first to form in a highly ordered crystal structure.

●5.1 γ_2

The γ_2-phase crystallizes in very thin plates, that is they have a very high **aspect ratio** (ratio of greatest to smallest dimension). As such, these usually appear as only thin lines in polished sections of amalgam. This appearance has very frequently led to a misinterpretation of the structure, so that it is described as forming 'needles'. Clearly, needle-shaped crystals, randomly oriented in space, if cut by an arbitrary plane will only produce a spotty pattern of cross-sections in that plane. Thin plates, equally, must produce a pattern of lines. If an amalgam is allowed to set against glass, for example, the γ_2 crystals can be readily identified on the surface from their morphology (Fig. 5.1), the narrow edges of the crystals corresponding perfectly to the lines of dark areas in the etched section (Fig. 5.2). As might be expected, only occasionally will a crystal form at the surface parallel to that surface (as at the centre of Fig. 5.1).

This characteristic 'ink-blot' pattern can also be found in polished sections when the plane of the section coincides with the plane of the crystal, as may be seen near the centre of Fig. 5.2. These thin plate-like crystals, interrupted frequently but with the different portions plainly aligned with each other, grow at random orientations. Inevitably they intersect each other from time to time, so that the pattern of seemingly disconnected grains in fact represents a good many inter-crystal contacts. It is too strong to describe this arrangement as a network but it is true that, if there is sufficient γ_2-phase present, a nearly continuous path throughout the entire piece could be found within that phase alone. This has an important implication for the corrosion of amalgam, a point to which we shall return (§9).

From the crystal habit it would seem likely that the γ_2-phase is in fact precipitating very early on in the setting process, before the γ_1 matrix has become sufficiently well-formed for the evident long-range structure of the γ_2 crystals to be prevented or obscured altogether. Were this timing even slightly different, isolated and uncorrelated near-isotropic crystals of γ_2 would be expected distributed evenly across any section. The interruptions that do occur in the line of a cross-sectioned crystal are probably due to fragments of γ-phase lying in the way when the crystal grows. That γ-phase may then have completely reacted and so not be detectable by the time the polished section is made.

●5.2   Copper

The appearance of polished and etched sections of amalgams prepared from **admixed** high-copper alloys is characterized by the presence of two types of particle outline (Fig. 5.3). The one is that of the conventionally-formulated γ-phase-based alloy, and the other more circular outline is marked by what has been described as a 'reaction halo', consisting principally of η'-phase surrounding the unreacted Ag-Cu eutectic. No γ_2 is usually detectable in such amalgams, at least not after a few days. The reaction may be summarized as:

$$Cu + \gamma_2 \Rightarrow \eta' \qquad (5.1)$$

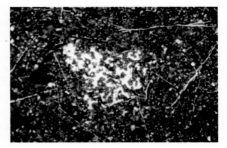

Fig. 5.1   Surface of low-copper 'conventional composition'[!] amalgam after setting against glass, showing γ_2-phase crystals mostly as edge-on plates. The crystal in the centre is flat on the surface. Note the very rough surface. ×500

Fig. 5.2 Polished and etched section of the same type of amalgam as in Fig. 5.1, showing γ_2-phase sites (black) as cut planes. The crystal near the centre was sectioned at a shallow angle. ×500

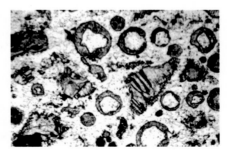

Fig. 5.3   Amalgam from 'Dispersalloy' alloy, sectioned, polished and etched. Spherical Ag-Cu eutectic particles surrounded by reaction 'halo' of η'-Cu_6Sn_5. Remains of Widmanstätten pattern in Ag-Sn particle to right of centre. ×500

where the η′ again is understood to contain Hg in solid solution. It is reported that a minimum of about 12 % Cu is required for there to be no detectable remaining γ_2-phase. Obviously, this figure is sensitive to the amount of mercury which reacts with the γ-phase to release Sn.

High-copper content amalgam alloys that use Ag-Cu eutectic particles are sometimes described as "disperse-phase", referring to those particles. This is an erroneous usage as the eutectic is, of course, two-phase.

The Cu-Sn phases are otherwise found dispersed as fine grains throughout both the core alloy and the reaction product matrix. Any original ε-Cu$_3$Sn that is exposed to the matrix is slowly converted to η′-phase, as described at reaction 1.4, which can be seen metallographically. Under other circumstances, most especially in high-copper amalgam, fine columnar grains of η′ can be precipitated throughout the γ_1 matrix and these are thought to contribute to the hardening of the amalgam (*cf.* 19§1.5) by inhibiting dislocation movement (note: they cannot act as is sometimes claimed as 'pins' between adjacent grains of γ_1 to prevent intergranular slip; this is not an available mechanism – precipitates cannot grow across grain boundaries in that fashion). This could be especially important in amalgam since the γ_1-phase does not show strain-hardening (nor indeed does amalgam overall). This property is probably attributable to the fact that the normal working temperature, *i.e.* in the mouth, is very close to the peritectic melting temperature of about 72 °C (§5.5, §7), so that the **homologous temperature**, $T/T_m \approx 0.90$ (using absolute temperature). In other words, recrystallization and diffusion to annihilate dislocations is easy (plastic deformation at normal temperatures is therefore best described as 'warm work').

The absence of strain-hardening also permits the **burnishing** of surfaces and margins without the detriment of increased brittleness or tendency to corrode. Such plastic deformation to reduce roughness and improve marginal adaptation, by rubbing with a polished, rounded instrument, results in rearrangement of material, cold-welding of overturned asperities, and thus the filling-in of surface porosity. Even so, burnishing is not a satisfactory means of rectifying poor work.

●5.3   Zinc

Zinc, if present, remains in solid solution in several of the phases already described. It requires an unrealistically large amount to be present for any separate Zn-based phase to appear. Its importance in the sense of amalgam constitution is restricted to very slightly changing the phase-field boundaries and melting temperatures of various phases, although (if the Zn has done its job properly, §2.4) these will be entirely negligible for most ordinary purposes. That said, it may have a role in self-sealing corrosion (§9.1).

●5.4   Heat effects

As has been indicated already, dental amalgam as it is normally prepared cannot be considered as being at equilibrium. A particular illustration of this is the persistence of the γ_1-phase in the presence of excess γ, when, according to the information of Fig. 1.2, reaction 1.5 is expected to occur. Being dependent on solid state diffusion, the rate of this reaction increases at higher temperatures. From about 60 °C or so, β_1-phase forms quite rapidly enough for the effects of its presence to be seen.

This order of temperature is not normally attainable in the mouth as a result of hot foodstuffs, but the friction during the polishing of an amalgam restoration may generate much heat, especially if inadequate or no cooling is used, or the polishing implement is applied continuously rather than intermittently. In the process, γ_2-phase will have reacted also:

$$\gamma + \gamma_1 + \gamma_2 \;\Rightarrow\Rightarrow\; \beta_1 \qquad (5.2)$$

This effect is readily demonstrated. Fig. 5.4 shows the corroded cross-section of a cut surface of an amalgam sawn with a high-speed circular saw under streaming water. Despite the copious supply of coolant, a layer of some 50 μm thickness has been very obviously altered as the characteristic etching pattern of the γ_2-phase is no longer detectable in that region. Electrochemical changes are also involved, and the corrosion behaviour modified. Fig. 5.5 similarly shows the corrosion (under different conditions) of a heat-transformed area (on the left), arising from the heat of cutting that

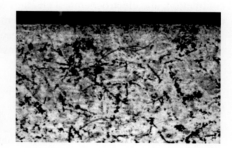

Fig. 5.4 Section of an amalgam through a cut surface. After polishing the section it was allowed to corrode, thereby showing a thermally transformed layer at the cut. ×150

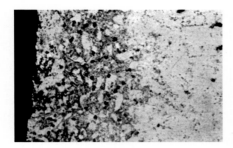

Fig. 5.5 Corroded area of polished amalgam corresponding to the overheated region. ×150

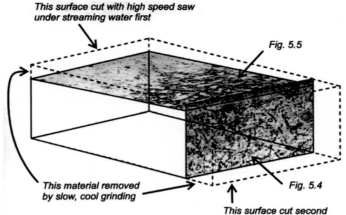

Fig. 5.6 Diagram of the orientation and procedure used to obtain Figs 5.4 and 5.5.

surface, while the deeper portion remains unaffected (see also Fig. 5.6). Clearly, amalgam polishing must be done in such a way as to minimize the local temperature rise, and avoid phase changes which may be deleterious. Likewise, if an existing amalgam restoration is to be repaired, requiring some cutting to be done, proper cooling is again essential.

•5.5   Peritectic melting

If the overheating is taken to extremes, further changes occur. Fig. 1.2 shows a peritectic reaction at about 127 °C:

$$\gamma_1 \quad \leftrightarrow \quad \beta_1 \; + \; \text{Liq} \tag{5.3}$$

In dental amalgam, where the presence of Sn, Cu, and possibly a small amount of Zn, is expected to alter the energy balance of all equilibria, the transformation temperature is found to be at about 72 °C. This temperature is very readily produced during abusive polishing with inadequate coolant. The melting is quite obvious. However, it is often incorrectly described as "bringing the mercury to the surface". This description is as unhelpful as it is inaccurate because it might be taken to imply a beneficial effect, whereas the converse is most definitely the case. There is no movement or migration of Hg, and the similarity of the appearance to that during proper packing, when Hg really is expressed, is quite superficial.

On cooling, of course, since the β_1-phase is expected to have a composition near to equilibrium (*cf.* Fig. 1.2: the isothermal tie-line at the peritectic temperature) the remaining Liquid, which is nearly pure Hg with only ~1 % Ag dissolved, will react with further γ-phase. Reaction of ε and γ_2-phases is also expected (reaction 1.4). These may appear to be advantageous changes, but the melting also implies mechanical disruption of the restoration with loss of contour, thermal expansion causing distortion followed by a non-compensatory contraction, and electrochemical changes which complicate the corrosion processes. Any one of these faults is enough to justify avoidance of the problem, quite apart from the challenge to the pulp of cooking it. However, as is seen in Fig. 5.4, the γ_2 phase also reacts, leading to the β_1 containing a substantial proportion of Sn. This may be the reason for the enhanced corrosion of the matrix when this has been overheated (Fig. 5.5).

§6. Dimensional Changes

Dimensional changes on setting have been alluded to above, and these are important from the point of view of an adequate marginal seal (*i.e.* not excessive shrinkage) and avoidance of pain or splitting of the tooth (not excessive expansion). A typical amalgam setting dimensional change curve is shown in Fig. 6.1. Immediately after packing a rapid contraction may be observed, followed by a slower expansion, and then a slight, and slow, contraction. It is necessary to appreciate why these dimensional changes occur.

●6.1   Shrinkage

The efficiency of packing of metal atoms in a crystal lattice depends on the lattice itself, whether it is close-packed or a more open structure, very regular or distorted. On the other hand, the packing density depends on the sizes of the atoms concerned. We may estimate the packing efficiency by dividing the mass density, ρ (kg/L), of the material by the average atomic molar mass, $\bar{A}$, to obtain the number of moles of atoms per litre (mol/L) – the 'molar' density. As the mole is defined in terms of a number of atoms, we could use this figure directly. However, slightly more convenient is the reciprocal of this, the molar volume (mL/mol). A list of values for some of the phases of interest is given in Table 6.1.

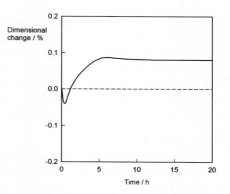

Fig. 6.1 Typical dimension changes occurring on setting of a dental amalgam.

Table 6.1 *Structure data for amalgam phases* *(some data are unavailable)*					
phase	**ρ**	**average atomic weight, $\bar{A}$**	**$\rho / \bar{A}$**	**molar volume**	**crystal structure**
	g/mL	**g/mol**	**mol/L**	**mL/mol**	
β-Sn*	7·3	118·7	61·5	16·3	b.c.tetragonal
Ag	10·5	107·9	97·3	10·3	f.c.c.
Hg	13·5	200·6	67·3	14·9	liquid
Ag_3Sn	9·8	110·6	88·6	11·3	orthorhombic†
Cu_6Sn_5	8·3	88·6	93·6	10·7	h.c.p.
Ag_4Hg_5	13·6	159·4	85·3	11·7	b.c.c.
$Sn_{7·6}Hg$	~8·0	~128·2	62·4	16·0	h.c.p.
Cu_7Hg_6	13·1	126·8	103·7	9·6	cubic
Cu	8·9	63·6	140·5	7·1	f.c.c.
* see box, §1.4			† similar to h.c.p.		

If we consider, say, the basic setting reaction (reaction 4.1), and the same simplified stoichiometry (4.2):

$$8Ag_3Sn + 31Hg = 6Ag_4Hg_5 + Sn_8Hg \tag{6.1}$$

the total relative volumes of reactant and product can be calculated:

$$8 \times 4 \times 11\cdot3 + 31 \times 14\cdot9 \Rightarrow 6 \times 9 \times 11\cdot7 + 1 \times 9 \times 15\cdot9 \tag{6.2}$$

or

$$823\cdot5 \Rightarrow 774\cdot9 \tag{6.3}$$

i.e. a decrease of ~5.9 % by volume, or ~2.0 % by length. This is, of course, a very large shrinkage and would not, on its own, be clinically acceptable (most amalgam standards specifications set the limit on overall acceptable dimensional change as "±20 μm/cm", *i.e.* ±0.2 %). This shrinkage can be seen to be due largely to the effect of 'solidifying' the Hg and, despite the lack of information concerning some other phases, similar effects are expected for the relevant reactions as well. However, it can be expected that the faster stages of the reaction (with β-Ag-Sn or β_3-Sn for example) will have occurred first, during the mixing and packing process. The reaction with γ-Ag_3Sn is slower, but still a proportion of that reaction will also have occurred by the time that packing is complete. Therefore, a large proportion of the shrinkage will be unobservable except by using special techniques. There is ordinarily a considerable delay in preparing a test piece for the experimental determination of dimensional changes, as indeed is true for a real restoration. A large portion of the shrinkage will thus merely be unobserved – which is fortunate. Were it not for this timing, amalgam would probably not be a viable restorative material. That part of the shrinkage occurring before packing is completed simply does not matter.

It may also be noted that in mixing amalgam mechanically, and appreciable rise in temperature must occur (15§2.2). However, given the small mass involved, and the length of time taken to pack, and the general circumstances of use, cooling will occur well before the restoration is completed such that thermal contraction is of no concern. On the contrary, the temperature of the tooth in an open mouth must be below normal, and thus after completion some thermal expansion must occur. The thermal expansion coefficient is about 25 MK^{-1} (Fig. 3§4.14), so that assuming a rise from 20 °C to 37 °C we have a total of about 0.04 %, which on its own would be significant, but this occurs while the material is still very soft (in much less than an hour) and so probably does not contribute effectively to the overall expansion on setting.

●6.2   Loss of gloss

Nevertheless, there is one very important manifestation of the Hg shrinkage: an amalgam can never seal the cavity into which it is placed, even if a laboratory test shows an overall expansion. The reason is similar to the **loss of gloss** phenomenon of gypsum products. Shrinkage due to reaction will initially cause a decrease in bulk volume, but once contacts between alloy particles or new crystals interfere with this there is no choice but for the Hg to be withdrawn into the mass, leaving behind the outline of the alloy particles (Fig. 6.2). The crinkly surface thus produced clearly cannot be in continuous contact with the cavity wall, and a leakage path therefore exists (*cf.* Fig. 5.1). This leakage path survives even if there is a subsequent expansion (whether thermally or otherwise) because the forces generated are unable to cause enough plastic deformation of the now quite solid amalgam, and especially not of the original alloy particles that are sticking out at the cavity wall surface.

Fig. 6.2 The failure of any amalgam to provide a marginal seal is due to the shrinkage of the liquid mercury on reaction and surface tension.

The effect does in fact lead to loss of gloss, as can be readily observed: a freshly mixed pellet of amalgam (if containing a clinically useful amount of mercury) is very smooth and shiny – metallically wet-looking, but as setting proceeds it acquires a 'frosted' or sand-blasted appearance. This effect can also be seen in amalgam set against glass (Fig. 5.1). It should be obvious that coating the cavity wall with varnish can have no effect on this process. But if one is concerned about the sensitivity of the tooth to the effects of marginal leakage, the use of a dentine bonding agent is probably the best solution. It will not prevent or change the leakage around the amalgam, but the living tissue will at least be isolated.

There is a second aspect which is illustrated in Fig. 6.2: the effect of surface tension. While it may be possible, with the application of a sufficient pressure to cause an unset amalgam paste to conform reasonably well to a cavity's detail, there will still remain a problem. This will be evident in particular at the corners between faces and edges, because the liquid mercury has a contact angle of 180° on tooth tissue. The situation is similar to that of casting metals in investment moulds (Fig. 18§3.7) – complete filling is impossible. Thus one can expect a leakage path along all such angles, and this may in fact be even more severe than that due to loss of gloss.

●6.3   Expansion

As in gypsum products, the impingement of growing crystals one on another will cause outward forces which will result in some (small) expansion. This is **crystal growth pressure** again (2§5.1). Superimposed upon this will be the slower settling down (*i.e.* shrinkage) of the crystal structure as disorder is reduced (the disorder expected because of the rapidity of the precipitation of the product phases and the 'warm-work' being done on the already precipitated crystals during packing). Again superimposed are further small changes in volume as other much slower reactions occur, such as with the Cu-Sn phases, but these are of unknown sign and magnitude. In addition, it must be presumed that the loss of gloss (§6.2) leads to a contraction force opposing expansion as in setting gypsum, a force that could be more substantial because the surface tension of mercury is so much higher than that of water (Table 18§2.1), given that mercury wets the reactant and reaction product phases. However, it is not known how the timing and magnitude of this affects the setting dimensional change curve. The situation would be complicated by the availability of excess mercury at the surface (*cf.* "hygroscopic" expansion, 2§5.2) in wet packing techniques.

Overall, the crystal growth pressures seem to dominate the observed macroscopic dimensional change. The effect of increasing packing pressure, for example, is to reduce the expansion (Fig. 6.3).[21] This can be understood in part by the increased plastic deformation of the still relatively soft γ_1-phase, although the effect of squeezing out excess mercury leading to a lesser extent of reaction must play a major part. The overall expansion due to crystal growth pressure is expected to produce a small amount of sub-microscopic porosity, which would not be distinguishable from the porosity due to the mixing and packing stages. Even so, too little is known about the system and it is impossible to sort out all of the components of the process exactly; only broad empirical indications of effects are possible.

For example, increasing mixing time leads to less expansion, and perhaps ultimately contraction (Fig. 6.4). Even particle size, because it affects the rates of the reactions, will give variations in dimensional change (Fig. 6.5).[22] And while the expansion due to excess Sn in the alloy (over the Ag_3Sn amount) is discernible from the data of Table 6.1, that of Cu in excess of 6 % by mass is not. Obviously, the effect of solid solution elements, such as Sn in γ_1, needs to be taken into account. There are nevertheless other manufacturing variables, largely in terms of heat treatment, that can affect setting dimensional change, and it seems safest overall to determine the amount experimentally for each alloy and set of mixing and packing conditions. Even so, it is clear that guidelines can be given:
- use fine grain alloy,
- do not overmix; and
- use the maximum packing pressure possible.

This last is probably largely due to the fact that the residual mercury will be minimized, and the extent of reaction therefore limited.

●6.4 Delayed expansion

There is another potential source of expansion quite apart from the reactions of the setting process, namely 'delayed' or moisture contamination expansion (Fig. 6.6).[23] This has caused much concern clinically, so much so that zinc-free alloys have been produced (and continue to be marketed). This effect is due to a straightforward corrosion reaction arising with Zn-containing alloys. Hydrogen may be produced if water (especially with dissolved salts as in sweat or saliva) is incorporated in the mix, and the Zn-content is high enough:

$$Zn + 2H_2O \Rightarrow Zn(OH)_2 + H_2(g) \qquad (6.4)$$

The pressure due to the hydrogen eventually becomes large enough to cause flow, distortion of the restoration, and possibly pain.

However, if normal good clinical practice is followed, there should never be any moisture contamination

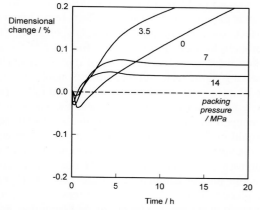

Fig. 6.3 The effect of packing pressure on setting dimensional change.

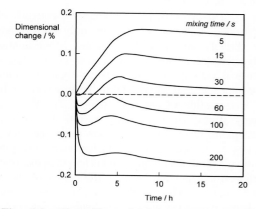

Fig. 6.4 The effect of mixing time on setting dimensional change.

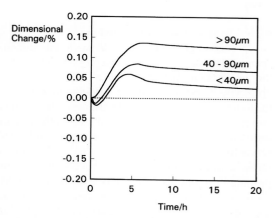

Fig. 6.5 Effect of alloy particle size on setting dimensional change.

and so no problem arising from this source. Indeed, it seems to require extraordinary efforts to demonstrate the effect, such as trituration with saline solution. This is unlikely to occur in the clinic. But even if a non-zinc alloy were used, the inclusion of *any* water during the packing of the restoration could reasonably be expected to be disastrous because corrosion would follow (even if without hydrogen production) and the strength of the amalgam would be severely affected because increments would not adhere to each other so well. The crevices remaining would facilitate corrosion in the mouth anyway. Another implication of this is that amalgam should not be handled with bare skin (as if mercury hygiene were not a strong enough reason). In fact, there is positive evidence in favour of Zn-containing amalgam (§2.4).

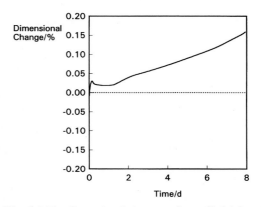

Fig. 6.6 The dimensional changes of so-called delayed expansion.

●6.5   Extrusion

However, of the many cases frequently cited of 'extruded' amalgam restorations, only a few, if any, may be directly attributed to moisture contamination and 'delayed' expansion. Extrusion is most often seen in cervical restorations where masticatory loads would place a compressive stress parallel to the free surface, because of deformation of the tooth (Fig. 6.7). When applied many times, such stressing results in creep of the amalgam, which thus becomes squeezed out of the cavity.[24] The deflections of the tooth structure, given that Young's modulus for amalgam is less than that of enamel, are also exaggerated because the cervix of the tooth is undermined. Also, while amalgam shows flow, tooth substance does not at the same loads. It does not require much extrusion for the edge of the restoration to catch a probe tip, but delayed expansion is unlikely to be the cause.

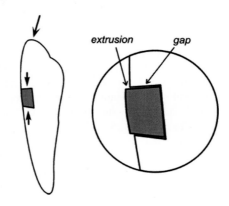

Fig. 6.7 Extrusion of a cervical amalgam restoration by deflection of the undermined tooth by masticatory forces.

●6.6   Matrix-band compression

Although nothing to do with the dimensional change of the amalgam, there is a problem which may give the impression that it is. An MOD amalgam will ordinarily be placed by using a steel matrix band around the tooth to form the proximal surfaces. Clearly, this must closely conform to the shape of the tooth to avoid the formation of ledges, and the band may be tightened by a screw to achieve this. However, the band also produces forces acting on the crown in a bucco-lingual direction, pushing those walls of the cavity together (Fig. 6.8). The

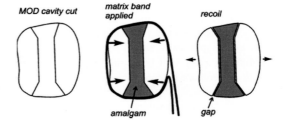

Fig. 6.8 An overtightened matrix band will compress the tooth and guarantee excessive leakage for an MOD restoration.

deeper the cavity, the greater the effect. The amalgam may be placed perfectly, but when the band is removed the highly-elastic enamel will recover its original shape, and there will now be no contact between tooth and amalgam along those walls. The gross leakage that is then present may suggest an untoward contraction of the amalgam, which is therefore blamed, whereas the fault is in the technique. The fact that the deformation of the tooth is not easily visible leads to the assumption that it is quite rigid, but one must recognize the effects of material properties whether or not they are immediately detectable. Thus, as little force as possible (preferably none) should be used in tightening the band, but good wedging will achieve the desired effects.

It should then be apparent that these remarks apply equally to any direct restorative material, whether filled-resin, glass ionomer, or anything else. The elastic recovery of the tooth will generate a force great enough to break any adhesive bond or mechanical key that may be achieved, and if not immediately then the stress will hasten its breakdown, mechanically and hydrolytically.

§7. Mechanical Properties

The contribution of the presence of the γ_2-phase to static creep has been mentioned (§1.6), but there are other factors. Creep is the phenomenon whereby at loads well below the yield point continuous plastic deformation occurs (1§11). The process is very similar to that depicted for Maxwell-type behaviour when the rheological properties of more fluid substances are discussed (4§4). The creep (or flow) shown by metals is just at a somewhat lower rate than, say, polymers. The initial part of plot of strain against time for a given load is an exponential curve, and is completed quite quickly with most metals. The straight line portion of the plot, or **secondary creep**, is the feature of general interest and the one studied for amalgam. This kind of creep is essentially due to diffusive processes. These become significant when $T/T_m > \approx 0{\cdot}3$ and because diffusion is an activated process, there is an Arrhenius Law dependency on temperature.[25]

The creep rate of amalgam is dependent on its mercury content (Fig. 7.1). This shows principally the effect of reducing the volume of the rigid, unreacted γ-phase particles, as the rate becomes extreme when there is none remaining. Creep is also dependent on the amount of γ_2. In fact, control of creep has been used as a major justification for the use of high-copper amalgam alloys because of the consequent elimination of γ_2. It should be noted that while the creep value for such amalgam may be much reduced, it cannot be eliminated, and a similar increase as the residual alloy is reduced should be expected. However, even though extrusion is observed in cervical amalgam restorations (which should no longer be very common), the clinical significance of creep remains obscure.

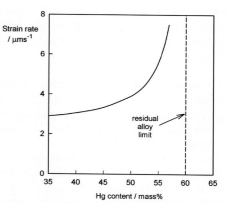

Fig. 7.1 Creep rate vs. mercury content for an ordinary low-copper alloy amalgam. Note the limit when liquid mercury remains.

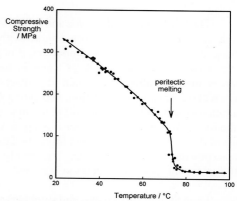

Fig. 7.2 Variation of amalgam strength with test temperature.

As mentioned earlier (§5.2), the normal operating temperature of dental amalgam (37 °C) is a large fraction of the peritectic melting temperature of the γ_1-phase, expressed in kelvin (§5.5), *i.e.* $T/T_m \approx 0{\cdot}9$. This means that, as for all material at such homologous temperatures, its strength is relatively low, and very temperature sensitive (that is, the slope against temperature is steep) (Fig. 7.2) (see also 10§3.8).[26][27] This provides another example of the need to match test conditions to service conditions if behaviour is to be understood properly.

§8. Porosity

As has been mentioned before, porosity severely reduces the strengths of materials (2§8), and this is no less true for amalgam (Fig. 8.1). While this is itself detrimental, it also contributes to the creep: the amount of creep after a fixed interval increases steadily with increasing porosity (Fig. 8.2). Likewise, the modulus of elasticity will be affected (9§4). It should be pointed out though that porosity in dental amalgam is quite unavoidable, there being four distinct sources.

●8.1   Sources of porosity

Firstly, during mixing the vigorous stirring and grinding together of the alloy powder and mercury encloses bubbles which

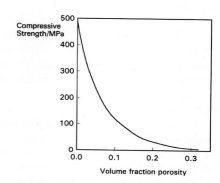

Fig. 8.1 Variation of amalgam strength with porosity at constant mercury content (*cf.* Fig. 2§8.3).

are unable to escape because of the high viscosity of the mixture, despite the very high surface tension of the still liquid mercury. Such bubbles can be seen in polished sections of amalgam pellets that have been allowed to set without disturbance after mixing, and these bubbles largely survive the packing process.

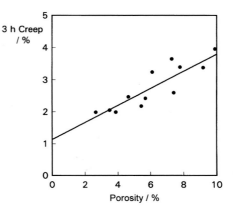

Fig. 8.2 Variation of creep rate in amalgam with porosity at constant mercury content.

Secondly, the packing process introduces more. Each increment of mix that is carried to the cavity has a more or less rough surface that traps air as it is pressed onto the surface of the amalgam that is already there, whether the technique employed is 'wet', 'dry', or anything in between. Such porosity appears in well-defined layers between each increment, very obvious in a polished section perpendicular to the packing direction (Fig. 8.3). Further, for each thrust of the condensing point that penetrates the surface, or folds over a ridge of displaced mix, further air is entrapped (Fig. 8.4). While good technique can reduce the amount of porosity, and high forces help, it cannot be eliminated.

The third source is the oxide and sulphide coating that is usually present on any amalgam alloy arising from contact with moist air; in other words, the tarnish. This is the single most important, if not the only, reason for the need for vigorous mixing, for clean Ag, Cu and Sn are spontaneously wetted by Hg, whereas amalgam alloy can remain in contact with Hg for many minutes or hours without appreciable reaction, depending on the amount of tarnish. The trituration process rubs or grinds off this film to permit reaction. The existence of the oxide film is detectable after mixing as a very fine grey coating on the walls of the capsules, sometimes in large enough amounts that it can be seen as a loose powder free in the capsule. The freshly mixed pellet may even appear dull grey for this reason when it too is coated.

Fig. 8.3 Porosity due to incremental build-up of an amalgam. (Sagittal section through cylindrical specimen, 3 mm diameter, 8 mm high.)

Even though such oxides and sulphides are not wetted by Hg (*i.e.* metal on ceramic), the particles may still be mixed into the amalgam, when they will be segregated in pores, exacerbating the overall porosity and interfering with the welding of layer to layer. However, it is where the oxide has not been lifted off the surface of the alloy that the most damage is done, for there simply will be no bond between matrix and core at that point, and thus the fundamental criterion for a successful composite structure will have been violated. Furthermore, that non-wetted region is in fact a pore, a real gap between matrix and alloy, possibly exaggerated by the mercury shrinkage discussed above (§6.1). It is porosity not readily detected by optical means, but corrosion behaviour can reveal its presence very clearly.

The potential fourth type is microscopic porosity due to crystal growth pressure expansion and mercury consumption shrinkage. The scale of this is likely to be very small. Its significance has not been investigated.

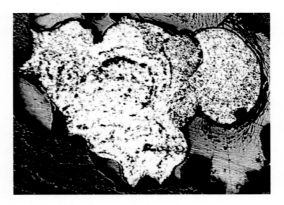

Fig. 8.4 Porosity in an amalgam due to poor packing technique. (Polished transverse section of a very large occlusal restoration.)

●8.2   Strength

The variation of amalgam strength with mercury content (Fig. 8.5)[28][29] can now be explained. It follows the usual understanding of a core-matrix composite (6§1.13), with the addition of the effect of porosity (2§8, Fig. 8.1). Thus, at low Hg there is insufficient liquid to fill the pores – the **minimum mixing liquid criterion** is not met. When the pores are just filled there is a maximum in strength because the volume fraction of residual alloy is also at a maximum (*cf.* Fig. 2§8.2). After that, increasing mercury steadily reduces the strength as original alloy is replaced by reaction product phases, which are weaker, but at some point excess liquid mercury remains when the stoichiometric point has been passed (Fig. 8.6). Then the strength falls very quickly because this liquid phase has no strength (*cf.* Fig. 1.6) and behaves as porosity (2§8.3). This may be compared with the effect on creep (Fig. 7.1).

As with gypsum products, the mixture corresponding to maximum strength cannot be used in practice because it is too viscous and has too high a yield point (the material is plastic dilatant, 4§7.7) and therefore difficult to pack – porosity would be introduced. Some dilution is required here also. But the attempt is still made during packing to express as much mercury as possible to increase strength (see also §4.2).

It is stressed that the minimum mixing volume here has nothing whatsoever to do with the mercury equivalence discussed in §4.2. The former is a purely physical concern, the latter purely chemical, exactly parallel to the situation with gypsum products (2§2.4) and all similar systems.

●8.3 Pins

Amalgam has been used to create relatively large restorations, and there have been concerns that its strength is inadequate in such contexts, and particularly when retentive tooth preparation (macroscopic key) is poor or not possible. In an attempt to overcome this limitation, pins of various kinds (plain and threaded) have been used, inserted in holes drilled in the tooth. Embedded thin 'dog-bone' strips and other approaches have also been tried. The fundamental requirement for any such reinforcement is the same as for any composite structure: matrix constraint through bonding (6§2). This cannot happen with stainless steel, which is commonly used, because there is no wetting and no reaction. As a result, the structure is weakened as the insert or pin acts as a void: cracking will occur at that interface. Gold- and silver-plating are ineffective because the gold and

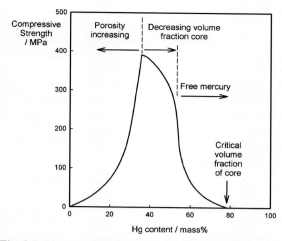

Fig. 8.5 The strength of a conventional Ag_3Sn-based, low-copper amalgam as a function of mercury content. Note the high cut-off due to excess mercury, the low cut-off due to porosity. Similar variation would occur for all amalgam products.

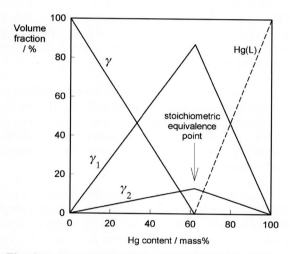

Fig. 8.6 Variation in phase proportions for a typical conventional Ag_3Sn-based, low-copper amalgam. Note the presence of liquid mercury after the equivalence point. (Note: the equivalence point depends on alloy composition.)

silver react with the mercury of the fresh mixture, and the steel still does not wet, but indeed gold could make the corrosion worse (13§4.7). Similar arguments apply many other systems, such as silver-plated sapphire (Al_2O_3) whiskers: the silver reacts and the aluminium oxide does not wet. However, silver pins, which will wet, can be used as retentive device. In this case, cracking does not follow the interface, although the strength overall cannot be improved (silver itself is too soft).[30] In addition, adaptation of amalgam to threaded pins is difficult, and many voids are left – a problem for both crack initiation and corrosion. Broadly, the case for pins and other kinds of "reinforcement" does not exist.

§9. Corrosion

The multiphase structure of dental amalgam offers several possibilities for the establishment of galvanic corrosion cells (13§1.2). The γ_2-phase is the most electropositive phase ordinarily expected in an amalgam, and thus the phase which is anodic and prone to corrode. Fortunately, the rate at which this corrosion ordinarily goes on a free, exposed surface is quite slow because the galvanic potential is not very large. However, if there is a surface which is in a crevice, such as the (unsealable) margin (§6.2), the oxidation of Sn^{II} to Sn^{IV} by dissolved oxygen means that the potential difference between crevice and free surface becomes large enough to drive the steady dissolution of Sn from the γ_2-phase. Such a corrosion reaction would be expected to come to a stop as the concentration of Sn in solution builds up but, as already mentioned (10§7.2), Sn^{II} hydrolyses readily, causing a lowering of the pH:

$$3Sn^{2+} + 4H_2O \;\;\rightrightarrows\;\; [Sn_3(OH)_4]^{++} + 4H^+ \tag{9.1}$$

and Sn^{IV} does likewise. The precipitation of hydroxides and related compounds containing phosphate and chloride follows rapidly, ensuring that the back e.m.f. due to Sn^{++} remains negligible. There is therefore continuous corrosion of the γ_2-phase.

This is one of the problems arising from γ_2-containing amalgams. If the γ_2-phase corrodes, as is expected from the electrochemical viewpoint, the pore that is left when it has dissolved is extensive, with a high probability of communicating with other interpenetrating γ_2-crystals (§5.1). The corrosion occurring is thus exaggerated and maintained by the ensuing oxygen concentration cell deficiency deep in the pit, irrespective of whether there is γ_2 present anymore or not. Other phases may then be attacked. Porosity from the other sources (§8) ensures that, even if a γ_2 grain is not connected to another, a path can usually be found somewhere. Corrosion throughout the restoration is thus possible, and has been observed frequently in practice. Polished sections of old amalgams show this very clearly, the corrosion tracking especially along the lines of porosity at increment boundaries.

The steps needed to reduce the extent of corrosion in this way are obvious: low residual mercury, and low porosity, both aided by good packing. It must be borne in mind that, while the amount of mercury used controls directly the amount of γ_2 produced, its amount must not be reduced at the mixing stage so far as to cause porosity by falling below the minimum mixing volume, or even so far as to make packing more difficult. The burnishing of margins (§5.2) may assist by reducing the diffusion from the crevice of corrosion product ions, thus encouraging precipitation.

It must be stressed that crevice corrosion is inevitable for any amalgam just as the crevice is inevitable (§6.2). It is sometimes said that a layer of cavity varnish (9§11) can prevent corrosion, but it should be obvious that it can have no effect on the formation of the crevice. However, it may prevent or limit contact between the acidic contents of the crevice when this is an active corrosion site and the tooth, so preventing its dissolution. A corrosion product-sealed crevice has the same effect, while an alkaline liner or base, *e.g.* Ca(OH)$_2$, may neutralize the acid from the reaction.

●9.1 High-copper amalgam

High-copper amalgam alloys avoid the difficulties due to the presence of the γ_2-phase, by eliminating it. There remains a problem even so. As mentioned above, it is impossible to obtain a marginal seal with amalgam on setting as packed. The percolation of oral fluids in the gap at the cavity walls is expected, and crevice corrosion must ensue. The rate of corrosion, though, depends on whether there is γ_2-phase present: if it is, in a low-copper amalgam, the rate is high and corrosion products rapidly fill the crevice, sealing the margin and eliminating further corrosion. Such corrosion is thus self-limiting in that there is no path available through to the surrounding electrolyte, the saliva. This limitation depends of course on the balance of diffusion against precipitation, and grossly porous amalgams and wide marginal gaps, caused by an excessively contracting amalgam or bad technique (see §6), will not self-seal. Amalgam's basic corrosion problem provides the means for it to be overcome. High-copper amalgams, lacking γ_2-phase, would not be expected to corrode at such a high rate, and diffusion therefore dominates in the removal of corrosion products from the anode site. Marginal leakage is observed to persist for long periods with high-copper amalgams, many months longer than for 'conventional' amalgams, although they may be affected beneficially (*i.e.* to corrode and seal more rapidly) when zinc is present.[31]

It should be noted that the phase which is most electropositive in a high-copper amalgam is η'-Cu$_6$Sn$_5$, essentially because of the Sn content. One of the pieces of evidence invoked to demonstrate the resistance of this phase to corrosion over the γ_2-phase is the observation that high-copper amalgams stay bright and shiny in the mouth for a long time. This means that the reactions otherwise expected to occur with sulphur-containing foodstuffs, to form AgS, CuS and possibly HgS (all black[5]), are not occurring. Such a reaction requires something like:

$$2\,Ag\ +\ S^{2-}\ \rightarrow\ Ag_2S\ +\ 2\,e^-$$
$$O_2\ +\ 2\,H_2O\ +\ 4\,e^-\ \rightarrow\ 4\,OH^-$$

$$(9.2)$$

Hence, if the surface is already negatively polarized, the first reaction is inhibited.

The reason may be found by considering a corrosion cell. The cathode is where, if anything, metal ions will be reduced to metal, so that oxidation (to sulphides) could not possibly occur. It may be deduced therefore that a shiny free surface is cathodic, and therefore that active corrosion is going on somewhere – and this usually means in the marginal crevice. Thus it is quite the converse of the usual interpretation that is true: high-copper amalgams stay brighter longer because they corrode actively for longer. It is therefore a reassuring sign if an amalgam is *not* bright and shiny. It shows that the margin has sealed and that the amalgam is no longer corroding where it cannot be seen. Of course, this does not prevent corrosion restarting if the restoration is cracked, a margin becomes chipped, or caries is allowed to develop on an adjacent tooth. This will not remove any accumulated discolouration. However, such defects should be easily detectable. Equally, repolishing an old, self-sealed amalgam – to make it look better – is likely to be a short-lived improvement. Of course, repolishing itself may cause damage to the seal if there is undue heating or stress generated, although peritectic melting (§5.5) would be much more damaging: adequate cooling and intermittent, light loading remain necessary.

As it turns out, high-copper amalgams do not appear to have an advantage in terms of strength or the effect of corrosion on strength.[32]

●9.2 Effect of microporosity

This phenomenon has been demonstrated in another manner. The microporosity that is inevitably present in an amalgam, because of the incomplete removal of oxide from the alloy, means that there will be present many active corrosion cells in the surface of a polished amalgam. Such amalgam stays bright for a fair time when exposed to a saliva-like medium, the surface being cathodically-protected again. However, if the alloy is first washed with dilute HCl to remove the oxide, rinsed and thoroughly dried, the amalgam then made becomes tarnished very quickly indeed.[33] The surface is no longer cathodically-protected, and the oxidation proceeds normally.

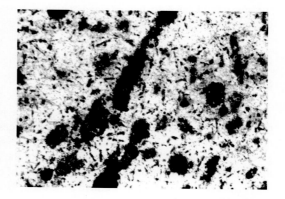

Fig. 9.1 Amalgam prepared from high-copper alloy where the admixed Ag-Cu-eutectic has not wetted because of tarnish. Some particles lie free in the macroscopic porosity. ×30

Some high-copper amalgam alloys, particularly those containing Ag-Cu eutectic, suffer more obviously from this effect.[34][35] The copper-rich phase in the spheres of eutectic, for example, becomes oxidized on exposure to air, producing a characteristic reddish colour. Once coated in this way it is not easy to remove the film at mixing and many eutectic particles may remain unwetted, segregating as a separate powder in the mixing capsule, and remaining within porosity in the mix (Fig. 9.1). Such amalgams are very porous, and such alloy should not be used. However, the oxidation of the copper may not be quite so obvious; what happens then is that wetting does not occur uniformly around the periphery, and a fine crevice is left here. This is enough, when exposed at the surface by polishing the amalgam, to form a corrosion cell and keep the surface bright by polarizing it cathodically. Such a lack of wetting can be seen from the irregularity of the 'reaction halo' around the Ag-Cu particles in Fig. 5.3 – in places it is missing altogether.

The presence of zinc in the Ag-Sn alloy is beneficial (§2.4) in a similar way. Although amalgams from Zn-containing alloys darken more rapidly,[36] this is probably because the microporosity is reduced or eliminated.

---

[5] HgS is black when ordinarily formed from aqueous solution, β-HgS. The mineral cinnabar, α-HgS, is red.

Amalgam alloys should be stored in dry conditions, and exposed to the air as little as possible if purchased as bulk powder. Even predosed encapsulated materials must be treated with care if not of the completely-sealed variety, and only relatively small stocks held to keep turnover regular and quick. The financial advantage of buying in bulk may not be so certain, at least from the patient's point of view.

●9.3   **New on old**
Sometimes it might be appropriate to place new amalgam in contact with old, and the question arises as to whether this represents a corrosion risk (*cf.* 13§4.9). In itself, there is no reason to suppose that there is. There are several situations that might be considered. Clearly, if the products are the same (and therefore the same phase description) there can be no potential difference between old and new. Even different brands of the same kind of amalgam alloy inevitably produce a similar set of phases on setting, so again no effect, as has been confirmed by experiment (Fig. 9.2). If high-copper amalgam is adjacent to a low-copper type, then the absence of γ_2 in the one means that all the activity is concentrated in the other part, yet this can be no more active than otherwise could be the case because η'-Cu-Sn will still be present to some extent and so the potential difference is unaffected. This would be similar to the situation where the old amalgam has already lost all the accessible γ_2. It would only be a problem if there to be such a substantial difference in the phase description or the minor constituents that the such that the electrode potentials were affected. This is considered to be a very remote possibility.

There is a problem, however, with the technique in that great care would have to be taken to avoid excessive porosity at the junction of the two (Fig. 9.2): a bright metal surface would have to be obtained which was then well-wetted by the new unset amalgam, that is, by careful packing.[37]

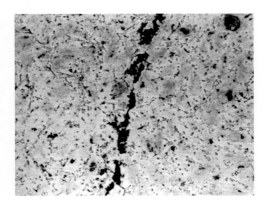

Fig. 9.2 New amalgam (left) set against old (right). The porosity (centre, black), due to poor wetting, was filled with resin for the corrosion test to avoid crevice corrosion. No galvanic corrosion effect is detectable. ×30

§10. Quality of Work

It will be apparent from the above that the long-term success of an amalgam restoration depends very much on the quality of its packing: how well excess mercury and porosity are reduced to a minimum. The demand therefore is that an adequate packing pressure be used, which can only mean the maximum possible given the strength of the operator, the ability of the patient to tolerate it, and the need to avoid the condensing instrument punching through the mixture instead of compacting it. Obviously, this last point depends in part on the correct selection of the instrument to be used. However, the first is not so easy. Recommendations vary, but a typical figure of about 40 N may be cited as an indication of what can be expected to be necessary for lathe-cut and similarly resistant alloys.

As a matter of principle, every thrust for every increment, every restoration, every patient, every week and every year must be the best possible. Despite Monday morning blues after a fine weekend, despite hunger towards lunchtime, despite the need to get the next patient into the chair or to avoid rush-hour traffic, despite those particular 'condensing' muscles aching, and despite the languor that may set in as the weekend once again approaches, that consistent effort must be made. Every patient expects the best possible treatment whether they are first or last in the appointment book, indeed, whether they are your very first or last ever patient.

In fact, of course, such observations regarding the demands for consistent high-quality work apply equally well to cavity preparation, endodontic treatment, impression taking and so on. It can be summed up by saying that no matter how good the materials are, the weakest link in the chain is the operator. There lies the ultimate responsibility for the success or failure of a treatment.[38]

References

[1] Brandes EA & Brook GB (eds). Smithells Metals Reference Book. 7th ed. Oxford, Butterworth-Heinemann, 1992.

[2] Darvell BW. Stoichiometry of the amalgamation reaction. J Dent 5(2): 149 - 157, 1977.

[3] Vrijhoef MMA, Vermeersch AG & Spanauf A. Dental Amalgam. Quintessence, Chicago, 1980.

[4] Darvell BW. Development of strength in dental silver amalgam. Dent Mater 28: e207 - e217, 2012.

[5] Darvell BW. Some studies on dental amalgam. Part 3. The constitution of amalgam and alloy. Surf Technol 4: 95 - 106, 1976.

[6] Jørgensen KD. Amalgame in der Zahnheilkunde. Hanser, München, 1977.

[7] Jensen SJ. Copper-tin phases in dental silver amalgam alloy. Scand J Dent Res 80: 158 - 161, 1972.

[8] Johnson LB. X-ray diffraction evidence for the presence of β(Ag-Hg) in dental amalgam. J Biomed Mater Res 1: 285 - 297, 1967.

[9] Johnson LB. Confirmation of the presence of β(Ag-Hg) in dental amalgam. J Biomed Mater Res 1:415-425, 1967.

[10] Marshall SJ & Marshall GW. Time-dependent phase changes in Cu-Rich amalgams. J Biomed Mat Res 13: 395-406, 1979.

[11] Youdelis WV. US Patent 3305356, 1967. http://www.google.co.uk/patents/US3305356

[12] Jensen SJ, Andersen P, Olsen KB & Utoft L. On the solubility of zinc in the γ-phase of the silver-mercury system. Scand J Dent Res 84: 338 - 341, 1976.

[13] Marshall SJ & Marshall GW. Dental amalgam: The materials. Adv Dent Res 6: 94 - 99, 1992.

[14] Johnson LB & Paffenbarger GC. The role of zinc in dental amalgams. J Dent Res 59(8):1412-1419, 1980.

[15] de Freitas JF. A survey of the elemental composition of alloy for dental amalgam. Austral Dent J 24 (1): 17 - 25, 1979.

[16] Jensen SJ, Vrijhoef MMA. Phases in preamalgamated silver amalgam alloy. Scand J Dent Res 84:183 - 186, 1976.

[17] Okabe T & Hochman RF. Amalgamation reaction on mercury-plated dental alloy (Ag₃Sn). J Biomed Mater Res 9: 221 - 236, 1975.

[18] Jensen SJ. Maximum contents of mercury in dental silver amalgams. Scand J Dent Res 93 (1): 84-88, 1985.

[19] Mahler DB & Adey JD. The influence of final mercury content on the characteristics of a high-copper amalgam. J Biomed Mat Res 13:467 - 476, 1979.

[20] Hirayama C et al. Metals in Mercury in: IUPAC Solubility Data Series Vol 25, Pergamon, Oxford 1986, 139 - 141.

[21] Ward ML & Scott EO. Effects of variation in manipulation on dimensional changes, crushing strength and flow of amalgams. J Amer Dent Assoc 19(10): 1683 - 1705, 1932.

[22] Jarabak JR. The effect of particle size on dimensional change in dental amalgams (1). J Amer Dent Assoc 29 (4): 593 - 605, 1942.

[23] Phillips RW. Skinner's Science of Dental Materials. 7th Ed. Philadelphia, Saunders, 1973.

[24] Dérand T. Creep in amalgam class V restorations. Odont Revy 27: 181 - 186, 1976.

[25] Ashby MF & Jones DRH. Engineering Materials 1. Pergamon, Oxford, 1981.

[26] Gray AW. Metallographic phenomena observed in amalgams. I Crushing strength. Nat Den Assoc J 6: 513 - 531, 1919.

[27] Caul HJ, Longton R, Sweeney WT & Paffenbarger GC. Effect of rate of loading, time of trituration and test temperature on compressive strength values of dental amlagam. J Amer Dent Assoc 67: 670 - 678, 1963.

[28] Mahler DB & Mitchem JC. Transverse strength of amalgam. J Dent Res 43(1): 121 - 130, 1964.

[29] Darvell BW. The Corrosion of Dental Amalgam - a Photometric Method. MSc Thesis. University of Birmingham, 1971.

[30] Moffa JP, Going RE & Gettleman L. Silver pins: Their influence on the strength and adaptation of amalgam. J Pros Dent 28 (5): 491 - 499, 1972.

[31] Mahler DB, Pham BV & Adey JD. Corrosion sealing of amalgam restorations in vitro. Oper Dent 34: 312 - 320, 2009.

[32] Darvell BW. Effect of corrosion on the strength of dental silver amalgam. Dent Mater 28: e160 - e167, 2012.

[33] Darvell BW. Some studies on dental amalgam. Part 5. Corrosion behaviour - effect of conditions. Surf Technol 7: 55 - 69, 1978.

[34] Darvell BW. Strength of Dispersalloy amalgam. Brit Dent J 141: 273 - 275, 1976.

[35] Darvell BW. Deterioration of disperse-phase amalgam alloy. Brit Dent J 144: 181 - 184, 1978.

[36] Darvell BW. Some studies on dental amalgam. Part 6. Corrosion behaviour - market survey. Surf Technol 7: 71 - 80, 1978.

[37] Jørgensen KD & Saito T. Reparation af amalgam. Tandlaegebladet 72(6) : 498 - 507, 1968.

[38] Darvell BW. The contribution of dental materials science to long-term success in dental care. in: Kurer PF (ed). The Kurer Anchor System. Quintessence, Chicago, 1984.

Chapter 15 · Mixing

The proper mixing of all materials supplied as two or more components is a crucial factor in obtaining the expected properties and the avoidance of problems. While many dental materials are mixed by hand, dental amalgam in particular is mixed using a specialized machine.

*An analysis of the energetics of the process involved in machine mixing of dental amalgam shows that the rate of the process of consolidating the mixture into a coherent body is dependent on, and very sensitive to, the oscillation **frequency** and the **amplitude** of the shaking motion, as well as the mass of the mixture. A critical factor is the ratio of the internal length of the **capsule** to that oscillation amplitude, and some capsules simply will not work on some machines. The **pestle** is an unnecessary part of the mixing apparatus, except possibly when pelleted alloy is used.*

These factors allow the selection of an amalgam capsule product and mixing machine that will work efficiently as well as allowing the identification of the reason for the failure of a given combination.

The proportioning of alginate impression materials presents problems because of the nature of the powder: it does not flow well and its bulk density is very variable. Recognition of the sources of error is important to consistent properties in the set impression. Similarly, products supplied in tubes present difficulties of proportioning that must be acknowledged to achieve accuracy.

A variety of other factors are involved in controlling the proportions of various other powder-liquid systems. Seemingly trivial, these factors have important effects, and are not ignored with impunity.

*A theoretical view of the general process of mixing yields some surprising insights into our ability to discern adequate mixing through the **Baker's transformation**. Great care is required to achieve a sufficiently intimate mixture for best effect.*

Materials Science for Dentistry
https://doi.org/10.1016/B978-0-08-101035-8.50015-8

Many dental materials, such as plaster, cements, and impression materials, must be mixed before use. The reason is obvious enough: the mixing of the two parts initiates a chemical reaction which develops the desired properties; that is, they set. The primary demand of mixing is also fairly obvious: the mixture must be intimate for even and complete reaction. But, given that, the time taken must be short to keep the working time long, *i.e.* keep the extent of reaction at the time of use low. Often consigned to an assistant as if it were a menial task, proper mixing is, however, usually critical to the success of a material and attention to detail is essential. In order to train an ancillary (assuming they are not already professionally-trained), or simply to ensure that standards are maintained, the demands and limitations of the various mixing processes used in dentistry should be understood. The peculiarities of zinc phosphate cement have already been dealt with (9§5), but we proceed now with some further aspects of mixing, which has more to do with machines and product selection than training, but all of which affect the quality of the outcome. Little effort is required, only attention to detail.

§1. Amalgamation

Liquid mercury reacts rapidly with several metals, wetting and spreading quickly over a large area as it reacts. Gold jewellery such as rings and watches are particularly prone to being rendered silvery by contact with stray droplets of mercury whether in the clinic or in a laboratory, and for that reason one is well-advised not to wear such adornment when mercury is around – advice that should be passed onto one's dental assistant. Any treatment to remove the mercury from the gold is necessarily drastic, and the piece may not survive unscathed. Silver, as has been seen (14§1.1), is also reactive to mercury, but ordinarily may not appear to be very much so because of the presence of a fine film of sulphide (Ag_2S) from exposure to hydrogen sulphide in the air and to other sulphurous materials. This may prevent immediate metal-to-metal contact and so delay reaction until some rubbing breaks the film, in much the same way that the extremely rapid oxidation of amalgamated aluminium cannot be demonstrated until the oxide coat is broken under the mercury. In the case of silver, the sulphide coat is not very strong and usually not very thick, so that little effort is required to initiate reaction.

●1.1 Need for trituration

Certainly, when amalgam fillings were being placed by the showmen of the travelling medicine shows in North America in about the 1840s, the powder from freshly-filed silver dollars reacted readily enough that amalgam could be mixed effectively (if a little alarmingly, given current awareness) in the palm of the hand. When tin, and then copper, were incorporated into silver-based alloys that were delivered to the dentist as a ready-prepared powder, this changed. Firstly, the powders were exposed to the air long enough that sulphide tarnish could develop, but secondly, the less noble copper and tin are much more readily oxidized.[1] The bulk metals steadily acquire an oxide coat and no less in their alloys, especially tin, being the more reactive of the two. Thus SnO_2, CuO and ZnO may be found in addition to Ag_2S. These corrosion products prevent contact between mercury and alloy and thus prevent reaction (Fig. 1.1). Amalgam alloy had therefore to be mixed a little more vigorously, and this was achieved with a mortar and pestle (Fig. 1.2). This process has commonly been referred to as **trituration**, and involves a grinding action intended to disrupt surface coatings and even break or plastically deform the alloy particles, which actions would also expose fresh metal to the mercury as well as create dislocations that would raise the energy and thus the reactivity. With the finding that alloys should be 'aged' to reduce and stabilize their reactivity (14§3) to obtain a more consistent and sensible setting rate, the problem was exacerbated. The ageing was done at an elevated temperature in air or water, both of which are conducive to the oxidation of tin.

The fact that the main purpose of trituration was the abrasion of the tarnish coat to expose clean metal is readily demonstrated. If a commercial amalgam alloy powder has a droplet of mercury dispensed onto it, nothing will happen. Maybe after some minutes or hours it will have been absorbed

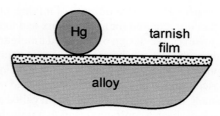

Fig. 1.1 Metal oxides, sulphides and the like that occur in tarnish films are not wettable by mercury.

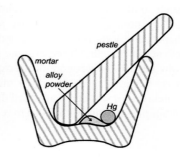

Fig. 1.2 Cross-section of pestle and mortar as formerly used for mixing amalgam; these were usually made of glass.

by the powder, but on a timescale of interest to a dentist essentially nothing changes. However, if the alloy is first washed in dilute hydrochloric acid to dissolve oxides and so on, thoroughly rinsed and dried and the test repeated, the result is dramatically different. The instant that the mercury droplet is delivered, wetting and reaction start. Whilst alloy powders could be prepared oxide free, and could be stored dry and sealed against the air, extraordinary precautions would be necessary and these would probably be considered uneconomic. It is therefore always necessary in practice to apply sufficient friction to disrupt the oxide coat. Again, the presence of that oxide is easy to demonstrate, and in a very simple manner. Merely inspect the pellet of amalgam produced from a current mixing machine (for it should no longer be necessary to use hand mixing) and one may notice a fine grey powder, perhaps covering the pellet, or loose in the capsule. Since it is not wetted by mercury it tends to get excluded from the consolidating amalgam, becoming visible evidence of the tarnish (see box). The older the alloy, and the poorer the conditions of storage, the more oxide.

> The exclusion of oxides and so on from liquid mercury is primarily driven by the density difference. The density of mercury is so high (~13.5 kg/L) that the buoyancy (4§9.3) of almost everything else is too great for it to stay immersed. This would not apply in mixed amalgam, where the effective viscosity of the paste is too high, and vigorous mixing can lead to entrapment in voids (14§9.2). These effects would also be seen to apply to fragments of plastic from capsules and other contaminants. Even so, since the reaction with mercury would lead to it creeping beneath oxides *etc.*, these would be lifted off with no drive for the mercury to close over the top – a straightforward surface tension effect: there is no significant wetting.

Hand-mixing is a relatively slow process, and of low efficiency. As a result it required the use of extra mercury which then had to be squeezed out of the mix, through gauze or chamois leather, or later on with a special press, at the chairside prior to condensation. Inevitably there were problems of non-uniformity arising from the difficulty of a consistent squeeze, as well as from the mixing process itself (to say nothing of the problems of mercury 'hygiene' – personal contact and workplace contamination; 28§6). Even so, the remarkable efficacy of amalgam as a restorative material is underlined by its success, measured by its durability of many years, despite these difficulties.

The introduction of mechanical mixing was with the aim of obtaining greater speed and uniformity of mixing, but this also reduced the amount of mercury required for a plastic mix and thus eliminated the need to squeeze. However, the design of so-called **amalgamators** was on a pragmatic rather than theoretical basis and consequently their development has been unscientific. For example, many machines were designed to work by an oscillating motion causing the alloy and mercury to be thrown about with a small, often metallic object, again called a pestle (Fig. 1.3). There appears to have been a conscious effort to mimic the old mortar and pestle (Fig. 1.2), in the belief that as the mix was thrown against the ends of the capsule, the pestle followed and did the 'trituration'.[1] It led to such claims as that the heavier the pestle, the better the mixing. This culminated in a short rod of solid tungsten weighing 8·5 g being sold as a pestle. The main effect of the use of this was the smashing of the ends of the capsule and the premature wear of the mechanism of the mixer. The danger of the bullet-like escape of such an object from a disintegrating mixing capsule is not to be dismissed, to say nothing of the mercury spread all over the room (28§6). The haphazard designs of machine and capsule also lead to such disconcerting events as the total failure of some products to mix at all on some machines, whether disposable or pre-dosed capsules. To overcome this, faster machines were designed. At least, their mechanisms were designed to move more quickly, as it was again assumed that only higher velocities were necessary. A simple analysis of the system reveals the fallacies of these designs.[2]

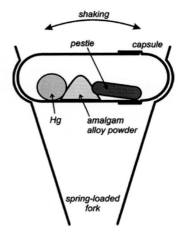

Fig. 1.3 General pattern of capsule used for machine mixing dental amalgam (NB: not to the same scale as Fig. 1.2).

---

[1] In the late 1800s, amalgam was sometimes mixed by shaking the components, vigorously, by hand, in a long tube; no pestle.

§2. Amalgam Mixer Analysis

Mechanical amalgam mixers essentially consist of a small cylindrical capsule into which is put the amalgam alloy powder and the mercury, and which is then shaken, typically end to end. The mechanism for producing this motion also resulted in various other motions, but these are relatively insignificant. So, ignoring the pestle for the time being, the kinetic energy, e_i, of each individual impact of the mix with the capsule end wall can be written:

$$e_i = \tfrac{1}{2}mv^2 \tag{2.1}$$

where m is the mass of the mix, and v the instantaneous collision velocity. While the mix is still powdered, all of this energy may be absorbed, i.e. this will be an inelastic collision. The mix as a whole will not bounce since interparticle collisions will rapidly randomize their directions of motion. It would be like throwing a handful of sand at a wall - single grains may bounce, but the whole assembly does not. Since v will be directly proportional to the frequency, f, and the amplitude, a, of the capsule motion ($s^{-1} \times$ m), no matter what the oscillation waveform (although it is usually close to sinusoidal), we may write (since we do not care about the constant of proportionality):

$$e_i \propto (f.a)^2.m \tag{2.2}$$

Since an impact will be expected at each end of the capsule, i.e. twice per cycle, the number of impacts per second is expected to be twice the oscillation frequency, so that multiplying expression 2.2 by f gives the relative rate of energy supply (t is time):

$$\frac{de}{dt} \propto f^3.a^2.m \tag{2.3}$$

The rate of energy supply per gram of mix is therefore given by:

$$\frac{de}{dt}.\frac{1}{m} \propto f^3.a^2 \tag{2.4}$$

The majority of this energy will appear as heat. If then we assume that there are no losses and that the rates of the amalgamation reactions obey an Arrhenius-type model for their temperature dependence (since they depend primarily on diffusion) (cf. equation 3§3.3), the relative rate factor ρ will be dependent on the temperature:

$$\rho \approx e^{-1/T} \approx T \tag{2.5}$$

for modest temperature rises from around 300 K. The temperature rise would be given by:

$$\Delta T \propto \frac{de}{dt}.\frac{1}{m}.\frac{t}{c_p} \tag{2.6}$$

where c_p is the specific heat capacity (J/K/g). Since the relative rate factor may be taken as 1 at room temperature, it is therefore about $(f^3.a^2)$ times greater at the end of the mixing process. However, we need the average relative rate factor for the mixing time to determine the overall effect. The geometric mean is appropriate to an exponential function; the average relative rate factor $\langle\rho\rangle$ will then be:

$$\langle\rho\rangle \propto \sqrt{1 \times f^3.a^2} = f^{1.5}.a \tag{2.7}$$

Hence the actual overall rate of amalgamation will be given by (2.4) × (2.7):

$$\langle\rho\rangle.\frac{de}{dt}.\frac{1}{m} \propto f^{4.5}.a^3 \tag{2.8}$$

Now the progress of amalgamation is observable because the mix changes steadily in appearance from a powder to a single pellet (Fig. 2.1).[3] This latter stage is very easily identifiable. We define the time taken for it to appear as the **coherence time**, t_c. This being a time, it is inversely proportional to

Fig. 2.1 An amalgam mix goes through a series of stages with time, culminating in a single pellet, on a mechanical mixing machine.

the rate (s⁻¹) for the process, so at constant mass:

$$t_c^{-1} \propto f^{4.5}.a^3 \qquad (2.9)$$

This shows the quite unexpected sensitivity of the rate on two design aspects of the machine: its frequency of oscillation, and the amplitude of that movement (the distance between the extremes of motion of, say, the centre point of the capsule). It does, however, confirm the general assumption that increasing the speed of the machine is beneficial. It follows also that, for consistency in operation, the electric motor of an amalgamator must have a very stable running speed, unaffected by changes in supply voltage, temperature, load, or age. Equally, the amplitude of the oscillation must be stable - meaning that the springs do not soften or spread with use. The high powers in relation 2.8 indicate high sensitivity to small changes.

●2.1 Power rating

Several other things emerge from this analysis. Firstly, the *design* of the mixing machine is characterized by relation 2.4. From this we can define a **Power Rating**, PR, for the machine:

$$PR = f^3.a^2 \qquad (2.10)$$

> It is always worthwhile checking the dimensions of an expression:
>
> W/kg = (J.s⁻¹)/kg
>   = (N.m)/(kg.s)
>   = (kg.m.s⁻².m)/(kg.s)
>   = m²s⁻³
> and Hz = s⁻¹

with the units of W/kg (see box) or, equivalently, mW/g, to put it on a more relevant scale. That is, the maximum power available for the mixing process is proportional to this value, independently of all other factors. PR is therefore an objective measure of the capability of mixing machines. Whether this capability is usable is a matter to be determined.

●2.2 Heating

Secondly, relations 2.4 and 2.6 show that:

$$\Delta T.s^{-1} \propto f^3.a^2 \qquad (2.11)$$

which implies that the rate of temperature rise is dependent only on the PR of the machine (apart from some efficiency factor hidden in the constant of proportionality). In other words, the mass of the mix is quite irrelevant since this only controls the rate of **energy** delivery, and the specific heat capacity (J/K/g) from mix to mix will be essentially constant. To put it another way, every mix of amalgam *must* get warm as a necessary consequence of being mixed. The warmth of the pellet when it emerges from the capsule cannot be used to judge whether the amalgam has been overmixed or not, as is sometimes suggested. 'Overheating' of mixes has been blamed for various faults from time to time, but clearly this heating effect cannot be avoided, unless the mixing rate is low enough that heat can be conducted away; meanwhile, setting reactions continue.

●2.3 Performance

Now whether or not the above analysis is sufficiently representative of the true state of affairs can be tested, since the formation of the single pellet is so clearly-defined as a stage in the mixing process. After this pellet has formed, so long as it stays plastic, the energy calculations will be similar but there is no longer any easily monitored attribute of the mix to use to follow the changes. If the pellet hardens, and the collisions become more elastic, the physics may be different and the calculations much more difficult. However, all of the factors which may reasonably be expected to control the efficiency of mixing, such as alloy particle size, shape, cleanliness and reactivity; capsule size, shape and wall friction; alloy : mercury ratio; mass of mix; and mixing machine details such as capsule path and the velocity profile along that path, can all be taken into account by one measurement of the coherence time. This therefore is a performance measure for the entire alloy-capsule-machine system.

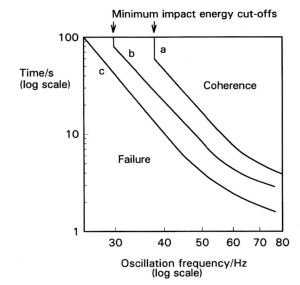

Fig. 2.2 Variation of coherence time (solid lines) with oscillation frequency for three values of amplitude; log-log plot. **a**: 15 mm, **b**: 18 mm, **c**: 22 mm. 'Coherence' means attainment of coherence, above the respective line; 'Failure' the converse, below the line.

Experimentally, the effect of variation in frequency and amplitude on coherence time is shown in Fig. 2.2. Although some deviations from the theoretical slope of -4·5 do occur at high frequency and high amplitude, the analysis above is seen to be satisfactory over a substantial region. The complete failure below a certain frequency on curves a and b (the vertical portions of the plots) illustrates the need for a minimum impact energy for amalgamation to occur, corresponding to an activation energy. If the impact is too gentle, abrasion of the oxide will not occur.

The effect of variation in the mass of the mix is also found to be broadly as predicted from relation 2.8, *i.e.*:

$$\text{rate} \propto f^{4\cdot5}.a^3.m \propto t_c^{-1} \qquad (2.12)$$

(Fig. 2.3). Strangely, this contradicts the advice usually to be found with amalgam products that the

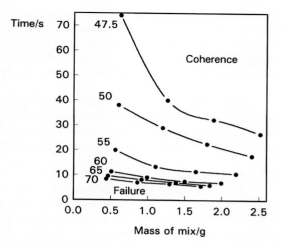

Fig. 2.3 Variation in coherence time with mass of mix for various mercury contents (in mass%). A low PR machine was used to show the effects more clearly.

larger doses (double 'spill' capsules) require longer mixing time. There is no physical basis for this advice, and no experimental evidence that it is necessary.

Figure 2.3 also shows that mixing to coherence occurs faster the more mercury is present, but that there is always a mercury content below which coherence will not be attained, no matter how high the machine's PR. This of course has more to do with the mechanics of the pellet than the reaction with mercury and so must be product-dependent. With little liquid the pellet will be very porous and so very weak; impact will fracture it rather than cause consolidation. More importantly, Fig. 2.3 also shows that the mercury : alloy ratio can have a marked effect on the mixing rate, particularly at the lower ratios appropriate to modern alloys and techniques. This means that, for consistent results, the proportioning must be done accurately, when this in done in the clinic for reusable capsules. Furthermore, it sets limits on the acceptable variation in dosing or, perhaps more significantly, the completeness of delivery of the mercury from its reservoir, in preproportioned (disposable) capsules. Many such products are found to have crevices or other features that trap part of the mercury, in a poorly reproducible fashion, and much apparent inconsistency in the behaviour of the mixture can be traced to this - on top of the actual variation in both alloy and mercury actually put into the capsule.

●**2.4 Capsule**

The capsule itself is an important component of the system and must be considered as part of the machine.[4] The crucial factor is the ratio of the internal length of the capsule to the amplitude (as defined above): the length : amplitude ratio (L/a). Because it is an oscillating system an harmonic analysis is appropriate. However, there are so many extra factors controlling the efficiency of mixing that this becomes impractical. Capsule geometry and both oscillation path and waveform are quite variable between brands, and the detail of their effects and interactions not well known. Hence a direct experimental determination of the combined effect of these factors is more effective (Fig. 2.4), and in the long run much more meaningful as a direct statement of performance. The optimum L/a value can be seen to be only broadly defined (which is fortunate as this gives some freedom of design), but very clearly the worst efficiency occurs for L/a ≥ π/2, that is ≥ ~1·6, even at large amplitude (curve c). It is quite simply impossible for amalgam to be mixed in some capsules

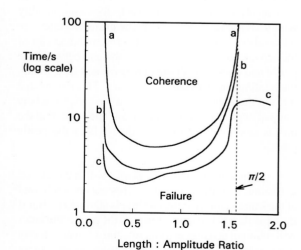

Fig. 2.4 Variation in coherence time with capsule length/amplitude ratio. Key as for Fig. 2.2. The capsule was adjustable to vary the internal length. (The lower limit is simply due to the volume of the capsule becoming too small for the mixture to move.)

on some machines – nothing happens (curves a, b). It is important to appreciate that this failure to mix has no direct bearing on the quality of an amalgam product or of a machine as such, they can be entirely satisfactory otherwise. It simply indicates a mismatch of parameters – an unserviceable combination. As can be seen from the figure, very short coherence times can be obtained by a suitable choice of capsule and machine. Most commercial machines have amplitudes of the order of 15 ~ 18 mm.

This behaviour can in part be understood by considering the impacts of the mixture with the capsule end. Unless the capsule motion takes the one end past the centre, overtaking the mixture, there will be no impact and it will not be accelerated. In other words, a long capsule can oscillate around the mixture without it ever being disturbed from the centre, just guided to remain a central cloud of particles.

●2.5   Mixing time

It must be stressed that the coherence time is used here only as a measure of the rate of the mixing process. A pellet which has been mixed just as far as coherence is *not* suitable for clinical use. A mixing time of about 5 t_c has been shown[5] to be near optimum for a variety of products, but this does depend on the alloy itself, and indeed of the definition of optimum. It is by no means clear how setting time, strength and dimensional change should be weighted in making such an assessment, for these properties are not closely correlated and a compromise is inevitable.[6]

●2.6   Pestle

So far, all of the discussion has been in terms of a capsule used without any pestle; the impact of the mixture with the capsule end walls has done all the work. Evidently, perfectly satisfactory amalgam mixing can be obtained without one. The assumption that a pestle is essential to the process is simply wrong. However, given that some designs of predosed, disposable amalgam capsule release a pestle which has been acting as a barrier between the amalgam alloy and the mercury for storage, and that some products include a pestle anyway, it is as well to consider their behaviour.

The mass of the pestle has very little, if any, effect.[7] In Fig. 2.5, the solid circles represent the coherence times obtained with a series of steel balls in one particular capsule on one machine. Their masses vary as d^3, yet it is evident that t_c decreases steadily with increase in d. Moreover, the projected intercept on the abscissa corresponds to the internal diameter of the capsule, d_i. Thus, the smaller the gap between the pestle and capsule wall the more efficient the mixing; the relationship is linear: $t_c \propto (d_i - d)$. This means that it is not the impact of the pestle on the mix, somehow trapped between it and the end wall, but rather that the mix is squeezed against the side walls as the pestle passes.

An object such as a pestle will have elastic collisions with the end walls. This means a rebound at the same velocity if the capsule is stationary. But if the capsule end is moving towards the pestle, the latter will rebound at a higher velocity, and a lower velocity if it is moving away. Thus, the pestle velocity cannot stay in step ("in phase") with that of the capsule; essentially it is moving independently – chaotically, in fact. This will therefore keep it out of step with the movement of the mix. Thus, often the pestle and mix will be moving in different directions and pass each

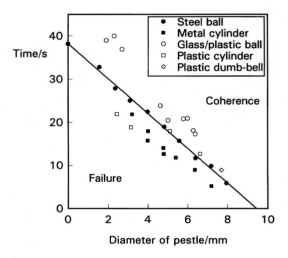

Fig. 2.5 The effect of various pestles on coherence time. The straight line is fitted through the steel ball data; it intersects the abscissa at the internal diameter of the capsule. L/a = 1.51. The cylinders may tilt, and appear of larger effective cross-section. The momentum of the low-density balls is too low and they tend to be swept along with the mix.

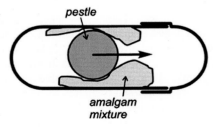

Fig. 2.6 The independent motions of mixture and pestle mean that the mixing process is achieved by squeezing the material against the capsule walls.

other in the capsule (Fig. 2.6). Otherwise they will be moving at different velocities and one will overtake the other, giving more squeezing against the walls. (The full explanation of the role of pestle diameter requires some further background; see §7.3.)

That the mass and length of the pestle have little contribution to make to the process is seen by the small scatter for the many other pestles shown in Fig. 2.5. This is emphasized by the result for the largest metal cylinder (solid square), the tungsten one referred to earlier, which has a mass some 15 times that of the largest plastic pestle (open diamond): the improvement observed in coherence time is not at all in proportion. It is also clear that the diversity of pestles that have been commercially supplied (Fig. 2.7) arises from a lack of understanding just how they work. The relative inefficiency of the smallest glass beads is probably due to their very low mass: their momentum must be so low that they are carried with the amalgam mix rather than oppose it. The greater efficiency of the cylinders than would be expected is probably due to them often lying at an angle in the capsule – not neatly aligned with the centre axis, so that the average gap is smaller than the diameter of the pestle would suggest.

Fig. 2.7 Some examples of pestles supplied for machine-mixing of silver amalgam. (Scale: large cylinder, bottom left: 15 mm long.)

The effect of the pestle is also dependent on the L/a ratio (Fig. 2.8). For a relatively short capsule there may be some deterioration in efficiency at small sizes, and no improvement may be attainable until a very large pestle is used. With relatively long capsules a large pestle is essential to get the combination to work at all. Obviously, if the development of mixers was done with such a capsule, the assumption that a pestle was essential would not seem unreasonable. But, had the L/a ratio fortuitously been optimized, around 0.7 say, the conclusion would not have been so obvious. However, a pestle is probably essential if pelleted alloy is used (weighed portions of alloy formed into a tablet under some pressure, to simplify dispensing into reusable capsules). The disintegration of the pellet may require squeezing against the side wall for it to be done efficiently, or at all. Again though, it should not require a heavy pestle for this to be effective, simply one that is large enough to create a small enough gap.

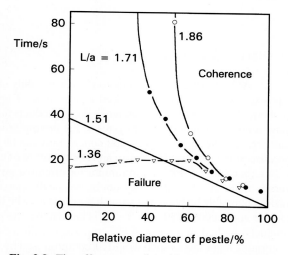

Fig. 2.8 The effect on t_c of steel ball pestles of varying diameter in capsules of differing L/a. The line for $L/a = 1.51$ is as in Fig. 2.5.

On a directly practical note, it is easy to check the compatibility on an intended purchase of capsules with an existing mixer, or *vice versa*, by simply noting the L/a ratio. When a capsule's external dimensions have been designed to fit an existing machine, which was itself arbitrarily dimensioned, and the thickness of the end wall is allowed to vary by several millimetres, the utility of such a check is plain. Equally, if a product fails to work as expected, it is a simple matter to determine whether the machine and capsule are compatible or not.

●2.7 Proportions

Many amalgam products are available in disposable capsules with pre-measured quantities of alloy and mercury. The effects of the mercury content have already been discussed (14§4, Fig. 14§8.5) but the assumption is that using predosed capsules avoids problems. This is false. Unfortunately, the design of many capsules (if not all) is such as to provide a risk of trapping some mercury (in particular) or alloy powder in various crevices. This means that occasionally the resulting pellet of mixed amalgam will be noticeably dryer or wetter than usual. If it is a dry mix it is probably better to discard it and use another capsule; if it is wet then extra care with condensation is necessary to get rid of the excess mercury. It is also worth noting that the capsules are machine-filled and that occasionally there will be larger errors than the normal slight scatter in the mass of the components. The effects and solutions are the same.

The trapped mercury has another implication: pollution. The design of some capsules is such as to allow some mercury to be shaken out during mixing thus contaminating the workplace,[8][9] but any residue (including mixed amalgam) represents a problem whether the discarded capsule is incinerated (air pollution)[10] or ends up in landfill (groundwater contamination).[11] Indeed, this applies to any amalgam waste as well. It is issues such as these that will lead to the discontinuation of the single most effective restorative material ever rather than any hazard (well-demonstrated to be negligible[12]) from amalgam restorations in the mouth (28§6.6).

●**2.8   Mulling**

It has sometimes been recommended that, after machine mixing with a pestle, a brief further period of shaking be given, the pestle having been removed; this was called **mulling**. Some mechanical mixing machines even went so far as to have a **mull** button (at additional cost), which gave a timed (2 ~ 3 s), lower-speed operation. This seems to have arisen as a carry-over from the practice when a pestle and mortar was used of gathering the mixed amalgam together in a more convenient, consolidated pellet. This was done in the palm of the hand (! 28§6), or later in a piece of rubber dam or a rubber finger-stall.

It seems never to have been demonstrated that this treatment had any effect, although some users claimed a change in handling properties. It does, of course, continue the mixing process, allowing further setting to occur before use, but it may have simply been easier to handle a coherent body rather than separate pieces. With the advent of sealed, predosed capsules it became impossible to remove any pestle and return the device to the machine. However, a lower-speed run may have assisted in gathering the mixture into a single pellet. The practice seems to have fallen out of use.

§3. Other Capsule Products

Although the above discussion has been solely in terms of amalgam, there are a number of other product types which are presented predosed in capsules. They are intended to be mixed on a machine of the sort generally known as an 'amalgamator'. Such products include zinc oxide-eugenol, calcium hydroxide liner, and glass ionomer cements. The physics of the mixing process is, of course, independent of the chemistry of the mixture, and the same analysis and conclusions can be reached in respect of these other materials. However, because of the increased viscosity and sticky nature of the mixture, many such materials may require machines with a much higher PR as a minimum for effective mixing (*cf.* Fig. 2.2 and the remark about the cut-off for curves a and b). Indeed, it may not be physically possible to mix such materials at the powder : liquid ratios easily obtainable by hand and so such products may be limited to cements and so on whose fluidity is important to their application (9§9). In fact, it is known that the powder : liquid ratio may be deliberately reduced for a product supplied in a capsule in comparison with the recommendation for the hand-mixed version. Direct restorative materials as such would be compromised by the demands of the mechanical mixing process, and the capsule product may only be usable for luting where the fluidity requirement is dominant.

The question of the completeness of the delivery of the liquid component assumes greater importance in some of these products than even in amalgam capsules. The high viscosity of, for example, the polycarboxylic acid solution of glass ionomer cements means that a greater time is required for the flow of the liquid. Inadequate time and pressure during the 'activation' of the capsule will result in a bad mixing ratio and poor properties when set, even if mixing proceeds properly. In this case, as in all others, it is important to follow the instructions given by the manufacturer. There is also a greater risk of dry powder being trapped in a capsule with a delivery nozzle; care should be taken that this is not the first material delivered to the target site, as occasionally does happen.

The viscosity, stickiness and low surface tension of the liquids of these other types of material are also the cause of a further problem: bubbles. While porosity in amalgam does arise from the mixing process, in these other materials it is more significant perhaps because it occurs in greater quantities but also because in weaker materials the effects of flaws in the Griffith sense (1§7) are less tolerable. Certainly, for one resin-modified glass ionomer material it was found to be unacceptable and the use of capsules discontinued. In all of this it must not be forgotten that a chemical reaction has been initiated: over-mixing will result in higher temperatures, thus faster reaction, hence higher viscosity and less working time, if not damage to the intended structure.[13]

§4. Alginate Proportioning

Alginate impression materials (7§9) are supplied as a powder which is ordinarily to be dispensed with the aid of a scoop (Fig. 4.1). This is to be filled by dipping it into the bulk powder, 'cutting' or 'striking off' the excess by means of a spatula drawn over the rim to obtain a powder surface level with the rim. This is supposed to be a sufficiently accurate and reproducible process to ensure the correct properties of the set material. Unfortunately, considerable care is required even to approach this intended result.

●4.1 Powder

First of all, the powder is rather fine and contains much low density material – the alginate salt – as well as the filler. This means that it does not settle well, so that the volume fraction of pores in the bulk powder is high. Worse, the particles do not slide well over each other (they tend to stick slightly, probably due to hydrogen bonding, 10§4, at contact points; *cf.* cellulose in paper, 27§2.2) so that further packing is easily obtained by vibration (as occurs in transport or normal handling of the container) or by compression, such as by pressing with a spatula or scoop. Such sliding is also inhibited by the humectants of so-called 'dustless' materials (7§9). Accordingly, the amount of powder collected by the scoop depends on how well the powder was settled beforehand and how much force was used to fill it. Thus it is important to note that the powder packing density is also affected by tapping the scoop, such as against the side of the bulk container or with a spatula. It is also affected by the angle at which the spatula is held with respect to the rim of the scoop when cutting off: any downward component of force tends to leave more behind by compressing the powder.

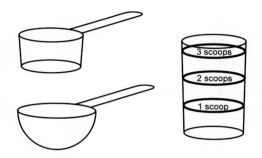

Fig. 4.1 Two types of powder scoop and a water measure for alginate impression materials.

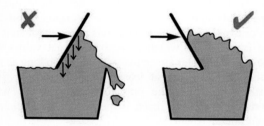

Fig. 4.2 In 'striking off' the excess, the spatula must be held with a positive rake angle (right) to avoid compressing the powder (left).

This behaviour underlies the usual instructions to 'fluff up' or loosen the powder by tumbling the bulk container a few times (but see 7§9.3), and then to avoid digging too deep in filling the scoop, followed by 'striking off' with the spatula leaning back somewhat (a positive rake angle, 20§3.1) (Fig. 4.2). Tapping is sometimes recommended to avoid large voids, which are particularly easy to obtain with scoops having flat, sharp-cornered bases (Fig. 4.1, top) since the powder does not flow well into such angles (*cf.* sand), but this runs the risk of extra consolidation as the amount of vibration to apply is hard top specify and harder to control. Such sharp-angled measures should perhaps not be used for powders that are not free-flowing. Even so, considerable variation is found in the amount dispensed.

It is common now, however, for alginates to be supplied in bags rather than rigid containers, and 'fluffing up' is more or less impossible. It is probably best to decant the powder into a large rigid container (with a good, air-tight seal) for this purpose.

●4.2 Water

Of course, the mixing water must also be measured. This is usually done by means of a cylindrical vessel with one or more calibration lines. The normal means of reading such a measure is at eye level (to avoid **parallax** error, Fig. 4.3) and using the lowest point of the meniscus as the reference level. The diameter of such measures is typically about 32 ~ 35 mm. This means that 1 mm depth is equivalent to about 0.8 ~ 1.0 mL water, or about 4 ~ 6% of that required for a 1 scoop portion of powder. It does get proportionately less as an error if the measure is used for 2 or 3 portions (*i.e.* the measure has 2 or 3 calibrated lines, see Fig. 4.1), but the error can be compounded if the

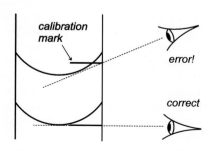

Fig. 4.3 Volumetric measures should be read with the line of sight horizontal, reading to the bottom of the meniscus when the contact angle θ < 90°.

measure is used twice, *e.g.* a 3-scoop portion of powder with a water measure only marked to 2 portions (see §4.4). All this indicates that care is required to achieve accuracy.

However, although some water measures have the calibration marks moulded in (and thus are probably reasonably accurate) some are only screen-printed, and then necessarily with a fairly thick line – around 0.5 ~ 0.7 mm seems commonplace. These therefore have a greater scope for variation – *i.e.* error – in the calibration on the one hand and of reading on the other: does one use the top or bottom of the line?

●4.3 Combination

It would be a mistake to think that these errors are entirely separate issues, even if they are independent of each other: the powder and water have an intended *ratio* in the mixture. Thus the combined distribution of errors must be considered. A typical standard specification requires that the measures provided with the product yield a powder : water ratio not more than 7.5% away from the manufacturer's specified value.[14] Given this, the **tolerance band** can be drawn (Fig. 4.4). From this it is apparent that it is of no consequence whether the exact target values for powder and water are actually achieved, so long as the ratio is correct, or at least within tolerance. The difficulty is that the operator cannot know with certainty what has been dispensed without check-weighing.

But this is neglecting scatter. There will be variation in each measurement, so if a point is plotted corresponding to the mean masses of each component associated error bars can be drawn. However, since the combination of measured powder and water is random (and independent), an ellipse must be drawn through the standard deviation (s) limits to show the expected range of variation (Fig. 4.5, *A*). This may be called a **confidence ellipse** (CE) and represents how the probability of values lying in their range is concentrated (Fig. 4.6). The interpretation is that some 68% of the time a mixture will lie within that ellipse if it is drawn through the ±1 s limits. Similarly, the 95% CE corresponds to ±2 s, and the 99% CE to ±3 s.[2] Looking at point *A* (Fig. 4.5), it can be seen that while strictly speaking accuracy in each component is not necessary if the ratio is right, there is still a risk of a result outside the tolerance band. Indeed, achieving accuracy in either powder (*B*) or water (*C*) alone is unacceptable. But even if only a 5% error on average is obtained in both components in different directions, most of the probability lies outside the tolerance band (*D*). Thus the risk of obtaining an out-of-specification mixture should be assessed, not just the mean accuracy of the component masses. (These ideas of course also

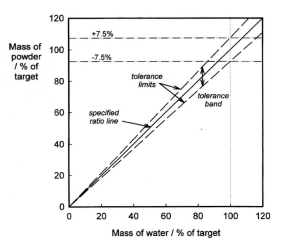

Fig. 4.4 The manufacturer's specified mixing ratio is satisfied along the diagonal of the plot of the two component quantities. The construction of the tolerance band for ±7·5% of powder:water is also shown.

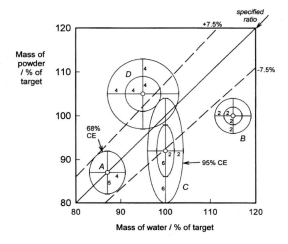

Fig. 4.5 Some confidence ellipses for the expected powder : water ratio given the independent dispensing errors for the two components. The values of the standard deviations are shown against the error bars.

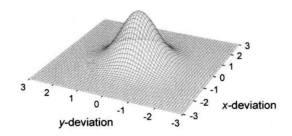

Fig. 4.6 Probability density for a joint normal error distribution (correlation zero). The axis scales are standardized in terms of the standard deviation. If these are unequal, the distribution is stretched in one direction or the other appropriately. The confidence ellipses of Fig. 4.4 represent the total density inside that contour.

---

[2] These are, of course, rough approximations. They ignore the question of sample size effects as well as other details.

apply to any other system in which two components are to be mixed: for example cements, amalgam and casting investments.)

Overall, it would seem worthwhile to 'calibrate' the operator to ensure that the correct mixture is obtained reliably for any given product. If this cannot be done, the only recourse is to weigh the powder and use a more accurate water dispenser. In fact, some alginates are now available in portion packets. It is not known what their accuracy is, but the water must still be dispensed accurately.

Whatever the outcome for the clinical situation it is clear that for any experimental work all proportioning should be done by mass, even if it is the clinical effect that is being investigated.

> Error bars: the uncertainty associated with a plotted value may be indicated by the length of line segments drawn from that point in the direction parallel to the corresponding axis. Conventionally, such bars correspond to small multiples (1 ~ 3) of the standard deviation (or standard error) of the data point. If there is uncertainty in two (or three) dimensions, the additional error bars may also be drawn. Points of constant probability then lie on an ellipse (or ellipsoid).

●4.4   Propagation of errors

In systems in which there are errors from separate sources (and when the variables are statistically and physically independent), it is of interest to calculate the error in a calculated quantity. For both the sum and difference of a pair of values, the **additivity of variance** rule applies.[15] What this means is that if we have two quantities with an associated error, say x with standard deviation σ_x and y with standard deviation σ_y, then for their sum or difference, $z = x \pm y$:

$$\sigma_z^2 = \sigma_x^2 + \sigma_y^2 \qquad i.e. \qquad \sigma_z = \sqrt{\sigma_x^2 + \sigma_y^2} \tag{4.1}$$

This extends in the obvious way to the sum of any greater number of values. In particular, if the individual variances are of the same size, the standard deviation of the sum increases as does the square root of the number of items in the sum:

$$\sigma_z^2 = n.\sigma_i^2 \qquad i.e. \qquad \sigma_z = \sqrt{n}.\sigma_i \tag{4.2}$$

Thus, in this case, taking two or more scoops of alginate increases the standard deviation of the total mass (that is, on repeating the exercise again and again more scatter would be seen) and the risk of an out-of-tolerance value increases. Contrary to the common view, the proportioning does not get more accurate by taking more scoops of powder, or indeed more measures of water, even if we know the *average* scoop contents better for the larger sample size. In practical terms, we may substitute the sample estimate s^2 for the population value σ^2 in the above equations.

§5. Tubes

Elastomeric impression materials such as polysulphide, the silicones and polyether (Chap. 7) are supplied in collapsible tubes (much like toothpaste). These are to be proportioned for mixing according to the length of the extruded 'rope', either of one part when the other reactant is in a liquid to be dispensed by counting drops, or for both parts in 'paste-paste' systems. The basic difficulty here is that the width of the extruded material depends on both the rate of extrusion (how hard one squeezes the tube) and on the speed of movement of the tube. It is notoriously difficult to achieve the requisite balance. It is not helped by the fact that the viscosities of the two components are often different, requiring a different amount of effort. The tube nozzles are usually of different sizes when the volume proportions are not 1 : 1, *i.e.* the rope thickness is supposed to be adjusted to give equal lengths when the ratio is correct. This may make judgement more difficult because of the conceptual conflict between volume and length.

If the intention is to achieve accuracy in proportions to achieve the best properties for the product, care must be taken to ensure that the task is done properly. Weighing remains the best option in case of doubt, and again is essential for experimental work. Similar difficulties arise in the case of putty materials: filling the scoop accurately is very difficult. In contrast, dual-syringe dispensers with disposable mixing nozzles would seem to avoid accuracy problems as well as the difficulties of mixing with a spatula, albeit at a price and with increased waste (§7.2).

For putty-like materials in particular, the co-reactant may be presented as a liquid in a small tube or plastic squeeze-bottle. The same considerations apply for dispensing these liquids as for those for cements discussed below (§6.1). The mixing of impression materials has in part been covered in 7§12.5.

One problem may be apparent with tube contents, particularly when there is a dense powder suspended in the liquid vehicle: settling (4§9.3). That is, if allowed to stand for some time, the denser material will form a more compact layer at the gravitational 'bottom', leaving a layer of the vehicle – usually identifiable as being a markedly less viscous or even clear liquid – on the top. If such a liquid emerges first, whether by being nearer the nozzle or simply squeezed out first by being less viscous, and the solid material was the reactant, the mixing proportions will clearly be wrong, even if the material is weighed. There is not much that can be done about this. 'Massaging' the tube may have some effect, but it cannot be efficient, and simply inverting the tube will not have the desired effect of even redistribution. Avoidance by firstly keeping stocks low, that is, with rapid turn-over, is the best that can be suggested, but if lengthy storage is unavoidable, secondly by turning over the containers from time to time might be effective, but this is clearly not very practical. In any case, any evidence of separation in this way means that extra care is required in the mixing process to ensure homogeneity at least. Even so, no control over the storage of such products in the wholesale and retail distribution chain is possible: packs are usually kept one way up the whole time.

§6. Other Powder-Liquid Products

Various other materials used in dentistry require mixing from powder and liquid. Some aspects of gypsum products have been dealt with earlier (Chap. 2), but there are a number of general issues that need to be recognized even though some have already been mentioned.

●6.1 Proportioning

Many cements as well as acrylic powder products are provided with measuring scoops. While acrylic powders tend to flow very freely, and settle easily with little vibration, many powders do not. Similar care to that required for alginate impression material (§4) is therefore necessary (since all the concerns over tolerances and errors are just as relevant here), including shaking the bottle to "fluff up" the powder (before each use), scooping gently without compressing the product into the scoop, and striking off level, again using a positive rake angle for the spatula (Fig. 4.2). In case of doubt, it is better to repeat the attempt than risk an error. Dispensing the powder before the liquid is also sensible as the latter may evaporate in the meantime sufficiently as to cause problems. Then while dispensing the liquid it must be kept apart from the powder to prevent premature reaction.

Liquids for cements (and elastomeric impression materials) present their own challenges. Although it is a considerable convenience that drop size for liquids dispensed from pipettes, droppers and squeeze-bottles can be remarkably constant, this constancy depends on several factors:

- angulation: ensure that the device is always held vertically[3] so that the geometry is fixed – the drop is supported *via* the contact line, and the length of this depends on the geometry;
- vibration: avoid shaking the drop free, let it fall away freely when its weight can no longer be supported by surface tension (see below, equation 6.1) – indeed, a steady hand is required to avoid any such extra forces;
- rate: dispense slowly so that each drop is separate and dynamic forces are minimized;
- free fall: do not touch the tip or the drop to the pad, let the drop fall freely so that its weight is the determining factor for breakaway;
- bubbles: ensure that the dispensing tip is completely filled the whole time, an ejected bubble will spoil the geometry and may causing splashing;
- cleanliness: for squeeze-bottles, ensure that the outside of the nozzle is clean before dispensing, again to ensure the correct drop geometry, but also to avoid contamination or the inclusion of more concentrated or dried material.

The analysis of hanging drops is rather complicated in detail[16] although in the limiting case of a fine capillary, radius *r,* and a relatively large drop, total volume V_h, we have from the balance of forces:

---

[3] However, read the instructions: at least one product specifies that the nozzle be held at 45°.

$$V_h \Delta\rho g \;=\; 2\pi r\gamma$$

gravitational   surface
acceleration   tension

(6.1)

Fig. 6.1 Water droplet on GI cement liquid dispenser tip, 0.9 mm diameter.

where $\Delta\rho$ is the difference in density between the liquid (of surface tension γ) and the air, and g is the local gravity (*cf.* equation 10§2.5); the dependence on the length of the supporting line is clear (Fig. 6.1). Of course, this does not take into account the wetting of the nozzle material, which would tend to increase the radius of the contact circle unless the tip were very sharp-edged, and such wetting is helped by old dried liquid. Thus, although the exact equation will depend on the particular nozzle design, through some correction terms, in principle it is a rather precisely determined physical quantity regardless of the detail of the tip geometry. For example, for water in a glass Pasteur pipette, with r = 1 mm, equation 6.1 gives ~46 µL, hence the common (but crude) idea of 20 drops to the millilitre. Whether by accident or design, many ordinary dispensers have tips of about that size. (Equation 6.1 also ignores viscosity – the timescale of dispensing of viscous liquids matters[17]: forcing a drop of a viscous liquid by increased dispensing pressure means that it will be oversized.)

Acrylic such as for denture bases has its own approach, relying on the free-flowing powder to give a very closely-constant volume fraction porosity in the bulk, so that the gauging liquid (2§2) is similarly constant. In addition, the liquid has very low viscosity and wets the polymer very well. Accordingly, the scoop quantity is not critical and liquid does not need precise measurement. Powder can be added in a stream into a portion of liquid until excess is present, this excess can then merely be tipped off from the saturated (2§2.5) powder beneath. Such a technique has the advantage that if the powder is added sufficiently slowly, air bubbles will not be trapped. A variation of this is used for gypsum products and casting investments. Both water and powder should be measured accurately, but the powder should then be added to the water in the mixing bowl in a steady stream ("sifted" in) to allow wetting and avoid the entrapment of large islands of dry powder which would present more difficulties for incorporation. The nature of the powder is such, however, that the saturation of the powder does not occur as easily as with acrylic. Sometimes a few seconds are then allowed for the wetting process to proceed, permitting capillarity (10§2.9) to eject much of the air.

For those powders or liquids that are measured out in a separate container before mixing, whether a scoop as for alginate impression materials or cements, a balance pan or measuring cylinder, the point is not so much what is put in but what is then tipped out. Powder may adhere or become stuck in corners (especially in small cement scoops), liquid drops may remain, or the film of liquid which wets the surface will not drain – all are in effect subtracted from the quantities actually mixed. Precise work will consider what is delivered, not what was measured out. A similar problem exists with prepackaged 'unit' doses for materials such as casting investments. Care must be taken that the packet is emptied completely enough, that material is not left in a corner or lost when it is cut or torn open. Related effects have already been noted for silver amalgam capsules (§2.7) and other capsule products (§3).

We may make two further observations. If such a system is mixed with less than the gauging volume (2§2.4), it necessarily has a non-glossy surface, but from that amount to the minimum mixing volume it may go matte during mixing (even if only temporarily) because of the mixture's dilatancy (*cf.* loss of gloss, 2§5.1, Fig. 10§2.12). Likewise, if reaction proceeds too far because of slow mixing or too high a temperature, for example, then a loss of gloss will also occur. It is worth stressing the implications of the incorporation of air bubbles due to this, and thus the effect on strength (1§7) – once present, they can hardly be removed. Loss of gloss, even if this occurs because of change of volume from reactants to reaction products, may also have an effect on setting expansion (10§2.2).

●**6.2   Combining**
For cements that are dispensed onto a mixing pad, it is sometimes a considerable challenge to gather and wet all the powder into a paste, especially because the microscopically-rough surface of the powder in bulk commonly prevents wetting and capillarity absorption of the usually quite viscous liquid into the mass. Although not required for reasons of special chemistry as with zinc phosphate cements (9§5), it is frequently easier to combine a small portion of the powder with the liquid first, then gradually incorporate the remainder. Fine powders can easily be lost through 'splashing' with the mixing implement and blowing away. Proceeding cautiously to start with allows better control.

●6.3 Mixing

Although efforts may be made to dispense the components of a cement or similar material accurately, it is still necessary to mix them properly. This requires:

- completeness: remnants of powder or liquid, or incompletely mixed material, must not remain anywhere on the mixing surface or spatula;

- thoroughness: uniformity of the mix is critical to the intended mechanical properties being attained, and layering, excess liquid or powder anywhere, or much variation in the effort expended on the various regions of the mass will be detrimental;

- no bubbles: consolidation of the mass and expulsion of air bubbles – Griffith flaws (1§7) – must be assiduously pursued – this is difficult, but their size can be minimized;

- promptness: as liquid may suffer evaporation, and thus affect the powder : liquid ratio, mixing should not be delayed after dispensing;

- speed: the moment contact between components occurs, reaction starts. It is essential to complete the mixing process before the reaction has proceeded far enough that detrimental results will follow – observe the manufacturer's instructions (and remember reaction allowance, 2§2.4);

- planning: once mixed, a product should be used immediately if so instructed, delay is normally detrimental as it encroaches on the working time available and further manipulation irreversibly breaks up the structure that is forming.

To achieve good mixing, repeated gathering of the material together, then squashing and spreading with the flat of the spatula, is most efficient. Simply stirring alone will not achieve the desired outcome because it is difficult to involve all the material without a deliberate effort. This applies on flat pads as well as in mixing bowls. For plaster and similar materials, the spatula is deliberately curved to ensure that all material at the walls can be gathered up.

Mixing is a skill. No amount of theoretical understanding of the principles and process can be a substitute for practice, although that understanding is, as elsewhere, important for appreciating the critical nature of this step no less than any other.

§7. Theory

Whilst the concepts of a mixture and the process of mixing may superficially seem trivial, it is worth exploring some of the implications in a little more detail. Initially, we may imagine two components to be mixed as being completely separate (Fig. 7.1 a), while the target of an ideally mixed system may be represented as in Fig 7.1 c, if each small square represents a component unit, whether a particle or a molecule. For comparison, intermediate states, or poor mixing, are inhomogeneous on a scale larger than the component unit scale (Fig. 7.1 d), while all that can actually be achieved in practice is random mixing (Fig. 7.1 b).

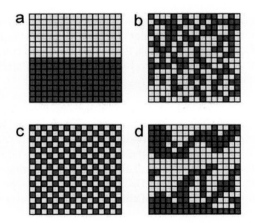

Fig. 7.1   Conditions of mixing: (a) unmixed, (b) random, fully mixed; (c) perfect mixing; (d) poorly mixed.

●7.1 Turbulence

The primary point to notice is that all mixing is based on shear: any stirred system involves flow, and all flow is by shear (Chap. 4).   There are two regimes to consider: **turbulent** and non-turbulent flow. Turbulent flow is a very difficult area of study (by some accounts, *the* most difficult), but we can dismiss this here because it is of very little concern in dentistry (with perhaps the exception of casting, Fig. 18§4.4). Turbulence means the creation of eddies or vortices in which shear continues by virtue of the momentum of the fluid: think of the swirling water in a river as it passes a bridge pier. Mixing in turbulent systems is very rapid: think of adding milk into stirred coffee. Non-turbulent flow is described as **laminar** – the sliding of layers (Fig. 4§3.8) – or **streamline flow** and is the basis of all the rheological discussion in Chap. 4 and elsewhere. The boundary between the two regimes of behaviour is identified by reference to the (dimensionless) **Reynolds number**, *Re*, of the system:

$$Re = \frac{\rho . u . r}{\eta} \qquad (7.1)$$

where ρ is the density, η the viscosity, u the velocity, and r a linear scale measure.[18] This expresses the balance between inertial (numerator) and viscous (denominator) forces. The scale measure is the most difficult part, that is, to identify precisely what this means in any given system, but for example it would be the radius of the ball in Stokes flow (4§10), and the diameter of the tube in Poiseuille flow (4§11); essentially, it is approximately the scale of the affected body of the flowing fluid and may be taken as representing the largest eddy – as the entity having kinetic energy – that can exist in the system. If this value *Re* is greater than about 2300 instability sets in, > 4000 the flow is definitely turbulent. To find the worst case, we may note that the scale of any system in dentistry is no greater than about 0.1 m, the density of impression materials and the like is no more than about 3 g/mL (*i.e.* 3×10^3 kg/m^3, to keep the units consistent), and the viscosity of the most fluid impression plaster is about 2 Pa.s; with these values the velocity must exceed ~15 m/s to encounter instability, and > 27 m/s for turbulence. Going to smaller scale, higher viscosity and lower density increases the critical velocity. For comparison, alginate when mixed has a viscosity of some 20 Pa.s, minimum, and the fastest that a plaster or alginate mixture could be stirred with a spatula is about 4 turns/s, or about 0.75 m/s. Since the maximum value of *Re* is therefore ~100, we can probably safely discount turbulent flow from our considerations.

●7.2   Shearing

Turning then to laminar flow, it is clear that disregarding chemical reaction effects, mixing must be at constant volume.[4] If then the block of two components (Fig. 7.1a) is drawn out (Fig. 3§4.10), shearing must be occurring such that the area of the interface increases while the thickness of the layers decreases. If the elongation was to double the length, and the material folded over, this would double the number of layers[5]. Repeated application of this **Baker's transformation**, named in direct reference to what bakers do in handling dough, results in **striation thinning**. Similar effects are obtained by cutting and stacking (Fig. 7.2). Foodstuffs such as flaky pastry, mille feuille, and baklava illustrate the outcome very well.

The question then is how many times is this necessary to achieve adequate mixing? The ultimate scale is that of the size of the molecules involved, D_M. This has to be taken in relation to the starting scale, D_0, which we can take to be half the size of the combined body – the depth of a layer in Fig. 7.1a. We can then write:

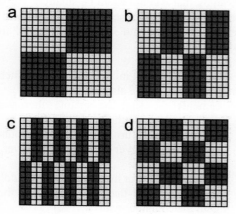

Fig. 7.2   Some possible stages of mixing by repeated cutting and stacking from the condition of Fig. 7.1a.

$$D_M = \frac{D_0}{2^n} \qquad (7.2)$$

D_M obviously varies somewhat, but we can take 1 nm (about 6 or 7 carbon atom diameters) as being a representative enough lower limit. This can be rearranged to allow us to determine the relevant **division number**, *n*:

$$n = \log(D_0 / D_M) / \log 2 \qquad (7.3)$$

where we can see, as might be imagined, that the relative magnitude of the start and finish scales is the key factor. Then, if we consider the maximum starting scale, D_0, to be nominally about 35 mm, we get $n \sim 25$. That is to say, that for a 1-dimensional mixing, some 25 foldings – as is done with flaky pastry – are necessary to get down to the molecular scale. Any further folding cannot make the layers any thinner, but merely continue to jumble the molecules – they cannot be mixed any more intimately. Mixing silicone putty impression material is done in this kind of manner. The range of scales involved in some dental materials is illustrated in Fig. 7.3.

In fact, since the requirement for reaction is simply that of contact between reactants, subdivision to the

---

[4]   Solvation effects generally lead to some change of net volume on mixing dissimilar liquids; these effects would be small for similar polymers.

[5]   As described by O. Reynolds, 1894.

scale of one molecule thickness is not necessary. Thus, the condition represented by Fig. 7.1c is quite unnecessary, while that of the top (or bottom) of Fig. 7.2c would appear to be quite enough: each component unit is in contact with its counterpart. n can therefore be reduced by 1 from implied by D_M.

Clearly, this is only a one-dimensional process, and three times as many folds would be necessary to create a three-dimensional subdivision to the same scale. This is obviously not necessary if the goal is just to bring reactants together, which the one-dimensional process achieves. Thus, the extra (horizontal) cut of Fig. 7.2c is actually unnecessary. There will also be variations due to factors such as unequal quantities of components, diffusion that is occurring while the mixing is going on, as well as the fact that diffusion will occur after mixing anyway, when the random walks of the diffusing molecules or chain segments will contribute to effective distribution of reactants, meaning that n could again be smaller. However, variation in the thickness of the layers will also reduce efficiency and should be allowed for by adding some steps.

The kinds of mixing process to which this analysis is applicable also include:
• cements – using the spatula alternately to gather and spread the mixture;
• alginate – each turn of the spatula around the bowl spreads a layer over a previous layer, then thins it out again;
• vacuum mixers for investments – the paddle can be seen to shear the mixture through the gap with the container wall.
Even an idealized simple rotation of a viscous body can be seen to have a similar effect by shearing and thinning out the layers (Fig. 7.4). It can be seen that the recommended 'stropping' motion for alginate mixing (see box), primarily said to be to remove large bubbles and make sure no powder is left unwetted, is in fact creating the appropriate shear conditions for efficient mixing.

When the components are strongly and contrastingly coloured, as are various silicone impression materials, for example, it is easy to see the streaking, layering and thinning going on during the process. However, one of the possible instructions is to mix until no more colour streaks can be seen. Since the visible light resolution limit is at about 0.5 μm, we can put this figure in instead for D_M in equation 7.3. This gives $n \sim 16$, implying that beyond that no change would be detectable optically, even with a microscope, and the target of "molecular" scale mixing would not have been obtained if the process were stopped at that point. Worse by far, however, is the limit imposed by our visual acuity, dependent in part on the spacing of cones in our eye's fovea. At a distance of about 30 cm, the very best we can do is resolve lines about 0.07 mm apart (acuity may be, and commonly is, substantially worse than this). Setting $D_M = 0.07$ mm gives $n \sim 9$! In other words, it would appear that we should continue to mix for some further time, even if

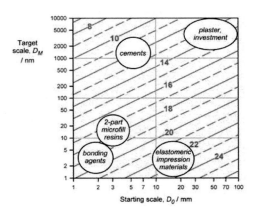

Fig. 7.3   Division number contour map for the range of scales relevant to dental materials' mixing, with some examples of possible value regions.

It is commonly said that alginate impression material should be mixed with a **stropping** motion. This action derives from the technique of cleaning up a **whetstone**-sharpened cut-throat razor, by dragging the blade edge 'backwards' in alternating directions on a strip of cow-hide leather called a "strop". The purpose then was to remove a very thin strip of steel called a 'wire' from the edge (left behind by the grinding process) – by fatiguing it – leaving the razor very sharp for shaving. Most of these concepts are archaic and thus the relevance not at all obvious to many in the present context (*cf.* 'brush-heap' [Chap. 7]). "Stropping" is often explained as a 'swiping' or wiping motion in which the mixture is squeezed – fairly forcibly – between the spatula and the wall of the mixing bowl. This is said to help to remove bubbles (presumably by forcing them to be exposed at the surface where they burst), but it would also break up any remaining dry powder-containing bubbles – which action would be crucial to obtaining a complete, homogeneous mixture. This can then be seen to be operating in the same manner and according to the same principle as applies to mixing on a pad (Fig. 7.7). A similar process is used in the mixing of dental plaster and stone, and even of casting investment materials, if this is done by hand.

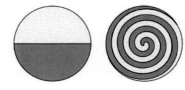

Fig. 7.4   Simple viscous rotation generates multiple layers that effect mixing.

Fig. 7.5   An impression material mixing nozzle screw: each segment is rotated 90° and has opposite sense with respect to its neighbour, to divide each stream, and recombine alternately.

homogeneity seems to have been achieved. For comparison, we may note the design of an impression material mixing tip (Fig. 7.5), a device known as a **Kenics**[6] mixer.[19][20] This example effectively subdivides the flow 11 times, as the baffles are arranged to cut each stream in two and recombine them in complementary pairs. This process can be followed by cutting the nozzle into slices corresponding to each segment (Fig. 7.6), when the

Fig. 7.6 Successive stages in the mixing of an impression material in a Kenics mixer nozzle.

successive subdivision is seen. Thus, the device should take the mixing just beyond the point at which streaking is visible (allowing perhaps for some inefficiency, which close visual inspection shows is present). As far as the user is concerned, the job appears to be done. (Some earlier devices have had 12, and even 13 flow-splitters.[7]) Other than speed and simplicity of use, a major advantage of this approach is supposedly the avoidance of bubbles incorporated during mixing. However, the blades of the mixing nozzle screw have sharp, square edges and this means that boundary layer separation must occur, especially if extrusion is rapid (23§10) (Fig. 7.7). This means that the first portion must have many bubbles as these are swept out, and only the later material is as desired (Fig. 7.8). Of course, any bubbles already present in the components will remain. (It is evident in Fig. 7.8 that the mixing nozzle had too few flow-splitters.)

Fig. 7.7 Bubbles form on the edges of the mixer screw blades because of boundary layer separation. Photo: Tom Smethurst

The homogeneity problem is not quite so bad with granular materials such as amalgam alloy powder, cements and gypsum products where reaction between the powder and liquid is involved. Assuming that the particle size is of the order of 1 μm, mixing beyond $n \sim 13$ or 15, depending on starting scale, will not achieve anything more (Fig. 7.3). However, it can still be noted that the absence of any strong colour contrast means that inhomogeneity is very hard (if not impossible) to detect, for example in cements, but the problem is especially acute with two-paste filled-resin luting materials (6§4.7), which are also much more viscous and therefore difficult to mix. In this case the problem is to achieve adequate mixing of the two matrix components so that they can react properly, not to disperse the powdered filler in the matrix (which is the manufacturer's problem). Accordingly, the target scale is somewhat smaller than that of the filler particles (Fig. 7.3).

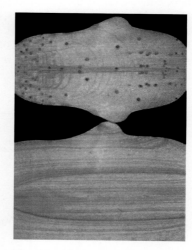

Fig. 7.8 The initial extrusion (top) has many bubbles, the later material (bottom) is free of them. Inadequate mixing streaks are also visible in this example. Photo: Tom Smethurst

With materials that are mixed on a pad or slab, there is some gain in efficiency from the method used. The calculation of equation 7.3 is based on an exact halving at each step, the **characteristic number** for the Baker's transformation is 2. Obviously, if the materials are spread by the spatula to cover, say twice the area at each step, the process will proceed twice as fast, assuming the gathering together of the mass after each spreading is effectively folding the sheet up. In fact, the characteristic number may have been increased to 4 or more. Thus, repeatedly spreading thinly then gathering into a compact mass gives very effective mixing, assuming that all material is gathered up at each step (Fig. 7.9). Careful use of the spatula to mix cements and impression materials of this type will indeed minimize the mixing time and extent of reaction at the

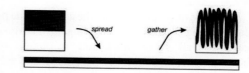

Fig. 7.9 Mixing on a pad may be more efficient if spreading is thin and gathering is complete.

[6] This is a trade name: http://www.flexachem.com/kenics-km-static-mixers/
[7] It is to be hoped that the decision to reduce the number of steps was not based on the visual evidence.

point of use. However, care has to be taken if two pastes are of differing viscosity that the more viscous one does not squeeze out the other.

In any case, the edges of the mass of components will not be incorporated as efficiently in such a scheme – it is not a mathematically precise operation – and extra attention will need to be paid to ensure that inhomogeneities do not arise from this source.

It still needs to be borne in mind that we have the following argument:
• greater intimacy of reactants gives better reaction
• reaction creates structure (as is required, of course)
• continued shear – *i.e.* mixing – breaks down that structure,
• mechanical properties are then expected to deteriorate
Clearly, there is a compromise necessary between these conflicting effects, and there is expected to be a cross-over point where structure damage begins to exceed the benefit of extra intimacy. Here, the rate of mixing in relation to the rate of reaction (*i.e.* the timescale of the rate constant) becomes relevant as more subdivisions can be achieved before that critical point when mixing is faster, but (of course) greater intimacy leads to a greater rate of reaction in the sense of the bulk conversion of the reactants (that is, not the rate in the chemical kinetics sense). Slowness, delay and interruptions are necessarily detrimental (Fig. 7.10). Over-mixing is shown to have an effect on the extreme right of Fig. 7.11 for compression set, but at an appreciably lower n for modulus of elasticity (Fig. 7.12).

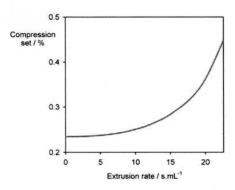

Fig. 7.10   Slow mixing can damage developing structure. Here, if the extrusion through a Kenics nozzle is too slow, the compression set of a silicone impression material is made worse.

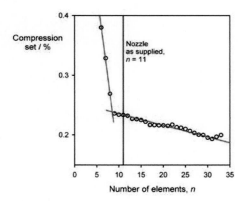

Fig. 7.11   The minimum number of Kenics mixing elements required for a silicone impression material is shown by mechanical properties, but also damage from over mixing.

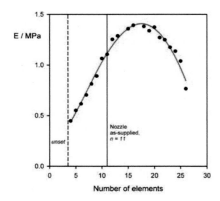

Fig. 7.12   Structural breakdown from over-mixing is also seen in modulus of elasticity, but with a different optimum n. (Cantilever beam self-weight deflection results, 23§4.1)

●7.3   Amalgam capsules

It is now apparent that the action of the pestle in the capsule (Fig. 2.6) is one of shearing the mixture against the capsule wall (very much like a plunger-type tissue homogenizer, Fig. 7.13[21]). The model is the Newtonian dashpot (Fig. 4§3.6), where the rate of shear depends on the annular gap. But now, clearly, the thinner the layer that is created, the greater the rate of mixing in the sense of §7.2, explaining the effect of ball diameter in Fig. 2.5 and illustrating the need for both of the two issues to be dealt with effectively: oxide removal and actual mixing. Even

Fig. 7.13   A Dounce tissue homogenizer: the close-fitting plunger leaves a narrow annular gap that causes a very high rate of shear. Photo: Lily_M

so, very few oscillations would be necessary if the system were designed appropriately. By implication, such a system would mix other encapsulated dental materials more efficiently: better, faster.

●7.4   Diffusion

For a full analysis of the mixing process, as suggested above, diffusion of the reactants must be taken into account. Accordingly, the temperature of the system is critical, and so is the viscosity of the medium. This involves not just the diffusion of small molecules, but also that of chain segments in polymeric systems (3§3). Some related points have been made in relation to filled-resin materials (6§4.5) and impression materials (7§11.2). The key issue is the mean diffusion distance over the time allowed from the end of mixing to the point at which the material is considered set and thus handled, supposedly safely. This is because the poorly-mixed regions are the ones most critical to the failure of the material, and the mixing time is generally very much shorter than the time allowed for setting (even though diffusion rates will be greater in the unreacted material). Thus, in Fig. 7.9, there is a clear transition at $n = 9$ from an inadequately mixed domain to one that is only slightly affected (in this one respect) by further mixing. At this point $D_M \sim 10$ μm (equation 7.2, for $D_0 = 5$ mm – the effective internal diameter of the nozzle), which suggests the possible useful diffusion distance in this case. However, it should be noted that this is already smaller than the scale where the stripes can be resolved by eye, emphasising that reliance on visual inspection is unsatisfactory.

●7.5   Wetting

In systems where wetting of the powder by the mixing liquid is possible, capillarity can drive the intermingling of the components (10§2.9) such that wetting out (2§2.5) is spontaneous, treating the powder mass as one body. In this context, the mixing process can be seen to be much more efficient, and the division number formally required could be very small. However, such penetration takes time (and reaction will occur while waiting), can lead to the trapping of bubbles, does nothing to break up any agglomerates (lumps) in the original powder, and does not ensure homogeneity. Some mixing is therefore still required, and the use of reduced pressure may be necessary when bubbles could be critical (18§1).

§8.   Spatulas

Mixing devices may not receive very much attention, but they are important in the sense that they are commonly the means by which the requirements indicated above are met. Operative textbooks explain design features such as the need for flexibility in the spatula used for mixing impression materials such as polysulphide, or the shape of an alginate- or plaster-mixing spatula. However, some thought also needs to be given to the material from which they are made.

Stainless steel (21§2) is the commonest material, for the simple reason of its relatively good resistance to corrosion (13§5.3), low cost, and generally adequate mechanical properties. However, as indicated in 20§3.5, it is not hard enough to resist abrasion by the fillers used in restorative materials, whether filled resin or glass ionomer. Hence, if discolouration would be important, such spatulas should not be used for hand-mixed materials. Attention should also be paid to its chemistry. As indicated earlier (9§6.2, 9§8.7), its oxide coating means that polycarboxylate-based cements will adhere if allowed to set in place, but that oxide is also susceptible to dissolution by both acid and alkali if strong enough, as well as fluorides (*cf.* 22§2.2), such that corrosion follows.

One approach to overcome some of the problems that has been suggested is the use of spatulas made of tantalum, a very hard, although dense (~16.7 g.cm⁻³) grey metal. It is almost completely immune to chemical attack at temperatures below 150 °C except by acidic solutions containing fluoride ion (such as HF) and sulphur trioxide – which latter is not encountered in dentistry – and only slowly by strong alkalis. Unfortunately, it is an expensive metal, partly because of its high melting point (3017 °C). It is also expected to adhere to polycarboxylate cements because of the passivating oxide coat (Fig. 13§5.1).

Agate is an extremely finely crystalline form of silica whose consequent hardness and lack of chemical reactivity (except strong base) has led to it being used for dental spatulas (Fig. 8.1). These have been recommended for use in mixing porcelain powders, primarily because any abraded

Fig. 8.1   One end of an agate spatula.

particles are chemically-compatible and cause no discolouration, as would happen with stainless steel (noting the strong effects of oxides, Table 25§7.1). They are, however, necessarily brittle, and not particularly strong, being prone to chipping if handled roughly and break if dropped. They must therefore be made rather thicker in the blade than is otherwise common. Formerly, they were recommended especially for silicate cements (9§7), where again abraded particles would not cause a problem. Sintered zirconia (25§11) is now being used for the blades of spatulas for porcelain work, where the much higher flexural strength allows a much thinner blade, while the hardness limits abrasion.

An alternative approach is to use (organic) polymeric materials, which if chosen carefully can be chemically- and solvent-resistant, such as polyethylene, polypropylene and polytetrafluoroethylene (ptfe). Although their stiffness may be low, section thickness can be increased to compensate. Certainly, most dental materials will not stick to such tools, but poor abrasion resistance would mean that incorporated abraded particles would effectively be flaws in the set material, even if not visible. (Strangely, there are some disposable plastic spatulas for sale for dentistry described as 'agate' – care is required.) Now-obsolete materials include tortoiseshell (actually hawksbill turtle shell) and ivory, both of which have long since been banned, but abraded ivory particles – elephant dentine – would of course be well-bonded in polycarboxylate cements, and neither would adhere to the usual polymerizable resin matrices. Overall, it is worth paying attention to the materials of mixing implements to avoid problems.

References

[1]   Grenga HE, Carden JL, Okabe T, Hochman RF. Auger analysis of surface films on Ag_3Sn. J Biomed Mater Res 9: 207 - 11, 1975.

[2]   Darvell BW. Efficiency of mechanical trituration of amalgam. II. Effects of some variables. Austral Dent J 26: 25-30, 1980.

[3]   Darvell BW. A performance criterion for amalgamators, capsules, pestles and alloys. Austral Dent J 25: 146-147, 1980.

[4]   Darvell BW. Efficiency of mechanical trituration of amalgam. I. Optimum capsule length. Austral Dent J 25: 325-332, 1980.

[5]   Brockhurst PJ & Culnane JH. Optimization of the mixing of dental amalgam using coherence time. Austral Dent J 32: 28 - 33 1987.

[6]   Watts DC & Combe EC. Early strength and adaptability of amalgam in relation to coherence time. Dent Mater 9: 74 - 78, 1993.

[7]   Darvell BW. Efficiency of mechanical trituration of amalgam. III. Practical comparisons. Austral Dent J 26: 236-243, 1980.

[8]   Anderson MH, Kuhl LV, Schloyer DD. Mercury leakage during amalgam trituration. Oper Dent 13: 185-190, 1988.

[9]   Wilson SJ, Wilson HJ. Mercury leakage from disposable capsules. Brit Dent J 153: 144-147, 1982.

[10]   Darvell BW. Mercury Hazard: Current Notes No. 57. Aust Dent J 24: 385, 1979.

[11]   Stone ME et al. Residual mercury content and leaching of mercury and silver from used amalgam capsules. Dent Mater 18: 289-294, 2002.

[12]   Osborne JW. Mercury, its impact on the environment, its biocompatibility. Oper Dent Suppl 6: 87-103, 2001.

[13]   Prentice LH, Tyas MJ Burrow MF. The effect of mixing time on the handling and compressive strength of an encapsulated glass-ionomer cement. Dent Mater 21: 704-708, 2005.

[14]   Specification for dental elastic impression materials Part 2. Alginate impression material. BS 4269 : Part 2 British Standards Institution, London, 1980.

[15]   Meyer SL. Data analysis for scientists and engineers. Chap. 10. Wiley, New York, 1975.

[16]   Boucher EA, Evans MJB. Pendent drop profiles and related capillary phenomena. Proc Roy Soc London A, Math Phys Sci 346 (1646): 349-374, 1975.

[17]   http://en.wikipedia.org/wiki/Pitch_drop_experiment

[18]   Reynolds O. An experimental investigation of the circumstances which determine whether the motion of water shall be direct or sinuous, and of the law of resistance in parallel channels. Phil Trans Roy Soc 174: 935 - 982, 1883.

[19]   Armeniades CD, Johnson WC & Thomas R. Mixing device. https://www.google.com/patents/US3286992

[20]   http://adlittlechronicles.blogspot.co.uk/2008/07/disposable-motionless-mixer.html

[21]   http://commons.wikimedia.org/wiki/File:Tissue_glass_teflon_Dounce_homogenizer-02.jpg, modified

Chapter 16   Waxes

The use of waxes in dentistry is inescapable in many applications because of their special combination of properties: weak, mouldable, low-melting, combustible, non-toxic. However, these benefits come at the price of considerable disadvantages. The successful use of waxes must therefore be with a full understanding of their behaviour.

*Waxes are typically moderately-large, essentially **paraffin-like** hydrocarbons with weak intermolecular forces. Thermal expansion is very large, and flow is easy. Even so, some of the important practical advantages include the ability to be carved without leaving rough surfaces and being moulded without cracking.*

*The **flow** of waxes is critical in the sense that they are frequently moulded when in a softened condition by the application of heat and pressure. Accurate moulding to the shape desired is important. This is assisted by its pressure-sensitive viscosity. However, waxes are viscoelastic and will continue to deform under applied stress even when cooled, and show retarded recovery of stresses frozen in during the moulding process. Both diminish the accuracy attainable and so require prompt processing to minimize the effects.*

*For metal casting, it is necessary to ensure that the wax is entirely **combustible** if untoward effects are to be avoided. Filled waxes, to control flow, thus have not found general application except in impression compound.*

Although waxes are indispensable in dentistry, they have severe problems of dimensional stability. With an understanding of their properties and limitations, these difficulties can be overcome for effective precision casting and other work.

Materials Science for Dentistry
https://doi.org/10.1016/B978-0-08-101035-8.50016-X

Waxes have a wide variety of types of application in dentistry, ranging from taking impressions, making patterns for lost-wax casting and acrylic denture baseplates, for bite detection, blocking out undercuts on models, the masking of regions to be protected in electroplating, and the temporary cementing of models. Even given this range of applications, the selection of waxes to do all of these jobs is of course coincidental, essentially relying on just one or a few of the characteristic properties of this class of materials.

§1. Chemistry

Waxes are not defined chemically but rather they take their name in allusion to beeswax, which in view of the presumed antiquity of its knowledge may fairly be called the archetype of all waxes. Beeswax is largely a collection of long chain esters, and it will be shown that this chemistry is associated with a kind of structure that leads to the same general properties arising both in a range of pure substances as well as in their mixtures.

The simplest waxes are the alkanes (C_nH_{2n+2}), whether straight chain or branched, and these **paraffins** may be considered as waxes when they are solid at room temperature or thereabouts. Thus, for n = 17, the melting point is about 22 °C. There is no effective upper limit to n. Polyethylene, named more from its manufacturing route than what for it is, an alkane, may also be considered as a wax, where n is then counted in the thousands or tens of thousands (it has a characteristic waxy feel). Indeed, it may be used in some dental products. However, a typical practical limit to dental usefulness for alkanes is with n ~ 60.

The alkanols ($C_nH_{2n+1}OH$) and alkanoic acids ($C_nH_{2n+1}COOH$) are also waxes if n is large enough, but in fact they are not very common as constituents of the natural products employed in dental wax manufacture. On the other hand, esters are very common (see beeswax, above), especially as plant products. In the general formula for saturated esters, $C_nH_{2n+1}.COO.C_mH_{2m+1}$, m and n may both be in the range 15 - 30. Similar length chains are found in the unsaturated esters, and in all similar compounds that exhibit beeswax-like properties, i.e. are waxy.

●1.1 Melting

The melting point of the straight chain alkanes is primarily dependent on the number of carbon atoms (Fig. 1.1) as might be expected, but there is a subtle effect associated with whether that number is odd or even. This is most noticeable in the lower members of the series, the odd-numbered chain having a slightly lower melting point than expected (or vice versa). This illustrates the subtlety of the factors involved in the crystallization of such compounds, that is, the neatness, if you will, of the packing. Branches in the chain can have much more dramatic effects. For example, for the series n = 4 to 10 (at least), the addition of a methyl group in the 2-position typically lowers the melting point by 18 ~ 24 K, despite the increase in molecular weight (Fig. 1.1). The position at which the substituent has been introduced is also extremely important (Table 1.1), but the effect is largely unpredictable. Clearly, it is difficult to arrange such molecules in fully-regular arrays, and crystallization becomes more difficult; thus the melting point is lowered.

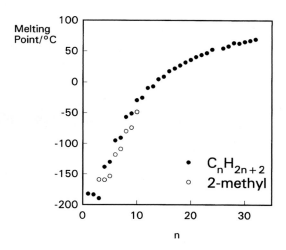

Fig. 1.1 Melting points of the n-alkanes and the 2-methyl alkanes (where n refers to the straight chain).

Table 1.1 *Melting points of substituted hexanes.*

		m.p. / °C
C_8	2,2- dimethylhexane	-121
	2,5- "	-91
	3,3- "	-126
	"1,6-" (= n-octane)	-57
C_9	2,2,4-trimethylhexane	-120
	2,2,5- "	-106
	2,3,3- "	-117
	2,3,5- "	-128
	2,4,4- "	-123
	3,3,4- "	-101
	n-nonane	-51

Natural waxes are likely to consist of a mixture of a great many compounds: mixtures of types (alcohols, esters, *etc.*) and mixtures of straight and branched chains, all of which will have different physical and mechanical properties, although chemically the properties will be dominated by the preponderance of the alkane parts. Although one or a few compounds may account for the majority, they never approach purity. Even in refined paraffin waxes derived (by distillation) from petroleum, there will be a range of sizes present, probably concentrated around some central value, but with tails in both directions. Under these circumstances it is not surprising to find that there is no single, sharp melting point for any of these natural products, but rather there is a pronounced **melting range**. Under these circumstances, the "melting point", T_m, can really only refer to the upper limit of the melting range. A mixture of two metals has already been discussed in the context of the phase diagram and cooling curve (12§1), and mixtures of two chemically different waxes (which are themselves mixtures) can be studied in a similar manner. Generally the mixture has a wider melting range than either component wax (Fig. 1.2).[1] But when the mixture consists of many compounds, with strong chemical similarities, it may become impossible to obtain precise values for the top and bottom of the melting range merely by studying ordinary cooling curves (Fig. 1.3) (see 11§2.7). In these cases, microscopic observation of a few shavings on a heating stage may be necessary to detect changes in appearance, but with an appropriate technique it is possible to demonstrate not only the onset and completion of solidification on cooling but also the presence of many other features that may then be related to mechanical properties (Fig. 1.4).[2]

●1.2　**Structure-property correlation**

Waxes can be described as being weak solids, having (apparently) low yield points at moderate rates of strain, with a tendency to creep even under low stresses, low melting points, and as being hydrophobic (*i.e.* showing poor wetting by water, contact angle > ~100°). It is not difficult to account for these properties.

First of all, being non-polar or nearly so (paraffinic chains dominating), intermolecular forces are almost entirely of the van der Waals type and so very weak (10§3). This means that strength is low, and the expected yield point much lower. It also fixes the low melting point, essentially because it does not require much energy to separate or slide such molecules over each other. Furthermore, it immediately accounts for the poor wetting as there are few or no groups to interact with the water. Waxes, with their typically long polymethylene

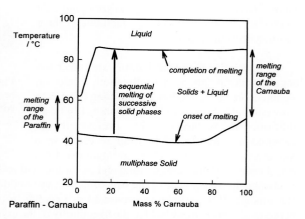

Fig. 1.2 Melting range for mixtures of two natural wax mixtures.

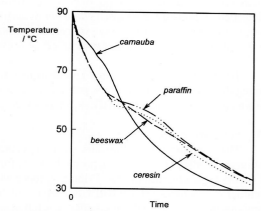

Fig. 1.3 Typical cooling curves for natural wax mixtures. Little detail is discernible.

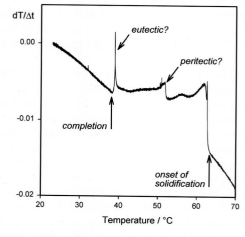

Fig. 1.4 Differential cooling curve for beeswax (rate of change of temperature with time) showing the complexity of the mixture. Broad bands indicate progressive freezing of multiple components. This graph is to be read from right to left, *i.e.* for cooling. Evolution of latent heat of solidification is indicated by the curve deviating upwards (slower cooling). $dT/\Delta t > 0$ means a rise in temperature, here because of the supercooling of the eutectic.

chains, $-(CH_2)_n-$, commonly crystallize in lamellar structures;[3][4] such layering facilitates shear deformation (Fig. 4§3.4).

Secondly, the chain lengths are rather too long to permit good crystallization, faults in alignment are easy, yet they are not long enough to allow entanglements as in high polymers (3§3). In addition, it is essentially impossible to accommodate the carboxyl group in esters in a regular array (Fig. 1.5). We might note that for a

Fig. 1.5 A straight-chain paraffin ester crystallizes less well because the molecule is kinked (in lowest energy conformation) and has a bulky carbonyl group. The molecule is also nearly reversible.

typical ester, with moderately long chains for the acid and alcohol, reversing the molecule (end to end), would not change the appearance of the structure, the two would be quite similar except perhaps for the position of the ester linkage. If it could be incorporated in a crystal structure at all, orientation would not matter very much energetically, but it would add to the irregularity of the crystal. Add to this the fact that most waxes are mixtures of very similar molecules, homologues differing by one or two carbon atoms or maybe by having an unsaturated carbon-carbon bond, and the strength-related properties are even more compromised. However, the chain lengths are great enough that some high polymer-like properties are observed. Glass transition temperatures (3§4) can be identified, with clearly brittle behaviour below and plastic behaviour above. There is therefore the expectation of strain-rate sensitivity, hence there is creep at low stresses. In amorphous regions, as in polymers, plasticizer effects will be observed if smaller molecules are present.

To all this must be added the kinetic aspect: rapid cooling will not allow such large molecules to form good regular crystallites, even if it is possible from solution. However, it does imply that such material may be far from equilibrium under normal conditions of use in dentistry. This aspect needs further consideration.

●1.3   Solid structure
The dissimilarities of the constituent molecules are great enough that continuous solid solutions, as are found in metal alloys (11§3.7), are probably never formed in commercial products. Rather, the microstructure of a wax is typically multiphase. Primarily, there will be a mixture of very small crystals of the nearly pure individual components which will have separated from the cooling melt. These will be distributed in an amorphous matrix of everything else that was left over, whether by not having enough time to crystallize out on existing crystals, or by being uncrystallizable. The relative amounts of matrix and crystalline core thus depends on the thermal history. Over time, solid waxes change their properties (both mechanical and optical) as solid-state diffusion permits further crystallization, the rate depending on temperature – we observe **ageing**. Here again we may notice that the working temperatures in dentistry, say 20 ~ 40 °C, are very close to the melting temperatures, T/T_m being as much as 0.95. Thus, strength is necessarily low and diffusion rates high (allowing for the fact that the molecules are typically fairly large). The ageing process can be observed optically because the translucency increases as the proportion of amorphous material between crystals decreases (3§4.1).

The picture is potentially more complicated than this. Since there is a solidification range as an obvious consequence of there being a melting range (Figs 1.2, 1.4), over a large temperature range there will be liquid present even if the mass appears to be solid. Such liquid, even in small amounts, could have profound effects on the mechanical properties by acting as a zero-strength dispersion (2§8), or even as a continuous three-dimensional network which will allow and lubricate the relative movement of crystallites. In other words, it is paste of weak solid in liquid. This is consistent with what one observes of wax handling properties. Such liquid provides a path of easy diffusion and so suggests that ageing effects may be quite rapid. Further, there is no guarantee that all fractions of the wax will solidify at any 'normal' working temperature, say down to 18 °C (see Fig. 1.1). Purification of such mixtures is difficult and expensive, and hardly worth the trouble for such non-critical substances. As an illustration of this, consider the odour of 'pure' waxes (that is, there are low molecular weight volatiles), the slow 'bleeding' of wax into plain paper to leave an oily mark, the fact that some waxes may show a 'bloom'[1] of crystalline material that results from migration of some component(s) to the surface on long standing. Similarly, some even show clear traces of liquid. In addition, many waxes tend to acquire a gloss spontaneously, as well as holding a glossy surface when used in domestic 'wax polish', both of which indicate high diffusivity and self-levelling (10§1.10) on a very small scale (*cf.* §2.6). Couple this with the general lowering of melting points in mixtures and it becomes more than likely that traces of liquid are important determinants of wax properties.

---

[1]   A dust-like or 'cloudy' coating, like the yeasts on grapes, and very much like the effect of refrigerating chocolate.

§2. Properties

From the above, waxes (where the fact that these are mixtures is taken as read) may be formulated to exhibit a wide range of properties such as melting range. But in order to achieve a more practically useful suite of properties for dental applications other additives must be used. For example, many waxes are not very tough, and flake or crumble when cut, more or less exhibiting brittle behaviour on a small scale due to high rates of strain. This would leave a carved surface rough and difficult to smooth. On that basis, slow carving of small slivers at a time will give a better result than attempts to remove large portions quickly. Plasticity is also important, for shaping a pattern without cracking. Other substances are therefore included. To the list of natural and synthetic waxes which are potential ingredients may be added carbohydrate gums, oils (*i.e.,* liquids at ordinary temperatures) and other fatty substances, and resinous materials.

●2.1   Colour

Colouring agents, which must be oil-soluble dyes (24§6.6), are also generally included in waxes, almost entirely to distinguish between product applications or brand. They have no appreciable effect on physical or mechanical properties. However, inlay and other casting waxes are usually intensely coloured to provide good contrast with tooth or die material and allow feather edges and other undesirable extensions of the pattern to be easily detectable. There seems to be no real reason, though, why baseplate waxes should be pink, and so on, except for easy recognition and a sense that it is the 'proper' colour. One brand did include temperature-sensitive (**thermochromic**) dyes that were intended to guide the user in obtaining the correct temperature for a given manipulation. However, temperatures change very rapidly for the kind of size or section thickness common in dentistry, so feel and experience are probably much more important than a prescriptive approach. In any case, because waxes are poor conductors of heat, edges cool appreciably faster than other areas and there must then be a variety of colours corresponding to the temperature gradients and the internal colour will, of course, not be visible. The utility of this is not clear and the formulation idea has not been generally adopted.

●2.2   Melting range

The melting range is of primary importance in designing a commercial wax product because this, and particularly the lower limit, controls the applicability of a given wax formulation in a particular function. Thus, a wax intended for use at 'room temperature' may be within the melting range at mouth temperature and so have little or no mechanical stability. For oral use the onset of melting would appear to be necessarily above 37 °C (or whatever 'open mouth' temperature is decided to be), but not so far as to risk burning the patient or otherwise traumatizing their tissues. Inflammation starts at least as low as 42 °C, and 45 °C is painful if applied for more than brief periods. Conversely, an impression wax needs to have considerable flow in a short time at mouth temperature to function. The width of the melting range is also important for allowing greater latitude in operation, avoiding complete melting for slight overheating and permitting moulding to the correct shape without it becoming too rigid too soon as it cools a little.

However, it may be noticed that 'room temperature' is not in practice very well defined. In the lower latitudes ordinary working temperatures may be above 30 °C, when products sold for more temperate climates, with working temperatures perhaps below 20 °C, would be unserviceably soft. Conversely, 'tropical' products would be too stiff and brittle under low-temperature conditions. Clearly, it is not even as simple as that because the air-conditioning may be adjusted for high or low temperatures in any climate – or non-existent, as may be the more normal condition in many places. The only appropriate advice would seem to be to find a wax product that performs properly under the local working conditions – and stay with it.

●2.3   Thermal expansion

Waxes typically have very high coefficients of thermal expansion (Fig. 2.1), particularly around the obvious melting range, with values for linear expansion typically in the range 300 ~ 600 MK^{-1} but even up to ~1500 MK^{-1} in some cases. Compare PMMA and other polymers at about 70 MK^{-1} below their T_g, or even 200 ~ 300 MK^{-1} above the T_g (see Fig. 3§4.2, 3§4.14). But, given the liquid+solid

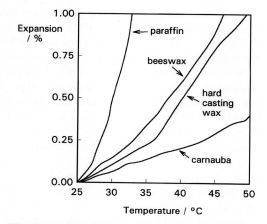

Fig. 2.1 Thermal expansion for a variety of waxes.

model for waxes over most of the ordinary temperature range of use, and the progressive melting that occurs on raising the temperature, it is apparent that the expansion coefficient is a complicated mixture of the ordinary thermal expansion contributions from both solid and liquid, plus that due to the progressive change of state as components melt – which expansion is normally a relatively large, discontinuous event, as seen clearly for pure substances (water is an exception). However, some of the persistent increases in slope on heating will be due to passing through the T_g for individual phases (Figs. 3§4.1, 3§4.2), particularly for waxes that are decidedly brittle at room temperature (such as sticky wax and hard casting waxes). The exceptionally high values recorded for the mixture of substances of a typical wax is therefore an artefact of the constitution, and is not really true thermal expansion, although it remains a convenience to refer to it as such from a macroscopic, phenomenological point of view.

A wax pattern for a lost-wax casting is expected (or, rather, intended) to be an accurate representation of the shape and size of the desired device. Whether or not that intention is to be realized obviously depends on the temperature at which the pattern was created and the temperature at which it was invested. The correct compensation through the use of investment expansion must take such factors into account if the failure of the casting through lack of fit is to be avoided (17§1). Thus, for example, the techniques used must distinguish between direct and indirect inlay patterns. However, if there is a change of dimension which is successfully part of the overall compensation for metal cooling shrinkage, then constancy of conditions and technique is evidently important for clinical success.

Low-melting paraffin waxes are also used for impressions, largely for soft tissue-related purposes but also for bite-registration. Generally, these techniques are not dimensionally critical and the waxes are considered to provide convenience of use. However, it should be recognized that the shrinkage on cooling from mouth temperature to that of the laboratory is substantial for these materials, probably well over 1% (Fig. 2.1), and adjustment or correction may well be necessary at some stage.

●2.4   Modulus of elasticity and flow

During the investment of wax patterns, and in other procedures, various forces are applied to the wax. It is evidently desirable that distortion by simple elastic deflection be minimized to preserve the accuracy of the procedure. In other words, the Young's modulus must be sufficiently high to resist appreciable deflection. This property is also strongly temperature-dependent (Fig. 2.2). But, because of the nature of these materials – weak intermolecular forces – the values are always rather low (10§3.6). This is a major limitation of which to be aware. We are referring of course to elastic behaviour, and permanent deformation through flow (creep) is perhaps of greater concern.

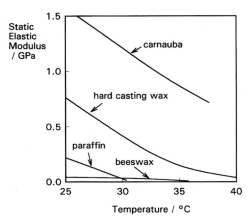

Fig. 2.2 Variation in static elastic modulus with temperature for a variety of waxes.

The stresses generated by an investment when it expands on setting, as it usually must, are more than sufficient to distort the pattern if there is any constraint acting on the investment to spoil the intended isotropic expansion. This is exacerbated by the investment's setting temperature rise, or the technique of 'hygroscopic' expansion in a water bath at an elevated temperature. The chemical nature of waxes is such as to leave no possibility of any substantial modification of this behaviour, given the other property demands. All techniques employing waxes in the sense of a dimensionally-accurate representation of an object, be it a metal casting, baseplate or impression, must take due recognition of this severe limitation. The same features that cause the plasticity and low melting temperatures, *i.e.* weak intermolecular forces and short chains, also guarantee plasticity under other forces. This necessitates a compromise that must be heeded.

●2.5   Residue

One other feature of casting waxes is essential to their efficacy: the residue after the burning out of the investment must be very small, in other words, the products of combustion should all be gaseous. Such a requirement ought to be self-evident, as the mould space clearly should not have any foreign material in it, nor should the permeability of the investment be impaired (17§7). This demands a very low inorganic content for the wax: silicone waxes, for example, could not be satisfactory ingredients of dental casting waxes no matter what other advantages they might have, and fillers could not be used to modify properties unless they too were completely combustible (although no examples of this are known). The effects on the casting may range from gross defects due to occlusion by the debris, to increased roughness or contamination by particular materials which may make polishing difficult or corrosion worse (see also §4).

●2.6   Thermal conductivity

Waxes have very low values of thermal conductivity, similar to those of polymers at around 1/3000th of the value for silver (which is about 425 W.m^{-1}.K^{-1}), and about 1/10th of the value for tooth enamel. The practical implications of this are clear: heat slowly. The problem arises because the temperature range over which substantial changes in viscosity occur, and then melting, is very small – typically only about 20 K. Thus, the surface can melt and start running away while the material only a millimetre thickness away is still at room temperature. Using a gas flame, as is commonly done, it may be impossible to avoid the very surface melting. However, intermittent exposure to the heat source with intervals allowed for the heat to spread more evenly will achieve a better result when a uniformly softly-plastic mass is required for moulding.

The converse of this situation also needs to be taken into account. It takes an appreciable time for wax to cool enough that flow and stress-relaxation do not occur to distort the work. It is not enough to wait until the surface appears cool and solid enough, time must be allowed for the whole mass to cool because the distortions will probably not be visible.

Some products have been designed in an attempt to deal with this issue. Thus, 'bite-wafers' containing copper and baseplate wax containing aluminium powder have been sold. The metal would, of course, act as a filler and both stiffen the cold, solid object and increase the viscosity of the softened material (Fig. 4§9.2). However, the main effect must be to ease the even warming of the wax before use as the two metals have very high thermal conductivities (Cu: 397 W.m^{-1}.K^{-1}; Al: 238 W.m^{-1}.K^{-1}), by far the highest values (apart from gold) after silver. They are sufficiently cheap metals for this to be a practical proposition.

Of course, the poor thermal conductivity is an advantage when a carved wax surface is to be 'polished'. A brief exposure to a flame as it is passed quickly over the surface achieves the necessary melting so that the liquid can very rapidly '**self-level**' (*cf.* 4§7.4), driven by surface tension (10§1.10).

§3. Flow

The flow of waxes is clearly important, not only as part of the deliberate moulding process but also as an undesirable aspect after the pattern or impression has been made. Rheology in general was dealt with in Chap. 4, and all of the principles remain valid in wax systems. The difficulty is essentially that wax viscosities are so very high at normal working temperatures ($10^8 \sim 10^{12}$ Pa.s) that macroscopic rheometric techniques are frequently quite inappropriate; yet the deformations are too large for the kind of approach that would be taken with creep metals in metals, for example.

Because of these practical difficulties, in dentistry the usual approach has been rather crude: the relative change in length of an arbitrary sample under an arbitrary static compressive load after an arbitrary time (American Dental Association Specification No. 4 [ADA4], which dates from a method of about 1931[5]) (Fig. 3.1). The inclusion of three arbitrary factors is less than scientific and although some useful comparisons can perhaps be made (Fig. 3.2),[6] the flow curve is non-linear because of the changing shape of the sample (Fig. 3.3), from a relatively narrow cylinder to a very thin disc. This results in a more or less abrupt change from very little to very much deformation as the test

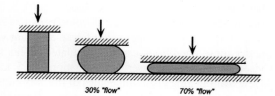

Fig. 3.1 Principle of the wax flow test of ADA4.

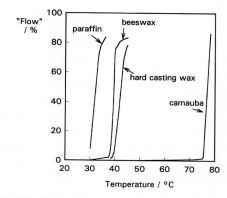

Fig. 3.2 Flow at 10 minutes of some representative waxes according to ADA4.

temperature is raised. No useful property value is derivable from this approach, and little information is obtainable about behaviour at other timescales or temperatures. Thus, although the 50% point may appear to offer a characterization of the wax, the temperature at which this is obtained depends on load, size and time. In particular, the kinds of deformation which would be important in terms of dimensional accuracy cannot be studied with an appropriate sensitivity. This situation is altogether rather odd for a class of materials that might be considered central to a number of high precision tasks.

In principle, it would be far better to determine the stress σ on the test piece (F/A), observing the **rate of change** of dimension as a function of temperature. Flow rate would then be expressed in terms analogous to a viscosity. Thus, the strain rate would be given by:

$$\dot{\varepsilon} = \frac{d\varepsilon}{dt} = \frac{\Delta L}{L} \cdot \frac{1}{t} \qquad (3.1)$$

where L is the original length and t the time taken. Then the 'flow modulus', D, would be given by:

$$D = \sigma/\dot{\varepsilon} \qquad (3.2)$$

The parallel with the definition of Newton's viscosity coefficient is evident (equation 4§3.5). If the deformation were kept small, so that the geometry could be considered constant, this would be an effective measure because it is a material property (1§1.1). The crude test would have been turned into something more sensitive and useful.

●3.1   Pseudoplasticity
An alternative approach to wax viscosity which has the advantage of more constant conditions is the 'falling ball' test derived from Stokes' Law (4§10).[7] Here, a steady-state condition for the ball velocity can be achieved (Fig. 3.4), and so a proper value for an apparent viscosity under these conditions can be calculated. This apparent viscosity, η_a, may be inserted in equation 3§3.5, taking the view that the long chain molecules of the wax constituents are somewhat similar to low polymers. Rearrangement, supplying a constant of proportionality (B), and taking logs, yields the following expression:

$$\ln(\eta_a/T) = B + E_a/RT \qquad (3.3)$$

where T is the absolute temperature, R is the gas constant (8.3145 J.K^{-1}.mol^{-1}), and E_a is the apparent activation energy for the flow. Thus, plotting $\ln(\eta_a/T)$ vs. 1/T gives a straight line of slope E_a/R. Some examples are shown in Fig. 3.5. This provides a very sensitive test and reveals that the viscosity of waxes can change steadily with temperature, quite unlike the impression gained from plots such as those in Fig. 3.2. However, what is more striking about such results is the stress-dependency of the viscosity: the higher load giving substantially lower viscosity.

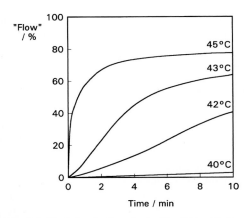

Fig. 3.3 Variation in flow (as in Fig. 3.2) with time for various temperatures with an inlay wax. The distortion due to the changing geometry is clear.

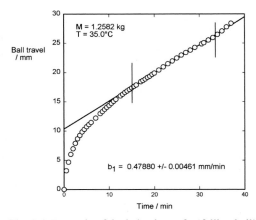

Fig. 3.4 Example of the behaviour of a 'falling ball' test. A steady-state is easily attained. (The initial high rate of travel is because the ball starts on the wax surface; see Fig. 4§10.3.)

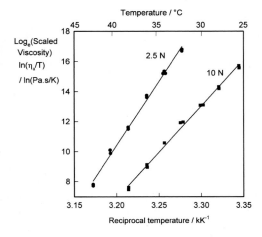

Fig. 3.5 Some experimental results for the 'falling ball' apparent viscosity of a baseplate wax.

Taking this further, and looking at the variation of ball velocity (which is proportional to shear strain rate) with the load on the ball, results such as are shown in Fig 3.6 are generally obtained. The first thing to notice is that there is no sign of a yield point (as is the perception described in §1.2): these are not plastic materials in the sense of 4§7.3. That is, even though the wax is seemingly entirely solid at 25 °C, it still behaves as a viscous liquid: 10^{11} Pa.s is a very high viscosity, so high in fact that it gives the impression of it being an ordinary glassy solid. One consequence of this is that there is no 'safe' temperature below which deformation may be ignored, and likewise, no safe load. That is, under any load, waxes deform continuously no matter what the temperature: it is only the rate that is affected. This is not unreasonable given that $T/T_m > 0.90$ for a 50 °C m.p. material at 20 °C, especially since lower m.p. solids will always be present, and some are necessarily liquid anyway (§1.3). A low load for a long time will result in distortion that could be significant as detrimental to the fit of a casting or denture base, or the accuracy of a bite registration.

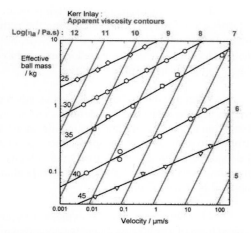

Fig. 3.6 An example of the load-dependency of the falling-ball velocity for a casting wax for a range of temperatures: 25 ~ 45 °C. The marked departure from Newtonian behaviour is clear. The steeper contour lines show the actual effective viscosity variation.

Secondly, it is clear that the slope of the plot at each temperature is markedly less than unity, meaning that the wax is non-Newtonian, and more specifically, strongly **pseudoplastic**. (4§7.4).[8] In other words, an equation of the following form applies:

$$\eta_a = a.x^b \tag{3.4}$$

for each temperature, where x is the velocity, a is a scale factor, and the exponent b is the slope of the log-log plot (as in Fig. 3.6) – and this may be termed the pseudoplasticity parameter. This can range from zero (instant collapse at any load) to unity (Newtonian behaviour). (A value greater than unity would mean that the material is dilatant.) These values can then be plotted against temperature, as on the right in Fig. 3.7. Apart from some remarkably large jumps, there is only a slight general effect detectable. However, remembering that many properties that are dependent on overcoming intermolecular forces, *i.e.* have an activation energy, such as diffusion (*cf.* 11§6.6), depend not so much on temperature but the proximity of that temperature to the material's melting point, we can replot.

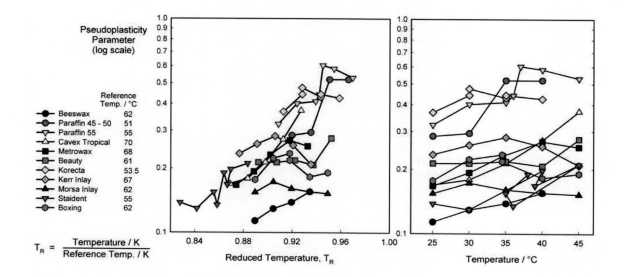

Fig. 3.7 Temperature dependence of the pseudoplasticity of dental waxes.

Thus, taking as the reference temperature that of the highest melting component (*i.e.* that of the onset of solidification, Fig. 1.4), and calculating the **reduced temperature**, T_R, we find a more informative pattern: a general trend towards Newtonian behaviour, b = 1, at T_R = 1. Conversely, the lower the reduced temperature, the greater the tendency to pseudoplastic behaviour. There are one or two oddities, but this might not be so surprising in such complex mixtures.

What this means in practice is that when moulding wax under the pressure of a tool or thumb, the harder one pushes, the easier the wax deforms, and strongly disproportionately so. The lower the working temperature in relation to that of complete melting, the more strongly will the effect be felt. This would apply therefore more strongly to an inlay wax, as opposed to say, a baseplate wax. This behaviour is obvious in Fig. 3.6: doubling the load results in about ten times the velocity. What this feels like is that the wax 'gives' under pressure, moulding responsively, but then apparently stopping altogether when the load is reduced. In other words, the material is "pseudo" plastic.

This behaviour may be explained by considering the expected typical form of a non-crystallized wax component: a miniature version of the polymer random coil, as in Fig. 3§2.3 (Fig. 3.8). Under shear, such a molecule would then be elongated – drawn – and oriented in the direction of the flow. This reduces intermolecular interference in itself, but also removes any (slight) entanglements between them that would make the effective molecular weight greater (see 3§3) and raise the viscosity. Nevertheless, thermal agitation means that the random coil would soon be re-established once shearing was stopped, albeit at a temperature-dependent rate; this is a diffusive process. In a sense, then, waxes appear to show some thixotropy (4§7.11), but in a directional manner rather than involving three-dimensional structure (4§7.10). Even so, the rate of relaxation would become sufficiently great at moderate temperatures that the return to the high-viscosity state would be so fast as to make this hard to detect.

Fig. 3.8 A wax molecule in a random coil conformation (*cf.* Fig. 1.5).

Turning to crystalline components, it is obvious that their orientation will be random initially with respect to the shear direction. That shearing will cause a reorientation of the lamellae, when the slip between layers will become easier. Thus there may be a range of crystallite 'yield points' causing a graduated decrease in the viscosity as the stress is raised.

There is yet another complication. For some waxes, in certain temperature ranges, a discontinuity may be found in the velocity plot (Fig. 3.9). That is, for a very small change in load, a very large increase in velocity occurs. In the examples shown here, the changes are of the order of ten- or one hundred-fold. This may be due to some component of the mixture suffering what may be called **stress-melting**, presumably of some definitely-crystalline component, which would be dispersed as fine crystals throughout the rest of the multiphase matrix. Conceivably, the extra liquid lubricates or provides a medium for the movement of more solid material by squeezing through interstices, approaching more a slurry-like condition. Whatever the exact cause, the outcome in practical terms is to exaggerate the pseudoplastic 'give' under load, and likewise the prompt return to the solid-feeling, high viscosity condition when the user stops pushing. Conceivably, though, such sensitivity might be disadvantageous when finer control is desired.

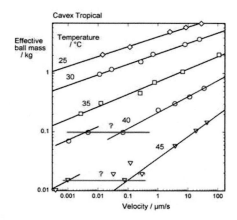

Fig. 3.9 Discontinuities in falling-ball velocity plots due to stress-melting.

It would therefore appear that the extraordinary usefulness of waxes for making patterns has a three-fold explanation. Firstly, the weak intermolecular forces allow slippage at low stress, especially since the working temperatures are close to the complete melting point (high T_R). Secondly, the amorphous condition of at least a large proportion of the material causes the pseudoplasticity (through molecular alignment), which allows the feeling of a pressure-sensitive mouldable material. Thirdly, this pressure-sensitivity is made even greater when the composition and temperature permit a stress-melting effect. Of course, this means that one must develop a feel for the particular wax in use; there are very clear differences between products so that training on one might not be immediately useful on another.

A number of other rheological behaviours are also exhibited by waxes. They do, for example, show marked viscoelasticity, so that in addition to showing Hookean and Newtonian compliances the retarded compliance component can be very strong, and this can be attributed to recovery from the shear deformation of the random-coil molecules. Thus, the attempts at the determination of a static (Young) modulus of elasticity (Fig. 2.2) will show results that are strain-rate dependent, as the relaxation times of the various molecular rearrangements must be taken into account. These are also strongly temperature-dependent. In fact, the chemical similarity of waxes to polymers (Chap. 3) in general means that many of the remarks made there are directly applicable to waxes. Thus, the interrelationships of temperature, strain rate, plasticizers (for such are many of the additives, in effect) and modulus of elasticity, as well as the interpretation and existence of glass and other transition temperatures in terms of molecular movements, are applicable here also. Very little that is specific to waxes can be added.

●3.2   **Residual stress**
Even so, one particular manifestation of the viscoelasticity of waxes is of great importance: that of warpage due to residual stresses (although this effect is not, of course, restricted to this class of material). That is, after moulding has been completed, and the pattern or impression is considered finished, some slow deformation occurs even in the absence of external stresses or elevated temperatures. To explain this we refer to the model of viscoelasticity (4§6).

If a stress is applied to such a body, then the primary response we expect to see is the gross flow which corresponds to the moulding process that is the intention. However, simultaneously, some strain energy is being stored in that part of the system that corresponds to the Kelvin-Voigt behavioural element, the retarded compliance component, *i.e.* some form of molecular deformation (which we now see is the shear-generated drawing and alignment of the molecules). At the usual moulding temperatures, that is, somewhat elevated above room temperature, the component elastic moduli and viscosities are all lowered. Deformation is easy, because molecular relaxation is easy, so that the stresses imposed during moulding tend to be relaxed quite quickly (and the pseudoplastic condition is shown). However, this recovery process slows down markedly as the wax cools, and this it must be doing continuously during the moulding. The Hookean elastic recovery (which is very small) of course occurs immediately on removing the stress (a finger), while recovery of the retarded elastic deformation is not. But because the wax has now cooled, both the elastic moduli and viscosities again have high values and the material appears solid. Even so, the remaining stored energy is unaffected – it is still stored. The rate of relaxation is thus low, but the driving force is undiminished. This is therefore an example of entropic recoil (7§2.3).

Room temperature may, for some waxes, be sufficiently low in comparison with the moulding temperature that no observable change will occur on ordinary timescales (say, up to a few hours), but equally for other waxes this may not be true and relaxation deformation will occur before an investment or model can be cast. Worse, should the wax be warmed in the meantime, such as by being left in the sun or too close to a bench light, or even during the investment process by the heat of the setting reaction or immersion water bath, the recovery will occur much more quickly, potentially with critical effect: failure of the device to fit.

Nevertheless, the longer the wax pattern is left before being invested, the greater the distortion that must result. In common, then, with most other impression or pattern materials, the next stage (such as pouring a model) should be done promptly to minimize the effects of such distortion. For safety, probably not more than about 30 min should be allowed before pouring such models or investments. A simple demonstration is possible: bend a stick of inlay wax when it is just warm enough to bend slowly, chill in cold water while under stress (*i.e.* still being bent), then float the piece on warm water. The bend will be seen to recover, straightening the stick. This demonstration is obvious, but the damage done in practice will not be visible to the eye; yet it will be great enough to be

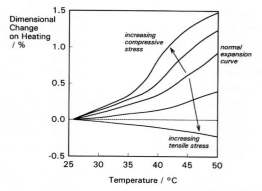

Fig. 3.10 Variation in apparent linear thermal expansion due to strain relief on warming for a typical wax.

detrimental. Residual strains also affect the apparent thermal expansion, as this represents a strain also (*i.e.* a change in dimension), and the two will be additive. That is, any residual strain from compression will result in exaggerated (one-dimensional) thermal expansion as the compressive strain is relieved, while tensile strain may reduce or even reverse the dimensional change on warming (Fig. 3.10). The complications these provide for casting accuracy are clear. The complementary tension and compression effects are at right angles (29§1), no matter what the imposed load, and this means that these effects are always anisotropic (1§2.7) (we can expect the changes to be essentially constant-volume, allowing for temperature). While annealing can remove these effects, the strain relief this allows causes dimensional changes anyway.

This situation is complicated by the high thermal expansion coefficient of waxes (§2.3) in the face of low thermal conductivity (§2.6). That is, simply heating one part of a wax body generates a stress that may then be frozen in when cooling occurs, at least in part. The thermal gradients will be complicated, as will the partial melting (§1.1): differential heating causes differential expansion, which generates the stress (this may be compared with the deliberate generation of stress in porcelain, 25§5.1 ~ 5.3). Accordingly, all wax that has been treated in such a fashion is subject to deformation through stress-relaxation on standing, and this is accelerated by raising the ambient temperature, even in the absence of moulding stresses.

§4. Burn out

The removal of the wax from a denture base mould prior to packing is a relatively straightforward process: it is easy to melt and flush away the wax from the open mould with boiling water – even though some dissolution of the mould must occur (2§9). The elimination of the wax from an investment mould is more involved.

Ignoring for the time being the chemical changes in the investment itself (Chap. 17), a combination of physical and chemical processes are required to ensure that the mould space is clear and the investment porous, and that there is no contamination of the investment or the casting. When the invested wax pattern is first placed in the burn-out furnace it is arranged 'upside down' – crucible former below to enable the majority of the wax to run out *via* the sprue as it melts. (It will also be tilted, *i.e.* on edge, to allow oxygen better access – see below.) However, the low viscosity at the same time permits some of the molten wax to soak into the porous investment (a question of capillarity, 10§2), where it will be held (Fig. 4.1(a)). As the temperature rises some of the wax will simply volatilize and diffuse out of the investment mass. But in that process some wax will begin to decompose, possibly catalysed by the hot investment, depositing carbon throughout the system of pores as a finely-dispersed particulate material – soot, in the 'vapour zone' (Fig. 4.1(b)). Some wax constituents may have high enough boiling points that they will decompose before evaporating, leaving a greater concentration of carbon in those places that the molten wax penetrated, essentially a zone of a few millimetres thickness around the mould space where the wax has charred (Fig. 4.1(a)). The removal of this carbon now depends on the diffusion of oxygen from the free surfaces of the investment mass (which are somewhat limited if a casting ring is used) and a temperature high enough for reaction to occur at an appreciable rate. Since the investment will tend to be hotter on the outside (at least initially), and the concentration of oxygen is also higher there, the oxidation proceeds from the outside in. Frequently a clear 'front' can be seen in a sectioned investment mass at this stage, a sharp boundary between the colour of the investment (dominated by that of oxidised iron, which may present as a contaminant) and a grey zone where carbon remains (Fig. 4.1(c)), where the iron is (partially) reduced. Oxygen will also diffuse relatively more quickly through the sprue or any risers, and a second front may be found commencing from the mould space surface.

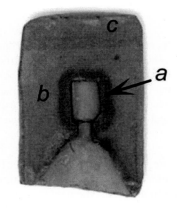

Fig. 4.1 Zones in a partially-burnt out casting investment mould. (a) infiltrated zone, (b) vapour zone, (c) oxidized zone.

These observations establish four important conditions for successful burn out:
1)   that the investment is porous, with the pores continuously connected to provide a diffusion path;

2) that sufficient oxygen is available (a tightly closed, unventilated furnace with much wax present may not meet this demand);

3) that the temperature is high enough for an adequate rate of oxidation; and

4) that time is allowed for that process to go to completion.

There are occasions when acrylic or other polymer materials may be used for specialized patterns or sprue-formers, whether for convenience or greater dimensional accuracy, resistance to deformation or special shapes. If these are used it must be remembered that they will not melt and run out of the investment mould as readily as waxes, if at all, and that charring is more likely. If used as a sprue they will also inhibit the running out of the wax itself, allowing more to soak into the investment mass; such sprue-formers should therefore be removed before burn out. In any case, more time for burn out might have to be allowed to obtain appropriate results.

It should go without saying that waxes for casting patterns should leave negligible residue on burn out, *i.e.* when the carbon has been removed. Nothing with involatile oxides is permitted; indeed, nothing involatile. This kind of residue would tend to block investment porosity, contribute to dimensional inaccuracy by simply being in the mould space, may become embedded in the metal surface, and may reduce the refractoriness of the investment by reacting with it (17§3.4). This latter has to do with the lowering of the melting point (eutectic temperatures) of silicates and other compounds that may be made from the contaminants and the components of the investment itself. Such melting would lead to mould distortion.

§5. Impression Compound

The plasticity of waxes at normal room and body temperatures limits their applications as impression materials to special tasks in prosthetic dentistry such as bite registration (although much more stable impression materials are available for this). However, in common with other materials discussed previously, various properties may be modified by the inclusion of a fine particulate filler. Modulus of elasticity, viscosity and hardness can all be increased.

So-called impression compounds are mixtures of waxes and resins with softening or melting ranges above body temperature, but heavily filled to limit their flow when heated and to confer rigidity when cooled. Common fillers include such cheap and easily comminuted materials as talc, chalk or pumice. Because they are composite materials, all that has been said about this class of materials, and about waxes, is also true for 'compound'. However, they have limited capacity for detail reproduction because their viscosities are high at their working temperatures, and they cannot be used for undercut impressions because of their rigidity. They are insufficiently brittle (that is, they appear to show a yield point) for the reassembly technique used with impression plaster in such circumstances to be applicable. They are also a little too strong to be easily broken whilst in the mouth. In addition, good dimensional accuracy cannot be expected, despite the presence of the filler, because the thermal expansion coefficient remains high. Enough filler to achieve this goal simply cannot be incorporated.

§6. Other Applications

Waxes do appear in a variety of other dental applications, for example wax pencils, mounts for sets of denture teeth and so on. The common feature is that each relies on the weak-solid properties of the wax to facilitate shear and transfer. There are no critical issues in these contexts. In some applications, however, there are some specific materials properties of importance.

●6.1 Sticky wax

Sticky wax is a mixture intended to be quite rigid and brittle (as waxes go) at room temperature to facilitate the temporary holding in position of objects while other processes are applied, such as parts of a bridge prior to investing for soldering or brazing. It can even be used as a repair agent for broken plaster models or impressions. For this to be effective the (full) melting point is necessarily rather higher than most other dental waxes at 65 °C or so. Even so, it is required to have a very low residue on boiling or burning out in order that it can be removed cleanly when required. It should be recognized that the brittleness of sticky wax is not absolute, but merely indicative of the strain-rain sensitivity: glassy at relatively low temperature (T/T$_m$), and at relatively high strain rates – as with hard casting waxes, it appears to have a clear glass transition, giving a glassy 'conchoidal' fracture when cold.

●6.2   Articulating paper

Waxes can be used effectively to identify those locations where teeth (whether natural, artificial or repaired) make contact during function, primarily to facilitate the adjustment of the occlusion by marking high spots. Obviously, the film used to carry the wax needs to be as thin as practically possible to avoid interfering with the occlusion, but the low strength and stickiness of the wax will only allow the transfer if the receiving surfaces are dry, because of the hydrophobicity of wax in general. Equally obviously, such waxes must be very intensely coloured to be detectable when transferred to tooth surfaces in very small amounts; dark red and dark blue are common.

●6.3   Floss

Dental floss is generally available in plain and waxed versions. The role of the wax is essentially to be a lubricant, to allow sliding over the tooth surface with less catching and wear on natural roughness and edges of various kinds that fray the floss and break it. However, by making the floss more hydrophobic and smoother (by at least partially filling in spaces between fibres), the ability to pick up and remove plaque is lessened. Either way, the effort required is about the same, although replacing breakages might be more annoying.

●6.4   Petroleum jelly

This is an ill-defined mixture of paraffinic and aromatic hydrocarbons (typically C$_{12}$~C$_{85}$, mostly above C$_{25}$; mol. wt. 375 ~ 630)[9] that forms a semi-solid paste-like material, also called **petrolatum**. It therefore has much of the chemical nature of waxes, indeed it contains substances that on their own are waxes, but it does not crystallize at ordinary temperatures even in the long term. It does not have any of the structure or characteristics of a gel (7§8, box), so the label 'jelly' is unhelpful. Commercial products vary somewhat in their composition and thus their (full) melting point. Some have components that melt a little below body temperature and so are more unctuous when applied. As a benign lubricant it finds a number of uses in dentistry, ranging from the provision of a water-repellant surface to easing the stretching of the labial commissure for easier access to the rear of the mouth.

References

[1]   Craig RG, Powers JM & Peyton FA. Differential thermal analysis of commercial and dental waxes. J Dent Res 46: 1090 - 1097, 1967.

[2]   McMillan LC & Darvell BW. An improved cooling curve technique as applied to waxes. Meas Sci Technol 10: 1319 - 1328, 1999.

[3]   Dorset DL. The bridged lamellar structure of synthetic waxes determined by electron crystallographic analysis. J Phys Chem B 104: 4613 - 4617, 2000.

[4]   Dorset DL. From waxes to polymers—Crystallography of polydisperse chain assemblies. pp 91 - 99 in: Hargittai I & Hargittai B (eds). Science of Crystal Structures: Highlights in Crystallography. Springer, Cham: 2015. [Structural Chem 13 (3/4): 329 - 337, 2002.]

[5]   Paffenbarger GC & Sweeney WT. Dental casting technic: Theory and practice. J Dent Res 11: 681 - 701, 1931.

[6]   Craig RG, Eick JD & Peyton FA. Properties of natural waxes used in dentistry. J Dent Res 44: 1308 - 1316, 1965.

[7]   Darvell BW & Wong NB. Viscosity of dental waxes by Stokes' Law. Dent Mater 5: 176 - 180, 1989.

[8]   McMillan LC & Darvell BW. Rheology of dental waxes. Dent Mater 16 (5): 337 - 350, 2000.

[9]   Screening Assessment - Petroleum Sector Stream Approach - Petrolatum and Waxes. Environment and Climate Change Canada, Health Canada June 2016. [Appendix B] http://www.ec.gc.ca/ese-ees/E78B1855-EE77-40D6-BBC7-964AF3A8E3EB/ ... FSAR_PSSA4-Petrolatum%20and%20Waxes_EN.pdf

Chapter 17 Casting Investments

Casting investments represent an important step in the process of producing cast metal devices such as partial denture frameworks, bridges, crowns and inlays. As such, the stage is crucial in ensuring that the device actually fits the teeth to which it will be attached. The concern here is that the hot cast metal must cool down to the correct size. Therefore the mould must be oversize at the time of casting. Ensuring this condition is a major goal of investment design and handling procedure.

*Available materials that can set hard to create a mould all shrink on heating. Providing compensation therefore requires the use of a second substance, the **refractory**, which can expand. Various forms of silica permit this, and reliance is placed on a unique type of crystal structural change to obtain sufficient expansion. The setting material is then only a **binder** holding the refractory together. Detailed consideration is given to the structural changes of silica.*

*The binders which may be used also interact with the refractory through crystal growth pressures, which offer a further means of increasing and controlling expansion. However, the **decomposition** of the binder on heating leads to some shrinkage which offsets the expansion and therefore must be understood if proper allowance is to be made. Gypsum-, phosphate- and silica-bonded types are discussed.*

*The success of casting also depends on the absence of mechanical constraint on the expanding mould so that if a casting ring as a container for the investment is used, a compressible **liner** is necessary. In addition, the air in the mould space must escape to permit filling by the molten metal, and the controlled **porosity** of the mould material is essential for this.*

*As the final size-determining step in the preparation of a cast dental device, the investment is crucial. It also permits the correction of size **errors** that may have accumulated through the impression, model and pattern stages. It is undesirable merely to follow instructions; an understanding of the mechanisms and means of control ensures reliably high quality work – that fits – by informing intervention and adjustment as well as the recognition of the sources and means of correction of problems.*

Materials Science for Dentistry
https://doi.org/10.1016/B978-0-08-101035-8.50017-1

The process of casting a metal restorative device requires a mould, one that can be formed around and to the exact shape of the wax pattern, in what is called the **lost-wax** process. The process of enveloping the pattern in a material which will become the mould is called **investing**, and the material doing the job an **investment**. Industrial castings may use a similar process but any precision surfaces will be machined afterwards, thus quite avoiding problems of distortion, dimensional errors, and roughness. There is, therefore, rather little concern with the accuracy or detail of the pattern, of the mould, or of the fresh casting. However, it must be said that the precision surfaces required in such contexts usually are of simple, mathematically defined shapes, certainly not the nearly randomly irregular types of surface associated with tooth cavity or crown preparations or tooth anatomy. In dentistry, in great contrast, machining is not yet a practical proposition and the general aim is for the casting to fit, exactly, first time (although some ceramic inlays are now being machined directly from stock pieces by a process of copy-milling, with some severe limitations). The difficulties inherent in the casting process, and means of overcoming them, are discussed in this chapter.

The lost-wax or *cire perdue* process of casting has been in use for many thousands of years. In antiquity all major civilizations on all continents have used it for jewellery, ornaments and decorated objects of many kinds, some extremely elaborate, in bronze, gold and other metals. In no case, however, has there been any concern with dimensional accuracy. Quite apart from the fact that measuring with the necessary resolution would not have been feasible to observe the effects, there simply has been no need for accuracy. In modern times the jeweller's *cire perdue* techniques have been borrowed and adapted for dental use, essentially paying most attention to the dimensional accuracy of the product.

§1. Requirements

Because the investment mould will be used to cast metal alloys with high melting temperatures, the first demand is that it survive those temperatures without mechanical or chemical breakdown. In other words, the investment must be **refractory**, here meaning resistant to change at high temperature. That is, it should not break, melt, decompose or react in the process. Indeed, to avoid premature chilling and freezing of the metal, and thus obtain very fine section castings without imperfections, it is necessary to preheat the mould to a temperature maybe only a few hundred degrees below the solidus temperature. But this overlooks the fact that the metal being cast must start out substantially above its liquidus temperature, for precisely the same reason – avoiding premature solidification. This hot metal must then heat the adjacent investment to the same temperature. The requirement for lack of mechanical breakdown, whether due to chemical change or thermal shock, is thereby made more severe.

The refractory itself needs to be held together in some fashion in order to create a mould. That is, the powder requires a **binder**, a material that can form a matrix to contain the core of the refractory particles. Thus, the principal kinds of dental casting investment are said to be gypsum-, phosphate- or silica-**bonded**. This should not be taken to imply the kinds of chemical bonding discussed in 10§3~6, but rather the concept is of the **bond** of a brick wall: the mortar forms the matrix, which by virtue of mechanical keying holds the bricks in place. The matrix itself may be held together purely mechanically (as in gypsum products) or with contributions from a more or less continuous cementitious mass (as in phosphate cements). Real interfacial bonding is unlikely to be obtained until very high temperatures are reached and sintering or melting occur.

At the high temperatures required for casting metals, there should be no reaction with the alloy or with any of the oxides that may form on its surface, the investment must be porous to allow the escape of air from all parts of the mould so that it fills quickly and completely, and yet it must have sufficient strength to withstand the forces of casting. In addition, the surface reproduction must be accurate, the unset material easily adaptable to the shape of the pattern without excessive force, the material compatible with the wax, and removable from the casting at the end of the procedure.

Furthermore, and possibly most demanding, the final casting must be of just the right size. In creating a dental metal casting there are many stages to the process in which dimensional accuracy and stability are of paramount importance in ensuring that the device will fit. The list might include the following:

- setting and removal of impression material
- cooling from mouth to room temperature (including the impression tray)

- setting of model material
- cooling of wax pattern when fabricated
- heating of wax pattern by setting investment
- investment setting expansion
- investment expansion on heating
- thermal contraction of metal on cooling to mouth temperature.

Table 1.1 *Linear thermal expansion coefficients for some metals and refractories* (25 °C).

Substance	α / MK⁻¹
WC (6% Co)	4·9
SiO$_2$ (fused)	0·5
Al$_2$O$_3$	6·2
Pt	9·0
Au	14·1
Ag	19·1
Al	23·5
Hg (liquid)	61·0

The most important of these in terms of magnitude is the last-mentioned. In fact, but for one happy accident, it would be a serious problem because, in general terms, the class of materials which can provide useful refractories in the sense required here is that of the ceramics. Ceramics typically have lower thermal expansion coefficients in comparison with the metals they are to contain (Table 1.1). It would therefore be impossible to gain enough expansion on heating from room temperature to the casting alloy's solidus temperature to compensate for the metal shrinkage. Obviously, the mould cannot be heated above the solidus temperature to any useful effect. The 'happy accident' is the existence of a phase transition in crystalline silica that, on heating, leads to a large expansion.

In addition, ceramic systems that can be made to 'set' in order to create the casting mould around the pattern will show some kind of decomposition on heating that results in shrinkage. Thus, while a gypsum product such as plaster of Paris is sufficient to satisfy some of the requirements at casting temperatures up to about 700 °C, it has the significant and fatal defect of great shrinkage on heating above ~300 °C, due to the conversion of the dihydrate to hemihydrate and then to anhydrite by dehydration (Fig. 1.1) (see 2§1). Since alloy shrinkage on cooling may be of the order of 2% (linear), it is clear that something else must be done.

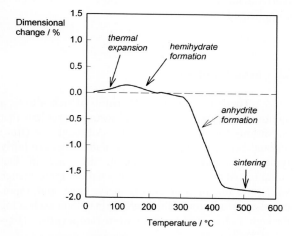

Fig. 1.1 Dimensional changes of pure gypsum plaster on heating showing the various causes.

In principle, noting the thermal expansion behaviour of composites (6§2.1), the shrinkage of a material such as gypsum could be reduced, and even turned into an expansion, by incorporating an inert filler with a suitably-large thermal expansion coefficient. Again, however, no ordinary refractory ceramic can be used because simple thermal expansion is unequal to the task and the required compensation cannot be attained, at least with materials which have a high enough melting point and are cheap enough to be appropriate.

There will also be variations in technique, ambient temperatures, types of impression and model materials, and of course variations in alloy properties, especially solidus temperature, to say nothing of errors in the calibration of furnaces. To accommodate such variation (maintaining control of the accuracy of the final product), the investment should have its expansion adjustable by some means, thus permitting some direct intervention at the last opportunity to do so, for mechanical adjustment is almost certain to be imprecise and ineffective.

●1.1 Thermal expansion

The dimensional changes of materials on heating is primarily due to the increase in the amplitude of thermal vibration – the stretching and bending of bonds – as more energy is supplied. This is known as **thermal expansion**. It was defined in volumetric terms in equation 6§2.1 over some temperature interval ΔT. However, the variation in interatomic distances is only very rarely even approximately linear with temperature, and a single, simple statement of that kind is not necessarily very helpful. In effect, it is a **secant** or average value that is being reported. Indeed, given the underlying reasons, it could not be expected (10§3.7). It is therefore sometimes more useful to use the differential form, corresponding to a **tangent** value at the temperature of interest (*cf.* 1§13, 4§7.5). Thus,

$$linear: \quad \alpha = \frac{1}{L} \cdot \frac{dL}{dT}$$

$$volumetric: \quad \beta = \frac{1}{V} \cdot \frac{dV}{dT}$$

(1.1)

Because such expansion rates are small, $\beta \approx 3\alpha$ (*cf.* equation 1§2.14, where quadratic and cubic terms are similarly dropped).

Thermal expansion depends on two principal factors: the strength of the bonds between atoms and molecules, and the type of structure. The stronger the bond, the lower its expansion coefficient. For example, that for the covalent bond Si-O itself is about 2.2 MK^{-1},[1] but for van der Waals forces, the values are much larger (*cf.* waxes, Chap. 16). Clearly, if a totally covalent system were to retain its structure completely except for scale (that is, expand completely homogeneously, and therefore isotropically), the covalent bond expansion would be reflected in the bulk value. Thus for diamond, in which there is no other effect operating, we have an overall value of ~1.2 MK^{-1}. However, if the detail of structure were to change, by twisting or other rearrangements, this would not follow. It will be seen that this kind of effect makes the behaviour of silica more complicated.

§2. Chemistry of Silica

There is a group of minerals which allows a solution to the above dimensional accuracy problem, that is, various forms of silica, SiO_2. Although the (true) thermal expansion values of these substances are fairly typical of ceramics in general, they are in fact complicated by a unique kind of crystallographic change which results in a relatively very large and rapid, microscopically-reversible increase in volume on heating through a characteristic **transition temperature** (Figs 2.1, 2.2).[2]

The stabilities and interconversions of the various forms may be explained in terms of their free energy *vs.* temperature curves (Fig. 2.3), the lowest curve at any temperature representing the most stable form.[3] It will be noticed that fused silica (a glass, and therefore amorphous) has a very low coefficient of thermal expansion, ~0.5 MK^{-1},[4] and therefore the chemistry of SiO_2 as such is not responsible for the large changes observed for the crystalline materials. Essentially, the thermal expansion coefficient of that material (Table 1.1) can only be due to the lengthening of the Si-O bonds, but even some of that change is 'lost' because of the free volume in the glassy structure.

Clearly, we must seek the cause of the larger-scale effect elsewhere, and it does in fact have a geometrical explanation. We need therefore to look at the structure of silica in some detail.

Silicon is four-valent and ordinarily tetrahedrally-coordinated. In each kind of silica and all derivatives the silicon atom is surrounded by four oxygen atoms (Figs 2.4, 2.5). If the

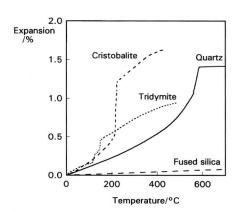

Fig. 2.1 Expansion on heating of the various silica allotropes and fused silica.

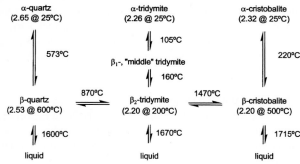

Fig. 2.2 The thermal phase transitions for silica. Horizontally, crystal form; vertically, displacive transformation. Densities (g/mL) are shown in parenthesis.

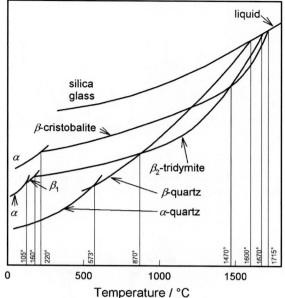

Fig. 2.3 Free energy diagram for silica.

silicon atom were ignored, the oxygen atoms can be seen to be spaced at just about the sum of their radii (Fig. 2.6), so that the silica tetrahedron has the size of a set of four just-touching oxygen atoms – even though they are not, of course, bonded to each other. This allows us to represent silica and silicate structures in a simple manner (Fig. 2.7).

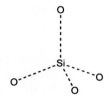 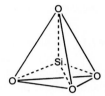

Fig. 2.4 The tetrahedral geometry of silica.

The actual molecule SiO_2 does not exist under normal conditions, and the oxygen atoms are shared between two adjacent silicon atoms, giving each tetrahedron shared apices (Fig. 2.8). To preserve the composition at SiO_2 in the silica minerals, every oxygen must be shared between two silicon atoms. Thus the empirical formula SiO_2 is the average over all the atoms of what is really a **giant molecule**.

There are several ways of connecting up these silica tetrahedra in three-dimensional networks which preserve this stoichiometry. In fused, glassy silica the connections are random (Fig. 2.9), but in the crystalline

 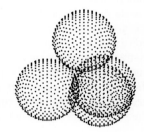

Fig. 2.5 A space-filling model of an SiO_4 group superimposed on a ball-and-stick model, using standard atomic radii.

Fig. 2.6 If the silicon atom is deleted, it is as if the oxygen atoms are nearly touching. (Slightly rotated view from that in Fig. 2.5 for clarity.)

cristobalite and **tridymite**[5] the tetrahedra are arranged in planes, each tetrahedron oriented alternately 'up' and 'down' (Fig. 2.10). In effect, there is a single plane of oxygen atoms in horizontal rings, and the silicon atoms are alternately just above and below it (Fig. 2.11). As was encountered in the discussion of the metal close-packed structures, h.c.p. and f.c.c. (11§3), such layers can be stacked in two distinct repeating patterns, although of course the silica structures are *not* close-packed.

Fig. 2.7 For convenience, the silica tetrahedra can be drawn as if they were a set of just-touching oxygen atoms with a silicon atom – not to scale! – embedded at the centre.

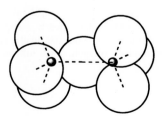

Fig. 2.8 In building networks, oxygen atoms are shared.

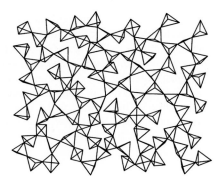

Fig. 2.9 Random linkage gives the network typical of glassy structures. NB: This is a two-dimensional representation of a three-dimensional structure.

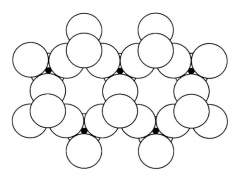

Fig. 2.10 Regular linkage can give a pattern such as this, one ring layer from tridymite or cristobalite.

●2.1   Tridymite

Tridymite consists of an alternating sequence of such layers, each plane being the mirror image of the one above and below, opposing apices of course corresponding to shared oxygens. The mirror plane then passes through those shared oxygen atoms. In this arrangement, which may be described as A,B,A,B..., each tetrahedron (silicon atom) is immediately above one in the plane below, and this is true for every layer (Figs 2.12 and 2.13). This is an hexagonal structure.[6] In this arrangement it can be seen that within the plane of alternating tetrahedra the six-membered rings are in the 'chair' conformation - much like cyclohexane's preferred arrangement. There are other six-membered rings to be found, but these are all in the 'boat' conformation (Fig. 2.14). The significance of this will be explained shortly.

Fig. 2.11 The tetrahedra in the plane of Fig. 2.10 are alternately 'up' and 'down'. Above, plan view; below, edge view of a portion of a layer. The junction points of the tetrahedra are the oxygen atoms; the silicon atoms lie at the centre of each (see Fig. 2.4).

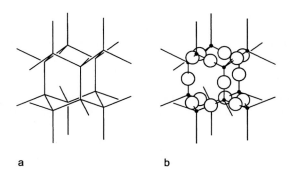

a   b

Fig. 2.12 The structure of tridymite (hexagonal). a. framework model, looking nearly along the planes of alternating tetrahedra. b. with the positions of the atoms shown: large circles – oxygen, dots – silicon.

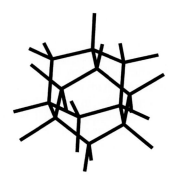

Fig. 2.13 A different view of a frame model of tridymite, showing the relative positions of the rings in successive planes. In this model, only the silicon atoms are shown. The oxygen atoms lie midway along each link (see Fig. 2.12).

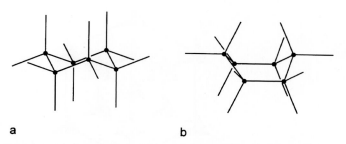

Fig. 2.14 The 'chair' (left) and 'boat' conformations of silica. Only the silicon atoms are shown; the larger oxygen atoms would lie on the joining lines.

Tridymite has been long believed to be an allotrope of silica, but there is now evidence which suggests that this is a mistake. It usually contains a small proportion of metal ions, making it in fact a *silicate* mineral, and it is these that cause the characteristic structure. In the present discussion this uncertainty can be ignored, since the thermal effects and structural reasons are real, and the practical outcome unaffected.

●2.2 Cristobalite

If on the other hand, in building structures with the basic plane of Fig. 2.10, each successive layer were positioned as if by lifting the previous plane and displacing it along one of the three directions indicated in Fig. 2.15 by exactly the distance between silicon atoms (*i.e.* one tetrahedron unit), then the downward facing oxygen atoms can again be made to match with upward facing ones. But this means that a third layer must be added in the same manner before the relative position of the first layer can be reproduced - at the fourth layer. The structure therefore becomes of the A,B,C,A,B,C... type. This makes it face-centred cubic (Figs 2.16, 2.17), silicon atoms being found on the corners and at the centre of each face of the unit cell. It can then be seen that the unit cell contains four six-membered rings in the chair conformation. These rings effectively lie on the faces of a tetrahedron, apices on alternating corners of the unit cell, edges on the face diagonals. There are thus four equivalent directions in this structure (Figs 2.18, 2.19): when viewed along the directions of each of the four cube diagonals the structure looks exactly the same. This is the structure of the mineral **cristobalite**.

Fig. 2.15 The displacement operation notionally required to generate the next layer of the cristobalite structure from the network in Fig. 2.11

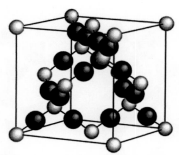

Fig. 2.16 The structure of cristobalite (face-centred cubic) in relation to the unit cell. The planes of rings, as in Fig. 2.15, are perpendicular to the cube diagonals. Here the silicon atoms are light, the oxygen dark.

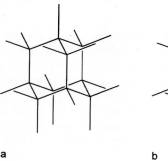 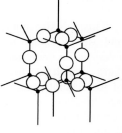

a   b

Fig. 2.17 The structure of cristobalite (f.c.c.). a. framework model, looking nearly along the planes of alternating tetrahedra. b. with the positions of the atoms shown: large circles – oxygen, dots – silicon.

Fig. 2.18   Framework model of cristobalite viewed from a similar angle to a plane of tetrahedra as Fig. 2.13 to show the relationship of successive layers, one tetrahedron lying over a ring below. (This is, in fact, with the oxygen atoms omitted, the diamond structure.)

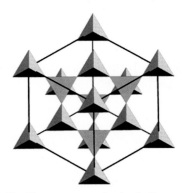

Fig. 2.19 The structure of cristobalite as viewed along a cube diagonal to show the relative orientations of the SiO₂ tetrahedra.

●2.3   **Quartz**

The structure of quartz, the third polymorph of silica of interest, is somewhat more complicated (Figs 2.20, 2.21): it lacks both any plane of symmetry and a centre of symmetry. It can be viewed as containing helices of linked tetrahedra with three per turn (Fig. 2.22, right) (that is, the fourth silicon is directly over the first). At the same time, these tetrahedra can be traced as belonging to two intertwined spirals with six per turn. (Fig. 2.22, left) Naturally, this structure is strongly anisotropic and also, because the helices may be right- or left-handed, optically active.

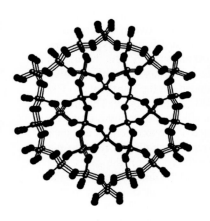

Fig. 2.20 Plan view of the structure of quartz. The 3-per-turn spirals of tetrahedra correspond to the apparent rings of three units, the 6-per-turn double spirals make the star-shaped channels.

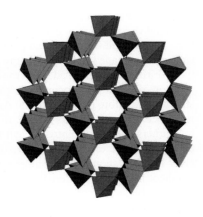

Fig. 2.21   An alternative model of the quartz structure, corresponding to the view in Fig. 2.20, indicating the SiO₂ tetrahedra.

Quartz is also highly anisotropic in its expansion, and this is put to good use – although not in dentistry. By cutting thin plates of a quartz crystal in one orientation, these may be used as piezo-electric oscillators which are nearly temperature-independent and are thus useful for accurate radio transmitters, clocks and watches. Cut in another direction, to utilize the large thermal expansion coefficient, such crystal plates can be used as thermometric devices, as the natural oscillation frequency will then vary markedly with temperature.

●2.4   **Nature of the transition**

The structures as discussed so far are in fact the high temperature or β-forms in each case, characterized by the regularity of the hexagonal arrays of tetrahedra in the planes (at least, in tridymite and cristobalite). The notably higher density of the low temperature or α-forms is due to a puckering of the rings (Fig. 2.23). This is due to slight rotations of each of the tetrahedra, changing their orientations with respect to one another, brought about by a slight change in the Si-O-Si bond angle.[7] The same effect occurs in quartz,[8] although it is harder to illustrate.

Fig. 2.22 Left: two intertwined 6-per-turn spirals in quartz; right: one of the 3-per-turn spirals. Viewed from the side of Fig. 2.21.

We can note a general geometrical result to help visualize the effect of this. Suppose we take a regular hexagon: the area enclosed is automatically a maximum, there is no way to arrange it to increase that area. Conversely, any distortion, treating the apices as hinges but keeping the sides the same length, must result in a decrease in enclosed area. A similar thing can be said about the enclosed volume of three-dimensional frameworks – consider the flattening of a cardboard box. Since the puckering in silica reduces the enclosed area of each ring in projection in the plane of the layer, the volume enclosed in a three-dimensional cell in such a lattice must also decrease. Therefore, the expansions observed on heating are explained by the reverse change, irregular to regular (Fig. 2.2).

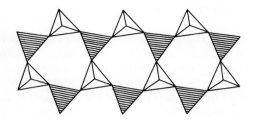

Fig. 2.23 Ring-puckering of the α-form of tridymite and cristobalite within the planes of tetrahedra. A similar effect would be seen viewed along the quartz spirals.

Fused silica on the other hand (Fig. 2.9), even though the Si-O-Si bond angle change may still be occurring, cannot adopt a 'puckered' higher density conformation at low temperature because the random chain structure with all its entanglements cannot change in any coordinated fashion. It therefore shows only a very low expansion on heating. Now, from the data of Fig. 2.1, it can be seen that cristobalite shows substantially more expansion than tridymite. We need, if possible, a structural explanation of this. The key difference would seem to be the six-membered rings that have the 'chair' conformation (Fig. 2.14a). These exist parallel to one plane only in tridymite. Regular puckering seems only to be able to occur in that sense, and the expansion on passing through the transition temperature must be anisotropic in this form. Cristobalite, on the other hand, can pucker in four planes simultaneously (that is, the planes of the faces of the tetrahedron, §2.2), and this would account for the greater volumetric change accompanying the α,β transition for this form.

●2.5   Interconversion

It will have been noticed (Figs 2.2, 2.3) that interconversion of the various polymorphs of silica is possible. Indeed, thermodynamically, this presents no problem. However, kinetically, as in many other systems discussed in this book, there are severe limitations. The reason is simple: the conversion of one silica crystal form to another requires the breaking of many covalent bonds. Si-O bonds are very stable (370 kJ/mol), and a very high activation energy is thus associated with such a **reconstructive transformation**. In practice then, such reactions do not occur on the timescales or at the temperatures of interest now, and there is no evidence that they are of any importance in dentistry. On the other hand, the change of bond angle causes only a slight shift in the relative positions of atoms and this therefore is called a **displacive transformation**. It has a negligible activation energy, occurs extremely quickly and completely reversibly, and at well-defined temperatures.

●2.6   Terminology

The volumetric change which occurs as a result of the displacive process is thus very clearly not to be identified with 'thermal' expansion, which is the result of a generally-increased amplitude of thermal vibration of atoms. The 'true' thermal expansion coefficients of the silica minerals is actually quite high for ceramics (Fig. 3§4.14), as may be judged from the slopes of the plots when the effect of the displacive transformation is subtracted (Fig. 2.24), effectively some 12 - 25 MK^{-1} (up to ~400 °C). However, the abruptness of the phase transitions themselves is sufficient to put this process in a separate class of action. This requires that the distinction be made very carefully between thermal expansion and **crystallographic transformation expansion**. Unfortunately, this is not usually done in the dental literature. Nevertheless, the extra expansion is of great importance to successful accurate castings in dentistry. Accordingly, it would be better to refer to **heating expansion** for those materials, where this is understood to be the combined effect of thermal and displacive expansions.

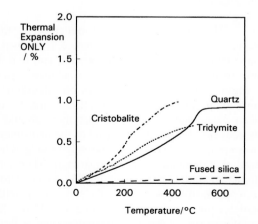

Fig. 2.24 The hypothetical thermal expansion component of the expansion on heating of the allotropes of silica (*cf.* Fig. 2.1). The displacive transformation expansion has been subtracted numerically for the sake of this illustration.

§3. Gypsum-bonded Investments

The silica minerals can thus be utilized to provide an unusual expansion on heating the investment. However, the effect of tridymite is somewhat smaller than the other two, and quartz requires higher temperatures, so that cristobalite is the principal refractory filler of many commercial investments. The effects of some such mixtures can be seen in Fig. 3.1.[9] It will be noticed, however, that the magnitude and abruptness of the expansion could lead to unacceptable stresses, as the heating of a such a poorly-conducting material could lead to steep temperature gradients. Cracked investment moulds lead to dimensional inaccuracies and extra metal fins that would need to be removed. The blending of some tridymite and quartz with the cristobalite would make the net effect rather more gradual and reduce the stresses, perhaps making the investment more tolerant of greater heating rates on burn out, or at least a greater margin of safety. Of course, the anisotropic expansions of quartz and tridymite are of no great consequence since the powder will be randomly oriented in the mass.

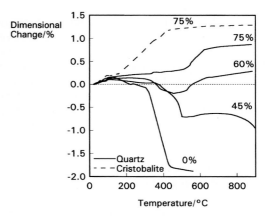

Fig. 3.1 Dimensional changes on heating various plaster-silica mixtures. The shrinkage is due to the dehydration of the gypsum (2§1.1).

●3.1 Strength

Calcium sulphate hemihydrate, whose reaction product with water is loosely termed 'gypsum' (even though this is the name for the natural mineral) was one of the first binders used for dental casting investments, probably because of its cheapness and familiarity as plaster of Paris. As might be expected, the incorporation of the inert silica material as refractory filler in the gypsum 'binder' results in changes in the mechanical properties and characteristics of the setting mass. Thus, raising the silica content (and it does not matter which crystal form this is) is said to result in a slightly longer time available for manipulation and a greatly extended setting time (Fig. 3.2),[10] as measured by a test such as the Gillmore needle (2§11).

However, it should be remembered that the presence of the silica does not change in any way either the reaction or the crystallization rate, and that the test only reveals how long it takes to achieve a particular strength. Such claimed results are therefore rather misleading, if not completely irrelevant, and it should be taken as a matter of course that there should be no unnecessary delay in the pouring of the investment once mixed properly. The point of this may be judged from the steady but dramatic decrease in strength when set as the amount of silica is increased (Fig. 3.3). Since there can be no reaction of the silica with the gypsum on setting, its particles may act partially as voids (*cf.* 2§8). There is therefore a risk of the investment cracking, or fragments being detached and becoming incorporated in the casting, neither of which is desirable for accurate and usable castings. Increasing the net expansion on heating by increasing the silica content therefore involves a compromise.

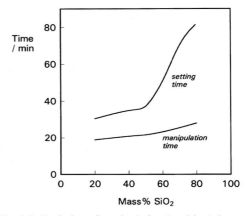

Fig. 3.2 Variation of manipulation (working) time and setting time against silica content for a gypsum-bonded investment.

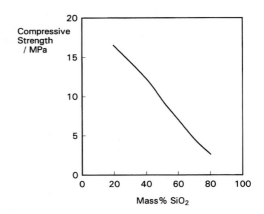

Fig. 3.3 Compressive strength against silica content for a gypsum-bonded investment.

●3.2   Factors for expansion

There is an alternative or at least adjunctive method for achieving mould expansion which has been discussed under gypsum products (2§5.2): 'hygroscopic expansion'. It remains true, of course, that it is not properly hygroscopic in the usual chemical sense. The addition of extra water during setting results in an exaggerated expansion for the same reasons as with, say, plaster alone (Fig. 3.4), but the effect is much greater than with pure plaster. This may be traced to the fact that in plaster adjacent crystallite clusters, the spherulites, can interpenetrate to some extent before individual needles impinge on each other and generate the expansion (Fig. 3.5, left). However, when there is an inert filler present no interpenetration whatsoever can occur at the filler-spherulite interface (Fig. 3.5, right) and all of the crystal growth after contact results in **crystal growth pressure**, and therefore in expansion.

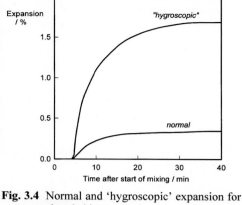

Fig. 3.4 Normal and 'hygroscopic' expansion for a gypsum-bonded investment.

Fig. 3.5 The setting expansion of plaster (left) is exaggerated by the presence of a filler (right).

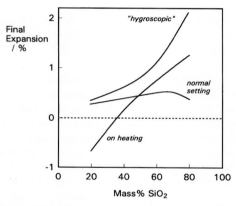

Fig. 3.6 Normal and 'hygroscopic' setting expansion, compared with expansion on heating, as a function of silica content for a gypsum-bonded investment.

Up to a point, the greater the filler content the greater the ordinary setting expansion (Fig. 3.6). (The decline at very high silica content is probably due to the dihydrate crystals not even reaching the silica: they just sit in the spaces in the packed powder.) Hygroscopic expansion, relying on avoiding capillary forces opposing the expansion, is thus to be expected to be strongly dependent on silica content. The expansion on heating, of course, increases markedly with increasing silica content. But it also follows from the interfacial interaction mechanism that the finer the silica powder the greater the expansion (whether normal setting or 'hygroscopic') for the same overall content: since the surface area of the interface is greater, the amount of ineffective interpenetration of spherulites is correspondingly less (Fig. 3.7), and therefore the sooner the pressure is exerted and the greater the total expansion possible.

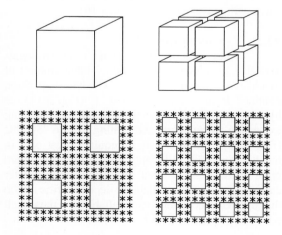

Fig. 3.7 The particle size effect on setting expansion. Subdividing to half the width doubles the total area, and so increases the area of interaction by the same amount.

Although 'hygroscopic' expansion has been discussed in terms of total immersion, by controlling the amount of water available to the investment the amount of expansion can also be controlled (Fig. 3.8).[11] However, it would seem unlikely to be a method of practical

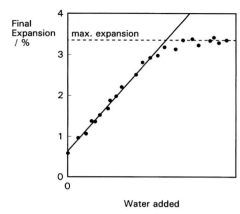

Fig. 3.8 Variation of 'hygroscopic' expansion with the amount of water added to the top surface of the poured gypsum-bonded investment.

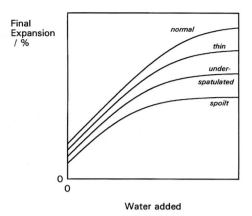

Fig. 3.9 'Hygroscopic' expansion is technique and condition sensitive.

importance as it is evidently quite sensitive to the accuracy of the water volume added, and this must be a function of the volume of investment concerned.

Changing the size of casting ring, pattern and crucible former, or even the amount of slurry actually poured into the ring, will add to the uncertainty, quite apart from requiring the water to be dispensed with an accuracy of about 0·1 mL. Of course, the magnitude of the expansion obtained from this source (as with any other) is sensitive to the treatment given to the investment (Fig. 3.9). Excess water, by 'diluting' the solid present in a larger volume, reduces the 'hygroscopic' expansion, as does insufficient spatulation (a smaller number of crystal nuclei), or the use of an old and humidity-affected investment. The water : powder ratio itself is critical in controlling the expansion realized on heating: the more water initially present, the less expansion, because there is less interaction of silica with gypsum crystals (Fig. 3.10).

●3.3   Expansion strategies

There are therefore two basic strategies available for obtaining the desired large amount of mould expansion: normal setting with large expansion obtained from heating to ~700 °C (Fig. 3.11), or hygroscopic expansion with moderate heating expansion, using quartz and heating only to ~500 °C (Fig. 3.12). Quite clearly it requires great care in handling and in following instructions to obtain reliably the same desired expansion, casting after casting. Even so, this reliability depends critically on the supplied wax pattern being reliable, and thus on all prior steps in the procedure.

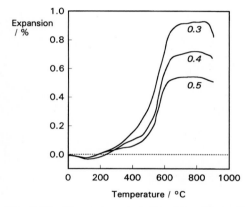

Fig. 3.10 The water : powder ratio affects the expansion on heating of gypsum-bonded investments.

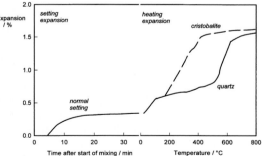

Fig. 3.11 The expansion of investments containing cristobalite and quartz.

There is one other approach that has been used. Some products have incorporated a small amount of starch. This reacts with water (*i.e.,* cooks) when the investment is heated to dry it, swelling appreciably, with the intention of generating more mould expansion. This appears to be dangerous, as the local inhomogeneity of that expansion must generate stresses in the matrix which might lead to cracking, as is done intentionally for impression plaster (*cf.* 27§2.4). Of course, any such material will burn out in the normal way, but it might require slightly longer to achieve because of the increased mass of combustible material and the rate-limiting step of the diffusion of oxygen into the mass. A small increase in porosity will result as a consequence.

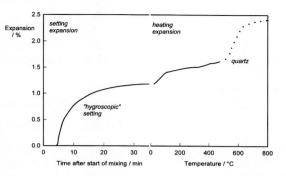

●3.4 Breakdown reactions

The calcium sulphate dihydrate of gypsum-bonded investments must dehydrate on heating (2§1.1). This results primarily in shrinkage, but also a weakening of the structure.[12]

Mould temperatures of up to ~700 °C are adequate for gold-based alloys, whose liquidus temperatures lie in the range of up to about 1050 °C. In contrast, the liquidus of Co-Cr alloys may be at up

Fig. 3.12 The expansion of a quartz-containing gypsum-bonded investment in the hygroscopic expansion technique.

to 1500 °C. These latter alloys require higher temperatures for the investment mould to avoid premature chilling and incomplete castings. However, if any carbon remains in the mould from an incomplete burn out of the wax pattern (and this may occur easily in regions remote from both the outer free surfaces of the investment and the mould surface itself), when the temperature rises above ~700 °C (which it will do anyway adjacent to the cast metal, even if gold-based) the following reactions will occur:[13]

$$CaSO_4 + 4C \Rightarrow CaS + 4CO \tag{3.1}$$

$$CaS + 3CaSO_4 \Rightarrow 4CaO + 4SO_2 \tag{3.2}$$

Several effects may result from this. The evolved sulphur dioxide causes severe embrittlement of gold alloys, making them quite unserviceable. The presence of sulphide may, in an exchange reaction with, say, copper oxide which may form on the casting surface, result in a blackening of the casting which is more difficult to remove:

$$CaS + CuO \Rightarrow CaO + CuS \text{ (black)} \tag{3.3}$$

The reaction may also result in a breakdown of the mould surface, and increased roughness of the casting. Particulate inclusion in the casting may follow.

These reactions certainly make it essential to avoid overheating the investment for gold alloys, but similarly they will cause excessive breakdown if Co-Cr alloys are cast. Thus, decomposition of anhydrite occurs from about 900 °C:

$$CaSO_4 \Rightarrow CaO + SO_3 \tag{3.4}$$

and this may be accompanied by reaction with the silica:

$$CaSO_4 + SiO_2 \Rightarrow CaSiO_3 + SO_3 \tag{3.5}$$

which becomes significant above about 1200 °C, and this temperature will certainly be reached locally, if not from preheating then from the adjacent liquid metal. Further breakdown of the mould will occur, as calcium silicate has a low melting point (~540 °C) (it is therefore formed as a liquid), but the extremely corrosive sulphur trioxide will damage the casting also (to say nothing of the surroundings and lungs). The formation of low-melting substances appears to occur anyway, from the presence of various contaminants and additives, as the mould material softens and may deform plastically in contact with hot metal, even if a gold alloy. Overheated metal therefore presents a risk of ill-fitting castings.

For higher-melting alloys, it is therefore necessary to eliminate gypsum products as the binding agent for the investment. This can be done in two ways.

§4. Phosphate-bonded Investments

Developed to be used with the high-melting Co-Cr alloys, phosphate-bonded investments are widely used. The chemistry of setting and burn out is, however, rather complicated.[14][15]

●4.1   Setting

All commercial dental products rely on the reaction of $NH_4H_2PO_4$ and MgO when mixed with water:

$$NH_4H_2PO_4 + MgO + 5H_2O \Rightarrow NH_4MgPO_4{\cdot}6H_2O \qquad (4.1)$$

The very soluble acid phosphate dissolves rapidly, lowering the pH and dissolving MgO, the magnesium ions from which precipitate promptly as a colloidal mixed phosphate coating around the MgO, reducing reaction rates by limiting diffusion (contrast gypsum), which over-mixing would tend to disrupt. The salt may also crystallize somewhat, heterogeneously-nucleated on the MgO, and this would also reduce the rate of reaction by occluding it more strongly. Overall, phosphate in solution, MgO and solid mixed phosphate can co-exist in the set material, even when there is excess MgO – another instance of the kinetic limitation of the approach to equilibrium. Some amorphous $Mg_3(PO_4)_2$ can also form directly:

$$2NH_4H_2PO_4 + 3MgO \Rightarrow Mg_3(PO_4)_2 + 2NH_4^+ + H_2O + 2OH^- \qquad (4.2)$$

and from similar reactions both amorphous $MgHPO_4$ and crystalline $MgHPO_4{\cdot}3H_2O$ may also be present. These reactions may be driven more at the free surface by the loss of ammonia during setting, which can be smelt, as this loss affects pH and, of course, solution concentration of NH_4^+. However, the setting reaction is strongly exothermic (peak temperatures from 50 to 90 °C are common). Thus, and in any given set-up, because of the poor thermal conductivity of the material, the temperature and the thermal history varies across the mass. The formation of other possible phases will vary correspondingly, and also between products and preparation conditions, but always with a thermal history-dependent spatial distribution in the mass. The size and shape of the cast investment will therefore also have an effect, as well as the nature of the casting ring (metal or removable plastic), and also immersion and immersion temperature, if this approach is used. This means that a definite description of the set material cannot be given, even for a single product.

The relatively high temperatures encountered during setting must mean that wax patterns suffer first appreciable expansion, considerable softening with a risk of distortion, then melting. This suggests that harder, higher-melting waxes are appropriate, but of course the expansion cannot be changed, becoming part of the system behaviour that needs to be taken into account.

●4.2   Burn out

On heating, the main reaction appears to be progressive decomposition of the mixed phosphate, first by partial dehydration:

$$NH_4MgPO_4{\cdot}6H_2O \Rightarrow NH_4MgPO_4{\cdot}H_2O + 5H_2O\uparrow \qquad \text{(from ~50 °C)} \quad (4.3)$$

Note that this process could start during setting itself, depending on local conditions (if water can be lost, such as at the free surface), further complicating the set material description. Further decomposition then occurs to leave amorphous, polymeric pyrophosphate, $(Mg_2P_2O_7)_n$:

$$2NH_4MgPO_4{\cdot}H_2O \Rightarrow Mg_2P_2O_7 + 2H_2O\uparrow + 2NH_3\uparrow \quad \text{(over by ~300 °C)} \quad (4.4)$$

– which reaction accounts for the characteristic strong smell of ammonia during burn out. As the temperature increases above ~690 °C, this glassy material begins to crystallize. Thermal decomposition of the mixed phosphate can also produce a mixture of meta- and ortho-phosphates:

$$4NH_4MgPO_4{\cdot}H_2O \Rightarrow Mg_3(PO_4)_2 + Mg(PO_3)_2 + 4NH_3\uparrow + 6H_2O\uparrow \qquad (4.5)$$

The metaphosphate, $Mg(PO_3)_2$, does not crystallize under these conditions, and remains glassy. Thus, pyro-, ortho- and meta- phosphates can all occur in the process, depending on local conditions.

Any ammonium dihydrogen phosphate that remained in solution after setting decomposes to form P_2O_5 (water having long since evaporated), and this can react with major and minor setting reaction products, the net effects being:

$$NH_4H_2PO_4 + NH_4MgPO_4 \Rightarrow Mg(PO_3)_2 + 2NH_3\uparrow + 2H_2O\uparrow \qquad (4.6)$$

$$2NH_4H_2PO_4 + 2MgHPO_4 \Rightarrow 2Mg(PO_3)_2 + 2NH_3\uparrow + 4H_2O\uparrow \qquad (4.7)$$

At high burn-out temperatures both pyro- and orthophosphate may remain, but metaphosphate disappears because

$$Mg(PO_3)_2 \;+\; Mg_3(PO_4)_2 \;\Rightarrow\; 2\,Mg_2P_2O_7 \tag{4.8}$$

In addition, and especially after the metaphosphate has been consumed, orthophosphate will be formed by the reaction of pyrophosphate with excess MgO, if present (which is usual):

$$Mg_2P_2O_7 \;+\; MgO \;\Rightarrow\; Mg_3(PO_4)_2 \tag{4.9}$$

so that for the usual burn-out temperature (1000 ~ 1300 °C), some magnesium pyrophosphate may remain in the structure or be replaced totally by magnesium orthophosphate, depending on the original composition.

In all such thermal decomposition reactions, the kinetic limitations need to be recognized. Reactions can commence slowly at low temperatures, and accelerate as the temperature rises. But, again because thermal conductivity is low, the temperature profile cannot be uniform across the body. Thus, we may have the situation of reactions proceeding at different rates in different places (and the Arrhenius rule of thumb of doubling a rate for a 10 °C rise is usefully indicative), on top of variation of setting reaction products, followed by differing reactions according to those variations. We therefore cannot expect uniformity in chemistry at any stage, or indeed uniformity in mechanical properties, and especially not uniformity in setting and heating expansion when anything else is allowed to vary, such as mixing ratio, mixing temperature, mould shape and size. Much inconsistency in the literature may have arisen from such factors not being recognized, and it certainly can account for inter-laboratory variation when standardization tests are made.

There are a number of other ingredients in these products (Zn is present in at least one to a large proportion), but the reasons are at present largely commercial secrets and unpublished. It can be expected that since the Mg and Zn phosphates form continuous solid solutions under most conditions (9§5.14), it is largely a matter of reactivity and rate. The various colours of the phosphate-bonded investments are due to small amounts of coloured, transition metal oxides (e.g. Fe_2O_3, Cr_2O_3), but these seem to have no other function.

The loss of ammonia between ~300 - 400 °C during burn out results in an appreciable shrinkage (as well as the unmistakable smell), but there is, of course, an overall expansion on heating due to the silica refractory filler. For mixing, instead of plain distilled water, a colloidal suspension of silica, known (unhelpfully) as 'Special Liquid', is generally used to increase the expansion, although the reasons for this effect are not yet clear. Fine control of the amount of expansion is therefore possible, both in heating and 'hygroscopic' expansion (Fig. 4.1).[16] The silica refractory of the powder, usually ~80% of the bulk, and typically cristobalite, still shows the same expansion behaviour on heating (Fig. 4.2).[17]

Given that there are a number of sources of variability in all the dimensional change processes, and notwithstanding the usual injunction to follow the manufacturer's instructions here as elsewhere, once a production process has been tuned to give castings that fit, consistency becomes the overriding concern.

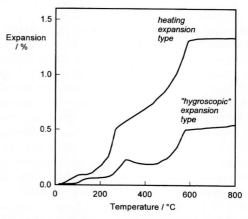

Fig. 4.1 Expansion *vs.* temperature for 'heating' and 'hygroscopic' expansion types of phosphate-bonded investment.

● **4.3 Slag**

There is one curious issue: the melting point of $Mg_3(PO_4)_2$ is 1184 °C while that of $Mg_2P_2O_7$ is 1383 °C. Given that the liquidus temperatures of many cobalt-chromium and similar casting alloys lie in the range 1200 ~ 1450 °C, and that for good castings heating a further 50 K or so is necessary, it seems to be inevitable that some melting of the binder will occur on casting. It is to be expected that these quoted temperatures are in fact upper limits because the effect of mixtures and contamination in general is to lower fusion temperatures. Reaction with silica will then occur, with further lowering of fusion temperatures, forming glassy materials on cooling. This melting need not be a problem from the point of view of dimensional accuracy as long as the investment mould is not overheated in the first place, and that freezing of the metal occurs reasonably promptly. However, it does seem likely that the slag which coats such castings will in part at least be due to such melting.

Related to this is the fact that because cobalt-chromium and the like are not oxidation-resistant alloys at their processing temperatures, their constituents will oxidize during casting (mostly the chromium). Some oxides can also react with the silica of the investment to form silicates (an acid-base reaction), contributing to the formation of slag. A possible example is:

$$2CoO + SiO_2 \Rightarrow Co_2SiO_4 \qquad (m.p. \sim 1345 \text{ °C}) \qquad (4.10)$$

although Cr_2O_3 does not appear to react to give an equivalent silicate, not even melting until over 2200 °C. Reactions with magnesium oxide and phosphates are also possible. It is, however, incorrect to say that the *metal* reacts with the silica: this is thermodynamically impossible.[1]

In any case, overheating of the investment (and metal) is likely to prove disadvantageous. At the least it will make cleaning the casting that much more difficult. At the worst the melting might be severe enough to cause distortion of the mould while the metal was still molten, and thus of the casting itself.

§5. Silica-bonded Investments

Silica-bonded investments do not have a setting reaction in the conventional sense: they rely on the drying of a silicic acid gel to provide the binder. A **sol** (colloidal solution) is first formed by hydrolysing ethyl silicate in aqueous solution with dilute hydrochloric acid:

$$Si(OC_2H_5)_4 + 4H_2O \Rightarrow Si(OH)_4 + 4C_2H_5OH \qquad (5.1)$$

When this is mixed with the investment powder, which contains magnesium oxide simply to neutralize the acid, the sol is converted to a gel. The silicic acid undergoes a condensation reaction (eliminating water) to form a three-dimensional network of high polymer (*cf.* the intermediate forms in Fig. 9§7.2; and condensation silicone impression material, 7§6.1):

$$2Si(OH)_4 \Rightarrow (OH)_3Si-O-Si(OH)_3 + H_2O \qquad (5.2)$$

and so on, progressively. This gel is, however, very weak and fragile and needs to be dried at an elevated temperature (~100 °C) to drive off both the alcohol and the water. This results in a small, so-called 'green' shrinkage (Fig. 5.1). As the investment powder is nearly all silica, the expansion on heating is proportionately larger than for other investments, depending of course on the choice of allotrope or mixture (and because there is no shrinkage due to a decomposition reaction). But while higher mould temperatures may be tolerated by silica-bonded investments (up to ~1200 °C), compared with ~1000 °C for phosphate-bonded), there is a tendency to sinter at high temperature, and this may offset some of the expansion on heating.[18]

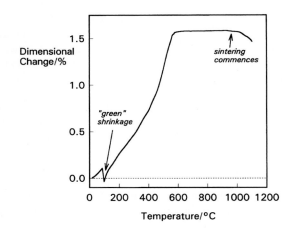

Fig. 5.1 Expansion on heating of a quartz-based, silica-bonded investment.

Unfortunately, there are a number of disadvantages associated with this type of investment, and although they can be overcome with extremely careful technique the disadvantages probably account for the lack of popularity. The silica sol is unstable and needs to be freshly prepared regularly. The investment mould is necessarily very weak, much weaker than other investments, and has to be handled extremely carefully. The silica powder (usually cristobalite) has to have a carefully-controlled particle size distribution to improve the strength, but settling and segregation occurs in transit due to vibration, and careful remixing is essential before use. Heating, likewise, has to be done very carefully, firstly to dry the gel, and then to avoid cracking on

---

[1] The main exception to this in dentistry is for titanium. The reaction to reduce the silica,

$$Ti + SiO_2 \Rightarrow TiO_2 + Si$$

is allowed thermodynamically and thus would lead to investment breakdown. Refractories such as ZrO_2 and MgO must be used; these do not react in this way.

taking it up to the casting temperature. This type of investment is rather prone to cracking because of the large expansion on heating. In addition, vacuum investing (ordinarily used to eliminate air bubbles, especially around the pattern) is not permissible because the ethanol would readily boil under the reduced pressure and disrupt the structure. In contrast, phosphate-bonded investments are simple to use and suffer none of these drawbacks.

§6. Ring Liner

The investment casting mould is usually prepared in a steel **casting ring**, initially to act as a mould for the investment itself when poured, but later to provide a strong casing that may be handled with tongs without damaging the investment when it is hot. Such a stiff casting ring quite clearly would not allow any investment setting expansion (whether ordinary or 'hygroscopic', and even on heating), at least in directions perpendicular to its axis. Any expansion along the axis would then cause severe distortion of the pattern, and a useless casting would result. Consequently, provision must be made for this essential investment expansion. This is usually done by using a **ring liner** (Fig. 6.1).

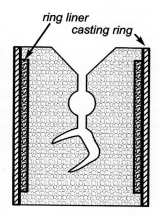

Fig. 6.1 A ring liner allows (some) expansion of the investment mould.

●6.1   Setting
A ring liner must have sufficient thickness to provide room for the expansion yet provide no appreciable resistance to lateral expansion, while having sufficient strength to avoid being crushed under the hydrostatic head of the investment slurry. In addition, there must be no undesirable effects on burning out the investment.

A loosely formed mineral fibre paper, similar to thick blotting paper or felt, satisfies these requirements. Asbestos was a natural candidate because of its fibrous crystal habit, but its use is now deprecated (if not banned outright by legislation) because of its carcinogenic properties.[19] Other mineral 'wools' have now been substituted but even so, it would be wise to avoid inhaling the dust from these as similar doubts about carcinogenicity have also been raised (and all silicate dusts, such as talc, are suspect as capable of causing pneumoconiosis and other diseases[20]). An appropriate thickness of such paper is therefore wound around the inside of the casting ring prior to pouring the investment slurry, avoiding a few millimetres at one, maybe each, end to allow the set investment to lock into the ring and not fall out. However, the paper is necessarily very porous and absorbent, and much of the investment slurry water would be absorbed, leaving possibly insufficient for reaction, but certainly causing **loss of gloss** (2§5.1) to occur very quickly and much reduced setting expansion as a result. The ring liner must therefore be wet before pouring the investment slurry, and this probably results in a little hygroscopic expansion on its own as it is essentially impossible to judge how much water to put in the liner to achieve no effect on balance.

It is certainly important to have a sufficiently thick layer to accommodate the anticipated expansion, but the approach is not obligatory. 'Ringless' casting techniques are sometimes used, relying on the strength of the set investment. This approach would still require a mould, of course, but a heat resistant type is then unnecessary as it is removed before burn out, and this mould can be made to allow for some expansion (see §6.3).

●6.2   Heating
A similar problem arises on heating, only now the expansion of the casting ring itself must also be taken into account. Thus, as shown in Fig. 6.2, the steel expands smoothly and ultimately has sufficient expansion to suit this particular investment, at least. However, there are two regions where the investment expands much more rapidly

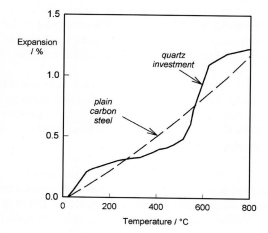

Fig. 6.2 Expansion on heating of a quartz-based, gypsum-bonded investment compared with that of a plain carbon steel casting ring.

and these pose a threat of stresses that would crack the investment by being squeezed by the ring. Thus, and particularly with 'heating expansion' method investments, a ring liner is necessary to accommodate the mismatch in expansion rates.

●**6.3   Ringless casting**

Reconsideration of Fig. 6.1 shows that such arrangements are not entirely satisfactory. The fact that the investment is in contact with the casting ring at top and bottom has undesirable effects. Firstly, axial setting and heating expansion is inhibited because the set investment is frictionally locked to the ring. Secondly, the radial constraint at top and bottom must be at least partially transferred to adjacent regions, so the radial expansion at the level of the mould space cannot be considered entirely free. The mould is therefore always distorted to some extent. The casting ring, in fact, has little part to play except as a mould for casting of the investment slurry (for which no great robustness is required) and to protect the set investment from rough handling (which should not occur anyway).

Given these considerations, the tendency is to dispense with steel casting rings, using either a cardboard tube (which must be soaked in similar manner to ring liner), which is stripped off after setting, or a plastic, removable split ring, both of which allow expansion more freely. Phosphate-bonded investment in particular is strong enough to survive casting without a supporting ring.

§7. Porosity

As remarked above, one of the requirements for an investment is that it be sufficiently porous to be permeable to the air (and possibly other gases) which would otherwise be trapped in the mould space by the molten metal when cast. An impermeable mould material would allow air bubbles to be trapped unless there was an elaborate series of vents from (gravitational) 'high' points on the pattern to the top of the investment mould. Indeed, for very large or complicated castings such arrangements are essential anyway, but under the right conditions this is not necessary for small castings – the majority of cases in dentistry. (See also 18§2.2).

The porosity of gypsum products arises from the extra water required for mixing, over and above that required for reaction, due to **gauging**, **dilatancy** and **dilution** considerations (2§2). Precisely the same considerations lead to the porosity of dry investments. However, although the reaction requirement for gypsum- and phosphate-bonded types may be relatively low because the volumetric proportion of refractory is so high, the other demands are not diminished. In the silica-bonded materials there is an extra problem: the nature of the silica gel binder is such as to have a naturally low permeability, even though the porosity may be high. Dried silica gel has a low density due to molecular-scale porosity, but naturally this would provide great resistance to the passage of gas. This lower permeability may be offset by the addition to the investment of **blind vents**, not communicating directly with the mould space but coming sufficiently close to it that the path length over which the gases must travel to reach a low impedance route to the atmosphere is drastically shortened. Such a technique does not add to the amount of metal which must be cast, nor to the difficulties of cleaning up the casting afterwards by adding parts which need to be cut off. In addition, such vents need not emerge only on the 'top' (sprue end) of the investment: since there should be no metal in them, vents opening on the 'gravitational' bottom of the casting ring may be more convenient for some types of pattern.

Porosity is also of concern in burn out schedules. The diffusion of oxygen to the wax residues is the limiting factor, such diffusion being facilitated by short path lengths from large diameter paths such as sprues and vents (16§4). Unless those residues are removed, in addition to reactions noted above (§3.4) and later (19§2.4), the permeability of the investment is also seriously reduced, and this may compromise the quality of the casting finally made.

It should go without saying that porosity due to air bubbles in the investment slurry should be avoided. A bubble lying adjacent to the mould space may be open or easily broken into during casting, and the resultant extra metal **bleb** may cause the device to fail to fit but will anyway require time-consuming effort to remove. So-called "vacuum-mixing" is the means to reduce the risk of this problem (see also 18§1.2).

References

[1]   Tucker MG, Dove MT & Keen DA. Direct measurement of the thermal expansion of the Si–O bond by neutron total scattering. J Phys: Condens Matter 12: L425–L430, 2000.

[2]   Paffenbarger GC & Sweeney WT. Dental Casting Technic: Theory and Practice. J Dent Res 11: 681 - 701, 1931.

[3]   Wyatt OH & Dew-Hughes D. Metals, Ceramics and Polymers. An Introduction to the Structure and Properties of Engineering Materials. Cambridge UP, 1974.

[4]   Souder W & Hidnert P. Measurements of the thermal expansion of fused silica. Scientific Paper of the Bureau of Standards, No. 524, Washington, 1926.

[5]   Cotton FA & Wilkinson G. Advanced Inorganic Chemistry, 5th ed. Wiley, New York, 1988.

[6]   Wells AF. Structural Inorganic Chemistry, 3rd. ed. Oxford UP, 1967.

[7]   Hatch DM & Ghose S. The α-β phase transition in cristobalite, SiO_2. Phys Chem Minerals 17: 554 - 562, 1991.

[8]   Dolino G. The α-inc-β transitions of quartz: a century of research on displacive phase transitions. Phase Transitions 21: 59-72 (1990)

[9]   Volland RH & Paffenbarger GC. Cast gold inlay technic as worked out in the cooperative research at the National Bureau of Standards and applied by a group of practicing dentists. J Amer Dent Assoc 19: 185 - 205, 1932.

[10]   Earnshaw R. Inlay Casting Investments. in: O'Brien WJ & Ryge G. An Outline of Dental Materials and their Selection. Philadelphia, Saunders, 1978.

[11]   Craig RG (ed). Restorative Dental Materials. 6th ed. St Louis, Mosby, 1980.

[12]   Luk HWK & BW Darvell BW. Effect of burnout temperature on strength of gypsum-bonded investments. Dent Mater 19 (6): 552 - 557, 2003.

[13]   Phetrattanarangsi T et al. The behavior of gypsum-bonded investment in the gold jewelry casting process. Thermochim Acta 657: 144 - 150, 2017.

[14]   Scrimgeour SN, Chudek JA & Lloyd CH. The determination of phosphorus containing compounds in dental casting investment products by 31P solid-state MAS-NMR spectroscopy. Dent Mater 23 (4): 415-424, 2007.

[15]   Scrimgeour SN, Chudek JA, Cowper GA & Lloyd CH. ^{31}P solid-state MAS-NMR spectroscopy of the compounds that form in phosphate-bonded dental casting investment materials during setting. Dent Mater 23(8): 934-943, 2007.

[16]   Zarb GA, Bergman B, Clayton JA & MacKay HF. Prosthodontic Treatment for Partially Edentulous Patients. Mosby, St Louis, 1978.

[17]   Luk HWK. High Temperature Strength of Dental Phosphate Bonded Investment. PhD, The University of Hong Kong, 1995.

[18]   Earnshaw R. Cobalt-chromium alloys in dentistry. Brit Dent J 101: 67 - 75, 1956.

[19]   Sichletidis L, Spyratos D, Chloros D, Michailidis K, & Fourkiotou I. Pleural plaques in dentists from occupational asbestos exposure: A report of three cases. Amer J Ind Med 52: 926 - 930, 2009.

[20]   Ghio AJ, Kennedy TP, Schapira RM, Crumbliss AL, Hoidal JR. Hypothesis: is lung disease after silicate inhalation caused by oxidant generation? Lancet. 336(8721): 967 - 969, 1990

Chapter 18   Casting

The process of the casting of a fluid material into a mould clearly depends on the viscosity of that fluid. But if the demands include the accurate reproduction of the details of the mould, and overall dimensional accuracy, additional factors must be taken into account.

*In the case of gypsum and investment products, the **flow** itself is affected by the special rheological properties of slurries, while conforming closely to the mould or pattern requires that the surfaces be wetted. **Wetting** is not possible with molten metals on ceramics and extra measures must be taken to obtain a good result.*

*In essence, accurate metal casting depends on supplying just sufficient **casting pressure**. This may be obtained hydrostatically, assisted by centrifugation, pressurized air, or the use of vacuum. However, excessive casting pressure is detrimental. Obtaining the correct balance depends on the understanding and fine control of the various factors.*

*A variety of **defects** can arise in the casting of metal, in part from the risks associated with the challenging conditions of high temperature and high speed. Faulty castings are not just a waste of time and money, but may also be detrimental to the life of the device if it is pressed into service. This is a treatment failure.*

Casting is the final critical stage in the construction of complex metal prosthetic and restorative dental devices. A variety of factors interact with each other to affect the outcome, and fine control of the process requires a proper understanding.

Materials Science for Dentistry
https://doi.org/10.1016/B978-0-08-101035-8.50018-3

Casting is the process of pouring a fluid material into a mould where, by some process (physical or chemical), that material hardens to produce a replica of an original pattern. The mould necessarily is a 'negative' or reversed three-dimensional image of the shape of the pattern. In the course of restoring tooth form for function, or in the provision of a larger scale prosthesis, both the reproduction of the shape of tissues for working models as well as the creation of the final functional device rely on casting in one sense or another. In the long chain of steps leading to the fabrication of a final device, the casting process itself is at least as important as the materials that are used for the attainment of dimensional accuracy and completeness of structure.

We are drawing a distinction here between systems in which flow is spontaneous (or, at least, occurring at a rate great enough to be useful) and those that require some assistance to flow in the form of a raised stress. Examples of the latter are commonplace in dentistry: all cements, denture base acrylic dough, most impression materials such as silicone and polysulphide rubbers, amalgam, filled resin restorative materials, waxes ... the list goes on. Making such a material conform to the shape of some surface is described as **moulding**. A little thought will show that the distinction is rather artificial, there being no sharp, true demarcation but rather a continuum again. It is more a case of the timescale on which the desired effect operates that determines whether, and how much, force is to be used. Indeed, all of these processes involve flow, and flow is time-dependent.

§1. Models and Investments

Dental plaster and artificial stone models and dies are prepared by casting a slurry of hemihydrate and water into the impression material mould. Several factors are of direct importance in controlling the accuracy of the replica produced. Firstly, the slurry must be sufficiently fluid that it may flow into all of the fine details of the mould completely. This is the purpose of the mixing water and the inclusion in it of a proportion which we called the dilution allowance (2§2), the particle shape and size distribution playing a part in controlling the viscosity. Coupled with this is the effect of the surface tension of the slurry: it must not be great enough to cause rounding of sharp corners by preventing penetration. Ordinarily no special steps are taken to reduce this as, in such aqueous mixtures, the surface tension is apparently low enough for few problems to arise. But in this context the wetting of the mould material by the slurry is the primary consideration, and so the penetrativity (10§2.4) becomes relevant here (as in fissure sealants) when fine detail is required. Aqueous gel impression materials, agar and alginate, naturally offer least difficulty but some silicone materials, being hydrophobic, can cause problems. Accordingly, polymers have been designed to be 'hydrophilic', meaning wetted by water, to be used in these systems with a lower risk of trapping air bubbles in critical places.

●1.1 Vibration

Even so, and with all impression materials, care must be taken to avoid the entrapment of air bubbles in fine detail or narrow spaces. To this end vibration is routinely employed to aid the displacement of bubbles, which must be expected to be present as a matter of course. Firstly, this is because mixing by hand cannot avoid the inclusion of bubbles in the mix, and mixing under reduced pressure ultimately only reduces the size of those bubbles. Secondly, in pouring viscous media it is very difficult to avoid creating bubbles, whether due to trapping air against a wall or by 'folding' into the liquid.

Vibration both improves the flow of the slurry and eliminates bubbles from within its bulk that were incorporated at the mixing stage. This depends on the phenomenon of **shear-thinning** (4§7.9): the solid particles are prevented from mechanically interlocking with each other by the vibration supplying the activation energy for relative movement. The same phenomenon can be observed in dry powders. Tilting a container of dry powder such as plaster of Paris may require a rather large angle before the powder moves (and then it collapses unhelpfully dramatically). However, tapping gently permits more controlled and gradual flow. This effect is used to good advantage when weighing powders accurately from a spatula: tapping (or a vibrating spatula) may help dispensing small increments with greater control. Even if a mixture is already fluid enough to flow spontaneously, vibration will reduce the viscosity. But bear in mind that, for gypsum products and the like, we wish to work as close to the minimum mixing liquid as we can in order to maximize strength. Shear-thinning can allow us to get closer to that ideal.

It does not require bubbles to be trapped at a critical surface for them to be detrimental. Even if a bubble were to be entirely surrounded by slurry, if it were to end up sufficiently close to the surface of the set material the thin wall may easily be broken in subsequent handling, causing a deficiency of possible later concern, such as at a cusp or fitting surface. In any case, the setting rate of the gypsum product, or whatever it is being cast, must not be so great as to compromise the above considerations.

●1.2 Investing

The production of an investment mould for metal casting is itself also a casting operation: the investment slurry is cast into the space between the wax pattern and the investment ring. Similar criteria as for pouring models must therefore apply for accuracy to be obtained. But, as the wetting of wax by water or watery mixtures is very poor, waxes necessarily having low energy surfaces, surface active agents (*i.e.* detergents or surfactants, 10§8.3) are generally applied to the surface of the pattern to avoid incomplete wetting and the trapping of air bubbles. The presence of sub-surface air bubbles is again of concern because the investment may break down under the impact of the molten metal (§3.8). Again, air may be eliminated by vibration, although care must be taken to avoid bubbles coming to rest underneath the pattern, so the proper orientation of the pattern on the sprue is crucial.

Even so, the application of vacuum (or, rather, reduced pressure) is a more reliable adjunct procedure. The reduction of pressure means that any bubbles that are present must expand (Boyle's Law: PV = constant). This means that their total **buoyancy**, the force acting to lift the bubble through the slurry, is increased. The buoyancy is simply the weight (mass × gravitational acceleration) of the slurry displaced by the bubble (we can ignore the weight of the contained air). If we were to reduce the pressure to, say, 25 mm Hg, the volume, and therefore the buoyancy, of a bubble would be increased about 30 times (normal atmospheric pressure being about 760 mm Hg). Under such conditions bubbles tend to be expelled readily. Any bubbles remaining, or fragments left behind from the other escaping bubbles, will then collapse to about one-thirtieth of their volume (or about one-third their diameter) when atmospheric pressure is restored. Notice that the pressure could not be lowered much further at normal room temperatures without the water in the slurry boiling, since its vapour pressure is then about 25 mm Hg

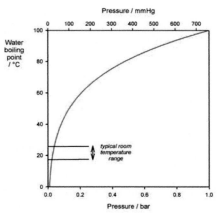

Fig. 1.1 Variation of boiling point of pure water with pressure. Dissolved substances will raise the b.p. slightly.

(Fig. 1.1). However, if this could be tolerated, it would have the effect of quickly flushing out the air, the vapour bubbles then quickly collapsing completely so that an essentially bubble-free casting would be obtained. The original air bubbles would form good nuclei for that boiling.

The use of a duplicating material such as agar, as for example is used to prepare the refractory model for a cast metal framework, involves pouring the fluid sol over the master model. This therefore is also a casting process. In this application, too, bubbles must be avoided. Although shear-thinning is not required, and wetting of the gypsum model by an aqueous sol should not be a problem if there is no stray wax on its surface, careless work could easily create problems of this kind.

§2. Metal Casting

The casting of molten metal at high temperatures into an investment mould has similar requirements as the above two operations as regards flow and detail reproduction, but these are made more critical because of the high speed and temperature involved. However, the casting of metal has its own special problems.

●2.1 Surface tension

Clearly, the alloy must be at a temperature above the liquidus when it first enters the mould, otherwise solid will be present which could occlude detail at a surface or even block channels. An incomplete casting must then result. The viscosity of molten metals is usually quite low and so this presents no problems with flow as such, but the surface tension is certainly very high (Table 2.1) and wetting of the investment may be taken as impossible (contact angle θ = 180°). Most ceramics simply are not wetted by metals, *cf.* mercury on glass (the notable exception to this is cobalt on certain carbides, 21§3). This effect will prevent sharp, detailed castings unless there is sufficient pressure to force the liquid into corners and fine sections. For the scale of dental castings it is unlikely that this could be realized by simple pouring, relying on the hydrostatic pressure under a long sprue, even allowing for the high density of gold alloys, for example. Thus, some mechanical assistance must be provided for the purpose, *i.e.* with centrifugal, pressure or vacuum casting machines. We return to this point below.

●2.2　Porosity

Even so, and quite unlike pouring gypsum products and investment casts, the molten metal is admitted to the mould space through a rather narrow sprue. Although a large amount of air may in fact escape through the sprue, bubbling back against the metal flow, not all of it may do so because of the shape of the mould space. This would further retard or even curtail the completion of the casting. However, investments are prepared from a powder with a relatively large quantity of mixing water, certainly in excess of the amount required for the setting reaction. This excess, of course, leaves a very porous structure when it has been evaporated, and this in turn allows trapped air to escape. Obviously, such porosity must be **connected** for flow through the mass to occur (Fig. 2.1). Even then, for casting complicated shapes and large volumes, vents may be added to the larger structural elements of the pattern, remote from the sprue, to permit ready escape of a large volume of the trapped air (17§7).

For uniform spheres the maximum packing density corresponds to a volume fraction of about 74%. Irregular particles decrease this figure considerably and careful selection of the particle size distribution of the refractory silica is essential if a dense, and therefore strong, structure is to be obtained. But, clearly, care must be taken not to go too far and impede the escape of the air through that structure. The pore size must be small enough to put a satisfactory upper limit on the roughness of the resultant casting. But even though the viscosity of air is low, because of the high speed necessary to get the metal into the mould without premature freezing, this factor is critical. The failure of the air to escape creates 'back-pressure' defects, bubbles trapped between investment and metal.

Table 2.1　*Surface tensions and viscosities of some molten metals.*

	Temp. / °C	Surface Tension / mN.m^{-1}	Viscosity / mPa.s
Ag	961	903	3.88
Au	1063	1140	5.0
Co	1493	1873	4.18
Cr	1875	1700	-
Cu	1083	1285	4.0
Hg	25	484	1.53
Ni	1454	1778	4.90
Pt	1769	1800	-
(H$_2$O	25	72	1.00)
(Eugenol	25	-	~7.4)

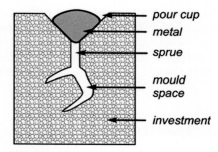

Fig. 2.1　Air must escape from the mould space through the investment material itself.

●2.3　Oxidation

Air has another connotation in metals casting: it is an oxidizing agent. The problems with this are two-fold: there will be a resultant change in alloy composition because the elements vary in their tendency to oxidize and the rate of their oxidation. Excessive amounts of oxides on the melt (known as **dross**) may possibly get carried over with the metal and be included in the structure, thus weakening it (these oxides, again, are not wetted by the metal); although under normal conditions most, if not all, should remain in the crucible. Even then, the failure to wet would usually mean that they would be ejected from the melt and only lie on the surface of the casting (*cf.* oxides on amalgam alloy, 15§1).

When a flame is used to melt the metal, reducing conditions are employed (see §5). Such conditions may be attempted within the investment to prevent oxides forming on the surface of the casting so as to reduce the work required for cleaning it up after divesting, but also to allow a smoother surface to the cleaned work. Powdered graphite or copper has been used in this way, dispersed in the investment powder, *i.e.* as easily oxidizable material which, by using up oxygen already present and as it continues to diffuse in, reduces the exposure of the metal. However, copper would dissolve in gold alloys, changing their composition, especially at the surface (which could make them more prone to tarnishing), while graphite is prone to being burnt out along with the wax residues and could not be used at high temperatures because of reaction with calcium sulphate (17§3.4). An alternative has been the use of a graphite crucible, but this may not be used for alloys that dissolve carbon because of the embrittlement that the extra carbide formed would cause. In principle, any reducing agent that does not react with the alloy could be acceptable. Thus, while boron and aluminium have been suggested for gold alloys,[1] Al has appreciable solid solubility in Ag, Au and Cu (and thus presumably in their alloys), and AuB$_2$ may form under such conditions; the options are limited (*cf.* 19§1.14).

The speed of heating obtained with induction casting machines greatly reduces the problem of air oxidation and such measures become less critical, but the price of such machines is likely to mean the continued use of flames in many countries for a long time yet. The more obvious procedure of doing the casting under an inert atmosphere (such as argon) is even more costly because of the specialized equipment required. However, it must be used for titanium because of its reactivity (28§1).

●**2.4   Cooling**
The need to 'burn out' an investment mould before casting has already been discussed from the point of view of removing the pattern materials (16§4) and achieving mould expansion by heating (Chap. 17) (see also §4). There is, however, a third requirement: avoidance of premature chilling. Although molten metal velocities in the mould space are known to be of the order of $0.5 \sim 1$ ms^{-1}, the metal probably then takes only about $0.1 \sim 0.2$ s to cool to the liquidus. In simple designs, such as crowns and inlays, this may not present a problem, but for large frameworks and mesh it is a much more serious matter. The relationships involved are complex and not fully worked out yet, but several factors are of obvious importance. Firstly, the 'temperature excess',[1] how far above the liquidus that the metal has been heated, controls how much heat is available to be lost before solidification commences. Secondly, the temperature of the investment then establishes the temperature gradient and thus the rate of loss of heat. The specific heat capacities of metal and investment, and the conductivity of the latter, are also involved but not amenable to control in practice. Clearly, it is not just a matter of simply raising both metal and investment temperatures. Mould space expansion, investment breakdown (also in contact with the hot metal), and alloy component oxidation and volatilization also need to be considered as critical issues in casting accuracy and resultant properties. Part of the problem, though, is the driving force for the hot metal into the mould. It is this we now consider.

§3. Physics of Casting

The near-total exclusion of air (or other gas) by casting in a vacuum would of course avoid both interference from effects due to porosity and also from oxidation, but this approach cannot of itself assist in obtaining complete and sharp castings as surface tension remains unaffected: some extra assistance is required. Centrifugal casting machines are commonly used to provide such assistance and the effect may be analysed in the following way.

●**3.1   Rotation**
Newton's Second Law of Motion states that

> **the rate of change of momentum is proportional to the applied force, and takes place in the direction in which the force acts.**

After integration, this leads to the expression

$$F = m.a \qquad (3.1)$$

where F is the force, m the mass, and a the acceleration. In terms of the units this is

$$N = kg \times m/s^2 \qquad (3.2)$$

which in fact provides the definition of the **newton** as the SI unit of force. For a body rotating uniformly about a fixed point, as for example a ball on a string, the equivalent of equation 3.1 is

$$F = m.v^2/R \qquad (3.3)$$

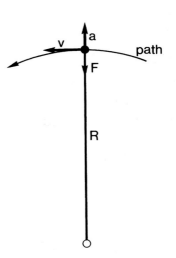

Fig. 3.1 Acceleration and restraining forces acting on a body in circular motion.

---
[1] This is sometimes, and quite erroneously, referred to as "superheat" in the dental technology context. The term should only be applied to steam heated above the normal boiling point, or to a substance being heated above a phase transition without the transition occurring.

where R is the radius of the path of the centre of gravity of the (small) body and v is its velocity. What this means is that, despite the rotation being uniform, there is still an acceleration of v^2/R acting, and this is outwards, away from the centre of rotation (hence 'centrifugal') (Fig. 3.1). This is due to the body being constrained to follow the circular path by the force F. The velocity that is to be measured is taken along the tangent to the path of rotation (which in fact shows that the body would move further away from the centre of rotation if it were not constrained) and so may be expressed by:

$$v = 2\pi Rn \qquad (3.4)$$

where n is the number of revolutions per second (i.e. the units of n are s^{-1}). Hence, substituting from 3.4 into 3.3, we have:

$$F = m.(2\pi Rn)^2/R \qquad (3.5)$$

$$i.e. \qquad a = 4\pi^2 n^2 R \qquad (3.6)$$

$$hence \qquad F = 4\pi^2 m n^2 R \qquad (3.7)$$

The units remain newtons, i.e. kg.m/s², of course. The force *restraining* the body, and preventing it from continuing on the tangential straight line path, is termed **centripetal** (*towards* the centre) when the body is properly restrained.

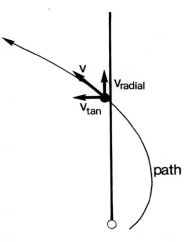

Fig. 3.2 Possible path in space of molten metal travelling from crucible to mould when constrained as if in a tube or channel when it starts to rotate. The radial and tangential components of the instantaneous velocity are shown.

Initially, of course, the molten alloy is not restrained, except by the side of the crucible. Thus it accelerates outwards (Fig. 3.2), tracing a spiral path in space. Under ordinary, simple conditions the velocity outwards (radially) would be calculated from

$$v_r = v_0 + at \qquad (3.8)$$

where t is the time and v_0 is the initial velocity (here it can be taken as zero) and a could be calculated as before. But the casting arm itself is accelerating after release, whether spring or motor driven; that is, n in equation 3.7 is not constant. Taking into account the path that the melt must take to get out of the crucible, the calculation is in reality much more complicated. However, it still means that the molten metal is travelling at a non-negligible velocity when it enters the mould space, even over the short distance from crucible to the inner end of the sprue. In fact, sprue velocities in the region of 1 m/s have been measured for ordinary casting conditions.[2]

●3.2   Centrifugation

An alternative view of the entire centrifugal casting procedure is provided by considering the effective local gravity. We can rewrite equation 3.1 as

$$F + (-ma) = 0 \qquad (3.9)$$

illustrating equilibrium conditions. The second term (-ma) can be interpreted as the **inertia force**, which is acting in a particular direction. In the case of the centrifuge this would be outwards. This force is indistinguishable from the inertia force any body experiences at rest in a gravitational field (acting 'downward'). Hence the concept of 'weight'. In a centrifuge the local gravitational field is not cancelled but has the centrifuge 'gravity' added to it. Because gravity has direction, it is a vector addition that is required: the effective 'down' direction is the direction of the resultant of the two 'gravities' (Fig. 3.3). We can therefore plot the 'down' direction in the centrifuge as a function of rotation rate, referred to the normal, at rest, down direction (Fig. 3.4). The effect of this is very clearly seen in the position taken up by the buckets of a laboratory centrifuge: as the speed increases they swing out to lie nearly horizontally (Fig. 3.5).

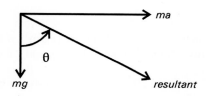

Fig. 3.3 Vector diagram for the effective gravity acting in a centrifuge. *mg* is the weight of the object, *ma* is the inertial force.

In a centrifuge the effective gravity acting on the body may be, indeed is intended to be, very much greater than that of the earth and consequently is normally expressed in terms of multiples of g_0,

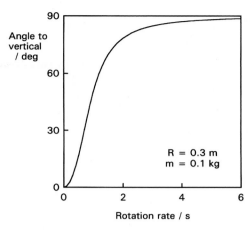

Fig. 3.4 Variation in the direction of 'down' with centrifuge speed, calculated for the mass and radius shown.

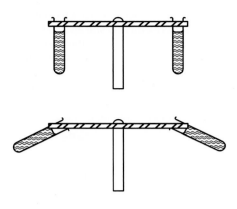

Fig. 3.5 Effect of change in effective gravity on the position of laboratory centrifuge buckets. Top: at rest; bottom: turning.

the (standard) acceleration due to gravity, which has the exact value of 9.80665 ms^{-2} (see box).

Relative Centrifugal Force (RCF)
From equation 3.6, since the usual units employed in the context are cm and rpm (revolutions per minute), the numerical scale factor is:

$$A = \frac{4\pi^2}{100\times 60^2 \times g_0} \approx 11.18\times 10^{-6}$$

so that   RCF = A × R.n^2
For example, 50 × g might be typical for a casting machine. Note: this is not a force, it is an acceleration, m.s^{-2}.

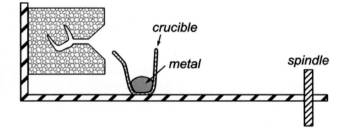

Fig. 3.6 Layout principle of the centrifugal casting machine: rotate the page anticlockwise to see the effect of changing the effective gravitational 'down' direction from the point of view of the molten metal.

In the case of the dental centrifugal casting machine, which has a rigid arrangement of crucible and investment mould, and seen from the point of view of the metal, it is as if the whole machine arm were tilted, eventually just pouring the molten metal into the mould (Fig. 3.6).

●3.3   Casting pressure

But it is not enough merely to deliver the metal to the mould. We need to provide some force to enable the attainment of sufficient detail, assuming that the metal is still fully liquid.

We start by noting that when the mould has filled, the metal is still subject to the centripetal acceleration; in other words, equation 3.9 applies. We need, then, to consider the **hydrostatic pressure**, P_h (in Pa), exerted by the melt; this is simply expressed:

$$P_h = \rho g h \tag{3.10}$$

where ρ is the fluid density (kg/m^3), h the height of the fluid column (measured in the direction of the local gravitational 'up' and 'down'), and g the local effective gravity.

The effect of this is best shown by a numerical example. Assuming that the height of the molten metal is 25 mm, and that it is a gold alloy of density 15 g/cm^3 (= 15000 kg/m^3), we have $P_h = 375 \times g$ (Pa). Then from equation 3.6, for a typical arm length of 250 mm, $a \approx 10\ n^2$ m/s^2. We then have $P_h = 3.75\ n^2$ kPa. Were the alloy to have been a cobalt-chromium type, with $\rho \approx 8$ g/cm^3, we obtain $P_h = 2.0\ n^2$ kPa. The corresponding pressures for a static casting, that is, upright on the bench, are 3.7 and 2.0 kPa respectively. Casting machines may operate at rotational rates of around 7 or 8/s, giving $a \approx 500$ m/s$^2 \approx g$, and P_h(gold) ≈ 200 kPa and P_h(Co-Cr) ≈ 100 kPa. Notice that ordinary atmospheric pressure, 1 bar, is about 101 kPa. The hydrostatic pressure of the molten metal in the mould is consequently much greater than in a static casting, for even modest rotational rates (by about 50 times in these examples, the scale factor for RCF – see box) and the metal may therefore be forced into fine detail and corners. Clearly, the tensile strength of the investment must be high enough for it to be able to contain such pressures without fracture or collapse of the porous structure. Such failures could either create undesired extensions to the casting ('fins'), cause the complete loss of the metal (a dangerous event) or simply extend the shape in an unexpected fashion.

The hydrostatic pressure actually required for a good sharp casting may be estimated from the surface tension of the molten metal. The pressure difference ΔP generating a capillary rise or depression is given by (equation 10§2.4):

$$\Delta P = \frac{2\gamma . \cos\theta}{r} \tag{3.11}$$

We can evaluate the magnitude of this pressure if we adopt a minimum value of 1000 mN/m for the surface tension of the molten metal at the casting temperature (see Table 2.1), making some allowance for the fact that alloying elements and higher temperatures (than the m.p.) will lower the value. We also make the reasonable assumption that the contact angle of metal on investment is $\theta = 180°$, *i.e.* $\cos\theta = -1$ (no wetting). We then have:

$$\Delta P = \frac{-2000}{r} \text{ mN/m}^2 = -2/r \text{ Pa} \tag{3.12}$$

Thus for a mesh for a framework of, say, 0.5 mm diameter wire, $r = 0.25$ mm and $\Delta P = 8$ kPa. This is the hydrostatic pressure required for the metal to penetrate such a mould space against surface tension.

But a 0.5 mm mesh is a relatively coarse feature of a casting: what about fine detail, or sharp corners? The same equation may be used to estimate that for a radius of, say, 20 μm to be obtained on a casting of a truly sharp mould edge, $\Delta P = 100$ kPa (*i.e.* approximately 1 atmosphere) (Fig. 3.7). This radius is now, however, about that of the investment particles themselves and so no further improvement to detail makes sense. Note that it is a minimum value: Co-Cr alloys may well require double the pressure for that degree of detail because their surface tensions are expected to be higher by about that amount (Table 2.1).

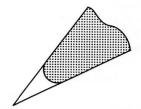

Fig. 3.7 The sharpness of cast edges is limited by surface tension.

●3.4   Roughness

There is a limit to this 'higher is better' calculation of a much more important kind. The other feature of the investment mould surface, in addition to the roughness represented by the refractory and binder particles, is the porosity. That is, there are pores between adjacent particles. Clearly, if the casting pressure is high enough, the metal can be forced into that porosity. Once in, all further pores in the metal's path will be of similar size, and thus there will be little impediment to further entry. This would result in a metal sponge, embedding the investment. There would be no way to rescue such a casting for use. Dissolving the investment with hydrofluoric acid would serve no purpose, and grinding it back until the investment was no longer detectable would be too imprecise. There is no choice but to discard that casting and start again. A knowledge of the size of the pores would allow a calculation of the upper safe limit to the casting pressure to be used, but if it happens, the only remedy is to reduce that pressure, whether by changing sprue length, casting arm length or, more generally, maximum rotation rate (assuming that we are referring to a centrifugal machine). This then implies that the investment particle size, porosity, and therefore the roughness of the investment, must be small enough to suit this pressure.

●3.5   Air pressure

In the alternative approach to metal casting of using air-pressure to force the melt into the mould, the required pressure can be calculated from the same equation 3.11. We obtain a figure of 1 ~ 2 bar, depending on the alloy and the detail required. This may be utilized simply by applying that air-pressure over the melt, which must of course be covering the sprue hole. It still requires the air in the mould space itself to diffuse through the pores of the investment. This takes time and a sufficient, albeit short, duration of application of pressure (as must be used in a centrifugal machine also) allows for this to happen. The fine detail of the mould space can therefore be reproduced before cooling below the solidus has occurred.

●3.6   Vacuum

The obvious way to eliminate this air flow resistance as a problem is to eliminate the air from the mould space. By evacuating a container holding the investment mould, and then applying air-pressure over the melt as before, an improvement in speed may be obtained, and thus improve the reproduction of detail. Oxidation of the metal is also limited this way, as are 'back-pressure' defects (see Fig. 4.4), but difficulties in transferring the investment mould after burn out while maintaining a proper mould temperature may then arise.

●3.7   Reheating

There are risks associated with variations in the temperature of the investment which have a bearing on the practical problems of a casting procedure. Specifically, reheating after cooling is quite unacceptable. As might be imagined, the relationships of the particles of binder and refractory are threatened by the chemical breakdown of the binder (*e.g.* gypsum) and the structural transition of, say, cristobalite (17§2). Although overall the expansion on heating would be uniform (isotropic), on a local scale it is highly anisotropic and even varies in sign (*i.e.* some shrinkage occurs). In addition, there will be the processes of solid state sintering, and even perhaps some partial melting, that will permit some consolidation of the structure, dependent on the total time-temperature history. In any case, on cooling again, the original relationships of one particle with another cannot be regained and disruption occurs, such that on reheating less expansion would be observed (Fig. 3.8). In fact, considerable shrinkage is observed for the completely cooled investment

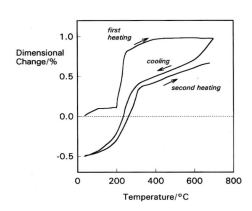

Fig. 3.8 The dimensional change curve of an investment is not retraced on reheating. (NB: this is not an example of hysteresis.)

due to this kind of effect. Some cooling must occur on taking the investment mould out of the burn-out furnace. Hence, return for reheating because of a problem with the casting machine or the alloy to be cast, is inappropriate.

●3.8   Metal velocity

The velocity with which the metal enters the mould is of some concern since the impact with the investment may be great. Since the impact energy is given by:

$$E = \tfrac{1}{2}mv^2 \tag{3.13}$$

it is clear that the investment must also have substantial impact 'strength' (1§10.3). As was mentioned above, in a centrifugal casting machine the metal may enter the sprue at about 1 m/s. This velocity is high enough to guarantee that the flow is turbulent (15§7.1). In addition, regular-section castings are very rarely required in dentistry and efforts must be made in the sprue design to minimize the extra turbulence in the high speed metal stream as it enters the mould due to sharp changes in direction. Great turbulence may generate porosity in the casting, particularly if the temperature is very close to that of the liquidus (when freezing may occur too soon), as well as provide greater stresses on the investment.

§4. Dimensional Considerations and Defects

In addition to the dimensional changes which were to be compensated by the investment (17§1), there are five further potential sources of variation which must be considered. They are:
• volumetric shrinkage of the alloy above the liquidus,
• change-of-state volume change on freezing,
• investment cooling before casting,
• casting ring expansion matching,
• cooling stresses.

The increased disorder associated with a molten metal over the solid (as with most substances) results in a discontinuous decrease in volume on freezing (Fig. 3§4.7). Only a few metals show an expansion on solidification, due to the peculiarities of their crystal structure, as indeed water shows such an expansion.

Unfortunately, none of these alloys can be applied to dentistry, mostly because of lack of strength. For dental alloys, then, there is a substantial decrease in volume during freezing as well as prior to that temperature (Fig. 4.1). The deficit must be made up eventually, by the time the metal is solid, if the casting is to be successful. This is achieved by casting rather more metal than is actually required for the intended cast object itself. The shrinkage above the liquidus should therefore be entirely of no consequence.

●4.1 Reservoir

But mere quantity is not enough. The extra metal must freeze last of all. This extra metal may simply be that in the sprue and button, and a large excess is ordinarily employed both for this reason and to obtain sufficient hydrostatic head. (After all, the requirements for detail reproduction must apply at the top of the casting as well as the bottom.) As molten metals (usually) wet perfectly their solid counterparts at the freezing point, then the cohesion of the two is sufficient to ensure that more molten metal will be drawn in to compensate for the freezing shrinkage if, that is, molten alloy is available.

If the regions most remote from the sprue freeze first (Fig. 4.2, left), then successive frozen increments draw on the still molten metal, ultimately drawing on the sprue and the button itself. However, if any part of this path is occluded by solid metal before all more remote parts have frozen, then voids will appear (Fig. 4.2, right). This is called **solidification cracking** or **supersolidus cracking**, indicating that it is the volume changes occurring between the liquidus and solidus temperatures that are the problem. It can therefore be seen that alloys with long freezing ranges are at particular risk of this kind of event.

Avoidance of this defect in a casting depends on a complex combination of mould and alloy temperature, sprue design and quantity of metal cast, so that the sprue and button can be understood as providing not just extra metal but a reserve of heat as well: the greater **thermal mass** in that region maintains the essential temperature gradient. To this end, the sprue is attached to the pattern at the thickest portion. However, there may also be included a spherical body near the sprue's midpoint called (appropriately enough) a **reservoir**, to enhance the effect (Fig. 4.3). This works because being surrounded by investment it must cool more slowly than the well-exposed button. But even the absence of a visible surface defect may not mean complete absence from the casting. Radiographs of castings frequently show translucencies due to such voids and, while perhaps merely a nuisance when test pieces for strength determinations are made, they are a serious source of premature fractures in service, often by being at critical points such as in the root of a clasp or other union of two sections (Fig. 4.4). If it occurs at the root of the sprue, as in Figs 4.2, 4.4, when this is cut off a large hole will result. In any case, it is important to maintain the casting pressure until solidification is complete, as well as casting enough metal that the reservoir be filled (*i.e.,* there is a button present).

The heating and 'hygroscopic' expansion of the investment mould will have been carefully determined to match the thermal contraction of the solid alloy from the solidus to mouth temperature to allow an accurate reproduction of the

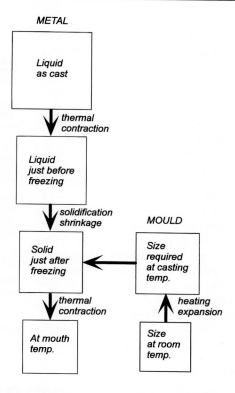

Fig. 4.1 Thermal volumetric considerations for the alloy on casting.

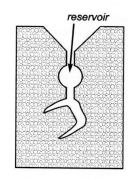

Fig. 4.2 The addition of a reservoir provides a source of heat and metal.

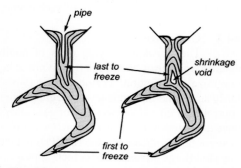

Fig. 4.3 Patterns of solidification in a freezing casting. Left: proper sequence, right: premature freezing in the sprue leads to a shrinkage void.

pattern. It should be obvious that this cannot take into account the reduction in volume accompanying the freezing of the metal.

It should be recognized that the design of the entire process includes the transfer of the investment mould from the burn-out furnace to the casting machine. That is, it is inevitable that some cooling occurs before the metal is cast. Accordingly, the mould space will shrink (Fig. 3.8). The effect is minimized by holding off transfer until the last possible moment, aided by the low thermal conductivity of the investment material. Nevertheless, since this is a normal event, the overall accuracy of the system has already effectively made allowance for its effect. It follows that a consistent procedure, times and temperatures, once set up and working is more important than a theoretical ideal.

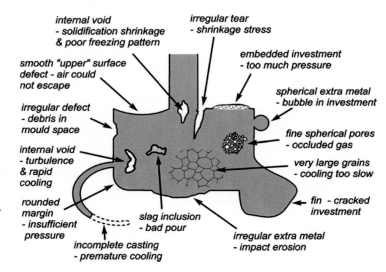

Fig. 4.4 The casting from hell: several types of casting defect which may be encountered due to various technique errors.

●**4.2   Thermal stress**
The very large dimensional changes associated with the refractory inversion (17§2), coupled with the unavoidably low thermal conductivity of such ceramic materials, is a source of risk in the initial heating. As the outside of the investment heats up first because it is then expanding faster than the interior position, appreciable tensile and shear stresses between layers will result. In other words, the thermal gradient between outside and inside results in a similar stress gradient (*cf.* Fig. 20§7.5). If the heating rate is too great then these stresses will exceed the strength of the material, and the mould will crack. If the mould is not completely destroyed, molten metal which entered the cracks will appear as fins on the casting, making it difficult to clean up for use, especially when fitting surfaces are involved (Fig. 4.5). However, the problem is in fact rather more serious: since the crack has a finite width, the dimension of the casting in that direction near the site of the fin will be increased by the same amount. This may be sufficient on its own to render the casting completely useless. Fragments of investment may even become detached in the process of such cracking, perhaps aided by or because of metal impact and turbulence, and these will cause gross defects in the casting. The heating rate must always be carefully controlled to minimize the risk of such events.

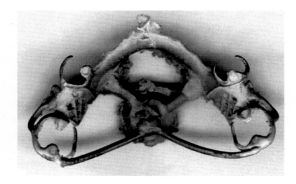

Fig. 4.5 A framework with bad fins due to investment cracking. Note also the infill of the mesh segments.

●**4.3   Grain size**
The effect of varying rates of cooling from the melt has already been described, particularly in macroscopic terms and in relation to phase equilibria (12§1), that is, the goal is primarily to avoid (or at least minimize) segregation. Fortunately, in most dental casting alloys, the gap between the liquidus and solidus is fairly small and the tendency for segregation reduced. The most important aspect in the present context is the control of grain size (especially for gold alloys) by cooling quickly enough. Single grains of Co-Cr alloys may have diameters of more than 1 mm, *i.e.* about the width or thickness of a clasp. This is disadvantageous in terms of strength, as indicated by the Hall-Petch relationship (11§4.1). Even so, the rate of cooling must not be so great that **shrinkage porosity** and defects appear. Microscopic distributed shrinkage porosity has been noted in gold alloys which have been cooled too quickly.

●4.4   Included gas

One other source of defects is noteworthy. Many metals and alloys dissolve gases from the surroundings when molten, and show a marked drop in this solubility when solid. Silver is the most well-known example: it may dissolve so much oxygen from the air when molten that it 'spits' vigorously on freezing, and this behaviour is carried over to some extent to its alloys (19§1.12). Obviously, not all of this gas could emerge from a dental casting, and this will appear as a finely disseminated porosity throughout large volumes of the solid. Such 'included gas' porosity might not be detectable until the casting is polished, when it would appear as a series of pits in the surface, and these will be sites of corrosion, leading to mechanical failure. The problem is usually one of over-heating, allowing time for the oxygen to dissolve, but it could also be overcome by melting under an inert atmosphere, such as of argon. The spitting is also a hazard: eye-protection should in any case be worn.

●4.5   Other casting defects

Other defects in the casting due to faulty investment of the pattern should be self-evident. For example, air-bubbles and contamination by extraneous materials are matters for careful technique and need not be elaborated upon. Such problems and others are summarized in Fig. 4.4. Indeed, it should not be overlooked that faults in the original impression, model and pattern, including various kinds of distortion, will all be carried through and these too must be recognized and dealt with appropriately. These have been discussed elsewhere. To this list may be added the error of 'short casting', which may arise from simply having insufficient metal in the crucible, or failing to melt it completely. Of course, the investment may fracture if it is subjected to excessive casting pressure or is weak, such as by having a thin wall or base (*i.e.*, from poor positioning of the pattern in the mould).

Most dental casting is done in air, and so any oxidizable metals will do so continuously if the occasion allows, despite using a reducing crucible or reducing investment. This will be true whilst the metal is in flight from crucible to mould space, and tiny particles of oxide will be formed. These can be enfolded in turbulent metal and, if the cooling is rapid enough to prevent ejection, will result in small oxide inclusions. These must have several effects: ductility will be slightly reduced, strength may be reduced because the voids containing the particles will act as crack nuclei (Griffith flaws), and corrosion may be enhanced if subsequent grinding and polishing reveals them at the surface (they may become active corrosion pits). Nevertheless, it is likely that the effect on the final product is generally not so great as to warrant an inert gas environment (titanium alloys, for example, are rather different [28§1]).

●4.6   Expansion of holes

The discussion of investment dimensional changes will not make much sense unless it is appreciated that a hole in a body expands and contracts at the same rate as the surrounding material. In other words:

the thermal expansion coefficient of a hole is the same as that of the material in which it is found.

There are a number of images that may be invoked to explain this. Consider, then, a thin circular ring which is cut, straightened out, then heated (Fig. 4.6, top). There will be a change in length from L to L+ΔL. If the ends are then rejoined, the length of the circumference of the circle so formed must be the same as this new length, *i.e.*

$$L + \Delta L = 2\pi(R + \Delta R) \qquad (4.1)$$

from which it is obvious that the diameter of the ring changes proportionately to, and with the same *sign* as, the linear expansion of the material. It will not matter whether the ring consists of a solid substance, or a string of particles whose size or spacing (or both) may change: the relationship is the same (Fig. 4.6, centre). Another model: a rectangular frame consists of beams pinned only at the corners (Fig. 4.6, bottom). If this is heated, the change in length of the side is the same as the change in length of the hole.

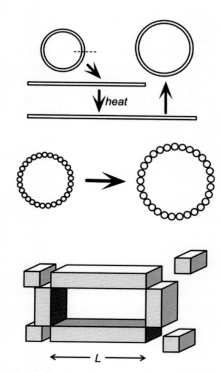

Fig. 4.6 Three demonstrations of the fact that, if unconstrained, holes expand thermally at the same rate as the body in which they are found.

The corner pieces, if now added, expand at the same rate as the adjacent material, so there is no shear at their junctions with the beams. Hence, holes have the same coefficient of thermal expansion as the material in which they are found. The only exception to this result arises if the expansion of the material is somehow constrained (17§6).

Two non-dental uses of this effect illustrate the point clearly. Wooden cartwheels have an iron hoop "tyre" fitted by heating the undersized hoop to fit over the assembled wheel. On cooling it compresses all the parts to hold it together as well as provide a hardwearing coating. Aluminium alloy car engines are not wear-resistant enough to withstand the effect of piston rings and so the cylinders are fitted with steel lining sleeves. These are inserted after chilling (in, say, liquid nitrogen). On rewarming they fit snugly in the prepared bore. In these two cases, circumferential **hoop stresses** are generated for good reason, but in the context of casting investments they are undesirable if excessive because they will crack the mould (§4.2, 20§7, 25§5.4).

●4.7   Expansion matching

This consideration is also relevant to the use of casting rings and ring liners (17§6.2). The thermal expansion coefficient of steels lies in the range $10 \sim 20\ MK^{-1}$, but mostly at the lower end of that range. It would require heating to 1000 °C to obtain 1% overall expansion in that case, well above the normal burn out temperatures of gypsum-bonded ($500 \sim 700$ °C) and phosphate-bonded investments ($800 \sim 900$ °C). Gold alloy cooling shrinkage from the solidus is of the order of $1.25 \sim 2\%$, while for Co-Cr alloys the figure is larger at $2 \sim 2.4\%$. Obviously, it might be possible to arrange for the correct total expansion, perhaps by choosing an appropriate grade of steel, but this would be to miss the critical point that the expansion of the investments themselves is not smoothly continuous, due to the form of the displacive transition expansion of silica (Fig. 17§2.1). In particular, cristobalite expansion would outpace the ring expansion by a factor of five or six at about 200 °C. This could only lead to mould damage. A ring liner or other technique (such as 'ringless' casting) is essential to avoid such damage.

There is a related issue in connection with the cooling shrinkage of the metal casting with respect to the investment. Unless the two dimensional changes are matched (which is essentially impossible), stresses will be generated in both metal and investment. Metal at high temperature (T/T_m) is weak, and may deform, which could spoil the fit of crowns and bridges and partial dentures alike, including both the overall span and clasps. However, if the tensile deformation caused by shrinkage cannot be compensated by the ductility of the alloy, then the metal will tear in what is known as **hot cracking** (Fig. 4.7), which is distinct from supersolidus cracking (§4.1). This is promoted by a high metal thermal expansion coefficient, but also by the casting being constrained by the investment, as here by the high palatal vault.

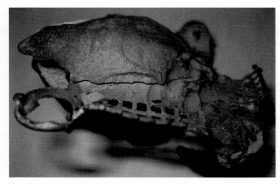

Fig. 4.7 A cobalt-chromium casting showing a hot crack in the palate.

Such stresses may be offset by the weakening of the investment on being heated to the casting temperature, and then beyond by the molten metal. Inlays may be exempt from this effect, there being neither thin sections nor long lever distances, but large frameworks and the like with an appreciable arch will be more affected where a greater absolute error will be more noticeable. Even a crown can show such cracks. Likewise, uneven cooling of the investment after casting may generate such stresses. The necessarily low thermal conductivity means that there will be a considerable temperature gradient in the investment, while the metal stays more uniform because of its much higher conductivity. Dropping cast investment moulds into water would be expected to exaggerate any such effects. This must be viewed in the context of §4.3, that is, with regard to avoiding grain growth.

§5. Flame

While many casting machines now employ electrical heating methods, there are still occasions when the use of a flame may be necessary, in soldering for example. Indeed, bunsen burners are commonly employed for heating the wax knife and for other laboratory and clinical procedures. For these reasons it is of importance to consider the structure of the flame and its influence on the material being heated. Even so commonplace a laboratory or clinical 'device' as a flame needs to be understood to be used properly.

●5.1   Types of flame

Typically, a flame in these circumstances represents the continuous oxidation of a fuel, usually hydrocarbon, the oxidant being the oxygen of the air. Such oxidation leads to the release of large amounts of heat, and so the flame itself and its (invisible!) effluent gases become very hot. A flame is a controlled explosion: the rate of flame propagation (towards the source of fuel) is equal to the flow velocity of gases from the burner. We may distinguish two types of flame. The first is where the gaseous fuel is ignited at a simple orifice or jet where the oxygen is available only through the agency of **diffusion** or turbulence (Fig. 5.1a). The second is where air is **premixed** with the fuel by entrainment (as in a bunsen burner) (Fig. 5.1b) or where air or oxygen is supplied from a metering valve to the gas stream and mixed by turbulence (a gas 'torch'). In both premixed cases diffusion is also very important for supplying more oxygen from the surrounding air. Burners which rely on wicks, such as the methylated spirits burner, paraffin (kerosene) lamp, candle, or even a match, create diffusion-controlled flames unless a blow-pipe is used to inject air more rapidly, when a small premixed type of flame is created.

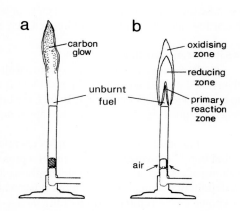

Fig. 5.1 (a) Diffusion and (b) Premixed flames on a bunsen burner.

●5.2   Diffusion flames

It should be apparent that the first type of flame is much less efficient than the second. The same rate of fuel delivery will produce a much larger 'soft' flame, which is therefore cooler because the same amount of heat is generated over a larger volume (assuming complete combustion). The dependence on diffusion and turbulence to mix in air from outside means that there is always an excess of fuel or partial combustion products, *i.e.* a lack of oxygen over a large volume of the flame. Such flames are therefore always **reducing**, except at the very edge. Under these conditions the chemistry of the fuel becomes important. For a hydrocarbon the first stage in the combustion process is the removal of hydrogen, which produces water, then successively various short-lived and partially and completely dehydrogenated paraffinic and aromatic species, the limit being carbon.

A cold surface introduced into such a flame will collect a black and tarry deposit of, mostly, carbon ("lamp black") with various other complicated aromatic molecules (Fig. 5.2) which give the greasy feel to such material (and which compounds are toxic or carcinogenic!).

The oxidation of hydrogen leads to the emission of blue light, but that of fine particles of carbon gives the characteristic white or yellowish glow of, for example, 'soft' bunsen burner flames. The initial oxidation product of carbon is of course carbon monoxide, and this in turn burns with a blue light to carbon dioxide. Outside of the luminous zone of a soft flame can be seen a shell of blue where this same reaction is predominant. The implication is that such a flame is inappropriate if contamination is to be avoided: the presence of carbon in the form of

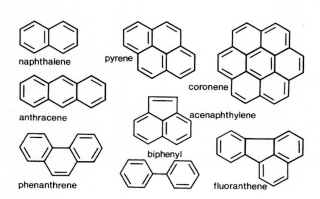

Fig. 5.2 Some of the species identified in lamp black obtained from hydrocarbon flames. There are many others. Successive dehydrogenation and ring fusion creates more and more graphite-like structures.

fine particles means that this element can be incorporated into metal that is being heated or simply make a wax knife dirty. In a bunsen burner the air supply should be adjusted until the yellow flame just disappears, but of course this cannot be done with a spirit burner or candle (naturally, a candle should have a luminous zone if it is to be an effective illuminant.

With some fuels, such as acetylene and paraffin, it is easy to have a fuel supply in excess of the rate that oxygen can diffuse into a diffusion flame. There will therefore be carbon-rich material left over by the time the flame temperature has fallen too low for oxidation to continue. In other words, we get smoke. Such smoke is more than just a dirty nuisance, it is a health hazard. Not only are the substances found in it likely to be toxic and carcinogenic in their own chemical senses, but there will also be a variety of free-radical species present, because the combustion process has more or less been interrupted in mid-stream. Free radicals are known to be hazardous and are implicated in lung cancer, as is well-known now for tobacco smoke.

●5.3   Premixed flames
In premixed flames there will be a well-defined **flame front**, usually conical in shape. Nearer the burner than this the gases are quite cold, and there is clearly no point whatsoever in allowing this section of the flame to contact the part being heated. In fact, the hottest part is just above the tip of the cone, just outside the primary reaction zone (Fig. 5.1b), where the rate of energy release is greatest.

Flames are extremely complex chemically, and clearly will depend in this sense very much on the fuel being burnt. However, two features of a premixed flame can be identified which are of direct relevance to the heating of casting alloys. If the air supply is adjusted properly, the hottest part of the flame (just above the cone) is definitely deficient in oxygen but it has a high concentration of oxidizable radicals and other species. This is known as the **reducing zone** of the flame. Here, metal oxides will form only slowly or actually be reduced to metal if present, depending on the metal. Casting alloys would thus be protected from composition changes. Additionally, oxide dross would not be present to accompany the metal into the mould, thus avoiding potential flaws that would cause rejection or premature failure of the casting.

The outer region of all flames will be oxygen rich, and especially near the tip of a premixed flame, as diffusion and turbulence puts oxygen in. This then is the **oxidizing zone**. Metals would oxidize rapidly if exposed to the gases here. If the flame is held correctly over the casting metal button, the oxidizing zone can be kept from contact. In soldering, however, it is inevitable that some part of the workpiece will be exposed to an oxidizing region. This is a further reason for minimizing time and temperature during such an operation.

●5.4   Hazards
The obvious hazards of working with flame and hot metal perhaps do not need much elaboration, but it is worth stressing that appropriate eye, face and other protection should always be worn, and guards and shields should always be in place. That investment moulds can crack, releasing molten metal, albeit rarely, is reason enough for caution. However, there is a further more insidious hazard: **thermal cataract**. Whilst this has been documented for prolonged and repeated exposure to furnace radiation, and it is not entirely clear what the mechanism might be,[3] it is an established industrial disease. It would be wise to minimize exposure as it is cumulative damage that is involved, not a threshold. Dark or infra-red-absorbing eye-protection is advisable.

References

[1]   Nakai A, Kakuta K, Goto S, Kato K, Yara A & Ogura H. Development of casting investment preventing blackening of noble metal alloys Part 2. Application of developed investment for Type 4 gold alloy. Dent Mater J 22(3): 321 - 327, 2003.

[2]   Luk HWK & Darvell BW. Casting system effectiveness – measurement and theory. Dent Mater 8: 89 - 99, 1992.

[3]   Vos JJ & van Norren D. Thermal cataract, from furnaces to lasers. Clin Exp Optom 87 (6): 372 - 376, 2004.

Chapter 19   Casting Alloys

The range of alloys that may be used in the oral environment is limited primarily by the need to avoid corrosion. This leads to the use of two major groups: corrosion-resistant or precious metal alloys, and non-precious but passive alloys. The first is exemplified by gold-based products, the second by cobalt-chromium and similar alloys. Each has advantages and disadvantages as well as special handling considerations that must be taken into account in order to obtain the desired outcome.

Corrosion resistance *also (usually) requires single (metallic) phase alloys, yet sufficient hardening and strengthening must be obtained for the cast devices to function without permanent deformation under service stresses. Detailed consideration is therefore given to all available **mechanisms for hardening**, and their appropriateness in the two groups of alloys, in order that their design and handling be understood. In the case of some gold and related alloys the hardening can be reversible, depending on the crystal structure, and controlled by the thermal treatment.*

Composition *is critical in determining mechanical properties, and especially in the cobalt-chromium types. This arises because the composition may lie dangerously close to that at which a brittle structure is obtained. Thermal history is also critical in that carbide grain growth is affected; this can either strengthen or weaken the metal.*

The selection of alloys for cast devices must be based on a knowledge of the properties of those alloys and their sensitivities to procedural and handling variables. Failure to recognize their individual limitations will result in treatment failure, for it is still a matter of compromise in balancing good and bad aspects of behaviour.

Materials Science for Dentistry
https://doi.org/10.1016/B978-0-08-101035-8.50019-5

Whilst the casting process has a long history, and the skills needed for successful casting can be learned by trial and error, it is essentially the properties of the alloys that control the outcome. The market these days is characterized by an enormous and potentially bewildering array of very similar products, even from a single manufacturer. To begin to comprehend the reasons for design choices it is necessary to consider some of the alloy systems in greater detail. However, it is also necessary to point out that in many cases the alloys themselves have often been designed by trial and error, and a comprehensive explanation of their behaviour in suitably scientific terms is simply not possible because the essential data have not been gathered. Nevertheless, it is not too difficult to outline the generic properties of the major classes.

When choices of casting alloy are to be made it is primarily the application that provides the guide in that strength, stiffness, density (as it affects the mass of a suitably strong or stiff framework, for example), ductility, suitability for porcelain bonding, and so on must first be considered. A second level of importance is that of the experience of the dental laboratory in handling that alloy. As has already been made plain, the outcome of the entire casting process depends on the combination of investment and alloy, with all the many handling and processing variables that this entails. But, despite the theoretical matching of these factors, it still requires a demonstration that the entire system works to give an acceptable outcome. Part of the system is the skill of the technician in running a process consistently, and being able to calibrate it, as it were, to tune critical steps to ensure success.

§1. Gold Alloys

●1.1 Pure gold

Pure gold is the outstanding restorative material. Because of its peculiar combination of properties: tarnish resistance, ductility with work hardening, and the ability to be cold-welded by pressure alone, it is usable as a direct filling material in the form of foil or powder (28§4) (although the skill, effort and expense of this process has led to it being more or less abandoned now). Aided by the deep-rooted and ancient mystical associations for the metal, its colour does not present problems for many people, despite the great contrast with tooth material. As has been implied elsewhere (Chap. 24), the patient's perceptions are at least as important as more fundamental criteria for material selection.

However, in its pure state, gold is not very strong for applications other than inlays, *i.e.* for crowns, bridges and removable partial dentures. The strength advantages of alloys are required, but these require casting rather than direct fabrication techniques. This, of course, precludes work hardening as a means of increasing rigidity because dimensional stability requirements are severe if fit is to be maintained. Recourse to hardening and strengthening by suitable alloying elements must be made.

●1.2 Solid solution hardening

The disorder introduced into crystal structures by elements in solid solution has previously been discussed in terms of the ease or otherwise of slip along certain planes of atoms (11§5). The effect is only moderate when there is a close match of atomic radii, as is exemplified by the Ag-Au system (radius ratio 144/143 = ~1.01) (Fig. 1.1). But even so, at an atomic ratio of ~1:1, when the disorder might be expected to be at its greatest, there is an appreciable peak in the strength. By way of contrast, with Al-Cu α-solid solution alloys the effect is much more

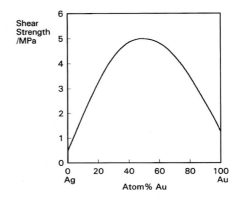

Fig. 1.1 Solid solution hardening in Ag-Au. The peak occurs near an Ag:Au ratio of 1:1.

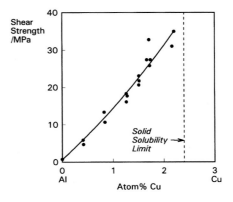

Fig. 1.2 Solid solution hardening in Al-Cu.

pronounced because the radius ratio (143/128 = 1.12) is very close to the general practical limit for a continuous solid solution, and the distortion is much greater (Fig. 1.2).[1] Indeed, the solubility limit for Cu in Al is 2.5 at% at 548 °C. The ratio for Au-Cu (144/128 = 1.13) is also close to that limit, and while a continuous solid solution is formed above ~400 °C, *i.e.* at all compositions, distinct new phases in fact form below that temperature (Fig. 1.3).[2] Of particular interest are those corresponding to stoichiometries around AuCu₃ and AuCu, but it is the latter which is the phase of principal interest in dentistry. A peak in solid-solution hardening is again observed at this atomic ratio (*i.e.* 1:1).

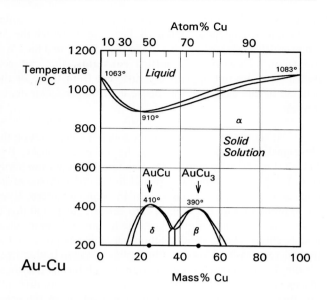

Fig. 1.3 The Au-Cu system equilibrium diagram and some associated crystal structures. The narrow gap between liquidus and solidus is important for avoiding appreciable coring in these and similar alloys when cast as there will be little time for segregation.

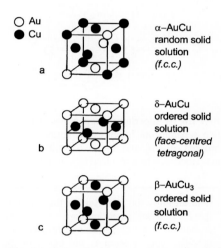

Fig. 1.4 Unit cells for the various phases of the Au-Cu system. Notice that in the δ-phase (the one of interest in dentistry) the unit cell height (c-axis dimension) is shorter than the a and b dimensions.

The Au-Cu constitutional diagram has a feature known as a **solution minimum** at about 20 mass% Cu. Here the reaction on heating is

α → Liq

that is, the solidus and liquidus coincide at this point. It is therefore a type of **point reaction**. Both liquidus and solidus have horizontal tangents at this point. This reaction should be distinguished from that of the melting of a pure component (see box, 12§1.2).

The crystal structure for the whole Au-Cu system is essentially f.c.c. (11§3) and, in common with many other solid solutions, the α-phase is one of random substitution (Figs 1.4a, 1.5a). That is, at any point in the lattice of an α-phase alloy of composition AuCu, the probability of finding either kind of atom is exactly 0.5.

●1.3   Ordered phases

In δ-phase AuCu the structure is **ordered**. The Au and Cu atoms alternate regularly, which is equivalent to alternating layers of Au and Cu atoms (Figs 1.4b, 1.5b), in a pattern that repeats over long distances. In contrast, a random substitutional solid solution does not have long-range order for the occupancy (element identity) of each lattice site, although the probability of finding a particular kind of atom is fixed at the overall atomic proportion for the phase. Such a structure arises because the layers fit together just a little more compactly, increasing the regularity of the structure, thus lowering its energy. This rearrangement results in a slight distortion from a perfect cubic lattice to a **tetragonal** one, in which one of the three unit cell dimensions is smaller than the other two (the vertical direction in Fig. 1.4b), the ratio being about 0.935. This structure can be seen to consist in effect of two interlaced lattices, in this case of identical pattern: one of Cu only, the other of Au only (Fig. 1.6). This type of structure is known as a **superlattice** and is characteristic of ordered solid

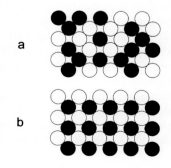

Fig. 1.5 Example planes in (a) random and (b) ordered 50:50 solid solution crystal structures.

solutions.[3] Alloys of the composition $AuCu_3$ produce a similar ordered, low-temperature phase (β, Fig. 1.4c) in which the Au atoms are placed at the corners of the f.c.c. lattice and the Cu at the faces. However, the **lattice constants**, the unit cell dimensions, are not appreciably different from those of the α-phase in this case, and this phase is of no particular dental interest.

It should be noted that in the random solid solution the close-packed planes (normal to the cube diagonals, Fig. 11§3.23) necessarily contain the two kinds of atom randomly (Fig. 1.4a), whereas the same planes in the ordered phases are populated with the two kinds of atom in a regular pattern. Even so, in terms of slip, these planes are still irregular, and this is made worse when slip takes them out of their ideal juxtaposition. Thus, solid-solution hardening is preserved in the ordered phases. However, the δ-phase is taken further from ideality by the **tetragonal distortion**, which means that these planes are not quite close-packed, making slip more difficult. For this reason alone δ is harder than α-phase at the same composition, an effect that does not operate for β-phase.

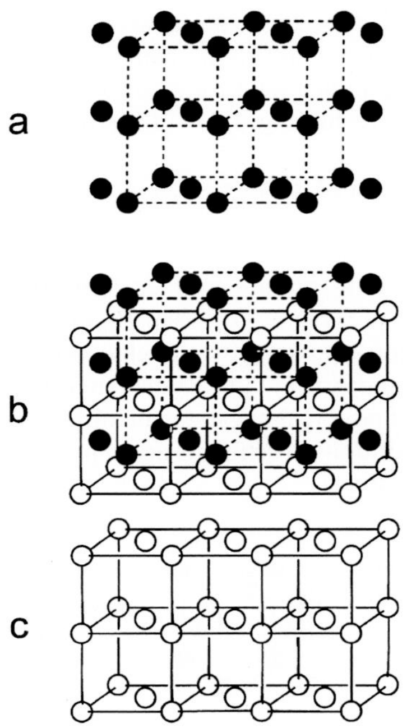

The significance of the ordered solid solution δ-phase, and its importance to dentistry, lies in the fact that it produces a structure markedly harder than just α-phase. While this makes such an alloy much more suitable for use in the dental context, it also allows a special kind of heat-treatment to be used to develop the degree of hardness required for any particular application or, conversely, to permit softening for adjustment which is then followed by rehardening. This process is enhanced by a further source of hardening.

●1.4   Coherency strain hardening

If an alloy with the composition AuCu is cast, rapid cooling or **quenching** to around room temperature will preserve the high-temperature, random, solid solution structure of the α-phase. There will be little segregation (12§1) because the liquidus and solidus are very close together. But if the temperature is then raised again and held below that of the upper limit of the boundary of the δ-phase field (at, say, about 400°C), equilibration will now occur by diffusion, and the superlattice structure will tend to form. The higher the temperature the greater the diffusion rate, because diffusion is a process requiring an activation energy. But because there is a change in unit cell dimensions, notably a reduction in the direction of the c-axis, with concomitant slight increases in the other two directions, the transformation produces a strain in the lattice.

The initiation of the α to δ phase change will take place at a large number of separate locations within any grain randomly and independently. This will necessarily be with random orientation of the tetragonal c-axis along any of the three axes of the original cubic structure. These regions of differing c-axis orientation within a given grain are called **domains**. Taken over a large enough region the net strain in each grain will average out to zero, and there will be very little volume change overall. But the variation from one orientation to another from place to place within the grain superimposes on those primary grains a microscopic **lamellar** or **granular** structure, and it is this that makes the heat-treated material hard. There are now effectively very many more grain boundaries, *i.e.* between domains, so dislocations travel much shorter distances before their movement is inhibited. There is less slip. The formation of β-phase does not cause a similar hardening because there is no associated axial dimensional change to induce strain. The δ-phase hardening process is entirely reversible: reheating above ~400 °C again produces the α-phase, which may again be preserved by quenching.

From another point of view, we can see that the **periodicity** of the structure in the δ-phase may also vary. The crystal lattice does not repeat indefinitely in any direction, but every so often skips a step, as it were, to

Fig. 1.6 A superlattice appears as if there are two separate but intermeshed lattices. This is ordered AuCu, and the two lattices may be described as " base-centred tetragonal".

If in an alloy the positions of the different species of atom are not random – that is, the probability that of a site being occupied by a specific type of atom is not equal to the atomic fraction of that kind of atom in the alloy – then it is said to be ordered, although that ordering may not be perfect. In addition, such ordering can be short- or long-range.

create a layering mismatch for a given *c*-axis orientation. The boundaries between areas of opposite **phase** (in the wave periodic sense, not the structural-constitutional sense) are strained, since the spacing between layers and along rows differs slightly across a domain boundary. This then is known as **coherency strain** across an **anti-phase boundary**. Slip through such boundaries is therefore more difficult, that is, there is **coherency strain hardening**. Fig. 1.7 shows what a random section through the junction between four such domains might look like. Since the spacing of the atoms in each kind of domain is different from that in its neighbour, the mismatch generates the strain.

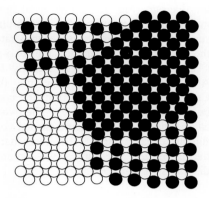

Fig. 1.7 An atom map showing how domains result in coherency strain in δ-AuCu ('atoms' roughly to scale).

Although δ-phase is expected to be a little harder because of the tetragonal distortion, on top of what would be expected from the substitutional effects, this on its own is not very great. Most of the hardening is due to the coherency strain; the superlattice by itself does not confer the observed change, which is largely due to the domains forming with random orientations.

Although there is strain energy associated with the domain boundaries, because the domains are coherent, the excess 'surface energy' of the interface is relatively low, which means that there is little driving force for domain growth. The structure tends therefore to be stable and not coarsen appreciably on continued annealing. Indeed, the domain structure can be compared to a foam: planar boundaries between two domains, three domains meeting along a line, four at a point, where the domains are distinguished by both orientation and phasing (*cf.* Fig. 1.7).

Current gold alloys for dentistry may consist of 6 or more major constituents: Au, Cu, Ag, Pt, Pd, Zn; several other minor additions may also be made. Clearly, the multi-dimensional phase diagram for such a system is going to be extremely complicated, and full details are not at present known for any such system. (For example, there are further aspects to the ordered structure in AuCu[4], although it is not clear if they apply in dental alloys in the presence of such other elements.) Detailed explanation of the purpose and effects of any single alloying element in such contexts cannot be given, even though sometimes rather vague statements may be made. However, certain aspects and broad principles can be summarized, and these are applicable no matter which alloy is under consideration.

●1.5   Precipitation hardening

When a second phase is allowed to separate from (usually) a high-temperature solid solution (see Fig. 12§3.9), those new precipitated crystals form randomly throughout the original grains of the alloy. They therefore get in the way of dislocation movement, limiting slip. Such alloys are described as being **precipitation hardened**. The example of the Ag-Cu system has already been discussed (Fig. 12§3.1). There, for example, the separation of the β-phase (solid solution of Ag in Cu) from a high temperature, Cu-rich α-phase (solid solution of Cu in Ag) at a temperature below the solvus results in just such a very fine-grained precipitate. Conversely, a heat treatment above the solvus will redissolve the precipitate to form a solid solution again. Reactions similar to these will play a part in the overall hardenability of such alloys. Time and temperature are the two important aspects of any heat-treatment: to determine the extent of any change, and the direction and rate of that change.

A continuous solid solution characterizes Ag-Au alloys (Fig. 12§1.1) as it does Au-Cu at high temperature, but in the ternary system Ag-Au-Cu below ~400 °C this is restricted to the very edges of

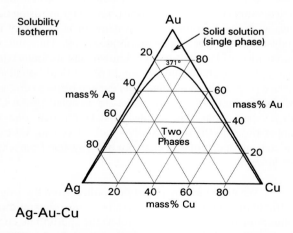

Fig. 1.8 The solid solubility limit in Ag-Au-Cu alloys at 371 °C. Compare Fig. 1.3, and Figs 12§1.1, 12§3.1. (The detail of the ordered phases is omitted).

the diagram (Fig. 1.8). (We shall ignore for the moment the formation of β and δ-phase Au-Cu superlattices.) However, as the temperature is raised the gold-rich boundary of the two-phase field (Au,Ag) + (Au,Cu) moves steadily towards the Ag-Cu boundary; at the same time the Ag- and Cu-rich solvi move away from the corresponding pure metal corners (Fig. 1.9). The two-phase field can be visualized as a dome-shaped volume in the full phase diagram (Fig. 1.10). Note that the faces of the prism are the three, two-component phase diagrams (*cf.* 14§3.7).

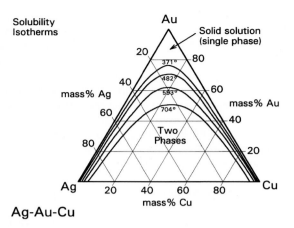

Ag-Au-Cu

Fig. 1.9 Several solid solubility isotherms of the Ag-Au-Cu system mapped on the same diagram. This is a plan view of the triangular prism of Fig. 1.10; the isotherms are therefore equivalent to the height contours of that surface.

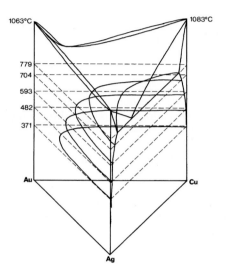

Fig. 1.10 A view of the three-dimensional reconstruction of the solid solubility limit surface of the Ag-Au-Cu system.

It should be clear from this that an alloy with a composition corresponding to the two-phase field (*i.e.* at low-temperature equilibrium) will, if quenched from the melt, as in a normal casting, maintain the high-temperature, single-phase solid-solution structure. Subsequent **ageing** at temperatures below the single-phase limit will cause the decomposition of the solid solution into two new solid solution phases.

●1.6 Ternary system phase-composition

The compositions of the conjugate phases present in the Ag-Au-Cu system cannot in fact be stated as the necessary analytical work does not yet seem to have been done. As explained in 8§4.2, this cannot be done in ternary systems simply by reference to the phase diagram, using tie-lines as is the case for binary systems; it is necessary to analyse each pair for a well-equilibrated alloy. Thus, unless the position of the plait point is known, and its locus with change in temperature on the single phase field boundary surface (Fig. 1.10), it is impossible even to begin to discuss the compositions of the two phases in an aged Ag-Au-Cu alloy, apart from the simple expectation that one is copper-rich and the other is silver-rich. Even so, we can understand aspects of the behaviour of the system from what we do know.

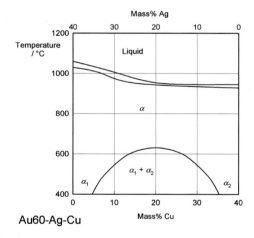

Au60-Ag-Cu

Fig. 1.11 Section through Fig. 1.10 at 60% Au. Note that isothermal tie-lines cannot be drawn in the $\alpha_1 + \alpha_2$ field of a pseudo-binary diagram of this kind – in general they lie at an angle to the plane of the page (*cf.* 8§4.2)

What we can say is that on cooling from the single-phase field, a second phase will precipitate in the matrix of the parent, or more generally a phase-separation (8§4.1) akin to that of eutectoid formation (Fig. 1.11), a minimum diffusion path to establish the appropriate compositions in least time. Of course, grain growth will occur if allowed to remain at a high-enough temperature.

●1.7   Reversible hardening

Thus, the hardenability of dental gold alloys is probably mostly due to the decomposition of the high-temperature solid solution into a fine-grain two-phase structure. The effect can be traced in Fig. 1.12, which shows the Vickers Hardness (1§8.2) for alloys in three conditions:

(1)   alloys that have been quenched after a **solution heat treatment**. This involves holding the metal above the temperature at which a single solid solution begins to be stable for long enough that diffusion can occur to re-equilibrate the structure, and so followed by **quenching**, rapid cooling to room temperature such as by sudden immersion in water while hot.

(2)   alloys that have been annealed in the two-phase region (**aged**), similarly to re-equilibrate the structure. This **age hardening** is perhaps more accurately described as **precipitation hardening**.

(3)   those that have been allowed to air-cool after casting, which represents an intermediate cooling rate through the transition temperature.

The effect of varying the gold content is shown. The quenched alloys show only solid solution hardening, while the aged alloys show the effect of the two-phase structure in addition. The air-cooled condition, *i.e.* which was slow enough to allow some development of the two-phase structure, lies somewhere between.

In addition, the hardening effect of the formation of δ-phase AuCu is clearly seen at 75 % Au (Figs 1.12, 1.13). Available evidence suggests that even though a third element (Ag) is present, superlattice structures of the AuCu-type may still form, and even when the Ag- and Cu-rich two phases are present. The hardenability of the dental Ag-Au-Cu alloys is therefore due to solid solution, precipitation and superlattice effects occurring together. Indeed, this is also known to be the case with Ag-Au-Cu-Pt alloys (*cf.* Figs 1.15 and 1.21 with Fig. 1.3).

●1.8   Alloy colour

The addition of Ag to Au-Cu alloys provides the benefit of precipitation hardening, but it also allows a certain amount of variation in the **colour** of the alloy (Fig. 1.14). This can be used to offset the reddening effect of the

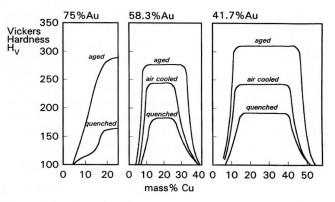

Fig. 1.12   Variation in indentation hardness as a function of composition for various heat treatments over three constant Au-content sections of the Ag-Au-Cu system. (See Fig. 1.13)

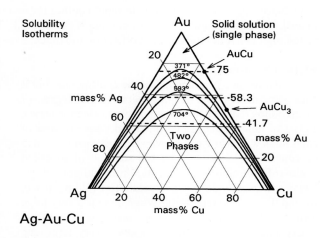

Fig. 1.13   The positions of the sections of the Ag-Au-Cu system shown in Fig. 1.12.

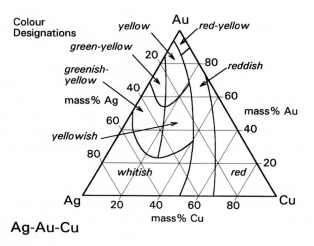

Fig. 1.14 Alloy colour *vs.* composition in the Ag-Au-Cu system. The names are those conventionally used in the jewellery trade. Silver whitens, copper reddens the alloy.

presence of the Cu. If 'white gold' is required, it is more likely to be alloyed with Pd, at about 5%, which has quite a strong effect; Pt has only a slight effect. (Ni is a particularly powerful 'whitener' of gold alloys for jewellery but because of its known hypersensitizing action is likely to be less used in dentistry). Obviously, such colour adjustments are not concerned with matching tooth tissue, but some people may find one colour more pleasing than another, or there may be a need to match the appearance of two alloys which have different mechanical properties because of different functional requirements.

> In some countries, "white gold" is used as the name for platinum and platinum alloys (whether or not they actually contain any gold), and even sometimes for rhodium and its alloys. This has to do with the unfamiliarity of platinum as a jewellery metal, and the higher status afforded the word "gold". Care is therefore necessary in some contexts to be very precise about what is being discussed.

●1.9 Grain size

The Hall-Petch relationship has previously been mentioned (11§4.1) as indicating the effect of decreasing grain size on the strength. Grain size is a function of the cooling rate: faster cooling gives more and smaller crystals. In addition, perfect crystals require time to grow. Rapid cooling does not permit the annealing out of spontaneous stacking faults. Errors will be covered over and locked into the structure by the solidifying metal on top. Thus, the faster the cooling, the more stacking faults there will be, in parallel with the formation of smaller grains. Stacking faults, *i.e.* dislocations, create strain fields that inhibit slip. There will therefore be some hardening effect simply due to rapid cooling for two reasons: the Hall-Petch effect and stacking faults. Indeed, it could be argued that slow cooling is detrimental. Thus, the investment mould temperature must not be too high to avoid inhibiting solidification too much.

●1.10 Grain-refining

Grain size is a function of the cooling rate, but it does vary from alloy to alloy because the degree of supercooling required for homogeneous nucleation varies, and so the number of nuclei formed under any given set of conditions will also vary. However, this aspect can also be controlled deliberately by the addition of so-called **grain-refining** alloying elements. In dental gold alloys this is now usually done with **iridium** (Ir) (although Ru, Re and Rh are also reported). When there is a very large difference between the melting points of two metals, their binary phase diagram tends to show a very large gap between the liquidus and solidus (Au-Pt is one such example which has been studied, Fig. 1.15), in great contrast to the Au-Cu system (Fig. 1.3). In other words, the solubility of the grain-refining element is low. This in fact means that the nucleation of grains occurs very readily. Referring to Fig. 1.15, it can be seen that even at 80 % Au at 1350 °C the first solid to form is about 75 % Pt.

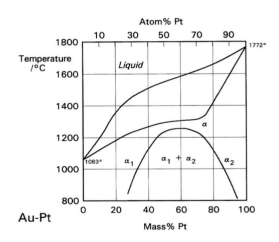

Fig. 1.15 The equilibrium diagram for the system Au-Pt. Such a diagram indicates a strong tendency to coring - note the great width of the α + Liq field above about 1350°C.

Iridium, which has a melting point of 2454 °C, shows such an effect, only much more exaggeratedly. Its solubility in gold is very low (less than 0.005 % by mass) and so can be expected to readily precipitate from the melt. Indeed, it is said that there is always this solid present at the point of casting. It would therefore initiate heterogeneous nucleation (11§2.5) of the Au-alloy, for which process the activation energy is much smaller than for homogeneous nucleation. In other words, the degree of super-cooling required is very small. Grain-refining thus ensures that the distance a dislocation can travel is small, thereby adding to the hardening of the alloy.

In passing, we can note that the phase diagram for **Au-Pt** has a two-phase field of two solid solutions (α₁ + α₂) enclosed within the general solid solution field which is otherwise continuous (*i.e.* stretching between the two terminals). Such a system is exactly parallel to the two-liquid systems discussed in 8§4.1, as it shows an upper critical solution point (*cf.* Fig. 8§4.1), and similar to the two-phase field shown in Fig. 1.11. Thus, this system would also show phase-separation on cooling into that region. Au-Pt is the basis of some alloys used for porcelain-fused-to-metal devices (25§6).

●1.11 Work hardening

Although work hardening is important in several metals' contexts, it is not an option applicable to cast alloys in general – the deformation involved clearly would spoil the dimensional accuracy of any such device. Occasionally, however, work hardening is significant. Thus, in the adjustment of clasps to compensate for residual errors in the casting, the deformation will inevitably harden and therefore embrittle the metal. There is then an increased risk of fracture on any further adjustment. Bear in mind that manual adjustment is in effect a 'strain-controlled' action (as opposed to load-controlled) (1§3.1) since the target outcome is a certain amount of deformation, but in doing this by hand it is rather difficult to control the amount. In practice this means that bending should be done in one direction only, gradually approaching the correct shape. Avoidance of work hardening is a good reason for using a reversibly hardenable alloy. But work hardening can also be beneficial when the margins of a crown or inlay have to be burnished, *i.e.* made to conform very accurately to the adjacent tooth. The hardening limits the possibility of deformation later in service.

●1.12 Related alloys

There are, as pointed out above, many other alloying elements possible for dental 'golds', and some indeed form alloy systems of their own that have found application in dentistry. We may briefly explore a few examples to illustrate further the general principles.

The **Ag-Pd** system is one of simple solid solution over the whole composition range (Fig. 12§1.2) (radius ratio: 1.05). No further hardening after the effects of solid-solution is possible except from cold-work, which annealing can then remove. After annealing, such alloys are quite weak and soft (Fig. 1.16), although they still show very clearly the general effect of hardening by solid solution disorder (*cf.* Ag-Au, Fig. 1.1). Unfortunately, these alloys have a strong tendency when molten to dissolve oxygen, which then appears as disseminated porosity in the cast metal (Fig. 18§4.4).

Au-Pd also forms a continuous solid solution (Fig. 1.17) (radius ratio also 1.05) and the ternary system **Ag-Au-Pd** is similarly one of a continuous solid solution. Although some variation in hardness is apparent in that system (Fig. 1.18), although with a complicated pattern, all of these alloys are still relatively soft and not amenable to age or precipitation hardening.

The **Cu-Pd** system (radius ratio 1.07) shows superlattice formation (Fig. 1.19), just as in Au-Cu (although not exactly at the expected 50-50 composition). As a result the **Ag-Cu-Pd** system shows essentially the same features as does

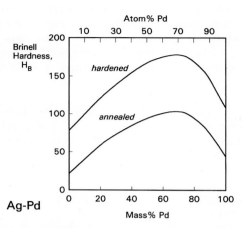

Fig. 1.16 The effect of work-hardening on the Brinell indentation hardness in the system Ag-Pd.

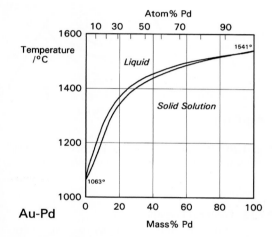

Fig. 1.17 The equilibrium diagram for the system Au-Pd. The narrow gap between liquidus and solidus suggests little coring in these and similar alloys when cast.

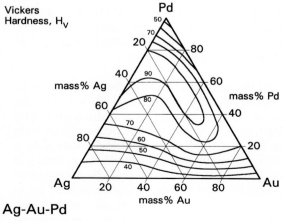

Fig. 1.18 Variation in Vickers indentation hardness with composition in the system Ag-Au-Pd (annealed).

Ag-Au-Cu, with age-hardening due to both CuPd superlattice formation and a precipitation of two solid solution phases. Substantial improvement in mechanical properties can be made this way (Fig. 1.20). Notice in particular the peak hardness near the centre of the diagram where most disorder would be expected.

The **Cu-Pt** system (radius ratio 1.09) also shows superlattice formation (Fig. 1.21), and age-hardening due to the reversible formation of CuPt is possible.

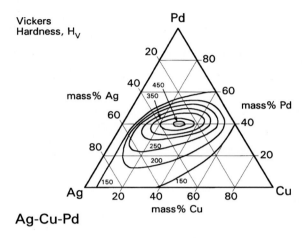

Fig. 1.20 Variation in Vickers indentation hardness with composition in the system Ag-Cu-Pd (annealed).

●1.13 General considerations

What all this illustrates is that there is considerable scope for the design of alloys, even using metals that are very closely related and have very similar chemical properties. The subtlety and complexity of these systems defeats any attempt at memorizing property values, or even compositions. In fact, this is quite pointless since small changes may cause profound changes in behaviour. It does emphasise the need to consider manufacturers' data and instructions very carefully when making product selections, or changing product.

One such aspect of all of these alloys that must be taken into account in casting them is that the liquidus temperature may vary considerably with composition. Some illustrative examples are given in Figs 1.22 - 1.24.[5] Care must be taken that the alloy is heated sufficiently above the liquidus that casting is successful. Different commercial alloys may show appreciable variation in the casting temperature due to the inclusion of further alloying elements and this must be checked. A second point is the **melting range**, the vertical (temperature) distance between the liquidus and solidus. The wider this gap the greater the

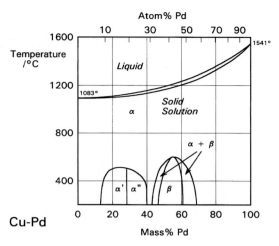

Fig. 1.19 The equilibrium phase diagram for the Cu-Pd system. The formation of ordered phases is comparable to that in the Au-Cu system.

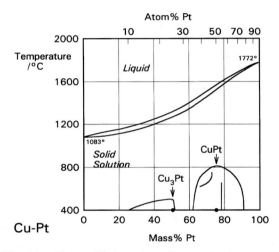

Fig. 1.21 The equilibrium phase diagram for the system Cu-Pt. Again, the formation of ordered phases is comparable to that in the Au-Cu system.

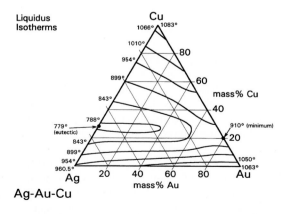

Fig. 1.22 The liquidus temperature contour map for the system Ag-Au-Cu.

likelihood of **coring**, the first freezing metal having a grossly different composition to that freezing later (12§1). **Homogenization annealing** then becomes necessary. This is particularly true when platinum forms an appreciable proportion of gold alloys (*cf.* Fig. 1.15); it is not normally present to more than about 4 mass% because of the tendency to coring that it conveys.

Two-phase alloy systems are likely to be less corrosion-resistant than are single-phase alloys, although both phases may be sufficiently **noble** or electronegative for this to be quite unimportant in alloys with high enough content of elements such Au and Pt. Even so, the requirements of mechanical, chemical (*i.e.* corrosion) and handling (*i.e.* casting) properties need to be carefully balanced for optimum clinical results. But with the emphasis turning more and more to alloys of lower intrinsic cost, the difficulties of maintaining corrosion resistance will increase, and polyphase structures will be more prone to suffer in this way.

Gold alloys suffer from sensitivity to the market price of the gold content, but they are also handicapped by limited strength and modulus of elasticity compared with the demands made on removable appliances. Other difficulties include their high density, which thus causes high weights for devices, possibly causing problems in stability and comfort. These kinds of factor justify consideration of other alloy systems.

●1.14 Scavengers

Because some of the alloying elements of gold alloys are oxidizable in air, copper in particular, and such oxides may form inclusions in the cast metal which have effects such as acting as flaws and embrittling the alloy, there is a need to limit this action with a **deoxidizer**. Zinc, and sometimes indium (In), may be added to the original formulation to provide a readily-reacting scavenger of dissolved oxygen. The resulting oxides then appear as part of the dross in the crucible. The process and outcome are similar to those in the case of amalgam alloys (14§2.4).

Of course, a scavenger is meant to be consumed, and such that little is intended to be present in the actual cast alloy (again, *cf.* Zn in silver amalgam alloy). Re-use of old buttons is problematic in that the composition must have changed by virtue of such oxidation, and not just by loss of scavengers: Sn, Ga, Si and others are also susceptible to reaction or volatilization.

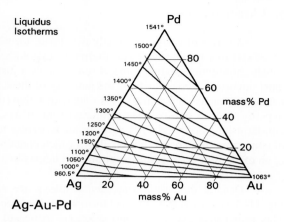

Fig. 1.23 The liquidus temperature contour map for the system Ag-Au-Pd.

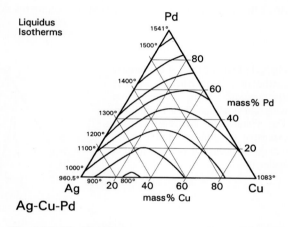

Fig. 1.24 The liquidus temperature contour map for the system Ag-Cu-Pd.

§2. Chromium Alloys

A series of alloys, originally developed for high strength at elevated temperatures for industrial use, called **stellites**, have provided an alternative to gold-based alloys for dental castings. Commonly referred to as cobalt-chromium, cobalt-chrome, chrome-cobalt or just chrome alloys, chromium is the nominal principal – and most important – constituent of the majority of these so-called 'base metal' alloys. Like aluminium, it is an extremely reactive metal that forms a tenacious, thin (thin enough to be colourless; Cr_2O_3 is green in bulk), and impermeable (at ordinary temperatures) film of oxide which limits further reaction, making it essentially non-tarnishing in air and many aqueous media of near neutral pH. It possesses the further remarkable advantage of conferring these properties on alloys which contain an appreciable proportion of it. This ready production of an oxide film is called **passivation**, and it has obvious interest in dentistry (13§5).

Despite this usefulness, dental alloys are not *structurally* based on chromium (which is b.c.c., Fig. 11§3.3); Ni, Co and Fe in fact form the greatest proportions of the elements present in commercial products. This is because extensive solid solutions (α-phase) are formed in the Co-Ni and Co-Fe systems at high temperatures; these have the **austenitic** structure, *i.e.* f.c.c. Only limited solid solubility is shown in other systems, such as Cr-Ni and Co-Cr itself. Even so, in the ternary and quaternary systems, the resulting f.c.c. solid solutions have very wide ranges of possible compositions (Figs 2.1, 2.2). That this occurs may be readily understood from the atomic radii of the metals in such alloys (in pm): Fe: 126, Co: 125, Ni: 125, Cr: 129, Mo: 140. The radius ratios for the first four elements do not exceed ~1.03, and Mo tends to be used up to only

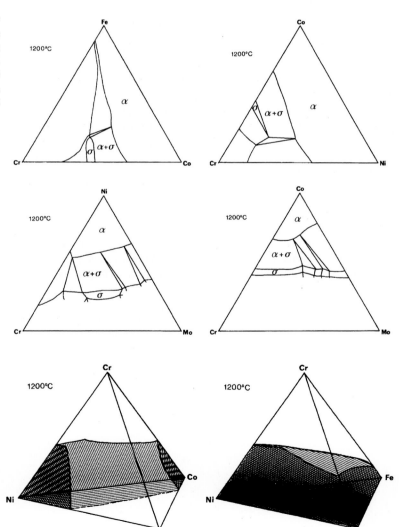

Fig. 2.1 Parts of four of the ternary isothermal equilibrium phase diagrams of interest in connection with 'chrome' dental casting alloys. The features of concern are the extensive α solid solution regions, and the adjacent α + σ phase field in each case.

Fig. 2.2 Two of the possible quaternary isothermal equilibrium phase diagrams for the chrome alloy series, showing only the α-solid solution regions (shaded).

about 5 mass%. Curiously enough, because these are all f.c.c. structures, the mechanical properties of ductility, softness and low strength are common to all such alloys with relatively little variation – as they stand they are still in fact entirely unsuitable for dentistry.

However, the intention is to create, if possible, an alloy with mechanical and chemical properties suitable to the task. The preference is therefore a single-phase alloy, to avoid galvanic effects, passivating, to reduce reaction rates and thus corrosion, while maximizing the disorder in the solid solution to increase the yield point.

In order to make such alloys functional in the sense of carrying large stresses over long spans with small cross-sections (in other words, to fit in the mouth), further hardening of such alloys may be achieved in a number of ways. The differences, as well as the similarities, between this set of possibilities and that for gold-based alloys should be noted.

●2.1   Other metallic alloying elements
The effect of the disorder produced in a solid solution by atoms of different sizes in a lattice has already been discussed. Suitable elements in the present context include Mn (~127 pm) and W (~140 pm). It is important to notice that these alloying elements are required to dissolve in the α-phase solid solution to avoid galvanic corrosion. Manganese can also play a role as a deoxidizer (as does Si in these kinds of alloy).

There are in fact many other alloying elements that are used in a bewildering variety of commercial products for dentistry, some of which products are predominantly one metal, some another. There are too many choices available. It is beyond the scope of this book to detail the possibilities and it is probably beyond the ability of any practitioner to understand the resulting plethora of property and behaviour data, usage options and prices. Accordingly, here – as for the similar situation with respect to gold and related alloys – only some broad principles are given.

●2.2   Metallic phase precipitation
Aluminium is added to at least one dental alloy which contains much Ni; a fine-grained precipitate of the compound $AlNi_3$ is formed which is believed to contribute greatly to the modulus of elasticity and strength of the alloy. Although this looks as though it may be a serious corrosion risk, the fineness of the precipitate may mean that it is soon dissolved when exposed at the surface, and no great roughness results. However, the frequency of occurrence of nickel sensitivity may be enough to preclude its use in biological contexts if only as a prudent precaution. The absence of serious mechanical problems does not adequately offset the risk.

●2.3   Dissolved carbon
The solid solutions discussed earlier (§1, §2; 12§1) were **substitutional**, that is, an atom of one kind was replaced by one of another kind, in other words a substitution at the exact same position in the unit cell. However, there is a further kind of solid solution possible when the solute atom is sufficiently small, and that is **interstitial solid solution**. This possibility arises because in some lattices the packing of the layers leaves enough space to accommodate, with some distortion of the surroundings, atoms such as B, C, N and O (Fig. 2.3). That is, even in close-packed crystal structures there are necessarily holes – **interstices** – between the layers of atoms (11§3) (not to be confused

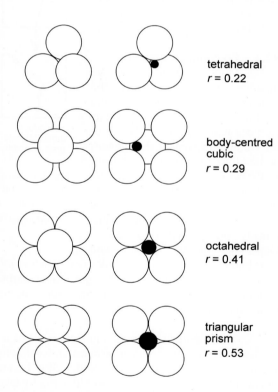

tetrahedral
$r = 0.22$

body-centred cubic
$r = 0.29$

octahedral
$r = 0.41$

triangular prism
$r = 0.53$

Fig. 2.3 Geometry of holes in the common metal crystal structures. The tetrahedral and octahedral holes occur between the close-packed planes of the h.c.p. and f.c.c. structures. The body-centred cubic hole is similar to the tetrahedral, but the tetrahedron is distorted by separating the atoms of two of the possible pairs slightly, making the hole a bit bigger. The triangular prism is not a normal crystal structure but is related to h.c.p. r here is the relative radius of the largest sphere that can fit in the space, compared with the parent atoms. In the right-hand column, the front-most atoms on the left have been removed to show the fitting interstitial atom.

with vacancies, 11§6.1). The resulting distortion of the crystal lattice itself is enough to inhibit slip, similar to the effect of the tetragonal distortion of δ-AuCu. In fact, it goes much further than this. The movement of dislocations past an interstitial atom is greatly inhibited because it physically gets in the way of the necessary metal atom movement.

The maximum radius ratio of an assumed hard, spherical atom that can be fitted into an f.c.c. lattice without distorting it is ~0.41 in an 'octahedral' hole (Fig. 2.3). Carbon is the commonest important solute. Its atomic radius, ~75 pm, is sufficiently small that it can be squeezed into the octahedral holes of the f.c.c. lattice of Co-Cr alloys, albeit with considerable distortion; the relative radius is about 0.60 that of the metal atoms involved. However, atoms are not hard spheres, they are slightly 'squashy' and their shapes also get distorted. Other non-metallic dissolved elements such as boron and silicon may also be used in certain alloys deliberately for similar effects, although these are also effective deoxidizers. Others, such as nitrogen and oxygen, may be unavoidable contaminants under ordinary processing conditions but which nevertheless add to hardness or reduce ductility (*cf.* amalgam alloy, 14§2.4, and titanium, Fig. 28§1.1) (but see §2.1). Note that substitution by such atoms is not possible because it destabilizes the lattice.

Thus, the inclusion of some carbon can provide a substantial increase in the yield point of these Co-Cr-based alloys. Even so, the maximum solubility of carbon is only of the order of 0.2 at%. Any excess results in the precipitation of carbides.

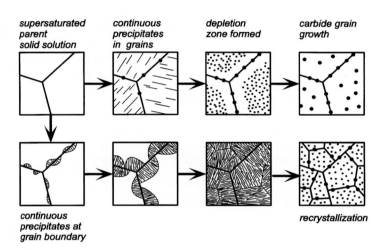

●2.4   Precipitated carbides
More carbon may be dissolved in the melt than in the solid (see also Steel, Chap. 21), and in the freezing process various carbides may separate. The faster the cooling the smaller the amount, and the finer and more dispersed will be the carbide grains. Slow cooling, or annealing at high temperatures, will cause the precipitation of more carbide and growth of the grains already precipitated (Fig. 2.4).[6] Depending on the amount and

Fig. 2.4 The range of types pattern of carbide precipitate that may be obtained in the α solid solution matrix according to the thermal history - time and temperature - of the casting. Note the variety, and the fact that while these changes all affect strength and ductility, none is reversible.

distribution of these brittle carbides (*e.g.* whether int<u>ra</u>granular or int<u>er</u>granular), the alloys may be relatively ductile or brittle, high or low strength. Large carbide grains are decidedly detrimental by providing longer continuous brittle paths for crack propagation. The majority of such alloys are therefore two-phase: α f.c.c. solid solution and carbide.

The important distinction in these systems is indeed *where* the carbide is to be found. There are two kinds of site. The intragranular carbides harden the alloy because they inhibit dislocation movement. In contrast, the intergranular carbides embrittle the alloy because there are no dislocations to inhibit here. The grain boundary is the path that a crack would normally take, and the brittle and unbonded carbide makes it easier for crack propagation.

Carbon is in fact the key element in these alloy systems and, as such, any variation in the amount will result in changes in properties which are almost certainly detrimental. Thus, both oxidizing and **carburizing** conditions, whereby carbon is removed or added (for example, from melting with an oxyacetylene flame), will clearly be disadvantageous. In fact, alloys intended for oxyacetylene flame melting are deliberately supplied with a lower carbon content than might be expected because of the tendency to pick it up. Conversely, alloys for induction melting tend to have slightly more carbon than finally desired because of the tendency to lose it.

In like manner, remelting such alloys from previous sprues and buttons is not recommended because of the likely loss of carbon. Conversely, graphite crucibles or carbon-based investment deoxidizers again are not to be used. While special heat treatment is possible for some alloys to redistribute or modify the carbides (Fig. 2.3), in general the properties are so sensitive to such changes, and to the exact initial conditions, that it is not recommended that it be attempted for dental work. Welding or soldering is particularly risky because of the localized heating and consequent unevenly distributed changes to be expected. Residual carbon in the investment from an incomplete burn-out could also be dissolved in the casting surface, making attention to that particular detail especially important.

It should be noted that alloys containing precipitated phases, whether metallic or carbides, have composite structures; their mechanical properties may be understood in part from this fact alone.

●2.5 Sigma phase

Extended heat treatment or mixing of these α-phase alloys is not recommended for another reason. It can be seen in Fig. 2.1 that a second phase has been marked for notice: the σ-phase. This phase has a complicated tetragonal structure (nominally FeCr, $c/a \sim 0.52$, 30 atoms to the cell) and nucleates readily at grain boundaries. Its significance, not having close-packed slip planes, and in marked contrast to the parent f.c.c. structure, is that it is in fact extremely brittle and would cause the rapid failure of the appliance if present. Some dental alloys have compositions that lie very close to the α / α + σ phase fields common boundary. Under some circumstances, the right conditions for σ-phase formation may arise. This risk is especially increased if the composition of the alloy is altered by mixing different products. The problem also arises for repeatedly melted metal, when the effects of the unavoidable differential oxidation will be felt. These effects are in addition to the problems of the loss or gain of carbon. Notice that oxides and so on that might be adhering to old, used metal will segregate on melting – they will not dissolve or get suspended in the melt (18§2.1). In any case, common sense dictates that metal to be reused would be cleaned up by sand-blasting first, although the risk of composition changes remains.

It is the same problem that precludes stainless steel (21§2) from use as a casting alloy. There is a very high risk of σ-phase forming, and consequent embrittlement, unless the melt is quenched very quickly indeed (working is impossible and further heat treatments are impractical). Clearly, this cannot be done for the lost-wax process as employed in dentistry (Chap. 18), even though this is considered to result in "rapid" cooling.

The 'base metal' alloys are generally understood to be cheaper than the gold-, silver- and platinum-containing 'noble' alloys. This is a major selling point. However, it should be noted that the higher casting temperatures required demand the use of specialized equipment (furnaces and casting machines) as well as investments capable of working under these more demanding conditions. This can make the total cost of working with such alloys comparable with that for the 'dearer' alloys. Yield point, modulus of elasticity and density must also be taken into account when selecting an alloy for a given application.

●2.6 Grain refining

The chromium-based and related dental casting alloys tend to form rather large grains despite the normally fairly rapid cooling encountered in dental casting. Grain refining is thus essential to attain the best properties in the cast metal, especially in thin sections and narrow clasps (see §1.9, §1.10). Molybdenum (Mo) confers this on an alloy of the present type, and this can be seen to be related to its relatively low solid solubility, which in turn arises from its large radius ratio.

Beryllium also has this desirable effect, as well as tending to lower the liquidus of the alloy by as much as 100 K, and of lowering the surface tension of the melt and thus improving the castability, when present in small quantities. Unfortunately, beryllium is a very toxic metal (28§7) and is present in the fumes produced by the casting process and in the dust arising from the cutting, grinding and polishing of its alloys. The special fume-extraction equipment and other precautions required to prevent harm to laboratory technicians working with it has meant that it is now less-used and alternative alloys have been found.

Table 2.1 *Possible means of hardening dental casting alloys.*

Source	Gold alloys	Chromium alloys
Solid solution	✔	✔
Rapid freezing	✔	✔
Grain refining	✔	✔
Metallic phase ppt.	✔	✔ *
Tetragonal distortion	✔	✗
Coherency strain	✔	✗
Interstitials (C)	✗	✔
Carbide	✗	✔
Cold work **	!	!

*   risk of corrosion
** not practicable; risk factor

The available sources of hardening for dental casting alloys of the two major classes are summarized in Table 2.1. These approaches are also applicable in principle to any other alloy system, in existence or yet to be invented. Whether any particular one or combination is used depends on a number of factors: alloy composition, device purpose, processing conditions. Yet the primary purpose always has to be the inhibition of dislocation movement – slip, whether by raising the activation energy for movement or by shortening the available path.

References

[1]   Wyatt OH & Dew-Hughes D. Metals, Ceramics and Polymers. An Introduction to the Structure and Properties of Engineering Materials. Cambridge UP, 1974.

[2]   Brandes EA & Brook GB (eds). Smithells Metals Reference Book. 7th ed. Oxford, Butterworth-Heinemann, 1992.

[3]   Stoloff NS & Davies RG. The mechanical properties of ordered alloys. Prog Mater Sci 13: 1 - 84, 1968.

[4]   Van Tendeloo G, Amelinckx S, Jeng SJ, Wayman CMThe initial stages of ordering in CuAu I and CuAu II. J Mater Sci 21 (12): 4395 - 4402, 1986.

[5]   Lyman T (ed). Metals Handbook. Cleveland : American Society for Metals, 1948

[6]   Lyman T (ed). Metals Handbook Vol. 8. Amer. Soc. for Metals, Ohio, 1973.

Chapter 20   Cutting, Abrasion and Polishing

This Chapter deals with the set of procedures involved in surface reduction. Whether this is the preparation of a tooth to take a filling or other device, the adjustment of the size and shape of that filling or device, or simply the final finishing of a surface to be smooth and shiny, the processes usually depend on the interaction of the mechanical properties of the tool or abrasive and the workpiece.

*The distinction is often made between bonded and loose **abrasives**, but this is superficial. Loose abrasives can only work as a cutting tool if held by some support, however temporarily. The behaviour of abrasives and cutting tools alike depends on the geometry of the interaction and the relative hardness with respect to the workpiece.*

Blasting employs loose particles, but these need not be sharp or hard to be effective, although variations here are important. The kinetic energy at impact is a major factor in determining the effect, as well as the direction of delivery. In this and ordinary abrasion, thermal effects may be very important, especially on polymers. There are also implications for vital teeth subject to cutting or polishing.

***Electrolytic polishing** is only applicable to metals but offers a means of polishing otherwise awkward shapes. It is affected by the previous mechanical history of the surface, and this provides insight into the processes of mechanical abrasion.*

*Consideration of surface reduction, and the need for it, is connected with the question of **roughness**. The expression of the idea of roughness depends in part on the techniques used to measure it. The terms are explained to facilitate discussion and interpretation of advertising and other literature.*

Frequently neglected, the techniques for the preparation and finishing of surfaces are of evident importance in determining the quality, even the success, of a treatment. Understanding the factors involved will give better control, more efficient operation, and reduced risk of faults in all the many areas in dentistry where surface reduction is employed.

Materials Science for Dentistry
https://doi.org/10.1016/B978-0-08-101035-8.50020-1

In the course of dental procedures there are many tasks which require materials to be shaped and their surfaces modified, most immediately teeth themselves, to take restorations. Then there are the restorative materials themselves: silver amalgam, filled resins, glass ionomer cements amongst the direct filling materials to be handled in the mouth, and acrylic, cobalt-chromium, gold, ceramics and so on to be cut, trimmed and polished in the laboratory, sometimes at the chairside. Other techniques, such as grit-blasting,[1] toothbrushing, electropolishing, and dental 'prophylaxis' are similarly under the heading of shaping and finishing – finishing in the sense of creating a visually smooth surface, *i.e.* polishing. In view of the frequency with which such operations are carried out, controlling not just the success but the very execution of a procedure, it is worthwhile giving some consideration to the factors which influence such processes. To these deliberate actions must be added the undesired or unintentional effects of abrasion, *i.e.* wear. Tooth on tooth, tooth on restoration, bur on tooth, abrasive components or contaminants of foodstuffs; each of these and others have a detrimental effect. Some may be minimized by a proper choice of materials, some may be reduced by proper design; none may be avoided altogether. All must be recognized and taken into account.

The three terms that form the title of this Chapter may appear to be separate ideas, even though the purpose, shaping and finishing, is more general. In fact, each involves surface reduction in some fashion: removal of material. This provides the common theme and the reason for the grouping. What is needed then is an understanding of the basis of the choices to be made, and the processes involved, such that the user can take control of the outcome. Nevertheless, this subject matter does not stand alone. It is intimately connected with the topics of Chapter 10 – surface creation or modification, and dependent on behaviours discussed in Chapters 1 and 4, *i.e.* cracking, strength and flow.

§1. Background

It is noteworthy that relatively little is understood about the processes involved in cutting and wear. This is despite the evident economic considerations of doing these kinds of task efficiently, and not least in the industrial context where milling and turning are routine manufacturing processes. Because of this, a vast amount of effort has gone into studies of cutting and wear, but the theoretical basis remains weak. Questions such as which mechanisms of those proposed in an enormous literature are actually operating or important cannot easily be answered in a direct and theoretically-informed way, although there is much data and much experience to guide decisions.

This situation is probably due to three factors. Firstly, as classified by chemical nature and mechanical properties (especially viscoelasticity and strength), the range of types and behaviours of material is very large. To this must be added the complications of composite materials, where very many combinations are possible and increasingly are being applied. Secondly, many of the range of several possible mechanisms for the removal of material from a substrate may be operating simultaneously. They will then vary in their relative contribution to the overall effect with changes in temperature, pressure, speed and time in any one operation, possibly continuously throughout an operation, and particularly so if the tool is itself wearing (which, of course, it always is). In addition, changes in material properties, most importantly work hardening but also temperature-related effects, add a further dimension of variability. Thirdly, what may be learned about the behaviour of a material under one reduction regime may provide little or no guide to behaviour under a different procedure. We shall rely greatly, then, on the results of careful observation, for it may be that for surface reduction processes, taken as a whole, there may never be a general explanatory theory. In common, then, with many other fields, cutting and abrasion in dentistry rely on empirical knowledge, but this does not preclude attempts at better understanding of the kinds of process that occur and the possible roles of the numerous factors.

Many of the materials themselves have already been discussed and what follows will relate only to the several possible factors of interest in order to give a broad overview of the subject, rather than attempt to assess their importance under specific conditions and on specific materials. However, it does assume that the structure and general behaviour of these materials is understood.

---

[1] Commonly referred to as 'sand'-blasting, it should be noted that sand as such is never used except in large-scale industrial contexts. Broadly, sand is too variable and ineffective to be worth using, even if cheap.

§2. Bonded Abrasives

An abrasive is a material which is understood to wear away the substrate to which it is applied, with some relative motion between the two. It does this by means of many small particles, whether sharp or otherwise, but which are not organized in any regular fashion. Often this wear is described as the result of friction, but we distinguish abrasion from the effects of merely rubbing (§3.2) because the former involves the generation of scratches, rough grooves which have been gouged in the surface by the passage of the abrasive. Repeated scratching results in the detachment of particles of the substrate and thus the reduction of the surface.

Abrasives are employed in two major forms in dentistry: bonded, as in abrasive papers, interproximal finishing strips, cut-off discs, grinding stones, diamond points, and model trimmers; and loose, as in toothpaste and various other kinds of prophylaxis[2] and polishing pastes, as well as for grit-blasting and similar operations. However, with the exception of the blasting materials, the loose abrasives can be usually treated as if they were bonded because in use they will be caught up in the bristles of brushes, the fibres of fabric mops and felt pads, or the pores of foamed materials, and thereby held, even if temporarily, to be dragged across the substrate. Indeed, unless they are caught up in this way they would tend only to roll across the surface and have relatively little effect. Thus the surface texture of the carrier material is important. For example, natural bristle toothbrushes provide greater abrasivity than nylon ones because the bristles of the former have sculptured surfaces, those of the latter are very smooth. We shall return to blasting materials later (§4). The term 'loose' abrasive also applies to products supplied in a waxy or viscous medium as well as dry powders because the essential feature is that the particles are permitted, indeed expected, to move relative to one another. Partly this is to allow the treatment of complicated surfaces, such as the occlusal form of a tooth, without the restriction of access imposed by a fixed shape tool or the damage that would ensue to carefully carved contours. Partly, too, it is to allow replenishment of the abrasive.

●2.1 Rate of removal

So, if we consider a generalized bonded abrasive particle in its interaction with a substrate (Fig. 2.1), there are a number of factors which are of interest as having an influence on the outcome.[1] Since the abrasive particle works by gouging a scratch, the first requirement is that the particle be pushed into the surface to some extent. Clearly, the load, F, acting on the particle controls the depth of penetration, p, as this must be related to the indentation hardness (1§8) of the substrate material. If the particle is now to move, the rate of doing the work of deformation obviously depends on the speed of the particle, u, and for strain rate-sensitive systems (such as polymers) this will be of great importance. However, the amount of deformation imposed on the emerging **swarf** is controlled by the

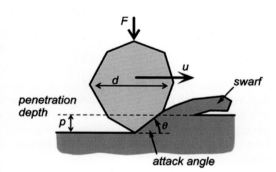

Fig. 2.1 A generalized abrasive particle being moved from left to right.

geometry of the system, in particular as measured by the included or 'attack' angle, θ, between the advancing face of the particle and the substrate surface.

If for the moment we ignore angle and velocity, we can commence an analysis by assuming (reasonably) that the cross-sectional area, A, of the groove produced by a single particle is on average indicated by:

$$A \propto p^2 \tag{2.1}$$

for particles of similar shape; the constant of proportionality is independent of particle size. If we treat the individual particles as hardness test indenters, a hardness value, H, may be represented by a relationship of the form (*cf.* 1§8):

$$H \propto \frac{F}{d_i{}^2} \propto \frac{F}{p^2} \tag{2.2}$$

because the indentation diameter d_i will be proportional to p, again on average. Now, the maximum number of particles that can contact unit area of the surface is indicated by:

---

2 Prophylaxis is the preventative treatment of disease. Thus 'dental prophylaxis' is a somewhat debased term because it is neither implicit that abrasive cleaning of teeth is involved nor necessary. A mouthwash could just as well be 'prophylactic'.

$$n \propto d^{-2} \qquad (2.3)$$

But the mean load per particle in contact with the surface is:

$$F \propto \frac{\sigma}{n} \qquad (2.4)$$

where σ is the applied load per unit area overall, the nominal stress. Thus, substituting from relations 2.2 and 2.4 in 2.1:

$$A \propto \frac{\sigma}{nH} \qquad (2.5)$$

If we further assume that the material ploughed from the groove is not merely pushed aside but broken free, the volume removed by each particle, δV, is proportional to the length of traverse of the particle, the length of the scratch, L, *i.e.*

$$\delta V \propto AL \qquad (2.6)$$

Therefore, for the *n* particles in contact,

$$V \propto nAL \qquad (2.7)$$

where V is the total volume. Hence, from relation 2.5, substitution for A gives

$$V \propto \frac{L\sigma}{H} \qquad (2.8)$$

If we then address the *rate* of removal, the volume per unit time is found by substituting velocity, *u*, for L:

$$\frac{dV}{dt} \propto \frac{u\sigma}{H} \qquad (2.9)$$

Already we have some justification for the intuitive view that the faster we move the abrasive, the harder that we press, and the softer the workpiece, the more rapidly we will indeed remove material. Conversely, this immediately gives us control of the process: light pressure and low-speed application of the abrasive permits more gradual removal of material, and therefore less risk of unintentionally going too far. Once material has been removed the workpiece can rarely be rebuilt. To this we must add intermittent application with frequent inspection, for safety.

In practice, the constants of proportionality in the above will depend on a number of factors, and vary with time. Thus, wear or breakdown of the abrasive particles themselves will change their shape and so their efficiency may deteriorate. For example, so-called gamma alumina is a milder abrasive than the alpha crystal form (corundum) because it is a weaker solid and breaks down during use. Thus the scratches it produces get smaller as the polishing process proceeds. Conversely, the alumina grinding stones used in dentistry ("white stones") should not break down for continued rapid reduction of hard substrates. Obviously, sharp particles will produce higher stresses beneath their points than would rounded ones, and penetrate further for a given load, while smooth rounded profiles will tend to slide more easily over the substrate and not plough up any material. Similarly, removal of abrasive particles from the work area as their bond to the carrier breaks down will reduce the values of *n* and A: abrasive paper wears out.

●2.2   Particle size

It might appear from the above that the effect of an abrasive is expected to be independent of particle size. This can be largely true under some circumstances in practice, but there are limits. One is clearly implied by relation 2.2: if $F > k.d^2H$ (where *k* is the unknown constant of proportionality); no further increase in the value of A can follow as no greater width of penetration than *d* is possible. Nor is greater depth of penetration possible if the particles are randomly oriented and roughly equiaxed, as is reasonable and commonly the case for abrasives. In fact, a bonded abrasive particle obviously does require at least part of its depth embedded in some binder: on average about half can be expected; this therefore limits the maximum penetration possible even further. However, from relation 2.4, for larger particles the load per particle is greater and thus the depth of the scratch produced is greater for a given overall stress, σ.

It is also difficult to ensure that the load on each particle is accurately controlled to be more or less uniform. This is because it is essentially impossible for all abrasive points to lie in the same plane, as is implicitly assumed in the above development, even with carefully selected abrasives; there must be a size *distribution* (a factor range for maximum and minimum of 1.5 ~ 3 and 0.7 ~ 0.3 is commonplace). Uniformity is certainly impractical with commercial abrasives (for reasons of cost if nothing else) and these generally consist of a range of sizes laid down on a smooth backing. It is also very difficult to control particle orientation, to ensure that the cutting points are all near an optimum orientation. This kind of effect is especially relevant to items such as hard-bonded grinding wheels and 'rubber' wheels which rely on wear to expose fresh, sharp abrasive.[3] In fact, a good abrasive must be brittle if it is to do its job without becoming uselessly distorted by plastic deformation, and brittle fracture cannot be expected to occur in a regular manner across the abrading surface. Thus, the maximum depth of scratch possible will depend on the irregularity of the abrasive layer, while the number of particles in contact will therefore vary with the pressure applied.

These two issues lead naturally to the general practice of using finer and finer grades of abrasive successively, not just to control the rate of material removal but to control the size of scratch remaining in the surface. Each stage should be aimed at eliminating all scratches from the previous stage. This means in part that less effort is required to complete the following stage because the amount of material it is necessary to remove is less, although the rate of removal is also lower, by the principles of §2.1. It also means that the particle size designation, the grade of **grit**, in each abrasive system that is available in the surgery or laboratory should be properly characterized if time and effort are not to be wasted by 'polishing' with a similar or (worse) coarser grit than that just used. A typical and efficient sequence involves using grits whose sizes

> The concept of polishing is to reduce the size of scratches, that is, the roughness, such that scattering of (visible) light does not occur, in other words, so that they cannot be seen (< ~0.5 µm). By definition then, polishing has that aim. Any process that leaves visible scratches is therefore not polishing as such, although it may be a step on that path. Even so, invisibility does not mean absence, and scratches will still be seen with an SEM.

> Because particle-sorting by size was (and still is) commonly done using sieves, the designation of abrasive powder grades by **mesh** is also used – that is, effectively the size of the holes through which the grit could pass, using a label such as "–200" to indicate, where the number refers to the number of holes per inch (!) in the sieve, and the negative sign means "to pass" – although this is often omitted. However, since the sieve mesh was often of woven wire (and so not giving 'flat' holes, which complicates the geometry), the thicker the wire the smaller the hole, so particle size cannot be directly deduced from mesh. In addition, because such holes are more or less square, the maximum diameter of an irregular particle that can pass the sieve is closer to the length of the diagonal, not the length of the side of the square hole. Furthermore, very fine particles also pass through such a mesh, so a size range would be indicated by adding another number, the "retained on" mesh, *e.g.* +300. Unfortunately, there are several systems in use, so that for detailed work great care is required to determine exactly what is meant. Size range, of course, does not mean size distribution.

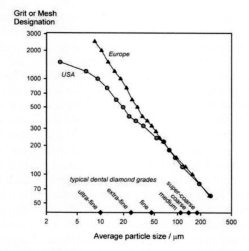

Fig. 2.2 Comparison of two standardized systems of abrasive particle designation, with an example of a dental manufacturer's labelling for diamond instruments.

differ by factors of ~1.5[4] to 2. Unfortunately, there are several systems in use for designating particles sizes and the correspondence between them may be poor (Fig. 2.2). Manufacturers may use different schemes or criteria, and these can vary between types of product so that the numbers not comparable, even within one manufacturer's range. Worse, few dental products give any indication of grit size, being coded by colour or just given crude

---

[3] An ordinary eraser is simply an abrasive powder embedded in a rubber matrix. The coarser the abrasive the faster and more destructive to the paper the action will be. There is no selectivity whatsoever for the deposited medium.

[4] Manufacturers often use a standard factor of √2 ≈ 1.4 as the coarsest definition of successive grades, but may use more closely spaced values in a kind of 'Preferred Number' sequence, as developed by Charles Renard (c. 1870). This kind of (near-)logarithmic sequence is a combination of inventory convenience and perceived differences (Weber-Fechner Law, 26§5.2).

labels ('coarse', 'fine'). These designations clearly vary between products, sometimes with very confusing results.

Notice that very fine finishes indeed are possible with coarse abrasives, as in industrial surface grinding of metal. Here, a high speed grinding wheel has its spindle fixed in the machine, and the workpiece is rigidly attached to a movable table. The workpiece is then moved back and forth, scanning slowly sideways, ensuring that any high point is eventually struck by the highest point on the abrasive wheel. That is, the relationship of abrasive to workpiece is very tightly controlled, *i.e.* constant separation between the rotation axis of the wheel and the base of the workpiece. This condition does not apply to any dental procedure. Normally, in dentistry, a rotating instrument is held freehand against the substrate and control even of the load applied is very difficult. Control of geometry in the workpiece is therefore usually poor unless great care is taken, but it can never be considered to be very good.

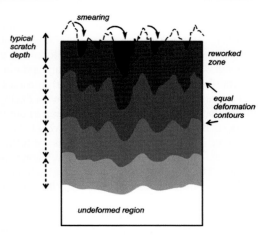

Fig. 2.3 The deformation of a substrate extends deeply into the body whenever the surface is abraded or machined.

●2.3 Effects of abrasion

In abrading a surface the material from the scratch is clearly displaced, and sometimes this just fills previously made scratches. But there may also be considerable distortion in the regions adjacent to the scratch, depending on how ductile or brittle is the substrate. Remembering that the effect of a hardness test indentation extends over several diameters into adjacent material (1§8.5), both sideways and down into the workpiece, the plastic flow and deformation produced may result in a considerable altered zone (Fig. 2.3).[2] Consequently, there may be work-hardening, especially with metals and sometimes polymers, which may change the efficiency of material removal as the grinding proceeds. The scale of these effects depends on the depth of the scratch and therefore the size of the abrasive particle under most conditions (Fig. 2.4). It is the depth of the affected zone that leads to the rule of thumb in metallographic polishing (12§2) to polish for *twice* as long as it takes to remove the scratches from the previous stage. That is, a proportion – say 50% – of the previously altered material must also be removed, but then without extending the depth of the deformed zone because the grit size is now smaller. The point of this is to allow the study of unaltered material, but it obviously has implications for the corrosion behaviour of a surface in practice if there is much strained material remaining. The goal then would be to reduce that to a practical minimum, limiting any consequent effects.

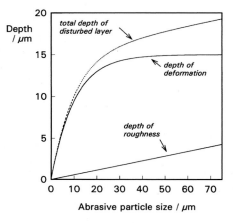

Fig. 2.4 Relationships of deformation and roughness to particle size during abrasion and polishing.

●2.4 Polymers

The properties of polymers are strongly dependant on temperature (Chap. 3) so that considerable variation in abrasion rates may be expected as friction warms the surface and the work of plastic flow is delivered (Fig. 2.5). The thermal conductivity of polymers is rather low, so that heating in this way is very rapid. The glass transition temperature is typically not far from normal working temperatures. As might be anticipated, property changes are most marked at temperatures around the glass transition, and any aspects of structure and the presence of plasticizers which change this value will produce changes in abrasion behaviour. But it is also apparent that, due

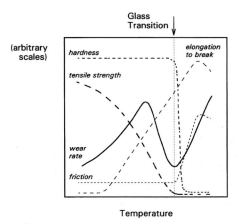

Fig. 2.5 Temperature dependence of some properties of typical polymers.

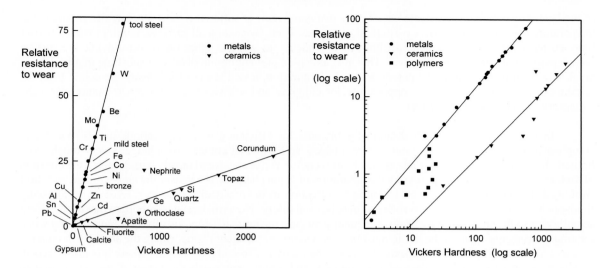

Fig. 2.6 The wear resistance of some materials plotted against their indentation hardness. Left: linear scales, for metals and ceramics. Right: same data on logarithmic scales, and including data for a range of polymers.

to the inverse equivalence of strain rate and temperature, the *speed* of the abrasive particle over the surface will also influence the *rate* of abrasion beyond that otherwise expected.

Fig. 2.5 shows how complicated are the interrelationships of the mechanical properties of such a material to produce the typical wear curve shown. The need to use diamond instruments at high-speed to trim filled-resin restorations before they are fully set is apparent from this, but equally the use of coolant and intermittent application of the abrasive device as a general rule when handling polymers is evidently of considerable importance. The most obvious example of what happens when this is not taken into account is in the final polishing of a PMMA denture base: if the temperature rises above about 70 °C, the internal strain relief (stress relaxation) that occurs may be enough to destroy the fit. But even if it does not the acrylic may become sufficiently softened that it will be deformed in the hand.

Relation 2.8 is supported by observations on metals (Fig. 2.6) which indicates a fairly strict dependence of wear resistance to hardness.[1] A similar relationship, although with greater scatter, is indicated for ceramic materials, but the differing slope is probably due to a different mode of failure, *i.e.* essentially brittle with little if any plastic deformation possible. The polymers seem to occupy an intermediate position, but with much greater scatter, reflecting the greater complexity of the processes involved for these materials.

●2.5   Clogging

There is a further source of diminishing efficiency with fixed abrasives: clogging. This is the filling of the roughness between abrasive particles by debris from the substrate. It has the effect of gradually reducing the apparent size of the abrasive, thereby reducing the scratch depth that it can achieve, and thus reducing the rate of material removal. The softer the substrate the greater the tendency for this to happen. It also depends to a large extent on the ability of that material to weld to itself under heat and pressure. Certainly, because of the work being done in friction and deformation, local temperatures can rise very high during abrasion, and melting is common, even if for a short time. Examples of such clogging are resinous woods in abrasive paper, aluminium or amalgam on stones, and nearly any polymer on any fixed abrasive. In fact, the chips arising from any abrasive action will be held for a while, even if they subsequently fall clear, because separation between the abrasive body and the substrate must first occur for this to happen. Thus, the penetration of abrasive particles into the substrate and the rate of removal of material are self-limiting as the chips fill the roughness, even if clogging as ordinarily understood does not occur. Some fixed abrasives are made with less than 100% coverage in order to leave more room for chips and reduce clogging.

●2.6   Lubricants

Such inefficiency is wasteful of time and materials, but is generally easily overcome by the use of a liquid lubricant. This will have functions besides the obvious reduction of ineffective friction (which would, on its own, achieve little in the way of surface reduction). The liquid will behave as a coolant, carrying away much of the heat, preventing general temperature rises that could melt or soften the substrate, but also reducing the temperature rise under the abrasive particles themselves. Then, by coating all surfaces, it will provide a barrier between particles of debris which will limit their tendency to weld together (as it were, deliberate surface contamination) or stick to the abrasive. This also assists in preventing the displaced substrate that has yet to break free from re-welding to the parent material. In addition, it will tend to flush away the debris from the grinding site, further reducing the opportunity for clogging and self-welding, although it cannot affect the temporary 'roughness-filling' effect.

In industrial contexts, the design of lubricants for efficiency is a very serious matter, and many complicated formulations exist to suit different tasks. In the dental context, the choice is very much more limited. In the mouth, naturally, water is the obvious choice as it is cheap and mostly harmless, but the question of toxicity or hazard is also important in the laboratory or at the chairside because of the close contact and short working distances normally involved: isolation or protection are not practical alternatives. Even so, occasionally fluids such as glycerol offer better results and may be used safely. Nevertheless, it is important that the lubricant actually gets to the site of active cutting. Frequently this just means that a copious supply is required, but when rotating instruments are used the liquid must be supplied on the 'upstream' side, where the abrasive or cutter is approaching the workpiece. This means that it will be carried or forced into the area where it is most needed. It will be spun off a high-speed tool very quickly and simply not reach the target if the jet is applied 'downstream'.

There is a further role of some importance: dust control. Many, if not most, of the materials that may be cut in dentistry would be undesirable if not an outright hazard if inhaled. Silicates, heavy metals, beryllium (28§7), resins[3] and others present problems of this kind. Given that it is often impossible to know exactly what is being cut in the mouth, liberal water spray and high-volume suction are essential.[4] The same reasoning would, of course, apply to potentially infectious material. In laboratory work, dust-control measures will need to be much more elaborate.

●2.7 **Zeta potential**
In this general context it is worth noting that the mechanical properties of the substrate depend in part on surface energy. The link here is that much new surface is being created, and the work of its creation must be supplied (1§7, 10§1.2). Surface energy is related to a factor known as the **surface** or **zeta potential**. This is the potential which is spontaneously developed due to charge separation across a solid-liquid interface. There are obviously some similarities with the idea of electrode potential (13§1), but it applies to all solids, not just metals. Zeta potential is strongly dependent on the characteristics of the solution, especially concentration and pH, and can be made to be zero by appropriate adjustments in the solution chemistry. These zeta potential variations are accompanied by large changes in the measured surface hardness, strength and, most noteworthy, abrasion and cutting rates. This has great economic importance in such tasks as drilling for oil. It may be that, with suitable non-toxic additives to the cooling water, tooth material cutting may be made much more efficient with less tool wear, although this is still some way off yet as a practical clinical procedure (more directly chemical dissolution methods may preempt this approach). In the laboratory, too, it may become possible to improve efficiency and the finish obtained by an appropriate choice of coolant.

However, despite all of the appropriate precautions being taken, fixed abrasives do sometimes still clog. With abrasive papers or interproximal finishing strips there is no economic choice but to use a new piece. With industrial grinding wheels it is usually necessary to 'dress' the wheel by using a single mounted diamond to strip away a layer and expose a fresh abrasive surface. Occasionally it might be possible to dissolve the clogging material with a solvent or acid, but this is rarely worth the trouble and expense. But if the clogging were due to improper procedures, it would be far better to use the correct abrasive for the job in the correct manner – cheaper and faster.

§3. Cutting Tools

The simple relation 2.8 was derived for a particle with an essentially conical shape at the contact point, but this configuration is not very efficient in terms of material removal because the principal action is one of 'ploughing': much material is displaced to the sides. Only if the deformation is sufficient to cause fracture will material be removed instead of merely being redistributed. The difference between abrasion and a true cutting action (Fig. 3.1) is very clear.

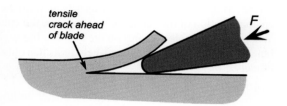

Fig. 3.1 The geometry of a generalized cutting tool. The blade is driven by the force F at a given depth in the material. A leading tensile crack is formed.

The terminology applied to industrial machine tools such as lathes and shapers – which use a single bladed cutter, and milling machines – which use multibladed rotary cutters, is directly applicable to dental instruments such as enamel chisels and the multibladed cutting tools ('burs') used in dental handpieces. It does not matter which part is fixed and which is moving, it is only the relative motion that is important. However, there is a most important difference between the usage in the two fields, as with abrasives. In the former the relative position of the tool with respect to workpiece is very rigidly controlled, whereas in dentistry the tools are hand-held and so subject to movement and consequently give less precisely shaped surfaces. There is a further difference: the use of the term 'blade' means a shape to the working part that is an extended edge, as opposed to the point of an abrasive particle. This is the essence of the distinction being made in Fig. 3.1: there is little or no sideways displacement of substrate, little scope for rewelding of substrate, and potentially much faster rates of removal.

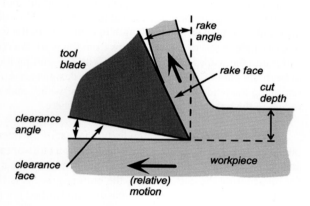

Fig. 3.2 Some terminology of cutting tools. The tool illustrated could be a lathe tool, where the workpiece rotates, or one blade of a rotary tool, where the substrate is stationary.

●3.1 Cutter design

In all cases, however, the **rake angle** of the blade (Fig. 3.2) is a critical factor. It is related to the '**included**' or '**attack**' angle θ mentioned earlier (Fig. 2.1), being given by $\theta - 90°$, *i.e.*, measured from the normal to the workpiece surface. It can be positive or negative (Fig. 3.3). The positive direction is that of the movement of the workpiece relative to the tool.

Large positive rake angles do not require much plastic deformation of the removed material (the chip) and thus might be expected to require less work per unit volume of material removed. However, while it is geometrically

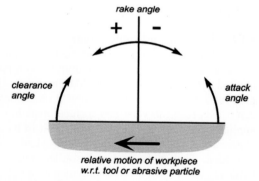

Fig. 3.3 Terminology and sign convention for the tool-substrate geometry.

difficult to achieve such angles on a multibladed tool, as the angle increases so does the risk that the chip will not clear the tool but clog it, filling the space between the blades, reducing its efficiency to zero. Spiral **flutes** (*i.e.* spiral blades) may help chip clearance by tending to direct the chips somewhat sideways, allowing them to flow along the flute until they clear – much like the action of twist drill. More serious is the problem that the blade will tend to dig in. This causes the work to come to an abrupt stop as the machine is stalled when it cannot deliver enough power to correspond to the rate of consumption by deformation and the work of creation of new surface. Such blades are also relatively weak because of the small amount of metal supporting the edge and so tend to bend and fracture on impact.

Negative rake angles, on the other hand, impose so much deformation on the chip that it may readily fragment, making clearance easier. This requires much more power to be expended and results in a rougher surface as the material tends to be torn away from the substrate rather than being cut. The compromise involves an appreciation of the substrate's properties, the cutter's strength, the power available, and the finish required. What is controllable depends on the system. Industrially, depth of cut and feed rate are the important variables, but, again, in dentistry with hand-held tools it is only the force of application that can be adjusted, and then only crudely.

To some extent the design of dental cutting instruments has taken this into account, presumably to achieve the smoothest operation even if not the most efficient that can be envisaged. Spiral flutes are also helpful in this respect. A straight blade would tend to strike the substrate simultaneously all along the operating portion of its length and hammer the substrate. The impact of the blade causes severe vibration and increases the risk of it being fractured as well as being unpleasant for the operator (with the risk of nerve and blood-vessel damage in finger tips) and extremely unpleasant for the patient. A spiral blade would commence contact at only one point, or over a very short length, and cutting would start there smoothly. Furthermore, the next blade would come into contact before the first had left, and possibly a third and a fourth, depending on the pitch of the spiral, the number of blades, and the length in contact with the workpiece. However, the spiral must be in the correct sense (handedness) to clear the chips from the work area (*i.e.* toward the handpiece) and this tends to screw the cutter into the workpiece, particularly if an end-cutting tool is used. More care is therefore required with these to keep the path of cut as intended. Some designs are effectively of two spirals of opposite handedness, thus removing the tendency for the instrument to 'walk', making it easier to control where the cutting is occurring, reducing the risk of unwanted damage elsewhere..

A further possible design feature is that of the interrupted blade. That is, a spiral flute is not continuous but has a series of small gaps along it. This is a **chip-breaker** design. Particularly on ductile materials (metals and polymers), the action of a spiral flute is to generate a long chip that has a greater risk of clogging the bur. By interrupting the cutting edge, the cutting process is limited to short stretches, and the chips are therefore small. They can be cleared more easily, and flushed out more readily by coolant.

There are further differences between products, and between designs intended for different purposes. Unfortunately, there is little systematic information about performance related to design for these hand-held cutters, and we are left only with these general remarks. Nevertheless, it remains important in principle to select an appropriate cutter for the job in hand. Experience must be the best guide at present.

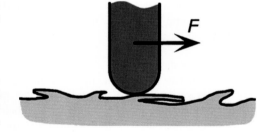

Fig. 3.4 The wear effects of a smooth object on a smooth polymer surface. The normal load, pressing the indenter into the surface, is usually omitted but it must not be forgotten.

●3.2   Rubbing and rolling wear

It may be thought that if the rake angle were $-90°$, which would correspond to a rotating smooth mandrel being applied to the surface, there would be no wear. However, the pure rubbing of a smooth blunt object on a smooth substrate can in fact result in wear and surface breakdown. The load applied normal to the surface will result in its deformation, elastically certainly, but locally the yield point may be exceeded. That is, the loading is similar to that of the Brinell hardness test (1§8.1). In a polymer-based material, plastic deformation and fracture may then cause microscopic lips and tears in the surface (Fig. 3.4),[5] exacerbated by the friction between the two bodies (1§5.5) applying tensile forces to those features. Subsequent passes may well then remove the projections. In the case of metals, plastic deformation arising from the local pressure means slip, and locally this must result in an irregular surface (*cf.* Fig. 11§5.4) which then will have fragments broken off by subsequent passes (Fig. 3.5). There may also be some partial welding of the two bodies, and of the

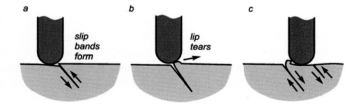

Fig. 3.5 The generation of roughness by a smooth object on a smooth metal surface. The pressure causes indentation and slip (a) and elastic-plastic deformation causes the lip to be raised. Some sticking of tool and substrate occurs to tear the lip (b), and the process is repeated (c).

fragments generated to either body, during this process. Of course, this type of wear would not be used as a practical process for surface reduction, but it does account for wear in service when there is no obvious abrasive action. One example of such a situation is a ball- or roller-bearing: the raceway surface is subject to wear of just this type. The most important example of this in dentistry is the ball-bearing of a high-speed air-turbine handpiece (Fig. 3.6). It should also be recognized that the effect is symmetrical – the roller or sliding object also experiences similar stresses, deformation and wear (Fig. 3.7).

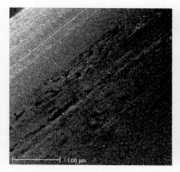

Fig. 3.6 Rolling damage to an air-turbine handpiece bearing raceway. (Image: Wei Min)

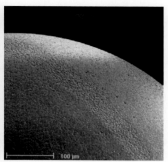

Fig. 3.7 Rolling damage to an air-turbine handpiece bearing ball. (Image: Wei Min)

●3.3　Wax carving

The carving of wax is an important procedure for many dental purposes. However, attempting to use a knife blade as in Fig. 3.1 only leads to the blade digging in because the depth of the cut cannot be controlled by hand. An equivalent system which would suffer in the same way is wood, as any boy with a stick and penknife knows (or soon learns). However, one approach here is the use of a plane (Fig. 3.8): the blade still has a very large positive rake angle, but the depth is controlled by the sole plate of the plane.

Fig. 3.8 The depth of cut of a plane is controlled by the sole plate, so a large positive rake angle can be used (attack angle > 90°).

This, of course, is not applicable to dental wax carving because the desired surface is not flat. Hence the blade is ordinarily held with a large negative rake angle (Fig. 3.9). The load applied to the blade is now easily controlled, and the depth that it penetrates into the wax is then limited in much the same way that a hardness indentation is limited. Drawing the blade over the surface then shaves off a layer by allowing it to flow up the blade – much as one would remove a portion of butter from a slab with a knife (any attempt to do this with a positive rake angle is doomed). The action of carving is actually slightly more complicated. The blade is moved in the desired direction with very little force applied initially, but this is then increased gradually, forcing the blade slightly into the surface. As it continues to be moved the sliver of wax is carved off, diminishing in thickness as the load is then reduced to nothing to approach the end of the stroke. Throughout, it is important to maintain a steady *negative* rake angle to ensure control. The same kind of action is used on many similar tasks: carving dental plaster and stone, amalgam, and PMMA, for example. From this list it can be seen to be applicable to the brittle materials as well as to the plastic.

Fig. 3.9 A wax knife cannot be depth-controlled by hand unless a negative rake angle is used (attack angle < 90°).

●3.4　Critical attack angle

To explore further the question of attack or rake angle we may return to bonded abrasives. Because of the random nature of the orientation of irregular abrasive particles themselves, the frequency distribution of attack angles will be spread over the full range (Fig. 3.10). The critical point for the change from "cutting" to "non-cutting"

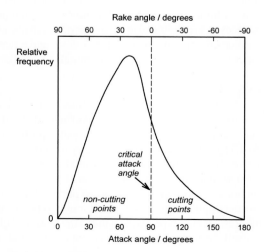

Fig. 3.10 The distribution of attack angles in a random sample of bonded abrasive particles.

(but nevertheless still abrading) points is 90°. It is in this sense that the distinction between the cutting action of dental burs and the abrasion of diamond powder points can be made.

When a cutting blade wears, the rounded and chipped edge presents a more and more negative rake angle. This blunting results in a decrease in efficiency, corresponding to the transition between θ > 90° and θ < 90°, due to the increase in ploughing action and greater energy loss in plastic deformation and friction. But more than this, the negative rake means that the blade tends to climb up (as opposed to digging in) so greater force must be applied to try to continue the cutting, thus increasing tool wear. Rapidly, the point is reached where nothing useful is happening. The only option now is to change the cutter. Continued use otherwise would lead to overheating of the workpiece and extra wear in the bearings of the handpiece, to say nothing of the waste of time. Clearly, changing the cutter earlier rather than later is preferable.

As the cutting and abrasion of the workpiece depends in part on its hardness and strength, its elasticity and strain hardening properties, it can be seen that similar criteria must apply to the wear of the abrasive and of cutting tools such as chisels and burs. Extending the conclusions of §2, the rate of wear of the substrate in practice depends on its hardness relative to that of the abrasive, H/H_a (Fig. 3.11). Intuitively, the abrasive must be harder than the substrate to do its job. Thus, when this ratio approaches 1, *i.e.* the hardness of the substrate is nearly the same as that of the abrasive, a rapid deterioration in substrate wear rate occurs. However, it does not become zero (as measured by the ability to scratch at all) even at $H/H_a = 1$, but at some point beyond, at a value around 1.2 ~ 1.5.

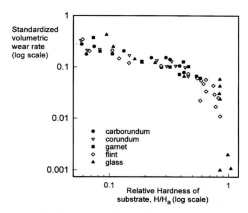

Fig. 3.11 Wear rates obtained with various abrasives as a function of relative indentation hardness (H_a is the abrasive hardness).

This observation accounts for the mutual slow wear of opposing similar materials, such as tooth on tooth, even in the absence of other abrasive materials. By symmetry, even with $H/H_a < 1$, the abrasive or cutter itself is necessarily being worn (even diamonds) and the useful life of a cutting tool or abrasive implement will be limited by this. It can thus be deduced that for efficiency and cutter life the cutter must be at least twice as hard as the intended substrate. Steel burs are of little use on Co-Cr alloys, for example, and tungsten carbide tools need to be used.

There are a variety of other contexts where abrasion will occur that is undesirable. Obvious examples are porcelain crowns and inlays against the opposing dentition, filled-resin restorative materials likewise, especially when these have been eroded somewhat (Fig. 4.1). Writing and printing paper and the like are commonly 'filled', made smooth, opaque, white and dense, with clay (kaolin, 25§2.1), chalk (calcium carbonate), aluminium hydroxide, titanium dioxide or similar substances. These are all abrasives, which accounts for the wear of fountain pen nibs and technical drawing pens. Such paper should therefore be kept away from critical surfaces. Care should also be taken with such mundane materials as tissue paper, which frequently contains talc, another silicate and also abrasive. Lenses and other optically-important surfaces should not be cleaned with such paper (filler-free 'optical' tissue should be used).

●3.5 Contamination
A similarly undesirable abrasion occurs with filled resin restorative materials when they are being inserted into a cavity. The filler is sufficiently hard that particles of metal may be readily removed from stainless steel instruments. This was a particular problem when two-paste products were in common use for fillings (6§4.7) and they had to be thoroughly mixed with a spatula. Agate or polymer instruments (15§8) are preferable to avoid contamination which would leave dark specks in what is meant to be an 'invisible' repair of the tooth. Even so, questions about the effects of any contamination arise since particles of metal, mineral or polymer will not bond to the matrix and so act as pores – flaws in the Griffith sense (1§7). The same idea is applicable to any powder-liquid cement system, although to be fair the amount of glass from a slab is likely to be very small, and stainless steel is expected to bond to the matrix of glass ionomer cement (9§8.7). Discolouration is not a problem in itself for materials used to cement crowns and inlays since they are out of sight.

§4. Erosion

As was mentioned above (§2), the effects of loose abrasive powders that are applied in slurries or pastes and carried to the work on mops, brushes and the like do not differ in their primary behaviour substantially from bonded abrasives. This is because the trapping and dragging of the particles by the carrier make them appear as temporarily bonded. However, some abrasion may be effected even by rolling the abrasive particles. Such particles will still behave as indenters when under load. If the contact area is small enough (as with sharply angular particles) the stress will be high enough that the yield point is exceeded, or even the ultimate strength. Flow and fracture must ensue. However, this mode of abrasion, better called **erosion**, may be critical in circumstances where the structure of the substrate is composite. Such multiphase systems have inevitable variation of hardness from site to site. As can be seen from Fig. 3.11, there will therefore be variation in the wear rate, depending on the relative hardness of substrate and abrasive for the different regions. Further, because the abrasive is loose, it may be pressed down into the worn areas to continue the process. This **differential erosion** may lead to rougher surfaces being produced than existed initially (*cf.* Fig. 12§2.3). These effects would be encountered especially when using brushes and mops whose flexibility allows the abrasive particles to follow the surface detail more closely, even though these are necessary for polishing irregular surfaces such as of the occlusal surface of a molar. Even resilient backings for fixed abrasives are a source of this difficulty.

Such an effect is found with filled-resin restorative materials (Fig. 4.1) where the abrasive must be particularly hard to abrade the filler particles (hence the use of diamond instruments for finishing), and tightly bonded to the matrix to prevent erosion of the resin. Of course, such erosion in the mouth is unavoidable with foodstuffs and toothpaste, and this is one reason for the limited service life of such restorations (and hence the attempts to use 'micro-filled' resin). The roughened surface will more readily trap debris which may discolour, but the exposed edges of filler particles will more readily allow them to be caught and torn out of the surface, tending to cause larger depressions and a rougher surface again. Importantly, this allows resin erosion to continue.

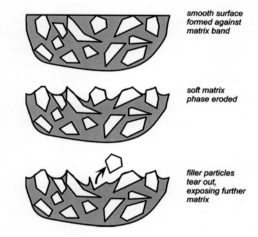

smooth surface formed against matrix band

soft matrix phase eroded

filler particles tear out, exposing further matrix

Fig. 4.1 Pattern of differential wear observed on a typical composite structure of hard core in softer matrix.

As can be seen in Fig. 4.1, the removal of a filler particle depends in part on the matrix having been eroded to reduce the area of the bonded interface and therefore the force required to break it free. It also depends partly on the lack of interference by adjacent particles: if contact is present, movement of the one requires another particle or particles also to move, raising the dislodging force required substantially. The threshold for this behaviour to become dominant would appear to lie at the random loose-packed limit (Fig. 4§9.2), when continuity of contact effectively means that most particles are held in place by contacts with neighbours. Of course, this is also the threshold for dilatant behaviour, and so in ordinary, direct-filling resins good wear resistance is not attainable. For the so-called packable types (6§3.4), which cross the line into the dilatancy domain, then improved wear-resistance might be expected, although again this is to be traded against deterioration in various other properties.

●4.1   Grit blasting

Grit blasting is an erosion procedure whereby the abrasive is suspended in or propelled by a stream of fluid. It is particularly suited to work on irregularly shaped workpieces. The stream of fluid, usually gas but it can be liquid, can be readily directed into places where other tools could not be used, or at any rate not so easily. On the other hand, it is more difficult to restrict the area affected. Grit- and shell-blasting are used primarily for cleaning up castings and acrylic respectively, but several applications are found in porcelain work: cleaning metal and preparing the surface of the porcelain as well as trimming excess, or indeed complete removal for remaking. A variety of grits are used for these different jobs and in order to understand their choice the influence of some of the relevant variables need to be considered.

The effectiveness of blasting erosion may be measured in a number of ways, but the commonest are in terms of the mass or volume of substrate removed per unit mass of abrasive delivered. Since the particles are delivered by virtue of their velocity, v, rather than by a force being applied to each directly, the energy of the impact of the grit particle is the natural starting point for enquiry on the grounds that work needs to be done to remove material from the substrate. The kinetic energy, E, of a particle of mass m is given by the familiar expression

$$E = \tfrac{1}{2}mv^2 \qquad (4.1)$$

If, therefore, we assume that the erosion rate, ε, is proportional to this, allowing for variable efficiency of material removal, we may as a first step guess:

$$\varepsilon \propto mv^2 \qquad (4.2)$$

The effect of impact velocity has been tested extensively (Fig. 4.2), and while the exponent takes the value 2 for very fine particles, this represents a lower limit; generally it has a value of about 2·3. This indicates that the efficiency of **energy deposition** or utilization is itself velocity dependent, that is, increasing with velocity. There is, however, always a minimum impact energy for a given system, and thus a minimum grit velocity, for erosion to occur at all. This limit corresponds to the point below which the deformation of the substrate is purely elastic – the particles just bounce. Clearly, in the absence of plastic deformation, $i.e.$ the yield stress of the material is not being exceeded, no erosion can occur.

The mass of the particle in relation 4.2 can be expected to be proportional to the cube of the diameter, d, but experimentally this is not often approached and a relationship nearer:

$$\varepsilon \propto d^2 \qquad (4.3)$$

is often found, particularly for brittle substrates, although little regularity is found.. This suggests that the area of substrate affected is more important. Even then, there may be an upper limit or 'saturation' rate of removal which may be reached for ductile materials (Fig. 4.3). Clearly, we are not yet in a position to explain erosion even in terms of the elementary variables.

From fixed abrasive studies we might expect that the hardness of the abrasive is of great importance, since if it deformed plastically on impact it would appear unlikely to achieve anything. Experimentally, a relationship of the form

$$\varepsilon \propto H_a^{2·3} \qquad (4.4)$$

has been found to hold approximately for a number of conditions (Fig. 4.4) (the value 2·3 for the exponent is presumably a coincidence). Similarly, the hardness of the substrate is important but while a number of metals show behaviour that parallels that for fixed abrasives (Fig. 2.5), there are some important departures (Fig. 4.5). The steels shown had their hardness modified by heat treatment with negligible effect on erosion rate.

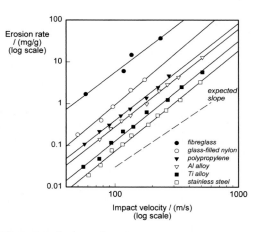

Fig. 4.2 Influence of grit impact velocity on erosion rate (mg substrate removed per g of grit delivered).

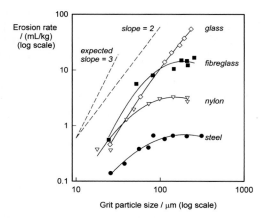

Fig. 4.3 Influence of grit particle size on erosion rate (volume of substrate removed per kg grit delivered).

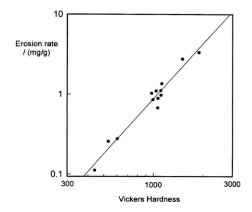

Fig. 4.4 Influence of grit indentation hardness on erosion rate.

Even so, soft materials can in fact erode hard ones. This can be understood in terms of the energy deposited at a site exceeding the capacity of the material to absorb it elastically, causing breakdown rather than simple mechanical cutting.

We can summarize the empirical knowledge with the following relationship

$$\varepsilon \propto d^2 . H_a^{2.3} . v^{2.3} \qquad (4.5)$$

over some ranges for some systems. At least this indicates that sensitive control of the process can be effected by the choice of grit and by adjusting the velocity. This latter is achieved by adjusting the flow rate of the carrying fluid, but this is usually indicated (rather vaguely) by the supply pressure (but see §4.7).

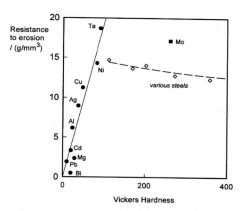

Fig. 4.5 Breakdown of hardness of substrate effect in some materials under grit blasting.

●4.2   Angle of impact

A further factor is the angle of impact of the grit, but the effect of this depends very much on whether the workpiece is ductile or brittle. For ductile materials, a glancing impact is more likely to gouge out a fragment, more like true abrasion, than in a direct, 90° impact when a predominantly plastic indentation deformation might be expected. In contrast, a brittle material is more likely to show fracture of the surface and the loss of fragments at high angles of impact when the strain in the surface would lead to cracking.

As a first step in an analysis of this type of system, and to distinguish the two kinds of effect, the particle velocity is resolved into the vertical, v_v, and horizontal, v_h, components (*i.e.*, referred to the substrate surface):

$$v_v \propto \sin \alpha \quad \text{and} \quad v_h \propto \cos \alpha \qquad (4.6)$$

where α is the angle of impact measured from the horizontal. The kinetic energy of the particles is constant:

$$E = \tfrac{1}{2}m(v_v^2 + v_h^2) \qquad (4.7)$$

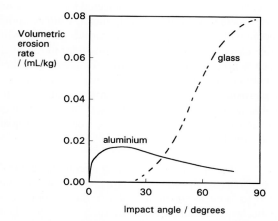

Fig. 4.6 The difference in behaviour of a ductile (Al) and a brittle (glass) material under grit blasting.

Since a gouging action depends on a combination of penetration and movement to be effective we may take it that at 0° and 90° it has zero efficiency. We then assume that maximum efficiency for gouging occurs at about 45° and model this by setting its angular dependence proportional to sin 2α. We therefore obtain the following relationship:

$$\varepsilon \propto A . \cos^2\alpha . \sin 2\alpha + B . \sin^2\alpha \qquad (4.8)$$

where A and B are weighting factors dependent on material properties to indicate the relative importance of ductile and brittle erosion mechanisms respectively (A + B = 1). A good example of such a contrast in behaviour is the distinction between aluminium and glass. These materials obey the above simple relation well enough for the principle to be illustrated (Fig. 4.6). This result shows that there is a further element of control for grit blasting depending on the nature of the material to be removed: high angle shatters brittle materials, low angles gouge ductile ones.

An additional complication to this is, of course, particle shape. Spherical particles will be extremely inefficient at gouging, yet may still transfer energy effectively at high impact angles. On the other hand, sharply angular fragments will produce ploughing and cutting effects at low angles, as do ordinary fixed abrasives. Bead blasting can therefore increase the distinction between the response of ductile and brittle materials at high angles, offering an extra degree of selectivity.

●4.3   Energy deposition

The excess energy delivered to a surface must be dissipated (after allowing for plastic deformation and fracture, that is). This excess must appear as heat and, at the extremely high rates of delivery occurring under impact, localized very high temperatures and possibly melting can be expected. Metal erosion rates in particular have been shown to be inversely correlated with the product of specific heat, c_p, and the difference between the test temperature and the melting point of the metal, ΔT_m :

$$\varepsilon \propto (c_p \Delta T_m)^{-1} \qquad (4.9)$$

This makes sense in terms of the general decrease in strength which is observed for metals as their melting point is approached (*cf.* Fig. 14§7.2). It should therefore in principle also be an important factor with other materials. Certainly, the heat generated by abrasion proper affects the local melting and flow of metals, polymers, and indeed ceramics, if the heating is sufficient. Even hydroxyapatite, as in teeth, melts when being cut by a high speed rotary device; this is the origin of the smear layer which is the focus of so much attention in bonding studies and treatment.

●4.4   Choice of grit

The overall effects of any particular grit and substrate under a particular set of conditions are probably best determined experimentally. Exact prediction is not yet possible from the available theory, but the principles outlined above do permit broad decisions to be made. They do indeed explain the choices that have been made in practice, probably on the basis of experience, for specific tasks. For example, in removing the very brittle porcelain from a facing that is rejected for some reason, large smooth particles are most effective in breaking up the porcelain without removing metal, which can therefore be reused. Glass beads are therefore the blasting medium of choice. While silicon carbide powder (which is sharp) could be used to remove investment and oxide from cast chromium alloys with low rates of removal of the metal beneath (because of the metal's toughness), such a grit would be disastrous for cleaning adhering mould plaster from denture base acrylic. Here, a combination of low density (*i.e.* low mass) particles with low strength and hardness are required to concentrate the erosion on brittle gypsum remains rather than the flexible and ductile polymer. Such an abrasive is the pulverized nut-shell commonly used for this job. It can be seen therefore that the incorrect choice of a blasting medium could be disastrous for the workpiece. Equally, cross-contamination of abrasive materials may lead to undesirable damage.

In all grinding and cutting operations, control of the waste is important. In grit blasting the difficulties are increased. Special containment and dust collection equipment is necessary, certainly to avoid the mess but also to prevent inhalation of the dusts, some of which (silicates, for example) are extremely harmful. The nature of the workpiece must also be considered here (*cf.* Be in Ni-Cr alloys, 19§2.6).

There is a further concern. High-velocity impact can result in particles becoming embedded in the substrate, especially if this is ductile. Thus, alumina can be embedded in titanium (Fig. 4.7). This can have physiological consequences if such a surface is exposed to living tissue, such as for an implant.[6] Should the surface be required to work against another, such as in an artificial joint, the embedded particles would abrade to the opposing surface causing wear debris to accumulate, and this in turn may be undesirable in terms both of toxicity and service life. Thus care in the choice of abrasive, or clean-up procedure if this is feasible, is required.

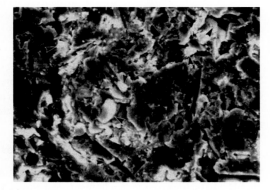

Fig. 4.7 Alumina abrasive particles (6 arrowed) embedded in titanium by grit blasting (frame width: 50 μm).

●4.5   Liquid honing

A variation of the erosion process is seen outside dentistry in the use of so-called **liquid honing**. This technique has the grit suspended in a liquid, typically water. It has the advantage of making dust control much easier, although the mechanisms of erosion would be expected to be somewhat modified by the suspension medium acting as lubricant and coolant. It has now been applied to dental 'prophylaxis' in the form of a slurry of a benign, preferably poorly-soluble substance: sodium bicarbonate ($NaHCO_3$; ~100 g/L, 23 °C) is commonly used, but glycine (NH_2CH_2COOH; ~25 g/L, 20 °C) and even erythritol ($HOCH_2(CHOH)_2CH_2OH$; ~470 g/L, 25 °C) are employed. This approach combines relative erosion mildness and ease of disposal of the waste with a lack of biological compatibility problems. Water, of course, is benign in the oral context. However, the efficiency of the process is such that great care must be exercised in its use if excessive unwanted loss of tooth tissue is not to occur. As will now be realized, there can be no absolute cut-off or selectivity between dental calculus on the one hand (which is largely calcium phosphates) and dentine, especially, on the other. The dentist's intent as such here is quite irrelevant: damage to tooth tissue is a serious risk.

●4.6   **Air abrasion**

There is, however, the also need to remove carious tooth tissue, and grit blasting can be used quite effectively for this. Presumably, the rather odd name for this process, **air abrasion**, arises from avoidance of the rather industrial-sounding term of 'grit blasting', but there is no practical difference – just the application has changed; the principles discussed above still apply.

There are a number of appreciable advantages with such an approach. Primarily it is very quiet in comparison with an air-turbine driven cutting instrument, with no vibration. It is also said to be more precise, but since there is no tactile feedback, as with a handpiece, caution and frequent inspection are essential. The low mechanical impact is said to avoid generating microscopic fractures in enamel (*i.e.* Griffith flaws) as well as making local anaesthesia unnecessary for shallow cavities. The absence of a water spray is considered by some an advantage. Certainly, the rounded cavity profiles created – absence of sharp line angles – mean that crack-inducing stress-concentrations are avoided, especially if the microscopic fracturing is eliminated.

The method is sometimes said to generate no heat, but this is incorrect. All mechanical interactions of this kind must involve the conversion of kinetic to thermal energy – work is being done, and it is quite possible to raise the temperature of tooth tissue substantially, with its usual adverse consequences (§7). The characteristic smell of overheated protein is said to be absent, but this is not a sufficient indicator of the temperatures attained. It may be that the air flow can provide sufficient cooling to avoid damage, or at least reduce the perception of heating, but this does not remove the need for intermittent use. The injection of some water into the air stream to enhance cooling has also been used.

There are a number of severe problems with the technique which must be heeded. The spent abrasive is spread widely, requiring a very good exhaust system. Although the abrasive used, alumina, is ordinarily believed to be generally biocompatible, this is not necessarily a sufficient condition (see §4.4). Eye protection, and protection against inhalation, is still required for all persons involved. This problem might be at least partially mitigated by water in the air stream. In addition, the particle jet is extremely aggressive to soft tissue, which therefore requires assiduous protection with, for example, rubber dam (*cf.* the heavy rubber gloves used in grit-blasting cabinets, including those in dental laboratories). While the method can in part discriminate carious tissue from sound, removing it more rapidly because it is softer, this is not an absolute distinction, and over-zealous use can remove large volumes of sound tissue. Adjacent teeth also need to be carefully protected. Likewise, it cannot be used to remove harder or tougher inlay materials, and it should not be used on silver amalgam because of the vaporization of mercury that occurs, due to the localized heating under each abrasive particle impact.

The greatest risk, however, lies in surgical emphysema: the injection of air into tissue. Whilst this can easily occur with an air-turbine handpiece, the pressure used there is typically only about 2 ~ 3 bar. Air abrasion requires pressures of 3 ~ 11 bar, with concomitantly increasing risk with increasing pressure. Soft tissue must therefore be protected, but also the risk applies to cutting in dentine: with a pulp exposure, air would be injected into the pulp chamber (destroying the pulp) and thus systemically. The risk is then one of air embolism. Note that the presence of abrasive is not essential for this to occur: compressed air is a hazard in itself, capable of flaying soft tissue and injecting gas through cuts, under finger nails, or wounds created by the air itself. Note well: compressed air can be, and has been, lethal. Much care is essential.

A similar but less aggressive device, called an **air brush**,[5] has been used to slightly roughen the surface of restorations so that the wax of articulating paper (16§6.2) can more easily adhere.

●4.7   **Specification**

In describing grit-blasting processes in dental contexts, the conditions are very nearly always specified only in terms of the type and particle size of the grit – which are reasonable, and the air pressure used. However, it should be apparent from the discussion at §4.1 - 4.2 that the air supply pressure is not a fundamental controlling variable as such, and that the erosion conditions are therefore insufficiently specified. With that information alone, the treatment cannot be reproduced by others unless identical equipment is used, a rather

---

[5]   Not to be confused with the device of the same name used for artwork, retouching photographs, and the like.

unlikely general circumstance. What principally is required instead is the nozzle velocity, which depends on the flow resistance of the nozzle – that is, the diameter and 'throat' shape, and the pressure difference between the attached supply line and the outside. Since the pressure gauge is often positioned at some point remote from the nozzle, the flow resistance in the apparatus (hoses and grit delivery system) must also be taken into account as this causes a pressure drop in a manner entirely analogous to electrical resistance causing a voltage drop. What shows on the gauge is not what is applied to the nozzle. If the system is badly designed, there will also be an appreciable pressure drop at the gauge when the air is flowing due to the resistance in the supply line to that point from the source such that the relevant reading is to be obtained only while the gas is flowing. Thus, without knowing as well the exact system details in all relevant respects, it is quite impossible to reproduce the conditions from a gas pressure statement alone.

It then needs to be assumed that the grit particles are in fact travelling at the gas velocity as they emerge, but of course this cannot be uniform across the section of the nozzle (Fig. 4§11.3). In addition, as the gas stream spreads after emergence from the nozzle, the net velocity must drop, and more so because the stream must be turbulent. Accordingly, the impact velocity of the particles must also depend on the distance of the workpiece from the nozzle – which value should also be controlled and reported (the momentum of small particles will be insufficient for them to carry very far independently of the gas velocity, and their shape and density will affect their behaviour). As indicated in §4.2, the angle of impact, at least at the centre of the stream, must also be controlled as appropriate.

It is evident that a full analysis of the flow is not a practical proposition for a statement to enable reproducible blasting conditions. Accordingly, a specification of the nozzle by type and catalogue number may be adequate in this respect. However, an additional concern is wear of the nozzle. Clearly, high-velocity abrasive moving through the nozzle must abrade those contacting surfaces, and in particular the throat. The flow characteristics will therefore change so that even specifying the nozzle by part number implies that it must be unworn. It cannot be assumed that behaviour is constant over time. Because wear is such a problem, hard materials such as alumina, tungsten carbide (21§3) and boron carbide (B$_4$C) are commonly used for nozzle liners. Even then, wear is appreciable and regular replacement is essential, even in non-critical contexts. Obviously, the amount of grit delivered needs to be specified since the number of impacts is the primary controlling factor for rate of material removal, and thus – integrated over time – the actual amount removed. Hence, there needs to be a means of metering and quantitating the delivery of grit into the gas stream if control is to be had and reporting complete.

It may be noted in passing that there is an effective upper limit to the amount of grit that a fluid stream can carry, essentially that of the random loose-packed limit for that grit shape and particle size distribution (*cf.* Fig. 4§9.2), although the viscosity of the composite stream will then rise, the flow rate must therefore fall (depending on how far the back from the nozzle the grit is added), and the impact velocity decline. (The relevant equations for a compressible fluid, a gas, are more complicated than for an essentially incompressible liquid, and these therefore lie beyond the present scope.)

It is clear, then, in a research context, or for an application demanding precision of process, air pressure alone is an entirely unsatisfactory means of specification of blasting conditions, even though it is commonly the only control variable available. For example, **tribochemical coating** is critically dependent on the impact energy of the particles being sufficient to create conditions where reaction can occur. Unless all factors are as required, the treatment must fail to achieve the intended effect.

●4.8   Microabrasion

In cosmetic dentistry, a treatment has arisen motivated by the desire to remove discolouration that is resistant to bleaching but which is not merely superficial, being deeper into the substance of the tooth enamel. In essence, this entails using pumice powder as an abrasive in the presence of a fairly concentrated solution of phosphoric acid (sometimes even hydrochloric acid).

Pumice is a foamed glass, formed when molten rock (lava) is depressurised rapidly on ejection from a volcano, causing dissolved gases to come out of solution (**exsolve**), but then rapidly cooled before the bubbles have chance to coalesce, burst and the gas escape. Ground to a powder, the particles have many sharp edges because of the fractured bubbles, and it is therefore convenient as an abrasive, especially because it is very cheap. Its weak, brittle, glassy nature also gives the tendency to fragment during use, making the particles smaller so that the finish obtained may be progressively finer, although this is not used as a working principle particularly deliberately.

The acid, of course, simply dissolves tooth tissue mineral – essentially hydroxyapatite. As this means etching in the same sense as is used for bonding filled resins, for example, it will leave a highly-sculptured topography that will be amenable to rapid abrasion by the pumice, as the peaks are broken off easily. The overall effect is therefore an accelerated and macroscopic reduction of the surface of the enamel.

The name of this treatment is therefore grossly misleading: it is certainly not 'micro-' in any sense, and while the intention is sometimes said to be to remove no more than ~0.2 mm of enamel, for example (and this is a significant proportion of enamel thickness), it is difficult to see how this can be controlled with any precision, bearing in mind that the etching must occur to a greater depth than the gross tissue removal if nothing else is done. It is certainly a more aggressive process than with pumice alone, as is used for "prophy" paste. There is also some hazard, particularly with the use of hydrochloric acid, as spatter is unavoidable, and full protection for operator and patient, especially of eyes, is essential. Adjacent teeth should not be exposed incidentally. Working under dam is therefore essential, and emergency treatment (sodium bicarbonate slurry) must be to hand. The etching effect also means that the remaining surface is left somewhat porous, and may be left more susceptible to further staining than before.

§5. Electrolytic Polishing

Even though blasting may be able to clean up a casting from adhering investment and oxide scale, there remains the problem of polishing the intricate shapes that are frequently required for dental devices. **Electrolytic polishing**, sometimes called **brightening**, offers a means of dealing with this kind of work relatively efficiently. It may be viewed as accelerated corrosion as the intention is to remove metal through a dissolution process (13§7). The piece to be brightened is made the anode of an electrolytic cell, $i.e.$ by applying a large enough positive voltage to it, so that metal dissolves readily though being oxidized to metal ions. The exact behaviour observed depends on the chemistry of the metal, of the electrolyte, and on the voltage applied.

Many metals and alloys show a consistent pattern of behaviour with respect to voltage. If the impressed voltage is too low then etching effects dominate, maintaining or increasing the roughness. A little higher and solid reaction products tend to coat the workpiece ($cf.$ the 'anodizing' of aluminium, 13§5.2). On the other hand, if the voltage is too high then evolution of oxygen dominates (through the electrolysis of water), which is pointless and therefore to be avoided. Within a certain range, however, electrolytic polishing can occur (Fig. 5.1). Only this region, where neither oxygen evolution nor reaction product layer formation occurs, is suitable. The composition, pH and temperature of the solution are the main controlling factors. Even so, the principal effect of electrolytic polishing can be traced to two factors inherent in a rough surface, and the fundamental idea of surface energy.

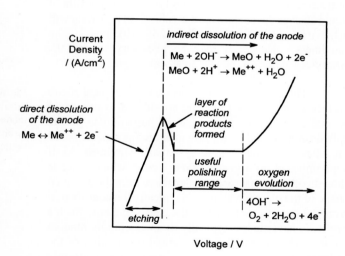

Fig. 5.1 Typical current-voltage plot for the electrolytic behaviour of a metal and the associated chemistry. The central region is associated with polishing effects.

The role of surface energy in affecting corrosion was discussed in 13§6.3. In essence, surface energy is in part a function of the local radius of curvature of the surface (10§2). Hence, for a rough surface, the radius of any feature may be locally very small and therefore have an associated high surface energy. In addition, any surface produced by an abrasion, cutting or erosion process implies a worked layer of relatively poor crystallinity and high residual stress (Figs 2.2, 2.3); this is associated with a higher surface energy simply because of the departure from equilibrium.

Such a surface is therefore subject to stress corrosion (13§6.1). Thus, atoms in positions of high surface energy (which necessarily are more easily oxidized) are removed first, and the roughness is expected to be reduced, eventually giving a single smooth surface of uniform curvature, without cold-worked material. The limitation is that the grain boundaries are themselves regions of disorder and can also be etched preferentially. This is the basis of a metallographic procedure to show the microstructure of metals (Fig. 12§2.6), but clearly it is undesirable in the context of polishing as the structure may be weakened from the surface cracks that the etched boundaries could represent. The very same factors are, of course, the reasons for the enhanced corrosion characteristics of rough and deformed surfaces.

§6. Roughness

The aim of polishing is to produce a "smooth" surface, but this begs the question of what it is that constitutes smooth. The analysis of surface texture provides descriptive measures to enable the state of the surface to be judged. Since these measures are used both in the literature to describe the outcome of a process and as a means of specifying a desired finish, it is necessary to understand their basis and the terminology.

Notice that the visual assessment of polish has to do with the quality of the reflection of light, *i.e.* whether or not it is mirror-like (24§5.7). This appearance is often achieved in domestic contexts by applying wax 'polish', a thin film that fills in roughness and which by rubbing becomes smooth and reflective. Clearly, nothing has been done to the underlying surface of the object and this kind of treatment is irrelevant to dentistry. However, a film of liquid such as the coolant used in (abrasive) polishing will achieve precisely the same effect (24§5.10), masking the roughness. Hence, it may be necessary to dry the workpiece surface to be able to assess progress.

The profile of a cross-section of a typical rough surface (Fig. 6.1, top) may be considered to consist of several components.[7] Thus, small variations in profile called **roughness** (Fig. 6.1) are imagined as superimposed on what might be a more systematic **waviness**, itself superimposed on a general **form error**, which latter is thought of as the macroscopic departure from the designed shape of the piece. Waviness is typically the contribution from the shaping tool, such as the blades of a bur. Such a profile is nevertheless typically of an essentially random nature. This means that the vertical distribution of the ordinates (that is, the heights) of successive, uniformly spaced points tends towards a Normal (or 'Gaussian') distribution (Fig. 6.2). This kind of distribution can be described by just two parameters, location and scale. Location is interpreted as the average height or mean position of the surface, which is of interest as a measure of overall dimensional accuracy, but this is beyond the scope of roughness as such. The scale parameter expresses the spread of the values about the mean, and the commonest such measure generally is the standard deviation; the scale parameter is thus a proper measure of roughness.

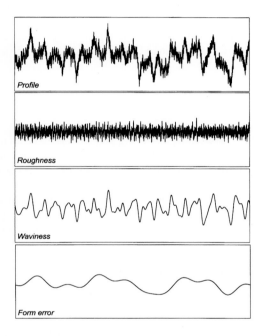

Fig. 6.1 Conceptual breakdown of the kinds of surface profile deviation that constitute 'roughness'. The random profile (top) has been filtered into three (arbitrary) wavelength band components. (NB: the vertical scale is much exaggerated.)

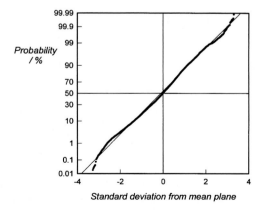

Fig. 6.2 Typical distribution of spot heights on a 'Normal' randomly rough surface. Data from the profile shown in Fig. 6.1.

•6.1 Centre line average roughness

Historically, however, it has been the **average deviation**, designated R_a, and sometimes called the **centre line average** or CLA, which has commonly been used to express roughness. It is actually defined by the expression

$$R_a = \frac{1}{L}\int_0^L |y - \overline{y}|.dL \qquad (6.1)$$

where $\overline{y}$ is the average height. *i.e.* the position of the **centre line**, and L is the distance over which the measurements are being made. Thus the centre line corresponds to that section through the profile which cuts off equal areas above and below it (Fig. 6.3). Notice that it is the mean of the *absolute* values of the deviations that is used.

In practice a strict integration cannot be performed, but it is approximated by making a series of height determinations at small but discrete intervals (Fig. 6.4). This then leads to the working equation:

$$R_a = \frac{1}{n}\sum_{i=1}^{n}|y_i - \overline{y}| \qquad (6.2)$$

where n is the number of points at which height measurements are made.

In a machine to measure roughness, a stylus is drawn over the surface (Fig. 6.5). The stylus is free to move vertically, and its position in relation to a rigid reference bar (which is called the internal machine straight line datum) is determined from the transducer to which the stylus is attached. The R_a value can be obtained from a simple electronic circuit or through a microprocessor program. In the latter case more detailed calculations of the surface characteristics can also readily be made.

•6.2 RMS Roughness

The 'root mean square' or RMS roughness measure, R_q, is equivalent to the standard deviation of the profile about a centre line and is defined by:

$$R_q^2 = \frac{1}{L}\int_0^L (y - \overline{y})^2.dL \qquad (6.3)$$

where $\overline{y}$ is now defined as the value giving a minimum sum of squares of deviations from the line. This is also

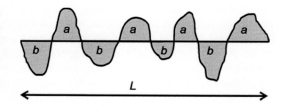

Fig. 6.3 Graphical definition of the Centre Line for the purposes of calculating a roughness: $\Sigma a = \Sigma b$.

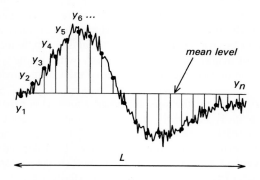

Fig. 6.4 Practical implementation of a Centre Line Average roughness calculation. The y values are the heights sampled at regular intervals (black spots) from the profile.

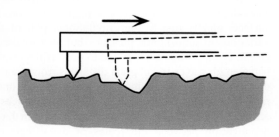

Fig. 6.5 Typical method of tracing a surface profile using a stylus. A position transducer provides height data relative to the machine's traverse datum.

called a 'least squares' fitted line (as in ordinary statistical practice), and it is necessarily not identical to the R_a centre line. Again, because the exact integration cannot be done, as any machine can only take a sample of heights of the surface, the practical form of the calculation is:

$$R_q^2 = \frac{1}{n}\sum_{i=1}^{n}(y - \overline{y})^2 \equiv \frac{1}{n}\left[\Sigma y^2 - \frac{(\Sigma y)^2}{n}\right] \qquad (6.4)$$

(This expression is, of course, strictly that for a population variance, but if the number of points sampled is large enough the difference will be negligible in the present context.) While the value of R_q is more useful in theoretical terms as a measure of roughness, R_a has until recently been easier (and cheaper) to compute by machine and consequently has wider currency and literature presence. The availability of cheap microprocessor-based instruments is expected to change this usage greatly. Both R_a and R_q have the dimensions of Length and may be expressed in convenient units; micrometers is usual. However, as the algebra shows, there is no way to interconvert the two; they are separate measures.

There are several other measures of roughness that can be obtained from modern profilometers. They each have their use as a descriptor, and some may be associated with external mechanical, physical and chemical effects more than others: it very much depends on the purpose which might be most appropriate. Since they are all derived from the same vertical displacement data, they are all highly correlated with each other; they are not independent measures.

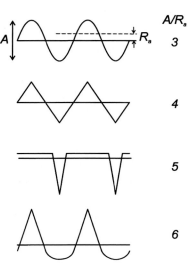

●6.3   Profile

The same value of R_a can, however, be obtained from a wide range of different profiles and textures (Figs 6.6, 6.7). This also applies to R_q (but is more difficult to illustrate). In other words, both R_a and R_q are insufficient to describe a surface. Further measures are necessary for a better description of the form of the roughness, one of which concerns the average separation of successive peaks or the **average wavelength**. Similar analyses can be made of wavelength as of height, but it is as necessary with this as with surface height variation to decide on the scale of the variation which is of interest.

Fig. 6.6 Because of the definition of centre line average roughness, R_a, many different profiles may have precisely the same value of R_a.

A profile was described above as consisting of roughness, waviness and form error. Clearly though, there can be no absolute boundaries separating the scopes of these labels, and the distinctions between them are entirely arbitrary. They depend on (a) the desired accuracy and precision of the surface, (b) the preparation method - *i.e.* the tool marks that are unavoidable (these are often all about the same size), (c) the magnification, and (d) the application.

The surface of the ocean provides a good analogy. To start with, the average surface is not spherical because of the uneven distribution of mass and the rotation of the earth; this is a form error. The tides too produce form errors, major deviations from the average surface, mostly due to the gravitational pull of the moon, as do variations in atmospheric pressure, but on a smaller (vertical) scale. Ocean swell ('waviness') is due to powerful storms, and these long wavelength waves travel great distances from their source. Coming down the scale, local winds, shipping, bathers and rain produce a range of successively smaller scale disturbances which may be equated with roughness. Nevertheless, it is plain that there is no sharp demarcation between the scales of the effects of these various phenomena or even the interpretations placed on them. The scale of disturbances ranges from extremely small to extremely large, although at any one place one particular magnitude of effect may be dominant. This is true whether we consider the vertical deviations or the wavelengths of the effects.

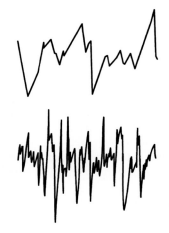

Fig. 6.7 Again, because of the definition of centre line average roughness, R_a, precisely the same value is associated with profiles of varying wavelength (or frequency).

A more useful view is that a **spectrum** of wavelengths characterizes the profile, that is to say the variation of amplitude (intensity) as a function of wavelength (scale). We can obtain such a spectrum by a Fourier analysis of profile data (for which a computer is needed). This method was used to separate the components of the profile shown in Fig. 6.1. In practice, for the simpler types of roughness machine, a high-pass filter is applied to the surface profile signal before analysis, so that features of a scale too *large* to be of interest are eliminated from the calculated roughness. In other words, if we are not concerned about the curvature of a specimen, the filter can be set to ignore it. On the other hand, if we are concerned about tracing a profile to study form error, we should not filter the signal.

●6.4   Problems

It should be apparent from Fig. 6.5 that any such attempt to measure the roughness of a surface corresponds to that of Fig. 2.1: a scratch is inevitable, and one that tends to reduce the magnitude of the peak excursion and increase the depth of the trough record, and to an extent depending on the load applied – zero is not possible. On relatively hard and relatively rough surfaces, this may not matter very much, but it is clearly a 'destructive' technique, and very often the scratch is visible to the naked eye, at the right viewing and lighting angle. The ability to trace the troughs also depends on the narrowness of the stylus tip, essentially the included angle of the point, but also the radius of curvature (r_c) of the tip. Small angles and small r_c increase the pressure and the depth of the unwanted scratch, while large angles prevent tracing by resting on (and deforming) adjacent high spots. It is for these reasons that **non-contact profilometry** is preferable, and this can be done by various optical techniques.

§7. Heating

Local heating due to friction and deformation has been mentioned at several points. In addition to changes in the materials themselves, this may be critical in the case of cutting tooth material, or in grinding or polishing a restoration in place, where the comfort of the patient and the possible destruction of a vital pulp is concerned. Practical studies and computer simulations have shown that, in the absence of coolant, normal cutting techniques result in large temperature rises (Figs 7.1 - 7.3), sufficient indeed to cook the pulp.[8]

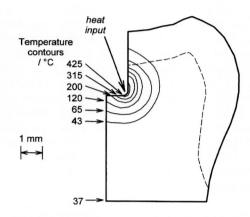

Fig. 7.1 The kind of temperature distribution expected from cutting a tooth without cooling.

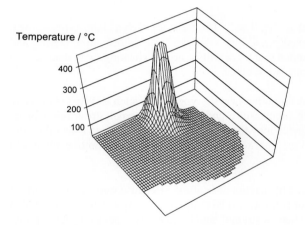

Fig. 7.2 Same data as in Fig. 7.1, replotted to show the temperature gradients.

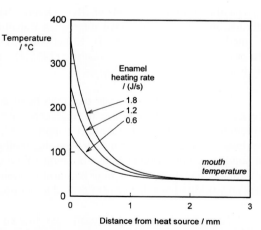

Fig. 7.3 Expected temperature profiles for different rates of heat deposition in enamel.

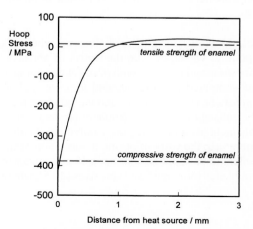

Fig. 7.4 Possible thermal stress pattern in enamel as a function of distance from heat source.

The non-uniformity of the temperature rise results in varying strains due to the variation of thermal expansion from point to point, and these generate stresses which conceivably could exceed the strength of the tooth materials (Fig. 7.4) and cause cracks to form. These cracks may not result in immediate failure, but they will certainly weaken the structure of the tooth and may permit fracture under occlusal loads more readily. The liberal use of coolant and light, intermittent, cutting techniques are thus essential for prevention of trauma and damage to the tooth (as well as being beneficial for the cutting tool and handpiece; see below). It is perhaps sufficient to note that inflammatory changes are pronounced at pulp temperatures of only 42 °C.[9] For this reason alone other proposed treatments which involve high temperatures, such as those for bleaching which require 80 °C or so, are suspect on vital teeth.

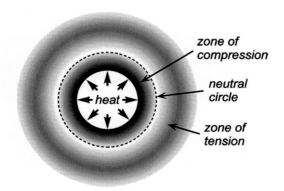

It is worth exploring the effect of heating like this in a little more detail. Referring back to Fig. 18§4.6, it is apparent that heating the material at the periphery of a hole must throw that material into compression as it tries to expand but is constrained by the surrounding, cooler body (Fig. 7.5). Equally, the body itself is being stretched by the piece that is trying to expand, *i.e.* it is in tension (the mechanics are of the kind discussed in 6§2.2). As shown in Fig. 7.4, there is a smooth transition from compressive to tensile stresses, passing through zero – at the neutral circle (*cf.* 23§2.6). Curiously, the failure may initiate in tension

Fig. 7.5   The state of stress in the material surrounding a hole whose walls are expanding on being heated. (The depth of shading indicates roughly the magnitude of the hoop stress at any point.)

at some distance from the point of application of the heat, and so might not be visible unless the crack propagates to a free surface. A larger volume of material may be subject to a large tensile stress and thus, on the Griffith criterion (1§7) run proportionately greater risk of a critical flaw being found there. Even so, it must be born in mind that there is no such thing as "compressive strength" (1§6.3) and that failure in the compression zone must be in shear (*cf.* 25§5.4).

Some abrasive and polishing products are sold as "heatless". The basis for such a claim is not known. However, it should be realized that this is not a physical possibility. All abrasion, deformation, cutting and friction must generate heat as work is done. If it is assumed that there will be no such heating, only damage can result. Always proceed with the appropriate precautions being taken.

Other materials may also suffer. Denture base and other acrylic-type resins, such as filled restorative materials, may be heated above their glass transition temperature and so allow internal stress relief or be deformed by external forces (§2.4). Either way, the device may thereby be rendered useless as it will then fail to fit properly. Amalgam will transform (14§5.4) and melt peritectically (14§5.5); glass ionomer cement will dehydrate.

It is not just the substrate that may be affected by frictional heating. Because cutting edges are of necessity of low bulk, their ability to conduct away heat is limited. The temperature may rise high enough that the melting point of the material may be approached, when its strength and hardness is expected to fall rapidly. Steel burs are readily blunted in this way; diamond grits are oxidized by the oxygen of the air as they get hot (diamonds are not 'forever'); the bonding of abrasive grits in general, which bonding may be a resin, electroplated metal or rubber, may be burnt or melted. Proper treatment of these tools will result in greatly extended working lives, better results, and lower operating costs.

There is a further aspect easily overlooked. In general, the binder for a bonded abrasive, whether a 'stone', a powder-coated instrument, a more dilute 'rubber' point or wheel, or even the matrix of abrasive-impregnated brush fibres, must also come into contact with the substrate and its chips, both *in situ* and as abraded fragments. Since the binder is nearly always relatively very soft compared with the intended abrasive, it plays an altogether negligible abrasive role. However, relative motion between such materials and the substrate must result in the generation of frictional heat, and this alone can be substantial (1§5.5). Rubbery binders can be expected to show this effect in an exaggerated fashion (and this includes otherwise glassy polymeric binders above their T_g – which soon happens in use anyway). Thus, more than the work of plastic deformation of the substrate is involved, re-emphasizing the need for adequate coolant to be supplied continuously as a matter of routine. It also provides a further reason for avoiding clogging of bladed tools and fixed abrasives (§2.5), where such friction must become dominant.

§8. Diamond

Diamond powder-coated rotary abrasive instruments are used in a variety of common tasks in dentistry, primarily because of diamond's extreme hardness (H_K ~7000 - 10,000, depending on quality and purity, *i.e.* 70 GPa or higher) (for its structure, see Fig. 17§2.18). Several other properties also contribute to its effectiveness as an abrasive. As might be expected, it is very brittle although strong, and this allows the formation of very sharp edges that show great wear resistance. In addition, it exhibits low friction and so does not produce as much heat as other abrasive materials, which property coupled with its relatively high thermal conductivity (1000 ~ 2000 $Wm^{-1}K^{-1}$, *cf.* Cu: ~400 $Wm^{-1}K^{-1}$; Al_2O_3: ~40 $Wm^{-1}K^{-1}$) offsets to some extent its instability at high temperature – it oxidizes in air to carbon dioxide over about 700 °C – and allows it to be used at high cutting speeds. Its reactivity otherwise is quite low, and shows no adhesion to most metals, unlike some metallic cutting tools, for example.

However, diamond does have some less than beneficial aspects, not least its cost (despite diamond abrasives being artificially manufactured). Its brittleness means that under impact the crystalline material may cleave easily, and although sharp edges are again generated, this is a source of wear. There is also a tendency to react with the carbide-forming transition metals of groups IVA to VIII, most importantly Ti, Ta, Zr and W, while it dissolves readily in Fe, Co, Mn, Ni, and Cr under high-temperature conditions, that is, under the working abrasive particle. This contributes to the blunting of edges in addition to the effect of oxidation, but also contamination of the workpiece that is undesirable in critical applications.

Noting that there are some essential compromises in the use of this material, it is appropriate to stress that adequate cooling water and intermittent use with light loads will prolong useful service.

The carbide-reaction and dissolution problems can be avoided by using cubic boron nitride, which is the next hardest material known (~50 GPa), generally chemically-inert, and stable in air up to about 1400 °C. Unfortunately, it is several times more expensive than diamond at present, and has yet to make an appearance in dental contexts.

References

[1]   Scott D (ed). Treatise on Materials Science and Technology. Vol. 13. Wear. Academic, New York, 1979.

[2]   Petzow G. Metallographic Etching. Amer. Soc. for Metals, Ohio, 1978.

[3]   Cokic SM, Duca RC, Godderis L, Hoet PH, Seo JW, Van Meerbeek B & Van Landuyt KL. Release of monomers from composite dust. J Dent 60: 56 - 62, 2017.

[4]   Schmalz G, Hickel R, van Landuyt KL & Reichl F-X. Nanoparticles in dentistry. Dent Mater 33(11): 1298 - 1314, 2017.

[5]   Eirich FR. The role of friction and abrasion in the drilling of teeth. in: Pearlman S (ed) The Cutting Edge - Interfacial Dynamics of Cutting and Grinding, pp 175 - 198. DHEW Pub. No. (NIH) 76-760, US Dept. of Health, Education and Welfare, NIH, 1976.

[6]   Darvell BW, Samman N, Luk HWK, Clark RKF & Tideman H. Contamination of titanium castings by aluminium oxide blasting J Dent 23: 319 - 322, 1995.

[7]   Dagnall H. Exploring Surface Texture. Rank Taylor Hobson, Leicester, 1980.

[8]   Lloyd BA, Christensen DO & Brown WS. Energy inputs and thermal stresses during cutting in dental materials. in: Pearlman S (ed) The Cutting Edge - Interfacial Dynamics of Cutting and Grinding, pp 175 - 198. DHEW Pub. No. (NIH) 76-760, US Dept. of Health, Education and Welfare, NIH, 1976.

[9]   Plant CG, Jones DW & Darvell BW. The heat evolved and temperatures attained during setting of restorative materials. Brit Dent J 137: 233 - 238, 1974.

Chapter 21   Steel and Cermet

Steel is the single most important engineering alloy, not least because the iron which forms its basis is cheap. The general lack of corrosion resistance, however, precludes applications in dentistry in both the oral environment and where sterilization is necessary, as in hand instruments. Consequently, stainless steel is the alloy of choice.

*The mechanical properties of steel depend on crystal structure, and especially as this is modified by the presence of **carbon** – which defines the alloy as steel. Control of the amount, distribution and form of the carbon is critical, and this is effected through combinations of thermal and deformation history. Equally, further thermal treatment and deformation can change those properties; such changes may be detrimental to the service behaviour of the material.*

Stainless steel *belongs to the passive group of corrosion resistant alloys, relying on the presence of a thin coating of chromic oxide to isolate it from the environment. The formation and maintenance of this oxide coat is a function of composition; heat treatment may spoil it. The mechanical properties have to be adjusted by adding other alloying elements to offset the effects of including the chromium.*

The proper care of stainless steel, and its application in dentistry, depends on a recognition of the factors that effect its structure. A basic understanding of its metallurgy is therefore necessary.

Cermets *are a class of materials that utilize the extreme hardness of carbides for cutting tools such as enamel chisels and burs for use on cobalt-chromium alloys. The difficulties of fabrication in manufacture, and the extreme brittleness in service, are overcome by using a matrix of metallic cobalt. Understanding the structure of these materials permits their effective use without damage.*

Materials Science for Dentistry
https://doi.org/10.1016/B978-0-08-101035-8.50021-3

Steel in several forms is used in the fabrication of many dental instruments: burs, scalpel blades, probes, and so on. As wires and bands, stainless steel finds applications in orthodontics and as matrices in restorative dentistry, and as small-bore tubing for injection needles. Other steels will probably form a number of parts of handpieces, retention pins, and such items as rubber dam clamps.

The topic of steel is of vast scope, forming the basis as it does of one of the world's major industries and having applications in nearly all possible fields. Its low cost, strength, ease of fabrication and general durability are the basis of this. The diversity of application arises from an equally diverse range of properties, depending on the particular treatment and composition. It is not proposed to deal with steel-making or its development in any detail, and we cannot help but oversimplify the subject in trying to extract some basic principles whereby the main ideas may be understood. Nevertheless, these ideas will serve to point out the applications - and the limitations - of steel in dentistry.

§1. Carbon Steel

Pure iron is a soft and malleable metal that has little structural use. A simple **steel**, an alloy consisting of iron and a small amount of carbon, is harder, tougher and stronger. The importance of the carbon arises firstly from the fact that it is a small enough atom that it can form **interstitial** solid solutions (19§2.3). The sensitivity of the properties of steel to the dissolved carbon is very great, as with the Co-Cr alloys. Even so, the diversity of steels is due to not just this possibility but is also related to the **dimorphism** of iron, that is, it exists in two distinct crystal forms. Pure Fe at low temperatures has a body-centred cubic (b.c.c., 11§3.4) structure. This converts to face-centred (f.c.c.) on heating to 906 °C, and then back again to b.c.c. at 1401 °C. This last structure then persists up to the melting point of ~1535 °C.

While the maximum radius ratio of an assumed hard, spherical atom that can be fitted into an f.c.c. lattice is ~0·41 in an 'octahedral' hole, in a b.c.c. lattice without distorting it is rather smaller at ~0·29 (Table 1.1; Fig. 19§2.3). For carbon and iron the radius ratio is actually ~0·60 so some considerable distortion of the iron lattice is expected if it contains any carbon at all. Indeed, C in b.c.c. Fe results in severe tetragonal distortion. This distortion is mainly the cause of the hardening effect (*cf.* δ-AuCu, 19§1.3). It follows from this that the solubility of carbon in iron is expected to be limited. Further, the solubility is expected to be much less in the case of b.c.c. iron than in f.c.c. If, then, there are to be structural transformations, the fate of the carbon must be identified.

Part of the C-Fe constitutional diagram is shown in Fig. 1.1,[1] and the processes involved can be traced here as with other alloys (Chap. 12). Note that because the molar mass of carbon (12) is much smaller than that of iron (56), a small mass proportion (lower scale) in fact corresponds to a large atomic proportion (upper scale) by very roughly a factor of four. For structural matters, we should of course think in terms of atomic proportions. For example, at 1·7 mass% C the proportion C : Fe is 1 : 12, or 1 C atom per 3 unit cells (11§3.3) – a rather easier concept. However, in common with the usual practice, all compositions given here will be in mass percent.

Table 1.1 *Relative sizes of interstitial holes in the common metal structures.*

hole type:	octahedral	tetrahedral
b.c.c	0.155	0.291
f.c.c., h.c.p.	0.414	0.225

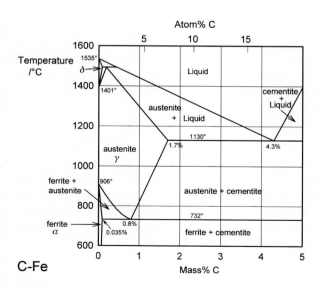

C-Fe

Fig. 1.1 Part of the C-Fe working constitutional diagram (non-equilibrium). NB: the right-hand edge is not Fe₃C, which would appear at ~6.7 mass% C, 25 at% C.

Steel has had a long history, and many special names have been introduced in the literature associated with various kinds of structure and composition. These names are frequently used as a kind of shorthand to refer to these forms. Thus, **ferrite** is the name given to the solid solution of C in b.c.c. Fe (also known as α-iron); this has a maximum solubility of about 0·035 % at 732 °C (Fig. 1.1) (*i.e.* C : Fe = 1 : 612!). **Austenite**, on the other hand, C in f.c.c. Fe (γ-iron), shows a much greater maximum solubility of ~1·7 % at 1130 °C (C : Fe = 1 : 13); this phase is only stable at high temperatures.

●1.1   Cementite

One other aspect of the system is of great importance: Fe forms a carbide, Fe_3C, called **cementite**, which corresponds to 6·7 % C. Because of the formation of this compound, interest in the C-Fe phase diagram is restricted to the Fe-rich end. However, cementite is in fact only **metastable**, the true equilibrium is between iron and graphite. This means that while cementite is a persistent species, this is for kinetic reasons (the decomposition to Fe and graphite is very slow). It is therefore not actually a thermodynamically equilibrium phase in the C-Fe system. Ordinarily, constitutional diagrams show only stable phases, that is, they are equilibrium diagrams. Although graphite occurs extensively in cast irons (2 ~ 4 % C), it is usually difficult to obtain this equilibrium phase in steels (0.03 ~ 1.5 % C). Thus, the relative stability of cementite is such that it is an important constituent of many iron products and so must be taken into account, being relevant to most steels in practice   (recall that dental amalgam does not attain thermodynamic equilibrium, and its working constitutional diagrams too are non-equilibrium; 14§1). This is another example of the working compromise between kinetics and thermodynamics.

In the process of smelting iron from its oxide ores with coke there is a naturally a great excess of carbon present. Up to 4·3 % of that C will dissolve in the molten Fe, corresponding to the composition of the eutectic of Fe and Fe_3C: this liquid freezes at 1130 °C (Fig. 1.1):

$$Liq \Rightarrow \text{austenite} + \text{cementite} \tag{1.1}$$

That is, the eutectic solid is austenite + cementite. Ordinary 'pig-iron', that is, cast from the original melt, has about that composition. If pig-iron is processed to remove all but the last traces of impurities, including C, the result is 'wrought iron', essentially pure Fe. Now carbon steels in practice may contain up to about 1·5% C, although 'mild' steel contains from only ~0·1 to 0·5% C. Thus either the addition or removal of carbon is involved in steel production, depending on whether wrought iron or pig-iron is (or is considered to be) the starting stock. This adjustment of composition is one of the critical factors determining the properties of the steel.

The presence of the carbide, cementite, is responsible for further hardening, depending on grain size and distribution as before (*cf.* Fig. 19§2.4), by inhibiting dislocation movement.

●1.2   Austenite

Austenite (f.c.c.) is a simple interstitial solid solution in which the C atoms are randomly arranged throughout the available octahedral holes (of which there is one per Fe atom), there being insufficient C for a regular structure. The solubility of C in austenite drops steadily from ~1·7% at 1130 °C to ~0·8% at 732 °C. Thus, when austenite containing more than ~0.8% C (the **eutectoid** composition, C : Fe = 1 : 27, 1 C atom per 6.75 unit cells) is cooled slowly the first reaction to occur is the separation of cementite at grain boundaries, to give a two-phase structure. Below 732 °C the remaining austenite is no longer stable and decomposes into the eutectoid mixture of ferrite + cementite:

$$\text{austenite} \Rightarrow \text{ferrite} + \text{cementite} \tag{1.2}$$

This mixture is called **pearlite** because of the fine-grained banded or lamellar structure which gives a pearly lustre in polished sections (this is due to differential erosion during polishing, 20§4, and the resulting effective formation of a diffraction grating-like surface). However, because it is mostly nearly pure Fe, pearlite overall is in fact not very hard. But because of the lamellar structure dislocations cannot travel very far in the ferrite and cracks would be deviated by the cementite. Hence, this material is nevertheless quite tough and fairly strong. It is important to note that pearlite is not a phase but a two-phase mixture, *i.e.* it is itself a composite.

●1.3   Martensite

If, instead of slow cooling, the austenite is quenched (cooled very rapidly, at 200 K/s or more) say by dropping into water, the transformation from f.c.c. to b.c.c. still occurs, but the decomposition into ferrite + cementite cannot happen because the necessary diffusion is inhibited by the lack of activation energy and lack of time. The f.c.c. structure is nevertheless too unstable to exist – there would be too much strain – so the structure distorts nearly instantaneously in a non-diffusive, displacive fashion into b.c.c. This kind of transformation has zero activation energy and therefore called **athermal**. It results, therefore, in a supersaturated solution of C in b.c.c. iron; this is called **martensite** (actually it is distorted enough that it is better called body-centred tetragonal when there is carbon present in the cell). This material is extremely hard and brittle because of the great strain in the lattice, but also because there are no close-packed planes in the structure slip is difficult (see Fig. 11§5.3). It is therefore of limited use as it stands. Note, however, that since the maximum solubility of C in austenite is about C : Fe = 1 : 13 atoms, on transformation this corresponds to 1 C per 6.5 or so b.c.c. unit cells. The strain is therefore rather localized, but severe, being essentially between the unit cells that are with and without a carbon atom. There is a further effect. About one-third of the carbon atoms are trapped on the face of the b.c.c. cell, which is still an octahedral hole (Fig. 19§2.3), but smaller (Table 1.1), hence the strain. Even so, this is rather less than the strain caused by the two-thirds of the carbon atoms trapped at an edge (again, of course, it was an f.c.c. octahedral hole), where the strain is very severe – hence the tetragonal distortion: the lattice cannot relax to cubic symmetry.

On such rapid cooling, the transformation is triggered on reaching what is called the **martensitic start temperature**, M_s (about 550 °C), when the strain becomes to great to sustain (that is, prior to that, the system is **metastable**). This transformation is described as **military** or **concerted** because of the systematic nature of the wave of distortion which travels at about half the speed of sound (*i.e.* of an elastic wave) in the material. That is, there is a one-to-one correspondence of the lattice positions in the two structures (Fig. 1.2). Note that because the transformation is purely displacive, the two lattices are coherent (*cf.* 11§6.1). The transformation continues progressively on further cooling and ceases at the **martensitic finish temperature**, M_f (about 350 °C for plain carbon steel), although there is usually some small amount of untransformed austenite which remains. The displacive nature of the transformation also means that it is completely reversible: reheating will return the original austenite, with all atoms in the exactly the same place (assuming that no time for diffusion is allowed).

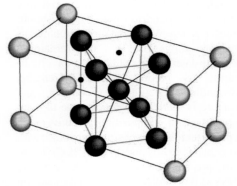

Fig. 1.2 How the atoms of the b.c.c. unit cell of martensite can be identified in the parent f.c.c. lattice (some face atoms are omitted for clarity). A carbon atom, if present, would occupy either an edge- or face-centred (octahedral) position in the b.c.c. unit cell (two of the 6 possible are shown as solid dots) because it does not diffuse from its octahedral hole position in the f.c.c. unit cell.

A further factor contributing to the strain of the system is that there is an associated transformation expansion. The specific volume of b.c.c. Fe is about 2 ~ 4% greater than that of f.c.c.,[2][3] depending on carbon content, and this expansion has to be accommodated in the parent matrix of the transforming grain, putting the martensite into compression, and the surrounding material into tension (6§2.2). It arises because, as can be seen in Fig. 1.2, while the *c*-axis (vertical) of the soon-to-be body-centred unit cell would need to shrink, or the the *a*- and *b*-axes would need to expand, to make the cell cubic rather than tetragonal. Thus there is a coherency strain between the transformed and untransformed regions in the grain (*cf.* 19§1.4). The presence of an interstitial carbon atom makes this worse, and in a concentration-dependant fashion. Furthermore, the presence of the carbon in an edge position prevents full transformation to a symmetrical cubic structure, adding another source of strain. [4] (Notice that although the b.c.c. lattice lies at 45° to that of the f.c.c., crystallographic axes are an artificial scheme of classificatory convenience – our point of view; no physical rotation is involved in this case. Incidentally, this shows that the f.c.c and b.c.c. lattices are rather closely related, despite the apparent strong distinction; 11§3.4.) Furthermore, since the *c*-axis of the soon-to-be b.c.c. unit cell can, randomly, lie in the *a*-, *b*- or *c*-directions of the parent lattice, the resulting tetragonal strain direction varies. This leads to domains in the same sense as for the gold alloys (19§1.4), adding to the coherency strain and thus the hardening.

All of this must be considered in the light of the effect of ordinary thermal expansion. On cooling, lattice dimensions decrease as atomic vibration decreases (10§3.7), but atom sizes do not change. Thus, the size of interstitial holes must decrease, and so also the solubility of interstitials (assuming nothing else is involved). This can in part be seen by the decrease in the solubility of C in f.c.c. γ-Fe from 1.7% to 0.8% on cooling from 1130 °C to 732 °C (Fig. 1.1). A similar effect is present for α- and δ-Fe. Thus, if cooling is fast enough to prevent diffusive phase changes and separations, the strain increases with decreasing temperature. It is this strain that triggers the martensitic transformation, and since the local critical value is a random variable, the transformation occurs progressively over a range of temperature.

●1.4   Tempering

The process of **tempering**, by which steels with more useful properties are made, involves the carefully-controlled reheating of martensite to 200 ~ 300 °C. As the transformation to ferrite + cementite is relatively slow (*i.e.* diffusion-limited), the duration of reheating and the temperature may be used accurately to control the amount of conversion. The newly converted eutectoid material is much coarser than pearlite and is known as **bainite**. The presence of the mixture bainite + martensite greatly increases the toughness of the steel (hard core, ductile matrix). Steel in general, then, has a composite structure, primarily because carbide is present (as in Co-Cr alloys). However, if the idea is continued, all alloys with more than one phase are also composite.

If the original austenite contained less than 0.80% C (that is, the eutectoid composition), ferrite is the first phase to separate as crystals on slow cooling, but now in the form of dendrites within the austenite grains (*cf.* the Widmanstätten structure in amalgam alloys, 14§3). The remaining austenite later (below 732 °C) converts to pearlite as before, and this remains surrounding the ferrite dendrites; the structure is therefore characteristic of such **hypoeutectoid** compositions. Conversely, **hypereutectoid**[1] compositions are those with more than 0·80% C and up to 1·7% C. These terms are sometimes used to describe the alloys as they indicate the amount of carbon

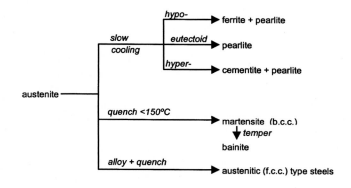

Fig. 1.3   An outline of some steel transformations.

present in relation to the eutectoid composition, but they cannot be used to accurately deduce structure (and therefore properties) except in a very limited sense. This is because the extent of the various transformations depends on the rate of cooling and subsequent treatment in terms of time and temperature, *i.e.* the **thermal history**. In general terms, the transformations that are possible are summarized in Fig. 1.3.

●1.5   Properties

There are many other details and complications to the science of steel and steel-making into which there is no need to go here. However, it is critically important to recognize that because all such transformations are time and temperature dependent, any subsequent heat treatment may be expected to modify, or even totally destroy, the properties for which the steel had been chosen and prepared. In particular, the hardness of the martensitic structure will be lost, and extremely soft metal will result from overheating. Wax knives, for example, should not be overheated if they are to remain useful. The flexibility of steel instruments such as endodontic files and reamers will be lost if they are exposed to too high a temperature, and the associated softening means that there will be very rapid wear of the cutting edges.

Carbon steels can have a vast range of property values and combinations of properties, depending on carbon content and both mechanical and heat treatments (11§6). Only a brief

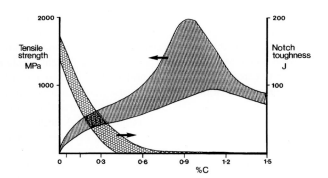

Fig. 1.4 The range of tensile strength and Izod notch toughness which may be obtained in plain carbon steels against carbon content. Arrows indicate which axis applies to which plot. Broad ranges of values are shown because of the variation possible due to thermo-mechanical history.

---

[1] These terms are relative not absolute, and depend on the orientation of the diagram – which element is on the left. In the case of C-Fe, it is conventional that the Fe terminal is on the left, and thus 'hypo-' means to the left of the eutectoid point, and 'hyper-' to the right. Likewise with hypo- and hypereutectic compositions. Care must be taken in any system where such terms are used to be sure of the sense intended.

indication of the possibilities can be made. The typical variations in tensile strength and notch toughness on impact ('Izod', see Fig. 1§10.1) are shown in Fig. 1.4, which indicates that there is a certain compromise to be made. Overall, toughness declines rapidly (*i.e.* brittleness increases) as the carbon content rises; simultaneously the strength can be made very high. It is this kind of variation that is the basis of the selection of steel grades for particular jobs. Some illustrative applications are given in Table 1.2.

Table 1.2 *Carbon steel grades and their applications.*

Grade	% C	*Some typical uses*
Commercial 'pure' iron	0·02 - 0·04	Electrical and magnetic circuits
Mild steel: 'dead soft'	0·05 - 0·15	Consumer goods: car bodies, refrigerators, *etc.*
Mild steel	0·15 - 0·3	Boilers, girders, bolts, rivets, concrete reinforcing bars
Medium carbon steel	0·3 - 0·5	Machine parts, axles, camshafts, gears
Tool steels		
low carbon	0·6 - 0·9	Springs, railway wheels & rails, woodworking tools
medium carbon	0·9 - 1·2	Chisels, saws, milling cutters, drills, dental burs
high carbon	1·2 - 1·5	Scalpel blades (now superseded by martensitic stainless)

●1.6 Alloy steels

The high temperature f.c.c. austenite structure is desirable because of the ductility associated with it. However, it can be stabilized and retained indefinitely at low temperatures by the use of a variety of alloying elements, particularly manganese and nickel, broadly because they cannot form solid solutions with b.c.c. Fe. This is related to the use of such elements in dental casting alloys (19§2).

The scope for alloying in steels is extremely wide, including the use of non-metals such as nitrogen and silicon. While such alloys may be seen in use in such ordinary tools as screwdrivers (Cr-V) and spanners (Cr-Mo) in connection with dental equipment, there is no need for such detail here because of one over-riding factor: rust. For use in the mouth, or in circumstances that involve frequent sterilization (whether hot or cold), or simply wet work such as in the laboratory with gypsum products and the like, the rusting of steel is a generally unacceptable property. This is why much emphasis is placed on the use of the so-called stainless steels in such contexts.

§2. Stainless Steel

Stainless steel[5] is defined essentially by the chromium content of what are otherwise Fe-based alloys: a minimum of ~13% Cr is required to obtain the passivity so useful in many applications. This passivity results from the thin, adherent, chromic oxide layer, coherent with the underlying metal lattice, which is essentially impermeable and insoluble under ordinary conditions (see 13§5.3). Fe-alloys with more than about 30% Cr are used only for high temperature applications and are outside the scope of 'steels' as such and are certainly not currently used in dentistry. However, as will be seen from the compositions of the Co-Cr dental casting alloys (19§2), strict dividing lines are again impossible to identify. It depends in part on one's point of view. However, the commonest grade in dentistry, as well as in many other uses, has 18% Cr with 8% Ni, so-called 18-8 stainless steel (the reason for the Ni is discussed below). There are three broad types of stainless steel, named for the principal structures to be found in them.

●2.1 Austenitic

As has been mentioned, the high-temperature form of Fe (austenite, f.c.c.) can be preserved at low temperatures by the presence of alloying elements such as Ni. Unfortunately, Cr is not a stabilizing element in this sense. The so-called **austenitic stainless steels** have that structure, but must contain anything from 3.5% to 22% Ni to achieve this, depending on other factors. However, there is a price to pay for the passivity due to the Cr. It is essential that any annealing that is done is at relatively high temperatures to ensure that the carbon stays in solid solution and does not begin to precipitate as chromium carbide, $Cr_{23}C_6$. In addition, cooling must be rapid to prevent transformation of the structure to a martensitic type, but then not so quickly that the fully soft condition is obtained. Unlike conventional plain carbon steels (which harden on quenching), austenitic stainless is said to be **quench annealed**. Even so, the rate of grain growth (and therefore of weakening and deteriorating surface finish) is high, and the time spent at high temperatures must be carefully limited. These austenitic steels cannot be sharpened effectively as they are not hard enough to take an edge, but are commonly used in the work-hardened condition as wires, surgical implants and swaged denture bases (28§3).

●2.2   Martensitic

Martensitic stainless steels, which are low in Ni, behave similarly to plain carbon steel on cooling from the high temperature austenitic condition. That is, they are hardened by the **martensitic transformation** to b.c.c., maintaining the carbon in supersaturated solid solution. Tempering is possible, as before, but the extreme hardness and a relative lack of corrosion resistance means that there are limited applications in dentistry. The exception is for tools requiring a sharp cutting edge such as scalpel blades and hypodermic needles. For example, because this type of steel has minimal ductility, such needles may not be straightened successfully once bent. The work hardening process will cause fracture. (Bent needles should be discarded and no attempt made to use them: removing a fragment from deep tissue is not a recommended exercise.)

●2.3   Ferritic

Commercially available **ferritic stainless steels** do not differ systematically from the martensitic types in any one compositional respect, although the particular combinations of C and Cr, coupled with the other trace elements present, result in a structure incapable of holding excess C in b.c.c. supersaturated solid solution, so that carbide separates on cooling. As a result, no heat-treatment other than for annealing is possible, and then principally to relieve stresses, especially those from work hardening. These steels are relatively cheap as well as corrosion resistant and so are commonly used for decorative purposes.

All three types of stainless steel work harden. Ferritic stainless does so the least but becomes rapidly unworkable. Austenitic types, however, work harden most rapidly but remain workable even when much harder than a ferritic stainless could be. Martensitic types, when hard, are also very brittle and work harden very quickly so that little work can be done before failing.

It might seem from the descriptions and classification given above that the types of stainless steel are very distinct and quite separate. In fact, it is possible for each type of structure to be represented in the same piece at room temperature. Cold-work alone will induce structural transformations of one sort or another, quite apart from those due to heat treatment. Thus, austenitic stainless steel, which is chosen for dental use because of its corrosion resistance, capacity for cold-work, and its mechanical properties of strength and ductility, tends to form some martensite along grain boundaries on being worked. This contributes to the work hardening effect. These structural labels, therefore, are only broad classifications. Mechanical and heat-treatment properties likewise merge imperceptibly over the full range available. Even so, the basic phase description of the unworked metal is predictable.

●2.4   Alloying elements

As with the carbon steels, many other alloying elements are possible. The effects of these elements, including C, must all be allowed for in considering the structure of stainless steels. Because, in general, they have two broad tendencies with respect to the stability of the austenite structure, exemplified by the effects of Cr and Ni, their effects are usually summarized

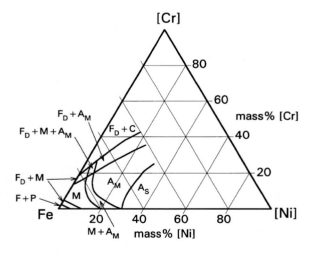

Fig. 2.1 Diagram to show the possible constitutional descriptions of quenched alloy steels in terms of the nickel-equivalent [Ni] (stabilizing) and chromium-equivalent [Cr] (destabilizing) alloying proportions.

A_M	Austenite, metastable
A_S	Austenite, stable
F	Ferrite, from austenite
F_D	Ferrite, high temperature form
M	Martensite
P	Pearlite
C	Carbide

by calculating their Cr- or Ni-equivalent, and a diagram such as Fig. 2.1 may be drawn. This shows the phases present after rapid cooling from the austenite temperature range. As such, it can be used to assess the likely structure and behaviour of cast, welded and annealed alloys.

The nickel and chromium equivalents are given by the following expressions, which have been derived from the practical examination of many alloys rather than any theoretical basis:

Austenite Stabilizing: $[Ni] = \% \, Ni + 30(\% \, C) + 0.5(\% \, Mn)$
Austenite Destabilizing: $[Cr] = \% \, Cr + \% \, Mo + 1.5(\% \, Si) + 0.5(\% \, Nb)$

where % refers to the actual amount of the element present. The details of these equations are not important in themselves here, but rather the fact that they demonstrate the kind of controlling influence and relative importance of alloy components is to be noted. Carbon in particular is seen to be especially effective. Alternatively, it may be said that the properties of the steel are (again) very sensitive to the carbon content. The remark made above about the labels austenitic and so on being somewhat vague is emphasized by a feature of Fig. 2.1 that cannot readily be shown: the boundary lines drawn have some uncertainty or diffuseness. They are very much dependent on the exact composition, starting temperature and cooling rate of the piece.

For comparison, the room temperature equilibrium phase diagram for the same Fe-Cr-Ni system is shown in Fig. 2.2. The considerable differences between this and Fig. 2.1 are indicative of the many changes that can occur on further heat treatment of such steels. This again emphasizes the care that must be taken if untoward effects on mechanical properties are to be avoided. Also of note is the appearance of the σ-phase at high Cr-content (*cf.* 19§2.5). This provides an upper limit to the amount of chromium that may be tolerated.

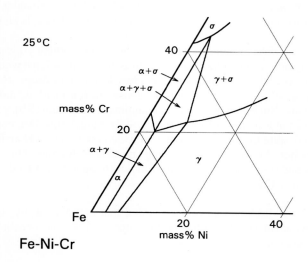

Fe-Ni-Cr

Fig. 2.2 Part of the room temperature isothermal equilibrium phase diagram for the system Fe-Cr-Ni. α : ferrite, γ : austenite.

●**2.5   Sensitization**
The precipitation of $Cr_{23}C_6$ referred to earlier is known as **sensitization** because the corrosion resistance of the alloy is thereby markedly reduced.[6]   This arises from the depletion of the Cr in the outer zone of every grain as it diffuses to the grain boundary where the carbide is forming (Fig. 2.3) (see also 19§2.4). Nucleation is very easy at this site, and the grains of carbide tend to be large. The carbon, being smaller, diffuses more rapidly than the chromium, so that even if present in small quantity it will be available to react to form carbide. Only a few seconds at 650 °C, for example, are required for that to occur, and this is one of the reasons why stainless steel is not used as a casting alloy (see also 19§2.5). The atomic ratio Cr:C in the carbide also indicates a high efficiency in the process of removing the chromium: about 4:1. When the local Cr-content falls below ~13%, the ability to form a continuous oxide coat is lost, and so therefore is the passivity. Intergranular corrosion then readily occurs.

It should be noted that the corrosion occurs *after* damage to the pre-existing oxide coat (Fig. 2.4), effectively exposing metal that cannot

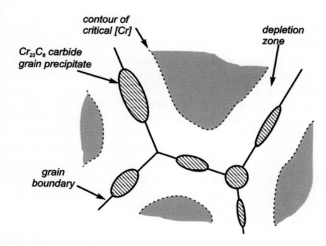

Fig. 2.3 Carbide precipitation causes corrosion sensitization in Cr-containing alloys (schematic cross-section).

then repair the damage by reforming a protective layer. This damage is commonly done in the cleaning-up process, usually with abrasive materials or tools.

This problem may be overcome by a solution heat treatment to redissolve the carbide in austenite or, at a somewhat lower temperature (which caused the sensitization in the first place), a much longer heating period to allow Cr to diffuse from within each grain to restore passivity (assuming that there is in fact sufficient excess Cr present for the 13% target figure to be attained for the metal grains). However, these procedures are generally impracticable because the state of the alloy cannot be determined without metallographic

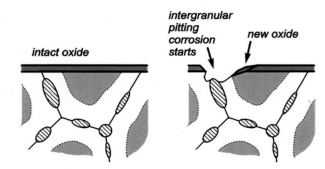

Fig. 2.4 If the oxide film is interrupted in a sensitized alloy corrosion may begin where the oxide cannot reform (schematic cross-section normal to the free surface).

analysis and therefore the amount of treatment required to rectify the situation will be quite unknown. Note, however, that the original structure cannot be regained by such a process. Grain growth, for example, will continue, and the mechanical properties of the object inevitably affected by the carbide now present (see 19§2.4). The same effect occurs in the **heat-affected zone** around a weld or a solder joint, and the resulting change in properties, especially the loss of corrosion resistance, is known as **weld decay**.

The problem is, of course, the carbon content. It is possible, in principle, to create alloys which do not rely on carbon for hardening, but to avoid sensitization it would have to be reduced to below 0.03%; such grades are available, although with a slightly lower yield stress. However, by the addition of other alloying elements in small proportions, which are chosen to form carbide more readily than does Cr, a **stabilized** alloy can be made. Such elements include titanium (Ti), niobium (Nb), vanadium (V) and tantalum (Ta), which are effective even at less than 0.1 mass%, but oddly enough this does not appear to be done in dental stainless steels. This is despite the various heatings to which the wires and bands of orthodontic appliances may be subjected as a matter of course in spot welding and silver soldering. Thus, all that can be said in using these procedures with dental stainless steel is that the time and temperature should both be kept to the minimum necessary. A stabilized or low-C alloy should be used if possible.

It should be borne in mind that any heating by way of normal processing or even a solution heat treatment, should this be feasible, must also result in annealing of the metal with consequent change of properties and grain growth (11§6.6).

●2.6   Pitting

A similar problem is the cause of stainless steel's usual tendency to be subject to pitting corrosion (13§5.5). Due to the common presence of manganese and sulphur as contaminants of steels, manganese sulphide (MnS) is usually present as small inclusions (about 1 μm across) distributed throughout the mass (the manganese scavenges the sulphur [14§2.4, 19§1.14], which otherwise would make the steel brittle). Depletion of the chromium in a narrow zone around the inclusion due to reactions that occur on cooling means that if such an inclusion is exposed at a surface, no coherent oxide will form (Fig. 2.5) and pitting can therefore start in this unprotected zone if the exposure conditions are appropriate.[7] Because the zone is narrow the crevice formed is also narrow, and the corrosion cell, once it has formed, is very active. Indeed, this pitting corrosion may be auto-catalytic,[8] stimulating the activation of adjacent pits, meaning that once started continuation is more likely. Very low-sulphur stainless steel is needed in order to have significantly improved corrosion resistance in this respect. Nevertheless, it is worth emphasizing that such alloys are still really only "corrosion-resistant".

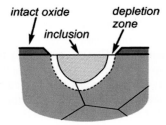

Fig. 2.5 Manganese sulphide inclusions are common in stainless steel, but cause local depletion of the Cr-content. This allows pitting corrosion to occur.

●2.7   Sterilization

To sterilize stainless steel endodontic files and other instruments, dry-heating immersion in a 'bath' of molten metal (28§8.2), salt, glass beads, or steel balls has been used.[9] Some 10 s at 213 ~ 218 °C has been reported to be necessary. Given that the recovery temperature for 18-8 stainless steel is ~200 °C ($T_R > 0.27$)

(11§6.6), and that some glass-bead baths may operate at 233 °C, allowing for the fact that temperature uniformity is not likely (with deviations up or down: thermostats may not be accurate, glass beads conduct heat poorly), it is clear that if the hardness of the instrument as-supplied depends on work-hardening this can be expected to be spoilt even under normal use. Indeed, one manufacturer at least recommends a maximum of 180 °C. Longer immersion, 'to be on the safe side', will exacerbate the deterioration. Note that the typical 'wet sterilization' (autoclave) temperature is 134 °C, too low to be a problem here.

§3. Cermet

Although steels are used for many dental instruments, under the severe conditions at the working edges of cutting instruments their life is very limited. A class of materials known as **cermets**[2] offers distinct advantages in this kind of application. Cermets are defined as a composite consisting of a disperse phase or filler of a crystalline ceramic in a matrix of metal. The most common example (and the one commonly used in dentistry, as well as for the ball of many ball-point pens) is tungsten carbide (WC) in cobalt.[3] The usual practice in dentistry is to refer simply to 'tungsten carbide' instruments. However, it is not a commercial proposition to prepare such devices from pure carbide; it is too hard and refractory. This simplistic labelling should not cause the special structure to be ignored.

Some of the demands made on dental cutting tools, especially burs, are:
- high strength in compression strength, particularly at the cutting edge;
- high tensile strength at points away from the edge;
- abrasion resistance, bearing in mind that indentation hardness may be only a rough guide to resistance;
- maintenance of properties at high temperatures – a problem in cutting even tooth materials; on hard alloys (such as Cr-alloy castings) the problem is exacerbated. Rotating instruments may generate local temperatures of several hundred degrees (by "local" is meant perhaps over distances of a few micrometres).

To illustrate the utility of cermet cutting tools we can note that a modern 'high-speed' tool steel, approximately Fe - 18%W - 4%Cr - 1%V - 0·7%C, has a 'cutting speed' (a measure of the usefulness of the tool) of up to 30 m/min. Because of their wear resistance WC-Co cermets may go up to 300 m/min – a tenfold improvement.

A typical WC-Co cermet contains 6 - 20 mass% Co. Tungsten carbide is very hard (Knoop hardness ~1800; *cf.* high-speed steel at ~740 and diamond ~7000), while Co is very ductile (h.c.p. at < 422 °C; f.c.c. above that temperature, 28§8.1). The important property of the Co here is its ability to wet the carbide and thus form a strong adhesive bond, something generally not possible between metals and ceramics. Although such a cermet is still brittle, the plastic deformation and energy absorbed before fracture (*i.e.* toughness) is at least 10 times greater in the composite than in the carbide alone, which is enough to make such tools a practical proposition.

To make a cermet tool a fine powder (typically with a grain size ~1 μm) of mixed WC and Co is either compacted and sintered or **hot pressed** (*i.e.* sintered under pressure). **Sintering** is the process of heating a powder below the melting point of any component to permit agglomeration and welding of particles by solid-state, mostly surface, diffusion. However, so-called **liquid-phase sintering** involves some melting, which then facilitates the processes involved by increasing diffusion rates and facilitating the **densification** under pressure (*cf.* the packing of silver amalgam). For WC-Co, this sintering is done at ~1400 °C, when a reaction occurs between the two components. Notice that the melting points of these are WC : 2600 °C (peritectic reaction) and Co : 1495 °C; these are not approached.

●3.1 Sintering reactions

For simplicity, the cermet can be considered as a pseudo-binary alloy system (Fig. 3.1).[10] This diagram (an **isopleth**) is a vertical section through the C-Co-W – temperature diagram, but at constant W:C ratio, *i.e.* corresponding to WC itself. We may understand the manufacture of these materials in much the same way that we studied zinc phosphate cement: by a thought experiment to consider the slow addition of WC to Co.

---

[2] Cermets are often referred to as "hard metal" materials, particularly in engineering and manufacturing contexts.

[3] Some other transition metals, such as Ni, can also be used in special contexts; the metallurgy is similar.

Thus, at the sintering temperature of about 1400 °C, WC diffuses into (*i.e.* dissolves in) the solid γ-phase Co. Thus the composition of this phase moves from right to left along the isothermal (Fig. 3.1). When the dissolved (solid solution) WC-content of the γ-phase cobalt has increased to roughly 10%, melting starts, as the composition enters the γ + Liquid phase field. This melting is completed by the time the composition of the combined liquid and γ-phase reaches ~30% WC, when the Liquid (only) phase field is entered. Further additions of WC would mean the composition of the Liquid would steadily change until it reached about 50% WC, at which point it is saturated. No more can dissolve, and the only effect of further additions would be to increase the volume fraction of carbide in the final product. The limit to what can be achieved in this direction is set by the gauging volume of the WC core after processing (2§2.2).

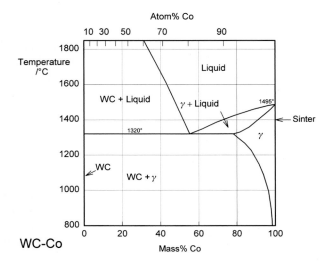

Fig. 3.1 The pseudo-binary constitutional diagram for Co-WC.

In practice, during the sintering process the Co-WC liquid which is forming wets the remaining WC particles and is drawn into the interstices by capillary action (10§2), the consolidation of the mass being aided by pressure. Under pressure the porosity of the original bulk powder is being eliminated, and up to 50% decrease in the volume of the part being moulded occurs. Commonly, the initial WC powder particles will be polycrystalline, but during the sintering process grain boundary attack will occur and this tends to break up the particles into single crystals, reducing the average grain size. The heating stage of the process is now complete and no further change can occur because the liquid is saturated with respect to WC (the solubility of Co in WC is nil). On cooling, some of the dissolved WC is precipitated from the solution, probably by growth on the remaining WC particles. This continues until the eutectic point is reached (1320 °C), at which temperature all of the remaining liquid solidifies as the two-phase eutectic mixture, γ + WC. The eutectic structure consists of finely-dispersed carbide in the γ-Co solid solution matrix. This disperse

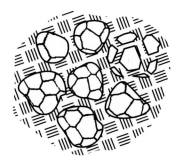

Fig. 3.2 Schematic diagram of the structure of a WC-cermet: WC core in a matrix of eutectic.

phase therefore toughens the matrix which embeds the remaining unreacted carbide particles (Fig. 3.2). As the cooling continues further precipitation of carbide occurs, in a solid state reaction, because the solubility of the carbide in Co decreases steadily with temperature, as can be seen from the markedly curved solvus at the Co end of the diagram. This too is likely to occur on pre-existing WC (original and precipitated) since nucleation within the γ-phase will be energetically unfavourable, but not impossible since diffusion rates will be low.

Although this description of the process is somewhat simplified, the main feature emerges clearly: the composite structure. The hard core resists the plastic deformation of the matrix, but this matrix inhibits crack propagation through the brittle carbide phase by being ductile and causing the crack tip to become blunt and diverted around grains of carbide. Thus the core particles which appear at the working edge (which must be shaped, of course, by a harder abrasive) may effectively be used as the actual cutting edges. This improved cutting performance is still at the expense of considerable brittleness, and great care must be taken to avoid impact or bending which would shatter the cermet. Such tools should, therefore, only have stresses applied to them in the sense in which they were designed to take them: compressive. However, the use of a cermet permits materials otherwise practically unworkable to be formed into complex shapes at very low cost, and their hardness utilized, whilst ameliorating the worst disadvantages.

Since deformation is restricted to the cobalt matrix, the finer the WC powder the greater the hardness and bending strength because the average available path length for dislocations to travel has been reduced. However, the impact resistance and toughness are reduced. Likewise, increasing the Co content lowers the hardness, but increases bending strength and toughness. The modulus of elasticity follows well the expectations for a composite with good matrix constraint due to the good wetting (Fig. 3.3).[11]

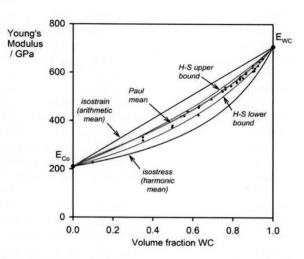

The great difference in the hardness of the two phases allows the use of such cermets as **non-bladed abrasive** materials, as opposed to a shaped cutter. When such a surface is applied to a hard substrate (harder, that is, than the cobalt matrix) there will be immediately some differential erosion (20§4), exposing many very fine cutting edges on the carbide particles. This will then act as an abrasive itself. Such bladeless instruments have been used as a very fine finishing tool for restorations.

It can be seen that in the present sense, any carbide-containing alloy could be considered as a cermet, cobalt-chromium alloys in particular, any steel with cementite present, but also sensitized stainless steel. Indeed, such systems represent a dilute composite, and of course are affected in similar ways in terms of mechanical properties, wear resistance – and causing wear, just not as extreme as WC-Co.

Fig. 3.3 Elastic modulus of WC-Co composite (*cf.* Fig. 6§2.12). H-S: Hashin-Shtrikman bounds; Paul mean: see 29§4.1.

●3.2 Carbide chemistry

There are some aspects of the WC-Co system that may seem puzzling. Firstly, how it can dissolve in a metal; secondly, how it can be wetted by a metal, when ceramics in general are not. We can start to understand this by noticing that chemically there are three main classes of carbide.

Salt-like carbides: When the electropositivity of the metal is very high, ionic substances are formed, and many of these contain C_2^{2-} anions in a crystal structure similar to that of NaCl. A common example is CaC_2. These hydrolyse with water to evolve hydrocarbons, leaving the metal hydroxide. Electrical resistivity tends to be very high at normal temperatures since ionic mobility is exceedingly low; effectively no electron conduction occurs. No metal wetting occurs, the chemical dissimilarity is too great.

Covalent carbides: Silicon and boron are similar to carbon in electronegativity and their carbides form giant covalent network structures. While such carbides are very hard (and are used as abrasives), and do not react with water, they also do not get wetted by metals. These too have extremely high electrical resistivity (SiC: > 10^6 Ωcm, B_4C: 10 Ωcm, depending on purity).

Interstitial carbides: In these materials the carbon occupies all available interstitial sites (*cf.* Fig. 19§2.3) in an essentially metallic lattice. The small size of the carbon atom means that the metal lattice is hardly affected, and there is no ionic character to the bonding. Such carbides therefore do not hydrolyse in water, and exhibit a metal-like electrical conductivity (Fig. 3.4).

Tungsten carbide is of this third kind, and thus shows substantial metallic character. The structure is h.c.p. (Fig. 3.5), with the W and C atoms forming separate layers – it is a superlattice structure (19§1.3). But while the W and C atoms are bonded covalently through *d-p* orbital

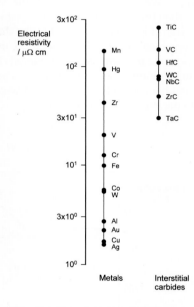

Fig. 3.4 Electrical resistivity of some metals and interstitial metal carbides. (Note scale: μΩcm)

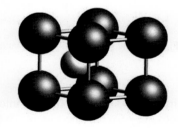

Fig. 3.5 'Open' view of the h.c.p. unit cell of WC (*cf.* Fig. 11§3.6). The carbon atom is in the 'prismatic' position, with six W near-neighbours in a triangular prism.

hybridization, there is also considerable bonding of a more metallic kind between the adjacent W atoms. Notice that the C atom is small, so the W layers are very close together; the *c/a* axis ratio for the unit cell is about 0.97 (Fig. 3.6), close to the "hard sphere" expectation of unity for the surrounding W atoms to be touching. The relative size of the 'hole' in the prismatic position is r ~0.53 (Fig. 19§2.3) while the C : W radius ratio is ~0.54; distortion is therefore expected to be slight. Hence, WC (along with other such carbides) is odd in that formally it is a ceramic in the sense of it being a distinct chemical compound between a metal and a non-metal, yet it has metallic properties, which accounts for the wetting, by which, of course, is meant that there is metallic bonding between the matrix and the core in the cermet. Thus the interface between the phases has strength, and does not behave as a flaw in the Griffith sense (*cf.* 11§6.8). In this light, the presence of carbides in cobalt-chromium casting alloys (19§2.4) and in steel (§1.1) as effective hardening agents does not seem so strange.

It is stressed that tungsten carbide is not a molecular compound: there are no discrete molecules with the formula WC or any multiple of this. We can then notice that the high-temperature solid solubility of carbon (interstitially) in γ-cobalt, which is f.c.c., is somewhat less than in γ-iron (Fig. 1.1), accounted for by the slightly smaller Co atom, but still appreciable (Fig. 3.7). In addition, the solubility of tungsten in cobalt at the same temperature is substantial (Fig. 3.8). This, of course, will be a substitutional solid solution. The combined effect of this is a simultaneous solubility of C and W of about 4 at% each (Fig. 3.9), close to the WC isopleth (Fig. 3.1).

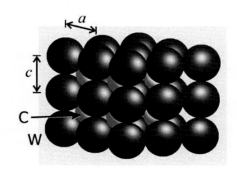

Fig. 3.6 The packing of WC, to scale, showing W layers in contact.

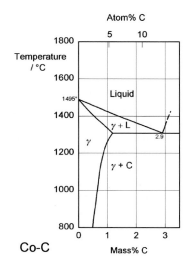

Fig. 3.7 The Co-rich end of the Co-C equilibrium system.

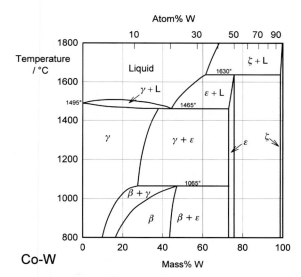

Fig. 3.8 The Co-W equilibrium system.

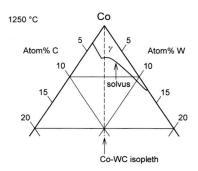

Fig. 3.9 The Co-apex of the 1200 °C isothermal section of the C-Co-W equilibrium system showing the extent of the γ-Co solid solution.

We can therefore interpret the solubility of WC in Co in the following way. Dissolution involves the simple dissociation of the WC into component atoms, which therefore behave independently. This applies to both the γ-Co and the Liquid. WC is more stable than W + C, or anything else, in contact with Co, so co-precipitation occurs on cooling to form the crystalline substance WC again. In effect, the reaction is a dissociation equilibrium:

$$WC \rightleftharpoons W + C \tag{3.1}$$

– there is a "vapour pressure" of W and C over the solid in that Co solution, whether this is solid or liquid. This may be compared with the peritectic melting of WC itself:

$$WC \rightleftharpoons C + Liq \tag{3.2}$$

Here, the C is in the form of graphite, and the liquid is a solution of ~41 at% C in W. Again, the non-molecular, non-ionic nature of WC is apparent. Thus, the apparently anomalous behaviour in those two respects, wetting and solubility, can be explained when the underlying nature of the system is appreciated.

One further point may be made. Clearly, both the interstitial carbon and the substitutional tungsten must cause hardening of the cobalt matrix by introducing some disorder (*cf.* 19§2.1, 19§2.3). The metallic 12-coordination radii are Co: 125 pm, W: 141; ratio 1.13. This further improves the composite's properties in one sense, but at the cost of reduced matrix ductility.

●3.3   Tool assembly

There remains only to fix the cermet tool to a handle, such as for an enamel chisel, or to a shank for a rotary instrument. The Co matrix, being metallic, can be expected to be wetted by at least some other liquid metals, and this provides the means for fixing the working part into a more easily fabricated, cheaper handle such as austenitic stainless steel. This process is soldering (22§1.2), but two things should be noted immediately. Firstly, the solder must melt at a low temperature to avoid changing the structure of either the cermet or the stainless steel, and that therefore overheating the tool in use could cause the solder to melt and the cermet working part to drop off. However, such a low-melting solder is necessarily weak as T/T_m is large. Secondly, the mixture of metals will almost inevitably result in corrosion at the interface, further weakening it, so that mistreatment may subsequently cause the cermet part to drop off when under load.

●3.4   Glass ionomer

Glass ionomer cement (9§8) has limitations of application because of its relatively poor strength, in particular in comparison with dental silver amalgam. In an attempt to overcome this, it has been suggested that it could be mixed with a powdered, much harder material. The suggestion was that amalgam alloy powder, normally readily available, satisfied not only the requirements of strength and hardness but also of corrosion resistance. The combination has been described (and sold) as 'miracle mix', which may be overstating the case, but the adhesiveness, strength, abrasion resistance, radio-opacity and lack of irritancy towards vital dentine are all apparently helpful. The material has also been described as a 'cermet'.

This idea has been taken further with the marketing of a similar material. Here, a fine pure silver powder is heated with the usual kind of glass, which is then ground to a powder in the usual way. It is not clear that the silver, which is described as 'fused' to the glass, is in fact bonded in any way at that stage because silver is rather inert and unreactive.[12] Indeed, it is less clear that, even if it is initially bonded in any way, it can remain so. The reaction of the glass with the acid to produce the gel matrix must undermine any such localized attachment, assuming that it occurs in the first place.

Thus, while these and similar materials might be called cermets in the inverted sense that they are a composite of metal in a ceramic matrix (albeit aqueous), the lack of bonding is a problem. That is, the primary requirement for a functional composite structure (6§2) is missing. In any case, these so-called **cermet cements** have lost a primary advantage of the glass ionomer system: a tooth-like appearance. The compromise between appearance (often complained about for silver amalgam) and performance is evident (see 9§8.8 and 9§8.13).

References

[1]   Brandes EA & Brook GB (eds). Smithells Metals Reference Book. 7th ed. Oxford, Butterworth-Heinemann, 1992.

[2]   Gulyaev AP & Zel'bert BM. Diagram of volume constitution phases in steel. Fiz Metal i Metalloved 6: 843 - 848, 1958.

[3]   Moyer JM & Ansell GS. The volume expansion accompanying the martensitic transformation in iron-carbon alloys. Metall Trans A 6A: 1785 - 1791, 1975.

[4]   Hulme-Smith CN *et al.* Further evidence of tetragonality in bainitic ferrite. Mater Sci Technol 31 (2): 254 - 256, 2015.

[5]   Zapffe CA. Who discovered stainless steel? The Iron Age, Oct 14, 1948 [ftp://ftp.wargamer.com/pub/TechSpec/Manufacture%20&%20Production/TechSpec%20Metalurgical%20Reviews/US%20Who%20Discovered%20Stainless%20Steel%201948.pdf]

[6]   Hanks RW Materials Engineering Science. An Introduction. Harcourt Brace & World, New York, 1970.

[7]   Ryan MP *et al.* Why stainless steel corrodes. Nature 415: 770 - 774, 2002.

[8]   Punckt C *et al.* Sudden onset of pitting corrosion on stainless steel as a critical phenomenon. Science 305: 1133 - 1136, 2004.

[9]   Koehler HM & Hefferen JJ. Time-temperature relations of dental Instruments heated in root-canal instrument sterilizers. J Dent Res 41(1): 182 - 195, 1962.

[10]   Wyatt OH & Dew-Hughes D Metals, Ceramics and Polymers. An Introduction to the Structure and Properties of Engineering Materials. Cambridge UP, 1974.

[11]   Doi H, Fujiwara Y, Miyake K & Oosawa Y. A systematic investigation of elastic moduli of WC-Co alloys. Met Mater Trans B 1(5): 1417 - 1425, 1970.

[12]   Sarkar NK Metal-matrix interface in reinforced glass ionomers. Dent Mater 15: 421 - 425, 1999.

Chapter 22   Soldering and Welding

Since it is not always possible to create a metal device in one piece, such as by casting, it is necessary to assemble it from separately prepared parts. In dentistry, the fabrication techniques used are then either soldering or welding. These approaches are used most often in the context of orthodontic appliances, where wires and bands may need to be joined.

Soldering is distinguished from **welding** by the use of a third body, the solder, between the two workpieces, but without melting either of them. The conditions for effective joins of these types are discussed.

Unfortunately, metals may oxidize sufficiently to interfere with wetting by the solder, whether this is because the metal is prone to corrosion or naturally-coated by a passivating oxide layer. The use of **fluxes** is therefore necessary to obtain a good joint. The chemistry of these is discussed, and the means of limiting their influence.

The main features of solder design are outlined, especially from the point of view of selection of the correct grade for the task in hand. Factors to be considered include melting point and toxicity.

Soldering and welding almost inevitably are detrimental to some aspect of the device being fabricated, such as hardness, strength and corrosion resistance. However, by recognizing the problems related to the structure and behaviour of the workpiece materials, it is possible to make appropriate compromises and to limit the damage in order to obtain functional appliances.

Materials Science for Dentistry
https://doi.org/10.1016/B978-0-08-101035-8.50022-5

Although casting provides a means of making complicated shapes in metals and other materials, there remain types of device where this is insufficient. For example, if alloys with differing properties, such as due to differing amounts of annealing or work hardening, are required to function in different parts of a device then because casting cannot give that variation, some joining method is essential. This is clearly so when alloys with differing compositions are involved. Examples of this include the assembly of orthodontic appliances from stainless steel wires and bands, assembly of bridges, and the fixing of precision attachments to cast devices. In addition, on occasions it may be necessary to repair a fractured device, although this is usually not a satisfactory permanent solution. Although the use of rivets or nuts and bolts may be appropriate fabrication techniques in many circumstances, for intra-oral use there will be severe disadvantages or impracticalities. Thus, there is usually little metal present in which to drill a hole, and this would cause a weakening of the object, even if there were space and soft tissue trauma could be avoided, which is unlikely. The risk of corrosion is also increased with the numerous crevices which are thereby created. The use of adhesives, epoxy resin and the like, is precluded because of relatively low strength, small interfacial area, and poor fatigue-resistance under wet, oral conditions. What are left, then, are soldering and welding as the only generally viable techniques.

§1. Definitions and Conditions

•1.1 Welding

Welding is the joining of two bodies brought into contact by almost any other means than soldering. For example, two perfectly clean gold surfaces (or, indeed, many other metals) placed in vacuum where the adsorbed gases and water may evaporate (perhaps helped by warming), and then placed in contact, will weld together instantly (28§4). The driving force for this is, of course, the reduction in total surface energy by the reduction of the free surface area but it is effected by the diffusion of atoms at the interface. This phenomenon is employed, with the aid of pressure, in the condensation of a direct gold restoration, when a vacuum is not necessary but it is still contingent on the lack of contamination of the surface of the gold. The term welding is usually taken to apply to metals, but it can also be used in the context of some polymers (a polyethylene bag has at least one seam welded), and even some ceramics. In addition, so-called **solvent welding** may be applied to some polymers: the surfaces to be mated are softened with a solvent and pressed together. When the solvent has evaporated, the entangled chains hold the workpieces together. However, what follows will be restricted to the context of metals.

Industrially, welding often means the melting of the parent metal of both bodies into a **weld pool**, with or without the addition of 'filler' metal (not to be confused with the filler of composite materials), which then freezes to obtain the union. As this technique is of relatively low precision, and the amount of heat required is large, it is better suited to large workpieces, and in fact because of this finds no place as such in dentistry. But welding can be achieved by any means that permits the close approximation of the surfaces to be joined, and this usually entails the use of pressure. **Spot-welding** does just that: while the workpieces are squeezed together by a pair of electrodes an electric current is passed through them to heat resistively the area to be welded. As the strength of the material decreases as its melting point is approached, the workpieces are plastically deformed between the electrodes, bringing them into close contact over the area of the 'spot', and allowing the weld to develop, aided by the rapid diffusion that can then occur. While a small amount of localized melting often occurs, it is not in fact necessary for effective union of the parts. Indeed, it is recommended that melting be avoided, since the low viscosity of the melt means that it will be squeezed or run out of place, reducing the amount of metal at the crucial site. Similarly, too much pressure applied by the electrodes can result in an unacceptable thinning of the workpieces, or even a hole being punched, which thus weakens the structure and fails to achieve an effective weld.

In spot welding the amount of energy delivered is given by the usual electrical relationship:

$$J = I^2.R.s \tag{1.1}$$

where R is the resistance of the workpiece. The temperature actually attained depends on the bulk of the metal present, its thermal conductivity and its heat capacity. Thus, because resistance will vary depending on area of contact, roughness, oxide coats, and alloy composition, for any particular job the current and its duration as well as the pressure applied must be carefully chosen. Too much energy and metal will melt; too much pressure and a hole could be punched in the workpiece. Different conditions are therefore required for different alloys and different thicknesses.

As has been mentioned before (21§2.5), the heating cycle is crucial to the maintenance or alteration of the mechanical properties and corrosion resistance of stainless steel fabrications, and this puts further constraints on the welding conditions chosen, *i.e.* to minimize the duration of the heating cycle, the temperature attained, and the volume of metal affected. It is, however, impossible to avoid at least some change.

●1.2   Soldering

This is the joining of two metallic items by an alloy which melts fully, *i.e.* has a liquidus at a lower temperature than the solidus of either of the pieces to be joined. This is a necessary condition when it is required that the integrity of the pieces to be joined be preserved.

Soldering thus sets out to achieve the same mechanical effect as welding, *i.e.* the union of two workpieces (which may or may not be of the same alloy), but it does this by the interpolation of a third body – the **solder** – which is joined to both workpieces. Apart from the basic requirement of a lower melting point, it is evident that the solder alloy must be capable of wetting the parent alloy(s); that is, it may be said to be compatible with them. This certainly means that chemical similarity has a bearing on the matter, but more that there is the potential ability (not necessarily realized) to form an alloy with the substrate metals.

Similar factors as were discussed under casting become of relevance here also. The surface tension of the molten solder and the interfacial tension between the solder and substrate clearly dominate the wetting of the latter and the attainment of good adhesion by close atomic approach. Similarly, the surface tension and the wetting control the ability of the solder to run into small spaces by capillary action, although again the penetrativity (dependent on viscosity, meaning that it takes time) (10§2.4) may be a more relevant figure of merit. The filling of crevices is important from at least two points of view: mechanical strength by reduction of the size of Griffith flaws; and avoidance of crevice corrosion – which would reduce strength still further.

The flow of the solder is clearly a crucial factor and the role of roughness must be considered. The advance of the liquid front through a rough-surface gap is unlikely to be uniform (*cf.* 10§2.9), especially when the gap cannot in general be controlled to be of constant width, and the trapping of bubbles is a common observation. Poor technique, applying solder at several locations in succession, and without providing enough to fill the gap unidirectionally, is more likely to trap large bubbles. Subsequent grinding or polishing may reveal these, and they will act both as surface flaws (stress concentration) and as crevices for initiating corrosion promptly; stress corrosion will be likely.

All of this also depends on the cleanliness of the substrate metal. Oxides, sulphides and many other possible contaminants cannot be wetted by solders (*cf.* mercury and amalgam alloys; casting alloys and investments). It is essential therefore to ensure that both surfaces to be soldered (or welded, for that matter) are scrupulously clean, initially by degreasing and then abrasion or filing. Even so, surface oxidation may be so rapid when the workpieces are heated as still to preclude effective wetting or wetting over a sufficient area. This is where **fluxes** come in.

§2. Fluxes

The word **flux** means flow, and in the present context means flow of the molten solder being achieved by allowing or facilitating the wetting of the substrate. The function of fluxes therefore is twofold:
(1)   to remove any oxides or other compounds present on the surface, and
(2)   to prevent further oxidation by excluding oxygen from the site, purely as a physical barrier.
In general it is not possible to reduce oxides to metal at low temperatures because the oxides are thermodynamically so very stable. Another chemical approach must be taken, which is to dissolve the oxide. Formerly, in workshop contexts, this was done by using a solution of hydrochloric acid or similarly acid substance. This, however, did not provide any protective action.

Boron now provides the basis of many commercial fluxes and the constituents for 'home-made' recipes in the form of borax or sodium tetraborate, $Na_2B_4O_7.10H_2O$, and boric acid, $B(OH)_3$. This latter compound dehydrates readily:

$$B(OH)_3 \; \underset{H_2O}{\overset{169°}{\rightleftharpoons}} \; HBO_2 \; \underset{H_2O}{\overset{300°}{\rightleftharpoons}} \; B_2O_3 \quad \text{(boric oxide)} \tag{2.1}$$

The melting point of metaboric acid, HBO_2, is 236 °C, while that of B_2O_3 is 450 °C. Borax also dehydrates in several stages:

$$.10H_2O \; \underset{H_2O}{\overset{100°}{\rightleftharpoons}} \; .5H_2O \; \underset{H_2O}{\overset{150°}{\rightleftharpoons}} \; .H_2O \; \underset{H_2O}{\overset{320°}{\rightleftharpoons}} \; \text{anhydrous} \tag{2.2}$$

and finally melts at 741 °C. As the large amount of water removed on heating may cause splattering, fluxes are usually compounded using the anhydrous salt and applied as a paste mixed with alcohol. The evolution of water can also be a problem with boric acid. All fluxes must be kept in tightly-sealed containers if they are designed to be used in dry form or alcohol suspension so that the water of crystallization is not taken up from the atmosphere, with consequent working difficulties. Nevertheless, it is a common practice to apply a flux as a paste mixed with water. This should be avoided.

Both of the above compounds form glasses when they have been fused, and are quite miscible with each other. The melting point of the mixture has a minimum for an addition of about 10% boric oxide in anhydrous borax, although this is only about 20 K lower than the melting point of the borax (Fig. 2.1). The boric oxide is said to allow the flux to "cling" to the metal better, which may mean that lowering the contact angle is more important than melting temperature.

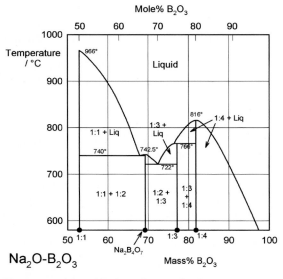

Fig. 2.1 Partial equilibrium diagram for the system Na_2O - B_2O_3. This shows two eutectic subsystems, while the 1:3 salt undergoes a peritectic melting at 766 °C.

The glasses formed are very similar to silica glasses in their structure. Boric oxide forms three-dimensional networks of $[BO_3]$ groups, sharing all oxygen atoms with adjacent bonded boron. This network may be described as having a **connectivity** of 3, since each node (the boron atoms) is attached to precisely three others. In contrast, and by similar reasoning, fused silica glass has a connectivity of 4.[1] Fused silica is therefore expected to have a much higher viscosity than fused boric oxide because more bonds must be broken for plastic deformation to occur. However, it is the lower melting temperature of the boric oxide glass that makes it useful as the basis for a flux: this must be below the substrates' solidi also. The low viscosity is sometimes a disadvantage. This

> The phase diagram of Fig. 2.1 has a new feature, arising from the presence of four distinct stoichiometric compounds in the composition range considered. The diagram can be understood as being the juxtaposition of three separate binary systems, whose terminals are coincident: $Na_2O.B_2O_3$ – $Na_2O.2B_2O_3$, $Na_2O.2B_2O_3$ – $Na_2O.3B_2O_3$, and $Na_2O.3B_2O_3$ – $Na_2O.4B_2O_3$, reading from the left. The boundaries between these systems are the intermediate compounds, and these show sharp melting points. Note that the 'phase field' for a stoichiometric compound has no width – hence the appearance of the corresponding vertical lines for each. That is, each 'binary' has a terminal phase field that shows no solution solid range.

may be overcome by the inclusion of a small amount of silica. Substitution of $[SiO_4]$ for $[BO_3]$ in the network occurs readily, again sharing oxygen atoms. This increases the connectivity, greater 'cross-linking' being obtained, and so the viscosity is higher, thus helping to keep the flux on the site required. The tetraborate group $[B_4O_7]^{2-}$ has a peculiar structure (as indeed is true of many boron compounds, defying simple description) but it too can also be incorporated into these glassy melts. The resulting excess negative charges are balanced by the Na^+ ions from the salt.

Boric oxide glasses (being acidic) have the property of dissolving many metal oxides, forming borates in the process, the metal ions adopting the same role as does Na^+ above. This is an acid-base reaction. So readily does this occur that it used to form the basis of the 'borax bead' test for identifying metals. This depended on the colour of the glass produced when borax and the unknown metal compound are fused together; many of the transition metals in particular were identifiable with the aid of this test (*cf.* 24§6.4). Thus, metal oxides and other compounds can be removed from the surface of the workpiece by being dissolved in the fused boric oxide mixture. The viscous fused glass then stays in position because it wets the now high-energy surface of the metal, and so excludes oxygen. The introduction of molten solder to the site, with a little disturbance of the flux, results in wetting of the solder by the flux and then the wetting of the workpiece by the solder, whence the solder flows or 'runs' to coat the available surface. The chemistry here is similar to the formation of silicate slag on cast alloys (17§4.3).

●**2.1   Antiflux**

The tendency for the solder to run is, of course, desirable and its promotion is the whole aim of fluxing. However, it would mean also that the solder could spread to undesirable locations if the workpiece metal were hot enough and clean enough. Some barrier to the operation of the flux is therefore necessary, one with which the flux cannot react, and which neither the flux nor solder wet. Such materials are sometimes referred to as **antifluxes**. Suitable barriers are ordinary graphite pencil marks (which also include a proportion of clay), or a deposit of calcium carbonate ('whiting'), or ferric oxide ('rouge'). These last two are obtained from an alcohol suspension of the powder which is painted around the site. By preventing the spread of the flux, so therefore is the spread of the solder. Ferric oxide (Fe_2O_3) is, of course, expected to dissolve in the flux, and therefore may not be a good choice. However, by putting on a relatively thick layer, the rate of dissolution becomes the limiting factor, and time is available for the work to be completed.

●**2.2   Chromium alloys**

Although fluxes based on borate glasses are applicable to many metals, there remain some whose oxides are soluble only with difficulty or not at all. Stainless steel and cobalt-chromium alloys present such problems, as the chromic oxide of the protective film falls in that group. The difficulty can be overcome by incorporating a fluoride salt, such as KF or KHF_2, which enables the dissolution of the chromic oxide, probably through the formation of the hexafluoro-complex which is known to be very stable:

$$Cr_2O_3 + 12KF + 6B_2O_3 \Rightarrow 12K^+ + 2[CrF_6]^{2-} + 3[B_4O_7]^{2-} \qquad (2.3)$$

The tetraborate ion, of course, will become incorporated in the glassy structure as before. Thus the unusually tenacious oxide film on stainless steel can be dissolved and prevented from reforming, permitting soldering to be performed.

There are a variety of other ingredients which may be found in commercial fluxes, which will have been optimized in some respect for particular tasks. Phosphates are one such type of additive, which also form low fusing glasses with alkali metals. The principles remain the same, however: the cleaning and protection of the surfaces to be joined. Nevertheless, a flux cannot be expected to work efficiently if the contamination is too great. Workpieces should be free of obvious foreign matter such as grease, and if heavily oxidized cleaned up by abrasion first. This would perhaps also be beneficial in providing a high energy surface and an additional mechanical key to supplement the strength of the bond.

§3. Solder Design and Selection

The choice of composition for the solder alloy depends on a number of factors. Clearly, as has been mentioned, the upper point (*i.e.* liquidus) of the melting range of the solder must be below the solidus of the workpiece alloy(s), but the relationship of the two ought to be a little clearer. Even though the workpiece needs to be preheated to the solder liquidus to avoid premature chilling (*cf.* the preheating of an investment mould for casting) and then allow wetting and flow, in purely practical terms it must be heated some way above that temperature for reliable soldering. This arises really only because of the difficulty of controlling temperatures accurately when heating with a flame, and the conductivity of the substrate ensures that there will be steep temperature gradients. The solder liquidus must therefore be some 50 or 100°C below the substrates' solidus temperature for safety and efficacy.

One of the common causes of corrosion is the galvanic effect of dissimilar metals in contact (13§1, 13§4), and this is indeed a strong motivation for using spot-welding. It is helpful, therefore, from the point of view of corrosion if the solder alloy composition is close to that of the workpiece to minimize electrochemical potential differences. Thus, for gold alloys, a gold-based solder is desirable. In fact, gold solders can be taken from within the same alloy system as dental gold casting alloys, *viz.* Ag-Au-Cu (19§1) and this permits such things as a closer matching of mechanical properties as well as colour (see Fig. 19§1.14), a very useful aspect for a good appearance in the finished appliance, as well as guaranteeing compatibility. However, the viscosity and surface tension of these basic compositions tend to be rather high, as indicated by the data outlined in Table 18§2.1, so that additions of metals such as In, Sn and Zn are made

Table 3.1 *Properties of molten gold solder additives (cf. Table 18§2.1).*

	Melting Temp. / °C	Surface Tension / mN.m^{-1}	Viscosity / mPa.s
In	156	556	1·89
Sn	232	544	1·85
Zn	419	782	3·85

(Table 3.1) to lower both property values. An additional advantage is the concomitant lowering of the solidus and liquidus temperatures, which thus satisfies the primary criterion mentioned above. However, it is obvious that the electrode potentials of the Zn and Sn will tend to be disadvantageous. Again, a compromise is evident.

●3.1   Choosing a gold solder

There are a number of designations used for gold solders according to either the gold-content of the solder itself or the gold-content of the metal with which the solder is 'suitable' for use. Neither is of any use in practice since, as is apparent from diagrams such as Fig. 19§1.22, the liquidus – and hence the solidus – temperature depends on all the constituents, not merely the gold-content. The use of additives such as Zn not only improves the flow and wetting but also substantially lowers the melting point at precisely the same gold-content. Likewise, the solidus temperature of the parent metal may even be lower than the liquidus temperature of a solder under some conditions, even though the 'carat' designations match. It is apparent that the only way to ensure applicability is to know both the solidus temperature of the parent metal and the melting range for the proposed solder. No other approach is feasible.

●3.2   Nickel

When the parent metal is a 'white' gold, that is, although there may be an appreciable concentration of gold the colour is not yellow (*cf.* Fig. 19§1.14), a 'white' solder is more appropriate. This may be achieved by the addition of some nickel to the composition, as Ni is a powerful 'whitener' of gold alloys (19§1.8). These solders are probably less satisfactory as the solid solubility of Ni in Au at low temperatures is limited and a two-phase structure, one Ni-rich, readily forms. This leads to poorer corrosion resistance. If it is the Ni which dissolves (a possibility, but not known to have been checked), then in contact with mucosa nickel sensitivity can be expected to develop. It is well known that nickel plating on watch bands and spectacle frames readily causes a contact dermatitis in the adjacent skin.

●3.3   Silver solder

White solders are usually required for stainless steel and chromium-based casting alloys for colour matching. Their formulation is very similar to gold solders, with the omission of the gold. Thus, the basic alloy will be Ag-Cu (Fig. 12§3.1) with, again, additions of Zn (Fig. 3.1),[2] Sn and In to modify and control the properties, primarily to lower the liquidus temperature. They are therefore known as **silver solders**. Outside of dentistry, 'silver solder' usually contains a substantial quantity of cadmium. However, this corrodes too readily, releasing cadmium, which is very toxic. Such solder must not be used in any biological context.

The constitution of silver solder can be understood by examining some isopleth sections of the diagram. We can first note that the Ag-Zn system, that is, the Ag-Zn face of Fig. 3.1, shows an extensive solid solution of Zn in Ag (Fig. 3.2). Likewise, Zn shows a similarly-extensive solid

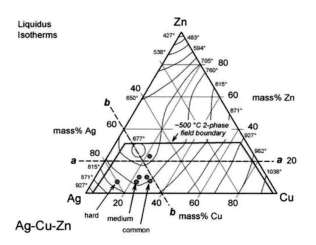

Fig. 3.1 Liquidus isotherms for the silver solder system: Ag-Cu-Zn. Marked are the compositions of some reported general and dental products with some designations. The sections marked a-a and b-b are shown in Figs 3.4 and 3.5 respectively.

solution in Cu (Fig. 3.3). Chemical similarity and radius ratio account for this; we have Ag: 144 pm, Cu: 128 pm, Zn: 137 pm, ~5% and ~7% difference respectively. Turning to the isopleth for 20 mass% Zn (Fig. 3.4), it can be seen that this solid solubility carries over for both parents in the ternary system so that there is a very large two-phase field (Ag) + (Cu). Indeed, the diagram shows a great similarity to that of Fig. 12§3.1, although of course it is not now a eutectic system – there is an additional field

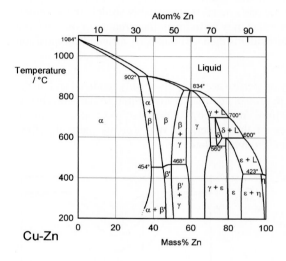

Fig. 3.3 Phase diagram for the Cu-Zn system. (Various alloys with compositions in the range ~15 ~ 50 % Zn are called brass.)

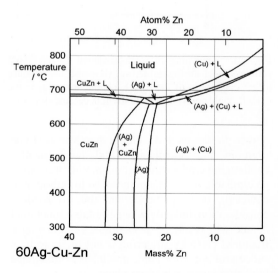

Fig. 3.5 Phase diagram for the system 60Ag-Cu-Zn, section **b** - **b** in Fig 3.1. This is drawn as viewed from the Ag apex of that diagram, for convenience.

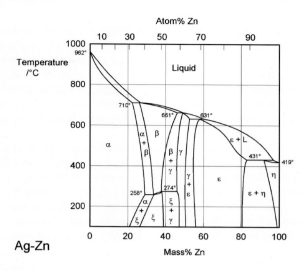

Fig. 3.2 Phase diagram for the Ag-Zn system.

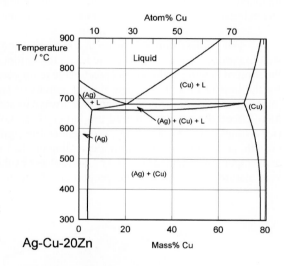

Fig. 3.4 Phase diagram for the system Ag-Cu-20Zn, section **a** - **a** in Fig 3.1. A phase marked in parenthesis, *e.g.* (Ag), indicates the solid solution corresponding to that parent metal.

corresponding to (Ag) + (Cu) + Liquid. The extent of this two-phase field, the projection marked in Fig. 3.1 for ~500 °C, can also be judged from the section shown in Fig. 3.5, that for 60 mass% Ag. It would appear that these alloys have as primary grains the (Cu) solid solution. However, the normally rapid cooling that solders are subjected to means that diffusion for phase separation can only occur over short distances and that the resulting metal will have a fine-grain, two-phase structure very similar to that obtained for the Ag-Cu eutectic itself. Obviously, a narrow freezing range is advantageous in this kind of context, preventing liquid and solid separating and forming a porous structure, aided by surface tension.

●**3.4   Thermal reactions**

The effects of heat on the structures and properties of a variety of alloys have previously been discussed under several headings, whether the heat arises in annealing, welding, soldering or some other process. In the case of soldering other detrimental effects are possible, whether these are through heating to too high a temperature, for too long, or with an inappropriate solder. It has been said that the solder 'bond' is due to the close approach of the various metal bodies and the atomic interactions thus providing 'adhesion' over a wide area, that is, metallic bonding is present, with the implication that chemical reaction does not occur. Rather, chemical reaction should not be allowed to occur to any appreciable extent. There are several reasons for this.

If the solder or any of its constituents reacts with the basis metal, the reaction is more likely to occur along grain boundaries as these offer good paths for diffusion. Then, because the resulting alloys are in all probability weaker than the grain boundaries they have replaced, intergranular cracking and failure of the piece under load is more likely (*cf.* Fig. 11§6.13). Conversely, if the basis metal is appreciably soluble in the solder, its bulk will be reduced, and if of high strength alloy the net effect may be to lower the strength of the soldered part well below that which could be obtained otherwise. The solder is then, of course, modified in composition and this may lead to undesirable structures or properties.

The elements added to solder to obtain the necessary properties are typically readily oxidized and, in the case of Zn, volatile. It is probably not possible in practice to protect adequately the molten solder with flux as a large excess is undesirable: it may become included in the joint and thereby weaken it. Overheating – time or temperature – is therefore liable to alter the composition of the solder. It will lose its particular flow characteristics, change the melting range and become difficult to manipulate.

All of these things indicate that any soldering or welding process should be conducted with the minimum duration of heating at the lowest practicable temperature for the best results. Even then, the limitations of such joints should be appreciated. There will always be some **heat-affected zone** with altered properties (Fig. 3.6)[3] (for example due to annealing, recrystallization, grain growth, or carbide precipitation), depending on the time and temperature attained (Fig. 3.7)[4], and solders by their very nature cannot match the properties of the basis

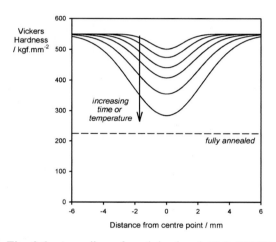

Fig. 3.6 Annealing of work-hardened 18-8 ("316") stainless steel during soldering: time and temperature affect magnitude of the effect, conduction widens the extent. Yield point and tensile strength decline similarly.

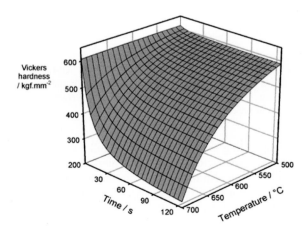

Fig. 3.7 Effect of time and temperature on the Vickers hardness of work-hardened 18-8 stainless steel. The ultimate tensile strength plot has exactly the same form.

alloys. Soldering and welding therefore are to be used in the chemically- and mechanically-challenging environment of the mouth only when unavoidable (*e.g.*, attaching hooks to orthodontic devices), or expedience is the over-riding consideration, but only when a suitable functional device will result. For example, re-attaching a fractured clasp by soldering is not likely to be a successful operation since it will have broken in a high stress area in the first place.

●3.5   Brazing

One of the problems with the silver solders described above is the zinc content as this renders them liable to more rapid corrosion, especially because such alloys are two-phase. It has been suggested that this can be avoided by using alloys such as Ag-30Cu-10Sn (liquidus 718 °C) (*cf.* Fig. 14§1.7) or Ag-28.5Cu-2.5Ni-6Sn (802 °C), both of which are three-phase. Such alloys are commonly called **brazing** alloys, although the process and principle are identical to those of soldering. It is, however, commonly understood that brazing is done at higher temperatures than soldering, although there can be no definite boundary between the two, and comparison with Fig. 3.1 shows overlap with alloys which are usually known as solders.

There are nevertheless other factors to take into account. Tin-containing silver 'brazing' alloys run and wet the parent metal less well, possibly increasing joint porosity, while the presence of nickel is now considered to be a major sensitization risk (§3.2, *cf.* 19§1.8, 19§2.2). Furthermore, if through intrinsic reasons (higher liquidus temperature) or behaviour (poorer wetting) such an alloy requires longer or higher heating, one can expect the effects on the parent metal to be exacerbated (§3.4).

§4. Amalgamation

The process of the amalgamation of dental amalgam alloy has much in common with soldering, although the intent is of course the reaction of the mercury with the basis alloy. But, to obtain the reaction, the mercury must first wet the alloy particles. This is usually inhibited by the presence of a very thin film of oxide (and probably including sulphide) from exposure to the air – which is usually humid and acidic (CO_2, SO_2, NO_x and H_2S are usually present). As before, these oxides and sulphides have high energy surfaces and will not be wetted or displaced easily by mercury. Mechanical abrasion is required to expose clean alloy, hence the use of the vigorous mechanical mixers now employed. Removal of the film by washing the alloy with very dilute HCl usually results in a very rapid wetting with little mechanical assistance, although this is not a practical process for clinical purposes. The model, however, remains a good one for a process that ordinarily occurs at high temperature and is not amenable to easy study. (See also 14§8.1, 14§9.2 and 15§1.)

References

[1]   Ray NH. Inorganic Polymers. Academic Press, London, 1978.

[2]   Lyman T (ed). Metals Handbook. American Society for Metals, Cleveland, 1948.

[3]   Wilkinson JV. Some metallurgical aspects of orthodontic stainless steel. Amer J Orthod 48(3): 192 - 206, 1962.

[4]   Wilkinson JV. The effect of high temperatures on stainless steel orthodontic arch wire. Austral Dent J 5: 264 - 268, 1960.

Chapter 23   Mechanics

*Much of the discussion so far has centred on the properties of materials as materials, but even more important is the application of those materials in real objects. It is the behaviour of those objects as **engineering** structures that is the ultimate goal of those studies. This chapter contains an outline of the mechanical behaviour, essentially the force-deflection characteristics, of some simple structures as encountered in dental applications. This is focused primarily on orthodontic contexts, but is generally applicable to a wide variety of other areas.*

*A number of foundation concepts are discussed, such as work and moment, and the means of **mapping** shear stresses and bending moments in a small group of elementary examples. An outline is then given of the principles underlying the development of the equations to describe behaviour, with some examples which may be applied to simple cases.*

*These ideas lead to the use of **flexural rigidity** as a descriptor for the behaviour of a beam as an object, combining the effects of material and shape. This is an important element of the design of structures.*

*A **case study** is then presented of the application of the above ideas to an orthodontic example, to illustrate the application of the principles and the information that may be deduced concerning the means of modifying designs for specific purposes.*

*Materials science in dentistry is not to be viewed in isolation. It is its application to real dental products and systems as a means to understand their behaviour that is important. In the case of mechanics it is the engineering principles which control the design and function of devices. This is true whether we are dealing with an orthodontic appliance, a bridge, a denture or an amalgam restoration, or even **dental hand instruments**.*

*All elastic systems can **oscillate**, and this may be problematic. Examples affecting radiography, the articulated arm of an intra-oral X-ray set, and rotary devices (air turbines) are explored. In addition, there may be **buckling** instability in slender columns, and this affects injection needles in particular.*

*Fluid systems are very common in dentistry, and consideration of **fluid mechanics** is then appropriate. The accuracy of **impression taking** and the efficacy of **endodontic irrigation** are areas of special concern.*

Materials Science for Dentistry
https://doi.org/10.1016/B978-0-08-101035-8.50023-7

      Dental treatment involves a great variety of different kinds of activity, ranging from the relatively minor interventions of topical medicaments to the complex engineering of orthodontic appliances, although these latter are rarely presented from that point of view. In fact, engineering principles are important in the design of many kinds of dental device: from small fixed bridges, through removable partial dentures, full dentures, the design of impression trays, MOD inlays and direct restorations, to crown posts and beyond. It is at this level that the mechanical properties of materials are put to use, not as tabulated numbers but in the active design process, ensuring that the device will survive and function as intended by choosing the correct dimensions, shape, cross-section, and so on. The purpose of the present chapter is to allow some insight into the behaviour of structures through examples of direct application in orthodontics. The treatment is unavoidably mathematical, but it is largely restricted to algebraic manipulation and all steps are laid out for clarity. Even so, it is the principles of the approach that are of concern, rather than the detail.

      In orthodontics various kinds of tooth movement are desired, from bodily displacement (translation) to rotation, in any or all planes, and about any axis. These movements are obtained from the forces applied through stored elastic energy in devices of various types, all of which rely upon a spring in one form or another; the work done in deforming the spring elastically (which is termed 'activation' in this area) is recovered as work done on the tooth and supporting structures. Of course, the movements are the result of the active biological remodelling of those supporting structures and not of the simple passage of the tooth through a viscous medium. The speed of response, therefore, is beyond the scope of materials science as such, although the stresses involved in eliciting the physiological response are addressable. In particular, we shall only be concerned with static systems, *i.e.* at equilibrium, assuming that any motions are so slow as to be safely ignored.

§1. Work and Moment

      The equation which defines Young's Modulus, E, for a body under uniaxial stress, σ, was given as:

$$\sigma = E.\varepsilon \tag{1.1}$$

(equation 1§2.4 a) where ε is the axial strain. This equation was obtained to be deliberately independent of the geometry of the system being tested in order to examine material properties in the intensive sense. But in order to study the deformation of real bodies, that is, not in any abstract sense but rather in engineering terms applied to whole structures, we are obliged to consider dimensions. So we retrace our steps slightly. Thus, if the length, L, and cross-section, A, of the test piece are put back, we obtain for the force:

$$F = E.A.\frac{\Delta L}{L} \tag{1.2}$$

where ΔL is the change in length. Recalling that work done, W (J $\equiv$ N.m), is given by the product of force and the distance, *s*, moved through, we have:

$$W = F.s \tag{1.3}$$

As the behaviour is assumed to be elastic and linear (*i.e.* Hookean) the average force (or, equally, the average distance moved) is given by dividing the sum of the extreme values by two. Thus:

$$W = F.\Delta L \ / \ 2 \tag{1.4}$$

This result for the whole object parallels that obtained for the material property of resilience (1§4). This type of analysis is applicable to simple springs of regular shape in compression, tension, shear or bending, irrespective of the form and material, with the sole condition that the device remains Hookean in its deformation characteristics over the range of interest. (If it were not, a more complicated calculation would be required, *i.e.* of the area under a force-deflection curve – but the principle remains the same.)

      In dealing with the sideways deflection of spring elements the idea of **torque** or **(turning) moment**[1], M, has to be introduced. Moment, the tendency of a force to rotate the body to which it is applied, is defined as the product of the force applied and the perpendicular distance *d* of its line of action to the axis of rotation (Fig. 1.1):

$$M = F.d \tag{1.5}$$

---

[1]   Where the context is understood, the qualifier "turning" is often dropped although the word "moment" has a more general meaning of a quantity × distance product.

This is seen to be similar to the definition of work, indeed it has the same dimensions – although always expressed as "newton-metres" or some equivalent, but the difference must be carefully noted. In fact, this similarity can be understood in the following sense: for the force to exist in static equilibrium there must be elastic strain in the system, thus work will have been done in reaching the final condition from an initial state of F = 0.

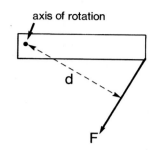

Fig. 1.1 The definition of a moment uses the perpendicular distance to the line of action of the turning force.

The above definition of moment, however, seems to require that the axis of rotation be defined. This is not strictly necessary. If instead we consider a pair of equal but opposite forces acting along parallel directions, separated by the distance d (Fig. 1.2), then for any point in the plane containing those forces the net moment is a constant, given by equation 1.5, taking into account the sign of the rotation about that point. Such a pair of forces together are said to constitute a **moment couple** or, simply, a **couple**.

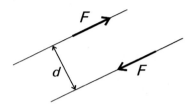

Fig. 1.2 The geometry of a moment couple.

●1.1 Sign conventions

To reach that result we had to establish the sense of rotation. Conventionally, now, anticlockwise rotation is defined as positive and clockwise as negative; the angles moved through in these directions have the corresponding signs (Fig. 1.3). This is the same sign convention as is used for the ordinary definitions of trigonometry.

Note that Newton's Third Law of Motion says that to any force there is an equal and opposite (in sign or direction) force, *i.e. Action = –Reaction*. Thus, for any moment to exist in an equilibrium system there must be a counteracting moment to establish that equilibrium.

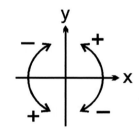

Fig. 1.3 The sign convention for rotation.

A sign convention also applies to forces, and is implicit in diagrams such as Fig. 1§12.1. Usually, forces acting upwards (+y direction) are given a positive sign, downwards are negative. Likewise, tension is considered positive while compression is negative.

It is also necessary to introduce a convention for the sign of the curvature produced in a bent object. Simply, this takes the sign of the radius of curvature (or, equivalently, the position of the centre of curvature) as indicated by the y-axis, taking the fixed point as the origin (Fig. 1.4). Concave upwards is taken as positive curvature, concave downwards as negative. By quadrant, these are the signs of the sine function (for positive rotation). Effectively, since curvature arises from an applied moment (F × d), we need to multiply the sign of the resultant rotation by the sign of the (initial) distance from the origin in the x-axis to get the sign of the moment and thus the sign of the curvature. Notice that the sign of the curvature is unaffected by the choice of origin for the curved element, which is only reasonable.

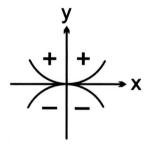

Fig. 1.4 The sign convention for curvature.

§2. Simple Beams

Orthodontic appliances, dentures and denture frameworks are complicated structures involving many parts, many cross-sections, and many points of contact with the dentition (and perhaps soft tissues). To understand the way in which they function, and therefore their design, we must begin by considering far simpler systems based on beams, or bars of uniform cross-section, and of material which is homogeneous and isotropic.

●2.1   Cantilever beam

The simplest system is a **cantilever** beam, one that is considered to be 'built in' to the support at one end and so cannot rotate there when the free end is loaded (Fig. 2.1a).[1]   The application of a force to some arbitrary point on the upper surface of the beam results in a clockwise moment at the support as before:

$$M_1 = F.d \qquad (2.1)$$

But for equilibrium this moment must be opposed by an equal and opposite moment:

$$M_2 = -M_1 \qquad (2.2)$$

This necessary condition is expressed by the **equation of moments**:

$$\Sigma M = 0 \qquad (2.3)$$

Similarly, the load F must be balanced by a reaction in the support of the same magnitude, but opposite direction:

$$R = -F \qquad (2.4)$$

This similarly essential condition is expressed by the **equation of statics**:

$$\Sigma F = 0 \qquad (2.5)$$

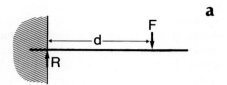
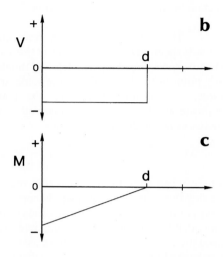

Fig. 2.1 The variation in shearing force, V, and bending moment, M, for a singly loaded, weightless, uniform section cantilever beam. Both V and M are negative because the resulting rotation of the beam end with respect to the left hand support is negative (clockwise).

These named equations of mechanics are alternative versions of Newton's Third Law. The point is that if equilibrium did not exist the unbalanced forces or moments would cause an adjustment of the geometry, a movement that would establish that equilibrium.

The application of the force F perpendicular to the axis of the uniform section beam means that it is in the plane of a cross-section. It generates a shear between that plane and the adjacent plane of atoms (working in the direction of the support). This is the situation discussed in 1§2 (Fig. 1§2.10). The shear force is transmitted undiminished to the next plane, and so on, all the way to the support. Shear force is constant over the entire length between the point of application of the load and the support (Fig. 2.1b). In addition, for constant cross-section, the shear stress is similarly constant. We give this force a negative sign because it is acting downwards:

$$V = -F \qquad (2.6)$$

It is emphasized that there is no shear stress operating between the point of application of the load and the free end. Because that end is free to move there is by definition no opposition to the deflection, therefore no reaction force, thus no shear.

It is worth ensuring that the implications of this are understood. Thus, in Fig. 2.2, the force F creates a shear action *A* at the interface between blocks **a** and **b**. However, since this a system in static equilibrium, there is necessarily a reaction *R* equal

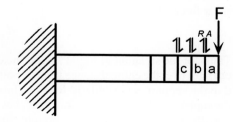

Fig. 2.2   Transmission of shear undiminished in a uniform beam.

in magnitude to, but opposing *A* at the same plane. This means that the full force of F is transmitted through block **b** to the interface between **b** and **c**, where a similar argument applies, and so on through the whole beam to the support. By the principle of uniformity (1§2.1) this must be true for all infinitesimal slices of the beam. Thus the shear stress in a uniform cross-section beam is everywhere the same.

The bending moment given at equation 2.1 was specifically calculated as at the support. However, we may also calculate the moment at any point in the beam due to the same applied force. Thus, measuring from the point of application of the load, the bending moment increases as the distance increases:

$$M = \int_0^x V.dx = V.\int_0^x dx = V.x \quad (2.7)$$

> **Constancy of geometry**
> It is worth emphasizing one thing. The response of a system depends on the magnitudes and directions of the applied forces (*i.e.* these are vectors). Those forces may result in deflections and rotations that change the relative orientations of parts of the structure and therefore of those vectors themselves with respect to the structure. The equations that are developed here are all on the assumption that all strains are sufficiently small that they cause no appreciable change to the geometry of the system. This is an extended version of the limitation of Hooke's Law (1§2, 1§13): "If a deformation is small enough ...". We require as a condition of validity the **constancy of geometry**. Of course, real systems do deform more than this and any calculations that fail to take this into account are necessarily only approximations: compare the discussion of true modulus of elasticity and true strain (1§2.2, 1§2.4). The difficulty here is that the deflections of beams – as with the deflections of orthodontic springs – may be very large in proportion to their size. Caution is required as deviations from the supposed behaviour must then be observed.

where *x* can vary from 0 to *d* (the reason for writing this in integral form will become apparent below). In other words, within the beam the moment is not constant from point to point but varies directly in proportion to the distance over which it is acting (Fig. 2.1c). Note that the starting point for this integration, the zero distance, is at the point of load application. We have used the force V because we have assigned a sign to this: negative. Thus the moment is also negative. The mechanical interpretation of this is that the centre of curvature is below the beam. Given that the moment can be represented as an integration of the shear force, in the other sense we also have

$$\frac{dM}{dx} = V \qquad\qquad (2.8)$$

●2.2   Additivity

Since the purpose of this type of analysis is to be able to handle real systems, as opposed to observing material properties in deliberately simplified mechanical tests (Chap. 1), we now proceed to the case of more than one load applied to the beam. Now, in working with Hooke's Law (1§2) it is implicit that each increment of load behaves as does any other increment, so that the responses of the system are likewise consistent: we would not have proportionality otherwise. Thus, in Fig. 2.1, it is of no consequence that F is applied as several small loads or as one: the shear force acting at a point is the sum of the shear forces generated by each small load. The response to each load is independent of the responses to other loads (assuming constancy of geometry).

Irrespective of the distribution of the loads on a beam, the total shear force at any point is simply the sum of the shear forces due to all sources:

$$V_{TOT} = \Sigma V_i \qquad\qquad (2.9)$$

Thus, for the situation of Fig. 2.3a, the shear force due to F_1 operates over the entire distance from its point of application, d to the support, and similarly for F_2. Thus, between d and the support both shear forces are acting, but between d and e only the shear force due to F_2 is present, since this region is 'outside' d. From e to the end of the beam there is no shear. This argument applies equally to bending moments, so that at any point we may write:

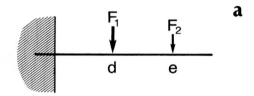

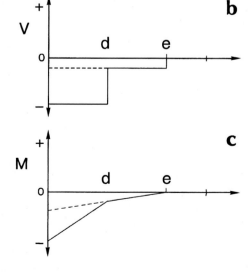

Fig. 2.3 Shearing force and bending moment in a doubly loaded cantilever beam. Superposition says that the effects of the two forces are additive.

$$M_{TOT} = \Sigma M_i \qquad (2.10)$$

We thus obtain the diagrams of Figs 2.3b and 2.3c by the simple addition of the respective diagrams, in the form of Figs 2.1b and 2.1c, for each of the two forces applied. In fact, if we also assume that the beam is of one uniform material, *i.e.* constancy of elastic modulus, deflections are also additive:

$$y_{TOT} = \Sigma y_i \qquad (2.11)$$

This simple additivity of the different types of response to the applied loads is called the **principle of superposition**. That is, assuming constancy of geometry (including magnitudes and directions of the forces applied), the full description of a system can be found as the summation of the effects of each element of load. This is in effect a version of the **principle of uniformity** (1§2).

●2.3   Self-weight

We can explore the implications of the principle of superposition with a real example. So far the beam has been treated as having no weight, *i.e.* ignoring the forces felt by the beam due to gravity. The effect of this so-called **self-weight** of the beam is that of a uniformly distributed load (Fig. 2.4 a). Therefore, for a constant weight per unit length of the beam, the **load intensity**, q, at any point is given by:

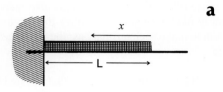

$$\frac{dF}{dx} = -q \qquad (2.12)$$

so that the shear stress at a given point depends on the total weight 'outside' that point:

$$V = \int_0^x -q.dx = -qx \qquad (2.13)$$

Thus, taking the integration route of equation 2.7, we have simply:

$$M = \int_0^x V.dx = -qx^2/2 \qquad (2.14)$$

where x is measured from the free end. This produces the conditions shown in the diagrams of Figs 2.4 b and 2.4 c. Again, if point loads were to be applied to such a beam, as in Figs 2.1 a or 2.3 a, the separate effects of self-weight and imposed loads can be superposed to find the net effect. The point is that complicated systems can be dissected into their elements to perform the analysis in stages, and the individual results assembled to give the overall picture.

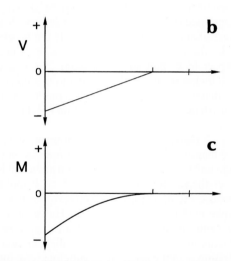

Fig. 2.4 The shearing force and bending moments due to self-weight are equivalent to a uniformly loaded beam. The effects are shown for a cantilever.

The cantilever is only one form of beam. Another common type is that of the **simply supported beam**; that is, both ends rest on some (frictionless) support. By the arguments given for the cantilever, the portions outside of the supports (assuming that no forces are applied there) have no influence on the results and will be ignored.

●2.4   Three-point bending

Consider first the situation in Fig. 2.5 a. A centrally placed load generates reactions in the supports, each of which is one half of the applied load. This is known as three-point loading and is a common type of mechanical testing arrangement (see Fig. 1§1.3; 1§6.4). Now, looking first at the beam to the left of the central load point, the shear force is acting so as to rotate each element of the beam here *clockwise* from its original position with respect to the left-hand point of support adopted here as origin. The shear force is therefore negative in sign in this region. Between the centre and the right hand end, however, the shear is acting in the opposite sense, tending to rotate each element of the beam *anticlockwise* about the centre point and thus must also be acting in the same sense about the adopted origin. The shear force here is therefore positive. The pattern of shear forces is thus as in Fig. 2.5 b.

We may dissect the system a little further to clarify this outcome. If we were to take away the right hand support, and thus its reaction, the load to be considered is just F. We then have precisely the load system of a cantilever beam as in Fig. 2.1, with no load, and therefore no shear, to the right of the centre point. On the other hand, if we were to take away the left hand support, and move our reference coordinate system origin to the right hand support, it is a cantilever beam again, with a similar unloaded portion to the left of the centre point, but with the shear in the opposite sense between the centre and the right hand support. We then add the two diagrams together to determine the net effect. It is therefore necessary to observe carefully the sign convention which has been adopted for a consistent picture to be obtained. It is not just a case of identifying 'up' and 'down', but the sense of rotation as well. This also means that it does not actually matter where we identify the origin of our coordinate system to make such an analysis: the outcome will always be the same.

By reasoning similar to that applied for the cantilever beam (§2.1), the shear acting between successive sections of a beam in three-point bending can be shown to be uniform in magnitude (Fig. 2.6). From this it is very clear that the sense of shear changes at the mid-point. If the blocks are imagined to slide, the clockwise motion of **a** with respect to **b**, of **b** with respect to **c**, and so on, is plain. Similarly, **a** moves in an anticlockwise sense with respect to **b'**, and so on. It is the need to distinguish between such conditions that requires the consistent use of a sign convention (§1.1).

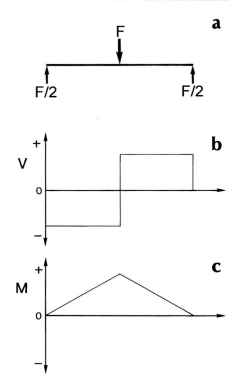

Fig. 2.5 Shear force and bending moment diagrams for a simply supported, centre loaded beam. This is termed 3-point loading.

Now, to get the bending moment, we apply the integration of equation 2.7 to the shear forces mapped in Fig. 2.5 b, from the right and towards our chosen origin at the left hand support. This gives the result shown in Fig. 2.5 c. It looks odd at first sight because the moment is everywhere positive, in contrast to the situation in the cantilever beam. However, it is correct because the centre point of the beam must be depressed so that its centre of curvature lies above it, *i.e.* the curvature is positive, the bending is in the sense of an anticlockwise turn about the origin for all points along the beam. (We would, in fact, achieve the identical result if we chose the right hand support for the origin,

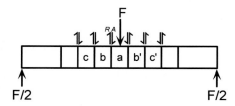

Fig. 2.6 Shear between sections of a beam in three-point bending.

because the distance to be used in the integration would then be negative.) We can therefore see that the curvature is the consequence of the imposed system of loads, the outcome and not the causation.

Thus, the overall result for a three-point beam is a shear stress which is uniform in magnitude everywhere except exactly at the centre point (where it is zero), and maximum bending moment, $M_{max} = FL/4$ (*i.e.* force × distance), at that centre point. Now, if we were doing a mechanical test, if failure occurred it would be due to a flaw initiating the crack. Such flaws would ordinarily be distributed randomly over the test zone, and have a random distribution of size and orientation, so there would be no guarantee that failure would occur at any particular location. This means that the details of the loading at the failure site would vary from test to test because the bending moment varies. However, a slight modification allows a clearer picture to be obtained for some purposes. This involves separating the centre load into two equal portions, symmetrically placed about the centre, to give what is termed the four-point loaded beam (1§6.4).

●2.5 Four-point bending

In a four-point beam the reaction forces are necessarily identical to those in the three-point system (Fig. 2.7 a). Following through the same kind of analysis, the shear forces are now found to be restricted to those regions at the ends of the beam between the load application points and the corresponding support, wherein the reaction is seen (Fig. 2.7 b). However, considering the force acting on **a** (Fig 2.8), while it is transmitted to the left and ultimately to the support, it is similarly transmitted to the right to the interface with the central block. But this block is also subject to an identical force – in both magnitude and direction – on the right at the interface with **a'**. Hence, although being pushed down, the forces acting on the central block are exactly balanced, therefore there is no shear whatsoever.

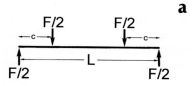

The striking thing here is then the central region in which there is no shear force acting. Notice that, in the limit, if the two upper forces F were moved out to act directly on the supports there could be no shear between the supports, and equally, in the three-point situation (Fig. 2.5 a), there is no shear under the coincident pair of forces, each F/2.

The integration for the bending moment is as above, and yields the result of Fig. 2.7 c: between the load points it is constant, $M_{max} = Fc/2$, corresponding precisely to the zone of no shear. This region experiences what is therefore known as **pure bending**.

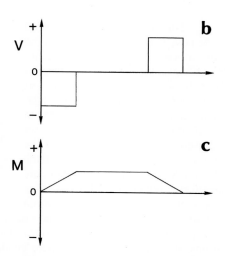

These examples of beam bending are clearly special cases, being symmetrical, although convenient for mechanical testing (1§6.4), but are illustrative of the main features of and the approach to be taken for more general systems. But shear force and bending moment are not the only effects operating. We must determine what other stresses and strains may be present. For example, it is evident that for any body to be bent, meaning that one surface is made convex and the other concave, the outer convex surface must be relatively longer than a corresponding portion of the inner surface.

Fig. 2.7 Shear force and bending moment diagrams for a simply supported beam in 4-point bending.

Imagine bending a beam into a full circle such that the ends were butted together: the outer surface has a larger radius of curvature; it is therefore longer than the inner surface. Whether this is due to the one being stretched or the other compressed is immaterial. For a relative change in length (*i.e.* strain) there must be a force acting, and the shear forces identified cannot be the source since shear alone does not change dimensions in that sense. We shall therefore stay with the case of pure bending to investigate this. For either stretching or compression (or both) of

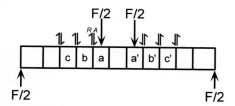

Fig. 2.8 Shear between sections of a beam in four-point bending. The central block suffers no shear stress with respect to its neighbours.

those surfaces of the beam to occur means that there are forces acting along its length, perpendicular to the cross-section. Thus, to determine the internal forces operating over any arbitrary cross-section of the central portion of Fig. 2.7 a, the deformation of the beam must first be considered.

●**2.6   Geometry of bending**
For a rectangular beam it is found experimentally that two adjacent parallel lines, *mn* and *pq*, on its sides, perpendicular to the length of the beam, remain straight during bending; but also that they rotate (about an axis perpendicular to the page) and remain everywhere perpendicular to the longitudinal 'fibres' of the beam (Fig. 2.9 a). The theory of bending then assumes this to be a universal result and also, by extension, that the entire section perpendicular to the bending axis remains plane and that every fibre remains perpendicular to that plane. It follows that transverse sections such as correspond to *mn* and *pq* (Fig. 2.9 a) must rotate as a whole with respect to each other during bending, so that they are no longer parallel.

This result can be seen to be expected from an alternative view. Since there is only a bending moment acting on any section of the beam (in the pure bending region), we can represent it as in Fig. 2.10 a which shows that, relatively speaking, the upper surface is under compression and the lower under tension, but the magnitudes are as yet unknown. Let us then consider just the extreme outmost fibres (Fig. 2.10b): if the elastic modulus is symmetrical about zero (as is reasonable for small deflections), the deflection of the top fibre must be identical to that of the bottom fibre – but opposite in sign. This means that the net length of the two fibres is unchanged, and that there is therefore no net force acting to extend or compress the pair – which are required conditions for pure bending. The couple is therefore acting to rotate the plane containing the ends of the two fibres about a point midway between the two.

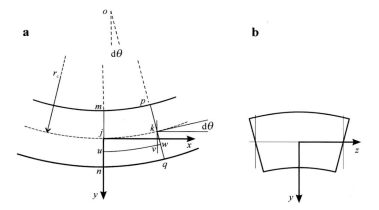

This argument can be extended to the next pair of fibres, just below the two surfaces, to obtain the same result. But, because these fibres are closer to the axis of rotation, the deflection of their ends is proportionately less. Continuing this process we end up with the situation of Fig. 2.10 c. The arrows represent both the magnitudes of the forces acting to compress or extend the fibres and their direction: they are therefore force vectors. Equally, they represent the corresponding strain vectors. Notice that the sum of the forces must be precisely zero (taking into account their sign), and the sum of the strains likewise, for this to be pure bending. The axis of rotation is necessarily through a point at the centre of the beam section – in the longitudinal section that we have taken of this. There is at this point no stress and no strain. It is called a **neutral point**. The stress acting on any fibre, and also the strain it suffers, is therefore proportional to its distance from the neutral point.

Fig. 2.9 a: The geometry of pure bending. **b**: Poisson distortion in the transverse section. The orientation of the *y*-axis is the same as in **a**. Notice the resulting **anticlastic** upper and lower surfaces: bending in opposite senses in the *x*- and *z*-axes (when the Poisson ratio is positive, 1§2.2).

The longitudinal fibres on the convex side of the beam are therefore under tension and those on the concave side are under compression, each with corresponding strains. This applies uniformly to every section through the beam in the pure bending region (the **principle of uniformity** again, 1§2). Since it must also be true over the entire cross-sectional area, there is also a line of neutral points on that section where there is no force acting. Correspondingly, it follows from this that there must be a surface, the locus of all those points, in which it is true. This is called the **neutral surface**, and its intersection with any (*xy*-)longitudinal cross-section is called the **neutral axis** of that section. Such an axis is defined more generally as the line through the **centroid**, *i.e.* the **centre of gravity**, of each transverse cross-section (*cf.* the **neutral circle** in Fig. 20§7.5).

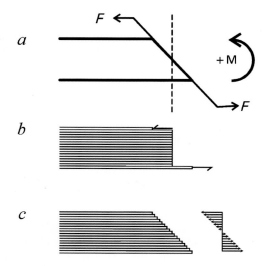

Fig. 2.10 A beam in pure bending:
a: The couple acting on the beam cross-section.
b: The effect of an element of the same couple acting on just the outermost fibres, assuming independence.
c: The stress (or strain) vectors acting on each fibre of the beam. The arrows represent the magnitude and the direction of the stress, *viz*: left = compression, right = tension.

It should be noted, however, that the 'layers' of such a beam are not independent. The strain in one is affected by the different strain in the adjacent fibre: there is constraint (*cf.* 6§2) because they cohere (*i.e.* are bonded to each other), and so there must be a shear stress operating between them.

•2.7   Stresses and strains
What remains is to calculate the magnitudes of the stresses and strains. In Fig. 2.9 a the neutral axis is the (dashed) line on which points *j* and *k* lie. Lines *mn* and *pq*, when extended, meet at *o*, the centre of curvature for the beam. Hence, $r_c = oj = ok$ is the radius of curvature of the neutral axis of the beam. For any initial fibre of length *uv* at distance *y* from the neutral surface, the lengthwise strain ε_x is defined by

$$\varepsilon_x = \frac{vw}{uv} \qquad (2.15)$$

Constructing the line kv, parallel to ju, by similar triangles $\Delta jok \equiv \Delta vkw$ (assuming that the curvature is small) we find approximately (Fig. 2.11):

$$\varepsilon_x = \frac{vw}{uv} = \frac{vw}{jk} = \frac{kw}{ok} = \frac{ju}{oj} = \frac{y}{r_c} \qquad (2.16)$$

As expected, the strain is proportional to the distance y of the fibre from the neutral surface. It follows from Hooke's Law that:

$$\sigma_x = E . \frac{y}{r_c} \qquad (2.17)$$

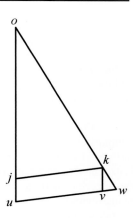

Fig. 2.11 Similar triangles argument for strain in a bent beam. Letters correspond to those in Fig. 2.9a

      This analysis obviously depends on the approximations implicit in equation 2.16 being valid. But, as before (Chap. 1 and above), we may use the results as being sufficiently accurate so long as the strains are small. To illustrate this we consider the Poisson strain arising from the strains in the beam 'fibres'. On the inner, concave surface the compression is maximum, and here is found the maximum Poisson expansion. Conversely, on the outer, convex surface the Poisson contraction is greatest in response to the axial tension. The difference in sign of these two effects means that the cross-section is also curved – as if a bending moment were applied across the width of the beam, and in the opposite sense to that causing the axial bending (Fig. 2.9 b).[2] This has little practical effect in stiff materials such as metals and ceramics, and can safely be ignored in the same way that 'engineering' results were found acceptable (1§2.4). However, it is readily observed in softer materials such as rubbers; bars of elastomeric impression materials and alginate ($v \sim 0.5$), for example, show it clearly. It can be demonstrated clearly with a rectangular soft rubber pencil eraser (Fig. 2.12).

Fig. 2.12 Anticlastic curvature of a bent beam (*cf.* Fig. 2.9).

      Although this has been developed for the case of pure bending, similar results can be derived for other cases, taking note of the principle of superposition. That is, the effects of bending are independent of those of shear.

●2.8   Pre-stress

      There is another implication of the principle of superposition which can be used to great advantage. The technique that uses it is called **pre-stressing**, and is of great importance in dentistry in the context of porcelain (25§5). A brittle material which has a strength of σ_t in tension obviously ordinarily fails when the applied tensile stress reaches that same value, $\sigma_{applied} = \sigma_t$ (Fig. 2.13, line a). Suppose, though, that before testing a compressive stress, σ_c, is first imposed on the piece by some means; this is the pre-stress. That pre-existing compressive stress must therefore first be overcome by the applied stress (point b) before a tensile stress can be generated in the piece. Only when the net stress attains the value corresponding to the strength of the material will failure occur: $\sigma_{applied} - \sigma_c = \sigma_t$ (numerically). Rearranging this, we have:

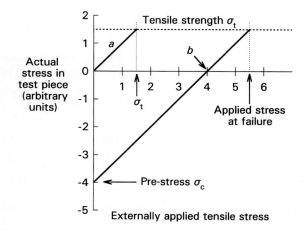

Fig. 2.13 Diagram illustrating the principle of pre-stressing.

---

[2] NB: this is only true for 'normal' materials with a positive Poisson ratio.

$$\sigma_{applied} = \sigma_t + \sigma_c \qquad (2.18)$$

as the condition for failure. Again, this illustrates additivity. The strength of the material will thus appear to have been increased.

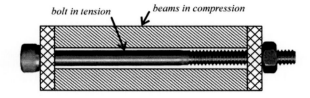

Fig. 2.14 Pre-stressing always involves balanced forces.

Of course, it is necessary that the compressive stress itself is not so large as to cause failure. But since collapse in compression typically is expected to occur at stresses some 8 times larger than those required in tension in the same material (1§7.2), the apparent increase in strength from compressive prestressing can be substantial, the limiting failure stress being about $9\sigma_t$. The only problem with this approach is that the prestressing stores a large amount of elastic energy in the piece. If failure does occur, the stored energy must be released. Cracking will then be dramatic and uncontrolled, resulting in the fragmentation of the piece. Car windscreens are a good example of this. The available mechanisms for applying this to dental porcelain are discussed later (25§5). Even so, it should be recognized while some benefit may be seen in the compressed element of a structure, the complementary tensile forces elsewhere may put that member at risk unless it is sufficiently strong (Fig. 2.14).

Effectively, pre-stressing also occurs in any material where there is a differential dimensional change. For example, polymerization shrinkage and water-sorption in filled resins (6§2.12), and in any composite or structure where there are phases of differing thermal expansion coefficient (5§5.5). It should be recognized that the outcome may be advantageous or disadvantageous, depending on the circumstances.

§3. The Second Moment of Inertia

So far we have assumed rectangular beams, but this is plainly not a general enough system to be of use in understanding even orthodontic wires. In particular, the summation of the forces considered in connection with Fig. 2.10 above was on the assumption of symmetry about the z-axis. This is not an essential condition. What is needed is a summation over the entire area of an arbitrary cross-section, element by element, taking note of the stress acting on each. Thus, from equation 2.17, the element of force dF acting on an area element dA is given by:

$$\frac{dF}{dA} = E \cdot \frac{y}{r_c} \qquad (3.1)$$

since dF/dA is a stress. Hence, the moment of that force, from equation 1.5, is

$$\frac{dM}{dA} = \frac{dF}{dA} \cdot y = E \cdot \frac{y}{r_c} \cdot y \qquad (3.2)$$

where y is still the perpendicular distance from the neutral surface. Integrating over the entire cross-section, equating that integral to the known (total) moment, M, of the external forces, we get:

$$M = \int \frac{dM}{dA} . dA = \int \frac{E}{r_c} . y^2 . dA = \frac{E}{r_c} . \int y^2 . dA \qquad (3.3)$$

This rearrangement has effectively separated the elements of equation 3.2 into two terms which are respectively material-dependent and material-independent. The integral

$$\int y^2 . dA = I_z \qquad (3.4)$$

is known as the **second moment of inertia**[3] (or the **moment of inertia of the area**) of the cross-section with respect to the z-axis, which is taken to pass through the centroid of the section (see Figs 2.9b and 3.1a). It is sometimes referred to as the **shape factor** for the section; tables of these can be found in engineering texts for a wide variety of patterns. It can be explicitly calculated in the case of many regular sections. Thus, for a rectangular section (Fig. 3.1b), height h (that is, measured in the direction of the load axis), breadth b, we have

---

[3] 'Second' moment because the square of distances is involved.

$$I_z = 2\int_0^{h/2} y^2 .b.dy = \frac{bh^3}{12} \qquad (3.5)$$

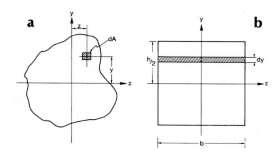

Similarly, for a circular cross-section, radius r, the integration gives:

$$I_z = \frac{\pi r^4}{4} \qquad (3.6)$$

Thus, from equation 3.3, for any beam we can write:

$$M = \frac{EI_z}{r_c} \qquad (3.7)$$

Fig. 3.1 The calculation of the second moment of inertia, a: for an arbitrary cross-section, b: for a rectangular section. y is the load axis.

so that the product EI_z emerges naturally as a parameter descriptive of the particular beam (*i.e.* section plus material). This is called the **flexural rigidity** of the beam and may be interpreted simply as the **resistance to bending** of the section. We thus return with a result that satisfies the requirement that we consider real structures rather than materials in isolation. We now have a means of investigating design specifically.

One point can be made immediately: the material far from the z-axis contributes more to I_z than does the closer material. Hence, if the material is redistributed such that most material is far away from the z-axis, for the same total area and therefore for the same amount of material, the structure can be made stiffer. Thus, an I-beam is very stiff to bending in the vertical plane compared with a solid bar (Fig. 3.2), and a circular tube is much stiffer than a rod of the same material, for a fixed material cross-sectional area. In the major connector of a bridge, for example, the roughly elliptical section gives very different stiffnesses in the planes corresponding to the major and minor axes of the ellipse (see §3.2).

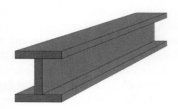

Fig. 3.2 An I-beam is stiff to bending in the plane of the central 'web' because material is concentrated in the flanges at a large distance from the centre line.

•3.1 Additivity

We have already seen that the application of the principle of superposition relies on the additivity of forces, bending moments and deflections. We may now extend this list by adding the second moment of inertia. Thus, if two beams are simply supported (as in Fig. 2.5) in parallel and loaded simultaneously at the centre point with the same load (Fig. 3.3), as by a cross-bearing member, the deflection observed is accounted for by using just the sum of the individual I_zs (that is, about their own neutral axes, not the centre line of the pair). In fact, this result applies no matter what the relative disposition in space of the two, or more, beams might be.

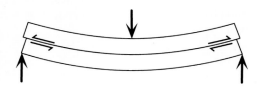

Fig. 3.3 Unless the shear which occurs between beams or layers is constrained by friction, adhesion or pinning, the deflection of the two is independent, and simple additivity of stiffness applies.

Hence, a pair of similar planks laid one on top of the other would only be twice as stiff as one (ignoring friction between the two). Yet, from equation 3.5, somehow gluing or clamping the two together would make the structure *eight* times stiffer (effectively doubling h) – which is the point of the remark at the end of §2.6. The point is that, when they are unbonded, the expansion of the lower surface of the top plank is not opposed by the compression of the upper surface of the other. This can be seen by duplicating the beam of Fig. 2.10c, one on top of the other. The maximum tension of the top piece is adjacent to the maximum compression of the lower part; there must therefore be sliding here during bending. Yet, if this junction were to be fixed, it would become the new neutral surface at which there is no strain, and no relative motion. It is the freedom of individual structural elements to move past each other that confers flexibility on multi-element structures of this kind.

The additivity rule is also useful when considering more complex arrangements, so long as the structure can be broken down into simpler pieces whose second moment about a given axis (not necessarily lying through the piece itself) can be found. This also applies to holes: the 'contribution' is subtracted, which makes some kinds of structure much easier to deal with. Thus, for example, for a circular tube, from equation 3.6, we can write simply:

$$I_z = \frac{\pi r_2^4}{4} - \frac{\pi r_1^4}{4} = \frac{\pi}{4}\left(r_2^4 - r_1^4\right) \tag{3.8}$$

where r_2 and r_1 are the outer and inner radii respectively.

●3.2   Practical implications

This effect is utilized in many ways. Braided wires are sometimes used in orthodontics, for example, containing three strands. Although an analysis in this case is rather complicated, the flat braid would tend to be stiffer for deflections in the wide plane, and less stiff in the perpendicular direction, thus offering some control of the action of the device. Electrical cables ("flex") are a more common example, where the conductor consists of many fine strands twisted together (the twisting in fact makes the cable even more flexible – see §6). There is a need here to balance the current-carrying capacity against flexibility. The resistance of the conductor varies as does $1/A$ (A = area), but I_z is proportional to A^2 (equation 3.6) if this were in one piece; strength, proportional to A, may also be of interest. But by using small conductors, I_z for the assemblage of wires becomes proportional to A. Single solid conductors may be used in fixed installations; however, where movement is expected they would quickly work-harden and fracture, but in any case be very stiff. Ships' mooring cables, the shrouds holding the mast on a yacht, and the drive wire for the pen of a chart recorder are all steel, made up from many smaller wires in order to achieve sufficient flexibility. Glass fibre woven fabric, which can be used for curtains, for example, is evidently made from a very stiff and brittle material but is soft and flexible. Even casting ring liner (17§6), being made of a mat or felt of many small ceramic fibres, is flexible, but only because the fibres are capable of moving past each other.

In such structures there is a further advantage: premature failure of a single element due to a fatal flaw (Griffith criterion, 1§7) is not likely to cause failure of the structure in terms of the load carrying capacity of that strand, because that which was carried by the now broken fibre will only be a small proportion of the total. In addition, the crack cannot be transmitted to adjacent strands, no matter how brittle the material. This effect is turned to great advantage in fibrous composite materials, where the embedding of the core in a bonded matrix makes the structure much stiffer than the unsupported fibres (29§4) but, more importantly, much tougher as well.

In addition, savings can be made in material, or weight, for a given stiffness by using a tube rather than a solid bar. Thus, for a fixed material cross-sectional area A, the second moment of area for a square tube is given by:

$$I_z = \frac{a^4 + 2a^2b^2}{12} \tag{3.9}$$

where a is the width of the equivalent solid bar (A = a^2), and b is the width of the (square) lumen (Fig. 3.3). Similarly, for a circular section tube, with a and b the corresponding diameters, the equation is:

$$I_z = \frac{\pi}{64}(a^4 + 2a^2b^2) \tag{3.10}$$

(Fig. 3.4). The effect is quite dramatic, and accounts for the use of steel tubing or bamboo (which has hollow stems – evolutionary design optimization) as scaffolding, and also for the design of long bones, to use the material efficiently without unnecessary weight. However, it must be realized that there is a practical limit to how far the diameter of the lumen can be pushed, given that thin walls buckle under low loads, and the yield point of the material becomes important in setting this limit.

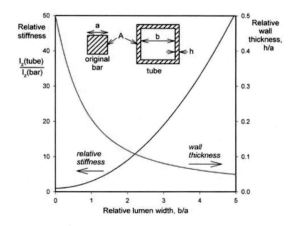

Fig. 3.4 For a given amount of material, cross-sectional area A, the stiffness rises rapidly as the tube lumen is increased.

Another means of control is to vary the section shape. Thus the stiffness may vary according to the direction of bend. For example, for an elliptical cross-section, we have

$$I_z = \frac{\pi a^3 b}{4} \qquad I_y = \frac{\pi a b^3}{4} \qquad (3.11)$$

where a is the semi-diameter normal to the z-axis, and b is the semi-diameter normal to the y-axis. It is easy to see that the ratio of the two, I_z/I_y is $(a/b)^2$, in other words, the square of the axial ratio, for a constant cross-sectional area, πab (Fig. 3.5). Thus, an approximately elliptical section casting, such as for the major connector of a partial denture framework,

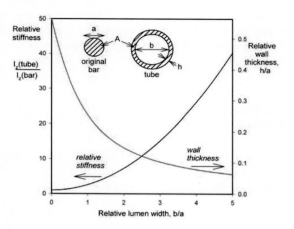

Fig. 3.5 Similarly, for a circular section tube, although slightly less efficiently than for the square section.

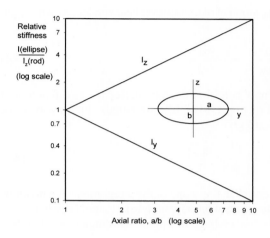

Fig. 3.6 Stiffness can be controlled by section shape. This is for an ellipse of constant cross-sectional area. Note the logarithmic scales.

which is primarily that shape for anatomical reasons, can be expected to show a substantial difference in flexibility according to the sense of the bending. Such matters must be taken into account in the design if untoward effects are to be avoided. This kind of behaviour is also seen in semicircular sections such as are commonly used for clasps, when out-of-plane bending of the arc is not desirable, but flexibility in opening is required for insertion and removal.

Incidentally, for the same cross-section, a square wire is a little stiffer than a round one. In equations 3.5 and 3.6, setting the areas to be equal:

$$A_{square} = bh = a^2 = A_{circ} = \pi r^2 \qquad (3.12)$$

we find the required wire radius to be:

$$r = \frac{a}{\sqrt{\pi}} \qquad (3.13)$$

Therefore:

$$\frac{I_{square}}{I_{circ}} = \frac{a^4/12}{\pi r^4/4} = \frac{a^4/12}{a^4/4\pi} = \frac{4\pi}{12} = 1.047... \qquad (3.14)$$

– which is a slightly more efficient use of material.

§4. Deflections of Loaded Beams

We have now reached the stage of being able to describe the state of stress in beams of various kinds (§2), and have deduced a characteristic parameter, the flexural rigidity, for the beam (§3). This is not enough, as a major intention of many structures is that they maintain their shape. It is necessary then to consider the deformation under load to complete the picture given in Figs 2.4, 2.5 and 2.7, for example. Although this was suggested in equation 2.17, the value of r_c was left unidentified. We proceed through an example to illustrate how the standard equations of deflections which appear in the tables for the various beam designs are obtained, concentrating on the principles rather than the details of the calculus. We shall consider the self-weight deflection of a cantilever beam (Fig. 2.4).

To calculate the deflections of each point along such a loaded beam we start by considering local geometry, and for the effect of the bending moment alone. (The effect of the shear forces on curvature can usually be ignored – if the deflections are small.) Thus, if in Fig. 2.9a we take points j and k to be separated only by the small interval dx, and taking the small angle jok to be $d\theta$, we may write:

$$dx = r_c.d\theta \quad \text{and} \quad \frac{1}{r_c} = \frac{d\theta}{dx} \tag{4.1}$$

Now, since the deflection is very small, we may make the following approximation, where $d\theta$ is also the angle of slope of the beam at that point:

$$d\theta \simeq \tan d\theta \simeq \frac{dy}{dx} \tag{4.2}$$

where dy is the deflection of the point k. Hence,

$$\frac{1}{r_c} = \frac{d\theta}{dx} = \frac{d^2y}{dx^2} \tag{4.3}$$

So that, from equation 3.7, and the expression for the bending moment of a cantilever beam under its own weight given at equation 2.14, we have for this particular case

$$M = \frac{EI_z}{r_c} = EI_z.\frac{d^2y}{dx^2} = -\frac{qx^2}{2} \tag{4.4}$$

where q is the load intensity (equation 2.12). The object now is to find the total deflection as a function of position along the beam, which requires successive integration to get rid of the differentials. First we integrate equation 4.4 with respect to x to the point of support ($x = L$) to give

$$\int_x^L M.dx = EI_z.\frac{dy}{dx} = -\frac{q}{2}\left(\frac{L^3}{3} - \frac{x^3}{3}\right) \tag{4.5}$$

Inspection of the boundary conditions is important here to establish the scope of the integration. We do know that, because it is built-in and unable to rotate, the *slope* of the beam at the support must be zero, *i.e.* when $x = L$, the length of the beam (because we are measuring x *from* the free end). At this point, rearranging, we therefore have an equation for the *slope* of the cantilever beam under its own weight as a function of the distance from the free end:

$$\theta = \frac{dy}{dx} = -\frac{q}{6EI_z}\left(L^3 - x^3\right) \tag{4.6}$$

The negative slope here is consistent with the convention for the deflection of the end of the beam downwards. Setting $x = 0$ tells us the slope at the free end itself.

Integration of equation 4.6 (*i.e.* equation 4.4 a second time), again with respect to x, and again to the point of support, therefore gives:

$$EI_z.\int_x^L \theta.dx = EI_z.y = -\frac{q}{6}\left(L^4 - \frac{L^4}{4} - L^3x + \frac{x^4}{4}\right) \tag{4.7}$$

This time we note that the *deflection* at the support must be zero, that is when $x = l$, $y = 0$. This expression gives

$$EI_z.y = -\frac{q}{6}\left(\frac{3L^4}{4} - L^3x + \frac{x^4}{4}\right) \tag{4.8}$$

i.e. $$y = -\frac{q}{24EI_z}\left(3L^4 - 4L^3x + x^4\right) \tag{4.9}$$

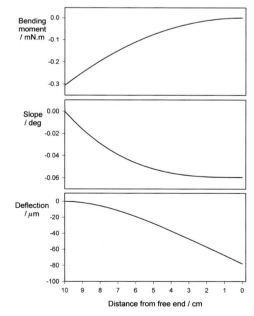

Fig. 4.1 Self-weight effects in a cantilever beam: 100 mm length of 1mm diameter stainless steel wire, E = 200 GPa, ρ = 7·93 kg/L. Note that the built-in end is at the left. The bending moment curve has the same shape as Fig. 2.4 c.

This, then, is the equation of the curve assumed by a cantilever beam under its own weight. The negative sign remains consistent with the convention that the force acting downwards is negative and gives a negative deflection.

These results are illustrated by an example in Fig. 4.1. It can be seen that the premultiplier term in q/EI_z for the slope and deflection amounts to a scale factor such that exactly the same shape curves would be obtained for any such cantilever – subject to the scale adjustment for the graph, and again assuming that the deflections were small – note the approximation involving the angle of slope at equation 4.2.

●4.1   Cantilever beam

Whilst neither the detail of the above equations nor the integrations themselves are of immediate interest, the procedure is important as it illustrates how, in general, any system may be analysed to find moments, slopes, and deflections, all of which depend inversely on the flexural rigidity. This conclusion similarly and inevitably arises in all other beam systems, whether as structural elements of some functional device, in a test for determining the mechanical properties of the beam material, or even as the working part of a mechanical test device.

For example, in the cantilever beam the deflection at the point of application of a **single load** (Fig. 2.1) is given by:

$$y \; = \; \frac{FL^3}{3EI_z} \tag{4.10}$$

(disregarding direction). This, of course, is the deflection required for a simple orthodontic finger spring. One of the features of interest in such a device is the sensitivity of the deflection to the load applied, what we may call the **flexibility** of the spring (*i.e.* as an object, not the material property: *cf.* equation 1§2.4 c). If this is defined as y/F, the deflection per unit load, we have from equation 4.10

$$\frac{y}{F} \; = \; \frac{L^3}{3EI_z} \tag{4.11}$$

This is in fact the **compliance** of the beam (with a similar remark being made for the equations of all beams). This gives immediate information of direct value in the design of appliances, showing the trade-off between the diameter and the length of the wire used.

Sometimes in dentistry, in orthodontics especially, the inverse of this quantity is of interest and is called, rather oddly, the **rate of force delivery**, *i.e.*

$$\frac{F}{y} \; = \; \frac{3EI_z}{L^3} \tag{4.12}$$

Effectively this is the **lateral load deflection stiffness** of the spring in the Hooke's Law sense, the force required for unit deflection[4] (*cf.* equation 1§2.1). In engineering contexts, this quantity, sometimes labeled k, is called the **equivalent stiffness** of the beam.

For very low elastic modulus materials, such as alginate impression materials, the **self-weight deflection** may be a convenient variable for monitoring behaviour. So, substituting $x = 0$ in equation 4.9, and rearranging, we obtain (again dropping the negative sign for convenience):

$$E \; = \; \frac{qL^4}{8yI_z} \tag{4.13}$$

Fig. 4.2 Toothbrush bristles are cantilever beams.

where q is the load per unit length, calculated from the mass of the beam beyond the support, and where the deflection measurement is made at the free end at the level of the neutral axis. This avoids the need to measure very low forces in a test, but it also indicates that such materials must be properly

---

[4] NB: the quantity EI/L which is sometimes quoted as the 'rate of force delivery' is seen to be quite wrong: it has the dimensions of work or turning moment – force × distance, but it is not clear how it should be interpreted. Likewise, the general confusion of force (F) and moment (F.d) in the orthodontic literature and textbooks is erroneous and unhelpful.

supported by the impression tray if distortion of the model is to avoided: the weight of the hemihydrate slurry will be more or less uniformly distributed on any overhanging material. Even if this overhang corresponds only to soft tissue, the error could be important in the context of a free-end saddle.

The 'stiffness' of a toothbrush (Fig. 4.2) (or indeed any other brush) can in this manner now be understood as arising from the effects of the length and flexural rigidity of the individual bristles as cantilever beams (equation 4.11), treated additively (§3.1). Thus, longer bristles make a softer brush as the cube of the length, but stiffness is only proportional to the number being deflected.

●4.2   Three-point bending

Following similar reasoning, the deflection of the **centre point** of a simply supported beam in a three-point (symmetrical) bend (Fig. 2.5) is given by:

$$y \; = \; \frac{FL^3}{48EI_z} \tag{4.14}$$

This may be used to determine the value of E in very stiff materials. Using a long span means that relatively large deflections can be obtained easily (bearing in mind the constancy of geometry requirement, §2.1). Otherwise, very delicate and sensitive strain-gauge extensometers have to be used, as would be the case in uniaxial tensile tests using specimens such as are shown in Fig. 1§6.1. This is sometimes called a **centre-point loading** test.

●4.3   Four-point bending

The simply supported beam in a four-point (symmetrical) bend test (Fig. 2.7) shows a deflection at the **centre** given by:

$$y \; = \; \frac{Fc(3L^2 \; - \; 4c^2)}{48EI_z} \tag{4.15}$$

where F is the total load, L is the overall length between the outer supports, and c is the distance of each load application point from the nearest support. This equation reduces to equation 4.14 when $c = L/2$. Commonly, the test is conducted with $c = L/3$, in which case it is sometimes called **third-point loading** (when care has to be taken to distinguish the usage from "three-point" bending), and the deflection equation reduces to:

$$y \; = \; \frac{23\,FL^3}{1296\,EI_z} \tag{4.16}$$

It is worth remembering at this point that the principle of superposition still applies in respect of deflections (equation 2.11). That is, complicated force arrangements, as are commonly the case in orthodontic appliances, can be analysed in a piecemeal fashion to understand overall behaviour. This will now be demonstrated through a common example.

§5. The Finger-Spring Analysed

The deflection behaviour of the simple cantilever beam is described by equation 4.10. But this is, in some circumstances, an insufficient device for orthodontic purposes in terms of the space required or the flexibility attainable. Thus, a coil may be included which then introduces several new factors: position, size and number of turns. As has already been stressed, the principle of superposition allows the calculation of the effects of several forces separately, combining the results by simple addition. Another implication is that complex structures may be analysed as a number of separate components and the effects added in the same way. This approach is therefore applicable to a composite structure such as the finger-spring illustrated in Fig. 5.1 (top).[2]

This device may considered as being assembled from three separate components (Fig. 5.1: 1,2,3), each having a 'built-in' end and each amenable to treatment as above. In dealing with this example some further deflection equations are introduced without derivation or explanation as it is the general method which is of interest, although they will have been obtained in similar fashion. They are nevertheless important to the comprehension of the outcome.

Thus, for component 1, which is a simple cantilever, the end deflection has already been given at equation 4.10:

$$y_1 = \frac{Fl_1^3}{3EI_z} \qquad (5.1)$$

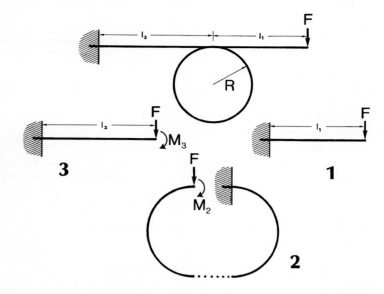

The free end of the loop (component 2) will suffer a vertical deflection under the influence of the shearing force F, which is transmitted unmodified throughout the device (see Fig. 2.1). This is given by:

$$y_2 = \frac{\pi FR^3}{3EI_z} \qquad (5.2)$$

Component 3 is again just a simple cantilever beam. However, this time there is, in addition to the shear force, a bending moment acting at its end. From Fig. 2.1c it can be seen that

Fig. 5.1 The finger spring containing a coil (top) is a form of cantilever. It may be broken down into three parts for the sake of analysis.

at the 'built-in' end of component 1 there is a moment, and this must have the value Fl_1. This acts on component 2, which therefore transmits it to component 3. Because the two ends of the loop are at the same distance from the free end of component 1, $M_2 = M_3 = Fl_1$. The addition of the moment M_3 leads to an additional deflection at the 'free' end of component 3:

$$y_3 = \frac{Fl_3^3}{3EI_z} + \frac{M_3 l_3^2}{2EI_z} = \frac{F}{EI_z}\left[\frac{l_3^3}{3} + \frac{l_1 l_3^2}{2}\right] \qquad (5.3)$$

The extra term is derived in a similar fashion as was the basic deflection equation. Note that the moment M_2 acting on the end of component 2 does not result in a vertical deflection; the equivalent term to that in equation 5.3 has the value zero.

In a weightless cantilever the portion outside of the point of load application must remain straight, *i.e.* of constant slope, because it has neither force nor couple acting on it. However, the deflection of the free end is magnified because of the lever-like effect (Fig. 5.2) (we can temporarily ignore the loop). The *slope* of the beam at the point of load is given by:

$$\theta' \simeq \frac{dy}{dx} = \frac{Fl_3^2}{2EI_z} \qquad (5.4)$$

(an analysis similar to that of §4 would be used to obtain this). In addition, there would be some extra turning because of the applied moment; this extra deflection angle is:

$$\theta'' \simeq \frac{M_3 l_3}{EI_z} = \frac{Fl_1 l_3}{EI_z} \qquad (5.5)$$

So that for both F and M_3 applied simultaneously,

$$\theta = \theta' + \theta'' = \frac{Fl_3^2}{2EI_z} + \frac{Fl_1 l_3}{EI_z} \qquad (5.6)$$

The extra deflection y_1' of the free end (Fig. 5.2), due to the lever effect, is simply

$$y' = l_1\theta \qquad (5.7)$$

i.e. from equation 5.6:

$$y' \;=\; \frac{F}{EI_z}\left[\frac{l_1 l_3^2}{2} + l_1^2 l_3\right] \qquad (5.8)$$

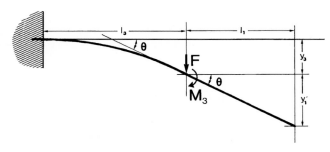

Returning to component 2 of the spring, the moment M_2 acting on its 'free' end is such as to cause a rotation of that point about the centre of the loop. This angular deflection of the 'free' end of the loop is given by:

$$\varphi \;=\; \frac{2\pi RNM_2}{EI_z} \;=\; \frac{2\pi RNF l_1}{EI_z}$$

$$(5.9)$$

where R is the radius of the coil and N is the number of turns in the coil.

Fig. 5.2 The deflection of the end of a cantilever magnifies the deflection of the load point.

The 'free' end of component 1 therefore suffers an extra deflection due to this angular displacement (Fig. 5.3). This is therefore given (approximately) by:

$$y'' \;=\; \varphi l_1 \;=\; \frac{2\pi RNF l_1^2}{EI_z} \qquad (5.10)$$

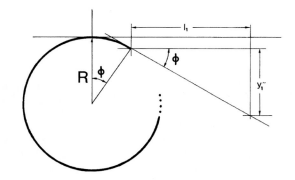

Having identified all of the mechanical and geometrical effects operating, we then obtain the total deflection of the point of application of the load simply by the addition of the results of equations 5.1, 5.8, 5.3, 5.2 and 5.10 (in that order):

Fig. 5.3 The deflection of the end of the coil causes a deflection in the attached arm.

$$
\begin{aligned}
y_{TOT} \;&=\; y_1 + y_1' + y_3 + y_2 + y_1'' \\[4pt]
&=\; \frac{F}{EI_z}\left[\frac{l_1^3}{3} + \left(\frac{l_1 l_3^2}{2} + l_1^2 l_3\right) + \left(\frac{l_3^3}{3} + \frac{l_1 l_3^2}{2}\right) + \pi R^3 + 2\pi RN l_1^2\right] \qquad (5.11) \\[4pt]
&=\; \frac{F}{EI_z}\left[\frac{l_1^3}{3} + l_1^2 l_3 + l_1 l_3^2 + \frac{l_3^3}{3} + \pi R^3 + 2\pi RN l_1^2\right]
\end{aligned}
$$

The first four terms inside the brackets are recognizable as the expansion of $(l_1 + l_3)^3/3$. Taking $L = l_1 + l_3$, which is the *overall* length of the spring (and not the total length of wire), the end deflection can thus be expressed as:

$$y_{TOT} \;=\; \frac{F}{EI_z}\left[\frac{L^3}{3} + \pi R^3 + 2\pi RN l_1^2\right] \qquad (5.12)$$

After all that we can check that the result is consistent by setting $R = 0$ (*i.e.* no coil) when we return to the basic equation 4.10 for the deflection of a cantilever. This also clearly demonstrates the utility of the principle of superposition inasmuch as the device can be considered as a straight wire plus a coil, whose effects are calculated separately.

We defined the flexibility of a spring as the deflection obtained per unit applied force (y/F) (equation 4.11). It is then evident from equation 5.12 that the flexibility of the finger spring may be increased by:

1)   increasing the overall length, L
2)   increasing the radius of the coil, R
3)   increasing the number of turns, N
4)   increasing l_1 which, for a given L, means moving the position of the coil towards the support, the limit being $l_3 = 0$

in addition to the influences implicit in EI_z, *i.e.* the choice of wire. This enables considerable freedom of design, taking into account the space available and the forces required to act on the tooth.

The weight of small orthodontic springs is usually small enough to be neglected in determining the deflection unloaded. Even so, it does not affect the force-deflection characteristics of the appliance which are, as a matter both of principle and theory, superposed on the starting condition, whatever it might be.

The above type of process of analysis is applicable to any type of orthodontic appliance, partial denture, and so on. The difficulties arise when considering the three-dimensional nature of the structures, loading and deflections where the geometry is more easily found to be affected. Also, the uneven or graduated cross-sections that are used (as in clasps, for example) and lack of symmetry in the cross-section (elliptical or D-shapes) all require more detailed investigation.

§6. Coils & Braiding

Sometimes long coils are used, in the form of a helical spring. Apart from the ability to generate forces axially (that is, in the sense of Hooke's Law applied to a spiral spring, 1§2), the lateral deflection may also be of interest. The appropriate equation is:

$$y = \frac{2\pi RNFl^2}{3EI_z}\left[1 + \frac{v}{2} + \frac{3R^2}{2l^2}\right] \qquad (6.1)$$

where N is the number of turns of radius R, and v is the Poisson ratio (1§2.2), l being the overall length of the spring.[3] This is, not surprisingly, very similar to the deflection of a simple cantilever beam (equation 4.10), with one important modification: instead of l^3 we have $2\pi RNl^2$. Notice that the total length of the wire in the spring is given by $2\pi RN$.

This equation may therefore be read as showing that the deflection (or the flexibility) increases simply in proportion to that total length, keeping the other dimensions constant. In other words, we may achieve greater flexibility in a given space by using a spiral. This is one reason why the strands in rope, electrical flex, steel cables, sewing yarn, as well as multistrand orthodontic wires and so on are twisted, apart from helping to hold the bundle together (see §3): each strand is a separate helix. The same kind of effect is operating in a braided material, whether steel or polymer, even though the pattern is more sinusoidal than spiral, so long as the fibres are free to slide past each other. Retraction cord, for example, needs to be very flexible to pack readily into sulci, and braiding assists this. Other examples include braided nylon or polypropylene sutures and floss tape.

§7. Hand Instruments

The manual skills of the dentist are central to the execution of many procedures, but of course those tasks are undertaken through the agency of some instrument, such as a probe, chisel or air-turbine handpiece. The design of such tools may, at first sight, appear to be strange, involving odd angles and curves (Fig. 7.1). However, there is a single underlying definite purpose: to bring the working point to lie on the projected axis of the handle or grip of the instrument. There are two principal reasons for this.

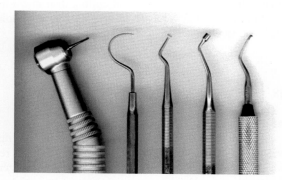

Fig. 7.1 Dental hand instruments are designed for accurate remote proprioception.

Firstly, we have learned to use tools as an extension of our hands and fingers in such a way that effectively extends our sense of proprioception to operate remotely. This is obvious through the use of such straight tools as pencils and paintbrushes where the grip employed is adjusted to suit. When the working direction of the tool is not along the main axis itself, but at some angle to it, it seems natural to design the working part to reflect that change of orientation. However, simply bending the tip to create that new orientation is inadequate, for now we tend to lose that proprioceptive sense – the working part is no longer where we feel it ought to be. The answer to this is to use a double bend or crank such that the second turn allows the working point to be returned to lie on the handle axis. This can be seen to be approximately the case in the examples of Fig. 7.1. In handpieces and hand instruments this is known as a 'contra-angle', but it really only serves that alignment function, no matter the actual orientation of the cutting tool itself. The same principle applies to tweezers, forceps, pliers and many other tools.

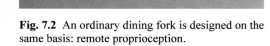

While it is possible to learn to use 'non-aligned' tools, they require more effort in practice because the remote proprioception is spoilt, and the device seems more awkward. A good example of this is an ordinary dining fork (Fig. 7.2), which shows exactly the same principle in operation. One gives no second thought to transferring food to one's mouth – which action is essentially unguided by sight, and manages to do this without injury, at speed, and frequently – even in the dark. However, one is easily disconcerted if the fork is bent (or made) 'out of true',

Fig. 7.2 An ordinary dining fork is designed on the same basis: remote proprioception.

where the line of the tine tips lies off the main axis: eating suddenly becomes very awkward (and very difficult with one's eyes closed). This may be easily rectified by rebending the neck so that it is realigned.

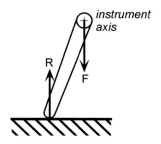

The second reason for the design is the avoidance of rotation when appreciable forces are involved, such as for packing amalgam or using an enamel chisel or excavator. Slight misalignment of the action force vector with the plane containing a curved or bent working part results in a moment couple as the reaction vector is no longer in the same plane (Fig. 7.3). The system is then quite unstable, as the grip required by the fingers to prevent rotation becomes very large (a small radius means a large turning force is required, the torque – equation 1.5). When the instrument axis passes through the point of contact such a moment vanishes (Fig. 7.4), as can be seen in Fig. 1.1 when the axis of rotation lies on the line of action.

Fig. 7.3 Misalignment of action force results in a turning moment for an off-axis point.

Thus, from this point of view, the working quality of hand instruments can be assessed from the accuracy with which the nominal alignment is achieved. Measurements on Fig. 7.1 show that there may be small discrepancies, and the value of the design may not be fully appreciated by the manufacturer even though it is supposed to be standardized. There are, of course, variations between items of the same product because of manufacturing variation – many of such instruments are hand-made. Selection, therefore, sensibly includes such a check. Of course, with devices such as handpieces, the choice of cutting instrument length, and its accurate location in the chuck, has an effect. The wrong choice or incorrect fitting leads to instability.

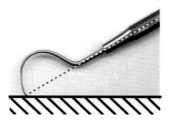

Fig. 7.4 Alignment of axis with working tip results in no turning moment when force is applied in any direction.

§8. Oscillation

●8.1 Linear

If an elastic body such as a cantilever beam is displaced, the work done is stored (§1) – it is now potential energy. If the displacing force is removed, the body recoils, converting the potential energy to kinetic energy, which reaches a maximum when the displacement is zero. However, the body will continue moving, converting the kinetic energy back to potential energy when the displacement reaches its maximum in the opposite direction. In that sense, the system resembles the pendulum of an impact tester (1§10.1). Assuming no losses (say, through internal friction or air resistance), this oscillation would continue indefinitely. This is then an example of simple harmonic motion, if the deflection is small enough that Hooke's Law (1§2) is a good approximation. Through processes similar to those used in §4, equating the potential and kinetic energies, for a massless beam with a simple (point) mass at the end (Fig. 8.1a), the natural oscillation (fundamental) frequency, f_1 (in hertz) is given by:

$$f_1 = \frac{1}{2\pi}\sqrt{\frac{k}{m}} \qquad (8.1)$$

where k is the Hookean spring constant (in the direction of displacement) and m is the mass. In the case of a rectangular section cantilever beam, k is given by equation 4.12, so that

$$f_1 = \frac{1}{2\pi}\sqrt{\frac{1}{m}\left(\frac{3EI_z}{L^3}\right)} \qquad (8.2)$$

For a simple real beam, with mass per unit length μ (Fig. 8.1b), the equation becomes more complex, although on the exact same principle as equation 8.1, leading to:

$$f_1 = \frac{1}{2\pi}\sqrt{\frac{c_1}{\mu L}\left(\frac{3EI_z}{L^3}\right)} \qquad (8.3)$$

where $c_1 \approx 4.12$.[5][4] Thus we can see that the equation is parallel to that for equation 8.2, but with an effective equivalent end-mass m_{eff} defined by:

$$m_{eff} = \frac{\mu L}{c_1} \qquad (8.4)$$

That is, m_{eff} is roughly one quarter of the total mass of the beam.

Since the effects of such masses are subject to superposition, for the combination of a real beam (*i.e.* one having mass) with an added point mass at the free end (Fig. 8.1c), we need only write:

$$f_1 = \frac{1}{2\pi}\sqrt{\frac{1}{m+m_{eff}}\left(\frac{3EI_z}{L^3}\right)} \qquad (8.5)$$

Although these equations are for the fundamental frequency, there are of course many higher vibrational modes possible (Table 8.1). These are of lesser importance in general. Other beam sections can be dealt with simply by substituting the appropriate expression for k in equation 8.1, taking care to use the correct expression for I_z according to the direction taken in the case of non-symmetric beams.

This kind of oscillating beam system occurs in the case of an intraoral dental X-ray set consisting of the generator head on a folding arm (26§8.3) (Fig. 8.2). Of course, the geometry is more complicated than a straight beam, but the potential for oscillation (with multiple modes) after repositioning remains as an important source of image blurring.

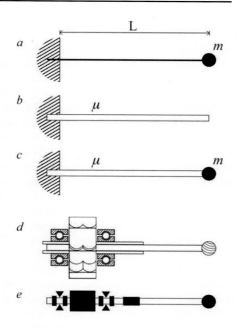

Fig. 8.1 Beam systems to be considered for vibration. *a*: simple, end mass. *b*: real, with mass. *c*: real with end mass. *d*: schematic of rotary instrument in handpiece turbine. *e*: equivalent doubly-supported shaft with several masses (triangles indicate supports).

Fig. 8.2 An intra-oral dental X-ray set consists of a tube head on the end of an articulated arm – a long cantilever beam.

Table 8.1 *Effective mass constant c for the first six vibrational modes of a real cantilever beam (equations 8.3, 8.4). The relative frequency with respect to the fundamental is also shown.*

Mode	c_i	f_i/f_1
1	4.1	1
2	161.8	6.3
3	1268.8	17.5
4	4872.4	34.4
5	13314.6	56.8
6	29711.8	84.9

5　The solution requires the roots, b_i, of the equation $\cos(x).\cosh(x) - 1 = 0$. Thus $b_1 \approx 1.875104$. Then $c_1 = b_1^4/3$.

●**8.2   Rotary**

There is another dental context where these calculations are relevant: rotating systems, as in a rotary instrument in an air-turbine handpiece. That is, taking the instrument – shaft plus working part – as a cantilever beam with an end mass (as in Fig. 8.1c), when the rotation rate corresponds to a natural transverse vibration frequency then the shaft bends off the centre line to create the condition knowing as **whirling**. This is the result of resonance between the two frequencies. While a perfectly balanced shaft will not whirl at any speed, in practice the centre of mass axis (the locus of the mass centroids of all sections) does not exactly correspond to the axis of rotation. This may be due to a lack of straightness or the instrument being 'out of round', whether due to manufacturing imperfection, wear or damage. As the shaft is rotated there are necessarily centrifugal forces (18§3.1) which result in bending moments which tend to deflect the shaft off axis, and this results in increased centrifugal forces, proportional to the radius for the centre of mass (that is, its deflection) and the square of the rotation rate, in a positive feedback process.

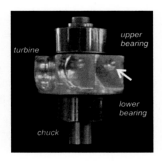

Fig. 8.3 A dental air turbine assembly, showing the drilled balancing dimples in the turbine itself (one is arrowed).

When the rotation rate becomes equal to the natural fundamental frequency of the system of shaft plus mass, the out-of-balance condition causes the shaft to vibrate, bending out of true, and so it begins to whirl. In this resonant condition, the extreme deflections of the shaft occur at the same points in each turn. The system is now in an unstable state, generating high stresses and vibration, acting against the reaction mass of the rest of the handpiece. Damage can then result if the shaft is maintained at this speed. If the speed is now further increased the shaft deflection may increase but the instability will cease as the system settles into a mass-axis rotation rather than the nominal geometric axis. Shaft whirling only occurs when there is a resonance between the rotation rate and a natural frequency of the shaft system. This is called **forced vibration**. Yet further increase of speed can then reach another resonant frequency, and so on.

A dental air-turbine handpiece is more complicated than a cantilever beam (Fig. 8.1d,e). Consisting of a turbine on a shaft supported by two bearings, the mass of that turbine plus those of the inner bearing races and the chuck assembly, as well as of the rotary instrument, are all important. Some care in manufacture is taken to balance the turbine (Fig. 8.3) but, as suggested above, this cannot be perfect. The effect of such imbalance can be seen in the free-running speed of a turbine (Fig. 8.4, 8.5) (that is, with no load on the instrument). In the second case, three resonances are found, the first and third of which are in the right proportions for the first and second cantilever modes (Table 8.1) (higher modes would seem to be inaccessible). The second resonance suggests that there is another beam system mode, probably due to the support of the bearings (Fig. 8.1e). Vibration from such resonances will cause extra wear of the bearings, one of the reasons for the need to replace these assemblies from time to time.[5] The effect is also seen when the rotary instrument is under load (*i.e.* cutting), showing that it is rotation rate that matters in exciting the resonance.

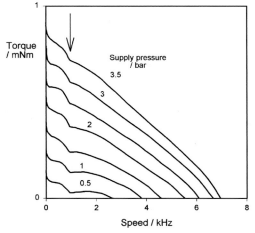

Fig. 8.4 An air-turbine handpiece showing a resonance effect (arrowed) at the same frequency irrespective of air-supply pressure and load.

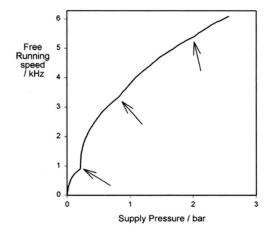

Fig. 8.5 An air-turbine handpiece showing resonance effects (arrowed) at various rotational rates using an unbalanced instrument.

Obviously, the use of damaged instruments (such as a chipped blade or tooth), or even one that is bent, would exacerbate the problems – with the risk of projectile injury.[6] Worse, however, might be the hammering effect on the substrate that whirl would cause: enamel, for example, might be cracked. It would certainly be unpleasant vibration for the patient.

§9. Buckling

There are some circumstances in dentistry when a beam (or some equivalent object) may be loaded end-on. This offers special risks of collapse and thus hazard, in particular in the case of hypodermic needles.

The compressive axial loading of short columns was mentioned earlier (1§6.3) when the point of interest was collapse through rupture in an attempt to understand the strength of a material. There are circumstances where such rupture does not occur, that is, when the column is long enough to **buckle**: a sideways deflection representing the path of least work. Here, the stored elastic energy in that axial compression exceeds the work required for that bending, so there is an abrupt switch from a straight column to the buckled one: that is, it had become unstable with respect to the second state. If the system were perfect, then axial compression could continue indefinitely – but the system would be in **unstable equilibrium** because the slightest disturbance would trigger the switch. Dimensional irregularity, non-coincidence of load and column axes, vibration, or even any inhomogeneity in the column material, would be enough to do this – a catastrophe of the kind discussed under Griffith's criterion (1§7): a switch from an unstable to a stable state when the

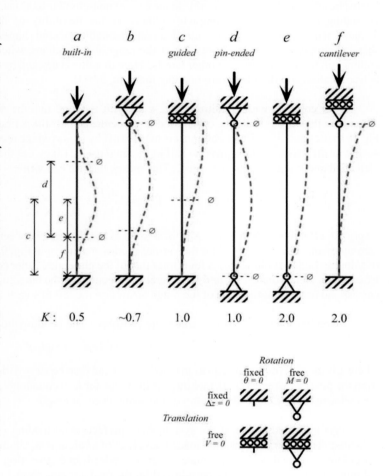

Fig. 9.1 Buckling modes for long columns according to the end constraint conditions. The buckled shape is indicated by the broken line. K is the effective length factor for the mode, and ⊘ indicates the points of zero bending moment, $M = 0$, as for the frictionless pin. z is the lateral displacement direction, that of the axis for the least value of I_z. V is the end shear, θ the end rotation.

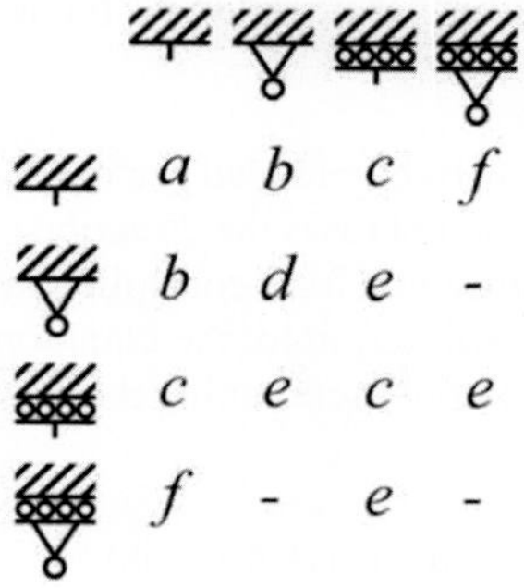

Fig. 9.2 Possible end-constraint combinations for columns (key letters refer to Fig. 9.1; "-" means that the column would slip and rotate, and stay straight).

stored elastic energy can pay for another process and lower the total energy of the system.

For a column subjected to axial loading, the kind of behaviour that occurs depends on the end constraints (or lack of them). The possibilities are of four types: fixed or built-in (§2.1); pinned or simply supported, where only rotation is permitted (*cf.* Fig. 2.5); guided, where only lateral movement – translation – is allowed, that is, allowed to slide; and free to both slide and rotate.

This leads to just six sensibly distinguishable cases (Fig. 9.1). Other combinations of end-constraint are possible, but these are either reflections of, or equivalent to, others, or trivial (Fig. 9.2).

The starting point for such studies is the so-called Euler[6] column or 'pin-ended' type *d* in Fig. 9.1. Again following a procedure similar to that of §4, the critical load for collapse, *i.e.* buckling, is determined to be given by:

$$F_{cr} = \frac{\pi^2 EI_z}{L^2} \tag{9.1}$$

where L is the length of the column.[7] Notice that again the flexural rigidity, EI_z, controls the behaviour. Here, I_z must be taken as the minimum value in the case of non-circular sections. For a load-controlled system (1§3.1), exceeding F_{cr} must lead to complete collapse as the flexibility of the system increases with the lateral displacement, while for a strain-controlled system a stable buckled shape is attained. For the Euler column, that shape is simply a sine curve, $\sin(x)$, over the range $0 \sim \pi$ radians, where *x* is the relative position along the column, taking L as the unit, in other words. Note that this is defined by the pivot points, the pins, where there is free rotation and therefore zero bending moment, $M = 0$.

For columns with conditions *a, c, e* and *f,* it is found that the deflection shapes are still sine curves, although with various scalings, offsets and displacements, but still defined by the points where $M = 0$. Thus in Fig. 9.1*a* the equivalent Euler column *d* is marked off by $\oslash - \oslash$ (which are the inflections in the buckling curve), so that its effective length is just one half of that in *d* itself. Likewise, the buckling shapes for *c, e* and *f* can be identified and marked off as indicated. This means that we can write:

$$F_{cr} = \frac{\pi^2 EI_z}{L_{eff}^2} = \frac{\pi^2 EI_z}{K^2 L^2} \tag{9.2}$$

where the **effective length** $L_{eff} = KL$, is defined as the distance between those inflection points, the points of zero moment. The values of *K* for the various modes are shown in Fig. 9.1. For example, the load for collapse of a doubly built-in column (*a*) is four times that of the pin-ended Euler column of the same physical length (*d*). Effectively, conditions *c, d, e,* and *f* are merely subsections of the buckling for condition *a*: their end-constraints correspond to those at the ends of the range arrows on the left of Fig. 9.1.

Condition *b* (Fig. 9.1) is a little more complex. The bending mode equation takes the form[8]:

$$y = \sin(\lambda x) - x.\sin(\lambda) \tag{9.3}$$

where λ is the first root of the equation $\tan(\lambda) = \lambda$. It can then be shown that the inflection is at a point ~ 0.6992 L from the pin (*i.e.,* where $M = 0$), making this the value for *K* in equation 9.2. Such a column has then about twice the collapse load of the Euler column of the same physical length.

As indicated above, if the system were perfect, axial loading could continue past the critical values indicated. However, dimensional inaccuracy, lack of straightness, alignment problems, or inhomogeneities in the column material, must precipitate collapse. Indeed, in engineering terms, for the design of load-bearing columns, a safety margin is always allowed for the load, ranging from 5 to 30% according to the mode. Worse than that, the end-constraints can rarely be ideal as depicted, and this produces further uncertainty, but always towards greater *K* with greater freedom.

Notice that the direction of buckling displacement is indeterminate for a perfectly loaded perfect column when this has any symmetry, subject only to the direction of minimum second moment (as is the direction of any sliding). The amplitude of the displacement is also left undefined, firstly because in a load-controlled system progress to collapse is prompt, but secondly because the Hookean constraint is still required: the equations are only valid for small deflections (see box, Constancy of geometry, §2.1). The point of ascertaining the buckling shape has been to identify the critical load F_{cr} from the value of L_{eff}.

To return to the case of the hypodermic needle, it can be seen that in ideal use it would appear to

[6] Pronounced "oiler".
[7] This and the other conditions also have other buckling modes that correspond to 'harmonics' of the wave shape of the deformation. Since these occur at higher loads, 'least action' says that only the primary mode is important, so only these solutions are discussed.
[8] This has several equivalent but less than obvious trigonometric identities, any of which might be found in the literature.

approximate mode *b* (Fig. 9.1): the built-in end is nominally at the connector, while the embedded point roughly corresponds to a pinned end in that it has little restraint to rotation (at least initially). Being hand-held, maintaining axial loading is clearly critical to avoid buckling, but the ability to hold the syringe steady in the lateral sense is difficult, meaning that the condition in actual practice is closer to condition *e* – pinned-guided (Fig. 9.2), when the value of *K* is 2, and the critical load is one-eighth of that otherwise to be expected: $0.7^2/2^2$. There are, then, two implications: firstly that as much support as possible should be given to the hand holding the syringe – mid-air is very dangerous, and secondly that the force applied for tissue penetration should be the minimum necessary, given that effectively it is into a viscous medium, and travel takes time (*cf.* Fig. 4§10.2) – there will be viscous drag from the immersed length that increases with penetration. The sharpness of the cutting facets at the tip, of course, plays a major part in limiting the penetration resistance.

What is not clear from Fig. 9.1 and the equations for F_{cr} is how the relative length of the column in relation to its lateral size affects the outcome. For this we need a dimensionless quantity, one that is therefore applicable in all circumstances, called the **slenderness ratio**. But first we have to define what we mean by 'lateral size', and this done with a quantity called the **radius of gyration**, ρ. This is derived by considering that all of the mass of a cross-section is concentrated at a radius ρ from the neutral axis of the column such that it has the exact same second moment as the actual cross-section of the column for the chosen direction:

$$A\rho^2 = I_z; \quad \rho = \sqrt{\frac{I_z}{A}} \tag{9.4}$$

The slenderness ratio is then simply given by:

$$s = \frac{L}{\rho} \tag{9.5}$$

Substituting from equation 9.4 in 9.2, and then from 9.5, yields

$$F_{cr} = \frac{\pi^2 E I_z}{K^2 L^2} = \frac{\pi^2 E}{K^2}\left(\frac{\rho^2 A}{L^2}\right) = \frac{\pi^2 EA}{K^2 s^2} = \sigma_{cr} A \tag{9.6}$$

where now σ_{cr} is the effective critical "strength"[9] of th column:

$$\sigma_{cr} = \frac{\pi^2 E}{K^2 s^2} \tag{9.7}$$

Here it is very obvious that the only material property that governs the resistance of the column to buckling is its modulus of elasticity, but crucially it also very dependent on the slenderness ratio, given the end constraints. However, buckling is not the only possible mode of failure: if the slenderness ratio is small enough, the compression stress in the material of the column will reach the yield stress in that mode, imposing an upper limit that is now material strength-dependent.

We might also note that for sections whose second moment varies with orientation, such as for an ellipse (§3.2), buckling will occur in a direction corresponding to the least value of I_z, and it is this value that must be used for the calculation. This is why, for example, one cannot bend a square section by hand along a diagonal, but also why I-beams (Fig. 3.2) must be designed to avoid lateral buckling by having sufficient material in the flanges.

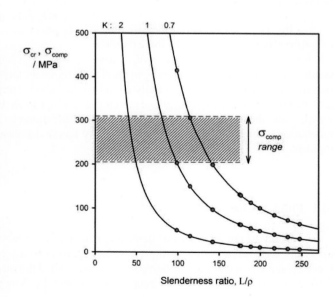

Fig. 9.3 Critical buckling stress for type 304 stainless steel hypodermic needle tubing, E = 200 GPa. For dimensions, see Table 9.1. The upper bound depends on the exact steel used; the possible range for the compressive collapse stress is shown.

---

[9] Better, perhaps, to say "bearing capacity" – *cf.* 1§6.3.

Table 9.1	Hypodermic needle examples for the calculations of Figs 9.3 and 9.4.						
	OD: outside diameter, ID: inside diameter						
Gauge	OD / mm	ID / mm	wall / mm	area / mm²	I_z / mm⁻⁴	ρ / mm	lengths / mm
25	0.514	0.260	0.127	0.155	5.19×10^{-3}	0.183	21, 26, 36.5
27	0.413	0.210	0.102	0.099	2.16×10^{-3}	0.148	21, 26, 32, 36.5
30	0.311	0.159	0.076	0.056	0.70×10^{-3}	0.112	11, 21, 26

We can now apply these ideas to the specific instance of hypodermic needles (with I_z given by equation 3.8). Taking a range of typical commercial dental needles as examples (Table 9.1), the critical stress is plotted in Fig. 9.3 for the ideal theoretical condition *b* (*K* = 0.7), the Euler column (*d*, *K* = 1), and the conservative condition *e* (*K* = 2) for a variety of typical lengths ranging from 11 to 36.5 mm. We may notice how the use of slenderness ratio reduces all results to a single curve in each constraint condition. The yield strength of the typical steel used (stainless type 304) has a range of possible values, but even taking the lowest, the buckling limit is likely to be reached before yield in compression occurs. Imperfections in material (wall thickness, and lumen circularity, smoothness and straightness are especially difficult to control in tube-making) or alignment will further depress the limiting line in both respects.

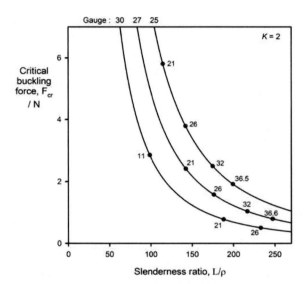

Fig. 9.4 Critical buckling force for condition *e* (Fig. 9.1) for type 304 stainless steel hypodermic needle tubing for some typical combinations of gauge and length. Each point is identified by the length of the tube in millimetres.

It is, however, informative in operational terms to consider the force required for buckling, taking the most conservative condition *e* (Fig. 9.4). Again bearing in mind that alignment faults (*i.e.* off-axis loading) will decrease the values required, and also that it is essentially a load-controlled system, it can be seen that in practice very low forces are essential. Since the needle gauge selected is a compromise between the achievable flow rate (4§11) – large lumen – and the minimization of pain – small outside diameter, the length used should be the least necessary. Of course, as the needle penetrates, the effective length decreases, making the initial puncture the most critical moment.

§10. Fluid Mechanics

Many of the processes of dentistry, whether in the laboratory or the mouth, involve the flow of fluid materials, from metal casting to impression taking, from pouring models to endodontic irrigation. In many cases these dynamic processes are straightforward, and beyond the rheological characteristics (Chap. 4) offer no great difficulties. However, there are circumstances where the flow process itself is critical.

●10.1 Bernoulli equation

For steady laminar flow (see 15§7.1) in an incompressible fluid (that is, one of constant density, ρ), in the absence of frictional losses, for points along a **streamline**, on the principle of the conservation of energy, the **Bernoulli equation** has the following form:

$$p + \frac{\rho v^2}{2} + \rho gh = \text{constant} \tag{10.1}$$

where p is the pressure, v the velocity, and ρgh is the hydrostatic pressure (equation 10§2.5).[10] The absence of frictional losses, *i.e.* due to the fluid's viscosity, means that this refers to what is known as **inviscid flow**. A streamline is simply the path that a small particle would take in that flow. The 'no frictional losses' condition is well enough approximated under many circumstances for the equation to be a useful guide to behaviour, and especially for non-turbulent flow. In dental contexts, the hydrostatic head is ordinarily negligible, and so the term can be dropped now (the pressure of the environment is taken to be constant and can be ignored, so these are effectively **gauge pressures**). Hence, for an example such as in Fig. 10.1, we can write:

$$\Delta p \;=\; p_1 - p_2 \;=\; \frac{1}{2}\rho\left(v_2^2 - v_1^2\right) \qquad (10.2)$$

Conservation of mass means that the amount of fluid passing the two points must be the same, so we have the **continuity equation**:

$$\rho A_1 v_1 \;=\; \rho A_2 v_2 \qquad (10.3)$$

Thus, it follows that if

$$A_2 < A_1 \;\Rightarrow\; v_2 > v_1 \;\Rightarrow\; p_2 < p_1 \qquad (10.4)$$

Obviously, the behaviour is symmetrical: reversing the flow has no effect on these relationships.

The Bernoulli equation has some interesting consequences. For example, if a stream of liquid impinges squarely onto a plate through which it cannot pass (Fig. 10.2), half must go one way, half the other (in a two-dimensional view – radial symmetry is expected). The dividing line is called the **stagnation streamline** because it terminates at the plate, where of course there is no flow: $v_0 = 0$; this is the **stagnation point**. Thus, taking as the reference the conditions far upstream, we find:

$$p_0 \;=\; p_\infty + \frac{\rho v_\infty^2}{2} \qquad (10.5)$$

– the highest value anywhere in the system. In words, this is the sum of the static pressure, p_∞, and the 'dynamic pressure' due to the velocity (imagine the force exerted by the jet from a hosepipe).

●10.2 Boundary layer separation

Were real fluids to be inviscid, flow through ducts of rapidly or abruptly changing section, or past obstacles of any kind or shape, would be continuous, following the implications of the Bernoulli equation everywhere. Thus, if the plate of Fig. 10.2 were not indefinitely extended, and immersed in a stream much wider than that, flow would occur around the ends and rejoin on the far side in a perfectly symmetrical fashion, including a trailing stagnation point. However, real fluids are viscous, and because in general there is no slip at the boundary the velocity there is zero (4§3). We therefore have to modify the Bernoulli equation (at least conceptually) to take this into account. Hence, for liquid flow in a tube (4§11), the pressure varies across the section as the square of the velocity, being lowest on the centre line for established flow. The flow vectors in Fig. 10.1 therefore should be modified accordingly. Nevertheless, the pressure effect is quite general, even when there is a boundary layer – indeed, it operates across the boundary layer.

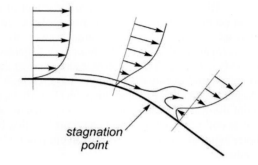

Fig. 10.1 Flow of a viscid stream of fluid past a large increase in cross-section.

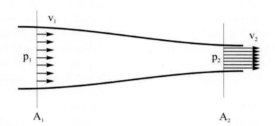

Fig. 10.2 Incompressible laminar flow through a duct of changing cross-section. Absence of boundary friction is assumed here, for convenience.

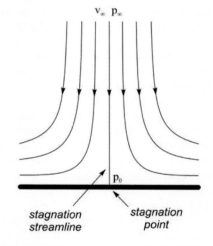

Fig. 10.3 Flow of a stream of liquid squarely onto a plate.

---
[10]   As an exercise, show that the three terms on the left have the same dimensions.

Given that, we can explore the effect of change in cross-section in more detail, in particular for large and rapid increase. The increase in area causes the pressure downstream to rise along the boundary because the flow velocity is falling. The rise in pressure opposes the flow, so to speak, and eventually becomes large enough to stop it altogether, creating a stagnation point (Fig. 10.3). After that point the flow is in reverse, and this creates an eddy (NB: this is not turbulence). This behaviour is called **boundary layer separation**, and arises from an unfavourable pressure gradient.

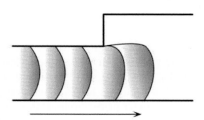

Fig. 10.4 A viscous fluid is extruded from a narrow duct rather than fill the expanded region.

Now, this situation arises in a fully-immersed system. If instead the fluid was being injected into such duct (and for the situation where the effect of gravity is negligible, in any orientation), the fluid front would take up roughly the shape of the velocity profile, and then, at about the location of the stagnation point, it would part company from the duct wall: there would be no driving force to fill that space (Fig. 10.4). Water flowing though the stem of an inverted funnel does not fill the funnel, much less honey or a material such as toothpaste. Consequently, in a situation such as taking an impression of a crown preparation, where a distinct change in the effective duct cross-section occurs at the outer margin of a shoulder – even if bevelled or 'feathered' – this design causes outward flow and separation (Fig. 10.5). This would lead to the model being oversized below the shoulder, and thus a crown being made oversized because the outer surface contour is made smoothly continuous with that of the model.

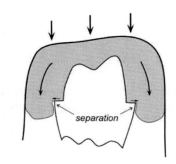

Fig. 10.5 Flow of impression material over a prepared shoulder may lead to separation.

A similar effect would occur further down the tooth if it were particularly tapered or its angulation in the jaw was marked, that is, similar to the situation in Fig. 10.3, assuming that contact was re-established after the shoulder. In any such case, the higher the velocity of the impression material, the greater the effect. For example, a narrow space between tooth and tray and higher seating force would both increase extrusion velocity, as well as the use of wax to dam the flow of the material, or even blocking the vents in a perforated tray.

The separation phenomenon is made worse by the low viscosity of air. That is, air will tend to flow more readily into the low pressure zone created by the high velocity, effectively opposing the desired flow of the impression material.

It can be seen, therefore, that the emphasis sometimes placed on generating 'hydraulic pressure', supposedly to improve the accuracy of an impression, is actually counterproductive because this must increase fluid velocity and therefore increase the magnitude of separations.

Fig. 10.6 Low-viscosity materials and air (on the right) are preferentially extruded at high velocity (model system in a wide-vent syringe). Photo: Tom Smethurst

•10.3 Viscosity effects

For a given seating force, the lower the viscosity of the impression material the faster it will flow since

$$\dot{\gamma} = \frac{\tau}{\eta} \tag{10.6}$$

(4§3.2). Plainly, separation effects will then be exacerbated. This would also apply in dual-viscosity methods, so-called 'putty-wash' and similar techniques in which both materials are unset, where the lower viscosity material is extruded first at high velocity (Fig. 10.6).

A similar effect would be seen with entrapped air, which in principle would seem to be a good thing: rapidly displaced to yield a good impression of the surface. This might, however, not work as intended. If a perforated tray is used, it is possible for air trapped between the impression material and the tooth on first inserting it to be displaced towards the nearest perforation, forming a pipe (Fig. 10.7). This may persist if the volume displaced is insufficient: because the flow of the impression material out through the perforation is easiest from near the hole, this will be dominant (principle of least work). Thus, the deeper part of the pipe will be more stable and persistent. This can be seen by reversing the flow on the stream lines in Fig. 10.2 – the stagnation point is at the bottom of the pipe in Fig. 10.7. As with the blebs formed by small trapped bubbles, the pipe requires the model to be cleaned up before fabricating the wax pattern. Obviously, blocking the holes, or using an unperforated tray, would not help.

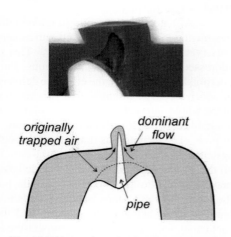

Fig. 10.7 Formation of a pipe from trapped air with a perforated tray. Photo: Tom Smethurst

If the impression material were pseudoplastic (4§7.4), or viscoplastic (which is more commonly the case) (§7.3), that is, where shear thinning occurs (§7.9), or even (really) thixotropic (§7.12), then the falling viscosity causes the shear strain rate – *i.e.* the velocity – to rise, that is, with positive feedback, and this would exacerbate the risk and magnitude of separation. Even if there were to be appreciable slip at the interface, that is approaching the assumption of inviscid flow, such as by the presence of a film of water, then while the Bernoulli equation would more closely be followed it would not affect the outcome: separation would still occur. Indeed, any water present would make it worse by flowing back, as would occur for air.

The principal determinant of behaviour is in fact the Reynolds number of the flow (15§7.1). If $Re \leq \sim 1$, the behaviour

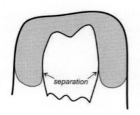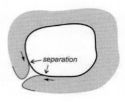

Fig. 10.8 Separation can occur at any flow-divergent surface. Photo: Tom Smethurst

is called **creeping flow**. In this condition, surfaces will be followed most exactly, and while rapid expansions of ducts might not be filled (as in Fig. 10.4), separation as in Fig. 10.5 will not be expected to occur. However, for $Re > \sim 10$, the streamlines will separate from even smoothly curved surfaces (Fig. 10.3). This suggests two things, firstly that the slower the seating of the impression tray the better, to minimize the velocity of the material. This also implies that the closer the fit of the tray, and the lower the viscosity of the material, the slower should be the seating. The second point is that separation may occur even over the normal convexity of a tooth if the fluid velocity is too high. This would also apply to flows that converged, such as around the circumference of a tooth (Fig. 10.8). Even so, the principle is plain enough. It follows then that tilted teeth present a much greater hazard in this respect, as mentioned above. Unfortunately, precise recommendations cannot be made here because the fluid velocity depends on too many factors: gap size in three dimensions and its rates of change, material viscosity, and pressure applied. However, coating all surfaces of interest with the (low viscosity) material before the tray is inserted would appear to be the minimum required preventative or ameliorative measure – that is to eliminate all air. In similar fashion, all water must be eliminated from all critical surfaces because this encourages separation, including before that low-viscosity material is applied

It should be noted that defects arising from flow separation may be very difficult to see in the impression itself, and especially if one is not actively looking for them. They are much more apparent on the model, and while this might not be inspected by the dentist, the technician will see these defects (Fig. 10.9). Overhanging crowns, however, will be noticeable to the dentist, if not the patient as a source of irritation, food entrapment, and a catch for floss, to say nothing of the risk of caries at such a site.

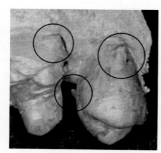

Fig. 10.9 Separations are quite evident in the cast model. Photo: Tom Smethurst

●10.4 Endodontic irrigation

A major consideration in endodontic treatment is the attainment (and subsequent preservation) of root canal sterility. This is addressed though the use of various anti-microbial agents (30§1), the effective delivery of which would be essential, that is, exposure of the whole internal surface to the solution at a suitable concentration for an adequate period of time. The difficulty is that a flow-through system is not possible, so that contra-flow is all that is usable. That is to say that delivery may be made through an inserted cannula, but the return flow must then be in the contained annular space between the cannula and the root canal wall. It is, however, imperative to avoid extrusion of the irrigant through the apical foramen because of the damage done to the peri-radicular tissues caused by endodontic irrigants such as hypochlorite.

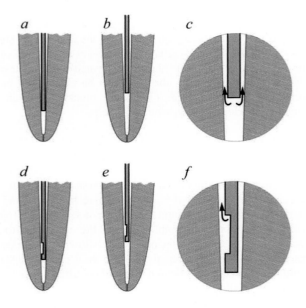

Fig. 10.10 Endodontic irrigation cannulae – and the unflushable apical static volume – in an idealized simple canal.

There are a number of limitations. If the cannula is of larger diameter than the prepared canal towards the apex, there is clearly a limit to its insertion (Fig. 10.10a), at which point the close fit to the walls – amounting to a seal if the canal is of circular section – means that the pressure rises in the enclosed apical volume, driving the liquid through the apex. If the cannula is backed off (*i.e.* withdrawn slightly) such that there is sufficient annular space for a reasonable flow rate (Fig. 10.10b), or the canal is sufficiently non-circular, that flow will necessarily only occur backwards, with a depth of penetration into the apical volume only about equal to the diameter of the cannula's lumen (Fig. 10.10c). The remainder of the liquid in that apical volume will be static (turbulent mixing will not occur). There can be no pressure gradient that drives fluid through the depth of the static volume from a cannula of similar dimension to that of the canal – no jet can be created. However, if the rate of delivery is increased by raising the pressure in the syringe, the pressure must rise in the apical volume, raising the risk of apical extrusion. Side-opening cannulae are intended to avoid raising the apical pressure and eliminating the risk of apical extrusion. However, if such a device is used (Fig. 10.10d), quite obviously there will be no flow into the apical region if it is inserted as far as it will go, but equally clearly there can be no flow in the region of most interest even if it is withdrawn (Fig. 10.10e, f). In both cases, the usual curvatures of the canal compound the problem by restricting insertion and raising the risk of the cannula buckling (§9), and of its fracture if pushed too far, especially if smaller gauge cannulae are used.

If somehow it is possible to remove fluid in the apical volume, say by absorbent paper point (27§2.2), the next irrigant will probably trap a bubble. This could not be displaced by the new fluid using a cannula, in just the same sense as if liquid were there.

The intention is generally understood to require minimal root-canal preparation because of the numerous risks that are entailed otherwise: awkward section shapes, side perforations, weakening and subsequent root fracture. It follows then that obtaining a large enough canal to allow full insertion of a cannula, and thus minimal static volume, is not appropriate as a routine. It might be thought that reversing the flow, using the cannula to suck out liquid provided in excess coronally, would be beneficial. To the extent that a raised apical pressure is completely avoided this is true, but the flow is necessarily fully-reversed and the streamlines would again not extend into the static volume unless near full insertion were made.

Consequentially, exposure of the walls of the apical static volume to the irrigant must rely on diffusion from liquid further away, and this means that adequate time must be allowed. Equally, flushing out the remnant is impossible in the absence of flow, such as for the alternation of hypochlorite and EDTA (30§1.1, 30§2.1) These remarks apply with even greater force to accessory canals and their interconnections (simple form is rarely to be expected), and indeed to the dentinal tubules themselves.[7][8]

References

[1]   Timoshenko S. Strength of Materials. Part I. Elementary Theory and Problems. Krieger, New York, 1958.

[2]   Waters NE. The mechanics of finger and retraction springs of removable orthodontic appliances. Arch Oral Biol 15: 349 - 363, 1970.

[3]   Timoshenko S. Strength of Materials. Part II. Advanced Theory and Problems. Krieger, New York, 1958.

[4]   Stanek FJ. Free and forced vibrations of cantiever beams with viscous damping. NASA Technical note D-2831, 1965.

[5]   Wei M, Dyson DE & Darvell BW. Factors affecting dental air-turbine handpiece bearing failure. Oper Dent 37 (4): E1 - E12, 2012.

[6]   Hailu K, Lawoyin D, Glascoe A & Jackson A. Unexpected hazards with dental high speed drill. Dent J 5(1): 10, 2017.

[7]   Hess W. Zur Anatomie der Wurzelkanäle des menschlichen Gebisses mit Berücksichtigung der feineren Verzweigungen am Foramen apicale. Zürich, Buchdruckerei Berichthaus, 1917. http://www.endoexperience.com/documents/Hesstext.pdf

[8]   Gutman JL. Apical termination of root canal procedures—ambiguity or disambiguation? Evidence-Based Endod 1(4), 2016. DOI 10.1186/s41121-016-0004-8

Chapter 24   Light and Colour

Increasing emphasis is being placed on cosmetic dentistry, essentially the invisible repair of teeth. This 'invisibility' requires that the optical properties of colour, gloss and translucency of tooth substance be reproduced in the restorative material. The factors concerned in this, and the conditions under which it may be obtained, involve understanding the nature of colour.

*Colour itself is a product of the interpretation by our brain of physical stimuli. Discussion of the factors and conditions relevant to colour matching involves defining the **terminology** and the nature and dimensions of colour. The roles of the absorption of light by the object and the spectrum of the illumination are explained, as well as the interactions between them. Understanding of these basic principles is necessary to deal with the issues affecting colour **matching** in practice, as applied to the real problems of shade selection.*

*Other physical attributes of materials also affect the optical appearance of materials. The most important of these is **refractive index**, and the effects operating at boundaries. This has implications for the design of materials, and the effects of faults and surface conditions, especially roughness.*

*The **chemical** basis of colour is also outlined in terms of the absorption (and emission) of light. This bears on the means of colouring various types of material, and also their colour stability over time.*

Colour is another topic that is generally neglected but yet is an important issue, affecting as it does the conditions under which shade matching is undertaken. Comprehension of these issues is necessary to avoid the wastage of effort in high-value cosmetic dentistry.

Materials Science for Dentistry
https://doi.org/10.1016/B978-0-08-101035-8.50024-9

Whilst a major role of dental treatment is to restore form and function to the dental apparatus, rehabilitation, the vanity of man is such that merely functional restoration is often insufficient and the appearance of the prosthesis, be it anywhere from fissure sealant to full denture, assumes great importance. 'Aesthetic dentistry' is a greatly overworked phrase, for it surely depends on the eye of the beholder and opinions differ: black denture teeth found favour in at least one country, gold work is often prominently displayed, and a diamond has been seen worn in at least one upper central incisor. As appearance (or, at least, intended appearance) is the wearer's choice, the term 'cosmetic dentistry' is perhaps more honest. This includes 'tooth whitening' - noticing that natural teeth are not truly white - and which often represents an attempt to change appearance to match some perceived but unrealistic ideal. A quick glance at a dental shade guide will convince you of this. However, both terms are frequently interpreted as meaning only reproduction of the appearance of natural tissue so that the restoration is undetected by onlookers in day to day life. Clearly, if that is the case, neither term is being used properly. If, however, we accept that the matching of the colour and other optical attributes of a restoration to those of the replaced or surrounding tissue is a worthwhile goal, which seems not to be a difficult proposition, then the nature of colour, its generation, and the factors that influence its perception, should be understood. This is to enable the practitioner to achieve several goals.

Firstly, the 'taking of a shade' is a critical process upon which patient acceptance depends. Secondly, limitations to the quality of the match that can be attained need to be understood to inform the selection of material. Thirdly, the patient needs to receive advice on what can be achieved and what is to be expected, especially when a material may show age-related changes in appearance. A related problem occurs in connection with tooth whitening when restorations of any type are present: it is extremely unlikely that there could be any change in the restorative material to match that in the teeth. This would leave the restorations rather obvious, presumably requiring replacement, which could be a very expensive proposition. Fourthly, appearance is affected by the surface condition (scratches) and internal structure of a material (*e.g.* porosity) and is therefore under at least some control in terms of finishing treatments and perhaps mixing.

§1. The Sensation of Colour

'Light' is generally understood to be electromagnetic radiation which may be detected by the receptors of the human eye, although no actual physical distinctions can be made to identify that particular portion of the continuous spectrum which ranges from extremely low frequency radio waves to gamma rays (Fig. 1.1). It is of course 'visible' only because of the simple chemical fact of the energy of the radiation being able to reach those receptors and be absorbed at the sensitive organelle. Indeed, no distinct boundaries between portions of the visible

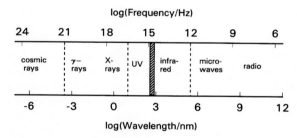

Fig. 1.1 Part of the electromagnetic spectrum. The visible region is the central, shaded part.

spectrum (*i.e.* 'colours') can be drawn sharply anywhere. In fact, what wavelength range is visible varies widely between species. All of the many designations are essentially phenomenological; that is, the labels are applied according to the spread of wavelengths associated with particular kinds of physical effects. Visible light then is no more special in this regard than any other region.

Our brain can utilize the information provided by the eyes to build up information about the disposition and nature of the immediate environment, coupled with other sensory input. As far as light is concerned, however, we may also distinguish a set of independent characteristics which are interpreted as the **psychophysical sensation** of **colour**. The term psychophysical means that colour is not, of course, a real physical attribute of radiation in the sense that frequency and wavelength are. Rather, it is an interpretative artefact of the brain, the psychological response to the physical stimulus. Thus, when we refer to the 'colour' of light, this interpretation will be taken as understood.

The present discussion is restricted to vision dependent on the retinal cone cells, **photopic** vision, as the rods do not permit colour perception. Those cone cells comprise, in 'normal' colour vision, three distinct types, sensitive to different regions of the spectrum; the peak sensitivities lie respectively in the yellow, green and blue regions (Fig. 1.2). However, it must be appreciated that they are each sensitive to a range of wavelengths; for example, we see red because the 'yellow' peak cones[1] have a response that extends to long wavelengths, and thus provide a visual signal in that region, while the other two have negligible response at those wavelengths. The brain[2] interprets such a combination of signals in the three channels as 'red'. Notice that it is the *relative* intensities of the responses (output signals) that provide the colour (*i.e.* wavelength) information for a single wavelength light. There are, therefore, only two degrees of freedom for assigning 'colour' since of the three comparisons, yellow *vs.* green, green *vs.* blue, and blue *vs.* yellow, only two need to be specified to determine the third, *i.e.* only two are independent (we are ignoring total intensity) (*cf.* the equivalent discussion of independent equations in 8§3.4). This is not the same, of course, as saying that only two signals are required by the brain, since there is a further dimension to consider, as we shall see below.

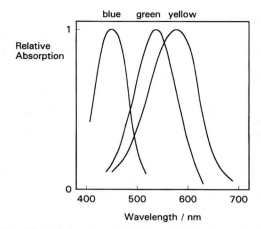

Fig. 1.2 Relative absorption spectra for the three human eye cone photopigments (peak sensitivities adjusted to unity - "normalized").

●1.1   Colour gamut

It is well known that a range of colours, the (visible) spectrum (the colours of the rainbow), may be obtained from white light by using a prism (Fig. 1.3) or a diffraction grating. However, this range is also very evidently only a very small part of the total **colour gamut**, the complete range of colours that can be created or perceived by humans. In fact, about 10,000,000 detectably different colours exist for anybody with 'normal' colour vision,[3] and it is immediately obvious that most of these are not spectral, meaning that they do not correspond to the colour of any single wavelength. What is not so obvious is how these other colours arise, and it is now necessary to explore some aspects of colour science in order to understand the implications for dentistry. We shall do this by describing the results of some simple experiments.

We shall of necessity have to restrict the following discussion to normal human trichromatic vision, or **trichromacy**. There are, of course, various types of colour vision defect[1], and even rare instances of **tetrachromacy** (which also occurs in various other species). What follows can be extended easily in a natural way to cover such circumstances, as no principles are critically dependent on trichromacy as such, should the need arise. Even so, it would be wise to bear in mind the possibility of defective colour vision ('colour blindness') in the context of cosmetic dentistry and colour matching in general, for the patient certainly but especially for the dentist, given that (depending on region) as much as 10% of the population may be affected in some way (males : females ~ 20 : 1).

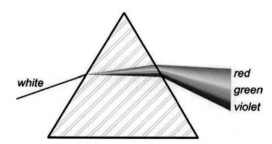

Fig. 1.3 White light may be dispersed by a prism or a diffraction grating to show the visible spectrum.

●1.2   Visual matching

Suppose that we illuminate a portion of a white[4] screen by a lamp, which can be arbitrarily chosen but for example an ordinary tungsten filament bulb, such as can be done with a slide projector or theatrical spotlight. Three other such lamps are then set up to be able to illuminate (only) the adjacent portion of the same screen. These other three lamps have widely different colours (for example, but not necessarily, red, green and blue), the brightness of each of which is independently adjustable (as by an iris diaphragm - as will be explained below, changing the voltage on a filament lamp will change the colour). These lamps can have their colours determined

---

[1]   Commonly referred to as the "red" cones.

[2]   This is a simplified story, and ignores the effective signal processing that occurs in the retina.

[3]   This may be compared with the 16,777,215 (*i.e.* 2^{24} - 1) colours supposedly obtainable with 24-bit computer colour systems.

[4]   Which we define as reflecting diffusely *all* incident radiation – see §3.10.

by selectively absorbent filters or any other appropriate means. The set-up is illustrated in Fig. 1.4. By trial and error it will be found that a perfect colour match can be produced, *i.e.* the light in the second portion of the screen will look as 'white' as that from the tungsten bulb; indeed it will be quite indistinguishable - by eye. Any small adjustment in the brightness of any of the three coloured lights will result in a distinct colour in the illuminated area. This experiment indicates that, as would be expected from the discussion above, 'white' is not really a property of the light as such but a particular response (on our part) to a particular set of stimuli. Furthermore, it does not matter how the lamp colours were established: the match can always be made perfect.

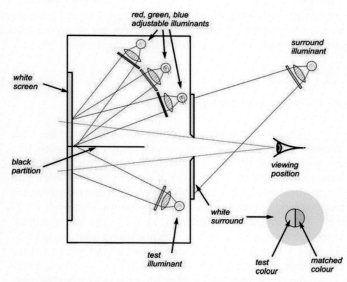

Fig. 1.4 The experimental set-up for colour matching experiments. The lamps in the upper part of the apparatus are independently adjustable in brightness. For the simple colour-matching experiment, the external lamp is turned off so that the surround is effectively black.

If now one of the three coloured lights is turned off altogether (say the blue), adjustments of the other two will be found to produce only colours varying between green and red, with yellow as an intermediate colour. With the presently assumed lamps, or **illuminants** - to make it more general, under no conditions will white be produced, and certainly blue will not be approachable. Equivalent remarks may be made for the other two pairs possible. The sensation of colour is thus said to depend on three stimuli; the total illumination of the test screen is certainly three-dimensional. However, if we were very careful (and it is a difficult experiment to perform), two illuminants could be found which would allow white to be produced, that is, with an appropriate ratio of intensities. This requires, in addition, for an arbitrary first illuminant, a very precise selection of the colour of the second. This is equivalent to the situation described above: having found a perfect match using any three illuminants, only one is then allowed to vary in intensity, the other two establish in effect a single but mixed-colour illuminant. More than three coloured lights could be used to obtain a match, but *in general* no less than three will do.

This outcome can be related directly to the fact that there are three separate wavelength-sensitive kinds of detector in the eye (Fig. 1.2). Three signals are generated, and information on the relative strengths of these three must be necessary to define a colour.

●1.3   Mixture diagram

What we are really dealing with is a three-component mixture, and diagrams of the type shown in Fig. 8§1.6 can be used to plot the results of such mixtures.[2] These diagrams are generally called a colour-mixing triangle (Fig. 1.5). Notice that in such diagrams it is the *proportions* of the components which are relevant, not the total intensity. A wide range of colours is capable of being reproduced by such a system. Indeed, it is the method employed in colour television and computer screens to produce the image, as close inspection will show (it often comes as something of a surprise to see that the three coloured dots are red, blue and *green* – get up close with a magnifying glass; §3.11). Photographic colour film uses exactly the same principle, but with coloured dyes which act as filters.

This diagram also indicates some pairs of **complementary colours**, *i.e.* those which in appropriate intensity ratios will generate white. These are indicated by the ends (as well as intermediate points) of all lines which may be drawn through the white point.

However, and significantly, there are always some colours which cannot be reproduced by *any* such system. This leads to

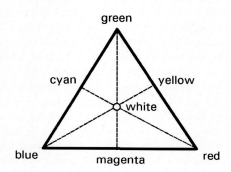

Fig. 1.5 The colour mixing triangle for showing the effect of mixing light as in Fig. 1.4. This is also called the 'additive' colour diagram.

limitations on the image quality of both televisions and photographs, no matter what the advertising says about true-to-life colours. Rainbows in particular are always disappointing because they are spectral colours and cannot be reproduced. This failure needs explanation.

§2. The Chromaticity Diagram

The **tri-stimulus** nature of the colour response has been stressed, so that the representation graphically of any particular combination of lights at specified intensities requires a three-dimensional plot. However, if we are concerned only with the proportions of each component, we can restrict our interest to the one plane where the sum of the photometric intensities $R + G + B = 1$, the so-called **unit plane** (Fig. 2.1). The 'composition' of the colour can be determined from the unit plane triangle in precisely the way that the composition of mixtures of three chemical components can be determined (8§1.2), relying on the idea of the perpendicular distances from the sides of the triangle representing proportions (Fig. 2.2). In other words, this is a colour mixing triangle for the chosen set of lights.

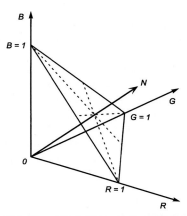

Fig. 2.1 The unit plane is constructed such that in it the total photometric intensity is constant and has the value 1, in any convenient units. N represents the neutral colour vector (white).

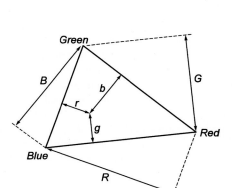

Fig. 2.2 The relative proportions of each stimulus in the mixture are given by the relative heights of the plotted point above the corresponding zero-intensity base.

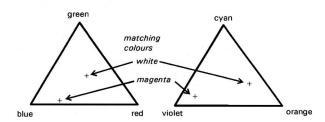

Fig. 2.3 Two separate colour mixing triangles in which corresponding matching colours have been identified.

Now, supposing we do exactly the same thing with a new set of different coloured lights (say, orange, cyan and violet) we would then have created a second colour mixing triangle. Then, using the set-up of Fig. 1.4, and replacing the lower lamp by the second set of coloured lights, we may be able to find a colour - for example, white – using the first set that we can match with the second set. The corresponding points can be plotted in the two diagrams. We go on to seek a second colour using the first set, say, magenta, that we can match with the second set. We again may plot the corresponding point in each diagram. This situation is then similar to that shown in Fig. 2.3. However, we know that these colour points in the two diagrams are matched. By rotation and scaling the second diagram we

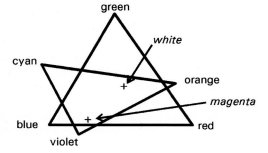

Fig. 2.4 By rotation and scaling, the colours of the right triangle of Fig. 2.3 (OCV), can be made to coincide exactly with the corresponding points in the RGB triangle. All overlapping points then match.

can arrange for the two matched points to be superimposed. We would then find that all points common to the two triangles – that is, lying in the overlap area – correspond to colours that can be made by mixing either set of illuminants. Furthermore, all such points coincide in the new diagram (Fig. 2.4).

•2.1 Spectral locus

If we were to continue creating and overlaying colour mixing triangles, using as wide a variety of illuminants as possible, we would gradually build up a map covering more and more of the colour gamut (Fig. 2.5). If we included in our list of tested illuminants the pure spectral colours, single wavelength lights (such as from a laser), after overlaying to match colours we would find that the spectral colours formed the outer boundary of the available colour gamut. This boundary is called the **spectral locus**, and the wavelengths of the spectral colours will be found to lie in order along it, although not on a uniform scale. Bear in mind that we are dealing with the psychophysical effects of light, and strict 'calibration' of our perceptions in wavelength – or frequency – could not be expected. No matter what illuminants we test, none will be found that will lie outside that boundary after the colour mixing triangle has been matched and overlain properly (Fig. 2.6).

•2.2 Chromaticity coordinates

Now, quite obviously, no single set of three illuminants can correspond to a colour mixing triangle that will give the entire colour gamut; no amount of scaling will change the shape of Fig. 2.6 to a triangle and maintain the proportionality requirements for plotting in that area. In other words, there are always colours that cannot be reproduced using any single set of three arbitrary 'primaries', even if these primaries are themselves pure spectral colours, *i.e.* single wavelength light which is by definition pure.

Obviously, if the spectral locus is to be plotted in a plane, there is a region outside it totally inaccessible through the mixing of any real illuminants. Subject to the condition of existence, therefore, colours may be identified in terms of rectangular coordinates.

What this in effect means is that an arbitrary mixing triangle for three *imaginary* illuminants (which do not and cannot exist), can be drawn to enclose the spectral locus (and this is in fact done in some in some colour definition systems). Even so, it is far easier to specify any point within the colour gamut by rectangular coordinates, conveniently chosen.   These are then called the **chromaticity coordinates**. (In fact, the diagram is scaled better to match human perceptions of the 'distances' between colours. The resulting, arbitrary coordinates are labelled v' and u'.)

The line which closes the **colour circuit**, joining the red and blue ends of the spectral locus, is known as the **purple line** (Fig. 2.7). Colours along this section range from red through magenta to purple and violet, following the rules for a mixture line (8§1.1). However, these are not spectral colours, and can only be created as mixtures. Clearly, since the ends of the spectral locus are the limits of attainable spectral illuminants, the purple line is also an unpassable boundary: no colour exists outside it. This then completes the **chromaticity diagram** (Fig. 2.7).[3]

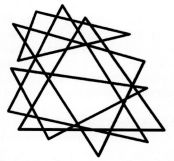

Fig. 2.5 Overlaying many colour mixing triangles includes more and more of the colour gamut.

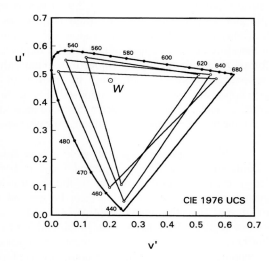

Fig. 2.6 The entire colour gamut can be built up from many overlapping tristimulus triangles, with the outer boundary defined by the spectral locus (the numbers on this are the corresponding spectral wavelengths in nm). *W* is the 'white point'.

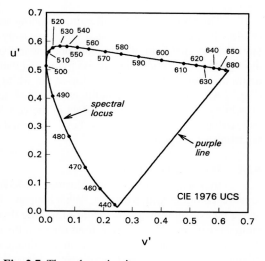

Fig. 2.7 The colour circuit.

Note that although it is possible to specify coordinates that lie outside the colour circuit, they do not correspond to realizable colours. Even so, it is noteworthy that the range of wavelengths said to be visible is not clearly defined. Estimates of the long wavelength limit vary between about 680 ~ 740 nm, while the short wavelength limit is given variously between about 380 and 430 nm. This clearly depends on individual sensitivities, some may see further into the red or blue, or both, than others. Thus, the purple line may assume different positions for these people. Nevertheless, the remainder of the diagram and all the observed effects and phenomena will remain unaffected by this.

●2.3   Colour map

The common named colours can now be mapped onto this diagram (Fig. 2.8), where conventional boundaries are identified by spectral wavelength. This can be seen to be insufficient for anything other than elementary work, and more detail can be added to take into account finer distinctions (Fig. 2.9). Even so, this is still not very precise, but it would rapidly become unwieldy if the process of subdivision were continued. It therefore becomes apparent that what is required is a formal means of specifying colour in a manner that is reasonably easy to interpret in terms of the psychophysical sensation, as opposed that is to the arbitrary scales of the CIE diagram. Some of the systems that have been developed are outlined in the next section.

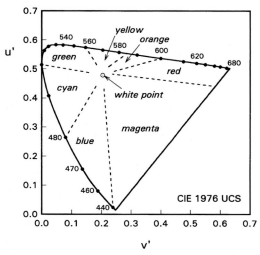

Fig. 2.8   The regions of the chromaticity diagram labelled according to a simple list of perceived colours.

No phosphor or dye can yield a pure spectral colour, so approach to the spectral locus is limited. However, the general convexity of the spectral locus ensures that no spectral colour can be created as a mixture of any other two, except perhaps in the red to yellow region. This means that, in general, a mixture line between two such pure colours cannot reproduce the appearance of intervening spectral colours, except where the locus is itself quite straight. These unavailable colours are said to be outside the gamut of the three particular stimuli chosen in the kind of experiment illustrated in Fig. 1.4. In the case of Fig. 1.5, for example, certain strong yellows, a range of greens, and violet to purple colours simply could not be produced.

The situation of two illuminants being used to create white can now be mapped more precisely. For any arbitrary first light, there is only one line that passes through the white point; the second light must therefore lie on this line, but on the other side of the white point (Fig. 2.6). The special conditions attached to this possibility do not diminish the effect of the conclusion that colour as we sense it is best described as a **tri-stimulus** phenomenon.

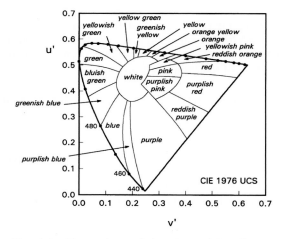

Fig. 2.9   The regions of the chromaticity diagram labelled according to a more detailed list of perceived colours.

●2.4   Weighted mixtures

In reaching the above conclusions we have only considered taking illuminants three at a time to create separate colour mixing triangles. What happens if we have four - or more - illuminants? Remember that if we take a pair of illuminants the result of mixing them is represented as a point on the mixture line joining their plotted positions. This is a **weighted mean** of the two illuminants, weighted by their intensity proportions. But this new point is itself equivalent to an illuminant, and we may repeat the process to calculate the corresponding mixture of this with the next illuminant. Ultimately, this process repeated will produce a weighted mean for any number of illuminants acting together, including the situation where all wavelengths are represented. We could, of course, have taken the illuminants in threes, using mixture triangles, to achieve exactly the same result. There are then many ways in which 'white' can be created, including the particular case of all wavelengths being represented uniformly. It also follows that any colour (except a spectral colour) can be made using a mixture line – that is, of just two illuminants, of any orientation or length (providing it lies within the colour circuit), that passes through that colour point. For example, white can be made by a suitable mixture of blue and yellow, or indeed of many other such pairs.

§3. Colour Specification

●3.1 Saturation

The coordinate system of the chromaticity diagram is not very intuitive: the use of a coordinate pair does not lead to a good conception of the colour involved despite the technical advantages of a cartesian (rectangular) coordinate system map and, as we have seen, the use of word labels lacks precision. Several alternative approaches have been offered. They all take as the natural point of reference or origin the **white point** which represents, by definition, an absence of perceived colour. On the other hand, the strongest colours (what we may consider the most extreme illuminants physically possible) are those which lie on the colour circuit, the spectral colours themselves or the purple line mixtures. Thus, there is a gradation from white, through pastel shades, to what is termed a **saturated** colour. White is therefore the least saturated colour possible; indeed, its 'saturation' is defined as zero. The measure of saturation for a colour at a given point is therefore the relative distance along a line drawn from the white or **neutral point** to the colour circuit through that point of interest (Fig. 3.1). This is, in itself, a mixture line, where the two components are white and the saturated colour.

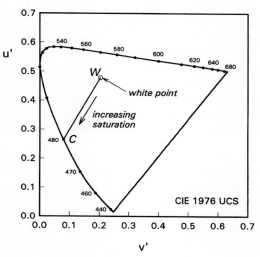

Fig. 3.1 A line of constant hue, drawn from the white or neutral point, W, to the spectral locus at C, where the colour is said to be fully saturated.

●3.2 Hue

To specify saturation is not enough. Since we are working in the plane, two values need to be specified, and in this case it is the direction to be taken in moving away from the neutral point. This corresponds to specifying wavelength (or frequency) on the spectral locus. We already have common names for the spectral colours: red - orange - yellow - green - cyan - blue - violet (*cf.* Fig. 2.8). Adding to this list, in order anticlockwise around the colour circuit, purple and magenta, we can specify the type of colour or **hue** in a natural manner that is easier perhaps to comprehend than wavelength. This kind of mapping is similar to a polar coordinate system: distance from centre and direction. Thus, in Fig. 3.1, W is the neutral point, C is a saturated colour (in this case, blue), while all points in between are whiter or paler or less saturated versions of the same hue.

●3.3 Grey scale

Notice that the **photometric intensity** of the illumination is not discussed (except to say that it is constant). This is because, except at extremely high levels when discomfort is involved and at low levels when **scotopic vision** starts to take over (*i.e.* the rods are doing the work), the psychophysical sensations of colour change very little, if at all. In addition, we have only discussed the appearance of the illumination of an assumed perfect diffusing surface, *i.e.* a 'white' screen. However, it is evident that 'absence of colour' has another dimension in that we are aware of shades of grey, ranging from a totally absorbent, non-reflective black to a totally diffusing pure white. Again, this is without reference to the actual amount, the photometric intensity of the light falling on the object.

Fig. 3.2 The grey scale.

It is necessary then to introduce a further dimension to complete the statement of the colour of objects: the **lightness** or **grey scale value** (Fig. 3.2). (Plainly this cannot apply to the illuminants themselves – 'grey' light does not exist.) If we have plotted hue and saturation in a plane, the lightness dimension is normal to it (Fig. 3.3). This means that we are working in what is called a **cylindrical coordinate system**: direction and distance from the centre in a plane, and the height above or below the plane, and we now have a three-dimensional reference frame for colour specification. Since therefore the range of the neutral colour is from black to white, we may represent this in terms of the overall **reflectance** of the object, which value can then range from 0 to 100% of all that impinging on it. (This property, as a proportion, 0 ~ 1, is also sometimes called the **albedo** of a surface or body, especially in astronomy.) It is therefore an absolute, not subjective, measure. In this sense 'grey' is a physical property statement when applied to a body, implying less then perfect reflectance, but then not total absorption.

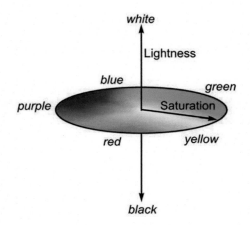

Fig. 3.3 The lightness or grey scale axis is orthogonal to the chromaticity diagram.

Notice, however, that for us to judge that something is grey requires an object for comparison (as in detergent advertising). Imperfection in whiteness can only be detected by comparison with a whiter object. In low light conditions, for example, this page may appear 'grey'. But what is the colour of a shadow on 'white' paper?[5] The absence of returned light is an absolute: this is black.[6][4] At the other end of the scale we have 100% reflectance as a definition of white; this is perfect behaviour,[7][5] but clearly it is not dependent in any way on the photometric intensity of the illuminant – it still holds as a *definition* in complete darkness. We are dealing, then, with the interpretation of grey as being a relative measure of the returned light as a proportion of that incident on the body. It is automatically scaled in our perception according to the current lighting conditions, typically by (unconscious) reference to the lightest object in the field of view. Figure 3.2 is unaffected whether viewed under floodlights or a single, white, low-power LED lamp, or even starlight. Of course, shadows are not perfect black when there is any other source of illumination, whether diffuse or direct. Hence, in snow, a shadow from the sun may appear blue when there is a clear sky because of its blueness, while the snow itself is 'white' because it is illuminated by the sun. Such effects are often clear in stage lighting. White light remains white no matter its photometric intensity, while a 'grey' object, in the sense of its reflectivity being less than 100%, may appear white if it is the lightest object in view, and especially if more brightly lit than the surroundings. Confusion with photometric intensity is common: a 'grey' day is one that has low light (because of the density of the cloud cover) compared with a sunny day, although the light may be spectrally uniform.

●3.4   Differential absorption

We have assumed, indeed required, for the experiments illustrated by Fig. 1.4 that there is no selective or **differential absorption** of the various wavelengths of the light impinging on the screen - that is why it is 'white'. When we consider shades of grey the requirement is that all wavelengths are uniformly absorbed, which also means uniformly reflected. In this way the saturation remains zero and the colour neutral for white illumination, no matter what its spectral make-up. But, it is evident that many objects are coloured even though they are illuminated by white light.

Let us, for the moment, assume that the phrase "white light" means that all wavelengths are equally represented, *i.e.* uniform photometric intensity. The weighted mean of these in the chromaticity diagram therefore lies at the neutral point. But if the object illuminated absorbs any wavelengths preferentially, in general the weighted mean of the reflected light will no longer lie at the neutral point. This means that the saturation will have increased in the direction of some hue, and typically away from the hue of the light which has been absorbed. The colour of the object under white light illumination is therefore determined by the relative proportions of the different wavelengths which are reflected. Thus, the colour of objects in general arises from differential absorption. How this may work in practice can be seen in plots of reflectance *vs.* wavelength, such as might be obtained from a scanning spectrophotometer, so-called **reflectance spectra** (Fig. 3.4), where the

---

[5]   Assume white light, but it actually depends on the surroundings – see §4.12.

[6]   Although this is impossible to achieve, a very close approach has now been made: Vantablack® – see the next reference.

[7]   Again, impossible to attain, but a good approximation can be made: Spectralon® – see the next reference.

reflectance indicates the proportion of incident light at each wavelength that is not absorbed. Thus, a 'red' object tends to absorb yellow to blue wavelengths, a 'green' object both blue and red wavelengths, and so on.

In summary, the colour of an object – hue and saturation – depends on the weighted mean (§2.4) of the reflected light, as plotted in the chromaticity diagram, while the grey scale value or lightness depends on the overall reflectivity of the object, without regard to wavelength.

The colour mixing triangle of Fig. 1.5 may be seen in effect to operate as an 'additive' colour diagram, the addition being of the lights impinging on the screen. In a complementary sense there are also 'subtractive' diagrams, which are used for paints, artists' colours and pigments in general to indicate what might be expected when they are mixed (Fig. 3.5). These two diagrams differ, in part, in having respectively white and black at their centres as the neutral colour. However, although pigment formulation may require such concepts, and this could be relevant to the colouring of dental porcelain for example, as far as the eye is concerned it remains the total wavelength mixture that is received which is important, as is apparent from Fig. 3.4.

Similar subtractive processes are operating in filters, where we would speak of **transmission** spectra, the essential similarity between filters and pigmented opaque objects (Fig. 3.6) being the spectral composition of the light received by the eye. In each case it is the intensity weighted mean of all the wavelengths present that determines what is perceived as the colour.

There are several other ways of describing colour.[6] These have been developed for particular purposes but are necessarily still based on the ideas of illuminant mixing and so on described above. Since some of them at least may be encountered in dental contexts, they are described briefly.

●3.5 Munsell system

One of these colour specification schemes is the proprietary **Munsell system**, an elaboration of the scheme shown in Fig. 3.3. This takes into account the sensitivity of the perception of saturated colours at both high and low lightness. Clearly, any colour that is very light cannot simultaneously be very saturated, because this would require that the albedo is higher than the proportion of, say, red light in the white illuminant. Likewise, something that is very dark, not returning much light from the white illuminant, cannot also appear to be saturated – it simply is not seen. The Munsell **colour space** is therefore a kind of distorted double cone (bases together), the apices being white and black respectively. The direction relative to the black-white axis is still termed **Hue** (although, oddly, labelled in a clockwise direction), but saturation is now called **Chroma**, while lightness is termed **Value** (*i.e.* grey scale value) (Fig. 3.7). The notation is brief and easily

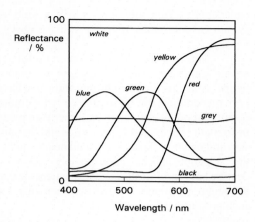

Fig. 3.4 Some examples of typical spectral reflectance curves for variously coloured objects.

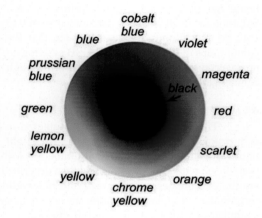

Fig. 3.5 An example of a subtractive colour diagram. Such a diagram can only ever be approximate since it is based on the assumption that the spectral absorption curves for each pair of opposing colours are exactly complementary, leading to total absorption. Normally, the result of mixing like this with pigments is a muddy colour because each surface particle must scatter some light. Filters are a more effective demonstration.

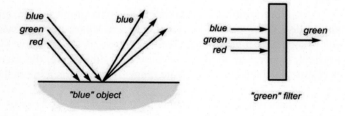

Fig. 3.6 The effects of opaque objects (left) and filters (right) on the light seen are necessarily similar.

understood. This therefore is still a kind of cylindrical coordinate system.

The Munsell system is very useful for colour matching purposes as sets of colour 'chips' with specific hue, chroma and value have been prepared. These are used directly for colour determination by eye matching rather than by an instrument. For example, they are useful for hair, skin and mucosa. The latter would be relevant to the selection of an appropriately matching denture base acrylic (although it is noticeable that the range of colours offered commercially for such products is very small).

A variant of this system is called the **HSB model**, that is Hue, Saturation, Brightness. Hue is described by the angle around the circle, $0°$ being red, proceeding through yellow at $60°$. Saturation is dealt with as a percentage from 0 to 100, as is brightness. It can be seen that the colour maps of the Munsell and HSB systems do not overlay on each other exactly, even though conceptually similar.

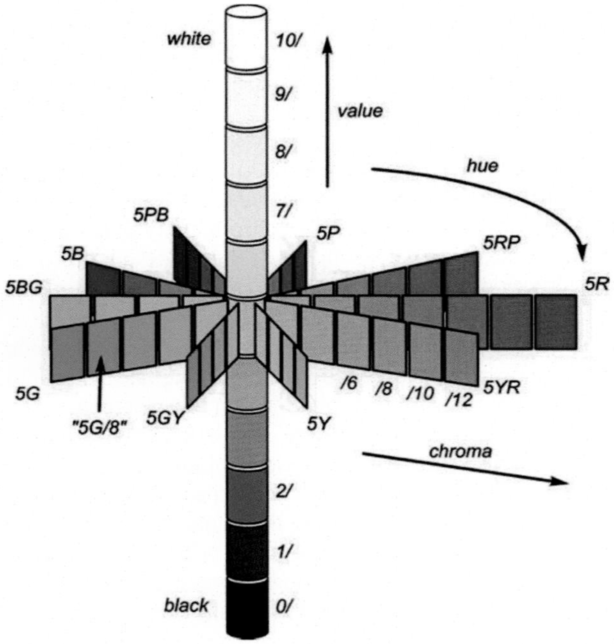

Fig. 3.7 The Munsell colour diagram and labelling system. Hue is represented by an abbreviation (RP = red-purple, Y = yellow, ... *etc.*), grey scale value by the first digit (*e.g.* 7/), and saturation or chroma by the second number.

●3.6   CIE L*a*b* system

This is more likely to be encountered in connection with colour meters. The light returned from the test object on illumination with a white light (such as is obtained from a photographic flash) is analysed with a set of filters and then described by three numbers: **L*** is the lightness value, while **a*** and **b*** are the coordinates on red-green and blue-yellow axes respectively. This therefore is a three-dimensional rectangular coordinate system for the colour space, the origin of which lies at the neutral black point. There are a number of variants of this kind of scheme.

●3.7   RGB system

This is normally used for defining the colour to be displayed on television and computer monitors, and similar light-generating equipment. As mentioned above, a colour monitor has three light emitters, red, green and blue (corresponding to the RGB of the name). The intensities of these three primary colour components determine the overall colour, through a mixing triangle such as Fig. 1.5, as well as the overall intensity. However, because no triangle can cover the entire colour gamut, as explained above, there remain colours which cannot be specified in such a fashion – they are, anyway, incapable of being created by those emitters.

●3.8   CMYK system

Colour printing, on the other hand, in its simplest form, employs dots of varying size of **C**yan, **M**agenta and **Y**ellow inks (see Fig. 1.5) (with blac**K** to adjust grey scale value, so-called four-colour printing), as can be easily seen with a hand lens. This, of course, is a subtractive system, but the same limitations on the gamut available apply. In addition, because it is subtractive, the seemingly odd selection of 'primary' colours is necessary to obtain an appropriate range: red plus green ink makes black, not yellow.

There is a significance in both RGB and CMYK systems for dentistry. On the one hand, there could be errors in the colours that would be found if comparison were made of a real object with a colour monitor image, if it was desired to demonstrate colour matching with such equipment in the surgery. This would not be because the unsaturated colours appropriate for teeth and so on could not be made, but because it would demand accurate calibration of the screen (*i.e.* of the intensities of the output of the electron guns or LEDs, for which purpose special devices and software are available), and this would also depend on ageing effects in the light emitters and driving electronics (requiring regular checks). An indication of the problem can be gained from displays of televisions in shops: rarely do two have the same colour balance! A further problem is that with LED displays the perceived colour depends on viewing angle, sometimes quite markedly, and for only small angular changes.

On the other hand, there are difficulties in making an accurate printed shade guide because of the high accuracy required for the amount of ink in each dot. High quality colour printing is expensive (and in fact may employ more than three colours to achieve good saturation and range of hues), but paper is a relatively fragile medium that does not lend itself to sterilization. In addition, printing ink dyes are **fugitive**, fading in time (§6.3).

Each of these colour systems is, of course, dealing with the specification of colours from the same overall set. Conversion between them is therefore possible, using various formulae, but this has no immediate dental relevance. Ultimately, dentistry is mostly concerned with the problems of colour matching, as exemplified by the use of a 'shade guide', such as is used for restorative materials.

●3.9   Colour space

Although there are three coordinates in, for example, the Munsell system, which may be specified independently, there are combinations which have no meaning. This excluded region has already been mentioned (§3.5), but it is worth emphasizing that this is a general condition, not relevant to only the one system. Just as there is an 'inaccessible' region outside the colour circuit (Fig. 2.7), so there are non-existent colour specifications outside the double-cone of the full realizable **colour space** (Fig. 3.8), the three-dimensional extension of the colour gamut (§1.1). It should be clear that a colour cannot simultaneously be fully-saturated and have either maximum or minimum lightness, indeed it follows that any deviation in the lightness direction for a fully-saturated colour is disallowed, remembering that lightness is not photometric but relative intensity. Notice that since white is made up of all wavelengths, reducing the proportions of any wavelengths must simultaneously

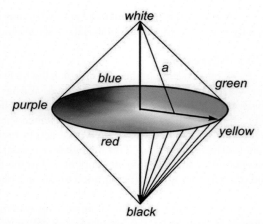

Fig. 3.8   The nature of the limits of the available, realizable colour space.

introduce a hue, increase the chroma, and reduce the grey-scale value. Reducing, then, the value for a single wavelength takes the colour towards black. But it is, of course, the same black regardless of the starting colour, and therefore this must be a single, unique point in the colour space, and the same as the lower end of the grey scale.

The effect can be envisaged by considering the double-cone of the colour space to be defined by the full series of mixture lines from the apices to the saturated circuit, for example white-purple, or black-yellow. Saturation therefore decreases as lightness increases. Similar mixture lines for unsaturated colours can also be considered, such as line **a** in Fig. 3.8.

●3.10   Black, white and grey

It is worth reviewing the meaning of 'white' at this juncture as it has three distinct senses in the development above, according to context:
*   *White light* – an illuminant with all wavelengths equally represented, *i.e.* a flat spectrum.
*   *White object* – diffusely and uniformly scattering all incident visible radiation, *i.e.* no absorption.
*   *White perception* – the end result of the eye receiving light whose weighted mean lies at the neutral point, *i.e.* which has zero saturation.

These aspects are respectively a spectral description of an illuminant, a physical property of a material body, and the psychophysical outcome of seeing a particular spectral profile. Thus we may perceive as white a non-white object under a non-white illuminant when the weighted mean of the returned light is neutral.

We may contrast these definitions with those for 'black':
*   *Black object* – absorbing all incident visible radiation, *i.e.* no reflection or scattering.
*   *Black perception* – the end result of the eye receiving no light. This may be for
    – a black object under any illuminant (albedo zero)
    – a coloured object absorbing all incident radiation, *e.g.* a 'red' object under blue light[8]
    – the absence of any incident visible radiation

Of course, 'black' light does not exist (despite the term being applied to that of ultraviolet lamps).

---
[8]   That is, when there is no overlap whatsoever of the illuminant and reflection spectra.

For completeness, we may add definitions for 'grey':

- *Grey object* – diffusely and uniformly scattering a fixed proportion less than 100%, but not zero, of all incident visible radiation, *i.e.* no differential absorption.
- *Grey perception* – the end result of the eye receiving light whose weighted mean lies at the neutral point, *i.e.* which has zero saturation, but by reference either
    - to an object which is perceived to be of greater lightness, and which thereby becomes a standard for 'white', or
    - to the same object under more intense illumination, effectively putting the first area in shadow.

Again, there is no such thing as 'grey' light, that attribute of dawn notwithstanding.

Clearly, it is important to be sure when any such words are used (and indeed other colour terms), what exactly is meant: illuminant, material, or sensation.

●3.11 LED colour

Consideration of the way in which the LEDs of computer screens and the like work to generate colour is informative. We have three separately-controlled RGB emitters, which may be set at any value between 0 and 100% output (Fig. 3.9). Since at normal viewing distance the individual emitters are not resolvable by eye because of the visual acuity limitation (15§7.2), the emitted light is effectively mixed. This then gives a system as shown in Fig. 2.1 with the exception that there is now an upper bound to the intensity of each component, and of course we are not then limited to the unit plane.

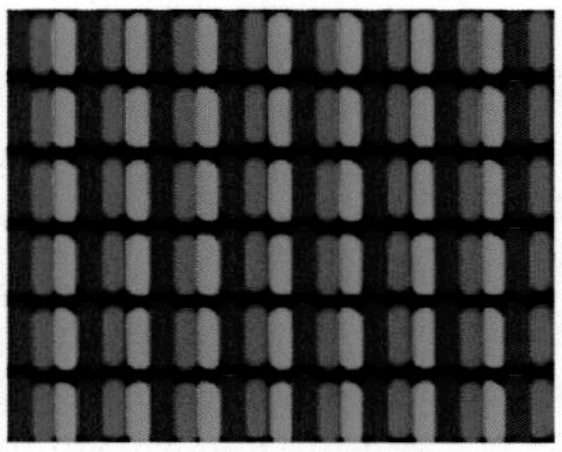

We therefore have the colour-mixing system shown in Fig. 3.10.[9][7] At the coordinate position (0,0,0) lies black – all off, at (100,100,100)[10] lies white – all on full. Yellow (100,100,0) is obtained with red and green full on, cyan (0,100,100) is green + blue, and magenta (100,0,100) is red + blue.

If the view point is changed to look (almost) down the cube diagonal from (100,100,100) to (0,0,0), and removing a number of plotted combinations for clarity (Fig. 3.11), the grey scale is revealed as lying along that diagonal, while the triangle drawn between the pure colour (100%) corners can be seen to correspond to the unit plane, and is of course now at constant

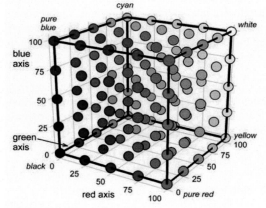

Fig. 3.9 Typical computer screen LEDs, as seen magnified when illuminated (use a hand-lens on any such screen).

Fig. 3.10 Three-component colour space mixing diagram.

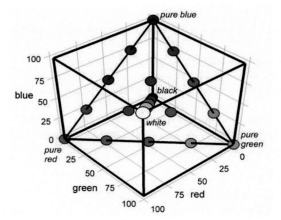

Fig. 3.11 Cube-diagonal view to show the grey scale and unit plane.

photometric intensity: the sum of the coordinates is 100 everywhere on it – this is the unit plane. Now, viewing just the visible (nearest) cube faces in Fig 3.11 gives Fig. 3.12. This is a kind of additive colour wheel, where the spokes correspond to mixture lines for the pure component at one end with the '100,100' mixture at the other, or equally, mixture lines of the saturated colours at the periphery with white at the centre. Conversely, looking at the other three faces, from the opposite end of

---

[9] Such a system was anticipated in the 13th century by Robert Grosseteste.

[10] In most computer programs, the RGB coordinates are specified on an 8-bit scale: 0 ~ 255, hence the figure in footnote 1: $(2^8)^3$.

the grey-scale diagonal (Fig. 3.13), we get a subtractive colour wheel similar to that shown in Fig. 3.5 Now we have mixture lines as diagonals showing in effect the behaviour of complementary colours again, whilst radiating from the centre are effectively mixture lines with black.

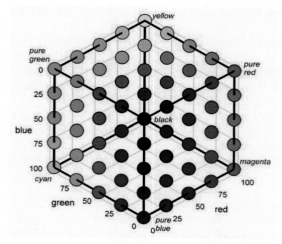

Fig. 3.13 The "subtractive" surfaces of the three-component colour space.

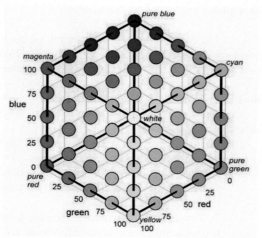

Fig. 3.12 The "additive" surfaces of the three-component colour space.

Rotating again, for a kind of 'equatorial' view, setting the grey-scale cube diagonal to be vertical (Fig. 3.14), we obtain the equivalent of Fig. 3.8, where primarily it can be seen why no colours can exist outside the double cone's surface, as delimited by the mixture lines radiating from the white and black poles, and that this must be true no matter what (or how many) light sources are involved. Secondarily, we can see that the colour gamut accessible to a three-light source system such as this is necessarily limited, and will fail to reproduce a rainbow properly, for example, as indicated by each of the mixture triangles in Fig. 2.6.

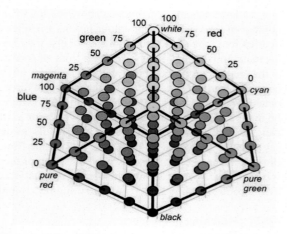

Fig. 3.14 The nature of the limits of available colour space. The grey-scale axis is vertical.

It is also possible to appreciate the absence of absolute photometric intensity in these considerations since the scale represented by 100% is not defined. All values can be multiplied arbitrarily by the same factor without changing the diagram or its import. It should be noted, of course, that the portion of real colour space covered by such a diagram is necessarily dependent on the emission spectrum of each LED (again, as in Fig. 2.6). This accounts for the often startling variation in the displayed colours in the many screens showing the same image seen in shops.

§4. Colour Matching in Practice

The discussion so far has centred on the background of the identification and mapping of colour. It is now necessary to consider the practical implications and use of this in the process of colour matching, especially as it relates to the use of shade guides in the selection of a restorative material.

•4.1 Principles of colour matching

The description above of the nature of the psychophysical sensation of colour can be summarized in three general principles.

    1. **Tri-stimulus**: We are capable of distinguishing only three kinds of variation in, or attributes of, colour: hue, saturation and lightness.

    2. **Weighted mean**: The perceived colour depends on the intensity weighted mean of the illuminants. Thus, if any one of the illuminants is steadily changed in intensity the colour perceived steadily changes, as the weighted mean varies.

    3. **Perceptive invariance**: Separate stimuli of the same colour (*i.e.* the same hue, same saturation and the same lightness - a perfect match) produce exactly the same perception in all mixtures of the two, regardless of the spectral composition of the individual stimuli.

This last principle is simultaneously the most important for colour matching, and the cause of all the problems encountered in this area.

The total colour gamut was shown earlier to be capable of physical expression only by overlaying a large number of three-component colour mixing triangles (Fig. 2.5). But, equally, there must be a large degree of overlap for many of these possibilities or else the idea of a colour *match* could not in general be realized. Thus, the matching colours may be made up from an endless series of combinations of pure spectral illuminants; that is the triangles may be drawn with their apices anywhere on the spectral locus so long as they enclosed the coordinates of interest. Any pair of these sets of three stimuli may be superimposed (that is, illuminate the same patch of white screen) without producing any difference in the perception of the colour. Similar remarks can be made for sets of stimuli which lie wholly or partly *within* the spectral locus. Now, given the idea of the weighted mean for more than three illuminants, as developed in §2.4, we may extend the above conclusion to say that the matching colours may be obtained by using any number of illuminants. In particular, those illuminants can be the spectral colours, and all of them at once, but with various intensities. Clearly, the same weighted mean can be attained in many ways, that is, there are many different spectra possible for the one colour. The spectra shown in Fig. 3.4 are very simple examples.

•4.2 Matching persistence

There is a fourth principle which must be mentioned: that of the **persistence of colour matches**. First, suppose that using the set-up of Fig. 1.4 a match is obtained for the lamp using the three separate illuminants in a suitable combination. If the illumination of the surround to the viewing aperture (the 'reduction screen') is changed from dark to light, or to any colour whatsoever using the external lamp, the colours that were obtained on the internal screen will continue to match.

However, the perceived actual colour of the two will appear to change, often quite markedly, as the surround changes in colour. This means that our *perception* of a colour also depends on the environment. It illustrates how the viewing conditions, the brightness and colour of adjacent objects, may affect the psychophysical *interpretation* of the colour of an object of interest. Thus a shade guide sample must be held adjacent to the reference tooth, for example, if the surroundings are to have no effect on the selection. This emphasises the fact that our judgement of a colour can be faulty, depending on circumstances. Man does not have an absolute sense of colour, and attempting to create or choose a colour without a reference object is risky.

•4.3 Metamerism

Now we are in a position to consider the combined effect of illuminant spectrum and object absorbance spectrum. Thus, if for a uniform perfect white screen, as in Fig. 1.4, using two sets of three coloured lights, we obtain an arbitrary colour on one side (*i.e.* using one set of lights), the Third Principle says that we can obtain a match on the other side, assuming only that the gamuts of the two three-stimulus groups overlap sufficiently. If now the colour of the screen itself is changed, that is, its saturation in respect of one hue is increased or decreased, in general there will no longer be a colour match for the two illuminants. By changing the colour of the screen we necessarily have differential absorption of the incident light. But that differential absorption means that there will be a greater or lesser effect on wavelengths represented more or less strongly in one illuminant but not the other. Thus, the colour *changes* of the two sides will be different. This is equivalent to saying that by adjusting the pigments in two coloured objects they can be made to match in colour under any arbitrary light source. However, they will not then, in general, match under any other source because light sources themselves differ very widely in their spectral makeup. This is the phenomenon of **metamerism**.

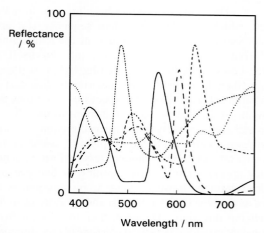

Fig. 4.1 A set of metameric spectral reflectance curves. Under one white light they are all a perfectly matched neutral grey.

There are in fact an essentially infinite number of spectral reflectance curves which will reproduce any given colour (hue, saturation, lightness) in an object under a given illumination. Thus, the objects corresponding to the spectral absorbance curves shown in Fig. 4.1 appear as the same neutral grey under white (spectrally uniform) illumination. There are many other such curves. But if the spectrum of the illumination were changed, however, then many of these **metamers** can be expected to show perceived colour changes. The implication of this is that if a colour match is obtained under one particular illumination there is no guarantee that the match will be maintained under a change of illumination. In fact it can be said more strongly than that: expect change.

It also follows from the colour matching principles (§4.1) that there is no information to be obtained about either the reflectance curve or the illuminant from the perceived colour, which can be understood from the myriads of sets of illuminants that can produce the same weighted mean (§2.4). We can perhaps better appreciate this by considering the individual cone responses (Fig. 1.2). There are only three signal channels, and each represents the sum of the response of that type of cone over all wavelengths, suitably weighted by the sensitivity curve. Such an integrated signal value can be created in many, many ways – *i.e.* combinations of intensities over all possible wavelengths – the incident spectra. All of that extra information is now completely lost, it therefore cannot be transmitted to the brain, and thus spectral information cannot be perceived.

This discussion, and especially from points made at §1.2 and §3.10, now leads immediately to the recognition of the existence of metamers in three distinct senses:
•   *Metameric illuminants* – having different spectral emittance curves that have the same weighted mean colour in the chromaticity diagram. (This is the basis of the illuminant colour matching of Fig. 1.4. using the white screen; §3.10.) Thus, any differentially-absorbing object would have a different perceived colour under each illuminant.
•   *Metameric objects* – having different diffuse spectral reflectance curves (or spectral transmission curves) for the one illuminant (*i.e.* of arbitrary spectrum) that have the same weighted mean colour in the chromaticity diagram (that is, as illustrated in Fig. 4.1). Thus, the match would be lost under a different illuminant.
•   *Metameric perceptions* – different spectra perceived by the one person as the same colour, due to
        (a) two illuminant + object combinations (neither of which components are metameric in themselves) yielding spectral reflectance (or transmission) curves that have the same weighted mean colour in the chromaticity diagram, or
        (b) limitations in the observer's visual response itself, whether due to colour-blindness (a defect in the visual apparatus) or dark-adapted vision, when the cones are not (fully-)functional.
In the last case, condition (a) is, of course, normal, and can be created in the apparatus of Fig. 1.4 by using different screens on the two sides, although it would be an unusual natural occurrence as it stands. However, it can now be seen that 'white perception' (§3.10) is in fact a metameric perception of this kind, using the concept or memory of lack of hue as an internal reference. Condition (b) is certainly a problem in low-light conditions, which is why lighting level is part of the specifications for workplaces where colour is important.

●**4.4   Daylight**
        Ordinarily, we take 'daylight' to be a nominal reference illumination. This is because of its naturalness: we suppose it to have a better quality in some sense. But if we were to take daylight to be a fixed illuminant we would be wrong: 'daylight' varies considerably (Fig. 4.2), as anyone who has been up at dawn or watched a spectacular sunset knows.[8]

The light of day emanates ultimately from the sun, which is a Type G, main sequence star and therefore yellow-white, having a surface temperature of some 5800 K.[9] However, the sky itself is well-known to be blue, because blue light is scattered more than the red end of the spectrum by fine dust in the atmosphere, shifting the spectrum – and thus the colour of the light – more and more to the blue at greater angles of view from the sun direction. Clouds, however, with their large droplets of water or ice scatter all wavelengths more or less uniformly; hence they appear white. Early morning or late evening direct sunlight, however, tends to be very yellow or red as the blue light has been scattered away from the viewing direction. Daylight, therefore, is far from constant in colour, depending on time of day, direction viewed, cloud cover and other atmospheric conditions. Even so, diffuse daylight (*i.e.* not direct sun) of one kind or another represents a common type of illumination, even indoors. Artists are well aware of the effects of such changes in illumination: the "cold northern light" is much preferred (at least in the northern hemisphere!) as it avoids much of the interfering variation.

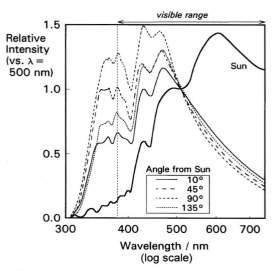

Fig. 4.2 Daylight varies in its spectral intensity curve according to the part of the sky that is delivering the light. The blueness of the sky is due to differential scattering by dust in the atmosphere; it is wavelength and angle dependent. Clouds tend to reflect or diffuse light which is closer to the sun's spectrum.

●4.5   Artificial lighting

Artificial lighting, on the other hand, is much more variable. The spectrum of a tungsten filament incandescent lamp, for example, is very dependent on the voltage at which it is run (Fig. 4.3). The filament of such lamps is resistively heated by the current flowing through it, and that current therefore varies with the applied voltage. The temperature attained by that filament then depends on the balance of the energy supplied with that radiated and conducted away. When objects are heated they are well known first to glow dull red, then a brighter red, yellow-white, then blue-white. This is because the relative emittance of radiation towards the blue end of the spectrum increases steadily with increasing temperature, making the light 'whiter'. The colours of stars show such a range for the same reason.

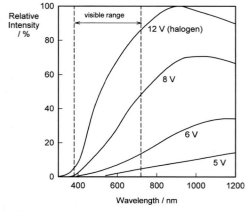

Fig. 4.3 Some spectral emittance curves for a tungsten filament run at different voltages and therefore different temperatures. (Halogen lamps can be run at higher temperatures than ordinary gas-filled types.)

●4.6   Black body emission

This effect is explained ultimately by **Planck's radiation law**. The emission power E, in W/m², of a black body, at a given wavelength λ, is given by:

$$E_\lambda = \frac{c_1 \lambda^{-5}}{e^{c_2/\lambda T} - 1} \tag{4.1}$$

where T is the absolute temperature, and c_1, c_2 are constants. A **black body**[11] is a theoretically-perfect, entirely passive emitter of radiation, emitting only by virtue of its temperature, without contributions from any other physical or chemical process or reaction. The nature of the emission curves is illustrated by Fig. 4.4, from which it can be seen that the peak emission moves steadily to shorter wavelengths, while the emitted power increases very rapidly, as the temperature rises.

---

[11] Strictly, this is defined as a body that absorbs all incident radiation, without reflection or transmission, independent of direction and wavelength.

The effect on the perceived colour of the emitted radiation can be seen by examining the region of the visible spectrum more closely (Fig. 4.5). Here, the emission curves have been normalized to the peak emission power (that is, scaled by dividing the values for a curve by the peak value for that curve) in order to avoid overall brightness affecting the picture. At 1000 K the emission is nearly all red, and this bias continues to 3000 K. After that there are substantial contributions from the blue end of the spectrum and the light becomes orange-red. By 5000 K, the power is fairly evenly spread with a peak around 550 nm, so that the colour is yellowish white. At 6500 K the peak is centred in the blue, corresponding to daylight's bluish white, after which the light becomes increasingly blue. For comparison, electrical discharges such as in arc welding and lightning reach temperatures of at least 10 000 K, hence their very blue-white, even violet appearance.

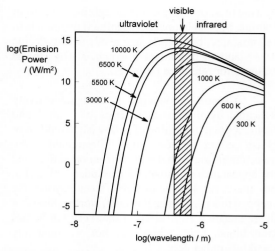

Fig. 4.4 A set of black body spectral emission curves according to Planck's radiation law. (Note the logarithmic scales.)

●4.7   Colour temperature

In fact, the wavelength, λ_p, of the peak emission of a black body is related to the absolute temperature, T, by the relationship:

$$\lambda_p = b / T \qquad (4.2)$$

where the constant $b = 2.898 \times 10^6$ nm.K. This is called **Wien's displacement law**. Hence the spectrum, and thus the colour, of the emitted radiation is very sensitive to the voltage applied to the filament. Because quartz-halogen[12] lamps (which are also tungsten filament lamps) can be run at a higher temperature, these produce simultaneously the brightest and whitest lights that can be obtained from this type of system.

Illuminants such as photographic floodlights and flashguns, and indeed any lamp where the colour of the light is important, have their output designated by their **colour temperature**. This is defined as the black body temperature that would give a colour of light identical to that obtained from the illuminant in

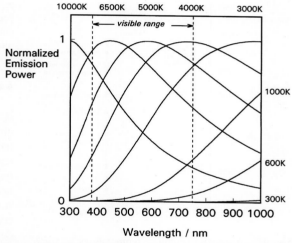

Fig. 4.5 The black body spectral emission curves for a range of temperatures (corresponding peaks indicated at the top). Compare the 300 K curve with that for the 12 V Halogen lamp in Fig. 4.3.

question. In the case of a filament lamp it is closely related to the actual temperature of the filament (whose behaviour is quite close to that of a perfect black body). The colour of such a lamp may be plotted at a variety of temperatures directly on the chromaticity diagram (Fig. 4.6). The locus of that plot, the **Planckian locus**,[10] is a smooth curve which passes very close to the neutral point at about 5500 K. It can be seen that this corresponds to the 'flattest' curve over the visible range (Figs 4.4, 4.5) – nearest to the ideal 'white' illuminant. It is the strong predominance of red in the light from ordinary filament lamps that led to the use of 'tungsten' as opposed to 'daylight' colour photographic film when such illumination was unavoidable; the sensitivities of the dyes have been adjusted to give a better approximation to a 'daylight' appearance. Indeed, the same principle applies to the 'white balance' settings of a digital camera.

Some idea of the range of colour temperatures that may be encountered can be obtained from Fig. 4.7. It is apparent that the variation in 'daylight' discussed above (§4.4) is indeed substantial, and that what we might

---

[12] A quartz envelope to prevent it melting at the high running temperature, and a small amount of a halogen (often iodine) included in the envelope to reduce the net evaporation of the tungsten filament. Note: m.p. of tungsten is ~3387°C, b.p. ~5555°C.

be used to as ordinary artificial lighting is far from natural. Special lamps are essential if daylight is to be simulated, but clearly one needs first to decide on a standard illuminant – which is why D65 is required for reference purposes.

Colour temperature is also subjectively indicated by the common terms of 'warm' and 'cold' applied to lighting of yellowish to reddish or bluish effect respectively. Presumably, the associations are with the glow from open fires and the appearance of snow and ice under a blue sky, but these are psychological effects beyond the scope of this book. It is, however, worth noting that the association is inverted in that very hot bodies emit bluish light, while 'red hot' is, relatively speaking, rather cool.

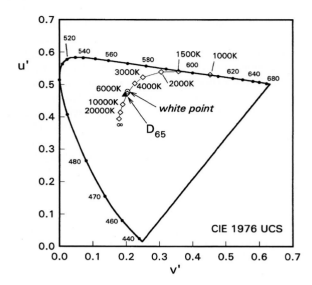

Fig. 4.6 The locus of the colour of the light from a tungsten filament at various temperatures (in kelvin) illustrating the idea of colour temperature. D65, 'standard daylight', is close to the black-body 6500 K point (▲). Notice that the light becomes decidedly blue at very high temperatures, when the emission spectrum also includes significant amounts of ultraviolet (Fig. 4.4).

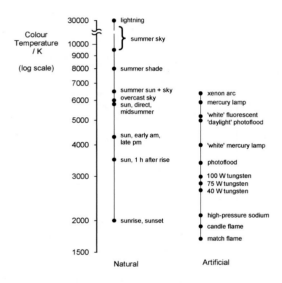

Fig. 4.7 Approximate colour temperatures of some common natural and artificial illuminants.

●4.8 Ultraviolet emission

Another consequence of Planck's radiation law is that as the temperature of the emitter rises so the quantity of ultraviolet (UV) radiation produced also increases. UV is damaging to the eyes, one of the reasons why UV-initiated polymerization in resin restorative materials was abandoned (6§5.9). This means that lamps running at a high enough temperature, such as quartz-halogen types, emit sufficient UV to be of concern. This has been recognized to be a hazard in the context of domestic lamps of that kind (such as desk lamps) because of the long exposure times involved, and they are now frequently sold with a UV (blocking) filter.

Such filters should therefore be installed in dental operating lights, which can otherwise be expected to be a risk because they have to be operated at a high enough colour temperature to allow good matching of shades under daylight-like conditions. Indeed, dental lights, and other lamps for high-quality colour work, are sold with specific claims of daylight-like illumination. This is not altogether unreasonable but, as should be clear from Fig. 4.2, it should not be interpreted as being capable of being strictly true. Nevertheless, such illuminants are effectively standardized and therefore useful. Notice from Fig. 4.6 that 'standard daylight' corresponds to a colour temperature of about 6500 K. (Compare the sun's surface temperature of ~5800 K.)

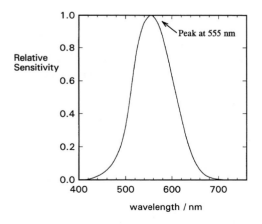

Fig. 4.8 Relative sensitivity against wavelength of the light-adapted human eye (photopic vision)

It should also be borne in mind that the overall sensitivity of the eye varies according to wavelength (Fig. 4.8). As was pointed out in 6§5.12, the intensity in the blue region from curing lights will be far greater than appears to be the case. The effect will be increased if other wavelengths are filtered out. Add this to the relatively intense blue emission from quartz-halogen lamps (Fig. 4.3), and the risk of damage can be seen to be significant. The use of the viewing filter is very important.

●4.9 Fluorescent lamps

The other commonly used source of artificial illumination is the so-called 'fluorescent' lamp or 'tube'. These operate by the conversion of ultraviolet light, from a mercury vapour discharge set up within the tube, to visible light through the use of particular chemical compounds known as **phosphors**, some examples of which are given in Table 4.1. The mercury vapour discharge produces what is known as a line spectrum (Fig. 4.9), which arises from the ionization and re-reaction with electrons of mercury atoms in the electric discharge.[11] (These electrons are in outer shells, *cf.* Fig. 26§1.1.) It is called a line spectrum because the electronic transitions are associated with specific precise wavelengths, and these are therefore very strongly represented.

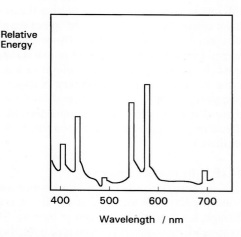

Fig. 4.9 The spectrum of a low-pressure mercury vapour discharge lamp, as used in fluorescent lamps. The bars indicate intense line emission.

Table 4.1 *Some typical phosphors which emit after stimulation by ultra-violet light or electron bombardment. The light emitted is due to electron transitions to lower energy states after excitation.*

Phosphor	Wavelength/nm of max. emittance
Calcium tungstate	440
Magnesium tungstate	480
Zinc silicate	540
Calcium halophosphate	590
Calcium silicate	610
Cadmium borate	615
Calcium-strontium phosphate	640
Magnesium arsenate	660

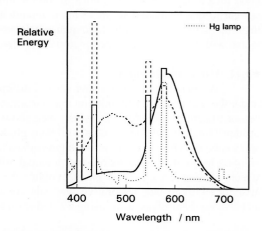

Fig. 4.10 Two typical spectra for fluorescent lamps showing the dominance of the Hg lines, and the degrees of success obtained in 'filling in' the spectrum with phosphors; compared with curve of Fig. 4.9 (dotted).

Inevitably, these lines will figure prominently in the output of the fluorescent lamp because they are not absorbed completely, even though much broadening and 'filling-in' of the spectrum can be done by the phosphors (Fig. 4.10). Phosphors also operate by emitting light in the process of electronic transitions to lower energy states, but the spectrum is more complicated and 'smeared out' because in the solid state, and with multi-atom compounds, there are many more possibilities for intermediate states (*cf.* the vibrational decay in polymerization photosensitizers, Fig. 6§5.2). The dramatic difference between the spectrum of tungsten filament and these fluorescent lamps is obvious.

By adjusting the mixtures of phosphors the spectral distributions of the light produced can be changed, but the strong lines are always present. It is therefore possible in principle to create the general *perception* of the colour of daylight ('D65', Fig. 4.11), even though the spectra are very

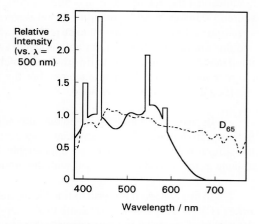

Fig. 4.11 The spectra of standard daylight (D65) and a typical fluorescent lamp show substantial differences.

different in general shape: by adjusting the intensities the weighted mean can be made to correspond with the target colour. So-called "daylight" tubes are available which are claimed to be good matches, and for which a premium price may be payable. However, depending on the purposes of the user, the 'cold' (rather bluish) or 'warm' (somewhat pinkish) light products may be preferred. That there are in fact large differences between products is easily seen by inspecting from a distance the colour of the light in the windows of a block of flats in the evening - and each occupant is under the impression that the light is natural. However, because of metamerism, many objects, paint and so on may take on 'peculiar' colours under fluorescent lights. Fluorescent illumination is therefore not recommended for high quality colour work, such as for dental porcelain. Peculiar effects are also possible with colour television cameras and photographic film because of their characteristic sensitivities.

●4.10 Ultraviolet emission (2)

There is a further issue of relevance with fluorescent lamps: the absorption of the UV by the phosphors is not perfect. The intensity of the UV is not great enough to have caused any concern over potential eye damage, but there are two effects of interest. Many paper products and domestic detergents contain fluorescent compounds that convert UV into blue light, thereby tending to make the treated object appear whiter or brighter (hence the term 'brightener' for these compounds). In comparison, an untreated but truly neutral colour will appear relatively yellowish under the same illumination. This is a distortion of perception because colour is psychophysical, not absolute. Secondly, many microorganisms are sensitive to UV. It is therefore not a very good idea to use fluorescent lamps in microbiological laboratories, where the ability to culture organisms is critical to diagnosis and so on. This unintended emission was also a reason for UV-cured resin restorative materials setting prematurely (and quartz-halogen lamps also emit UV, §4.8), but the problem is, of course, far worse with blue light-sensitized materials, so that any exposure to room or natural light is undesirable.

●4.11 Use of shade guides

We may therefore be using several different kinds of illumination, with widely varying spectra. It is under these changes of illumination that metamerism is problematic. It is possible, indeed common, to observe colour matches under daylight which then show striking differences under artificial lighting, whether under ordinary tungsten filament lamps with their predominantly red emission, or fluorescent lamps with their very spiky spectra. This is a common problem with paint, clothing and cosmetics, particularly with the more saturated colours; hence the very sensible and practical utility of making comparisons outside the shop (in 'daylight' – which, of course, itself varies, §4.4), avoiding the distortions of artificial lighting. In dentistry, such effects can arise for all 'tooth-like' restorative materials – resin, GI, porcelain – with respect to adjacent tooth tissue, as well as between themselves, as in using a filled resin to repair a broken porcelain device.

Thus, in dentistry, when identifying from a shade guide which shade of artificial tooth, restorative or other material to use, and appearance is important, it is good practice to check that the match is good at least under natural daylight, and preferably also under fluorescent tube illumination. This is to make sure that untoward effects do not occur. Checks with multiple illuminants are always necessary unless it is known with certainty that the shade guide material is identical in every respect to the actual material to be used, so that the absorbance spectrum is precisely the same. This is not likely to be the case with printed shade guides. (It also assumes that the manufacturer has already designed – and checked – the product to avoid problems from this source.) However, and worse, no dental material can possibly have precisely the same spectral absorbance spectrum as the tissue it is meant to be replacing. Metameric effects are therefore inevitable, given the sensitivity and discriminatory ability of our colour visual system. With care, these difficulties can be minimized. Again, it is a matter of compromise, to find the best match possible under a variety of lighting conditions. This is mostly the responsibility of the manufacturer, but the dentist still needs to recognize the nature of the problem if significant errors are to be avoided.[12]

In passing, it should be noted that although the number-letter shade designations ('A3', and so on) used for restorative materials by many manufacturers are uniform in style, it should not be assumed that they are indistinguishable. In fact, there are appreciable differences between the colours they represent. Accordingly, a shade guide should only be used with the product for which it is intended.

In this context it is worth noting the existence of a figure of merit for lamps called the **Colour Rendering Index** (CRI). This is a 0 ~ 100 scale that is supposed to indicate how accurately a light source will reproduce the colour of an object in comparison with a 'natural' illumination, such as a standardized 'daylight'(*e.g.* D65) or a blackbody source. The calculation is complicated (and negative values are possible!), and not a little controversial.[13] In essence, it is derived from the arithmetic mean shift in measured chromaticity coordinates

(**Euclidean distance**) between the two sources for a set of "representative" test colour swatches. Whilst high values may be reassuring, it is not sufficient to rely on the CRI for critical applications, and an actual visual check under daylight remains a sensible precaution.

●4.12 Effect of surroundings

A related but rarely recognized effect of the same kind can arise from the surroundings, not by influencing perception as such but by changing the effective illuminant. Thus, the light reflected from walls, ceilings and floors will have had its spectrum modified by the pigments in the paint and so on (Fig. 3.6). Red walls remove most of the blue from the light, leaving in effect a red illuminant to augment the direct illumination. The overall colour of the illumination thus moves towards the hue of the wall, and clearly will tend to give faulty shade matches (Fig. 4.12). Daylight reflected from hoardings outside the surgery window will also have a **colour cast** which depends on what is currently being advertised there, or affected by whether there are buildings, neon signs, street lighting, trees in leaf, or snow outside – which with a clear sky gives a very blue light, as mentioned earlier. Colour matching can thus be influenced by location, building orientation (*via* sun angle, §4.4), the weather, the seasons, and the time of day. In addition, the use of tinted windows changes the spectrum of the daylight entering. The wearing of brightly-coloured garments by the dentist will similarly modify the light illuminating the patient, as will an aquarium lit with a strong UV component for visual effect, or a TV or computer screen (which tend to be blue-dominant). Indeed, the patient's clothes will also have an effect and, if of strong as opposed to neutral colours, these should be covered with a white or neutral (grey) drape. There is therefore considerable sense in the notion that dental surgeries should be decorated in pale colours, that tunics should be similarly neutral, and that attention be given to possible external interferences if serious colour matching is contemplated, quite apart from using a suitable standard illuminant. Given the increasing emphasis on 'cosmetic dentistry', this seems to acquire greater importance. Of course, in other kinds of practice, orthodontic or paediatric, for example, this might not matter. In summary, the illumination of the object of interest is the sum of all sources of light (Fig. 4.13), direct and scattered, and so the colour of that overall illuminant is dependent, in ordinary colour-mixing fashion, on their colours.

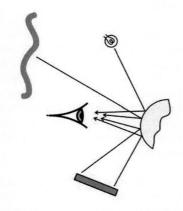

Fig. 4.12 Apparent colour is affected by all aspects of the surroundings and illuminants.

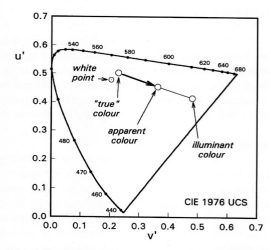

Fig. 4.13 The apparent colour of a tooth like-coloured object is shifted by the addition to the intended white illuminant of some light reflected from coloured surroundings.

These are dramatic effects. More subtle, so that it is more important that it be recognized, is the effect of adjacent soft tissue. Quite clearly the colour of the mucosa will also create colour cast effects. Matching must therefore be done under the normal circumstances of the illumination of the anterior teeth being modified in that way, that is, including that reflected light. Green rubber dam will not help, and even the usual grey will screen the light scattered from oral tissues. Obviously, cosmetic lip colour should be avoided as well. (Incidentally, such colour can become infiltrated into tooth tissue through inadvertent – and frequent – contact, and this can result in odd staining, which of course affects shade matching.)

In this discussion we are referring specifically to the effect of incident light. However, teeth are commonly sufficiently translucent that light transmitted through the tissue affects the perceived colouration, especially at incisal edges. While this is another reason for avoiding rubber dam (or other foreign materials) being present, it should also be recognized that if, say, a porcelain shade guide were not subject to the same retrograde illumination, and of appropriate thickness, it cannot result in a proper match.

●4.13　Perception

There is nevertheless a perception problem as well. We have learnt through long experience the colours of familiar objects: the blue of the sky, the colour of our own skin, the white of paper, and so on. Our brains have therefore adjusted the weight to be afforded each type of retinal cone's signal for a reasonable colour balance in our view of the world. This is known as **chromatic adaptation**.[14] But if, for example, the amount of red light that is being received is greater than usual for any length of time, this is interpreted as a defect in processing and the gain, as it were, of the red channel is decreased in an attempt to restore 'normality' to the scene (this is also due in part to the relative bleaching of the retinal pigments on excessive exposure to more intense colour). After a while we become less aware of the red excess. But if now the illumination is returned to a more usual balance, everything appears distinctly greenish, because effectively the red gain is relatively too low. This effect fades in a few minutes. A similar and common example might be to use a microscope with a green filter in order to improve the contrast of certain images. Each time that one looks up after some extended use the surroundings are suffused with a rosy glow. The wearing of tinted sunglasses produces similar effects. Similarly, under water (scuba diving) where red light is filtered out, the strong blue effect becomes less noticeable with time – photographs taken without flash or filter indicate just how strong this bias is, sometimes surprisingly so.

The point of this is that under a particular illumination, say tungsten filament light, which is rather deficient at the blue end of the spectrum and somewhat overstrong at the red end, the colour balance in our heads will have been adjusted appropriately for 'normality'. Adding this effect to the system compounds the problem of metamerism by changing the way we process the tristimulus information being transmitted by the cones. It is no longer just a matter of the light which is reaching the objects being colour matched, but the general colour balance of the surroundings now has a bearing. This circumstance must be distinguished from that in Fig. 1.4 where illuminants are being compared, but it does account for the colour-shifts discussed under 'Matching persistence' (§4.2).

Despite all of the above, it is important to recognize that there will be variation in shade selection due to observer preferences: nobody is really objective about their appearance. Our memory for colour is also extremely poor, and certainly not an absolute sense, so that attempting to select a well-matching colour without the reference is doomed. Furthermore, traditional and advertising emphases on the virtues of 'white' teeth (teeth are anything but white[15]) bias perceptions even more. The appropriate approach is to recognize the difficulties of the task and arrange the necessary circumstances for doing the best job possible.

●4.14　Aesthetic dentistry

The word 'aesthetic' appears very frequently in dentistry, even occurring in the titles of several journals. It is, however, essentially meaningless in many of these contexts, as it is never applied to the appreciation of beauty (a purely philosophical and personal matter), but rather nearly always to just the appearance of a restoration (and especially as the plural form: "aesthetics of ..."). However, an 'aesthetic' restoration, it would seem, is merely one which has achieved a high standard of similarity to the tooth tissue in or on which it sits. This confusion and abuse is unfortunate as it relates more to advertising than prowess, function, or material properties.[16]

Appearance may be considered from two points of view. Mimicry of tooth tissue, essentially 'invisible repair', is the ordinary common – and quite reasonable – goal, as this seeks to match adjacent tooth tissue in shade, translucency, and so on, and often with heed given to the localized non-uniformities of natural teeth (as in high-quality porcelain work). The undetectability of the restoration is the measure of goodness: **invisible dentistry** the aim. On the other hand, **cosmetic dentistry** (an honest term in itself) is concerned with the reconstruction of the dentition is terms of arrangement (orthodontics), shape (extension for closure of diastemata, edge regularity), or blemish obliteration or whiteness (bleaching; 30§1.3), albeit far too often towards some imagined (and unnatural) paradigm of perfection. This may be confused with (the perhaps simultaneous) repair of congenital or developmental defects, or the repair of function otherwise impaired, or the reconstruction required following surgery or the rectification of pathological conditions, but the purposes are quite distinct: vanity-driven mutilation or exaggeration as opposed to re-establishment of anatomical and functional adequacy related to absence of undesirable side-effects or to quality of life. In contrast, there have been attempts to address appearance from the point of view of aesthetics (properly) in both classical[17] and pragmatic[18] terms.

It should be borne in mind that standards of beauty are not universal, but culturally-determined. There remain societies in which, for example, anterior teeth are filed to sharp points or even removed (particularly in Africa), black teeth are thought to be desirable (in parts of Malaysia, the Philippines, and formerly Japan), societies in which demonstration of wealth is achieved through gold-capped or -inlaid anterior teeth (in China, even if the 'gold' is brass), inlaid diamonds are not unknown, and various other decorations are promoted. While people are usually vain (*i.e.* show appearance-consciousness), to a greater or lesser extent, and some account can legitimately be taken of this in prescribing a treatment, it remains inappropriate, misleading (and a source of bias) to describe any of this as "aesthetic" or involving "aesthetics". The usages are meant to impress, but are essentially hollow and pretentious; their use unthinking and uncritical. Thus, wherever the words appear in dentistry, "aesthetics" can normally be replaced by words such as "appearance", and "aesthetic" by "tooth (enamel)-like" or "cosmetic", with no loss – in fact, with clearer meaning.

Even so, it should still be recognized that the service performance of a material remains a major consideration of ethical dentistry. Thus, for posterior restorations, for example, strength and wear-resistance are extremely important, and may override the demand for tooth-like appearance altogether (silver amalgam, gold) or partially (machined ceramics). Thus, tolerance of non-tooth-like appearance, as for partial-denture clasps, is possible if not common. Cost is also pertinent.

§5. Physics of Light

Colour as such is clearly the most important aspect of matching for a 'tooth-like' restorative material, artificial tooth or denture base acrylic, but there are other aspects of appearance which need to be noted for the full simulation of natural tissue. In other words, there are other optical properties which contribute to the appearance of objects. These factors include **gloss, opacity** and the natural **fluorescence** of tooth substance. The first two items are concerned with the physical effects of the medium on the light, apart from spectral effects. There will also be contributions from the surface texture (20§6) and what may be termed the granularity or scale of internal structure. These are outside the present scope but cannot, ultimately, be ignored.

●5.1 Refraction

On meeting the boundary between two media of differing refractive index, a light ray is both refracted and reflected (Fig. 5.1) and the following familiar relationships apply:

Angle of Incidence = Angle of Reflection

i.e. i = f   (5.1)

and, for the ray crossing the boundary,

$$\frac{\sin i}{\sin r} = \frac{n_2}{n_1} \quad (5.2)$$

where n_1 and n_2 are the refractive indices of the two media (Table 5.1) and the angles are measured from the perpendicular to that boundary, the 'normal'. Thus a ray perpendicular to the surface is not **deviated**. The refractive index of a medium is in fact defined fundamentally by the following ratio:

$$n = \frac{\text{speed of light in vacuum}}{\text{speed of light in medium}} \quad (5.3)$$

In other words, any medium reduces the speed of light in it. Graphically, this can be expressed as a retardation of the wave front (Fig. 5.2). Refractive index varies with wavelength; an effect that is commonly seen to be operating in the **dispersion** of wavelengths by a glass prism to produce a spectrum (Fig. 1.3) or, of course, a rainbow by water droplets. The important point here is that at the boundary between two media (the interface) an incident

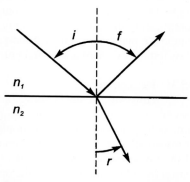

Fig. 5.1 Light is simultaneously reflected and refracted at a boundary between regions of differing refractive index. In this example $n_2 > n_1$.

Table 5.1 *Some examples of refractive index (23 °C; λ = 589.3 nm - sodium yellow light).*

Diamond	2.417
Zirconia	1.98
Alumina	1.76
Hydroxyapatite*	1.649, 1.643
Mylar (PET)	1.641
Quartz*	1.544, 1.553
Feldspar	1.52
Soda-lime glass	1.512
PMMA	1.495
Glycerol	1.474
Water	1.333
Ice (0 °C)	1.31
Air (1 bar)	1.00029

* these materials show double refraction

light ray is deviated from its original path, whether by reflection or by refraction, when the ray is not normal to that boundary.

•5.2   Total internal reflection

For a ray approaching a boundary from the medium with the higher refractive index, n_2, we calculate the angle of the refracted ray by rearranging equation 5.2:

$$\sin r = \sin i \cdot \frac{n_2}{n_1} \qquad (5.4)$$

(*i.e.* swapping n_1 and n_2). From this it is obvious that if the right hand side has a value greater than 1, $\sin r$ is meaningless, *i.e.* the limiting case is

$$\sin i = \frac{n_1}{n_2} \; ; \quad \sin r = 1 \qquad (5.5)$$

so that $r = 90°$. This means that the emergent ray lies in the plane of the interface or, in other words, is tangent to it. So what happens if the right hand side of equation 5.4 is greater than 1? The ray is totally internally reflected (Fig. 5.3). When equation 5.5 is true the angle of incidence, i, is called the **critical angle**. Obviously, such an effect could be important in complicating the path taken by a ray of light through composite materials such as filled-resin restoratives, glass ionomer cement and dental porcelain.

•5.3   Reflectance

However, as stated above, and except for the special case of total internal reflection, light approaching an interface is not simply either reflected or refracted: both usually occur, simultaneously. Part of the ray goes one way, part the other. This is the reason for reflections on water, the operation of 'beam splitters' as used in microscopes, the possibility of so-called 'head up' displays in cars and aircraft cockpits, and the annoying reflections on computer screens referred to as 'glare'.

The proportion, ρ, of light reflected at an interface as opposed to being transmitted, the **reflectance**, is also a function of refractive index and the angle of incidence. Thus, at $0°$ incidence (*i.e.* perpendicular rays) the proportion is given by:

$$\rho_0 = \left(\frac{n_2 - n_1}{n_2 + n_1} \right)^2 \qquad (5.6)$$

For example, for an ordinary glass with a refractive index of 1.5, given that the refractive index of air at standard temperature and pressure is close enough to 1, $\rho_0 = 0.04$. For other angles as i increases the proportion rises steadily to a value of 1, meaning complete reflection, at $i = 90°$, *i.e.* at grazing incidence (Fig. 5.4). There are a number of implications arising from this behaviour.[19]

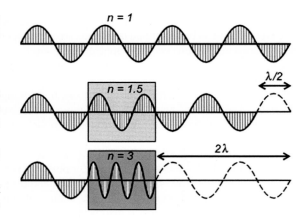

Fig. 5.2 The velocity of light is reduced from its vacuum value in a denser medium according to the refractive index. It is said to be retarded.

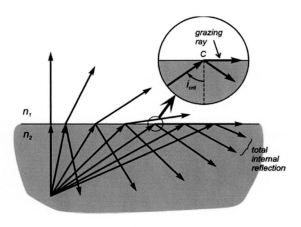

Fig. 5.3   Total internal reflection occurs when the angle of incidence from the medium with the higher refractive index exceeds the critical angle, *e.g.* at *C*.

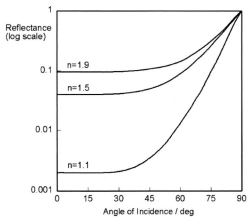

Fig. 5.4 Variation in reflectance of a ray incident on the boundary from air into a another medium with variation in angle of incidence and the refractive index of the medium. n = 1.5 is near the value of that of ordinary soda glass, as used for windows and mirrors, amongst other things.

It should be noticed that reflectance has nothing to do with the colour or transparency of the medium: this is purely a surface phenomenon (see 'gloss', §5.8, §5.9). Black glass is reflective.

●5.4 Multiple internal reflection

The first effect is that of **multiple internal reflection** (Fig. 5.5). Here, the word 'internal' refers to the medium with the higher of the two refractive indices. A light ray reaching the surface of, say, a sheet of window glass is divided into two parts: the reflected ray with relative intensity ρ, and the refracted ray $(1 - \rho)$. This latter will pass through the sheet of glass and suffer some absorption (for no medium other than the vacuum is completely transparent, showing 100% transmission), say by proportion T. Thus, what reaches the far interface is at $T(1 - \rho)$ of the original intensity.

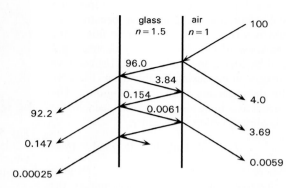

Fig. 5.5 Multiple internal reflections in a plain sheet of glass for near normal incidence. The percentage of the original intensity in each ray is shown assuming $T = 1$ (no absorption).

At that same interface partial reflection again occurs, so that the intensity of that 'internally' reflected ray is $T(1 - \rho)\rho$ of the original, while the emergent ray is $T(1 - \rho)^2$. The intensity of the internally reflected ray reaching the first surface again is then $T^2(1 - \rho)\rho$. The process repeats itself indefinitely, although it is evident that the intensities rapidly diminish. Nevertheless, it is responsible for the slight degradation in the clarity of any image seen through a window, essentially because of the displacement of the second transmitted ray (no matter how high the quality of the glass) and for the slight dimming of the view.

Such reflections also reduce the amount of light in the primary image in telescopes and camera lenses and degrade that image by creating 'stray' light; the more interfaces (lens elements) the worse it gets (and the matte black inside of the containing tube or body still reflects some slight, especially at near grazing angles, Fig. 5.4). This effect can be seen when looking through a stack of 50 microscope slides. The other point to notice here is the pair of *reflected* rays of about 4% of the original intensity. These images are most apparent looking at a window from within a lighted room at night.

●5.5 Mirrors

The same effect is, of course, operating in mirrors, where the highly reflective coating is applied to the second or rear surface (Fig. 5.6). This 'silvering' is indeed usually silver, and is therefore protected from sulphide tarnish on the critical surface by the glass (and by a special paint on the back). The same split of rays occurs at the 'front' surface as above, ρ reflected, $(1 - \rho)$ refracted. As now the reflection, R, at the rear surface is very efficient (for convenience we shall assume 100%, but see Fig. 28§5.1), the ray re-emerging from the front surface has the relative intensity $RT^2(1 - \rho)^2$. After a further reflection from the rear surface, the next emergent ray has the intensity $R^2T^4\rho(1 - \rho)^2$. There are therefore two subsidiary reflection images, each of about 4% of the intensity of the main image. These lie either side of the main reflection, at least when the incident ray is not normal to the surface. Notice that the sequence and values of the intensities are the same as in the plain glass sheet (Fig. 5.5), but 'folded' so as to appear on the same side (ignoring the effects of less than perfect glass transmission T and mirror reflection R).

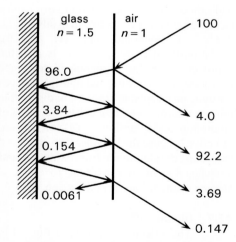

Fig. 5.6 Multiple internal reflections in a rear-silvered mirror for near normal incidence. Ray intensities as in Fig. 5.5; mirror reflectivity $R = 1$, transmission efficiency $T = 1$ (no absorption).

●5.6 Dental mirrors

This would not matter very much therefore in the case of a 'looking glass', when normally one can only view oneself from near normal incidence. However, in the case of dental mouth mirrors, where the object is to get a view of a region not directly visible and therefore larger angles of incidence are involved, the displacement effects are worse (Fig. 5.7). For detailed work, and when the contrasts in the objects may be small, the image degradation is an unacceptable interference. At small angles of incidence, the effect will appear to be a blurring of the main image, but at higher angles a clear separation will occur, and the first 'ghost' become very obvious (Fig. 5.8). A factor in this is the thickness of the glass, both by decreasing T and by increasing the spacing of successive images on increasing the thickness (Fig 5.8), for the necessary strength and rigidity. The successive image spacing d as seen in the viewing direction is given by:

$$d = \sqrt{2}\,t\tan[\sin^{-1}(\sin i / n)] \quad (5.7)$$

where t is the glass thickness. (Higher refractive index for the glass makes the separation less but the ghost image strengths greater.)

In a looking glass the question of the weight of the mirror and its strength determines that the thickness must be relatively large (cheap, large, thin mirrors can distort noticeably), although in a dental mirror of a diameter of about 20 mm this might not be so significant; clearly there is a practical limit. Even so, rear surface mirrors are not good enough here because the glass cannot be made thin enough without making them too fragile or at risk of distortion.

The solution is to use front surface mirrors, eliminating both absorption dimming and secondary images. Because the metal surface is now exposed, silver cannot be used and rhodium is the coating of choice (although it is less reflective than silver, Fig. 28§5.1), a moderately hard ($H_v \sim 120$), oxidation-resistant metal. This is also the approach taken for astronomical telescope mirrors where dimming and multiple images are quite unacceptable (28§5). For disposable mirrors a simpler solution exists: a coating of aluminium on Mylar film (27§4) as a rear surface mirror is effective because the polymer film can be very thin while protecting the aluminium from oxidation, and it can also be stretched taught over a frame to create a flat enough surface, if the frame is rigid enough.

•5.7   Surface effects

The reflection occurring in Fig. 5.1 is envisaged as being from a plane surface. Naturally, such reflection is a 'local' phenomenon, and on a rough surface each locality will present a different angle to the rays of an incoming beam. The reflected rays may therefore be at any angle with respect to the mean surface (*cf.* 20§6). Similarly, the refracted rays may be at a wide range of angles. On the one hand we have **specular** ('mirror-like') reflection and transmission through a **transparent** medium, in which there is no scrambling or confusion of a beam and images are coherent: the obvious example of transparency is plain

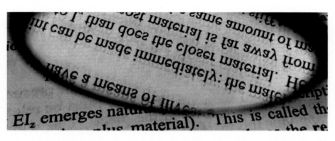

Fig. 5.7 Double image in a back-silvered glass mouth mirror. Variation in the strength of the first ghost with angle is evident. (The third image, above the main one, is much weaker because of the viewing angle – see Fig. 5.8.)

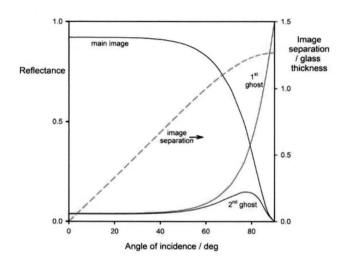

Fig. 5.8 The relative strengths of the ghost images depend on the viewing angle, with the first ghost being similar to the main image at ~80° incidence. The apparent separation is proportional to the glass thickness. (Calculated for $n = 1.5$.)

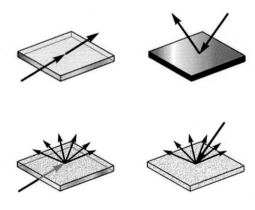

Fig. 5.9 On reaching a boundary, light may be transmitted (left) or reflected (right), either in a specular (top) or a diffuse (bottom) manner, or a wide range of intermediate conditions.

window glass. On the other hand we have **diffuse** reflection from a **matte** surface and transmission through a **translucent** material, in which much scattering of the light occurs (Fig. 5.9) and no image information is preserved, such as with etched or grit-blasted glass. This scattering may be from macroscopic detail, such as an irregular surface (Fig. 5.10), but it is also affected by small particles of other phases which bend the light one way or another (Fig. 5.11), as well as bubbles of gas or liquid within the body.

●5.8 Size limit

For scattering to occur requires that the object has a size greater than ~½ the wavelength of the light, otherwise it simply is not 'seen'. This does not prevent **absorption** by dye molecules, for example, which clearly are much smaller than that criterion. One of the implications of this limit is that while a rough surface will scatter light, if the scratches are made smaller than ~½ wavelength the surface will appear polished and show specular reflection or transmission. Since blue light has a wavelength of about 400 nm, or 0.4 μm, a surface with scratches of less than about 0.2 μm will appear perfectly polished, even under a microscope, if it uses light for imaging.

The same applies to 'micro-filled' resins (6§3.2), where the average particle size of the fumed silica filler used ranges from about 7 to 50 nm, so that there is no scattering of visible light. Thus, unless other particulate phases are included on a large enough scale, or agglomerations are present (Fig. 4§7.11), such materials would be transparent. The effect is also seen with bubbles in porcelain (25§4.3).

The reflectivity of smooth surfaces is known as **gloss**. It is an important property of paints and other decorative finishes, but it is also an important subjective means of assessing the quality of polishing, as in the finishing of dental restorations and prostheses. Gloss can be seen to be a mixed rather than fundamental property in that it combines the effects of lack of scattering due to roughness and refractive index, but it is also dependent on the angle of view (Fig. 5.4). If a 'glossmeter' is used for such assessments the angle of illumination and view must be specified.

●5.9 Chromaticity shift

For an apparently opaque object to return light to the eye, and for the effects of differential absorption to be seen, *i.e.* for it to have colour and not a metallic appearance, the light must in the first instance have penetrated the surface, and then be scattered back out. However, as described above, there is always some superficial simple reflection. The importance of this is that the light received by the eye from any given non-metallic surface is always a mixture of the internally scattered and the superficially reflected. The superficially reflected light has not undergone any spectral modification by absorption within the medium; it is spectrally identical to the incident light. This then is light added to or mixed with the 'coloured' light, representing the result of differential absorption, emerging from within the object. This mixture necessarily gives a less-saturated colour to the object

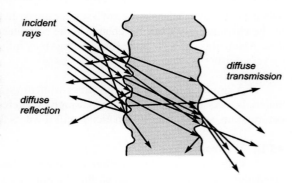

Fig. 5.10 Scattering may be due to an irregular surface ...

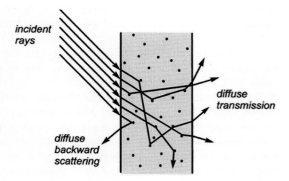

Fig. 5.11 ... or due to small particles. Compare the effect of dust in the air (Fig. 4.2).

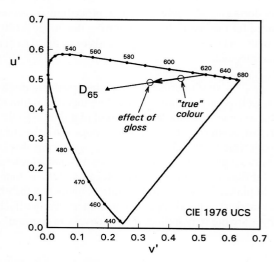

Fig. 5.12 The chromaticity shift that may be observed with daylight on a reflective red object.

than it would have in the absence of a surface gloss. In terms of the chromaticity diagram this effect is expressed by the **chromaticity shift**, the change in the coordinates towards those of the illuminant (Fig. 5.12), whatever it is. Again, this shows that the surroundings may influence the perceived colour of an object if it is glossy, but by yet a further mechanism.

A similar effect is seen with stress-whitening (5§5.4, 7§2.4) in what are otherwise relatively translucent materials. As incident light is increasingly scattered from within by the structural changes occurring on deformation, so there is a chromaticity shift towards the illuminant, which is of course usually "white".

●5.10 Wet roughness

Only a very small proportion of a rough surface will lie parallel to the nominal or mean surface. Thus, no matter what angle of incidence the illuminating rays may have to that mean surface, locally many will effectively have high angles of incidence. Therefore, no matter what the angle of view of such a surface, and no matter what the angle and direction of illumination, some superficial reflection in the view direction will occur. Under ordinary illumination conditions rough surfaces therefore tend to appear whitish, and not to show the real colour of the underlying bulk material. Hence the "frosty" appearance of etched enamel.

However, if a film of liquid which wets the solid is spread on it, the rough surface is now against a medium of much higher refractive index. For example, applying equation 5.6 to water on PMMA (see Table 5.1), ρ_0 = ~0.003 (compared with 0.04 before) so very little reflective scattering now occurs at that interface. On the other hand, for the air-water interface, which is necessarily smooth because of the surface tension effect (10§1), ρ_0 = ~0.02 and glossy reflection will dominate. However, and most importantly, it becomes easy to find a viewing angle that avoids glossy highlights obscuring the underlying material, and its colour becomes obvious. In addition, if the PMMA is **transparent**, not containing any scattering or absorbing particles, an image viewed through the rough but wet surface will be clear, while the dry surface will scatter so much as to destroy the image. The same applies to water on ground glass, and even oil on paper, which becomes much more **translucent** by reducing the scattering at each fibre. This was the basis of the old 'grease-spot' photometer of school physics classes.

In passing, it is just this air-polymer interface reflection that causes crazes in PMMA to be visible (§5.9, 5§5.4), even if they are very narrow. Similarly, the presence of bubbles in denture base resin through faulty mixing or processing causes less translucency, an opaque and milky appearance. Likewise, the elimination of the porosity of porcelain on sintering (25§4.1) converts the mass from an opaque to a translucent body. Note that even black glossy materials show roughened or scratched areas as grey or whitish because of such scattering of ambient light.

The same kind of effect is operating when wax 'polish' is applied to a surface, or a varnish is used: the roughness is filled in by a high refractive index medium that is then given – or generates – a smooth surface. It is also effective in the context of a glaze on dental porcelain, even though the primary purpose is strengthening (25§5.1). The converse effect must also be borne in mind. If the roughness of a surface is to be assessed by eye, as during an (abrasive) polishing procedure, it is essential that this be done on the dry surface, irrespective of the need for coolant or lubricant during that process (20§2.6, 20§7). The same applies when examining an etched tooth surface, for example; it needs to be dried to determine whether the extent and depth of etching are in fact as intended.

A similar effect is seen when teeth are dried. Tooth enamel is naturally slightly rough (through wear) as well as very slightly porous. The superficial film of saliva or water dominates the glossy appearance, and thus the associated chromaticity shift of highlights (§5.9). Thus, when this film is removed, although the gloss effect is lost, scattering at the roughness means that the chromaticity shift is more generalized, that is, the tooth appears both whiter and lighter. Further drying means that liquid in the porosity is lost, and progressively deeper. The extra scattering from the numerous air-enamel interfaces amplifies the whitening and lightening as the amount of differential absorption is reduced and the colour of the scattered light more nearly approaches that of the illuminant – assumed to be white. Obviously, the effect is similar to that of etching, mentioned above, but not as marked. This change in appearance clearly affects shade-matching (§4.11), and must be taken into account by allowing rehydration. Full recovery of 'natural' colour may take 20 ~ 30 minutes.

●5.11 Composite materials

The various phases present in translucent composite materials almost inevitably will have differing refractive indices. Indeed part of the design of restorative materials such as filled resins and porcelain takes this into account specifically to achieve the desired effects. This means that at every interface encountered there will be reflection and refraction. Since those deviated rays will themselves be subject to further deviation at successive interfaces, the net effect is for light to be scattered diffusely (Fig. 5.13). The path taken by any ray will therefore be complicated. In addition, there will be differential absorption if any phase is 'coloured'. The optical appearance of the object therefore depends on the refractive indices and absorption characteristics of each phase present.

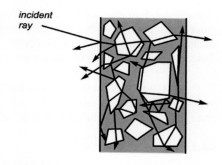

Fig. 5.13 Many light paths are possible in a translucent composite medium through reflection and refraction when the refractive indices are different.

This effect is also seen in single-component polymers that are not completely amorphous (3§4.1). Here the refractive index varies from place to place as the crystallinity of the polymer varies, and thus the density. This variation in density is gradual rather than abrupt, resulting in the light path bending rather than being kinked, but the overall effect is the same. Polyethylene in very thin films is evidently quite transparent ("cling" film), but as the thickness increases so first a slight image distortion may be detectable ("polythene" bags), then increasing cloudiness (bottles), finally becoming quite opaque in thick sections (chopping boards). Where clarity is important, a polymer that might otherwise tend to crystallize (say, on cooling slowly after melt processing) may be modified as a copolymer simply to increase the irregularity of the chain and better preserve the amorphous structure.

However, the refractive index of the matrix of filled resins depends on the polymerization shrinkage *via* degree of conversion. That is, it changes continuously during the curing process. Since it is the difference in refractive indices that affects the deviation and scattering, depending on how the filler is chosen, and the chemistry of the resin, the effect may increase or decrease the translucency. This then affects how underlying material affects the apparent shade (§5.12). This is in addition to the shade shift that occurs as the diketone is consumed (6§5.16). Hence, it is important to rely on the manufacturer' shade guide (§4.11) and ignore the initial appearance of the unirradiated material. However, there is one context where this is not so reliable.

If a light-cured resin is intended for luting a thin porcelain shell crown or veneer, using transillumination to effect the cure, shade selection is much harder as the combined effect of the three materials – tooth, lute and ceramic – is difficult to judge. Even so, trying the restoration in place with the proposed luting material cannot give the desired effect because its translucency and colour will change when cured. In any case, the sensitivity to ambient light is such that it would have partially set by the time removal was attempted (and making that removal very difficult). It has been suggested that a special try-in filled resin be used, one that cannot set on ambient irradiation (no diketone, or no amine) but which has the actual optical properties of the corresponding shade of set material (by adjusting the refractive index of the matrix). Thus, if a different shade needs to be checked it can be done readily because the try-in material is easily removed. Even so, if any slight remnants of this remain they will be bonded to the lute because the same polymerizable monomers are present. Thus, an elaborate clean-up is avoided (no solvents for set material), minimizing the risk of damage to the ceramic device.

The alternative approach of the so-called try-in "gel" (see box in 7§8; §5.14), based for example on glycerine and fumed silica, cannot replicate the full optical characteristics of a filled resin and may present problems in clean-up and drying. The silanation of the porcelain may be affected, requiring it to be redone.

●5.12 Chameleon effect

In this sense of translucent scattering, dental restorative materials are designed to mimic tooth tissue, whether enamel (whose hydroxyapatite crystals are embedded in a protein matrix, no matter how little – the interfaces still exist) or dentine, where the tubules normally contain a more watery medium or cell substance as well as having a mineral-protein composite part. Thus, teeth are not glass-like and transparent. On the other hand, the translucency does mean that light may reach deeper sites after passing through tooth or restoration and then be returned by scattering to emerge at the surface again. If those deeper sites are discoloured or stained then the effect will be seen as a general shift in the colour of the tooth or restoration. To some extent this chromaticity shift (§5.9) is thought to be beneficial in that a restoration "tends to take on" the colour of the adjacent tooth (and *vice versa*), perhaps compensating for a not quite exact shade match in the first place. (Needless to say, this should not be relied upon to excuse less care and attention.) It is sometimes referred to as the **"chameleon effect"** (notwithstanding the fact that chameleons operate in a quite different way! It should not be thought that this is other than a purely passive, entirely physical effect). However, it may also be detrimental if is the intention that a stain is to be masked. Thus, an opaque liner for a resin or glass ionomer restoration may be necessary, or an opaque layer in a porcelain restoration used to advantage. For porcelain on metal devices, this is essential.

If it were desirable to 'switch-off' the effect, so as to remove the influence of adjacent temporary, metallic or discoloured restorations, a mirror strip would need to be inserted so that all reflected light was the same as that emerging. This might be done with a 'space blanket' material, aluminium-coated Mylar film, for example.

●5.13 Glass ionomer ageing

A related effect is seen in glass ionomer cements. These, being composite, consist of a glass core and reaction product matrix, and clearly these will have different refractive indices (Fig. 9§7.6). Initially, when freshly set, these materials tend to be rather less translucent than they will eventually become when reactions are more complete. This extra scattering, and therefore relative opacity, is due in part to the magnitude of the difference. But, as the reaction proceeds, and more glass has been dissolved, the boundaries between the unaffected glass, its hydrated silica coat, and the matrix proper become blurred. There is then more a gradient of refractive index than a sharp interface. Reflection scattering therefore becomes less. It is thus necessary to ignore the fact that initially the match does not appear particularly good, and indeed to place one's confidence in the shade guide to indicate the final appearance correctly. The same problem existed with silicate cements (9§7).

●5.14 "Gel" toothpaste

Some toothpastes are described as "gels", apparently on the grounds that they are transparent. Such materials may well be plastic (4§7) in that they have a yield point, but there is no evidence that they are gels (7§8, 7§9) in the sense of having a network structure. This is a common use of the term gel, *i.e.* for viscous, possibly plastic, and transparent materials and it also turns up in cosmetic contexts. It is therefore an unhelpful usage. Nevertheless, the transparency of the toothpaste in the context of what must still by definition be an abrasive material requires consideration. It should now be obvious that all that is required is that the abrasive material itself is transparent and is suspended in a medium whose refractive index has been matched to that of the dispersed abrasive – which is therefore invisible as it does not scatter light by refraction.

Similar effects are found for dentine 'cleared' by soaking in methyl salicylate (*cf.* Fig. 9§10.1) when refractive-index matching makes demineralized tooth roots more or less transparent (Fig. 5.14).[20] Likewise, 'white spot lesions' of enamel can be masked by infiltrating with a polymerizable resin precursor,[21] reducing the scattering of ambient light by reducing the refractive index mismatch from that of hydroxyapatite against saliva. The spot is 'white' because of the chromaticity shift to be expected (§5.9). Of course, if the match were perfect the spot would become translucent, *i.e.* 'cleared', the ideal condition that of matching the refractive index of enamel matrix, not that of its mineral component.

Fig. 5.14 'Cleared' demineralized tooth roots – refractive index-matched with methyl salicylate. Photograph courtesy of S. Rosler.

●5.15 Curing lights

The reflectance effects described in §5.3 have a bearing on the efficacy of curing light irradiation for filled resins. Primarily, it is the loss at the upper surface of a Mylar or similar matrix strip that will be apparent, but if there is an air-bubble underneath, that interface will also contribute, along with the top surface of the resin itself. If, as is now common, a polyethylene sheath is used for infection

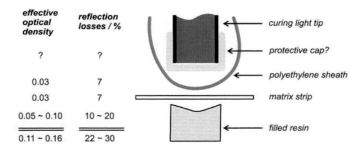

effective optical density	reflection losses / %	
?	?	← curing light tip
		← protective cap?
0.03	7	← polyethylene sheath
0.03	7	← matrix strip
0.05 ~ 0.10	10 ~ 20	← filled resin
0.11 ~ 0.16	22 ~ 30	

Fig. 5.15 Resin curing light reflection losses associated with the use of matrix strips and other accessory materials.

control, both of those surfaces must also be counted. Indeed, if a protective cap is used to protect the glass-fibre exit window of the curing light, this too will contribute. (Bear in mind that the final figures are the products of the transmissions, they are not additive except as effective optical densities; *cf.* 26§5.7.) The figures given in Fig. 5.15 are for normal incidence, but as it is likely that a greater angle will be involved in many cases, the losses may increase substantially (Fig. 5.4). If, however, an oxygen barrier material is used (6§6), that is, instead of the matrix strip, and care is taken to fill the space between the sheath and the resin without bubbles, at least two surfaces can effectively be eliminated because of the better refractive index matching (Table 5.1). In principle, using such a high-refractive index fluid under the cap (if any), and between the cap and the sheath, will all but eliminate those losses.

§6. Chemistry of Colour

As has been discussed above, colour commonly arises from differential absorption, assuming that we are given white (*i.e.* spectrally uniform) illumination. It is the range of mechanisms of that absorption that we now address. There are also light emission processes that sometimes also have to be taken into account. In summary there are the following classes of system:

- dyes
- pigments
- spectral effects – diffraction, interference (which, however, have little relevance to dentistry)
- emission – fluorescence.

Colour arises from a number of fundamentally different types of process at the atomic or molecular level, resulting in the selective absorption or emission of light at particular wavelengths. Electromagnetic radiation in general interacts with matter through a variety of mechanisms (Fig. 6.1).[22] These interactions are due to changes in nuclear and electronic spin, molecular rotation and distortion (bond angles and lengths, but in the sense of the amplitudes of vibration), or electron redistribution. Which of these may occur depends on the energy of the radiation concerned, because each change is associated with a particular scale of energies. In addition, radiation is quantized, and close matching of the incoming radiation to the energy of the transition concerned is necessary. In particular, the absorbed quantum must have *no less* than the necessary energy. But, as can be seen from Fig. 6.1, only transitions in outer shell electrons are associated with energies equivalent to visible and ultraviolet light. These affected electrons are the valence electrons, or electrons in molecular orbitals.

It should be noted that all of these processes are independent, meaning that in a mixture of several dyes or pigments, or both, each makes its own separate contribution to the perceived colour (unless, of course, there is a chemical interaction).

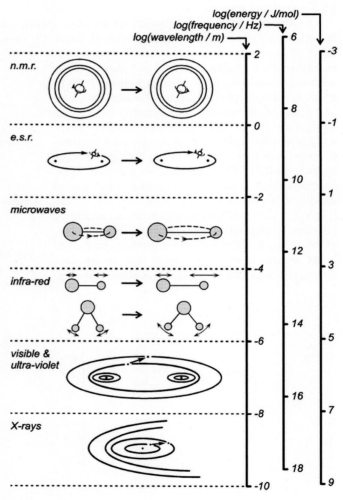

Fig. 6.1 The various possible atomic and molecular energy absorption processes for electromagnetic radiation. Only outer shell and molecular orbital electronic transitions are associated with colour. NB: the boundaries are not sharp and exact, as shown; there is considerable overlap at each, because while the processes are distinct their energies may have considerable range, depending on the chemistry.

●6.1   Electronic transitions

In absorbing a quantum of radiation, an electron is promoted to a higher energy level. In atomic or molecular terms, this means promotion to a higher energy orbital.[23] In organic materials, such as the dyes used in colouring polymers and fabrics, only certain kinds of transition are allowed (Fig. 6.2): σ-bonding electrons may be promoted to σ*-antibonding orbitals (Fig. 6.3, left), π-bonding electrons may be promoted to π*-antibonding orbitals (Fig. 6.3, right), and electrons in non-bonding orbitals, n, may be promoted either to σ* or π* (Fig. 6.2). Notice that the transition σ → σ* is represented here as the highest energy type of transition. Were this to happen it would in fact mean the rupture of a molecule (if that were the only bond), the σ-bond being the primary bonding between two atoms. On the other hand, n → π* transitions are typically of the lowest energy and such transitions may easily cause absorption in the visible range, causing 'colour', whereas σ → σ* typically requires the energies associated with ultraviolet light.

●6.2   Organic systems

The energy of π → π* transitions depends, amongst other things, on the extent of **delocalization** of the electrons.[24] Thus, in a series of fused rings (Fig. 6.4), where the number of energy levels available increases and their spacing decreases, transitions from π-bonding (lower half) to π*-antibonding (upper half) orbitals become both more likely and of successively lower energy (ΔE). The absorptions due to these transitions therefore move steadily into the visible range with increasing numbers of rings. Another example is the conjugated 'ene' system $Ph\text{-}(CH_2\text{=}CH_2)_n\text{-}Ph$ (Ph = phenyl): as n increases from 1 to 7 the absorption band moves steadily from the ultraviolet range into the visible (Fig. 6.5). Any side groups which can act as part of resonant conjugated systems, such as $-NO_2$, $>C=O$, $>C=S$, $-NH_2$, $-C\equiv N$, as well as many others, may act to lower energy levels still further and so increase absorption. Such groups are known as **chromophores**.

The energy E (J/mol) of e.m.r. quanta can be calculated from:
$$E = N_A.h.c\,/\lambda$$
where N_A is the Avogadro constant, h is Planck's constant, c the speed of light and λ is the wavelength. (See also 26§2.1)

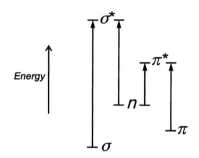

Fig. 6.2 The types of electronic transition available in an organic molecule.

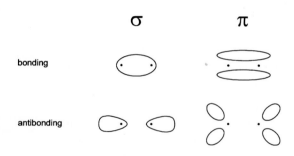

Fig. 6.3 Approximate shapes of bonding and antibonding molecular orbitals for σ (left) and π (right) bonds.

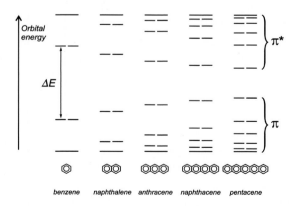

Fig. 6.4 Energy levels of the π and π* orbitals of a series of fused ring compounds. All π-bonding orbitals are filled in these compounds.

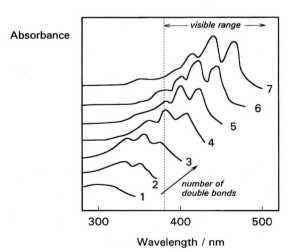

Fig. 6.5 Absorption spectra for the series of compounds with conjugated double bonds $Ph\text{-}(CH_2\text{=}CH_2)_n\text{-}Ph$, with n = 1 to 7.

These extensive **conjugated** systems are the basis, for example, of the many pH indicators as well as virtually the whole of the dye-stuffs industry. In particular, they are the colorants used in dental polymers such as for denture bases and restorations.

We may note here that such absorptions first occur at the blue end of the spectrum so that the colour of the substance tends to be yellowish. It is this effect that causes the yellowing discolouration of resin restorations, for example, when continued (but undesirable) reactions create more strongly absorbing molecules, such as with aromatic amines and eugenol (6§1.1, 9§2.3).

●6.3   Colour stability

An unfortunate effect of the absorption mechanism is that the excited, higher energy state so created is reactive. Although the extra energy may decay away thermally (*cf.* photosensitizers, 6§5.2) through collisions, there is always the possibility of reaction with another molecule. This may create the coloured substances referred to just now, but equally, if the absorbing molecule is large and complicated (as dyes tend to be in order to achieve their strong visible light absorptions) the special structure may be spoilt, bleaching it. This is especially so in the presence of oxygen (which is difficult to avoid in the dental context), so that colours may fade with time. The oxidation that occurs is usually of the unsaturated, and therefore reactive, chromophores. This is very obvious on advertising posters exposed to sunlight (with its ultraviolet component). Dyes which are prone to this kind of effect are called **fugitive**. Taken together, the colour stability of resin restorative materials is a considerable challenge.

In some polymer systems the problem of UV bleaching has been addressed by the incorporation of (colourless) efficient UV absorbing compounds. These dissipate the energy of the radiation without involving the breakdown or activation of the dye molecules. These compounds are known as **UV stabilizers**.

●6.4   Metal ion complexes

The generation of colour in metal oxides and similar systems again depends on changes in electron energy, but of a different type. The non-transitional elements, groups IA, IIA, and IIIB to VIIB, have electron configurations which may be written as: [(core), ns^y, np^z], where n is the first quantum number. The total number of valence shell electrons is $y + z$ and has the range 1 - 8. In transition elements the electron configuration is: [(core), $(n-1)d^x$, ns^y], $y = 1$ or 2, $x = 1 - 10$. When the d shell is incomplete there is a very much larger number of electrons available for bonding than in the non-transition elements, and many more orbitals are possible, with smaller energy differences between them. This increased scope of bonding capacity is expressed in the chemistry of these metals by the ready formation of **complexes**, compounds in which the metal ion is bound tightly to a group of **ligands**, often in an octahedral arrangement (*cf.* Ca^{++} in its chelates, Fig. 9§8.2), *i.e.* with a coordination number of 6. The ligands themselves may be either negatively charged ions or species such as H_2O and NH_3 which can donate electrons to the bonding. (The lone pairs of these molecules are the source and are highly polarizable, 10§3.3.) As these ligands provide or modify an electrostatic field around the metal ion, so the energies of the d-orbitals themselves, originally all equal (*i.e.* **degenerate**), are changed.

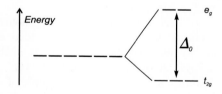

Fig. 6.6 Ligand field splitting of the energies of the d-orbitals in an octahedral field (6 coordinating ligands).

As an example: energy calculations based on the symmetry of octahedral complexes lead to the division of the five d-orbitals into two distinct groups (Fig. 6.6). The separation in energy between them, Δ_0, depends on the **electrostatic field strength** of the ligands (*cf.* equation 9§8.1), but typically it is rather small.[25] Ligands may be arranged in a **spectrochemical series**, indicating their relative strength in this regard:

$$I^- < Br^- < Cl^- < F^- < OH^- < H_2O < NH_3 < CN^- \quad (6.1)$$

This so-called **ligand field splitting** of the d-orbital energies results in absorptions in the visible and ultraviolet, as when these orbitals are incompletely filled an electron may be promoted to a higher level (Fig. 6.7). Thus, in a complex such as the hydrated ion $[Ti(H_2O)_6]^{3+}$

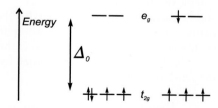

Fig. 6.7 For a d^4 transition metal ion the ground state (lowest energy) is as on the left. The promotion of an electron to a higher energy by the absorption of radiation leads to the configuration on the right.

in which there is only one d-electron, the absorption spectrum shows a band corresponding to the promotion energy from the lower to the higher level (Fig. 6.8), the peak absorption lying at about 500 nm (bluish-green). This complex is therefore in fact purple (red and blue wavelengths predominating).

The same principles apply in the solid state, and no less in glasses, irrespective of the lack of crystallinity. The environment of the dissolved metal ions (as in the porcelains) will tend to be similar to that in solutions and crystals, forming coordination compounds, and so similar absorptions may occur generating colour. The oxygen ions surrounding the metal ion in oxides and silicate glasses will have similar ligand field strengths to the OH⁻ and H_2O ligands in the spectrochemical series, but the colours will vary according to the metal and its oxidation state (i.e. the number of d-electrons). Small changes will be observed due to other factors such as the glassy nature of the matrix, the temperature, and whether other anions are present, such as F⁻, which may arise from fluxes and which may lead to various mixtures of ligands around the metal ions.

The same remarks apply to the colour of the corrosion products of dental and other alloys. With but few exceptions the components of these alloys are transition metals. Strong absorptions are therefore expected, given the presence of strong ligand fields due to sulphide (S⁼), oxide (O⁼) and hydroxide (OH⁻), modified by the presence of other ions found in saliva, such as chloride, sulphate, phosphate and so on. The colours of ferric oxide (Fe_2O_3), silver and mercuric sulphides (AgS, HgS) in particular, are well known and very strong.

●**6.5   Fluorescence**

As was discussed under photosensitizers (6§5.2), absorbed light may not be degraded entirely to thermal energy but may be re-emitted as visible light, although always at longer wavelengths (i.e. lower energy).[26] Anthracene (Fig. 6.4) is an example of a compound which fluoresces under some conditions. Colourless by absorption (Fig. 6.9), if the energy is not rapidly dissipated by collisions some energy may be emitted in the visible region, so long as the illumination has some ultraviolet or far blue light in it. This is the basis of the dyes, histological "stains", used for fluorescence microscopy. Similarly, some transition metal complexes may show fluorescence. The corresponding spectra for a solution of a deep red [Ru²⁺] complex are shown in Fig. 6.10. Notice that both the absorption and the emission occurs in the visible region, and so in this and similar cases the colour perceived will be modified by what we may call the self-luminance.

Hydroxyapatite is the basis of some commercial phosphors (see Table 4.1: Ca-phosphate and Ca-Sr-phosphate) which are characterized by the use of small amounts of transition and other metal ions substituted for Ca⁺⁺, as well as various anions substituted for OH⁻. These together give a variety of environments in which electronic transitions may be made and so a variety of absorptions, but notably there will also be fluorescent emission spectra.

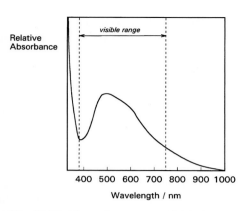

Fig. 6.8 The absorption spectrum for the aqueous ion $[Ti(H_2O)_6]^{3+}$.

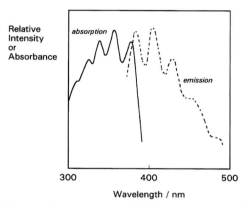

Fig. 6.9 The absorption and fluorescence emission spectra for anthracene. Note that the absorption is at wavelengths shorter than the visible range.

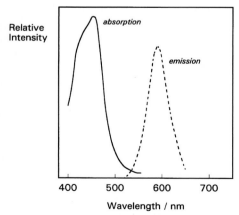

Fig. 6.10 Absorption and fluorescence emission spectra for an aqueous solution of a ruthenium complex.

Hydroxyapatite, as the basis of tooth substance, is not a pure substance. It will contain various contaminating ions (fluoride included) and is the natural counterpart of the synthetic phosphors. Its fluorescence therefore under normal lighting conditions, in particular in sunlight with its significant ultraviolet component, is important in modifying its appearance. As a result of this type of behaviour it is plain that, in any attempt at a close simulation of tooth material, its own natural fluorescence ought be taken into account. At one time uranium salts were used to lend fluorescence to dental porcelains. Not unnaturally, this practice came in for some criticism because of the associated radioactivity, and their use was dropped (25§7).

●6.6 Dental dyes and pigments

The distinction between a dye and a pigment is that the former is a coloured molecular structure, where in solution or bound, while a pigment is a dispersed particulate material which is essentially opaque. While there are many dyes used to colour, for example, dental waxes, their particular chemistry is of little importance except to note that they must all be 'oil soluble' or hydrophobic, firstly so that they can dissolve in such substances, but also so that they do not leach into the surroundings.

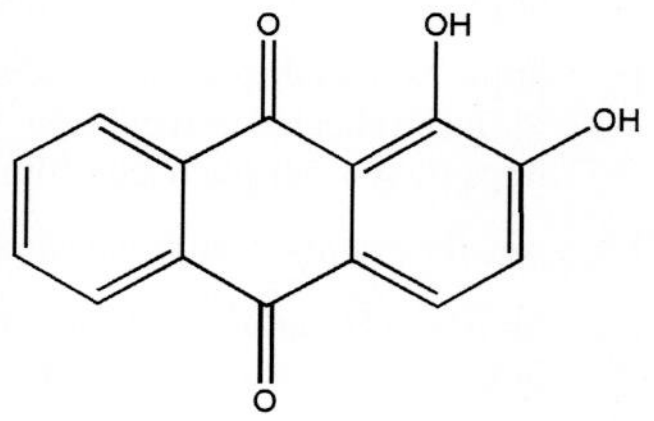

This requirement not to leach is also a concern when highly-polar or hydrophilic systems are involved. For example, dental plaster, stone and die-stones are frequently coloured so that they may be easily distinguished from each other in models, while impression plaster is commonly pink. This is done with a dye that reacts to become an insoluble substance known as a **lake**, by chelating with a polyvalent cation. Alizarin (Fig. 6.11) is the archetypal member of a very large family of such dyes, and widely used in dentistry, in which the key structural aspect is the adjacency of the hydroxyl and carbonyl groups. The calcium ions of gypsum products are functional in this respect (Fig. 6.12; cf. Fig. 9§8.1), the chelate being sufficiently strong that the colour does not bleed or run into adjacent material when the model is cast on the impression. Indeed, this class of dye is very commonly used as an histological calcification stain, where presumably the binding is to the surface of the calcification in much the same way as polycarboxylic acids (9§6, 9§8.7).

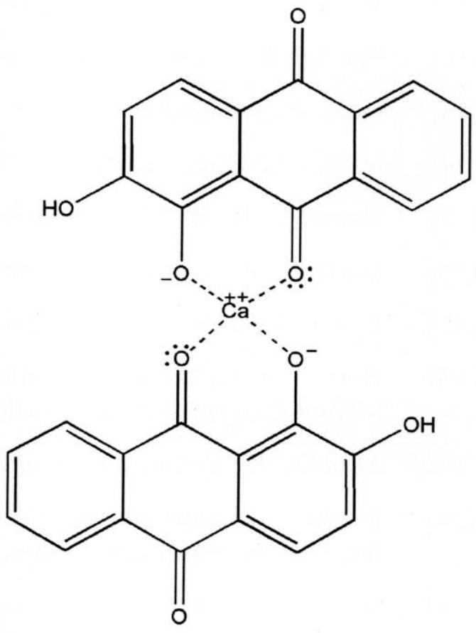

The multiple hydroxyl and carbonyl groups means also that such dyes bind strongly to polar materials such as the proteins and polysaccharides of dental plaque and so are the common basis of disclosing agents. However, there is some risk of such materials dissolving in the resin of restorative materials leading to a chromaticity shift that would spoil any colour match. One would also expect chelation binding reactions to occur on the surface of glass ionomer cement, where calcium and aluminium ions, in particular, are present, if the stereochemistry of the relevant groups was favourable. [27]

Fig. 6.11 Alizarin.

Fig. 6.12 Alizarin complex with calcium – the insoluble lake. Water molecules may also be coordinated to make such complexes octahedral.

References

[1]   http://en.wikipedia.org/wiki/Color_blindness

[2]   Maxwell JC. On the theory of compound colours, and the relations of the colours of the spectrum. Phil Trans 150: 57 - 84, 1860

[3]   Clulow FW. Colour: Its Principles and their Application. Fountain, London, 1972.

[4]   https://www.surreynanosystems.com/vantablack

[5]   https://www.labsphere.com/labsphere-products-solutions/materials-coatings-2/targets-standards/diffuse-reflectance-standards/diffuse-reflectance-standards/
        https://www.labsphere.com/site/assets/files/2553/a-guide-to-reflectance-materials-and-coatings.pdf

[6]   see, for example, http://www.colorsystem.com

[7]   Smithson HE *et al.* A three-dimensional color space from the 13th century. J Opt Soc Amer A 29(2): A346-A352, 2102
        http://www.grosseteste.com/cgi-bin/textdisplay.cgi?text=de-colore.xml

[8]   Henderson ST. Daylight and its Spectrum. 2nd ed. Hilger, Bristol, 1977.

[9]   Ridpath I & Tirion W. Collins Guide to the Stars and Planets. Collins, London, 1984.

[10]   https://en.wikipedia.org/wiki/Planckian_locus

[11]   Wyszecki G & Stiles WS. Colour Science: Concepts and Methods, Quantitative Data and Formulas. Wiley, New York, 1967.

[12]   Lee YK & Powers JM. Metameric effect between resin composite and dentin. Dent Mater 21: 971 - 976, 2005.

[13]   https://en.wikipedia.org/wiki/Color_rendering_index

[14]   https://en.wikipedia.org/wiki/Chromatic_adaptation

[15]   Joiner A. Tooth colour: a review of the literature. J Dent 32: 3 - 12, 2004.

[16]   Darvell BW. Esthetic Dentistry. Amer J Esthet Dent 3 (3): 167 - 168, 2013.

[17]   Levin EI. Dental esthetics and the golden proportion. J Prosthet Dent 1978;40:244–52, 1978.

[18]   Burke FJT, Kelleher MGD, Wilson N & Bishop K Introducing the concept of pragmatic esthetics, with special reference to the treatment of tooth wear. J Esthetic Rest Dent 23 (5): 277-293, 2011.

[19]   Judd DB & Wyszecki G. Colour in Business, Science and Industry. 3rd ed. Wiley, New York, 1975.

[20]   Rosler S. Transparent teeth: A powerful educational tool. Roots 4: 30 - 31, 2010.
        http://www.oemus.com/archiv/pub/sim/ro/2010/ro0410/ro0410_30_31_rosler.pdf

[21]   S. Parisa, F. Schwendicke, J. Keltscha, C. Dörfera, H. Meyer-Lueckel. Masking of white spot lesions by resin infiltration in vitro J Dent 41, Suppl 5, e28–e34, 2013.

[22]   Banwell CN. Fundamentals of Molecular Spectroscopy. McGraw-Hill, New York, 1966.

[23]   Dyer JR. Applications of Absorption Spectroscopy of Organic Compounds. Prentice-Hall, New Jersey, 1965.

[24]   Smith LO & Cristol SJ. Organic Chemistry. Reinhold, New York, 1966.

[25]   Heslop RB & Jones K. Inorganic Chemistry. A Guide to Advanced Study. Elsevier, Amsterdam, 1976.

[26]   Barrow GM. Physical Chemistry, 4th ed. McGraw-Hill, New York, 1979.

[27]   Hino DM, Mendes FM, De Figueiredo JLG, Gomide KLMN & Imparato JCP. Effects of plaque disclosing agents on esthetic restorative materials used in pediatric dentistry. J Clin Pediat Dent 29 (2): 143 - 146, 2005.

Chapter 25   Ceramics

The dental porcelains offer one means of addressing very effectively the demands for cosmetic dentistry. This arises from the special combination of mechanical, chemical and optical properties that they possess – properties that depend on composition, structure, technique and thermal history. There are, however, some disadvantages. To use these materials successfully demands a proper appreciation of each of these factors.

*The **composite** nature of porcelains is the key aspect. The glass matrix chemistry and structure is first described including some explanation of terminology and the relationship of dental porcelains to other types of product.*

*The formation of a porcelain structure arises from reactions that occur on heating a mixture of basic ingredients. The control of these reactions is a matter of **time and temperature**, and varying grades of material can be manufactured to allow the special incremental build-up technique of dentistry to be used successfully. The firing process also involves physical changes as partial melting occurs. The control of **porosity** in this process is very important to both appearance and mechanical properties.*

*Because dental porcelain usually needs to be used in thin sections, strength is of great significance, but the brittleness of ceramics in general is a disadvantage to this. Various means of **strengthening** and reducing the sensitivity of the structure to flaws and scratches are discussed. This includes the use of metal-ceramic combinations which attempt to get the best out of each type simultaneously, offsetting their individual drawbacks.*

The use of porcelain in dentistry is very demanding, from the design of the restoration through the many stages of the fabrication processes to the cementation of the finished device. A thorough comprehension of this class of materials is essential.

The inadequacy of porcelain, being very brittle, in meeting service demands has been addressed through a number of alternative ceramic systems and processes, most recently based on zirconia where advantage is taken of crystallographic changes to obtain toughening.

Materials Science for Dentistry
https://doi.org/10.1016/B978-0-08-101035-8.50025-0

Ceramics are capable of very high strength and stiffness (Figs 1§14.1, 2), and in general have low densities in comparison with metals. Their use may therefore offer considerable advantages. However, their often extreme brittleness (and thus flaw-sensitivity) limits their applications, and especially to circumstances where tensile loading – particularly through bending – is minimal or non-existent. Nevertheless, porcelain and similar ceramics are of great value in dentistry for one overriding reason: they can be made to resemble natural tooth materials extremely well indeed. Porcelain inlays, denture teeth and facings for metal work have low wear and high chemical resistance to oral conditions, while crowns can offer unsurpassable mimicry.[1] Even so, there are problems which need to be overcome.[2]

§1. Structure

Porcelains are composite materials. They consist of a silica-based glassy matrix embedding a core of various kinds of crystalline substance, primarily silica and silicates; but other minerals are possible and sometimes used. The utility of the composite structure lies in several distinct areas: strength, optical appearance, and fabrication. These are, of course, interrelated. Glasses on their own are very scratch- and flaw-sensitive, optically unlike tooth tissue, and would require casting to achieve the accuracy required for dentistry. This as a 'one-shot' technique has limitations, particularly severe when graded properties – mechanical or optical – are necessary. Casting can be used (see §9.1), but the end result is not a glass.

The strength and stiffness issues of composites have been dealt with before (e.g. 6§2), and are developed further below (§3); the optical issues have been explored in 24§5. The question of fabrication in dental contexts is a matter of adapting traditional techniques to the demands of dentistry.

●1.1   Historical background

Some background helps to understand aspects of the dental material. Pottery or earthenware is typified as *terra-cotta* ('cooked earth'), which means that natural clays have been moulded as a wet paste, dried, then heated to a temperature that causes the mass to bind together as a strong body unaffected by rewetting. The particles of the clay have partially reacted, partially fused to give a porous structure that corresponds to the **biscuit** stage of dental porcelain (§4). The porosity is, however, interconnected so that the structure is permeable. If pottery is to hold liquid it must be **glazed**; that is, a glassy and impermeable layer is added. Stoneware takes this a little further, by raising the firing temperature, which increases the strength and hardness. Raising the temperature further still leads to porcelain. In general, porcelain is described as **vitrified**, meaning that much melting has occurred to create a glassy matrix. The porosity is thereby much reduced, and the structure is typically not permeable. The extensive glassy matrix now makes the mass much more translucent, as opposed to the dense opacity of pottery. The fact that porcelain tends to be prepared from white ingredients, for reasons of appearance and the ease of further decoration, has no bearing on these structural definitions.

So-called *true* or *hard-paste porcelain* was developed in China, sometime in the Tang Dynasty (618 - 907 CE), from a paste prepared from pulverized partially-decomposed granite called petuntse (白土瓷). This was therefore a natural mixture of kaolin (高嶺), feldspar and quartz. Subsequently, there have been many varieties of porcelain in which the proportions of the minerals have varied as well as the firing temperature, and other substances introduced into the mixture. These do not affect the broad description of porcelain. However, it will be seen that the extent of the melting (proportion of glassy matrix), the remaining porosity and the proportions of the minerals – the composition, are all important variables in determining the properties.

●1.2   Dental porcelain

Dental porcelain is not, in these broad terms, significantly different. In the final fired condition, it consists of a glassy alumino-silicate matrix (with various cations; §2, §7) in which are embedded several crystalline phases, in particular feldspar (§2), mullite (§2.1), quartz – which is part of the initial formulation and survives unreacted, and perhaps alumina (§3). Various minor phases of pigments (§7) are also present. The key structural aspect is that the matrix wets and bonds directly to each phase of the core, which confers the strength of the system. The presentation and manner of preparation does differ from ordinary porcelain: the so-called **powder method**. The dry powder is mixed with a little water to make a slurry, so that, using a small spatula, the desired shape can be built up incrementally on a model (the **die**) of the tooth to be restored, with vibration to help consolidate the mass (cf. 2§2.2, 4§7.9), when water is said to be 'brought to the surface', essentially as the density of **powder compact** increases. Excess water can be removed by absorbent paper so that the body does not collapse or slump (see 4§7.8). This technique, rather than moulding a plastic paste like clay for pots, permits the variation of composition from place to place in the structure such as to vary the opacity, strength and colour of the porcelain after firing. The powders may have organic dyes included in them to permit identification once they are in place on the restoration as it is built up. Being organic, these are intended to burn away on firing. But before we consider the firing reactions (§2) we need first to discuss the chemistry of the glassy matrix.

● 1.3 Silica glasses

The random linking of [SiO$_2$] tetrahedra in fused silica has previously been described (Fig. 17§2.9), the links between the silicon atoms being the shared oxygen atoms. This glass can be considered as the parent structure from which other silica glasses are derived. Because each silicon unit is linked to four others we may describe this three-dimensional network as having a **connectivity** of 4 (22§2).[3] Alternatively, this structure may be considered from the point of view of the frequency of cross-linking, the **cross-linking density** having the value 2, the excess over the minimum connectivity of a (linear) polymer of 2.

If any of these links were to be broken it would be necessary to add an extra oxygen atom in order to satisfy the valence requirements of one of the silicon atoms, which would otherwise be exposed. This in turn would leave the structure with excess negative charges which must be balanced by the inclusion of metal ions such as Na$^+$. This is equivalent to reacting the silica with a metal oxide:

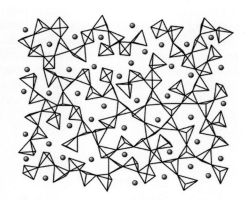

Fig. 1.1 Schematic diagram of the structure of a soda glass, Na$_2$O-SiO$_2$.

$$-\overset{|}{\underset{|}{Si}}-O-\overset{|}{\underset{|}{Si}}- \ + \ Na_2O \ \Rightarrow \ -\overset{|}{\underset{|}{Si}}-O^- \ 2Na^+ \ {}^-O-\overset{|}{\underset{|}{Si}}- \qquad\qquad (1.1)$$

Continuing this reaction further leads to a structure such as that represented in Fig. 1.1, which is now that of a plain 'soda' glass. As might be expected, any metal ions can be substituted for the sodium, and in endless mixtures, giving rise to a vast range of values for the properties. Some of this range was mentioned earlier (9§7), but other common varieties include 'soda-lime' glass, as used for windows and so on, which includes some calcium oxide; 'crown' glass, where potassium or barium oxide is substituted for the sodium; and 'flint' glass, which uses lead oxide. One result of such variation is that the refractive index changes. While this is useful for designing lenses, it also affects the behaviour of light in the composite body, exactly as in filled resins and glass ionomer cements (24§5.11).

The pure fused silica glass naturally does not possess an equilibrium structure. It would revert eventually to one of the crystalline forms if the activation energy for the conversion were available (Fig. 17§2.3). Similarly, fused mixtures of metal oxides and silica would separate out into distinct crystalline phases if allowed to do so (Fig. 1.2): at equilibrium all mixtures of "Na$_2$SiO$_3$" and SiO$_2$ would result in crystals of those two compounds only.[4] But this would require extremely slow cooling (geological timescales!). Under normal circumstances of moderate rates of cooling amorphous glasses (Fig. 1.1) are always produced.

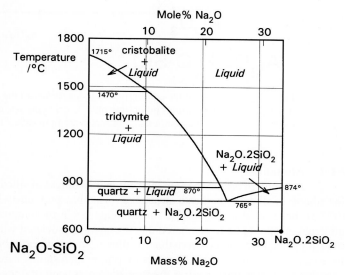

Fig. 1.2 Part of the constitutional diagram of the Na$_2$O - SiO$_2$ system Note the eutectic mixture of the two compounds. (N.B. this is an equilibrium diagram). Note that Na$_2$O·2SiO$_2$ and the like have polymeric anions, such as (SiO$_3$)$_n^{n-}$, which accounts for the difficulty of crystallization of all such substances. The ion SiO$_3^-$ does not exist.

This is because to form crystals requires covalent bonds to be broken followed by diffusion in a viscous medium. Here, then, is another example of kinetics limiting the approach to what the thermodynamics demands.

One circumstance where **devitrification** is encountered is in glassblowing, where repeated heating and plastic deformation of the viscous mass can cause nucleation, if there are sufficient cations present. The crystals then tend to grow quite quickly, rapidly making the glassware very weak as well as brittle, therefore useless.

There are two important consequences to arise from dissolving metal oxides in silica glass. Firstly, there is a substantial lowering of the melting point (*cf.* Fig. 1.2). Rather, since it is a glass, we should refer to the **softening point** (see §4.6). This enables working with materials that otherwise would be extremely refractory. Secondly, and obviously connected with the first point, the viscosity of the melt is also considerably lower because the presence of the metal ions has resulted in broken chains and lower network connectivity (*cf.* 22§2). More than this, the negatively charged 'oxide' oxygens may encourage bond exchange by further lowering the activation energy for bond breaking, attacking existing bonds. Such events are essential for a cross-linked network system to flow at all. This may be compared with the stress-relaxation mechanisms operating in rubbery impression materials (Chap. 7).

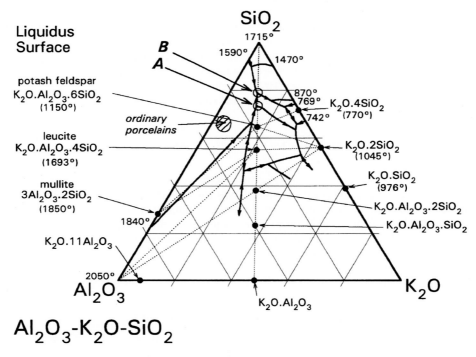

Fig. 1.3 Map of part of the equilibrium liquidus surface of the system Al_2O_3 - K_2O - SiO_2 showing some melting grooves (the arrows point 'downhill'), stable phases, and compositions of interest. Compound melting points are in parenthesis. The scales are in mass%. The points *A* and *B* refer to Fig. 2.1.

●1.4   Aluminium

Aluminium is an important element in the context of silica chemistry because of its ability to substitute for silicon in silica and silicate structures. This has previously been discussed in connection with cements (9§7.3), but it is also responsible for the huge range of compositions and structures in the aluminosilicate minerals. It should be remembered that the charge associated with the presence of the aluminium is different (one more negative charge) from that for silicon, so that an extra balancing metal cation charge must be present for every Al atom included if the structure is to remain simply substitutional. Otherwise, an oxygen atom must be removed, which thus requires the formation of a bond (the reverse of equation 1.1). An example of such a system is that of Al_2O_3 - K_2O - SiO_2 (Fig. 1.3). [5] This is of importance because it includes a range of compositions corresponding to those used in dental porcelains. Again, this is an equilibrium diagram, whereas it will be seen that normally dental porcelain is far from equilibrium.

The K₂O - SiO₂ system, corresponding to the right, upper edge of Fig. 1.3, is shown in Fig. 1.4. From this it is apparent that the system is a little more complicated than in Fig. 1.2, but the lowest eutectic temperature is even lower at about 742 °C. What this also illustrates is that a variety of compounds may be involved in the chemistry of the firing process for porcelains and similar materials, according to *local* conditions. That is, given a mixture of reactants, which can be expected to react slowly, reactions at interfaces may not represent an overall process or move towards overall equilibrium, at least initially. Time, temperature, diffusion rates and reaction rates, as well as thermodynamics determine the outcome – kinetics is extremely important.

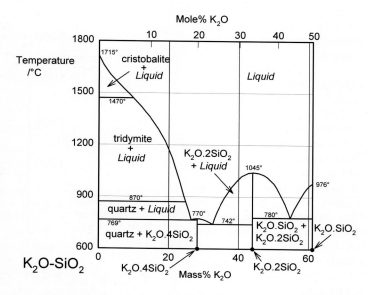

Fig. 1.4 Part of the equilibrium constitutional diagram of the K₂O - SiO₂ system.

●1.5   Labelling convention

A convention commonly used in this field needs introduction. In the phase diagrams discussed so far there have been only two kinds of product representation: either elemental solid solutions or definite stoichiometric compounds, such as Ag₃Sn or ZnHPO₄. The substances represented at the corners of the diagrams, *i.e.* at the ends of the composition lines, the components, are then seen as being combined in a natural manner.

The phase diagram of Fig. 1.4 is another example of juxtaposed binary systems: SiO₂ - K₂O·4SiO₂, K₂O·4SiO₂ - K₂O·2SiO₂, and K₂O·2SiO₂ - K₂O·SiO₂, reading from the left, where the boundaries between these systems are the intermediate compounds, each of which shows a sharp melting point. In this case, the mixtures are all eutectics. For comparison, Fig. 1.2 shows only one eutectic mixture of a kind similar to the binary systems here.

However, in ceramic work, where glasses are very often involved, quite clearly stoichiometry is very frequently not a relevant consideration except for the occasional crystalline substance. It is therefore not sensible to talk in terms of mixtures such as "Na₅AlSiO₆" as if they were compounds, which clearly have no chemical meaning.   Rather, it is convenient to speak in terms of multiples of the component oxides, *viz.* 5Na₂O·Al₂O₃·2SiO₂ for the above example, and even for good chemical compounds, whilst recognizing that the oxides do not exist as individual molecules in either the melt or the glass. This is essentially a matter of practical utility – it is easy to locate compositions in diagrams, and awkward sums are avoided. In addition, it ensures that the underlying stoichiometry of the oxides is preserved, since no variation in relative oxygen content is meaningful. Given this symbolic convention it is now easier to discuss some of the reactions of dental porcelain.

§2. Firing Reactions

It is apparent from Fig. 1.3 that, even with just the three components, many crystalline compounds are possible at equilibrium. However, certain regions (*i.e.* ranges of composition) are capable of producing quasi-stable glasses on cooling from the melt. Typical porcelain, as in art work and domestic ware, has an overall composition close to K₂O·5Al₂O₃·20SiO₂, rather richer in silica and potash (K₂O) than earthenware, which is based on clays. Dental interest centres around the mineral **potash feldspar**, K₂O·Al₂O₃·6SiO₂. This compound melts peritectically at 1150 °C to produce the compound **leucite**:

$$K_2O \cdot Al_2O_3 \cdot 6SiO_2 \Rightarrow K_2O \cdot Al_2O_3 \cdot 4SiO_2 + 2SiO_2 \tag{2.1}$$

This reaction is better understood by examining the vertical section (the isopleth) of Fig. 1.3 taken between the leucite point and the SiO_2 apex (Fig. 2.1). The co-existing liquid at the peritectic temperature has a composition corresponding to **A** in Fig. 2.1, approximately $K_2O \cdot Al_2O_3 \cdot 9SiO_2$, *i.e.* the silica 'ejected' from the structure may be thought of as dissolving a proportion of the leucite formed in that decomposition reaction. The most significant point to arise from this, as remarked above, is the dramatic lowering of the effective melting point of the silica in the presence of the other oxides. Indeed, silica in excess of that required to form feldspar actually melts at 985 °C, which is the 'eutectic' temperature of the pseudo-binary system (**B** in Fig. 2.1); the liquid here has a composition approximately $K_2O \cdot Al_2O_3 \cdot 12SiO_2$.

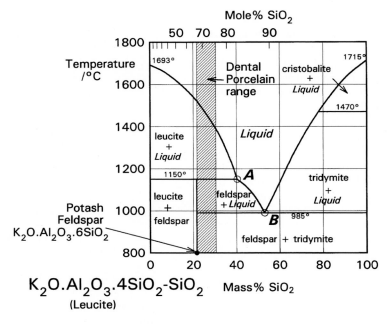

Fig. 2.1 The pseudo-binary system leucite-silica, an isopleth of the $K_2O\text{-}Al_2O_3\text{-}SiO_2$ system (Fig. 1.3). The points *A* and *B* refer to Fig. 1.3.

A certain proportion of sodium, a small amount of rubidium, and even occasionally some calcium, may be found in natural feldspar. These metal substitutions will tend to lower the eutectic and other transformation temperatures in this system. In addition, the sodium analogue of potash feldspar, **albite**, $Na_2O \cdot Al_2O_3 \cdot 6SiO_2$, may also be added to mixtures for dental porcelain, with similar effects.

●2.1 Kaolin

A typical dental porcelain will mainly consist of a mixture of feldspar with, say, about 15 ~ 20% powdered quartz and a few percent of **kaolin** as a **binder**. Kaolin is an hydrated clay mineral, $Al_2O_3 \cdot 2SiO_2 \cdot 2H_2O$, and is in the form of very fine hexagonal plates in the range 1 ~ 5 µm diameter and about 1 µm thick. Because of the presence of strong surface charges on these particles (which are mutually repellent) they form colloidal suspensions in water, binding large amounts of it, and can be used to provide a certain amount of coherence and plasticity to a porcelain powder slurry, facilitating the handling. The effect otherwise would be like wet sand: surface tension holds together a sandcastle (Fig. 2.2), when there is water only at the contact points – the many high curvature surfaces generating considerable tensile force (*cf.* loss of gloss; 2§5.1, 10§2.2), but it is a very fragile structure. The presence of some clay-like mud in the mixture makes it much tougher (but only when the powder is wet). The

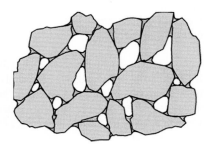

Fig. 2.2 Surface tension holds the mass of porcelain particles together during build-up.

amount of kaolin in dental porcelain must be limited, however, because of the resulting opacity in the product; tooth-like appearance depends much on translucency. The coherence of the porcelain powder, like the sandcastle, otherwise depends entirely on the capillary forces acting in the films between grains, and thus clearly depends on a minimum amount of water being present. Recall that gypsum and investment slurries only flow well when glossy; after 'loss of gloss' they are much stiffer. Excess water must therefore be carefully blotted away during the manual build-up of a porcelain restoration (which works because the affinity of cellulose for water is much greater, 27§2.2). The residual water must, of course, be carefully removed by drying slowly with gentle heat just before firing – and in the absence of vibration – to avoid wrecking the structure.

On firing a dental porcelain the first important reaction is the dehydration of the kaolin at about 450 °C:

$$Al_2O_3 \cdot 2SiO_2 \cdot 2H_2O \Rightarrow Al_2O_3 \cdot 2SiO_2 + 2H_2O(g) \tag{2.2}$$

The solid product is called **metakaolinite** (Fig. 2.3). This is in fact an unstable substance (not a true compound) which, by being finely divided and amorphous, is relatively reactive. Its composition lies in the silica – **mullite** two-phase field. On further heating to about 1000 °C or so, therefore, this material decomposes to mullite, $3Al_2O_3 \cdot 2SiO_2$ (which is a well-characterized, definite compound), and amorphous silica (Fig. 2.4). This latter remains reactive because it is still finely divided as well as having a non-equilibrium structure. The mullite will be well-crystallized, as it is above its recrystallization temperature (*cf.* 11§6.6) of 918 °C.

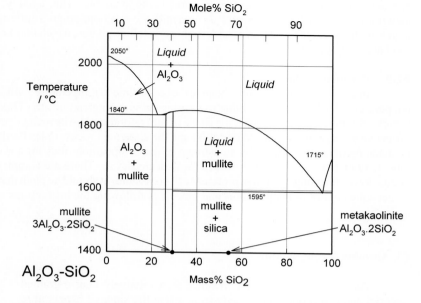

Fig. 2.3 The system Al_2O_3-SiO_2. This is one face of the diagram in Fig. 1.3. Notice that there is no appreciable solid solubility of either alumina in silica or *vice versa* at equilibrium: the SiO_2 and Al_2O_3 phase fields do not show. The silica polymorphs are omitted for clarity (see Fig. 1.2).

As the temperature is raised further still, the feldspar now starts to melt and decompose at ~1150 °C (reaction 2.1), and not at a lower temperature despite the presence of the free silica, because the reactions are so slow and equilibration can never be achieved on practical timescales. The liquid formed will flow and start to cause the consolidation of the powder particles because of its own capillary action (10§2). Reactions with the amorphous silica and mullite now start to occur, and metal ions from the feldspar start to react with the quartz powder particles originally included, but which up to now have not altered at all except for the α-β transition at ~573 °C (Fig. 17§2.2). On cooling this partially-melted mass, crystallization of the liquid does not occur because of the viscosity of the liquid phase and the high activation energy of crystallization. The final structure now is a glassy silicate matrix embedding a core of a mixture largely of quartz and unreacted feldspar, with a dispersion of the separated mullite.

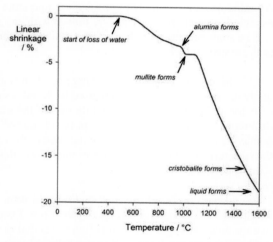

Fig. 2.4 Behaviour on heating of a pure kaolin powder body. The shrinkage after the formation of mullite is due to sintering and glass formation. In the presence of other substances, other reactions may occur.

●2.2 Processing temperature

The above scheme is applicable to the so-called **high-fusing** dental porcelains, where little if any glassy phase is present initially. However, the melting point of the mixture has been *lowered* by the partial reaction of the components, so that on *re*firing the onset of melting and further reaction occurs at a lower temperature, even though the overall composition is unchanged. The **low-** and **medium-fusing** dental porcelains are prepared in just this way, by controlled reheating of already-fired porcelain. After cooling, usually by pouring the mass into water to induce much cracking by thermal shock (steep stress gradients caused by rapid cooling of the low thermal conductivity material, §5.5), the mixture is ground again to a fine powder, which is then essentially the commercial product.

There are thus two ways of controlling the temperature required for processing a dental porcelain: the composition and the degree of reaction permitted in previous firings. The successive-increment procedures employed in porcelain restoration fabrication are dependent to some extent on the existence of a group of products with a range of fusion temperatures being available. But, equally, excessive refiring of any such material can be deleterious because this permits the reactions to go a little further towards equilibrium and therefore the possible formation of undesirable crystalline phases, which may be expected to alter the optical and mechanical properties.

●2.3 Pyroplastic flow

In addition to the chemical changes, excess 'time at temperature' will also result in the lowering of the temperature at which distortion due to sagging or slumping can occur. This is known as **pyroplastic flow**. Clearly, this needs to be avoided if the shape of the restoration is to be maintained. Nevertheless, there is a trade-off between resistance to this deformation and the sintering process (§4). Partly, some tuning of behaviour can be made by the manufacturer by adjustment of the composition: thus for a given total amount of alkali metal ions, K_2O increases the viscosity while Na_2O decreases it. There are complementary changes in the melting points: lower for using sodium in preference to potassium. It should be plain that the quality of the work depends on accurate control of temperatures and times, *i.e.* following the instructions.

§3. Alumina Strengthening

As the quartz originally included is a relatively coarse-grained material and reacts only very slowly (if at all) at the kinds of temperature used, this component remains as a distinct crystalline phase in the glassy vitrified matrix. Hence, as mentioned above, porcelains are composite materials and much of their strength is due to this structure. Quartz, however, is not a particularly strong material. Considerable improvement in porcelain strength has been obtained by using alumina in the same role. Alumina has an extremely high fusion temperature (Fig. 2.3) and can be expected to react only very slowly indeed with any matrix, although there will be some reactions on the surface which will tend to bond it covalently to the matrix, which is desirable.

Fig. 3.1 Dental feldspathic porcelain is a composite of several crystalline phases in a glass matrix.

Since such alumina-containing porcelain is also composite, we can apply the same principles as were used for filled resins (6§2.7). The value of E for alumina is ~380 GPa, while that of the glass is ~70 GPa. This difference is enough for us to expect substantial matrix constraint (if bonded) and therefore a substantial increase in the effective value of E for the structure, and thus also for the strength – *i.e.* the stress causing failure – in proportion (6§2.10).

Accordingly, **aluminous** or **high alumina** porcelain consists of about 40 ~ 50 mass% Al_2O_3 in a typical low-fusing glass matrix. The extraordinary strength of alumina (ten times that of feldspathic porcelain; Table 5.1) is thus utilized in the composite. However, care must be taken to match closely the thermal expansion coefficient of the matrix to that of the alumina (see §3.2); this will be done by adjusting the mix of metal oxides used. If the match is not good, thermally-induced shear stresses developed at the interface (on cooling after firing) may lead to a weakened bond between the two and a fracture path which follows the periphery of the particles. The bond is relatively weak because of the limited reaction referred to above. If the shear stresses can be avoided the fracture will be

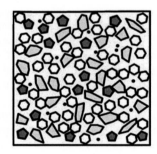

Fig. 3.2 Aluminous porcelain includes a very high stiffness and high strength material to increase matrix constraint and crack path convolution.

tend to be **transgranular** and the overall strength will depend more on the fracture energy of the alumina. Clearly though, the balance is a fine one as the total surface energy for a complicated fracture path may be only slightly more than a straight fracture through zones of sometimes higher surface energy of formation.

Unfortunately, this is still a 'powder method' material, and suffers from the same problems of porosity as ordinary porcelain (see §4). But, more importantly, it is rather opaque, and can only be used as a basement layer, the **aluminous core**, on which are built up further layers to represent dentine and enamel. Nevertheless, high-alumina porcelain represented a major advance in strength for such restorations.

●**3.1   Fracture energy**

We can try to understand this kind of system by writing out the overall specific fracture energy W_f in terms of the surface energies γ_g and γ_a of the components, glass and alumina. To start, we write down the fracture energy for the glass alone, which is effectively for a **transgranular** path (indicated by the symbol $\neq$) because there are no other phases (or grain boundaries) involved:

$$W_f(\neq) = 2\gamma_g \qquad (3.1)$$

This is the familiar expression of the work of creation of a crack (*cf.* equation 10§1.10); the factor of 2 is because we are counting both new surfaces (10§1.2). Since the proportions of each type of surface in a random plane section through an alumina-in-glass composite are the same as the volume fractions, φ, of each phase, we can then write down directly the fracture energy for a random plane transgranular crack:

$$W_f(\neq) = 2[\varphi_g\gamma_g + \varphi_a\gamma_a]$$
$$= 2\gamma_g + 2\varphi_a(\gamma_a - \gamma_g) \qquad (3.2)$$

From this it can be seen that if the work of fracture of the alumina is higher than that of the glass then the overall work must be higher – which is true. (The square bracketed expression is, of course, a simple **mixture rule**, *cf.* 6§2.6.)

Similarly, we consider a crack that crosses the glassy matrix directly but deviates around the alumina particles, that is, peripheral fracture (in a form of the Cook-Gordon blunting process, 6§2.13). This must be an **interfacial** failure (symbol =) between the glass and alumina (g|a). Thus, we have:

$$W_f(=) = 2[\varphi_g\gamma_g + \gamma_{g|a}.A_{g|a}] \qquad (3.3)$$

where $\gamma_{g|a}$ is the work of fracture of that interface, whose relative area, $A_{g|a}$ is for the moment unknown. However, for simplicity, we can assume that the alumina particles are spherical and that the crack is aligned with a diameter (the maximum effect). The interfacial area between a spherical particle and the matrix on one side ($4\pi r^2/2$) is then exactly twice the area of the diametral cross-section (πr^2) of that particle. If it can also be assumed that the energy of formation of the two new surfaces, *i.e.* glass and alumina, each have about the same energy whether from the homogeneous material or the interface, we have from equation 3.3:

$$W_f(=) \approx 2[\varphi_g\gamma_g + \varphi_a(\gamma_g + \gamma_a)]$$
$$= 2\gamma_g + 2\varphi_a\gamma_a \qquad (3.4)$$

Comparing this with equation 3.2, we find that there is a difference: extra work effectively due to the presence of the alumina causing an increased fracture surface area in the glass, thus:

$$W_f(=) \approx W_f(\neq) + 2\varphi_a\gamma_g \qquad (3.5)$$

Under these circumstances the perigranular fracture path will be distinctly unfavourable, requiring more energy. The material is therefore tougher because of the presence of the alumina. This is true whether the crack goes through (equation 3.2) or around (equation 3.5) those particles, assuming that there is a bond between the two phases. However, if the interfacial bond has already been broken, equation 3.4 reduces to

$$W_f(=) \approx 2\varphi_g\gamma_g \qquad (3.6)$$

(*i.e.* setting to zero the second term on the right, the energy for the breaking of the interface) so that the composite actually becomes weaker with increasing, **unbonded** alumina content. This is, of course, what would be expected, as the matrix would then appear to have holes in it (see 2§8, 6§2).

If the alumina particles are irregular (as they normally would be) the balance is pushed further toward transgranular cracking, as the ratio of interfacial area to cross-sectional area is increased compared with that for a sphere. However, this still requires more energy than plain glass (equation 3.2). Thus, the very high strength of the alumina is directly utilized in strengthening the porcelain, so long as it has been wetted by and become bonded to the matrix.

●**3.2   Thermal expansion matching**

The matching of thermal expansion coefficients is critical in this case because the matrix and filler are both so brittle: no plastic deformation whatsoever is possible, and failure occurs at very small tensile strains.

If the contraction of the alumina is greater than the matrix the interface is put in tension; if the contraction of the glass is greater it is the glass itself which is in tension, and therefore pre-stressed and prone to fail, even though the interface is in compression.

Alumina and alumina-silica based ceramics are quite feasible (Fig. 2.3) but the very high melting points in this system preclude their fabrication in ordinary equipment and hence by dental laboratories (but see §9). Great strength is obtainable from virtually pure alumina, and this material is used in commercially-prepared backings onto which are fused lower-melting porcelains. Again, it is essential to obtain wetting and covalent bonding if such inserts are to function properly as strengtheners.

§4. The Firing Process

The processing of dental porcelain involves the conversion of a collection of separate particles into one fused, continuous mass (Fig. 4.1). As indicated above, this involves only a partial melting of the constituents and the process is known as **sintering**. In fact, sintering does not require melting: solid state diffusion can be enough for bonds to form between particles (*cf.* 21§3), but in the case of porcelain the melting is a necessary and important part of the process. The driving force for sintering is the reduction in the total surface energy of the system by the reduction of the surface area, the glass of porcelain having a high surface energy.

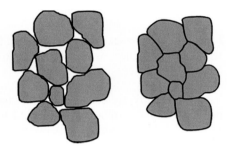

Fig. 4.1 The aim of sintering is to produce one continuous mass, pore-free.

Since the shape of the (unfired) restorative device has been built up from a porcelain powder, there is a second factor to consider: the porosity of the mass. The packed powder **compact** (Fig. 4.1, left) necessarily has a large volume fraction of porosity (2§2.1), and this will lead to shrinkage on firing, usually more than 30% but as much as 50% depending on particle size and how well it was compacted. The design of the device and the processing must take this into account.

Because the melting of feldspar-based glasses is sluggish, and the melt itself is quite viscous, the sintering process can be closely controlled by time and temperature. This is utilized to particular advantage in the dental procedure whereby different layers may be built up and fired in turn to a state of cohesion, although to avoid serious effects on previous increments the successive layers may need to have successively lower fusion temperatures (§2.2). This first stage in the sintering process is referred to as **low bisque** or **low biscuit**. This stage is entered when only the contact points between particles have fused together to achieve this (Fig. 4.2) and the porosity is virtually unchanged.

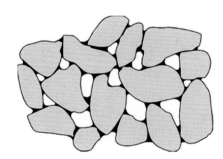

Fig. 4.2 The low biscuit stage of sintering gives a weak and porous structure, held together only at contact points.

●4.1 Bubbles

On further heating, when more flow can occur under surface tension effects in the viscous melt, air will be expelled from the interstices and gross shrinkage will be observed. Clearly though, at some point, as the coalescence proceeds, the interstices cease to be a continuous three-dimensional network. What was **connected porosity** now becomes discontinuous. Air would therefore remain trapped as bubbles distributed throughout the mass (Fig. 4.3), but initially with irregular shapes. This corresponds to leaving the low bisque stage and entering the condition known as **high bisque** (or **high biscuit**). As the firing continues, the bubbles assume a more and more rounded shape, ultimately becoming spherical, when the high bisque stage is left and "acceptable" firing has been achieved. The surface then has what is described as an 'egg shell sheen' – not quite matte, almost glossy,[6] like a hen's egg. Of course, if the heating is prolonged, melting and slumping may occur, which would be unacceptable as the shape of the restoration would be lost.

Porosity has several effects on the porcelain. Firstly, it is a source of weakness in that pores are inclusions of zero strength and thus provide paths of low resistance to crack propagation, quite apart from a general lowering of the overall Young's Modulus. The work of fracture would also be low. Secondly, the strength would be determined by the largest such pore (Griffith criterion, 1§7) (assuming uniform stress). Irregular porosity would be associated with lower strength because locally the radius of curvature may be very small and cause greater stress concentrations. Thirdly, any attempt at polishing such a material would leave a rough surface as the pores became exposed. This would enhance the accumulation of staining debris, and also provide sites of high stress concentration in the surface which would easily initiate a fracture under even a slight tensile stress, as might occur in bending. Fourthly, the optical properties of the material would be impaired by increasing the light scattering and therefore its opacity (24§5).

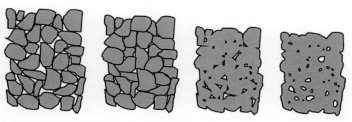

Fig. 4.3 Full densification of a sintered mass cannot be achieved because of trapped air. Left to right: 'green compact', entering low bisque, entering high bisque (porosity disconnected but irregular – note shrinkage), "acceptable" firing (porosity rounded).

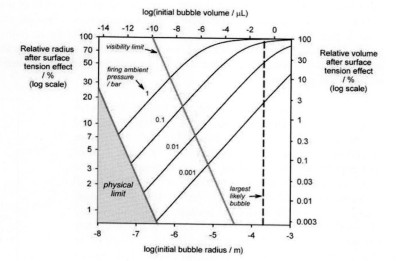

Fig. 4.4 The effect of Laplace bubble pressure on the size of equilibrium bubbles in a typical glassy matrix, assuming no dissolution. Read from the initial bubble size to a 'firing ambient pressure' line, then across to the relative bubble size due to the effect of surface tension. The physical limit is approximate and corresponds to the density of liquid air! Bubbles smaller than the visibility limit (*i.e.* after shrinkage) simply cannot be seen.

●**4.2   Air firing**

The surface tension of the liquid phase is expected to raise the pressure in such bubbles (P_b) above ambient (P_a) (10§2):[1]

$$P_b = P_a + \Delta P \qquad (4.1)$$

where the expected excess pressure, ΔP, is given by the Laplace equation (10§2.2) for that initial bubble radius. However, when the bubble first becomes formed, pinched off from the original network as an isolated site, it is a fixed mass of gas that has been enclosed, and that was at the ambient pressure. Surface tension then acts to compress that bubble, decreasing the bubble radius until equilibrium is reached (1 bar line, Fig. 4.4). This compression does not occur instantaneously because of the high viscosity of the melt, but eventually a balance would be obtained between the bubble radius and that extra pressure in it, ΔP. It can also be seen that the effect is very limited unless the bubbles are already quite small.

As described at 10§2.9, the effect of the increased bubble pressure and the operation of Henry's Law tends to drive the dissolution of the air in the glassy melt surrounding the bubble. The concentration gradient established from adjacent to the bubble to the external surface drives the diffusion of the dissolved air out of the mass. As the air dissolves and the bubble shrinks, so ΔP rises and the process tends to accelerate. However, although these effects are in the right direction the diffusion process is relatively very slow in this kind of system. This mechanism cannot be relied upon, therefore, as a practical means of eliminating porosity in porcelain, especially since time at temperature must be limited for other reasons, *i.e.* extent of reaction and sagging. Air-firing may leave porosity of the order of 5 vol%.

---

[1]   Strictly, hydrostatic head should also be taken in to account in such calculations, but since densities are low and fluid depths small this can be ignored without appreciable error in this context.

•4.3 Vacuum firing

Several techniques may be utilized to reduce the porosity while keeping the time short, the most common being so-called "vacuum" firing. Under reduced pressure, at the point when the bubble is first formed by the separation of the continuous porosity into discrete separate pores, the *mass* of air actually trapped in each is less than would be the case at normal atmospheric pressure, in proportion to the new pressure (Boyle's Law, 18§1.2), since the pore geometry, and thus its volume, is controlled by the powder particle packing only. However, the surface tension of the melt is hardly affected, so that the expected pressure difference across the bubble, ΔP, due to curvature is just the same. However, this means that the ratio between the expected bubble pressure P_b and the ambient pressure is increased as P_a is reduced:

$$\frac{P_b}{P_a} = \frac{P_a + \Delta P}{P_a} \tag{4.2}$$

Thus the initial surface-tension driven compression of the bubble is now greater, leading to an equilibrium bubble size rather smaller than that for firing at atmospheric pressure, since ΔP continues to increase and will reach higher values than before. Dissolution and diffusion of the air in the bubble is therefore encouraged even more. This effect can be traced in Fig. 4.4, where it can be seen that the shrinkage obtained just by reducing the amount (mass) of gas trapped in the bubble – by reducing the firing ambient pressure – leads to a much improved shrinkage at equilibrium.

After a given firing time the bubble will have become smaller. Then, if atmospheric pressure is restored before cooling commences, the bubble will be further compressed. Remaining bubbles are then expected to be of an insignificant size. Indeed, as the bubble shrinks the radius decreases at an increasing rate ($r \propto \sqrt[3]{V}$), because ΔP increases more and more rapidly ($\Delta P \propto 1/r$). Very small bubbles must vanish altogether because at very small radii of curvature the pressure becomes enormous and forces any gas into solution in the melt (Fig. 10§2.4).[2] There is an absolute minimum size of bubble than can exist, of course, set by the size of the gas molecule (150 ~ 200 pm radius) – although it would be quite wrong to describe just one molecule as a bubble when it is then already clearly 'dissolved' in the matrix. The particle size distribution of the porcelain powder and the surface tension of the liquid phase (which is temperature-dependant) are of importance here: the former because it determines the maximum size that a pore may have between touching grains in the original powder body (Fig. 2.2), the latter because it controls the Laplace bubble pressure. The only difficulty with this process is that the compression of the bubbles when atmospheric pressure is restored also takes time because of the viscosity of the melt. The temperature must therefore be maintained at least for a short time at atmospheric pressure. The kind of schedule that is used to accommodate this behaviour is shown in Fig. 4.5.

A typical sequence of events may be summarized as in Fig. 4.6. Suppose the firing pressure is 0·01 bar. The Laplace pressure effect takes the bubble size, here taken to be initially ~14 μm, to the point indicated by *a*. Then, applying normal atmospheric pressure means in effect a 100 : 1 compression ratio, tracing the effect of the after-shrinkage size at *b* to the compressed size at *c*.

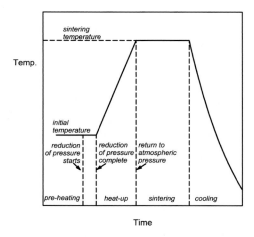

Fig. 4.5 Outline of a typical firing schedule for powder-method porcelain.

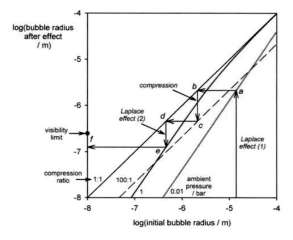

Fig. 4.6 Schematic of the effect on porosity of firing under a reduced-pressure schedule as in Fig. 4.5.

---

[2] Interesting things happen, however, at very small scales, at least in water, where nanoscopic bubbles persist.
See: Ball P. Big trouble over tiny bubbles. Chemistry World 9(9): 60 - 63, 2012.

However, a bubble of this size (*d*) is subject to further Laplace pressure shrinkage (*e*), giving a final radius some one-hundredth of the original value (*f*). Whether this is actually attained depends on whether there is time for the flow processes to occur, but it would be augmented in effect by dissolution and diffusion, as indicated. From such a sequence it can be seen that low-pressure firing is very effective.

It should be noted that while very small bubbles are expected under appropriate conditions, there is a limit to their optical detectability – about 0·25 μm radius. But, while such bubbles cannot be seen with visible light (even with a light microscope) and so do not affect the appearance of the porcelain (24§5.8), they will still act as Griffith flaws and affect strength. It should not be assumed, therefore, that invisibility means absence (*cf.* effect of visual acuity, 15§7.2).

The outcome of the firing of powder-method porcelains is summarized in Fig. 4.7. The reduction of the volume fraction of porosity is clearly seen to be time- and temperature-dependent, but attaining the desired low value must be traded against the general appearance, which depends on the flow of the matrix: too little and it appears chalky and opaque, too much and the surface glazes and edges start to melt and deform, the goal being an intermediate eggshell-like sheen. The acceptable zone corresponds closely to the formation of spherical pores (*cf.* Fig. 4.3), but clearly the temperature has to be high enough that the viscosity is low enough for the necessary flow to occur and realize the benefit of the Laplace bubble-pressure effects. Of course, the temperature of the porcelain is affected by its thermal conductivity and related properties, so there is a strong boundary on the left corresponding to the time taken to heat through.

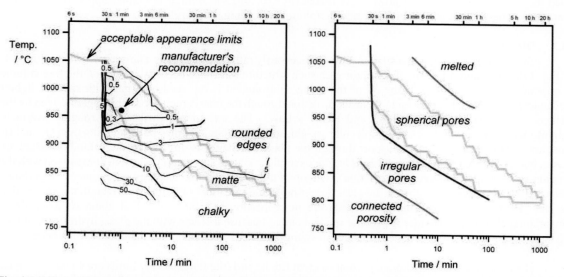

Fig. 4.7 Effect of time and temperature on the porosity and appearance of a typical powder-method porcelain. Left: some experimental contours for volume percentage porosity, with the limits of acceptable gross appearance (hair-pin staircase line); also showing the conditions recommended by the manufacturer. Right: porosity type domains.

The effect of vacuum firing can be seen to be dominated by the quality of the vacuum itself: the lower the pressure, the better, although some manufacturers have claimed an 'optimum' value at about 0.05 bar (the reason for this is obscure – it is clear from the above that the optimum is an absolute vacuum, assuming that no degassing can occur to generate porosity in any melted material). The factors to be taken into account are the speed and quality of the pump, the quality of the seal (given that it is a furnace that operates at high temperature), and therefore the rate at which the pressure can be dropped in comparison with the firing schedule. These affect the price. Keeping the seal in good condition, and clean, would be important for continued proper operation and thus porcelain quality.

It should be noted that although the firing process is at an elevated temperature, so that the density of the air trapped in bubbles is lowered according to **Charles' Law**:

$$\frac{1}{V} \propto \frac{1}{T}$$

(4.3)

and the mass of gas in a bubble correspondingly smaller, this only affects how much gas has to diffuse out for the bubble to shrink to nothing, which shrinkage therefore occurs somewhat faster. The pressure statements all remain valid as given since no appreciable shrinkage can occur on subsequent cooling (*i.e.* by Charles' Law, again) because the viscous matrix soon becomes solid, fixing the bubble size. The pressure in the bubble, however, then does fall, as indicated by **Amonton's Law**:

$$P \propto T \tag{4.4}$$

While this then puts the matrix under a slight compressive stress (*cf.* lower part of Fig. 6§2.7), it is of no particular significance here.

●4.4 Pressure

Cooling under pressure after air-firing would also reduce bubble size and be equivalent to firing in a vacuum at the outset. However, the use of high pressure is more problematic than vacuum. If, say, the vacuum furnace operated at 1/25th atmospheric pressure, a not unreasonable figure, restoration of atmospheric pressure before cooling would be expected to cause an initial collapse to 1/25th of the original bubble volume. However, for an applied pressure method to have a similar effect a pressure of 25 bar would be required. There is considerable hazard associated with such pressures, and the design of the furnace much more difficult. For comparison, air-turbine handpieces typically require air at no more than 3 or 4 bar.

There is a parallel here with the heat-processing of denture base acrylic in relation to air bubbles trapped when the monomer and polymer powder are mixed (5§2.6, 10§2.9). Similar principles apply, including the operation of the Laplace bubble pressure and dissolution according to Henry's Law.

●4.5 Diffusible gases

The network structure of a silicate glass is necessarily rather open – it has a large free volume (Fig. 1.1, *cf.* Fig. 17§2.9),[7] and diffusion of gases through such structural, intrinsic porosity is fast (*cf.* the free volume of organic polymers, 3§4.2). If the air of the firing environment is replaced by a gas with smaller, lighter molecules, such as He, H_2, or even H_2O, faster diffusion through the matrix melt is possible, again driven by the surface tension effect on the bubbles and Henry's Law. The rate of gas diffusion is roughly inversely proportional to the square root of the molecular weight, so He is about 2·8 times faster-moving than O_2, and H_2 four times. However, the bulk transport through the glass also depends directly on the solubility. Data are sparse, but He is about 4 times as soluble as N_2, for example, a simple size effect (that is, in comparison with the structural porosity, since there is no reaction). On top of this is the activation energy requirement for getting a molecule from location to another in the glass: larger molecules are more of a squeeze. Overall, this means that for He : H_2O : O_2 the diffusion coefficients are in proportions ~2500 : 30 : 1 at about ordinary firing temperatures.[8] The use of water is peculiar in this context because it reacts with the glass, hydrolysing it (see Fig. 8.1), and more or less permanent changes in strength, melt viscosity and other properties also occur.

However, the use of hydrogen presents a great fire hazard (the flammability limits are 4 and 75% in air at 20 °C, 1 bar, requiring just a 20 μJ spark – such as an invisible static discharge from clothes. It has an autoignition temperature in the region of 500 ~ 560 °C, well below porcelain firing temperature), to say nothing of the risk of explosion (the lower and upper explosive limits are 17 and 56%). On the other hand, helium is rather expensive, and would be an inessential waste of a diminishing resource. Although the water molecule is larger than either of the other possibilities, and its diffusion through the solid would be that much slower, it is still better than O_2 and N_2 in that respect (ignoring the hydrolysis problem). Even so, vacuum firing techniques remain the most attractive and technically simple, and properly applied are capable in practice of reducing the total porosity to about 0·1 vol%.

One further factor is involved: bubble surface area. Since the bulk rate of loss of gas, *i.e.* volume $\propto r^3$, from a bubble into the surrounding medium, the total flux, is proportional to its area, *i.e.* as r^2, the rate of change of radius for a constant driving force is constant. Since that driving force is the Laplace pressure $\Delta P \propto 1/r$ (assuming that the gas diffuses rapidly away from the surface), and obviously not constant, this remains as the principal measure of the rate of disappearance of the bubble:

$$-\frac{dV}{dt} = -\frac{4\pi}{3}\frac{dr^3}{dt} = -\frac{4\pi}{3}.3r^2\frac{dr}{dt} \propto 4\pi r^2.\frac{1}{r} \tag{4.5}$$

$$\rightarrow -\frac{dr}{dt} \propto \frac{1}{r}$$

Of course, the gas still has to diffuse away and out of the porcelain, and this takes time, modifying the effect.

●4.6 Glass transition

The rate of sintering and **densification** of the porcelain depends on the flow characteristics of the melt. As might be expected, the complex glasses that form the basis of the matrix in the normal low- to medium-fusing porcelains do not exhibit a sharp melting point and so no sudden change in viscosity. Instead, at the **softening point**, which is entirely analogous to the glass transition temperature of organic polymers, the rate of change of viscosity with temperature is at a maximum (Fig. 4.8).[9] Thus the expulsion of the air by surface tension effects is both time- and temperature-dependent. Clearly, though, gross deformation of the porcelain or sagging under its own weight must be avoided and the firing process must be very carefully controlled (§2.3). In addition, the amorphous phase must be a **strong liquid** and therefore show what is called **low melt fragility**, that is, have a broad softening range. A fragile melt would have the viscosity fall too quickly and allow too much flow on the outside before the inner regions have heated enough for densification to proceed. (Strong liquids are also less likely to suffer devitrification.) Again, a compromise is required to balance these conflicting demands. Nevertheless, the overall dimensional shrinkage arising from densification is inevitable. It depends on how well the powder has been condensed, as well as on the particle size distribution (the blend of sizes), but can be as much as 30 or 50 vol%. The incremental build-up technique is necessary in part to compensate for this. Dental porcelain fabrication is thus akin to sculpting, since precise moulding is not feasible, but this must be done while attempting to allow accurately for the firing shrinkage.

The glass transition in such materials also has other consequences similar to those in organic polymers. Thus, the coefficient of thermal expansion goes from a low to a substantially higher value, reflecting the more liquid-like state above the T_g (Fig. 4.9). As you also expect, there are differences in behaviour due to heating rate (Fig. 4.10) because of the possibility of relaxation effects (*cf.* Fig. 3§4.8). Accordingly, transient stresses on heating, and residual stresses when cold, are thermal history-dependent.

●4.7 Foil matrix

There is an obvious problem with the handling of porcelain powder for a full jacket crown: it has very little strength to start with so could not be removed from the die. It cannot be fired on its die because that would be destroyed if it were gypsum, and any other ceramic would itself react with and bond to the porcelain. This is overcome by using a thin platinum foil support or carrier for the powder. Unfortunately, this is also called a **matrix**, but it is clearly different from that of a composite structure. The demands on such a material are several and severe:

- It must have a high enough melting point to withstand the firing process.
- There must be no reaction with the porcelain so that it can be removed after firing.
- It must be ductile enough that thin foils are available and so that it can be adapted closely to the die of the prepared tooth.

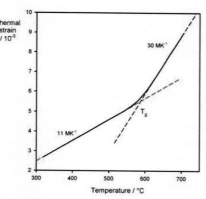

Fig. 4.8 Change of thermal expansion coefficient for a silicate glass on going through the glass transition (slow heating).

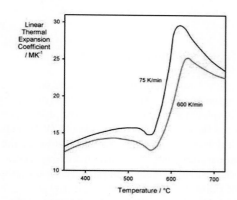

Fig. 4.9 Thermal expansion of porcelain depends on heating rate.

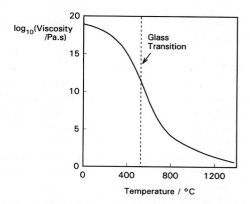

Fig. 4.10 Viscosity of glasses against temperature. Note the clear presence of a T_g, which is defined conventionally by $\eta(T_g) = 10^{12}$ Pa.s for all glass-forming liquids.

Platinum (m.p. 1769 °C; H_V 40, annealed) meets these requirements, and also has a low thermal expansion coefficient (9·75 MK⁻¹). Its high cost is offset by its high scrap value. The process of adapting and burnishing the foil to the die is, of course, cold work, and the metal will become work hardened (11§6) in the process. It is then necessary to anneal the foil by heating to cherry red (~1200 °C), T/T$_m$ ~ 0·72), such as in a bunsen flame, and then letting it cool in air. Burnishing can then proceed further without the risk of cracking.

The use of such a foil matrix has another benefit. By acting as a **spacer** between the die and porcelain, assuming that the die is accurate in the first place, allowance is made for the cement film (9§9). This gap must not be too small as the required seating force may then be high enough to fracture the crown. Clearly, the gap could, in principle, be adjusted by varying the chosen foil thickness, although foil of about 25 μm is usual. There may, as well, be a region of increased thickness on the lingual aspect from the lap joint that is made to complete the matrix. This may assist the extrusion of excess cement, although it is not a deliberate design feature for this purpose.

●4.8   Binders and burn-out

Since kaolin's primary role is as a binder, if the opacity that it causes is too great alternatives must be sought. Instead of water a solution of sugar or starch may be supplied by the manufacturer to be used as a slightly sticky medium for the build-up process. Naturally, on firing, this organic material will decompose and char, causing a blackening of the porcelain body. If this were to remain the work would be spoilt. Accordingly, the firing procedure is adjusted to include a preheating stage, in air, to allow such substances to be burnt out. The situation here is very similar to the burning out of wax from a casting investment (16§4). This is after the gentle heating necessary to evaporate the remaining water without it boiling violently and disrupting the fragile powder. In fact, despite the best of care, dust or fibres from the absorbent blotting paper may have been incorporated accidentally; these too would have similar effects. In any case, the organic dyes that are used to identify the different types and shades used in the build-up must also be burnt out. It is therefore appropriate to ensure that any blackening has disappeared before proceeding to the sintering stage.

§5. Strengthening Treatments

Under normal circumstances any porcelain surface will exhibit irregularities due to the powder particles originally used. The rough surface of the dry powder will only be slightly smoothed by the firing process. Complete melting is not part of the process, indeed it is undesirable, and the pores presenting at the surface will act as flaws in the Griffith sense (1§7). In addition, many microscopic cracks will be present. These result from local thermal stresses, arising because of the differences in thermal expansion coefficients, given the large temperature ranges over which cooling occurs after firing. Such a porcelain is then inherently very weak because crack initiation is easy, and stress concentration occurs around defects in response to external loads. To illustrate the point: in a 3-point bend test a glazed layer 1/80th of the overall depth of the bar increases the measured strength of a typical porcelain by some 60%. The roughness of an as-fired surface would also be detrimental to any soft tissue with which it came into contact, as well as accumulating discolouring debris very readily. In addition, such a surface will be extremely abrasive and, being harder than enamel, will abrade an opposing tooth very rapidly. A smooth **glazed** surface is therefore desirable, particularly if any surface grinding has been done before completion to adjust the occlusion. There are two ways of achieving this goal: overglazing and tempering.

●5.1   Overglazing

The application of a thin layer of low melting point glass powder, which on firing fuses completely and flows to a smooth, even surface (driven by a minimization of surface area by surface tension), is the technique of **overglazing**. This **glaze** is chosen to have a *lower* coefficient of thermal expansion than the underlying porcelain so that on cooling it is under compression (by being in effect an oversize layer). This pre-stress makes the product much more resistant to fracture (23§2.8). The compressive stress thus generated in the glaze has to be reversed before any tensile stress appears (principle of superposition, 23§2.2), to which porcelain and glasses in general are most sensitive.

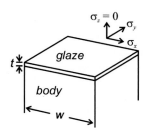

Fig. 5.1 Stress system of a glazed ceramic.

To understand this effect, we can extend the Generalized Hooke's Law equation (6§2.4) to include the strain due to the thermal expansion coefficient (simple additivity, superposition: 23§2.2):

$$\varepsilon_x = \frac{1}{E}[\sigma_x - \nu(\sigma_y + \sigma_z)] + \alpha\Delta T \tag{5.1}$$

Clearly, the strains (and thus the stresses also) in the x- and y- directions are the same (Fig. 5.1), and so must be the strain either side of the interface, in the glaze and the body of the ceramic (they are bonded, there is no slippage):

$$\varepsilon_x = \frac{1}{E_g}[\sigma_{xg} - \nu_g(\sigma_{yg} + \sigma_{zg})] + \alpha_g\Delta T = \frac{1}{E_b}[\sigma_{xb} - \nu_b(\sigma_{yb} + \sigma_{zb})] + \alpha_b\Delta T \tag{5.2}$$

Furthermore, the stress in the z-direction must be zero in both parts since this is at a free surface (unconstrained).

$$\frac{1}{E_g}[\sigma_{xg}(1 - \nu_g)] + \alpha_g\Delta T = \frac{1}{E_b}[\sigma_{xb}(1 - \nu_b)] + \alpha_b\Delta T \tag{5.3}$$

$$\frac{1}{E_g}[\sigma_{xg}(1 - \nu_g)] = \frac{1}{E_b}[\sigma_{xb}(1 - \nu_b)] + (\alpha_b - \alpha_g)\Delta T \tag{5.4}$$

Since the glaze is meant to be very thin, its thickness t in comparison with the body is very small, $t_g \ll t_b$, and we require a balance of forces (stress × area) acting parallel to the plane of the surface (equation 23§2.5):

$$w_b t_b \sigma_{xb} = w_g t_g \sigma_{xg} \tag{5.5}$$

(where w is the width of each part, and $w_b = w_g$), then the stress in the body can be considered negligible, and therefore so is that component of its strain. Hence, equation 5.4 reduces to:

$$\sigma_{xg} = \frac{E_g}{1 - \nu_g}(\alpha_b - \alpha_g)\Delta T \tag{5.6}$$

Inserting some typical numbers for a feldspathic porcelain, say $E_g = 70$ GPa, $\nu_g = 0.3$, $\alpha_g = 6$ MK^{-1}, $\alpha_b = 7$ MK^{-1}, for $\Delta T = 500$ K (i.e. cooling to mouth temperature), we have $\sigma_{xg} = \sigma_{yg} \approx -50$ MPa, that is, in compression. This can be compared with the 'compressive' strength of a such a material, ~150 MPa, while the ordinary tensile strength is ~25 MPa. By the argument of 23§2.8, an applied tensile stress of ~75 MPa would be required for failure.

If the relative magnitudes of the thermal expansion coefficients are reversed, a residual tensile stress in the skin will result on cooling, leading to tension cracking in an irregular pattern all over the surface. This is called **crazing**. (Note that the identically-named crazing of polymers is fundamentally different in character, even if caused by tension, 5§5.) In the extreme case **spallation** may occur, the loss of the layer in large sections. Although glazes that craze are used deliberately on pottery for visual effect, they are of no use in dentistry because they reintroduce the flaws in the surface that the glazing was supposed to eliminate. This effect, however, can be the result of cooling too quickly: the thermal contraction of the skin after it has solidified generates tensile stresses.

●5.2   Tempering

The second method may be called **tempering**. It depends on the poor thermal conductivity of glasses and of porcelain's constituents. If the furnace temperature is raised quickly above the normal firing temperature, the interior of the porcelain will remain at a lower temperature than the skin, which then melts almost completely to form a glaze. More rapid than usual cooling is then applied so that the glaze is solidified at the size of the underlying still-hot porcelain (but not cooling so far that crazing results). Subsequent slow cooling then provides the compressive stress in the skin to increase the fracture resistance as before (Fig. 5.2), and similar calculation as above can be applied, but allowing for the body stress. This technique is also used in the manufacture of so-called "tempered" glass for drinking tumblers, car windscreens, diving masks and the like. Although the strength is increased, when failure does occur it tends to be catastrophic, the stored elastic energy providing at

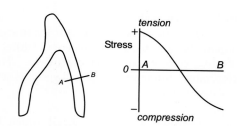

Fig. 5.2 Tempering is meant to result in a stress gradient through the thickness of the ceramic, but this puts the inner surface at risk of tensile failure.

least some of the work of fracture, often more than enough. The effect of this stored energy will have been seen in the many small fragments that result from a broken car windscreen.

●5.3   Ion-exchange

It is the same principle of pre-stressing that underlies the procedure of **ion-exchange**. By immersing the porcelain in a molten potassium salt (at a temperature well below the softening point of the matrix glass) sodium ions from the surface will tend to dissolve in the melt and be replaced by potassium. This is a concentration-driven effect and an equilibrium will eventually be established, not a complete exchange. However, potassium has an ionic radius of 133 pm, somewhat larger than that of sodium, at 95 pm. Thus, the potassium ions must in effect be squeezed into the glass and this generates a compressive stress in the skin, although this must be assisted by the thermal expansion at the temperature of the salt bath. Obviously, this could only work if there were a sufficient concentration of sodium in the porcelain initially. Subsequent cooling back to mouth temperature after the salt-bath treatment will increase the magnitude of the compressive stress. Note that this treatment does not affect the roughness of the surface, which must be dealt with separately and beforehand by some glazing process. It should be noted that this process is diffusion-limited and may only affect perhaps a layer of 100 μm thickness.

●5.4   Stresses

In any case, whichever means of strengthening is employed, the residual stresses must be finely balanced to avoid fracture, for an excessive compressive stress in the skin can generate shear stresses to exceed the strength of the material (1§6.3). The nature of any imbalance can be judged from the nature of the cracks (Fig. 5.3), if they form. Tensile stresses in the skin lead to perpendicular cracking, whereas compression stresses produce shear cracks at angles close to 45° to the surface. Of course, neither type of crack can be tolerated in service, both because of the low strength of the fabrication (the propagation of an existing crack would be easy) and the discolouration that would ensue.

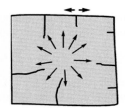

Fig. 5.3 The nature of the stresses causing cracks may be deduced from the orientation of the cracks. Left: tensile – *e.g.* from superficial cooling too rapidly; right: compressive skin stresses – *e.g.* by 'tempering' from too high an interior temperature.

Despite the benefits of the composite structure, use of alumina and post-fabrication strengthening, the inherent limitations of ceramics remain. The sensitivity to scratches and flaws is essentially unaffected, and this means that direct tensile stress is highly undesirable. The design of dental porcelain devices of whatever type must therefore pay particular attention to avoiding tensile loading and bending, and achieving adequate thickness. It is here that the critical nature of dimensional accuracy in impressions, models and dies become most apparent. For example, if a porcelain jacket crown is undersized such that placing it on the conical prepared tooth tends to widen a cervical diameter – wedging it open, failure is almost guaranteed because the inner surface of the crown is then in tension. This surface is left rough, unglazed, to facilitate cementation (retention will be through mechanical key). Similarly, the firing shrinkage generates stresses and the design of the tooth preparation should also take this into account to reduce the risk of fracture. However, it can be noted that since, in the Griffith sense, flaw-sensitivity depends on the sharpness of crack tips, anything that causing blunting will be beneficial. Etching of the inner surface may therefore attack those high-energy sites first and thus cause some crack-tip blunting. This may occur naturally to some extent with a strongly acid (*cf.* Fig. 8.1), but could better be done deliberately in advance with a stronger etchant (containing hydrofluoric acid – the use of which requires extreme care! See §10.).

●5.5   Thermal shock

The skin stresses generated for strengthening purposes have been discussed in terms of differences in effective size or expansion coefficient. However, temperature differences alone will generate stress that may cause failure.

Consider a bar of material which is fixed rigidly at both ends (Fig. 5.4a). If this is cooled from a higher to a lower temperature, instead of it shrinking a tensile stress will be developed equivalent to that

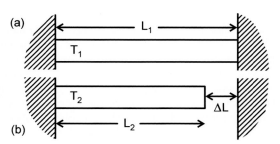

Fig. 5.4 Strain associated with thermal contraction. The end walls are considered as completely rigid. $T_1 > T_2$.

required to return it to its original length were it allowed to contract freely on cooling (Fig. 5.4b):

$$L_1 = (1 + \alpha(T_1 - T_2))L_2$$

$$or \quad \Delta L = \alpha.L_2.\Delta T$$

$$i.e. \quad \varepsilon = \frac{L_1 - L_2}{L_2} = \alpha.\Delta T \tag{5.7}$$

where $\Delta T = T_1 - T_2$, and α is the coefficient of linear thermal expansion. Now if E is the modulus of elasticity of the material, the tensile stress required for this much extension is given by equation 1§2.4 a:

$$\sigma = E.\varepsilon \tag{5.8}$$

Clearly, if σ reaches σ_t, the tensile strength of the material, failure will occur. Thus, combining equations 5.7 and 5.8 we get:

$$\Delta T_{max} = \frac{\sigma_t}{E.\alpha} \tag{5.9}$$

which is the maximum temperature drop that can be survived without cracking, for a constrained-size piece.

This equation also represents the maximum temperature drop that can be sustained at the surface of a hot body where the inner portion constrains the size of the skin (*cf.* Fig. 5.3, left). This, then, is the **thermal shock resistance** and can be seen to be defined as a material property. It may be envisaged as the maximum temperature which will be survived without cracking of a large object dropped into water, which is indeed how it is often tested in practice. Some examples are given in Table 5.1. Note how poor are dental porcelain, and especially enamel (assuming that its properties are correctly known) (see 20§7).

Table 5.1 *Thermal shock resistance of some materials.*

	Tensile strength / MPa	Young's modulus / GPa	Thermal expansion (linear) / MK^{-1}	Thermal shock resistance / K
diamond	1200	1050	1·2	1000
borosilicate glass	33	65	3·3	150
alumina	250	380	4·3	150
soda glass	30	74	8·5	50
porcelain (feldspar)	25	60	7	60
dental enamel	10	70	11·4	13
dentine	100	15	8·3	800

The effect depends however very much on the surface condition of the object since the crack initiation, which is in tension, will be affected by defects there (1§7). In other words, the shock resistance can be made worse by bubbles or scratches, and improved by glazing.

In practice, since temperature changes do not normally occur in a step-like abrupt fashion, time is part of the equation. That is, diffusion of heat affects the temperature profile through the section, therefore the temperature gradient, and thus the stress gradient, which in turn is affected by the shape and size of the object being tested or exposed. Accordingly, the **thermal diffusivity** of the material is also a factor affecting the outcome, and the ranking of the above figures would be modified somewhat. Then again, it depends on the fluid doing the quenching, as heat must be transferred across the interface, the effectiveness of which is measured by what is known as the **Biot**[3] **modulus**. Although there is no need to develop these ideas here it provides a further indication that experimental work needs to mimic service conditions if the results are to be useful in that context.

---

[3] "Be-oh"

§6. Metal-ceramics

One approach to the problems of ceramics is to use a 'porcelain fused to metal' (PFM) technique.[10] This relies on the strength of a cast metal device, and simply provides a **facing** or **veneer** of porcelain to give a tooth-like appearance to the labial surface; the lingual surface, being ordinarily unseen, can often remain plain metal. This approach primarily requires that the veneer be bonded to the metal, but it must also be opaque enough to mask it, and this is dealt with by a thin layer of very opaque porcelain. However, this leaves little room for the translucent material, and such restorations tend to appear less natural than full porcelain crowns.

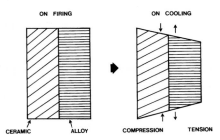

Similar considerations apply to these so-called 'metal-ceramic' (or 'ceramo-metal') devices as for other strengthening methods (§5) in the sense of residual and interfacial stresses. It is essential that the coefficient of thermal expansion of the ceramic is less than that of the alloy to ensure that the ceramic is under compression rather than tension (Fig. 6.1). Contrary to popular ideas, this cannot conceivably affect the actual strength of the metal-ceramic bond and indeed must in effect reduce it as a considerable shear stress is thereby applied to the interface. There is therefore a limit to the coefficient mismatch that can be tolerated if premature failure in shear (**spalling**) is to be avoided, and this maximum tolerable difference is of the order of 0.7 MK^{-1}, although 0.5 MK^{-1} is considered safe. This effect will be sensitive to flaws at the interface. However, the presence of a compressive stress in the

Fig. 6.1 It is intended that a porcelain veneer be under compression after cooling the work, although this results in the interface being subjected to shear.

ceramic again means that a much larger tensile stress must be applied (superposed, 23§2.2) before the net stress is large enough in tension (*i.e.* under bending) to cause failure in the ceramic itself. This, too, is a form of pre-stress (23§2.8). It may give the impression of a stronger *structure*, but the nature of the failure must be carefully identified for the behaviour to be understood. For the scheme to work it is necessary that a means of **stress transfer** across the interface exists. This is similar to the idea of the bonding required within particulate composites for the filler to exert an effect (6§2). The metal-ceramic structure here is still a composite, even if of only two layers. We now consider the requirements and means of bonding of the two types of material.

●6.1 Mechanical key

Molten glasses in general do not wet clean metal surfaces. This is similar to the case of resin-based restorative materials in contact with enamel, and very similar to soldering fluxes on clean metal (22§2): the systems are chemically dissimilar. As a result no true adhesion can be obtained from van der Waals forces. Mechanical retention must be utilized in the sense of obtaining a 'key' between the two, rather than large-scale dovetails or similar designs as are applied in the 'retention form' of amalgam restorations, for example. There are several methods of obtaining a key, all of which effectively

Fig. 6.2 Mechanical key may be created by adding metal buttons to the surface.

roughen the cast metal surface: abrasion, etching or by depositing small granules of metal (Fig. 6.2). Whichever method is used the aim is to obtain an effective and small-scale interlock with the ceramic. But, no matter how intimate such an interlocking may be, there must still be a crevice, and stress transfer from the one to the other must be through a relatively small number of points, and stress concentrations will result. This is exaggerated by the unavoidable difference in modulus of elasticity between the two materials: under stress they will show different strains. The situation can be improved upon.

●6.2 Oxide coats

As has been mentioned elsewhere, some metal oxides can be extremely strongly adherent to the substrate metal under certain conditions (see 13§5, 19§2, 21§2). Chromic oxide on Cr-containing alloys is one of the best known examples (19§2, 21§2). It is also evident that silicates may form from the reaction of metal oxides with silica and its glasses, and in fact the wetting of metal oxides and their direct dissolution in the melt occurs quite readily in many cases (*cf.* 22§2). The opportunity exists therefore for an

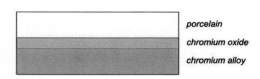

Fig. 6.3 The silicate glass of a porcelain matrix can wet and bond to metal oxides.

intermediate layer between the porcelain and the metallic substrate, *i.e.* an oxide coat to which both may adhere strongly (Fig. 6.3). When the metal naturally has such a coating this kind of adhesion may be effective directly. Otherwise, an intermediate metal or metals may have to be deposited. Mostly, however, special alloys have been developed for this technique to meet the severe demands:

- The solidus temperature must be high to permit the fusing of the ceramic without melting the metal.
- The thermal expansion of alloy and ceramic must be properly matched to generate the required compressive stress in the porcelain.
- The metal should not creep or sag at the firing temperatures (*i.e.* under its own weight); similarly, appreciable grain growth should not occur.
- Finally, the alloy surface must oxidize to be wetted by and react with the ceramic to form the bond.

This last requires the inclusion of alloying components that are intended to oxidize but whose oxides are strongly adherent to the metal beneath. The high quality of this bond is also essential to the elimination of microscopic crevices that could readily become active corrosion sites (13§4.8). No ordinary dental casting alloys are suitable.

Dental gold alloys as used for crowns and inlays are not suitable because of the copper content (19§1), the oxide of which dissolves readily in silica glasses to give a strong green colouration. Alloys based on Au-Pd and Ag-Au-Pd (see Figs 19§1.17,18) and related systems are used, but typically with relatively low gold content and with additions of several elements such as Fe, In and Sn, at around 0·5 mass%. These latter form the oxide coating. Ag-Pd alloys, again alloyed with Sn and In, can also be used but there is a tendency to produce a greenish-yellow discolouration which has reduced their popularity. This is thought to be due to colloidal silver, formed by diffusion into the porcelain from the alloy or vapour.[11] This may be partially offset by the presence of CeO_2 as an oxidizing agent. However, it is possible to tin-plate a gold alloy, heat to allow this diffuse partially into the metal, then oxidize the surface so that the porcelain may wet it (Fig. 6.4).

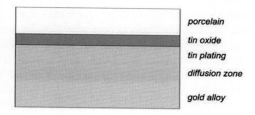

porcelain
tin oxide
tin plating
diffusion zone
gold alloy

Fig. 6.4 Gold alloys cannot hold an oxide coat, so an intermediate oxidizable metal coating is used.

Cr-Ni alloys are cheaper but much more complex, with many compositions possible and available. Frequently containing Al, these alloys depend on the chromic and aluminium oxides produced on heating for the formation of the bond. However, these alloys are also typically multiphase, which complicates the bonding but in addition increases the risks of corrosion (13§4.9). Since nickel sensitivity is a widespread and serious problem, care must be taken not to use such alloys in allergic patients. Even so, it must be borne in mind that a patient may be sensitized to nickel by the use of these alloys since some corrosion is likely.

Although oxide is required, the thickness must be carefully controlled: if it is too thick its crystallinity would make it structurally weak, and thus give a weak bond. Accordingly, to prevent further growth during subsequent firings, it is necessary to purge the furnace with nitrogen or argon before applying the vacuum (which is, of course, not perfect) to reduce the amount of oxygen present to a very low value.

●6.3   Platinum-bonding

One technique creates what amounts to a platinum-bonded full jacket crown.[1] A second, thin platinum foil 'matrix' is laid down over the first. It is then grit-blasted, tin-plated electrolytically, and this layer is then allowed to alloy with the platinum by heating in a vacuum. The heating is then continued in air to create an oxide coat. The build-up of the crown then follows in the normal way, allowing of course for the masking of the metal first with an opaque layer. On firing, the bond is developed by the SnO_2 partially dissolving in and bonding to the glass, although the grit-blasting assists by increasing the surface area for reaction and providing some mechanical key. The platinum foil which is thus incorporated in the crown has little strength in itself (being only 25 ~ 50 µm thick, and Pt is a soft metal), but it eliminates surface flaws and microscopic firing shrinkage cracks on the inner surface of the porcelain. The effective strength is therefore increased markedly. There is a second advantage. Since the Pt-foil is tin-plated and therefore oxidized on both surfaces, the strength of the cementation can be improved by using a polycarboxylate cement which will bond to that oxide as well as the tooth. The loss of mechanical key on this inner surface, compared with the rough unglazed porcelain, is therefore offset by the chemical bonding now possible. The fact that the coefficient of thermal expansion of Pt is ~9·75 MK^{-1}, similar to many porcelains, also helps to avoid stresses and distortions during repeated heating and cooling.

§7. Colouring

The solubility of metal oxides in glasses is important also for the colouration of the porcelain better to simulate natural dentition, as well as to account for both patient variability and irregularity of tint within even small areas of tooth. The principles have been dealt with elsewhere (24§6.4).

There are a wide variety of colours available, mostly from the oxides of transition metals (Table 7.1). The reactions are entirely analogous to those of the alkali metal oxides, except that the products are coloured. The colourations are usually quite intense and only very small amounts of the oxides are used for the shades required in dentistry. They are therefore typically provided as a dilute solution in a powdered glass, *i.e.* already reacted, and thus become part of the single-phase glassy matrix. Other metal oxides can have other effects: thus CeO_2, TiO_2, SnO_2, ZrO_2 and ZnO can all be used to increase the opacity of the glass. This works because the reaction to form the silicate, and thus show colour, only occurs very slowly even at high temperature and so remain as separate phases. They are therefore better described as pigments. Some elements also have a range of oxidation states and can be affected by the oxygen-status of the furnace as well as each other. The chemistry of porcelains can therefore be quite complex. Appearance can therefore be dependent on thermal history: time and temperature. This all suggests that care must be taken to avoid over-firing if undesired colours are to be avoided.

Table 7.1		Some oxides used for colouring dental porcelain.
Indium	In_2O_3	ivory
Praseodymium	Pr_2O_3	ivory
Iron	Fe_2O_3	brown
Nickel	NiO	brown
Cobalt	CoO_2	blue
Copper	CuO	green
Chromium	Cr_2O_3	green
Manganese	MnO_2	lavender
Vanadium	V_2O_5	brown, yellow-green
Cerium	CeO_2	yellow, pale green
Titanium	TiO_2	yellow-brown; opaque white
Tin	SnO_2	opaque white
Zirconium	ZrO_2	opaque white
Zinc	ZnO	opaque white

The oxides of lanthanides (including Ce) and actinides may be used to produce some fluorescence in an attempt to mimic the natural fluorescence of tooth substance. As might be imagined, however, the use of uranium salts has been discontinued as a source of fluorescence in this context because of the radioactivity it confers on the porcelain. However, to be fair, this was more by way of the safety of the technicians, because of the large quantities that they would handle over long periods, than of the patients whose dose would be very small indeed. Radiation arising from granite chips used in the concrete of buildings is a far more significant worry.

§8. Boric Oxide

One other oxide is noteworthy in the present context, that of boron. Boric oxide (B_2O_3) is one of the most difficult of all substances to obtain in a crystalline form, and it confers some of this resistance to **devitrification** on the silicate glasses in which it is incorporated. This may be of considerable value when several firings are required, and a better approach to equilibrium would otherwise be expected to be made. As the firings proceed finely-divided crystalline material would be produced. This would be distributed throughout the glassy matrix and result in greater opacity, because of light scattering at boundaries (where there are refractive index changes, 24§5), and weakness. Many silicate minerals have layer structures which **cleave** easily along the planes of the layers. In addition, boric oxide confers some toughness to the glass (*i.e.* it is a little less brittle) as well as a lower coefficient of thermal expansion. Unfortunately, it is rather more prone to hydrolysis and dissolution than silica, especially when much metal oxide is present, and so can only be used in limited proportions (compare the removal of boric oxide-based soldering fluxes from the workpiece by dissolving them in water, 22§2.)

In fact, resistant though silica itself is to hydrolysis, some etching by aqueous media does occur under normal conditions, and particularly in the presence of strong bases.[12][13] The weakening effects of that etching on the strength of glass fibres are well documented (Fig. 8.1) (see 1§7). Indeed, this effect is thought to account for the delayed fracture of glass under stress (so-called **static fatigue**), whether that stress is internal or externally applied.[14] It is assumed that this mechanism also plays a part in limiting the strength that can be achieved in practice with dental porcelain, where both service stresses and internal frozen stress are present. Etching is enhanced by the alkali metal content of glasses (which breaks up the network – see reaction 1.1 and Fig. 9§7.1), and the feldspar of porcelain contributes much to this in the dental context. The system provides a further example of how testing under service conditions, which include time of exposure, is necessary for a proper understanding of material behaviour.

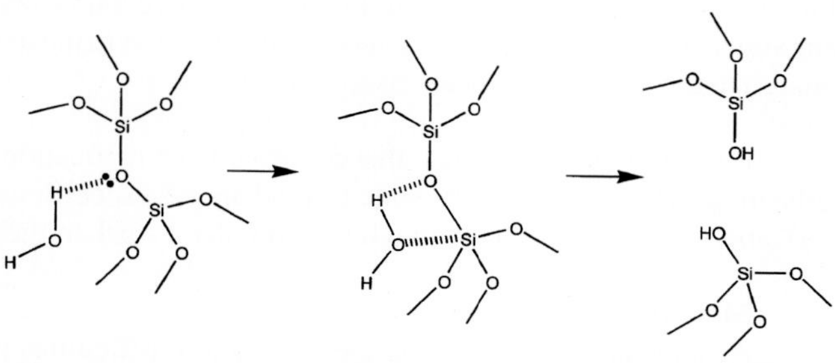

Fig. 8.1 Hydrolysis of Si-O-Si bonds by water. This process is aided by stress in the silicate and OH⁻ ions.

For comparison, borosilicate glass, as used in laboratory and domestic glassware, has a composition approximating to 80% SiO_2, 13% B_2O_3, 4% Na_2O, 2% Al_2O_3. The high boron content and low metal content result in considerable resistance to acid hydrolysis, very low thermal expansion (3·3 MK^{-1}), and good thermal shock resistance (Table 5.1). The low alkali content, however, raises the softening point considerably, and such a formulation would not be workable in a dental porcelain (especially with the thermal expansion mismatch that would occur with respect to crystalline phases).

§9. Other Approaches

Dental porcelain works as a restorative material because the extreme brittleness and weakness of glass has been offset by the presence of harder, stronger particles of some filler, such as quartz and alumina (§3). However, while porcelain denture teeth can be mass-produced by hot-pressing in metal moulds, the individual crown and inlay designs required for restoration work preclude such expensive procedures. Porcelain, as described above, being composite and having a thermal history-dependent structure, is not itself a castable material. Nevertheless, several other approaches have been designed.

●9.1   Castable ceramics

Glass can obviously be cast, using the lost-wax process (Chap. 17) but as such would be too fragile for service as well as being transparent or, if of the 'opal' type (*cf*. 9§8.1), of an unnatural appearance compared with tooth. However, glass compositions are possible which are sufficiently unstable that, on annealing, that is, a special heat-treatment, precipitation of other phases will occur in exactly the same sense that overfiring ordinary dental porcelain causes crystallization. This **controlled devitrification** is the process (oddly) named as **ceramming** by the manufacturers of dental castable ceramics (oddly, since from the materials science point of view the glass is a ceramic in the first place – the point seems to be that they are then composite materials). The phase precipitated in these products is a mica-type mineral in the form of very thin plates (Fig. 9.1). These are produced with random orientations and frequently intersect each other. This therefore forms a more or less continuous network throughout the mass, and has been called a '**house of cards**' structure despite its great irregularity (*cf*. the similar arrangement of γ_2-crystals in amalgam, Fig. 14§5.2) (Fig. 9.2). This precipitate, which can be as much as 55 vol%, by virtue of its ready cleavage in the plane of those plates (they have marked layer structure), increases the toughness of the ceramic by increasing the crack path length and by blunting the crack tip. The light-scattering these crystals produce in this case is beneficial in converting the transparent glass to a translucent composite.

Fig. 9.1 Precipitated mica-type crystals increase the toughness of the ceramic by forcing crack deviation through the cleavage planes of the mica.

Fig. 9.2 A 'house of cards'. Note that these do not intersect as do the structures so named.

Although this kind of material is free of the porosities of sintered porcelains (it is necessarily fully-dense to start with, as-cast) and so has a much simpler preparation technique, the opportunity has been lost for the detailed colour characterization that skilled dental porcelain workers can employ in creating good tooth mimicry. Some superficial colouring however could be applied, refiring as necessary, although tricky. However, this class of material appears to have been discontinued.

It should be noted that the controlled devitrification here should be contrasted with that due to overheating (§1.3, §8) as the crystals formed are of a specific form (plate-like), particular structure (cleavable mica), and limited extent – too much would indeed weaken the material.

●9.2   Machinable ceramics

As a means of overcoming the fabrication difficulties inherent in powder-method dental porcelains, machinable ceramics have been introduced to dentistry. These ceramics are formulated to be tougher (*i.e.* not crack and shatter) when they are shaped by machining with high-speed abrasive tools.   They owe this to the crack-stopping and blunting mechanisms arising from a two-phase structure, containing precipitated mica-like crystals, as with the castable ceramics (§9.1), but to a higher volume fraction of ~70% (Fig. 9.3). However, it is limited so far, by the nature of the copy-milling type of machine used, to the production of inlays. However, this is the very area where the powder method has most problems: firing shrinkage makes the attainment of an accurate fit very difficult. However, because the shape and dimensions of the machined inlay are determined by mapping from a video image which has been outlined by the dentist, the accuracy does not yet match that of a good metal casting. This **computer-aided machining** (CAM) also requires a rather expensive set of equipment.

Fig. 9.3 Machinable ceramics have a similar structure to that of heat-treated castable glasses, but to a higher volume fraction of mica.

The difficulties of the direct CAM copy-milling approach have been addressed in part by the use of a method similar to that used for lost-wax casting. A pattern is prepared using a light-cured modelling material, and this pattern is then copied mechanically using a surface-tracing system which is linked directly to the cutter. Accordingly, the prospects for success using this technique seem better in that the pattern dimensions are more easily controlled, even if in an intermediate material. In addition, the pattern can be either direct – prepared in the tooth, or indirect – prepared on a model.

Whichever technique is used, the restoration is necessarily of just one uniform colour and has rough surface after milling. Staining and glazing (in a furnace) to achieve both a better appearance and a smooth, flaw-free surface are required.

●9.3   Infiltration

As was said above, copy-milling has yet to be able to deal with crowns, yet powder-method porcelain suffers from the difficulty (amongst others) that porosity cannot be entirely eliminated. In addition, its strength is not high enough – even with alumina – in thin enough sections to offer a suitably conservative treatment. The infiltration technique addresses these problems quite effectively although at the expense of an elaborate procedure.

In this approach a deliberately porous (bisque stage) substructure is first prepared with pure alumina using the powder method. The little shrinkage that occurs at this stage is limited by keeping the time and temperature low.   This means that dimensional accuracy (fit) is much improved although the bisque itself is fragile. The outer (non-fitting) surface only is then coated liberally in a slurry of a glass powder and, after drying, the device is fired again. Now the glass melts and is **infiltrated** into the porosity of the alumina skeleton by capillarity (10§2.9) (note that the flow must be unidirectional to avoid closing the porosity – *cf.* Fig. 10§2.17: bubble flaws may be inevitable). Obviously, this requires sufficient wetting of the alumina and low enough viscosity for the glass (10§2.4). After cooling, the excess glass is trimmed away and the surface ground back and alumina-blasted (20§4).   At this point we have a randomly-connected **bicontinuous**, two-phase composite (Fig. 9.4) which is homogeneous (*i.e.* everywhere similar) and essentially bubble-free

Fig. 9.4 Infiltrated bisque structures have no simple crack path through one phase and so are tough and strong.

(although not perfect, 10§2.9). The strength is therefore raised substantially (Griffith criterion, 1§7). The intrinsic strength and stiffness of the alumina network is the other major aspect of interest. The glass infiltrate also controls the optical properties. Subsequently, the crown is coloured, veneered and glazed with further firing.

If the infiltration firing were performed at low pressure (in a 'vacuum' furnace), any bubbles that would be trapped by the irregularity of the infiltration front would contain less gas and shrink more due to Laplace bubble-pressure and dissolution as for ordinary porcelains (§4.3).

The advantage of this bicontinuous structure is that a crack cannot propagate through just one phase: it is obliged to cross alumina and glass roughly in proportion to their volume fractions because no path of avoidance exists, even if it becomes very convoluted. The material will therefore be tougher than ordinary porcelains as well as stronger, but dependent on the continuity and strength of the sintered contacts of the alumina.

The ordinary powder method is now replaced by **slip casting** for such materials, although it is really only a minor variation of the technique (derived from the industrial process used for mass-produced pottery). The die on which the device, the **core**, is built up is deliberately absorptive, so that as the **slip** – a liquid slurry of the powder – is applied the water is automatically absorbed such that the particles stay in place. This approach is also used to make cores from spinel, which is magnesium aluminate ($MgAl_2O_4$, *alias* $MgO \cdot Al_2O_3$), and from zirconia with 30 mass% alumina as a binder.

●9.4   Milling plus infiltration
One disadvantage of the infiltration method (§9.3) is that in order to avoid the expected shrinkage the alumina is only fired at about 1120 °C. This is a rather low temperature compared with its melting point of 2050 °C, *i.e.* $T/T_m \sim 0.60$. If the sintering were to be more extensive, and the structure therefore stronger, it would require higher temperatures or very much longer times but be accompanied by much greater shrinkage. This would make the method unusable. It would also require a special furnace. All this can be avoided by a combination technique in which the computer-aided milling of an oversized, low-density, partially-sintered alumina blank (the "green" body) from a stock piece is followed by sintering to a greater density, effectively to the target size, and then glass infiltration as before (and then colouring, veneering and glazing, of course). While it would have the additional advantage that the porosity of the alumina core would be more controlled, the strength would be substantially higher because the sintering would be more complete and more reliable. This avoids the problem of milling a dense alumina blank. As can be expected for a material used as an abrasive, it would itself be very hard and difficult to grind. This technique is also used for low-density zirconia blanks.

●9.5   Pressable ceramics
An alternative to the castable ceramics, which require a precipitation heat-treatment, is to include a second phase in the glass to start with. The problem, of course, is that the material is then very viscous even when the matrix has been melted, and so the only means available for the casting is to use high pressure moulding, and time must be allowed for flow. Given that, higher volume fractions of disperse phase may be used than might otherwise be considered (Fig. 9.5). Essentially zero porosity is ensured because this is the state of the initial, manufactured ingot. Such materials, of course, require specialized equipment.

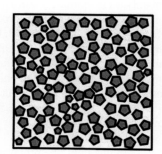

Fig. 9.5 High volume fractions of disperse phase can be used in pressable ceramics.

A variety of such products have appeared, relying on leucite (§2), (35 vol%), lithium disilicate ($Li_2O \cdot 2SiO_2$) (70 vol%), or lithium phosphate (Li_3PO_4) – which is soluble in water (0.4 g/L), although this is likely only to affect the cementation surface and could be an advantage by providing more key. One benefit of such a system is that the refractive index match between core and matrix can be better (24§5.11), conferring more translucency and tooth tissue-like appearance. Nevertheless, such materials are still monolithic and require characterization for realism.

Whatever the relative merits of the various alternative strategies outlined here, it is to be expected that dental ceramic strength will improve further. Thus the service lifetime of such devices may increase, and the treatment become more conservative as they can be made thinner. However, it is a key observation that despite a great improvements in formal strength, the failure rate of ceramic restorations has not declined. This indicates that the longevity is at least in part dependent on the design, and thus the skill and care of the dentist.

§10. Etching

A porcelain restoration needs to be cemented into place. Unfortunately, no direct chemical reaction is feasible. For example, the cation concentration is too low for polycarboxylate cement to be effective, although intermediary silanes can be used (10§5.1) if a filled-resin 'cement' is used (24§5.11). The usual approach is to grit-blast (20§4.1) the inside of all-ceramic crowns in order to obtain the necessary mechanical key (Fig. 10.1). This has the awkward side-effect of lowering the strength substantially because the roughness itself constitutes flaws in the Griffith sense, but also because the grit-blasting produces many minute cracks, and the inner surface is normally put into tension by the flexure arising from biting forces.

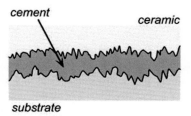

Fig. 10.1 Retention of cemented ceramic restorations depends on mechanical key.

Incidentally, the susceptibility to tensile failure from the inner surface of crowns (Fig. 5.2) is why glass-ionomer cement should never be used for all-ceramic crown restorations: the absorption of water and thus the expansion, even after setting, is sufficiently great to split the crown. Obviously, this does not affect inlays.

The alternative to grit-blasting is etching by a solution of hydrofluoric acid, in which the following type of reaction occurs rapidly:

$$SiO_2 + 4HF(aq) \Rightarrow SiF_4 + 2H_2O \tag{10.1}$$

This not only creates roughness through differential reaction rates on matrix and core phases, but highly-strained areas will be attacked more quickly (*cf.* §8), as will be cracks or other flaws present as a result of the firing process (§5). Thus, the advantages are that prestressed areas (23§2.8) are removed and flaws are blunted, reducing the stress-concentration effect, in the process of creating the key. It should be noted in this context that 'APF' (10§7.3) also etches porcelain, and such treatment should not be applied where there are porcelain restorations of any kind: the rough surface would stain as well as weaken the device.

After etching, porcelain can be silanized (10§5.1) to improve the short–term bond strength, but this will not work on alumina or zirconia cores – indeed, neither of these can be etched, their reactivity being too low.

The disadvantage is that HF is a considerable hazard, even if dilute, and requires handling with extreme care: all necessary protective gear must be worn and adequate ventilation provided (the vapour is also hazardous, as is the gas SiF_4 from the above reaction); spill clean-up kit is essential. All users should be properly trained. Not only is HF extremely corrosive to metals, oxides of many kinds and all silicates, it is described as a 'contact' poison, acting systemically. Because it is not a "strong" acid in the sense that it does not ionize fully as does HCl, for example (see Fig. 9§3.3), the diffusion of HF through tissue is rapid and burns are therefore fast and very deep, although typically not immediately obvious because nerves are damaged so quickly that there is no pain – at first. This allows further penetration before the need for treatment is apparent, by which time amputation is likely to be the only solution. Bone damage and heart problems follow. Although as a first step, rinsing with water is useful, it is not enough by any means. Workplaces where HF is used should be provided with the means of appropriate first aid – commonly a calcium gluconate gel. No exposure should ever be ignored.[15]

In this light, the proposal to use HF to etch porcelain in the mouth for repair is extraordinary.

§11. Zirconia

As will be apparent, there are deficiencies with all existing ceramics for dental prosthetic applications in terms of strength and toughness, especially when coupling the demands for appearance. Zirconium dioxide, ZrO_2, commonly known as zirconia, has been seen as offering a solution as it is said to be capable of offering a toughening mechanism along with very high strength. It is a very hard (Vickers: ~12.7 GPa), extremely refractory material, m.p. ~2690 °C, and has very high modulus of elasticity (E ~ 210 GPa).

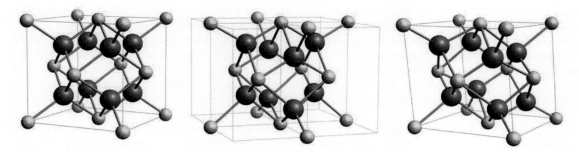

Fig. 11.1 Crystal structures of zirconia (left to right): cubic, tetragonal, monoclinic. Unit cells and orientations chosen to show equivalent atoms and the successive distortions (some atoms omitted from the outside of the tetragonal structure for clarity). Zr^{4+}: small; O^{2-}: large.

●11.1   Constitution

Zirconia shows **allotropy**, having three crystal structures (Fig. 11.1) according to temperature. Above ~2370 °C it is cubic (*c*), but tetragonal (*t*) below that temperature down to ~1170 °C, and below that it is monoclinic (*m*) (Fig. 11.2).[16] The transformations between these allotropes are martensitic, that is, diffusionless and reversible, but they are **athermal** and so are progressive over small temperature ranges (21§1.3).[17] The transformation of *c*-ZrO_2 to *t*-ZrO_2 on cooling involves the oxygen ions alternately being displaced up and down slightly, which leads to the tetragonal distortion: the axial ratio of the (effective) unit cell goes from c/a = 1.00 to ~1.02, with an overall increase of volume of ~0.8%, although the bulk (polycrystalline) change is higher at ~2.3%, implying that there are intergranular voids arising from the random orientation of the tetragonal c-axis. Bear in mind that there can be no preferred orientation for the crystal axes in polycrystalline cubic zirconia.

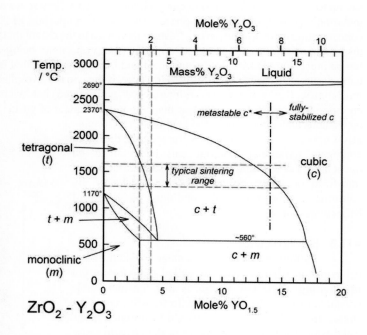

Fig. 11.2 Portion of the zirconia-yttria phase diagram. The approximate range for ~100% *t** is indicated by the vertical dashed lines.

However, the change from *t*-ZrO_2 to *m*-ZrO_2 is much more dramatic, with a crystallographic cell volume change of about 2%, but the distortion leads to a bulk volume change which can range from 3 to 5% or more, depending on the history of the material, as further intergranular voids are created by the random orientation of the shearing direction: there are four possibilities for tilting the original tetragonal c-axis. Such a magnitude of change cannot be accommodated in a ceramic, and massive cracking occurs. This is important because if sintering of a powder compact is to be used to create a fully-dense material for a prosthetic device, this has to be done in the region of 1400 ~1600 °C in order that diffusion occur at a useful rate (cation diffusion all but stops below ~1200 °C), well above the *m-t* transition temperature. Hence, on cooling, destruction of the device would inevitably occur.

The crystallographic transformations are affected by solid-solution lattice substitutions, and a number of oxides are known to have useful effects: MgO, CaO, CeO_2 ... but yttrium oxide, Y_2O_3, or yttria, has been found to be effective in the dental context and may be taken as the archetype. The Y^{3+} ion (radius 104 pm) is substantially greater in size than that of Zr^{4+} (86 pm), so must introduce appreciable distortion, but this is associated with stabilization of the cubic structure, as can be seen by the $c / c + t$ phase field boundary dropping to the right in Fig. 11.2. That distortion may be accommodated to some extent by the oxygen-deficiency (*i.e.* vacancies are present) due to the different stoichiometry of yttria: one vacancy per two Y^{3+} ions. (These vacancies can, of course, occur randomly at any of the oxygen sites, and are not 'associated' with the Y^{3+} ion as nearest neighbours, as is often indicated in drawings. There are no yttria 'molecules'. Indeed, the vacancies are more likely to occur one step further away – 'next nearest neighbour'. Indeed, it is said that it is the vacancies themselves that produce the stabilization effect.[18]) The stabilization also applies to *t*-ZrO_2, depressing the temperature for the damaging transformation to *m*-ZrO_2. If enough **dopant** is added, then **fully-stabilized zirconia (FSZ)** is obtained. Below that amount, the material is said to be **partially-stabilized (PSZ)**. The dopant can then be indicated, *e.g.* **Y-PSZ**, with a number prepended to indicate the molar-percentage, e.g. **3Y-PSZ**, where the percentage is usually in terms of Y_2O_3.

On its own, the effect of these transformations would not be enough to be useful – the spontaneous formation of monoclinic would still be expected at low temperature. However, the phases formed when only partially-doped are now **metastable**, that is spontaneous transformation of $c \Rightarrow t$, or $t \Rightarrow m$ does not occur. In effect, the martensite start temperature, M_s, has been depressed below ordinary temperatures (*cf.* Fig. 28§2.2). The transformation can nevertheless be triggered by stress, and this effect may be turned to good use.

Two strategies for the design of the ceramic are now possible. Firstly, at low concentrations of yttria, say up to ~3 or 4 mol%[4] sintering can yield only *t**-ZrO_2 (where we add the asterisk to emphasize that this is a metastable phase), or that same phase with a small proportion of *c**-ZrO_2 (see Fig. 11.2). At higher yttria concentrations, the proportion of c^* can be increased to the extent that the t^* is considered to be relatively minor dispersion within the c^* grains as a matrix (the proportions being indicated by the tie-line at the sintering temperature, Fig. 8§3.9). Fully-stabilized and homogeneous c, that is, crossing into the c field at the sintering temperature, is a further possibility. However, if the c in the two-phase $c + t$ field contains more than about 14 mol% yttria, that is, having been sintered (*i.e.* equilibrated) below about 1400 °C, it will persist as stable c on cooling. In contrast, for c with less than about that amount of yttria (*i.e.* sintered above ~1400 °C) it becomes metastable, and transforms: $c^* \Rightarrow t$, which t is then stable at low temperature. In that temperature region, the initial conjugate tetragonal material can be retained on quenching, *i.e.* it would then be metastable, but would transform on slow cooling: $t^* \Rightarrow m$. Clearly, by adjusting composition and conditions, a range of behaviours can be accessible.

●11.2 Strengthening

As indicated above, advantage is taken of this behaviour in what is generally called **transformation toughening**. The key to this is that the metastable crystal forms can be prompted into transformation by stress. Intergranular cracking would be initiated in the usual way at a flaw that is active in the Griffith sense (1§7). However, when such a flaw is loaded there is a stress concentration at the crack tip, depending on how sharp it is, but this may be five or more times the nominal stress. Under this stress, the metastable phase is caused to transform, but as has been seen above, this transformation is associated with a volume change. In the confines of the bulk material, this expansion is constrained, either closing the existing crack or

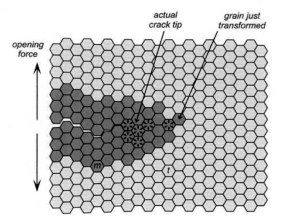

Fig. 11.3 The transformation field around a growing intergranular crack in metastable tetragonal (*t**) zirconia. The transformed material is monoclinic (*m*). The stress field ahead of the crack tip causes prompt transformation, which creates a compressive prestress, opposing crack opening.

---

[4] Great care must be taken with the interpretation of the composition specification. On a cation-substitution basis, Y^{3+} for Zr^{4+}, the yttria formula must be reckoned as $YO_{1.5}$, and much work has been reported in these terms. However, there is also much work that uses Y_2O_3 as the formula basis, numerically just half of the former for the same actual composition. In some cases, mass proportions have been used. All three scales are therefore shown in Fig. 11.2. Care must therefore be taken to be sure which is relevant.

generating a large compressive stress that opposes the opening tensile stress (Fig. 11.3). This is equivalent to a prestress in the material (23§2.8), requiring the stress to be applied for further crack growth to rise. That is, if a given applied stress initiates a crack, that crack is stopped in its tracks by the transformation. Note that the stress field at the crack tip extends some distance ahead of it, and thus the transformation is triggered before cracking has actually occurred in that volume. Since it is a martensitic transformation, the speed of the transformation front is about half the speed of sound in the material, and this is of the order of 5 or 6 km/s. Hence, crack growth would have to be very fast indeed for the transformation not to keep pace and stop that propagation. For such rapid cracking it would be fair to say that the grains ahead of the crack tip were in the process of transforming (as is often depicted), but not otherwise. The same process would apply whether the bulk was purely $t*$ or any proportion of $t*$ in c or $c*$ (when the $c*$ would also be liable to transform). Clearly, it is a matter of balancing the crack-closing effect against the detrimental effects of the volumetric expansion that make complete transformation on cooling the unstabilized material so disastrous.

> The speed of sound v in a solid is given by:
>
> $$v = \sqrt{\frac{K}{\rho}}$$
>
> The bulk modulus K for zirconia is about 200 GPa, and the density ρ is about 6 Mg/m³.

    In the highly-constrained circumstances of the stress-triggered formation of the monoclinic phase, which involves a shearing action, the uniform conversion of each grain is not favourable: the stress generated would be very high, and very large voids would be created, potentially themselves acting as crack nuclei. Instead, a twinned structure is usually formed, with alternating bands of opposing shear directions (*cf.* Fig. 28§2.4).

    It should be noted that the driving force for such transformations is the stored energy in the metastable crystals: it is a thermodynamic process of lowering the energy of the system by releasing it. Now, while it can be seen that the force required to generate the crack-opening stress must rise, this must result in more energy being stored elastically in the zirconia, to be released again on cracking – plus the extra from the transformation. That is, while the work of deformation rises, this is a perfectly resilient system (1§4) – there is no dissipative mechanism that would warrant the use of the term 'toughening': no plastic deformation, no crack-tip blunting, no extra crack path tortuosity. On this basis, it would appear to be better termed **transformation strengthening** in that the nominal stress for crack growth is raised, just as in ordinary prestressing. However, what is generally meant is that the fracture toughness is increased. Since this is not a work of fracture statement, but rather is related to the stress required to drive cracking, according to flaw size (1§7), it must be understood that 'transformation toughening' refers to fracture toughness.

●11.3   Variations

    Hafnium dioxide, hafnia (HfO_2) is **isomorphic** with zirconia, and the crystal ionic radius of Hf^{4+} (85 pm) is very similar to that of Zr^{4+} (86 pm), allowing complete solid solubility, with similar *c-t-m* allotropic behaviour (Fig. 11.4), and so behaviour similar to that of zirconia in respect of yttria and other dopants can be expected in such solid solutions. The slight difference in the ionic radius, however, does mean that the unit cell volumes vary in the same sense, meaning that the solid solution introduces some lattice distortion, presumably with the effect of increasing the strength and toughness. Hafnia, however, is about ten times the price of zirconia, but even so it is used in some dental products, to the extent of some 2 ~ 5 mass%.

    There are a number of problems with the use of zirconia in dentistry. For example, in bulk, the polycrystalline material is typically dead-white. This arises from the scattering at grain boundaries because of the high refractive index and the optical anisotropy

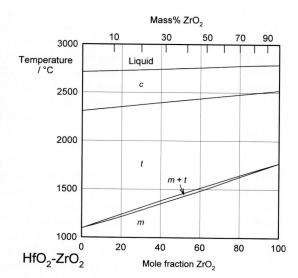

Fig. 11.4 Phase diagram for hafnia-zirconia.

in both monoclinic and tetragonal forms, as well as the mismatch that occurs in two-phase material. This may be addressed in part by using very fine-grained material. This also has effects on the stress-sensitivity of the phase transformations: there is a lower limit for the grain size that allows the transformation to m to occur, essentially based on the trade-off between surface and volume energies (*cf.* 11§2.2).[19] Indeed, there may also be a maximum grain size for the fracture toughness to be improved by transformation,[20] above which point the transformation occurs spontaneously. The problem of appearance has been addressed in some dental products by using FSZ, which is cubic and has much greater translucency because it is optically isotropic.

A small proportion of alumina (Al_2O_3) is commonly used in the formulation as well, although its solubility in zirconia is low because the ionic radius of Al^{3+} is rather small at ~68 pm (Fig. 11.5). This addition improves the densification on sintering, presumably by increasing ion mobility, and it also has useful effects on the fracture toughness if the amount is kept low.[21] Although there is some stabilizing effect, too much results in phase separation: dental products generally have well below 1 mass%. Silica can have a similar effect, but this must be kept at a minimum because the aluminosilicate glass formed tends to take up Y^{3+} from the PSZ, thereby destabilizing it. Indeed, mullite (§2.1) can also form, which is not stable in the wet environment of the mouth.

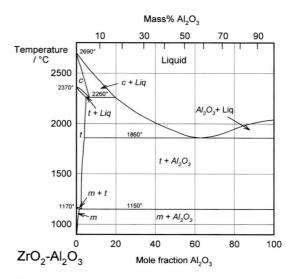

Fig. 11.5 Phase diagram for zirconia-alumina.

●11.4   Difficulties

The chemical properties of zirconia that lead to its strength and hardness are also responsible for its lack of reactivity. This means that despite presenting an apparently very high-energy surface, it has proven singularly difficult to achieve strong and stable chemical bonding to it, despite much effort.[22] Indeed, such claims seem never to have been verified. This may be due to the fact that the surface seems to react with water readily, with a coat of hydroxide groups, that lowers the energy substantially. Displacing this appears to be the problem. Accordingly, mechanical key by roughening in some manner is the major requirement,[23] whether or not silica coating and silanation (10§5.1) are used. However, the material cannot readily be etched – it is resistant to all available etchants, even HF is very slow.[24] The creation of an aluminosilicate glass phase, which would be etchable, is not appropriate either (see above). Grit-blasting has therefore been the general approach.

This unreactivity is also responsible for the difficulties of veneering. The general opacity, and inability to be coloured by the inclusion of transition metal oxides (§7), means that for reasonable tooth-tissue mimicry a more conventional porcelain veneer is required that may then be characterized in the usual fashion. The lack of bonding again means that grit-blasting is required first. Even then, delamination is a recurring problem. This is perhaps exacerbated by interfacial porosity if wetting is incomplete, and thermal expansion coefficient mismatch: Y-PSZ has values below about 10 MK^{-1},[25] while many porcelains have values in the range 12 ~16 MK^{-1} – such a difference presents risks (§6).[26] It might be noted that there is no solubility of silica in zirconia, while reaction to form zirconium silicate ($ZrSiO_4$) only occurs well above normal processing temperatures.[27]

Unfortunately, roughening by a grit-blasting process causes further difficulties in that the stresses involved can be sufficient to trigger phase transformation in PSZ, and this results in cracks that further compromise the outcome. Tight control of the blasting conditions (grit size, velocity; 20§4) is required to minimize these effects. However, it is possible to reconvert the *m* to *t* by a so-called 'regeneration' firing (> ~1000 °C or so for a useful rate). Roughening by laser blasting seems to avoid the phase transformation.[28]

Despite the general lack of reactivity of zirconia, it is susceptible to hydrolytic effects. That is, in a wet environment, the grain boundary is slowly attacked, presumably because the disorder that must be present there offers a means of lowering the energy. This might be due to the loss of yttrium destabilizing the *t*-phase.[29] The consequence is that spontaneous phase transformation occurs, the

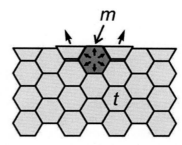

Fig. 11.6 Low-temperature degradation by hydrolytic attack at grain boundaries prompts phase transformation.

volume increase of which puts stress on adjacent grains which may then be lost from the surface (Fig. 11.6). This process may then continue steadily, and appears to be exacerbated by fatigue loading (1§12), *e.g.* by chewing, in what is termed **low-temperature degradation** ('LTD'), for which, as yet, there appears to be no solution (although a variety of factors are known to influence the rate). The flaws that are generated then make the piece more vulnerable to cracking and failure, notwithstanding transformation toughening.

●11.5 Fabrication

Given the strength and hardness of zirconia, and the fact that sintering requires rather high temperatures (given that there is no glassy phase involved), the usual processes for dental ceramics are not applicable. Accordingly, a specialized CAD-CAM approach is used.

In outline, this process involves the milling of a blank, which is in the form of a block of zirconia sintered to a low biscuit-like stage (§4.1). The milled form is then sintered for full densification to the final form. Obviously, because this involves considerable shrinkage, the milled form must be oversized very precisely so that the sintering produces the exact size of device required – no manual adjustment is feasible without difficulty and risk. This entails that the blanks are manufactured under very precisely controlled conditions to enable a precise, highly reproducible outcome. Even so, variation occurs in the density and isotropy of the shrinkage behaviour of these blanks such that each batch must be calibrated. The calibration, which may involve separate values for the xy-plane and the z-direction, may then be contained in a bar code which is to be read by the milling system computer such that the pattern measurements may be suitably scaled for the oversize milling, enabling the shrinkage to be accurately predetermined.

References

[1]   McLean JW. The Science and Art of Dental Ceramics (2 vols). Quintessence, Chicago, 1979.

[2]   O'Brien WJ. Dental Porcelain. in: Williams DF (ed) Concise Encyclopedia of Medical and Dental Materials. Pergamon, Oxford, 1990.

[3]   Ray NH. Inorganic Polymers. Academic Press, London, 1978.

[4]   Rosenthal D & Asimov RM. Introduction to Properties of Materials. 2nd ed. Van Nostrand Reinhold, New York, 1971.

[5]   Wyatt OH & Dew-Hughes D. Metals, Ceramics and Polymers: An Introduction to the Structure and Properties of Engineering Materials. Cambridge UP, 1974.

[6]   Cheung KC & Darvell BW. Sintering of dental porcelain: effect of time and temperature on appearance and porosity. Dent Mater 18: 163 - 173, 2002.

[7]   Malavasi G, Menziani MC, Pedone A, Segre U. Void size distribution in MD-modelled silica glass structures. J Non-Crystalline Solids 352 (3): 285-296, 2006.

[8]   Bansal NP & Doremus RH. Handbook of Glass Properties. Academic, Orlando, 1986.

[9]   Jones GO. Glass. London, Methuen, 1956.

[10]   Meyer J-M. Porcelain-Metal Bonding in Dentistry. in: Williams DF (ed) Concise Encyclopedia of Medical and Dental Materials. Pergamon, Oxford, 1990.

[11]   O'Brien WJ, Boenke KM, Linger JB & Groh CL. Cerium oxide as a silver decolorizer in dental porcelains. Dent Mater 14: 365 - 369, 1998.

[12]   Charles RJ. Static fatigue of glass I. J Appl Phys 29 (11): 1549 - 1553, 1958.

[13]   Michalske TA & Freeman SW. A molecular mechanism for stress corrosion in vitreous silica. J Amer Ceram Soc 66(4): 284 - 288, 1983.

[14]   Orowan, E. Fracture and strength of solids. Phys Soc Prog Reports 12: 185 - 232, 1949.

[15]   http://www.bt.cdc.gov/agent/hydrofluoricacid/basics/facts.asp
               http://ehs.whoi.edu/ehs/occsafety/HFsafetyGuideline.pdf
               http://www.chem.purdue.edu/chemsafety/Equip/hfmsds.pdf
               http://www.fap.pdx.edu/safety/hydrofluoric_acid/

[16]   Scott HG. Phase relationships in the zirconia-yttria system. J Mater Sci 10: 1527 - 1535, 1975.

[17]   Subbarao EC, Maiti HS & Srivavstava KK. Martensitic transformation in zirconia. Phys Stat Sol A 21: 9 - 40, 1974.

[18]   Fabris S, Paxton AT & Finnis MW. A stabilization mechanism of zirconia based on oxygen vacancies only. Acta Materialia 50: 5171 - 5178, 2002.

[19]   Evans AG, Burlingame N, Drory M & Kriven WM. Martensitic transformations in zirconia – particle-size effects and toughening. Acta Metallurg 29: 447 - 456, 1981.

[20]   Liang YM & Zhao JH. Effect of zirconia particle size distribution on the toughness of zirconia-containing ceramics. J Mater Sci 34: 2175 - 2181, 1999.

[21]   Vasylkiv O, Sakka Y & Skorokhod VV. Hardness and fracture toughness of alumina-doped tetragonal zirconia with different yttria contents. MAter Trans 44 (10): 2235 - 2238, 2003.

[22]   Della Bona A, Pecho OE & Alessandretti R. Zirconia as a dental biomaterial. Materials 8: 4978 - 4991, 2015.

[23]   Sanli S, Dündar Çömlekoglu M, Çömlekoglu E, Sonugelen M & Darvell BW. Influence of surface treatment on the resin-bonding of zirconia. Dent Mater 31: 657 - 668, 2015.

[24]   Lee MH, Son JS, Kim HKH & Kwon TY. Improved resin–zirconia bonding by room temperature hydrofluoric acid etching. Materials 8: 850 - 866, 2015.

[25]   Hayashi H, Saitou T, Maruyama N, Inaba H, Kawamura K & Mori M. Thermal expansion coefficient of yttria stabilized zirconia for various yttria contents. Solid State Ionics 176 (5-6): 613 - 619, 2005.

[26]   Tanaka CB, Harisha H, Baldassarri M, Wolff MS, Tong H, Meira JBC & Zhang Y. Experimental and finite element study of residual thermal stresses in veneered Y-TZP structures. Ceramics Int 42 (7): 9214 - 9221, 2016.

[27]   Suzuki M, Sodeoka S & Takahiro Inoue T. Structure control of plasma sprayed zircon coating by substrate preheating and post heat treatment. Mater Trans 46 (3): 669 - 674, 2005.

[28]   Liu D, Matinlinna JP, Tsoi JKH, Pow EHN, Miyazaki T, Shibata Y & Kan CW. A new modified laser pretreatment for porcelain zirconia bonding. Dent Mater 29: 559 - 565, 2013.

[29]   Pandoleon P et al. Aging of 3Y-TZP dental zirconia and yttrium depletion. Dent Mater 33(11): e385 - e392, 2017.

Chapter 26   Radiography

Radiography is a basic tool of clinical medicine and dentistry that is used for routine diagnostic purposes. In that context, restorative and prosthetic materials – even other foreign bodies – may well be present. It is appropriate therefore to consider the factors which influence the appearance of structures in radiographic images. The physics of this is discussed with a view to a better understanding of the images obtained as well as the control that may be exercised in the process.

*A brief indication of the aspects of atomic structure associated with the formation and absorption of X-rays is followed by an outline of the practical means of generating X-radiation and the type of **spectrum** thus obtained.*

*The critical issues centre on the **differential absorption** of X-radiation according to the elements present, the laws expressing these effects, and the need for **filtration**. The contributions of true absorption and scattering processes are distinguished and some of the practical implications are described.*

*Radiographic film has been the main means of recording such images, so it is necessary to explore the generation and properties of the image with respect to exposure as well as explain the terminology. Practical aspects of the handling and interpretation of the image (in a non-diagnostic sense) are presented. The utility and limitations of **intensifying screens** are outlined. However, alternative techniques are displacing film in favour of **digital images**.*

In terms of handling, image interpretation and radiological protection it is clearly important to understand the nature of X-radiation and its interaction with matter. It is this interaction that ties the subject to dental and biomedical materials. No distinction can be drawn in this sense between biological tissues and the foreign materials that may be present, but the discrimination of the two may be critical.

Materials Science for Dentistry
https://doi.org/10.1016/B978-0-08-101035-8.50026-2

Radiography is a procedure used both clinically and industrially to obtain information non-destructively about the internal structure of objects. While the anatomical interpretation of clinical radiographs is of no concern here, that interpretation depends ultimately on the interaction of radiation with matter, both in the object being investigated and in the recording medium. That brings radiography firmly within the ambit of materials science. It is therefore considered important that these interactions be considered as an adjunct to understanding the interpretation process, especially since non-biological materials will frequently be present in the field, *e.g.* if any prior restorative treatment has been done.

§1. Atomic Structure

As was noted in 24§6, electromagnetic radiation may interact with matter in a variety of ways, depending on the wavelength of that radiation. So-called X-rays[1][1] are associated with changes in the energy of inner electrons.

The electron configuration of an atom, that is, the distribution of the electrons among the various available orbitals 1*s*, 2*s*, 2*p* and so on, is governed by the **Aufbau Principle**. This effectively says that in filling the available set of atomic orbitals the electrons are placed in order of decreasing stability, with the added proviso of the **Pauli Exclusion Principle**. This states that no two electrons in one atom may have identical sets of quantum numbers. In other words, under normal conditions in the ground state the inner shells are all completely filled; there are no gaps.

However, any electron may be removed from its orbital by the provision of a quantum of sufficient energy. In general the energy required corresponds to what has become known as the X-ray region of the electromagnetic spectrum, and the deeper the shell from which the electron was removed the higher the energy state of the atom which remains. Having thus excited the atom, any outer electron may fall down into the vacancy, and in so doing it will emit a quantum of radiation of energy corresponding to the energy difference between the two states of the atom. This quantum is known as a **characteristic X-ray**. If the original ejected electron is from the lowest energy shell, the *K*-shell, and an electron from the next lowest shell, the *L*-shell, falls into it, the characteristic X-ray is known as *K*α radiation. If the *K*-shell vacancy is filled by an *M*-shell electron, the *K*β radiation is seen, and so on. A similar series of transitions to the *L*-shell results in *L*-series spectral lines. The processes involved are shown schematically in Fig. 1.1.

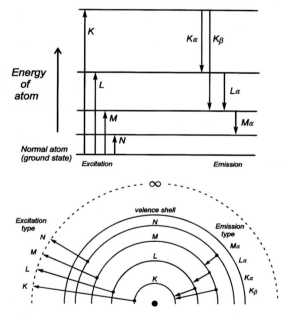

Fig. 1.1 Energy levels associated with X-ray absorption and emission.

It should be apparent that in attempting to remove a given electron (and thus, incidentally, ionizing the atom), unless enough energy is provided to correspond with the energy of the transition, or somewhat more, no absorption of that energy can occur. Thus, at a supplied energy just below this **critical absorption wavelength** that particular electron cannot be ejected, although others in higher shells may be. This is one of the consequences of the quantization of radiation. The energy of the critical absorption wavelength is enough to remove the electron to infinity, but then be at rest. Any energy in excess of that exact amount must appear as the electron's kinetic energy.

---

[1] "A piece of sheet aluminium, 15 mm thick ... allowed the X-rays (as I will call the rays for the sake of brevity) to pass, ..."
W.C. Röntgen (1896).

§2. X-ray Generation

What was described in §1 has to do with the atom-specific effects. In particular, for absorption, it presupposes that there is a source of radiation of appropriate energies for those electron transitions to occur. We now describe the means of generating such radiation, and this is in a non-specific manner.

●2.1 Energy

Free electrons absorb or emit radiation on acceleration or deceleration. On striking atoms they lose energy successively with each impact, and they may be accelerated or decelerated by an electric field. If an electron is accelerated over an electric field with voltage difference V between the start and end points, irrespective of the distance between those two points, the kinetic energy E of the electron is then given by:

$$E = eV \qquad (2.1)$$

where e is the electronic charge. This energy corresponds to electromagnetic radiation with wavelength λ:

$$E = hc / \lambda \qquad (2.2)$$

where h is Planck's constant and c is the speed of light. This is in fact the energy of the quantum that would be emitted if the electron were brought to an abrupt halt. The relationship between accelerating voltage and wavelength may conveniently be given by:

$$\lambda = \frac{hc}{e} \cdot \frac{1}{V} = \frac{12.40}{V} \qquad (2.3)$$

where λ is in Angstroms if V is in kilovolts (kV) (Fig. 2.1). X-ray wavelengths are traditionally expressed in Angstrom[2] units (Å). Currently this persists, even though it is not an S.I. unit. The equivalence is 1 Å = 0.1 nm. Note here that the energy unit, the **electron volt** (eV), commonly used in this kind of context, is defined by equation 2.1. Thus 1 keV is equivalent to passing through a 1 kV field, whatever distance this requires.

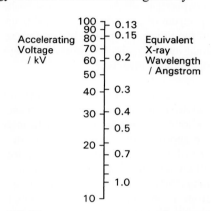

Fig. 2.1 Comparison of accelerating voltage and equivalent X-ray wavelength in the region of diagnostic X-rays.

●2.2 Spectrum

Normally, the total kinetic energy of the decelerated electron will only very rarely be converted into precisely one quantum of radiation. This is because that electron will undergo several collisions before being brought to rest. In effect, the incoming electrons ricochet off those of atoms, like snooker balls. Most of its energy will appear, therefore, as heat in the **target**, the body or area struck by the electrons. As this thermal energy is removed, and the kinetic energy of the electron decreases, the wavelength of the X-radiation capable of being emitted gets longer. The radiation produced by a beam of uniformly-energetic electrons hitting a target will therefore not be monochromatic, *i.e.* of a single wavelength, but will instead show a considerable spread. Even so, equations 2.2 and 2.3 are important because they give the *minimum* wavelength (maximum energy) of the radiation which may be observed for a given accelerating voltage – the so-called **short-wavelength limit**.

The continuous radiation spectrum produced in such a system is called '**white**' in this field, simply because all wavelengths up to the limit are represented, ignoring the fact that the intensity is not uniform. It is this kind of radiation that is used for radiography. Typical curves of emission intensity *vs.* wavelength for several accelerating voltages are shown for a tungsten target in Fig. 2.2.[2] Note the sharpness of the short-wavelength limit, the shift of the peak intensity to shorter wavelength, and the increase in area beneath the curves (total intensity) with increasing voltage.

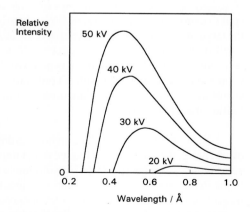

Fig. 2.2 Emission spectrum for a tungsten target at several accelerating voltages.

---

[2] The unit was named in 1905 for the Swedish physicist A.J. Ångström, pronounced approximately as "ong-strerm", who first used such a unit in his studies of the solar spectrum.

●2.3   Generation

A schematic diagram of an X-ray tube of the so-called Coolidge type[3] is shown in Fig. 2.3, and an actual tube in Fig. 2.4. The envelope is borosilicate glass and holds a high vacuum. Electrons emitted from a resistively-heated filament (the cathode) are accelerated by a high voltage to strike a metal target, the anode. As the number of such **thermionic** electrons emitted by the cathode depends only on its temperature, the **tube current** (that flowing between the cathode and the anode, usually expressed in milliamps [mA]) is independent of the accelerating voltage. Thus the spectrum of the radiation emitted is itself independent of the tube current, which therefore only controls the total intensity (the 'brightness' of the source), a great operational advantage.

Experimentally, it has been found that for the 'white' radiation produced under these conditions both the total radiation produced and the intensity of the peak are approximately proportional to the atomic number (Z) of the target material. Thus, for good emission efficiency the target should be of a heavy metal, and tungsten (Z = 74) is usual. Emission spectra as obtained from a tungsten target are illustrated in Fig. 2.2. It should be realized that these spectra are essentially those of decelerated electrons, as there are no element-specific effects discernible. The total radiation

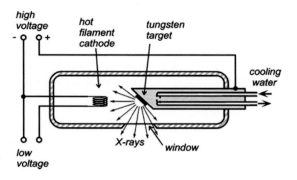

Fig. 2.3 Schematic diagram of the basic structure of an X-ray tube. The water cooling system would only be present in high-current machines.

Fig. 2.4 A dental X-ray tube, anode and copper heat-sink on the right, which passes through the envelope.

produced (the area under the emission curve) is proportional to the square of the applied voltage, for a given tube current. This voltage should then be increased as far as possible for efficient high intensity radiation. The total radiation produced is also directly proportional to the tube current, *i.e.* the number of electrons making impact, as might be expected.

The production of X-rays is extremely inefficient. The best that can be managed is about 0.2 % at 100 kV. Hence, for a (continuous) tube current of, say, 10 mA only 2 W of X-rays in total may be produced. However, nearly 1 kW of heat is being dumped into a target of only a few square millimetres. This means that high melting point metals – such as tungsten (m.p. 3422 °C, the highest value of all elements) – are advantageous, if not essential. The target may reach a temperature as high as 2700 °C and would glow white-hot (see Fig. 24§4.6). Large X-ray sets for high intensity or long exposures require cooling water to be pumped through the target, which is usually a plate of the chosen metal embedded in a massive copper base, for good heat conduction (Fig. 2.4). For even larger machines rotating anodes are used to increase the effective surface area of the target spot to many times the actual size. Even then, because the radiation will be emitted over a full sphere, some into the target itself and some into free space, the available power in the chosen output 'beam' direction will be a very small fraction of that actually produced. Note that there is no practical way that X-rays for diagnostic use can be usefully focused because there is no possible equivalent of the mirror and condenser lens of a spotlight. To control the irradiation of the subject only **collimation** by masks may be used to create a 'beam'. That is, highly absorbing casing and shutters are used to limit the angular extent of the emerging radiation, typically only just to illuminate the subject.

The intensity of the electron bombardment of the target means that to obtain sufficient output and avoid overheating, the area of the effective target must be large. However, a large target means that the image quality deteriorates by blurring (§8.3). The solution to this is a compromise: an elongated 'line target' placed at an angle to the intended beam direction, that is, at less than 90° to the electron beam axis. This allows the view of the target from the collimator to be foreshortened so that the source is apparently smaller, but the heat production is spread out. Obviously, the smaller the view angle (the **target** or **bevel angle**), the smaller the apparent source size in length (the width is unaffected). The angle cannot be zero because of internal absorption (§8.4). Normally, it may lie between about 10 and 20°. To achieve the necessary X-ray source shape, the cathode

---

[3] Invented by William Coolidge in 1913.

filament is an extended coil parallel to the beam direction, and lying in a U-shaped recess in the electrode (Fig. 2.5). The shape of this **focussing cup** modifies the electric field between the filament and the anode, guiding the electrons to the **line focus** (Fig. 2.6).

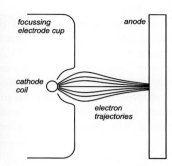

Fig. 2.5 Example of a cathode filament and focussing cup from a dental X-ray tube.

Fig. 2.6 Electrons from the cathode are focused onto a narrow strip on the target. The view is of a section of the device from the X-ray beam direction.

As the accelerating voltage applied to an X-ray tube is increased there comes a point when the energy of the electrons, corresponding to the short-wavelength limit (equation 2.3), is just sufficient to eject K-shell electrons from the target. For this and higher voltages, the characteristic X-rays for the refilling of those vacancies then appear in the emission spectrum for that target, superimposed on the 'white' radiation due to the impacting electrons' deceleration. This effect is shown for a molybdenum target operated at 35 kV in Fig. 2.7. In the case of tungsten the limit is 70 kV (hence the non-appearance of such peaks in Fig. 2.2). Of course, the same applies to L, M, N ... excitation at progressively lower energies (if the atom possesses these levels filled), but for reasons seen later these are not of interest in the present context. The usual range for clinical diagnostic X-rays is 30 - 100 kV.

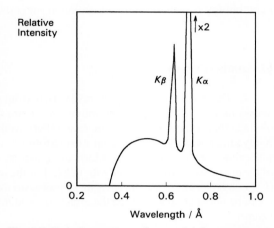

Fig. 2.7 Emission spectrum for a molybdenum target at a tube voltage of 35 kV. Mo targets are used for some medical imaging purposes, such as mammography.

●2.4   Moseley's Law

The variation in the energy of the characteristic X-rays was first studied by Moseley (one of Rutherford's former students), who came to the conclusion that

$$\sqrt{\nu} = Z \qquad (2.4)$$

where ν is the frequency of the radiation and Z is the atomic number of the element (and wavelength, of course, is inversely proportional to frequency) (Fig. 2.8). Although in fact this is only an approximation and not a strict relationship (the lines are curved), it is known as **Moseley's Law**. It is an expression of the increasing stability of the inner shells as the atomic number increases (in fact, Moseley was responsible for developing the idea of atomic number[3]). This behaviour therefore further emphasizes the need for heavy metals to be used as targets if characteristic X-ray emission needs to be avoided. As will be seen later, relatively intense emission at long wavelengths would create a filtration problem because such radiation is not of diagnostic radiographic value, and a very thick filter that would appreciable attenuate the diagnostically-useful radiation would be needed. If the characteristic radiation is at a useful energy, as for Mo (Fig. 2.7), or for W (at 69.5 keV), then it is only a matter of making appropriate allowance for the exposure (§5.1). Characteristic radiation can be as much as 10 to 30 % of the total.

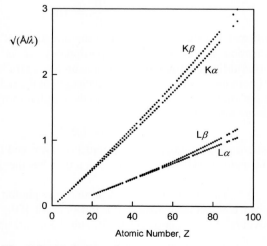

Fig. 2.8 Variation of the wavelength of K and L characteristic X-rays for the elements.

•2.5   Kinetic energy

It should be noted that the energy statement represented by equation 2.1 is not accurately equivalent to that of the usual (Newtonian) kinetic energy equation, $E_k = \frac{1}{2}mv^2$, for the voltages relevant to dental diagnostic X-ray generation. This is because the electron velocity is an appreciable fraction of that of light in a vacuum, c, (which is an upper bound) and relativity must be invoked. The velocity is then given by:

$$\frac{v}{c} = \sqrt{1 - \left(\frac{m_0}{m_0 + eV/c^2}\right)^2} \qquad (2.5)$$

where m_0 is the rest mass of the electron. This function is plotted in Fig. 2.9, and compared with the Newtonian result. It can be seen that the errors becomes noticeable above about 30 kV, noting that the electron velocity at impact on the target is around 0.4 that of the speed of light, but lower than the simple calculation suggests.

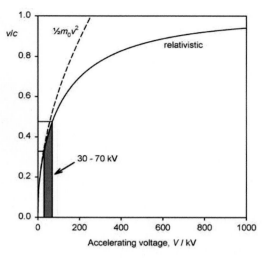

Fig. 2.9 Relative velocity, with respect to the speed of light, of electrons in X-ray tubes. The relativistic calculation starts to deviate appreciably from the Newtonian kinetic energy value above about 30 kV.

§3. Lambert's Law

The striking property of X-rays was, to Röntgen,[1] their penetrating power. Indeed, it is this property that allows their use for imaging the internal structure of objects that are opaque to visible light. However, it is obvious that if they were too penetrating there would be no information conveyed in the emerging radiation. X-rays must undergo **differential absorption** (as with visible light, to create colour, 24§3.4) or differential scattering, for there to be such information, that is, variation between materials is essential. The mechanisms of these processes will be dealt with in §4. For the moment only the macroscopic behaviour will be addressed. (For convenience we may discuss these processes as if they were due to absorption alone, for the moment, but it can be understood to include the effects of scattering as well: the result is the same in the line of the incident beam.)

X-radiation, like other electromagnetic radiation, has been found to obey **Lambert's Law**. That is to say, the intensity decreases by the same fraction on passing through each successive, constant thickness layer of the medium, *i.e.* in differential terms, the rate of decrease of intensity in the beam is proportional to the intensity (more properly called the **irradiance**) I at that point:

$$-\frac{dI}{dx} = \mu I \qquad (3.1)$$

where x is distance measured in the direction of the beam. In this equation the constant of proportionality, μ, is called the **linear attenuation coefficient**.[4] The value of this coefficient is characteristic of both the substance *and* the wavelength of the radiation. Integrating equation 3.1 with respect to x gives the familiar form of the exponential decay function:

$$I = I_0 e^{-\mu x} \qquad (3.2)$$

where x is the thickness of the object or layer[5] and I_0 is the intensity of the incident radiation (see 11§2.6); μ then has the units of cm^{-1}.

> In discussing attenuation in contexts such as this, in the absence of any explicit statement otherwise, it is implicitly assumed that a parallel beam of radiation is being used. That is, a divergent beam would entail a reduction in the irradiance with distance, even in a vacuum (see equation 6.5, §8.2). Since collimation is not a practical possibility, the assumption is therefore an approximation.

The risk of interaction of an X-ray photon with an atom of a given element is a constant. It is independent of the chemical state of the atom because chemical reactions only involve outer, valence electrons, and the e.m.r. absorption of these does not lie in the X-ray region of the

---

[4]   This is normally called an absorption coefficient, but as we shall be distinguishing between absorption and scattering components in the overall effect with X-rays, the phenomenological term of attenuation coefficient is preferable.

[5]   Usually expressed in cm (N.B. *not* consistent with S.I. usage).

spectrum. Therefore, if a given number of atoms lie in the beam, it does not actually matter what the path length is, *i.e.* the distance over which they are spread.

Thus, if a given mass of vapour of, for example, bromine were contained in a long tube with a movable (X-ray transparent) end wall as a plunger to compress the gas, the measured absorption would be constant no matter where that end wall was placed. Nor, indeed, would it matter if the bromine were condensed as a uniform film of liquid on the end wall. Again, freezing it would make no difference. To measure the total amount of bromine present per unit area of cross section, irrespective of its distribution in the tube, we must take the product ρx (which expresses the amount of matter per unit irradiated area: g/cm^2), where ρ is (average) density over the length of the tube x. If this is inserted in equation 3.2 we must also make a corresponding adjustment to the coefficient to maintain the exponent as a dimensionless quantity, thus:

$$I \ = \ I_0 e^{-(\mu/\rho).\rho x} \tag{3.3}$$

The term in parenthesis, μ/ρ, is called the **mass absorption coefficient**.[6] It expresses the fact that it is the *number* of absorbing atoms that matters rather than the distance over which they are spread. We can see this by modifying equation 3.1, taking the differential with respect to ρx:

$$-\frac{dI}{d(\rho x)} \ = \ -\frac{dI}{\rho.dx} \ = \ \frac{\mu}{\rho}I \tag{3.4}$$

Since ρ is a constant for the chosen substance and set-up (the material), this can be taken outside the differential, so the need similarly to divide the right-hand side of equation 3.1 becomes apparent.

Thus, while diamond is extremely transparent to visible light, and graphite is completely opaque (because of its metallic nature, 11§1), both forms of carbon have exactly the same mass absorption coefficient for X-rays. Similarly, water has the same value for μ/ρ in the solid, liquid and gaseous states; the same value in fact as a mixture of H_2 and O_2 in the stoichiometric proportions of water, 2 : 1, or even a 1 : 1 mixture of H_2 and H_2O_2.

It may be recognized that density is equivalent to concentration, since ρ/A = mol/mL (A = molar mass) and the mole is defined as a fixed number of entities, *i.e.* Avogadro's Constant. If x is held constant, equation 3.3 is therefore effectively a version of **Beer's Law**.[4] Taken together, equations 3.2 and 3.3 are sometimes known as the **Beer-Lambert Law**.

●3.1   Calculation

The last two examples above take the chemical independence of X-ray absorption of a given element a step further: that it is not affected by the other elements that are present. Alternatively, we can say that, in a given medium, the overall mass absorption is the sum of the mass absorptions for all elements present - simply additive. We have, of course, to take account of the relative amounts of the various elements over the complete path, so the summation is weighted by the mass fraction (M) for each of the n elements present, *i.e.* the overall mass absorption coefficient is given by:

$$\left(\frac{\mu}{\rho}\right)_o \ = \ \sum_{i=1}^{n}\left[\left(\frac{\mu}{\rho}\right)_i \times M_i\right] \tag{3.5}$$

where $\Sigma M_i = 1$ in the usual way. Then, to obtain the linear absorption coefficient, μ, for the real material as a whole, we simply multiply by its observed overall density. We can therefore deal with any material no matter what its physical state or its arrangement in space, whether it is solid, a foam, a loose packed powder, or just a stack of plates with gaps.

The use of μ/ρ follows the approach taken in other contexts where intensive properties of a material are preferred, such as modulus of elasticity: the effect of size or shape of the object has been eliminated. Thus, there are tables of mass absorption coefficient for many elements and X-ray wavelengths.[5] These can be used to calculate μ/ρ for any material in any condition. We may take PMMA as an example. Since it has an empirical formula somewhat similar to that of soft tissue overall, a consideration of its general behaviour is instructive.

Given the empirical formula: $C_5H_9O_2$, and formula weight: 101.126, for PMMA, and for $\lambda = 0.5609$ Å

---

[6] Usually expressed in cm^2/g, with density in g/mL.

(Ag-Kα radiation),[3] we have:

	Mass fraction	μ/ρ	product
C	0.5939	0.400	0.2375
H	0.0897	0.371	0.0333
O	0.3164	0.740	0.2342
Σ =	1.0000		0.5050

i.e. μ/ρ for PMMA is 0.505 cm²/g at that wavelength. Its density at room temperature is about 1.19 g/cm³, so that μ = 0.601 cm⁻¹. Similar calculations give values for μ/ρ for acrylic for a variety of wavelengths (Table 3.1, Fig. 3.1).

Table 3.1 *Variation in mass absorption coefficient with wavelength for PMMA.*

λ/Å	(μ/ρ)/(cm²/g)
0.5609	0.505
0.7107	0.820
1.5418	6.41
1.7902	9.87
1.9373	12.4
2.2909	20.2

●3.2 Filtration

Especially to be noted is the steady and rapid increase in μ/ρ with increasing wavelength, even though the elements concerned are very light. Soft tissue also shows this (Fig. 3.1), tending to absorb long wavelength radiation efficiently. This has two results:

1) such radiation is very inefficient for diagnostic radiographs because the usable beam intensity reaching the film would be extremely low, and

2) since the absorption of radiation must result in ionization and free radical formation, damage to DNA, proteins, and other cell components must ensue from the reactions such ionization would initiate (§4.5).

Thus, for diagnostic purposes, **filtration** of the X-rays of the incident beam is essential. Aluminium is frequently chosen as a filter medium because it is both a light element - and so absorbs radiation appropriately (which means letting enough of the short wavelength radiation through to be useful) – and reasonably cheap. In addition, it is easily fabricated into any desired shape. The effectiveness of such a filter can be calculated by rewriting equation 3.2 as:

$$T = \frac{I}{I_0} = e^{-\mu x} \qquad (3.6)$$

where T is the fraction of the radiation transmitted; this is known as the **transmittance** of the filter. The density of aluminium is about 2.70 g/cm³ and, using the values of μ/ρ given in the second column of Table 3.2, T can be calculated *vs.* λ. The long wavelength filtration is seen to be very efficient.

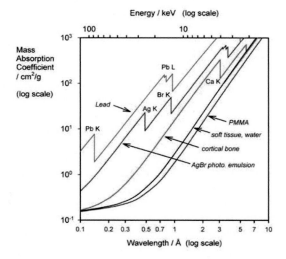

Fig. 3.1 Variation of mass absorption coefficient with X-ray wavelength for some important materials. (The discontinuities are absorption edges, see §4).

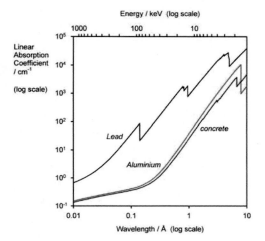

Fig. 3.2 Variation of linear absorption coefficient with X-ray wavelength for some common shielding materials.

Table 3.2 *X-ray filtration by aluminium.*

λ / Å	Energy / keV	(μ/ρ)_Al / cm²g⁻¹	T (2.5 mm Al)
0.5609	22.1	2.65	1.7×10^{-1}
0.7107	17.4	5.16	3.1×10^{-2}
1.5418	8.0	48.6	5.7×10^{-15}
1.7902	6.9	74.8	1.2×10^{-22}
1.9373	6.4	93.9	3.0×10^{-28}
2.2909	5.4	152.0	2.8×10^{-45}

Similar calculations can be applied to determine the thickness of shielding required around X-ray generators, using such materials as lead and concrete. Although the latter is much less absorbent, indeed rather similar to aluminium (Fig. 3.2), it is much cheaper and so can be used in great thickness if necessary.

●3.3   Spectral effects

So far we have only discussed the behaviour of monochromatic radiation, whereas the output of an X-ray tube is 'white'. Since the mass absorption coefficient is specific for each element at each wavelength, the total absorption would have to be calculated by reference to the incident spectrum, treating each wavelength independently and summing the results. While this is not in general a practical proposition, it does indicate some features of behaviour which should be recognized.

Firstly, in the context of radiation protection this means that successive increments of shielding are progressively less efficient. The long wavelength radiation may have been reduced to an insignificant intensity, but the short wavelengths – being more penetrating – require substantially more material to achieve similar reductions. The standard references take this into account. It suggests that it is protection against the most energetic X-rays (*i.e.* the highest voltage used in the generating tube) that is the dominating consideration.

Secondly, if an attempt is made to determine the linear absorption coefficient of a material by noting the intensity *vs.* thickness behaviour and using white radiation (even if filtered as for diagnostic use), the value may be seen to diminish with thickness. This is not a failure of Lambert's Law but the result of the spectrum of the radiation changing as it travels through the material because of the **differential absorption** of the various wavelengths. In any case, one could only report the apparent linear absorption (attenuation) coefficient for that particular source and for the thickness of the material approaching zero (the limiting value), as a guide value rather than a true material property.

3.4   Rectification

It has already been pointed out that the X-ray spectrum depends on accelerating voltage (Fig. 2.2). But we have been discussing this as though the voltage was constant for any given exposure; it is not. To understand the implications of this it is necessary to examine how the accelerating voltage is generated.

In general, domestic mains electricity is commonly supplied at some 100 or 200 Vac, and at a frequency of 50 or 60 Hz. Clearly, such a voltage is too low to be useful (equation 2.3) and must be stepped-up with a transformer, a pair of coils wound on a common magnetic circuit (typically of easily-magnetizable – "magnetically soft" – iron). The effect of this is to raise the output voltage in proportion to the ratio of the number of turns in the primary and secondary windings of the transformer. The variation of the current in the primary coil induces a magnetic field, the variation of which induces a current in the secondary coil, but now 90° out of phase with that in the primary. The phase, of course, is of no consequence to the production of X-rays, but the alternating voltage is: only one half of each cycle would have the correct sense to generate an appropriate tube current (see Fig. 2.3), the other would be wasted as heat (Fig. 3.3, top). A system that behaves like this is

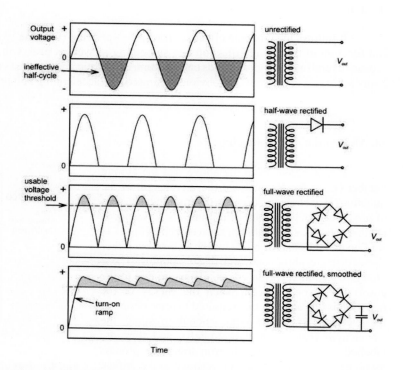

Fig. 3.3 Simple rectification and smoothing of X-ray tube accelerating voltage. The double winding of the transformer is shown symbolically at the left of each partial circuit diagram.

called **self-rectified**, as the behaviour is intrinsic. Accordingly, the voltage must be **rectified**, that is, made to be of just one polarity. The device which achieves this is known as a **rectifier**. This is based on an electronic device called a **diode**, whose characterizing property is that it conducts electricity in one direction very much more easily than in the other. Thus, applying an alternating voltage to a diode results in transmission of the voltage on every second half cycle (Fig. 3.3, second graph). This is then known as **half-wave rectification**, for the obvious reason. It is not very efficient in the sense that again half the power expended is wasted as heat in the transformer. A **full-wave rectifier** makes full use of both halves of the cycle (Fig. 3.3, third graph) by using the 'one-way valves' of the diodes to good effect. More elaborate circuits may be used.

Even so, it is apparent that the voltage still varies from zero to the peak and back to zero in the course of each half-cycle. If we notice that radiation corresponding to less than some accelerating voltage is of no diagnostic value (say, 50 kV), and would in any case be largely filtered out (Table 3.2), there is still considerable inefficiency in that much of the cycle is still wasted. Even so, to reduce the patient's unnecessary exposure, the filtration would have to be very heavy. This may be overcome by **smoothing**, the use of capacitors in the circuit such that the voltage is built up during each half-cycle, but then decays (exponentially) at a rate lower than the timescale of input voltage variation as current is drawn. The effect is to generate an output voltage waveform that **ripples** rather than oscillates over the full range (Fig. 3.3, bottom). By such a means, the voltage supplying the X-ray tube can be maintained above that which is minimally useful for diagnostic purposes. Modern equipment uses more efficient circuits: "switching" supplies which generate high frequency output – several kilohertz, and diode arrays called voltage doublers that can produce the necessary high voltage without a transformer. Rectification is still required, but the capacitors for smoothing can be much smaller, making the X-ray set itself much smaller and neater altogether.

None of this would be of any particular concern in dentistry were it not for one thing: the emitted X-ray spectrum depends on voltage (Fig. 2.2). Thus, during the course of a half-cycle, whether in a half- or full-wave (unsmoothed) circuit, the spectrum must vary dramatically. Such variation is considerably reduced by smoothing, and obviously the better the smoothing the better the constancy, and the closer will be the radiation spectrum to that suggested by the peak voltage (Fig. 2.2). Obviously, at the start of the exposure the voltage must rise from zero in the 'turn-on ramp', and likewise return at the end. Although the amount of non-useful radiation produced this way is small, as a proportion of the total in very short exposures (0.05 s or less, that is, just 2 or 3 a.c. cycles) it is reduced to insignificance in high-speed switching power supplies because the time spent in that region is very small.

It follows then that the effective irradiation spectrum (the time-weighted average) for patient and film is a function of type of rectification and smoothing as well as peak voltage and filtration. This means that because overall attenuation depends on spectrum, diagnostic images must vary somewhat according to the actual X-ray set in use, and not just the obvious settings of the accelerating voltage, milliamperage and exposure time.

§4. Absorption and Scattering

It is apparent from the above that the atomic number (Z) of the element and the wavelength of the radiation (λ) are both important in determining the value of μ/ρ. But a detailed plot of μ/ρ *vs.* λ is revealing in another sense. Using platinum as an example (Fig. 4.1) we find that the curve is by no means smooth but rather shows discrete, abrupt and large discontinuities. These are known as **absorption edges**. They are due to the absorption to be expected when the energy of the radiation is just equal to that required to remove one of the electrons (*i.e.* just large enough, reading the wavelength scale from right to left), and they are labelled according to the electron transition process involved. Absorption edges correspond to the critical absorption wavelengths mentioned earlier (§1).

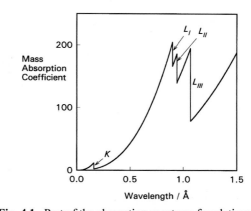

Fig. 4.1 Part of the absorption spectrum for platinum.

For each of these separate types of electronic transition, as the energy increases (λ decreases) so the efficiency of the interaction of electron and radiation decreases. Each such process is independent of the others so, dissecting out the separate absorption curves, we obtain a figure such as Fig. 4.2. The overall effect at any wavelength is therefore seen to be a summation of all possible absorptions (Fig. 4.3).

If ejection of the electron does occur the energy in excess of that required just to remove the electron will appear as its kinetic energy. The energies of these absorption edges therefore correspond to the excitation transitions of Fig. 1.1. What happens next is of clinical importance. Fast electrons are capable of initiating chemical reactions as a result of *many* interactions with outer (bonding) electrons, the collisions imparting some energy to the other electrons such that they are ejected from their molecular orbitals. This is why X-rays are described as **ionizing radiation**. The absorption of radiation that does occur is thus made more effective (and therefore damaging) in this sense than a one electron per photon calculation would indicate.

The actual chemistry of the formation of the image in a photographic or radiographic film 'emulsion'[7] is a complex matter that need not be discussed here; however, one feature can be noted. There are absorption edges for silver and bromine, *i.e.* the constituents of the sensitive silver bromide crystals in the emulsion of the radiographic film, at 0.486 and 0.920 Å respectively. This leads to intense absorptions of radiation by these elements at those, and shorter, wavelengths. Such radiation is therefore rather more effective in producing a latent image in the emulsion

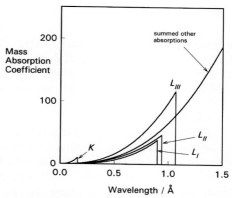

Fig. 4.2 The individual absorption spectra for several electronic transitions in platinum.

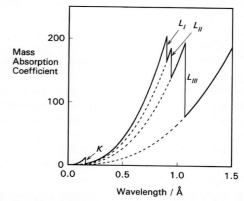

Fig. 4.3 The absorption spectrum for platinum dissected to show the additivity of the individual spectra for the overall effect.

than visible light, which can only affect the outermost electrons. The effect can be demonstrated by using a crystal as a diffraction grating to obtain a spectrum, much like visible light may be diffracted. It should be noted that radiation is indeed present but simply passing through the film, at energies corresponding to the gaps between the *L*-series lines (see Fig. 1.1). The image is lacking because the absorption by the film is very weak. However, in practice, the radiation corresponding to the region of the *L*-series characteristic X-rays of W would be heavily absorbed by an Al-filter and would not contribute to an ordinary clinical radiographic image.

●4.1   Scattering

Absorption has been discussed as if there were only the one process operating, as indicated at the beginning of §3. There are in fact two separate components to the reduction of transmitted beam intensity. **True absorption** is due to the transformation of X-rays into the kinetic energy of electrons, having first promoted them to the 'free' state as discussed above, hence the description of this process as **photoelectric absorption**. These ejected electrons leave vacancies into which other electrons may cascade (§1), thereby producing **fluorescent** radiation which will be emitted in all directions. As with other fluorescence processes (6§5.2, 24§4.9, 24§6.5), the wavelength of such secondary characteristic radiation cannot be shorter than the original absorbed radiation. It is therefore capable of being subsequently absorbed quite efficiently, and therefore adding to the damaging action of the primary beam, but over a larger volume.

Scatter, however, is due to a transfer of energy from the incident beam to secondary rays through interaction with the atoms of the medium, but not mediated by electronic transitions. The observed overall mass 'absorption' or attenuation coefficient can thus be separated into two components:

---

[7] A special usage of the word emulsion that does not signify the usual stable dispersion of immiscible liquid droplets in a second liquid, silver halides being crystalline.

$$\frac{\mu}{\rho} = \frac{\tau}{\rho} + \frac{\sigma}{\rho} \tag{4.1}$$

for true absorption (τ/ρ) and scattering (σ/ρ). This scattering coefficient is nearly constant, varying little with either λ or Z. But scatter is itself due to two entirely separate processes. (Electron-positron (e⁻, e⁺) pair production can be ignored for dentistry because the accelerating voltage used is generally not high enough, *i.e.* $< \sim 1$ MV, since 2×0.511 MeV, the rest mass-energy of each particle, is required for this.)

●4.2   Compton effect

Electromagnetic radiation imposes, as its name implies, oscillating electric and magnetic fields on the matter on which it impinges. Electrons, being charged, will then tend to oscillate in such fields. But an oscillator must itself re-radiate energy, so that the net effect is for energy to be removed from the incident beam and reradiated in random directions. Such scattering is therefore described as **incoherent** and **inelastic**. The radiation that is not scattered, continuing on in the beam, must then have a slightly longer wavelength, corresponding to its reduction in energy. This process only applies to free or loosely bound electrons such as in the outermost, bonding, shells where the available energy levels are very close together. Their occurrence is therefore essentially constant despite changes in Z. However, these electrons are *relatively* more important at low Z, as the proportion of the total number of electrons present. This scattering process is called the **Compton effect**.

●4.3   Coherent scattering

The second scattering process is due to the fact that the separations between atoms (centre to centre) are of about the same scale as the wavelengths of X-rays. There are therefore reflection and interference phenomena associated with this condition. As far as the determination of crystal structures is concerned, this **coherent** or **Rayleigh** scattering from well-organized layers of atoms as in crystals, otherwise known as **diffraction**, has been overwhelmingly important to our understanding of the structure of matter through **X-ray diffraction analysis**. However, the general irregularity (non-crystallinity) and complexity of matter, and especially soft tissue, as well as its thermal motion, results in a general 'background' of scattered radiation of *exactly* the same wavelength as the incident beam. That is, there is *no* absorption and reradiation of energy, and it may be viewed as 'pure' lossless scatter. For practical structure determination by diffraction a monochromatic X-ray source is ideal. With the polychromatic or 'white' continuum of radiographic X-rays the superimposition of all possible diffraction peaks, even from an obviously well-crystalline material such as hydroxyapatite, results in no structural information being decipherable.

The existence of coherent scattering means that a slight adjustment must be made to the thinking about attenuation coefficients as expressed by equation 4.1. Clearly, the crystallographic form affects the amount of scattering (the type of crystal symmetry and angular orientation of the crystal planes to the beam direction), but also the atoms involved (Z), the temperature (as it affects lattice spacing) and ionic charges are involved (Z in effect referring to the electron count, the nuclear charge having no relevance[8]). Thus, although crystallinity has an effect, it is physical rather than chemical in origin. Even so, for biological systems the magnitude of the effect is small and can be ignored.

These two scattering processes add further to the non-primary beam radiation to which the surroundings of the deliberately irradiated area are subject. This accounts, in part at least, for the use of 'lead' aprons and thyroid shields. It is also the reason for the lead foil included in radiographic film packets.

Although both Ag and Br have moderately high atomic number, Z, and the corresponding mass absorption coefficients are high, the rather small amount of AgBr present in the emulsion (of the order of a few mg/cm²) does not absorb more than a few percent of the incident beam. The radiation which penetrates will continue on to the film packet or cassette and, in the case of intraoral films, to tissue on the other side of the mouth. This transmitted radiation will produce secondary scattered radiation by interacting with that tissue or other materials. Since this radiation is emitted or scattered in all directions, some would come back to the film. This would also interact with the film emulsion and tend to 'fog' the diagnostic image (there can be no image information in that **backscattered** radiation). (There is also some scattering in the forward direction from all material in front of the film and this adds to the fog. This becomes significant for deep body sections.)

---

[8]   Since all chemical substances are charge-neutral overall, the effect of ionic charge on true absorption cancels out exactly. Hence it makes no difference overall to the true absorption how atoms are combined or ionized.

•4.4   Lead

Lead is the heaviest element ($Z = 82$) to have no natural radioactive isotopes. Bismuth ($Z = 83$), which has a half-life of $\sim 2 \times 10^{19}$ y and emits a very low-energy α-particle with a range in solids of only a few micrometres,[6] is stable enough to be useful (not a problem in HSCs, for example; 9§12.3). However, the latter is not very abundant (3.4×10^{-6} % of the earth's crust by mass) compared with lead (0.002 %) and about twice the price. Lead is also very ductile (f.c.c.) and does not work-harden (its recrystallization temperature is below room temperature, Fig. 11§6.10). It is therefore suited in all respects as a high efficiency X-ray absorber.

To minimize the effect of **backscatter** on the radiograph a thin lead foil is placed in the back of intra-oral film packets. Typically, this is between about 0.05 and 0.07 mm thick or about 60 to 80 mg/cm^2, and gives a total transmittance for 100 kV (white) radiation of about 0.30, somewhat less for lower accelerating potentials. However, the absorption efficiency is much greater for secondary radiation, which can be expected to have much longer wavelengths. Thus, the tissue irradiation on the far side of the film is reduced substantially, but returning secondary radiation is practically eliminated.

Because lead is so soft, foil and other forms should not be handled ungloved if possible. Some skin absorption can occur via transfer of metallic particles (lead marks paper, and used to be used for that, hence pencil "lead"; 鉛筆), and superficial oxide is also transferred. Ingestion from the hands is then feasible. While the effectiveness of migration to the blood is debated, and gut absorption is inefficient, it would be prudent to minimize intake of this known toxic heavy metal.

•4.5   Ionization

X-rays and other radiation are called **ionizing** because they interact with matter to generate ions. Thus, for water, we have the following, for example:

$$H_2O + hv \rightarrow H_2O^+ + e^-$$
$$H_2O + H_2O^+ \rightarrow H_3O^+ + \bullet OH$$
$$e^- + H_2O \rightarrow HO^- + \bullet H \tag{4.2}$$

However, it is not the ions themselves that are the problem. It is the two radicals that are generated that cause the problem: these can go on to many other reactions as in other contexts, for example to produce superoxide:

$$H\bullet + \bullet O_2\bullet \rightarrow HO_2\bullet \tag{4.3}$$

which is known to be damaging to living tissue. The enzymes to handle such chemistry (superoxide dismutase, in this case) may be overwhelmed, or not exist. Such reaction chains may continue for a long time because the free radical concentration may be low, and mutual annihilation therefore rare.

In fact, there is a further, more damaging mechanism.[7] The first ion formed in reaction 4.2 is actually high energy since it is an inner electron that has been ejected:

$$H_2O + hv \rightarrow *H_2O^+ + e^- \tag{4.4}$$

where the asterisk means an excited state. The characteristic radiation emitted in returning to the ground state is greater than the ionization energy of an outer electron and so can be absorbed by an adjacent water molecule:

$$*H_2O^+ \rightarrow H_2O^+ + hv$$
$$hv + H_2O \rightarrow H_2O^+ + e^- \tag{4.5}$$

so the net effect is

$$hv + 2\,H_2O \rightarrow 2\,H_2O^+ + 2\,e^- \tag{4.6}$$

The ions and electrons then may react as above, that is, double the effect because each can initiate the radical cascade. But first the two adjacent cations suffer a strong repulsion because of the like charges and give what may be called a **Coulombic explosion**, which itself can cause yet further damage, which for DNA would be serious.

Such ionization is also instrumental in creating free radicals that can initiate polymerization in filled-resin restorative materials. Thus, dental surgeries in which chairside X-ray sets are used have the potential for causing deterioration of such products in that room if they are not well shielded. They should be protected as is the stock of radiographic film, perhaps in a lead-lined box, even if they are kept in a refrigerator.

●**4.6   Emission spectrum (2)**

It will be noticed that the spectra shown in Figs 2.2 and 2.5 have a rather odd shape. The short-wavelength limit has already been explained (§2.2), but not the presence of the peak. It is clear that in order to be decelerated, and so emit radiation, an electron must enter the target. Accordingly, that emitted radiation is generated inside the target material, not right at the surface (Fig. 4.4). The theoretical shape of the actually-emitted **bremsstrahlung** or "braking" radiation is as in Fig. 4.5 a – a linear decrease to the short-wavelength limit when this is plotted against energy.[8] Obviously, much of this is in the infra-red, visible and ultra-violet (§2.3), and even only a few micrometres of tungsten are not transparent to the long-wavelength X-rays generated. The spectrum emitted is therefore necessarily heavily filtered, even before it reaches the envelope of the X-ray tube.

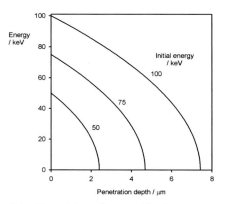

Fig. 4.4 Remaining electron kinetic energy as a function of depth of penetration. As energy is lost interaction increases and so the remaining energy falls, roughly linearly as its square: $E^2 = E^2_{max} - m.x$. Hence, the velocity falls roughly linearly as its fourth power – but see §2.5.

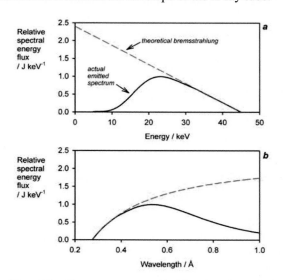

Fig. 4.5 For tungsten at 45 kV, the theoretical bremsstrahlung emission spectrum (broken line) and that actually observed outside the tube (solid line) plotted against energy (**a**) and wavelength (**b**). (For other accelerating voltages, in **a** the theoretical lines are parallel, passing through the peak voltage point.)

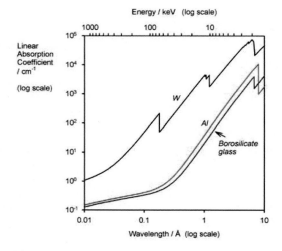

Fig. 4.6 Variation of linear absorption coefficient for common target (W), envelope (borosilicate glass) and filter (Al) materials.

Of course, the glass of that X-ray tube also absorbs X-radiation (Fig. 4.6), adding to the attenuation, and ordinarily so that what actually emerges from the tube (after blocking remaining UV and longer wavelengths, which in some tube designs is achieved with a thin beryllium shield, but see 28§7) is more as is shown by the solid line in Fig. 4.5 a. When that data is replotted against wavelength, it appears as in Fig. 4.5 b – the more familiar form. There remains, however, a substantial long-wavelength X-ray component that is not required for imaging (§3.2), and further filtration is applied with aluminium.

The variation in spectrum due to ripple (§3.4) can now be seen as a distortion of the shape of the emitted spectrum, biassed towards the low-energy side, as a time-weighted average.

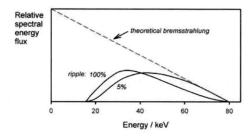

Fig. 4.7 Effect of non-constant accelerating voltage on the emitted bremsstrahlung spectrum. Ripple is measured as the proportion of variation from the peak value. (Characteristic radiation is omitted.)

§5. Radiographic Film Density

As was mentioned above, X-rays passing through the crystals of the silver bromide of a photographic (radiographic) emulsion may be absorbed directly, the electrons ejected creating the Ag^0 (metallic) atoms that constitute the so-called **latent image**, essentially defects in the crystal. Both types of scattered radiation, whether from within the film or from tissue, as well as the direct beam, contribute to that image, which becomes visible upon the chemical processing of the film called 'development'. In this process a halide crystal (or 'grain') that carries such a damage site is reduced entirely to metallic silver.

●5.1   Exposure

For a given film development schedule it might be expected - by analogy with visible light photographs – that the degree of blackening observed in the processed film is a direct increasing function of the **exposure**. Exposure, E, may be defined as the total radiation **dose** that has passed through the film and, for a constant radiation intensity, I, incident on the film, this may be expressed as the product of that intensity and time, t:

$$E = I.t \qquad (5.1)$$

Since for a given X-ray tube, target-film distance, operating voltage and filtration the intensity is proportional to the tube current, i, we may write instead:

$$E \propto i.t \qquad (5.2)$$

The same value of E (expressed in mA.s) clearly may be obtained through various combinations of i and t, governed by $i \propto t^{-1}$, but the blackening of the film will not necessarily be constant. This arises because of the (still imperfectly understood) nature of the chemistry of silver bromide emulsions. However, this **reciprocity failure** is not serious and the ideal **reciprocity law** (equation 5.2) holds well enough in radiography over intensity ratios as high as 1000 : 1. For more complete generality we should then write equation 5.2 as:

$$E \propto \int i.dt \qquad (5.3)$$

since tube current variations may occur for various reasons. (In X-ray sets that do not have smoothed accelerating voltage, that voltage goes up and down in step with that of the rectified mains supply. Since clinical exposure times are always rather longer than 0.01 s, the variation of tube voltage during a half cycle and the consequent spectrum variation, is of no practical consequence.)

●5.2   Optical density

Equation 3.5 defined the transmittance of a body with reference to X-rays in particular. However, the definition of transmittance as the ratio of transmitted to incident radiation is a perfectly general one where the absorbing or scattering centres are randomly distributed over both the area and thickness of the object. In the case of photographic or radiographic film the developed emulsion contains a very fine dispersion of the absorber - the precipitated particles of silver - in just such a manner. The idea of transmittance applies, therefore, equally well to the behaviour of *visible* light in a *developed* film.

Now, if we invoke Beer's Law for the absorption of visible light according to the concentration c of silver particles per unit area in the more or less fixed thickness of the emulsion, we have:

$$-\frac{dI}{dc} = \alpha I \qquad (5.4)$$

where the absorption coefficient is α. Integrating this with respect to c gives:

$$I = I_0 e^{-\alpha c} \qquad (5.5)$$

Thus, we similarly define the transmittance for this system:

$$T = \frac{I}{I_0} = e^{-\alpha c} \qquad (5.6)$$

However, these numbers rapidly become very small and inconvenient to use. In dealing with the transmission of light through objects it is common to use the **Optical Density**, D, as the measure. This is defined as the negative of the logarithm (to base 10) of the transmittance:

$$D = -\log_{10}(T) = \log_{10}\left(\frac{I_0}{I}\right) \quad (5.7)$$

In other words,

$$D = \log_{10}(e^{\alpha c}) \quad (5.8)$$

i.e. numerically,

$$D = 0.4343\alpha c \quad (5.9)$$

Optical density provides a measure of how dark we will perceive the developed film to be by eye. This may be said to be in 'Optical Density' units, OD. The usefulness of OD is that equal steps of apparent brightness are in a geometrical progression of the actual intensities, *i.e.* their logarithms are perceived to be equally spaced or nearly so (**Weber-Fechner Law** [9][10]).

●5.3 Exposure and film density

If now we assume that the number of halide grains made developable, *c*, is proportional to the exposure (this means really that the exposure is relatively low so that the proportion of all grains present rendered developable is small), we can write:

$$c \propto E \quad (5.10)$$

noting that efficiency of capture of X-ray quanta by atoms in the emulsion is very low (typically only about 4 %). It then follows immediately from equation 5.9 that:

$$D \propto E \quad (5.11)$$

This is a very convenient result.

The amount of blackening of a film is measured using an instrument known as a **densitometer**. This simply measures the intensity of the light transmitted (*i.e.* through the film) and expresses it according to equation 5.7. Typical plots of D *vs.* E are shown in Fig. 5.1. Note that variations must occur if the intensity *vs.* wavelength profile is altered, as for example by increasing the accelerating voltage. But, in general, the range of voltages used for clinical diagnostic radiography is relatively small, and the spectrum shape varies correspondingly little. The film response is not very sensitive to such moderate changes in the spectral composition of the X-rays in the region of interest.

●5.4 Fog

It will be noted from Fig. 5.1 that the plots do not pass through the origin. That is, in the absence of any deliberate exposure of the undeveloped film to X-rays, a developed film will nevertheless show some absorption of light. This is called **fog** and has three components. It is distinguished from that fogging due to fluorescence and scatter in the radiographed object (tissue) and the film since these must be proportional to the exposure. Partly, fog is due to **background** radiation. This has several sources: the natural radiation from rocks (granite in particular, which

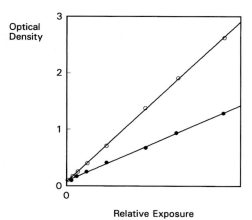

Fig. 5.1 Variation in film optical density after development against relative X-ray exposure for two high density film types.

The half-life of a radioisotope is defined by
$$T_{1/2} = \ln(2)/\lambda$$
where λ is the decay constant (s⁻¹) such that the number of atoms of that isotope at time *t* is:
$$N(t) = N(0).e^{-\lambda t} \quad (11§2.6)$$
The activity *A* (Bq) of a sample of N atoms is:
$$A = N\lambda = N.\ln(2)/T_{1/2}$$
For potassium, u ≈ 39.1, so given the total concentration [K] = 2.5 g.kg⁻¹ we have:
$$N(K^{40}) = N_A.[K]/u \times 1.18\times10^{-4}$$
atoms of K⁴⁰ per kg lean body mass. Inserting this in the activity equation yields
$$A(K^{40}) \approx 80 \text{ Bq.kg}^{-1}$$

may be used in building materials[11]), cosmic rays (about 2 cm⁻².min⁻¹), and man-made releases into the environment. Even natural potassium contains about 0.0118 % of K⁴⁰, which is a β- and γ-emitter with a half-life of 1.26 × 10⁹ y. Since there is a total of about 2.5 g/kg of potassium in a lean body, this means about 80 Bq.kg⁻¹ from this source (of the approximately 120 Bq.kg⁻¹ total radioactivity of the body, most of the rest being due to C¹⁴). Such radiation, of course, can be from both operator and patient.[12]

Another aspect is the **thermal degradation** of the emulsion. With time, even at 20 °C, film will gradually accumulate fog due to the inherent instability of the silver bromide. This is the reason for storing film stocks in a refrigerator. However, and even then, film will fog at 4 °C (consider the Arrhenius equation), so film stocks have a limited life. The management of X-ray facilities must take this into account in buying and using stock. Another source of fog is the inevitable development of some AgBr grains in the complete absence of any exposure of any kind. This problem is time- and temperature-dependent: that is, the longer and hotter is the developer the worse such fog will be. Of course, accidental but unknown exposure to X-rays in or around a radiography installation will also contribute to fog.

Fogging also occurs due to visible light exposure that may occur when film is removed from its packet or cassette. This light may arise from a bad choice of safelight in the darkroom, from leakage into supposedly light-tight processing equipment, or even static electricity. Exposure to gases such as ammonia, formaldehyde and hydrogen sulphide also causes fogging, as does mechanical damage, such as kinking the film.

●5.5   Base density

Indistinguishable from fog under normal conditions of measurement will be the optical density of the (processed) emulsion and the film base due to light absorption and scattering by their components, although usually this is small. However, radiographic film for dental use is usually sold with a blue- or grey-tinted base, which is said to improve the ease with which the image may be viewed and read by eye. This in itself will contribute an appreciable increment to the optical density, and this increment is known, appropriately enough, as the **base density**.

●5.6   Interface effects

One other aspect of the optical density of a film, usually overlooked, is the reflection of light at the interface between two media of different refractive index (24§5.3), irrespective of the direction in which the interface is viewed. Many organic polymers have a refractive index (n_2) of about 1.5. If we assume that film base has a similar value, and that of air (n_1) is 1.0, then for perpendicular incidence:

$$\rho_0 \;=\; \left(\frac{n_2 \,-\, n_1}{n_2 \,+\, n_1} \right)^{2} \;=\; 0.04 \tag{5.12}$$

So the overall transmittance for two such interfaces (ignoring absorptions) is given by:

$$T \;=\; \left(1 - \rho_0\right)^{2} \;=\; (1 - 0.04)^2 \;=\; 0.9216 \tag{5.13}$$

then

$$D \;=\; \log_{10} \frac{1}{T} \;=\; 0.035 \tag{5.14}$$

Real films may have a protective layer over the emulsion and an additional layer between the emulsion and the base to promote adhesion. If the film is also double-coated (emulsion both sides) to improve the sensitivity and maximum attainable blackening (see below), the number of interfaces is also (nearly) doubled. The apparent optical density due to reflection for a film of zero actual absorbance is not negligible.

In practice, these effects would be measured as part of the 'base' density and are inseparable from it. Even so, this analysis implies that if measurements of density are to be made it should be on the bare, unlaminated film otherwise spurious increments to density may confuse the analysis. Worse, if the laminating layer is not everywhere in intimate contact with the film it means that instead of the small difference in refractive index across that intended boundary, there will now be two large differences either side of the trapped air. This effect would be large and inconsistent from place to place on the film. Effects would also be expected for finger-grease: the change in glossiness of fingerprints is obvious (*cf.* 24§5.10).

●5.7   Additivity

It follows from equation 5.7 that optical density is **additive**. Thus, if we write:

$$D = \log_{10}\left(\frac{I_0}{I}\right)$$

$$= \log_{10}\left(\frac{I_0}{I_1} \cdot \frac{I_1}{I_2} \cdot \frac{I_2}{I_3} \cdot \frac{I_3}{I_4} \cdot ... \cdot \frac{I_{n-1}}{I_n}\right) \tag{5.15}$$

$$= \log_{10}\left(\frac{I_0}{I_1}\right) + \log_{10}\left(\frac{I_1}{I_2}\right) + \log_{10}\left(\frac{I_2}{I_3}\right) + \log_{10}\left(\frac{I_3}{I_4}\right) + ... + \log_{10}\left(\frac{I_{n-1}}{I_n}\right)$$

for n successive absorbers, as if they were in layers, we must therefore have:

$$D = D_1 + D_2 + D_3 + D_4 + ... + D_n \tag{5.16}$$

Thus, the contributions of various sources can be considered separately as they are independent.[9] We may therefore write, for a film,

$$D = D_{fog} + D_{reflections} + D_{base} + D_{emulsion} + D_{Ag} + ... \tag{5.17}$$

This enables us to work with optical densities in a very straightforward manner to analyse film response, for example.

●5.8   Film response

Returning to Fig. 5.1, it is plain that the anticipated linear relationship holds over an appreciable range for X-ray exposure, *i.e.*:

$$D = aE + f \tag{5.18}$$

where f is the density due to fog + base, and a is a constant, a measure of the sensitivity of the film. The greater the value of a the steeper the line, and the greater the blackening of the film for a given exposure. This equation, while useful over the range of exposures normally encountered in radiography, is nevertheless an approximation, as will be shown below.

It must be said that a direct comparison with the behaviour of visible light exposure of photographic films cannot be made because of a fundamental difference in the mode of interaction of the respective radiation with the film. Not because the film itself is very different (films for radiographic purposes have more AgBr in order to compensate for the low detection efficiency of the radiation) but because one X-ray photon is capable of generating many free electrons which guarantees the developability of the grain which has been struck. Visible light, in great contrast, can generate at most one promoted electron, and this may decay without making the grain developable. Usually several absorbed visible light quanta are required for one latent image site to form. This is why dim red 'safelights' can be used in a radiographic film processing laboratory without seriously affecting the fog. (Visible light films have special dyes called sensitizers added to the emulsion to increase greatly the sensitivity to red light.)

If one plots optical density against exposure, as suggested by the form of equation 5.15, and over a large enough exposure range, the curve is in fact an exponential approach to a limiting value (Fig. 5.2).

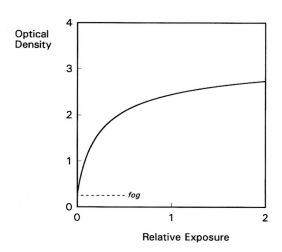

Fig. 5.2 The type of curve obtained by plotting simply optical density *vs.* exposure. Notice the absence of the spurious features identified in Fig. 5.3.

---

[9] This is true if there is no differential absorption, or only one type of such contributor. The only colour normally present is due to the film base, all other components being neutral, so eq. 5.17 is good.

Unfortunately, and due to an historical precedent set in 1890, it has long been the custom to plot D *vs.* log E when examining film response, resulting in a sigmoidal curve, such as in Fig. 5.3, called the 'characteristic curve'.[13] This plot has no basis whatsoever in the chemistry or physics of the system, but was based on the fact that our perceptions are logarithmic or nearly so (§5.2). That is, while we perceive evenly-spaced steps in the logarithm of intensity, *i.e.* optical density, this should not be applied to the causal stimulus as well in a physical chemical system. Even so, this has not prevented a number of terms being defined in terms of that curve's shape and used to describe supposedly the film's response for given exposure and development conditions. The 'shoulder' and the phenomenon of 'reversal', or at least a plateau at a maximum achievable density, depends of course on the limited density that must result from the limited amount of AgBr present in the emulsion, as seen in Fig. 5.2. These effects would be seen if the plots of Fig. 5.1 were extended sufficiently.[14]

The J-shape of the rest of the curve of Fig. 5.3 is entirely due to the logarithmic transformation applied to the E-axis (Fig. 5.4). Thus, several terms depend absolutely on this transformation and have <u>no meaning</u> in the context of the theory of the film response. Furthermore, because these terms all depend on this one aspect of the plot, they are themselves mutually interdependent and cannot increase the information available about the film. There are only two parameters for film response to X-rays: saturation optical density and sensitivity (*i.e.* slope in Fig. 5.1).

●5.9   Double coating

The maximum optical density that can be achieved with a given film depends only on the amount of silver halide in the original emulsion as this determines how much silver can be formed on development. This can be seen to increase the sensitivity of the film as well (Figs 5.5, 5.6) as the response curve is made steeper. There is, however, a practical limit to how much silver halide can be handled in one emulsion layer in the manufacture of the film. This corresponds to a saturation density, Q, in the region 3 to 4 OD. The answer to this is to use films which are **double-coated**. Q in the region of 7 to 8 OD is then easily achievable. This has the practical effect of reducing the exposure of the patient to X-rays: shorter exposure for the same developed film density.

●5.10   Useful range

Working range is said to be identified in the "characteristic curve" as the optical density range corresponding to the ends of the "linear portion". Unfortunately, there is in fact no linear portion, only the approximately straight region around the inflexion. However, we can define working range in a rather different way, one that is related to the actual conditions

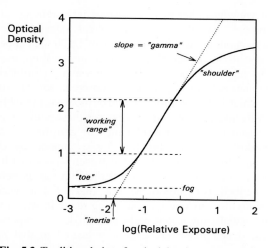

Fig. 5.3 Traditional plot of optical density of a developed film against log(exposure) to give the so-called "characteristic curve". This plot has no physical basis, and 'inertia', 'toe', 'gamma' and 'working range' are all spuriously defined and therefore invalid.

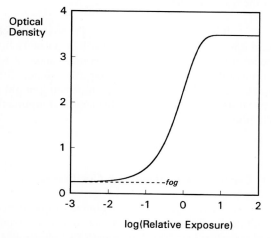

Fig. 5.4 The same data as in Fig. 5.2, replotted for a logarithmic exposure axis.

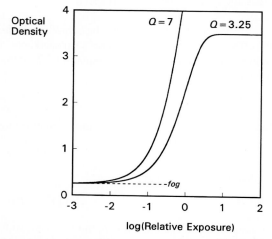

Fig. 5.5 The effect of using a double-coated film on the sensitivity and saturation optical density in the conventional form of plot.

of viewing rather than a spurious interpretation of a curve. To the eye, an optical density of 2, corresponding to a transmittance of 1/100, will appear very black. Evidently, no image information above this point will be useful for the viewer. Similarly, another arbitrary limit can be chosen at, say, D = 0.5 at which light absorption would be too weak to yield an easily interpretable image. Between these points might be said to be the **useful range** of densities, or the working range of *any* film for evaluation by eye (Fig. 5.6). This operational approach shows that it is not so much the behaviour of the film that is being described (or important) but that of our visual apparatus: it is our eye's working range in which we are actually interested, and emulsion design, exposure and development are expressly geared to that and nothing else.

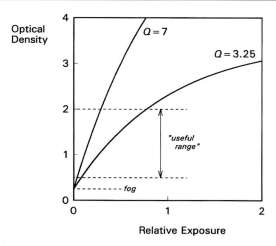

Fig. 5.6 The data of Fig. 5.5 replotted for a linear exposure axis, showing how 'useful range' is not a film property, and also that Q affects sensitivity.

Similarly, consideration of equation 5.15 and Fig. 5.6 shows that there is no possibility of defining both **sensitivity** and **contrast** in distinct ways. For X-ray exposure they are one and the same thing, the slope of the response line.

It is worth noting that the light-box on which a radiograph is viewed will affect the useful range. That is, if it is brighter, the upper limit will be pushed a little further, but equally the lower limit will be raised as less dense areas become too bright for comfort and easy reading. However, the light would have to be twice as bright to move the OD limits by 0.3, $\log_{10}(2)$. Hence, little is to be gained by adjusting this brightness. Even so, screening-off the surroundings, which being bright would otherwise close the pupil and make the film image appear darker, remains a sensible action. Largely, this is a physiological issue, not a fundamental property of the film.

§6. Experimental Attenuation Coefficient

The radiographic density or radio-opacity of dental materials is of interest when these have to be detected in clinical radiographs; for example, to recognize root-filled teeth or filled cavities. As was discussed in 6§1.10, and underlined by the calculation in §3.1 above, PMMA and similar polymers would be difficult to distinguish in an ordinary radiograph. The addition of absorbing fillers is one way of overcoming this. Thus, the experimentally determined value of a material's linear attenuation coefficient is of immediate application in assessing performance in this respect.

Equation 5.15 describes the behaviour of the film in terms of the radiation exposure. We now consider the effect of an absorber of thickness x placed between the radiation source and the film, *i.e.* as the object being radiographed. Substituting from equation 3.2 for the radiation intensity emerging from the absorber in equation 5.1 we have:

$$E = t.I_0 e^{-\mu x} \tag{6.1}$$

Substituting this in equation 5.18, and correcting for base and fog density, we get:

$$D - f = at.I_0 e^{-\mu x} \tag{6.2}$$

so that, taking logarithms,

$$\ln(D - f) = \ln(at.I_0) - \mu x \tag{6.3}$$

Of course, a and I_0 are unknown, but constant, so that if the exposure time and tube current are held fixed the effective linear attenuation coefficient for a material can be found from the slope of $\ln(D - f)$ *vs.* x. The values thus determined can only be valid under the chosen X-ray conditions (*i.e.* voltage and filtration, therefore spectrum), as opposed to data such as in Tables 3.1 and 3.2; but nevertheless it is obviously this very fact that makes it of direct clinical relevance. It may be necessary even so to determine the slope at $x = 0$ because of the differential absorption effect discussed in §3.3.

A further factor arises from the **inverse square law** because the 'beam' is divergent:

$$I \propto d^{-2}$$
(6.4)

If I_0 is taken as the value at the specimen surface, which is distance d from the tube target, then a further attenuation is present due to the slightly longer path through the test material, which may be accounted for by writing (instead of equation 6.1):

$$E = t.I_0 \frac{d^2 e^{-\mu x}}{(d + x)^2}$$
(6.5)

However, it can be seen that if $d \gg x$, the effect will be very slight (but see §8.2).

§7. Screens

In order to reduce the exposure of the patient to X-rays to minimize tissue damage, there are obvious advantages in using 'fast' – that is, sensitive – film. However, the efficiency of X-ray absorption by the film remains low. It is nevertheless possible to increase the apparent efficiency by converting some X-rays to visible light, to which AgBr emulsion is rather more sensitive. Such a system is the **intensifying screen**. This consists of a thin sheet of a material containing a layer of a salt, typically calcium tungstate, $CaWO_4$. This sheet is placed in close contact with the emulsion. Some rare earth salts are also used, with a wide variety of compositions and properties. The salt is typically chosen such that under X-irradiation it fluoresces strongly in the blue and ultra-violet as this is the wavelength band most effective for the purpose, for example, about 1000 photons emitted per X-ray photon at 50 kV. (Compare the use of $CaWO_4$ as a phosphor in fluorescent tubes; Table 24§4.1.)

The fluorescent radiation is produced in random (spherically symmetrical) directions, as might be expected, and so will not all be effective in exposing the film (Fig. 7.1). This also is the cause of some loss of image quality as there is then a small circular region effectively exposed for each point of irradiation, broadened by some scattering of the emitted light in both the screen and the film as well as X-ray scattering by the crystalline grains of the metal (typically, aluminium) in the cassette front face and in the screens themselves. It is important that the screen be in very close contact with the film to minimize image degradation by minimizing the area of film exposed. The improvement in image density obviously depends on many factors: attenuation of the X-rays in the front face of the cassette, the absorption efficiency of the fluorescent salt with X-rays, which will depend on the spectrum of the radiation; the light emission efficiency in the film direction, taking into account crystal size, layer thickness (which increases scattering and useless absorption) and the uniform spherical emission pattern; and the sensitivity of the film to the fluorescent radiation.

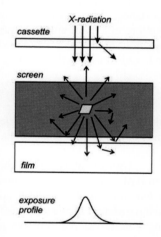

Fig. 7.1 Phosphors emit randomly, with losses due to scattering and reflections.

Losses of emitted light on the non-film side of the screen phosphor layer may be reduced by adding a reflector layer, for example titanium dioxide (Fig. 7.2). However, this can only increase the image degradation as the returned light is now scattered over a larger area. It also requires the phosphor layer to be thin and translucent enough to allow an appreciable quantity to be returned. Some X-ray scattering must arise in such a reflector layer.

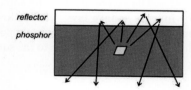

Fig. 7.2 A reflector layer can increase the film exposure from a screen, but the scatter is increased.

While the use of double-coated film (§5.9) and two screens can improve the sensitivity of the system, and thus reduce the exposure, there is a considerable loss of sharpness due to the fact that the efficiency of absorption of visible light by the emulsion is not perfect, a lot gets through to cause a **cross-over** exposure of the halide on the other side of the film (Fig. 7.3). As much as 40 % of the effective exposure may come from this **punch-through** effect. It also includes the effects of multiple internal reflections at all interfaces (§5.6). This may

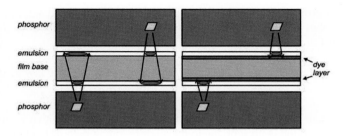

Fig. 7.3 Cross-over exposure with double-coated films between screens may be controlled by suitable absorbing dye layers.

be counteracted by including under the emulsion a dye layer that absorbs the phosphor emission. Of course, that dye must be dissolved or bleached when the film is processed in the normal way. Although the effective film sensitivity is reduced substantially, the image quality is much improved.[15]

Although intensifying screens can produce appreciable reductions in patient radiation dosage of the order of 1/50 or 1/100, their use is limited to non-intraoral film by the bulk of the cassette required to hold them.

§8. Image Properties

There are a number of properties of radiographic images which need to be appreciated to avoid errors of interpretation and measurement, as well as inherent limitations of image quality.

•8.1   Geometry
There are two geometrical points to be made concerning the behaviour of X-rays, and these are of great importance. Firstly, the radiation travels in straight lines. This is of course in common with visible light travelling in a medium of uniform refractive index. The implication is that all radiographs are direct linear projections of the object irradiated. The use of 'similar triangles' arguments easily shows that the ratio of image size to object is directly given by the ratio of the target-film and target-object distances:[10]

$$\frac{\text{Image width}}{\text{Object width}} = \frac{\text{Target-film distance}}{\text{Target-object distance}} \tag{8.1}$$

with the sole proviso that the object plane is *parallel* to the film plane, no matter that the ray path from focus to film is oblique to the plane of the latter ('target' here means the X-ray source) (Fig. 8.1).

Unfortunately, real bodies have a definite thickness in the direction perpendicular to the film plane. Even if of constant width, the magnification will vary for different parallel sections because of the variation in the denominator of the right-hand side of equation 8.1 (Fig. 8.2). Obviously, a section nearer to the tube target will have a larger image than one further away. However, there is *no* distortion of the image of thin plane objects lying parallel to the film, as has sometimes been claimed. Unfortunately, the images of objects or structures which are not lying parallel to the film plane may suffer considerable distortion (Fig. 8.3) and great care is required if measurements are intended.

The second point is the applicability again of the **inverse square law**. This follows from the first point because if lengths are in proportion to distance then areas are in proportion to the square of distance:

$$A \propto d^2 \tag{8.2}$$

Thus, for a constant irradiance of the object (*i.e.*

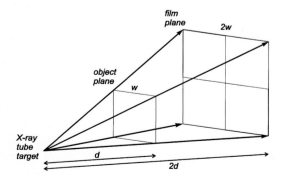

Fig. 8.1 The scaling of projected images.

---

10   FFD: film-focus distance; SID: source-image distance.

expressed as per unit area), the irradiance at the film plane will be in inverse proportion to its distance from the source (for uniform attenuation in the object):

$$I \propto d^{-2} \quad (8.3)$$

Then, for a constant optical density (D) in the film, the exposure (film dose) must be correspondingly increased:

$$E \propto d^2 \quad ; \quad D = \text{constant} \quad (8.4)$$

From equation 5.2 this is seen to be equivalent to:

$$i.t \propto d^2 \quad ; \quad D = \text{constant} \quad (8.5)$$

Therefore:

$$i_2.t_2 = i_1.t_1 \times \left(\frac{d_2}{d_1}\right)^2 \quad ; \quad D = \text{constant} \quad (8.6)$$

where d_1, d_2 represent the distances to the film in the two positions. Adjustments may therefore be made to either the tube current or the exposure time, or both, to compensate for variations in target-film distance.

●8.2   Beam parallelism

It has been implicit in earlier sections that the beam of X-rays traversing the material of interest is parallel, and in the above that the object is irradiated uniformly. However, as indicated earlier (§2.3), and indeed relied on above, the beam from an X-ray tube is necessarily divergent. Hence, by virtue of the inverse-square law (and assuming a point source), the irradiation of a flat object, normal to the beam axis, will not be uniform. The above exposure calculations can therefore only be an approximation. Since the actual target-object distance h varies according to $\cos \theta = \cos[\tan^{-1}(y/d)]$ (Fig. 8.4):

$$h = d/\cos \theta \quad (8.7)$$

so the actual irradiance varies across the irradiated plane object according to:

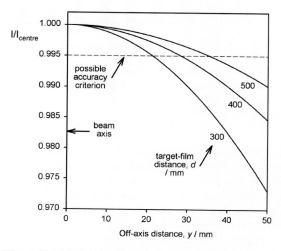

Fig. 8.5 Relative irradiance of a film or object as a function of off-axis position (equation 8.9).

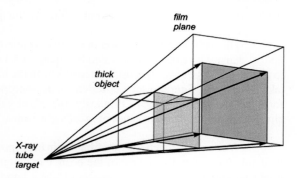

Fig. 8.2 Different sections of thick objects have different projected sizes.

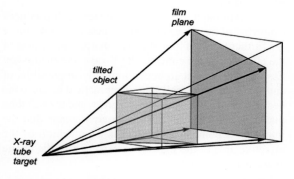

Fig. 8.3 Objects tilted with respect to the film will have distorted images.

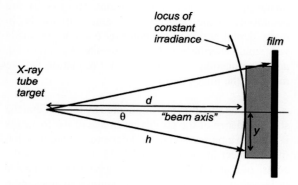

Fig. 8.4 Beam divergence means that the irradiation of flat objects cannot be uniform.

$$I \propto (d/\cos \theta)^{-2} = (d^2 + y^2)^{-1} \quad (8.8)$$

or, in relative terms:

$$I/I_{\text{centre}} = d^2/(d^2 + y^2) \quad (8.9)$$

(from Pythagorus' theorem). It follows that any attempt to measure the absorption characteristics of a system must use a target-film distance sufficiently large that this 'spherical aberration' is negligible over the width of the irradiated object. It also follows that for a sufficiently thick object, the actual path length through it will vary according to $\tan \theta$, and the image of the edge of the object will be markedly affected as the X-rays there emerge at some point before reaching the film.

The nature of the variation in relative irradiance is shown in Fig. 8.5. By virtue of the dependence of optical density on exposure indicated by equation 5.11, we may see that in the region of values of importance to the useful range (§5.10), a difference of 1 % in the irradiance corresponds to about 0.01 OD. Hence, we may take this kind of figure to set a criterion for an appropriate target-object distance (assuming the object is thin) for experimental work (§6). Primarily, the effect explains the observed small variation in density across a clinical film, although this is not of great significance for diagnostic purposes.

We can note that because a film is thin in proportion to the target-film distance, change in irradiance over its depth due to the divergence of the beam can be ignored (see box near equation 3.1). However, this effect might be appreciable for thick objects (*cf.* §6); large target-film distances are therefore appropriate for any detailed work of this kind.

●8.3   Blurring

Although the beam divergence itself produces a blurring of the edges of thick objects (Fig. 8.4), there is a further cause of loss of image sharpness: the size of the effective target emitting the X-rays, the focus size. Essentially, a point source is logically impossible, of course, but a very small spot is impractical because the heat concentration would be too great, and the fact then that X-rays are emitted from an area means that the image of a point in the object is a projection of the effective area of the focus (Fig. 8.6). Thus, the image size c is determined by the ratio

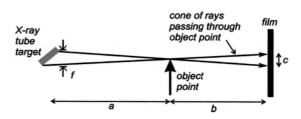

Fig. 8.6 Blurring of image by non-point X-ray source.

$$c/b = f/a \qquad (8.10)$$

Hence, the closer the film is to the object, the better the image quality. Focus dimensions (projected) are commonly in the range 0.4 ~ 2 mm. Obviously, in many dental and medical contexts b cannot be made very small, leaving as options increasing the value of a, and then needing to increase i or t (equation 8.5) to maintain the image density.

The situation is complicated by the fact that the shape and size of the projection image of the focus (as if through a pinhole 'camera') depends on the location of the object in the irradiated field (Fig. 8.7). At the centre of that field, the image is supposed to be a more or less equiaxed spot. But, at an angular displacement in the direction of the anode, on the centre line, equal to the target angle the image is reduced to a thin line. Towards the cathode, it is drawn out into a longer and longer image. Laterally, the image is sheared into a parallelogram as the distances of the two ends of the target are different. This effect is in addition to the blurring caused by intensifying screens (§7).

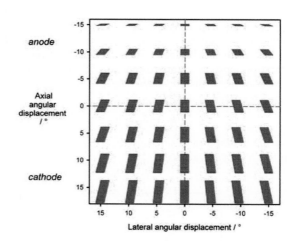

Fig. 8.7 Distortion of projection of focal "spot", the image of the anode target, with angular displacement from the centre of the intended 'beam'; 15° anode angle (exaggerated for clarity).

The above blurring is an unavoidable consequence of the geometry of the system, and arises essentially from design compromises and the nature of the object – the patient. There are two other causes to be considered: movement of the source, and movement of the object (with the film or sensor). The general assumption is that the elements of the system (Fig. 8.6) are in a fixed relationship, and in certain cases this may well be so. However, in a dental context, when the X-ray tube housing is on the end of a long and multiply-jointed arm, complete rigidity is not, of course, possible. Such a cantilevered arrangement must have a certain effective flexibility (23§4.1) because the various joints must have sufficient friction (1§5.5) to remain in place without sagging under the weight of the arm. That is to say, to reposition such a joint the coefficient of static friction is equivalent to a yield point, and the elastic deformation component represents stored energy. Thus, the arm is a spring, and on release after adjustment must recoil somewhat and oscillate (23§8). This is readily observable. Such movement gives the tube focus an apparent size (f in equation 8.10) rather greater than its real size. On top of this is blurring from the movement of the patient, the normal inability to hold perfectly still being exacerbated by discomfort from the film or sensor, especially for certain kinds of image, and the need to hold the position for an apparently (subjective) protracted period. A head-rest of some kind would greatly reduce the magnitude of this source, while sufficient time must be allowed for the decay of the cantilever arm's oscillation (against the sense of urgency often associated with taking an awkward image). Given the size of the tube focus, only very small movements would be necessary to degrade the resolution of an image, coupled with the fact that in photographic terms the typical exposure range of 0.04 ~ 0.8 s is not fast enough to 'stop motion'. For critical endodontic measurements, for example, this could be problematic. The use of hand-held (portable) X-ray sets can only be associated with additional blurring from the unsteadiness of the operator.

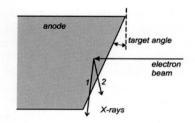

Fig. 8.8 Cause of the heel effect: surface X-ray generation is filtered by passage through the target material.

●8.4 Heel effect

In addition to the blurring there is a further effect of position in the irradiated field. As seen in Fig. 4.4, there is a filtration effect due the bremsstrahlung being generated inside the target material. However, the length of the path taken to emerge from the target depends on the angle of emission (Fig. 8.8): path 1 is longer in the tungsten than path 2. In accordance with Lambert's Law (§3), and the differential absorption effect for longer wavelengths (§3.3), the spectrum of the radiation that emerges depends on direction (Fig. 8.9). This is called the **heel effect**. At very low angles, although the image must be the sharpest possible for the width of the focus, the decline is very marked indeed. Usually the collimation for the irradiated field excludes this region altogether. There is also some effect in the lateral displacement sense, but this is less marked as the effective path length varies more slowly. The heel effect may be reduced by tilting the electron beam axis to point slightly away from the plane of the collimator (instead of being parallel to it) and so moving the exit window closer to the normal to the target. This, however, would offset the desired foreshortening of the focal spot image (Fig. 8.7), adding to blurring. The only answer to this is to make the focal spot much smaller; 0.5×0.5 mm^2 has been used, but the power density means that the acceptable current and exposure duration are much more restricted because of the heating.

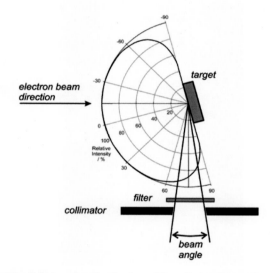

Fig. 8.9 Angular distribution of X-radiation for a point source.

The majority of the loss at low-angle emission is at longer wavelengths, and since the filtration that is normally present primarily affects such low-energy X-rays, the angular variation in the irradiated field is somewhat less than shown.

With use, however, the target **ablates**, showing a distinct 'burn' mark (Fig. 8.10). Despite the very high melting point, the tungsten slowly evaporates (Fig. 8.11), despite having the lowest vapour pressure of all metals (the result of this is visible as a blackening on the inner surface of the X-ray tube envelope, similar to that of an ordinary tungsten filament incandescent lamp). In addition, the very high alternating thermal stresses cause cracking and structural breakdown of the surface – bits fall off. This erosion of the surface means that X-radiation is emitted below the level of the surrounding target, as if in a ditch, and so must more and more pass through it, causing a steady fall in output, and an increase in the severity of the

Fig. 8.10 An old tungsten X-ray tube target showing the 'burn' of the focus. Note the heavy copper block into which the tungsten disc is let.

heel effect. Old tubes show this deterioration of the target very clearly (Fig. 8.12).

This effect is reduced (but not eliminated) by using a solid-solution alloy of some 3 ~ 15 mass% rhenium (Z = 75) in the tungsten (Fig. 8.13). Although this lowers the melting point, the ductility of the metal is greatly improved (W is b.c.c.), possibly by the ease with which **twinning** (28§1.2) then occurs, but in part this is due to a scavenging effect of Re for oxygen which otherwise segregates at grain boundaries and dislocations (*cf.* 14§2.4, 19§1.14); Re_2O_7 is volatile. There is a minimum in indentation hardness around W-5 mass% Re. The rhenium also stabilizes the grain structure. A further advantage is that the alloy is also more ductile at

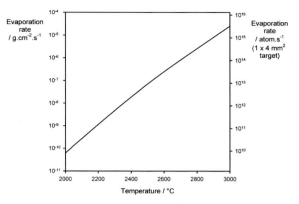

Fig. 8.11 Evaporation of tungsten at high temperature. Scale on the right is for a typical X-ray tube focus size.

low temperature, facilitating fabrication, as it lowers the temperature of tungsten's ductile-brittle transition. A disadvantage is the very high cost of rhenium – more than 100 times that of tungsten, but the trade-off for service life is evidently acceptable.

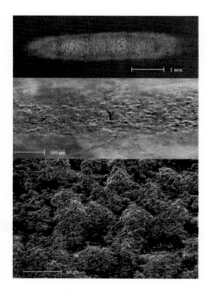

Fig. 8.12 SEM images of the 'burn' mark of Fig. 8.10. Note the cracking and pitting.

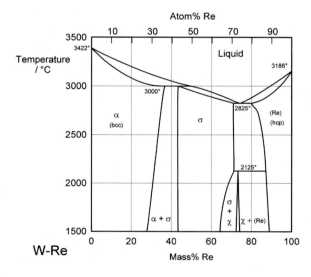

Fig. 8.13 Phase diagram for the W-Re system. X-ray tube targets are commonly α-solid solution.

§9. Alternative Imaging Techniques

Radiography using silver halide emulsion films has a number of drawbacks. Radiation exposure is higher than is desirable and there is a general desire to minimize this. The need on occasions for retakes, whether due to an initial bad exposure choice or position and movement errors, makes this worse. The silver of the film is relatively expensive, and recycling is not economically feasible. Processing is wet and requires large volumes of toxic chemicals whose disposal presents problems; it is often associated with artefacts in automated systems; needs constant monitoring to maintain (near) uniformity of outcome – both between films and across each film; and is relatively slow. The unexposed film itself cannot be stored for long periods without deterioration, and ideally needs refrigeration even then. Storage of images requires appreciable space, copies are awkward to make, and forwarding to others slow. Accordingly, there is much interest in alternative imaging systems that can offer improvements in some or all of these matters.

•9.1   Xeroradiography

The element selenium is a semiconductor. That is, the energy gap between the **valence band**, the set of highly-localized orbitals where bonding electrons normally lie (*cf.* 24§6.1) and the more metal-like, delocalized **conduction band**, is relatively small in comparison with insulators. In ordinary semiconductors at normal temperatures, a few electrons are naturally energetic enough to cross the **band gap** (Fig. 9.1), and they can then be moved under the influence of an electric field, that is, they conduct electricity. However, in selenium, the band gap is somewhat larger, lying somewhere between insulators and true semiconductors. In fact, the band gap is about 2.3 eV, which corresponds to green light of wavelength about 535 nm. Thus, if selenium is irradiated at this or higher energy, valence electrons are promoted into the conduction band and may be moved by an applied voltage. Selenium is therefore a **photoconductor**. This works just as well with X-radiation as it does with (short enough wavelength) visible and ultraviolet light, except that now electrons from deeper shells may be excited (Fig 1.1).

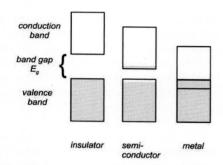

Fig. 9.1 The distinction between metals, semiconductors and insulators. At normal temperatures, there is insufficient energy to promote an electron to the conduction band of insulators, while in ordinary semiconductors a small number are available because the band gap is sufficiently small.

The problem now is to use this property to record image information. The technique is indirect. In the dark, selenium remains non-conducting. A thin coat of amorphous (glassy) selenium on an insulated aluminium backing-plate is exposed to a corona discharge from a high-voltage (~7 kV) device scanned across the sheet so as to lay down a uniform positive charge (to a potential of about 600 V) over the whole surface (Fig. 9.2, top).[16] Since the selenium is non-conducting, that electrostatic charge is stable (at least in the short term). Now, if irradiation occurs with sufficient energy to promote electrons to the conduction band, under the influence of the electrostatic field created by the surface charge, those electrons travel to the surface and neutralize the surface positive charges. The 'holes' created at the same time drift towards the base plate (they are effectively positively charged), and are filled by electrons from that plate. Thus the remaining positive charges on the selenium surface are distributed as an image: the more charge, the less irradiation occurred (Fig. 9.2, bottom).

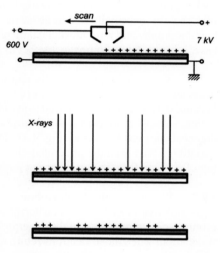

Fig. 9.2 Formation of an image in xeroradiography. The induced negative charges at the Se-Al interface are not shown.

The second step is to extract the information. There are three possibilities. In the process that is most like photocopying, to create a paper image, the irradiated sensitive surface is treated by exposure to a cloud of negatively-charged **toner** particles, very fine polymer beads with, for example, black pigment, *e.g.* carbon black. The toner powder adheres to the selenium in proportion to the charge there. This powder can then be transferred to paper (using a stronger electrostatic field to lift it off the selenium), then the polymer is fused to the paper (or a non-melting transparent polymer film) by running it over a hot roller. This is now a positive image: light where there was most irradiation, dark where little – the reverse of a radiographic film. This outcome means that the form of the image is unfamiliar to radiographers, and even if printed onto transparent film might remain inconvenient in this sense.

The second possibility for extraction of information is a direct reading of the charge with an electrometer device in so-called **selenium-plate** systems (although no different in principle from the direct method above). That is, the electrostatic field of the surface charges left after exposure is detected by a linear array of sensors which is scanned across the surface at very close range. Such a system generates analogue field intensity data which can be readily converted to digital form and so be stored, manipulated and displayed electronically, including image inversion to the conventional 'radiographic' negative form. Such a system is most suitable for use in large-scale radiography, where the selenium is on a drum which is therefore rotated for exposure and reading.

The third approach integrates the field sensors as a full array of conventional silicon-based transistors beneath and insulated from the selenium layer. Each transistor corresponds to a pixel (an image element), and so can be addressed in sequence to read the image information directly. Again, this permits digital data to be obtained. However, the large scale of the chip circuitry means that the cost is very high for such a system.

Although there are a number of advantages to such **xeroradiography** techniques, including an improved **dynamic range** (that is, between the noise and saturation levels), there is one ambivalent issue that arises from the nature of the process, depending as it does on electrostatic charges. That is, the image information is affected by the state of the adjacent selenium surface. Like charges repel, and there is a tendency for charge redistribution to occur such that areas of initially identical charge cease to be so (older photocopiers suffered from this badly). In fact, the edges of such regions tend to become more marked.

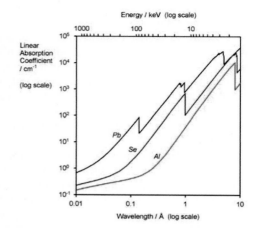

Fig. 9.3 Variation of linear absorption coefficient for selenium.

This so-called **edge enhancement** tends to improve the appearance of structure, especially on large scale images, and so appears to be favourable for chest radiography, mammography and musculoskeletal imaging. However, since it in fact implies greater radio-opacity than is the case, it is too strong an image artefact to allow use in dental applications, and has fallen out of favour.

Xeroradiography typically requires exposures similar to those of film, and rather more than would be necessary with intensifying screens (§7). However, screens cannot be used in conjunction with selenium plates because the close contact required to avoid further image blurring would dissipate the charge. A further problem is that, as with all elements, the absorption coefficient for X-rays falls rapidly with increase in energy in the diagnostically-useful region (Fig. 9.3). Thus, exposures must be adjusted in the same way as with film for such variation. Other than the integrated-chip, third approach described above, this system is substantially cheaper than for X-ray film, in particular not requiring a darkroom.

●9.2   Photo-stimulable phosphor

As has been described earlier, phosphors are substances that can absorb certain wavelengths of radiation, promoting electrons to a higher energy orbital to create an excited state, and re-emit radiation of a longer wavelength when those electrons return to the ground state (6§5.5, Table 24§4.1). Such materials are the basis of intensifying screens (§7). When the re-emission is prompt, the phenomenon is called fluorescence, when there is appreciable delay, phosphorescence. It is, however, possible to design materials which do not phosphoresce spontaneously, but which require stimulation of some kind to do so. When that stimulation is light itself, we have a **photo-stimulable phosphor** ("PSP"). The principle of the phenomenon is as follows: the ejection of highly-energetic electrons by X-rays (Fig. 1.1) allows them to move through the conduction band, even of insulators (Fig. 9.1). If there are sites that correspond to local energy minima, but with no permitted path for escape (*cf.* the triplet state of Fig. 6§5.4), loss of energy through collisions (**thermalization**) may see some of those electrons trapped in a relatively high-energy state. Even so, providing the activation energy to escape the trap allows the electron to return to the ground state, emitting the stored energy correspondingly as a stimulated phosphorescence.

The system in common use2016-10-24 is based on a mixed-halide crystal, barium fluorobromide, $BaFBr$.[17] This is **doped** with a small concentration of the rare earth element europium in the form of Eu^{2+} ions which replace some barium ions in solid solution. In addition, the crystal naturally has a proportion of lattice defects, vacancies (*cf.* Fig. 11§6.1), with one missing metal ion corresponding to two missing halide ions, to preserve charge balance. The particular transition of interest here is the X-ray ejection of an electron from the Eu^{2+} ion since the Eu^{3+} ion then formed is stable. The ejected electron may, of course, recombine with the Eu^{3+} ion, but may also be trapped at a halide vacancy site – literally in the position that the halide should occupy, as if it were an ion (Fig. 9.4[11][18]). This electron forms a **colour centre** (known as an **F-centre**[12]) as it has its own broad absorption spectrum in the visible range (and into the infra-red). Thus, if now illuminated with such

---

11 A Jablonski diagram indicates schematically the successive electronic states of a system.

12 'F' from the German 'Farbe' = colour.

F-centres are a general phenomenon. One kind is seen in zinc oxide on heating (9§3.1), but is also seen in the glass envelopes of old X-ray tubes (Fig. 2.4) – they become quite yellow with use because electrons are continually being ejected and some get trapped.

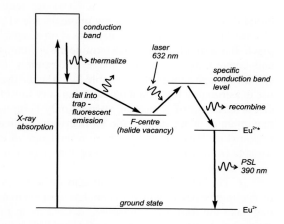

Fig. 9.4 Jablonskí diagram for the mechanism of photostimulated luminescence (PSL). Electrons in F-centre traps represent the image information, the activation energy for their escape is provided by the laser irradiation.

radiation, and typically a red He-Ne laser at 632 nm may be used, the trapped electron is once again promoted to the conduction band whereupon it reacts by recombining to form an excited-state Eu^{2+} ion:

$$Eu^{3+} + e^- \Rightarrow Eu^{2+}*$$

which then spontaneously decays fluorescently, in less than 1 µs, emitting **photostimulated luminescent** ("PSL") radiation centred at about 390 nm (Fig. 9.5). (The fluorescent radiation that must be emitted in recombination is filtered out from detection.) The F-centre trapped electrons are fairly stable, decaying spontaneously (*i.e.* without laser irradiation) and exponentially by only about 1 % per day at 20 °C, so the image information is better preserved than on a selenium plate.

In practical terms, fine crystals (~5 µm) of the salt are bound in a thin layer of polymer matrix on a support sheet, and coated with a protective layer. The X-ray irradiated imaging plate is scanned with the laser in a **raster** pattern. A photomultiplier detector, with a blue-pass filter to prevent the laser light being detected, then provides an analogue signal to be digitized. However, because the stimulated emission is not perfect, some residual image information remains which would interfere with the next exposure. This residual image is conveniently erased simply by exposure to a bright white light source, such as a film-viewer, for 2 or 3 minutes.

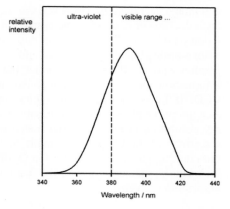

Fig. 9.5 Spectrum of photostimulated luminescence (PSL) from BaFBr:Eu^{2+}.

This kind of device has a linear dynamic range of the order of 10^4 : 1, the ratio between the maximum image-useful exposure and that required to be detectable about background noise, in photons per pixel. This can be compared with about 10^2 : 1 for film radiography. One problem, however, is that in scanning for image-detection the laser light will be scattered within the necessarily translucent matrix because of the refractive index mismatch (24§5.11), stimulating emission over a larger area than intended and reducing the sharpness (the resolution) of the image. Even so, the resolution can be much higher than for other devices, and 600 dpi or ~24 pixels/mm is easily achieved. The stimulated light is emitted in all directions and also scattered, of course, and some may be lost because of total internal reflection at the interface of sensitive layer and coating, as well as coating and air (24§5.2), so the PSL detection efficiency is necessarily limited (*cf.* emission from screens, §7). Similarly, a compromise has to be accepted between crystal size and the thickness of the sensitive layer (~100 ~ 250 µm) on the one hand, and the quantum-efficiency and sharpness of the image obtained on the other: small and thin are good for sharpness, but bad for efficiency.[19] However, major advantages of this system are that the plates can be made very large without special problems and are flexible, similar to film. Nevertheless, care is necessary to avoid scratching the plate in handling if the image quality is not to be degraded with repeated use.

●**9.3   Charge-coupled device sensors**

The third approach is based on a more sensitive detection of the light generated by X-rays in phosphor screens (§7), on the basis that the weak link there is the relatively low sensitivity of the film to the visible light produced.

Charge-coupled devices (**CCDs**) are semiconductor-based detectors which rely on the photoelectric effect, but the electrons produced are accumulated as if on a capacitor at each pixel. It is a simple matter to determine the accumulated charge and convert that to digital image information. CCDs have extremely high sensitivity to visible light (hence their common use in astronomy for distant objects), but very low efficiency in direct detection of X-rays of interest, only being useful to a maximum of about 10 keV, too far below the clinically-useful region (Fig. 9.6). They also have a tendency to cumulative damage from X-radiation, shortening their working life.

The lack of sensitivity to diagnostic X-rays is overcome by coating the device with a layer of luminescent material. This may take the form of a phosphor, as used in an intensifying screen. A common type is $Gd_2O_2S:Tb$, a gadolinium oxysulphide, which is doped with another rare earth (terbium, in this case). This emits rapidly in the green (peak at 545 nm) (Fig. 9.7), and has high X-ray absorption because of the high Z of Gd (64). The problems with phosphors in this application are similar to those in screens (§7): they scatter, they absorb. Hence, there is a trade-off between X-ray absorption, efficiency and image degradation. However, the required thin phosphor layer does not provide adequate protection of the CCD from X-radiation damage. The solution to this problem is to interpose a **fibre-optic plate** (Fig. 9.8), an assembly of short parallel optical fibres that conduct the image information coherently from the phosphor to the CCD. If this is made sufficiently thick and using an appropriate glass composition, satisfactory X-ray absorption occurs. A better match with the sensitivity of the CCD can be made by using a red-emitting phosphor, such as $Gd_2O_2S:Pr$ (Fig. 9.7), but this cannot overcome the scattering and absorption problems.

Loss of efficiency may be overcome by using a **scintillator**, that is, a luminescent material that is transparent at the emitted wavelength. One common type is thallium-doped caesium iodide, CsI:Tl, effective because all elements have high Z (55, 53, and 81, respectively). This has a broad emission spectrum, although unfortunately not in the most sensitive region of CCDs (Fig. 9.7). Although absorption and scattering are much reduced, the problem of the spherically-symmetric emission pattern remains: much light is lost and a circular region of the CCD is exposed. Image resolution is therefore degraded with increasing thickness. As with phosphors, the thin layer required absorbs X-radiation inadequately to protect the CCD, and a fibre-optic plate is still necessary. A further improvement can be obtained by growing the caesium iodide as columnar crystals on the fibre-optic plate (Fig. 9.9). This then gives a greater optical depth for the scintillator, improving the light emission efficiency, but also a larger proportion of that light is channelled forward as the crystals themselves then act as optical fibres. Since CsI is water-soluble, and hygroscopic, the whole device must be well-sealed.

"CCD" actually refers to the means by which the charge information in the array is read for each pixel – sequentially, element by element, line by line, by physically moving the charge of each in the sense of a 'bucket relay'. The details of this reading process are of no particular concern here.

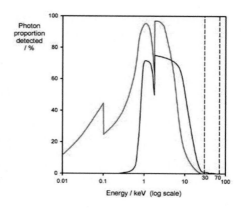

Fig. 9.6 Direct X-ray detection efficiency of CCDs for two kinds of device. Note the clinically-useful region (30~70 keV).

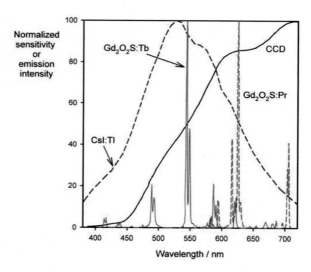

Fig. 9.7 Typical visible-region absorption spectrum for a CCD, and the X-radiation-stimulated emission spectra for doped caesium iodide scintillator and for 'red' and 'green' emitting gadolinium phosphors.

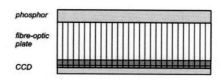

Fig. 9.8 Schematic section of a phosphor CCD assembly for X-ray imaging.

All of this means that intraoral CCD devices are necessarily rather thick, and of course inflexible. The need to be connected by a cable to an external device (for direct or indirect access to the image information) might also be a disadvantage. A significant problem with CCDs, however, is that they are extremely expensive. Typically (so far), devices cannot be made larger than about the size of an intra-oral film (up to about 25 × 30 mm^2) and so are not generally available at the kinds of size useful for extra-oral dental radiography (although it is possible to create 'tiled' arrays, the price goes up very rapidly).

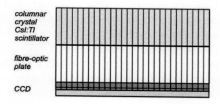

Fig. 9.9 Schematic section of a scintillator CCD assembly for X-ray imaging.

If larger image fields were required, an image-size reduction arrangement is required to use them at all. A typical structure is illustrated in Fig. 9.10. Here, the key feature is that the fibre-optic plate is replaced by a **fibre-optic taper**, a device made from a similarly image-coherent bundle of fibres, each of which is now tapered so that light arriving at the large area front end is delivered to a much smaller area at the back, which is in contact with the CCD. Thus the image is **demagnified**. However, while it is possible to have CCD pixel sizes of 20 - 50 µm, and thus potentially good image resolution, light is lost in the fibre-optic taper as the square of the demagnification. Hence, common tapers of say 2.5:1 or 4:1 result in only 1/6th or 1/16th of the light reaching the CCD, requiring correspondingly

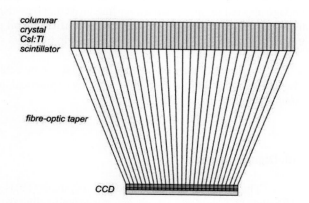

Fig. 9.10 Schematic section of a demagnifying scintillator CCD assembly for X-ray imaging.

longer exposure for the same image quality, and thus longer X-ray exposure – which sharply reduces their applicability to dental and medical diagnostic applications.

There are a variety of other techniques available for transferring an image in a scintillating or fluorescent screen to a CCD, essentially optical devices – cameras, or for collecting ejected electrons and focussing these electrostatically onto a smaller detector – image intensifiers. A variety of problems, such as inefficiencies and image distortion, bulk, expense, and short-lifetimes mean that none has achieved great popularity in dentistry.

•9.4 Thin-film transistor devices

As indicated earlier, conventional transistor arrays, which are built from crystalline silicon, are impractical for reasons of scale and cost. However, it is possible to fabricate transistors and other semiconductor devices using amorphous silicon in what are called **thin-film transistor** or **TFT** devices. The technique used is **chemical vapour deposition**, **CVD**, in which large areas can be processed on relatively cheap substrates such as glass. Again, there are many possibilities for the circuitry and materials which can be employed, but they all rely on the same general principles: conversion of X-rays to light, capture of the light, storage as electrons, programmed reading of the charge on the array to create the image information. In one sense, the resulting **flat-panel sensor** is similar to an LCD monitor, with a photodiode (sensor) rather than a light-emitting diode. The key part is the TFT switch which allows accumulation of charge in the photodiode (Fig. 9.11). The light may be generated in a layer of columnar CsI:Tl, deposited directly on the TFT layer; no fibre-optic plate is required because the TFT device is not as prone to damage by X-radiation (Fig. 9.12). The scintillator emission peak matches fairly well the peak sensitivity of the amorphous silicon photodiode, which is itself quite efficient (Fig. 9.13), giving high

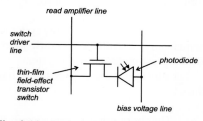

Fig. 9.11 Basic circuitry of one pixel of a thin-film transistor flat panel X-ray sensor.

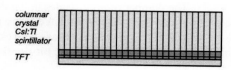

Fig. 9.12 Schematic section of a scintillator TFT device for X-ray imaging.

efficiency overall. Note, however, that the overall spectral sensitivity of the combination depends on the convolution of the two spectra – see 6§5.14. Even so, this does not take into account scintillation efficiency (X-rays to light) or the collection efficiency (channelling of the light to the detector). Such factors need to be taken into account in all systems to determine overall sensitivity and thus the exposure dose required for an image.

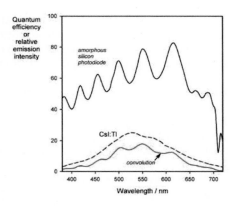

For an exposure, the TFT switches are off. X-radiation generates light in the scintillator, which in turn generates charge which accumulates on the diodes. At the end of the exposure, a voltage applied to the switch driver line turns on all switches in the row, and the charge flows out to the amplifiers for all pixels in that row simultaneously. Each row is read in turn.

Fig. 9.13 Emission spectrum for CsI:Tl as an X-ray scintillator compared with the capture and conversion efficiency of amorphous silicon for visible light.

A difficulty may be the pixel size that can be achieved. The fabrication techniques presently set a lower practical limit of around 100 μm.

§10. Digital Images

The dynamic range of the non-film systems (§9) is generally much greater than that of film. For the PSP, CCD and TFT devices, which are intrinsically digital, this can be turned to distinct advantage. Despite experience and the routine of much radiography, judgement is still required as to the correct exposure: there remains some trial-and-error. This is critical for film: under- or over-exposure renders the image unusable. However, when the dynamic range is great, the image that is displayed on the monitor can be adjusted so that diagnostic information lies in the **useful range** (§5.10), which is defined by our own visual apparatus and cannot be altered. That is, the raw pixel exposure values of the acquired image are scaled so that the upper and lower bounds of the raw image are made to correspond to the upper and lower bounds of a suitably-defined grey scale. Primarily, this involves adjusting the effective slope of the response (the sensitivity), but it may also include adjustment of the zero point (the offset). Sometimes more elaborate algorithms are used which modify the linearity of the image information, and even from place to place in the image in an attempt to 'level' extremes.

Hence, by adjusting the image gain, both under- and over-exposed images (as conventionally interpreted) may be brought into a visually-workable range. Of course, if the under-exposure is too great, the noise in the image may become unacceptable, and if the over-exposure is severe, when the sensor is saturated, image information will be lost. Even so, considerable latitude is available.

Commonly, digital X-ray systems incorporate a software function that attempts to make such adjustments automatically, so called **automatic gain control (AGC)** or **automatic exposure control (AEC)**. The latter should be distinguished from systems that measure separately the amount of radiation received and terminate the exposure when the intended dose has been attained. In any case, the need for repeat exposures is greatly reduced.

There are other advantages in that images may be further manipulated digitally after the event for cropping, enlargement, contrast adjustment, enhancement and special effects such as edge-detection, smoothing and false colour, as well as quantitative measurements.

●10.1 Grey scale

The number of grey shades in the digital image is given by the number of binary digits (bits) used to define the numerical value associated with a pixel. Depending on the system, the grey scale values may be from 8- to 16-bit. The darkest grey shade ('black') is usually defined by the value zero while the lightest ('white') has the value 2^n -1, where n is the **bit-depth**. Thus, for an 8-bit system the maximum value is 255.

The implication of bit-depth is the coarseness of the binning process in converting actual photo-electron counts to image information. As has been described (§5.2), the perception of brightness or darkness is logarithmic so grey-scale value (G) is nearly equivalent to optical density, but in a complementary sense (opposite scale direction), for example: D ≃ $k(255 - G)$ for an 8-bit image, where k is some scale factor, bearing in mind that optical density is an open-ended scale, while G is bounded on both sides. Thus it is the logarithm of the photo-electron count at each pixel that is finally stored, but the numerical representation is then limited in its resolution by the bit-depth. This is illustrated in Fig. 10.1 for $n = 3$, giving just 8 steps in the grey scale. Up to some threshold value, G must be zero; no information can be recorded. Likewise, above some upper limit, there can be no further information (even if the detector is not saturated). Obviously, the greater is n, the finer the granularity of the binning, and the higher the grey-scale resolution for the image. This in turn means that there is greater scope for adjustment according to under- or over-exposure. This is complicated by the fact that common image-file formats are only 8-bit, no matter the internal representation of the original image data.[13] Conversion to the image-file bit depth must therefore occur after any 'exposure' adjustment.

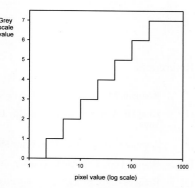

Fig. 10.1 Grey-scale granularity (for a 3-bit image) (arbitrary scale factor, k).

Consequently, it is not possible to recover from an output image any more information than is encoded therein. Unless the raw data are recorded separately it is not possible to recreate or reanalyse the original data.

In a parallel with the occurrence of the base and fog density of film (§5.4, §5.5), digital systems may suffer from non-image data from background noise or so-called **dark-current** effects, where counts occur in the absence of X-rays. For digital systems therefore, equation 6.3 may be rewritten as:

$$\ln(g - G) \;=\; \ln(at.I_0) \;-\; \mu x \qquad\qquad (10.1)$$

where g is the background, no-exposure value. Thus, the attenuation coefficient can be found from the slope of a plot of $\ln(g - G)$ *vs. x*, in the same sense as for film.[20] However, since the variability of film-processing is not involved, the quality of the calculated value is much improved. Even so, because of the distortions that may be introduced by AGC, this process can only work with image data that have been subject to no more than an overall gain (slope) adjustment.

---

[13]   12- and 14-bit data, for example, are 'padded' (with zeros) to be written to 16-bit file formats, or scaled (= truncated) for 8-bit.

References

[1]   Röntgen WC. On a new kind of rays. Nature 53 (1369): 274 - 276, 1896.

[2]   Klug HP & Alexander LE. X-Ray Diffraction Procedures for Polycrystalline and Amorphous Materials. New York, Wiley, 1974.

[3]   Asimov I. Biographical Encyclopedia of Science and Technology. Pan, London, 1975.

[4]   Barrow GM. Physical Chemistry 4th ed. McGraw-Hill, New York, 1979.

[5]   Brandes EA & Brook GB. Smithells Metals Reference Book 7th ed. Butterworth-Heinemann, London, 1992.

[6]   de Marcillac P, Coron N, Dambier G, Leblanc J & Moalic J-P. Experimental detection of alpha-particles from the radioactive decay of natural bismuth. Nature 422, 876-878, 2003.

[7]   Mucke M *et al*. A hitherto unrecognized source of low-energy electrons in water. Nature Physics 6: 143 - 146, 2010.

[8]   Birch R & Marshall M. Computation of bremsstrahlung X-ray spectra and comparison with spectra measured with a Ge(Li) detector. Phys Med Biol 24(3): 505 - 517, 1979.

[9]   http://en.wikipedia.org/wiki/Weber%E2%80%93Fechner_law

[10]   The neural basis of the Weber – Fechner law: a logarithmic mental number line. Trends Cog Sci 7(4):145 - 147, 2003.

[11]   Pavlidou S, Koroneos A, Papastefanou C, Christofides G, Stoulos S & Vavelides M. Natural radioactivity of granites used as building materials in Greece. Bull Geol Soc Greece 36: 113 - 120, 2004.

[12]   Asimov I. The radioactivity of the human body. J Chem Educ 32 (2): 84 - 85, 1955.

[13]   Darvell BW. Kinetic models for the development of density in radiographic and photographic film. J Chem Soc Faraday Trans 81: 1647 - 1654, 1985.

[14]   Darvell BW. A method for calibrating non-screen radiographic film. Austral Dent J 33: 27 - 31, 1988.

[15]   Doi K, Loo LN, Anderson TM & Frank PH. Effect of crossover exposure on radiographic image quality of screen-film systems. Radiol 139:707 - 714, 1981.

[16]   Schaffert RM. Electrophotography. Focal Press, London, 1965.

[17]   Amemita Y & Miyahara J. Imaging plate illuminates many fields. Nature 30(3): 89 - 90, 1988.

[18]   Jabłoński A. Efficiency of anti-Stokes fluorescence in dyes. Nature 13: 839-840, 1933.

[19]   Cowen AR, Workman A & Price JS. Physical aspects of photostimulable phosphor computed radiography. Brit J Radiol 66: 332 - 345, 1993.

[20]   Nomoto R, A Mishima A, Kobayashi K, McCabe JF, Darvell BW, Watts DC, Momoi Y & Hirano S. Quantitative determination of radio-opacity: equivalence of digital and film X-ray systems. Dent Mater 24 (1): 141 - 147, 2008.

Chapter 27 More Polymers

Any treatment of the subject of dental materials naturally emphasizes the 'most important', whether this is because they are used in great bulk, are the basis of a common procedure, or are critical in some sense to the success or quality of outcome of a process. However, there is diverse group of auxiliary or adjunct materials whose properties are no less significant, but whose routine or everyday nature may obscure that significance. Some of these have already been dealt with. It is the purpose of this chapter to outline the chemistry and mechanical consequences for a group of such materials which are polymeric.

Rubber dam *and* ***gutta percha*** *are closely related chemically, but the marked differences in properties are traced to a simple difference in the structure of the polymer chain. Such structural differences also underlie the great variety of properties that may be found for* ***polysaccharides****: paper, starch and related materials are also discussed in those terms.*

The ***polypeptides*** *are a major group of compounds in the sense of biological chemistry, but the mechanical properties of these are also controlled by the details of chain structure. The major use of such materials in the present context is for surgical sutures. Here, the compromise includes considering biocompatibility and resorbability. These factors can also be addressed through synthetic suture materials, and an example is discussed.*

The design, selection and handling of materials for use in and around dentistry is essentially based on structure - property relationships. This applies even to 'minor' materials that may be overlooked. The ideas of earlier chapters are therefore brought to bear to explain those relationships. It is no less important to understand these materials than it is the others for competent and comprehending use in the pursuit of better dentistry.

Materials Science for Dentistry
https://doi.org/10.1016/B978-0-08-101035-8.50027-4

Polymers exhibit a very wide range of properties, depending on the chemical nature and structure of the repeating units. The major kinds in use in dentistry have already been discussed, but there remain a number of polymers with specialized but miscellaneous applications. They are of a kind in that they illustrate the links between structure and function as well as the reasons for their selection. However, they have generally received little attention in this field, although of quite different types to those used for impressions, prostheses and restorations.

§1. Polyisoprene

Isoprene or, more formally, 2-methyl-1,3-butadiene (Fig. 1.1) is the notional starting point for an enormously wide range of natural products, the terpenes, which range from essential oils (such as that of ginger) to vitamin A_1 and the steroids. The present interest lies in its naturally-produced high polymers: (natural) **rubber**,[1] also known as caoutchouc, and **gutta percha**.[2] These two differ only in their stereochemistry, that is the relative **configuration**[3] of successive units of the polymer chain. Balata ('rubber')[4] is similar to gutta percha. (Notice that there is no polymerization reaction involved in any dental context – this has been done by the plant.)

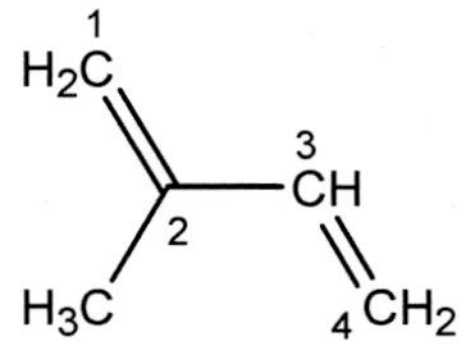

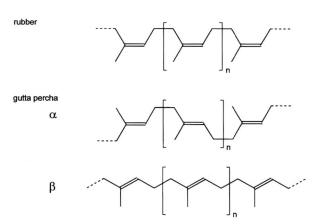

Fig. 1.1 Isoprene, the notional monomer of the terpenes, as well as rubber and gutta percha.

●1.1   Structure comparisons

Rubber, as is used in the manufacture of the elastic bands of orthodontics or rubber dam, is essentially pure *cis*-1,4-polyisoprene (Fig. 1.2, top), whereas gutta percha, as used in root canal filling points, is nearly pure *trans*-1,4-polyisoprene (Fig. 1.2). The non-equivalence of these **geometric isomers** derives, of course, from the lack of rotation about the double bond. The stereochemical effect on properties is profound: fresh pure rubber (molecular weight, MW ~10^6) is soft, highly flexible and tacky; fresh pure gutta percha (MW ~3 x 10^4), however, is relatively tough, hard and horn-like, showing comparatively little flexibility.

Fresh rubber is normally a gum with little tendency to crystallize but with a 'melting point' of about 27°C, when it becomes very sticky. Gutta percha, on the other hand, is usually about 60% crystalline, 40% amorphous.

Fig. 1.2 The structure of the natural polyisoprenes. top: rubber; centre: gutta percha in the low temperature, compact conformation; bottom: gutta percha in the high temperature extended, conformation.

Both polymers have the possibility of free rotation about the -CH_2-CH_2- bond, but in gutta percha this leads to two distinct crystal forms which melt at ~64 and ~74 °C, respectively. At high temperature the chains of gutta percha tend to be in the extended **conformation**[3] in the amorphous, liquid-like state. Rapid cooling, as might be expected, tends to preserve this more extended form in the crystalline β-phase (Fig. 1.2). The more stable and slightly more dense α-phase is formed slowly from the β-phase on slow cooling. The transition from β- to α-phase depends on thermal history, time and temperature, as it relies on the availability of the activation energy for rotation of chain segments to change the conformation; yet another case of kinetic limitation.[1] The density change due to this transformation (~1%) would tend to spoil the seal in a filled root canal even if nothing else happened.

---

[1]   Obtained from the sap (latex) of certain tropical trees, chiefly *Hevea brasiliensis*.

[2]   Obtained from the latex of several trees of the genus *Palaquium*; the name is an adaptation of the Malay for "gum of the percha tree". ['Percha' is pronounced as 'percher' – unvoiced 'r'.]

[3]   Note the difference in meaning of the terms of *conformation* and *configuration*: *conformational isomers* are interconvertible by bond rotation at normal temperatures; *configurational isomers* cannot be so interconverted (see 3§2).

[4]   Obtained from *Mimosops balata, Manilkara bidentata*.

●1.2   Gutta percha

Clinical use of gutta percha requires a compromise between the rigidity of the natural product and the mouldability of the plasticized material. Waxes are apparently the additives used to achieve the latter property, while about 80 mass% (~42 vol%) zinc oxide is used to control the viscous component of its viscoelastic nature, although this filler would also confer the radio-opacity valuable in identifying the extent, or even the presence, of a filled root canal. Temperature is therefore an important aspect of the use of gutta percha, and moulding to the root canal is thus possible with warmed instruments. Indeed, this moulding occurs slowly at body temperature against soft tissue when it is used as a "functional" impression material in cleft palate patients. Even so, temperatures up to 100 °C are required for fluidity, although some decomposition occurs. It follows that appreciable thermal shrinkage is to be expected and this must jeopardize any seal formed in a canal using heat, whether direct or frictional from a rotating instrument. The amorphous material has a typical polymer thermal expansion coefficient of about 75 MK^{-1},[2] while the usual partially-crystalline material has a lower value at around 54 MK^{-1} (Fig. 3§4.14). Since the adaptation obtained would depend on flow, much elastic recovery would also be expected under normal clinical 'compaction' procedures. Consequently, some sealing cement is ordinarily required.

Gutta percha dissolves in, and is also plasticized by, a good many solvents: chloroform and eucalyptol (the chief constituent of eucalyptus oil, Fig. 1.3) are typical. These have been used to soften the surface of gutta percha points to allow better adaptation, although such improvement is necessarily only transitory. When solvent dissolves in the polymer there must be a volumetric expansion. The shrinkage that must then inevitably ensue on evaporation or migration of the solvent after placement will prevent any possibility of a seal being maintained, even if initially obtained.

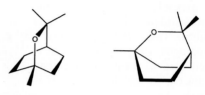

Fig. 1.3   Two views of the structure of eucalyptol, a small, compact, non-polar molecule.

The viscoelasticity of gutta percha has already been mentioned and its slow continuous deformation under stress is of course stress-relaxation. Pure natural rubber is highly elastic, but there is still a sufficient viscous component to allow appreciable stress-relaxation as, again, there is no cross-linking between chains. Indeed, partial crystallization may be induced through the alignment of the molecules (*cf.* Fig. 3§4.10) and will contribute to permanent deformation. The natural product, then, is *not* a rubber in the sense of a good elastomer (7§2). There is no three-dimensional network.

●1.3   Vulcanization

Both polyisoprenes have one double bond per chain unit remaining from the original two of the supposed monomer. As in other molecules containing unconjugated organic double bonds, these remain reactive. It was discovered[5] that heating natural rubber with sulphur at about 200 °C gave a product that was more elastic and which did not become tacky on warming. Indeed, it stayed flexible at lower temperatures. This reaction between the sulphur and the double bonds results in cross-links being formed. These establish a true elastomeric structure which inhibits crystallization under stress and lowers the glass transition temperature. The process is called **vulcanization** and employs typically about 1 or 2 mass% sulphur, although now many additives and modifiers are used with a wide range of processing conditions to produce many varied products such as tubing and tyres. Dental applications include rubber dam, where extreme extension is required but, unlike for impression materials, the small remaining viscous component is of little consequence. It is also used as the binder for the abrasive in dental 'rubber wheels', but perhaps most significantly in protective gloves. Such vulcanized natural rubber is often called **latex rubber**, to indicate its natural origin.

Viscous flow stress-relaxation, probably through a bond-exchange mechanism similar to that operating in polysulphide impression material (7§4), accounts for the steady decline in the force applied by latex rubber elastic bands in orthodontics. Thus their frequent replacement is required if the force is to maintained. The relaxation is also aided by the plasticizing effect of absorbed water if they are used intra-orally. Modern elastomeric polymers such as polyurethanes may reduce this flow effect.

Vulcanization was taken to the extreme with very much more sulphur for the denture base material called **Vulcanite** where the proportions by mass of rubber and sulphur were 2 : 1, very close to the molar weight ratio of the notional monomer unit and sulphur of 68 : 32. Thus, one sulphur atom was provided for each double bond

---

5   By Charles Goodyear, in 1839.

in the polymer. How complete the reaction actually was is not known (and hardly relevant now), but clearly the cross-linking was at least very extensive.

The processing of rubber at high temperature for long periods is not very convenient (and may cause damage) for products such as gloves, and alternative approaches have been developed. To promote the cross-linking of the rubber at temperatures as low as 100 °C, and more quickly, compounds such as zinc dithiocarbamates (Fig. 1.4) are used either directly or as a precursor with zinc oxide, to accelerate the reaction with sulphur. In a reaction whose details are complex and poorly understood,[3] but which is remarkably similar to that of the cross-linking of polysulphide rubber (Fig. 7§4.2), an allylic hydrogen (that is, one that is activated by being on a carbon adjacent to the double bond) is attacked and substituted by the accelerator-sulphur system. The resulting polysulphide side chain can then react with another of its kind to form a cross-link. Subsequent reactions may reduce the number of sulphur atoms to 1 or 2, to stabilize the structure. Many more reactions and types of cross-link are possible, even those based on direct reaction with the double bond, as occurs in the simple process of heating with sulphur alone.

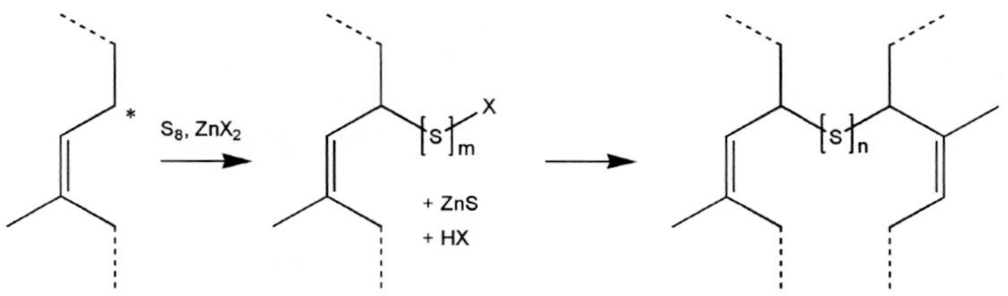

Fig. 1.4 Zinc dithiocarbamate. R-can be methyl-, ethyl, *etc.*, or form a loop -R-R- in a cyclic compound.

Fig. 1.5 Accelerated vulcanization of natural rubber. X = dithiocarbamate or similar compound; m, n are small numbers, 3 ~ 8. The allylic carbon is marked *.

Gloves and similar products are made by **dip-moulding** from a **latex**, an emulsion of the polymer in water, into which has been mixed the cross-linking reactants, by
- immersing a shaped former into a solution of coagulant
- drying the coagulant
- immersing the coated former into the latex mixture
- withdrawing the former to leave it coated with coagulated latex
- drying in a hot air oven
- vulcanizing the resultant film at an elevated temperature.

Three factors may represent a hazard to the user. Some accelerators may have carcinogenic potential: they are all meant to be highly reactive, and there will be some residues left in the processed rubber.[4] Secondly, natural latex contains proteins from the parent plant which may be allergenic. A third issue is glove powder: natural latex rubber is tacky, and getting the gloves on is made easier by coating with a powder. Talc (a hydrous magnesium silicate) is no longer used because it causes granulomas if introduced into wounds, but even corn starch-coated gloves (see §2.4) also have problems in surgical contexts,[5] where powder-free gloves are recommended. The dust from these gloves easily becomes airborne and so is respirable, which carries both allergenic proteins and vulcanization accelerators into the lungs. There is yet another problem with latex gloves containing accelerator residues – interference with the setting of addition silicone materials (7§6.3).

To overcome the problem of the tackiness and hydrophobicity of natural latex rubber, and avoid the use of powder, a second layer of a hydrophilic polymer (*cf.* 5§6.1) may be added.

●1.4 Oxidation
Ordinary vulcanized rubber obviously contains many (if not most) of the original double bonds, while those of gutta percha remain entirely, and these bonds remain vulnerable to attack. In particular, reaction with the oxygen of the air occurs steadily, especially in the presence of ultraviolet light which will produce the $\pi \to \pi^*$ transition (24§6.1) and facilitate the reaction with the ubiquitous diradical. Such oxidation causes cross-linking and other reactions which cause both materials to become brittle, and eventually even to depolymerize. Elastic (natural rubber) bands are well-known to become unserviceably stiff and inelastic with time, and then to be reduced to a sticky mess. Old gutta percha, that is, having been stored for too long, is thus less likely to be serviceable and will be prone to breakage during handling, with consequent difficulties if this occurs in the root canal during placement. In fact, gutta percha in service in a tooth also shows such ageing: an old root filling may be very difficult to retrieve because it has become brittle due to oxidation. The deterioration of elastic (rubber)

bands is due to the same reaction, particularly if these are under stress, when the decrease in energy due to stress relaxation as a result of the then available rotation of the bond may be seen to help drive the reaction.

●1.5   Chicle

Another polyisoprene natural product, also collected as a latex, is chicle[6]. This has both *cis*- and *trans*-bonds, in the ratio ~1 : 2, which presumably inhibits crystallization. This is used as the basis of chewing gum, of course without cross-linking, and can be used as a means of delivering therapeutic agents to the mouth, although this is only by formulation as a simple mixture so that release is gradual as the mixture is masticated with saliva. This gum, too, shows the effects of oxidation as do gutta percha and natural rubber, although not in any sense that affects its use in practice, although there are implications for discarded gum – a public nuisance in many places.

§2. Polysaccharides

Two polysaccharides have been discussed in the context of impression materials (7§7 ~ 9), but another – **cellulose** – is used in endodontics in the form of absorbent paper points for drying canals and the application of medicaments.[6] This material has many of the general properties to be expected of polysaccharides and therefore can be understood in much the same terms.

●2.1   Cellulose

Cellulose (Fig. 2.1), a major constituent of plant cell walls, is an unbranched condensation polymer of β-D-glucose (Fig. 2.2) using the 1,4- linkage. It consists of **fibrils** of about 3.5 nm in diameter containing about 40 molecules, each with up to 10,000 sugar residues in wood and 15,000 in cotton. These elementary fibrils are organized into larger bundles, eventually forming the familiar macroscopic fibres that we observe in cotton and paper.

Cellulose is entirely insoluble in water because of the presence of highly organized regions in which the concerted hydrogen bonding is very strong, *i.e.* there are many bonds close together. However, because of the numerous hydroxyl groups also available in the intervening amorphous regions, which therefore have access to water, the fibrils will absorb large quantities and thus swell somewhat in the process. In this regard the water binding may be compared with that in alginates and agar, although disassembly of fibrils cannot be achieved without using more drastic treatment, such as strong base. In addition, the surface has many 'exposed' hydroxyl groups, leading to good *ad*sorption of water. The wetting of such a material is therefore good and the penetrativity high (10§2.4), aided by its porosity allowing it to hold a relatively large volume fraction (10§2.9), making it 'absorbent' and useful in such products as paper towels, the rolled-up slivers called paper 'points' for drying prepared root canals, and even the paper used in preparing a porcelain restoration (25§1.2).

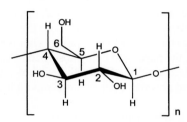

Fig. 2.1 The saccharide repeating unit of cellulose – anhydroglucose – in the more stable 'chair' conformation which is adopted.

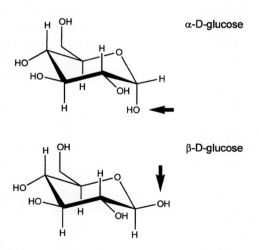

Fig. 2.2 The monosaccharide D-glucose exists in two cyclic forms, α and β. In the β-form all hydroxyl groups and the hydroxymethyl group are in the equatorial position, whereas the hydroxyl at C_1 (arrowed) is axial in the α-form. In aqueous solution, D-glucose forms an equilibrium mixture of about 36% α, 64% β, *via* the open-chain aldehyde sugar (the free-energy difference is about 1.4 kJ/mol).

---

[6] Obtained from *Achras sapota, Mimusops globosa, Sapota zapotilla.*

●2.2 Paper

Paper relies for its structure and integrity less on fibre entanglement than inter-molecular hydrogen bonding, these bonds forming at contact points between fibres. Water interferes with this very noticeably. Compare the stiffness of a sheet of dry writing paper with that when it is wet, or even when it is merely exposed to a high-humidity atmosphere. Consequently, the **wet strength** of paper is said to be low. While this may be rather irritating when blowing one's nose, fibres can easily be shed from endodontic paper points which may cause problems of tissue irritation if they remain, especially at root apices (where even a few fibres may be problematic as intense, chronic and persistent inflammation may occur[7]), and interfere with and prevent a seal being created between the root-filling and the wall. The wet strength can be improved considerably by a variety of chemical treatments that create permanent cross-links between fibres and fibrils. The problem is that, in so doing, all such treatments reduce the wetting and thus the absorbency of the paper. This is yet another compromise. Care must be taken therefore to avoid the disintegration of a point in use.

A similar problem arises in the use of paper for dental bibs, a seemingly trivial matter but one which must be of concern to the patient. Spray and spillage must be absorbed, yet the bib must not collapse or allow the liquids to soak through to the clothing beneath. Any treatment which strengthens the paper to prevent collapse increases the chance of the spillage simply rolling straight off before wetting the paper and being absorbed. One compromise is to use a bib with a thin film of polyethylene hot-pressed to the underside of untreated absorbent paper, simply avoiding 'soak-through'.

●2.3 Carboxymethylcellulose

This is a derivative of cellulose that has the group -CH₂-COO- attached to the oxygen on C₂ (Fig. 2.3). The effect of the addition of this highly polar group is to push the formation of sols somewhat further: they remain stable at lower temperatures. The substance is commonly used to increase the viscosity of various aqueous preparations at low concentrations, which it does by binding water molecules. However, as this is an anionic material (the sodium salt is usual), the negative charges on the chains prevent aggregation by mutual repulsion. Presumably, chelation (in the style of alginate) is prevented or delayed by using a monovalent cation. Even so, at high concentrations this substance does form gels.

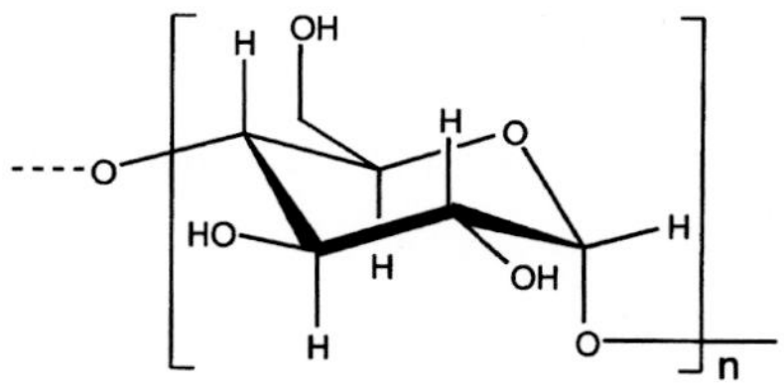

●2.4 Starch

In prosthetic dentistry in particular, impression plaster may be used. The impression will then be used to prepare a stone model in the usual way, but this may not be very easy or convenient to remove after the model has set, despite the use of a separating medium and their extreme brittleness. To help overcome this some brands of impression plaster have incorporated a proportion of **starch**. This substance, which forms the main food reserve of many plants, consists essentially of two polysaccharides: amylose (Fig. 2.4), MW: 2 ~ 10 × 10⁵, and amylopectin (Fig. 2.5), MW: 1~ 10 × 10⁶. Both are 1,4-condensation polymers of glucose, as is cellulose, but have the α- rather than β-linkage.[8] In addition, amylopectin is very highly branched at carbon 6, amylose only slightly. These quite subtle changes result in profound differences in their behaviour with water as compared with cellulose.

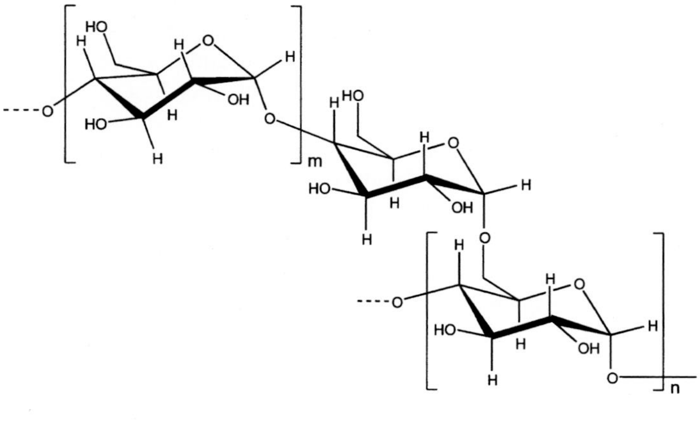

Fig. 2.3 The modified repeating unit of carboxymethyl cellulose.

Fig. 2.4 The repeating saccharide unit of amylose.

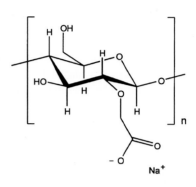

Fig. 2.5 A partial structure for amylopectin.

In aqueous suspension the naturally formed starch granules will swell by the absorption of water, and eventually burst at temperatures of about 60 ~ 80 °C (depending on the source), then partially dissolve, forming the viscous colloidal dispersion (a **sol**) generally called (starch-)**paste**. On cooling, the viscosity increases markedly and a gel is eventually formed due to the aggregation of linear amylose chains, held together by hydrogen bonds. This process, very similar to crystallization, has obvious parallels with agar impression material (7§9). The disruption and swelling process in fact forms the basis of the cooking of many starchy foodstuffs, to make the starch digestible. However, in dentistry, the swelling of the starch granules included in plaster creates sufficient stress that the plaster disintegrates, making its removal from the set model a relatively simple matter. The stress generated is a measure of the strength of the hydrogen bond and the mutual affinity of water and polysaccharide.

Corn (maize) starch is also used as a glove powder (§1.3), for which purpose the particles are surface treated to partially cross-link the chains (Fig. 2.6) to make the material less prone to swelling and less soluble when wet, whether by tissue fluid or sweat (and especially if steam sterilized), yet remain resorbable if left in a wound.

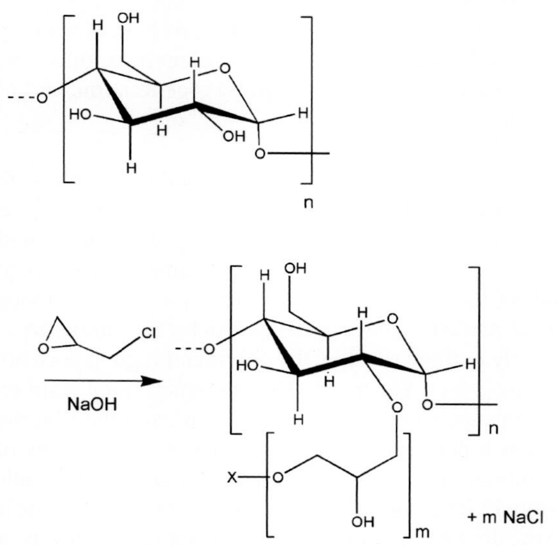

Fig. 2.6 Starches can be cross-linked with epichlorohydrin under alkaline conditions. Reaction is restricted to the surface of granules by using mild conditions (0.5% NaOH, 25 °C) to limit swelling. All hydroxyl groups are susceptible to attack, only one example is shown. m is a small number; X can be the next glucose unit or H.

Dextran, an α-1,6 polymer of glucose (as forms part of the matrix of dental plaque), can be cross-linked in a similar fashion. This yields the column-packing materials used in gel filtration (or permeation) chromatography for separating macromolecules such as proteins.

●2.5 Regenerated cellulose

Although cellulose is entirely insoluble in water, it can be treated to make it so in alkali. By reaction with sodium hydroxide solution, the cellulose swells and is converted to a sodium salt, and also partially depolymerized (by hydrolysis). By adding this still-insoluble material to carbon disulphide (CS_2), the hydroxyl group on C_6 (Fig. 2.1) is converted to an anion called a xanthate, $-CH_2-O-CS_2^-$ (this salt is an analogue of carbonate, $-OCO_2^-$). This material is then soluble in alkali, and the resulting solution is called **viscose**. The reaction is reversible by treatment with acid, thus regenerating cellulose itself (by the elimination of the CS_2). Commercially, there are two types of product employing this chemistry: fibre and film.

Viscose may be **extruded** under pressure directly into an acid bath through a set of fine orifices in a device called a **spinneret** (after the spider's organ of the same name and function, but looking more like a shower head). The acid bath reaction precipitates the cellulose, which is therefore reconstituted in the form of fine filaments. These are taken up by rollers which turn at a rate far faster than the viscose is extruded so that the filaments are fully drawn (3§4.7). This makes them narrower, but very stiff in the axial direction, with a high yield point, and little capacity for further plastic deformation (drawing). The chain alignment facilitates the reformation of the interchain hydrogen bonds, adding to the mechanical effects. This process is called **wet spinning**.

These filaments are then twisted into fibres, and the fibres into a yarn called **rayon**. Rayon fabrics are of very wide application and may appear in many medical contexts, including dentistry, because they have good handling and draping properties (relatively low lateral stiffness because of very small diameter filaments), are absorbent of water (hydroxyl groups), and can be sterilized at high temperature. Thus, rayon is only affected by decreasing strength above about 150 °C (well above the common wet autoclave temperature of 134 °C), and does not char or decompose until well over 170 °C. Similarly, it does not melt or stick (self-diffusion welding) at such elevated temperatures – it is not a thermoplastic material. For these reasons it is often used for surgical drapes and gowns. Note, however, that these properties also apply to cotton, which is a natural cellulose fibre. In a similar fashion, chopped fibres of both kinds may be used as surgical packs and other absorbing products, such as dental "cotton" rolls and cleaning cloths – so-called **non-woven** products. The water-absorbency of cellulose is not much affected by regeneration, and rayon's wet strength is also appreciably poorer than when it is dry.

Viscose may also be extruded through a slit-like die to form a film – a thin sheet of the regenerated cellulose called **cellophane**. The same process applies: acid-bath treatment precipitates the cellulose, which is then fully-drawn by rollers operating at surface speeds above the extrusion rate. The resulting **extruded film** has excellent clarity because recrystallization is kept to a minimum so there are no domains of varying refractive index (3§4.1, 24§5.11) to scramble the view. Although it is used widely for food and other packaging, it finds use in dentistry as a separating film for the temporary closure of a denture base mould, to enable it to be reopened cleanly without spoiling the polymer dough: it does not react with or dissolve in methyl methacrylate. Similarly, it can be used as a matrix strip to which filled resin restorative materials will not stick. Again, the cellulose of cellophane is highly water-absorbent, and the film therefore swells and is weakened when wet. However, this makes it permeable to water, although not to many other substances, and it may be used as a semi-permeable membrane for dialysis. Even so, in the dry state, cellophane film is much stiffer in bending than, for example, polyethylene, for the same cross-section, because the interchain hydrogen bonding, aided by the chain alignment in the drawn condition, means that chains do not slip easily over each other. Thickness, of course, can be tailored to give the required stiffness by taking note of the second moment of inertia (23§3.2).

●2.6   Cellulose acetate

Cellulose can also be treated with acetic anhydride in a process that ends up with, on average, about two out of the three hydroxyl groups on each sugar residue (Fig. 2.1) converted to acetate ester. This material is soluble in acetone and can be turned into fibre in a process similar to that for viscose (§2.5) except that the solvent is removed by evaporation in a stream of air; this is **dry spinning**. The product is known as **cellulose acetate** (or diacetate), or simply 'acetate'. In addition, it can also be cast into film, when the solvent is again evaporated. Here, because the polymer is thermoplastic (very few hydrogen bonds), it can be rolled (**calendered**) to the final thickness. Such film tends to be more isotropic than drawn extruded products.

The bulky side-groups, the irregularity of the sites of esterification, and the reduction in the number of possible hydrogen bonds between chains means that cellulose acetate is much less stiff in flexure than cellophane because chains can slide past each other. The ideas are similar to those of 23§3.1 ~ .2 where shear-stress transfer is the crucial condition for stiffening. Remove that interaction, and sliding occurs (see §3.5).

Cellulose acetate is commonly encountered as overhead projection 'foil' (although this is going out of fashion). It is used in dentistry for matrix strips, but also for temporary crown shells for which moulded forms are supplied which can be easily trimmed with scissors. A similar application is as a mould for a temporary crown in a filled resin where the transparency permits light-curing. The crown form is then easily stripped off when the resin is hard.

Cellulose acetate is also commonly used in dental research for making (inverse) replicas of surfaces such as of teeth and (non-polymeric) restorations for examination with an SEM. Brief exposure of one side of a slip of acetate film to acetone results in a softened surface that may be pressed onto the area of interest. When the acetone has evaporated, the film may be carefully stripped off, sputter-coated (to make it conducting) and examined.

●2.7   Cellulose nitrate

Another cellulose ester system is of historical interest but of only slight current relevance – the nitrates. While between two and two and half nitrate groups per sugar residue yields a commercial explosive (nitrocellulose), full esterification (three nitrates) yields a dangerously unstable material. However, with between one and two nitrate groups the resulting material, called **pyroxylin**, was the basis of the first commercially important plastic, **celluloid** (which was once a trade-name). Unfortunately, its inflammability was problematic, while its long-term instability causes problems today for film archives. It seems that the only general application currently is for table-tennis balls.

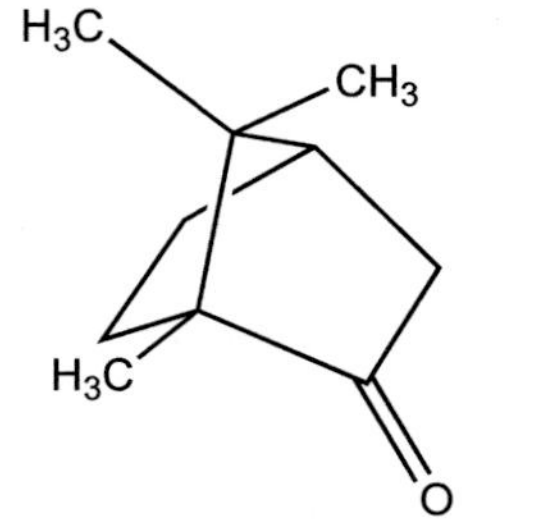

Fig. 2.7 Camphor – an isoprene-derivative natural product. (*cf.* camphorquinone, Fig. 6§5.8)

Cellulose nitrate, solid pyroxylin, is brittle and not easy to work, but if mixed with camphor (Fig. 2.7), a workable thermoplastic material is obtained. Camphor has a melting point about 180 °C, but under both heat and pressure would melt and dissolve in the pyroxylin, plasticizing it before the cellulose nitrate could decompose. In fact, the camphor acted as a solvent under those conditions, but one that remained in place when the piece was cold (*cf.* cast cellulose acetate). This was once tried for denture bases, but the problems of its unpleasant taste and dimensional instability, because of both the slow relaxation of frozen-in stresses due to the moulding conditions and the steady loss of the slightly volatile camphor, meant that it was rapidly superceded by vulcanite (§1.3). The behaviour and molecular shape of camphor here may be compared with those of eucalyptol (§1.2).

Pyroxylin is soluble in a variety of solvents; a solution in a mixture of diethyl ether and ethanol is known as **collodion**. This has been used in dentistry as a separating agent for plaster impressions (§2.4, 2§3.1) as the film formed when the solvent evaporates seals the surface preventing ingress of liquid from the slurry of the model material poured against it. Similar solutions might be encountered as an "instant skin" wound dressing or adhesive for skin electrodes (*e.g.*, EMG, ECG) (as well as a theatrical make-up aid). In using such materials, extreme care to avoid sources of ignition in the vicinity must be taken because ether-air mixtures are explosive (the upper and lower explosive limits are 1.7 and 48% in air at 25 °C, where the vapour pressure is ~0.7 bar). Cellulose nitrate (and in a mixture with cellulose acetate) is also used in microbiological filters in the form of membranes with pores of a near-uniform size, as well as various other biological research contexts.

●2.8   Chitin, chitosan

Chitin is poly(β-(1-4)-N-acetyl-D-glucos-2-amine) (Fig. 2.8), in other words another substituted cellulose, acetylamine at position 2. It is abundant in nature as the principal organic component of the exoskeletons of crustacea – crabs, lobsters, shrimps *etc.* – as well as insects. It is finding medical use as a facilitator of wound-healing, and so appears in wound and burn dressings where its resorbability (see §3.1) means that they can be left in place without the need for removal, likewise as sutures. Many other such applications are being investigated.

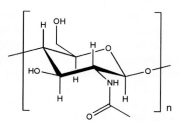

Fig. 2.8 The repeat unit of chitin.

Chitosan is prepared from chitin by partial deacetylation, which process is necessarily random, and so it may be described as a random copolymer of β-(1-4)-linked D-glucos-2-amine and N-acetyl-D-glucos-2-amine. The importance of this chemical change is the variation of the charge, becoming more basic, although it remains essentially insoluble, presumably by virtue of extensive interchain hydrogen bonding. But, as a result, it strongly promotes blood clotting and is also incorporated in wound dressings, particularly for where bleeding is severe. It remains as highly biocompatible and biodegradable as chitin.

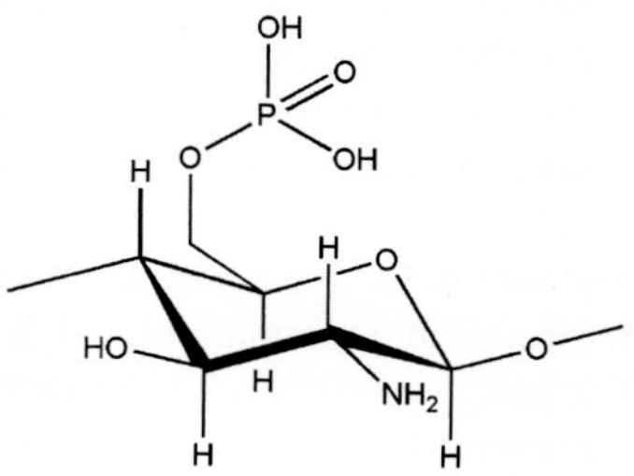

Treatment of chitosan with phosphorus pentoxide produces the phosphorylated derivative, with random replacement of hydroxyl groups by phosphate (Fig. 2.9). This material is now water-soluble because the phosphate groups readily ionize, and this also permits complexation with metal ions (especially calcium) in a fashion similar to alginate (7§9) and so can be used to make gels and membranes. These are being studied for application in tissue engineering, especially bone, as it may induce calcification by facilitating hydroxyapatite precipitation.

Fig. 2.9 One possible repeat unit of phosphorylated chitosan.

§3. Polypeptides

The mechanical stabilization of the approximated edges of a wound, whether this wound is deliberate or accidental, is a significant factor in the process of wound healing. Indeed, in reconstructive surgery it attains an enhanced importance. The properties and behaviour of the materials used in such an invasive manner are of great importance, both from the mechanical and biological points of view.

With the exception of some stainless steel and other wires with specialized uses, **sutures** for wound closure are made from polymeric materials of one kind or another as these offer the required properties of flexibility and toughness.[9][10] Whilst there are aspects such as 'hand' or 'feel', and knot-security which, although important clinically are not easy to quantify, the most important polymers offer some valuable insights.

●3.1   Collagen

'**Catgut**' or, simply, '**gut**' sutures are prepared from animal intestines. The collagen of the submucosal connective tissue is the structural material utilized after this has been stripped of essentially all extraneous material.

The protein **collagen** occurs as a fibrous structure with three levels of organization.[2] First is the protein chain itself. Secondly, three of these chains are assembled in a characteristic 'triple helix' molecule called **tropocollagen**. The smallest of the amino acids, glycine, occurs at every third position in the sequence of each chain so that it may lie on the inside of the triple helix. It is this special feature which allows the close approach of the three chains. This aspect of the stereochemistry is also critical for the stability of the structure, which is held together by hydrogen bonds. At the third level these molecules are then assembled into **fibrils**, held together by secondary forces in what is termed a **quarter-stagger** pattern (each tropocollagen molecule being offset by one-quarter of its length from its neighbour). In addition, in the so-called "type IV" collagen, there are cross-links at end chain segments. This ensures efficient transfer of shear stress within the fibril from molecule to molecule, essentially giving collagen its strength.

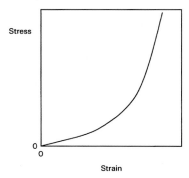

Fig. 3.1 The stress-strain behaviour of a typical collagen-containing tissue such as skin.

The arrangement of fibrils in membranes is complex, there being both oriented and random regions, some linear, some crimped, which together provide the unique properties of the tissues in which they are found. The general behaviour may be summarized by the characteristic stress-strain diagram (Fig. 3.1). A region of very great extensibility at low stress is followed by a region of very much higher modulus - higher by perhaps a factor of hundreds. This transition is due to the uncrimping and alignment of fibrils, which process requires relatively little stress until all the deformation must be taken by primary bonds, when the modulus becomes large. However, it is possible now to 'draw' the material containing the aligned molecules, reducing the diameter in the same way as for other polymers (see Fig. 3§4.10). This deformation is permanent and produces the filament used as a suture. The process will be aided by drawing through a die of the appropriate diameter in the presence of water which, because of the extremely polar nature of the protein, will be absorbed strongly and plasticize the material; this facilitates the rearrangement of the fibrils. Under moderate stress the presence of water would improve the flexibility of the filament by allowing some movement of fibrils past each other lengthways. This absorption of water is one reason for the early loss of strength of collagenous sutures, and for the observed marked swelling in contact with tissue fluids.

Such a suture material has one important property: it is resorbable.[7] The chemical nature of the collagen is common to a wide range of organisms and is readily broken down by available enzymes, phagocytosis being the removal route. However, a strong inflammatory response is expected, although this is likely to be due to remnants of other proteinaceous material from the source animal. In some circumstances this high degradation rate may not be desirable and **chromic gut** sutures may be used instead. In a process very much akin to the processing of 'chromate' leather, the protein is denatured and fixed (in the histological sense) by treatment with an acidic solution of a chromate salt (CrO_4^-). This produces cross-links both between molecules and between

---

[7] A suture is generally said to be resorbable if it has lost 50% of its initial dry strength after 60 days of placement, and non-resorbable otherwise. Evidently, this arbitrary division obscures a continuum of behaviour, from very short to more or less permanent.

the chains of the triple helix, making the material much more resistant to enzymatic degradation. The greater longevity is bought at a price: greater tissue irritation. Chromium is a toxic metal and, even though relatively small quantities are present, a distinct inflammatory response is to be expected from this source.

Using special techniques it is possible to solubilize collagen from tendons ('type I' collagen) with dilute salt or acid solutions as this collagen lacks cross-links. It can then be reprecipitated as an extruded solid filament after a more thorough purification. The extruded filament would then be drawn down to the required size, during which process the molecules would again be aligned for the maximum elastic modulus in tension. The longevity of 'chromic' sutures from this source may be doubled compared to plain gut, but on balance the tissue response is hardly affected.

●3.2   Gelatine

If the normally insoluble collagens ('types II - IV') are treated to hydrolyse them partially, for example by boiling in water, the result is a colloidal solution of a material known as **gelatine**. On cooling this sol below about 35 - 40 °C a hydrogen-bonded gel results, and in this form is frequently used in foodstuffs. Drying the gel can produce a brittle, transparent solid and this forms the basis of photographic 'emulsions': a thin layer of the sol with suspended AgBr crystals is coated onto film base and then cooled and dried. It remains extremely absorbent to water because polypeptides are highly polar, and on immersion it swells appreciably, becoming very soft. The water absorption allows the developing and fixing chemicals to diffuse into the emulsion and subsequently to be washed out. Wet film is readily damaged as the soft emulsion may be easily removed from the base, even though modern emulsion gelatine may be 'hardened' by chemical treatment.

●3.3   Protein structure

Proteins are distinguished from 'ordinary' polymers by the fact that they are made from many types of monomer – the naturally-occurring amino acids, 20 or so common ones, with a variety of others occasionally encountered – and so their structure can be correspondingly complex. This structure is described in terms of four hierarchical levels. **Primary structure** is simply the connection sequence, *via* covalent bonds, of the entire list of amino acids: what is joined to what. **Secondary structure** describes the local conformation, residue by residue, to gain a sense of the folding pattern, for example whether a helix or a sheet, on a scale of a few to modest numbers of units. Steric effects and hydrogen bonding between near-neighbours in the chain are important determinants at this level. **Tertiary structure**, however, is about the overall folding pattern and thus the shape of the whole molecule, for example whether rod-like or globular, looking at the relationships of sections such as local helices to one another. This takes into account hydrogen bonding between chain regions. **Quaternary structure** describes the patterns and connections between whole molecules in clusters and more complicated systems.

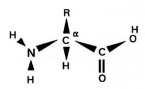

Fig. 3.2 The structural pattern for all amino acids.

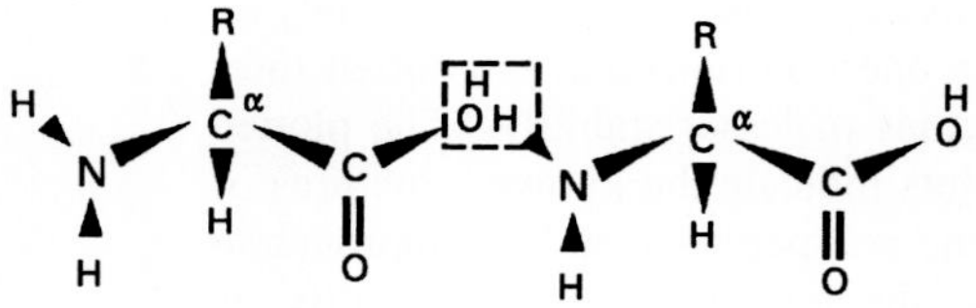

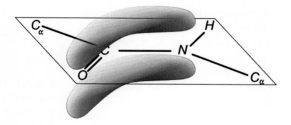

Fig. 3.3 The condensation which forms the peptide linkage.

To understand the secondary and tertiary structures of proteins, that is the conformation of the chains about each bond in turn, and the relationships of chains to each other and to themselves, it is necessary to examine the peculiarities of these polymers.

The monomers of proteins are all α-amino acids (Fig. 3.2), *i.e.* the carbon adjacent to the carboxyl group bears the amino functional group.[8] Proteins are, of course, condensation polymers, as water is eliminated in forming the peptide linkage (Fig. 3.3). Despite the formal depiction of a single bond at this

Fig. 3.4 Delocalization at the peptide linkage prevents rotation at the C-N bond giving coplanarity of the six atoms of the amide plane.

---

[8]   Strictly speaking, proline and hydroxyproline are imino acids since the side-chain is attached at both ends, one to the nitrogen, to give a cyclic structure. This does not affect the argument.

point, resonance stabilization occurs due to the lone pair on the nitrogen and the electronegativity of the carboxyl oxygen. This gives considerable double-bond character to the C-N amide bond, stabilizing the conformation about it. In fact, this constrains six atoms to be coplanar – the so-called **amide plane** of polypeptides (Fig. 3.4). Note that the arrangement of bonds around the nitrogen would otherwise be pyramidal. Some details of the geometry of the peptide linkage are shown in Fig. 3.5.

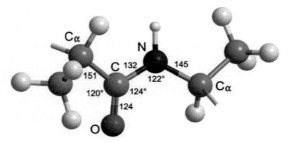

Fig. 3.5 Details of the geometry of the peptide linkage. Bond lengths in pm.

●**3.4   Ramachandran plot**

Because of the rotational constraint at the C-N amide bond there remain only two rotational degrees of freedom for each residue (amino acid monomer unit), *i.e.* about the αC-C and αC-N bonds. Because of this, determination of these two angles (labelled ψ and φ respectively) at each residue determines absolutely the conformation of the entire protein chain. Recalling the conformational analysis of n-butane (3§2.5), it is apparent that not all values of ψ and φ will be equally favoured and, depending on the size of the prosthetic groups R (Figs 3.2, 3.3), the range available may be even further restricted.

It is possible to compute the relative steric interactions around the two critical bonds of polypeptides and thus the relative energy as a function of conformation (Fig. 3.6). From such calculations, a stability map or **Ramachandran** plot may be drawn (Fig. 3.7).[11][12] The solid lines enclose areas of greatest stability (based only on this one criterion) and the dotted lines regions of lesser stability. The plotted points indicate the known structures of some polypeptides, and the approach to the theoretical areas of stability is obvious. The position for collagen is marked, as is the most common 'α-helix' which is found in many proteins. While the steric considerations for the conformations about these two bonds are of clear importance, thermal effects would certainly tend to scramble the overall form were it not for one thing. Stabilization of these structures ultimately depends on the formation of hydrogen bonds, holding sequences in position at points separated by one or several peptides (Fig. 3.8). There are also, of course, covalent cross-links of

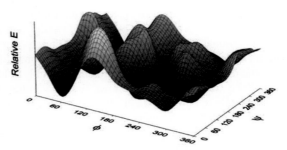

Fig. 3.6 Typical kind of energy surface for the rotation of a dipeptide. Image: Greg Chass.

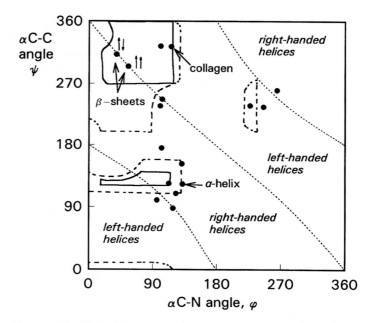

Fig. 3.7 The Ramachandran plot for polypeptides. Solid lines indicate regions of theoretically high stability, dotted lines regions of slightly lesser stability, based only on steric factors. Plotted points correspond to known polypeptides.

various kinds possible; these greatly stabilize the tertiary structure of proteins and their effect may dominate the steric energy. These extra factors account for the discrepancies between the plotted points and the steric-only stability regions in Fig. 3.7.

●3.5   Silk

Another polypeptide structure of interest is the "antiparallel β-pleated sheet" (Fig. 3.7) characteristic of the structural protein of silk,[9] **fibroin**. The primary structure of this protein is quite special, being essentially an alternating sequence of glycine and alanine (or serine); these are the three smallest and least polar amino acids. By arranging these chains in sheets, the sequences running alternately in opposite directions, the side chains pack together very compactly and the most stable hydrogen bonding can be established, which has the -N-H···O=C- group in a nearly straight line (Figs 3.9 ~ 3.11). A vector diagram shows the relevant forces both within and between chains in such a structure (Fig. 3.12). This indicates that the fibre will be stiff in tension along the chains (because bond angles would have to be changed), but flexible in 'out-of-plane' bending because the sheets can move relative to one another since there are only weak forces holding them together.

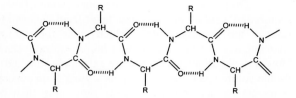

Fig. 3.8   Hydrogen bonds stabilize polypeptide conformation.

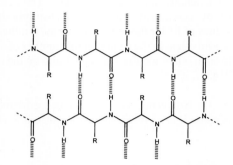

Fig. 3.9   Schematic diagram of the hydrogen bonding occurring within the sheets of fibroin. R = methyl.

The behaviour arising from the lack of constraint to shear between layers is similar to the bending of a bundle of sheets of paper: relative motion must follow (*cf.* 23§3.2). Notice, however, that bending must involve some bond rotation within the chains and therefore some conformational change – but this is relatively easy because the steric hindrance due to the peptide linkage and inter-chain hydrogen bonding is offset by the lack of substituents on the carbonyl carbon and the amino nitrogen. A further comparison may be made with graphite (11§1.2), in which bonding between planes is weak and non-localized van der Waals-type, while in-plane the sheets are stiff and strong. Out of plane bending is, however, much harder because no conformational change is possible, and all strain is then in bond angles.

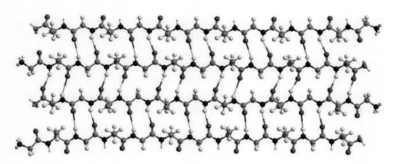

Fig. 3.10   Plan view of the hydrogen bonding occurring between adjacent chains within the sheets of fibroin.

The structure of silk is interesting because it is essential crystalline – it has very great regularity in both chain extension and chain packing in the other two directions. This comes about because the initially amorphous secretion is highly drawn on deployment by the 'silk worm' such that the crystallization from the solution is spontaneous. Obviously, no further drawing of the chains is possible because they are already fully extended, but in that condition

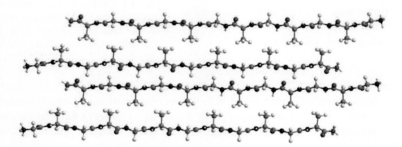

Fig. 3.11   Edge-on view of the sheets of chains in fibroin. The way in which the methyl groups of the alternate alanine residues fit together suggests less steric hindrance to sliding between every second pair of sheets.

rotation of segments to allow the hydrogen bonding to form is easy and highly favourable energetically. Thus the sheets of fibroin automatically form into stacks of parallel planes.

---

[9]   A secretion of the silk-moth *Bombyx mori* in the form of a very long, very fine thread from which the pupation cocoon is wound.

The properties of silk are, of course, some of those required of a suture material and, apart from the need to remove the gummy protein which coats the natural filament, silk can be used more or less in its 'raw' state. However, the filaments of silk are extremely fine and a usable suture has to be made up as a **yarn** from several filaments twisted together. Such yarns may then be **braided** for higher strength and resistance to untwisting. This has the important effect of increasing the strength (in direct proportion to the cross-sectional area) without a great penalty in increased stiffness. The flexural rigidity, EI, of such a multistrand system is simply the *sum* of the

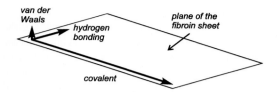

Fig. 3.12 A vector diagram indicating the relative strengths of the bonding forces in fibroin. The 'up' direction refers to the between-sheets interaction.

values of EI for the individual strands, rather than depending on the fourth power of the effective radius which is equivalent to the square of the number of strands. This arises because adjacent strands may slide with respect to each other and exert no constraint on the deformation (*i.e.* bending) observed, except for a little friction (see 23§3.2, 23§6).

Silk is not considered to be resorbable but a slow degradation may still be expected, with phagocytosis, because it is a protein material and so it is vulnerable to enzymes.

§4. Polyesters

A synthetic resorbable suture material which produces little inflammatory response is **poly(glycolic acid)**, a condensation polymer of glycolic acid: $HO-CH_2-COOH$. The mechanism for that polymerization, **transesterification**, is an unusual one in the dental context. The starting point would be, for example, the methyl ester of the acid. The hydroxyl group can then attack the carboxyl group carbon on another molecule and eliminate methanol after some rearrangement (Fig. 4.1). As the methanol has a low boiling point, this elimination product can be continuously removed by distillation, thus driving the reaction to completion. R and R′ can be either monomer or polymer, both hydroxyl and methyl ester groups remaining active in the sense of this reaction irrespective of the length of the polymer chain(s) to which they are attached. This is therefore not a chain reaction polymerization in the normal sense.

Poly(glycolic acid) turns out to be a rather stiff material: because of its small side groups it is likely to have a high degree of crystallinity when spun and drawn into fibres. Accordingly, it is essential to form sutures of this material from many very fine fibres in a braid in order to attain sufficient flexibility and strength (see silk, above).

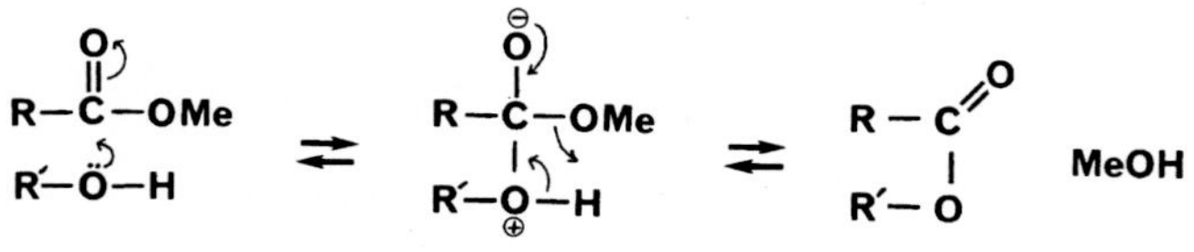

Fig. 4.1 Outline of the transesterification reaction. Substitute water for methanol and read from right to left to indicate hydrolysis. NB: the other ends of both of the groups R, R′ always have a reactive functional group available for further chain growth.

The manufacturing route for the material is, of course, of little direct interest to the user except that the degradation of the polymer *in vivo* is the simple hydrolysis corresponding to the reverse reaction, although regenerating glycolic acid itself rather than its ester. There is no phagocytosis involved. Pathways exist in the mammalian metabolism to deal with glycolate. This is a normal intermediate in some reactions, and elimination either as glycolate, glyoxylate or oxalate (the two successive oxidation products) can occur. Accordingly, little untoward tissue inflammation would be expected arising from the use of this material, and indeed this is just what is observed.

Transesterification is also used in the production of poly(ethylene terephthalate), which is known by the trade names **Terylene** or **Dacron** in fibres, and **Mylar**, a matrix film used for filled resin restorative materials. These, however, are much more resistant to hydrolysis.

§5. Epoxy Resins

Named for the strained three-membered ring of the **epoxide** reactive group (more formally, **oxirane**), epoxy resins[13] have found many uses as adhesives commercially and domestically, as embedding materials for specimens for transmission electron microscopy, to make surface replicas for scanning electron microscopy, as dental die materials, as pit and fissure sealants, and proposed as a hardener for gypsum dies (2§12.1).

The general setting reactions are shown in Fig. 5.1. The epoxide ring is opened by a primary ($-NH_2$) or secondary ($>NH$) amine through the nucleophilic attack of the nitrogen lone pair on the methylene ring member ($<CH_2$), with the transfer of hydrogen to the oxygen. Of course, the secondary amine generated in this way from a primary amine is itself reactive in the same way, and in this way extensive branching is developed. Furthermore, the hydroxy group is sufficiently nucleophilic as to also open the oxirane ring, especially at elevated temperatures (60 ~70 °C is commonly used for tissue embedding), adding to the extent of branching.

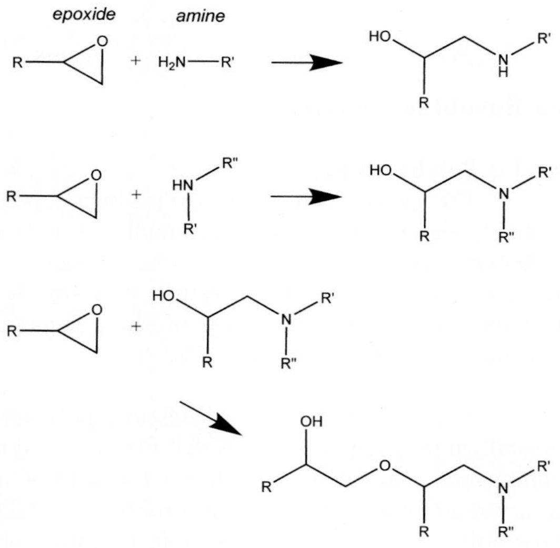

Fig. 5.1 Reactions involved in the setting of epoxy resins.

The strength and hardness of set epoxy resins are further increased by the common use of difunctional monomers and amines. There is in fact a vast number of such monomers and amines (and indeed a variety of other nucleophilic "activators") in use in various combinations for commercial purposes, with an equally vast range of properties, so much so that there is no value in specific examples here. A single example will suffice as an archetype for the monomer, that based on bisphenol-A (6§4.3), which is indeed commonly used.

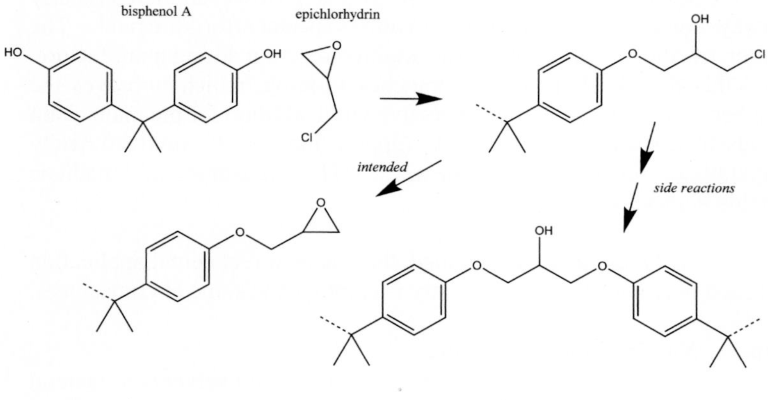

Fig. 5.2 Occurrence of other compounds in the manufacture of epoxy resins.

Through a series of steps, epichlorhydrin is reacted with bisphenol-A to the bis-epoxy (Fig. 5.2). However, a variety of side-reactions are possible, such that dimers and higher oligomers (as far perhaps as 25 units), as well as branched species, are inevitably present, adding to the complexity and viscosity of the 'base' resin. Using a difunctional amine in the correct proportions then yields a very highly cross-linked resin which typically is thermally-stable to quite high temperatures (120 ~ 230 °C) and resistant to attack by many solvents and reagents.

The free-volume loss expected for the addition reaction of setting is partially compensated by the gain from the oxirane ring opening, the polymerization shrinkage found with epoxy resins is generally somewhat less than found with vinyl polymerizations (5§2.4), and in some formulations is essentially negligible, even though the degree of conversion of the epoxide can exceed 90%. Values of around 0.1 ~ 0.3 % or so are found for dental die materials. Such dies are more robust than those of gypsum (2§12), but because the oxirane ring is attacked by water (Fig. 5.3)

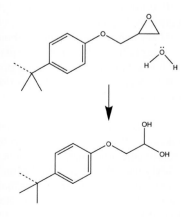

Fig. 5.3 Reaction of the oxirane ring with water.

they cannot be poured into alginate or agar impressions. However, the setting is relatively slow for most such systems, typically several hours but as long as several days for effective cessation, if no heating used.

§6. Butadiene Copolymers

●6.1   Polybutadiene

The 'parent' of isoprene (§1) is buta-1,3-diene (also called 1,3-butadiene, or just plain butadiene), and can readily be polymerized to industrially-important rubbers, principally for car tyres. The polymerization is a simple addition reaction, but it may take any of three separate forms, depending on conditions (Fig. 6.1).

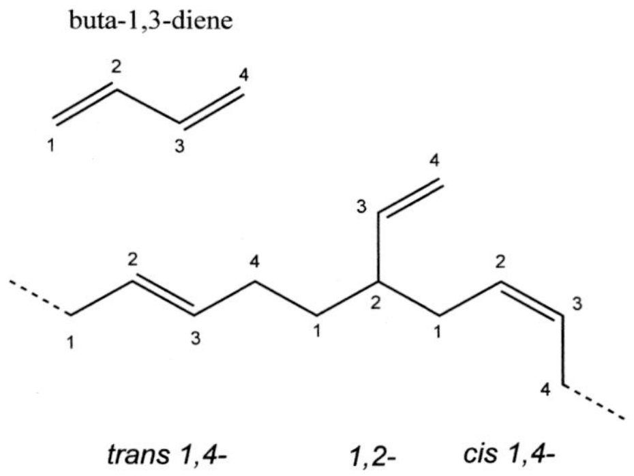

The *trans* 1,4-addition leads to a polymer that can crystallize readily (*cf.* gutta percha, Fig. 1.2) and so has no commercial value. However, the *cis* 1,4-addition product is a very good rubber (*cf.* natural latex rubber, Fig. 1.2). A third possibility is vinyl- or 1,2-addition, in which one of the original double bonds remains as a side-chain (the attachment point, carbon 2, is then chiral [3§2.2]). The proportions of the three structures can vary widely according to the catalyst used, but most interest focuses on so-called "high-*cis*" material, which has the best rubbery properties ($T_g < -100$ °C). However, there is nearly always some *trans* and *vinyl* present unless special efforts are made. The latter, of course, is amenable to an addition reaction during manufacture, and this then leads to longer branches forming, which improves the rubbery properties, although excessive vinyl addition for the main chain leads to a steadily increasing T_g (approaching 0 °C) and ultimately crystalline "high-vinyl" polymers.[14][15] The 1,2-isomer of butadiene yields high-vinyl polymer.

As far as can be ascertained, there is no direct dental application of such polymers, but the chemistry is employed in some circumstances.

●6.2   Nitrile butadiene rubber

Butadiene rubber is not resistant to oils and solvents in general and so is not a useful material for protective gloves, despite its low modulus of elasticity. However, copolymerization with acrylonitrile (15 ~ 50 %) confers considerable such resistance (but *not* to polar solvents and reagents), increasing steadily with its proportion, in what is now called nitrile butadiene rubber ("NBR") (Fig. 6.2). It does however at the same time make the modulus of elasticity higher. Again, *trans*-, *cis*- and *vinyl* additions will occur, but still with the *cis*- being preferred for rubbery behaviour. (Some 1,2-butadiene may be included in the reaction mixture to generate branches.) Even so, the plain copolymer is inadequate as an elastomer and needs to be cross-linked. This may be done in a variety of ways, but the commonest approach is to use sulphur-containing agents as for natural latex rubber (§1.3). These react with a small proportion of the remaining double bonds to create the necessary three-dimensional network.

buta-1,3-diene

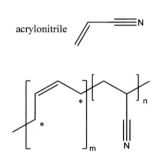

trans 1,4-   *1,2-*   *cis 1,4-*

Fig. 6.1 Butadiene and its polymerization.

acrylonitrile

Fig. 6.2 Structure of acrylonitrile and nitrile butadiene rubber. The allylic carbons are marked *. The nitrile group is highly polar and readily forms hydrogen bonds with relevant solvents, allowing them to diffuse through the polymer.

For gloves, this means that there is a practical limit to how far the proportion of acrylonitrile may be taken, at the expense of some hand fatigue in flexing against that stiffness. However, there is a great improvement in strength and puncture resistance as well with this kind of formulation, so such elastomers are widely used for medical and dental gloves. The occasional sensitivity experienced to the residues of sulphur-containing cross-linkers means that alternatives are desirable. Peroxides have been found effective. Generating HO· radicals on heating, these can abstract the allylic or α-hydrogen adjacent to a double bond (Fig. 1.5), and that free radical can then go on to attack other double bonds (whether in the chain or of the vinyl type), or mutually annihilating another free radical, effecting the cross-linking.[16]

•6.3   Styrene butadiene

As has been indicated earlier (3§6.2, Chap. 6), the desirable properties of hardness, strength, stiffness and so on in polymers is achieved through cross-linking. This prevents chain slippage, but makes the material brittle (3§6.2). The problem is to make glassy polymers damage-tolerant and tough. The primary method of doing this is to find a way of increasing the work for crack-propagation (6§2.14) or ease of crack-tip blunting – which on the face of it is contradictory with the idea of cross-linking. However, the effect can be achieved by establishing a dispersion of rubbery particles in the glassy matrix: a propagating crack encountering such a particle is then blunted and its work of propagation through the particle made high because the elastomeric chains draw readily (7§2.2). But this can only be achieved if the rubbery phase is covalently-bonded to the matrix, else the crack would propagate around it at near zero cost (25§3.1) and nothing would be gained – indeed, the material would be more easily cracked, with the toughness lowered. Creating the dispersion by simple mixture does not work, but it can be done by arranging for a **phase separation** (8§4.1) to occur whilst maintaining a complete, single, three-dimensional network.

The answer is to create a **diblock copolymer** where the two kinds of polymeric region are mutually immiscible. That is, not a random copolymer, but linked chains of the two kinds. One common example is that of poly(styrene-butadiene) (Fig. 6.3). The two kinds of chain are sufficiently different chemically for the system to phase-separate – that is, go to a lower-energy state – creating two kinds of domain: one being essentially pure butyl rubber in a matrix of essentially pure polystyrene. This is the basis of 'high-impact polystyrene' or 'HIPS', meaning that it can survive impact without shattering, as pure polystyrene would (atactic, $T_g \sim$ 90 °C). (It is this type of polymer that was used for scale model kits, which often took considerable abuse, but is still widely used for many consumer products.)

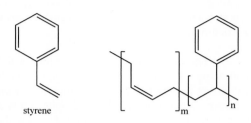

styrene

Fig. 6.3 Styrene, and the diblock copolymer with butadiene. The two kinds of block domain phase-separate under the right conditions.

Of course, the phase separation is constrained by the fact that it consists of very long chains with branches and cross-linked in the poly(butadiene) regions because of the occasional vinyl side group, but also by the proportions of the two components and the sizes of the blocks. Processing conditions, such as temperature and extrusion can also effect the outcome.

Manufacture of this kind of system typically involves dissolving the (relatively low molecular weight) butadiene polymer in styrene, and then polymerizing the latter (noting that a simple copolymer of butadiene and styrene would not lead to the formation of blocks and therefore there would be no phase separation). As the polymerization proceeds, which increases the polystyrene block size, at some point complete miscibility is lost and a second phase starts to separate (Fig. 6.4).[17][18] If the proportions of the two components are extreme (that is, one is present at low concentration), then a fine dispersion of droplets is formed. If the proportions are made less extreme, then separation tends to generate threads or cylinders of the one in the other, whilst for more equal proportions a lamellar structure is formed. The boundaries between these behaviour domains depend on

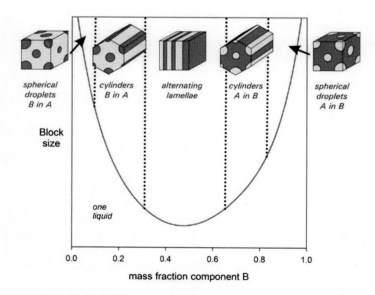

Fig. 6.4 Styrene, and the diblock copolymer with butadiene. The two kinds of block domains phase-separate under the right conditions. (Simplified scheme: more complex patterns can occur.)

many factors, including according to the polymer systems involved, and may involve the occurrence of **bicontinuous** phases between the two cylinder and lamellar conditions.[19] Parallels in the phase separation patterns can be seen with the formation of eutectics, *e.g.* Ag-Cu (12§3.3), pearlite (21§1.2) and Co-WC (21§3.1). In HIPS and the like kind of system, the proportion of the rubbery material (say, component B in Fig. 6.4) is kept relatively low, leading to the droplet or thread dispersion of that phase in the continuous component A phase. The stiffness of the copolymer is therefore closely related to that of the pure A polymer, polystyrene in this example (*cf.* Fig. 6§2.12, with the relative values of the moduli reversed).

It will be recalled that metallic eutectics are tough because crack deviation must occur frequently at successive grain boundaries (12§3.4). A toughening effect also operates in HIPS and similar polymer systems, but this time because a crack tip must be blunted as it reaches the almost liquid-like rubbery inclusion, where chain drawing is relatively easy (Fig. 6§2.18) and this therefore consumes energy as the path of least work. In other words, catastrophic crack propagation does not occur, despite the brittle matrix. Furthermore, these kinds of structure are craze-resistant. That is, the drawing of matrix chains (5§5) is inhibited because the path length between successive disperse phase droplets is small (micrometre-scale) and the deformability of the rubber great. The stress required to cause fibril formation is therefore harder to attain, and thus it, and therefore fibril rupture, are deferred.

The actual polymerization process to make HIPS may show a structural progression (Fig. 6.5). Following the implications of Fig. 6.4, low-molecular weight poly(butadiene) is soluble in styrene. But as the styrene is polymerized, the polymer now formed cannot dissolve the poly(butadiene), so the polystyrene separates out as droplets. These droplets grow as polymerization proceeds, increasing the concentration of the rubbery component in the solution, but now those droplets coalesce to form a continuous phase, leaving the poly(butadiene)-rich solution first as a second continuous phase (that is, the system is at that stage bicontinuous) but then soon as separate droplets when the volume fraction of the polystyrene phase becomes high enough.

However, within those poly(butadiene) droplets will be trapped, as well as newly-formed as the polymerization proceeds, droplets of the solid polystyrene, creating what is called the "salami" structure. Depending on the circumstances – with active mixing, for example – the structure may be even more complicated than illustrated here, with more layers and nesting of domains. Nevertheless, what has been formed still behaves as a craze- and crack-resistant material.

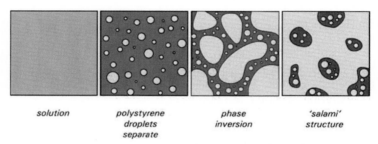

solution | polystyrene droplets separate | phase inversion | 'salami' structure

Fig. 6.5 Typical changes occurring during the polymerization to form HIPS. The continuous phase is initially solution, growing more concentrated in poly(butadiene), but after the phase inversion is solid polystyrene.

Despite the appearance of the simplification of the phase-separated material here, there cannot be a sharp boundary between the two phases because entanglement, complicated by branching and the cross-linking between blocks means that separation would be at best kinetically limited if not physically impossible. This means that the two phases are in effect – if not actually – bonded across the extended transition zone that represents the

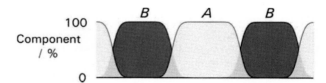

Fig. 6.6 Phase-separated block copolymers have transitions rather than sharp interfaces between regions because of entanglement, conferring matrix constraint.

interface (Fig. 6.6), thereby providing the matrix constraint that confers the toughening behaviour. Without this, the rubbery phase would simply occupy holes and weaken the material (*cf.* Fig. 6§2.15).

Phase separation has a further consequence: the material becomes less than transparent (which it was to start with). If the domains are large enough, the difference in the refractive index causes light-scattering, as it will in any composite material (24§5.11). Accordingly, depending on proportions and the exact chemistry of each phase (variations are possible for both), the material may appear anywhere from misty to opaque. In some contexts, a compromise between mechanical properties and appearance is again required.

Despite the commercial importance of such HIPS, it is not a suitable system for the fabrication of denture bases, where the powder-method process for PMMA remains convenient. Injection moulding is feasible, but difficult to justify when each denture is unique, requiring an individual 'prototype' mould, while special mould systems are required to handle the high pressures necessary. The cost is therefore high.

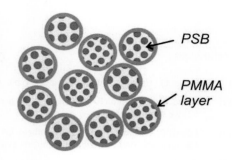

Fig. 6.7 Concept of PMMA-coated poly(styrene-butadiene) [PSB] beads for high-impact denture-base PMMA.

•6.4   High-impact poly(methyl methacrylate)

The benefits of the HIPS concept can be applied to denture-base acrylic. This has taken two paths. In the first, suspension-polymerized (5§2.1) beads of butadiene-polystyrene block copolymer are grown on by polymerizing a shell of PMMA in a similar process, the PMMA being in effect a third block covalently-linked to the HIPS core. This creates a powder (Fig. 6.7) that may be handled and processed in the usual fashion (5§2.2). The drawback is that the greater complexity of the production process means that the bulk price of the coated beads is nearly three times that of the simple PMMA powder, although this increase will be a very small proportion of the total cost of the production of the denture, especially since typically only some 5% is used admixed with plain PMMA beads.

In some versions of this approach, the initial suspension polymerization also includes some methyl methacrylate, which is likely to be incorporated as a random copolymer in the polystyrene blocks.

The second approach is simply to include particles of poly(butadiene) rubber in the PMMA powder,[20] but it is clearly important to ensure that, one way or another, there is covalent bonding to ensure stress transfer and matrix constraint.[21]

References

[1]   Fisher D. Crystal structures of gutta percha. Proc Phys Soc B 66 (1): 7 - 16, 1953.

[2]   Mandelkern L, Quinn FA & Roberts DE. Thermodynamics of crystallization in high polymers: gutta percha. J Amer Chem Soc 78 (5): 926-932, 1956.

[3]   Ghosh P, Katare S, Patkar P, Caruthers JM *et al*. Sulfur vulcanization of natural rubber for benzothiazole accelerated formulations: From reaction mechanisms to a rational kinetic model. Rubber Chem Technol 76 (3): 592 - 693, 2003

[4]   Tinkler J, Gott D & Bootman J. Risk Assessment of dithiocarbamate accelerator residues in latex-based medical devices: Genotoxicity considerations. Food Chem Toxicol 36: 849 - 866, 1998

[5]   Anon. Medical Glove Powder Report. Center for Devices and Radiological Health, US Food and Drug Adminstration, September 1997.

[6]   Edwards RO & Bandyopadhyay S. Physical and mechanical properties of endodontic absorbent paper points. J Endo 7(3): 123 - 127, 1981.

[7]   Brown DWP. Paper points revisited: risk of cellulose fibre shedding during canal length confirmation. Int Endod J 50 (6): 620 - 626, 2017.

[8]   Coultate TP. Food - The Chemistry of its Components. 2nd ed. Roy Soc Chem, London, 1984.

[9]   Swanson NA & Tromovitch TA. Suture Materials, 1980s: Properties, Uses and Abuses. Int J Dermatol 21: 373 - 378, 1982.

[10]   Taylor TL. Suture material: A comprehensive review of the literature. J Amer Podiatry Assoc 65: 649 - 661, 1975.

[11]   Ramachandran GN, Ramakrishnan C & Sasisekharan V. Stereochemistry of polypeptide chain configurations. J Mol Biol 7: 95 - 99, 1963.

[12]   Vincent JFV. Structural Biomaterials. Macmillan, London, 1982.

[13]   Castan P. Swiss Patent CH211116 (A) 1940-08-31, submitted 1938.

[14]   Bhowmick AK & Stephens HL. Handbook of elastormers. 2nd ed. Dekker, New York, 2001.

[15]   Halasa AF & Massie JM. Polybutadiene *in:* Kirk-Othmer Encyclopedia of Chemical Technology. Wiley, 2000. DOI: 10.1002/0471238961

[16]   Liu XF, Zhou T, Liu YC, Zhang AM, Yuan, CY & Zhang WD. Cross-linking process of cis-polybutadiene rubber with peroxides studied by two-dimensional infrared correlation spectroscopy: a detailed tracking. RSC Adv 5: 10231 - 10242, 2015.

[17]   Pethrick RA. Polymer Structure Characterization. 2nd ed. RSC Publishing, Cambridge, 2014.

[18]   Vonka M & Kosek J. Modelling the morphology evolution of polymer materials undergoing phase separation. Chem Eng J 207-208: 895 - 905, 2012.

[19]   Wang XJ, Goswami M, Kumar R, Sumpter BG & Mays J. Morphologies of block copolymers composed of charged and neutral blocks. Soft Matter 8: 3036 - 3052, 2012.

[20]   Stafford GD, Bates JF, Huggett R & Handley RW. A review of the properties of some denture base polymers. J Dent 8: 292 - 306, 1980.

[21]   Rodford R A. Further observations on high impact strength denture-base materials. Biomaterials 13: 726 - 728, 1992.

Chapter 28   More Metals

Titanium *and its alloys are of increasing interest in medical and dental applications because of their biocompatibility. However, these metals have properties that differ significantly from normal expectations, and recognition of this is important to their understanding. Key to this is the existence of allotropic forms and **deformation twins**.*

*Nickel-titanium goes further in having two very special behaviours: **pseudoelasticity** and **shape-memory**, dependent on a **martensitic transformation**. These allow a range of novel types of application.*

*Where casting is not appropriate, various other methods of forming metals are available, mostly mechanically. There are various advantages and disadvantages with these, depending on factors such as ductility and work-hardening. **Superplastic forming**, however, relies on a new mode of deformation: **grain switching**.*

Pure gold *for direct filling may be little used these days, and while relying on **cold-welding**, it can be understood in familiar terms: annealing and strain-hardening.*

Dental mouth mirrors *provide a nice case-study in the selection of materials according to several properties.*

Mercury *provides several challenges in its handling, storage and usage because of its toxicity, both directly to the user and environmentally. These matters are set out for reference. In this context, the obsolete restorative material **copper amalgam** is described fully in order to understand its demise. A related toxicity issue has to do with **beryllium**, a metal that should no longer be used in dentistry.*

*Dentistry employs many other other alloys for various purposes: **Elgiloy** is related to cobalt-chromium casting alloys, but also has features similar to titanium, as well as a further mechanism for hardening through a heat-treatment.*

Materials Science for Dentistry
https://doi.org/10.1016/B978-0-08-101035-8.50028-6

In addition to the common classes of metals and alloys that have already been covered, there remain some of more specialist use which have aspects of particular interest. It is the purpose of the present chapter to explore some of these systems. Again, it is broadly a matter of the relationship of properties and behaviour to the requirements of an application that control use, but in order to achieve the desired outcome recognition of the principles and limitations is necessary.

§1. Titanium

Titanium has long been known for its valuable combination of high strength with low density. Coupling this with remarkable corrosion resistance (it is unaffected by both acids and alkalis at room temperature), its application in diverse fields is not surprising,[1] and its use in medicine and dentistry has been assured for some time. It is used in various forms: commercially-pure metal (§1.1), β-phase alloys (§1.4), and nickel-titanium (§2). It is a relatively expensive metal in part because of the need for special processing conditions using vacuum or an inert atmosphere; this arises from its extremely high reactivity with oxygen (a fracture surface ignites spontaneously in oxygen at 25 bar, and the molten metal just in air). However, it is this reactivity that leads, in the presence of atmospheric oxygen, to the immediate formation on the metal at ordinary temperatures of an oxide layer (TiO_2) that is both adherent and stable, even in quite acidic conditions. This **passivation** (13§5) therefore inhibits further corrosion.

●1.1 Commercially-pure titanium

Titanium itself has two **allotropic** forms designated α and β. The α-phase has the h.c.p. structure (11§3.4) and is stable up to the **transus** temperature of 882 °C, at which it converts to the b.c.c. β-phase. This process is similar to the transformations occurring in iron (21§1). The α–phase is ductile, as is to be expected from the structure, but this is adversely affected markedly by dissolved (interstitial) oxygen (Fig. 1.1), and so the choice of grade of commercially-pure (CP) titanium used for a given application can be important.

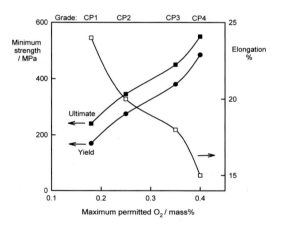

Fig. 1.1 Effect of dissolved oxygen on the mechanical properties of commercially-pure titanium. (For elongation, see 1§3.3.)

The effect of oxygen is so marked that it is commonly said that CP-Ti is better considered as an α-phase oxygen alloy. Carbon and nitrogen are also soluble interstitially and have similar effects on mechanical properties, such that one can write the oxygen equivalent thus:

$$[O] = \% O + 2(\% N) + 0.67(\% C) \qquad (1.1)$$

where each 0.1 % O equivalent increases the strength by 100 ~ 120 MPa, but with a decrease in toughness (*i.e.* increase in brittleness).

Many metallic elements are soluble to a greater or lesser extent in titanium, giving many alloying possibilities, although few of these have yet found applications in dentistry. Many of these additions have an effect on the transus temperature and accordingly are commonly classified as follows:
- α-stabilizers (*e.g.* Al, O, N, C)
- β-stabilizers, in turn divided into two groups:
    - isomorphous (*e.g.* V, Mo, Nb, Ta, Re)
    - eutectoid (*e.g.* Fe, Mn, Co, Cr, Ni, Si, H, Cu, Ag, Au)
- neutral (*e.g.* Zr, Hf, Sn)

An α-stabilizer raises the transus temperature, a β-stabilizer does the reverse, allowing the β phase to be stable at low temperature, but it may destabilize the solid solution such that it undergoes a eutectoid reaction (12§3.6). A neutral solute in principle has no effect on the transus, although in practice this is a matter of degree rather than an absolute statement, there being only relatively slight changes. These effects are shown schematically in Fig. 1.2.

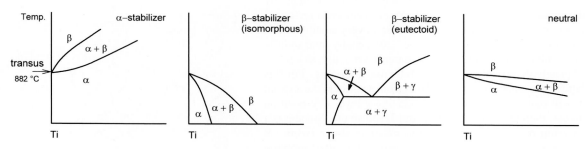

Fig. 1.2 Types of effect of alloying elements on the phase diagram layout.

The significance of the existence of the α-β transus lies in the behaviour of the cast metal: the initially-solidified metal is necessarily β-phase, because this is the stable phase at high temperature, and this must therefore transform on cooling. This transformation occurs in two stages, by a mechanism which is already familiar: nucleation and diffusional growth. Thus, at a temperature depending on the alloy and the rate of cooling, α-phase first nucleates at β-grain boundaries, producing a layer over all β-grains, but then further nucleation occurs allowing plate-like crystals of α to grow into the β-grains. Nucleation completely within β-grains can also occur, especially if the cooling rate is increased. This produces a **Widmanstätten** structure (Fig. 1.3)[1][2] (*cf.* Fig. 12§3.10), and eventually would, in pure Ti, be expected to result in a full conversion to α-phase. However, the existence of iron as a common contaminant, although of strictly controlled amounts in CP-Ti, as well as other elements, means that a small proportion of β-phase

Fig. 1.3 Widmanstätten structure in CP-Ti, in which β has been transformed to α+β, dominated by the growth of α-Ti in the remnant matrix of β-phase.

normally remains (some 2 ~ 5 vol% is typical), stabilized by the iron, even after annealing. This effect is enhanced by the partitioning of solute elements: β-stabilizing elements migrate into the remaining β-phase, while α-stabilizers become enriched in the α-phase, reinforcing the stability of the α+β two-phase system. In addition, the α-phase grain size is limited during the recrystallization of annealing by the presence of these remnants of the β-matrix. All this means that cast "pure" Ti is nothing like as ductile as otherwise would be expected: the fine grain, incoherent two-phase structure provides appreciable resistance to slip (*cf.* grain refining, 19§1.10). It also means that what is commonly described as "α" material is in fact more accurately interpreted as two-phase. Thus, although the alloying elements described above may not be used deliberately in 'titanium' as such in dentistry, the presence of any of them as impurities affects the behaviour of "pure" titanium. Clearly, it is of great importance to avoid contamination in casting titanium, much more so than perhaps with gold alloys, for example.

●1.2   Twinning

It will be recalled from 11§3.5 that the pattern of stacking of close-packed planes in the face-centred cubic (f.c.c.) structure may be characterized as A,B,C,A,B,C ... , referring to the relative position of atoms in successive planes with respect to those in an (arbitrary) reference layer. As can be seen from Fig. 11§3.20, it can be visualized as a regular stepping displacement of each layer in a particular direction as a result of this. But, as is apparent from Fig. 11§3.17, the h.c.p. structure is obtained when this displacement occurs in alternating directions 60° apart for successive layers. Now consider just one alternation of this kind in the f.c.c. sequence, giving for example: A,B,C,A,B,**C**,B,A,C,B,A,C ... . From the underlined "C" onward it is still the f.c.c. pattern, but in a different directional sense and so the letter sequence is inverted (Fig. 1.4). Notice that these ABC labelled atoms (Fig. 11§3.20) constitute the close-packed diagonal of a unit cell face. Thus, if the sequence undergoes such a switch the diagonal direction must thereby be changed, and so the crystallographic axes themselves have a different orientation. The structure is crystallographically identical, and the boundary is perfectly coherent, but it is as if it has been reflected across a mirror plane. Such a crystal is said to be **twinned**.

---

[1]   Image courtesy of M. Brezner and T. Okabe, Baylor College of Dentistry, Dallas, TX.

Twinning can arise in two ways: firstly, accidentally during solidification or recrystallization, because the energy of this most perfect of grain boundaries is so very small that this kind of stacking error has little energetic penalty (and so little thermodynamic driving force to change); secondly, as a result of plastic deformation – giving **deformation twins**. For this latter case, it may be envisaged that from a normal f.c.c. crystal a twin may be generated by displacing a layer one position across, then the second, and the third, and so on. Since it is a shearing process it is also said to generate **shear twins**. It occurs in some metals and alloys as a normal result of applying tensile or compressive stresses – recall that slip occurs along close-packed planes even though these may be at some angle to the principal stress direction. It is worth re-emphasising that, despite the simplistic description given here, such slip is

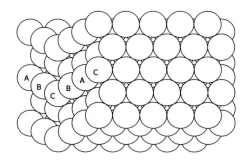

Fig. 1.4 Twinning in an f.c.c. structure by reversing the stacking sequence.

not homogeneous but must occur atom by atom (11§5, 6). Nevertheless, the key point in such a process is that it is **non-diffusive**: every atom retains precisely the same set of nearest neighbours and they are said to move "in concert" (at least, as viewed after the event). In addition, although such rearrangements are driven by changes in lattice dimensions or orientations, because the crystal structure type is unchanged, there is no change in volume except for the accumulation of voids at grain boundaries as the deformation in adjacent grains cannot match exactly (Fig. 11§6.8).

Twinning can occur in several crystallographic systems, including types in which a crystallographic lattice rotation occurs as opposed to a reflection, although the f.c.c. example is perhaps the simplest. Thus, pure α-Ti – h.c.p. – readily shows deformation twinning. However, this mode of response to applied stress is suppressed by the presence of solute atoms such as aluminium and oxygen (α-stabilizers), and so only occurs in low-oxygen CP-Ti. Even so, the ductility of very pure α-Ti (if it were obtained) would be due to this response in addition to the more usual slip by the movement of dislocations (11§6). Single-phase β-Ti alloys (b.c.c.) also exhibit deformation twinning, a behaviour which in fact enhances the ductility of what is ordinarily understood to be a relatively brittle crystal structure. In addition, interstitial oxygen does not cause the great lattice distortion that it does in b.c.c. iron and steel (metallic radius is much larger – Ti: 147 pm, Fe: 126), so slip remains relatively much easier even in the presence of this common contaminant.

Thus we have the curious situation that in practice α-Ti is expected to be ductile but is not because of an actual two-phase structure, grain boundary effects and distortion by interstitial contaminants; on the other hand, β-Ti is expected to be relatively brittle but is instead ductile because of twinning and lack of effect of interstitials.

●1.3   Casting

As indicated above, titanium is extremely reactive and great precautions must be taken in the process of casting the metal. In part this is due to the high m.p., 1670 °C, substantially above that of Co-Cr alloys, and special equipment is therefore necessary. In addition, this must provide an inert atmosphere, such as with argon or vacuum. Even then, care must be taken to exclude oxygen after casting because above ~550 °C oxygen diffusion through any oxide coating and into the metal is very fast, leading to severe embrittlement as well as an excessively thick oxide coating. The embrittlement arises because the solid solubility of oxygen in Ti is very high (~14.5 mass% = 33 at%) and forms an **oxygen-diffusion zone** beneath the actual oxide layer. This "α-case" forms even on β-alloys, and may be a source of cracking in tension even if the underlying metal is ductile. The oxide commonly appears dark blue or other colours, partly through interference effects, but also because it is anion-deficient, creating colour centres (*cf.* ZnO, 9§3.1). This anion-deficiency is the source of the high diffusion rate of oxygen through the oxide – there are vacancies at the oxygen sites.

Cooling of the casting well below that critical temperature of ~550 °C must therefore be allowed to occur in the same inert atmosphere to avoid these effects. A further reaction problem arises with the investment if this contains silica, as indicated earlier (17§4): titanium can reduce silica to silicon.

The low density of the pure metal (4.5 g/mL) also causes difficulties. Centrifugal casting machines (18§3.2) are impractical because of the controlled-atmosphere requirements and gas-pressure (18§3.5) is commonly used in Ti-casting machines. Even so, turbulence porosity is often a major problem that requires careful control of spruing and casting conditions, especially in thicker sections. A further difficulty arises from the effects of other elements on the mechanical properties, as described above. Accordingly, casting with titanium must be done under scrupulously clean conditions to avoid contamination by other metals.

●1.4   β-Titanium

Elements that stabilize the β-structure have a curious effect: they can reduce the modulus of elasticity markedly when present in modest concentrations, to perhaps to 60 or 70 % of the value for α-Ti at around room temperature (that is, from about 105 GPa). This is curious because ordinarily modulus of elasticity is little affected by anything other than temperature (assuming no constitutional changes) because it is a measure of interatomic forces – bond strength – and in chemical terms the transition elements involved are not that dissimilar. Thus it seems that β-stabilizing elements disturb that bonding. This effect is seen in such alloys as Ti-10Mo (meaning 10 mass% molybdenum), where there is enough molybdenum to stabilize the β-structure at around room temperature, and likewise for alloys near Ti-15V-3Cr and a number of others. Such bond-weakening might also represent a further contribution to the ductility of this type of alloy, although heat-treatment after cold-work is likely to result in detrimental metallographic changes.

Some β-phase alloys are not fully stable at low (normal working) temperatures and can tend to produce the so-called ω-phase, a very fine grain (2 ~ 4 nm) precipitate, yet again by a diffusionless shear-transformation, as a hexagonal or distorted b.c.c. structure which has a diffusely-coherent grain boundary with the β. This is not itself an equilibrium phase, but forms on nucleation in the transition from α to β. Such β-alloys are said to be **metastable**, and the presence of ω-phase causes severe embrittlement. Thus, heating of such alloys may be detrimental to their structure and thus their properties. Conditions of use must be carefully observed.

In dentistry, a particular titanium-molybdenum alloy, Ti-11Mo-7Zr-4Sn, with the trade name **TMA**, is used for orthodontic wires. This has the β-Ti structure, principally stabilized by the molybdenum content, but the presence of the zirconium and tin (otherwise nearly neutral with respect to the transus temperature, §1.1, but see §1.6) inhibits the formation of the brittle ω-phase. It is therefore weldable without detriment. The lack of Ni is an advantage for intraoral use.

●1.5   Corrosion

Although titanium and its alloys are corrosion-resistant, as with all passive alloys they are susceptible to crevice corrosion attack and all the normal remarks apply (13§5). However, there is an additional factor for CP-Ti and α-phase alloys, especially with aluminium, and that is **hydrogen embrittlement**. This generally arises because the cathode process

$$2H^+ + 2e^- \rightleftharpoons H_2(g) \tag{1.2}$$

(13§3) can occur readily enough to cause a problem: hydrogen is sufficiently soluble in α-Ti that the mechanical properties can be affected, in particular a loss of ductility, but it can be driven to supersaturation when the brittle phase TiH_2 may form by precipitation. This precipitation is aided by the presence of tensile stress (reducing the constraint on the hydride's crystal growth), and so stress-concentration by notches becomes a factor of relevance in focusing the embrittlement on just the region when crack growth needs to be suppressed by crack-tip blunting, *i.e.* where ductility would provide toughness.

The type of effect known as **internal hydrogen embrittlement** only occurs when hydrogen is present over and dissolves in the molten metal, which therefore becomes supersaturated with hydrogen on solidification. This is not normally likely to occur in ordinary dental laboratory contexts but might be a problem at manufacture. **Environmental hydrogen embrittlement**, on the other hand, results from hydrogen absorption by solid metal. In dentistry, this might occur during electrolytic processes or, more likely, from corrosion reactions. This seems to be the case in the presence of fluoride, and can result in changes in the properties of α-phase orthodontic wires. Since these are meant to be stressed in use, delayed fracture (**static fatigue**) may occur in service.[3] While β-phase wires do absorb hydrogen, this stays in interstitial solution and may have only relatively minor mechanical effects.[4] Similarly, etching of castings, for example with strong sulphuric acid, whether for obtaining a mechanical key for cementation or to improve the biocompatibility of the surface of an implant, may generate much TiH_2.

•1.6   Zirconium & Hafnium

Both zirconium and hafnium are isomorphous with titanium, and show the same α-β transition, the temperature of which is only slightly affected in their alloys (hence their description as 'neutral'; §1.1, Fig. 1.2) (Fig. 1.5). The metallic radii are Zr: 160 pm and Hf: 159 pm, giving radius ratios with Ti (147 pm) of 1.09 and 1.08 respectively. By the Hume-Rothery rules (Table 11§3.2) the complete solid-solubility that is observed is expected. In addition, both Zr and Hf are spontaneously passive (13§5), and the ability to form the coherent oxide coat, as has Ti, is present in the alloys. The radius ratios are still sufficient to generate appreciable solid solution hardening, reaching a maximum at proportions around 50-50,[5] also as expected, without employing alloying elements that pose a biological risk, such as Al, Ni, and Cr. The only significant problem is the high price of both metals, that of Hf being much greater than that of Zr. Commercially-pure grades are used for surgical implants.[6] Zr can also be half-substituted in Ti-alloys (such as Ti-6Al-4V) for improved hardness and strength.

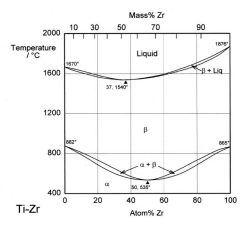

Fig. 1.5 Phase diagram for Ti-Zr.

§2. Nickel-Titanium

Nickel-titanium alloys have two main areas of application in dentistry, but with a third distinct possibility. These areas are in orthodontic appliances, endodontic instruments, and surgical correction of bony anatomy. Each depends on properties that are commonly only encountered with this group of alloys (although others exist outside dentistry), but to understand these it is first necessary to explore some of the underlying structure.

•2.1   Transformations

Nickel and titanium form a series of intermetallic compounds: Ni₃Ti, NiTi and NiTi₂. It is NiTi that is of relevance here, the equimolar alloy, or nearly so. But far from being a sharply defined 'compound' as are both Ni_3Ti and $NiTi_2$, it can be seen from the phase diagram (Fig. 2.1) that the compositional limits for NiTi can be quite broad above 630°C in the **eutectoid** phase field. Notice that below this temperature there is a two-phase field that stretches from about 25 to 66 % Ti, Ni_3Ti + $NiTi_2$. However, cooling NiTi itself below that temperature does not result in that phase separation, but the preservation of NiTi as a metastable phase.[7] Cold working aids the reaction.

Also known as Ni-Ti ("nye-tye") or "Nitinol" (from the US Naval Ordnance Laboratory where it was first investigated), nickel-titanium alloys approximating the equimolar composition NiTi are remarkable

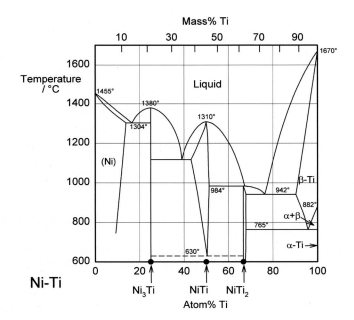

Fig. 2.1 Phase diagram for the system Ni-Ti.

for exhibiting a range of unusual properties, in particular the shape-memory effect and pseudoelasticity. These arise from certain reversible crystallographic changes.

NiTi at relatively high temperatures has an ordered b.c.c. structure which is known in this system as **austenite**. On cooling, one observes the expected thermal contraction, more or less linearly, until at a particular temperature the rate of change of dimension with temperature increases. This is caused by a spontaneous but progressive (*i.e.* with respect to temperature, not time) shear transformation to a **monoclinic** structure called **martensite**.[2] Such a change is therefore called a **martensitic transformation**, simply because crystallographic shear is involved. Hence, the temperature at which this process commences *on cooling* is called the **martensitic start temperature**, M_s (Fig. 2.2). At some lower temperature, this process is completed: this is the **martensitic finish temperature**, M_f, where the proportion of martensite is now 100 %. Below M_f, ordinary cooling shrinkage occurs as before. Notice that in this process, all neighbours are preserved (no diffusion), and therefore the monoclinic structure is also ordered.

If now this martensitic NiTi is reheated, the reverse process occurs, eventually completely reforming the original austenite. That is to say, the shear transformation is progressively reversed so that all atoms return to their original positions in their original crystallographic arrangement. However, the **austenitic start temperature**, A_S, is somewhat higher (typically by about 20 K) than M_f, and likewise the **austenitic finish temperature**, A_f, is somewhat higher than M_S. These transformations therefore show **hysteresis**, which occurs over the **transformation temperature range** ("TTR"). This whole process is more or less indefinitely reversible. Such a thermally-controlled crystallographic change is a normal consequence of the thermodynamic imperative of adopting a lower energy state, there being a simple path to allow it to occur and no limitation arising from diffusion. What is peculiar is the existence of the hysteresis; it is a **second-order** transition, unlike the **first-order** transition of ferrite to austenite and *vice versa* for iron (Fig. 21§1.1), for example. Thus, the change is not microscopically-reversible, as would be expected for a normal equilibrium process.

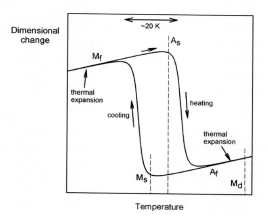

Fig. 2.2 Thermal transformations in NiTi.

A_s	Austenite start	(on heating)
A_f	Austenite finish	(on heating)
M_s	Martensite start	(on cooling)
M_f	Martensite finish	(on cooling)
M_d	Martensite deformation start	(on cooling)
	(relative position variable, ~50 K > A_f)	

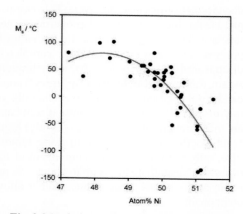

Fig. 2.3 Variation in the value of the martensitic start temperature, M_s, with composition for NiTi alloys. Note the scatter in reported values, as thermomechanical history and contamination would have some effect.

The progressive nature of the transformation may in part be explained like this. It is clear that if the composition departs from exactly 50:50 Ni:Ti in atomic proportions, it is not possible to maintain complete ordering over a whole grain (assuming that a second phase does not appear). However, this disorder will itself be randomly distributed. The energy conditions required for the transformation therefore vary slightly from place to place according to statistical fluctuations in the degree of disorder, and hence the temperature at which it occurs will also vary over some range. Further, even at exactly NiTi, stacking faults may occur in which Ni and Ti occupy the wrong sites, a further source of slight disorder affecting the strain energy of the lattice. The effect that this has can be seen by the variation of the value of M_s with composition (Fig. 2.3). In addition, remembering that grains will have randomly-oriented crystallographic axes, the transformation will be slightly constrained by the stress generated as neighbouring regions and grains cannot exactly accommodate the strain. The driving-force needed therefore continues to increase slightly to overcome this resistance.

An extra complication to this behaviour is that some NiTi alloys also show a third phase, the so-called R-phase, after its rhombohedral crystal structure (Table 11§3.1). Generally, this only appears on cooling as an

---
[2] Note that this usage does not relate to the terminology for steels.

intermediate stage before the full martensitic transformation. Again, however, the formation of this involves only shear and so does not interfere with, but rather becomes part of the continuous process of transformation. We may ignore such details in favour of the larger picture.

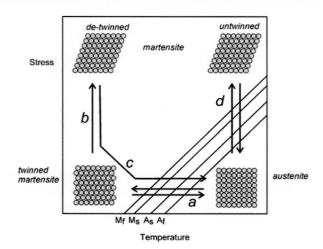

Austenitic NiTi shows quite ordinary mechanical behaviour at temperatures well above A_f and has no special attributes; it is quite hard, and resembles α-Ti (§1.1). However, the martensite that is reversibly formed thermally, *i.e.* only on cooling, is heavily twinned (§1.2) in a special way: it may be imagined as similar to a pleated structure or a "concertina-fold" (Fig. 2.4, process *a*). It may be seen that the formation of this type of structure leads to a small change in volume (Fig. 2.2), but not of shape (we assume here that this is in a polycrystalline piece of metal, where the net strain is isotropic; *cf.* Fig. 1§2.12) as there is no systematic displacement from layer to layer. This twinned martensite does have special properties: when this is mechanically deformed, the twinning is progressively lost until it has completely

Fig. 2.4 Thermomechanical transformations in NiTi (schematic).

a	No stress, thermal	(reversible)
b	Deformation, low temperature	(not reversible)
c	Re-form austenite on heating	(shape memory)
d	Pseudoelasticity	(above A_f)

Note that the crystallographic structures of de-twinned and untwinned martensite are indistinguishable, only the path taken is different.

disappeared (Fig. 2.4, process *b*), again through a plain shearing mechanism, with all near neighbours preserved. Again, this is a **regimented** (or **'military'**) non-diffusive crystallographic change. In this way, very large strains can be obtained but at nearly constant, only very slightly increasing load. Furthermore, this strain too is not microscopically-reversible because it is not an *elastic* strain: atoms have moved from one energy minimum to another through, of necessity, a (slightly) higher energy state. This is quite unlike the elasticity arising from bond-length or bond-angle strain alone; removing the stress does not result in the strain returning promptly to zero, only the purely elastic component of the deformation is recovered. The nearly-atomic scale of shifts occurring mean that strains up to about 8 % may be obtained. This should be compared with the 0.1 ~ 0.2 % strain at the elastic limit of many other alloys, which is therefore about the magnitude of the pure elastic component here.

●2.2   Shape-memory

The deformed, **de-twinned** martensite is thus stable in the absence of any applied stress if kept at low temperature, relative that is to A_S. However, if it is now heated to above A_S it becomes unstable and spontaneously reverts to austenite (Fig. 2.4, process *c*). Yet again, because this is purely a shear transformation, all neighbours are still preserved. Consequently, the original dimensions of the piece are regained because all of the imposed strain is recovered. Under this regime, therefore, NiTi demonstrates **shape-memory**, and is therefore called a **shape memory alloy** ("SMA"), a property it shares with a variety of other alloys that are not used in dentistry. It is clear that work has been done in creating the deformed martensite with respect to the original austenite, and this work is stored in the structure. Hence, when heated, this work is recoverable and can be used to exert forces and thus in turn do work on an external system. It is similar to the work done by the energy stored (purely elastically) in an ordinary spring, such as in driving a clockwork mechanism. The difference here is that the release in NiTi is triggered by the imposed rise in temperature. Obviously, the deformation can be of any type: in tension, bending, or torsion.

The low-temperature martensitic phase is very plastic. A piece can be deformed in a manner rather similar to that of pure tin: it can repeatedly be bent, backwards and forwards, without strain hardening. This enables any necessary shaping to be done with little risk of fracture, unlike metals such as stainless steel and dental gold alloys. The reason for this is that the activation energy for the movement of the twin boundaries is low, and – again unlike other alloys – without the formation or involvement of dislocations, which through pile-up would otherwise lead to the formation of microscopic defects that can initiate fracture (11§6.3). The remarkable bonus is that the original shape can then be recovered just by mild heating.

For obvious reasons, the above shape-memory behaviour is called "**one-way**": a mechanical deformation must be imposed first, and this is undone on heating above A_S. The initial thermally-induced transition is, overall, dimensionally isotropic because the b.c.c. austenite crystal axis that becomes the c-axis in the monoclinic martensite is chosen randomly from grain to grain (*cf.* the formation of the ordered AuCu domains, 19§1.4). However, it is possible through elaborate combinations of thermal and mechanical treatments to make the martensitic transformation on cooling anisotropic, that is, obtain a selective orientation of the martensite c-axis. Thus, cooling through that transformation also causes a dimensional change without loading, and this is recoverable on reheating. This "**two-way**" shape-memory can therefore do useful work in either direction. Note that the thermal hysteresis is still present, although it may be reduced in extent. The details of such **shape training** are, however, beyond the present scope.

●2.3 Pseudoelasticity

There is a third special behaviour. Austenitic NiTi is said to be mechanically unstable over a narrow range of temperatures. That is, in this temperature range the martensitic transformation can be induced purely mechanically, by the imposition of a stress. This temperature range is marked above by what is termed the **martensite deformation start temperature**, M_d (typically up to about 50 K above A_f), and below by M_f, where there is no austenite remaining. For the moment, however, we shall restrict the discussion to temperatures above A_f, that is, where no martensite is present. Thus, for a piece of austenitic NiTi in that temperature range, the application of a load first causes a perfectly ordinary elastic deformation with an unremarkable modulus of elasticity (Fig. 2.5). However, at some point the **stress-induced shear transformation** commences and, if the load is increased slightly, more austenite transforms to untwinned martensite, and so on until all has been converted. The transformation is again progressive and second-order. Very large strains, again of the order of 8 %, can thereby be obtained with only very small increases in stress. This

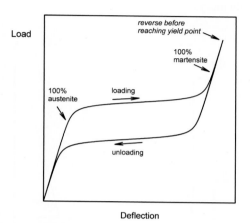

Fig. 2.5 Pseudoelasticity of NiTi for $M_d > T > A_f$.

is the phenomenon of **superelasticity**, also known as **pseudoelasticity**, which is perhaps the better, more accurately descriptive label. After the completion of the conversion the elasticity again becomes normal – the metal is stiff and elastic – until an ordinary yield point is reached. Remarkably, if the load is reduced before that (true) yield has occurred, the mechanical deformation is reversible, both the ordinary elasticity (of course) and the **pseudoelastic** deformation due to the martensitic transformation. This is represented in Fig. 2.4 by process **d**. It necessarily involves the direct formation of the untwinned martensite because of the imposed strain, and the structure is indistinguishable from that formed by stress at low temperature (except for ordinary thermal expansion). Lowering the temperature while under stress has no effect on structure. Again, there is hysteresis: the unloading curve sees the start and finish of the return to austenite at lower stresses than caused the martensite to form (Fig. 2.5); the loading and unloading curves in this region do not coincide.

As might be imagined, there is some temperature- and composition-sensitivity to the value of the stress that induces the transformation. In fact, M_d represents the highest temperature at which this may be achieved and it also requires the highest stress to initiate. This arises because above that point the critical transformation stress becomes inaccessible, being higher than the yield point for ordinary, irreversible plastic deformation of the austenite (that is, by the movement of dislocations), which is therefore the path taken, according to the principle of least work. At lower temperatures, the martensite becomes energetically progressively more favourable and the activation energy for the transformation, as represented by the stored (true) elastic energy, lower.

Although the stress to induce the transformation is temperature-dependent, the speed of the transformation is dependent on neither temperature nor strain rate. It is indeed a fast transformation, occurring at about half the speed of sound in the metal (that is, at about 2500 m/s), in other words, at half the speed of transmission of the elastic wave. Thus, for small pieces, the progressive transformation occurs essentially instantaneously. Similarly, the temperature-induced transformations are equally fast, and do not require any more time to complete. All of this arises from the non-diffusive nature of the crystallographic changes.

Now, if instead of starting at T > A$_f$, the loading occurs for A$_f$ > T > A$_s$, the material is, of course, 100 % austenite to start with, and can stress-transform fully to martensite. The loading curve lies below that for T > A$_f$ as the transformation is easier at lower temperature. However, on unloading, the reconversion to austenite cannot occur fully because T < A$_f$, and a proportion must therefore remain as martensite, as would be expected in heating to this temperature range under no load from below M$_f$ (Fig. 2.6). This permanent deformation can, of course, be undone completely by heating above A$_f$. Thus the extent of the pseudoelasticity is diminished by this now partial shape-memory.

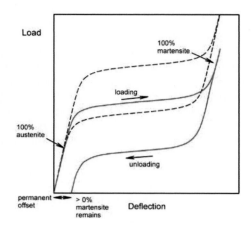

Fig. 2.6 Pseudoelasticity of NiTi, A$_f$ > T > A$_s$. The curves for M$_d$ > T > A$_f$ are shown broken.

Lowering the temperature to A$_s$ > T > M$_s$, full conversion to untwinned martensite still occurs as expected on loading, although again at a lower load than before, only now no recovery other than due to the elasticity of the martensite is now observed (Fig. 2.7), simply because T < A$_s$, where the shape-memory effect does not operate, and the permanent deformation is very large.

If the temperature is further lowered to the range M$_s$ > T > M$_f$ before loading commences, there now is a proportion of twinned martensite present, the amount depending on the temperature. Thus, on loading, the first process is that this twinned martensite is detwinned, at quite a low load. Subsequent to this, the remaining austenite is progressively transformed to untwinned martensite, as before (Fig. 2.8). Again, this process occurs at loads lower than for A$_s$ > T > M$_s$. But since the temperature is lower than A$_s$, no reverse transformation occurs on unloading and only the (Hookean) elastic recovery of the martensite occurs. The permanent deformation is again very large.

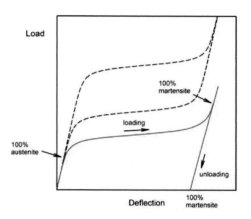

Fig. 2.7 Deformation of NiTi, A$_s$ > T > M$_s$. The curves for M$_d$ > T > A$_f$ are shown broken.

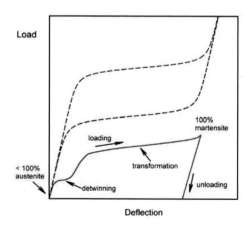

Fig. 2.8 Deformation of NiTi, M$_s$ > T > M$_f$. The curves for M$_d$ > T > A$_f$ are shown broken.

Lower still, for M$_f$ > T, the only permanent deformation available is from detwinning, and this at low load (Fig. 2.9). This is now in the full shape-memory domain, and heating above A$_f$ will recover the original shape, providing the yield point of the untwinned martensite is not exceeded. It can be seen that the curve of Fig. 2.8 has those of Fig. 2.7 and 2.9 as limiting cases, the relative size of the detwinning

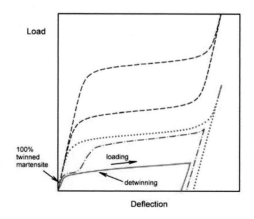

Fig. 2.9 Pseudoelasticity of NiTi, M$_f$ > T. The curves for M$_d$ > T > A$_f$, A$_s$ > T > M$_s$ and M$_s$ > T > M$_f$ are also shown for comparison.

step depending on the proportion of thermally-transformed, twinned martensite originally present.

It can be seen, therefore, that the mechanical behaviour is highly dependent on the exact temperature of the piece in relation to its particular values of M_f, M_s, A_s, A_f and M_d. Further complexity would arise according to the loading pattern and temperature sequence experienced. That is, the previous thermomechanical history is of great importance. Indeed, even if expected in certain regions, as implied here, complete reversibility may not be achieved indefinitely, some fatigue damage in particular would accumulate (although the fatigue life tends to be longer than usual). Indeed, the complexity of metallographic structure that may be obtained by combinations of alloying, heat-treatment and cold work is very great, and the scope of titanium and its alloys in general in medicine and industry, to say nothing of dentistry, is both very great already and increasing.

●2.4   Corrosion

Environmental hydrogen embrittlement seems to affect Ni-Ti as it does other Ti alloys (§1.5),[8] resulting in the formation of a hydride (NiTiH) in pseudoelastic alloys at least. Care would have to be taken to avoid electrochemical action (contacting dissimilar metals) and aggressive environments (fluoride) if the long-term action of shape-memory and similar devices were not to be compromised by hardening, embrittlement and fracture. Etching would also seem to be problematic, therefore.

●2.5   Applications

The central problem of orthodontics is the application of a force to a tooth in order to change its position with respect to the rest of the dentition or jaw. Biologically, this force must lie within certain limits: the force must be great enough that tissue remodelling occurs, yet not so great that damage ensues. Strictly, of course, it is the local stress that is the figure of relevance, and this is dependent on a variety of factors, not least tooth size and geometry. The magnitude of the force to be applied therefore is subject to a clinical assessment, while how it is to be applied depends on mechanics and physics. Ideally, once the mechanics has been worked out (see Chap. 23), covering questions such as direction, rotation and anchorage, the force should be constant until the desired movements have been attained.

However, given the essentially linear ramp of force *vs.* deflection for cantilever beams, three- and four-point bending (23§4), more elaborate designs (23§5), as well as for coils and braids (23§6), all of which are expressions of Hookean behaviour (1§2), it is apparent that there is no simple way of creating that constant force through a mechanical device. Thus, orthodontics has been greatly occupied with identifying that combination of alloy, wire cross-section, and spring design that will best achieve an approximation to this requirement. These enquiries in respect of the alloy translate primarily to finding a convenient modulus of elasticity, E. This is insufficient on its own because permanent deformation must be avoided – there is no value in exceeding the yield point when **activating** a spring, that is, deflecting it to its initial working condition (which usually involves some over-deflection to get it into position). It is clear, however, that when the effect of second moment of inertia, I_z (23§3), and spring design are taken into account (*e.g.* 23§5), even allowing for the physical limitations of the space within which the device must function, the system has several degrees of freedom: there is no unique solution. Other factors may therefore come into play, such as the capability of being welded or soldered, ease of working into the desired shape (ductility *vs.* work hardening), corrosion resistance, hardenability by heat treatment, and so on.

Under these circumstances, it clearly is not possible to formalize selection or design rules. Nevertheless, one is limited to working within the limits illustrated by Fig. 2.10. The force applied declines linearly as the tooth moves and the deflection of the spring from its equilibrium (zero force, zero strain) position decreases, but this is only usable down to the point at which the force becomes ineffective. Indeed, the effective range of deflection can be rather small in comparison with the full elastic range. The implications of this are small movements for each setting of the spring, and thus necessarily frequent adjustments or replacements of the device, the time and cost of which are significant to the patient.

It is against this background that the possibilities of pseudoelastic alloys must be viewed. Although the system remains multifactorial, as above, but with the additional variability of M_d, it now becomes possible to have a force application curve as in

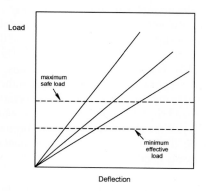

Fig. 2.10 Orthodontic force generation with a pseudoelastic spring. (Note: this must be the unloading curve, *cf.* Fig. 2.5.)

Fig. 2.11, with an extended, near-constant plateau. This is commonly understood to mean fewer adjustments and therefore longer intervals between appointments.

Since the wires in orthodontic use are employed in a bending mode (even in coils and other forms), the situation is slightly more complicated by the fact that part is necessarily in compression (23§2.6), rather than simple tension. This has the effect of reducing the working range of deflection somewhat, but it still offers a greater range than pure elastic devices.

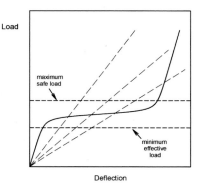

Fig. 2.11   Orthodontic force generation with Hookean springs.

§3. Metal Forming

Denture bases made from PMMA are limited by the low strength of the material, especially as it affects the palate where fractures are especially common unless it is made particularly thick. Metallic denture bases are attractive for their higher strength, fatigue resistance, thinness and thermal conductivity, but casting large thin sheets is technically demanding. This can be achieved for cobalt-chromium alloys, but it is more suited to partial dentures where clasps can be cast in the one piece. For full dentures, accordingly, **swaged** sheet metal offers an alternative approach, and stainless steel (which is not suited for casting) is the alloy of choice. Even so, a variety of other metals have been explored: Au alloys, Ag, Ag-Pd, Al, Ir-Pt, Pt, none of which appear now to be in use. In **swaging**, a thin sheet of the alloy is formed over the model. Whilst shaping metal sheet by hand by hammering over dies is commonplace for a variety of applications (car body panels, silver hollow ware) there are problems with this approach in dentistry. Generally, this is related to the precision of the work required, the difficulty of conforming the metal to steep contours, work-hardening, springback (equation 1§2.4 d), and avoiding a lengthy grinding and polishing process to obtain the required smooth surface.

●3.1   Press forming

The industrial technique of forming between metal dies using a press can be used, but the difficulty and expense of making the metal die and counter-die by meticulous machining preclude this for the one-off denture base, even though the dies would be hard and accurate. The dies have been made by casting phosphor bronze, Cu with 5~10 mass% Sn, but this is a foundry process as the casting temperature is over 1100 °C (see Fig. 12§5.1). Instead of this, dentistry has used a variety of **low-fusing** (low melting point) alloys. These can be cast into refractory moulds made from impressions in the usual way to create the positive die, the counter-die being made by casting against that die.

The primary consideration here is that the working temperature is low enough to be convenient, with alloys that are cheap. An added advantage is that the overall thermal contraction of the solidified metal is thereby reduced (Fig. 18§4.1), and thus the dimensional inaccuracy. Similar factors apply as for ordinary casting (Chap. 18). However, because there are extra steps involved, detail is apt to be lost. A bigger problem, however, is the strength of the alloy. By selecting low-melting mixtures, the **reduced** or **homologous temperature**, T/T_m, under normal working conditions, say 25 °C, necessarily has a large value, typically 0.6 ~ 0.8. This means that the strength is necessarily low, and deformation of the die and counter-die during pressing is likely. This is offset slightly by the fact that many of these alloys are multiphase, and with much solid solution hardening, but it cannot avoid damage altogether. There are many such alloys, primarily various combinations of Bi, Pb, Sb and Sn, but with Cd, In or Zn additions. Even those alloys with names (such as Wood, Babbitt, Molot and Rose) have reported composition ranges rather than precise formulations, and detailed tabulation has no great merit. One feature which seems to be general is associated with the inclusion of Sb: this causes the freezing metal to expand slightly (unlike most alloys, which show a marked freezing shrinkage). This assists in attaining a sharp copy of the mould, and hence its former use in "type metal", that is, for creating the letters of movable printing type. Toxicity concerns will be evident in the use of Pb and Cd in open systems where "fume" – oxidized vapour – needs good control.

Zinc has a much higher melting point (~419 °C) than those low-fusing alloys, and so has much more strength. Going further, alloys of Al-Cu-Mg, with yet higher melting temperatures are yet harder. However, the increasing complexity of the melting and handling indicates further compromise in addition to the cooling shrinkage issue as well as the logistics of melting handling large quantities of high-temperature metal. Normal dental casting is on a much smaller scale.

A further difficulty is that by forming the counter-die directly on the die (which is cold, and therefore already potentially undersize), there is necessarily no provision for the thickness of the metal sheet that will be pressed. Such relief is essential if tearing of the sheet – or destruction of the moulding surface – is to be avoided. This may have been the reason for using, in one technique, a pure lead counter-die, which was therefore soft, against a relatively hard zinc die.

In addition, there is no provision for the edges of the blank sheet to be clamped, and this leads to wrinkling before the press has closed fully. The edges of the sheet are drawn inward as the centre is pressed down, and this requires the periphery to occupy a smaller circle (consider a circle of cloth placed over a cup, then depress the centre). The only path for relaxing the resulting compressive hoop stress is for it to buckle (Fig. 3.1). These undulations would get flattened eventually, but if these folds were to appear in other than the area to be trimmed away they would spoil the denture base. Even with clamping, some wrinkling remains possible, depending on the exact properties of the steel. Work hardening of the steel is appreciable, and the process involved two or three successive stages of pressing, annealing between each. As such annealing needs to occur at elevated temperatures (1000~1100 °C), oxidation would be a serious problem, and a protective environment is essential. Even so, de-scaling and polishing are necessary.

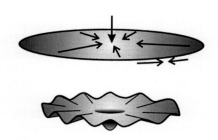

Fig. 3.1 Edge-buckling occurs readily if the periphery is not clamped.

●3.2   **Hydraulic forming**
One way of avoiding the die and counter-die issue is to only use the one half, the 'positive' model of the jaw. If then a hydrostatic pressure (through water or oil) is applied to a sheet of the metal placed over the model, and the pressure is enough to exceed the yield point, the metal will deform plastically to conform to the entire surface.[9] The concept is easily visualized with a water-filled balloon as a model: this immediately moulds itself to a container. A seal is effected by an O-ring on the steel (Fig. 3.2), which is to be clamped securely at the edges to prevent (or reduce) wrinkling. The space under the steel must be vented to avoid backpressure preventing full moulding. This need not be deliberate as the seal of the steel sheet to the base block may not be efficient.

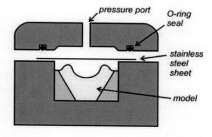

Fig. 3.2 Schematic diagram of the set-up for hydraulic forming of a denture base plate. The top plate is to be bolted down.

The model for this process has been made from a filled epoxy resin, cast directly from an impression, in order to be strong enough to withstand the pressures involved, typically up to about 50 MPa, but depending on the thickness of the sheet. According to the depth of the drawing, it may be necessary to do the forming in as many as four stages, again annealing appropriately between-times, as the work hardening would mean that the steel would tear rather than plastically deform further – its ductility would have been used up (11§6.8). Even so, the formability of austenitic stainless steel is inherently low, and the sheet tends to thin and fail (which affects die and counter-die press-forming as well), so that near-vertical surfaces would be very difficult or impossible to reproduce. Springback is still a problem.

●3.3   **Explosive forming**
At exceedingly high strain rates the mechanism for the plastic deformation of metals changes. As has been mentioned elsewhere, any yield process must occur when the stress for it is attained. In any metallic grain, there will be several potential slip planes, but only the one that first reaches the critical resolved stress will respond. That deformation will proceed until the stress falls, or work hardening causes the critical value to rise. It then takes a rise in the applied stress to cause further deformation, which may then occur at a different site, on a different slip plane. If, however, the stress is raised so quickly that it cannot be relieved by the deformation process – that is, faster than the dislocations can move (cf. Fig. 1§9.1) – we are no longer dealing with a fine balance between the applied stress and the consequent yield. The stress may now be higher than the critical resolved value on many of the available slip planes – all of which become active simultaneously. Accordingly,

plastic deformation occurs in many directions within each grain, with little if any preference. Under these conditions, work-hardening is not so severe, and much greater drawing is possible, even for austenitic stainless steel.

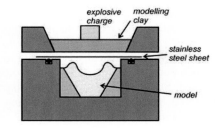

Such high stress rates can be achieved by explosive forming (Fig. 3.3).[10] The detonation of high explosive near the stainless steel sheet generates a very large pressure pulse, a **shock wave** (*i.e.* one travelling at greater than the speed of sound in the medium; sound is an elastic process, shock is not; Fig. 1§9.1) which is transmitted to the metal *via* a semifluid medium (such as modelling clay) or water.[11] The shock wave in the metal then achieves the conditions for multiple slip systems to be active. The steel then conforms to the model much more effectively, the grains tending to spread rather than be drawn out into elongated shapes, with little undesirable localized

Fig. 3.3 Schematic diagram of the set-up for explosive forming of a denture base plate. The steel sheet is to be clamped by the top ring.

thinning. Again, wrinkling is limited by clamping the edges tightly. After this process, there may be little or no springback because elastic deformation has essentially not occurred – there has not been time. Despite these changes in the system response, forming in two or three stages is necessary, again with annealing between-times. Backpressure still needs to be avoided, and the space over the model needs to evacuated, hence the O-ring seal. The model again needs to be of epoxy resin or similar strong and crush-proof material (gypsum is porous and too weak).

●3.4 Superplastic forming

As is apparent from the above, one of the problems of such work is the drawing behaviour of the alloy. Commonly, the response in yielding is necking (1§3.1), and this means that the deformation tends to be very localized and result in considerable thinning and even tearing of the sheet where the surface to be reproduced approaches being parallel to the forming direction. Such effects are clearly offset to some extent by work-hardening, where the stress for further yield rises with the accumulated strain (1§5.1, 11§6.4), but for most alloys this is insufficient. However, if the strain rate-sensitivity of the alloy is great enough, the difficulty can be avoided. That is to say, where the greatest plastic deformation is to occur is where the rate of strain is greatest, and that is where the control needs to be exerted. This can be expressed as follows, where σ is the actual yield stress, σ_0 the yield stress at very low strain rate (*i.e.*, the 'static' value), $\dot{\varepsilon}$ is the imposed strain rate, and m is the so-called **strain-rate sensitivity exponent**:

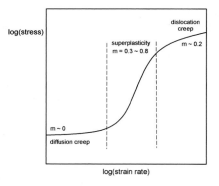

$$\sigma \propto \sigma_0 \dot{\varepsilon}^m \qquad (3.1)$$

m is also the slope of the log-log plot of stress against strain rate. If m is great enough, the resistance to the formation of necks and therefore to excessive thinning is great enough to be useful. In practice, if the alloy shows m > ~0.3 it is described as **superplastic** (Fig. 3.4). Clearly this is an entirely arbitrary criterion, but it arises as an **operational definition**, that is, one based on the practicality of its application.

Fig. 3.4 Stress-strain rate relation for superplasticity in susceptible alloys.

Ti-6Al-4V is one such type of alloy, but it needs to be drawn at a high-enough temperature (~875 °C in this case) that it is in effect continuously annealed, the diffusion rate being great enough to prevent accumulation of dislocations (11§6.6), yet not so high that recrystallization occurs. A significant advantage of this is that the yield stress σ is then so low, and the recovery so rapid (coupled with the low modulus of elasticity in the case of Ti alloys) that the springback on removing the forming load is negligible or non-existent, and high shape-accuracy is attainable in a single-step process.[12]

In superplastic forming grain-boundary sliding is a major feature and this represents a third mode of deformation in metals (*i.e.*, after dislocation movement and martensitic transformation). This is called **grain switching**. Mapping of grains in the surface of a piece has shown that their integrity is largely preserved, but that they are translated with respect to one another so that their neighbours may change, even to emerge (protrude) somewhat from that surface, moving from the interior. However, it is clear that if simple rearrangement of grains were to occur, spaces would open up in the metal between those grains (Fig. 3.5 b), there would be **cavitation**, but on a scale greater than produced by dislocation pile-up. There must be some further process operating. Part of the accommodation of these movements is due to grain rotation (as seen by scratches on their free surface) (Fig. 3.5 c), and this occurs randomly in direction and extent from grain to grain. However, the grains are nearly equiaxed in these alloys, and for rotation to occur only a small amount of ordinary plastic deformation (that is, mediated by dislocation movement) is required in a thin shell around each grain, the so-called **core and mantle** model (Fig. 3.5 d). (Notice that rotation alone would also lead to cavitation, as such a structure must show dilatancy (Fig. 2§2.2)). Diffusive mass transfer is also occurring, in part directed by the stresses at the grain boundaries, because the relative temperature (T/T_m) is high. Thus, the incipient spaces between grains during sliding and rotation are filled by plastically deforming the mantle, leaving the great bulk of the grain intact.

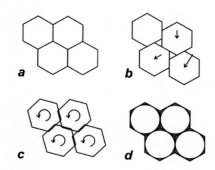

Fig. 3.5 Deformation in superplasticity. a) original grains, b) effect of grain sliding alone, c) effect of grain rotation, d) core-mantle model: shaded area represents overlap for rotation alone.

      As the process conditions for superplastic forming are in the region of around 900 °C for up to 2 h, and 5 ~10 MPa, the die material needs to be suitably resistant. Dental casting investments, both phosphate-bonded (17§4) and aluminous, have been found to be suitable.[13][14]

●3.5   Electroforming

      As was mentioned in 13§7.2, models can be plated electrolytically if they are first coated, in the appropriate pattern, with a conductive paint that forms the cathode. This process can be used to produce denture bases. In practice, this is not as straightforward as it seems since the plating can develop stress that cause it to distort when removed from the model. The stress depends on many factors: metal, temperature, current density, and solution chemistry. Hence, only some metals can be used for this purpose. Nickel has been proposed, but to avoid corrosion, this needed to be electroplated with gold.[15] The problems with this approach are many. Firstly, nickel is a known toxic heavy metal that is a strong sensitizer: very strong reactions can occur in people that have previously been exposed, or it will create that sensitivity. It is no longer appropriate for use in direct skin or mucosal contact. Secondly, the plating needs to be complete to avoid galvanic corrosion (13§4.9) and, as with other gold plating, it is vulnerable to damage (13§4.7). Thirdly, acrylic needs to be attached and this requires a bond to the underlying metal (as, indeed, it does for any metal base). The use of gold makes this unachievable since it does not oxidize, and a crevice is inevitable.

●3.6   Wire drawing

      The wires used in dentistry, as elsewhere, are manufactured by drawing down from larger diameter stock in a progressive process of area reduction to the desired size. While metals may neck under tensile load (Fig. 1§3.3), this is not a feasible route for producing wires of great length and uniformity. The wire diameter must be controlled. This is done by pulling the wire through a die (Fig. 3.6). This may be a very hard metallic block for soft wires, tungsten carbide (21§3) or diamond for harder materials. The die has a highly-polished throat with a special profile. The reduction in area that is achieved must be tailored to the force required to pull the wire through, the strength of the drawn wire, the amount of work-hardening that occurs, and the ductility of the alloy. Thus, several stages are required depending on the diameter of the original feed and the final diameter. To enable passage through the next die in the sequence, the drawn wire must be annealed (11§6.6) to re-establish ductility.

      The shear occurring during drawing results in the original equiaxed grains being distorted into an elongated form; the wire is said to acquire a **fibre texture**. On annealing, recrystallization destroys this completely. However, except for special purposes, fully-annealed wires are not commonly required and, especially for use as springs in orthodontics, the yield point must be raised. This is achieved by the final reduction in cross-sectional area being chosen to correspond with the desired degree of strain-hardening. Fully-hard wire corresponds to the reduction approaching the limit

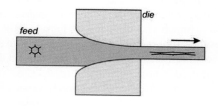

Fig. 3.6 Wire drawing through a die. Note the elongation of grains.

that the alloy will sustain. It is then, of course, very brittle. Notice that the elastic modulus is the metal is not affected by this process (11§6.8), only the yield point.

§4. Pure Gold

Pure gold is considered to be the most malleable of all metals, being capable of beaten into foil so thin (~0.08 μm) that it will visibly transmit some light. In addition, pure gold also shows the unusual property of 'self-welding' at ordinary temperatures, that is, for $T/T_m \sim 0.22$, where normally this requires substantially higher temperatures. This is critically-dependent on the cleanliness of the approximating surfaces: contamination readily inhibits the effect. This property has been turned to advantage in the form of direct-filling gold.

Such gold has been provided in a variety of forms: the foil itself, powder (atomized from the melt or precipitated from solution), and 'mat' gold. The latter is derived from a powder produced electrolytically as small dendritic crystals (cf. Fig. 14§3.1). The powder is then shaped and sintered at just below the melting point to a stage corresponding to the low bisque of porcelain (25§4). The porous material resulting is much easier to handle than the loose powder, being easily cut into convenient pieces. Handling convenience has also led to forms such as ropes and rolls of foil, laminated foil (several sheets stacked together and cut into convenient pieces, the edges welding in the process), corrugated sheet, parcels of powder in foil, compressed powder pellets, and so on. However, the actual form used makes little difference to outcome, the structure is much the same regardless. Operator preference is the sole determinant of the choice. In the past, a porous "sponge" gold in a solid form was produced by heating a gold amalgam; this has obvious drawbacks (§6).

Despite gold being unreactive as we would normally understand it, this does not prevent gases being adsorbed on its surface: oxygen, water, hydrogen sulphide, sulphur dioxide and others. The extent of this adsorption is enough that the self-cohesion of gold is normally not very noticeable. However, it has been found that in a vacuum gold parts in contact that were intended to move have seized up because the desorption occurring then has led to such clean surfaces that full contact is obtained; the ensuing diffusion causes the weld to form. Hence, even if a manufacturer carefully degasses direct-filling gold, and packages it to avoid contamination, it would require extraordinary precautions to prevent exposure at all to the atmosphere. It is therefore necessary to reheat each portion just prior to use, carefully avoiding melting it. Desorption is a diffusive process, so time and temperature are both important. Plainly, it is clear that contaminating contact must also be avoided both before and after heating: fingers, rubber gloves, plastics and other materials that might leave traces that are involatile or might char would be deleterious. However, sulphur-containing contaminant gases do not desorb on such heating.

Heating also has another effect. Those products that have been cold-worked, i.e. foils, will be somewhat strain-hardened, which would make subsequent deformation more difficult. Recovery, recrystallization, and grain growth may occur: the material is annealed (11§6.6). Since no work is done between this point and use, no annealing is done at the precautionary desorption reheating, although of course some grain growth could occur, depending on time and temperature. Proper annealing is important for the full ductility of the metal and easier compaction.

Some techniques of use of direct-filling gold are thought to be inhibited by the cohesiveness of the clean metal. For this purpose, so-called non-cohesive gold has been supplied. This has been exposed to ammonia, which adsorbs strongly but which is still removable by heating, should this be required.

The creation of a direct gold restoration involves the progressive build-up of a cold-welded mass of metal from the porous and malleable pellets or pieces. Similar considerations apply as to the packing of silver amalgam: the pressure under the instrument has to exceed the yield point to obtain deformation, but not so high as to punch a hole that will be hard to refill. A careful balance between the instrument face area and force applied is therefore required. The forces to be used are not easily applied by hand with an instrument large enough to be useful. This is complicated by the fact that the work-hardening of the gold in the process means that the force required is not consistent (Fig. 4.1). Accordingly, even well-prepared restorations are porous, laminated, and uneven in superficial hardness because of the variation in the condition of the material beneath.

Compaction was necessarily done with the aid of mallets or mechanical devices delivering a series of blows in order to attain the necessary pressure for welding. But despite best efforts, such restorations are always porous: complete densification cannot be attained, somewhere between 75 and 82 % being typical. There is therefore a risk of marginal leakage leading to caries unless special care is taken. However, emphasis has been placed on the total energy delivered in the compaction process, as if this were simply additive. This cannot be since the only criterion for deformation is that the yield point stress is exceeded. Repeated blows with, say, a large diameter instrument will achieve nothing unless yield occurs. It is pressure, not energy, that is important. Thus, while a powder may be compacted somewhat with the first blow of a given magnitude, when contact stresses at points will be large, the condition for yield will change. The second such blow will achieve much less, and so on. Soon there will be no effect whatsoever. It is also apparent that the effectiveness of the process depends on the thickness of the layer to be compacted: the stress will diminish rapidly with distance from the instrument contact. Compaction at some depth might require a blow that the tooth (or the patient) could not sustain.

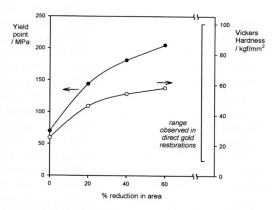

Fig. 4.1 Yield point and indentation hardness for cold-worked pure gold, compared with effective hardness of a direct-gold restoration (which at the low end also includes the effect of porosity). Note equation 1§8.1 and the remarks there and in 1§8.4.

Essentially, the energy available for plastic deformation is that remaining after all elastic deformations have occurred. This includes the tooth, periodontium, jaw, *etc.* Similarly, the strain-hardening of the gold is not a function of the energy deposited in any sense, but depends on the cold-work done, the amount of plastic deformation actually sustained. This has a limit (Fig. 4.1).

The design of the compaction tool is of interest in this context. The effect of area is obvious, but serrations or pyramid designs can only be effective to a depth of about their characteristic scale: local stresses under a point may be very large, but a few scale-lengths away it will have no effect. Even so, at the edge of any instrument, if this is square or sharp, there will be a 'singularity' – a stress so high that an indentation step will form that will add to porosity as it could not then be filled easily.

Similarly used was so-called 'platinum' foil in which a thin sheet of pure platinum was sandwiched between two gold foils. The procedure involved was much the same. Such metal was somewhat harder than pure gold, platinum itself being about twice as hard as gold (Fig. 5.2), because some alloying would occur in the process, but clearly greater effort would be required to achieve similar compaction. This seems to have been a compromise instead of using pure platinum, which was also applied in the same technique.

§5. Mirrors

In 24§5.6, some mention was made of the design of dental mouth mirrors. It is the purpose here to elaborate on the factors affecting the choice of metal for the reflective surface. Those factors include:

* uniformity of spectral reflectivity
* hardness
* corrosion resistance
* cost

Clearly, if a mirror is intended to reflect faithfully the appearance of the object, colour veracity is a major concern. This much is obvious for 'cosmetic' purposes, when to see ourselves as others see us is important. In the dental context, a mouth mirror is unlikely to be critical in that regard since the purpose has more to do with accessibility, and remote, otherwise unseen areas are hardly of importance with respect to shade matching.

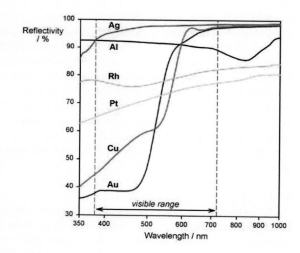

Fig. 5.1 Reflectivity of metals commonly used for mirrors.

Consideration of the spectral reflectivity of some common candidate metals shows several things (Fig. 5.1). Silver is by far the most efficient reflector over the range of the visible spectrum, and this accounts for its continued use in back-surface mirrors in domestic contexts; there is little colouration and the image is bright. Unfortunately, silver reacts with sulphides in the air very readily, and so quickly tarnishes, as the owners of silverware know all too well. Gold, of course, does not tarnish (although low-gold alloys may), but its distinct yellowness rules it out of common use, although as a reflectivity standard or for infra-red use it is valuable. Aluminium is second only to silver for colour-truth, but it is very soft (Fig. 5.2), indeed, hardly better than silver and gold. For disposable front-surface mouth mirrors this is tolerable, but not for anything intended to be more durable. Its softness is not an issue for use in astronomical optical telescopes, where a front-surface coating is important for avoiding image aberrations, but this also requires it being kept dry and chloride-free if corrosion is to be avoided. Increasingly, it is being used for back-surface domestic mirrors, where it can be protected from damage, because it is cheaper than silver and nearly as good a reflector. The softness and sulphide reactivity of silver means that such back-surface mirrors must still be protected from mechanical and chemical deterioration. This is done with special backing paints that must seal the edges of the silvering and remain intact.

Platinum is tarnish- and corrosion-resistant, but is not hard enough to be of general use; it is also expensive. Rhodium is similarly resistant to reaction, is more colour-faithful and slightly brighter than platinum, but is also very much harder, taking a high polish easily (which accounts for its use for plating in jewellery). In most respects it provides a good solution to the problem of the coating of a front-surface mirror.

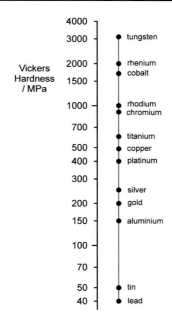

Fig. 5.2 Hardness of some pure metals in the fully-annealed condition. (Note: such values are very sensitive to conditions, reports may vary by ~ ±10 %, or more.)

§6. Mercury

Mercury is remarkable for being the only metal liquid at ordinary 'room' temperatures (m.p. −38.9 °C), and is capable of reacting with a number of metals to form alloys called amalgams. One such material is in common use in dentistry (Chap. 14). It is also remarkable for being readily available in very pure form[16] and most material as purchased is suitable for use in dental silver amalgam. In dry air, there is negligible reaction over many years. When contaminated with 'base' metals (*i.e.* ones more easily oxidized), mercury rapidly becomes dulled by a scum of oxides. Even with very small quantities of contaminants it '**tails**', an effect seen down to about 1 part in 10^7 by mass.[17] That is, if swirled in a clean flask, the liquid does not form a rounded droplet easily, but drags out into a tail that sticks to the glass.[18] This is sometimes called the "Wichers test". In air with some water vapour, slow oxidation occurs, again creating a scum, but this can be removed by running the mercury through a pinhole in a filter paper. It does not reform immediately, as it does on contaminated mercury.

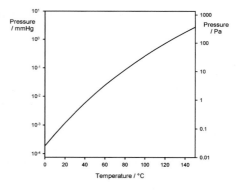

Fig. 6.1 Vapour pressure of mercury.

•6.1 Spillage

Although bulk mercury is becoming less common in dentistry as silver amalgam becomes less popular, spillage from broken or leaky capsules remains a possibility. The vapour pressure of mercury is surprisingly great, due to its low boiling point: 1.9×10^{-3} mbar (0.25 Pa) at 25 °C (Fig. 6.1). The corresponding mass concentration is ~15 mg/m³

Mass concentration (g/m³) is found from the fractional pressure, p_{Hg}/p_{tot}, the standard molar volume, V_{m0} (22.414 L/mol), the measurement temperature, T_m, and the molecular weight of the mercury, M_{Hg} = 200.59:

$$[Hg] = \frac{p_{Hg}}{p_{tot}} \cdot \frac{M_{Hg}}{V_{m0}} \cdot \frac{273}{T_m}$$

(see box). This should be compared with the WHO occupational exposure limit of 0.05 mg/m^3. Since mercury vapour is toxic and prolonged exposure is hazardous even at low concentrations, it is important to clean up any spill promptly. This is made difficult by the metal's density (13.55 g/mL), low viscosity and high surface tension (Table 18§2.1), which account for its tendency to splash and fragment into very small droplets, which cannot easily be swept up. It is possible to retrieve some using a suction device ("aspirator"), with a trap, or even a syringe and hypodermic needle, but much may escape attention. (NB: do not use a vacuum cleaner; mercury reacts with copper in the motor, and the heat from the motor means that much vapour will be emitted.) Commercial devices (for example, polymer sponges) for recovering spillage are available, but these are of low efficacy. Reaction seems to be the most effective means of rendering the mercury inactive or at least substantially lowering its rate of evaporation. A simple means of collecting a small number of droplets is contact with a strip of tin foil,[3] which reacts as in dental silver amalgam (14§1.2). An oxide coating may need to be rubbed off or abraded first (15§1.1).

Another source of spillage is broken fluorescent light tubes. These typically contain some $4 \sim 60$ mg metallic mercury, depending on brand and environmentally-benign design (at the low end of this range in modern products, including compact designs, up to ~5 mg). Liquid when cool, this mercury is vaporized in operation (24§4.9) and so presents a major inhalation hazard if the tube is broken while hot. Cold breakage is much less of a problem if disposal is prompt and proper. Mercury-in-glass thermometers are commonly used in and around dentistry. They contain about 1 g mercury, and should be handled with care and protected from breakage.

Skin absorption is rapid, so liquid mercury, including unset amalgam, should not be handled without gloves (the former technique of "palming" or mulling the freshly-mixed material in the hand would have led to substantial effects). It might also be noted that clinical face-masks offer no protection against the inhalation of mercury vapour: the atomic diameter (300 pm) is rather smaller than the size of pores of the fabric, through which, of course, one can breathe. Mercury vapour reacts with gold very readily, which forms the basis of an exposure dosimeter. The mercury can be vaporized by heating and assayed by atomic absorption or similar photometric means. Contact of mercury with gold jewellery is, for the same reason, not desirable. The amalgam is silvery-white, and while it can be removed by reaction with concentrated nitric acid, other damage is also likely. Heating is also not a good option. It is best if reactive metal jewellery is entirely avoided in any context where mercury or dental silver amalgam is present.

Even after a spillage has been cleaned up, it is likely that liquid mercury will remain in very small droplets somewhere, and these will continue to release vapour, although the oxide film that forms can reduce the rate substantially.[19] **Activated carbon** is a specially-treated material which has both a very great specific surface area (commonly $300 \sim 2000$ m^2/g) and a very high specific surface energy. It is used in many contexts for purification by adsorption, and mercury is one of the substances which can be adsorbed. Metallic mercury also reacts with halogens directly, which themselves can be adsorbed on activated carbon. Thus, iodine-treated activated carbon can immobilize mercury as the relatively involatile mercuric iodide:

$$Hg + I_2 \rightarrow HgI_2 \qquad (6.1)$$

Activated carbon of this kind can therefore be spread to act as a continued scavenger of mercury vapour from the remains of a spillage, including that of slow desorption from other surfaces which had been exposed to liquid or vapour and so had themselves absorbed or adsorbed mercury. Such surfaces would release the mercury steadily over time causing a continued hazard.

●**6.2 Sulphide formation**

Mercury has a remarkably strong affinity for sulphur and compounds containing it (including proteins). One procedure formerly in use and still recommended in places is to spread finely-powdered sulphur (**sublimated** or "flowers" of sulphur, which is pale yellow) liberally over the affected area, brushing it into crevices where the mercury may lie, and leaving it for some considerable time. Mercuric sulphide (black) is formed directly, if slowly:

$$Hg + S \rightarrow HgS \qquad (6.2)$$

The mixture, which is brown, can then later be swept up for proper disposal. A similar reaction occurs with calcium polysulphide (CaS_x), but this tends not to be complete. The layer of mercuric sulphide which forms is an effective barrier against further evaporation, but only if left undisturbed.

---

[3] Elemental Sn, not aluminium – which is what "tin foil" in ordinary speech may refer to. Amalgamated Al oxidises rapidly, but the mercury remains unreacted and therefore available.

●6.3   Zinc amalgamation

A rather faster method is to use powdered zinc. As can be seen from the phase diagram (Fig. 6.2). there are three possible intermetallic compounds that can be formed directly with mercury, α (Hg$_3$Zn), β (HgZn$_2$) and γ (HgZn$_3$), depending on temperature. Notice that we again invoke the Ostwald Rule of Successive Transformation (8§2.6) to use Fig. 6.2 by reading from the left, as if adding small quantities of Zn to dissolve in the Hg (the solubility of Hg in Zn is small here and can be ignored). Assuming that the temperature is above 20 °C, γ-phase will form, which requires ~50 : 50 by mass Zn : Hg. Since Zn (7.14 g/mL) has about half the density of Hg, about twice the volume (of metal) as a minimum is therefore required for complete reaction. Given that the bulk powder density will be less (at ~60 %, say, Fig. 4§9.2), three or four times the bulk volume of the Hg spilt would be safe. If the temperature were lower, less Zn is required for reaction to the other possible phases, so the above figure remains safe. Note that such reacted powder should not be collected with a vacuum cleaner, whose internal temperature can exceed the peritectic temperature of ~43 °C and so release mercury vapour.

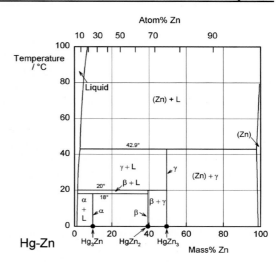

Fig. 6.2 Phase diagram for the system Hg-Zn.

Unfortunately, zinc is inflammable and is sufficiently reactive to oxidize slowly in air in the presence of any moisture. Indeed, powdered zinc can be explosive or burst into flames if allowed to get damp in storage. The oxide and hydroxide coating formed will prevent wetting by mercury (*cf.* amalgam alloy, 15§1.1) and thus reaction. Grinding ("trituration") is not feasible, so the oxide-hydroxide coat must be removed or at least breached by dissolving it in dilute acid; 5 ~ 10 % sulfuric (H$_2$SO$_4$), hydrochloric (HCl), acetic (CH$_3$COOH) and sulphamic (H$_3$NSO$_3$) acids have been suggested; the choice is a matter of convenience, it has no

The use of sulphamic acid (correctly, aminosulfonic acid) is due to the convenience for storage and transportation of the stable, non-hygroscopic, non-volatile powder in a low-bulk and reduced-hazard form for use in commercial mercury spill clean-up kits. It is still a moderately strong acid: pK$_a$ = 1 at 25 °C.

bearing on the application. The use of an acid is commonly described in commercial products as "activation", but clearly it has no more than a cleaning function. Likewise, it has been described as shifting the equilibrium in favour of the β-phase; it can do no such thing. Thus, after gross contamination has been removed, the affected area is sprinkled liberally with zinc powder, and this is then dampened with the acid. The resulting paste is scrubbed or rubbed into the affected area to encourage close contact with the mercury and hence reaction. The products are then swept up. (It is probably wise to neutralize any remaining acid, such as with sodium bicarbonate solution, and care should be taken in use to avoid contact, especially with eyes.) Disposal of the waste must then be through some route that will not put mercury into groundwater (*i.e. via* landfill) or the air (by incineration). Zinc-plated steel wire wool pads (also called mercury "sponges") are also used in a similar fashion; they commonly contain some acid powder (*e.g.* citric), and so must be wetted with water for use. Even so, the vapour pressure of Hg over the amalgam remains substantial unless reaction has gone to completion.

It will be recognized that there are several unsatisfactory aspects to the above procedure, but the seriousness of the contamination warrants the effort.

In passing, it can be noted that the powdered zinc is made by atomization (*cf.* 14§2.2) in a high-velocity stream of inert gas, to reduce the particle size.

●6.4   Waste storage

While the vapour pressure of mercury over set amalgam is negligibly low, at something in the range 10^{-6} to 10^{-10} mmHg[20] (otherwise it would be a problem in the mouth, which it is not[21][22]), because of the binding of the mercury in the intermetallic compounds, as indicated above, that of liquid mercury is appreciable. Mercury is also soluble in both water (0.06 mg/L) and oils. Accordingly, the vapour pressure of mercury over such a liquid "barrier" is unaffected, by the principles discussed in 8§3, and will simply diffuse through. However, metallic mercury can be oxidized to mercuric ion (Hg^{2+}) by solutions such as 0.2 % acidified potassium permanganate or 2 % ferric chloride (which is acidic by hydrolysis). Thus, waste with appreciable liquid mercury, that is, in excess of that required for reaction with the alloy present (14§4.2) would usefully be stored under such a solution, but there would be little point with fully-set amalgam.

A bigger problem is the free mercury remaining in mixing capsules (15§2.7). The volume of the waste poses problems, but immersion in an oxidizing solution is probably of much greater importance. Disposal of such materials remains poorly addressed in many places.

If nothing else is done, all such waste must be kept in tightly-closed containers, and all workspaces where mercury is handled maintained with good ventilation. Cutting, grinding or polishing amalgam can be expected to produce significant quantities of mercury vapour and particulate aerosol. The debris from placing, cutting or polishing amalgam in the mouth is increasingly being required to be removed from the waste-water stream with 'amalgam separators' for pollution control. The filter and the debris must also be treated as hazardous waste.

Another topic that is of concern is the fate of dental silver amalgam in cremations.[23][24] Although this is beyond the scope of dentistry as such to address, there are ramifications elsewhere.

•6.5 Copper amalgam

The general problem of a direct restoration is to find a material that is initially plastic but that then hardens to a serviceable condition. Early dentistry tried many things, copper amalgam was one, dating from at least 1840 (as "Sullivan's Cement"), although the history is very obscure.

As can be seen from the phase diagram (Fig. 6.3), there is a well-defined intermetallic compound (β) Cu_7Hg % Hg by mass, which is has silvery-white appearance and a peritectic melting temperature of 128 °C:

$$\beta \rightarrow \alpha + Hg(L) \tag{6.3}$$

The two phases produced are almost pure copper and almost pure mercury, as the solid solubility of Hg in Cu is very low (radius ratio 150/128 = 1.17), and similarly that of Cu in Hg; there is no appreciable solid solution at low temperature. The method of use involves heating a portion of the alloy, which is supplied in the form of pellets contain % by mass Hg, until liquid mercury appeared, *i.e.* some peritectic melting had

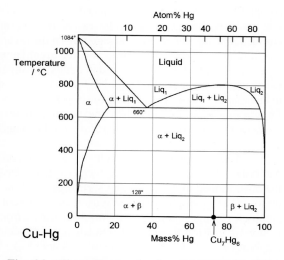

Fig. 6.3 Phase diagram for the system Cu-Hg. This system is an example showing an upper consolute point (8§4.1).

occurred. The resulting material is then triturated (with a pestle in a mortar, Fig. 15§1.2), as was formerly done for silver amalgam, until a pasty mass is obtained which can then be packed into a cavity. It has been described as more plastic and easier to use than silver amalgam. It has also been said that the trituration is unnecessary, and this can be deduced from the reaction: the α-Cu is necessarily in a very finely divided state, and unless allowed to separate by overheating would not require re-mixing. Even so, it is apparent that mercury released by the reaction must appear as globules on the surface as its volumetric expansion coefficient (183 MK⁻¹, 0 ~ 100 °C) is rather larger than that of copper (51 MK⁻¹) and so must be squeezed out (we can assume that the value for Cu_7Hg_6 is of the same order of magnitude as that of other solid metals). This is on top of the volumetric expansion of the mercury on going from the solid to the liquid state. Thus, using the data from Table 14§6.1 and the calculation method of 14§6.1, we have:

$$13 \times 9 \cdot 6 \rightarrow 7 \times 7 \cdot 1 + 6 \times 14 \cdot 9 \tag{6.4}$$

or
$$124 \cdot 8 \rightarrow 139 \cdot 1 \tag{6.5}$$

That is, a volumetric expansion of ~ %11.5, assuming no temperature change.

Some commercial alloys may contain up to 1.5 mass% Cd, as well as small quantities of any of Ag, Sn or Zn. These additions would be expected to lower the peritectic melting temperature (14§5.5), and a temperature of 96 °C has been reported for the appearance of liquid Hg.

The heating is done over an open flame in a test-tube or, more commonly, an iron spoon, open to the air, and as can be seen from Fig. 6.1, the vapour production is expected to be substantial. The occupational exposure hazard to the one doing the preparation is great. On cooling, which would be prompt, the reverse reaction,

$$\alpha + Hg(L) \rightarrow \beta \qquad (6.6)$$

is sufficiently slow (taking several hours, depending on how much peritectic melting occurred) that the cavity filling can be effected easily. Indeed, liquid mercury was typically still present at 24 h.

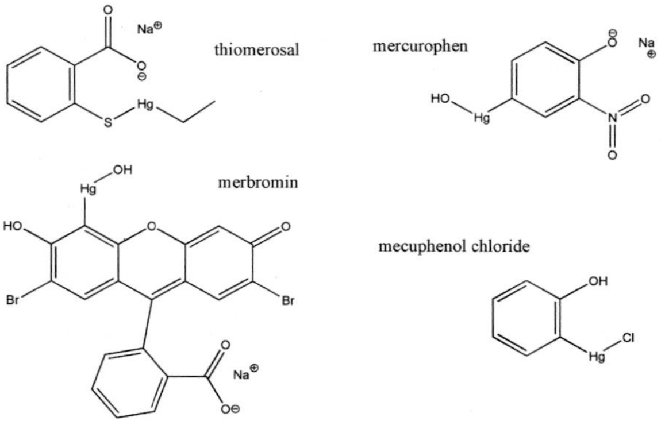

Copper amalgam used to be recommended for the primary dentition because of the antibacterial effect of the copper;[25][26] it was used especially

Fig. 6.4 Some mercury compounds used for antiseptic, disinfectant and preservative purposes.

when carious tissue could not, for whatever reason, be removed completely. Although cooling would have been complete when the mixture was packed, shrinkage on "setting" was marked, leading to much leakage (although this could be limited to a certain extent by controlling the amount of peritectic melting it could not be eliminated). The shrinkage is apparent from the reverse of equation 6.5: a volumetric shrinkage of ~ %10.3, or %~3.6 linear. However, despite this, recurrent caries was less frequent around such restorations than other types. It is evident that it was used in adults as well and such restorations might still be encountered: it was in extensive use at least until the mid-1990s in Norway, and yet later (if still not current) in Russia. It was banned in Denmark in 1984. The toxicity of copper (as well as cadmium) extends to people as well.[27] It was also used for retrograde root fillings, where the cytotoxicity was marked,[28] and as a die material in the dental laboratory – where again the occupational hazard would be major problem.

The amalgam corrodes badly in the mouth, releasing ions of these metals, and this is responsible for the bad taste. The corrosion is driven by an oxygen concentration cell (13§4.8) forming in any crevice, porosity in the mass (added to by the setting shrinkage), and the galvanic effect of excess mercury once any copper had dissolved as a two-phase system must then be present if it was not already. This would also be the case before complete setting. Inhomogeneity would also arise from the usual slight excess of copper, and the fact that setting is slow: corrosion would start immediately. Discolouration of restoration and tooth from the formation of sulphides was marked; both CuS and HgS are black. Teeth commonly could also go green (from other copper corrosion products).

●6.6   Mercury sensitivity

"Dental amalgam" has at various times been subject to criticism as the source of mercury which causes a long list of chronic problems. Despite a great deal of effort, no evidence for this has ever been found for dental silver amalgam.[29][30] Mercury in fish remains the largest source for many people. However, this long-running debate over the toxicity of "dental amalgam" may have been confused by the indiscriminating use of the phrase and the evident problems of copper amalgam, in comparison with which those of silver amalgam are negligible. The two seem to have been conflated in the popular press, especially because of the term "high copper" amalgam (14§9.1). They also look very similar when fresh.

Nevertheless, there are some matters of relevance of which a dentist should be aware. Repeated exposure to heavy metals can result in sensitization, and mercury is one such. The problem is that mercury and mercury compounds occur in many therapeutic contexts.[31] These are primarily antiseptics such as mercurochrome (merbromin), but formerly were prescribed for many things, *e.g.* mercuric benzoate for syphilis, and "blue mass" – finely divided metallic mercury in a carrier – for depression and much else. Many have been around for some hundreds of years, but the use of some substances is still widespread, such as thimerosal as a pharmaceutical preservative (Fig. 6.4). In addition, in many parts of the world, skin-whitening creams and other cosmetics are sold which contain calomel (mercurous chloride, Hg_2Cl_2) and other mercury salts. These are hazardous. "Mercury intoxication" is well-documented and needs to be treated promptly (with chelators such as 2,3-dimercapto-1-propanesulfonic acid, 'DMPS'[32], Fig. 6.5).

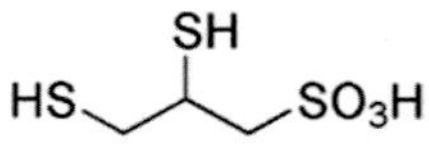

Fig. 6.5 DMPS – a heavy metal chelator for detoxification.

The point is that, while silver amalgam is not itself a problem, people who have previously used mercury-containing substances may have become sensitized and so respond quickly and badly to such a restoration being placed. Thus whilst silver amalgam may precipitate a reaction episode, it is the prior exposure to more active forms of mercury and its compounds that is to blame, and that exposure will probably not be recognized to have occurred.

§7. Beryllium

As mentioned earlier (19§2.6), beryllium may be present in casting alloys as a means of modifying the casting properties, especially by lowering viscosity and melting point the 'castability' was said to be improved, but also some grain-refining followed. As such, it appeared in alloys sold for use in crowns, bridges and partial denture frameworks. It was also found to increase the bond strength to porcelain (25§6), and to enable electrolytic etching, again to improve bonding. As its atomic radius is quite small (112 pm, *cf.* C: 86 pm), some limited interstitial solid solution is possible in some alloys, leading to lattice strain and therefore hardening. Larger concentrations would lead to precipitates being formed.

However, exposure to the vapour of beryllium or respirable particles is known to cause a number of debilitating diseases, principally **chronic beryllium disease** (CBD), a granulomatous lung condition, but it is also carcinogenic.[33] A contact dermatitis also occurs as well as sensitization. All of these diseases are incurable. The metal itself is probably not the problem, but the oxide BeO, which forms readily by reaction with the oxygen of the air on heating (and some scavenging activity may be present in casting alloys; 14§2.4, 19§1.14). Indeed, beryllium-containing ceramics are similarly exceedingly toxic, as are the metal's compounds. Whether there are problems from exposure to the metal in dental alloys in the mouth is not clear, although it can be released by corrosion.[34]

Accordingly, great precautions are essential in processing any Be-containing alloy, whether for casting, cutting, grinding or just polishing. This would entail working in a glove-box or fume-cupboard with good extraction, although the propriety of exhausting the fumes into the external environment must be questioned. The only advice now for dentistry is to avoid absolutely all Be-containing metals; alternatives exist. Other contact for dentists and paradental staff is rather unlikely: except for particle physics, the only other use at all likely to be encountered is for the exit window of X-ray tubes for mammography, although some dental X-ray tube designs do contain a beryllium filter (26§4.6). The brittleness of the pure metal and the fragility of those windows poses a special risk. An old tube containing a beryllium filter must be disposed of safely, and not in domestic waste. Even so, beryllium copper (2 ~ 2.5 mass% Be), which is precipitation hardened (β-BeCu), is useful for electrical spring contacts and may be widespread but unrecognized. Nevertheless, there is only one piece of advice in the event of a potential exposure: run away,[4] secure the area, then seek specialist decontamination of the site.

§8. Other Alloys

●8.1 Elgiloy

Cobalt, as has already been discussed (19§2), forms extensive solid solutions with a number of other transition metals, offering much scope for solution hardening. There is a further aspect to its metallurgy which is of interest. There is a **transus** at 422 °C corresponding to the transformation from the high-temperature f.c.c. α-phase to the h.c.p. β-phase, a martensitic phase change, that is, non-diffusive. This behaviour carries over into many alloys, and in parallel with the transus behaviour in Ti alloys (Fig. 1.2), the transition can occur at higher (*e.g.* with Cr, Mo, Pt) (Fig. 8.1) or lower temperature (*e.g.* with Mn, Ni, Pd), depending on composition. There is a very wide range of such alloys, generally called 'superalloys' by virtue of their strength and low creep at elevated temperatures ($T/T_m > \sim 0.7$), as well as corrosion resistance. The **stellites**, as represented in dentistry by the Co-Cr casting alloys (19§2), are members of this group.

In such alloys the transus is sufficiently low in terms of the homologous temperature, $T/T_m = 0.4 \sim 0.5$, that under normal circumstances ($T/T_m \sim 0.18$) the transition cannot occur for rapidly cooled material, that is,

---

[4] This is not a joke.

quenched. The f.c.c. α-phase is therefore metastable. However, deformation twinning occurs readily in such alloys, because the energy difference between the regular f.c.c. stacking and the reversed sequence (Fig. 1.4) is slight. Indeed, the h.c.p. β-phase is closely-related to a repetitively twinned structure. The result of deformation then consists, in effect, of multiple, thin, twinning lamellae, criss-crossing the grain. This twinning gives the equivalent effectively of reducing the grain size (Fig. 11§4.5). In addition, deformation can trigger the transus phase change, resulting in the appearance of the second phase as a fine-grain precipitate within the α matrix. This has the form of thin platelets, similarly crossing the primary grains. Thus, the result of cold-work is marked inhibition of the movement of dislocations, primarily by the precipitate boundaries but also by the twin boundaries. In other words, these alloys work-harden rather quickly. Nevertheless, the martensitic nature of the transformation, in addition to the twinning, means that substantial 'plastic' strain is associated with the transition, and such alloys appear relatively ductile.

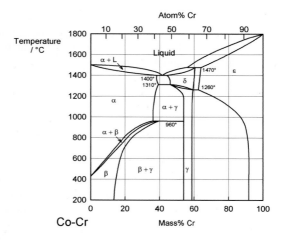

Fig. 8.1 Phase diagram for the cobalt-chromium system. Note the transus at 422 °C for Co.

Since the primary difficulty with precipitations of that kind is nucleation of the second phase, **age-hardening** can then allow these crystals to grow.[35] That is, raising the temperature, but keeping it below that of the transus, provides the activation energy for the transformation, further increasing the total length of the α-β grain boundary and thus increasing the resistance to dislocation movement. Solute partitioning may then also occur, diffusively, because the two phases have different most-stable compositions (corresponding to the tie-line in the two-phase field), and so decreasing the coherency of the boundary (*cf.* 19§1.4). It also follows that a **solution heat treatment** is possible (12§3.6), *i.e.* heating to above the transus temperature, to return the alloy to a fully α-phase condition, although, of course, recovery, recrystallization and grain growth also occur.

Table 8.1 Nominal composition specification of Elgiloy	
Co	40
Cr	20
Ni	15
Fe	16
Mo	7
Mn	2
C	0.15 max.
Be	0.01 max.

'Elgiloy' is the trade name of one such alloy, of a rather complex composition (see box),[5] which is used for orthodontic archwires. As can be seen from Fig. 8.2, the yield point of Elgiloy is markedly affected by cold-working, the value more than tripling at the maximum usable reduction %), that is, from the fully solution-treated (ST) state, on drawing the wire (§3.6) down to some 53 % of its original diameter, $\sqrt{(1 - 0.72)} \times 100$.[36] Although there is no hardening whatsoever on applying the age-hardening heat treatment in the ST condition, about 50 % more again of the increase due to work-hardening first applied is now added to the value of the yield point when maximally drawn.

For orthodontic use, wire of this alloy is sold in four "tempers" (designated by a colour code), presumably corresponding to various degrees of cold work. Accordingly, as-supplied, the user has a choice of initial ductility, and then, after heat-treatment, a corresponding range of yield points which permit various device designs. The actual values for the product, unfortunately, do not appear to have been published, but the tensile strength follows the same pattern.[37]

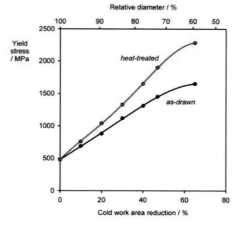

Fig. 8.2 Behaviour of Elgiloy wire according to cold work and subsequent heat treatment (527 °C, 5 h).

---

[5] The original composition included an appreciable amount of beryllium. This was changed some years ago, and no deliberate addition is now made, according to the alloy manufacturer.

●8.2   Lead-tin

The molten metal formerly used in instrument sterilizers (21§2.7) appears to have been something like a 50:50 alloy (by mass) of lead and tin (a formulation corresponding to what is called "soft solder"), at least in some implementations.[38] Such alloys have a eutectic temperature of 183 °C (Fig. 8.3), and although often described as melting at that point they would require a somewhat higher temperature (~240 °C at 50:50) to be fully liquid, which would have provided a reasonable indicator of efficacy (21§2.7).

Such a molten alloy pool, on being left exposed to the air, will develop a **dross** layer. This will consist of SnO_2, SnO coating finely dispersed Pb, and a transition layer of SnO plus Pb-Sn.[39] This material adhered to instruments as they were removed from the bath, frequently leaving contamination invisible to unaided visual inspection, even though gross contamination might be shaken off. Thus it can be expected that, for example, root canals became contaminated with metallic tin and lead as well as oxide. Such residues, and their effects, would persist indefinitely.

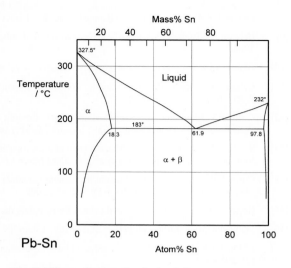

Fig. 8.3 Phase diagram for the lead-tin system.

§9.   Silver

In massive form, silver has limited use in pure or nearly pure form in dentistry, most obviously formerly for endodontic points. These usually contain a small amount of copper (0.1 ~ 0.2 %) and sometimes nickel in solid solution. Although a very ductile (f.c.c.) metal, the yield point is still far too high to allow good adaptation to the root canal wall and so could not seal on its own. Furthermore, while silver is ordinarily considered to be corrosion resistant, it is prone to sulphide tarnish (13§6.6), but more importantly here it is also susceptible to chloride corrosion (13§4.1). This is seen in endodontic points that need to be retrieved because the treatment has failed, and especially where apical leakage has occurred. The chloride concentration of tissue fluids is enough to drive this and can lead to marked pitting or more extensive loss and fracture.

Silver castings are occasionally produced, especially for orthodontic devices. To achieve sufficient hardness, the alloy is typically **sterling silver**, 92.5 % Ag, the balance being Cu. As can be seen from Fig. 12§3.1, such an alloy is expected to be **duplex**, *i.e.* two-phase. Although suitable for jewellery, it would not be satisfactory for long-term oral use. Indeed, even for the few months of the orthodontic application, sulphide tarnish is severe. There have been many attempts to create a tarnish-resistant high-silver alloy, judging by the patent literature; it would appear that none works in practice.

(For silver's nanoscopic use, see 30§1.7.)

References

[1]   Lütjering G & Williams JC. Titanium. Springer, Berlin, 2003.

[2]   Aoki T, Okafor ICI, Watanabe I, Hattori M, Oda M & Okabe T. Mechanical properties of cast Ti-6Al-4V-XCu alloys. J Oral Rehab 31(11): 1109 - 1114, 2004

[3]   Yokoyama K, Kaneko K, Miyamoto Y, Asaoka K, Sakai J & Nagumo M. Fracture associated with hydrogen absorption of sustained tensile-loaded titanium in acid and neutral fluoride solutions. J Biomed Mater Res Part A 68(1): 150 - 158, 2004.

[4]   Yokoyama K, Ogawa T, Asaoka K & Sakai J. Hydrogen absorption of titanium and nickel-titanium alloys during long-term immersion in neutral fluoride solution. J Biomed Mater Res B Appl Biomater 78B(1): 204 - 210, 2006.

[5]   Kobayashi E, Matsumoto S, Doi H, Yoneyama T & Hamanaka H. Mechanical properties of the binary titanium-zirconium alloys and their potential for biomedical materials. J Biomed Mater Res 29 (8): 943 - 950, 1995.

[6]   Grandin HM, Berner S & Dard M. A review of titanium zirconium (TiZr) alloys for use in endosseous dental implants. Mater (5): 1348 - 1360, 2012.

[7]   Murray JL. Ni-Ti (Nickel-Titanium). pp. 342-355 *in*: Nash P (ed). Phase diagrams of Binary Nickel Alloys, ASM International, Materials Park, Ohio, 1991.

[8]   Yokoyama K, Kaneko K, Moriyama K, Asaoka K, Sakai J & Nagumo M. Delayed fracture of Ni-Ti superelastic alloys in acidic and neutral fluoride solutions. J Biomed Mater Res Part A 69(1): 105 - 113, 2004.

[9]   Bahrani AS, Blair GAS & Crossland B. Slow rate hydraulic forming of stainless steel dentures. Brit Dent J 118(10): 425 - 431, 1965.

[10]   Blair GAS & Crossland B. The explosive forming of stainless steel upper dentures. Dent Practit 13(10): 413 - 419, 1963.

[11]   Bahrani AS, Blair GAS & Crossland B. Further developments in the explosive forming of stainless-steel upper dentures. Dent Pract 14(1): 499 - 505 (1964)

[12]   Curtis RV, Majo DG, Soo S, *et al*. Superplastic forming of dental and maxillofacial protheses. Chap 15, *in*: Curtis RV, & Watson TF (eds) Dental Biomaterials: Imaging, Testing and Modelling. Woodhead, Cambridge, 2008.

[13]   Soo S, Palmer R & Curtis RV. Measurement of the setting and thermal expansion of dental investments used for the superplastic forming of dental implant superstructures. Dent Mater 17: 247-252, 2001.

[14]   Curtis RV. The suitability of dental investment materials as dies for superplastic forming of medical and dental prostheses. Mat.-wiss. u. Werkstofftech.39 (4-5): 322-326, 2008.

[15]   Rogers OW. Electroformed metal palates for complete dentures. J Pros Dent 23(2): 207 - 217, 1970.

[16]   Wilkinson MC. Surface properties of mercury. Chem Rev 72(6): 575 - 625, 1972.

[17]   Elliott TA & Wilkinson MC. The effect of base-metal impurities on the surface tension of mercury. J Colloid Interface Sci 40 (2): 297 - 304, 1972.

[18]   Wichers E. Pure mercury. Rev Sci Instr 13(11): 502 - 503, 1942.

[19]   Benjamin DJ. The effect of gases and vapours on mercury evaporation. Mat Res Bull 19: 443 - 450, 1964

[20]   Wieliczka DM, Spencer P, Moffitt CE, Wagner ES & Wandera A. Equilibrium vapor pressure of mercury from dental amalgam *in vitro*. Dent Mater 12: 179 - 84, 1996.

[21]   Berglund A. Release of mercury vapor from dental amalgam. Swed Dent J Suppl 85, 1992.

[22]   https://www.bda.org/dentists/policy-campaigns/public-health-science/fact-files/Documents/amalgam_fact_file.pdf

[23]   https://no2crematory.files.wordpress.com/2010/09/2015-summary-of-references-on-mercury-emissions-from-crematoria.doc

[24]   Mari M & Domingo JL. Toxic emissions from crematories: A review. Environ Int 36 (1): 1312 - 137, 2010.

[25]   Leirskar J. On the mechanism of cytotoxicity of silver and copper amalgams in a cell culture system. Scand J Dent Res 82: 74 - 81, 1974.

[26]   Hyyppa T & Paunio K. The plaque-inhibiting effect of copper amalgam. J Clin Periodont 4:231-9, 1977

[27]   Gerhardsson L, Björkner B, Karlsteen M & Schütz A. Copper allergy from dental copper amalgam? Sci Total Env 290(1-3): 41 - 46, 2002.

[28]   Keresztesi K & Kellner G. The Biological effects of root filling materials. Int Dent J 16(2): 222 - 231, 1966.

[29]   Dental amalgam: Update on safety concerns. J Amer Dent Assoc 129: 494 - 503, 1998.

[30]   Lauterbach M *et al*. Neurological outcomes in children with and without amalgam-related mercury exposure. J Amer Dent Assoc 139: 138 - 145, 2008.

[31]   Budavari S (ed). The Merck Index. 12th ed. Merck, Whitehouse Station, NJ, 1996.

[32]   Casarett LJ, Klaassen CD & Doull J. Casarett and Doull's Toxicology: The Basic Science of Poisons. 7th ed. McGraw-Hill, New York, 2007; p. 817.

[33]   Report on Carcinogens. 12th ed. U.S. Department of Health and Human Services, Public Health Service National Toxicology Program. 67 - 70, 2011

[34]   Lopez-Alias JF, Martinez-Gomis J, Anglada JM & Peraire M. Ion release from dental casting alloys as assessed by a continuous flow system: Nutritional and toxicological implications. Dent Mater 22(9): 832 - 837, 2006.

[35]   Assefpour-Dezfuly M. & Bonfield W. Strengthening mechanisms in Elgiloy. J Mater Sci 19(9): 2815 - 2836, 1984.

[36]   http://www.matweb.com/search/QuickText.aspx?SearchText=elgiloy%20wire

[37]   Philip SM & Darvell BW. Effect of heat treatment on the tensile strength of 'Elgiloy' orthodontic wire. Dent Mater 32: 1036 - 1041, 2016.

[38]   Findlay J. Report on the efficacy of molten metal and ball bearings as media for sterilization. Brit Dent J 98: 318 - 323, 1955.

[39]   de Kluizenaar EE. Surface oxidation of molten soft solder: An Auger study. J Vac Sci Technol A 1(3): 1480 - 1485, 1983.

Chapter 29   More Mechanical Testing

Mechanical testing is the basis of much of the research and development of dental materials (as elsewhere). More elaborate service loading conditions require extension of the basic ideas already presented. It is the purpose of this Chapter to explore such ramifications and present methods of comprehending the behaviour of these more complicated conditions for further insight.

*Commonly overlooked, **shear** is a necessary consequence of any mechanical loading, and as a result is a determinant of the process of failure, brought out by study of **Mohr's circle**. Likewise, the **plastic-brittle** transition is a fundamental and universal behaviour, despite our perception 'window' making it appear to be uncommon.*

*Because **bend-strength** tests are commonplace some relevant equations are given for reference, but the basic analysis can be extended to cover **composite** and **laminated** beams, which would then apply to complex shapes and multiple material dental devices.*

*Earlier discussion of **filled resins** was restricted to the **rules-of-mixture** limits for particulate filler. Extending the analysis to drive an approximate solution allows firstly a better understanding of those systems, but this can then be extended to **fibre reinforcement**. The strength of such composites depends more particularly on the fibre properties, orientation and length.*

*The role of flaws has been emphasized in may places already, but now the nature of the **stress concentration** they cause is now described, and how this leads to the concept and measurement of **fracture toughness**, and thus to a measure of **brittleness**.*

Materials Science for Dentistry
https://doi.org/10.1016/B978-0-08-101035-8.50029-8

Chapter 1 discussed the major basic topics of the behaviour of bodies when tested to determine mechanical properties. A number of aspects have been developed subsequently, in Chapters 4, 20 and 23 in particular, but there are further topics that are of direct interest to dental systems, or which help to clarify certain relationships and behaviours.

§1. Shear

Shear is often treated as if it were a completely separate mode of loading from tension and compression. Indeed, for simplicity we have spoken in just such terms in several places (Fig. 1§1.2, 1§2.5, 23§2.1), although the fact that it arises in what may at first seem simple conditions of compression (1§6.2, 1§6.3) or tension (Figs 11§5.4, 11§6.8) might be taken as indicating something more complex. The caption for Fig. 1§5.2 is another such indicator, that is, in a photoelastic model, where the situation in a body in compression is mapped entirely in terms of shear stress. It is therefore appropriate to consider such a system in a little more detail.

●1.1   Equivalent forms

Taking as a starting point the shear in a solid body of square cross-section between parallel planes as in Fig. 1§2.8, or even of a fluid body (Fig. 4§3.3), it is evident that as drawn there is a turning moment (Fig. 1.1 a), as can be understood from Fig. 23§1.2. Obviously, in a static test in which such a loading is attempted, there is no continuous rotation. Newton's Third Law of Motion again decrees that there be an equal and opposite reaction, in this case an opposing moment couple. Whether or not this is arranged deliberately, it is clear that the loading device must provide the necessary constraint; the test machine is expected to be rigid enough to achieve this (the fact that for convenience we usually ignore such forces does not mean that they do not exist). We therefore have the situation depicted in Fig. 1.1 b.

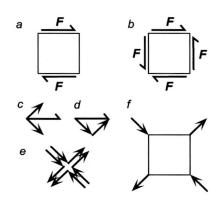

Fig. 1.1   Equivalence of pure shear and biaxial tension-compression.

Each of the surface **traction** forces F, for example that on the top surface, may be resolved into two equal, **orthogonal** components (*i.e.*, at right angles) (Fig. 1.1 c), symmetrically arranged with respect to F, the lower of the two being at -45° with respect to the horizontal direction. Thus, the magnitude of the component directed at +45°, f_x, is given by:

$$f_x = F.\sin \theta = F/\sqrt{2} \tag{1.1}$$

and the complementary f_y by

$$f_y = F.\cos \theta = F/\sqrt{2} \tag{1.2}$$

(because for $\theta = 45°$, and $\sin \theta = \cos \theta = 1/\sqrt{2}$). We may check the magnitude of the resultant for this special case of orthogonal vectors using Pythagoras' Theorem:

$$|\mathbf{f_x} \oplus \mathbf{f_y}| = F\sqrt{(\sin^2 \theta + \cos^2 \theta)} = F \tag{1.3}$$

where the symbol $\oplus$ means vector addition (not just arithmetic addition). Such resolved vectors may be redrawn as if acting individually at each end of the original vector (Fig. 1.1 d) without loss of meaning. This pattern of vector resolution may be repeated at each face in turn, resulting in paired sets of diagonally-directed vectors (Fig. 1.1 e) equivalently applied at each corner of the original body (Fig. 1.1 f). The net corner-applied vectors then have magnitude $2F/\sqrt{2} = F\sqrt{2}$, but with opposite signs from the point of view of the body: the one diagonal is subject to tension, the other to compression. Thus the pure, facially-applied shear of Fig. 1.1 b is found quite naturally to be equivalent to the application of identical tensile and compressive forces along the diagonals of the square, and *vice versa*.

●1.2   Mohr's circle

Another view may be taken by considering the equilibrium of a body in simple uniaxial tension (Fig. 1.2) in terms of the resolution in the x and y directions of the shear $(F_{\parallel})$ (*i.e.* parallel) and normal $(F_{\perp})$ (*i.e.* perpendicular) forces acting on a plane inclined at an angle θ to the load axis.

$$F_x = F_{\perp}.\sin\theta + F_{\parallel}.\cos\theta$$
$$F_y = F_{\perp}.\cos\theta - F_{\parallel}.\sin\theta \qquad (1.4)$$

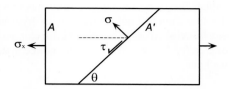

Fig. 1.2   Shear and normal stresses on an arbitrary plane of a body in uniaxial tension.

Given that $F_y = 0$, and $A = A'.\sin\theta$, dividing by the area of the plane, with a little rearrangement we have in terms of stresses

$$\sigma_x = \tau(\tan\theta + \cot\theta)$$
$$\sigma = \tau.\tan\theta \qquad (1.5)$$

which by substitution and some trigonometric identities lead to

$$\sigma = \frac{\sigma_x}{2}(1 - \cos 2\theta)$$
$$\tau = \frac{\sigma_x}{2}\sin 2\theta \qquad (1.6)$$

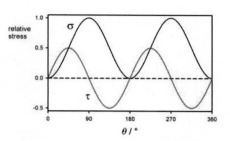

Fig. 1.3   Relative magnitudes of the shear and normal stresses in Fig. 1.2 as a function of the angle of the plane.

The stresses thus vary sinusoidally with θ (Fig. 1.3). Notice that since there is no distinction to be made between planes 180° apart, a natural symmetry, the right-hand half of such a plot is redundant. These two values (equations 1.6) can be plotted as a coordinate pair (σ,τ) as a function of θ, that is, in a **parametric plot**. This yields the plot shown in Fig. 1.4: a circle, tangent to the ordinate. This graphic expression of the relationship between the shear and normal stresses acting on a plane in a body is known as **Mohr's circle**.

Some properties and implications can be identified immediately. Even with simple uniaxial tension, the body is subject to shear at all angles except parallel to the load axis, the maximum of which is precisely half the value of the maximum tensile stress, and this occurs at $\theta = 45°$. The centre of the circle, whose radius $R = \sigma_x/2$, is therefore at $(\sigma_x/2, 0)$, while at $\theta = 0°$ there is neither tension nor shear acting. More generally, the state of stress (σ, τ) for a point within the body, and at any plane orientation through it, is found to lie on that circle.

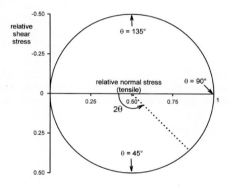

Fig. 1.4   Mohr's circle for uniaxial tension: a parametric plot of the data of Fig. 1.3.

If, now, a second tensile stress is applied to the body, orthogonal to the first, to create a state of **biaxial tension**, the same kind of analysis can be applied, allowing for the additional stress components acting on the plane (Fig. 1.5).

$$\sigma = \frac{\sigma_x}{2}(1 - \cos 2\theta) + \frac{\sigma_y}{2}(1 + \cos 2\theta)$$
$$\tau = \frac{\sigma_x - \sigma_y}{2}\sin 2\theta \qquad (1.7)$$

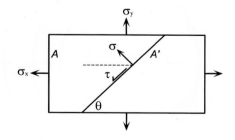

Fig. 1.5   Shear and normal stresses on an arbitrary plane of a body in biaxial tension.

which can be seen to reduce to equations 1.6 for $\sigma_y = 0$. The two terms for the normal stress are 180° out of phase, corresponding to the fact that the applied tensions are at right-angles. The effects of increasing the transverse tension are illustrated in Fig. 1.6: the maximum tensile stress is the greater of the two components, the minimum, the lesser of the two. It can be seen that the maximum shear stress depends on the difference between the two **principal stresses**. However, as the two tensions approach equality the maximum shear stress decreases, eventually to vanish at $\sigma_y = \sigma_x$; the circle diminishes to a point. This is the two-dimensional equivalent of hydrostatic loading, **equibiaxial stress**, when there is no change of shape, only of area (*cf.* 1§2.6). We therefore notice that, quite simply, we have:

$$\sigma_{max} = \sigma_x \qquad \sigma_{min} = \sigma_y$$
$$\tau_{max} = \pm\frac{\sigma_x - \sigma_y}{2} \qquad (1.8)$$

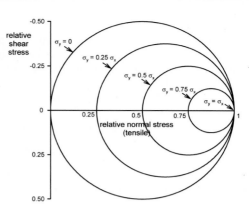

Fig. 1.6 Mohr's circles for biaxial tension: variation with relative magnitude of the transverse stress.

If instead of applying a tensile transverse stress it is made compressive instead, we use the same equations but treat the compression as a negative tension. This leads to the conditions illustrated in Fig. 1.7: the left-hand limit, the maximum transverse stress, moves leftward to correspond to the chosen value of σ_y as before, while the maximum relative shear stress naturally increases to ±1 when the axial stresses are equal in magnitude and opposite in sign, as indicated by equations 1.8. This corresponds exactly to the conclusion of §1.1, the state of pure shear.

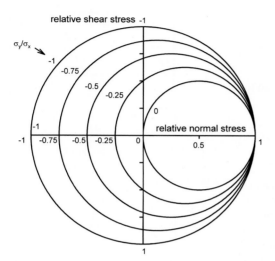

Fig. 1.7 Mohr's circles for biaxial tension plus compression: variation with relative magnitude of the transverse compressive stress. The innermost circle, for $\sigma_y = 0$, is the same as the outermost of Fig. 1.6. Note the state of pure shear for $\sigma_y = -\sigma_x$.

Were we to continue this process of varying the relative values of the two axial stresses it would be seen to be equivalent to reducing the value of σ_x towards zero. This leads first to a diagram (Fig. 1.8) which is the mirror-image of Fig. 1.7, and then on into biaxial compression, in a diagram which is a reflection of Fig. 1.6 (Fig.15.9). Again, we reach a circle of zero radius at the point where the two axial stresses are equal, this time in a state of biaxial hydrostatic compression, when there is no shear at all.

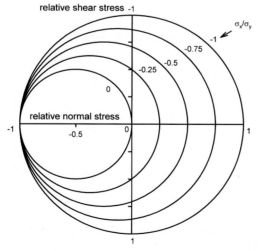

Fig. 1.8 Mohr's circles for biaxial tension plus compression: decreasing the relative magnitude of the tensile stress. The outermost circle corresponds to pure shear, the innermost to uniaxial compression.

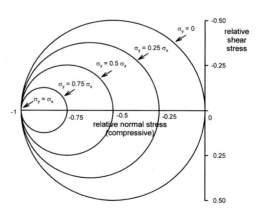

Fig. 1.9 Mohr's circles for biaxial compression: variation with relative magnitude of the transverse stress. The outer circle here is the same as the innermost of Fig. 1.8.

We are therefore able to represent graphically the state of stress of a body under any combination of axial stresses. What is not addressed is the state of such a body on which there is also an applied external shear stress. Of course, the pattern of Fig. 1.1 a is not at equilibrium, so we must consider the full balanced pattern of Fig. 1.1 b. As this pure shear is acting in the *xy*-plane, it is labelled accordingly: τ_{xy}, and is considered as **superposed** (23§2.2) on the axial stresses, *i.e.* treated as independent and additive in effect. However, since it is pure shear, this is equivalent (by virtue of equations 1.8, and the principle discussed in §1.1), to stresses of identical magnitude $\pm\sigma_s = \tau_{xy}$ applied at an angle of 45° to the axes of the body. Pure shear is represented in Figs 1.7, 1.8 by the circle centred on (0, 0). The equations for this are:

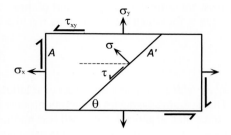

Fig. 1.10 Shear and normal stresses on an arbitrary plane of a body in biaxial stress with a superposed shear stress, τ_{xy}.

$$\sigma = \tau_{xy} \sin 2\theta$$

$$\tau = \tau_{xy} \cos 2\theta$$

(1.9)

So that, with superposition of equations 1.7 and 1.9, we get:

$$\sigma = \frac{\sigma_x}{2}(1 - \cos 2\theta) + \frac{\sigma_y}{2}(1 + \cos 2\theta) + \tau_{xy} \sin 2\theta$$

(1.10)

$$\tau = \frac{\sigma_x - \sigma_y}{2} \sin 2\theta + \tau_{xy} \cos 2\theta$$

The parametric plot of this remains a circle, and a great deal more analysis can be done using it.[1] However, here we will simply note some important properties (Fig. 1.11), using:

average normal stress, which defines the centre of the circle:

$$\sigma_{av} = \frac{\sigma_x + \sigma_y}{2}$$

(1.11)

and the *radius*:

$$R = \sqrt{\left(\frac{\sigma_x - \sigma_y}{2}\right)^2 + \tau_{xy}^2}$$

(1.12)

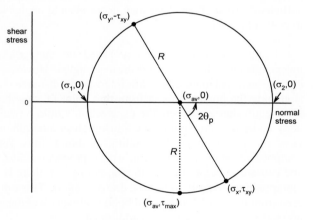

Fig. 1.11 General properties of Mohr's circle for an arbitrary stress system.

It is then obvious that this gives the value of the maximum shear stress, $\tau_{max} = R$, and so we also have the very important result

$$\tau_{max} = \frac{\sigma_1 - \sigma_2}{2}$$

(1.13)

where σ_1, σ_2 are the **principal stresses**, which themselves can be found from equations 1.11 and 1.12:

$$\sigma_1, \sigma_2 = \sigma_{av} \pm R$$

(1.14)

Thus, graphically, knowing σ_x, σ_y and τ_{xy} allows the end-points of a diameter to be plotted at (σ_x, τ_{xy}) and $(\sigma_y, -\tau_{xy})$, thus defining the whole circle (Fig. 1.11). However, looking at this circle it is apparent that there must still be an angle θ_p (*i.e.* as a θ in Fig. 1.10) for which there is no shear stress, only biaxial stress:

$$\tan 2\theta_p = \frac{2\tau_{xy}}{\sigma_x - \sigma_y}$$

(1.15)

Actually, this equation has two solutions, 90° apart, meaning that there are two mutually-perpendicular axes found by rotating the axis system of Fig. 1.10 by θ_p. These are termed the **principal axes** for the system, and these are normal to the **principal planes** corresponding to the principal stresses. Likewise, there must be an angle θ_s for which the shear stress is a maximum. This angle is found from

$$\tan 2\theta_s = -\frac{\sigma_x - \sigma_y}{2\tau_{xy}} \qquad (1.16)$$

again with a complementary pair of solutions, and it can be seen that these lie $45°$ away from the principal axes. That is to say, the planes of maximum shear stress are always at $45°$ to the principal planes. However, we should notice that, in general, there will be a normal stress acting on the plane of maximum shear stress, *i.e.* τ_{max} occurs for $\sigma = \sigma_{av}$. The only exception is the state of pure shear.

●1.3   Strength revisited

In Chapter 1, 'strength' was treated as a simple concept: force per unit area at collapse (or yield), in tension, although the essential nature of shear (*i.e.* slip) in the case of metals was brought out in Chapter 11. The dependence of strength on bond behaviour was brought out in 10§3.6. Since it is now apparent that even in the apparently simple case of uniaxial tension (Fig. 1.2) there is necessarily shear stress acting, one can enquire about the criteria for collapse – whether this will be in tension or shear. We can consider this in the context of Mohr's circle.

Thus, in Fig. 1.12, for an arbitrary tensile stress applied (as in Fig. 1.2, 1§3.1) there is simultaneously a maximum shear stress at $45°$. From equation 1.6 we see that $\tau_{max} = \sigma_x/2$. Since the yield point (which is necessarily a shear criterion) and the strict tensile strength (as a brittle failure – the separation of planes in a direction normal to them) have fixed values for that particular specimen, it is only a matter of which criterion is reached first as Mohr's circle grows in diameter. If the shear strength stress is reached first, the specimen is obliged to respond by shearing. If there are slip planes in the material that are sufficiently weak, then shear may occur even at angles less than $45°$(11§5.1).

Such an interpretation is based on the assumed independence of the tensile and shear strengths. However, it can be seen that both affect the interatomic distance – by stretching the bond – and so from that point of view have some equivalence. Hence, the expectation would be that a body under tension would have a lower shear strength, and *vice versa*. Conversely, putting a bond into some compression would provide some offset (pre-stress, 23§2.8), and cause the shear strength to rise, noting that there is no 'compressive strength' limit to be drawn (1§6.3). This can be seen from Figs. 10§3.4, 3.5: if a bond is already lengthened, no matter by what means, less additional force needs to be applied to cause it to rupture.

Thus, if the actual states of stress at the point of collapse of a series of bodies, loaded in various ways, could be determined, then we might end up with a **failure envelope** of the kind shown in Fig. 1.13, expressing the fact that there is some interdependence of the failure processes. In other words, collapse is not expected, in general, to occur to correspond exactly to either the maximum shear stress or the maximum tensile stress, as would be predicted from the idealized form of Fig. 1.12, but rather at a point which expresses least work in the face of

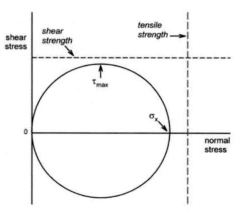

Fig. 1.12   Idealized strength criteria with respect to Mohr's circle for uniaxial tension.

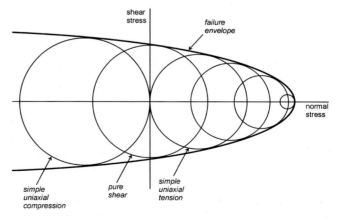

Fig. 1.13   Possible envelope of failure conditions showing the interaction of shear and normal stresses for actual collapse.

both. That is, the angle θ^* at which the shear failure surface develops is not θ_s, the angle of maximum shear stress, but its value depends on the normal stress. It follows that the locus of the failure envelope is everywhere tangent to the corresponding Mohr circle. Thus, during an actual test (or service loading of a body), the Mohr circle is envisaged to grow as the stress system is developed. Failure occurs when it touches the envelope – which is a complete representation of the actual failure behaviour of the material (for a given temperature, strain rate, and so on).

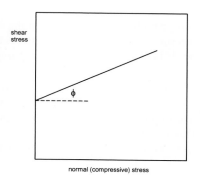

One of the consequences of this behaviour which has a more direct bearing on dental materials testing arises from the left-hand side of Fig. 1.13, that is, shear failure under compressive loading. We have already noted that (a) there is no such thing as failure in compression, and (b) that in such testing shear appears to provide the collapse process (1§6.3). That the shear stress to cause failure depends on the normal compressive stress has been expressed in the so-called **Mohr-Coulomb Law** (Fig. 1.14):

$$\tau = \tau_c + \sigma.\tan\phi \qquad (1.17)$$

Fig. 1.14 Form of the Mohr-Coulomb Law. (Note: this is reversed from the orientation of Fig. 1.13.)

Commonly used to describe the behaviour of soils, it can be seen to be an approximation to the envelope of Fig. 1.13 on the left-hand side, where the angle φ is clearly related to θ^*. What this means is that the shear surface that forms on collapse has an angle which depends on the **internal friction** of the material, φ being called the **friction angle**, essentially the sensitivity of the material to the effect of a normal stress on the shear failure plane. Thus, when a specimen of a cement or gypsum product, for example, is tested in compression, the angle of the shear cone that forms reflects the influence of this material property.

We can go further. As indicated in Fig. 1.10, we can draw the resultant Mohr circle for any arbitrary system of forces acting on a body, and now we can see that the stress at failure (and thus the nominal test load at that point) is a function of all such forces. Hence, if there are any accidental or **parasitic** stresses arising from, say, the difference between the elastic moduli of the specimen and the grips or platen, or the amount of friction between those parts – end constraint – we can expect to see an effect on the load at failure, and thus the strength as calculated from any of the simple formulae (essentially F/A). Thus, any mechanical test is at least in part dependent on the exact conditions employed: grip roughness, presence of lubricant substances, padding layers. Avoidance of such spurious effects, and the attainment of reproducible, defined loading conditions, are central to all mechanical testing, not least for comparability between laboratories. In other words, to understand and compare published results requires a careful consideration of the exact circumstances of the tests.

●1.4   Poisson strain in shear

While the Poisson strain in tension (and compression) is straightforward (Fig. 1§2.6) and simply defined, that occurring during shear is not so obvious.

From equations 1§2.21 and 1§2.22, where $\tan\theta = \gamma$ (Fig. 1.15), with the usual caveat about the strain being small:

$$\gamma = \frac{2(1+v)\tau}{E} \qquad (1.18)$$

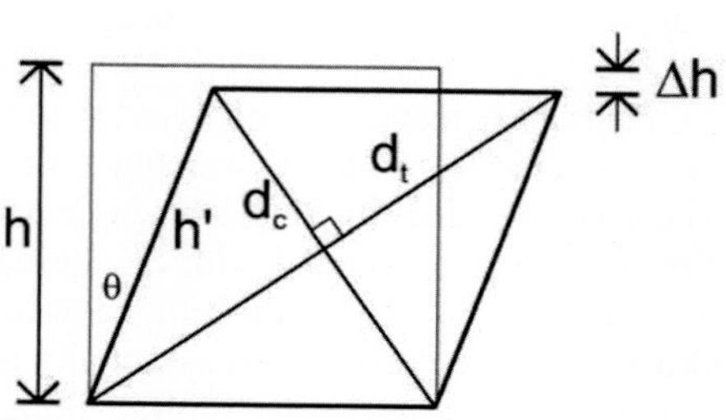

Fig. 1.15 Deformation of an element in shear.

Similarly, the tension and compression diagonal strains are:

$$\varepsilon_t = \frac{(1+v)\tau}{E} \qquad \varepsilon_c = \frac{-(1+v)\tau}{E} \qquad (1.19)$$

This means that, for the initial element side h = 1, the diagonal lengths become:

$$d_t = \sqrt{2}(1+\varepsilon_t) \qquad d_c = \sqrt{2}(1+\varepsilon_c) \qquad (1.20)$$

since the tension and compression stress have the same magnitude as the shear stress (see equation 1.1, for diagonal length d = √2) Since the diagonals must remain orthogonal (at right-angles), the new length h′ > h of the side of the element in shear is given by Pythagorus' theorem:

$$h' = \sqrt{\left(\frac{d_t}{2}\right)^2 + \left(\frac{d_c}{2}\right)^2} = \sqrt{1 + \frac{\tau^2}{4G}} \tag{1.21}$$

In a free body, the top surface has then dropped from its initial altitude:

$$\Delta h \simeq \frac{3(1+v)^2\tau^2}{2E^2} = \frac{3\tau^2}{8G^2} \tag{1.22}$$

(dropping higher order terms of the series expansion). This is the vertical strain. This then means that if that top surface is constrained (as by being fixed to parallel rigid shearing platens of fixed separation), the normal stress is then given (similarly approximately) by:

$$\sigma_n = E\Delta h = \frac{3(1+v)^2\tau^2}{2E} = \frac{3(1+v)\tau^2}{4G} \tag{1.23}$$

The implication of which is that pure shear is not possible if the shearing planes are constraining, or equivalently that an appreciable stress tending to separate a solid body from the shearing platens will be present, *i.e.* there is necessarily a normal force acting on the system for any $v > -1$.

§2. Plastic-Brittle Transition

We have become used to the idea of characterizing materials according to their mechanical behaviour in a rather definite manner (1§4.2), and primarily in terms of whether or not they are brittle. That this is not necessarily an absolute distinction became clear in discussing polymers, where strain-rate sensitivity meant that at one temperature the behaviour could range from hard glass – essentially (assumed to be) perfectly brittle – to liquid-like flow (Fig. 3§5.7). The size effect for strength has been discussed (1§7.1), but size also matters in terms of the mode of failure, whether plastic or brittle.

The effect is implicit in indentation hardness testing (1§8), indeed it is clear from the link to yield strength (equation 1§8.1). Obviously, for an indentation test to work at all for a solid material it is essential that the material be rearranged plastically, even when testing glasses and other ceramics. In general, such approaches are described as **microhardness** tests because of the scale of the indentation created. If the load is increased the indentation tends to become larger, as would be expected from the defining equations. However, there comes a point where cracking commences, from the corners of a Vickers indentation, for example, as may be seen in testing a glass microscope slide. This invalidates it as a hardness test, but more importantly it signals a size of indentation at which a transition from purely plastic to brittle behaviour occurs (see §5.3). No material could be exempt from this. Even diamond would show plastic-flow blunting of an initially sharp tip: infinite stress is not supportable, a single atom can bear only so much load before being displaced permanently (10§3.6).

Similar evidence can be found in the discussion of the Griffith criterion (1§7). At equation 1§7.6 it is shown that if the stored energy can pay for a crack to be created, it will happen. However, if the stress required for shear is exceeded, by the Mohr circle argument (§1), this mode of failure must supervene, since uniaxial tension is always associated with an equivalent shear stress – which is maximum at 45° to that axial load direction (or as modified by the failure envelope, §1.3). Now while shear strength under simple uniaxial loading conditions may be considered as essentially fixed – the load for failure scaling directly with the area under stress – this is simply not the case for brittle failure, where the critical stress varies as does $1/\sqrt{x}$, the scale of the body (equation 1§7.6). Hence it can be deduced that if the ordinarily brittle body is sufficiently small, plastic deformation must occur first. Conversely, a plastic or ductile material in a sufficiently large piece must become brittle. It is this kind of behaviour that results in the blunting of knives, hardness indenters, abrasives, and so on: atomically sharp edges and points cannot sustain large loads without deformation.

There are some examples in addition to the ceramic indentation observation above. It has, for example, been observed that even alumina particles deform plastically in compression on a scale below about 1 μm, and it is a common observation that sharp blades may lose their initial edge simply on contact with a modestly hard material, such as bone, no matter how hard the blade material. Lumps of doughy, ductile materials can crack at the periphery when squashed, such as freshly mixed dental silver amalgam, while the fully-set material may appear perfectly brittle in static tests at a scale of about 5 mm. Then again, under a microscope, grains can seem all too soft and plastic under the tip of a needle. Likewise, large scratches in many materials produce a shattered, irregular furrow, while fine scratches appear smooth and evenly deformed. Large steel bodies have a tendency to fracture in a brittle manner, despite laboratory tests (on small specimens) indicating adequate ductility for the purpose. In other words, what matters is the scale at which we are ordinarily used to working. Our perceptions and interpretation may be controlled by the range that we can see and handle easily, say from ~1 mm to ~10 cm, yet it is clear that our view is a very narrow one in terms of the possible scale of systems and objects: dentistry has a well-defined field of operation that establishes that scale. This is yet another reason to design mechanical tests on a scale appropriate to the service conditions of the material. This is a general phenomenon.[2]

What needs to be clear is that the Griffith criterion actually refers only to crack initiation. Should there be any mechanism for energy dissipation beyond creation of new surface at a rate faster than it can be supplied then the crack will not propagate catastrophically. Essentially, the distinction needs to be made between stable and unstable crack growth. If there is more elastically-stored energy than is needed for crack growth, with or without crack tip blunting or other dissipative mechanism, then the failure is prompt. If further strain is necessary, it is in the stable region. Crack initiation depends on the presence and nature of flaws but also on their numbers; the larger the volume tested, the greater the chance of a failure initiating at a lower stress. Thus, a three-point bend test has a small high-stress volume and tends to give higher strengths with slow crack growth, while a four-point test involves a much larger volume, with consequently greater stored strain energy, and thus lower strength is recorded on average, but with more explosive failure because it is now more brittle. For some materials we may not ordinarily see the transition because of the scale on which we work: the plasticity of glass becomes effective below that usual scale, while the brittleness of mild steel is only apparent at far greater than normal testing scales, but certainly on the scale of larger structures (Fig. 2.1).

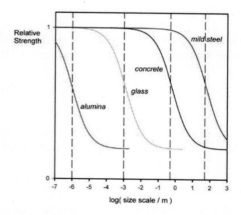

Fig. 2.1 The size effect for strength across the plastic-brittle transition on increasing scale for some common types of material (schematic).

§3. More Beam Mechanics

●3.1   Bend strength

In 1§6.4 some equations were given for flexure strength without derivation. Using the principles of Chap. 23, and especially noting Fig. 23§2.10, it can be shown that all such loading leads to an equation for the stress σ on a fibre in the beam of the form:

$$\sigma = -\frac{M}{I} \cdot y \qquad (3.1)$$

where M is the maximum moment, I_z the second moment of area (23§3), and y the distance from the neutral surface. This is the **flexure formula**.[1] Thus, for the most distant fibre of a doubly-symmetrical beam (that is, about y- and z-axes) we have

$$\sigma_{max} = \frac{M}{I_z} \cdot \frac{h}{2} = \frac{M}{S} \qquad (3.2)$$

where h is the depth of the beam (ignoring the sign, now). Often, the **section modulus**, $S = 2I_z/h$ is used to simplify the tabulation for various cross-sectional shapes. Effectively, this view relates the mechanics of the load system to the geometry of the beam cross-section.

Hence, since the moment is given by the *component* of the total force × distance from the support carrying it, for the three-point bending of rectangular beam specimens we have:

---

[1] Note the sign convention: above the neutral axis, $y > 0$, positive moment gives compression, which is taken as a negative stress.

$$\sigma_{max} = \frac{F}{2} \cdot \frac{L}{2} \cdot \frac{12}{bh^3} \cdot \frac{h}{2} = \frac{3FL}{2bh^2} \tag{3.3}$$

as at equation 1§6.4, while for four-point bending we have

$$\sigma_{max} = \frac{Fc}{2} \cdot \frac{12}{bh^3} \cdot \frac{h}{2} = \frac{3Fc}{bh^2} \equiv \frac{3F(L-L_i)}{2bh^2} \tag{3.4}$$

as at equation 1§6.5, where c is the horizontal distance of an upper load point from the line of a support (Fig. 1§6.5). The entirely equivalent version on the right-hand side is in terms of the span of the inner rollers, L_i, a form sometimes used. When this four-point test is conducted with $c = L/3$, known as the **third-point loading** test (23§4.3), the equation reduces to:

$$\sigma_{max} = \frac{FL}{bh^2} \tag{3.5}$$

while setting $c = L/2$ of course yields the right hand side of equation 3.3. Sometimes, one encounters **quarter-point** loading, with $c = L/4$; from equation 3.4 this gives:

$$\sigma_{max} = \frac{3Fc}{bh^2} \equiv \frac{3FL}{4bh^2} \tag{3.6}$$

For circular or elliptical sections, the appropriate substitutions for I_z can be made from equations 23§3.6 and 23§3.10, as appropriate. For example, for a circular rod we have in three-point bending:

$$\sigma_{max} = \frac{F}{2} \cdot \frac{L}{2} \cdot \frac{4}{\pi r^4} \cdot r = \frac{FL}{\pi r^3} \tag{3.7}$$

with I_z from equation 23§3.6, while for four-point bending we have

$$\sigma_{max} = \frac{Fc}{2} \cdot \frac{4}{\pi r^4} \cdot r = \frac{2Fc}{\pi r^3} \equiv \frac{F(L-L_i)}{\pi r^3} \tag{3.8}$$

as at equation 3.4. When this test is conducted with $c = L/3$, the third-point loading test (23§4.3), the equation reduces to

$$\sigma_{max} = \frac{2FL}{3\pi r^3} \tag{3.9}$$

Likewise, quarter-point loading of a circular bar must use:

$$\sigma_{max} = \frac{Fc}{2} \cdot \frac{4}{\pi r^4} \cdot r = \frac{2Fc}{\pi r^3} \equiv \frac{FL}{2\pi r^3} \tag{3.10}$$

Then, providing the specimen fails in tension, we can report the flexural strength, sometimes called the **modulus of rupture**. Again, it must be stressed that deflections must be small for these equations to be valid without correction terms (*e.g.* [3]): obviously, large deflections means that large shear strains are present, and this affects the calculations because it affects the load causing collapse (§1.3). However, given the large variety of test conditions – specimen section and loading pattern – care must clearly be taken to use the correct equation. Appreciable errors otherwise result.

Such calculations can be used in the design of a load-bearing in system, taking into account the material's strengths in both compression and tension, by setting σ_{max} = M/S to be the allowable stress, that is, with a sufficient safety margin.

●**3.2   Composite beams**
In Chapter 23, all beams, wires, tubes and so on were considered to be of one homogeneous, isotropic material, and in many systems in practice this is entirely sufficient. However, there are some situations in dentistry where the 'beams' are either not symmetrical about the z-axis, or of more than one material, or both.

To start, we need to be able to identify the location of the neutral axis, a point that was glossed over before as symmetry about the z-axis was assumed. If the section is symmetrical about the vertical y-axis, the neutral axis must pass though this. Then, the height above a reference baseline for the overall **centroid** for a section is found as the **first moment of area** with respect to that baseline, that is, as the area-weighted mean height $\overline{\overline{y}}$ of the separate component sections (Fig. 3.1), for as many as there are:

$$\overline{\overline{y}} = \frac{\sum_i A_i \overline{y}_i}{\sum_i A_i} \qquad (3.11)$$

where the weights $A_i = b_i h_i$, and it is through this point that the neutral axis must pass (but see equations 3.17 ~.18 for proof).[2] (The baseline is entirely arbitrary – any can be chosen, as is most convenient.) The second step is to calculate the total **second moment of area** (23§3) about the axis parallel to that baseline which passes through that same point:

$$I_z = \sum_i \left(\frac{b_i h_i^3}{12} + A_i \left(\overline{y}_i - \overline{\overline{y}} \right)^2 \right)$$
$$= \sum_i I_{zi} \qquad (3.12)$$

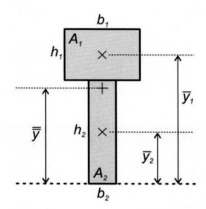

Fig. 3.1 A composite beam consisting of two sections of the same material. The heights of the centroids are measured from the (arbitrary) baseline (broken line).

correcting for the distance of each component's centroid from the neutral axis, using what is known as the **parallel axis theorem**. This can be read as saying that the overall value of I_z is simply the sum of the component I_zs (the I_{zi}) about the overall neutral axis (not about their own centrelines), which is another example of superposition, *i.e.* simple additivity (23§3.1). This therefore deals with the situation of arbitrary sections.

We move now to a beam of uniform width but composed of sections having differing elastic moduli, that is to say, a **laminated** beam (Fig. 3.2, left). We start by noting that the flexural rigidity of a section is given by the product EI_z (23§3), and this must apply no matter what is taken as the reference baseline, so that we have for the whole beam:

$$E_{ref}I_z = E_{ref}\sum_i \left(b_i \frac{E_i}{E_{ref}} \left[\frac{h_i^3}{12} + h_i d_i^2 \right] \right)$$
$$= \sum_i \left(b_i E_i \left[\frac{h_i^3}{12} + h_i d_i^2 \right] \right)$$
$$= \sum_i \left(E_i \left[\frac{b_i h_i^3}{12} + A_i d_i^2 \right] \right) \qquad (3.13)$$
$$= \sum_i E_i I_{zi}$$

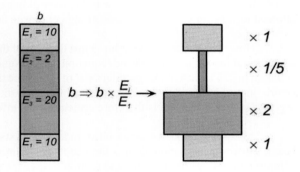

Fig. 3.2 A composite laminated beam consisting of different materials. The effective relative breadths of the sections are in proportion to their relative elastic moduli against any reference value. The overall scale factor is unknown.

Now, for each section, the value is proportional to the breadth of the section, b_i, as well as to the value of E_i. Thus, to keep the flexural rigidity of a section constant, doubling the breadth (for example) would require that the elastic modulus be halved, and *vice versa*. Therefore, to see the effect of various E_i, we can scale the breadth of sections in proportion to their elastic moduli with respect to a reference value, chosen for convenience (Fig. 3.2, right). Changing the elastic moduli of course means that the level of the centroid is changed, which therefore needs recalculating. So, expanding from equation 3.11, we can write:

$$\overline{\overline{y}}_L = \frac{\sum_i \left(\frac{E_i}{E_{ref}} b_i \right) h_i \overline{y}_i}{\sum_i \left(\frac{E_i}{E_{ref}} b_i \right) h_i} = \frac{\sum_i E_i A_i \overline{y}_i}{\sum_i E_i A_i} \qquad (3.14)$$

where the reference value cancels out and is therefore of no consequence in itself at this point. This is therefore

---

[2] If the beam is not laterally symmetrical, a similar process can be used to find the first moment about a perpendicular reference line, and therefore the intersection of the two locates the point through which the neutral axis must pass.

a weighted mean first moment, where the weights are the values of the $E_i A_i$. Naturally, if the E_i are all the same, equation 3.14 reduces to 3.11.

Applying the same scaling principle to the calculation of the overall value I_{zL}, we get:

$$I_{zL} = \sum_i \left(\frac{E_i}{E_{ref}} b_i \right) \left(\frac{h_i^3}{12} + h_i \left(\bar{y}_i - \bar{\bar{y}}_L \right)^2 \right) \tag{3.15}$$

using the new centroid position, although we do not at this point know the appropriate value of E_{ref} for this to make sense. However, if we multiply both sides by E_{ref}, we have:

$$\begin{aligned} E_{ref} I_{zL} &= E_{ref} \sum_i \left(\frac{E_i}{E_{ref}} b_i \right) \left(\frac{h_i^3}{12} + h_i \left(\bar{y}_i - \bar{\bar{y}}_L \right)^2 \right) \\ &= \sum_i \left(E_i \left[\frac{b_i h_i^3}{12} + A_i \left(\bar{y}_i - \bar{\bar{y}}_L \right)^2 \right] \right) \\ &= \sum_i \left(E_i A_i \left[\frac{h_i^2}{12} + \left(\bar{y}_i - \bar{\bar{y}}_L \right)^2 \right] \right) \quad \ldots * \\ &= \sum_i E_i I_{zi} \end{aligned} \tag{3.16}$$

with E_{ref} cancelled out on the right-hand side. But, by comparison with equation 3.13, we see that this is just the flexural rigidity of the whole laminated beam, $i.e.$ $E_{ref} = E_L$. However, the separate values for E_C and I_{zL} cannot be found. (The line marked * is included to show that the weights $E_i A_i$ which emerged at equation 3.14 remain the controlling weight factor in the overall behaviour.) It also follows that the widths b_i can also be allowed to vary for calculations concerning composite laminated beams, should the need arise.

However, as can be seen, this is just the sum of the component flexural rigidities referred to the overall, elastic modulus-weighted, centroid (equation 3.14). Again, this is a superposition. The difference here from the situation in 23§3.1 is that the 'parallel axis' offsets place the second moments about the correct overall, weighted centre line, as opposed to the individual centre lines, thus seeing the benefit of the components being bonded together (which condition is therefore critical for the effect).

At equation 3.12 we have a clear, independent statement of the effective second moment in a composite beam. However, we cannot do this for a composite laminated beam because, in effect, there are many ways of combining modulus of elasticity and section to give the same overall behaviour: a given value for the flexural rigidity is not uniquely defined in a such structure.

It is worth showing that the neutral axis does in fact pass through the beam centroid. For equilibrium, the net force, taken across the whole cross-section, must be zero (equation 2.5). We therefore have:

$$\sum_i \int_{A_i} \sigma_i . dA = \sum_i \frac{E_i}{r} \int_{A_i} y . dA = \sum_i E_i \int_{A_i} y . dA = 0 \tag{3.17}$$

That is, summing across the area A_i of each of the i components of the beam, where y is the distance from the neutral plane, and the radius of curvature is taken to be a constant value (see equation 23§3.1). The last integral ($\int y . dA$) is just the first moment of area of that section with respect to the level of the neutral axis. Hence, for the rectangular sections we have used here, for simplicity, this becomes:

$$\sum_i E_i b_i h_i \left(\bar{y}_i - \bar{\bar{y}}_L \right) = \sum_i E_i A_i \left(\bar{y}_i - \bar{\bar{y}}_L \right) = 0 \tag{3.18}$$

That is, the weighted first moments about the neutral axis sum to zero.

We can now proceed to determine the relative stress distribution in such beams. In any beam, for continuity, the strain must be linearly related to the distance y from the neutral axis, where the strain must be zero:

$$\varepsilon_x = \frac{y}{r_c} \tag{3.19}$$

where r_c is the radius of curvature, which is inversely proportional to the bending moment. But since $\sigma = \varepsilon E$

(equation 1§2.4a), the stress distribution is exactly similar (Fig. 3.3a). For a simple homogeneous beam, these distributions are symmetrical about the neutral axis. The similarity of strain and stress distributions applies also to the composite beam when the location of the neutral axis has been determined (Fig. 3.3b), but now they are not symmetrical. In a laminated beam, however, the same linearity applies to the strain, of course, but now not only is the correct neutral axis required, as above, but the stress must be calculated locally, corresponding to the value of E in each segment (Fig. 3.3c). This leads automatically to a discontinuity in the longitudinal stress field and thus an extra shear stress at the interface. A similar process may be used in the natural extension to examine a composite laminated beam where both the component breadths and elastic moduli vary, and for any number of components. (Again, it should be said that these simple calculations apply for small deflections only.)

Having calculated stresses, it is of course now possible to consider the strength of such structures under bending and to determine whether collapse (or plastic deformation) occurs in tension or compression, *i.e.* on which face does the failure stress get attained first.

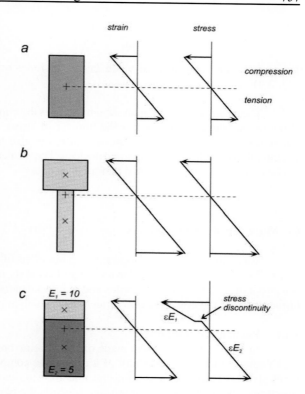

Fig. 3.3 The nature of the strain and stress distributions in a beam under bending: a) homogeneous, b) homogeneous composite, c) laminated. Component centroids: ×; overall centroid: +.

●3.3 Shear

As discussed in §2.1~2.5, there may be a shear force acting across the section of beam except in the case of pure bending. Assuming a uniform cross-section along the length of the beam, the vertical shear stress is simply given by $\sigma_{yz} = V/A$. However, it is apparent that in a beam where such a stress occurs there must also be a longitudinal (*i.e.* lengthwise) or **horizontal shear stress.** This can be seen by considering Fig. 23§3.3: if there is no slip, such as by gluing the pieces together, there must be a shear stress along that centre line. The derivation of the relevant equation is somewhat elaborate and serves no purpose here in itself, so is simply presented. The shear stress varies across the depth of the beam, and it must be zero at the top and bottom surfaces (as these are free surfaces no stress could be sustained). Thus we have:

$$\tau = \frac{VQ}{I_z b} \tag{3.20}$$

where V is the shear force, and b is the breadth as before. Q is a bit more complicated:

$$Q = \int_{y'}^{y_{max}} by.dy \tag{3.21}$$

where y' is the height of the level of interest with respect to the centroid of the section, *i.e.* above or below the neutral surface. For the simple case of a rectangular section, $Q = A' y'$, where A' is the area of the section outside y', *i.e.* from y' to y_{max}, but composite beams of any shape be can dealt with by the integration, noting that variation in b must be recognized in equation 3.21. However, in the simple rectangular case, Q is given by:

$$Q = \frac{b}{8}\left(h^2 - 4y^2\right) \tag{3.22}$$

Hence:

$$\tau = \frac{V(h^2 - 4y^2)}{8I_z} \tag{3.23}$$

which is a simple parabolic function, with a maximum at y = 0 given by:

$$\tau_{max} = \frac{3V}{2bh} = \frac{3V}{2A} \qquad (3.24)$$

that is, 1.5 times the average value based on force over area (Fig. 3.4).

The existence of horizontal shear underlines the importance of the union between laminae for the transfer of stress, both to realize the expected stiffness of the structure, but also to avoid delamination in built-up structures and devices. These calculations can be extended to composite and laminated beams.

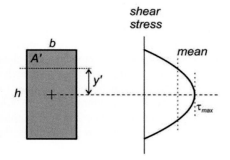

Fig. 3.4 Distribution of horizontal shear stress in a simple beam in bending.

§4. More Filled Resins : Fibres

While the mechanical properties of particulate-filled resins have been discussed at some length already (6§2), fibre reinforcement is used in a variety of contexts, *e.g.* denture bases, bridges, and endodontic posts. The behaviour of such systems is complex and requires a more elaborate explanation. What complicates the picture further then are the effects of fibre orientation, which must therefore be taken into account.

●4.1　Particulate composite

In 6§2.6, estimates were made of the extreme bounds of the (Young) modulus of elasticity of a particulate composite in terms of rules of mixtures, restricting the modelling to simple layer structures. We may take one more step towards the particulate system in the following fashion as a prelude to considering fibre reinforcement, but always with the assumption of good bonding at the interface, for stress transfer.

Consider a unit cube of the matrix material into which is set in one corner a smaller cube of the disperse phase, the filler (Fig. 4.1a). The volume fraction of the filler is then φ_f, the area fraction in the (top) face is α_f, and the depth or length fraction in a side is correspondingly λ_f. These values are related:

$$\phi_f = \lambda_f^3 \;\; ; \;\; \alpha_f = \lambda_f^2 \qquad (4.1)$$

So that

$$\lambda_f = \phi_f^{1/3} \;\; ; \;\; \alpha_f = \phi_f^{2/3} \qquad (4.2)$$

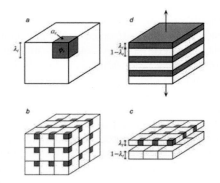

Fig. 4.1 Approximation to a particulate-filled composite structure. (a) The unit cube. (b) Unit cubes assembled. (c) Equivalent layer dissection. (d) Equivalent isostress model.

Then, assembling a body from those unit cubes (Fig. 4.1b), it can be seen that this may be dissected into layers of two kinds (Fig. 4.1c): one being composite, of thickness λ_f, the other being matrix only, of thickness $1 - \lambda_f$, and the whole body can therefore be viewed as a Reuss isostress system (Fig. 4.1d). The first kind of layer, however, is clearly a Voigt isostrain system in itself (*cf.* Fig. 6§2.9) and so its effective modulus of elasticity is given by the weighted arithmetic mean:

$$E_{c\lambda} = \alpha_f E_f + (1 - \alpha_f) E_m \qquad (4.3)$$

This value can then be inserted in the isostress harmonic mean equation:

$$\frac{1}{E_c} = \frac{\lambda_f}{E_{c\lambda}} + \frac{(1 - \lambda_f)}{E_m} \qquad (4.4)$$

where the weights are the layer thicknesses. Substitution and rearrangement then gives:

$$E_c = E_m \frac{\lambda_f^2 (E_f - E_m) + E_m}{(\lambda_f^2 - \lambda_f^3)(E_f - E_m) + E_m} \qquad (4.5)$$

as a kind of weighted mean of the Voigt and Reuss outer bounds.[4]3 This estimate may be termed the **Paul mean**. Obviously, the behaviour is isotropic on this basis as the layer dissections in the three directions are equivalent. That this is a reasonable working approximation can be seen from the results for the Co-WC system (Fig. 21§3.3).

Despite the manner in which the array of unit cubes was assembled, it is clear that, firstly, successive filled layers (in the load direction) do not have to be aligned vertically in any way for $E_{c\lambda}$ to be calculated, and secondly that there is no need for the particles to be regularly arranged in those layers, or indeed to be of uniform area α_f, if columnar shape[4], λ_f and total volume fraction are all preserved (with a sensible approach to homogeneity for the distribution). Thus, subject to those three constraints, random sizes, random (end surface) shapes and random distribution of particles within each filled layer yield the exact same solution. However, further generalization is not possible (that is, any relaxation of those constraints).

Comparison with the development in 6§2.9 shows that the shear modulus may be treated in exactly the same fashion, substituting the values of G for those of E in equation 4.5. Estimates of both the bulk modulus:

$$B = \frac{E}{3\left(3 - \dfrac{E}{G}\right)} \tag{4.6}$$

and Poisson ratio

$$\nu = \frac{E}{2G} - 1 \tag{4.7}$$

may then be made (see 1§2.5, 1§2.6) ... with caution – it is stressed that the Paul mean equations are simplistic (there are many more elaborate models in the literature).

•4.2   Long fibre reinforcement - elastic modulus

A common approach to such reinforcement is to use oriented bundles of long fibres. As is apparent from Fig. 4.2a, considered in the sense of the load axis, a plain Voigt isostrain model for the modulus of elasticity is appropriate, using $\alpha_f = \varphi_f$, and again we have

$$E_c = \alpha_f E_f + (1 - \alpha_f)E_m \tag{4.8}$$

It follows that this result is unaffected by the spatial arrangement, or fibre cross-sectional areas and shapes. To understand what can be achieved in this mode we need to consider how much fibre can be incorporated, as we did for the packing of spherical particles (Fig. 4§9.2).

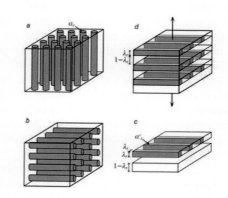

Fig. 4.2   Approximation to a fibre-filled composite structure. (a) Isostrain system. (b) Transverse structure. (c) Equivalent layer dissection. (d) Assembled structure.

The packing of fibres is an easier problem, equivalent to packing discs in a plane, but now this simple model is closer to real systems because most fibres in use are circular in section or nearly so, and generally more or less uniform in size in this respect. Simple geometry shows that for hexagonal close packing (Fig. 11§3.15) we have $\alpha_f = \phi_f = \pi/2\sqrt{3} \approx 0.91$, although this is very difficult to approach even with the best techniques (strictly speaking, the matrix then becomes discontinuous because the fibres are in contact along their length and there would be no bonding directly between them). The densest random packing of uniform, parallel fibres is found to give $\phi_f \approx 0.81$,[5] while the random loose-packed limit has been found to be $\phi_f \approx 0.55$.[6] Typically, however, the densest packing that can be achieved with normal processes is around $\phi_f \approx 0.60 \sim 0.65$, which is not a lot better than random loose-packing. It would not take many alignment defects in a supposedly parallel bundle to cause this. Fibres are obviously much more constrained as to possible arrangements than are particles.

For a load axis perpendicular to the fibres, transverse loading, we again have a more difficult situation (Fig. 4.2b). However, if the composite can be dissected into layers (Fig. 4.2c), as was done above in Fig. 4.1c, then we may proceed in like fashion, but now we must approximate the round fibres with square ones of the same

---

3   Paul derives an equivalent equation *via* a different route (energy theorems), without reference to Voigt or Reuss.

4   That is, uniform cross-section, as for an extrusion of the end-face area, with the axis normal to the plane of the layer.

cross-sectional area (to preserve the axial value of E_c), that is, $\lambda^2_f \propto \pi r^2$, where r is the fibre radius. Looking at the top surface, it can be seen then that the value of $\alpha'_f = \lambda_f$, so instead of equation 4.3 we have:

$$E_{c\lambda} = \lambda_f E_f + (1-\lambda_f)E_m \qquad (4.9)$$

The isostress equation remains as at equation 4.4 because the layer thicknesses are defined in the same way (and $\phi'_f = \lambda^2_f$), so that after substitution from equation 4.6 and rearrangement we have:

$$E_c = E_m \frac{\lambda_f(E_f - E_m) + E_m}{(\lambda_f - \lambda^2_f)(E_f - E_m) + E_m} \qquad (4.10)$$

(note the change of powers for the λs). Such a model (Fig. 4.2d) does not admit of much freedom in the fibres except in their regularity of lateral distribution, and in any case is a relatively crude representation. Nevertheless, it provides an indication of the main effects for an oriented fibre reinforcement, just as equation 4.5 is indicative. Now, however, because the fibre itself may be structured, such as a highly-drawn polymer, its properties will be anisotropic. Thus, for example, for a polyaramid (Kevlar) fibre with a longitudinal value of $E_{fl} = \sim 150$ GPa, the transverse value is only $E_{ft} = \sim 4$ GPa, leading to a dramatic difference in behaviour. Thus, in an epoxy matrix (E = ~3.5 GPa) these fibres have essentially no effect (Fig. 4.3) (and the elaborate calculation unnecessary). A similar outcome can be expected for the commonly used polyethylene fibres ($E_{fl} = \sim 100$ GPa, $E_{ft} = \sim 3$ GPa) in an acrylic resin. The contrast with the effect of glass fibre ($E_f = \sim 71$ GPa; isotropic) is marked, but even here the fibre loading must be very high to make an appreciable difference. Notice, though, that very high volume fractions are not possible.

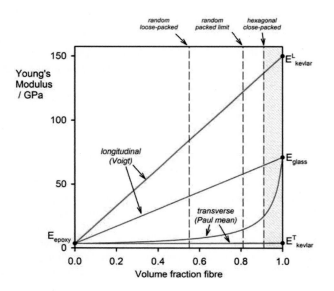

Fig. 4.3 Behaviour of Kevlar- and glass fibre-filled composites. (The hatched area on the right is inaccessible.)

As above, the shear modulus may be estimated by the appropriate substitutions; anisotropic fibre properties must again be taken into account, although the transverse values for drawn polymer fibres may not be available. However, since the composite structure itself is anisotropic, the simple estimates of the bulk modulus and Poisson ratio are not legitimate, even for isotropic fibres.

We may also note that the modulus of elasticity in bending (Fig. 23§2.10) for a composite in which the fibres run the length of the beam (Fig. 4.2b, extended in the fibre direction) is also an isostrain system and that equation 4.8 would be expected to apply. Likewise, the transverse stiffness would be estimated from equation 4.10, but again using the transverse E value for the fibre when this is anisotropic. Thus, diagrams similar to Fig. 4.3 would again be obtained for such systems. Broadly, lengthwise bending stiffness can be raised substantially, but the transverse value is hardly affected.

●4.3   Long fibre reinforcement - strength

In the matter of the effect of fibres in a composite, we are obliged to assume that the fibres are both stiffer and stronger in tension than the matrix else there could be no worthwhile effect. Then, because we are dealing with an isostrain Voigt model (Fig. 4.2a), and because it is strain ultimately that controls the strength (6§2.10, 10§3.6), we have two situations to consider:

(a) the strain for failure of the matrix is less than that for the fibre, $\varepsilon_m^* < \varepsilon_f^*$, and
(b) the converse, $\varepsilon_m^* > \varepsilon_f^*$.

Thus, for situation (a), looking at just the load-carrying capacity of the fibres, σ_f, that is, the strength of the bundle alone, this is just proportional to their total cross-sectional area:

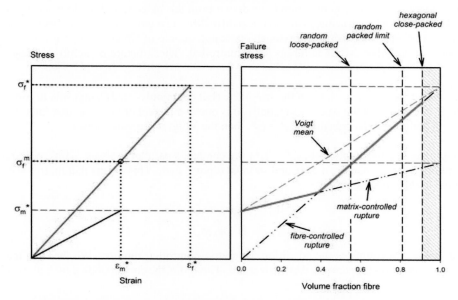

Fig. 4.4 Left: when the matrix fails first ($\varepsilon_m^* < \varepsilon_f^*$), the fibres are asked to carry the load. Right: The effect of volume fraction of the strength of an oriented long-fibre composite under that condition. The actual collapse stress is shown by the heavy line.

$$\sigma_f = \phi_f \sigma_f^* \qquad (4.11)$$

where σ_f^* is the failure stress for the fibre itself. This is plotted in Fig. 4.4 (right) as the condition for fibre-controlled rupture. On the other hand, the load-carrying capacity of the matrix is determined by its strain, which is limited by definition to match that of the fibres. Therefore, from this point if view the overall stress at failure, σ, is given by:

$$
\begin{aligned}
\sigma &= \phi_f E_f \varepsilon_m^* + (1-\phi_f) E_m \varepsilon_m^* \\
&= \phi_f E_f \varepsilon_m^* + (1-\phi_f) \sigma_m^* \\
&= \phi_f E_f \frac{\sigma_m^*}{E_m} + (1-\phi_f) \sigma_m^* \\
&= \phi_f E_f \sigma_f^m + (1-\phi_f) \sigma_m^*
\end{aligned}
\qquad (4.12)
$$

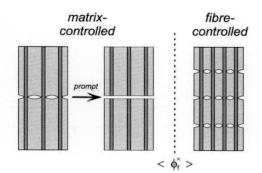

Fig. 4.5 Failure process for $\varepsilon_m^* < \varepsilon_f^*$. At low fibre content ($\phi_f < \phi_f^\times$), the fibres cannot carry the load when the matrix fails. Above $\phi_f^\times$ the matrix breaks down progressively, until the fibres fail, in a process called "slabbing".

where σ_m^* is the strength of the matrix material and $\sigma_f^m = E_m/\sigma_m^*$ is the stress in the fibres when the matrix fails (and because, of course, $\varepsilon_m = \varepsilon_f$). This stress is then plotted in Fig. 4.4 (right) as for matrix-controlled rupture. At low values of ϕ_f, the strain is obviously not enough for the fibres to break (Fig. 4.5). However, when the matrix fails the fibres are unable to carry the total load, and complete rupture occurs immediately. At high values of ϕ_f, although the matrix will have ruptured, and even in many places, the fibres can still carry the load. Thus the ultimate strength continues to rise with ϕ_f. The transition from matrix- to fibre-controlled rupture, where the two lines cross, is given by the usual coordinate geometry formula,[7] which reduces to:

$$\phi_f^\times = \frac{\sigma_f^m}{\sigma_f^* - \sigma_f^m + \sigma_m^*} \qquad (4.13)$$

Of course, in some circumstances it might not be very useful to have a composite with many cracks, but the benefit of a large proportion of strong fibres is clearly seen in that very much higher stresses can be carried that could the matrix on its own and the fibres serve to maintain the structure coherent if not intact. The breakdown occurs in a characteristic fashion called **slabbing**, which successively subdivides the (remaining) sections into equal portions until the stress on the fibres is great enough to cause them to fail.

Even so, there is a positive implication to be drawn from this behaviour: bubble-insensitivity. That is, the existence of bubbles in the matrix due to the practical difficulties of making up such fibre-reinforced systems, whether by infiltration *in situ* or using a previously-saturated bundle of fibres, 'pre-preg', does not in itself constitute a source of fatal flaws while failure is fibre-controlled. The structure is usefully damage-tolerant in that respect.

One can also note from Fig. 4.4 that the simplistic (but commonly-used) Voigt rule of mixtures model for the composite strength does not work. Essentially, it is one component or the other that fails. (However, if it happens that the failure strains are equal, $\varepsilon_m{}^* = \varepsilon_f{}^*$, then the Voigt mean will apply.)

Turning to the second situation (b, $\varepsilon_m{}^* > \varepsilon_f{}^*$), we conduct a parallel analysis. Thus, we notice that the fibres can tolerate less strain than the matrix and so must rupture first. This means that matrix is carrying the load proportional to its volume fraction:

$$\begin{aligned} \sigma_m &= \phi_m \sigma_m^* \\ &= (1-\phi_f)\sigma_m^* \end{aligned} \tag{4.14}$$

This is plotted in Fig. 4.6 (right) as the condition for matrix-controlled rupture. We then see that the load-carrying capacity of the fibres is determined by their strain. Therefore, the corresponding overall stress at failure, σ, is given by:

$$\begin{aligned} \sigma &= \phi_f E_f \varepsilon_f^* + (1-\phi_f)E_m \varepsilon_f^* \\ &= \phi_f \sigma_f^* + (1-\phi_f)E_m \varepsilon_f^* \\ &= \phi_f \sigma_f^* + (1-\phi_f)E_m \frac{\sigma_f^*}{E_f} \\ &= \phi_f \sigma_f^* + (1-\phi_f)\sigma_m^f \end{aligned} \tag{4.15}$$

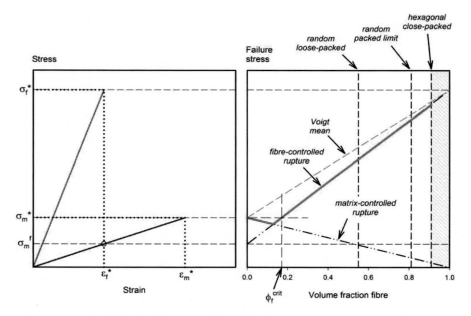

Fig. 4.6 Left: When the fibres fail first ($\varepsilon_m{}^* > \varepsilon_f{}^*$), the matrix is asked to carry the load. Right: The effect of volume fraction of the strength of an oriented long-fibre composite for that condition.

Again, $\varepsilon_m = \varepsilon_f$, necessarily, and $\sigma_m^f = E_f/\sigma_f^*$. We can then plot this in Fig. 4.6 (right) as for fibre-controlled rupture. It is now apparent that low values of ϕ_f cause the composite strength to fall, with the minimum occurring at the intersection given by:

$$\phi_f^{\times} = \frac{\sigma_m^* - \sigma_m^f}{\sigma_f^* - \sigma_m^f + \sigma_m^*} \tag{4.16}$$

Indeed, it is necessary to exceed a critical volume fraction value given by:

$$\phi_f^{crit} = \frac{\sigma_m^* - \sigma_m^f}{\sigma_f^* - \sigma_m^f}$$

(4.17)

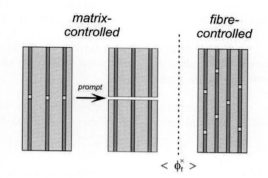

for the composite strength to be greater than the matrix alone. Thus, just a few relatively stiff fibres in a composite serve no useful purpose and indeed are detrimental. Below the minimum intersection, fibre failure leads to complete rupture (Fig. 4.7), but above that point the matrix can carry the load. That is, although the fibres are broken the matrix stays intact. Of course, not all fibres will in practice fail at the same moment as the 'weakest link' sequence must prevail (1§7.1), although it is likely that the spread in values may be small. In fact, further fractures of the same fibre may occur along its length as the conditions for rupture apply locally. In this situation, bubbles in the matrix may well constitute fatal flaws, and their elimination is therefore critical.

Fig. 4.7 Failure process for $\varepsilon_m^* > \varepsilon_f^*$. At low fibre content ($\phi_f < \phi_f^x$), the matrix cannot carry the load when the fibres fail. Above ϕ_f^x the fibres break down progressively until the matrix fails.

Again, a simple arithmetic averaging for strength does not work (Fig. 4.6), and indeed would be very misleading at low volume fractions of fibre, and more so as the strain at failure for the fibres falls. As above, the only way this model would work is if the strains at failure were identical: $\varepsilon_m^* = \varepsilon_f^*$, when in principle failure would be simultaneous. This would require a very highly cross-linked resin and a soft glass, for example.

●**4.4   Angulation**
So far, we have considered only on-axis loading of fibres for the strength of their composites, although it would be difficult to find situations in dentistry with such pure axial loading as the intended system. The behaviour in general now becomes very much more complicated. Not only does the possibility of shear failure arise, but also buckling (23§9) of the fibres and failure of the fibre-matrix interface bonding (which is required for matrix constraint, as in all composites) need to be considered. The detail will not be considered here, only some general trends outlined.

For a simple, oriented bundle system (Fig. 4.2a), of a given volume fraction of fibre, the theoretical strength (stress at failure) is indicated by:

$$\sigma = \frac{\sigma_1^*}{\cos^2 \theta}$$

(4.18)

where the asterisk refers to the reference value for failure in on-axis loading (subscript 1), and θ is the angle of the fibre axis to the load axis. Likewise, for transverse loading (subscript 2), we have:

$$\sigma = \frac{\sigma_2^*}{\sin^2 \theta}$$

(4.19)

Then, for in-plane shear (§5), we have:

$$\sigma = \frac{\tau_{12}^*}{\sin \theta \cos \theta}$$

(4.20)

where the shear occurs in a plane parallel to the fibre axis. The overall effect of this is shown in Fig. 4.8, where of course the principle of least work requires that the strength with the lowest value is observed (that is, the minimum value of those three equations). Thus, the manner of failure is strongly dependent on the orientation of the load axis with

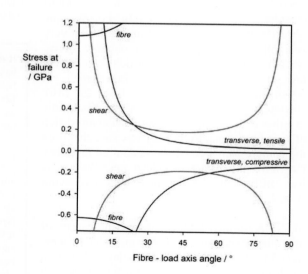

Fig. 4.8 Effective strength of a typical glass fibre-epoxy resin unidirectional composite according to load axis angle: in tension above, compression below.

respect to the fibre direction, but most importantly the strength falls dramatically for loading more than a few degrees off axis.

The picture is made more complicated by the fact that other failure criteria exist (e.g. maximum strain), given that Poisson strain and bonding (including its defects) need also to be taken into account, and that anisotropic fibres (drawn polymers) can dramatically change the behaviour, as well as the possibility of fibre buckling in compression. Ultimately, experimental work is required, even if designs are guided by theory.

Clearly, the behaviour in the presence of flaws such as bubbles in the matrix, or imperfect fibre bonding, for off-axis loading may show some sensitivity, although situations in dentistry may not be as critical as for aerospace applications.

An attempt has been made to describe the 'efficiency' of fibre reinforcement.[8] Taking axial loading as 100 %, and transverse as 0 % efficient, the following function was used to model the behaviour:

$$\eta = \sum_i \alpha_i \cos^4 \theta_i \qquad (4.21)$$

where α_i is the proportion of the fibres present with orientation θ_i with respect to the load axis, for i sets of fibres. Some results for this commonly-used **Krenchel** efficiency formula are shown in Fig. 4.9.

For a single orientation, the predicted fall-off is not as steep as actually occurs for glass in resin (Fig. 4.8), overestimating the benefit, but clearly the 0% efficiency at 90° is unrealistically extreme as some effect is still present in practice. For two orthogonal sets of fibres, the factor η varies between 0.5 and 0.25. For three or more sets of fibres, $\eta = 0.375$ identically at all angles (this is a mathematical result). Overall, the implication is that both the stiffness and strength of laminar long fibre composites in tension can be made more isotropic and, most simply, with three layers, although at the cost of lower values.

In bending, however, the situation is more complicated. Firstly, the variation in modulus of elasticity from layer to layer (due to orientation) must be taken into account (§3.2). For isotropic fibres, such as glass, this is relatively straightforward, but for drawn polymer fibres, where the lateral modulus of elasticity can be very low, the overall effect on stiffness depends on which layer is where (Figs. 4.10, 23§2.10). Obviously, the transverse fibres will contribute little in either case, but the value for axial compression of such fibres can also be very low such that if on the compression side of the neutral surface they may have little effect, or indeed make the lamina less stiff than the unfilled matrix material (Fig. 4.10).

A further problem arises in respect of multiple orientations and the attainable volume fraction. Where the structure is layered, with a single orientation in each, the typical value $\varphi_f \sim 0.6$ or so applies (§4.2), although at the interface there will be a slightly poorer packing, of course. For a

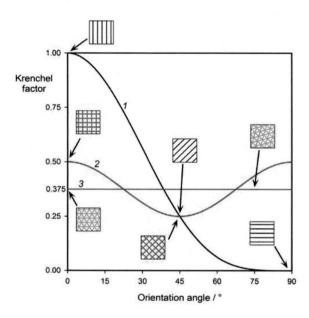

Fig. 4.9 Krenchel 'efficiency' factor for (1) a single set of parallel fibres, (2) two equal sets at right-angles, and (3) three equal sets at 60°, as a function of orientation angle. This does not take into account volume fraction effects.

It must be said that the derivation of equation 4.21 was based on the implicit independence of the fibre strain behaviour in an assumed symmetrically-loaded pair of fibre sets (as if in the self-same plane!) at an arbitrary angle. This was then applied as the 'efficiency' factor for the fibre term in the Voigt mean elastic modulus (equation 4.8), which – as has been shown above – cannot, in general, apply (Fig. 4.3). This Voigt mean is, however, very widely used, quite uncritically, in such calculations anyway (often with a series of various other *ad hoc* correction factors), and must therefore everywhere be in error except for the particular condition of axial loading (when the other corrections disappear). The slight resemblance of the shape of the curve to the net effect shown in Fig. 4.8 is a coincidence, as the actual failure behaviour switches modes according to angle (and this is clearly the most important aspect of the process), whereas equation 4.21 assumes fibre tensile failure uniformly, which obviously cannot happen.

woven fibre fabric, this figure drops to 0.45 ± 0.05 because of the packing loss at each fibre crossing. This is also subject to detrimental effects due to the 'crimp' of the fibres in the weaving, but also the effective modulus of elasticity in tension is reduced because pure axial loading is not possible (the flexural stiffness is also reduced, *cf.* 23§6). The packing inefficiency rises further for a random mat of fibres, *i.e.* non-woven two-dimensional fabric, when typically $\varphi_f \sim 0.20 \pm 0.10$ is obtained. For such a random mat, because the angles are approximately uniformly distributed, the Krenchel $\eta = 0.375$ would apply, were it in fact valid (see box). A similar calculation for a three-dimensional random fibre filler gives $\eta = 0.20$, but then the maximum φ_f is ~0.20,[9] which suggests a relatively small effect overall ($\eta.\varphi_f \nRightarrow \sim 0.04$). It can also be expected that surfaces and edges will have lower fibre content, especially if efforts are made to prevent exposure on trimming or polishing the piece, or even just accommodating the

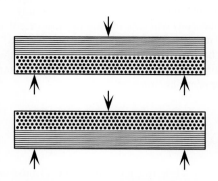

Fig. 4.10 The stiffness of drawn polymer fibre composites in bending depends on which fibres are in tension or compression.

possibility of wear (the so-called 'gel coat' on fibreglass boat hulls is just such an allowance). Exposed hard fibre ends, such as glass, will be abrasive and irritant to soft tissue.

●4.5   Critical fibre length

The discussion above has been in terms of 'continuous' fibres, that is, spanning the object. This is not always practical, but the question is how long must the item be to count as a fibre, as opposed to a particulate filler? A simple analysis uses the so-called **shear-lag** model in which it is envisaged that the tensile stress in the fibre is transferred to the matrix *via* the interfacial shear stress, τ_{fm} (Fig. 4.11), which transfer is taken to be linear (no load being carried on the fibre end-face). Then, balancing the forces, we obtain:

$$\sigma_f \pi r^2 \ = \ \tau_{fm} 2\pi r \frac{L_{crit}}{2} \qquad (4.22)$$

where it emerges that the tensile stress in the fibre is less than the maximum possible if the fibre is shorter than L_{crit}. If that maximum is the failure stress of the fibre, σ_f^*, then the **critical fibre length**, L_{crit}, is defined by:

$$L_{crit} \ = \ \frac{\sigma_f^* r}{\tau_{fm}^*} \qquad (4.23)$$

where τ_{fm}^* is the interfacial shear failure stress. This is better expressed in terms of the **axial ratio** of the fibre:

$$\frac{L_{crit}}{D} \ = \ \frac{\sigma_f^*}{2\tau_{fm}^*} \qquad (4.24)$$

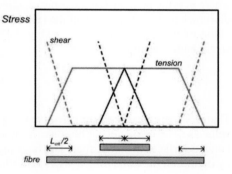

Fig. 4.11 Critical fibre length corresponds to the peak tensile stress being attained in the fibre.

where D is the fibre diameter, and this quite clearly depends very much on the quality of the interfacial bond (although if the matrix shear strength is lower and this value should be used instead). For glasses that may be silanized this ratio can be good, but for low reactivity materials such as polyethylene fibres there are serious challenges. No bond would mean no reinforcement, as before (6§2.10). Typical values for the axial ratio range from about 10 to 40, and obviously the smaller this value the better (*i.e.*, through strong bonding), especially when fibre fragmentation may occur during processing. What this then means for the second situation above (§4.3) (b, $\varepsilon_m^* > \varepsilon_f^*$) is that at some point the remaining fibre lengths reach or fall below L_{crit} and failure occurs through fibre pull-out if the matrix has not already failed in tension.

In a three-dimensional random, short-fibre filled material, the effects may be dominated by those fibres that are loaded transversely (Fig. 4.8). That is, crack initiation must occur at the weakest point, and this is probably the matrix interface for those transverse fibres. Again, the bonding is seen to be critical. Thus, the axially-loaded fibres must increase the modulus of elasticity enough that the strain for interfacial failure does not arise. The obvious remarks may also be made in respect of bubbles in the matrix.

The creation of a three-dimensional random, short-fibre filled object may be difficult in practice. In any system where flow is involved, such fibres will tend to align with the local shear direction (they experience a turning moment, Fig. 4§3.3). Thus, any such material that has to be packed into place (such as a resin restorative material) will tend locally to show orientation effects, assuming that the fibres do not get broken in the process. That is, the intended isotropy would be lost, with consequential variation in strength and stiffness characterization by direction, across the object. Whether this is good or bad clearly depends on the circumstances and the location, but it could also be turned to advantage if the flow could be controlled in an appropriate manner.

§5. Fracture Mechanics

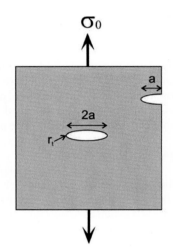

For the most part, the discussions of properties in relation to collapse or failure have been in terms of strength in one sense or another, with little regard given the processes or conditions relevant to the propagation of the crack apart from some references to the Griffith criterion (1§7) and polymer drawing (3§4.7, 5§5, 6§2.14). Fracture mechanics is the study of those processes and conditions, with a view to avoiding problems and improving the design of devices and materials.

●5.1   Stress concentration
For a crack lying perpendicular to the applied stress, the stress, σ_{max}, at the crack tip is found to be given by:[10]

$$\sigma_{max} = \sigma_0 \left(1 + 2\sqrt{\frac{a}{r_t}} \right) \qquad (5.1)$$

where σ_0 is the nominal (engineering) applied stress (1§2.4), a is the length of a crack at the surface, or half the length of an internal crack, and r_t is the radius of curvature at the crack tip (Fig. 5.1). We can therefore define the **stress concentration factor**, K:

Fig. 5.1 Cracks under stress at the surface and in the bulk.

$$K = \frac{\sigma_{max}}{\sigma_0} = 1 + 2\sqrt{\frac{a}{r_t}} \qquad (5.2)$$

That is, sharp cracks are more effective for crack initiation than are blunt cracks. (Obviously, the implied infinite value as $r_t \to 0$ cannot occur because the atomic scale is the limit; 1§7.) Hence, grain boundary defects, or non-bonded surfaces in composites, say, have high values of K. In the case of spherical bubbles, radius r_b, which is by definition the same as r_t: $a = r_b = r_t$, we have

$$K = 1 + 2\sqrt{\frac{r_b}{r_t}} = 3 \qquad (5.3)$$

that is, K now is not sensitive to size, although clearly the effect of such a flaw is a serious reduction in the nominal stress to cause cracking. Since bubbles are a common feature of dental materials – for example, porcelain (25§4.2) – an approach to theoretical strength is not possible, even if there are no other flaws, which is unlikely. It should be said that the rest of the shape of the crack is of no great concern – its convolution or 'roughness'– it is only the curvature at the tip at the maximum width in the direction normal to the load axis that matters.[ibid]

If there is no crack tip blunting mechanism, *i.e.* the material is brittle, the new crack is necessarily sharp, the value of K high, and the crack grows promptly without having to increase the applied stress – uncontrolled crack growth. However, if there is a blunting mechanism (*e.g.* as in the Cook-Gordon mechanism, 6§2.13), the crack does not grow spontaneously and the stress must be raised to drive it further – stable crack growth. Indeed, a deliberate blunting may be used to reduce the stress concentration dramatically and prevent (or at least delay) the onset of a crack (Fig. 5.2).

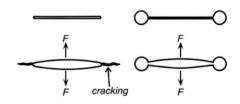

Fig. 5.2   Crack tip blunting prevents uncontrolled cracking.

The principle is used to advantage in various contexts. Thus the shoulders of tensile test pieces are rounded (Fig. 1§6.1), and buttonholes in garments may have a larger radius at one end, and especially when slits have to be made in cheap plastic straps, to prevent tearing when the stud is inserted or removed (Fig. 5.3). In fact, in some situations where cracks are part of the expected wear and tear, a hole may be drilled at the end of a (slowly) growing crack to arrest it (Fig. 5.4), preventing or at least delaying catastrophic rupture. Crack-stop holes are a well-known engineering fix in such fatigue contexts. Bear in mind that the original full crack length is still present, increased by the radius of the hole, but the end curvature has been reduced substantially: larger (diameter) holes are more effective, up to a point:

$$K \;=\; 1+2\sqrt{\dfrac{a+r_h}{r_h}} \;=\; 1+2\sqrt{1+\dfrac{a}{r_h}} \qquad (5.4)$$

Fig. 5.3 Examples of deliberate 'crack' blunting by design: buttonhole (top), camera strap (bottom).

where r_h is for the new hole. For example, if a 100 mm crack has a (rather large) tip radius of 10 µm (K = 201), and a 1 mm radius hole is drilled (K = 21.20), the stress concentration is reduced ~10-fold. Of course, this new hole must be cleanly cut and smooth if new flaws are not to be introduced or the old one continue (that shown in Fig. 5.4 is rather dubious in that respect). It follows that in preparing cavities in teeth, the junction of the floor with the walls should not be a sharp corner but smoothly rounded, to prevent cuspal fracture and so on (Fig. 5.5)[5]. Likewise, all changes of section in metal and ceramic bridges must be radiused to avoid such stress concentrations. On the other hand, notches can be used to control the location of deliberate fracture,[11] especially when such features are present in the service condition. In this case, the sulcal groove in an artificial denture, but the frenal notch is also a point of concern (1§10.2).

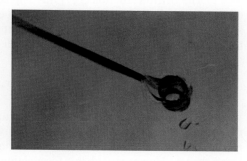

Fig. 5.4 A crack in an acrylic aircraft window can be arrested by drilling a hole, effectively blunting it.

The same principle applies to the root of screw threads. These must be cut such that they have a defined radius, R_t, (usually in terms of a proportion of the pitch) so as to minimise the stress concentration, given the load that the thread itself carries (Fig. 5.6). This is built into technical standards for such devices. While a perfectly sharp root cannot be cut it would of course be undesirable, but flat roots have two sharp corners that may initiate cracking. Such cracking may underlie many failures of the screws used to retain the superstructure on dental implants, although it appears not to be well-recognized as relevant.[12][13] A similar problem might affect the implants themselves, when these are threaded.

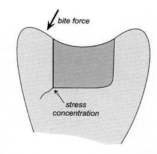

Fig. 5.5 Sharp internal angles should be avoided in stressed structures to prevent cracking.

●5.2   Fracture toughness

The Griffith equation (1§7.1) was based on the stress-concentration concept (§5.1) and can be rewritten to express the underlying energy balance:

$$2\gamma_s E \;=\; \sigma^2 \pi a \qquad (5.5)$$

separating the material properties from the crack conditions. Irwin approached the problem from a different direction, defining the **strain energy release rate**, G, *i.e.* the released energy that goes directly into forming the crack surfaces, that is, the rate as it grows, per unit area (specifically, not a rate in time).[14] This was explicitly for a "somewhat brittle" material, *i.e.* showing some plasticity local to the crack tip, therefore including this energy dissipation as well as that of the formation of new surface. Then, taking the geometrical **stress intensity**

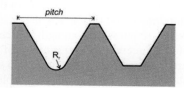

Fig. 5.6 The root of a screw thread should be radiused (left) to balance the strength of the thread itself against the risk of cracking from the root. Flat thread roots have sharp corners (right).

---

[5] It would follow that such cavities would also be a lot easier to fill (6§3.4).

factor, K, as given by:[15]

$$K = \sigma\sqrt{\pi a} \tag{5.6}$$

(care is needed because of the use of the same symbol as for stress concentration factor, §5.1, although ultimately they are closely-related expressions for the same kind of system). It can be shown directly[16] then that (with scaling by E) these terms are equivalent:

$$EG = \sigma^2 \pi a$$
$$i.e. \quad G = \frac{K^2}{E} \tag{5.7}$$

which is known as the Griffith-Irwin equation, where now it is explicitly about crack growth and the parallel with equation 5.5 is obvious.

The Griffith criterion (1§7) was emphasized to be relevant to only to brittle materials because the strain energy dissipation was restricted to that of the formation of new surface, as indicated by the inclusion of the surface energy, γ, that is, the specific work of formation of the new surface. Orowan[17] had already extended the idea to include the work done in plastic deformation[6], effectively giving:

$$\sigma_{crit} = \sqrt{\frac{2E(\gamma_s + \gamma_p)}{\pi a}} \tag{5.8}$$

where γ_p is that extra work done per unit area of the crack that forms, wherever that work is done (it is important to recognize that all this is on the basis of "fixed grips" – that is, the external force-applying system does not move, otherwise extra work is done that has no bearing on the problem). It is generally the case that γ_p is much larger than γ_s. However, with this formulation the application of the concept is much broadened, as most if not all materials in practice show some plasticity (ductility) on very small scales (§2) such that the strength predictions made using the unmodified Griffith equation can be seriously in error.

So far this discussion has been in terms of simple axial tension, and this is known as **Mode I**. Shear was shown earlier (§1) to be the only other means of causing failure, although that was implicitly what is termed **in-plane shear**, **sliding mode** or **Mode II**. There is also the possibility of a **tearing mode** of crack opening (as of a sheet of paper) which is labelled **out-of-plane** (or **antiplane**) **shear**, that is, **Mode III** (Fig. 5.7). There is no other way of driving a crack (except, of course, for combinations, or mixtures, of these three modes of loading), no matter what the external loading arrangements might be. Notice in particular that compressive loading can only lead to Mode II failure (see 1§6.3, §1.3).

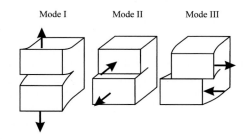

Fig. 5.7 The three fundamental modes of loading and fracture.

Given this, the stress-intensity factor and strain energy release rate must be associated with the mode of crack opening, and so must be subscripted: K_I, G_I, *etc*. Then, because the focus is on the actual initiation or progression of the crack, that is, the critical value, the subscript 'c' is added to indicate this, as for example in K_{Ic}. This last is then a measure of the resistance shown by a material to crack growth and is labelled **fracture toughness**, and thus this is in some respects a rather more important characterization of a material in a load-bearing context than strength (which is defined as if simultaneous, instantaneous, failure occurs across the entire stressed section). Determination of such values requires rather elaborate physical set-ups, but the key concerns are to ensure a very sharp crack to start with (rather than waiting for one to initiate), and then the obtaining of stable crack growth, while being able to measure the actual crack size for the stress acting. Because there is some sensitivity to the size of the crack (a) in relation to the test piece thickness (w), fracture toughness is determined according to an equation such as:

$$K_{Ic} = Y\sigma_c\sqrt{\pi a} \tag{5.9}$$

where Y is a rather complicated function of a and w, which function itself depends on the geometry of the test set-up.[18] Further elaboration is beyond the present scope.

---

[6] In a speculative afterthought footnote, p. 214 [ibid].

There is also another sensitivity to test piece thickness. The above has been developed implicitly for a thin test piece, where what is termed the **plane stress** condition applies, when the stress acting normal to the plate (z-direction) is essentially zero, or at least can safely be assumed to be so. However, if the test piece is a thick plate (say), the expected contraction in the z-direction must be allowed for as we now have what is termed **plane strain**, where the normal strain is assumed to be zero (or nearly so). This requires the Poisson ratio to be taken into account thus:

$$\sigma_{crit} = \sqrt{\frac{2\gamma_s E}{\left(1 - v^2\right)\pi a}}$$ (5.10)

Likewise, we have for that same plane strain condition:

$$G = \frac{K^2}{E'} = \frac{K^2}{E}(1 - v^2)$$ (5.11)

●5.3   Brittleness index

As has been seen (1§8), indentation hardness can be seen as a general measure of the yield stress of a material, and thus of the resistance to deformation. On the other hand, fracture toughness is a measure of the resistance to cracking. These may be seen as properties for which there is an essential compromise in any material: one cannot have both. It has been suggested that this essential trade-off may be expressed in the ratio of the two, called the Brittleness Index:[19]

$$BI = \frac{H_v}{K_{Ic}}$$ (5.12)

which has the units of $m^{-0.5}$, which is perhaps a little hard to interpret. However, looking at the square of the inverse, it is found to be related to a controlling maximum flaw size, c*:

$$c* = \mu \left(\frac{K_{Ic}}{H_v}\right)^2$$ (5.13)

while a controlling maximum stress, P*, can also be found:

$$P* = \lambda K_{Ic} \left(\frac{K_{Ic}}{H_v}\right)^3$$ (5.14)

(The scale factors λ, μ require elaborate determination.) That is, if the actual flaws in the Griffith sense are larger than ~c* then failure changes from hardness-controlled to toughness-controlled, and *vice versa*. This can be seen now to be the underlying criterion for indentation cracking (§2). Either way, P* and c* are interpreted as the maximum values that a body can carry without failure. Thus BI may provide a useful measure in evaluating materials for specific service conditions.

References

[1]   Gere JM. Mechanics of Materials. 6th ed. Thomson Brooks/Cole, London, 2004.

[2]   Glucklich J. Strain-Energy Size Effect. NASA Technical Report 32-1438, Jet Propulsion Laboratory, California Institute of Technology, 1970.

[3]   Musanje L & Darvell BW. Effects of strain rate and temperature on the mechanical properties of resin composites. Dent Mater 20: 750 - 765, 2004.

[4]   Paul B. Prediction of elastic constants of multiphase materials. Trans TMS-AIME 218: 36 - 41, 1960.

[5]   Williams DEG. Packing fraction of a disk assembly randomly close packed on a plane. Phys Rev E 57(6): 7344 - 7345, 1998.

[6]   Buryachenko VA. Micromechanics of Heterogeneous Materials. Springer, New York, 2007.

[7]   http://www.mathopenref.com/coordintersection.html

[8]   Krenchel H. Fibre reinforcement; theoretical and practical investigations of the elasticity and strength of fibre-reinforced materials. Thesis, Technical University of Denmark. Copenhagen: Academisk Førlag, 1964.

[9]   Bader MG. Polymer composites in 2000: structure, performance, cost and compromise. J Microsc 201 (2): 110 - 121, 2001.

[10]   Inglis CE. Stresses in plates due to the presence of cracks and sharp corners. Trans Institute Naval Architects 55: 219 - 241, 1913. ; http://www.fracturemechanics.org/ellipse.html

[11]   Chung RWC, Darvell BW & Clark RKF. The bonding of cold-cure acrylic resin to acrylic denture teeth. Austral Dent J 40 (4): 241 - 245, 1995.

[12]   Chang CL, Lu HK, Ou KL, Su PY & Chen CM. Fractographic analysis of fractured dental implant components. J Dent Sci 8(1): 8 - 14, 2013.

[13]   Morsch CS, Rafael CF, Dumes JFM, Juanito GMP, de Souza JGO & Bianchini MA. Failure of prosthetic screws on 971 implants. Braz J Oral Sci 14 (3): 195 - 198, 2015

[14]   Irwin GR. Analysis of stresses and strains near the end of a crack traversing a plate. J Appl Mech 24: 361 - 364, 1957.

[15]   Irwin GR. Crack-extension force for a part-through crack in a plate. J Appl Mech 29(4): 651 - 654, 1962.

[16]   Neal-Sturgess CE. A direct derivation of the Griffith-Irwin relationship using a crack tip unloading stress wave model. 2013. arxiv.org/pdf/0810.2218 14th Int Conf Fracture, Rhodes, Greece, June 2017

[17]   Orowan, E. Fracture and strength of solids. Phys Soc Prog Reports 12: 185 - 232, 1949.

[18]   Zhu XK & Joyce JA. Review of fracture toughness (G, K, J, CTOD, CTOA) testing and standardization. Eng Fract Mech 85: 1 - 46, 2012.

[19]   Lawn BR & Marshall DB. Hardness, toughness, and brittleness: An indentation analysis. J Amer Ceram Soc 62 (7-8): 347 - 350, 1979.

Chapter 30   More Chemistry

A large proportion of the materials and processes used in dentistry can fairly be described as depending on their chemistry quite specifically. This chapter gathers together a number of topics that do not sit easily under earlier chapter headings, yet because they are commonplace to the extent that they may easily be overlooked it is necessary to be aware of their significant aspects.

Oxidizing agents turn up in several distinct contexts: **sterilization** *and* **bleaching** *being most prominent. Hypochlorite is perhaps the most common, but various other agents are also used, many of which all depend ultimately on the effect of hydroxyl radicals. The key exception is silver, known for its effect (albeit not the chemistry) from antiquity. All demand care in use because they are potent.*

While endodontics uses such agemts, **dissolution** *of the mineral of dentine is also major component of root canal preparation, alongside protein maceration. Thus both acidic (or chelating) and alkaline properties are employed, as well as the biocide chlorhexidine.*

Aldehydes 'fix' protein and as such are biocidal but are generally inappropriate for contact with living systems, despite historical enthusiasm. **Formaldehyde** *is also a byproduct of other reactions.*

Calcium hydroxide, *generally considered to be beneficial, has disadvantageous aspects, as does* **zinc oxide**, *when ingested inappropriately.*

Sensitivity to cold provides a means, through **evaporative cooling***, of ascertaining the need for endodontic work in the first place. The physical chemistry of this allows a figure of merit to be defined.*

Materials Science for Dentistry
https://doi.org/10.1016/B978-0-08-101035-8.50030-4

§1. Oxidizing Agents

Sterility is considered paramount for endodontic treatment success, and various means of addressing this have been used, most often involving strong oxidizing agents. However, treatments of this kind are used in other dental contexts as well.

●1.1   Sodium hypochlorite

The principal component of household bleach, a solution of sodium hypochlorite is ordinarily used as an oxidizing agent, and therefore as a sterilant for endodontic irrigation. Because hypochlorous acid is a weak acid (pK_a ~ 7.5), a solution of the sodium salt is naturally of high pH, but it is also made higher (pH > ~11) by the deliberate addition of some sodium hydroxide, which stabilizes the solution against decomposition to chlorine:

$$OCl^- + 2 H^+ + Cl^- \rightleftharpoons H_2O + Cl_2 \qquad (1.1)$$

i.e. by keeping the concentration $[H^+]$ low (Fig. 1.1). The chloride ion is present from the manufacturing process, which is essentially the reverse process:

$$2\,NaOH + Cl_2 \rightleftharpoons 2\,NaOCl + NaCl + H_2O \quad (1.2)$$

The deliberate addition of chloride (as is sometimes done) would drive equilibrium 1.1 to the right. However, loss of chlorine gas on standing leads to the deterioration of the product at any pH as the equilibration is fast, meaning that long term storage is not appropriate.

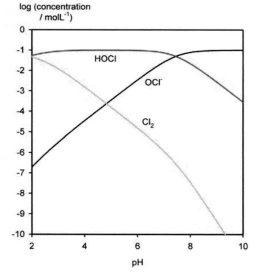

Fig. 1.1 The principal equilibria in sodium hypochlorite solution (0.1 molL^{-1}, ~7.4 gL^{-1}).

Dental products contain from 0.5 ~ 6 mass% NaOCl, at pH 9 ~ 13. The oxidizing activity of both the hypochlorite anion and chlorine (*i.e.* as electron acceptors) destroys organic material directly. However, the high pH solution also **macerates** tissue, in other words hydrolyses proteins, mucopolysaccharides and lipids (saponification). Thus, through these two chemical actions hypochlorite solutions are effective in removing the organic matrix of dentine and its remains in any debris from the canal preparation. Naturally, any micro-organisms present will be destroyed by the same processes – exactly as in domestic applications of 'bleach'. For the same reasons, contact with skin, mucosa and – especially – eyes is very damaging: this risk must be avoided clinically.

Hypochlorite solutions are also sometimes advocated for the deproteinization of etched tooth tissue in order to achieve better penetration of resins for bonding. However, in this context, as well as for free-radical polymerization endodontic sealants, the presence of residual hypochlorite interferes with the setting by reacting with those radicals:

e.g. $$HOCl + R\bullet \Rightarrow ROH + Cl\bullet \qquad\qquad\qquad (1.3)$$

Accordingly, treatment with reducing agents to remove or at least decrease the amount of residual hypochlorite may be used, for example with sodium thiosulphate ("hypo"):

$$4\,OCl^- + S_2O_3^{2-} + 2\,OH^- \Rightarrow 4\,Cl^- + 2\,SO_4^{2-} + H_2O \qquad\qquad (1.4)$$

or ascorbic acid (6§6.1).[1] Obviously, after such treatment the site must be well flushed with water.

The use of fresh solution is desirable because, in addition to reaction 1.1, successive self-oxidation can occur:

$$2OCl^- \rightleftharpoons ClO_2^- + Cl^- \qquad\qquad\qquad (1.5)$$

$$OCl^- + ClO_2^- \rightleftharpoons ClO_3^- + Cl^- \qquad\qquad\qquad (1.6)$$

$$OCl^- + ClO_3^- \rightleftharpoons ClO_4^- + Cl^- \qquad\qquad\qquad (1.7)$$

where reaction 1.5 is the rate-limiting step. This has implications of two kinds. Firstly, deterioration of the efficacy of the irrigant will occur with time. Secondly, because it is a second-order reaction, concentrated stock solutions deteriorate much faster, roughly as the square of concentration. This means that if the working solution is to be made up by dilution from a concentrated stock, this should be done early, not at the point of use. Storage cool is beneficial; the rate of reaction 1.5 increases ~3.5 times for a 10 K increase in temperature. (It should not be forgotten that 'storage' includes time spent in the supply chain; this could include periods at elevated temperatures.) At 25 °C, the 'half-life' of a 6% solution is probably no more than a year, so bulk purchases may be false economy. As indicated above, deterioration is also much accelerated at pH < ~11, as can easily arise if a concentrated stock solution is diluted without the addition of additional base. In addition, these reactions are catalysed (by 1 or 2 orders of magnitude) by transition metal ions.

A further reaction can also occur:

$$2OCl^- \Leftrightarrow O_2 + 2Cl^- \tag{1.8}$$

This is also catalysed by transition metals, especially nickel (as in stainless steel) and copper, even at very low concentrations. This can lead to elevated pressure in the container (because the solubility of oxygen is relatively low), and care should therefore always be taken in opening such to avoid spray or worse. In any case, eye protection should always be used.

The question remains why hypochlorite solutions are oxidizing, as indicated above. Remembering that oxidation is the removal of one or more electrons from a species, we have nominally at the root:

$$[O] + H_2O + 2e^- \Leftrightarrow 2OH^- \tag{1.9}$$

so firstly:

$$OCl^- + H_2O + 2e^- \Leftrightarrow Cl^- + 2OH^- \tag{1.10}$$

(under alkaline conditions; if acid then water is formed, not hydroxide), and then for each oxygen in the chlorates, another reaction 1.9 is added. Chlorine itself is an oxidising agent:

$$Cl_2 + 2e^- \Leftrightarrow 2Cl^- \tag{1.11}$$

Hence, chromophores (24§6.2) are decolourized – which is bleaching, and organic material in general tends to be destroyed.

●1.2   Photo-activated disinfection

Although hypochlorite is in principle effective enough as a disinfectant, it is not effective (at least, not on suitable timescales) against organisms embedded in biofilms, in which the extracellular matrix is a significant barrier.[2] Even so, there are a number of disadvantageous aspects to its use – it is too broadly biocidal as well as macerating dentine matrix (§1.1). An alternative approach has been found in the use of **photo-activated disinfection**. This relies on the absorption of light whose energy is then (in part) used to create several reactive oxygen species (ROS) which then go on to damage cells walls.

Often used is the common histological stain toluidine blue O (TBO), generally in the dental context called tolonium chloride (Fig. 1.2), and this has a convenient absorption spectrum (Fig. 1.3) in that light of the common red diode laser wavelength of 635 nm is strongly absorbed. In so doing, an excited state is created which, as is normal, may return to the ground state through vibrational decay and fluorescence (Fig. 6§5.2). An inter-system crossing to a triplet state is possible (Fig. 6§5.4), although this is a relatively infrequent event because the spin change is a

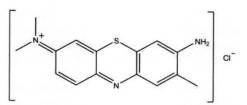

Fig. 1.2 Molecular structure of toluidine blue O.

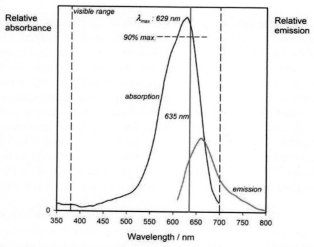

Fig. 1.3 Absorption spectrum for toluidine blue O in the visible region and part of the corresponding emission spectrum (fluorescence plus phosphorescence). NB: on different scales.

'forbidden' transition. This triplet state again may undergo some vibrational decay and then, through a phosphorescent emission, return to the ground state (Fig. 1.4)[3]. Neither of these two decay processes has any biological implications except for the slight warming possible for the infra-red component (Fig. 1.3) (although all absorbed light is ultimately converted to heat). However, whilst in that excited triplet state, reaction with molecular oxygen may occur through one of two processes. The first, (a), involves spin exchange whereby the oxygen molecule goes into the excited antiparallel spin singlet state, $^1O_2{}^*$, a very reactive species[4], rapidly generating peroxides which go on to generate many damaging radicals (*cf.* reactions 6§6.3 and 6§6.4). The second mechanism, (b), relies on the oxygen's reduction, abstracting an

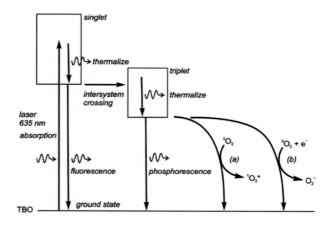

Fig. 1.4 A Jablonskí diagram for the photo-activated generation of reactive oxygen species by toluidine blue O.

electron from any convenient oxidizable source (which may be the dye itself). The result of this is the formation of the superoxide anion, O_2^-. This is not particularly reactive, as it happens, but one path available is disproportionation:

$$2H^+ + 2O_2^- \Rightarrow H_2O_2 + O_2 \qquad (1.12)$$

This peroxide is itself reactive (§1.4), but the bigger problem is that certain metal ions (Fe^{2+} in particular) can catalyse the **Haber-Weiss reaction**[5] with further superoxide:

$$2O_2^- + H_2O_2 \Rightarrow OH^- + {\cdot}OH + O_2 \qquad (1.13)$$

in which the very reactive and thus toxic hydroxyl radical, $\cdot OH$, is formed.[6]

We therefore have the situation that while atmospheric oxygen is available, irradiated TBO acts as a catalyst (that is, unchanged in the ground state) for the formation of two highly reactive species. These ROS are then available to attack any organic material to hand (which includes the sensitizing dye, of course), but bacteria especially are vulnerable to cell wall damage and thus death.

There are many similar dyes and other compounds available which would achieve the same effect, and likewise there is a corresponding range of irradiating wavelengths that would be effective. The absorption spectra of those dyes would, of course, vary, but providing the irradiating light corresponded to within, say, 90%, of the peak absorption, then satisfactory efficiency could be expected. Even so, using 690-nm light with TBO has nevertheless been reported to be effective.[7]

The efficiency with which ROS are formed under these circumstances does not appear to have been reported, but clearly good access to ambient air is required. From this it should be apparent that the power output of the laser need only be sufficient to obtain a population inversion in the dye, *i.e.* of the excited singlet state, to maximise the rate of intersystem crossing. Similarly, the total energy delivered (J) is not in itself a relevant figure of merit (as has been said), nor indeed the energy or power density (*e.g.* as J/cm^2 or W/cm^2), nor even the duration of irradiation, as the production of ROS depends on the availability of oxygen, *sine qua non* – it is essential. The type of dye (and thus its absorption spectrum and molar extinction coefficient), the concentrations of dye and oxygen, and both the wavelength and irradiance of the light are all to be considered along with duration. Indeed, the concentrations used have varied widely as have durations and irradiances. It has been reported that the temperature rise under such treatments is slight, implying that there is no risk of damage to soft tissue.[8] Even so, it might be wise to try the "burn test" before proceeding, applying the irradiating tip between pinched finger tips for the proposed duration, for example (*cf.* 6§5.18).

This kind of process is used for treating root canals, where sterility is believed to be of very high importance, but also for the disinfection of carious tooth tissue as part of an effort to reduce the amount of tissue removed in the course of restoration. In an extended field of use, such chemistry is referred to as **photo-dynamic therapy** (PDT) (and numerous other names) and is applied to the treatment of tumours as well as other kinds of lesion in addition to infections of various kinds.[9] In many of these uses the dye must be allowed to infiltrate tissue (which red light can penetrate relatively well), where the generally cytotoxic ROS are then effective, given that selectivity cannot be expected – any cells of any kind of organism so exposed will be damaged, as well as viruses.[10] Suitable time for diffusion is therefore critical.

●1.3   Hydrogen peroxide

Hydrogen peroxide, H_2O_2, is a very strong oxidizing agent (oxidizing potential[1] 1.8 V, acidic conditions), behind ozone (2.1 V) and fluorine (3.0 V), and much stronger than hypochlorite (0.89 V). It has been used in dilute solution as a topical disinfectant, *i.e.* on wounds, although this is now thought to be disadvantageous to healing because of the damage it does to soft tissue. Even so, solutions of hydrogen peroxide are used for endodontic irrigation because of its disinfectant behaviour. Again, recalling the definition of oxidation as the removal of electrons, many organic substances can suffer this very easily:

$$e^- + H_2O_2 \Rightarrow OH^- + \cdot OH \tag{1.14}$$

thus creating the extremely reactive hydroxyl radical, whereupon many other reactions follow. This can also be done by transition metals, such as iron, in the **Fenton reaction:**

$$Fe^{2+} + H_2O_2 \Rightarrow Fe^{3+} + OH^- + \cdot OH \tag{1.15}$$

after which a very extensive and complex set of reactions may ensue[11] (again, *cf.* reactions 6§6.3 and 6§6.4), in the course of which many organic substances may be destroyed, including keratin and collagen. In particular, the functional groups known as chromophores, which impart colour to organic substances (24§6.2), are good electron donors. Accordingly, their destruction by such means leads to the decolourization of the material containing them, hence hydrogen peroxide is a strong bleaching agent.

The activity of hydrogen peroxide is sometimes said to be due to 'nascent' oxygen. Chemically, this is now known to be meaningless. As can be seen from the above reaction, the active entity is the hydroxyl radical. The formation of gaseous oxygen is an irrelevant distraction arising because H_2O_2 is thermodynamically unstable:[2]

$$H_2O_2 \Rightarrow H_2O + O_2 \qquad -98 \text{ kJmol}^{-1} \tag{1.16}$$

(for comparison, the heat capacity of water is about 75 $Jmol^{-1}K^{-1}$ – the temperature rise can be substantial). This decomposition ordinarily is rather slow (the reactions involved are complex), and has a substantial activation energy (~170 $kJmol^{-1}$), which means however that the rate increases rapidly with temperature. Even so, decomposition can be catalysed by many things, including unreactive rough surfaces and dust. While the fizzing might be dramatic, and the bubbles are sometimes said to be useful in helping to flush the root canal, they have no more chemical significance than does the oxygen of the air.

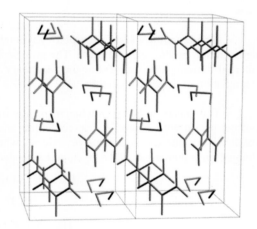

Fig. 1.5 Crystal structure (4 unit cells) of urea peroxyhydrate ("carbamide peroxide") showing separate rows of urea and hydrogen peroxide (in the *b*-direction). No new distinct molecular species is present and the material may be better described as a co-crystal. (Partial molecules are omitted for clarity.)

Given that much of the staining of tooth enamel is due to exogenous organic substances, which become incorporated in enamel because its matrix is permeable, efforts to reverse this are made through the application of 'whitening agents', principally hydrogen peroxide solution, or more usually a dispersion of the solid urea peroxyhydrate – commonly described as the 1:1 adduct[3] with urea, and frequently called "carbamide peroxide"[4] (Fig. 1.5)[12] (which is effectively the same in its solution chemistry – the hydrogen peroxide is simply released on dissolving, there are no complexes) (Fig. 1.6).[13] These products may take the form of viscous liquids, impregnated adhesive strips, or toothpaste.

---

[1]   This can be viewed as 'electron avidity'.

[2]   Care is always required in opening bottles or other containers where gas evolution is a risk.

[3]   An adduct results from the chemical combination of two species (without loss) to form a new molecular entity. There is no such new entity in this case.

[4]   This name is misleading, although in very common use: it is *not* a peroxide of urea (the preferred name, not carbamide). The hydrogen peroxide is equivalent to water of crystallization (which also does not form a new molecular entity).

The oxidation that occurs with hydrogen peroxide, however it is delivered (see also §1.8, §1.9), is not in any way selective. The matrix of tooth enamel contains keratin[14] and collagen[15][16] in addition to a variety of other enamel-specific proteins. Consequently, in addition to bleaching the undesirable colourants, the enamel matrix, on which depends its structural integrity, will be degraded. This will then lead to loss of mineral. This damage cannot be avoided, repaired or reversed.[17] Such damage will be exacerbated if the peroxide solution is made acidic (more mineral dissolution) or alkaline (more matrix destruction), or heated.[5] The lack of selectivity also means that if these materials come into contact with the mucosa a chemical "burn" may result, with irritation or pain. Pain can also result if penetration of the enamel as far as the dentine occurs. Indeed, contact with skin and eyes (especially) of any hydrogen peroxide solution is damaging and should be avoided. Eye-protection should be used.

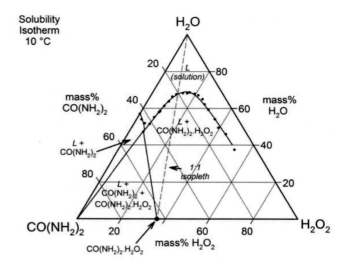

Fig. 1.6 Solubility diagram for the urea-hydrogen peroxide-water system. The solubility increases appreciably with temperature.

The common use of urea peroxyhydrate in whitening agents may have a secondary effect: protein denaturation by the released urea. The process is not fully understood, but it involves interference with the internal bonding that establishes the secondary and tertiary structure.[18] This could facilitate the oxidative degradation of the enamel matrix.

Care needs to be taken when disposing of excess hydrogen peroxide-containing products that they do not get mixed with waste acetone-containing material (10§9.7), particularly in the presence of acid. When oxidized, acetone forms acetone peroxides, which are a highly-unstable explosive compounds. Accidental detonation is then a risk. Hydrogen peroxide also reacts with hypochlorite:[19]

$$OCl^- + H_2O_2 \Rightarrow \cdot OCl^- + \cdot OH + OH^- \qquad (1.17)$$

$$\cdot OCl^- + \cdot OH \Rightarrow HO\text{-}OCl \qquad (1.18)$$

$$HO\text{-}OCl + OH^- \Rightarrow Cl^- + H_2O + O_2 \qquad (1.19)$$

where some of the dioxygen produced is singlet (and thus more reactive than atmospheric triplet O_2;[6] see §1.2).[7] The problem here is that this reaction can be violent if the concentrations are high enough. As with all reactive substances, disposal should be undertaken properly and carefully. Plainly, they should not be mixed deliberately.

●1.4 Ozone

Ozone, O_3, (trioxygen) is an allotrope of oxygen. As indicated above (§1.3), it is a strong oxidizing agent, and indeed much stronger than dioxygen, O_2. In aqueous systems, the electron acceptance reaction is

$$O_3 + 2e^- + 2H^+ \Rightarrow O_2 + H_2O \qquad (1.20)$$

In water on its own, ozone slowly decomposes through complex pathways that involve hydrogen peroxide, superoxide (HO_2) and hydroxyl radicals. However, it is otherwise a remarkably selective oxidant, preferentially attacking carbon-carbon double bonds, and olefins in particular (Fig. 1.7).[20] The result of the addition to the double bond is a very unstable 1,2,3-trioxolane that readily rearranges to the ozonide (a 1,2,4-trioxolane). This is then capable of numerous reactions with water to yield various carbonyl and other species, breaking the carbon

---

[5] The use of a resin-curing light to 'activate' the process is often suggested. There is no absorption in a colourless material, therefore no such activation can possibly occur. However, the heating consequent on irradiation would accelerate all reactions.

[6] The direct interaction of a triplet ground state with a singlet-state molecule is a spin-forbidden process; see 6§5.3, 5.4.

[7] NB: HO-OCl is a true peroxide and *not* the acid from the chlorite anion, ClO_2^-, which has the structure $O=Cl\text{-}O^-$.

backbone. The end result is the destruction of many molecules in living organisms, especially the unsaturated lipids of membranes. Cell death follows.

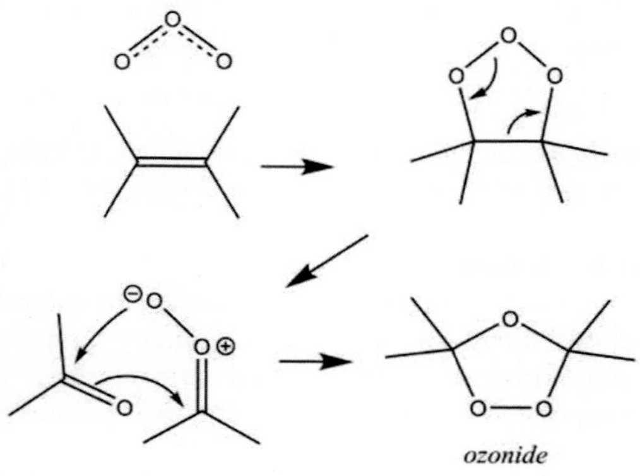

Ozone can be produced in a plasma induced by a high voltage, a so-called **silent** or **corona discharge**. Dioxygen molecules are broken to yield oxygen atoms, which then react with further dioxygen to form O_3. (This is the reason for the characteristic sharp 'electrical' smell around high-voltage apparatus and sparks, including lightning.) The discharge is arranged to occur in a flowing stream of air or oxygen, which gas can then be delivered to the treatment site. Concentrations of some $0.3 \sim 3$ % or so by mass ($3 \sim 30$ mg/L) are possible. Whether it is for the sterilization of root canals, carious lesions, other infections, or anything else, it is important to contain the ozone-containing gas such that only the target is exposed to it, else other tissues such as mucosa (pharynx, lungs)

Fig. 1.7 The addition reaction of ozone with unsaturated organic compounds. Ozonides are unstable in the presence of water.

and eyes will be damaged. Elaborate evacuation and destruction systems must be used to avoid the workplace being polluted and harm to both patient and operator.[21] Furthermore, traces of ozone are enough to cause the embrittlement and cracking of any elastomeric material where double bonds are present (27§1.4), for example polyisoprene (27§1.1) and polybutadiene (27§6). Prevention of leaks and good scavenging are essential.

Ozone reacts with peroxide anion in the so-called **peroxone process**, thought to be represented by:

$$O_3 + HO_2^- \Rightarrow\Rightarrow \cdot OH + O_2^- + O_2 \tag{1.21}$$

thus producing both hydroxyl radicals and superoxide anion, both of which are very reactive. This reaction scheme may, however, be simplistic.[22] Either way, the oxidation system is rather more aggressive than either component alone, and if used in dentistry, as is proposed,[23] it should be with appropriate care.

●1.5 Iodine

Iodine, I_2, being a halogen, is also an oxidizing agent:

$$I_2 + 2e^- \leftrightarrow 2I^- \tag{1.22}$$

although somewhat weaker than is chlorine. It has long been used as an effective topical antiseptic because of this, commonly in solution in ethanol. It has a further action which makes it effective in this regard, the iodination of lipids, simply and rapidly adding directly to carbon-carbon double bonds (Fig. 1.8).[8] This is

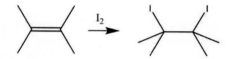

Fig. 1.8 Iodine adds directly to unsaturated carbon bonds.

sufficient to disrupt their role in cell membranes, especially because of the large size of the atom. However, because the dissociation of I_2 is weak, the uncharged molecule can also diffuse through membranes rather freely, reaching and disrupting other bacterial cellular mechanisms. It is also strong enough to oxidize thiols (–SH) to disulphide links (–S-S–), which interferes with proteins such as enzymes.

Unfortunately, iodine's action is strong enough to be an irritant and a sensitizer when used in simple solutions at the usual high concentrations. The discovery that poly(vinyl pyrrolidone) ('PVP') can bind iodine stably, with the equilibrium concentration of free iodine kept low, meant that iodine could be used with far fewer and weaker side-effects. However, reaction of the free iodine disturbs that equilibrium so that more is released:

$$PVP[I_2] \leftrightarrow PVP + I_2 \Rightarrow\Rightarrow \{\text{other reactions}\} \tag{1.23}$$

The PVP is thus said to be an **iodophore**. Commonly known as **povidone iodine** (PVP-I), this formulation is effective against a wide range of microorganisms.

Iodine may also be used in potassium iodide solution, where triodide, and possibly other polyiodides, may act as reservoirs in much the same way as PVP-I (*cf.* Fig. 8§3.3). Such a solution, commonly in the molar ratio KI : I_2 :: 3 : 1, which is ~1:2 by mass, is known as **Lugols's solution**, is commonly used as a topical

---

[8] Indeed, it is used for the quantitative determination of the degree of unsaturation of fats – expressed by the "iodine number".

antiseptic. However, in an endodontic context, where the mixture with calcium hydroxide may be used, iodine reacts:

$$6Ca(OH)_2 + 6I_2 \Rightarrow Ca(IO_3)_2 + 5CaI_2 + 6H_2O \qquad (1.24)$$

and while the iodate is still a fairly good oxidizing agent, its antimicrobial activity will be slight in comparison with that of the hydroxide's very high pH (>12).

●1.6 Iodoform

Iodoform (HCI_3) is an analogue of chloroform ($HCCl_3$) but with rather different properties. It is a pale yellow, somewhat oily solid at ordinary temperatures. It has long been used as an external disinfectant and wound dressing in general medical contexts,[24] although it is now uncommon for a variety of reasons. It is incorporated in a number of endodontic products. It has no internal medical value, despite old usage.

Its disinfectant properties may be due to reactions with olefins, and thus to unsaturated lipids, interfering with cell membranes as for iodine, but it is also a weak oxidising agent and so can affect many other cellular and microbial functions. Some reactions may release iodine, although this is not by hydrolysis. It is only slightly soluble in water (100 mg/L) but much more soluble in lipids, and so can cross cell membranes, and in particular the blood-brain barrier[25] where it is implicated in a range of neurological problems.[26] Indeed, it has a more general toxicity because its metabolism produces the toxic carbon monoxide (CO) via the iodine analogue of phosgene, OCI_2,[27] and which is itself very toxic. The mean fatal dose of iodoform is said to about 3 g.[28] Liberal use and ingestion (or applications where this will happen) are therefore contraindicated.

The endodontic use is commonly in a mixture with $Ca(OH)_2$, whose antimicrobial effects are well-documented (9§12.4). However, it does appear that this combination is essentially pointless, as the minor effect of iodoform is swamped.[29] The material is generally resorbable (for example if extruded beyond the root apex, but also intra-radicularly), i.e. diffuses into the blood and other tissues, from whence the toxic effects in liver and brain will arise.

●1.7 Silver

Ionized silver is a potent oxidizing agent:

$$Ag^+ + e^- \Rightarrow Ag^0 \qquad E_0 = 0.80 \text{ V} \qquad (1.25)$$

similar in strength of effect to hypochlorite, with no selectivity. The consequent production of metallic silver is the reason for the dense black stain that is produced by silver nitrate solution on organic matter, including soft tissue. The effect is strong enough that the wet solid[9] or a solution of silver nitrate is used for **chemical cautery**, effectively 'burning' unwanted soft tissue, stopping bleeding, and so on. As such, it is also destructive of micro-organisms, which has led to its being used on the primary dentition to arrest caries. The stain is not removable (except mechanically), hence the inapplicability of the treatment to the permanent dentition. Even so, great care has to be taken to prevent contact with non-target tissue. A solution of silver diammine[30] fluoride, $Ag(NH_3)_2F$, essentially made by dissolving AgF in ammonia solution, and hence is alkaline, is considered to be less damaging than the acidic nitrate solution, although its anticaries effect is entirely due to the action of the silver ions.[31] The blackening effect remains, albeit much more slowly because the actual solution concentration of Ag^+ is low, given the low dissociation constant for the complex:

$$Ag(NH_3)_2^+ \leftrightarrow Ag^+ + 2NH_3 \qquad K = {\sim}9.0 \times 10^{-8} \qquad (1.26)$$

Silver metal has long been known as an antibacterial agent[32] and has been used as such in many forms and contexts, and its applications continue to expand.[33] This effect is due to the presence of a low equilibrium concentration of silver ions in aqueous media in contact with the metal (13§1.1), whether this is massive solid or finely subdivided (nanoscopic particles). These ions will react as above, and disrupt the chemistry of cell membranes, DNA and any other metabolic process. In particular, biofilm production is inhibited.[34] It may be that the use of silver endodontic points (28§9) had this as an unrecognized benefit. Nevertheless, it should be recognized that silver ions must be toxic to all cells, eukaryotic or not. (Copper ions are also known to be similarly toxic and thus antimicrobial, which is why (with hindsight) it was used in the now-abandoned copper amalgam, 28§6.5.)

---

[9] In the form of a stick it was formerly known as 'lunar caustic'.

It may be noted that the oxidizing effect of Ag$^+$ results in the regeneration of Ag^0 – metallic silver atoms, which may therefore continue to generate ions spontaneously. In this sense, the effect is more or less permanent. However, there is a 'sink' reaction. Many proteins have thiol groups (–SH), as mentioned above, and these too may be oxidized by Ag$^+$. The eventual outcome here, though, is the formation of silver sulphide, AgS, which is black and has an exceptionally low solubility product, ~6 × 10^{-51}. The precipitation of this results in the slow development of a dark stain around silver or silver alloys embedded in soft tissue. In particular, this was the cause of the so-called **silver tattoo** around fragments of dental silver amalgam that were not removed before wound closure after the retrograde filling of root canal apices, or indeed particles embedded in the mucosa through trauma by rotary cutting instruments when amalgam fillings were being trimmed or removed. In each case, the fragments might not have been easily visible, but the stain would spread rather further and become obvious. Such sulphide is otherwise harmless.

•1.8   Sodium perborate

Hydrogen peroxide is unstable on storage while in high concentrations (> 12 mass%) it is a controlled substance,[35] hence the use of urea peroxyhydrate peroxide (§1.3). Sodium perborate (a true peroxide) offers a convenient alternative that is stable indefinitely in solid form (with a variable amount of water of crystallization). It is formed by the reaction of a solution of sodium borate with hydrogen peroxide, when the peroxide crystallizes out. The borate in solution is first formed from borax (2§7.4, 22§2) with alkali:

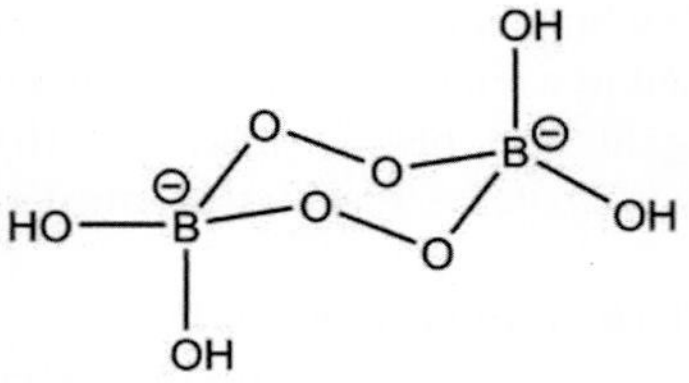

Fig. 1.9 The perborate ion in solution.

$$Na_2B_4O_7 + 2\,NaOH + 7\,H_2O \;\rightarrow\; 4\,Na^+ + 4\,B(OH)_4^- \tag{1.27}$$

Added hydrogen peroxide then reacts slowly (simplified scheme):

$$B(OH)_4^- + H_2O_2 \;\rightleftharpoons\; B(OH)_3 + H_2O + HO_2^- \;\rightleftharpoons\; HOOB(OH)_3^- + H_2O \tag{1.28}$$

The perborate anion is a cyclic structure in solution (Fig. 1.9). Since this is an equilibrium, in contact with water (tetrahydrate solubility ~37 g/L at 32 °C) hydrogen peroxide is released *via* the decomposition to the peroxide anion. However, in solution it is also capable of direct oxidation of electrophilic substances ("Ep"):[36]

$$HOOB(OH)_3^- + Ep \;\rightarrow\; HOO\text{-}Ep^- + B(OH)_3 \;\rightarrow\; O=Ep + B(OH)_4^- \tag{1.29}$$

that is, many biological molecules, including cell membrane lipids.

The hydrogen peroxide formed (from the reverse of reaction 1.28) itself will diffuse away and act as before, and this is then a relatively slow process. This has led to the use of the 'internal' or 'intracoronal' so-called "walking bleach technique", in which a paste of sodium perborate with water (or sometimes hydrogen peroxide solution) is packed into a prepared root canal for non-vital bleaching of a discoloured tooth, before root-filling and restoration, over a period of some days.[37] This bleaching can be accelerated by the use of a so-called 'activator', tetraacetylethylene diamine (TAED)[10][38] (Fig. 1.10), which is in fact an electrophile in the sense of reaction 1.29, whereby the acetyl groups are oxidized to peracetic acid:

$$HOOB(OH)_3^- + CH_3CO\text{-}R \;\rightleftharpoons\; CH_3CO\text{-}OOH + B(OH)_3 + \ldots \tag{1.30}$$

Peracetic acid is itself a strong oxidizing agent (E_0 = 1.76 V), *i.e.* nominally similar to hydrogen peroxide, but it is much more reactive, and therefore an irritant and capable of causing damage, *e.g.* if inhaled as vapour or extruded through the root apex – indeed, the latter is a problem with any bleaching agent. The pH of the paste is around 10 ~11, and so the bleaching effect will be assisted by the maceration of tooth matrix proteins. Sodium perborate is also used in some toothpastes.

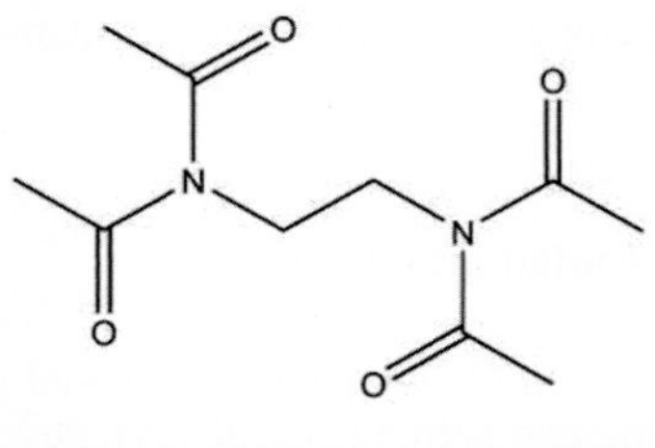

N,N,N',N'-tetraacetylethylenediamine

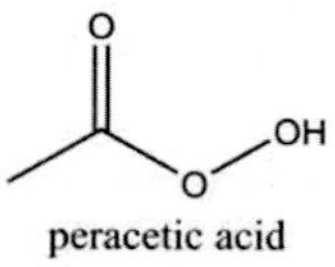

peracetic acid

Fig. 1.10 TAED and peracetic acid.

---

[10] Not to be confused with EDTA (§2.1).

●1.9   Sodium percarbonate

As with urea (§1.3), sodium carbonate will form a stable peroxyhydrate on crystallization from a solution containing hydrogen peroxide: $2Na_2CO_3 \cdot 3H_2O_2$ (Fig. 1.11), so the common name, again, is chemically incorrect[11]. On being dissolved in water (solubility ~140 g/L at 20 °C),[39] the peroxide is simply released, as with urea peroxyhydrate (§1.3), and accordingly is commonly used as a laundry bleaching agent (amongst other things), but is also used for intracoronal endodontic bleaching, much as is sodium perborate. It is also used in whitening toothpaste, and in mouthwash. Again the pH is high (> 10), so protein hydrolysis, *i.e.* maceration, must occur.

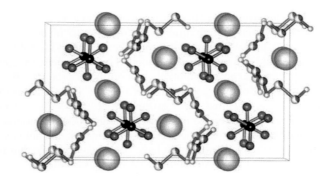

Fig. 1.11 Unit cell of sodium carbonate peroxyhydrate, "sodium percarbonate". (There is some disorder in the H_2O_2 positions, and not all sites may be occupied.) (Partial molecules are omitted for clarity.)

●1.10   Calcium peroxide

A true ionic peroxide, *i.e.* containing O_2^{2-} ions, CaO_2 in water is very reactive, being capable of oxidising all organic materials to carbonate, even at body temperature, and indeed is described as "dissolving pellicle" (*i.e.* destroys it by oxidation). Contact with soft tissue would have to be avoided. Reaction with water also yields hydrogen peroxide directly (the reverse of the manufacturing process), but in both cases it results in the formation of calcium hydroxide:

$$CaO_2 + 2H_2O \;\Rightarrow\; Ca(OH)_2 + H_2O_2 \tag{1.31}$$

and thus must lead to a high pH (§4). This substance is being used in toothpaste where, because of its reactivity, it appears to be coated or encapsulated, which coating is disrupted by the 'dry' brushing of the instructions to obtain direct contact with superficial stain (the solubility is low, 1.7 g/L at 20 °C). An alleged 'accelerator' for this treatment is merely urea peroxyhydrate (§1.3). It is also used as a coating on dental floss. Calcium peroxide slowly decomposes on standing by reaction with water as above (from vapour, for example, given that no coating can be a perfect barrier), and the hydrogen peroxide itself will then slowly decompose (reaction 1.16) (venting or expansion volume is required to avoid the container bursting), so the shelf-life of such products will be limited.

There are many variations on the theme of tooth whitening agents, yet they all have a common underlying chemistry, albeit with some more aggressive than others. The consequences remain the same: bleaching comes at the cost of damage to the matrix of tooth tissue.

§2. Endodontic Irrigants

In the course of root canal preparation, various means are used to modify the shape in order to allow filling and achieve a seal. Whatever the mechanical means, modification of the surface of the canal wall and production of debris occurs. It is considered desirable to remove the **smear layer**, that reworked and consolidated zone of the dentine, in part at least because of the inclusion of microorganisms, and also any loose particulate material, in order better to approach a seal of that canal against the ingress of fluids and other microorganisms.[40] As mentioned already, sterility is considered paramount for endodontic treatment success.

Various agents are used as irrigants in order to achieve the several goals: removal of both organic (see §1.1) and inorganic material, and sterilization (§1). Sometimes, these solutions are incorrectly described as **solvents**. A solvent simply dissolves a substance without essential chemical change (apart, that is, from ionization and solvation). Thus, water is the solvent for the active ingredients of these irrigants, but those active substances all effect irreversible chemical changes on their target materials.

---

[11] A percarbonate, properly a dioxidan-2-idecarboxylate, would have the CO_4^- ion, where -C-O-O⁻ is present.

•2.1   EDTA

By its nature, the commonly-used sodium hypochlorite solution has little effect on the mineral of tooth tissue – biological hydroxyapatite (HAp). Thus, the larger part of the debris of root canal preparation cannot be dissolved appreciably during such irrigation because the solubility of HAp is very low at neutral and high pH (hence the normally long survival of tooth enamel). Very simplistically:

$$Ca_5(PO_4)_3OH \leftrightarrow 5Ca^{2+} + 3PO_4^{3-} + OH^- \qquad (2.1)$$

which indicates that dissolution is suppressed when $[OH^-]$ is high. Likewise, if the value of $[Ca^{2+}]$ is low, dissolution is promoted.

EDTA (ethylene diamine tetraacetic acid – but it has a staggering number of synonyms![41]) is a common, (relatively) non-toxic, chelating agent for polyvalent cations, such as Ca^{2+}, because it is a **polydentate** species – not just the carboxylate groups being involved, but also the nitrogen lone pairs (Figs. 2.1, 2.2). This means that the binding or **sequestration** of such ions is very efficient, the equilibrium lying well to the right, in favour of the complex:

$$EDTA^{4-} + Ca^{2+} \leftrightarrow Ca\text{-}EDTA^{2-} \qquad (2.2)$$

– the complex formation constant K is about $10^{10.69}$. That being so, the concentration of the free calcium ion is driven to be very low, such that HAp continues to dissolve until essentially all of the EDTA is titrated to the complex.

The problem with this agent, however, is that the solubility of the acid itself is very low, and that of the disodium salt only becoming usefully high under neutral or slightly alkaline conditions (Fig. 2.3). Accordingly, the dental products are normally described as having had sodium hydroxide added (roughly equivalent to using the tri-sodium salt), and indeed this is essential to reach the concentration range of 15 ~ 24 % by mass which is used. This pH then has the effect of limiting the rate of dissolution of the HAp somewhat, but the pH is not so high as to macerate (as rapidly) the organic components of the tissue as do hypochlorite solutions (§1.1). Thus, alternating use of the two solutions is necessary if both organic and inorganic material is to be removed.

EDTA solution is frequently referred to as a "lubricant". That it is no such thing (beyond the effect of water alone) should be apparent. Any facilitation of the function of endodontic instruments is due to dissolution of HAp.

Chlorhexidine cannot be combined with EDTA in the same irrigant solution because an insoluble product is formed (§2.2).

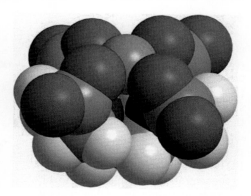

Fig. 2.1 A space-filling model of a calcium ion (top, centre) chelated by EDTA. Water would occupy the remaining open coordination position.

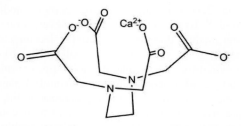

Fig. 2.2 A framework structure for the complex of Fig. 1.2, rotated about 60° to the right, to show the nitrogen involvement.

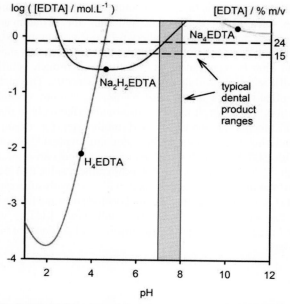

Fig. 2.3 Solubility plots of the acid, di-sodium, and tetra-sodium salts of EDTA, with the typical ranges of dental endodontic irrigation products indicated. (The tri-sodium salt solubility is above 2 mol.L^{-1}.)

●2.2   Citric acid

Citric acid is a moderately strong acid, with a pK$_a$ of about 3.13, and so can readily dissolve biological apatite. It is aided in this by that fact that, like EDTA (§2.1), it forms a number of complexes in solution with polyvalent cations, shifting the equilibrium even further in the direction of dissolution. On this basis it is sometimes used as an alternative to EDTA in the preparation of root canals, in concentrations ranging from 5 to 40% my mass, sometimes in combination with detergent and antibiotic. It has also been used for etching tooth enamel to improve restoration retention.

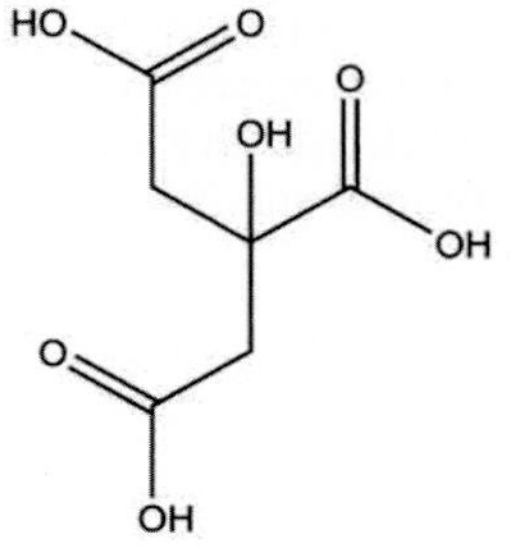

Fig. 2.4 Citric acid.

Etchants for enamel are frequently described as 'conditioners'. This is both quite uninformative and misleading. There is no chemical or physical modification involved, nor even just decontamination. Whilst it is true that the surface itself is modified in terms of its topography, the destructive nature of the treatment may be either unrecognized or ignored. Indeed, it has been said 'conditioning' is distinct from etching, and etching has been denied. With such agents this simply cannot be.

●2.3   Chlorhexidine

Rather than the overt destruction caused by oxidizing agents such as hypochlorite (§1.1), antimicrobial chemistry can take the form of therapeutic agents often called **antiseptics** or **disinfectants**.[42] Chlorhexidine (Fig. 2.5), a member of a large class of medicinal compounds derived from biguanide, is one such, widely used.[43] Its solubility in water, however, is very low at around 80 mg/L,[44] and that of most salts not much better, presumably due to substantial intermolecular hydrogen bonding in the solid (*cf.* silk, 27§3.5). However, polyhydroxy acids raise the solubility enormously to over 700 g/L, which can be attributed to multiple pairing hydrogen bonds. Accordingly, the digluconate is the common and preferred salt solution, although at high concentrations it becomes very viscous – hydrogen bonding again (*cf.* polysaccharides, 7§7); a 20 % solution (by mass) is the common formulation.

Chlorhexidine works by disrupting bacterial cell membranes, binding to key constituents, and then coagulating cell contents by cross-linking indiscriminately. A secondary benefit of its use is that it also binds to the endotoxins which are released on cell death, and thus the inflammatory response that these would otherwise induce is suppressed.[45] The polydentate nature of chlorhexidine also means that it can bind to many surfaces effectively (hydrogen bonding still), being released slowly from such reservoirs. This makes it effective as a mouthwash by having a more sustained action. However, it likewise also binds effectively to many constituents of foodstuffs and toothpaste, for example, rendering it ineffective. In the endodontic context,[46] it reacts with EDTA – which is polyanionic – to form an insoluble salt.[47] By implication, any chlorhexidine residue that is not flushed out, if this agent is used, will react with subsequent EDTA, leaving a coating that will occlude dentinal tubules or accessory canals. Likewise, following EDTA with chlorhexidine with inadequate flushing will produce the same effect.

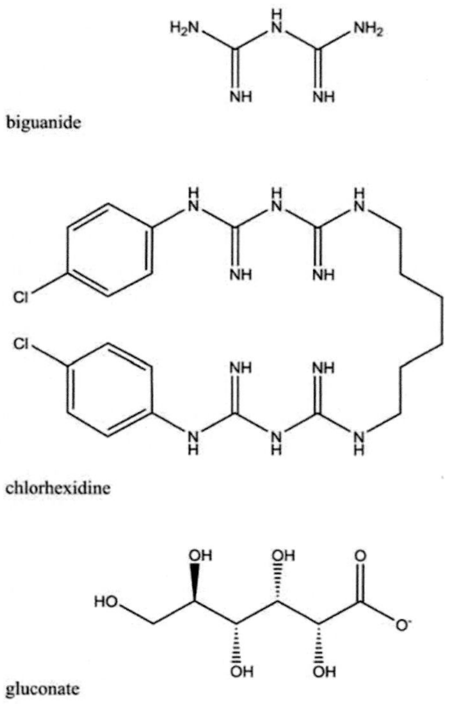

Fig. 2.5 Structures of biguanide, chlorhexidine and the solubilizing anion gluconate.

§3. Aldehydes

●3.1 Formaldehyde

Paraformaldehyde is the informal name of **polyoxymethylene**, a polymer of formaldehyde (also known by many other and confusing names, such as 'paraform', 'formagene', 'para', 'polyoxymethane'). It is slowly formed as a white precipitate by condensation from the predominant species **methanediol** (formaldehyde hydrate) in solutions of formaldehyde (which may also be called 'formalin', 'formal', or 'formalose') on standing, in an equilibrium (Fig. 3.1). The solution is predominantly of oligomers, but when *n* becomes large enough the material becomes sufficiently insoluble as to precipitate, when the condensation may still continue. The resulting solid may have *n* range from ~8 to 100, or more. The reaction is driven to the left, to release formaldehyde, by a low concentration of formaldehyde, and accelerated by acidic or alkaline conditions. Solid paraformaldehyde smells plainly of the monomer (b.p. −21 °C), so it is essentially a convenient means of delivering formaldehyde slowly.

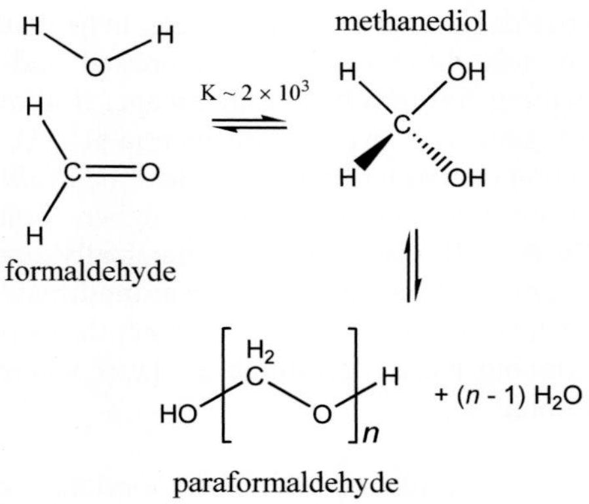

Fig. 3.1 Equilibria of aqueous formaldehyde.

Formaldehyde reacts (as the hydrate) with proteins, cross-linking them, by condensing with secondary amines at the peptide linkage (Fig. 27§3.5), and with the primary amines at the N-terminal or the side chains of arginine, histidine and lysine residues to create irreversible **methylene bridges** (Fig. 3.2).[48][49] Similar reactions also occur with the -SH group of cysteine residues, and the amines of DNA and RNA. Whilst such reactions are useful when tissue needs to be fixed for histological work, or simply museum specimens (or cadavers for dissection), clearly they are problematic in living systems. Indeed, formaldehyde is now known variously to be allergenic, generally toxic, extremely cytotoxic, mutagenic and carcinogenic, and so is increasingly under strict control in many contexts. Inhalation of the vapour must be avoided (the vapour pressure is high[50]). It is implicated as an asthma-inducer or exacerbator, and is a major component of house-fire smoke and photochemical smog. The use of paraformaldehyde, therefore, as an ingredient of endodontic cements[51] – to achieve so-called "mummification" (*i.e.* fixed tissue) – is now considered inappropriate (although not without controversy[52]). The inclusion of alkaline ingredients only serves to accelerate the hydrolysis and depolymerization. Sterilization would, of course, occur anyway. Formaldehyde itself is also used.[53] The acid hydrolysis of hexamethylene tetramine[54] (Fig. 3.3) (solubility in water ~850 g/L, 20 °C), an ingredient of one known endodontic product,[55] also yields formaldehyde.

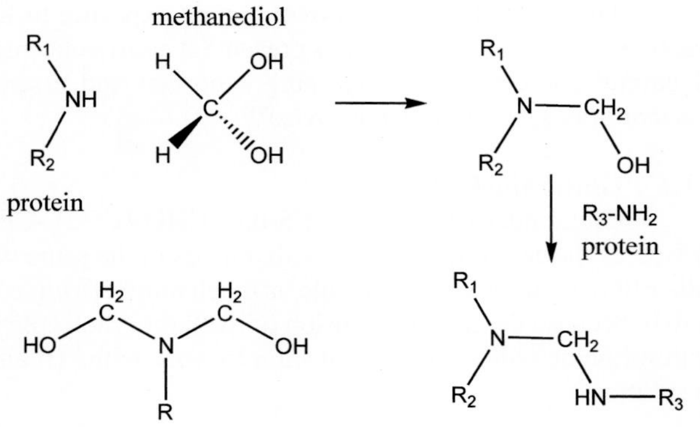

Fig. 3.2 Examples of reactions leading to the cross-linking ('fixing') of proteins. Double reaction with primary amines is also possible (bottom left).

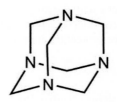

Fig. 3.3 Hexamethylene tetramine.

Formaldehyde was formerly used as a dentine desensitizing agent, and has even been included in toothpastes (with a pungent taste!) for the same supposed effect, although unsuccessfully.[56][57]

Formaldehyde is also formed as a by-product of free-radical polymerization of methacrylates (as in filled resins) in the presence of oxygen (6§6),[58] and in this context may be a contributory factor to adverse reactions,[59] being released slowly over a long period, presumably as the peroxides break down. This might also occur with acrylic denture bases as an irritant for denture stomatitis or 'sore mouth' in addition to residual MMA (5§2.7).[60] Since the precursor peroxides are not thermally stable, in heat-cured materials these will be decomposed and the resulting formaldehyde may escape, if allowed sufficient time. In cold-cure materials (5§3), this decomposition will not occur, and the available concentration will therefore be higher. Similar effects will occur in any chemically-similar system, such as so-called "resin-modified" GI cement (9§8.9).[61][62] This underlines the value of removing the oxygen-inhibited layer wherever possible.

Degradation of cyanoacrylates may proceed through depolymerization by hydrolysis (10§6.2), by the simple hydrolysis of the ester first, but another reaction occurs that also generates formaldehyde (Fig. 3.4).[63] Again, this may lead to the irritation of living tissue.

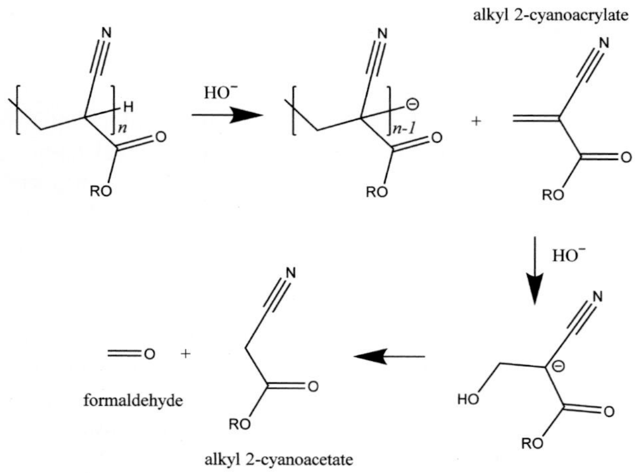

Fig. 3.4 One possible degradation route for poly(cyanoacrylate).

There is much concern over human exposure to formaldehyde because of the possibility of adverse reactions, despite the fact that it is present (at a low concentration, generally) in the environment from a number of natural sources as well as being a normal and essential physiological metabolite in man at very low concentrations, where it is not toxic.[64]

●3.2   Glutaraldehyde

Glutaraldehyde, pentane-1,5-dial, $CHO-(CH_2)_3-CHO$ (b.p. 187 °C), which is used as a sterilizing agent as well as tissue fixative, reacts with amines in the same way as formaldehyde, but being di-functional, and a rather longer but flexible molecule, is much more effective as a cross-linker – *cf.* bis-GMA (6§4.3), albeit more slowly because the rate of diffusion of the larger molecule is less. The much lower vapour pressure compared with formaldehyde also makes it safer to work with. (It also polymerizes on standing, with oligomers present in solution.)[65]

Glutaraldehyde is used in some topical dentine desensitizing products in about 5 % solution, and is also used on cut dentine as part of the procedure for bonding a restoration, after etching. It is supposed that the precipitated, denatured proteins occlude dentinal tubules, preventing the fluid flow that causes the pain, according to the 'hydrodynamic' theory of sensitivity. However, nerve endings must also be destroyed. Denatured proteins would, of course, degrade with time, compromising any seal. But, in addition, it would act as a local sterilization agent, killing all micro-organisms. Obviously, it is important to avoid contact of such solutions with the gingivae or tongue, or indeed any other vital tissue; working under rubber dam would seem to be essential (even so, it should be removed by suction, not with a water jet, to avoid splatter). Rubber gloves and eye protection are required for working with glutaraldehyde sterilants: they are described as caustic.

§4. Calcium Hydroxide

As discussed elsewhere (9§10), calcium hydroxide is commonly used on its own, or as the core of a setting cement, as a cavity liner or indirect pulp cap (9§10), relying on the high pH it causes to sterilize the site. Likewise, direct pulp capping (*i.e.* of exposed pulp) and temporary root filling also rely on that property, but in addition it is used to stimulate vital pulp into the formation of reparative dentine.[66] It is also a by-product of the setting reaction of hydraulic silicate cements (9§12), where it is similarly considered beneficial.

Unfortunately, the very high pH also macerates soft tissue (§1.1), and this effect would still operate on the matrix of dentine – it would be dissolved. This must result in a reduction in the strength of the dentine, with consequent root fracture a risk.[67] The reaction with phosphate and carbonate (9§12.4) may limit this in due course, but any damage already done cannot be reversed. It should be said that calcium hydroxide has no effect on the mineral of teeth, biological apatite. The macerating effect also means that extrusion of the material through the root apex, fractures or perforations may have very damaging consequences for surrounding tissue,[68] but it also represents a significant hazard for the operator: eye-protection should be worn.[69]

§5. Zinc Oxide

Denture retention pastes work on the basis of being sticky (that is wetting tissue and denture base acrylic), high viscosity fluids making the breakaway time sufficiently long (10§9.10). Whilst making a liquid viscous can only go so far, using solutions of substances such as carboxymethyl cellulose (27§2.3), it is apparent that a greater effect can be obtained using a filler (4§9). Accordingly, it used to be the practice that such pastes would be formulated with zinc oxide (9§3), presumably on the long-held assumption that this is benign. Indeed, its solubility is sufficiently low that absorption ordinarily would be negligible, as is perhaps indicated by its lavish and unproblematic (it would seem) use in sun-block creams. However, once ingested and under the low pH conditions of the stomach (mean pH ~2.6) it dissolves readily, and thus further on is easily absorbed.[70] Being a heavy metal, one would expect zinc to be toxic to at least some extent, but its effect is more complex: it blocks the uptake of copper, causing a deficiency of an essential element.[71] Neurological and haematological problems follow.

The problem arises when a denture base is sufficiently ill-fitting that instead of the infrequent application that would be expected, very large quantities have been used and much of the paste has been systematically swallowed. The recognition of this difficulty has led to the rapid reformulation of many products to be "zinc-free".

§6. Evaporative Cooling

The vitality of teeth, that is, of the pulp, may be tested by applying a variety of stimuli, one of which ordinarily used is cold. A common means of doing this is to apply a few drops of a substance such as ethyl chloride (C_2H_5Cl). The question is, how does this chilling occur? The mechanism is **evaporative cooling**.

Generally, in discussing the vapour pressure of liquids (such as at 5§2.6, 8§3.2, and 28§6.1), we are mostly concerned about equilibrium conditions, usually isothermally, where the energy accounting is of little consequence and therefore ignored. However, in open, non-equilibrium systems, the energy supply is important. Thus, equilibrium was defined in 8§3.3 in terms of the equality of the fluxes of atoms or molecules in the two directions across a phase boundary. Implicitly, on average, the kinetic energy transferred in the two directions is the same, and therefore there can be no temperature change, where we can note that temperature is essentially defined as a measure of the average kinetic energy of the molecules of the system. For a molecule to escape from the surface of a liquid, it needs to have sufficient kinetic energy to overcome the intermolecular forces that otherwise keep it in that condensed state (that is, have the activation energy for escape; *cf.* Fig. 3§3.2). But an open system means that the rate of evaporation far exceeds the condensation of molecules back into the liquid. This therefore means that more kinetic energy is taken into the vapour phase than is returned to the liquid: the liquid must cool as its total – and thus its average – kinetic energy is decreased because only the most energetic molecules can escape (Fig. 6.1).

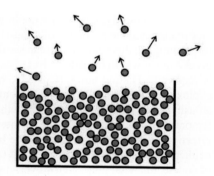

Fig. 6.1 Molecules escaping from a liquid (*i.e.,* evaporating) necessarily have higher average kinetic energy than those remaining, and therefore simply by escaping cause the liquid to cool, because they carry away as a minimum the escape activation energy.

It is then a matter of the rate of loss of heat (*i.e.* as kinetic energy) exceeding the rate of gain of heat from the surroundings by conduction and so on. Several factors are conducive to this. Firstly, the liquid boiling point: the lower the value, the greater the value of the equilibrium vapour pressure (or 'escaping tendency') of the liquid at a given temperature, and also the greater the actual temperature of the liquid the greater the vapour pressure will be. Secondly, the actual vapour pressure of the substance in the vapour phase is important in the sense that it is the difference between this and the equilibrium value, Δp_v, that controls the net flux across the boundary (as a simple matter of concentration). This difference can be maximised by ensuring that it is indeed an open system, that the vapour does not accumulate, and that there is a draught – blowing across the liquid surface increases the cooling rate. In

Table 6.1 *Some properties of liquids for evaporative cooling.*

	b.p. / °C (at 1 bar)	H_{evap} / kJ.mol^{-1}	vapour pressure / bar (at 35 °C)[1]	Q / Jm^{-2}s^{-1} (at 35 °C)
Water	100.0	41	0.06	20
Diethyl ether	34.6	28	1.01	223
Ethyl chloride	12.4	26	2.26	470
1,1,1,2-tetrafluoroethane	-26.6	22	8.54	1480
Dichlorodifluoromethane	-29.8	20	8.15	1297

[1] This is similar to the pressure that the container must sustain (the actual margin of safety may be somewhat larger) and, recalculated for the current room temperature, that which drives the substance out of the dispenser nozzle (in the case of the last three).

addition, if the equilibrium vapour presure is greater than atmospheric, the liquid must boil immediately. Vapour pressure can be calculated from the Clausius-Clapeyron equation (equation 5§2.1). Thirdly, the latent heat of vaporization, H_{evap}, is the measure of how much energy is carried away as a function of the amount of substance lost.

The maximum evaporative heat loss rate Q (Jm^{-2}s^{-1}) can then be estimated from an equation due to Langmuir:[72]12

$$Q = \frac{p_v H_{evap}}{\sqrt{2\pi MRT}} \tag{6.1}$$

where M is the molecular weight, R is the molar gas constant, T is the absolute temperature, H_{evap} the latent heat of evaporation, and p_v the vapour pressure (assuming no vapour already present). Of course, the actual temperature drop depends on how much liquid is present, its specific heat capacity, the size and thermal properties of the substrate, and several other factors, but Q gives a reasonable comparative measure of the effect, *i.e.* as a figure of merit. (Of course, the vapour pressure is a function of temperature [equation 5§2.1], so the cooling of the liquid reduces the value of Q.[73])

Accordingly, evaporative cooling occurs with all liquids, and most notably with water, hence the function of sweating and the problems of wind-chill: a wet body in the wind can rapidly cool to the point of being life-threatening. In this context, one can see the effect of the actual vapour pressure of water in the air: high humidity gives a lower wind-chill effect. But to maximize the effect for dental diagnostic purposes, a lower boiling liquid is required. Diethyl ether, $(C_2H_5)_2O$, would work, but is highly inflammable, and has the risk of forming explosive peroxides on standing in contact with oxygen; these can detonate on drying (which would not be amusing if this happened in the mouth). Ethyl chloride has a boiling point (b.p.) rather lower than body temperature (Table 6.1) and so evaporates very rapidly from a test pledget on teeth in the mouth. It is sufficiently effective that it is used to freeze tissue by topical application. Care should therefore be taken that freezing of neither the user's nor the patient's soft tissues occurs: necrosis follows – this is frostbite.[74] This applies also, of course, to a vital pulp.13 However, it should be noted that the use of this substance has been discontinued in many contexts because of its stratospheric ozone depletion potential. It is also flammable, toxic, anaesthetic, and a potential carcinogen. If it is used, it should be used sparingly and in a very well ventilated location. A variety of other substances and mixtures have been used for this purpose, and 1,1,1,2-tetrafluoroethane is replacing ethyl chloride as the major component of pulp-test refrigerants, being substantially more effective (much higher Q), but it also is non-inflammable and has no ozone depletion potential (although it is a 'greenhouse' gas) or known toxicity, but it remains anaesthetic: good ventilation is still required. Dichlorodifluoromethane was in common use formerly but, as it is an extremely potent ozone depleter and greenhouse gas, it has been discontinued and should not be used.

There is one further hazard. Fluorine-containing substances form a variety of toxic substances, including

12 By multiplying through by H_{evap}/M.

13 It should be noted that the converse test, of applying heat, is rather riskier as inflammation starts at ~42 °C and protein denaturation not much higher.

a high enough temperature (>~250 °C) in air, such as in or even just near a flame. Accordingly, open flames should not be present when such refrigerants are being used. Likewise polytetrafluoroethylene (ptfe) and pfte-coated instruments (such as spatulas and tweezers), pipe-sealing tape (as may be used on compressed air lines) and the like should not be overheated. The toxicity of inhaled HF is severe (25§10), while COF_2, the fluorinated analogue of phosgene, is acutely toxic, having a short-term threshold limit value (TLV) of just 2 ppm; it reacts immediately with water to form HF, *i.e.* as in the airway and eyes.

The same cooling effect occurs when a jet of compressed air is used to dry tooth surfaces. The compressed air is necessarily of low humidity (removed as condensate from the compressor) and so can not only displace liquid but cause evaporation. That evaporation causes cooling, as will be well-understood by anyone who has stood in the wind while wet. Even though this effect is relatively slight (Table 6.1), the cooled surface must immediately cause condensation on to it from the high humidity of the oral cavity (unless this is isolated by rubber dam), or even from the ambient air, depending on the dew point (9§5.13, footnote). Thus, it cannot be assumed that a surface so treated is in fact dry. Indeed, adsorption of a monolayer is inevitable (2§2.5). This complicates the wetting behaviour of the substrate for non-aqueous materials (10§1.11) such as vinyl polysiloxane for impressions (10§9.3).

References

[1]   Prasansuttiporn T, Nakajima M, Kunawarote S, Foxton RM & Tagami J. Effect of reducing agents on bond strength to NaOCl-treated dentin. Dent Mater 27(3): 229 - 234, 2011).

[2]   Bergmans L, Moisiadis P, Huybrechts B, Van Meerbeek B, Quirynen M & Lambrechts P. Effect of photo-activated disinfection on endodontic pathogens ex vivo. Int Endod J 41: 227 - 239 (2008)

[3]   Jebaramy J, Ilanchelian M & Prabahar S. Spectral studies of toluidine blue O in the presence of sodium dodecyl sulfate. Dig Nanomat Biostruct 4(4): 789-797, 2009.

[4]   Ogilby PR. Singlet oxygen: there is indeed something new under the sun. Chem Soc Rev 39: 3181 - 3209, 2010.

[5]   Haber F & Weiss J. The catalytic decomposition of hydrogen peroxide by iron salts. Proc Roy Soc London 147: 332 - 351, 1934.

[6]   Ingraham LL & Meyer DL. Biochemistry of Dioxygen. Plenum, New York, 1985.

[7]   Dörtbudak O, Haas R, Bernhart T & Mailath-Pokorny G. Lethal photosensitization for decontamination of implant surfaces in the treatment of peri-implantitis. Clin Oral Implants Res 12(2): 104 - 108, 2001.

[8]   Dickers B, Lamard L, Peremans A, Geerts S, Lamy M, Limme M, Rompen E, De Moor RJG, Mahler P, Rocca JP & Nammour S. Temperature rise during photo-activated disinfection of root canals. Lasers Med Sci 24: 81 - 85, 2009.

[9]   Konopka K & Goslinski T. Photodynamic therapy in dentistry. J Dent Res 86: 694 - 707, 2007.

[10]   George S & Kishen A. Advanced noninvasive light-activated disinfection: assessment of cytotoxicity on fibroblast versus antimicrobial activity against *Enterococcus faecalis*. J Endod 33: 599 - 602, 2007.

[11]   Petri BG, Watts RJ, Teel AL, Huling SG & Brown RA. Fundamentals of ISCO using hydrogen peroxide. Chapter 2 *in:* Siegrist RL *et al.* (eds.). In Situ Chemical Oxidation for Groundwater Remediation, Springer, 2011.

[12]   Fritchie CJ & McMullan RK. Neutron Diffraction Study of the 1:1 urea: hydrogen peroxide complex at 81 K. Acta Crystallographica B 37: 1086 - 1091, 1981.

[13]   Zhao HK, Tang C, Zhang DS, Li RR Wang YQ & Xu WL. Solubility prediction for the nonelectrolyte urea–hydrogen peroxide–water system and thermodynamic solubility product calculation for $CO(NH_2)_2 \cdot H_2O_2$ using the modified Pitzer model. Computer Coupling of Phase Diagrams and Thermochemistry 31: 281 - 285, 2007.

[14]   Duverger O, Ohara T, Shaffer JR, Donahue D, Zerfas P, Dullnig A,Crecelius C, Beniash E, Marazita ML & Morasso MI. Hair keratin mutations in tooth enamel increase dental decay risk. J Clin Invest 124(12): 5219 - 5224, 2014.

[15]   Açil Y, Mobasseri AE, Warnke PH, Terheyden H, Wiltfang J & Springer I. Detection of mature collagen in human dental enamel. Calcif Tissue Int 76(2): 121 - 126, 2005.

[16]   McGuire JD, Walker MP, Mousa A, Wang Y & Gorski JP. Type VII collagen is enriched in the enamel organic matrix associated with the dentin-enamel junction of mature human teeth. Bone 63: 29 - 35, 2014.

[17]   Al-Tarakemah Y, Darvell BW. On the permanence of tooth bleaching. Dent Mater 32: 1281 - 1288, 2016.

[18]   Das A & Mukhopadhyay C. Urea-mediated protein denaturation: a consensus view. J Phys Chem B 113: 12816 - 12824, 2009.

[19]   Castagna R, Eiserich JP, Budamagunta MS, Stipa P, Cross CE, Proietti E, Voss JC & Grec L. Hydroxyl radical from the reaction between hypochlorite and hydrogen peroxide. Atmos Env 42: 6551 - 6554, 2008.

[20]   Criegee R. Mechanism of ozonolysis. Angew Chem Int Edit 14 (11): 745 - 752, 1975.

[21]   Burke FJT. Ozone and caries: A review of the literature. Dent Update 39 (4): 271 - 278, 2012.

[22]   Merényi G, Lind J, von Naumov S & Sonntag C. Reaction of ozone with hydrogen peroxide (Peroxone process): A revision of current mechanistic concepts based on thermokinetic and quantum-chemical considerations. Environ Sci Technol 44: 3505 - 3507, 2010.

[23]   Al-Omiri MK, Hassan RSA, AlZarea BK & Lynch E. Effects of combining ozone and hydrogren peroxide on tooth bleaching: A clinical study. J Dent 53 : 88 - 93, 2016.

[24]   Kirk-Othmer Encyclopedia of Chemical Technology. 3rd ed., New York, NY: Wiley, 1984. p. 7(79) 803.

[25]   Lagarce L,Tessiera B, Harry P, Lelievrec B, Bourneaua D,Drabliera G & Lainé-Cessac P. The dangers of iodoform gauze : a retrospective study. 8ème Congrès de Physiologie de Pharmacologie et de Thérapeutique - Angers, 22 - 24 April, 2013. http://www.atout-org.com/p2t2013/abstract_display!fr!!!!65f6b2f8-8e84-1030-b866-9251dd645b9d!bc970a90-d673-1030-b866-925 1dd645b9d

[26]   Block SS (ed). Disinfection, Sterilization, and Preservation. 5th ed. Lippincott Williams & Wilkins. Philadelphia, 2001.

[27]   Stevens JL & Anders MW. Metabolism of haloforms to carbon monoxide—III: Studies on the mechanism of the reaction. Biochem Pharmacol 28 (21): 3189 - 3194, 1979.

[28]   https://toxnet.nlm.nih.gov/cgi-bin/sis/search/a?dbs+hsdb:@term+@DOCNO+4099

[29]   Estrela C, Estrela CRdeA, Hollanda ACB, Decurcio DdeA & Pécora JD. Influence of iodoform on antimicrobial potential of calcium hydroxide. J Appl Oral Sci 14(1): 33 - 37, 2006. https://www.ncbi.nlm.nih.gov/pmc/articles/PMC4327168/#B04

[30]   Lou YL, Botelho MG & Darvell BW. Erratum to "Reaction of silver diammine fluoride with hydroxyapatite and protein" [J. Dent. 39 (2011) 612–618] J Dent 40 (1): 91-93, 2012).

[31]   Lou YL, Botelho MG & Darvell BW. Reaction of silver diamine fluoride with hydroxyapatite and protein. J Dent 39 (9): 612 - 618, 2011. [see also: J Dent 40 (1): 91 - 93, 2012.]

[32]   Russell AD & Hugo WB. Antimicrobial activity and action of silver. Prog Med Chem 31: 351 - 370, 1994.

[33]   Lemire JA, Harrison JJ & Turner RJ. Antimicrobial activity of metals: mechanisms, molecular targets and applications. Nature Reviews Microbiol 11: 371 - 384, 2013.

[34]   Markowska K, Grudniak AM & Wolska KI. Silver nanoparticles as an alternative strategy against bacterial biofilms. Acta Biochim Pol 60 (4): 523 - 530, 2013.

[35]   https://www.gov.uk/government/publications/supplying-explosives-precursors/supplying-explosives-precursors-and-poison

[36]   McKillop A & Sanderson WR. Sodium perborate and sodium percarbonate: Cheap, safe and versatile oxidising agents for organic synthesis. Tetrahedron 51 (22): 6145 - 6166, 1995.

[37]   Attin T, Paqué F, Ajam F & Lennon ÁM. Review of the current status of tooth whitening with the walking bleach technique. Int Endod J 36 (5): 313 - 329, 2003.

[38]   E. U. Çelik EU, Türkün M & Yapar AGD. Oxygen release of tetra acetyl ethylene diamine (TAED) and sodium perborate combination. Int Endod J 41: 571 - 576, 2008.

[39]   www.inchem.org/documents/sids/sids/15630894.pdf

[40]   Violich DR & Chandler NP. The smear layer in endodontics – a review. Int Endod J 4: 2 - 15, 2010.

[41]   http://www.commonchemistry.org/ChemicalDetail.aspx?ref=60-00-4

[42]   Maris P. Modes of action of disinfectants. Rev Sci Tech Off Int Epiz 14(1): 47 - 55, 1995

[43]   Gomes BPFA, Vianna, ME, Zaia A, Almeida JFA, Souza-Filho FJ & Ferraz CCR. Chlorhexidine in endodontics. Braz Dent J 24(2): 89 - 102, 2013.

[44]   Senior N. Some observations on the formulation and properties of chlorhexidine. J Soc Cosmet Chem 24: 259 - 278, 1973.

[45]   Zork o M & Jerala R. Alexidine and chlorhexidine bind to lipopolysaccharide and lipoteichoic acid and prevent cell activation by antibiotics. J Antimicrobial Chemotherapy 62(4): 730 - 737, 2008.

[46]   Mohammadi Z & Abbott PV. The properties and applications of chlorhexidine in endodontics. Int Endod J 42: 288 - 302, 2009.

[47]   Rossi-Fedele G, Do gramaci EJ, Guastalli A, Steier L & Poli de Figueiredo JA. Antagonistic interactions between sodium hypochlorite, chlorhexidine, EDTA, and citric acid. J Endod 38(4): 426 - 431, 2012.

[48]   Metz B, Kersten GF, Hoogerhout P et al. Identification of formaldehyde-induced modifications in proteins: reactions with model peptides. J Biol Chem 279: 6235 - 6243, 2004.

[49]   Kiernan JA. Formaldehyde, formalin, paraformaldehyde and glutaraldehyde: What they are and what they do. Microscopy Today 00-1: 8 - 12, 2000.

[50]   Oancea A, Hanoube B, Focsa C & Chazallon C. Vapour-liquid equilibrium of the formaldehyde-water system. Journées Interdisciplinaire de la Qualité de l'Air (JIQA), 2008. http://www.jiqa.fr/doc/2008/Article/OANCEA.pdf

[51]   Spångberg LSW. Biologic effect of root canal filling materials: the effect on bone tissue of two formaldehyde-containing root canal filling pastes, N2 and Riebler's paste. Oral Surg Oral Med Oral Pathol 38: 934 - 944, 1974.

[52]   http://www.studiodentisticovenuti.it/2012/12/10/n2-and-sargenti/
         http://www.aesoc.com/
         http://www.dentalwatch.org/questionable/sargenti/overview.html
         http://www.sargentipaste.org/index.html

[53]   Lewis B. The obsolescence of formocresol. Brit Dent J 207: 525 - 528, 2009.

[54]   Hexamethylenetetramine [MAK Value Documentation, 1993]. The MAK Collection for Occupational Health and Safety.
         356–372, 2012. DOI: 10.1002/3527600418.mb10097kske0005

[55]   Spångberg LSW, Barbosa SV & Lavigne GD. AH 26 releases formaldehyde. J Endod 19: 596 - 598, 1993.

[56]   Smith BA & Ash MM. Evaluation of a desensitizing dentifrice. J Amer Dent Assoc 68: 639 - 647, 1964.

[57]   Addy M, Mostafa P & Newcombe R. Dentine hypersensitivity: a comparison of five toothpastes used during a 6-week treatment
         period. Brit Dent J 163(2): 45 - 51, 1987.

[58]   Øysæd H, Ruyter IE & Sjøvik Kleven IJ. Release of formaldehyde from dental composites. J Dent Res 67: 1289 - 1294, 1988.

[59]   Schmalz G, Arenholt-Bindslev D. Biocompatibility of Dental Materials. Springer, Berlin, 2010.

[60]   Ruyter IE. Release of formaldehyde from denture base polymers, Acta Odont Scand 38: 17 - 27, 1980.

[61]   Ruyter IE. Physical and chemical aspects related to substances released from polymer materials in an aqueous environment.
         Adv Dent Res 9: 344 - 347, 1995.

[62]   Geurtsen W. Biocompatibility of resin-modified filling materials. Crit Rev Oral Biol Med 11: 333 - 355, 2000

[63]   Dumitriu S. (ed) Polymeric Biomaterials. 2nd ed. Dekker, New York, 2002. pp. 719 - 721

[64]   IARC Monographs on the Evaluation of Carcinogenic Risks to Human, IARC Monographs, Volume 88. WHO, Geneva, 2006.
         http://monographs.iarc.fr/ENG/Monographs/vol88/mono88-6.pdf

[65]   Migneault I, Dartiguenave C, Bertrand MJ & Waldron KC. Glutaraldehyde: behavior in aqueous solution, reaction with proteins,
         and application to enzyme crosslinking. BioTechniques 37(5): 790 - 802, 2004.

[66]   Carrotte P. Endodontics: Part 9 Calcium hydroxide, root resorption, endo-perio lesions. Brit Dent J 197: 735 - 743, 2004.

[67]   Mohammadi Z & Dummer PMH. Properties and applications of calcium hydroxide in endodontics and dental traumatology.
         Int Endo J 44: 697 - 730, 2011.

[68]   De Bruyne MAA, De Moor RJG & Raes FM. Necrosis of the gingiva caused by calcium hydroxide: a case report. Int Endod
         J 33: 67 - 71, 2000.

[69]   Lipski M, Buczkowska-Radlińska J & Góra M. Loss of sight caused by calcium hydroxide paste accidentally splashed into the
         eye during endodontic treatment:Ccase report. J Can Dent Assoc 78: c57, 2012.

[70]   Lee HH, Prasad AS, Brewer GJ & Owyang C. Zinc absorption in human small intestine. Amer J Physiol - Gastrointestinal Liver
         Physiol 256 (1): G87 - G91, 1989.

[71]   Cathcart SJ & Sofronescu AG. Clinically distinct presentations of copper deficiency myeloneuropathy and cytopenias in a patient
         using excessive zinc-containing denture adhesive. Clin Biochem 50 (12): 733 - 736, 2017.

[72]   Langmuir I. The vapour pressure of metallic tungsten. Phys Rev 2 (5): 329 - 342, 1913.

[73]   Xu XF & Ma LR. Analysis of the effects of evaporative cooling on the evaporation of liquid droplets using a combined field
         approach. Sci Rep 5 : 8614, 2015.

[74]   https://en.wikipedia.org/wiki/Frostbite

Chapter 31   Equipment

A wide variety of equipment and devices are used in the support of dentistry and to execute various tasks. The design and operation of these machines is, of course, based on various principles of materials science, but their proper function affects the outcome of the systems on which they are used. It follows that an understanding of their principles and behaviour is condusive to their effective and efficient use as well as to their selection in the first place and proper maintenance subsequently.

The air-turbine handpiece is a central piece of equipment, the use of which defines dentistry for the public. Understanding its behaviour in relation to its supplied air pressure permits effective use, while the importance of lubrication for bearing working life is easily seen.

Semiconductor diodes have many applications, but apart from LED computer screens the most prominent application is for filled-resin curing lights. In that context, stability of output is critical for effective curing, while the ability to monitor output depends on similar physics but is equally prone to error.

Materials Science for Dentistry
https://doi.org/10.1016/B978-0-08-101035-8.50031-6

Both the clinical and the laboratory work of dentistry is dependent on the tools, instruments and equipment used to execute the various tasks. The design and other characteristics of these devices control what may be done and the quality of the outcome. In the same sense as applies to the more obvious materials, these items form an important part of the armamentarium of dentistry and are thus deserving of proper consideration, yet they are generally overlooked. Aspects of dental amalgam mixing machines have already been discussed (15§2), as have some features of cutting tools (20§3). However, there remain a variety of devices that may be used more or less frequently, the understanding of which will be of benefit in both selection and use.

§1. Air-turbine Handpiece

A great deal of dentistry involves the removal or shaping of tooth tissue, and the usual manner of achieving this is by a rotary cutter driven by an air-turbine handpiece. The history of the development of these tools[1] is of no particular concern here, but the principal goal of the design generally understood is of high speed such that cutting enamel requires little force and little time, at least in comparison with previous equipment. The questions to be asked then are

- what controls the speed of the device?
- is speed as such so important?
- what controls the efficiency of cutting?

Some answers to these questions will be offered. The last, however, is dealt with in large part in Chapter 20.

●1.1 Speed and torque

The basic principle of an air turbine as used in dental handpieces is illustrated in Fig. 1. The pressure drop between the inlet and the outlet ports results in a pressure difference across each blade of the turbine and thus a force tending to rotate it about its axis, that is, a **torque** is generated (23§1). This force is unbalanced, partly because of the momentum of the flowing air and the direction in which the inlet port is pointing, and thus the initially-stationary turbine accelerates in the usual Newtonian sense:

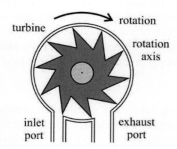

Fig. 1.1 Basic outline of the design of a circumferential flow air-turbine for a dental handpiece.

$$\tau_p = I_r \alpha \qquad (1.1)$$

where τ_p is the instantaneous torque arising from the applied pressure, α is the angular acceleration (rad/s), and I_r is the rotational inertia (kg.m^2) (see equation 1.9 below; *cf.* 18§3.1). However, there is friction (1§5.5) within the bearings of the turbine, as well as viscous forces arising from the bearing lubricant (oil). Thus, there comes a point where the acceleration due to the pressure drop is exactly balanced by the frictional deceleration (*cf.* equation 4§10.1). Alternatively, this can be viewed as meaning that the frictional torque now exactly balances the driving torque, so the net torque is zero. This condition is known as **free running**, because no external (useful) work is being done, and this is at the **free-running speed**. In the absence of other effects we might then expect a straight line speed-pressure relationship. This is not what happens however, although in the region of the usual supply pressures it is close enough in practical terms (Fig. 1.2). Partly, this is because gases flowing through an orifice in this fashion cannot exceed the speed of sound, an effect called **choke**. This may be understood by imagining an instantaneous pressure drop in a flowing gas when the wave of rarefaction, travelling upstream,

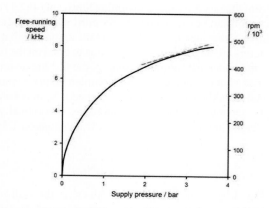

Fig. 1.2 Typical form of the dependence of turbine free-running speed on handpiece supply pressure. In the region of the typical recommended supply pressures (1.9~3.5 bar), the response is roughly linear.

In the context of turbine handpieces and compressors, supply air pressures are strictly to be referred to as **gauge pressures**, p_g, that is, the difference between the absolute pressure of the gas, p_{abs}, and that of the surroundings, *i.e.* atmospheric pressure, p_{atm}:
$$p_{abs} = p_g + p_{atm}$$
However, it is common to omit the word 'gauge'. Even so, it should be remembered that the gas laws are expressed in terms of absolute pressures.

corresponds to the cyclic rarefaction in a sound wave in that gas, and this corresponds in turn to gas flowing into the rarefaction. In addition, the speed of the tips of the turbine blades cannot exceed the speed of sound, as this would mean that the gas flowing there was also supersonic. In practice, this limit is approached **asymptotically** as the supply air pressure is increased. There is a further complication in that the air is expanding nearly **adiabatically** (it does not have time to equilibrate with its surroundings), and thus suffers some cooling, and this also affects the flow rate attainable. Overall, pressure-flow rate curves looks more like those in Fig. 1.3, with a clearly defined limit.[2]

If the turbine were to be prevented from rotating at all, when it is said to be **stalled**, there are no frictional or viscous forces acting to oppose the turning moment due to the flowing air, merely the applied restraining torque, and the torque experienced by the turbine due to the air pressure is at a maximum (Newtonian action and reaction, static equilibrium); this is the **stall torque**. It depends simply and directly on the air supply (gauge) pressure:

$$\tau_p = \Phi p_g \qquad (1.2)$$

(Fig 3.4). The constant of proportionality, Φ, the **stall torque coefficient**, depends on the exact design of the handpiece and its turbine, but cannot yet be predicted.

It follows then that if the restraining torque is reduced, the turbine rotation rate must increase until the frictional forces once again balance the accelerating force. The work done in opposing the retarding forces arising from the viscosity of a Newtonian lubricant (4§7.1) and of friction can be expected to be proportional to velocity (*i.e.* force × distance in unit time). Thus, assuming that other factors are negligible, the net torque exerted on the turbine – and hence the torque exerted by the turbine on any tool in its chuck – decreases in a complementary manner, as then does the rotational rate (*i.e.* speed, N, in revolutions per second) of the turbine increase, theoretically in a straight-line relationship, that is, from equation 1.2:

$$\tau_p = \Phi p_g \left(1 - \frac{N}{N_{fp}} \right) \qquad (1.3)$$

(Fig. 1.5). Note the limiting values: at the stall torque, there is no rotation; at the free running speed, N_{fp}, there is no (available) torque to drive a tool to do useful work because it is all used to overcome the retarding forces. If external work is done the speed must fall.

The 'useful work' in the present context is the cutting of the tooth tissue or other materials. Remember that the creation of new surface requires the input of at least the corresponding surface energy (1§7, 10§1.2), with the work of plastic deformation (1§4) and frictional work (1§5.5) providing additional sinks: cracking, tearing, deforming, heating. Thus the capacity of the rotary

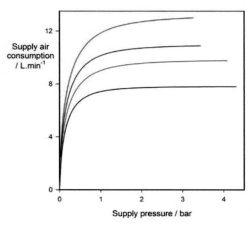

Fig. 1.3 Typical form of the dependence of the air used as a function of handpiece supply pressure showing the effect of choke. It is data such as this that determines the compressor specification.

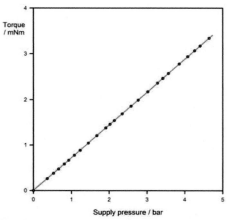

Fig. 1.4 Stalled turbine torque is proportional to the supply air pressure. For circumferential flow (Fig. 1.1), the position of the blade in relation to the supply orifice has a small effect that quickly averages out at very low speed, so the overall (weighted) mean is plotted here.

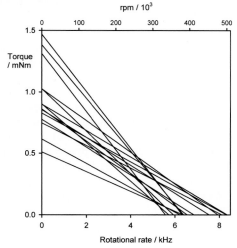

Fig. 1.5 Ideal torque behaviour for a series of commercial handpieces at their minimum recommended supply pressure.

system to do work is of fundamental importance. It can be seen that turbine speed is not itself important, it is the capacity to do external work that matters.

●1.2 Power

In a rotary system, (external) work rate or power, P, is calculated from:

$$P = \tau\omega \tag{1.4}$$

(Nm × s^{-1} = Js^{-1} = W) where ω is the rotation rate in radians/second:

$$\omega = 2\pi N \tag{1.5}$$

Thus, we have for a dental air turbine, from equation 1.4 substituting from equations 1.3 and 1.5:

$$P = 2\pi\Phi p_g N\left(1 - \frac{N}{N_{fp}}\right) \tag{1.6}$$

which is a quadratic equation in N. The variation of work rate with speed is therefore expected to be parabolic, with zero work at the two end points and a maximum at exactly half the free running speed (Fig. 1.6), and thus generating half the stall torque. This type of relationship is in fact true for many kinds of simple, fixed-input motor-driven system. So, substituting $N_{fp}/2$ for N in equation 1.6, we have

$$P_{max} = \frac{\pi}{2}\Phi p_g N_{fp} \tag{1.7}$$

as measure of the maximum possible power at the chosen supply pressure. If then the variation of free-running speed with pressure is taken into account, a curve of the form shown in Fig. 1.7 is obtained. This indicates that the maximum power varies nearly linearly with supply pressure in the region of the usual recommended pressure for such machines. Hence, care should be taken if quoted handpiece power is taken into account when selecting a product, as this can be increased by the manufacturer simply by an arbitrary increase in recommended supply pressure. Clearly, for objective comparison of devices, a standardized value, such as at p_g = 2 bar, should be used, irrespective of the actual recommended pressure.

Claims are sometimes made regarding 'head size', effectively the turbine rotor radius. However, it is clear that the situation is more complicated than that might suggest (Fig. 1.8), and no clear guidance can be given in respect of any obvious design feature. However, it is apparent that the effective size of the orifice has a major effect, as might be expected since this controls the delivery of air to the turbine.[2]

The form of equation 1.6 means that the most efficient use of the cutting tool in a handpiece is when it is loaded to reduce the speed just to half of the free-running speed. However, the parabolic shape of the work curve (Fig. 1.6) means that in practice one should try to stay on the high-speed side of the peak such that

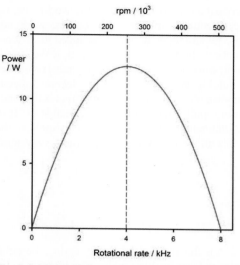

Fig. 1.6 Form of the dependence of turbine output power (effective work done) on rotational rate – the peak is at half the free-running speed.

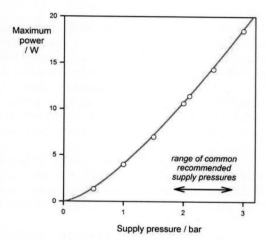

Fig. 1.7 Form of the dependence of peak turbine output power on supply pressure for a typical handpiece.

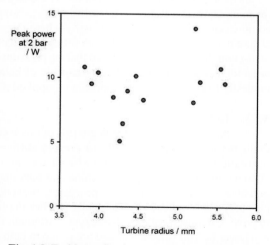

Fig. 1.8 Turbine radius is not the main determinant of turbine power: 14 examples of commercial handpiece.

increasing the load applied on the workpiece, while certainly decreasing the speed, actually increases the cutting work being done – which removes material and tends to allow the speed to increase – *i.e.* negative feedback (a situation allowing stable use). But, if the speed is reduced beyond the mid-point, work done (and thus cutting rate) is reduced, so material is not removed so quickly and this then tends to decrease speed further – **positive feedback** (an unstable situation). The net effect is that once the mid-point is passed, the cutter rapidly stalls, coming to a dead stop. It should be noted that the mass and diameter of the turbine are very small, so the stored (kinetic) energy, E_k, in this miniature **flywheel** is similarly very small. Thus,

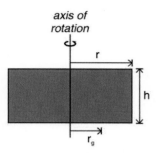

$$E_k = \frac{I_r \omega^2}{2} \qquad (1.8)$$

where

$$I_r = mr_g^2 \qquad (1.9)$$

and m is the mass of the object, and r_g, the radius of gyration, the weighted distance from its rotational axis, which for a simple cylinder is given by the first moment of area about that axis (*cf.* equation 29§3.11) (Fig. 1.9):

$$r_g = \frac{A.(\frac{r}{2})}{A} = \frac{rh.(\frac{r}{2})}{rh} = \frac{r}{2} \qquad (1.10)$$

Fig. 1.9 Radius of gyration – the effective position of the total mass of the turbine.

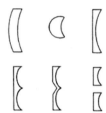

Fig. 1.10 Turbine blades may have a variety of shapes.

To allow for the radial and circumferential detail (Figs 1.1, 1.10) would require some effort, but assuming a large turbine (r = 5.5 mm, Fig. 1.8), effectively solid, h = 5 mm, and made of aluminium (ρ ~2.7 g/mL),we have m = ~0.4 g, so I_r = ~10 × 10^{-9} kg.m². At 8 kHz = 50 rad/s (*i.e.,* free running), E_k = ~12 J (this can be taken as an upper bound). This is more than enough to do damage to soft tissue, but not enough to sustain cutting in enamel, and especially not at, say, 2 kHz, when the stored energy would be 1/16th of that figure. This means that one cannot rely on the angular momentum of the rotating system to pay for the extra demands once into the positive feedback domain.

Even so, it is very difficult to judge the speed of the turbine, and a gentle, stroking motion, with little force applied, tends to be a more efficient approach, backing off before the speed drops too far. This, of course, also allows both the cutter and the substrate to cool, for the unavoidable frictional heating may damage both tooth and cutter (20§7), in part by better allowing the water spray into the site of cutting.[3]

As with (more or less) incompressible fluids flowing in a tube, air suffers a pressure drop along the length of the tubing in which it is flowing (4§11). Constrictions, roughness, irregularities and abrupt turns also inhibit flow and cause further pressure drops, as does the length of miniature tubing inside the handle of a handpiece, from the connector to the turbine chamber. Thus the pressure measured statically at the handpiece control valve on the dental unit may not correspond very closely to the actual pressure at the connector (which is presumably the point at which a manufacturer will specify a "supply pressure"), nor indeed to the pressure at the more remote compressor reservoir (which is usually where the upstream gauge is fitted). Care must therefore be taken to ensure that the pipeline and all of its fittings are adequate for the purpose, that is, of sufficiently large bore and well-enough designed and installed that the pressure drop, when air is flowing (*i.e.* being used), from compressor to handpiece is either negligible or capable of being compensated by raising the working reservoir pressure modestly to ensure that the handpiece is supplied as specified and expected.

●**1.3   Lubrication**

As was indicated at the beginning of this section, much of the use of air-turbine handpieces is for cutting tooth tissue, and this is mostly to prepare surfaces for bonding in one form or another, whether metallic restorations with cements or resin-based restorations *via* 'adhesive dentistry' (10§6). Such tasks requires uncontaminated surfaces if the etching and bonding are to work properly, and oils and greases are particularly potent in interfering with these processes, preventing wetting and forming a barrier to chemical reactions (10§2.7). The air used to clear debris and dry the field must therefore be clean, and thus the compressor used to provide it must be of the oil-free type. Although filters of various kinds can remove some if not most oil, they can never be 100 %-efficient and are prone to being rapidly overloaded and thus acting themselves as sources of oil spray. Note that breathing-air compressors may use non-toxic vegetable oils, but such air is not suitable for application in this kind of dentistry.

It is therefore rather odd that air-turbine handpieces are commonly lubricated with paraffin-based oils. This lubrication has two main functions: to reduce the wear in the bearings by reducing the friction between the moving parts (20§3.2), and to provide a barrier film against corrosion of steel bearings when the handpiece is sterilized in an autoclave – which involves water vapour at about 135 °C. Cleaning and lubrication involves flooding the head with the oil, and although much is drained or blown out during a deliberate short run before use, the remaining film provides a spray of oil for a considerable time, indeed for far longer than would be employed in treating a patient (Fig. 1.11),[4] thus risking contamination of all cut surfaces that neither water nor air sprays can remove. Acetone-based primers for adhesives may dissolve this oil but, unless a large excess is used and blown away, when the acetone has evaporated the oil will remain.

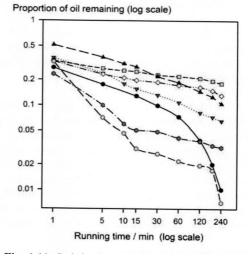

Fig. 1.11 Lubricating oil continues to be ejected from a handpiece for a very long time. Data for seven brands.

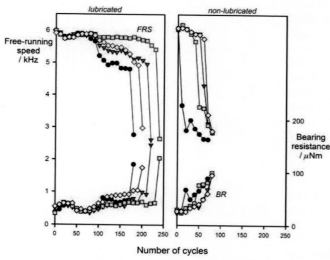

Fig. 1.12 In a simulated work cycle, with autoclaving, turbine bearings deteriorate rapidly if not lubricated. Each graph is for four replacement cartridges in the same handpiece.

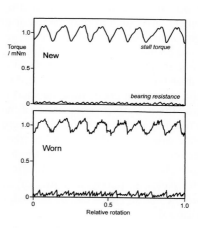

Fig. 1.13 A new turbine bearing assembly shows little resistance to rotation, but after wear becomes 'gritty', affecting stall torque as well as free-running speed.

Nevertheless, lubrication is important. The working lifetime of a turbine bearing assembly is severely shortened in its absence (Fig. 1.12).[5] This is seen in the decline in free-running speed but also in the increase in bearing rotation resistance, which can be felt when a cutting tool in place is turned with the fingers – a gritty feel that also affects the stall torque (Fig. 1.13).[6] Continuing to run a wearing bearing results in destruction of the cage though overheating and seizure of the bearing (Fig. 1.14).[7] Wear is exacerbated by cutting tools being out of balance, such as by being unevenly worn, chipped or bent (23§8.2).

Fig. 1.14 A turbine bearing ball cage after seizure. Note the fracture and the jammed ball. Photo: Wei Min

§2. Light-emitting Diodes

At present, light-emitting diodes (LEDs) only specific use in dentistry is as sources for the visible-light curing of photosensitized resins (6§5.14), and it can reasonably be expected that incandescent (quartz-halogen) lamps for this purpose will soon be obsolete if they are not considered so already, although it is likely that many will remain in service for some time yet.

LEDs are semiconductor devices, that is, the materials from which they are made are intermediate between insulators and metallic conductors (Fig. 26§9.1). Typically, such materials are non-metals or non-metallic compounds in which the bonding is essentially all covalent. Electrons are therefore localized, and are not easily free to move under the influence of an imposed electric field. However, the electrical conductivity is strongly influenced by impurities, and when such impurities (*i.e.* at low concentration) are introduced deliberately they are known as **dopants**. Such doping can be of two kinds: **n-type** and **p-type**, depending on whether they create an excess or a deficit of bonding electrons, and thus provide charge carriers which are electrons themselves (negative, n-type) or holes (positive, p-type). Whilst there are many kinds of semiconductor, the principles can be illustrated by reference to silicon, the archetype.

●2.1   Diode function

Silicon is tetravalent and has a diamond-type crystal structure (Fig. 2.1, left) in which every bond involves the usual two shared electrons, that is, covalently. Indeed, every valence electron from every atom is bound in such a way. However, if a silicon atom in this structure is substituted by a pentavalent element atom such as phosphorus or arsenic (*i.e.* one from Group VB), then only four of its valency electrons can be bound – one is left over (Fig. 2.1, top right), and is free to move around. Likewise, if a trivalent atom such as boron or gallium is included (*i.e.* from Group IIIB), one of the bonds must be short of one electron, hence there is effectively a notional 'hole' that should be filled to complete the bond. Nevertheless, this is a stable situation, and the 'hole' is free to move by an electron moving in. Despite these 'defects', such materials must of course be electrically neutral: the discrepancy is purely in terms of whether or not paired-electron bonds can be formed, that is, they are structural – not charge – defects.

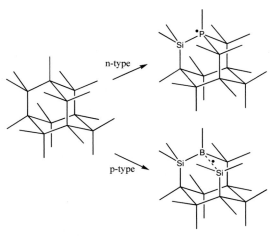

Fig. 2.1 Types of doping in silicon.

Now, as with metals, semiconductors have an electron affinity, and this is modified by the presence of dopants. Electron-deficient structures will tend to have their energy lowered by the addition of electrons (completing the bonds), and *vice versa*. However, such addition or subtraction then necessarily makes the material electrically-charged, raising its electrical potential, which therefore tends to oppose further change: there is an equilibrium to be established between the two opposing tendencies, thus there is here a **Volta potential**, as with metals (13§1.1). Thus, if a piece of n-type doped material is brought into contact with (or built on, as is required for the device to work properly) a piece of p-type, the n-type gives up some electrons to the p-type, thereby lowering the energy of both parts, but then creating an electrical potential difference between them, in what is called the **junction field**. The magnitude of this potential depends on the parent material and the nature of the doping, but the

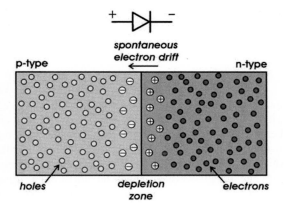

Fig. 2.2 Schematic formation of a depletion zone at the junction of n- and p-type doped regions in a semiconductor single crystal. The symbol for and polarity of the corresponding diode is shown above,

effect only operates over a thin layer at the contact, known as the **depletion zone** (Fig. 2.2). This name refers to the lower concentration, or absence, of charge carriers, due to their **recombination**. Accordingly, the electrical resistance of this region rises – it once again becomes an insulator. Typically, the concentration of holes in the p-type material falls by a factor of 10^{-4}, and likewise (of course) for the concentration of electrons in the n-type material, in the junction zone.

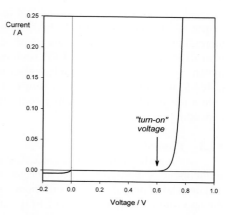

Fig. 2.3 Form of a silicon diode current-voltage characteristic curve. After 'turn-on', the current rises close to exponentially. (The very small reverse-bias current is due to processes which may be ignored here.)

Then, if a voltage is applied across the ends of the device, firstly such that the negative (electron-rich) connection is made to the n-type material – **forward biased**,[1] more electrons are driven first into the n-type material, cancelling the positive charges at the junction, and thus supplying more electrons to the p-type material. Of course, if the electrons are 'sitting in' the holes, *i.e.* paired or combined, they do not move. They need a high enough potential – the **turn-on voltage** – to promote them to the conduction band (Fig. 26§9.1), where there is little appreciable resistance – it becomes metallic-like conduction. Effectively, opposing the spontaneous junction (Volta) potential with a large enough voltage allows current to flow (in effect this allows the Volta potential to be measured, which of course cannot be done directly[2]). On the other hand, if the device is **reverse biased**, with the negative connection to the p-type material, the holes are caused to drift towards the connection. Likewise, electrons are drawn to the positive connection in the n-type material. The net result is that the junction zone is made thicker, and the insulating status is unchanged (at a high enough reverse bias breakdown and device destruction occurs). The current-voltage diode characteristic curve then looks something like that in Fig. 2.3. This, then, is diode behaviour: a one-way valve, so to speak, for electrical current. Such devices are in widespread use for converting alternating to direct current (26§3.4).

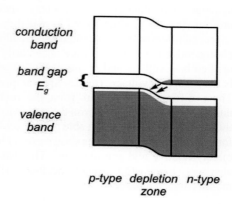

Fig. 2.4 Schematic band diagram for a diode junction under forward bias.

●2.2 LED function

What essentially is happening is that the electrons in the n-type material, which are at a higher energy than the holes of the p-type material, are driven to recombine in the depletion zone (Fig. 2.4). The energy difference ordinarily will appear as heat. However, if the difference is great enough, the energy is emitted as photons, whether this is infrared (as is used in television remote

Table 2.1	*Some common semiconductors used for LEDs*	
Aluminium gallium arsenide	AlGaAs[1]	IR, red
Aluminium gallium indium phosphide	AlGaInP	red, orange, yellow
Gallium phosphide	GaP	yellow, green
Indium gallium nitride	InGaN	green, blue, violet, UV

[1] Note that these are symbolic labels only, not chemical formulae.

controls, for example), or in the visible region, as used in computer screens (24§3.11), and most relevantly here, in LED resin curing lights (6§5.14). This is **electroluminescence**. Emission can now even extend into the ultraviolet (Table 2.1). Because of the nature of the band gap, that is, a nominally single precise value, the emission has a quite small wavelength range. Obviously, the device must be transparent (at least in a suitable direction) for this emission to be useful.

Even though light is being emitted, these devices are not perfectly efficient in that respect. Heat is generated anyway (they have a small but important resistance), but much of the emission may be 'inwards', so

---

[1] The difference between the "conventional current" direction, positive to negative, and electron flow, must be remembered.

[2] Volta potentials will also exist for metals and semiconductors in contact.

to speak, and this is then absorbed and thermalized. Thus, although the voltage characteristic suggests that output can be increased simply by raising the driving voltage (Fig. 2.3), the limit is set by the capacity for thermal dissipation: heat sinks are required for high-powered devices, and destruction is easily possible if this is insufficient. Nevertheless, LED sources are generally of much higher luminous efficiency, *i.e.* in terms of lumens per watt. That is, much less electricity is consumed for a given illumination than by tungsten filament (24§4.5) or fluorescent (24§4.9) lamps. However, the very steep voltage-dependence on output means that unless the supply circuit is tightly controlled, battery-driven curing lights can show a considerable variation depending on state of charge.[8][9]

●2.3 Photodiode function

As perhaps might be expected, if radiation of sufficient energy impinges on the depletion zone of a semiconductor diode, a current can be generated. In fact, electron- hole pairs can be generated anywhere in such a material, but only if they are sufficiently close to the junction can they separated, generating a current (and in the opposite sense to that for an LED). This is then the **photovoltaic effect**. If, though, the diode is reverse biased, the thicker junction zone increases the capacity to generate a current, and indeed makes the device more sensitive. It is then said to be in **photoconductive** mode.

These behaviours can then be used to provide a measure of the output of a curing light and are the basis of the **radiometers** sold for that purpose. There are however a number of problems associated with their use in this context. Spectral sensitivity is not uniform (see, *e.g.,* Fig. 26§9.13) and they are associated with appreciable temperature coefficients. In addition, for those that use photoconductive mode, they are necessarily very sensitive to the applied bias voltage and so are dependent on the stability of the battery supply voltage (unless special precautions are taken). These difficulties are apart from the problem of the lack of uniformity of the output across the curing light tip.[10] Thus, while there is some value in being able to monitor light output over time for a given device, the values shown cannot be precise or absolute, merely relative and approximate. The battery problem that can also apply to the LED curing lights themselves (§2.2) makes such monitoring even harder.

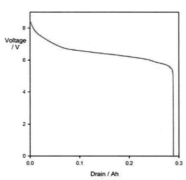

Fig. 2.5 Typical form of the voltage sag with use for an alkaline battery. Carbon-zinc are worse, NiCd and Ni-MH are similar. Only Li-cells have a nearly flat characteristic.

References

[1]   Dyson JE & Darvell BW. The development of the dental high speed air turbine handpiece. Part 1 Aust Dent J 38 (1) : 49 - 58, 1993. Part 2 Aust Dent J 38 (2) : 131 - 143, 1993.

[2]   Dyson JE & Darvell BW. Flow and free running speed characterization of dental air turbine handpieces. J Dent 27 (7): 465 - 477, 1999.

[3]   Leung BTW, Dyson JE & Darvell BW. Coolant effectiveness in dental cutting. NZ Dental J 108 (1): 25 - 29, 2012.

[4]   Pong ASM, Dyson JE & Darvell BW. Discharge of lubricant from air turbine handpieces. Brit Dent J 198(10): 637 - 640, 2005.

[5]   Wei M, Dyson JE & Darvell BW. Factors affecting dental air-turbine handpiece bearing failure. Oper Dent 37 (4): E1-E12, 2012.

[6]   Dyson JE & Darvell BW. Torque, power and efficiency characterization of dental air turbine handpieces. J Dent 27 (8): 573 - 586, 1999.

[7]   Wei M, Dyson JE & Darvell BW. Failure analysis of the ball bearings of dental air-turbine handpieces. Aust Dent J 58(4): 514 - 521, 2013.

[8]   AlShaafi MM, Harlow JE, Price HL, Rueggeberg FA, Labrie D, AlQahtani MQ & Price RB. Emission characteristics and effect of battery drain in ''budget''curing lights. Oper Dent 41 (4): 397 - 40-8, 2016.

[9]   Tongtaksin A & Leevailoj C. Battery charge affects the stability of light intensity from light-emitting diode light-curing units. Oper Dent 42 (5): 497 - 504, 2017.

[10]   Price RB, Labrie D, Kazmi S, Fahey J & Felix CM. Intra- and inter-brand accuracy of four dental radiometers. Clin Oral Invest 16 (3): 707 - 717, 2012.

Index

For common terms, only selected instances are shown, with the more important in bold.

Printed in the United States
By Bookmasters